Pseudo-reductive Groups

Second Edition

Pseudo-reductive groups arise naturally in the study of general smooth linear algebraic groups over non-perfect fields and have many important applications. This monograph provides a comprehensive treatment of the theory of pseudo-reductive groups and explains their structure in a usable form.

In this second edition there is new material on relative root systems and Tits systems for general smooth affine groups, including the extension to quasi-reductive groups of famous simplicity results of Tits in the semisimple case. Chapter 9 has been completely rewritten to describe and classify pseudo-split absolutely pseudo-simple groups with a non-reduced root system over arbitrary fields of characteristic 2 via the useful new notion of "minimal type" for pseudo-reductive groups.

Researchers and graduate students working in related areas such as algebraic geometry, algebraic group theory, or number theory will value this book, as it develops tools likely to be used in tackling other problems.

BRIAN CONRAD is Professor of Mathematics at Stanford University.

OFER GABBER is Directeur de Recherche CNRS at the Institut des Hautes Études Scientifiques.

GOPAL PRASAD is Raoul Bott Professor of Mathematics at the University of Michigan.

NEW MATHEMATICAL MONOGRAPHS

All the titles listed below can be obtained from good booksellers or from Cambridge University Press. For a complete series listing visit www.cambridge.org/mathematics.

1 M. Cabanes and M. Enguehard *Representation Theory of Finite Reductive Groups*
2 J. B. Garnett and D. E. Marshall *Harmonic Measure*
3 P. Cohn *Free Ideal Rings and Localization in General Rings*
4 E. Bombieri and W. Gubler *Heights in Diophantine Geometry*
5 Y. J. Ionin and M. S. Shrikhande *Combinatorics of Symmetric Designs*
6 S. Berhanu, P. D. Cordaro and J. Hounie *An Introduction to Involutive Structures*
7 A. Shlapentokh *Hilbert's Tenth Problem*
8 G. Michler *Theory of Finite Simple Groups I*
9 A. Baker and G. Wüstholz *Logarithmic Forms and Diophantine Geometry*
10 P. Kronheimer and T. Mrowka *Monopoles and Three-Manifolds*
11 B. Bekka, P. de la Harpe and A. Valette *Kazhdan's Property (T)*
12 J. Neisendorfer *Algebraic Methods in Unstable Homotopy Theory*
13 M. Grandis *Directed Algebraic Topology*
14 G. Michler *Theory of Finite Simple Groups II*
15 R. Schertz *Complex Multiplication*
16 S. Bloch *Lectures on Algebraic Cycles (Second Edition)*
17 B. Conrad, O. Gabber and G. Prasad *Pseudo-reductive Groups*
18 T. Downarowicz *Entropy in Dynamical Systems*
19 C. Simpson *Homotopy Theory of Higher Categories*
20 E. Fricain and J. Mashreghi *The Theory of H(b) Spaces I*
21 E. Fricain and J. Mashreghi *The Theory of H(b) Spaces II*
22 J. Goubault-Larrecq *Non-Hausdorff Topology and Domain Theory*
23 J. Śniatycki *Differential Geometry of Singular Spaces and Reduction of Symmetry*
24 E. Riehl *Categorical Homotopy Theory*
25 A. Munson and I. Volić *Cubical Homotopy Theory*
26 B. Conrad, O. Gabber and G. Prasad *Pseudo-reductive Groups (2nd Edition)*
27 J. Heinonen, P. Koskela, N. Shanmugalingam and J. T. Tyson *Sobolev Spaces on Metric Measure Spaces*
28 Y.-G. Oh *Symplectic Topology and Floer Homology I*
29 Y.-G. Oh *Symplectic Topology and Floer Homology II*

Pseudo-reductive Groups

Second Edition

BRIAN CONRAD
Stanford University

OFER GABBER
Institut des Hautes Études Scientifiques

GOPAL PRASAD
University of Michigan

CAMBRIDGE
UNIVERSITY PRESS

CAMBRIDGE
UNIVERSITY PRESS

University Printing House, Cambridge CB2 8BS, United Kingdom

Cambridge University Press is part of the University of Cambridge.

It furthers the University's mission by disseminating knowledge in the pursuit of education, learning and research at the highest international levels of excellence.

www.cambridge.org
Information on this title: www.cambridge.org/9781107087231

First published 2010
Second edition 2015

A catalogue record for this publication is available from the British Library

Library of Congress Cataloging in Publication data

Conrad, Brian, 1970– author.
Pseudo-reductive groups / Brian Conrad, Stanford University, Ofer Gabber, Institut des hautes études scientifiques, Gopal Prasad, University of Michigan. – Second edition.
pages cm. – (New mathematical monographs)
ISBN 978-1-107-08723-1 (Hardback)
1. Linear algebraic groups. 2. Group theory. I. Gabber, Ofer, 1958– author II. Prasad, Gopal, author. III. Title.
QA179.C667 2015
512′.55–dc22 2014029481

ISBN 978-1-107-08723-1 Hardback

Dedicated to Jacques Tits

Contents

Preface to the second edition

In addition to correcting minor errors/misprints and simplifying some proofs as well as improving some results, the major changes in this edition are: a complete rewriting of Chapter 9 (to obtain definitive results over all fields of characteristic 2), a simplified and improved exposition of Tits' results on unipotent groups given in Appendix B, and additional material C.2.11–C.2.34 in Appendix C to provide a version of the Borel–Tits relative structure theory (with relative root systems, etc.) in the pseudo-reductive case and beyond. Apart from Chapter 9, parts of §11.4, and the final displayed expression in §1.6, all numerical labels for results, examples, equations, remarks, etc. remain identical to the labels in the first edition (but the formulation of some results has been strengthened and new material has been added at the end of some sections).

We recall that pseudo-split pseudo-simple groups with non-reduced root systems can only exist over imperfect fields k of characteristic 2. In Chapter 9 of the original version of this monograph, we constructed such groups and explored their properties only when $[k : k^2] = 2$. In the revised Chapter 9, we introduce some new ideas (especially the property that we call "minimal type") and use them to eliminate the degree restriction, yielding a definitive treatment of such groups; when $[k : k^2] > 2$ there arise several new phenomena with no analogue when $[k : k^2] = 2$. For the convenience of the reader we have retained the hypothesis $[k : k^2] = 2$ for the classification results in Chapter 10, but the results in this new edition (especially in Chapter 9) are used in [CP] to study the automorphism groups of pseudo-semisimple groups and give a complete classification of pseudo-reductive groups (in the spirit of Chapter 10) without any restriction on $[k : k^2]$ when $\mathrm{char}(k) = 2$.

We thank Vikraman Balaji, Bas Edixhoven, Skip Garibaldi, Philippe Gille, Teruhisa Koshikawa, Andrei Rapinchuk, and Bertrand Rémy for their comments. B.C. and G.P. acknowledge support from the National Science Foundation, with grants DMS-0600919 and DMS-1100784 for B.C. and DMS-1001748 for G.P., and are grateful to the Institute for Advanced Study where much of the work on this revised version was carried out.

Introduction

Why go beyond reductive groups?

The theory of connected reductive groups over a general field, and its applications over arithmetically interesting fields, constitutes one of the most beautiful topics within pure mathematics. However, it does sometimes happen that one is confronted with linear algebraic groups that are not reductive, and whose structure may be rather mysterious.

Example: *forms of a variety*. Let X be a projective variety over a field k. Grothendieck constructed a scheme $\mathrm{Aut}_{X/k}$ classifying its automorphisms, but this is hard to understand in general. For example, finite generation of its component group is unknown, even if $k = \mathbf{C}$. Likewise, little is known about the identity component $\mathrm{Aut}^0_{X/k}$ apart from that it is a k-group scheme of finite type. Since $\mathrm{H}^1(k, \mathrm{Aut}_{X/k})$ classifies k-forms of X, there is arithmetic interest in $\mathrm{Aut}_{X/k}$ even though our knowledge of its structure is limited.

If $\mathrm{char}(k) = 0$ then by a structure theorem of Chevalley for smooth connected groups over perfect fields, $\mathrm{Aut}^0_{X/k}$ is an extension of an abelian variety by a smooth connected affine k-group. If k is imperfect then Chevalley's Theorem does not apply (even if $\mathrm{Aut}^0_{X/k}$ is smooth). Nonetheless, some general problems for connected k-group schemes G of finite type (e.g., $G = \mathrm{Aut}^0_{X/k}$) reduce to the cases of smooth connected affine k-groups and abelian varieties over k. We do not know any restrictions on the smooth connected affine groups arising in this way when $G = \mathrm{Aut}^0_{X/k}$.

Example: *local-to-global principle*. Let X be a quasi-projective (or arbitrary) scheme over a global field k. Suppose that X is equipped with a right action by a linear algebraic k-group H. Choose a point $x \in X(k)$. Does $H(k)$ act with only finitely many orbits on the set of $x' \in X(k)$ that are $H(k_v)$-conjugate to x for all v away from a fixed finite set S of places of k?

This question reduces to the finiteness of the Tate–Shafarevich sets $Ш^1_S(k, G)$ (of isomorphism classes of right G-torsors over k admitting a k_v-point for all $v \notin S$) as G varies through the isotropy group schemes H_x of all $x \in X(k)$. Such isotropy groups are generally disconnected and non-reductive (and non-smooth if $\mathrm{char}(k) > 0$), even if H is connected and semisimple. Over number fields k the finiteness of $Ш^1_S(k, G)$ for any linear algebraic k-group G was proved by Borel and Serre [BoSe, Thm. 7.1]. The analogous result over global function fields reduces to the case of smooth G (see Example C.4.3) and the results in this monograph provide what is needed to settle this case.

Imperfect base fields

Over fields of characteristic 0, or more generally over perfect fields (such as finite fields), it is usually an elementary matter to reduce problems for general linear algebraic groups to the connected reductive case. The situation is entirely different over imperfect fields, such as local and global function fields. The purpose of the theory of pseudo-reductive groups, initiated by Borel and Tits, is to overcome this problem over imperfect fields.

To explain the reason for the difficulties over imperfect fields, consider a smooth connected affine group G over an arbitrary field k. Recall that the unipotent radical $\mathscr{R}_u(G_{\overline{k}})$ over an algebraic closure $\overline{k}$ of k is the unique maximal smooth connected unipotent normal $\overline{k}$-subgroup of $G_{\overline{k}}$, and that G is *reductive* if $\mathscr{R}_u(G_{\overline{k}}) = 1$. If k is perfect then $\overline{k}/k$ is Galois, so by Galois descent $\mathscr{R}_u(G_{\overline{k}})$ (uniquely) descends to a smooth connected unipotent normal k-subgroup $\mathscr{R}_u(G) \subseteq G$. In particular, for perfect k there is an exact sequence of k-groups

$$1 \to \mathscr{R}_u(G) \to G \to G/\mathscr{R}_u(G) \to 1 \qquad (*)$$

with $G/\mathscr{R}_u(G)$ a connected reductive k-group. For imperfect k the unipotent radical $\mathscr{R}_u(G_{\overline{k}})$ can fail to descend to a k-subgroup of G (as we will illustrate with many examples), in which case G is not an extension of a connected reductive k-group by a smooth connected unipotent k-group.

The *k-unipotent radical* $\mathscr{R}_{u,k}(G)$ is the maximal smooth connected unipotent normal k-subgroup of G. For an extension field k'/k, the natural inclusion

$$k' \otimes_k \mathscr{R}_{u,k}(G) \hookrightarrow \mathscr{R}_{u,k'}(G_{k'})$$

is an equality if k'/k is separable but can fail to be an equality otherwise (e.g., for $k' = \overline{k}$ when k is imperfect). Over a global field k there are arithmetic problems (such as the ones mentioned above) that can be reduced to questions about smooth connected affine k-groups H for which the inclusion $\overline{k} \otimes_k \mathscr{R}_{u,k}(H) \hookrightarrow \mathscr{R}_u(H_{\overline{k}})$ may not be an equality when $\mathrm{char}(k) > 0$.

A *pseudo-reductive group* over a field k is a smooth connected affine k-group with trivial k-unipotent radical. For any smooth connected affine k-group G, the exact sequence

$$1 \to \mathscr{R}_{u,k}(G) \to G \to G/\mathscr{R}_{u,k}(G) \to 1$$

uniquely expresses G as an extension of a pseudo-reductive k-group by a smooth connected unipotent k-group. (For perfect k this is $(*)$.) To use this exact sequence, we need a structure theory for pseudo-reductive groups that is as informative as the one for connected reductive groups.

Structure theory

A useful structure theory for pseudo-reductive groups over imperfect fields cannot be developed by minor variations of the methods used in the reductive case, essentially because pseudo-reductivity is not a geometric property. More specifically, pseudo-reductivity can be lost under an inseparable extension of the base field, such as extension from an imperfect field to an algebraically closed field. Also, pseudo-reductivity is not inherited by quotients in general, as will be seen in many examples. It seems unlikely that there can be an interesting theory of pseudo-reductive groups over a base ring other than a field, in contrast with the theory of connected reductive groups.

Pseudo-reductive groups were the topic of courses by J. Tits at the Collège de France in 1991–92 and 1992–93; see [Ti2] and [Ti3]. In these courses, he studied pseudo-reductive groups over any field k in terms of pseudo-parabolic k-subgroups and a root datum, and he constructed "non-standard" examples of pseudo-reductive groups over separably closed fields of characteristics 2 and 3.

Most of Tits' results in his second course were announced by Borel and Tits in [BoTi2] without proofs. Tits' résumé [Ti3] sketched some of the proofs, and a more detailed exposition was given by Springer in [Spr, Ch. 13–15]. Tits also attempted to classify pseudo-reductive groups over separably closed fields of nonzero characteristic, but he ran into many complications and so did not write up all of his results. For example, the structure of root groups was shrouded in mystery; their dimension can be rather large. Inspired by [BoTi2], Springer obtained (unpublished) classification results in cases with geometric semisimple rank 1, and he made partial progress in characteristic 2.

We discovered a new approach to the structure of pseudo-reductive groups. If k is a field with $\mathrm{char}(k) \neq 2$ then our main result over k is a canonical description of all pseudo-reductive k-groups in terms of two ingredients: connected semisimple groups *over finite extensions* of k, and *commutative* pseudo-reductive k-groups. (These commutative groups turn out to be the Cartan k-subgroups, which are the maximal k-tori in the connected reductive case.) We also get a similar result when $\mathrm{char}(k) = 2$ if k is "almost perfect" in the sense that $[k : k^2] \leqslant 2$. This includes the most interesting cases for number theory: local and global function fields over finite fields. The case of imperfect fields of characteristic 2 involves new phenomena. Most of the hardest aspects of characteristic 2 show up in the study of G whose maximal geometric semisimple quotient $G^{\mathrm{ss}}_{\overline{k}}$ is SL_2 or PGL_2.

Main results and applications

There is a lot of "practical" interest in connected reductive groups (and especially connected semisimple groups), due to applications in the study of interesting problems not originating from the theory of algebraic groups. The reason for interest in pseudo-reductive groups is largely theoretical: they arise naturally in proofs of general theorems. Whenever one is faced with proving a result for arbitrary smooth connected affine groups over an imperfect field (e.g., reducing a problem for a non-smooth affine group to an analogous problem for a related abstract smooth affine group by the method of §C.4), pseudo-reductive groups show up almost immediately whereas reductive groups do not. More importantly, experience shows that the structure theory of pseudo-reductive groups often makes it possible to reduce general problems to the semisimple and solvable cases. This is why we believe that pseudo-reductive groups are basic objects of interest in the theory of linear algebraic groups over imperfect fields.

Our main discovery is that, apart from some exceptional situations in characteristics 2 and 3, all pseudo-reductive groups can be canonically constructed in terms of three basic methods: direct products, Weil restriction of connected semisimple groups over *finite extensions of* k, and modifying a Cartan subgroup (via a central pushout construction).

To explain how this works, first consider a finite extension of fields k'/k and a smooth connected affine k'-group G'. The Weil restriction $G = \mathrm{R}_{k'/k}(G')$ is a smooth connected affine k-group of dimension $[k' : k] \dim G'$ characterized by the property $G(A) = G'(k' \otimes_k A)$ for k-algebras A. When k'/k is separable, the functor $\mathrm{R}_{k'/k}$ has good properties and is very useful. For example, it preserves reductivity, and any connected semisimple k-group G that is k-simple and simply connected has the form $\mathrm{R}_{k'/k}(G')$ for a canonically associated pair $(G', k'/k)$ where k'/k is a separable extension and G' is *absolutely simple* and simply connected over k'.

The functor $\mathrm{R}_{k'/k}$ exhibits some unusual properties when k'/k is not separable: it does not carry tori to tori, nor does it preserve reductivity or perfectness of groups or properness of morphisms in general, and it often fails to preserve surjectivity or finiteness of homomorphisms. However, it does carry connected reductive groups (and even pseudo-reductive groups) to pseudo-reductive groups. The most basic examples of pseudo-reductive groups are products $\prod_i \mathrm{R}_{k'_i/k}(G'_i)$ for a finite (non-empty) collection of finite extensions k'_i/k and absolutely simple and simply connected semisimple k'_i-groups G'_i; the simply connectedness of G'_i ensures that thc k-groups $\mathrm{R}_{k'_i/k}(G'_i)$ are perfect, and such Weil restrictions are non-reductive whenever k'_i/k is not separable.

Cartan k-subgroups in pseudo-reductive groups are always commutative and pseudo-reductive. For example, if k'/k is a finite extension and G' is a nontrivial connected reductive k'-group given with a maximal k'-torus T' then $\mathrm{R}_{k'/k}(T')$ is a Cartan k-subgroup of $\mathrm{R}_{k'/k}(G')$ but it is never a k-torus if k'/k is not separable. It seems to be an impossible task to describe general commutative pseudo-reductive groups over imperfect fields, as we will illustrate with numerous examples. (That is, we do not expect there to be any description as convenient as the use of Galois lattices to describe tori.) The representation theory of connected reductive groups rests on the commutative case (tori), so it seems difficult to say anything interesting about the representation theory of general pseudo-reductive groups. The structure of the commutative objects turns out to be the only mystery, in the following sense. We introduce a class of non-commutative pseudo-reductive groups called *standard*, built as certain central quotients

$$\left(C \ltimes \prod_i \mathrm{R}_{k'_i/k}(G'_i)\right) \Big/ \prod_i \mathrm{R}_{k'_i/k}(T'_i)$$

with a commutative pseudo-reductive k-group C, absolutely simple and simply connected k'_i-groups G'_i for finite extensions k'_i/k, and a maximal k'_i-torus T'_i in G'_i for each i. (This also recovers the commutative case by taking the collection $\{k'_i\}$ to be empty.)

We give a canonical way to choose such data in this "standard" construction, and prove that if $\mathrm{char}(k) \neq 2, 3$ then *every* non-commutative pseudo-reductive k-group is standard. In this sense, non-commutative pseudo-reductive k-groups away from characteristics 2 and 3 come equipped with an algebraic invariant

$k' = \prod k'_i$ and can be obtained from semisimple groups over its factor fields along with an additional commutative group. One of the main ingredients in our proofs is Tits' theory of root groups, root systems, and pseudo-parabolic subgroups of pseudo-reductive groups; we give a self-contained development because we need scheme-theoretic aspects of the theory.

Building on some non-standard examples of Tits, we introduce a "generalized standard" construction in characteristics 2 and 3 as well as an additional exceptional construction in characteristic 2 (with a non-reduced root system). We prove that this gives a classification theorem in characteristic 3, as well as in characteristic 2 when $[k : k^2] \leqslant 2$. (Many of our proofs are characteristic-free, and in characteristic 2 we get quite far before we need to assume $[k : k^2] \leqslant 2$.) The general classification problem in characteristic 2 without restriction on $[k : k^2]$ is solved in [CP].

Our description of pseudo-reductive groups in terms of connected semisimple groups and commutative pseudo-reductive groups isolates the mysteries of the commutative case in a useful manner. For example, we discuss automorphisms of pseudo-reductive groups and we gain new insight into the large dimensions of root groups and the non-completeness of coset spaces G/P for non-reductive pseudo-reductive k-groups G and pseudo-parabolic k-subgroups P that are "not too big": the root groups are canonically identified with Weil restrictions of $\mathbf{G}_{\rm a}$ from finite extensions of k (apart from some exceptions when $\mathrm{char}(k) = 2$) and the coset spaces G/P are $\prod_i \mathrm{R}_{k'_i/k}(G'_i/P'_i)$ for semisimple G'_i and parabolic k'_i-subgroups P'_i for finite extensions k'_i/k (apart from some exceptional situations in characteristics 2 and 3). As an indication that the difficulties with characteristic 2 are not just a matter of technique, we mention another application of our structure theorem: if $\mathrm{char}(k) \neq 2$ then a pseudo-reductive k-group is reductive if and only if its Cartan k-subgroups are tori, and this is *false* over every imperfect field of characteristic 2.

In one of the appendices we provide the first published account with proofs of Tits' important work on the structure theory of unipotent groups over arbitrary fields of nonzero characteristic. This is an essential tool in our work and we expect it to be of general interest to those who work with algebraic groups over imperfect fields. In another appendix we use our results on root groups (valid in all characteristics, via characteristic-free proofs) to give proofs of theorems of Borel and Tits for arbitrary connected linear algebraic groups G over any field k. This includes $G(k)$-conjugacy of: maximal k-split k-tori, maximal k-split smooth connected unipotent k-subgroups, and minimal pseudo-parabolic k-subgroups. Another such result is the Bruhat decomposition for $G(k)$ (relative to a minimal pseudo-parabolic

k-subgroup of G). In Chapter 3 we prove the existence of Levi k-subgroups of pseudo-reductive k-groups in the "pseudo-split" case.

The most interesting applications of our structure theorem are to be found in number theory, so let us say a few words about this. In [Con2], this structure theorem is the key to proving finiteness results for general linear algebraic groups over local and global function fields, previously known only under hypotheses related to reductivity or solvability. Among the main results proved in [Con2] are that class numbers, Tate–Shafarevich sets, and Tamagawa numbers of arbitrary linear algebraic groups over global function fields are finite. We hope that the reader finds such arithmetic applications to be as much a source of motivation for learning about pseudo-reductive groups as we did in discovering and proving the results presented in this monograph.

Plan of the monograph

We assume the reader has prior familiarity with basics of the theory of connected reductive groups over a field (as in [Bo2]), but we do not assume previous familiarity with earlier work on pseudo-reductive groups. We develop everything we need concerning pseudo-reductivity.

In Part I we establish some basic results that only require the definition of pseudo-reductivity and we carry out many fundamental constructions. We begin in Chapter 1 with an introduction to the general theory, some instructive examples, and the construction of "standard" pseudo-reductive groups. In Chapter 2 we define (following Tits) pseudo-parabolic subgroups and root groups in pseudo-reductive groups over any field k (assuming $k = k_s$ when it suffices for our needs). In Chapter 3 we give the properties of pseudo-parabolic subgroups and root groups, and define the root datum associated to a pseudo-reductive group. We also establish the Bruhat decomposition and existence of Levi subgroups for pseudo-reductive k-groups when $k = k_s$ (to be used in our work in characteristic 2).

Part II takes up the finer structure theory of standard pseudo-reductive groups (in arbitrary characteristic). In Chapter 4, we rigidify the choice of data in the "standard" construction. In Chapter 5 we state our main theorem in general and record some useful results on splitting of certain central extensions. We also prove some invariance properties of standardness, including that standardness is insensitive to replacing the base field with a separable closure, and we reduce the task of proving standardness for all pseudo-reductive groups (over a fixed separably closed base field k) to the special case of pseudo-reductive k-groups G such that $G = \mathscr{D}(G)$ and the maximal geometric semisimple quotient $G^{ss}_{\overline{k}} = G_{\overline{k}}/\mathscr{R}(G_{\overline{k}})$ is a simple $\overline{k}$-group; such G are called (absolutely) *pseudo-simple*. For such G we establish a standardness criterion. This is used in Chapter 6 to prove the general classification away from characteristics 2 and 3, building on the rank-1 case that we handle using calculations with SL_2.

Part III introduces new ideas needed for small characteristics. In Chapter 7 we construct a class of pseudo-reductive groups (discovered by Tits) called *exotic*, which only exist over imperfect fields of characteristics 2 and 3. In Chapter 8 we study the properties of these groups, including their automorphisms and the splitting of certain central extensions. The structure theory in characteristic 2 involves new difficulties and requires additional group-theoretic constructions (known to Tits) resting on birational group laws. The absolutely pseudo-simple case is studied in Chapter 9, and the classification in characteristics 2 and 3 is completed in Chapter 10 using a "generalized standard" construction. In characteristic 2 we need to assume k is almost perfect (i.e., $[k : k^2] \leqslant 2$) in order to prove that the constructions we give are exhaustive. (If $[k : k^2] > 2$ then there are more possibilities, and these are classified in [CP].) Chapter 11 gives applications of our main results.

Part IV contains the appendices. Appendix A summarizes useful (but largely technical) facts from the theory of linear algebraic groups, with complete proofs or references for proofs. We frequently refer to this appendix in the main text, so the reader may wish to skim some of its results and examples (especially concerning Weil restriction) early in the process of learning about pseudo-reductive groups. Appendix B proves unpublished results of Tits [Ti1] on the structure of smooth connected unipotent groups over arbitrary (especially imperfect) fields with nonzero characteristic. This is crucial in many proofs of the deeper aspects of the structure theory of pseudo-reductive groups. Appendix C proves many interesting rational conjugacy theorems, and the Bruhat decomposition, announced by Borel and Tits in [BoTi2] for all smooth connected affine groups over arbitrary fields, and discusses root data, BN-pairs, and simplicity results in such cases too.

Acknowledgments for the first edition

We are grateful for the encouragement and comments from many people who read earlier versions of this work or saw presentations on it in seminars and workshops, especially Stephen DeBacker, Skip Garibaldi, Johan de Jong, and an anonymous referee. And most of all, we thank Indu Prasad for her patience and support in so many ways throughout our work on this project.

The work of the first author was supported by a grant from the Alfred P. Sloan Foundation and by NSF grants DMS-0093542 and DMS-0917686. The work of the third author was supported by the Humboldt Foundation, the École Normale Supérieure (Paris), and NSF grant DMS-0653512.

Terminology, conventions, and notation

Algebraic geometry. All rings are understood to be commutative with identity unless we indicate otherwise. For example, if B is such a ring then a B-algebra is understood to be commutative with the "same" multiplicative identity element. For a field k, we write k_s to denote a separable closure and $\overline{k}$ to denote an algebraic closure. If X is a scheme over a ring k and R is a k-algebra then $X_R = X \otimes_k R$ denotes the R-scheme obtained by scalar extension. We use similar notation $M_R = M \otimes_k R$ for k-modules M.

We often work with schemes and maps between them by thinking functorially. In particular, if k is a ring and X is a k-scheme then a *point* of X often means an element in $X(R)$ for a k-algebra R that we allow to vary. The context should always make clear if we instead mean points of the underlying topological space of a scheme (such as generic points). For a map of schemes $f: Y' \to Y$ and a locally closed subscheme $X \hookrightarrow Y$ we write $f^{-1}(X)$ to denote the scheme-theoretic pullback $X \times_Y Y'$ viewed as a locally closed subscheme of Y'.

Preimages and intersections of subschemes are always taken in the scheme-theoretic sense (and so may be non-reduced) unless we indicate otherwise. The underlying reduced scheme of a scheme X is denoted X_{red}. Beware that if k is an imperfect field and G is an affine k-group scheme of finite type then G_{red} can fail to be a k-subgroup scheme of G! Moreover, even when it is a k-subgroup scheme, it may not be k-smooth. See Example A.8.3 for natural examples. (A case with good behavior over any field k is that if T is a k-torus and M is a closed k-subgroup scheme of T then M_{red} is a smooth k-subgroup; see Corollary A.8.2.)

An extension of fields K/k is *separable* if $K \otimes_k \overline{k}$ is reduced. Note that K/k is not required to be algebraic (e.g., we can have $K = k(x)$, or $K = k_v$ for a global field k and place v of k).

In algebraic geometry over a field k, Galois descent is a useful procedure for descending constructions and results from k_s down to k. We sometimes need to descend through an inseparable field extension. For such purposes, we use faithfully flat descent. The reader is referred to [BLR, 6.1] for an introduction to that theory, covering (more than) everything we need.

The reader is referred to [BLR, Ch. 2] for an excellent exposition of the theory of *smooth* and *étale* morphisms of schemes. (For example, a map of schemes $X \to Y$ is smooth if, Zariski-locally on X, it factors as an étale morphism to an affine space over Y.) We sometimes need to use the *functorial criterion* for a morphism to be smooth or étale, so we now recall the statement of this criterion in a special case that is sufficient for our needs. If X is a scheme locally of finite type over a noetherian ring R then $X \to \operatorname{Spec} R$ is smooth (resp. étale) if and only if $X(R') \to X(R'/J')$ is surjective (resp. bijective) for every R-algebra R' and ideal J' in R' such that $J'^2 = 0$.

Algebraic groups. For an integer $n \geqslant 1$, a commutative group scheme is *n-torsion* if it is killed by n. We write $C[n]$ to denote the n-torsion in a commutative group scheme C (with $n \geqslant 1$), and $\mu_n = \mathrm{GL}_1[n]$. We also write $[n]$ to denote multiplication by n on a commutative group scheme.

For a smooth group G of finite type over a field k, its *derived group* is $\mathscr{D}(G)$ (Definition A.1.14).

We always use the notions of *kernel*, *center*, *centralizer*, *central*, *normalizer*, *subgroup*, *normal*, and *quotient* in the scheme-theoretic sense, though sometimes we append the adjective "scheme-theoretic" for emphasis; see Definitions A.1.8–A.1.11.

If S is a scheme (generally affine) and G is an S-group scheme then $\underline{\mathrm{Aut}}(G)$ denotes the functor on S-schemes assigning to any S' the group $\mathrm{Aut}_{S'}(G_{S'})$ of S'-group automorphisms of $G_{S'}$. We likewise define the functors $\underline{\mathrm{End}}(G)$ when G is commutative and $\underline{\mathrm{Hom}}(G,H)$ for commutative G and H. If we wish to consider functors respecting additional structure (such as a linear structure on a vector group; see Definition A.1.1) then we will say so explicitly.

For a smooth affine group G over a field k and a k-split k-torus $T \subseteq G$, $\Phi(G,T)$ denotes the set of *nontrivial* weights of T under the adjoint action of G on $\mathrm{Lie}(G)$ (i.e., the set of nontrivial k-homomorphisms $a : T \to \mathbf{G}_{\mathrm{m}}$ such that the a-weight space in $\mathrm{Lie}(G)$ is nonzero). This definition also makes sense for any group scheme G locally of finite type over k, but we only require the smooth affine case. (By using Lemma A.8.8, this definition carries over to smooth group schemes and split tori over any connected non-empty base scheme, but we will not need this generalization.)

PART I

Constructions, examples, and structure theory

1

Overview of pseudo-reductivity

1.1 Comparison with the reductive case

The notion of pseudo-reductivity is due to Borel and Tits. We begin by defining this concept, as well as some related notions.

Definition 1.1.1 Let k be a field, G a smooth affine k-group. The *k-unipotent radical* $\mathscr{R}_{u,k}(G)$ (resp. *k-radical* $\mathscr{R}_k(G)$) is the maximal smooth connected unipotent (resp. solvable) normal k-subgroup of G. A *pseudo-reductive k-group* is a smooth connected affine k-group G such that $\mathscr{R}_{u,k}(G) = 1$.

In the definition of pseudo-reductivity, it is equivalent to impose the stronger condition that G contains no nontrivial smooth connected k-subgroups U such that $U_{\overline{k}} \subseteq \mathscr{R}_u(G_{\overline{k}})$. To prove the equivalence it suffices to show that $U \subseteq \mathscr{R}_{u,k}(G)$, or equivalently that U is contained in a smooth connected unipotent *normal* k-subgroup. The smooth connected k_s-subgroup U' in G_{k_s} generated by the $G(k_s)$-conjugates of U_{k_s} is normal in G_{k_s} and satisfies $U'_{\overline{k}} \subseteq \mathscr{R}_u(G_{\overline{k}})$, so U' is unipotent. By construction, U' is $\mathrm{Gal}(k_s/k)$-stable, so it descends to a smooth connected unipotent normal k-subgroup of G that contains U, as desired. A consequence of this argument is that if N is a smooth connected normal k-subgroup of a smooth connected affine k-group G, so $\mathscr{R}_u(N_{\overline{k}}) \subseteq \mathscr{R}_u(G_{\overline{k}})$, then $\mathscr{R}_{u,k}(N) \subseteq \mathscr{R}_{u,k}(G)$. As a special case, if G is pseudo-reductive then so is N. In particular, every smooth connected k-subgroup of a commutative pseudo-reductive k-group is pseudo-reductive.

Pseudo-reductive k-groups are called *k-reductive* in Springer's book [Spr]. To define a related notion of pseudo-semisimplicity, triviality of $\mathscr{R}_k(G)$ (assuming connectedness of G) is a necessary but not sufficient condition for the right definition. The reason is that for any imperfect field k, there are smooth connected affine k-groups G such that $\mathscr{R}_k(G) = 1$ and $\mathscr{D}(G) \neq G$ (see Example 11.2.1). We will study pseudo-semisimplicity in §11.2.

Over a perfect field k, pseudo-reductivity coincides with reductivity for smooth connected affine k-groups G because $\mathscr{R}_{u,k}(G)_{\overline{k}} = \mathscr{R}_u(G_{\overline{k}})$ if k is perfect. In contrast, for any imperfect field k there are many pseudo-reductive k-groups G that are not reductive. We will see interesting examples in §1.3, but we wish to begin with an elementary commutative example of a non-reductive pseudo-reductive group to convey a feeling for pseudo-reductivity.

Our example will rest on the Weil restriction functor $\mathrm{R}_{k'/k}$ relative to a finite extension of fields k'/k that is not separable. Weil restriction is a familiar operation in the separable case (where it is analogous to the operation of viewing a complex manifold as a real manifold of twice the dimension), but it is not widely known for general finite extensions k'/k. We will use it extensively with inseparable extensions, so we first review a few basic facts about this functor, referring the reader to §§A.5 and A.7 for a more thorough discussion.

If $B \to B'$ is a finite flat map between noetherian rings (e.g., a finite extension of fields) and if X' is a quasi-projective B'-scheme, then the *Weil restriction* $X = \mathrm{R}_{B'/B}(X')$ is a separated B-scheme of finite type (even quasi-projective) characterized by the functorial property $X(A) = X'(B' \otimes_B A)$ for all B-algebras A. The discussion of Weil restriction in §A.5 treats the general algebro-geometric setting as well as the special case of group schemes. For any smooth connected affine k'-group G', the Weil restriction $G = \mathrm{R}_{k'/k}(G')$ is an affine k-group scheme of finite type characterized by the property

$$G(A) = G'(k' \otimes_k A)$$

functorially in k-algebras A. By Proposition A.5.11, the k-group $G = \mathrm{R}_{k'/k}(G')$ is smooth and connected. Its dimension is $[k' : k] \dim G'$, since $\mathrm{Lie}(G)$ is the Lie algebra over k underlying $\mathrm{Lie}(G')$ (Corollary A.7.6).

Remark 1.1.2 Beware that if k'/k is inseparable then the "pushforward" functor $\mathrm{R}_{k'/k}$ from affine k'-schemes to affine k-schemes has some surprising properties in comparison with the more familiar separable case. This will be illustrated in Examples 1.3.2 and 1.3.5.

Example 1.1.3 Now we give our first example of a non-reductive pseudo-reductive group. Let k be an imperfect field of characteristic $p > 0$, and let k'/k be a purely inseparable finite extension of degree $p^n > 1$. Consider the smooth k-group $G = \mathrm{R}_{k'/k}(\mathrm{GL}_1)$ of dimension p^n. (Loosely speaking, G is "$k'^{\times}$ viewed as a k-group".) This canonically contains GL_1 as a k-subgroup.

The smooth connected quotient G/GL_1 of dimension $p^n - 1 > 0$ is killed by the p^n-power map since $k'^{p^n} \subseteq k$, so it is unipotent and hence G is not reductive. (More explicitly, $G_{\overline{k}}$ is the algebraic group of units of $k' \otimes_k \overline{k}$, in which the subgroup of 1-units is a codimension-1 unipotent radical.) However, $G(k_s) = (k' \otimes_k k_s)^\times$ has no nontrivial p-torsion since $k' \otimes_k k_s$ is a field of characteristic p, so the smooth connected commutative unipotent k-subgroup $\mathscr{R}_{u,k}(G)$ must be trivial. That is, G is pseudo-reductive over k. The k-group G naturally occurs (up to conjugacy) inside of GL_{p^n}. A non-commutative generalization of this example is given in Example 1.1.11.

Remark 1.1.4 Let G be a smooth connected affine group over a field k. It is elementary to prove that $G/\mathscr{R}_{u,k}(G)$ is pseudo-reductive over k. In particular, G is canonically an extension of the pseudo-reductive k-group $G/\mathscr{R}_{u,k}(G)$ by the smooth connected unipotent k-group $\mathscr{R}_{u,k}(G)$. To use this extension structure to reduce problems for G to problems for pseudo-reductive k-groups, we need to know something about the structure of smooth connected unipotent k-groups.

A general smooth connected unipotent k-group U is hard to describe (especially when $\mathrm{char}(k) > 0$), but any such U admits a characteristic central composition series whose successive quotients are k-forms V_i of vector groups $\mathbf{G}_\mathrm{a}^{n_i}$ (by [SGA3, XVII, 4.1.1(iii)], or by Corollary B.2.7 and Corollary B.3.3 when $\mathrm{char}(k) > 0$). We have $V_i \simeq \mathbf{G}_\mathrm{a}^{n_i}$ when k is perfect (by [SGA3, XVII, 4.1.5], or by Corollary B.2.7 when $\mathrm{char}(k) > 0$) but V_i is mysterious in general if k is imperfect and $U \neq 1$. Nonetheless, such V_i are commutative and p-torsion when $\mathrm{char}(k) = p > 0$, so there are concrete hypersurface models for V_i over infinite k (Proposition B.1.13). This often makes $\mathscr{R}_{u,k}(G)$ tractable enough so that problems for general G can be reduced to the pseudo-reductive case.

There are interesting analogies between pseudo-reductive k-groups and connected reductive k-groups. For example, we will prove by elementary methods that the Cartan k-subgroups (i.e., centralizers of maximal k-tori) in pseudo-reductive k-groups are always commutative and pseudo-reductive; see Proposition 1.2.4. We do not know an elementary proof of the related fact that, when $\mathrm{char}(k) \neq 2$, a pseudo-reductive k-group is reductive if and only if its Cartan k-subgroups are tori. Our proof of this result (Theorem 11.1.1) rests on our main structure theorem for pseudo-reductive groups. It seems unlikely that an alternative proof can be found (bypassing the structure theorem), since the avoidance of characteristic 2 is essential. Indeed, we will show in Example 11.1.2 that over every imperfect field k with characteristic 2 there are non-reductive pseudo-reductive k-groups G whose Cartan k-subgroups are tori.

In view of the commutativity and pseudo-reductivity of Cartan subgroups in pseudo-reductive groups, a basic reason that the structure of pseudo-reductive groups is more difficult to understand than that of connected reductive groups is that we do not understand the structure of the commutative objects very well. For example, whereas k-tori are unirational over k for any field k, in Example 11.3.1 we will exhibit commutative pseudo-reductive k-groups that are not unirational over k, where k is any imperfect field.

Example 1.1.5 The property of pseudo-reductivity over arbitrary imperfect fields exhibits some behavior that is not at all like the more familiar connected reductive case. We mention two such examples now, the second of which is more interesting than the first.

(i) Pseudo-reductivity can be destroyed by an inseparable ground field extension, such as scalar extension to the perfect closure. This is not a surprise, since by definition it happens for every non-reductive pseudo-reductive group (of which we shall give many examples).

In contrast, by Proposition 1.1.9(1) below, pseudo-reductivity is insensitive to separable extension of the ground field. This will often be used when passing to a separably closed ground field in a proof. An arithmetically interesting example of a non-algebraic separable extension is k_v/k for a global function field k and a place v of k. Thus, a group scheme G over a global function field k is pseudo-reductive over k if and only if G_{k_v} is pseudo-reductive over k_v.

(ii) The second, and more surprising, deviation of pseudo-reductivity from reductivity is that it is generally not inherited by quotients. For example, consider a non-reductive pseudo-reductive group G over a field k with $\mathrm{char}(k) = p > 0$. By Galois descent, the subgroup $\mathscr{R}_u(G_{\overline{k}})$ in $G_{\overline{k}}$ is defined over the perfect closure k_{p} of k. Since k_{p} is perfect, the k_{p}-descent of $\mathscr{R}_u(G_{\overline{k}})$ has a composition series over k_{p} whose successive quotients are k_{p}-isomorphic to $\mathbf{G}_{\mathrm{a}}$ (Proposition A.1.4). But k_{p} is the direct limit of subfields $k^{p^{-n}}$, so if n is sufficiently large then $\mathscr{R}_u(G_{\overline{k}})$ descends to a nontrivial unipotent subgroup U of $G_{k^{p^{-n}}}$ such that U is $k^{p^{-n}}$-split. The p^n-power map identifies $k^{p^{-n}}$ with k carrying the inclusion $k \hookrightarrow k^{p^{-n}}$ inside of $\overline{k}$ over to the p^n-power map of k, so $G_{k^{p^{-n}}}$ is thereby identified with the target of the n-fold Frobenius isogeny $F_{G/k,n} : G \to G^{(p^n)}$ as in Definition A.3.3.

Hence, *every* non-reductive pseudo-reductive k-group G is a (purely inseparable) isogenous cover of a smooth connected affine k-group G' such that the smooth connected unipotent k-group $\mathscr{R}_{u,k}(G')$ is nontrivial and is a successive extension of copies of $\mathbf{G}_{\mathrm{a}}$ over k. Such examples are of limited interest since the structure of $\ker(G \twoheadrightarrow G')$ is not easily understood. But there are more interesting examples of pseudo-reductive k-groups G admitting quotients G/H that are not pseudo-reductive. This was illustrated in Example 1.1.3 in the commutative case with H a torus, and we will see examples using (infinitesimal) finite central multiplicative H (Example 1.3.2 for commutative

G and Example 1.3.5 for perfect G) as well as very surprising examples using perfect smooth connected H and perfect G (Example 1.6.4). These examples exist over any imperfect field.

An important ingredient in our work is the minimal intermediate field in $\overline{k}/k$ over which the geometric unipotent radical $\mathscr{R}_u(G_{\overline{k}})$ inside of $G_{\overline{k}}$ is defined. This is a special case of a useful general notion:

Definition 1.1.6 Let X be a scheme over a field k, K/k an extension field, and $Z \subseteq X_K$ a closed subscheme. For an intermediate field k'/k, Z is *defined over* k' if Z descends (necessarily uniquely) to a closed subscheme of $X_{k'}$. The (minimal) *field of definition* (over k) for $Z \subseteq X_K$ is the unique such k'/k that is contained in all others. (For a thorough discussion of the general existence of such a field, we refer the reader to [EGA, IV_2, §4.8], especially [EGA, IV_2, 4.8.11].)

Sometimes for emphasis we append the word "minimal" when speaking of fields of definition, but in fact minimality is always implicit.

Remark 1.1.7 The mechanism underlying the existence of (minimal) fields of definition of closed subschemes is a fact from linear algebra, as follows. Consider an extension of fields K/k, a k-vector space V, and a K-subspace $W \subseteq V_K$. Among all subfields $F \subseteq K$ over k such that W arises by scalar extension from an F-subspace of V_F, we claim that there is one such F that is contained in all others.

To see that such a minimal field exists, choose a k-basis $\{e_i\}_{i \in I}$ of V and a subset $B = \{e_i\}_{i \in J}$ projecting bijectively to a K-basis of $(K \otimes_k V)/W$. Then F is generated over k by the coefficients of the vectors $\{e_i \bmod W\}_{i \notin J}$ relative to the K-basis $B \bmod W$. We call F the *field of definition* over k for the K-subspace W.

The following lemma records the simple behavior of fields of definition under a ground field extension.

Lemma 1.1.8 *Let k be a field, $L'/k'/k$ a tower of extensions, and L an intermediate field in L'/k. Let X be a k-scheme and Y a closed subscheme of $X_{k'}$, and $F \subseteq k'$ the minimal field of definition over k for Y. The minimal field of definition over L for the closed subscheme $Y_{L'} := Y \otimes_{k'} L'$ in $X_{k'} \otimes_{k'} L' = X_L \otimes_L L'$ is the compositum FL inside of L'.*

In particular, if $L' = Lk'$ and $Y \subseteq X_{k'}$ has minimal field of definition over k equal to k' then the closed subscheme $Y_{L'} \subseteq X_{L'}$ has minimal field of definition over L equal to L'.

The most useful instance of this lemma for our purposes will be the case when k'/k is purely inseparable and L/k is separable (with $F = k'$) but no ambient L' is provided. In such cases $L \otimes_k k'$ is a field (as we see by passage to the limit from the case when k'/k is finite) and so we can take $L' := L \otimes_k k'$.

Proof We first reduce to the affine case, using a characterization of the minimal field of definition F/k that is better suited to Zariski-localization: if $\{U_i\}$ is an open cover of X (e.g., affine opens) then a subfield K of k' over k contains F if and only if the closed subscheme $Y \cap (U_i)_{k'} \subseteq (U_i)_{k'}$ is defined over K for all i. (This follows from the uniqueness of descent of a closed subscheme to a minimal field of definition.) In other words, F is the compositum of the fields of definition over k for each $Y \cap (U_i)_{k'} \subseteq (U_i)_{k'}$. Since the field of definition for $Y_{L'} \subseteq X_{L'}$ over L is the corresponding compositum for the closed subschemes $Y_{L'} \cap (U_i)_{L'} \subseteq (U_i)_{L'}$, if we can handle each U_i separately then passage to composite fields gives the global result. Thus, we now may and do assume X is affine, say $X = \operatorname{Spec} A$ for a k-algebra A.

The ideal of Y in $A_{k'}$ has the form $k' \otimes_F J$ for an ideal J in A_F, and F is the smallest intermediate extension in k'/k to which $k' \otimes_F J$ descends as an ideal. But descent as an ideal is equivalent to descent as a vector subspace (as the property of a subspace of an algebra being an ideal can be checked after an extension of the ground field), so F/k is the minimal field of definition over k for the k'-subspace $V = k' \otimes_F J$ inside of $A_{k'}$.

The proof of existence of minimal fields of definition for subspaces of a vector space (as in Remark 1.1.7) implies that F is also the minimal field of definition over k for the L'-subspace $L' \otimes_{k'} V \subseteq A_{L'}$. Let K be the minimal field of definition over L for this L'-subspace. Our aim is to prove that $K = FL$. Since the L'-subspace $L' \otimes_{k'} V$ is defined over FL (as it even descends to the F-subspace J in A_F), we have $L \subseteq K \subseteq FL$. The problem is therefore to prove that $F \subseteq K$. But $L' \otimes_{k'} V$ does descend to a K-subspace of A_K by definition of K/L, so by the minimality property for F relative to L'/k we get $F \subseteq K$. □

Proposition 1.1.9 *Let K/k be a separable extension of fields, and G a smooth connected affine k-group.*

(1) *Inside of G_K we have $\mathscr{R}_{u,k}(G)_K = \mathscr{R}_{u,K}(G_K)$. In particular, G is pseudo-reductive over k if and only if G_K is pseudo-reductive over K.*
(2) *Choose a k-embedding $\overline{k} \hookrightarrow \overline{K}$ of algebraic closures. The field of definition E_k over k for $\mathscr{R}_u(G_{\overline{k}}) \subseteq G_{\overline{k}}$ is a finite purely inseparable extension and $K \otimes_k E_k = E_K$ inside of $\overline{K}$.*

The analogue of Proposition 1.1.9(1) for the maximal k-split smooth connected unipotent normal k-subgroup $\mathscr{R}_{us,k}(G)$ lies deeper, and is given in Corollary B.3.5.

Proof First we prove (1). By Galois descent, if k'/k is a Galois extension then $\mathscr{R}_{u,k'}(G_{k'})$ descends to a smooth connected unipotent normal k-subgroup of G. Such a descent is contained in $\mathscr{R}_{u,k}(G)$, so this settles the case where K/k is Galois. Applying this to separable closures k_s/k and K_s/K (with K_s chosen to contain k_s over k), we may assume that k is separably closed. Since $\mathscr{R}_{u,k}(G)_K \subseteq \mathscr{R}_{u,K}(G_K)$ in general, to prove the equality as in (1) it suffices to prove an inequality of dimensions in the opposite direction when $k = k_s$.

More generally, if U is a smooth connected unipotent normal K-subgroup of G_K, say with $d = \dim U \geqslant 0$, we shall apply a specialization argument to U to construct a smooth connected unipotent normal k-subgroup of G with dimension d when $k = k_s$. By expressing K as a direct limit of its finitely generated subfields over k, there is such a subfield F for which U descends to an F-subgroup of G_F that is necessarily smooth, connected, unipotent, and normal in G_F. Thus, upon renaming F as K we may assume that K/k is finitely generated. Separability of K/k then allows us to write $K = \mathrm{Frac}(A)$ for a k-smooth domain A.

The normality of U in G can be expressed as the fact that the map $G_K \times U \to G_K$ defined by $(g,u) \mapsto gug^{-1}$ factors through U, and the unipotence can be expressed as the fact that for some finite extension K'/K the K'-group $U_{K'}$ admits a composition series whose successive quotients are $\mathbf{G}_\mathrm{a}$. Since K is the direct limit of its k-smooth subalgebras $A[1/a]$ for $a \in A - \{0\}$, by replacing A with a suitable such $A[1/a]$ we may arrange that $U = \mathscr{U}_K$ for a closed subscheme $\mathscr{U} \subseteq G_A$.

Let A' be an A-finite domain such that $A'_K = K'$. By "spreading out" of properties of $U = \mathscr{U}_K$ and $\mathrm{Spec}\, K' = (\mathrm{Spec}\, A')_K$ from the fiber over the generic point $\mathrm{Spec}\, K$ of $\mathrm{Spec}\, A$, upon replacing A with a further localization $A[1/a]$ we can arrange that $\mathscr{U}$ is an A-smooth normal A-subgroup of G_A with (geometrically) connected fibers of dimension d, and that there is a finite faithfully flat extension $A \to A'$ (with generic fiber K'/K) such that $\mathscr{U}_{A'}$ admits a composition series by A'-smooth normal closed A'-subgroups with successive quotients isomorphic to $\mathbf{G}_\mathrm{a}$ as an A'-group. Hence, all fibers of $\mathscr{U}$ over $\mathrm{Spec}\, A$ are unipotent. Since A is k-smooth, if $k = k_s$ then there exist k-points of $\mathrm{Spec}\, A$. The fiber of $\mathscr{U}$ over such a point is a smooth connected unipotent normal k-subgroup of G with dimension d. This proves (1).

For (2), we apply Lemma 1.1.8 to the tower $\overline{K}/\overline{k}/k$, the intermediate field K, and the k-scheme $X = G$ and $\overline{k}$-scheme $Y = \mathscr{R}_u(G_{\overline{k}})$ (for which

$Y_{\overline{K}} = \mathscr{R}_u(G_{\overline{K}})$). The assertion of the lemma in this case is that the field of definition E_K over K for $\mathscr{R}_u(G_{\overline{K}})$ is the composite field E_kK inside of $\overline{K}$. Provided that E_k/k is purely inseparable, the ring $E_k \otimes_k K$ is a field (since K/k is separable) and hence the surjective map $E_k \otimes_k K \to E_kK$ is an isomorphism. It therefore remains to prove that E_k/k is a purely inseparable extension of finite degree. By Galois descent, $\mathscr{R}_u(G_{\overline{k}})$ is defined over the perfect closure k_{p} of k, which is to say that it descends to a k_{p}-subgroup $U \subseteq G_{k_{\mathrm{p}}}$. Since the ideal of U in the coordinate ring of $G_{k_{\mathrm{p}}}$ is finitely generated, and every finite subset of k_{p} is contained in a subextension of finite degree over k, we can descend U to a k'-subgroup of $G_{k'}$ for some finite purely inseparable extension k'/k. Necessarily $E_k \subseteq k'$, so we are done. □

To make interesting non-commutative pseudo-reductive groups, we need to study the Weil restriction of scalars functor $\mathrm{R}_{k'/k}$ applied to connected reductive groups over finite extension fields k'/k. For later purposes, it is convenient to work more generally with nonzero finite reduced k-algebras k', which is to say $k' = \prod k'_i$ for a non-empty finite collection $\{k'_i\}$ of fields of *finite* degree over k. This generality is better suited to Galois descent (which we use very often): the functor $k_s \otimes_k (\cdot)$ carries nonzero finite reduced k-algebras to nonzero finite reduced k_s-algebras, but generally does not carry fields to fields. For such general $k' = \prod k'_i$, a quasi-projective k'-scheme X' is precisely $\coprod X'_i$ for quasi-projective k'_i-schemes X'_i, and $\mathrm{R}_{k'/k}(X') = \prod \mathrm{R}_{k'_i/k}(X'_i)$.

Proposition 1.1.10 *Let k be a field, k' a nonzero finite reduced k-algebra, and G' a k'-group whose fibers over* Spec k' *are connected reductive (or more generally, pseudo-reductive). The smooth connected affine k-group $G = \mathrm{R}_{k'/k}(G')$ is pseudo-reductive.*

Proof By Proposition A.5.2(4) and Proposition A.5.9, the affine k-group G is smooth and connected. Let $\iota : U \hookrightarrow G$ be a smooth connected unipotent normal k-subgroup of G. We must show that ι is the trivial map (i.e., $U = 1$). It is equivalent to prove triviality of the k'-map $\iota' : U_{k'} \to G'$ that corresponds to ι via the universal property of Weil restriction. The image $H' \subseteq G'$ of ι' is a smooth unipotent k'-subgroup of G' with connected fibers over Spec k', and we claim that it is normal. Once this is proved, then reductivity (or even just pseudo-reductivity) of the fibers of G' over Spec k' implies that ι' is trivial, so G is indeed pseudo-reductive over k.

To verify the normality of H' in G', we first observe that (by construction) ι' is the restriction to $U_{k'}$ of the canonical map $q : G_{k'} \to G'$ that corresponds to the identity map of $G = \mathrm{R}_{k'/k}(G')$ under the universal property of Weil restriction. Smoothness of G' implies that q is surjective

(Proposition A.5.11(1)), so normality of $U_{k'}$ in $G_{k'}$ implies that the k'-group $H' = q(U_{k'})$ is normal in G'. □

The k-group $G = \mathrm{R}_{k'/k}(G')$ for a k'-group G' with connected reductive fibers is especially interesting when G' has a nontrivial fiber over some factor field k'_i of k' that is not separable over k, because then the pseudo-reductive k-group G is *not* reductive. This follows from Example 1.6.1 (using Example 1.1.12). Here is a concrete instance of this fact.

Example 1.1.11 Consider a finite extension of fields k'/k and a central simple k'-algebra A' with $\dim_{k'} A' = n^2$. The k'-group G' of units of A' is a connected reductive k'-group (a k'-form of GL_n). The k-group $G := \mathrm{R}_{k'/k}(G')$ of units of the finite-dimensional simple (but non-central when $k' \neq k$) k-algebra underlying A' is pseudo-reductive, by Proposition 1.1.10. When k'/k is not separable, the unit group G is not reductive over k; this is a special case of Example 1.6.1.

Example 1.1.12 Let k'/k be a finite extension of fields and G' a connected reductive k'-group. Let k_1/k be the maximal separable subextension in k'/k (so k'/k_1 is purely inseparable), and k_s/k a separable closure. Since $\mathrm{R}_{k'/k} = \mathrm{R}_{k_1/k} \circ \mathrm{R}_{k'/k_1}$, the decomposition of $k_1 \otimes_k k_s$ into a finite product of copies of k_s gives rise to the decomposition

$$\mathrm{R}_{k'/k}(G')_{k_s} \simeq \prod_{\sigma} (\mathrm{R}_{k'/k_1}(G') \otimes_{k_1,\sigma} k_s), \tag{1.1.1}$$

with σ ranging through the k-embeddings of k_1 into k_s. This often reduces general problems for $\mathrm{R}_{k'/k}(G')$ to the special case of purely inseparable extension fields.

1.2 Elementary properties of pseudo-reductive groups

We now prove some general properties of pseudo-reductive groups that follow from the definition of pseudo-reductivity (and from general facts in the theory of reductive groups). To begin, we record a lemma that is extremely useful in many arguments with pseudo-reductivity.

Lemma 1.2.1 *Let G be a pseudo-reductive k-group, and X a closed subscheme of G such that $X_{\overline{k}} \subseteq \mathscr{R}_u(G_{\overline{k}})$ and $X_{\overline{k}}$ is reduced and irreducible. Then X is a k-rational torsion point in the center of G, trivial if* $\mathrm{char}(k) = 0$ *and with p-power order if* $\mathrm{char}(k) = p > 0$. *In particular, if X contains the identity then it is the identity point.*

Note that the center of G contains only finitely many k-rational p-power torsion points when $\mathrm{char}(k) = p > 0$. Indeed, the Zariski closure of the group of such points is a smooth unipotent central k-subgroup of G, so its identity component is normal in G and thus is trivial.

Proof We may and do assume that k is separably closed, so $X(k)$ is non-empty. To prove the weaker claim that X is a single k-rational point we may translate by x^{-1} for some $x \in X(k)$ to arrange that $X(k)$ contains the identity. Let $H \subseteq G$ be the smooth *connected* k-subgroup such that $H(\overline{k})$ is generated by $X(\overline{k})$. (The existence of such an H is [Bo2, I, 2.1, 2.2].)

Clearly $H_{\overline{k}} \subseteq \mathscr{R}_u(G_{\overline{k}})$. Likewise, the collection of $G(k)$-conjugates of H generates a smooth connected k-subgroup $U \subseteq G$, and U is normal in G because $G(k)$ is Zariski-dense in G (as k is separably closed). But $X_{\overline{k}} \subseteq \mathscr{R}_u(G_{\overline{k}})$, so $U \subseteq \mathscr{R}_u(G_{\overline{k}})$. In particular, U is unipotent. Since G is pseudo-reductive we conclude that $U = 1$, so X is the identity point as well.

In the original situation (prior to translation), we conclude that the geometrically reduced and geometrically irreducible k-scheme X is a single k-rational point. Obviously X must be the identity point if G is reductive, which includes the case of characteristic 0, so it remains to check that if $\mathrm{char}(k) = p > 0$ then $\Gamma := G(k) \cap \mathscr{R}_u(G_{\overline{k}})$ is finite and central with p-power order.

The Zariski closure of Γ in G is a smooth k-subgroup of G each of whose connected components is geometrically irreducible (as $k = k_s$). Thus, each of these components is a single k-rational point, by what we have already shown. It follows that Γ is finite. But Γ is stable by $G(k)$-conjugation, so it is normal in G since G is smooth and k is separably closed. By connectivity of G it follows that the finite constant group Γ is central in G. Since Γ is unipotent, it has p-power order. □

A useful application of Lemma 1.2.1 is:

Proposition 1.2.2 *Let $f, f' : G \rightrightarrows H$ be a pair of k-homomorphisms between smooth connected affine k-groups, with H pseudo-reductive. Assume $f_{\overline{k}}$ and $f'_{\overline{k}}$ carry $\mathscr{R}_u(G_{\overline{k}})$ into $\mathscr{R}_u(H_{\overline{k}})$, so they induce maps $f_{\overline{k}}^{\mathrm{red}}, f_{\overline{k}}^{\prime\mathrm{red}} : G_{\overline{k}}^{\mathrm{red}} \rightrightarrows H_{\overline{k}}^{\mathrm{red}}$ between maximal reductive quotients over $\overline{k}$. If $f_{\overline{k}}^{\mathrm{red}} = f_{\overline{k}}^{\prime\mathrm{red}}$ then $f = f'$.*

Any k-homomorphism f from a smooth connected affine k-group G to a commutative pseudo-reductive k-group is determined by its restriction to a maximal k-torus of G, and any action by a smooth connected k-group on a commutative pseudo-reductive k-group is trivial.

Proof Consider the k-scheme map $\phi : G \to H$ defined by $g \mapsto f'(g)f(g)^{-1}$. Over $\overline{k}$ this lands in $\mathscr{R}_u(H_{\overline{k}})$. For the schematic closure X of the image of ϕ over

k, $X_{\overline{k}}$ is the schematic closure of the image of $\phi_{\overline{k}}$. Since $1 \in X(k)$, Lemma 1.2.1 implies that $X = 1$, so $f = f'$.

For a commutative pseudo-reductive group C with maximal k-torus T, the natural map $T_{\overline{k}} \to C_{\overline{k}}^{\rm red}$ is an isomorphism. Thus, when H is commutative (so f factors through $G/\mathscr{D}(G)$, and a maximal k-torus in G maps onto one in $G/\mathscr{D}(G)$) then f is indeed determined by its restriction to a maximal k-torus in G.

If a smooth connected k-group $\mathscr{G}$ acts on a commutative pseudo-reductive k-group C then the action preserves the maximal k-torus T (as we may check at the level of rational points over k_s). Since the automorphism functor $\underline{\rm Aut}(T)$ of T is represented by an étale k-group and $\mathscr{G}$ is connected, the homomorphism $\mathscr{G} \to \underline{\rm Aut}(T)$ classifying the $\mathscr{G}$-action on T is trivial, so $\mathscr{G}$ acts trivially on T. Thus, the action by $\mathscr{G}(k_s)$ on C_{k_s} is trivial (due to triviality on T_{k_s}), so $\mathscr{G}$ acts trivially on C. □

The following result is an analogue of two well-known properties of connected reductive groups:

Proposition 1.2.3 *Let G be a pseudo-reductive k-group such that it either has a unique maximal torus over k_s or is solvable. The k-group G is commutative. In particular, a non-commutative pseudo-reductive k-group is non-solvable.*

Proof First assume that there is a unique maximal torus over k_s, so it is defined over k by Galois descent. Call this torus T. The uniqueness of the maximal torus implies that G is nilpotent [Bo2, 11.5(3)], so G is solvable.

It remains to prove that a solvable pseudo-reductive k-group G is commutative. The derived group $\mathscr{D}(G) \subseteq G$ is a smooth connected k-subgroup of G, and by the structure of smooth connected solvable groups [Bo2, 10.6(1)] this derived group is unipotent. Thus, $\mathscr{D}(G) = 1$ by the pseudo-reductivity of G over k, so G is commutative. □

In a connected reductive group, the centralizer of a torus is connected reductive and the Cartan subgroups are tori. We have the following useful extension to the pseudo-reductive case.

Proposition 1.2.4 *If G is pseudo-reductive over k then for any k-torus T in G the centralizer $Z_G(T)$ is a pseudo-reductive k-group. The k-group $Z_G(T)$ is commutative if T is a maximal k-torus in G (so Cartan k-subgroups of G are commutative and pseudo-reductive).*

Proof Note that $T_{\overline{k}}$ maps isomorphically onto its image in $G_{\overline{k}}^{\rm red}$ (Example A.1.7). Since G is a smooth connected affine group over a field k, the scheme-theoretic centralizer $Z_G(T)$ for the conjugation action by the k-torus T is a smooth connected k-group (Proposition A.2.5). Also, the image of $Z_G(T)_{\overline{k}}$ in the maximal reductive quotient $G_{\overline{k}}^{\rm red} = G_{\overline{k}}/\mathscr{R}_u(G_{\overline{k}})$ is $Z_{G_{\overline{k}}^{\rm red}}(T_{\overline{k}})$, due to the compatibility of torus centralizers with the formation of quotients of smooth connected affine groups (Proposition A.2.8).

The centralizer $Z_{G_{\overline{k}}^{\rm red}}(T_{\overline{k}})$ is a connected reductive group [Bo2, 8.18, 13.19]. Hence, any smooth connected normal unipotent k-subgroup $U \subseteq Z_G(T)$ satisfies $U_{\overline{k}} \subseteq \mathscr{R}_u(G_{\overline{k}})$. This forces $U = 1$, by Lemma 1.2.1. In other words, $Z_G(T)$ is pseudo-reductive over k.

Now assume that T is a maximal k-torus. The centrality of T in the pseudo-reductive k-group $Z_G(T)$ forces T_{k_s} to be the only maximal k_s-torus in $Z_G(T)_{k_s}$. Thus, by Proposition 1.2.3, $Z_G(T)$ is commutative. □

As a consequence of the commutativity of Cartan k-subgroups in pseudo-reductive k-groups, we get an analogue for pseudo-reductive groups of the fact that the derived group of a connected reductive group is its own derived group. First we need a general lemma concerning tori in pseudo-reductive groups.

Lemma 1.2.5 *Let M and N be smooth connected k-subgroups of a pseudo-reductive k-group G, T a maximal k-torus in G, and Z the maximal k-subtorus of T which commutes with N.*

(i) *If the images of $M_{\overline{k}}$ and $N_{\overline{k}}$ in $G_{\overline{k}}^{\rm red}$ commute then M and N commute.*
(ii) *Assume N is normal in G. For the maximal k-torus $S := T \cap \mathscr{D}(N)$ in $\mathscr{D}(N)$ there is an almost direct product decomposition $T = Z \cdot S$, and $Z_G(T) \cap N = Z_N(S)$ (so $Z_G(T) \cap N$ is a Cartan k-subgroup of N).*
(iii) *The correspondence $T \mapsto T \cap \mathscr{D}(G)$ is a bijection between the sets of maximal k-tori of G and $\mathscr{D}(G)$. In general, any maximal k-torus T is an almost direct product of $T \cap \mathscr{D}(G)$ and the maximal central k-torus in G.*

Note that S in (ii) is a maximal k-torus of $\mathscr{D}(N)$ (in particular, smooth and connected) by Corollary A.2.7 since $\mathscr{D}(N)$ is normal in G when N is.

Proof By Galois descent, the formation of Z commutes with extension of the ground field to k_s. Thus, we may and do assume $k = k_s$. Let $G' = G_{\overline{k}}^{\rm red}$, so the quotient map $\pi : G_{\overline{k}} \twoheadrightarrow G'$ carries $T_{\overline{k}}$ isomorphically onto a maximal $\overline{k}$-torus T' in G'.

The commutator subgroup (M,N) is a smooth connected k-subgroup of G, and since $M' := \pi(M_{\overline{k}})$ commutes with $N' := \pi(N_{\overline{k}})$ it follows that $(M,N)_{\overline{k}} \subseteq \mathscr{R}_u(G_{\overline{k}})$. Thus, by Lemma 1.2.1, we have $(M,N) = 1$. This proves (i).

To prove (ii), now assume N is normal in G, so N' is a smooth connected normal subgroup of G'. In particular, N' is reductive and so $T'N'$ is a connected reductive subgroup of G'. Let $Z' \subseteq T'$ be the maximal central torus in $T'N'$ and $S' := T' \cap \mathscr{D}(T'N') = T' \cap \mathscr{D}(N')$, so S' is a maximal torus in $\mathscr{D}(N')$ and $T' = Z' \cdot S'$ as an almost direct product. Since $\pi : T_{\overline{k}} \to T'$ is an isomorphism and $k = k_s$, there is an evident inclusion-preserving bijective correspondence between subtori of T' and k-subtori of T. Let S and Z be the k-subtori of T corresponding to S' and Z' respectively. Thus, $T = Z \cdot S$ as an almost direct product. Applying (i) with $M = Z$ then gives that Z commutes with N, and clearly $T \cap \mathscr{D}(N) \subseteq S$. But $T \cap \mathscr{D}(N)$ is a maximal k-torus in $\mathscr{D}(N)$ (Corollary A.2.7), so $T \cap \mathscr{D}(N) = S$. Since Z commutes with N and we have $T = Z \cdot S$, clearly $Z_G(T) \cap N = Z_N(S)$. This proves (ii).

Finally, taking $N = G$ (so Z is the maximal central k-torus in G), (iii) is an immediate consequence of (ii). □

Proposition 1.2.6 *Let G be a smooth connected affine group over a field k. For any Cartan k-subgroup C, we have $G = C \cdot \mathscr{D}(G)$. If G is pseudo-reductive over k then $\mathscr{D}(G)$ is perfect (i.e., $\mathscr{D}(G) = \mathscr{D}(\mathscr{D}(G))$).*

Proof We first carry out some general considerations with an arbitrary smooth connected affine k-group H and k-torus S in H. If $H \twoheadrightarrow H'$ is a quotient map of k-groups and S' denotes the k-torus image of S in H' then $Z_H(S)$ maps onto $Z_{H'}(S')$ (Proposition A.2.8). In particular, if H' is commutative then $Z_H(S)$ maps onto H'. Taking $H' = H/\mathscr{D}(H)$, we conclude that

$$H = Z_H(S) \cdot \mathscr{D}(H).$$

Taking S to be a maximal k-torus gives that a smooth connected affine k-group is generated by its derived group and any single Cartan k-subgroup. Applying this general result to the smooth connected k-group $S \cdot \mathscr{D}(H)$ gives that

$$S \cdot \mathscr{D}(H) = Z_{S \cdot \mathscr{D}(H)}(S) \cdot \mathscr{D}(S \cdot \mathscr{D}(H)).$$

Now let $H = G$ be pseudo-reductive over k and choose $S = T$ to be a maximal k-torus, so

$$G = Z_G(T) \cdot \mathscr{D}(G) = Z_G(T) \cdot (T \cdot \mathscr{D}(G)) = Z_G(T) \cdot \mathscr{D}(T \cdot \mathscr{D}(G)).$$

Let $\mathscr{D}^2(G) = \mathscr{D}(\mathscr{D}(G))$. By Lemma 1.2.5(iii), $\mathscr{D}(T \cdot \mathscr{D}(G)) = \mathscr{D}^2(G)$. Thus, $G/\mathscr{D}^2(G)$ is a quotient of the commutative $Z_G(T)$, so $\mathscr{D}(G) \subseteq \mathscr{D}^2(G)$. The reverse containment is clear, so we are done. □

Proposition 1.2.7 *Let G be a pseudo-reductive group over a field k, and H a smooth connected normal k-subgroup of G and N a smooth connected normal k-subgroup of H. If N is perfect then it is normal in G.*

In Remark 3.1.10 we will show that the perfectness hypothesis on N in Proposition 1.2.7 can be dropped (by applying Proposition 1.2.7 to the perfect $\mathscr{D}(N)$ and using Lemma 1.2.5 and the description of perfect smooth connected normal k-subgroups of G provided in §3.1).

Proof We may assume that k is separably closed, so $G(k)$ is Zariski-dense in G. Thus, the $G(k)$-conjugates of N generate a smooth connected normal k-subgroup of G that is contained in H. We may replace H with this subgroup so that the $G(k)$-conjugates of N generate H. Since N is perfect, H is also perfect. In this case we will prove that $N = H$.

Consider the smooth connected images $\mathscr{N}$ of $N_{\overline{k}}$ and $\mathscr{H}$ of $H_{\overline{k}}$ in $G_{\overline{k}}^{\rm red}$. Clearly $\mathscr{N}$ is normal in $\mathscr{H}$, which in turn is normal in $G_{\overline{k}}^{\rm red}$. The classification of smooth connected normal $\overline{k}$-subgroups in connected reductive $\overline{k}$-groups implies that $\mathscr{N}$ is normal in $G_{\overline{k}}^{\rm red}$. But $\mathscr{H}$ is generated by conjugates of $\mathscr{N}$, so $\mathscr{H} = \mathscr{N}$. This implies that $H_{\overline{k}}/N_{\overline{k}}$ is unipotent, so H/N is solvable. But H/N is perfect (since H is), so $H/N = 1$. □

1.3 Preparations for the standard construction

Many of the difficulties in the classification of pseudo-reductive groups are seen in the "simplest" interesting case: for an arbitrary field k, consider non-reductive pseudo-reductive k-groups G such that the maximal reductive quotient

$$G_{\overline{k}}^{\rm red} = G_{\overline{k}}/\mathscr{R}_u(G_{\overline{k}})$$

of $G_{\overline{k}}$ is a simple semisimple group (so G is non-solvable and $G_{\overline{k}}^{\rm red}$ coincides with the maximal semisimple quotient $G_{\overline{k}}^{\rm ss}$). For later purposes, we now consider any smooth connected affine group G over an arbitrary field k, but the reader may wish to keep in mind the above special case.

The quotient map $\pi : G_{\overline{k}} \twoheadrightarrow G_{\overline{k}}^{\rm red}$ has a smooth unipotent (scheme-theoretic) kernel, but if G is pseudo-reductive and non-reductive then this kernel fails to descend to a k-subgroup of G. Let k'/k be the field of definition (in the sense of Definition 1.1.6) for the $\overline{k}$-subgroup $\ker \pi \subseteq G_{\overline{k}}$. In Examples 1.3.3 and 1.3.5 we will compute the field of definition over k in some interesting cases.

Since the $\overline{k}$-subgroup $\mathscr{R}_u(G_{\overline{k}})$ of $G_{\overline{k}}$ descends to the perfect closure of k in $\overline{k}$ (by Galois descent from $\overline{k}$), the field of definition k' for $\ker \pi$ over k is purely inseparable of finite degree over k. The quotient G' of $G_{k'}$ modulo the k'-descent of $\ker \pi$ is a connected reductive k'-group. The quotient map $\pi' : G_{k'} \twoheadrightarrow G'$ over k' corresponds to a k-group map

$$i_G : G \to \mathrm{R}_{k'/k}(G'). \tag{1.3.1}$$

In Example 1.6.3 we will exhibit commutative pseudo-reductive G for which $\ker i_G \neq 1$, and Example 5.3.7 provides non-commutative pseudo-reductive G for which $G^{\mathrm{ss}}_{\overline{k}}$ is simple and $\dim \ker i_G$ is arbitrarily large. There are also interesting examples of pseudo-reductive G for which i_G is not surjective; see Example 1.3.3 for commutative G and Example 1.3.5 for non-commutative G.

Remark 1.3.1 Consider a non-reductive pseudo-reductive k-group G such that $G^{\mathrm{red}}_{\overline{k}}$ is a simple semisimple group. It follows from Theorem 1.6.2(2) that the k-group G arises as the Weil restriction of a connected reductive group over a finite extension of k if and only if i_G is an isomorphism. Our main classification theorem leads to a simple characterization of when i_G is an isomorphism for such G, provided that we require $G = \mathscr{D}(G)$ and moreover that $[k : k^2] = 2$ when $\mathrm{char}(k) = 2$: it is necessary and sufficient that the order of the fundamental group of $G^{\mathrm{ss}}_{\overline{k}}$ is not divisible by $\mathrm{char}(k)$, except possibly when $\mathrm{char}(k) = p \in \{2, 3\}$ and the Dynkin diagram of $G^{\mathrm{ss}}_{\overline{k}}$ has an edge of multiplicity p or is a single vertex when $p = 2$ (i.e., type G_2 for $p = 3$ and types B, C, or F_4 for $p = 2$). This is proved in Proposition 11.1.4 and Remark 11.1.5 (which also explicitly describe all exceptional cases) when G_{k_s} has a reduced root system. The case of a non-reduced root system rests on Theorem 2.3.10.

The construction of commutative examples of non-surjective i_G rests on Weil restriction through arbitrary finite extensions of fields. The following example shows that inseparable Weil restriction can convert an infinitesimal group into a (non-smooth) group with dimension > 0, and so can ruin the property of a map being an isogeny:

Example 1.3.2 Let k'/k be a nontrivial purely inseparable finite extension of fields with $\mathrm{char}(k) = p > 0$. By Example 1.1.3, $G := \mathrm{R}_{k'/k}(\mathrm{GL}_1)$ is pseudo-reductive over k, and G/GL_1 is nontrivial and unipotent. Thus, not only is G

not reductive, but we see that pseudo-reductivity can be lost after passage to the quotient by a smooth connected k-subgroup.

Let f be the p-power map on GL_1 over k' (an isogeny), so $\ker(f) = \mu_p$. Thus, $\mathrm{R}_{k'/k}(f)$ is the p-power map on the pseudo-reductive k-group $\mathrm{R}_{k'/k}(\mathrm{GL}_1)$, and this map fails to be surjective for dimension reasons since its kernel

$$\ker \mathrm{R}_{k'/k}(f) = \mathrm{R}_{k'/k}(\mu_p) = \mathrm{R}_{k'/k}(\mathrm{GL}_1)[p]$$

has dimension at least $p - 1 > 0$. To see this, first note that for any finite extension of fields L/K and integer $m > 0$, applying the snake lemma for fppf (*fidèlement plate de présentation finie*) abelian sheaves to the m-power endomorphism of the short exact sequence of K-groups

$$1 \to \mathrm{GL}_1 \to \mathrm{R}_{L/K}(\mathrm{GL}_1) \to \mathrm{R}_{L/K}(\mathrm{GL}_1)/\mathrm{GL}_1 \to 0$$

proves that the natural map $\mathrm{R}_{L/K}(\mu_m)/\mu_m = \mathrm{R}_{L/K}(\mathrm{GL}_1)[m]/\mu_m \to (\mathrm{R}_{L/K}(\mathrm{GL}_1)/\mathrm{GL}_1)[m]$ is an isomorphism. Now choose an intermediate field $k_0 \subseteq k'$ of degree p over k. Clearly $\mathrm{R}_{k'/k}(\mathrm{GL}_1)$ contains $\mathrm{R}_{k_0/k}(\mathrm{GL}_1)$, and

$$(\mathrm{R}_{k_0/k}(\mathrm{GL}_1)[p]/\mu_p)_{k_0} \simeq (\mathrm{R}_{k_0/k}(\mathrm{GL}_1)/\mathrm{GL}_1)_{k_0} \simeq \mathbf{G}_{\mathrm{a}}^{p-1} \tag{1.3.2}$$

by using the truncation of $\log(1+x)$ in degree $< p$ in characteristic p.

Thus, the isogeny f has Weil restriction $\mathrm{R}_{k'/k}(f)$ that is not an isogeny. Also, the quotient $\mathrm{R}_{k'/k}(\mathrm{GL}_1)/\mu_p$ by the finite multiplicative k-subgroup μ_p is not pseudo-reductive since its k-subgroup $\mathrm{R}_{k_0/k}(\mu_p)/\mu_p$ is nontrivial, smooth, connected, and unipotent by (1.3.2).

Here is a commutative example with non-surjective i_G.

Example 1.3.3 Let k be an imperfect field of characteristic $p > 0$, and let $k''/k'/k$ be a tower of purely inseparable extensions of degree p such that $k''^p \not\subseteq k$ (i.e., $k' = k(a^{1/p})$ and $k'' = k(a^{1/p^2})$ for some $a \in k$ that is not a p-power). Thus, the only intermediate fields in k''/k are k, k', and k''. Consider a smooth connected k-subgroup G of $C'' := \mathrm{R}_{k''/k}(\mathrm{GL}_1)$ that strictly contains $C' := \mathrm{R}_{k'/k}(\mathrm{GL}_1)$. (We will construct examples of such $G \neq C''$ at the end. The reason we focus on these G instead of smooth connected proper k-subgroups of C' that strictly contain GL_1 is that no G of the latter type exist when $p = 2$, as the codimension of GL_1 in C' is $[k' : k] - 1 = p - 1$.) Such a k-group G inherits pseudo-reductivity from the commutative pseudo-reductive C''.

Since $C''_{k''}$ is the unit group of $k'' \otimes_k k'' \simeq k''[x]/(x^{p^2})$, it is $\mathrm{GL}_1 \times U''$ where U'' is the group of 1-units. Hence, $G_{k''} = \mathrm{GL}_1 \times (U'' \cap G_{k''})$, so $U'' \cap G_{k''}$ must be nontrivial, smooth, and connected. It follows that G is non-reductive with k''/k containing the field of definition of its geometric unipotent radical, so this field of definition is either k' or k''.

We now check that the field of definition over k for the $\overline{k}$-subgroup $\mathscr{R}_u(G_{\overline{k}}) \subseteq G_{\overline{k}}$ is k'' by ruling out the possibility that it is k'. The k'-group $C''_{k'}$ is the unit group of

$$k'' \otimes_k k' = k'' \otimes_{k'} (k' \otimes_k k') = k''[x]/(x^p),$$

so $C''_{k'} = \mathrm{R}_{k''/k'}(\mathrm{GL}_1) \times U'$ for a smooth connected p-torsion k'-group U'. Observe that the projection $\pi : C''_{k'} \to \mathrm{R}_{k''/k'}(\mathrm{GL}_1)$ restricts to the identity on GL_1. Transitivity of Weil restriction provides another quotient map

$$q : C''_{k'} = \mathrm{R}_{k'/k}(\mathrm{R}_{k''/k'}(\mathrm{GL}_1))_{k'} \twoheadrightarrow \mathrm{R}_{k''/k'}(\mathrm{GL}_1)$$

via the construction in Proposition A.5.7 (with $X' = \mathrm{R}_{k''/k'}(\mathrm{GL}_1)$ there), and we claim that $\pi = q$. To prove this, observe that both are homomorphisms between the same pair of smooth commutative affine k'-groups and their restrictions to the maximal k'-torus GL_1 coincide (equal to the identity map). This forces the maps to be equal, by Proposition 1.2.2.

The equality $\pi = q$ ensures that π cannot carry $G_{k'}$ into GL_1. To explain this, observe rather generally that if X'' is a quasi-projective k''-scheme and $f : Y' \to \mathrm{R}_{k''/k'}(X'')$ is a k'-map from a k'-scheme Y' then (by the very definition of $q_{X''}$ in (A.5.3)) the composite map $q_{X''} \circ f_{k''} : Y'_{k''} \to X''$ corresponds under the universal property of Weil restriction to the given map f. Since $\pi = q$, the restriction of π to $G_{k'}$ therefore corresponds to the inclusion of G into $\mathrm{R}_{k'/k}(\mathrm{R}_{k''/k'}(\mathrm{GL}_1)) = C''$. Thus, if $\pi|_{G_{k'}}$ were to land in GL_1 then $G \subseteq \mathrm{R}_{k'/k}(\mathrm{GL}_1) = C'$ inside of C''. But we assumed that G strictly contains C', so this would be a contradiction.

Now we can show that k'/k is not the field of definition of the geometric unipotent radical of G. If it were, then $\mathscr{R}_{u,k'}(G_{k'})$ would be a complement to GL_1 inside of $G_{k'}$, yet π would have to kill it (as the target of π is commutative and pseudo-reductive over k'). Hence, $\pi|_{G_{k'}}$ would land in GL_1, which we have shown does not happen.

Since the field of definition over k for the geometric unipotent radical of G has now been determined to be k''/k, the universal property of Weil restriction makes i_G correspond to the evident projection from $G_{k''}$ onto its maximal reductive quotient GL_1. But this projection factors as the composite of $G_{k''} \to$

$C''_{k''}$ and the maximal reductive quotient map for $C_{k''}$. In other words, i_G factors as

$$G \hookrightarrow C'' \stackrel{i_{C''}}{\to} \mathrm{R}_{k''/k}(\mathrm{GL}_1).$$

The map $i_{C''}$ is the identity map (since the evident quotient map $C''_{k''} \to \mathrm{GL}_1$ coincides with the map $q_{\mathrm{GL}_1,k''/k}$ from Proposition A.5.7), so if $G \neq C''$ then i_G is not surjective.

It remains to exhibit nontrivial smooth connected proper k-subgroups of $U = C''/C'$. This k-group is smooth, connected, commutative, and p-torsion, so by Lemma B.1.10 it admits an étale k-isogeny onto a vector group V. Since $V \simeq \mathbf{G}_{\mathrm{a}}^{p^2-p}$ and $p^2-p > 1$, we can choose a nonzero smooth connected proper k subgroup of $\mathbf{G}_{\mathrm{a}}^{p^2-p}$, and the identity component of its preimage in U gives rise to the desired G.

Examples of the failure of surjectivity of i_G with *perfect* pseudo-reductive G rest on:

Proposition 1.3.4 *Let k be a field, k' a nonzero finite reduced k-algebra, and G' a reductive k'-group with connected fibers over* Spec k'. *For any k'-subgroup scheme $\mu \subseteq Z_{G'}$, the natural k-group map*

$$G := \mathrm{R}_{k'/k}(G')/\mathrm{R}_{k'/k}(\mu) \to \mathrm{R}_{k'/k}(G'/\mu)$$

is a normal subgroup inclusion with commutative cokernel. In particular, G is pseudo-reductive over k.

If $G' \to$ Spec k' *has simply connected semisimple fibers then $\mathrm{R}_{k'/k}(G')$ is perfect, so in such cases G is the derived group of $\mathrm{R}_{k'/k}(G'/\mu)$.*

Proof The k-group $\mathrm{R}_{k'/k}(G'/\mu)$ is pseudo-reductive (Proposition 1.1.10), so pseudo-reductivity of G will follow from the normality claim. To prove the normality, view the central extension

$$1 \to \mu \to G' \to G'/\mu \to 1$$

as a short exact sequence of fppf group sheaves over Spec k'. Pushforward along f : Spec $k' \to$ Spec k yields an exact sequence of low-degree higher direct image sheaves

$$1 \to \mathrm{R}_{k'/k}(\mu) \to \mathrm{R}_{k'/k}(G') \to \mathrm{R}_{k'/k}(G'/\mu) \stackrel{\delta}{\to} \mathrm{R}^1 f_*(\mu). \qquad (1.3.3)$$

The connecting map δ is a homomorphism due to the centrality of μ in G'.

It follows that $R_{k'/k}(G')/R_{k'/k}(\mu) = \ker\delta$, so it is a normal k-subgroup of $R_{k'/k}(G'/\mu)$ in the sense of fppf group sheaves over $\mathrm{Spec}\, k$. Thus, it is also a normal closed k-subgroup in the sense of k-group schemes (see Proposition A.2.1). When G' has simply connected fibers, the perfectness of $R_{k'/k}(G')$ is Corollary A.7.11. In such cases $G = \mathscr{D}(G)$, so G is the derived group of $R_{k'/k}(G'/\mu)$ since δ has a commutative target. □

Proposition 1.3.4 provides examples of non-commutative pseudo-reductive groups G for which i_G is not surjective, as we show below. The justification rests on results to be proved in §1.6, and also yields examples of inseparable isogenies f between connected semisimple groups such that the Weil restriction of f through a field extension is not an isogeny:

Example 1.3.5 Let $G = R_{k'/k}(\mathrm{SL}_p)/R_{k'/k}(\mu_p)$ for k'/k a nontrivial finite purely inseparable extension with $\mathrm{char}(k) = p > 0$. By Proposition 1.3.4, G is pseudo-reductive and perfect, and by Proposition A.5.11(1),(2) the maximal geometric semisimple quotient $G_{\overline{k}}^{\mathrm{ss}}$ maps isomorphically to the simple semisimple group $\mathrm{SL}_p/\mu_p = \mathrm{PGL}_p$.

We will show that i_G is not surjective. Granting this, it then follows (via Proposition A.5.11(1),(2)) that G is not a k-isogenous quotient of $R_{K/k}(H)$ for any finite extension K/k and connected reductive K-group H. Thus, G is "far" from Weil restrictions of connected reductive groups.

The first issue to settle in the proof of non-surjectivity of i_G is the determination of the (minimal) field of definition over k for the geometric (unipotent) radical of G. We claim that this field is k'. That is, relative to the evident k-structure on $G_{\overline{k}}$, we claim that the scheme-theoretic kernel of the quotient map

$$\pi : G_{\overline{k}} \twoheadrightarrow G_{\overline{k}}^{\mathrm{ss}} = \mathrm{PGL}_p$$

has (minimal) field of definition over k equal to $k' \subseteq \overline{k}$.

This field of definition over k is contained in k' since a k'-descent of π is given by the canonical map $\pi' : G_{k'} \twoheadrightarrow \mathrm{PGL}_p$ induced by the composite map

$$R_{k'/k}(\mathrm{SL}_p)_{k'} \xrightarrow{q} \mathrm{SL}_p \to \mathrm{PGL}_p$$

that kills $R_{k'/k}(\mu)_{k'}$ (due to functoriality of (A.5.3) with respect to the inclusion $\mu_p \hookrightarrow \mathrm{SL}_p$). An application of Theorem 1.6.2(1) with $G' = \mathrm{PGL}_p$ and $\widetilde{G}' = \mathrm{SL}_p$ implies that the field of definition over k for $\ker\pi$ cannot be a proper subfield of k', so it is equal to k'.

The identification of k'/k as the field of definition of $\ker \pi$ implies that the k-map

$$i_G : G \to \mathrm{R}_{k'/k}(\mathrm{PGL}_p)$$

corresponds to π' via the universal property of Weil restriction. For the inseparable k'-isogeny $f : \mathrm{SL}_p \to \mathrm{PGL}_p$ we have $\mathrm{R}_{k'/k}(f) = i_G \circ h$, where

$$h : \mathrm{R}_{k'/k}(\mathrm{SL}_p) \to \mathrm{R}_{k'/k}(\mathrm{SL}_p)/\mathrm{R}_{k'/k}(\mu_p)$$

is the canonical projection. Left-exactness of $\mathrm{R}_{k'/k}$ implies that

$$\ker(\mathrm{R}_{k'/k}(f)) = \mathrm{R}_{k'/k}(\mu_p) = \ker h,$$

so we conclude that $\ker i_G = \{1\}$. However, i_G fails to be surjective for dimension reasons. (Proposition 1.3.4 even identifies its image with $\mathscr{D}(\mathrm{R}_{k'/k}(\mathrm{PGL}_p))$.) In particular, although f is an isogeny, the map $\mathrm{R}_{k'/k}(f) = i_G \circ h$ is not surjective.

This example exhibits an additional feature, as follows. By Proposition 1.1.10, the k-group $\mathrm{R}_{k'/k}(\mathrm{SL}_p)$ is pseudo-reductive over k. Also, the central quotient G of $\mathrm{R}_{k'/k}(\mathrm{SL}_p)$ by a connected k-subgroup of dimension > 0 is pseudo-reductive over k (Proposition 1.3.4). However, the central quotient $\mathrm{R}_{k'/k}(\mathrm{SL}_p)/\mu_p$ of $\mathrm{R}_{k'/k}(\mathrm{SL}_p)$ by a finite multiplicative k-subgroup is not pseudo-reductive over k. Indeed, this central quotient contains the p-torsion central k-subgroup $\mathrm{R}_{k_0/k}(\mu_p)/\mu_p$ for any subfield $k_0 \subseteq k'$ of degree p over k, and this subgroup maps isomorphically onto the (unipotent) smooth connected k-group $\mathrm{R}_{k_0/k}(\mathrm{GL}_1)/\mathrm{GL}_1$ of dimension $p-1$ (see (1.3.2)).

Remark 1.3.6 In Example 1.6.4 we will give a subtle variant on Example 1.3.5: a perfect pseudo-reductive k-group G containing a smooth connected perfect normal k-subgroup $N \subseteq G$ such that G/N is not pseudo-reductive. In particular, the preimage N' of $\mathscr{R}_{u,k}(G/N)$ under the quotient map $G \twoheadrightarrow G/N$ is a smooth connected normal k-subgroup of G such that $N' \neq \mathscr{D}(N')$ (as N' has a nontrivial unipotent quotient).

Note the contrast with the theory of connected semisimple groups. Indeed, if G is a connected semisimple group over a field then for all smooth connected normal subgroups N of G both N and G/N are semisimple (and in particular, N is perfect). This illustrates some of the difficulties in any attempt to develop a good notion of pseudo-semisimplicity.

Let k'/k be a nonzero finite reduced k-algebra. Consider a k'-group G' whose fibers over $\operatorname{Spec} k'$ are connected and reductive. Let T' be a maximal k'-torus in G', so T' is a Cartan k'-subgroup. (Explicitly, $T' = Z_{G'}(T')$.) By Proposition A.5.15, the maximal k-torus T in $\mathrm{R}_{k'/k}(T')$ is a maximal k-torus in the pseudo-reductive k-group $G = \mathrm{R}_{k'/k}(G')$ and the Cartan k-subgroup $C := Z_G(T)$ is equal to $\mathrm{R}_{k'/k}(T')$. Likewise, the center Z_G is $\mathrm{R}_{k'/k}(Z_{G'})$. We will be interested in smooth connected k-groups equipped with a left action on G that restricts to the identity on C. One example is the conjugation action on G by the commutative C, but not even the action of C/Z_G on G explains all such actions (especially when k' is not étale over k):

Example 1.3.7 Let k'/k be a nontrivial purely inseparable finite extension of fields of characteristic $p > 0$, and T' a maximal k'-torus in SL_p with image $\overline{T}'$ in the central quotient PGL_p. Let T be the maximal k-torus of $\mathrm{R}_{k'/k}(T')$.

The action of PGL_p on SL_p over k' makes $\overline{T}'$ act on SL_p with trivial action on T'. Applying $\mathrm{R}_{k'/k}$ then defines an action of $\mathrm{R}_{k'/k}(\overline{T}')$ on $G = \mathrm{R}_{k'/k}(\mathrm{SL}_p)$, and this action fixes the Cartan k-subgroup $\mathrm{R}_{k'/k}(T') = Z_G(T) =: C$ pointwise. However, the natural inclusion $C/Z_G \to \mathrm{R}_{k'/k}(\overline{T}')$ is *not* surjective, as $\dim C = \dim \mathrm{R}_{k'/k}(\overline{T}')$ but $\dim Z_G > 0$ (Example 1.3.2). Theorem 1.3.9 below implies that it is the action of $\mathrm{R}_{k'/k}(\overline{T}')$ (and not the action of C/Z_G!) on G that is universal for the property of fixing C pointwise.

In general, for a smooth affine k-group G, a maximal k-torus T in G, and the Cartan k-subgroup $C = Z_G(T)$ in G, we are interested in the following functors on k-schemes:

$$\underline{\mathrm{Aut}}_{G,T} : S \rightsquigarrow \{g \in \mathrm{Aut}_S(G_S) \mid g \text{ restricts to the identity on } T_S\},$$
$$\underline{\mathrm{Aut}}_{G,C} : S \rightsquigarrow \{g \in \mathrm{Aut}_S(G_S) \mid g \text{ restricts to the identity on } C_S\}.$$

Observe that if H is a k-group then to give a k-homomorphism of group functors $H \to \underline{\mathrm{Aut}}_{G,C}$ is the same as to define a left action of H on G that restricts to the identity on C.

Example 1.3.8 In an evident manner, $\underline{\mathrm{Aut}}_{G,C}$ is a subgroup functor of $\underline{\mathrm{Aut}}_{G,T}$. If G is pseudo-reductive over k and F/k is a separable extension field then the inclusion $\underline{\mathrm{Aut}}_{G,C}(F) \subseteq \underline{\mathrm{Aut}}_{G,T}(F)$ is an equality. Indeed, separability ensures that G_F is pseudo-reductive over F, so we may rename F as k to reduce to showing that a k-automorphism ϕ of G that restricts to the identity on T also restricts to the identity on C. By Proposition 1.2.4, C is a commutative pseudo-reductive group. The automorphism ϕ of G restricts to an automorphism of $C = Z_G(T)$ that is the identity on the maximal k-torus T in C. Thus, by the final part of Proposition 1.2.2, ϕ is the identity on C as well.

In Theorem 2.4.1 we will prove that $\underline{\mathrm{Aut}}_{G,C}$ is represented by an affine k-group scheme of finite type (denoted $\mathrm{Aut}_{G,C}$). This k-group contains a maximal smooth k-subgroup $Z_{G,C}$ (via the construction in Lemma C.4.1), and we will prove that if G is pseudo-reductive then $Z_{G,C}$ is commutative with pseudo-reductive identity component. The following result makes $Z_{G,C}$ explicit in a special case. We record it here solely for the purpose of motivating the "standard" construction of pseudo-reductive groups in §1.4; it is not logically necessary at this stage of the theory.

Theorem 1.3.9 *Let $\mathscr{G} = \mathrm{R}_{k'/k}(G')$ for a nonzero finite reduced k-algebra k' and a k'-group G' with connected reductive fibers over* Spec k'. *Choose a Cartan k-subgroup $\mathscr{C}$ in $\mathscr{G}$, and let T' be the unique maximal k'-torus in G' for which $\mathscr{C} = \mathrm{R}_{k'/k}(T')$.*

Using the left action of $\mathrm{R}_{k'/k}(T'/Z_{G'})$ on $\mathscr{G}$ induced by the natural action of $G'/Z_{G'}$ on G', the natural map $\mathrm{R}_{k'/k}(T'/Z_{G'}) \to \mathrm{Aut}_{\mathscr{G},\mathscr{C}}$ is an isomorphism onto the maximal smooth k-subgroup $Z_{\mathscr{G},\mathscr{C}}$. In particular, $Z_{\mathscr{G},\mathscr{C}}$ is connected and the action of $\mathrm{R}_{k'/k}(T'/Z_{G'})$ on $\mathscr{G}$ is final among all smooth k-groups equipped with a left action on $\mathscr{G}$ that restricts to the identity on $\mathscr{C}$.

The correspondence between the choice of $\mathscr{C}$ in $\mathscr{G}$ and the choice of Cartan k'-subgroup T' in G' as used above is part of Proposition A.5.15(3).

Proof We may and do assume $k = k_s$. The natural map $\mathrm{R}_{k'/k}(T'/Z_{G'}) \to \mathrm{Aut}_{\mathscr{G},\mathscr{C}}$ uniquely factors through a k-homomorphism $f : \mathrm{R}_{k'/k}(T'/Z_{G'}) \to Z_{\mathscr{G},\mathscr{C}}$ between smooth affine k-groups, and we need to prove that this is an isomorphism. It suffices to prove that f is birational. In general, if k is a field and $h : X' \to X$ is a k-morphism between integral k-schemes with respective generic points η' and η, and if h is injective on $k(\eta')$-valued points and is surjective on $k(\eta)$-valued points then h is birational (in the sense that $h(\eta') = \eta$ and the induced map $k(\eta) \to k(\eta')$ is an isomorphism). Indeed, surjectivity on $k(\eta')$-valued points implies that the canonical map $\mathrm{Spec}\, k(\eta) \to X$ factors over k as $\mathrm{Spec}\, k(\eta) \xrightarrow{j} X' \xrightarrow{h} X$, so h is dominant and hence $h(\eta') = \eta$. The resulting k-map $\mathrm{Spec}\, k(\eta') \xrightarrow{h} \mathrm{Spec}\, k(\eta) \xrightarrow{j} X'$ coincides with the canonical k-map $\mathrm{Spec}\, k(\eta') \to X'$ since both have the same composition with h (which is assumed to be injective on $k(\eta')$-points). Hence, $j(\eta) = \eta'$ with j inverse to h between function fields. We conclude that it suffices to prove f induces a bijection between F-valued points for every separable extension field F/k (for $F = k = k_s$ this will ensure that $Z_{\mathscr{G},\mathscr{C}}$ is connected, hence integral, so the above reasoning may be applied).

For any separable extension field F/k, the finite F-algebra $k' \otimes_k F$ is reduced. Thus, we may rename F as k (and again pass to k_s) to reduce to proving the bijectivity property on k-points. In view of how the maximal smooth k-subgroup $Z_{\mathscr{G},\mathscr{C}}$ is constructed when $k = k_s$, this is the assertion that the map $(T'/Z_{G'})(k') \to \mathrm{Aut}_k(\mathscr{G})$ is an isomorphism onto the subgroup of k-automorphisms of $\mathscr{G}$ that restrict to the identity on the Cartan k-subgroup $\mathscr{C} = \mathrm{R}_{k'/k}(T')$. We want to pass to factor fields of k' to reduce to the case where k' is a field, so we first need to check that a general k-automorphism f of $\mathscr{G} = \mathrm{R}_{k'/k}(G')$ that restricts to the identity on $\mathscr{C}$ is built from corresponding automorphisms arising separately from fibers of (G', T') over the factor fields of k'. Here is the result:

Lemma 1.3.10 *Choose $f \in \mathrm{Aut}_{\mathscr{G},\mathscr{C}}(k)$. There is a unique k'-automorphism f' of G' that restricts to the identity on T' and satisfies $\mathrm{R}_{k'/k}(f') = f$. That is, the map $\underline{\mathrm{Aut}}_{G',T'}(k') \to \underline{\mathrm{Aut}}_{\mathscr{G},\mathscr{C}}(k)$ induced by Weil restriction is bijective.*

Proof First we treat the case where k' is a field. Consider the natural map $q_{G'} : \mathscr{G}_{k'} \to G'$ as in (A.5.3), so $q_{G'}$ is a smooth surjective map by Proposition A.5.11(1). Since k'/k is purely inseparable, it follows from Proposition A.5.11(2) that $\ker q_{G'}$ is connected and unipotent, so G' is the maximal reductive quotient of $\mathscr{G}_{k'}$. Hence, $f_{k'}$ induces an automorphism f' of G' that restricts to the identity on the image of $\mathscr{C}_{k'}$ in G'. But this image is T' (by functoriality of $q_{T'}$ and $q_{G'}$ with respect to the inclusion of T' into G'), so $f' \in \underline{\mathrm{Aut}}_{G',T'}(k')$. Functoriality of $q_{G'}$ with respect to f' then implies that f and $\mathrm{R}_{k'/k}(f')$ are k-automorphisms of $\mathscr{G}$ whose associated k'-automorphisms of $\mathscr{G}_{k'}$ induce the same automorphism f' of the maximal reductive quotient G'. This forces $f = \mathrm{R}_{k'/k}(f')$ by Proposition 1.2.2.

For general k', let $\{k'_i\}$ be the set of factor fields of k', and G'_i the k'_i-fiber of G'. Let $\mathscr{G}_i = \mathrm{R}_{k'_i/k}(G'_i)$, so $\mathscr{G} = \prod \mathscr{G}_i$. Observe that if T'_i denotes the k'_i-fiber of T' then $\mathscr{C} = \prod \mathscr{C}_i$, where $\mathscr{C}_i$ is the Cartan k-subgroup $\mathrm{R}_{k'_i/k}(T'_i)$. It suffices to prove $f = \prod f_i$ for $f_i \in \underline{\mathrm{Aut}}_{\mathscr{G}_i,\mathscr{C}_i}(k)$, as then the problem is reduced to the settled case where k' is a field. For each i, consider the centralizer of $\prod_{j \neq i} \mathscr{C}_j$ in $\mathscr{G}$. This is clearly $\mathscr{G}_i \times \prod_{j \neq i} \mathscr{C}_j$, and so f must restrict to a k-automorphism of this subgroup (as f restricts to the identity on $\prod_{j \neq i} \mathscr{C}_j$). Hence, f restricts to an automorphism of its derived group, which is $\mathscr{D}(\mathscr{G}_i)$. By Proposition 1.2.6, $\mathscr{G}_i = \mathscr{C}_i \cdot \mathscr{D}(\mathscr{G}_i)$. Thus, it suffices to check that f restricts to an automorphism of $\mathscr{C}_i$ for each i. Even better, f restricts to the identity on $\prod \mathscr{C}_i = \mathscr{C}$. □

Continuing with the proof of Theorem 1.3.9, the description of f as a Weil restriction in Lemma 1.3.10 reduces the problem to a separate argument over each factor field of k'. That is, we can now assume that k' is a field. We have to prove that any k'-automorphism f' of G' that restricts to the identity on T' is induced by the action of a unique point $t' \in (T'/Z_{G'})(k')$. The effect of f' on the root datum for (G', T') is trivial since f' restricts to the identity on T', so the Isomorphism Theorem for the split pair (G', T') over the separably closed field k' (Theorem A.4.6) gives that f' is induced by the action of a unique $t' \in (T'/Z_{G'})(k')$. □

1.4 The standard construction and examples

We now construct pseudo-reductive groups going beyond the examples provided by Proposition 1.1.10 (and motivated by Theorem 1.3.9). These will be the *standard pseudo-reductive groups*. The ubiquity of this construction will be explained by our main classification result for pseudo-reductive groups (stated in a weak form in Theorem 1.5.1). To construct standard pseudo-reductive groups, we need a procedure to modify a given pseudo-reductive group by either passing to a central quotient or expanding a Cartan subgroup without losing pseudo-reductivity.

To explain the idea, consider a pseudo-reductive k-group $\mathscr{G}$ and a maximal k-torus $\mathscr{T}$ in $\mathscr{G}$. Let $\mathscr{C}$ be the Cartan k-subgroup $Z_{\mathscr{G}}(\mathscr{T})$. According to Proposition 1.2.4, $\mathscr{C}$ is commutative and pseudo-reductive. There is a natural left action of $\mathscr{C}$ on $\mathscr{G}$ via conjugation, and this restricts to the identity on $\mathscr{C}$. Suppose there is given another commutative pseudo-reductive k-group C with a left action on $\mathscr{G}$ which fixes $\mathscr{C}$ pointwise, and a k-homomorphism $\phi : \mathscr{C} \to C$ compatible with the actions on $\mathscr{G}$. In other words, we are given a factorization diagram of group functors on k-schemes

$$\mathscr{C} \xrightarrow{\phi} C \xrightarrow{\psi} \underline{\mathrm{Aut}}_{\mathscr{G},\mathscr{C}}.$$

Suppose that $\underline{\mathrm{Aut}}_{\mathscr{G},\mathscr{C}}$ is represented by an affine k-group scheme of finite type, and let $Z_{\mathscr{G},\mathscr{C}}$ denote its maximal smooth k-subgroup, so we get a factorization diagram

$$\mathscr{C} \xrightarrow{\phi} C \xrightarrow{\psi} Z_{\mathscr{G},\mathscr{C}}.$$

In §2.4 we will prove that the representability always holds and that $Z_{\mathscr{G},\mathscr{C}}$ is commutative with pseudo-reductive identity component. (What follows is merely motivation for Definition 1.4.4, so there is no risk of circularity.)

Note that $\ker\phi$ acts trivially on $\mathscr{G}$ and so is central in $\mathscr{G}$. We wish to create another pseudo-reductive group in which C replaces $\mathscr{C}$ as a Cartan k-subgroup. A side-effect of this procedure will be to kill $\ker\phi$, so the center may be replaced with a quotient even as the Cartan k-subgroup may get bigger.

Here is the most important example.

Example 1.4.1 Suppose $\mathscr{G} = \mathrm{R}_{k'/k}(G')$ for a nonzero finite reduced k-algebra k' and a k'-group G' with connected reductive fibers. Let T' be a maximal k'-torus in G', and let $\mathscr{C}$ be the Cartan k-subgroup $\mathrm{R}_{k'/k}(T')$ of $\mathscr{G}$, and $\mathscr{T}$ be the maximal k-torus of $\mathscr{C}$. Since $G'/Z_{G'}$ acts on G', the quotient torus $T'/Z_{G'}$ acts on G'. Hence, $\mathrm{R}_{k'/k}(T'/Z_{G'})$ naturally acts on $\mathscr{G}$, and the natural map $\mathscr{C} = \mathrm{R}_{k'/k}(T') \to \mathrm{R}_{k'/k}(T'/Z_{G'})$ is compatible with the actions on $\mathscr{G}$. Observe that this map may not be surjective, as we illustrated in Example 1.3.7. Also, its kernel is $\mathrm{R}_{k'/k}(Z_{G'})$ due to left-exactness of Weil restriction, and this is the entire center of $\mathscr{G}$ (Proposition A.5.15(1)).

By Theorem 1.3.9 (which is conditional on §2.4), the natural map $\mathrm{R}_{k'/k}(T'/Z_{G'}) \to Z_{\mathscr{G},\mathscr{C}}$ is an isomorphism. Thus, to give a pair (C,ϕ) consisting of a commutative pseudo-reductive k-group C equipped with a left action on $\mathscr{G}$ which fixes $\mathscr{C}$ pointwise and a k-homomorphism $\phi : \mathscr{C} = \mathrm{R}_{k'/k}(T') \to C$ respecting the actions on $\mathscr{G}$ amounts to a factorization

$$\mathrm{R}_{k'/k}(T') \xrightarrow{\phi} C \xrightarrow{\psi} \mathrm{R}_{k'/k}(T'/Z_{G'}) \tag{1.4.1}$$

of the natural map $\mathrm{R}_{k'/k}(T') \to \mathrm{R}_{k'/k}(T'/Z_{G'})$ of k-groups (with C a commutative pseudo-reductive k-group). Since Theorem 1.3.9 is presently conditional on §2.4, we note that it makes sense to consider factorization diagrams (1.4.1) even without knowing Theorem 1.3.9, since ψ provides a C-action on $\mathscr{G}$ that restricts to the identity on $\mathscr{C}$. (The point of Theorem 1.3.9 is that the construction resting on (1.4.1) gives *all* possibilities for pairs (C,ϕ).)

Using the C-action on $\mathscr{G}$, we get a semidirect product group $\mathscr{G} \rtimes C$. The pair of homomorphisms

$$\phi : \mathscr{C} \to C, \quad \iota : \mathscr{C} \hookrightarrow \mathscr{G}$$

defines the twisted product map

$$\alpha : \mathscr{C} \to \mathscr{G} \rtimes C \tag{1.4.2}$$

via $c \mapsto (\iota(c)^{-1}, \phi(c))$. This is readily checked to be an isomorphism onto a central subgroup. Consider the resulting smooth connected affine quotient group $G := (\mathscr{G} \rtimes C)/\mathrm{im}(\alpha)$.

Remark 1.4.2 The k-group $\mathscr{C} \rtimes C$ is a Cartan k-subgroup in the semidirect product $\mathscr{G} \rtimes C$. Hence, its image in G is a Cartan k-subgroup (Proposition A.2.8). But this image is exactly C with its natural homomorphism into G (that has trivial kernel). This procedure therefore replaces $\mathscr{C}$ with C, as desired. Explicitly, the commutative k-group C is the centralizer of the image of $\mathscr{T}$ in G, as well as of the maximal k-torus T of C (which in turn must be maximal in G). Note that T is the unique maximal k-torus of G containing the image of $\mathscr{T}$, due to the commutativity of C.

We claim that the k-group $G = \mathrm{coker}(\alpha)$ is pseudo-reductive. It is convenient to work more generally, without mentioning Cartan subgroups. Consider the following generalization: $\mathscr{G}$ is a pseudo-reductive k-group, $\mathscr{C}$ is a commutative pseudo-reductive k-subgroup such that $\mathscr{C} = Z_{\mathscr{G}}(\mathscr{C})$ (for example, a Cartan k-subgroup), C is a commutative pseudo-reductive group equipped with a left action on $\mathscr{G}$ that is trivial on $\mathscr{C}$, and $\phi : \mathscr{C} \to C$ is a k-homomorphism compatible with the actions on $\mathscr{G}$. The k-scheme map

$$\alpha : \mathscr{C} \to \mathscr{G} \rtimes C$$

via $c \mapsto (c^{-1}, \phi(c))$ is easily checked to be a central k-subgroup inclusion, so we can form the smooth connected affine k-group

$$G = \mathrm{coker}\,\alpha = (\mathscr{G} \rtimes C)/\alpha(\mathscr{C}).$$

Proposition 1.4.3 *The k-group $G = (\mathscr{G} \rtimes C)/\alpha(\mathscr{C})$ is pseudo-reductive under the above conditions.*

Proof Let $\mathscr{H} = \mathscr{G} \rtimes C$ and $\mathscr{Z}$ denote the smooth connected central k-subgroup $\alpha(\mathscr{C})$; the subgroup $\mathscr{Z}$ is isomorphic to the commutative pseudo-reductive group $\mathscr{C}$, and by definition $G = \mathscr{H}/\mathscr{Z}$. It is easily seen that $\mathscr{Z} \cdot C = \mathscr{C} \times C$ and the multiplication map $\mathscr{Z} \times C \to \mathscr{Z} \cdot C = \mathscr{C} \times C$ is an isomorphism. Since the centralizer of $\mathscr{C}$ in $\mathscr{H}$ is $\mathscr{C} \times C$, clearly $Z_{\mathscr{H}}(\mathscr{C} \times C) = \mathscr{C} \times C$, so the center of $\mathscr{H}$ is contained in $\mathscr{C} \times C$.

We will prove that $U := \mathscr{R}_{u,k}(G)$ is trivial; i.e., G is pseudo-reductive. Consider the preimage $\mathscr{N}$ of U under the smooth quotient map $\mathscr{H} \to \mathscr{H}/\mathscr{Z} = G$. This is an extension of U by $\mathscr{Z}$ ($\simeq \mathscr{C}$), so $\mathscr{N}$ is a smooth connected solvable normal subgroup. We assert that the smooth connected normal subgroup $(\mathscr{H}, \mathscr{N})$ of $\mathscr{H}$ is unipotent. This is a general fact for smooth connected affine groups and smooth connected solvable normal subgroups, having nothing to do with pseudo-reductivity, so to prove it we may assume $k = \overline{k}$. By replacing $\mathscr{H}$ with its maximal reductive quotient, we are reduced to showing that if $\mathscr{H}$ is connected and reductive then $(\mathscr{H}, \mathscr{N})$ is trivial. By reductivity, the smooth connected normal solvable subgroup $\mathscr{N}$ is a torus. But a normal torus in a smooth connected affine group is always central (as the automorphism scheme of a torus is étale), so indeed $(\mathscr{H}, \mathscr{N}) = 1$.

Returning to our initial situation over a general ground field k, the semi-direct product $\mathscr{H} = \mathscr{G} \rtimes C$ is pseudo-reductive since $\mathscr{G}$ and C are pseudo-reductive. Thus, the smooth connected unipotent normal k-subgroup $(\mathscr{H}, \mathscr{N})$ is trivial, which is to say $\mathscr{N}$ is central in $\mathscr{H}$. But as we observed above, the center of $\mathscr{H}$ is contained in the subgroup $\mathscr{C} \times C$, so $\mathscr{N} \subseteq \mathscr{C} \times C$. Now since $\mathscr{N}/\mathscr{Z} \hookrightarrow (\mathscr{C} \times C)/\mathscr{Z} \simeq C$, and $\mathscr{N}/\mathscr{Z}$ is isomorphic to the smooth connected unipotent k-group U, the pseudo-reductivity of the commutative k-group C implies that $\mathscr{N}/\mathscr{Z} = 1$; i.e., $\mathscr{N} = \mathscr{Z}$ and hence U is trivial. □

Definition 1.4.4 A *standard pseudo-reductive k-group* is a k-group G that is k-isomorphic to the pseudo-reductive k-group $\mathrm{coker}(\alpha)$ with α as in (1.4.2) in the setting of Example 1.4.1. That is,

$$G \simeq (\mathrm{R}_{k'/k}(G') \rtimes C)/\mathrm{R}_{k'/k}(T') \tag{1.4.3}$$

for a nonzero finite reduced k-algebra k', a k'-group G' with connected reductive fibers, a maximal k'-torus T' in G', and a factorization (1.4.1) with C a commutative pseudo-reductive k-group.

We say that G *arises from a 4-tuple* $(G', k'/k, T', C)$ (suppressing explicit mention of the factorization diagram (1.4.1), which is always understood from the context).

In the preceding definition, we only use the description (1.4.3) and not the universal property of $\mathrm{R}_{k'/k}(T'/Z_{G'})$ from Theorem 1.3.9 (which was conditional on §2.4). At present, this universal property merely motivates the definition of the standard construction. Later it will be useful in the proof of standardness of all pseudo-reductive groups, apart from some exceptional situations in characteristics 2 and 3.

Two special cases of standard pseudo-reductive k-groups are $\mathrm{R}_{k'/k}(G')$ (take ϕ to be the identity map in (1.4.1)) and any commutative pseudo-reductive k-group C (take $G' = 1$). In particular, any solvable pseudo-reductive group is standard because it is commutative (Proposition 1.2.3). Also, if K/k is a finite extension of fields and G is a standard pseudo-reductive K-group then $\mathrm{R}_{K/k}(G)$ is a standard pseudo-reductive k-group; this follows immediately from the fact that $\mathrm{R}_{K/k}$ preserves short-exactness for smooth affine groups (Corollary A.5.4) and preserves pseudo-reductivity (Proposition 1.1.10).

We emphasize (see Remark 1.4.2) that in the standard construction we have $C = Z_G(T)$ for the unique maximal k-torus T of G that contains the image in G of the unique maximal k-torus of $\mathrm{R}_{k'/k}(T')$.

Remark 1.4.5 The construction of standard pseudo-reductive groups is an instance of a general non-commutative pushout construction, as follows. The general setup to consider, for abstract groups, group sheaves, or group schemes of finite type over a field, is a triple of group objects N, G', and G'', equipped with left actions of G'' on the groups G' and N, and G''-homomorphisms $f'' : N \to G''$ and $f' : N \to G'$, where G'' acts on itself on the left by conjugation. Assume

$$f''(n).g' = f'(n)g'f'(n)^{-1}$$

in G' for all $n \in N$ and $g' \in G'$. (An interesting example is N a normal subgroup of G'' and f' a G''-equivariant quotient.)

Consider the semidirect product $G' \rtimes G''$, defined using the left G''-action on G'. The map $f : N \to G' \rtimes G''$ defined by

$$f(n) = (f'(n)^{-1}, f''(n))$$

is a homomorphism that centralizes the normal subgroup $G' \subseteq G' \rtimes G''$ and intertwines G''-conjugation on $G' \rtimes G''$ with the given G''-action on N. In particular, the image of N under f is normal, so we can form the quotient $Q = (G' \rtimes G'')/f(N)$ that we view as fitting into a pushout diagram

$$\begin{array}{ccc} N & \longrightarrow & G' \\ \downarrow & & \downarrow \\ G'' & \longrightarrow & (G' \rtimes G'')/f(N) \end{array}$$

If G'' acts trivially on N (as in the construction of standard pseudo-reductive groups) then $f(N)$ is central in $G' \rtimes G''$. If all three groups are commutative and G'' acts trivially on both N and G' then the preceding pushout construction is the classical pushout construction as in any abelian category. Finally, if N is a normal subgroup of G'' and f' is a G''-equivariant quotient then G' is a normal subgroup of the pushout Q, with $Q/G' = G''/N$.

In the setting of the construction of standard pseudo-reductive groups, consider the central k-subgroup scheme

$$Z := \ker\phi \subseteq \mathrm{R}_{k'/k}(Z_{G'}).$$

The quotient $\mathrm{R}_{k'/k}(T')/Z$ is a smooth connected k-subgroup of the commutative pseudo-reductive k-group C, so it is pseudo-reductive. If we replace C with this k-subgroup in (1.4.3) then we get a k-subgroup inclusion

$$Q := \mathrm{R}_{k'/k}(G')/Z \hookrightarrow G.$$

By Corollary A.7.11, this k-subgroup coincides with $\mathscr{D}(G)$ when the fibers of $G' \to \operatorname{Spec} k'$ are semisimple and simply connected (a harmless hypothesis on G', as we will see in §4.1).

It seems difficult to describe all k-subgroups $Z \subseteq \mathrm{R}_{k'/k}(Z_{G'})$ for which $\mathrm{R}_{k'/k}(T')/Z$ is pseudo-reductive. Likewise, the quotient $\mathrm{R}_{k'/k}(G')/Z$ for a general k-subgroup Z of $\mathrm{R}_{k'/k}(Z_{G'})$ can fail to be pseudo-reductive over k, even if G' is semisimple and simply connected. We saw counterexamples in Example 1.3.5 with $G' = \mathrm{SL}_p$ and $Z = \mu_p$.

Remark 1.4.6 In general, when G' is semisimple, Proposition 1.3.4 gives a special class of $Z \subseteq \mathrm{R}_{k'/k}(Z_{G'})$ for which $\mathrm{R}_{k'/k}(G')/Z$ is pseudo-reductive. Let us now check that a central quotient $G := \mathrm{R}_{k'/k}(G')/Z$ is standard *whenever* it is pseudo-reductive.

Pick a maximal k'-torus T' in G', so $Z_{G'} \subseteq T'$. The smooth connected commutative k-group $C = \mathrm{R}_{k'/k}(T')/Z$ makes sense and it is pseudo-reductive over k because it is a Cartan k-subgroup in the pseudo-reductive k-group G and Cartan k-subgroups of pseudo-reductive groups are always pseudo-reductive (Proposition 1.2.4). There is an evident isomorphism

$$G := \mathrm{R}_{k'/k}(G')/Z \simeq (\mathrm{R}_{k'/k}(G') \rtimes C)/\mathrm{R}_{k'/k}(T') \tag{1.4.4}$$

in which the right side is coker(α) as in (1.4.2), so G is indeed standard over k.

We now give a proof that in some cases the standard pseudo-reductive groups as in Proposition 1.3.4 are not themselves the Weil restriction of a connected reductive group:

Example 1.4.7 Let k'/k be a nontrivial purely inseparable finite extension of fields with $\mathrm{char}(k) = p > 0$. Let $G = \mathrm{R}_{k'/k}(G')/\mathrm{R}_{k'/k}(\mu)$, where G' is a connected semisimple k'-group that is absolutely simple and simply connected and μ is a non-étale subgroup of the finite center $Z_{G'}$. (For example, we can take $G' = \mathrm{SL}_p$ and $\mu = \mu_p$.) Observe that $Z := \mathrm{R}_{k'/k}(\mu)$ has dimension > 0 (because we may take k to be separably closed, so μ contains μ_p, and in Example 1.3.2 we saw that $\mathrm{R}_{k'/k}(\mu_p)$ has dimension > 0). Also, by Proposition 1.3.4, G is pseudo-reductive and perfect (the latter using the simply connected hypothesis on G').

We claim that the k-group G is not a k-isogenous quotient of any k-group of the form $\mathrm{R}_{K/k}(H)$ for a nonzero finite reduced k-algebra K and a K-group H whose fibers over $\mathrm{Spec}\, K$ are connected and reductive. This fact, which will never be used in our development of the theory of pseudo-reductivity, rests on Theorem 1.6.2(2) below, as we now explain.

We argue by contradiction, so assume G is a k-isogenous quotient of $\mathrm{R}_{K/k}(H)$ for some K and H, and seek a contradiction. First we claim that K must be a purely inseparable field extension of k. If K is a nontrivial product or is a field not purely inseparable over k then $G_{\overline{k}}$ has maximal reductive quotient that is not a simple semisimple group. However, the maximal reductive quotient of $G_{\overline{k}}$ is isogenous to the simple semisimple group $G' \otimes_{k'} \overline{k}$ (Proposition A.5.11). This also shows that $\dim H \geqslant \dim G'$.

With K/k a purely inseparable field extension, we see that the maximal reductive quotient of $G_{\overline{k}}$ is an isogenous quotient of $H_{\overline{k}}$, so $\dim H = \dim G'$. But $\dim G < [k' : k] \dim G'$ since $\dim Z > 0$, so if k' admits a k-embedding into K then

$$\dim G < [k' : k] \dim G' \leqslant [K : k] \dim H = \dim \mathrm{R}_{K/k}(H),$$

yielding a contradiction for dimension reasons. It follows that k' is not contained in K over k.

Let $L \subseteq \overline{k}$ be the field of definition over k for the subgroup $\mathscr{R}_u(G_{\overline{k}}) \subseteq G_{\overline{k}}$. Since the $\overline{k}$-subgroup $\mathscr{R}_u(\mathrm{R}_{K/k}(H)_{\overline{k}})$ in $\mathrm{R}_{K/k}(H)_{\overline{k}}$ is defined over K, and by hypothesis there is a k-isogeny $\mathrm{R}_{K/k}(H) \twoheadrightarrow G$, we deduce that $L \subseteq K$ over k. We have seen that k' is not contained in K over k, so to get a contradiction from the assumed existence of $(K/k, H)$ it suffices to show $L = k'$ as purely inseparable extensions of k. This determination of the field of definition L/k is the goal of the remainder of the argument.

Obviously $L \subseteq k'$. The field of definition over k for the unipotent radical of the $\overline{k}$-fiber of a smooth connected affine k-group is unaffected by an étale k-isogeny, so we can replace Z with Z^0. That is, we may assume that

μ is infinitesimal, so $\mu \subseteq Z^0_{G'}$. By Galois descent we may assume that k is separably closed. In this case G' descends to a connected semisimple k-group $\mathscr{G}$ (Corollary A.4.11), and $\mathscr{G}$ is necessarily absolutely simple and simply connected. Also, $\mathscr{G}$ is naturally a k-subgroup of $\mathrm{R}_{k'/k}(G')$, so $Z \cap Z^0_{\mathscr{G}}$ makes sense.

Since $Z^0_{G'} = (Z^0_{\mathscr{G}})_{k'}$ and $Z^0_{\mathscr{G}}$ is a finite group of multiplicative type over the separably closed field k, $Z \cap Z^0_{\mathscr{G}}$ is the unique (infinitesimal) k-subgroup $M \subseteq Z^0_{\mathscr{G}}$ such that $M_{k'} = \mu$ in $Z_{G'}$. But the k-subgroup

$$Z/M = Z/(Z \cap Z^0_{\mathscr{G}}) \subseteq \mathrm{R}_{k'/k}(Z^0_{G'})/Z^0_{\mathscr{G}}$$

is a connected *central* unipotent subgroup scheme in $\mathrm{R}_{k'/k}(G')/(Z \cap Z^0_{\mathscr{G}})$, so its $\overline{k}$-fiber is contained in the unipotent radical of $(\mathrm{R}_{k'/k}(G')/M)_{\overline{k}}$ even though Z/M is generally not smooth. (Here we use that the center of a connected reductive group is of multiplicative type in the sense of Definition A.1.5 and so has no nontrivial unipotent subgroups, by Example A.1.7.) Thus, the geometric unipotent radicals of $\mathrm{R}_{k'/k}(G')/M$ and $G = \mathrm{R}_{k'/k}(G')/Z$ have the same field of definition over k. It therefore suffices to show that the unipotent radical of $(\mathrm{R}_{k'/k}(G')/M)_{\overline{k}}$ has field of definition k' over k.

The composite quotient map

$$\mathrm{R}_{k'/k}(G')_{k'} \to G' \to G'/\mu = (\mathscr{G}/M)_{k'} \tag{1.4.5}$$

over k' kills $M_{k'} = \mu$, and the resulting quotient map $(\mathrm{R}_{k'/k}(G')/M)_{k'} \twoheadrightarrow (\mathscr{G}/M)_{k'}$ over k' descends the maximal reductive quotient over $\overline{k}$. Thus, the field of definition over k for the geometric unipotent radical of $\mathrm{R}_{k'/k}(G')/M$ is the same as the field of definition over k for the kernel of (1.4.5). By Theorem 1.6.2(1) this latter field of definition over k is equal to k', so we are done.

Remark 1.4.8 Consider a standard pseudo-reductive k-group G and a 4-tuple $(G', k'/k, T', C)$ from which it arises. Let $\overline{G}' := G'/Z_{G'}$ be the maximal adjoint semisimple quotient of the reductive k'-group G' with connected fibers, and $\overline{T}' = T'/Z_{G'}$. From the construction, there is a natural left action of G on $\mathrm{R}_{k'/k}(G')$ lifting its action on the central quotient $\mathrm{R}_{k'/k}(G')/(\ker\phi) \subseteq G$ that is normal in G. We claim that there is a natural map $f : G \to \mathrm{R}_{k'/k}(\overline{G}')$ with central kernel which respects the left actions of both terms on $\mathrm{R}_{k'/k}(G')$.

Using the compatible maps $G' \to \overline{G}'$, $T' \to \overline{T}'$, and $C \xrightarrow{\psi} \mathrm{R}_{k'/k}(\overline{T}')$, we get a map of k-groups

$$f : G = (\mathrm{R}_{k'/k}(G') \rtimes C)/\mathrm{R}_{k'/k}(T') \\ \to (\mathrm{R}_{k'/k}(\overline{G}') \rtimes \mathrm{R}_{k'/k}(\overline{T}'))/\mathrm{R}_{k'/k}(\overline{T}') \simeq \mathrm{R}_{k'/k}(\overline{G}'),$$

and this respects the actions on $\mathrm{R}_{k'/k}(G')$. The kernel of f is $Q/\mathrm{R}_{k'/k}(T')$, where Q is the group of pairs (g, c) such that $g \in \mathrm{R}_{k'/k}(G')$ maps to some $\bar{t} \in \mathrm{R}_{k'/k}(\overline{T}') \subseteq \mathrm{R}_{k'/k}(\overline{G}')$ and $c \in C$ maps to $\bar{t}^{-1}$. Since the preimage of $\overline{T}'$ in G' is T', $\ker f$ is identified with $\ker(\psi : C \to \mathrm{R}_{k'/k}(\overline{T}'))$. This is central in G due to the definition of how C acts on $\mathrm{R}_{k'/k}(G')$.

Example 1.4.9 We now give a non-commutative counterpart to Example 1.3.2. Our aim is to construct a non-commutative standard pseudo-reductive k-group G and a smooth connected normal k-subgroup N such that G/N is not pseudo-reductive.

Let G be as in Definition 1.4.4, with $G' \to \mathrm{Spec}\, k'$ a semisimple k'-group whose fibers are connected, absolutely simple, and simply connected. Let $N \subseteq G$ be the image of $\mathrm{R}_{k'/k}(G')$ in G, so this is a smooth connected normal k-subgroup of G. Observe that N is perfect, by Corollary A.7.11. The quotient G/N is identified with the commutative $\mathrm{coker}\, \phi$, where $\phi : \mathrm{R}_{k'/k}(T') \to C$ is the map in (1.4.1), so $N = \mathscr{D}(G)$.

Take $C = \mathrm{R}_{k'/k}(T'/Z_{G'})$, so ϕ restricts to an isogeny between maximal k-tori. Hence, $\mathrm{coker}\, \phi$ is unipotent, so G/N fails to be pseudo-reductive when $\mathrm{coker}\, \phi \neq 1$. By dimension reasons, $\mathrm{coker}\, \phi \neq 1$ precisely when $\dim(\ker \phi) > 0$. This occurs when k'/k is a nontrivial purely inseparable field extension and $Z_{G'}$ is not k'-étale. For example, if k is any imperfect field of characteristic $p > 0$ then we can take $G' = \mathrm{SL}_p$.

Such examples of the failure of G/N to inherit pseudo-reductivity from G are due to G/N being unipotent. It would be more interesting to find examples in which G/N is not pseudo-reductive but is its own derived group. In Example 1.6.4 we will construct such examples, even with perfect G. In any such example, the preimage N' of $\mathscr{R}_{u,k}(G/N)$ in G is a smooth connected normal k-subgroup of G such that $N' \neq \mathscr{D}(N')$. This is surprising, since in the connected reductive case, if G is perfect then so are all of its smooth connected normal subgroups.

1.5 Main result

Modulo the intractable case of commutative pseudo-reductive k-groups, the pushout construction of standard pseudo-reductive groups leads to a description of all pseudo-reductive k-groups aside from some exceptional situations in characteristics 2 and 3.

Our main result is Theorem 5.1.1, and setting aside some complications in characteristics 2 and 3 the theorem asserts the following:

Theorem 1.5.1 *Let k be a field, and let G be a pseudo-reductive k-group. If k is imperfect and $p = \mathrm{char}(k) \in \{2, 3\}$ then assume that the Dynkin diagram of the maximal semisimple quotient $G_{\overline{k}}^{\mathrm{ss}} := G_{\overline{k}}/\mathscr{R}(G_{\overline{k}})$ over $\overline{k}$ has no edge with multiplicity p, and no isolated vertices when $p = 2$.*

The pseudo-reductive k-group G is standard.

In this formulation, the simple factors of $G_{\overline{k}}^{\mathrm{ss}}$ ruled out from consideration over imperfect fields of characteristic $p \in \{2, 3\}$ are type G_2 when $p = 3$ and types B, C, and F_4 when $p = 2$. In Theorem 5.1.1 we make no such restrictions, but we do assume $[k : k^2] \leqslant 2$ when $\mathrm{char}(k) = 2$.

In Chapter 4 we will prove (in arbitrary characteristic) that the data $(G', k'/k)$ entering into the construction of standard pseudo-reductive groups in Definition 1.4.4 can be uniquely chosen to satisfy some additional properties. In this sense, we get a classification of pseudo-reductive k-groups up to k-isomorphism in the spirit of the classification of connected reductive groups except that (i) we have to treat the commutative case as a black box (corresponding to the unknown structure of the commutative pseudo-reductive centralizers of maximal k-tori), (ii) we have to carry along a new invariant, a nonzero finite reduced k-algebra k' (akin to k' in Example 1.3.5) as part of the classification, and (iii) over imperfect fields of characteristics 2 and 3 we need special constructions to handle exceptional phenomena related to type G_2 in characteristic 3 and types B, C, and F_4 in characteristic 2. (Briefly, in these low characteristics some commutation relations among root groups degenerate or some roots are divisible by p in the character lattice.)

Example 1.5.2 Let us explain how connected semisimple groups G over a field k are described from the viewpoint of the pushout construction of standard pseudo-reductive groups. This is a variant on the classical description of G as a quotient $\widetilde{G}/\mu$ of its simply connected central cover $\widetilde{G}$ modulo a finite central (hence multiplicative) k-subgroup scheme μ.

Pick any maximal k-torus $\widetilde{T} \subseteq \widetilde{G}$. The central μ is contained in $\widetilde{T}$. The k-torus image $C = \widetilde{T}/\mu$ of $\widetilde{T}$ in G acts on the left on $\widetilde{G}$ by conjugation, and the natural multiplication map

$$(\widetilde{G} \rtimes C)/\widetilde{T} \to \widetilde{G}/\mu = G$$

is an isomorphism, where $\widetilde{T}$ embeds as a central k-torus in $\widetilde{G} \rtimes C$ via

$$\alpha : t \mapsto (t^{-1}, t \bmod \mu).$$

The quotient presentation $(\widetilde{G} \rtimes C)/\widetilde{T}$ of G is what is provided by the general procedure to describe pseudo-reductive k-groups in the proof of Theorem 1.5.1, due to the fact that $\widetilde{G} = \mathrm{R}_{k'/k}(G')$ for a nonzero finite étale k-algebra k' and k'-group G' whose fibers are absolutely simple and simply connected (Proposition A.5.14). Note that we can choose $\widetilde{T}$ to be the maximal k-torus $\mathrm{R}_{k'/k}(T')$ for any maximal k'-torus T' in G'. Also, $C := \widetilde{T}/\mu$ is a maximal k-torus in G and hence is a Cartan k-subgroup of G. It is the viewpoint of Cartan k-subgroups that indicates the analogue of C in the general pseudo-reductive case.

Due to the appearance of the inversion in the first component of the definition of

$$\alpha : \widetilde{T} \to \widetilde{G} \rtimes C$$

(rather than in the second component), the natural map $\mathrm{Ext}(G, \mu) \to \mathrm{Ext}(G, \widetilde{T})$ of abelian groups of central extensions (arising from the inclusion of μ into $\widetilde{T}$) carries the extension class

$$1 \to \mu \to \widetilde{G} \to G \to 1$$

to the *negative* of the extension class

$$1 \to \widetilde{T} \xrightarrow{\alpha} \widetilde{G} \rtimes C \to G \to 1.$$

We prefer to use the embedding α of $\widetilde{T}$ into $\widetilde{G} \rtimes C$ as defined above rather than the embedding $t \mapsto (t, t^{-1} \bmod \mu)$ obtained from α via inversion on $\widetilde{T}$. The reason is that α is the embedding that works in more general situations in which $\widetilde{T}$ is replaced with a non-commutative group and we use a semidirect product structure against a left action (as opposed to a right action); cf. Remark 1.4.5.

1.6 Weil restriction and fields of definition

In order to prove that a pseudo-reductive k-group G arises from the standard construction, at least away from certain exceptional situations in characteristics

2 and 3, we need a method to use G to reconstruct the k-algebra k' appearing in such a standard construction. The place to look is in fields of definition over k for certain subgroups of $G_{\overline{k}}$. Thus, we now undertake a preliminary study of fields of definition associated to Weil restrictions.

The most basic example of a field of definition over k that naturally arises in any attempt to classify pseudo-reductive groups over a field k is the field of definition over k for the $\overline{k}$-subgroup $\mathscr{R}_u(G_{\overline{k}}) \subseteq G_{\overline{k}}$. This is interesting when G arises as a purely inseparable Weil restriction:

Example 1.6.1 Let k'/k be a finite purely inseparable extension of fields, and let G' be a smooth connected non-unipotent (resp. non-solvable) k'-group. An example of interest is when G' is nontrivial and reductive (resp. nontrivial and semisimple). Consider the Weil restriction

$$G = \mathrm{R}_{k'/k}(G').$$

By Proposition A.7.8(2), the (minimal) field of definition over k for $\mathscr{R}_u(G_{\overline{k}})$ (resp. $\mathscr{R}(G_{\overline{k}})$) as a subgroup of $G_{\overline{k}}$ contains k'. In particular, if G' is nontrivial and reductive then the k-group G is not reductive when $k' \neq k$, although it is pseudo-reductive (by Proposition 1.1.10).

A natural first attempt to describe a general pseudo-reductive k-group G, at least when $G_{\overline{k}}^{\mathrm{red}}$ acts irreducibly on its Lie algebra (to avoid G being a Weil restriction from a nontrivial product of finite extensions of k or from a separable nontrivial finite extension of k), is as follows. Let G be any smooth connected affine k-group. Let $k' \subseteq \overline{k}$ be the field of definition over k for the $\overline{k}$-subgroup $\mathscr{R}_u(G_{\overline{k}}) \subseteq G_{\overline{k}}$. The extension k'/k is finite and purely inseparable.

Let $\mathscr{R}_u(G_{k'})$ denote the descent of $\mathscr{R}_u(G_{\overline{k}})$ to a smooth connected unipotent normal k'-subgroup of $G_{k'}$. For the connected reductive k'-group $G' = G_{k'}/\mathscr{R}_u(G_{k'})$ there is a natural homomorphism

$$i_G : G \to \mathrm{R}_{k'/k}(G') \tag{1.6.1}$$

arising from the quotient map $\pi_G : G_{k'} \to G'$ via the universal property of Weil restriction; we studied this map in some examples in §1.3. Concretely, i_G is the composition of $\mathrm{R}_{k'/k}(\pi_G)$ with the canonical map $G \to \mathrm{R}_{k'/k}(G_{k'})$. It follows from Theorem 1.6.2(2) below that such a G arises as the Weil restriction of a connected reductive group over a purely inseparable extension of k if and only if i_G is an isomorphism. By the universal property of Weil restriction, the closed k-group $\ker i_G$ has $\overline{k}$-fiber contained in $\mathscr{R}_u(G_{\overline{k}})$ and it contains every closed k-subgroup of G with that property.

Our further study of i_G will make extensive use of the natural map $q_{X'} : \mathrm{R}_{k'/k}(X')_{k'} \to X'$ from (A.5.3) for quasi-projective k'-schemes X' (especially smooth affine k'-groups). We refer the reader to Proposition A.5.7 for an explicit description of $q_{X'}$ at the level of points valued in k-algebras, and to Proposition A.5.11 for its properties when X' is k'-smooth (e.g., $q_{X'}$ is then a smooth surjection).

The following theorem is the source of all fine control that we can exert on minimal fields of definition over k, a crucial aspect of many of our later proofs concerning the classification of pseudo-reductive groups.

Theorem 1.6.2 *Let k'/k be a finite extension of fields, G' a nontrivial smooth connected k'-group, and $G = \mathrm{R}_{k'/k}(G')$.*

(1) *Let $\widetilde{G}' \twoheadrightarrow G'$ be a surjective k'-homomorphism from a smooth connected k'-group $\widetilde{G}'$. Assume that the induced map of Lie algebras is nonzero. Then the scheme-theoretic kernel of the natural surjective composite k'-map*

$$\widetilde{\pi} : \mathrm{R}_{k'/k}(\widetilde{G}')_{k'} \xrightarrow{q_{\widetilde{G}'}} \widetilde{G}' \longrightarrow G'$$

has field of definition over k equal to k'.

(2) *If k'/k is purely inseparable and G' is reductive then the natural quotient map $q : G_{k'} = \mathrm{R}_{k'/k}(G')_{k'} \to G'$ is surjective with smooth connected unipotent kernel (so $G' = G_{k'}/\mathscr{R}_{u,k'}(G_{k'})$). The $\overline{k}$-subgroup $\mathscr{R}_u(G_{\overline{k}}) \subseteq G_{\overline{k}}$ has field of definition over k equal to $k' \subseteq \overline{k}$ and the map i_G in (1.6.1) is an isomorphism.*

(3) *Assume G' is semisimple and absolutely simple over k'. Let $q : G_{k'} \twoheadrightarrow G'$ be the natural quotient map. The kernel $\ker q \subseteq G_{k'}$ has field of definition over k equal to k'.*

As σ' varies through all k-embeddings of k' into $\overline{k}$, the quotients $Q_{\sigma'} = G' \otimes_{k',\sigma'} \overline{k}$ of $G_{\overline{k}}$ are (without repetition and up to unique isomorphism) the maximal simple semisimple quotients of $G_{\overline{k}}$. In particular, the subgroup $\ker(G_{\overline{k}} \twoheadrightarrow Q_{\sigma'})$ in $G_{\overline{k}}$ has field of definition over k equal to $\sigma'(k') \subseteq \overline{k}$.

The case of most interest in the first assertion is when G' is semisimple and $\widetilde{G}' \to G'$ is a central isogeny (especially the simply connected central cover of such a G'), in which case the non-vanishing hypothesis always holds. Note that in (2) there is no ambiguity concerning the identification of k' with the field of definition over k in $\overline{k}$ for $\mathscr{R}_u(G_{\overline{k}})$ since k'/k is purely inseparable (so k' has a unique k-embedding into $\overline{k}$).

Proof We begin by proving the first assertion in which G' can be any smooth connected k'-group (not necessarily connected reductive, nor even affine). Since $\widetilde{q} := q_{\widetilde{G}'}$ is a smooth surjection (Proposition A.5.11), it follows that $\widetilde{\pi}$ is nonzero on Lie algebras. Thus, $V = \mathrm{Lie}(\ker \widetilde{\pi})$ is a proper k'-subspace of $\mathrm{Lie}(\mathrm{R}_{k'/k}(G')_{k'}) = k' \otimes_k \mathrm{Lie}(G')$ that contains

$$\mathrm{Lie}(\ker \widetilde{q}) = \ker(\mathrm{Lie}(\widetilde{q})) = \ker(k' \otimes_k \mathrm{Lie}(G') \to \mathrm{Lie}(G')).$$

The field of definition over k for $\ker \widetilde{\pi}$ certainly contains the field of definition over k for the k'-subspace V in $k' \otimes_k \mathrm{Lie}(G')$, so by Lemma A.7.7 (with $K' = k'$) this field is k'.

Turning to (2), assume that k'/k is purely inseparable. According to Proposition A.5.11 the natural quotient map $q : G_{k'} \to G'$ is surjective with smooth connected unipotent kernel. Thus, the map q over k' descends the maximal reductive quotient of $\mathrm{R}_{k'/k}(G')_{\overline{k}}$, so $(\ker q)_{\overline{k}} = \mathscr{R}_u(G_{\overline{k}})$. By Proposition A.7.8 the field of definition over k for $\mathscr{R}_u(G_{\overline{k}})$ is k'/k, so by definition i_G corresponds (under the universal property of Weil restriction) to the k'-descent q of the projection $G_{\overline{k}} \twoheadrightarrow G_{\overline{k}}^{\mathrm{red}}$ onto the maximal reductive quotient. But the definition of q is that it corresponds (under the universal property of Weil restriction) to $\mathrm{R}_{k'/k}(\mathrm{id}_{G'})$, so i_G is an isomorphism since $\mathrm{id}_{G'}$ is an isomorphism. This completes the proof of (2).

Finally, we consider (3), so now k'/k is any finite extension. Let K/k be the separable closure of k in k', so by applying (2) to k'/K we see that the geometric unipotent radical of the K-group $\mathrm{R}_{k'/K}(G')$ has field of definition over K equal to k'. (This identification is unambiguous because k' has a unique K-embedding into an algebraic closure of K, as k'/K is purely inseparable.) The geometric fiber $G_{\overline{k}} = G \otimes_k \overline{k}$ is

$$(\mathrm{R}_{K/k}(\mathrm{R}_{k'/K}(G')))_{\overline{k}} = \prod_{\sigma} \mathrm{R}_{k'/K}(G') \otimes_{K,\sigma} \overline{k}, \tag{1.6.2}$$

with σ ranging through the k-embeddings of K into $\overline{k}$. Each such embedding σ uniquely lifts to a k-embedding $\sigma' : k' \to \overline{k}$, so the maximal semisimple quotient of $G_{\overline{k}}$ is $\prod_{\sigma}(G' \otimes_{k',\sigma'} \overline{k})$. This is a product of $\overline{k}$-simple semisimple groups, so any choice of maximal $\overline{k}$-simple semisimple quotient $G_{\overline{k}} \twoheadrightarrow \overline{H}$ is uniquely isomorphic to the projection $\Pi_{\sigma'}$ from $G_{\overline{k}}$ onto some factor $G' \otimes_{k',\sigma'} \overline{k}$. The projection $\Pi_{\sigma'}$ is $q \otimes_{k',\sigma'} \overline{k}$, and the determination of the field of definition of $\ker q$ over k is a special case of Proposition A.7.8(1) (taking $K' = k'$ and $H = \ker q$ there). □

The criterion in Theorem 1.6.2(2) yields some commutative pseudo-reductive groups G in characteristic $p > 0$ with the interesting property that their finite subgroup of rational p-torsion points is nontrivial and $\ker i_G \neq 1$.

Example 1.6.3 Let k'/k be a nontrivial purely inseparable finite extension of fields with $\mathrm{char}(k) = p > 0$, and assume $k' \subseteq k^{1/p}$. Let $p^n = [k' : k]$. Let $C = \mathrm{R}_{k'/k}(\mathrm{GL}_1)$. By Example 1.3.2, C is a pseudo-reductive k-group and $U := C/\mathrm{GL}_1$ is a smooth connected unipotent k-group of dimension $p^n - 1 > 0$. The isogenous quotient C/μ_p of C fails to be pseudo-reductive over k because it is commutative and contains the k-group scheme $\mathrm{R}_{k'/k}(\mu_p)/\mu_p$ that maps isomorphically onto U (since U is p-torsion, as $k'^p \subseteq k$, and the snake lemma argument in Example 1.3.2 describes $U[p]$).

Loosely speaking, C is $k'^{\times}$ viewed as a k-group. Choose a p-basis $\{a_1, \ldots, a_n\}$ of k' over k, so $a_i^p \in k$ is not a p-power in k and $k' = k(a_1, \ldots, a_n)$ with $[k' : k] = p^n$. Thus, we have a k'-algebra isomorphism

$$k' \otimes_k k' \simeq k'[t_1, \ldots, t_n]/(t_1^p, \ldots, t_n^p),$$

where t_i corresponds to $1 \otimes a_i - a_i \otimes 1$. In other words, C is a k-descent of the unit group $(k'[t_1, \ldots, t_n]/(t_i^p))^{\times}$ viewed as a k'-group and U is a k-descent of the subgroup of 1-units. We claim that $U_{k'} \simeq \mathbf{G}_{\mathrm{a}}^{p^n-1}$ as k'-groups. To prove this, first note that the maximal ideal $\mathfrak{m}$ of $k' \otimes_k k'$ is killed by the p-power map. Hence, for $x \in \mathfrak{m}$ the truncation $\exp_{<p}$ of exp in degrees $< p$ defines a k'-homomorphism $\mathbf{G}_{\mathrm{a}} \to U_{k'}$ by $t \mapsto \exp_{<p}(tx)$. By forming the product of such maps for the members of the k'-basis of $\mathfrak{m}$ given by the nonzero monomials in the t_i, we thereby obtain a k'-homomorphism $\mathbf{G}_{\mathrm{a}}^{p^n-1} \to U_{k'}$ that is an isomorphism (as is checked by using the $\mathfrak{m}$-adic filtration on $U_{k'}$). We conclude that the group $\mathrm{Ext}_{k'}(U_{k'}, \mathbf{Z}/p\mathbf{Z})$ of commutative extensions over k' is isomorphic to $\mathrm{Ext}_{k'}(\mathbf{G}_{\mathrm{a}}, \mathbf{Z}/p\mathbf{Z})^{p^n-1}$, so there exist *connected* commutative k'-group extensions E' of $U_{k'}$ by $\mathbf{Z}/p\mathbf{Z}$ (via Artin–Schreier theory).

Choose such an extension E', so we can view E' as a finite étale cover of $U_{k'}$ (with a canonical trivialization over the identity) and view $E' \times E'$ as a finite étale cover of $U_{k'} \times U_{k'} = (U \times U)_{k'}$. The group law on E' corresponds to a map

$$E' \times E' \simeq m_{U_{k'}}^*(E')$$

over $(U \times U)_{k'}$ satisfying associativity, where $m_{U_{k'}} : U_{k'} \times U_{k'} \to U_{k'}$ is the multiplication map. Since k'/k is purely inseparable, for any k-scheme S (e.g., $\operatorname{Spec} k$, U, or $U \times U$), it follows from [SGA1, IX, Thm. 4.10] that pullback along $S_{k'} \to S$ induces an equivalence of étale sites. The extension structure E' thereby canonically descends to a commutative extension E of U by $\mathbf{Z}/p\mathbf{Z}$ over k. (This makes explicit that $\mathrm{H}^1_{\text{ét}}(U, \mathbf{Z}/p\mathbf{Z}) = \mathrm{H}^1_{\text{ét}}(U_{k'}, \mathbf{Z}/p\mathbf{Z})$.) In particular, E is a smooth connected unipotent commutative k-group.

The fiber product $G = E \times_U C$ is an extension of E by

$$\ker(C \twoheadrightarrow U) = \mathrm{GL}_1,$$

so G is a smooth connected commutative affine k-group for which $\mathscr{R}_u(G_{\overline{k}}) \simeq E_{\overline{k}}$. Also, $G \to C$ is an étale isogeny with covering group $\mathbf{Z}/p\mathbf{Z}$, so G is pseudo-reductive over k (since C is) and the field of definition over k for the subgroup $\mathscr{R}_u(G_{\overline{k}}) \subseteq G_{\overline{k}}$ is the same as that of $\mathscr{R}_u(C_{\overline{k}}) \subseteq C_{\overline{k}}$, which is to say k' (by Theorem 1.6.2(2)). Via the isomorphism

$$G_{k'}/\mathscr{R}_u(G_{k'}) \simeq C_{k'}/\mathscr{R}_u(C_{k'}) = \mathrm{GL}_1,$$

the map i_G is the natural map $G \twoheadrightarrow C$ that is an isogeny with kernel $\mathbf{Z}/p\mathbf{Z}$, so $\ker i_G \neq 1$.

It also follows that G cannot be k-isomorphic to $\mathrm{R}_{k'/k}(G')$ for a nonzero finite reduced k-algebra G' and a k'-group G' with connected reductive fibers. Indeed, otherwise the commutativity of $G(k) = G'(k' \otimes_k k_s)$ would force G' to be a k'-torus, which contradicts that $G(k)[p] \neq 1$.

Example 1.6.4 Let k be an imperfect field with $\mathrm{char}(k) = p > 0$. We now make an example of a perfect pseudo-reductive k-group G and perfect smooth connected normal k-subgroup N such that G/N is not pseudo-reductive. As we pointed out at the end of Example 1.4.9, such a G contains a smooth connected normal k-subgroup N' that is not perfect, which never happens in connected semisimple groups.

Let k'/k be a nontrivial purely inseparable finite extension such that $k' \subseteq k^{1/p}$. Let G' be a connected semisimple k'-group that is absolutely simple and simply connected such that $Z_{G'}$ contains a k'-subgroup μ' that is a k'-form of μ_{p^2}. (Some such G' with $Z_{G'} = \mu_{p^2}$ are $G' = \mathrm{SL}_{p^2}$ for any p and the spin double covering of SO_{2n} for odd $n > 3$ when $p = 2$.) Let T' be a maximal k'-torus in G', and define the k-subgroup $Z \subseteq \mathrm{R}_{k'/k}(T') \times \mathrm{R}_{k'/k}(T')$ to correspond to the k-subgroup inclusion

$$\mathrm{R}_{k'/k}(\mu') \hookrightarrow \mathrm{R}_{k'/k}(T') \times \mathrm{R}_{k'/k}(T')$$

defined by $\zeta \mapsto (\zeta, \zeta^p)$.

Define $G'' = G' \times G'$ and $T'' = T' \times T'$. By Proposition 1.3.4, the k-groups $G = \mathrm{R}_{k'/k}(G'')/Z$ and $C = \mathrm{R}_{k'/k}(T'')/Z$ are pseudo-reductive and G is perfect. Since C is commutative and acts on $\mathrm{R}_{k'/k}(G'')$ in the evident manner, we get a standard presentation of G via the natural isomorphism

$$G \simeq (\mathrm{R}_{k'/k}(G'') \rtimes C)/\mathrm{R}_{k'/k}(T'').$$

Let $N \subseteq G$ be the smooth connected normal k-subgroup that is the image in G of the first factor $\mathrm{R}_{k'/k}(G')$ of $\mathrm{R}_{k'/k}(G'')$, so N is perfect. We shall prove that the quotient G/N is not pseudo-reductive over k.

In view of the construction of G and N, there is a k-group isomorphism

$$G/N \simeq (\mathrm{R}_{k'/k}(G') \rtimes \overline{C})/\mathrm{R}_{k'/k}(T') \tag{1.6.3}$$

with $\overline{C} = \mathrm{R}_{k'/k}(T')/M$, where $M = [p](\mathrm{R}_{k'/k}(\mu')) \subseteq \mathrm{R}_{k'/k}(T')$. This presentation of G/N looks like the standard construction but beware that we have not addressed whether or not $\overline{C}$ is pseudo-reductive over k (and it will not be).

We shall now use (1.6.3) to exhibit a nontrivial smooth connected central unipotent k-subgroup of G/N (even contained in $\overline{C}$), thereby proving that G/N is not pseudo-reductive over k. The k-subgroup $M \hookrightarrow \mathrm{R}_{k'/k}(T') \subseteq \mathrm{R}_{k'/k}(G')$ is central, as it is the (scheme-theoretic) image of the p-power map on the central k-subgroup scheme $\mathrm{R}_{k'/k}(\mu')$. The natural map $\mathrm{R}_{k'/k}(G') \rightarrow G/N$ has kernel equal to M and so gives rise to the central k-subgroup $Q = \mathrm{R}_{k'/k}(\mu')/M$ in G/N. It therefore suffices to prove that Q contains a nontrivial smooth connected unipotent k-subgroup. If such a k-subgroup exists then there is a unique maximal one containing all others, so by Galois descent we may and do assume $k = k_s$. In particular, $\mu' = \mu_{p^2}$.

It remains to prove that the quotient $Q = \mathrm{R}_{k'/k}(\mu_{p^2})/[p](\mathrm{R}_{k'/k}(\mu_{p^2}))$ contains a nontrivial smooth connected unipotent k-group. Since $k'^p \subseteq k$, the geometric unipotent radical of $\mathrm{R}_{k'/k}(\mathrm{GL}_1)$ is killed by p. That is, the p-power map on $\mathrm{R}_{k'/k}(\mathrm{GL}_1)$ lands inside of the unique GL_1-subgroup. More specifically, we have a canonical isomorphism of $\overline{k}$-groups $\mathrm{R}_{k'/k}(\mathrm{GL}_1)_{\overline{k}} = \mathrm{GL}_1 \times U$ with $pU = 0$. Hence, for any $n \geqslant 1$ the p^n-torsion $\mathrm{R}_{k'/k}(\mu_{p^n})_{\overline{k}}$ of $\mathrm{R}_{k'/k}(\mathrm{GL}_1)_{\overline{k}}$ is $\mu_{p^n} \times U$. Taking $n = 2$, the scheme-theoretic image of

$\mathrm{R}_{k'/k}(\mu_{p^2})_{\overline{k}}$ under the p-power map is the canonical subgroup μ_p over $\overline{k}$, so likewise the k-subgroup μ_p inside of $\mathrm{R}_{k'/k}(\mu_{p^2})$ is the scheme-theoretic image of the p-power map. It follows that $Q = \mathrm{R}_{k'/k}(\mu_{p^2})/\mu_p$. This contains $\mathrm{R}_{k'/k}(\mu_p)/\mu_p$, and we saw in Example 1.6.3 that this k-group is isomorphic to the k-group $\mathrm{R}_{k'/k}(\mathrm{GL}_1)/\mathrm{GL}_1$ that is nontrivial, smooth, connected, and unipotent.

2

Root groups and root systems

The category of pseudo-reductive groups over a field k has weak stability properties under basic operations on group schemes, such as extension of the base field and passage to quotients (even by finite central multiplicative or smooth normal closed subgroups). Fortunately, pseudo-reductive groups do satisfy many properties generalizing those of connected reductive groups, and this chapter establishes such properties for convenient use later on.

Following Tits, we use an interesting construction of unipotent groups to construct a theory of root systems and root groups for pseudo-reductive groups over any separably closed field, or more generally for *pseudo-split* pseudo-reductive groups, i.e., the ones which contain a split maximal torus (over an arbitrary ground field). Tits developed a version of this theory without requiring a split maximal torus, but it will suffice for our needs in the main body of this book to have this theory for pseudo-split groups. In Appendix C we give a detailed exposition of k-root systems, relative Weyl groups, and k-root groups, as well as construct a Tits system in $G(k)$ for any smooth connected affine group G over any field k; see C.2.13 to C.2.34.

In contrast with the connected reductive case, root groups in pseudo-split pseudo-reductive groups over a field k may have dimension larger than 1, though they turn out to always be vector groups over k (i.e., k-isomorphic to a power of $\mathbf{G}_{\mathrm{a}}$). Our structure theorem for pseudo-reductive groups explains *a posteriori* why root groups can have large dimension: they turn out to naturally be Weil restrictions of 1-dimensional vector groups over a finite extension of k if $\mathrm{char}(k) \neq 2$, and we prove a related result when $\mathrm{char}(k) = 2$.

2.1 Limits associated to 1-parameter subgroups

The construction of root groups rests on a generalization to pseudo-reductive groups of the theory of parabolic subgroups and their unipotent radicals in connected reductive groups. Recall that a k-subgroup P in a smooth connected affine group G over a field k is called *parabolic* if P is smooth and closed and the quotient G/P is k-proper. This definition turns out not to be so useful in the pseudo-reductive case, essentially because Weil restriction through a nontrivial purely inseparable finite extension of fields *never* preserves parabolicity (for proper subgroups), as it never preserves properness for smooth quasi-projective schemes of positive dimension; see Example A.5.6.

The viewpoint that works well in the pseudo-reductive case is a variant on the classification of parabolic k-subgroups of connected reductive groups in terms of relative root systems and centralizers of k-split tori. This variant involves choosing a 1-parameter k-subgroup $\lambda : \mathrm{GL}_1 \to G$ (i.e., a k-homomorphism) and studying the limiting behavior of $\lambda(t)g\lambda(t)^{-1}$ as $t \to 0$ in an appropriate sense. Actually, we can consider more generally the case of an abstract left action of GL_1 on the k-group G and study the limiting behavior of $t \mapsto t.g$ as $t \to 0$ in an appropriate sense. This general case with abstract actions can be studied in more concrete terms using conjugation by 1-parameter subgroups in the semidirect product $\mathrm{GL}_1 \ltimes G$ defined by the given abstract action. For clarity of exposition, we therefore initially carry out the development for limits associated to 1-parameter subgroups, and in Remark 2.1.11 we address the generalization to limits associated to abstract left actions by GL_1.

The basic features of the theory of such limits are best illustrated by two examples (each of which will be useful in later general considerations).

Example 2.1.1 Let k be a ring, and $G = \mathrm{GL}(V)$ for a finitely generated projective k-module V. Let $\lambda : \mathrm{GL}_1 \to G$ be a k-homomorphism. By Lemma A.8.8, there is a canonical decomposition

$$V = \bigoplus V_{\chi_i},$$

where $\{\chi_1, \ldots, \chi_m\}$ is the finite set of pairwise distinct weights for the GL_1-action on V via λ (and $m \geqslant 0$). In particular, the V_{χ_i} are nonzero finitely generated projective k-modules and $\chi_i(t) = t^{c_i}$ for a unique $c_i \in \mathbf{Z}$. By working Zariski-locally on $\mathrm{Spec}\, k$ for now, we may assume that $V \neq 0$ and the V_{χ_i} are free k-modules. We also may and do arrange the χ_i so that c_i is strictly decreasing in i.

Consider the parabolic k-subgroup $P_G(\lambda)$ consisting of automorphisms of V respecting the increasing filtration by subspaces $F_j = \bigoplus_{i \leqslant j} V_{\chi_i}$. Choose an ordered k-basis $\{v_1, \ldots, v_n\}$ of V adapted to this filtration: each v_r is in a weight space $V_{\chi_{i_r}}$ and $i_1 \leqslant \cdots \leqslant i_n$. Relative to such a basis we have $G = \mathrm{GL}_n$ with

$$\lambda(t) = \mathrm{diag}(t^{e_1}, \ldots, t^{e_n})$$

for $e_r = c_{i_r}$ and $e_1 \geqslant \cdots \geqslant e_n$.

If $g = (a_{rs})$ in $\mathrm{GL}(V) \simeq \mathrm{GL}_n$ then g is a point of $P_G(\lambda)$ if and only if $g(V_{\chi_j}) \subseteq F_j$ for all $j \geqslant 1$, which is to say that $a_{rs} = 0$ whenever $i_r > i_s$. But it is elementary to compute that for any g,

$$\lambda(t) g \lambda(t)^{-1} = (t^{e_r - e_s} a_{rs}) = (t^{c_{i_r} - c_{i_s}} a_{rs}), \tag{2.1.1}$$

so the rs entry has negative t exponent precisely when $i_r > i_s$.

Now returning to the case of general k, V, and λ (without assuming the existence of bases), we conclude that $P_G(\lambda)$ represents the functor of points $g \in G(R)$ (for any k-algebra R) such that the R-map $(\mathrm{GL}_1)_R \to G_R$ defined by $t \mapsto \lambda(t) g \lambda(t)^{-1}$ extends (necessarily uniquely) to an R-map $\mathbf{A}^1_R \to G_R$. We summarize this latter condition by saying that

$$\lim_{t \to 0} \lambda(t) g \lambda(t)^{-1}$$

exists, and the image of 0 in $G(R)$ is considered to be the limit.

This limit procedure functorially characterizes the parabolic k-subgroup $P_G(\lambda)$ entirely in terms of the 1-parameter k-subgroup $\lambda : \mathrm{GL}_1 \to G$ without any mention of the coset space $G/P_G(\lambda)$ (or its properties). This paves the way for a viewpoint that will be useful when G is replaced with a general smooth affine k-group whose fibers over $\mathrm{Spec}\, k$ are connected.

We can give a similar functorial description of the "k-split unipotent radical" $U_G(\lambda)$ of the parabolic k-subgroup $P_G(\lambda)$: it consists of automorphisms acting trivially on the graded space

$$\mathrm{gr}_\bullet(V) = \bigoplus (F_j / F_{j-1}) \simeq \bigoplus V_{\chi_j}$$

for the filtration $\{F_j\}$ of V. Indeed, by (2.1.1) we see that for any k-algebra R, if $g \in P_G(\lambda)(R)$ then $\lim_{t \to 0} \lambda(t) g \lambda(t)^{-1}$ lies in the subgroup $\prod \mathrm{GL}(V_{\chi_j})$ and is the automorphism induced by g on $\mathrm{gr}_\bullet(V)$. k-subgroup $U_G(\lambda)$ represents the functor of points of $P_G(\lambda)$ with trivial action on $\mathrm{gr}_\bullet(V)$, so $U_G(\lambda)$ is the functor of points g of G such that $\lambda(t) g \lambda(t)^{-1} \to 1$ as $t \to 0$. This functorial description of $U_G(\lambda)$ only uses the 1-parameter subgroup λ and makes no reference to unipotence (or to $P_G(\lambda)$).

Example 2.1.1 can be pushed further via the reciprocal 1-parameter k-subgroup $-\lambda : \mathrm{GL}_1 \to G$ defined by $t \mapsto \lambda(1/t)$. The scheme-theoretic intersection $P_G(\lambda) \cap P_G(-\lambda)$ is the subgroup of block matrices $\prod \mathrm{GL}(V_{\chi_j}) \subseteq G$ that is exactly the scheme-theoretic centralizer $Z_G(\lambda)$ of λ in G. By inspection, the natural multiplication map

$$Z_G(\lambda) \ltimes U_G(\lambda) \to P_G(\lambda)$$

is an isomorphism of k-groups; i.e., $Z_G(\lambda)$ serves as a quotient $P_G(\lambda)/U_G(\lambda)$.

Before moving on to the next example, we directly prove the following result without reference to the structure theory of reductive groups (such as Bruhat cells):

Proposition 2.1.2 *Let k be a ring and V a finitely generated projective k-module. For $G = \mathrm{GL}(V)$ and a k-homomorphism $\lambda : \mathrm{GL}_1 \to G$, the multiplication map of k-schemes $\mu : U_G(-\lambda) \times P_G(\lambda) \to G$ is an open immersion.*

In Proposition 2.1.8(3) this will be proved for any smooth affine group over any ring, and any affine group of finite type over any field.

Proof The intersection $U_G(-\lambda) \cap P_G(\lambda)$ is trivial by inspection, so μ is a functorial injection (i.e., monomorphism). By [EGA, IV_4, 17.9.1], an open immersion is the same thing as an étale monomorphism, so to prove that μ is an open immersion it suffices to prove that it is étale. Since each of $U_G(-\lambda)$, $P_G(\lambda)$, and G are smooth, μ is étale if it is an isomorphism on tangent spaces at all rational points of $U_G(-\lambda) \times P_G(\lambda)$ in geometric fibers over $\mathrm{Spec}\, k$. Hence, we can assume that k is an algebraically closed field, and by translation considerations it suffices to check the tangential isomorphism property at a single k-point of $U_G(-\lambda) \times P_G(\lambda)$. We choose the identity point.

At the identity, the desired tangential isomorphism property says that the addition map

$$d\mu_{(1,1)} : \mathrm{Lie}(U_G(-\lambda)) \oplus \mathrm{Lie}(P_G(\lambda)) \to \mathrm{Lie}(G)$$

is an isomorphism of k-vector spaces. By inspection, $\mathrm{Lie}(U_G(-\lambda))$ is the span of the negative weight spaces in $\mathrm{Lie}(G)$ for the GL_1-action induced by

$g \mapsto \lambda(t)g\lambda(t)^{-1}$. Likewise, $\mathrm{Lie}(P_G(\lambda))$ is the span of the nonnegative weight spaces for this GL_1-action. Hence, $\mathrm{d}\mu_{(1,1)}$ is an isomorphism. □

To indicate how Example 2.1.1 will carry over to pseudo-reductive groups, we consider a special case.

Example 2.1.3 Let k'/k be a finite extension of fields, and let $G = \mathrm{R}_{k'/k}(\mathrm{SL}_2)$. This naturally contains SL_2 as a k-subgroup. Let $\lambda : \mathrm{GL}_1 \to G$ be the k-homomorphism that is the standard 1-parameter diagonal subgroup of this SL_2; that is, define $\lambda(t) = \mathrm{diag}(t, 1/t)$.

Inside of the k'-group SL_2, let D' be the diagonal k'-subgroup, B' the upper triangular Borel k'-subgroup, and U' the unipotent radical of B'. Note that if k'/k is purely inseparable then the k-subgroup SL_2 in G is a Levi k-subgroup and $\lambda(\mathrm{GL}_1)$ is a maximal k-torus in G.

Consider the subgroups $D = \mathrm{R}_{k'/k}(D')$, $B = \mathrm{R}_{k'/k}(B')$, and $U = \mathrm{R}_{k'/k}(U')$ in G. If k'/k is a nontrivial purely inseparable extension then the quotient $B/U = D$ is pseudo-reductive but not reductive (so $U = \mathscr{R}_{u,k}(B)$ is not a k-descent of $\mathscr{R}_u(B_{\overline{k}})$) and $G/B = \mathrm{R}_{k'/k}(G'/B') = \mathrm{R}_{k'/k}(\mathbf{P}^1_{k'})$ is not proper over k (see Example A.5.6). However, for any k'/k and any k-algebra R we have

$$\lambda(t)\begin{pmatrix} x & y \\ z & w \end{pmatrix}\lambda(t)^{-1} = \begin{pmatrix} x & t^2 y \\ t^{-2} z & w \end{pmatrix}$$

in $G(R) = \mathrm{SL}_2(k' \otimes_k R)$ for any $t \in R^\times = \mathrm{GL}_1(R)$. Thus, in general D is the centralizer of λ in G, B represents the functor of points g of G such that $\lambda(t)g\lambda(t)^{-1}$ has a limiting value as $t \to 0$, and U represents the functor of points g of G such that this limit exists and equals 1. In this sense, B is analogous to a parabolic subgroup, U is analogous to its unipotent radical, and D is analogous to a Levi subgroup of B.

Motivated by Examples 2.1.1 and 2.1.3, we seek to study limits $\lim_{t\to 0} t.v$ for the action of GL_1 on an affine k-scheme V for any ring k. Since GL_1 is "dual" to $\mathbf{Z}$, by Lemma A.8.8 the coordinate ring $k[V]$ of V admits an associated weight space decomposition

$$k[V] = \bigoplus_{n\in\mathbf{Z}} k[V]_n \tag{2.1.2}$$

with $k[V]_n$ consisting of the $f \in k[V]$ such that $f(t.v) = t^n f(v)$ in R for all k-algebras R, $v \in V(R)$, and $t \in \mathrm{GL}_1(R) = R^\times$; such an f is called an *n-eigenvector* for the GL_1-action on V.

Lemma 2.1.4 *Let k be a ring and V be an affine k-scheme equipped with a GL_1-action. Define the ideal*

$$J := \sum_{n<0} k[V] \cdot k[V]_n$$

in $k[V]$. Let $V' = \mathrm{Spec}(k[V]/J)$ denote the zero scheme of J in V.

The closed subscheme V' in V represents the subfunctor of V that associates to any k-algebra R the set of points $v \in V(R)$ such that $\lim_{t\to 0} t.v$ exists in the sense that the orbit map $\alpha_v : (\mathrm{GL}_1)_R \to V_R$ over R defined by $t \mapsto t.v$ extends (necessarily uniquely) to an R-morphism $\mathbf{A}^1_R \to V_R$. If K is any k-algebra then V'_K represents the corresponding subfunctor of V_K on the category of K-algebras, and the formation of V' commutes with passage to any GL_1-stable closed subscheme $Z \subseteq V$ (i.e., $Z' = Z \cap V'$).

For $v \in V'(R)$, the image of 0 in $V(R)$ under the extended map $\mathbf{A}^1_R \to V_R$ is denoted $\lim_{t\to 0} t.v$ in what follows.

Proof For $f \in k[V]$, let f_n be its component along $k[V]_n$ (so all but finitely many f_n vanish). The R-map α_v corresponds to the k-algebra map $\alpha_v^* : k[V] \to R[t, 1/t]$ defined by $f \mapsto \sum_{n\in\mathbf{Z}} t^n f_n(v)$. Thus, α_v extends to an R-morphism $\mathbf{A}^1_R \to V_R$ if and only if α_v^* has image contained in the subring $R[t] \subseteq R[t, 1/t]$. This containment says exactly that for all $n \in \mathbf{Z}$ we have $n \geqslant 0$ whenever $f(v) \neq 0$ for some $f \in k[V]_n$.

Equivalently, the necessary and sufficient condition is that if $n < 0$ then $f(v) = 0$ for all $f \in k[V]_n$. In other words, J vanishes at $v \in V(R)$, which is to say that $v \in V'(R)$. This establishes the functorial characterization of V'.

By the functorial characterization (or the construction via J), it follows that the formation of V' commutes with any extension of the base ring. The compatibility with passage to GL_1-stable closed subschemes is handled similarly. (In concrete terms, the coordinate ring of such a subscheme is a quotient of $k[V]$ compatible with the $\mathbf{Z}$-grading.) □

Let $V = G$ be an affine group scheme over a ring k and let $\lambda : \mathrm{GL}_1 \to G$ be a k-group homomorphism. Endow G with a GL_1-action via $t.g = \lambda(t)g\lambda(t)^{-1}$. By Lemma 2.1.4, there is a closed subscheme $P_G(\lambda) \subseteq G$ representing the functor of points g of G for which $\lim_{t\to 0} t.g$ exists in G. Moreover, the lemma ensures that the formation of $P_G(\lambda)$ is compatible with any extension of the base ring and that if $H \subseteq G$ is a closed k-subgroup scheme through which λ factors then $H \cap P_G(\lambda) = P_H(\lambda)$ (using λ to also denote the homomorphism $\mathrm{GL}_1 \to H$).

Following standard conventions with coroots in connected reductive groups, we write $-\lambda$ to denote the reciprocal homomorphism $t \mapsto \lambda(1/t) = \lambda(t)^{-1}$.

Lemma 2.1.5 *The closed subscheme $P_G(\lambda) \subseteq G$ is a k-subgroup scheme, and*

$$Z_G(\lambda) := P_G(\lambda) \cap P_G(-\lambda)$$

is the scheme-theoretic centralizer in G for the GL_1*-action via conjugation by* λ *(i.e., its R-points for a k-algebra R are the elements $g \in G(R)$ that centralize the R-homomorphism λ_R).*

The subfunctor of G defined by

$$R \rightsquigarrow \{g \in G(R) \mid \lim_{t \to 0} t.g = 1\}$$

is represented by a closed normal k-subgroup scheme $U_G(\lambda) \subseteq P_G(\lambda)$, and if G is of finite type over k then $U_G(\lambda)$ has unipotent fibers over Spec *k. The formation of $U_G(\lambda)$ commutes with any extension of the base ring, and if $H \subseteq G$ is a closed k-subgroup scheme through which λ factors then $H \cap U_G(\lambda) = U_H(\lambda)$.*

If G is finitely presented over k then so are $P_G(\lambda)$, $U_G(\lambda)$, and $Z_G(\lambda)$.

Proof To check that $P_G(\lambda)$ is a subgroup scheme we compute on R-points for any k-algebra R, as follows. Choose $g, g' \in G(R)$ such that $t.g \to g_0$ and $t.g' \to g'_0$ as $t \to 0$. That is, there are (necessarily unique) R-morphisms

$$f, f' : \mathbf{A}^1_R \rightrightarrows G_R$$

whose restrictions to $(\mathrm{GL}_1)_R$ are respectively $t \mapsto t.g$ and $t \mapsto t.g'$, and they satisfy $f(0) = g_0$ and $f'(0) = g'_0$ in $G(R)$. Use the group law on G to define a product map

$$f \cdot f' : \mathbf{A}^1_R \to G_R$$

over R. This carries 0 to $g_0 g'_0$, and its restriction to $(\mathrm{GL}_1)_R$ is $t \mapsto (t.g)(t.g') = t.(gg')$. Hence, $gg' \in P_G(\lambda)(R)$. A similar argument shows that $P_G(\lambda)$ is stable under inversion and contains the identity section, so it is a k-subgroup of G.

This argument also shows that $g \mapsto \lim_{t \to 0} t.g$ is a multiplicative k-map $P_G(\lambda) \to G$. Such a map is necessarily a k-group homomorphism, and its scheme-theoretic kernel is exactly the desired $U_G(\lambda)$. The compatibility of $U_G(\lambda)$ with extension of scalars is clear. Explicitly, the ideal of $U_G(\lambda)$ is generated by the n-eigenvectors $f \in k[G]$ (i.e., $f \in k[G]_n$) for $n \leqslant 0$ such that $f(1) = 0$ (i.e., f lies in the augmentation ideal of $k[G]$).

The equality $H \cap U_G(\lambda) = U_H(\lambda)$ holds because $H \cap P_G(\lambda) = P_H(\lambda)$. To check unipotence of the fibers of $U_G(\lambda)$ over $\operatorname{Spec} k$ when G is of finite type, we may assume k is a field. In this case we can choose a k-subgroup inclusion $j : G \hookrightarrow \mathrm{GL}_n$, so $U_G(\lambda) = G \cap U_{\mathrm{GL}_n}(j \circ \lambda)$. Thus, the unipotence of $U_G(\lambda)$ is reduced to the special case $G = \mathrm{GL}_n$, where it was seen in Example 2.1.1.

Consider $Z_G(\lambda)$ in general. To show that it is the centralizer of the GL_1-action, we have to prove that for any k-algebra R and $g \in G(R)$, g centralizes the map $\lambda_R : (\mathrm{GL}_1)_R \to G_R$ if and only if $g \in Z_G(\lambda)(R)$. Centralizing λ_R means precisely that $t.g = g$ for all functorial points t of $(\mathrm{GL}_1)_R$, and the membership $g \in Z_G(\lambda)(R)$ means precisely that $\lim_{t \to 0} t.g$ and $\lim_{t \to 0} t^{-1}.g$ both exist. Clearly if g centralizes λ_R then both limits exist (and equal g). Conversely, if both limits exist then the maps $t \mapsto t.g$ and $t \mapsto t^{-1}.g$ glue to define an R-morphism $\mathbf{P}^1_R \to G_R$ from the projective line to an affine R-scheme. The only such map is a constant R-map, which forces $t.g$ to be independent of t, so $t.g = g$ for all functorial points t of $(\mathrm{GL}_1)_R$.

If G is finitely presented over k then G and λ can be descended to a noetherian subring of k. Hence, the compatibility of the formation of $P_G(\lambda)$, $U_G(\lambda)$, and $Z_G(\lambda)$ with scalar extension on k reduces the finite presentation claim to the case of noetherian k, where it is clear. □

Example 2.1.6 Let k be a field. It is a well-known fact that for a nonzero finite-dimensional k-vector space V, the parabolic k-subgroups of $\mathrm{GL}(V)$ correspond to flags of subspaces of V. Thus, by Example 2.1.1, in $G = \mathrm{GL}_n$ the groups $P_G(\lambda)$ as λ varies are exactly the parabolic k-subgroups, with $U_G(\lambda)$ its unipotent radical and $Z_G(\lambda)$ a Levi k-subgroup of $P_G(\lambda)$.

For $G = \mathrm{R}_{k'/k}(\mathrm{SL}_2)$ and the standard λ as in Example 2.1.3, the k-subgroup $P_G(\lambda)$ is the Weil restriction of the standard upper triangular Borel subgroup of the k'-group SL_2 and the k-subgroup $U_G(\lambda)$ is the Weil restriction of the unipotent radical of this Borel subgroup. Also, $Z_G(\lambda)$ is the Weil restriction of the diagonal k'-subgroup of SL_2.

Remark 2.1.7 In the setup of Lemma 2.1.5, for any integer $n \geqslant 1$ we have $P_G(\lambda^n) = P_G(\lambda)$ and $U_G(\lambda^n) = U_G(\lambda)$. Indeed, suppose that $\lambda(t^n)g\lambda(t^n)^{-1} \to g' \in G(R)$ as $t \to 0$ for some k-algebra R and $g \in G(R)$. That is, we assume that the R-map $(\mathrm{GL}_1)_R \to G_R$ defined by $t \mapsto \lambda(t^n)g\lambda(t^n)^{-1}$ extends to an R-map $\phi : \mathbf{A}^1_R \to G_R$, with $g' = \phi(0)$. The problem is to show that $\phi = \widetilde{\phi}(x^n)$ for some R-map $\widetilde{\phi} : \mathbf{A}^1_R \to G_R$ such that $\widetilde{\phi}|_{(\mathrm{GL}_1)_R}$ is $t \mapsto \lambda(t)g\lambda(t)^{-1}$ and $\widetilde{\phi}(0) = g'$.

First we address the factorization of ϕ through the nth-power map. By descent theory, it is necessary and sufficient that $\phi(x) = \phi(y)$ whenever $x^n = y^n$ (for functorial points x and y of the affine line over R). The residue classes $t, t' \in R[T, T']/(T^n - T'^n)$ of T and T' respectively are not zero divisors, so it suffices to consider only points x and y of the multiplicative group instead of the affine line, in which case the hypothesis on ϕ gives the desired equality.

Since $\widetilde{\phi}(x^n) = \phi(x)$, faithful flatness of the nth-power map on $(\mathrm{GL}_1)_R$ implies that $\widetilde{\phi}(t) = \lambda(t)g\lambda(t)^{-1}$ for points t of $(\mathrm{GL}_1)_R$. We also have $\widetilde{\phi}(0) = \phi(0) = g'$.

The next task is to relate $P_G(\lambda)$, $Z_G(\lambda)$, and $U_G(\lambda)$ to a GL_1-action on the tangent space to G at the identity. We first review some elementary facts related to such tangent spaces for general group schemes over any ring. (We will only need the case of affine group schemes.) Let G be a group scheme over a ring k, with $e : \mathrm{Spec}\, k \to G$ the identity section. Since sections to scheme maps are locally closed immersions, the section e is the zero locus of a unique quasi-coherent ideal $\mathscr{I}_G$ near e, so $\mathscr{I}_G/\mathscr{I}_G^2$ corresponds to a k-module $\mathrm{Cot}_e(G)$ (of finite presentation if G is locally of finite presentation over k), the *cotangent space* to G along e. Its formation commutes with any base change on k. For the linear dual $\mathfrak{g} = \mathrm{Tan}_e(G) := \mathrm{Hom}_k(\mathrm{Cot}_e(G), k)$, there is a natural linear base change morphism

$$k' \otimes_k \mathrm{Tan}_e(G) \to \mathrm{Tan}_{e_{k'}}(G_{k'}) \tag{2.1.3}$$

that is an isomorphism whenever G is locally of finite presentation over k and either $k \to k'$ is flat or $\mathrm{Cot}_e(G)$ is k-flat (e.g., k a field or G smooth over k). When (2.1.3) is an isomorphism for every k-algebra k', we say that *the formation of* $\mathrm{Tan}_e(G)$ *commutes with any base change on* k.

As in the classical case over a field, there is a natural bijection of sets

$$\mathfrak{g} = \mathrm{Tan}_e(G) \simeq \ker(G(k[\epsilon]) \to G(k))$$

that is functorial in G, and by working locally along e it is easy to check that this is a homomorphism and carries the k-action on the left over to the functorial action through k-scaling on ϵ on the right. It identifies the k-linear map $\mathrm{Tan}_e(G) \to \mathrm{Tan}_{e_{k'}}(G_{k'})$ underlying (2.1.3) with the map induced by $G(k[\epsilon]) \to G(k'[\epsilon])$ for any k-algebra k'. We generally write $\mathrm{Lie}(G)$ to denote $\mathrm{Tan}_e(G)$ because in §A.7 we endow it with a natural Lie algebra structure over k. However, in this section we will never use the Lie algebra structure.

If the formation of the k-module $\mathrm{Lie}(G)$ commutes with any base change on k then the conjugation action of G on itself induces an action of $G(k')$ on $\mathrm{Lie}(G_{k'}) = \mathrm{Lie}(G)_{k'}$ functorially in k-algebras k'. This is the *adjoint representation* of G on $\mathrm{Lie}(G)$.

Proposition 2.1.8 *Let k be a ring, and G an affine k-group scheme of finite presentation. Let $\lambda : \mathrm{GL}_1 \to G$ be a k-homomorphism, and let GL_1 act on $\mathfrak{g} = \mathrm{Lie}(G)$ through the derivative of the action*

$$t.g = \lambda(t)g\lambda(t)^{-1}$$

on G. Define $\mathfrak{g}_+ = \bigoplus_{n>0} \mathfrak{g}_n$ where $\mathfrak{g}_n := \{v \in \mathfrak{g} \,|\, t.v = t^n v\}$ for $n \in \mathbf{Z}$, and define $\mathfrak{g}_-$ and $\mathfrak{g}_0$ similarly.

(1) *We have $\mathrm{Lie}(Z_G(\lambda)) = \mathfrak{g}_0$, $\mathrm{Lie}(U_G(\pm\lambda)) = \mathfrak{g}_\pm$, and $\mathrm{Lie}(P_G(\lambda)) = \mathfrak{g}_0 \oplus \mathfrak{g}_+$ inside $\mathfrak{g}$.*
(2) *The multiplication map $Z_G(\lambda) \ltimes U_G(\lambda) \to P_G(\lambda)$ is an isomorphism of k-groups.*
(3) *Assume k is a field. The multiplication map $U_G(-\lambda) \times P_G(\lambda) \to G$ is an open immersion of k-schemes and if $j : G \hookrightarrow G'$ is a k-subgroup inclusion into an affine k-group G' of finite type then*

$$G \cap (U_{G'}(-\lambda') \times P_{G'}(\lambda')) = U_G(-\lambda) \times P_G(\lambda),$$

where $\lambda' = j \circ \lambda$.

The open immersion property holds in general over any ring k if G is k-smooth, in which case $P_G(\lambda)$, $U_G(\lambda)$, and $Z_G(\lambda)$ are all smooth.
(4) *The fibers of $U_G(\lambda)$ over $\mathrm{Spec}\, k$ are connected, and the same holds for $P_G(\lambda)$ and $Z_G(\lambda)$ if G has connected fibers.*

Note that Proposition 2.1.8 only uses the underlying k-modules of the Lie algebras. Also, since GL_1 is k-flat, the GL_1-action on $\mathfrak{g}$ makes sense, and the existence of the weight space decomposition of $\mathfrak{g}$ is a consequence of Lemma A.8.8.

Proof Since $Z_G(\lambda) = P_G(\lambda) \cap P_G(-\lambda)$ as subschemes of G, we have the same relation on Lie algebras inside of $\mathfrak{g}$. That is,

$$\mathrm{Lie}(Z_G(\lambda)) = \mathrm{Lie}(P_G(\lambda)) \cap \mathrm{Lie}(P_G(-\lambda)).$$

Hence, to prove (1) it suffices to compute the Lie algebras of $U_G(\lambda)$ and $P_G(\lambda)$. We shall compute using points valued in the k-algebra of dual numbers $k[\epsilon]$.

Step 1. By the construction of $P_G(\lambda)$, the set $P_G(\lambda)(k[\epsilon])$ consists of the points $g \in G(k[\epsilon])$ that kill $k[G]_n$ for all $n < 0$. But

$$\mathrm{Lie}(P_G(\lambda)) = P_G(\lambda)(k[\epsilon]) \cap \mathrm{Lie}(G)$$

inside $G(k[\epsilon])$, so $\mathrm{Lie}(P_G(\lambda))$ consists of the maps $k[G] \to k[\epsilon]$ of the form

$$f \mapsto f(1) + D(f)\epsilon$$

that kill $k[G]_n$ for all $n < 0$, with $D : k[G] \to k$ a k-linear point derivation at the identity. Since $1 \in P_G(\lambda)(k)$ (i.e., $f(1) = 0$ for all f vanishing on $P_G(\lambda)$), the necessary and sufficient condition is that D kills $k[G]_n$ for all $n < 0$. Under the weight space decomposition of $\mathfrak{g}$ for the GL_1-action, the weight-m subspace $\mathfrak{g}_m$ (consisting of D such that $t.D = t^m D$ for all points t of GL_1) consists of exactly those D killing $k[G]_n$ for all $n \neq m$. Hence, $\mathrm{Lie}(P_G(\lambda)) = \mathfrak{g}_0 \oplus \mathfrak{g}_+$, as desired.

Next, we compute $\mathrm{Lie}(U_G(\lambda))$ by using the functorial interpretation of $U_G(\lambda)$. For any $v \in \mathrm{Lie}(P_G(\lambda))$ with corresponding weight decomposition $v = \sum_{n \geqslant 0} v_n$ we have $t.v = \sum_{n \geqslant 0} t^n v_n$. Thus, the map $\mathbf{A}^1_{k[\epsilon]} \to G_{k[\epsilon]}$ defined by

$$x \mapsto \sum_{n \geqslant 0} x^n v_n$$

is the unique extension of the action map $t \mapsto t.v$ over $k[\epsilon]$. In particular, $\lim_{t \to 0} t.v = v_0$ inside of $\mathfrak{g} \subseteq G(k[\epsilon])$. The property $v \in \mathrm{Lie}(U_G(\lambda))$ therefore says exactly that $v_0 = 0$, or equivalently $v \in \mathfrak{g}_+$. This completes the proof of (1).

Step 2. For the proof of (2), recall the k-group homomorphism $f : P_G(\lambda) \to G$ defined by $g \mapsto \lim_{t \to 0} t.g$ as in the proof of Lemma 2.1.5, where the kernel was identified with $U_G(\lambda)$. We claim that this k-homomorphism is a projector onto the k-subgroup $Z_G(\lambda)$, from which the desired semidirect product decomposition is immediate.

The map f clearly restricts to the identity on $Z_G(\lambda)$, so we just have to check that it factors through $Z_G(\lambda)$. Consider any k-algebra R and $g \in P_G(\lambda)(R)$, and define $g' = \lim_{t \to 0} t.g$ in $G(R)$. We want $g' \in Z_G(\lambda)(R)$. For any R-algebra R' and $t' \in \mathrm{GL}_1(R')$ we have $t'.g' = \lim_{t \to 0}(t't).g$ because the GL_1-action on itself by translation extends to an action of GL_1 on the affine line preserving the origin. This limit clearly equals g', so $t'.g' = g'$ for all t'. That is, $g' \in Z_G(\lambda)(R)$ as desired.

STEP 3. The deduction of (3) lies deeper. We first prove the open immersion property in general over a field by using the crutch of GL_n, and then (in Step 4) we will use a more sophisticated method to handle the case of smooth G over any ring k (which gives another proof when k is a field).

With k now assumed to be a field, by Proposition A.2.3 there is a closed k-subgroup inclusion $G \hookrightarrow \mathrm{GL}(V)$ for a nonzero finite-dimensional k-vector space V. By Example 2.1.1 and Proposition 2.1.2, (3) holds for $\mathrm{GL}(V)$. Thus, it suffices to treat the case where there is a subgroup functor inclusion $j : G \hookrightarrow G'$ such that G' satisfies the open immersion property in (3) relative to $\lambda' := j \circ \lambda$. We will show that the open subscheme

$$G \cap (U_{G'}(-\lambda') \times P_{G'}(\lambda')) \subseteq G$$

is identified with $U_G(-\lambda) \times P_G(\lambda)$ via multiplication. Consider a k-algebra R and points

$$u'_- \in U_{G'}(-\lambda')(R), \quad u'_+ \in U_{G'}(\lambda')(R), \quad z' \in Z_{G'}(\lambda')(R)$$

such that $u'_- u'_+ z' = g \in G(R)$. It suffices to prove that $u'_+, u'_-, z' \in G(R)$ since we have already proved (2) for G.

By Proposition A.2.4, there is a finite-dimensional k-vector space V, a k-homomorphism $\rho : G' \to \mathrm{GL}(V)$, and a line L in V such that G is the scheme-theoretic stabilizer of L in G'. Let $v \in L$ be a basis element, so $\rho(g)(v) = cv$ in $V_R = R \otimes_k V$ for a unique $c \in R^\times$. Since $g = u'_- u'_+ z'$, we get

$$\rho(u'_+ z')(v) = c\rho((u'_-)^{-1})(v) \tag{2.1.4}$$

in V_R.

For any point t of GL_1 valued in an R-algebra R', the point $\lambda'(t)$ of $G'(R')$ lies in $G(R')$ and so acts on v (through ρ) by some $R'^\times$-scaling. Hence, we can replace v with $\rho(\lambda'(t)^{-1})(v)$ on both sides of (2.1.4). Now act on both sides of (2.1.4) by $\rho(\lambda'(t))$, and then commute $\rho(\lambda'(t)^{-1})$ past $\rho(z')$ (as we may, since $z' \in Z_{G'}(\lambda')(R)$) to get the identity

$$\rho((t.u'_+)z')(v) = c\rho(t.(u'_-)^{-1})(v) \tag{2.1.5}$$

as points of the affine space $\underline{V}_R$ over R covariantly associated to V_R.

Viewing the two sides of (2.1.5) as R-scheme maps $(\mathrm{GL}_1)_R \to \underline{V}_R$, the left side extends to an R-map $\mathbf{P}^1_R - \{\infty\} = \mathbf{A}^1_R \to \underline{V}_R$ and the right side extends to an R-map $\mathbf{P}^1_R - \{0\} \to \underline{V}_R$. By combining these, we arrive at an R-map $\mathbf{P}^1_R \to \underline{V}_R$ from the projective line to an affine space over R. The only such map is a constant R-map to some $v_0 \in \underline{V}_R(R) = V_R$, so both sides of (2.1.5) are independent of t (and equal to v_0). Passing to the limit as $t \to 0$ on the left side and as $t \to \infty$ on the right side yields $\rho(z')(v) = v_0 = cv$. We have proved that z' carries v to an $R^\times$-multiple of itself. Thus, the point $z' \in G'(R)$ is an R-point of the functorial stabilizer of L inside of V. This stabilizer is exactly G, by the way we chose ρ, so z' is an R-point of $G \cap Z_{G'}(\lambda') = Z_G(\lambda)$.

Since $\rho(z')(v) = cv$, by cancellation of c on both sides of the identity (2.1.5) we get

$$\rho(t.u'_+)(v) = \rho(t.(u'_-)^{-1})(v)$$

with both sides independent of t and equal to $c^{-1}v_0 = v$. Taking $t = 1$, we see that $u'_\pm$ lies in the stabilizer G of L, so $u'_\pm$ is an R-point of $G \cap U_{G'}(\pm\lambda') = U_G(\pm\lambda)$, as required. This completes the proof that $U_G(-\lambda) \times P_G(\lambda) \to G$ is an open immersion when k is a field.

The connectedness of $U_G(\lambda)$ in such generality follows from the fact that every point u of $U_G(\lambda)$ is joined to the identity by the morphism $\mathbf{A}^1 \to U_G(\lambda)$ that extends $t \mapsto t.u$ on GL_1. If G is connected then it is irreducible (as may be checked over $\overline{k}$), so the open subscheme $U_G(-\lambda) \times P_G(\lambda)$ is connected for such G (using (3)). Thus, if G is connected then $P_G(\lambda)$ is connected, and so by using (2) we deduce that $Z_G(\lambda)$ is also connected. This proves (4) by passing to fibers.

Step 4. Now assume that G is smooth over an arbitrary ring k, and we need to establish the open immersion property as in (3) and the smoothness of $U_G(\lambda)$ and $Z_G(\lambda)$ (from which the smoothness of $P_G(\lambda)$ follows via (2)). By construction, $U_G(-\lambda)$ is cut out by the n-eigenvectors of the augmentation ideal of G with $n \geqslant 0$, and $P_G(\lambda)$ is cut out by the n-eigenvectors with $n < 0$. Together these span the augmentation ideal as a k-module, so $U_G(-\lambda) \cap P_G(\lambda)$ is the identity section. Thus, the multiplication map $\mu : U_G(-\lambda) \times P_G(\lambda) \to G$ in (3) is a monomorphism. By [EGA, IV_4, 17.9.1], an open immersion is the same thing as an étale monomorphism, so to prove that μ is an open immersion it suffices to show that it is étale.

Granting for a moment that $U_G(-\lambda)$ and $P_G(\lambda)$ are smooth, let us prove that μ is étale. We may assume k is strictly henselian with an algebraically closed residue field, and only need to check étaleness at rational points of the special fiber. But these lift to k-points by smoothness, so translations reduce the problem to étaleness at the identity. The k-smoothness hypothesis reduces étaleness at the identity to the isomorphism property on tangent spaces along the identity section, which in turn is just the weight space decomposition for the Lie algebras as in (1).

It remains to prove that $U_G(\lambda)$ and $Z_G(\lambda)$ are smooth when G is smooth. Smoothness of the torus centralizer $Z_G(\lambda)$ is a special case of Proposition A.8.10(2) (taking $H = \mathrm{GL}_1$, $Y = G$, and $Y' = \mathrm{Spec}\, k$), or alternatively is a special case of [SGA3, XI, 5.4]. For $U_G(\lambda)$, we may again assume that k is strictly henselian with algebraically closed residue field, and it suffices to prove smoothness at rational points in the special fiber. Since $U_G(\lambda)$ has geometrically irreducible fibers (by (4)), if V is any open subscheme of $U_G(\lambda)$ around the identity section then the multiplication map $V \times V \to U_G(\lambda)$ is surjective. Thus, if there is such a V that is smooth then $U_G(\lambda)(k)$-translates of V cover the special fiber, so smoothness of $U_G(\lambda)$ will follow. In other words, to prove smoothness of $U_G(\lambda)$ it suffices to do so along the identity, and for this we may assume that k is local noetherian, and even complete. Thus, the completion of the local ring of G at the identity point of the special fiber is identified with the I-adic completion A of the coordinate ring of G, where I is the augmentation ideal.

Consider the cotangent space I/I^2 along the identity. This is a finite free module (dual to $\mathrm{Lie}(G)$), and as such it has a linear GL_1-action that is induced by the GL_1-action on the entire coordinate ring of G via conjugation through λ. The weight space decomposition of I/I^2 therefore lifts to I, which is to say that we can find GL_1-eigenvectors $h_1, \dots, h_d \in I$ that lift a k-basis of I/I^2. Thus, there are integers $a_i \in \mathbf{Z}$ such that $t.h_i = t^{a_i} h_i$ for all points t of GL_1 (i.e., $h_i \in k[G]_{a_i}$ for all i), so the local k-algebra isomorphism

$$k[\![x_1, \dots, x_d]\!] \simeq A$$

defined by $x_i \mapsto h_i$ is GL_1-equivariant modulo each power of the augmentation ideal when using the action $t.x_i = t^{a_i} x_i$. Since the ideal of $U_G(\lambda)$ in $k[G]$ is generated by the n-eigenvectors in I for $n \leqslant 0$, and weight space decompositions are compatible with equivariant quotients (such as the projection $k[G] \twoheadrightarrow k[G]/I^{r+1}$ for each $r \geqslant 0$), the completion of $U_G(\lambda)$ along the identity section is the quotient of A by the ideal generated by monomials $\prod x_i^{e_i}$ (all $e_i \geqslant 0$ and some $e_i > 0$) such that $\sum a_i e_i \leqslant 0$. Any such monomial must have a coefficient $e_{i_0} > 0$ for some i_0 such that $a_{i_0} \leqslant 0$, so the ideal of $U_G(\lambda)$ in the completion A is generated by the x_i for which $a_i \leqslant 0$. This proves the smoothness of $U_G(\lambda)$. (The same method proves the smoothness of $Z_G(\lambda)$ near the identity, using the ideal generated by x_i for which $a_i \neq 0$, but does not suffice to prove smoothness of $Z_G(\lambda)$ away from its fibral identity components.) □

We now record how the formation of the k-subgroups $P_G(\lambda)$, $U_G(\lambda)$, and $Z_G(\lambda)$ behaves with respect to surjective k-homomorphisms $\pi : G \to G'$ between smooth connected affine groups over a field k. Since a surjective homomorphism $G \to G'$ between smooth connected k-groups is automatically flat (by [Mat, 23.1]), it is natural to avoid smoothness hypotheses on G or G' but to assume that π is flat:

Corollary 2.1.9 *Let $\pi : G \to G'$ be a flat surjective homomorphism between connected affine groups of finite type over a field k, and let $\lambda : \mathrm{GL}_1 \to G$ be a k-homomorphism. For $\lambda' := \pi \circ \lambda$, π restricts to flat surjective maps $P_G(\lambda) \to P_{G'}(\lambda')$, $U_G(\lambda) \to U_{G'}(\lambda')$, and $Z_G(\lambda) \to Z_{G'}(\lambda')$. In particular, $\pi(P_G(\lambda)) = P_{G'}(\lambda')$ and similarly for $U_G(\lambda)$ and $Z_G(\lambda)$.*

Proof It is clear from functorial considerations (and calculation of limits as $t \to 0$) that π carries $P_G(\lambda)$ into $P_{G'}(\lambda')$, $U_G(\lambda)$ into $U_{G'}(\lambda')$, and $Z_G(\lambda)$ into $Z_{G'}(\lambda')$. Since $\pi : G \to G'$ is flat, the induced map

$$\begin{aligned} U_G(-\lambda) \times Z_G(\lambda) \times U_G(\lambda) &= U_G(-\lambda) \times P_G(\lambda) \\ &\to U_{G'}(-\lambda') \times P_{G'}(\lambda') = U_{G'}(-\lambda') \times Z_{G'}(\lambda') \times U_{G'}(\lambda') \end{aligned}$$

between open subschemes is flat. Hence, the maps between the factors are flat. But a flat map between k-schemes of finite type is open, and any homomorphism between affine k-groups of finite type has closed image. By Proposition 2.1.8(4) the groups $P_G(\lambda)$, $U_G(\lambda)$, and $Z_G(\lambda)$ are connected (since G is connected), and likewise for G', so the maps $P_G(\lambda) \to P_{G'}(\lambda')$, $U_G(\lambda) \to U_{G'}(\lambda')$, and $Z_G(\lambda) \to Z_{G'}(\lambda')$ are surjective. □

In the theory of connected reductive groups over a field k, it is an important fact that the unipotent radical (over $\overline{k}$) of any parabolic k-subgroup is defined over k as a subgroup and is k-split in the sense that it admits a composition series over k with successive quotients k-isomorphic to $\mathbf{G}_{\mathrm{a}}$. We have an analogous result for $U_G(\lambda)$:

Proposition 2.1.10 *For a smooth affine group G over a field k and k-homomorphism $\lambda : \mathrm{GL}_1 \to G$, the smooth connected unipotent k-group $U_G(\lambda)$ is k-split.*

Proof By Theorem B.3.4, any smooth connected unipotent k-subgroup U contains a unique k-split smooth connected normal k-subgroup U_{split} such that U/U_{split} is k-wound in the sense of Definition B.2.1 (i.e., admits no non-constant k-morphism from $\mathbf{A}^1_k$), and moreover the formation of U_{split} commutes with separable extension on k. To prove that $U := U_G(\lambda)$ is k-split, we will show that $U = U_{\text{split}}$. Now we can replace k with k_s and assume that $k = k_s$. For any nontrivial element g in $U(k)$, the map $t \mapsto \lambda(t)g\lambda(t)^{-1}$ extends to a morphism $\mathbf{A}^1_k \to U$ carrying 0 to 1. The induced morphism $\phi : \mathbf{A}^1_k \to U/U_{\text{split}}$ is constant since U/U_{split} is k-wound. This implies that $g \in U_{\text{split}}(k)$. Thus we have shown that $U(k) \subseteq U_{\text{split}}(k)$. From the Zariski-density of $U(k)$ in U we conclude that $U = U_{\text{split}}$. □

Remark 2.1.11 The theory of $P_G(\lambda)$ and $U_G(\lambda)$ can be generalized to the case of an abstract GL_1-action on G, not just action through conjugation by a 1-parameter k-subgroup. Indeed, consider an affine group scheme G of finite presentation over a ring k, and a left action $\lambda : \mathrm{GL}_1 \times G \to G$ over k. We write $-\lambda$ for the composition of λ with inversion on GL_1.

For any k-algebra R we define

$$P_G(\lambda)(R) = \{g \in G(R) \mid \lim_{t\to 0} \lambda(t, g) \text{ exists}\},$$

$$U_G(\lambda)(R) = \{g \in G(R) \mid \lim_{t\to 0} \lambda(t, g) = 1\},$$

and also define $Z_G(\lambda)(R)$ to be the set of $g \in G(R)$ that are invariant under the action. Observe that if G' is a closed k-subgroup of G stable under the GL_1-action then $P_{G'}(\lambda) = G' \cap P_G(\lambda)$, $U_{G'}(\lambda) = G' \cap U_G(\lambda)$, and $Z_{G'}(\lambda) = G' \cap Z_G(\lambda)$ as subfunctors of G.

Let $H = \mathrm{GL}_1 \ltimes G$ denote the semidirect product defined by the action λ, so H is k-smooth (resp. has connected fibers over $\mathrm{Spec}\, k$) whenever the same holds for G. For the canonical inclusion $\mu : \mathrm{GL}_1 \to H$ viewed as a 1-parameter k-subgroup, we have (as subfunctors of H)

$$P_H(\mu) = \mathrm{GL}_1 \ltimes P_G(\lambda), \;\; Z_H(\mu) = \mathrm{GL}_1 \ltimes Z_G(\lambda), \;\; U_H(\mu) = U_G(\lambda).$$

In particular, $P_G(\lambda) = G \cap P_H(\mu)$ and $Z_G(\lambda) = G \cap Z_H(\mu)$, so each of $P_G(\lambda)$, $Z_G(\lambda)$, and $U_G(\lambda)$ is represented by a closed k-subgroup of G whose formation commutes with any scalar extension on k. The conclusions in Proposition 2.1.8, Corollary 2.1.9, and Proposition 2.1.10 continue to hold for $P_G(\lambda)$, $U_G(\lambda)$, and $Z_G(\lambda)$ by applying those results to $H = \mathrm{GL}_1 \ltimes G$ and μ.

Proposition 2.1.12 *Let G be a connected affine group scheme of finite type over a field k, and $\lambda : \mathrm{GL}_1 \times G \to G$ a left action of GL_1 on G. Assume that $(G_{\overline{k}})_{\mathrm{red}}$ is solvable, and define $P_G(\lambda)$ and $U_G(-\lambda)$ as in Remark* 2.1.11.

(1) *The open immersion $U_G(-\lambda) \times P_G(\lambda) \to G$ defined by multiplication is an isomorphism of k-schemes.*

(2) *Let H be a k-subgroup scheme of G that is stable under the* GL_1*-action. Then the multiplication map* $U_H(-\lambda) \times P_H(\lambda) \to H$ *is an isomorphism, and H is connected if and only if* $Z_H(\lambda)$ *is connected.*
(3) *For any action on G by a k-torus T and any closed k-subgroup scheme S in T, the centralizer* $Z_G(S)$ *is connected.*

Proof We may assume k is algebraically closed, so for (1) the problem is to show that the map of sets $U_G(-\lambda)(k) \times P_G(\lambda)(k) \to G(k)$ is surjective. The closed subscheme $G' = G_{\mathrm{red}}$ is a smooth k-subgroup of G since $k = \overline{k}$, and λ factors through G', so $P_{G'}(\lambda) = G' \cap P_G(\lambda)$ and $U_{G'}(-\lambda) = G' \cap U_G(-\lambda)$. But $P_{G'}(\lambda)$ and $U_{G'}(-\lambda)$ are smooth since G' is smooth, so $P_{G'}(\lambda) = P_G(\lambda)_{\mathrm{red}}$ and $U_{G'}(-\lambda) = U_G(-\lambda)_{\mathrm{red}}$. We may therefore replace G with G' so that G is smooth and solvable.

If G is commutative then the image of $U_G(-\lambda) \times P_G(\lambda)$ is an open subgroup of G (Proposition 2.1.8(3)) and hence by connectedness it equals G, so (1) is clear in such cases. To prove (1) in general, we induct on $\dim G$ and may assume that G is not commutative. The derived series of G is stable under the GL_1-action, so if D denotes the final nontrivial term of this series then the result holds for the commutative normal subgroup D, and by induction on dimension the result also holds for G/D, with the induced GL_1-action. Applying Corollary 2.1.9 to $G \twoheadrightarrow G/D$ gives that

$$G(k) = U_G(-\lambda)(k) \cdot P_G(\lambda)(k) \cdot D(k) = U_G(-\lambda)(k) \cdot D(k) \cdot P_G(\lambda)(k).$$

But $D(k) = U_D(-\lambda)(k) \cdot P_D(\lambda)(k)$ by the settled commutative case, so (1) is proved.

For (2), we first observe that by part (1) and Proposition 2.1.8(3) (and Remark 2.1.11) applied to $H \hookrightarrow G$ we have

$$H = H \cap (U_G(-\lambda) \times P_G(\lambda)) = U_H(-\lambda) \times P_H(\lambda)$$

via multiplication. But $U_H(\pm\lambda)$ are connected and $P_H(\lambda) = Z_H(\lambda) \ltimes U_H(\lambda)$, so clearly H is connected if and only if $Z_H(\lambda)$ is connected.

Finally, consider a k-torus T acting on G, and a closed k-subgroup scheme S in T. To prove that $Z_G(S)$ is connected we may assume $k = k_s$, so T is split. Choose a 1-parameter k-subgroup $\lambda : \mathrm{GL}_1 \to T$ that is not annihilated by any nontrivial T-weight on $\mathrm{Lie}(G)$. The inclusion $Z_G(T) \subseteq Z_G(\lambda)$ is an equality on Lie algebras, so the T-action on $Z_G(\lambda)$ is trivial on the Lie algebra. But $Z_G(\lambda)$ inherits connectedness from G (by Proposition 2.1.8(4), in the generalized setting of Remark 2.1.11), so by Corollary A.8.11 the T-action on $Z_G(\lambda)$ is trivial. That is, $Z_G(\lambda) = Z_G(T)$. By part (2), $H := Z_G(S)$ is connected if and only if the k-group $Z_H(\lambda)$ is connected. Since

$$Z_H(\lambda) = H \cap Z_G(\lambda) = H \cap Z_G(T) = Z_G(T)$$

and as $Z_G(T) = Z_G(\lambda)$ is connected, we are done. □

Proposition 2.1.13 *Let $k \to k'$ be a finite locally free map of rings, G' an affine k'-group scheme of finite presentation, and $G = \mathrm{R}_{k'/k}(G')$. Let $\lambda' : \mathrm{GL}_1 \to G'$ be a k'-homomorphism, and $\lambda : \mathrm{GL}_1 \to G$ the corresponding k-homomorphism via the universal property of Weil restriction. (Explicitly, λ is the composition of $\mathrm{R}_{k'/k}(\lambda')$ with the natural inclusion $\mathrm{GL}_1 \hookrightarrow \mathrm{R}_{k'/k}(\mathrm{GL}_1)$.) Then*

$$P_G(\lambda) = \mathrm{R}_{k'/k}(P_{G'}(\lambda')), \quad U_G(\lambda) = \mathrm{R}_{k'/k}(U_{G'}(\lambda')),$$
$$Z_G(\lambda) = \mathrm{R}_{k'/k}(Z_{G'}(\lambda')),$$

as k-subgroup schemes of G.

Proof By the universal property of Weil restriction, the set of k-morphisms $f : \mathbf{A}^1_k \to G$ is in natural bijective correspondence with the set of k'-morphisms $f' : \mathbf{A}^1_{k'} \to G'$, and the element $f(0) \in G(k)$ corresponds to the element $f'(0) \in G'(k')$ via the natural equality $G(k) = G'(k')$. Thus, λ extends to some such f if and only if λ' extends to some such f', so the closed subschemes $\mathrm{R}_{k'/k}(P_{G'}(\lambda'))$ and $P_G(\lambda)$ inside G have the same k-points. Via base change to k-algebras R it follows likewise that they have the same R-points for every R, so these closed subschemes coincide. By the same reasoning with the additional requirements $f(0) = 1$ and $f'(0) = 1$ we obtain that $\mathrm{R}_{k'/k}(U_{G'}(\lambda')) = U_G(\lambda)$. Since $Z_{G'}(\lambda') = P_{G'}(\lambda') \cap P_{G'}(-\lambda')$ by Lemma 2.1.5, and likewise for $Z_G(\lambda)$, we deduce that $Z_G(\lambda) = \mathrm{R}_{k'/k}(Z_{G'}(\lambda'))$. □

2.2 Pseudo-parabolic subgroups

The following definition is a useful generalization to arbitrary smooth connected affine groups of the notion of a parabolic subgroup of a connected reductive group.

Definition 2.2.1 Let G be a smooth connected affine group over a field k. A k-subgroup of G is *pseudo-parabolic* if it has the form $P_G(\lambda)\mathscr{R}_{u,k}(G)$ for some k-homomorphism $\lambda : \mathrm{GL}_1 \to G$.

Since $\lambda(\mathrm{GL}_1)$ is contained in a maximal k-torus T of G, so $T \subseteq Z_G(\lambda) \subseteq P_G(\lambda)$, the maximal k-tori in a pseudo-parabolic k-subgroup of G are also maximal k-tori of G. In Proposition 2.2.9 we will relate pseudo-parabolic subgroups and parabolic subgroups in the connected reductive case, but beware that in general if G is pseudo-reductive over k then G/P may not be k-proper for pseudo-parabolic k-subgroups P of G (see Example 2.1.3 with $P = B$ there). Also, it can happen that P is a pseudo-parabolic k-subgroup of G but $P_{k'}$ is not a pseudo-parabolic k'-subgroup of $G_{k'}$ for an inseparable extension of the base field k'/k (see Example 2.1.3 with k'/k a purely inseparable nontrivial extension). Behavior of pseudo-parabolicity with respect to separable extension of the base field works out nicely, as we will see in Proposition 3.5.2.

Example 2.2.2 If G is pseudo-reductive over k then the pseudo-parabolic k-subgroups are the ones of the form $P_G(\lambda)$ for a k-homomorphism $\lambda : \mathrm{GL}_1 \to G$. When G is not pseudo-reductive over k then it is important to remember that $\mathscr{R}_{u,k}(G)$ is put inside of pseudo-parabolic k-subgroups by definition.

For example, consider the standard semidirect product $G = \mathrm{GL}_1 \ltimes \mathbf{G}_\mathrm{a}$ (which is not pseudo-reductive over k). Letting $\chi : G \to \mathrm{GL}_1$ be the standard projection $(t, x) \mapsto t$, we see that $P_G(\lambda) = G$ when $\langle \chi, \lambda \rangle \geqslant 0$ and that $P_G(\lambda) = \lambda(\mathrm{GL}_1)$ when $\langle \chi, \lambda \rangle < 0$. Hence, $P_G(\lambda)\mathscr{R}_{u,k}(G) = G$ for all λ, so there is no proper pseudo-parabolic k-subgroup of this G.

Lemma 2.2.3 *Let G be a smooth connected affine group over a field k.*

(1) *There exists a proper pseudo-parabolic k-subgroup of G if and only if $G/\mathscr{R}_{u,k}(G)$ contains a non-central k-split k-torus.*
(2) *If G is pseudo-reductive over k then a pseudo-parabolic k-subgroup $P_G(\lambda)$ is a proper k-subgroup of G if and only if $\lambda(\mathrm{GL}_1)$ is not central in G.*

Example 2.2.2 shows the necessity of the pseudo-reductivity hypothesis in the second part of the lemma.

Proof We first address part (2). If $\lambda(\mathrm{GL}_1)$ is central in G then clearly $P_G(\lambda) = G$. Conversely, if $P_G(\lambda) = G$ then $U_G(\lambda)$ is a smooth connected unipotent normal k-subgroup of the pseudo-reductive k-group G. In such cases $U_G(\lambda)$ must be trivial, so $Z_G(\lambda) = P_G(\lambda)$. This says that $\lambda(\mathrm{GL}_1)$ is central in G.

Now assume that G admits a proper pseudo-parabolic k-subgroup P. By the definition of pseudo-parabolicity, P contains $\mathscr{R}_{u,k}(G)$. Hence, $P/\mathscr{R}_{u,k}(G)$ is a proper k-subgroup of the pseudo-reductive k-group $G' := G/\mathscr{R}_{u,k}(G)$, and by Corollary 2.1.9 this proper k-subgroup is a pseudo-parabolic k-subgroup. In view of the settled part (2), it follows that G' contains a non-central k-split torus.

Conversely, assume that G' contains a non-central k-split torus. Such a torus is generated by copies of GL_1 over k, so there must be a non-central GL_1 subgroup in G'. Let $H \subseteq G$ be the preimage of a fixed non-central $\mathrm{GL}_1 \subseteq G'$ under the quotient map $G \twoheadrightarrow G'$. This H is a smooth connected affine k-group that is an extension of GL_1 by $\mathscr{R}_{u,k}(G)$. Since the formation of maximal k-tori in smooth connected affine k-groups is compatible with passage to quotients (Proposition A.2.8), a maximal k-torus T in H must map isomorphically onto the quotient GL_1. In particular, $T = \mathrm{GL}_1$ and this is non-central in G since it maps onto a non-central GL_1 in the quotient G'.

We have proved the existence of a non-central k-homomorphism $\lambda : \mathrm{GL}_1 \to G$ such that the composite map $\lambda' : \mathrm{GL}_1 \to G'$ is also non-central. The proof of part (2) implies that $P_{G'}(\lambda')$ is a proper k-subgroup of G'. Since $P_G(\lambda)$ maps into $P_{G'}(\lambda')$, clearly $P_G(\lambda)\mathscr{R}_{u,k}(G)$ is a proper k-subgroup of G. This is a pseudo-parabolic k-subgroup of G, so we are done. □

There is an alternative equivalent definition of pseudo-parabolicity that is useful, so we now present it and prove the equivalence of the two definitions. Let $\mathscr{R}_{us,k}(G)$ denote the maximal k-split smooth connected unipotent normal k-subgroup of G. (This exists due to dimension reasons and the normality requirement. In Corollary B.3.5 we give another characterization of it in terms of Theorem B.3.4 applied to $\mathscr{R}_{u,k}(G)$.) For any k-homomorphism $\lambda : \mathrm{GL}_1 \to G$, let $P'_G(\lambda)$ be the smooth connected k-subgroup of G generated by the smooth connected k-subgroups $M \subseteq G$ that are normalized by $\lambda(\mathrm{GL}_1)$ and for which the weights a of $\lambda(\mathrm{GL}_1)$ on $\mathrm{Lie}(M)$ satisfy $\langle a, \lambda \rangle \geqslant 0$. By Proposition 2.1.8(1) this weight condition says precisely that $\mathrm{Lie}(M) \subseteq \mathrm{Lie}(P_G(\lambda))$. The following result implies that an equivalent definition of pseudo-parabolic k-subgroups of G is that these are the k-subgroups of the form $P'_G(\lambda)\mathscr{R}_{us,k}(G)$ for some k-homomorphism $\lambda : \mathrm{GL}_1 \to G$.

Proposition 2.2.4 *For any smooth connected affine k-group G and k-homomorphism $\lambda : \mathrm{GL}_1 \to G$, we have $P_G(\lambda)\mathscr{R}_{u,k}(G) = P_G(\lambda)\mathscr{R}_{us,k}(G)$ and $P_G(\lambda) = P'_G(\lambda)$ inside of G.*

Proof The description of $\mathrm{Lie}(P_G(\lambda))$ in Proposition 2.1.8(1) implies that $P_G(\lambda) \subseteq P'_G(\lambda)$ by definition of $P'_G(\lambda)$. To prove that this inclusion of smooth connected k-subgroups of G is an equality, it suffices to prove that $P_G(\lambda)$ contains every smooth connected k-subgroup $M \subseteq G$ that is normalized by $\lambda(\mathrm{GL}_1)$ and for which all weights a of $\lambda(\mathrm{GL}_1)$ on $\mathrm{Lie}(M)$ satisfy $\langle a, \lambda\rangle \geqslant 0$, as such M generate $P'_G(\lambda)$ by definition.

Since $\lambda(\mathrm{GL}_1)$ normalizes M, the smooth connected k-subgroups $P_M(\lambda)$, $U_M(\pm\lambda)$, and $Z_M(\lambda)$ make sense by Remark 2.1.11. By the hypothesis on $\mathrm{Lie}(M)$ we see that $\mathrm{Lie}(U_M(-\lambda)) = 0$, so $U_M(-\lambda) = 1$. Proposition 2.1.8(3) applied to M (via Remark 2.1.11) therefore gives that $M = P_M(\lambda) \subseteq P_G(\lambda)$, as desired. This completes the proof that $P_G(\lambda) = P'_G(\lambda)$.

The desired equality

$$P_G(\lambda)\mathscr{R}_{u,k}(G) \stackrel{?}{=} P_G(\lambda)\mathscr{R}_{us,k}(G)$$

means that $P_G(\lambda)\mathscr{R}_{us,k}(G)$ contains $\mathscr{R}_{u,k}(G)$. Let $G' = G/\mathscr{R}_{us,k}(G)$ and let $\lambda' = \pi \circ \lambda$, where $\pi : G \twoheadrightarrow G'$ is the canonical quotient map. By Corollary 2.1.9, the natural map $P_G(\lambda) \to P_{G'}(\lambda')$ induced by π is surjective. We may therefore replace (G, λ) with (G', λ') to reduce to the case where $\mathscr{R}_{us,k}(G) = 1$, which is to say (by Corollary B.3.5) that $\mathscr{R}_{u,k}(G)$ is k-wound in the sense of Definition B.2.1.

It remains to show that $\mathscr{R}_{u,k}(G) \subseteq P_G(\lambda)$ if $\mathscr{R}_{u,k}(G)$ is k-wound. By normality in G, conjugation by $\lambda(\mathrm{GL}_1)$ defines an action on $\mathscr{R}_{u,k}(G)$ by the k-torus $\lambda(\mathrm{GL}_1)$. There is no nontrivial action of a k-torus on a k-wound smooth connected unipotent k-group (Proposition B.4.4), so $\mathscr{R}_{u,k}(G) \subseteq Z_G(\lambda) \subseteq P_G(\lambda)$. □

Corollary 2.2.5 *With notation as in Proposition 2.2.4, for the pseudo-parabolic k-subgroup $P := P_G(\lambda)\mathscr{R}_{u,k}(G)$ we have*

$$\mathscr{R}_{u,k}(P) = U_G(\lambda)\mathscr{R}_{u,k}(G), \quad \mathscr{R}_{us,k}(P) = U_G(\lambda)\mathscr{R}_{us,k}(G).$$

In particular, if G is pseudo-reductive then $\mathscr{R}_{us,k}(P) = U_G(\lambda) = \mathscr{R}_{u,k}(P)$.

Proof Since the k-split property of smooth connected unipotent k-groups is inherited by quotients (Proposition A.1.4), and the k-group $U_G(\lambda)$ is a smooth connected unipotent normal k-subgroup of $P_G(\lambda)$ that is k-split (Proposition 2.1.10), it follows that $U_G(\lambda)\mathscr{R}_{us,k}(G)$ is a k-split smooth connected unipotent k-subgroup of G. Also, $P = P_G(\lambda)\mathscr{R}_{us,k}(G)$ by Proposition 2.2.4, so $U_G(\lambda)\mathscr{R}_{us,k}(G)$ is normal in P and hence is contained in $\mathscr{R}_{us,k}(P)$. A simpler variant of this argument shows that $U_G(\lambda)\mathscr{R}_{u,k}(G) \subseteq \mathscr{R}_{u,k}(P)$. We have to prove the reverse inclusions.

Let $G' = G/\mathscr{R}_{u,k}(G)$ and $P' = P/\mathscr{R}_{u,k}(G)$, and let $\lambda' : \mathrm{GL}_1 \to G'$ be the composition of λ with the quotient map $G \twoheadrightarrow G'$. Corollary 2.1.9 implies that $P' = P_{G'}(\lambda')$ and that $U_G(\lambda)$ has image $U_{G'}(\lambda')$ in G'. Since G' is pseudo-reductive, so $Z_{G'}(\lambda')$ is pseudo-reductive, we have $U_{G'}(\lambda') = \mathscr{R}_{u,k}(P') = \mathscr{R}_{u,k}(P)/\mathscr{R}_{u,k}(G)$. Hence, $\mathscr{R}_{u,k}(P) = U_G(\lambda) \cdot \mathscr{R}_{u,k}(G)$.

To prove that the inclusion $U_G(\lambda)\mathscr{R}_{us,k}(G) \subseteq \mathscr{R}_{us,k}(P)$ is an equality, we may replace G with $G/\mathscr{R}_{us,k}(G)$ so that $\mathscr{R}_{us,k}(G) = 1$. Thus, $\mathscr{R}_{u,k}(G)$ is k-wound. In particular, as we saw in the proof of Proposition 2.2.4, $\mathscr{R}_{u,k}(G) \subseteq Z_G(\lambda)$, so $P = P_G(\lambda)$ and $U_G(\lambda) \cap \mathscr{R}_{u,k}(G) = 1$. It follows that the product group $\mathscr{R}_{u,k}(P) = U_G(\lambda) \cdot \mathscr{R}_{u,k}(G)$ is a semidirect product. Therefore, the k-split $U_G(\lambda)$ is normal in $\mathscr{R}_{u,k}(P)$ with quotient $\mathscr{R}_{u,k}(G)$ that is k-wound. Hence, $U_G(\lambda) = \mathscr{R}_{us,k}(P)$ by Corollary B.3.5. □

Before we explore pseudo-parabolicity further, it is convenient to recall some useful notions from the general theory of root systems.

Definition 2.2.6 Let Φ be a root system. A subset $\Psi \subseteq \Phi$ is *closed* if $a+b \in \Psi$ for any $a, b \in \Psi$ such that $a + b \in \Phi$. If moreover $\Phi = \Psi \cup -\Psi$ then Ψ is called *parabolic*.

An element $a \in \Phi$ is *multipliable* if $2a \in \Phi$ and *divisible* if $a/2 \in \Phi$. We define the notions of *non-multipliable* and *non-divisible* similarly.

The reason for restricting to doubling and halving in the definitions of multipliability and divisibility is that the only **Q**-multiples of a root a apart from $\pm a$ that can be roots are $\pm 2a$ or $\pm a/2$ (and not both).

Closed sets of roots admit the following useful characterization:

Proposition 2.2.7 *Let Φ be a root system spanning a **Q**-vector space V, and Ψ a subset of Φ. Then Ψ is closed if and only if $\Psi = \Phi \cap A$ for a subsemigroup $A \subseteq V$. In such cases, Ψ is uniquely determined by its subset*

$$\Psi_{\mathrm{nd}} = \{a \in \Psi \mid a/2 \notin \Psi\}.$$

The empty set in V is considered to be a subsemigroup.

Proof If Ψ is closed then the elements of Ψ not in Ψ_{nd} are precisely those $a \in \Phi$ such that $a/2 \in \Psi_{\mathrm{nd}}$. Hence, in the closed case Ψ is uniquely determined by Ψ_{nd}. It is clear that if $\Psi = \Phi \cap A$ for a subsemigroup A in V then Ψ is closed. For the converse, we argue similarly to the proof of [Bou2, VI, §1.7, Prop. 22], as follows. Without loss of generality, we can require A to be the subsemigroup generated by Ψ. Thus, the problem is to show that if $a_1, \ldots, a_q \in \Psi$ $(q \geqslant 1)$ and

$$b := a_1 + \cdots + a_q \in \Phi$$

then $b \in \Psi$.

The case $q = 1$ is trivial, and the case $q = 2$ is the definition of closedness. In general we proceed by induction on q. We can assume $b \neq a_i$ for all i. Let $W = W(\Phi)$ be the Weyl group of the root system, and choose a W-invariant $\mathbf{Q}$-valued inner product $(\cdot, \cdot)$ on V. Since $(b, b) > 0$, we have $(b, a_{i_0}) > 0$ for some i_0. Hence, $b - a_{i_0} \in \Phi$ by [Bou2, VI, §1.3, Thm. 1, Cor.]. By induction on q, we then have $b - a_{i_0} \in \Psi$, so the identity $b = (b - a_{i_0}) + a_{i_0}$ forces $b \in \Psi$ by closedness. □

The structure of parabolic sets of roots is given by:

Proposition 2.2.8 *Let Φ be a root system spanning a $\mathbf{Q}$-vector space V, and Ψ a subset of Φ. The following conditions are equivalent:*

(1) *The subset Ψ is parabolic.*
(2) *The subset Ψ contains a positive system of roots Φ^+ in Φ and a unique subset I of simple positive roots such that $\Psi = \Phi^+ \cup [I]$ where $[I]$ denotes the set of roots that are $\mathbf{Z}$-linear combinations of elements in I. In particular, $\Psi \cap -\Phi^+$ is the set of roots in $[I]$ whose coefficients with respect to I are $\leqslant 0$.*
(3) *There is a linear form $\lambda : V \to \mathbf{Q}$ such that*

$$\Psi = \Phi_{\lambda \geqslant 0} := \{a \in \Phi \mid \lambda(a) \geqslant 0\}.$$

The subset $\Phi_{\lambda > 0} = \{a \in \Phi \mid \langle a, \lambda \rangle > 0\}$ is contained in a positive system of roots in Φ. In condition (2), such a unique I exists for any positive system of roots contained in Ψ when the equivalent conditions hold.

Proof The equivalence of (1) and (2) is [Bou2, VI, §1.7, Prop. 20]. The uniqueness of I given Φ^+ in (2) is obvious, and the existence of I for any positive system of roots $\Phi^+ \subseteq \Psi$ is [Bou2, VI, §1.7, Lemma 3]. It is clear from the definition that sets $\Phi_{\lambda \geqslant 0}$ as in (3) are parabolic. Conversely, suppose Ψ is as in (2). Since the simple positive roots are a basis of V, we can choose λ to vanish on I and to be positive on all other simple positive roots. It is then clear that $\Psi = \Phi_{\lambda \geqslant 0}$. If $\mu : V \to \mathbf{Q}$ is a linear form non-vanishing on Φ and close to λ in $V^*_{\mathbf{R}}$ then $\Phi_{\lambda>0} \subseteq \Phi_{\mu>0} \subseteq \Phi_{\lambda\geqslant 0}$ and $\Phi_{\mu>0}$ is a positive system of roots. □

To justify the terminology "pseudo-parabolic", we prove the following result that will be quite useful later on.

Proposition 2.2.9 *Let k be a field and G a connected reductive k-group. A k-subgroup $P \subseteq G$ is pseudo-parabolic if and only if it is parabolic. That is, the parabolic k-subgroups of G are precisely the subgroups $P_G(\lambda)$ for k-homomorphisms $\lambda : \mathrm{GL}_1 \to G$.*

Proof We first treat the problem over an algebraically closed field, so assume $k = \overline{k}$. To show that any subgroup $P = P_G(\lambda)$ is parabolic, we pick a maximal torus T of P containing the image of λ (as P contains $Z_G(\lambda)$, T is maximal in G) and let $\Phi = \Phi(G,T)$. As noted in Proposition 2.2.8, the set $\Psi = \Phi(P,T) = \Phi_{\lambda\geqslant 0}$ contains a positive system of roots $\Phi^+ = \Phi_{\mu>0} = \Phi_{\mu\geqslant 0}$ for a cocharacter $\mu \in \mathrm{X}_*(T)$ that is not orthogonal to any roots. The smooth connected subgroup $P_G(\mu)$ containing T also contains the smooth connected subgroup $P_G(\mu) \cap P_G(\lambda) = P_{P_G(\lambda)}(\mu)$ which has the same Lie algebra as $P_G(\mu)$, so these groups coincide; i.e., the subgroup $P = P_G(\lambda)$ of G contains $P_G(\mu)$. Thus, to show that P is parabolic it suffices to show that $P_G(\mu)$ is a Borel subgroup of G. By [Bo2, 13.18(5)], there is a Borel subgroup $B \supseteq T$ such that $\Phi(B,T) = \Phi_{\mu>0}$. The inclusion $P_B(\mu) = B \cap P_G(\mu) \subseteq B$ between smooth connected subgroups of G is an equality on Lie algebras and hence an equality of subgroups, so $B \subseteq P_G(\mu)$. But this is also an equality on Lie algebras, so $P_G(\mu) = B$ is a Borel subgroup of G.

Conversely, let P be a parabolic subgroup of G. Choose a Borel subgroup B of G contained in P and a maximal torus T of B (so T is maximal in G). Let $\Phi = \Phi(G,T)$, $\Phi^+ = \Phi(B,T)$ (a positive system of roots in Φ), and $\Psi = \Phi(P,T)$. We claim that Ψ is a closed set of roots. Granting this, we prove the pseudo-parabolicity of P as follows. Since Ψ is a parabolic set of roots, it equals $\Phi_{\lambda\geqslant 0}$ for some $\lambda \in \mathrm{X}_*(T)$ (by Proposition 2.2.8). Thus, the inclusion $P_G(\lambda) \cap P = P_P(\lambda) \subseteq P$ between smooth subgroups of G containing T is an equality on Lie algebras and hence is an equality of groups (as P is connected). The resulting inclusion $P \subseteq P_G(\lambda)$ is also an equality on Lie algebras and so $P = P_G(\lambda)$ since $P_G(\lambda)$ is connected (Proposition 2.1.8(4)). Hence, we just have to show that Ψ is closed in Φ.

For any $c \in \Psi$ such that $-c \in \Psi$ we claim that Ψ is stable under the reflection $r_c \in W(G,T) = W(\Phi)$. More specifically, we claim that P contains a representative $n_c \in N_G(T)$ of r_c. Since $B \cap Z_G(S)$ is a Borel subgroup of

$Z_G(S)$ for any torus S in B, the smooth group $P \cap Z_G(S) = Z_P(S)$ is a parabolic subgroup of $Z_G(S)$ for any torus $S \subseteq T$. Hence, $P \cap \mathscr{D}(Z_G(S))$ is a parabolic subgroup of $\mathscr{D}(Z_G(S))$ for any torus $S \subseteq T$. Thus, letting $S = T_c := (\ker c)^0_{\mathrm{red}}$, the 3-dimensional semisimple subgroup $G_c = \mathscr{D}(Z_G(T_c))$ meets P in a parabolic subgroup of G_c containing both weight spaces $\mathfrak{g}_{\pm c}$ for the maximal torus $c^\vee(\mathrm{GL}_1)$. But G_c has PGL_2 as a central quotient, so it has only two proper parabolic subgroups containing this torus and the Lie algebra of neither contains both $\mathfrak{g}_c$ and $\mathfrak{g}_{-c}$. Hence, P contains G_c, whence it contains an element in $N_G(T)$ representing the reflection $r_c \in W(G_c, c^\vee(\mathrm{GL}_1))$.

The problem is now combinatorial: if Φ is a reduced root system and Ψ is a subset containing a positive system of roots such that Ψ is stable under the reflection r_c whenever $\pm c \in \Psi$ then we claim that Ψ is closed in Φ. For any $a, b \in \Psi$ such that $a + b \in \Phi$, necessarily a and b are linearly independent, so the problem of showing $a + b \in \Psi$ reduces to the case of the reduced root system $\Phi' = \Phi \cap (\mathbf{Q}a + \mathbf{Q}b)$ of rank 2 and its subset $\Psi' = \Psi \cap \Phi'$ that satisfies the same hypotheses as Ψ does relative to Φ. Thus, the problem reduces to the case where Φ is a reduced and irreducible root system of rank 2, and this is easily settled by inspection of the three such root systems.

Now consider a general ground field. Consider a pseudo-parabolic k-subgroup P in G, so $P = P_G(\lambda)$ for a k-homomorphism $\lambda : \mathrm{GL}_1 \to G$. Since $P_{\overline{k}} = P_{G_{\overline{k}}}(\lambda_{\overline{k}})$ is pseudo-parabolic in $G_{\overline{k}}$, it is also parabolic. But parabolicity is a geometric property, so P is parabolic in G. Conversely, let P be a parabolic k-subgroup of G, and let T be a maximal k-torus in P, so T is also maximal in G and $P_{\overline{k}} = P_{G_{\overline{k}}}(\lambda')$ for some $\lambda' \in \mathrm{X}_*(T_{\overline{k}})$ (by the argument used over $\overline{k}$). We just need to find such $\lambda' : \mathrm{GL}_1 \to T_{\overline{k}}$ that is defined over k.

Let k'/k be a finite Galois extension that splits T and let $\Gamma = \mathrm{Gal}(k'/k)$. The homomorphism λ' descends to a k'-homomorphism $\mathrm{GL}_1 \to T_{k'}$ (again denoted as λ'), so $P_{k'} = P_{G_{k'}}(\lambda')$. The key point is that the set of roots

$$\Phi(G_{k'}, T_{k'})_{\lambda' \geqslant 0} = \Phi(P_{k'}, T_{k'})$$

is Γ-stable inside of $\mathrm{X}(T_{k'})$ since $(P_{k'}, T_{k'})$ descends to (P, T). Hence, for any $a \in \Phi(G_{k'}, T_{k'})$, whether $\langle a, \lambda' \rangle$ is $\geqslant 0$ or < 0 is the same across all members of the Γ-orbit of a. But $\langle \gamma(a), \lambda' \rangle = \langle a, \gamma^{-1}(\lambda') \rangle$ for all $\gamma \in \Gamma$, so if we define

$$\mu = \sum_{\gamma \in \Gamma} \gamma(\lambda') \in \mathrm{Hom}_k(\mathrm{GL}_1, T)$$

then $\langle a, \mu \rangle \geqslant 0$ if and only if $\langle a, \lambda' \rangle \geqslant 0$. Thus, the inclusion of smooth connected k'-groups

$$P_{k'} \cap P_G(\mu)_{k'} = P_{P_{k'}}(\mu)_{k'} \subseteq P_{k'} = P_{G_{k'}}(\lambda')$$

containing $T_{k'}$ is an equality on Lie algebras and hence of groups, so $P \subseteq P_G(\mu)$. Again comparing Lie algebras (over k') gives $P = P_G(\mu)$. □

The theory of pseudo-parabolic subgroups is primarily of interest in the pseudo-reductive case, due to the fact that $\mathscr{R}_{u,k}(G)$ is always contained in pseudo-parabolic k-subgroups (by definition). The following result makes the emphasis on the pseudo-reductive case more precise.

Proposition 2.2.10 *Let G be a smooth connected affine group over a field k, and let*

$$\pi : G \twoheadrightarrow G' = G/\mathscr{R}_{u,k}(G)$$

be its maximal pseudo-reductive quotient over k. Then

$$P \mapsto \pi(P), \quad P' \mapsto \pi^{-1}(P')$$

are inverse bijections between the sets of pseudo-parabolic k-subgroups of G and G'. If P and P' correspond under this bijection then the natural map $G/P \to G'/P'$ is an isomorphism.

Likewise, if T' is a maximal k-torus of G' then $\pi^{-1}(T')$ contains a maximal k-torus T of G, and P contains T if and only if P' contains T'.

Proof By definition, a pseudo-parabolic k-subgroup P in G has the form $P = P_G(\lambda)\mathscr{R}_{u,k}(G)$ for a 1-parameter k-subgroup $\lambda : \mathrm{GL}_1 \to G$. For the induced 1-parameter k-subgroup $\lambda' = \pi \circ \lambda : \mathrm{GL}_1 \to G'$, we have

$$\pi(P) = \pi(P_G(\lambda)) = P_{G'}(\lambda')$$

by Corollary 2.1.9, and this is a pseudo-parabolic k-subgroup of G' since G' is pseudo-reductive over k.

Conversely, let $P' \subseteq G'$ be a pseudo-parabolic k-subgroup, so $P' = P_{G'}(\lambda')$ for a 1-parameter k-subgroup $\lambda' : \mathrm{GL}_1 \to G'$. We claim that $P := \pi^{-1}(P')$ is a pseudo-parabolic k-subgroup of G. If $\lambda' = 1$ then $P' = G'$, so there is nothing to do. Now suppose $\lambda' \neq 1$, so $\lambda'(\mathrm{GL}_1)$ is a 1-dimensional k-split torus in G'. In this case $H = \pi^{-1}(\lambda'(\mathrm{GL}_1))$ is an extension of $\lambda'(\mathrm{GL}_1)$ by $\mathscr{R}_{u,k}(G)$. Since $H \twoheadrightarrow \lambda'(\mathrm{GL}_1)$ has unipotent kernel and must carry maximal k-tori (isomorphically) onto maximal k-tori, H is a smooth connected solvable k-group whose maximal k-tori are all k-split of dimension 1. Pick $\lambda : \mathrm{GL}_1 \hookrightarrow H$ such that $\lambda' = \pi \circ \lambda$. Then $P' = P_{G'}(\lambda') = \pi(P_G(\lambda))$, so the k-subgroup $\pi^{-1}(P')$ of G equals the pseudo-parabolic k-subgroup $P_G(\lambda)\mathscr{R}_{u,k}(G)$. This clearly reverses the first construction, and the isomorphism between coset spaces is obvious.

Finally, consider the claim concerning maximal k-tori. Let T' be a maximal k-torus in G', so $\pi^{-1}(T')$ is an extension of T' by $\mathscr{R}_{u,k}(G)$. Hence, $\pi^{-1}(T')$ is a smooth connected solvable k-group and every maximal k-torus T in $\pi^{-1}(T')$ maps isomorphically onto T'. If P' contains T' then certainly $P := \pi^{-1}(P')$ contains T. Conversely, if P contains T then its image $P' = \pi(P)$ contains $\pi(T) = T'$. □

We conclude our initial discussion of pseudo-parabolic subgroups by addressing the issue of how we can change λ without changing $P_G(\lambda)$.

Proposition 2.2.11 *Let G be a smooth connected affine group over a field k, and let $\lambda : \mathrm{GL}_1 \to G$ be a k-homomorphism. Let $P = P_G(\lambda)$. For any $g \in P(k)$, the k-homomorphism $g.\lambda : t \mapsto g\lambda(t)g^{-1}$ satisfies $P_G(\lambda) = P_G(g.\lambda)$, and if T is a maximal k-torus in P then there exists $g \in U_G(\lambda)(k)$ such that $g.\lambda$ has image contained in T.*

Proof For any point x of P, the conjugate

$$\lambda(t)(g^{-1}xg)\lambda(t)^{-1} = g^{-1}((g.\lambda)(t))x((g.\lambda)(t))^{-1}g$$

has a limiting value as $t \to 0$ since $g^{-1}xg$ is a point of P. Hence, $P_G(\lambda) = P \subseteq P_G(g.\lambda)$. In particular, $g^{-1} \in P_G(g.\lambda)(k)$. We can apply the same argument to g^{-1} and $g.\lambda$ to get the reverse inclusion. To find $g \in U_G(\lambda)(k)$ so that $g.\lambda$ is valued in T it is equivalent to find a $U_G(\lambda)(k)$-conjugate of T containing the image of λ. Let $U = U_G(\lambda)$ and define $\lambda' : \mathrm{GL}_1 \to P/U$ to be the composition of λ with the quotient map $P \to P/U$. The image of λ' is central in P/U since $Z_G(\lambda) \to P/U$ is an isomorphism, so $\lambda'(\mathrm{GL}_1)$ lies in every maximal k-torus of P/U (as we may check over k_s). In particular, $\lambda'(\mathrm{GL}_1)$ is contained in the maximal k-torus image T' of T in P/U, so the k-torus $\lambda(\mathrm{GL}_1)$ inside P is contained in the k-subgroup $H = T \ltimes U$ of P. By [Bo2, 19.2] all maximal k-tori in H are $U(k)$-conjugate to each other because H is a solvable smooth connected affine k-group with unipotent radical U, so there is a $U(k)$-conjugate of T that contains the image of λ. □

Proposition 2.2.12 *Let $f : G \to \overline{G}$ be a surjective homomorphism between smooth affine groups over a field k, and assume* $\ker f$ *is central in G.*

(1) *For every maximal k-torus $\overline{T} \subseteq \overline{G}$, the scheme-theoretic preimage $f^{-1}(\overline{T})$ is commutative and contains a unique maximal k-torus T. This k-torus is*

maximal in G, and $f(T) = \overline{T}$. *The map* $\overline{T} \mapsto T$ *defines a bijection between the sets of maximal k-tori of* $\overline{G}$ *and G, with inverse* $T \mapsto f(T)$.

(2) *Assume G is connected and perfect. The schematic center* $Z_{\overline{G}}$ *is equal to* $Z_G/(\ker f)$. *In particular, if* $\overline{G} \to \overline{\overline{G}}$ *is another central quotient map then the composite surjection* $G \to \overline{\overline{G}}$ *has central kernel and* G/Z_G *has trivial center.*

(3) *If G and* $\overline{G}$ *are pseudo-reductive then* $P \to f(P)$ *is an inclusion-preserving bijection between the pseudo-parabolic k-subgroups of G and pseudo-parabolic k-subgroups of* $\overline{G}$, *with inverse* $\overline{P} \mapsto f^{-1}(\overline{P})$.

Proof To prove (1), first note that by dimension considerations, in any (not necessarily smooth) commutative finite type k-group scheme there exists a k-torus that contains all others. Thus, it suffices to prove more generally that for any k-torus S and central extension

$$1 \to Z \to E \to S \to 1$$

by a commutative affine k-group Z of finite type, E is necessarily commutative and its maximal k-torus maps onto S. To prove that E is commutative, we observe that the commutator morphism of k-schemes $E \times E \to E$ factors through a bi-additive pairing $b : S \times S \to Z$, so it is enough to show that the only such b is the trivial pairing. For that purpose we may extend scalars to arrange that k is separably closed (so $S(k)$ is Zariski-dense in S). For each $s \in S(k)$, $b(s, \cdot)$ must factor through the maximal k-torus in Z, so b does as well. The case when Z is a torus is well known (due to the étaleness of the Hom-scheme between tori). This proves that E is commutative.

Since E is commutative, by Lemma C.4.4 the formation of its maximal k-torus commutes with any extension of the ground field. Hence, to show that this maximal torus maps onto S we may assume k is algebraically closed and then replace E with $E^0_{\rm red}$ so that E is smooth and connected. By commutativity, E is then a direct product of a torus and a unipotent group, so (1) is proved.

In the setting of (2), we may assume $k = \overline{k}$. Any perfect smooth connected affine k-group H is generated by its maximal k-tori (since the smooth connected subgroup N generated by such tori is normal with H/N having no nontrivial tori, so H/N is unipotent yet perfect and thus trivial). Hence, the scheme-theoretic center of any such group is defined by the condition of centralizing all maximal tori. In other words, the center is the intersection of all Cartan k-subgroups. It therefore suffices to show that the map $C \mapsto \overline{C} := C/(\ker f)$ from the set of Cartan k-subgroups of G to the set of Cartan k-subgroups of $\overline{G}$ is bijective. Injectivity is clear, and surjectivity follows from (1).

Finally, to prove (3) we assume G and $\overline{G}$ are pseudo-reductive. Every pseudo-reductive k-subgroup P of G has the form $P_G(\lambda)$ for a 1-parameter k-subgroup $\lambda : \mathrm{GL}_1 \to G$, so $f(P) = P_{\overline{G}}(f \circ \lambda)$ (Corollary 2.1.9) and hence $f(P)$ is a pseudo-parabolic k-subgroup of $\overline{G}$. To prove that every pseudo-parabolic k-subgroup $\overline{P}$ of $\overline{G}$ arises in this manner, let $\overline{\lambda} : \mathrm{GL}_1 \to \overline{G}$ be a 1-parameter k-subgroup of $\overline{G}$ such that $\overline{P} = P_{\overline{G}}(\overline{\lambda})$. Applying part (1) to a maximal k-torus of $\overline{G}$ containing the image of $\overline{\lambda}$, we obtain a 1-parameter k-subgroup $\lambda : \mathrm{GL}_1 \to G$ such that $f \circ \lambda = \overline{\lambda}^n$ for some integer $n > 0$. Since $\overline{P} = P_{\overline{G}}(\overline{\lambda}^n)$ (Remark 2.1.7), we have $\overline{P} = f(P)$ for $P = P_G(\lambda)$. The k-group P contains the Cartan k-subgroup $Z_G(T)$ for any maximal k-torus T in P, so P contains the schematic center of G (and therefore contains all central closed k-subgroup schemes). Hence, $f^{-1}(\overline{P}) = P$. □

Proposition 2.2.13 *Let k be a field and k' a nonzero finite reduced k-algebra. Let G' be a k'-group with pseudo-reductive fibers over the factor fields of k', and let $G = \mathrm{R}_{k'/k}(G')$. Let $\pi : G_{k'} \to G'$ be the natural surjective homomorphism. There is a natural bijective correspondence between the set of pseudo-parabolic k-subgroups of G and the set of fiberwise pseudo-parabolic k'-subgroups of G'.*

Explicitly, every pseudo-parabolic k-subgroup P of G has the form $P = \mathrm{R}_{k'/k}(P')$ with $P' := \pi(P_{k'})$ a fiberwise pseudo-parabolic k'-subgroup of G', and conversely if Q' is a fiberwise pseudo-parabolic k'-subgroup of G' then $\mathrm{R}_{k'/k}(Q')$ is a pseudo-parabolic k-subgroup of G.

Proof This is essentially an immediate application of Proposition 2.1.13. In more detail, by writing $k' = \prod k'_i$ for fields k'_i and G'_i for the k'_i-fiber of G' we have: $G = \prod \mathrm{R}_{k'_i/k}(G'_i)$, $G' = \coprod G'_i$, and π over k'_i is the composition of the ith projection $G_{k'_i} \to \mathrm{R}_{k'_i/k}(G'_i)_{k'_i}$ and the natural map $\pi_i : \mathrm{R}_{k'_i/k}(G'_i)_{k'_i} \to G'_i$. Consequently, since the pseudo-parabolic k-subgroups of a direct product $H_1 \times H_2$ of pseudo-reductive groups H_1 and H_2 are precisely the subgroups $P_1 \times P_2$ for pseudo-parabolic k-subgroups P_j of H_j (because 1-parameter k-subgroups of $H_1 \times H_2$ are precisely the maps $(\lambda_1, \lambda_2) : \mathrm{GL}_1 \to H_1 \times H_2$ for 1-parameter k-subgroups $\lambda_j : \mathrm{GL}_1 \to H_j$), the general case reduces to the case where k' is a field.

Now assume k' is a field. Any pseudo-parabolic k'-subgroup of G' has the form $Q' = P_{G'}(\lambda')$ for a 1-parameter k'-subgroup λ' of G', so by Proposition 2.1.13 we have $\mathrm{R}_{k'/k}(Q') = P_G(\lambda)$ for the 1-parameter k-subgroup $\lambda : \mathrm{GL}_1 \to G$ corresponding to λ' via the universal property of Weil restriction. A further application of the same proposition gives that for any pseudo-parabolic

k-subgroup $P = P_G(\mu)$ of G, necessarily $P = \mathrm{R}_{k'/k}(P_{G'}(\mu'))$ for the 1-parameter k'-subgroup μ' of G' that corresponds to μ. Thus, by naturality, the restriction of π to $P_{k'}$ is the composition of the natural quotient map $P_{k'} \twoheadrightarrow P_{G'}(\mu')$ and the inclusion of $P_{G'}(\mu')$ into G'. Hence, $\pi(P_{k'}) = P_{G'}(\mu')$ is a pseudo-parabolic k'-subgroup of G'. □

2.3 Root groups in pseudo-reductive groups

The constructions $P_G(\lambda)$, $U_G(\lambda)$, and $Z_G(\lambda)$ will now enable us to define and prove the uniqueness of "root groups" in pseudo-reductive groups over a separably closed field. We first set forth some terminology and notation.

Let k be a field, and G a smooth connected affine k-group. In Theorem C.2.3 we shall prove that all maximal k-split k-tori in G are $G(k)$-conjugate. The proof of this result rests on the structure theory of root groups that is developed below.

Definition 2.3.1 The k-group G is *pseudo-split* (over k) if it contains a maximal k-torus T that is k-split. (If G is connected reductive then this coincides with the usual notion of being k-split.) We call any such (G, T) a *pseudo-split pair.*

Now let T be an arbitrary k-split k-torus in G. Since T is k-split, $\mathrm{Lie}(G)$ equipped with its adjoint action by T decomposes as a direct sum of weight spaces for k-rational characters of T (Lemma A.8.8). The set of nontrivial characters of T that occur in $\mathrm{Lie}(G)$ will be denoted by $\Phi(G, T)$.

Example 2.3.2 Let k' be a nonzero finite reduced algebra over a field k, and let G' be a smooth affine k'-group with connected fibers. Thus, $G := \mathrm{R}_{k'/k}(G')$ is a smooth affine connected k-group. Assume that all factor fields of k' are purely inseparable over k (as always happens when $k = k_s$).

Let T be a maximal k-torus in G, so by Proposition A.5.15(2) it is the maximal k-torus in $\mathrm{R}_{k'/k}(T')$ for a unique maximal k'-torus T' in G'. Thus, if $\{k'_i\}$ is the set of factor fields of k' and (G'_i, T'_i) is the k'_i-fiber of (G', T'), then $G = \prod G_i$ and $T = \prod T_i$ where $G_i := \mathrm{R}_{k'_i/k}(G'_i)$ and T_i is the maximal k-torus in $\mathrm{R}_{k'_i/k}(T'_i)$. The pure inseparability hypothesis on k'/k implies that T is k-split if and only if T' is k'-split (i.e., T'_i is k'_i-split for all i), so G is pseudo-split if and only if G' is pseudo-split (in the sense that it admits a k'-split maximal k'-torus, which can be checked on fibers over $\mathrm{Spec}\, k'$). When this condition holds, the k-split maximal k-tori of G are in natural bijection with the k'-split maximal k'-tori of G'.

Assume T is k-split, so $\Phi(G,T) = \coprod \Phi(G_i, T_i)$ compatibly with the identification $X(T) = \prod X(T_i)$. To relate $\Phi(G_i, T_i)$ and $\Phi(G_i', T_i')$, we may rename k_i' as k' so that k' is a field. Since k'/k is purely inseparable, the natural map $T_{k'} \to T'$ is an isomorphism. This identifies $X(T)$ with $X(T')$. Under the identification of $\mathrm{Lie}(\mathrm{R}_{k'/k}(G'))$ with the Lie algebra over k underlying $\mathrm{Lie}(G')$ (Corollary A.7.6), we then see that the root spaces for T acting on $\mathrm{Lie}(\mathrm{R}_{k'/k}(G'))$ are the underlying k-vector spaces of the root spaces for T' acting on $\mathrm{Lie}(G')$. This identifies $\Phi(G,T)$ with $\Phi(G',T')$ inside of $X(T) = X(T')$.

Returning to the general setup, we fix a k-split k-torus T in a smooth connected affine k-group G, and for each nonzero $a \in X(T)_{\mathbf{Q}}$ let T_a be the corresponding codimension-1 k-subtorus $(\ker na)^0_{\mathrm{red}}$ for any integer $n \geqslant 1$ such that $na \in X(T)$ (the choice of which does not matter). The centralizer $Z_G(T_a)$ is a smooth connected k-subgroup of G (Proposition A.2.5), it is pseudo-reductive when G is pseudo-reductive (Proposition 1.2.4), and its Lie algebra is the subspace of T_a-invariants in $\mathrm{Lie}(G)$ (Proposition A.2.5). Thus, for any nonzero rational multiple of a in $X(T)_{\mathbf{Q}}$ that is contained in $\Phi(G,T)$, the associated root space in $\mathrm{Lie}(G)$ is contained in $\mathrm{Lie}(Z_G(T_a))$.

Consider the adjoint action by the (k-split) $\bar{k}$-torus T of $Z_G(T_a)$ on $\mathrm{Lie}(Z_G(T_a))$. The codimension-1 subtorus T_a acts trivially, and the weight spaces in $\mathrm{Lie}(Z_G(T_a))$ apart from $\mathrm{Lie}(Z_G(T))$ are the root spaces in $\mathrm{Lie}(G)$ for the rational multiples of a which lie in $\Phi(G,T)$. To separate the positive rational multiples of a occurring in $\Phi(G,T)$ from the negative ones, we will use the following device: choose a k-homomorphism

$$\lambda_a : \mathrm{GL}_1 \to T$$

such that $\langle a, \lambda_a \rangle > 0$ (so $\lambda_a(\mathrm{GL}_1)$ is an isogeny complement to T_a in T).

Lemma 2.3.3 *The k-subgroup $U_{(a)} := U_{Z_G(T_a)}(\lambda_a) \subseteq G$ is independent of λ_a, and* $\mathrm{Lie}(U_{(a)})$ *is the span of the root spaces for T in* $\mathrm{Lie}(G)$ *for all roots in $\Phi(G,T)$ that are positive rational multiples of a in $X(T)_{\mathbf{Q}}$. In particular, $U_{(a)} \neq 1$ precisely when some positive rational multiple of a lies in $\Phi(G,T)$.*

The k-group G is generated by the k-subgroups $Z_G(T)$ and $U_{(a)}$ for all $a \in \Phi(G,T)$.

Proof Since $\lambda_a(\mathrm{GL}_1)$ is an isogeny-complement to T_a in T, the composite map

$$\lambda_a' : \mathrm{GL}_1 \xrightarrow{\lambda_a} \lambda_a(\mathrm{GL}_1) \to T/T_a$$

is an isogeny. Conjugation by T on $Z_G(T_a)$ factors through an action of the quotient T/T_a, so $U_{(a)}$ represents the functor of points

$$R \rightsquigarrow \{g \in Z_G(T_a)(R) \mid (\lambda'_a)_R(t).g \to 1 \text{ as } t \to 0 \text{ in } (\mathrm{GL}_1)_R\}.$$

Hence, to prove that $U_{(a)}$ is independent of λ_a, it suffices to show that replacing λ_a with λ_a^n for an integer $n > 0$ does not affect the definition of $U_{(a)}$. This invariance follows from Remark 2.1.7.

By Proposition 2.1.8(1), $\mathrm{Lie}(U_{(a)})$ is the span of the $\lambda_a(\mathrm{GL}_1)$-weight spaces in $\mathrm{Lie}(Z_G(T_a))$ with weight > 0. But $\mathrm{Lie}(Z_G(T_a))$ is the T_a-centralizer in $\mathrm{Lie}(G)$ (Proposition A.2.5), so $\mathrm{Lie}(U_{(a)})$ is the span of the root spaces in $\mathrm{Lie}(G)$ for the elements c in $\Phi(G,T)$ that kill T_a and satisfy $\langle c, \lambda_a\rangle > 0$. Any $a' \in \Phi(G,T)$ killing T_a is a nonzero rational multiple of a, and if $a' = ca$ in $\mathrm{X}(T)_{\mathbf{Q}}$ with $c \in \mathbf{Q}^\times$ then $\langle a', \lambda_a\rangle = c\langle a, \lambda_a\rangle$ has the same sign as c. Thus, $\mathrm{Lie}(U_{(a)})$ has the desired description.

The k-group G is generated by $Z_G(T)$ and the $U_{(a)}$ for all $a \in \Phi(G,T)$ since these groups are all smooth and connected and $\mathrm{Lie}(G)$ is spanned by $\mathrm{Lie}(Z_G(T))$ and the $\mathrm{Lie}(U_{(a)})$ (due to the weight space decomposition for $\mathrm{Lie}(G)$ relative to its T-action). □

Definition 2.3.4 For $a \in \mathrm{X}(T)_{\mathbf{Q}} - \{0\}$ and any λ_a as above, $U^G_{(a)}$, or $U_{(a)}$ if G is clear from the context, is the smooth connected unipotent k-subgroup $U_{Z_G(T_a)}(\lambda_a)$ in G (with $T_a = (\ker na)^0_{\mathrm{red}}$ for any $n \geqslant 1$ such that $na \in \mathrm{X}(T)$).

Remark 2.3.5 The k-subgroup $U_{(a)}$ in G is normalized by $Z_G(T)$ since $Z_G(T)$ centralizes λ_a. Due to its definition, $U_{(a)}$ only depends on a up to a positive rational multiple in $\mathrm{X}(T)_{\mathbf{Q}}$.

Remark 2.3.6 Let G be a smooth connected affine k-group and T a k-split k-torus of G. For any smooth connected k-subgroup $H \subseteq G$ that contains T, and any nonzero $a \in \mathrm{X}(T)_{\mathbf{Q}}$, since $Z_H(T_a) = H \cap Z_G(T_a)$ for $T_a = (\ker a)^0_{\mathrm{red}}$ we see (via Lemma 2.1.5) that $U^H_{(a)} = H \cap U^G_{(a)}$ (where $U^G_{(a)}$ is the group $U_{(a)}$ in Lemma 2.3.3 associated to (G,T,a) and similarly with $U^H_{(a)}$ and (H,T,a)). In particular, $H \cap U^G_{(a)}$ is smooth and connected.

Consider a quotient map $\pi : G_{\overline{k}} \twoheadrightarrow G'$ over $\overline{k}$, and let $T' = \pi(T_{\overline{k}})$, so $\pi|_{T_{\overline{k}}}$ induces an embedding of $\mathrm{X}(T')_{\mathbf{Q}}$ into $\mathrm{X}(T)_{\mathbf{Q}}$ that we use to identify the former with a subspace of the latter. From Corollary 2.1.9 we see that for any nonzero $a \in \mathrm{X}(T')_{\mathbf{Q}}$, π carries $(U^G_{(a)})_{\overline{k}}$ onto $U^{G'}_{(a)}$ (note that Proposition A.2.8 implies that $\pi(Z_G(T_a)_{\overline{k}}) = Z_{G'}(T'_a)$). In particular, if $a \in \Phi(G',T')$ then since $U^{G'}_{(a)} \neq 1$ we have $U^G_{(a)} \neq 1$ and hence some positive rational multiple of a lies in $\Phi(G,T)$. Note that $N := (\ker \pi)^0$ is normal in $G_{\overline{k}}$ since smoothness of

G ensures that such normality can be detected using conjugation just against $G(\overline{k})$, so if N is solvable then $\mathscr{R}_u(N) \subseteq \mathscr{R}_u(G_{\overline{k}})$. Let U be a smooth connected unipotent k-subgroup of G such that $U_{\overline{k}} \subseteq N$. If N is solvable then $U_{\overline{k}} \subseteq \mathscr{R}_u(N) \subseteq \mathscr{R}_u(G_{\overline{k}})$, forcing $U \subseteq \mathscr{R}_{u,k}(G)$ by Lemma 1.2.1 (and thus $U = 1$ if G is pseudo-reductive).

Now assume G is pseudo-reductive and N is solvable, so π cannot kill $(U^H_{(a)})_{\overline{k}}$ for $a \in \Phi(H,T)$. Consequently, for $H' = \pi(H_{\overline{k}})$ and $a \in \Phi(H,T)$ the k-group $U^{H'}_{(a)}$ is nontrivial, so a cannot completely "disappear" in $\Phi(H',T')$. That is, some positive rational multiple of a lies in $\Phi(H',T')$ when G is pseudo-reductive and N is solvable. The same assertion holds if instead of assuming N is solvable we assume that $T_{\overline{k}}$ centralizes N. Indeed, by definition $U^H_{(a)} = U_{Z_H(T_a)}(\lambda_a)$ for a suitable $\lambda_a \in \mathrm{X}_*(T)$, so $Z_G(T) \cap U^H_{(a)} \subseteq Z_H(T) \cap U^H_{(a)} \subseteq Z_{Z_H(T_a)}(\lambda_a) \cap U_{Z_H(T_a)}(\lambda_a) = 1$.

Example 2.3.7 An interesting class of such π that arise in the classification of pseudo-reductive groups is central quotients $\mathrm{R}_{K/k}(\mathscr{G}) \to \mathrm{R}_{K/k}(\mathscr{G})/Z$ for a nonzero finite reduced k-algebra K, a semisimple K-group $\mathscr{G}$ with connected fibers, and a k-subgroup scheme $Z \subseteq \mathrm{R}_{K/k}(Z_{\mathscr{G}}) = Z_{\mathrm{R}_{K/k}(\mathscr{G})}$ (see Proposition A.5.15(1)). To prove that $(Z^0_{\overline{k}})_{\mathrm{red}}$ is unipotent it suffices to treat the case $Z = \mathrm{R}_{K/k}(Z_{\mathscr{G}})$, so the problem reduces to proving the unipotence of $\mathrm{R}_{A/k}(\mu_n)^0_{\mathrm{red}}$ for an algebraically closed field k, a nonzero finite k-algebra A, and any $n \geqslant 1$. Since $\mathrm{R}_{A/k}(\mu_n)$ is the n-torsion in $\mathrm{R}_{A/k}(\mathrm{GL}_1) = \mathrm{GL}_1 \times U$ for a smooth connected unipotent k-group U when A is local, and $(\mu_n)^0_{\mathrm{red}} = 1$, the result is clear.

We will prove shortly (in Theorem 2.3.10) that if $\mathrm{char}(k) \neq 2$ and (G,T) is a pseudo-split pair with G pseudo-reductive then $\Phi(G,T)$ is naturally identified with the *reduced* root system $\Phi(G^{\mathrm{red}}_{\overline{k}}, T_{\overline{k}})$ in $\mathrm{X}(T_{\overline{k}}) = \mathrm{X}(T)$, so $\mathrm{Lie}(U_{(a)})$ is the a-root space in $\mathrm{Lie}(G)$. When $\mathrm{char}(k) = 2$ and k is imperfect this can fail (i.e., $\mathrm{Lie}(U_{(a)})$ may support more than one T-weight), as we will see when we study the classification of pseudo-reductive groups in characteristic 2.

Lemma 2.3.8 *Let U be a smooth connected commutative unipotent group over a field k, and if* $\mathrm{char}(k) = p > 0$ *then assume that U is p-torsion. Let T be a split k-torus equipped with an action on U such that the trivial character of T is not a weight for the T-action on* $\mathrm{Lie}(U)$.

The k-group U is a vector group, and it admits a T-equivariant linear structure (in the sense of Definition A.1.1). Such a linear structure on U is unique if T acts on $\mathrm{Lie}(U)$ *with a single weight.*

The uniqueness can fail when $\mathrm{char}(k) = p > 0$ if the T-action on $\mathrm{Lie}(U)$ has more than one weight. For example, if $T = \mathrm{GL}_1$ acts on $V = \mathbf{G}_\mathrm{a} \times \mathbf{G}_\mathrm{a}$ via $c.(x, y) = (cx, c^p y)$ then $(x, y) \mapsto (x, y + x^p)$ is a non-linear k-automorphism of V that is T-equivariant.

Proof If $\mathrm{char}(k) = 0$ then the algebraic exponential isomorphism with the Lie algebra equips U with a functorial linear structure, so it is T-equivariant. If $\mathrm{char}(k) = p > 0$ then since T acts on $\mathrm{Lie}(U)$ with only nontrivial weights, it follows from Theorem B.4.3 that U is a vector group and admits a T-equivariant linear structure. This settles the existence in all characteristics.

To prove the uniqueness of the linear structure when T acts on $\mathrm{Lie}(U)$ with a single weight $\chi \neq 1$, we may assume k is algebraically closed. Any T-equivariant linear structure on U must have χ as its unique weight, since the Lie algebra can detect the weight space decomposition relative to the linear structure. For any $c \in k^\times$ we may write $c = \chi(t)$ for some $t \in T(k)$, so the linear structure on U must satisfy $cu = \chi(t)u = t.u$ for all $u \in U(k)$. Since $t.u$ does not depend on the linear structure, we are done. □

Proposition 2.3.9 *Let G be a pseudo-split pseudo-reductive group over a field k, and let T be a k-split maximal k-torus in G.*

For each $a \in \Phi(G,T)$, the k-group $U_{(a)}$ is a vector group and admits a T-equivariant linear structure. If $a' \in \Phi(G,T)$ is such that $qa+q'a' \notin \Phi(G,T)$ inside of $\mathrm{X}(T)_\mathbf{Q}$ for all $q, q' > 0$ in $\mathbf{Q}$ then $U_{(a)}$ and $U_{(a')}$ commute with each other.

Moreover, there is a unique positive rational multiple of a in $\mathrm{X}(T)$ that lies in $\Phi(G^{\mathrm{red}}_{\overline{k}}, T_{\overline{k}})$.

Proof By Lemma 1.2.1, to prove that the commutator $(U_{(a)}, U_{(a)})$ vanishes (i.e., $U_{(a)}$ is commutative) it suffices to show that $(U_{(a)}, U_{(a)})_{\overline{k}}$ has trivial image in $G^{\mathrm{red}}_{\overline{k}}$. By applying Remark 2.3.6 to the quotient map $G_{\overline{k}} \twoheadrightarrow G^{\mathrm{red}}_{\overline{k}}$ with unipotent kernel, some positive rational multiple $\overline{a}$ of a occurs in the reduced root system $\Phi(G^{\mathrm{red}}_{\overline{k}}, T_{\overline{k}})$ and $(U_{(a)})_{\overline{k}}$ has image in $G^{\mathrm{red}}_{\overline{k}}$ equal to the $\overline{a}$-root group. This root group is commutative, so $U_{(a)}$ is commutative.

We conclude that for any $a \in \Phi(G,T)$ we can define $q_a \in \mathbf{Q}$ to be the unique rational number > 0 such that $\overline{a} := q_a \cdot a \in \Phi(G^{\mathrm{red}}_{\overline{k}}, T_{\overline{k}})$ and $(U_{(a)})_{\overline{k}}$ maps onto the $\overline{a}$-root group in $G^{\mathrm{red}}_{\overline{k}}$.

For linearly independent $a, a' \in \Phi(G,T)$, if

$$q_a \cdot a + q_{a'} \cdot a' \notin \Phi(G^{\mathrm{red}}_{\overline{k}}, T_{\overline{k}})$$

then the root groups in $G_{\overline{k}}^{\rm red}$ corresponding to $q_a \cdot a$ and $q_{a'} \cdot a'$ commute. These latter root groups are the respective images of $(U_{(a)})_{\overline{k}}$ and $(U_{(a')})_{\overline{k}}$ in $G_{\overline{k}}^{\rm red}$. By Lemma 1.2.1, we conclude that

$$q_a \cdot a + q_{a'} \cdot a' \notin \Phi(G_{\overline{k}}^{\rm red}, T_{\overline{k}}) \Rightarrow (U_{(a)}, U_{(a')}) = 1.$$

This applies when $qa + q'a' \notin \Phi(G,T)$ for all $q, q' \in \mathbf{Q}_{>0}$ because $\Phi(G_{\overline{k}}^{\rm red}, T_{\overline{k}}) \subseteq \Phi(G,T)$ inside of $\mathrm{X}(T_{\overline{k}}) = \mathrm{X}(T)$ (as $\mathrm{Lie}(G_{\overline{k}}^{\rm red})$ is a $T_{\overline{k}}$-equivariant quotient of $\mathrm{Lie}(G_{\overline{k}}) = \mathrm{Lie}(G)_{\overline{k}}$).

Finally, we prove that $U_{(a)}$ is a vector group admitting a T-equivariant linear structure. Observe that $U_{(a)}$ is a smooth connected commutative k-group, and the T-action (through T/T_a) on $\mathrm{Lie}(U_{(a)})$ has no occurrence of the trivial weight (due to Proposition 2.1.8(1)). Thus, in view of Lemma 2.3.8, we just need to check that if $\mathrm{char}(k) = p > 0$ then $U_{(a)}$ is p-torsion. This goes as the proof of commutativity: the root group image of $(U_{(a)})_{\overline{k}}$ in $G_{\overline{k}}^{\rm red}$ is p-torsion, so $p \cdot (U_{(a)})_{\overline{k}} \subseteq \mathscr{R}_u(G_{\overline{k}})$. Hence, $p \cdot U_{(a)} = 1$ by Lemma 1.2.1. □

For a pseudo-split pseudo-reductive k-group G and a maximal k-torus T, an element $a \in \Phi(G,T)$ is *divisible* if $a = nb$ for some $b \in \Phi(G,T)$ and integer $n > 1$.

Theorem 2.3.10 *Let G be a pseudo-split pseudo-reductive group over a field k, and let T be a k-split maximal k-torus in G. If* $\mathrm{char}(k) \neq 2$ *then* $\Phi(G_{\overline{k}}^{\rm red}, T_{\overline{k}}) = \Phi(G,T)$ *inside of* $\mathrm{X}(T) = \mathrm{X}(T_{\overline{k}})$*; in particular,* $\Phi(G,T)$ *has no divisible elements. If* $\mathrm{char}(k) = 2$ *then*

$$\Phi(G_{\overline{k}}^{\rm red}, T_{\overline{k}}) \subseteq \Phi(G,T) \subseteq \Phi(G_{\overline{k}}^{\rm red}, T_{\overline{k}}) \coprod \frac{1}{2} \cdot \Phi(G_{\overline{k}}^{\rm red}, T_{\overline{k}})$$

inside of $\mathrm{X}(T) = \mathrm{X}(T_{\overline{k}})$*; in particular, all divisible elements of* $\Phi(G,T)$ *lie in* $\Phi(G_{\overline{k}}^{\rm red}, T_{\overline{k}})$.

When $\mathrm{char}(k) = 2$*, the case* $\Phi(G,T) \neq \Phi(G_{\overline{k}}^{\rm red}, T_{\overline{k}})$ *can only happen when k is imperfect and $G_{\overline{k}}^{\rm red}$ has a connected simple semisimple normal subgroup that is simply connected of type* C_n *with $n \geqslant 1$ (i.e., isomorphic to* Sp_{2n}*).*

Proof As we saw in the proof of Proposition 2.3.9, $\Phi(G_{\overline{k}}^{\rm red}, T_{\overline{k}}) \subseteq \Phi(G,T)$. In view of that proposition, our problem is to prove that for $a \in \Phi(G_{\overline{k}}^{\rm red}, T_{\overline{k}})$ the only positive rational multiples of a that can occur in $\Phi(G,T)$ are a and $a/2$, and that if $a/2 \in \Phi(G,T)$ then

(i) k is imperfect with $\mathrm{char}(k) = 2$,

(ii) $G^{\rm red}_{\overline{k}}$ has a simple smooth connected and simply connected normal subgroup of type C_n for some $n \geqslant 1$.

The assertion concerning simply connected type C_n, granting the rest, is due to the fact that in the classification of reduced and irreducible root data, this is the only case for which there is a root that is divisible (by 2, in fact) in the character lattice. For the remaining assertions we only need to consider elements of $\Phi(G,T)$ that kill the codimension-1 torus $T_a = (\ker a)^0_{\rm red}$. Hence, in view of Proposition A.4.8 it is harmless to replace G with $Z_G(T_a)$. This brings us to the case where there is a codimension-1 torus T' in T that is central in G. This is the maximal central torus, since T acts on Lie(G) with the nontrivial weight a.

The quotient $G/\mathscr{D}(G)$ is commutative, so its Lie algebra has no nontrivial T-weight. It follows that Lie($\mathscr{D}(G)$) contains the root spaces in Lie(G) relative to T and (by Lemma 1.2.5(iii)) that the k-split maximal k-torus $S = T \cap \mathscr{D}(G)$ of $\mathscr{D}(G)$ acts nontrivially on all of these root spaces. Hence, by Proposition A.4.8 there is no loss of generality by replacing (G,T) with $(\mathscr{D}(G), S)$. By Proposition 1.2.6, we also now have that $G = \mathscr{D}(G)$. In particular, $G^{\rm red}_{\overline{k}} = G^{\rm ss}_{\overline{k}}$.

Since a central torus of $G_{\overline{k}}$ has trivial image in $G^{\rm ss}_{\overline{k}}$, we conclude that $G^{\rm ss}_{\overline{k}}$ has a 1-dimensional maximal torus. Thus, this semisimple group is of type A_1 (i.e., it is SL_2 or PGL_2). Also, since $T_{\overline{k}}$ is a maximal torus in $G^{\rm red}_{\overline{k}} = G^{\rm ss}_{\overline{k}}$, it follows that $\dim T = 1$. Since $a \in \Phi(G^{\rm ss}_{\overline{k}}, T_{\overline{k}})$ we see that $\Phi(G^{\rm ss}_{\overline{k}}, T_{\overline{k}}) = \{\pm a\}$.

Fix a T-equivariant linear structure on the vector group $U_{(a)}$ and consider the weight space decomposition

$$U_{(a)} = \bigoplus V_{\chi_j}$$

according to this linear structure. Choose the unique isomorphism $\lambda : GL_1 \simeq T$ such that $\langle a, \lambda \rangle > 0$. There is an isomorphism of $G^{\rm ss}_{\overline{k}}$ onto SL_2 or PGL_2 carrying $T_{\overline{k}}$ over to the diagonal subgroup D such that λ is carried to the isomorphism $t \mapsto \mathrm{diag}(t, 1/t)$ when $G^{\rm ss}_{\overline{k}}$ is simply connected and is carried to the isomorphism $t \mapsto \mathrm{diag}(t, 1)$ when $G^{\rm ss}_{\overline{k}}$ is adjoint. Hence, by Corollary 2.1.9 applied to the quotient map $G_{\overline{k}} \twoheadrightarrow G^{\rm ss}_{\overline{k}}$, we see that $(U_{(a)})_{\overline{k}}$ is carried onto the upper triangular unipotent subgroup U^+ of $G^{\rm ss}_{\overline{k}}$ via the chosen isomorphism with SL_2 or PGL_2. This is equivariant with respect to the isomorphism $T_{\overline{k}} \twoheadrightarrow D$.

Let π be the composition of the quotient map $G_{\overline{k}} \twoheadrightarrow G^{\rm ss}_{\overline{k}}$ with the chosen isomorphism onto SL_2 or PGL_2. Since $a \in \Phi(G^{\rm ss}_{\overline{k}}, T_{\overline{k}})$, for any points t of $T_{\overline{k}}$

and u of $(U_{(a)})_{\overline{k}}$ we have

$$\pi(tut^{-1}) = a(t) \cdot \pi(u)$$

via the canonical linear structure on the 1-dimensional vector group U^+ (using any choice of $\overline{k}$-group isomorphism $U^+ \simeq \mathbf{G}_a$). To deduce information about positive rational multiples of a in $\Phi(G,T)$ aside from a, we shall study the restriction of

$$\pi : (U_{(a)})_{\overline{k}} \to U^+$$

to the $T_{\overline{k}}$-weight spaces $(V_{\chi_j})_{\overline{k}}$ relative to the linear structure we have chosen on $U_{(a)}$ (making T act linearly).

The additive map $\pi : (V_{\chi_j})_{\overline{k}} \to U^+$ satisfies

$$\pi(\chi_j(t)v_j) = a(t) \cdot \pi(v_j)$$

for all t in $T_{\overline{k}}$ and v_j in $(V_{\chi_j})_{\overline{k}}$. The map π cannot vanish on $(V_{\chi_j})_{\overline{k}}$. Indeed, if it does then the quotient map $G_{\overline{k}} \twoheadrightarrow G^{\mathrm{ss}}_{\overline{k}}$ would kill $(V_{\chi_j})_{\overline{k}}$, which is to say $(V_{\chi_j})_{\overline{k}} \subseteq \mathscr{R}(G_{\overline{k}}) = \mathscr{R}_u(G_{\overline{k}})$. By Lemma 1.2.1 it would follow that $V_{\chi_j} = 0$, a contradiction.

Choose $v_j \in (V_{\chi_j})_{\overline{k}}$ such that $\pi(v_j) \neq 0$. We therefore obtain a nonzero $T_{\overline{k}}$-equivariant map $\mathbf{G}_{\mathrm{a}}(\chi_j) \to \mathbf{G}_{\mathrm{a}}(a)$, where $\mathbf{G}_{\mathrm{a}}(\chi)$ denotes $\mathbf{G}_{\mathrm{a}}$ equipped with a $T_{\overline{k}}$-action through scaling via a $\overline{k}$-rational character χ of T. Since $a \neq 1$, this endomorphism of $\mathbf{G}_{\mathrm{a}}$ must be $x \mapsto c_j x^{d_j}$ with $c_j \in \overline{k}^{\times}$ and $d_j = p^{s_j}$ for some integer $s_j \geqslant 0$ and p equal to the characteristic exponent of k. Moreover, $a = \chi_j^{d_j}$. This shows that $\chi_j = a$ when $\mathrm{char}(k) = 0$, and that $\{\chi_j\}$ is a set of p-power roots of a when $\mathrm{char}(k) = p > 0$. In particular, we now may and do assume $\mathrm{char}(k) = p > 0$.

If some $\chi_j \neq a$ occurs then $\chi_j^{p^s} = a$ in $\mathrm{X}(T) = \mathrm{X}(T_{\overline{k}})$ with some $s \geqslant 1$, so the root $a \in \Phi(G^{\mathrm{ss}}_{\overline{k}}, T_{\overline{k}})$ is divisible by p^s in the character group $\mathrm{X}(T_{\overline{k}}) \simeq \mathbf{Z}$. But the roots for PGL_2 are generators of the character group and the roots for SL_2 are twice generators of the character group. Hence, additional positive rational multiples of a can only occur in $\Phi(G,T)$ when $p = 2$, and then only $a/2$ can occur. Moreover, in this case k must be imperfect because otherwise pseudo-reductive k-groups are reductive and so $\Phi(G,T)$ cannot contain a pair of distinct linearly dependent elements aside from those which are equal up to sign. □

For a pseudo-split pair (G,T) over k, we call $a \in \Phi(G,T)$ *multipliable* if $2a \in \Phi(G,T)$, so by Theorem 2.3.10 the set of non-multipliable elements of

$\Phi(G,T)$ is $\Phi(G_{\overline{k}}^{\text{red}}, T_{\overline{k}})$. In §3.2, the pair $(\mathrm{X}(T), \Phi(G,T))$ will be endowed with a natural structure of root datum (possibly non-reduced when $\mathrm{char}(k)=2$), compatible with the natural root datum structure on $(\mathrm{X}(T_{\overline{k}}), \Phi(G_{\overline{k}}^{\text{red}}, T_{\overline{k}}))$.

Although the k-groups $U_{(a)}$ for $a \in \Phi(G,T)$ were constructed using 1-parameter k-subgroups of G, they admit a characterization in terms of the T-action on G without any reference to 1-parameter subgroups of G. This will be a consequence of the following result.

Proposition 2.3.11 *Let G be a pseudo-split pseudo-reductive group over a field k, and let $T \subseteq G$ be a k-split maximal k-torus. For any $a \in \Phi(G,T)$ there is a unique T-stable smooth connected k-subgroup $U_a \subseteq G$ such that* $\mathrm{Lie}(U_a)$ *is the a-root space in* $\mathrm{Lie}(G)$ *when a is non-multipliable and is the span of the a-root space and the $2a$-root space when a is multipliable.*

The k-subgroup U_a is a vector group, and it admits a T-equivariant linear structure. Such a linear structure is unique when a is non-multipliable.

If a is multipliable then the quotient U_a/U_{2a} has a unique T-equivariant linear structure (and it makes T act with weight a), but U_a only has a non-canonical T-equivariant linear structure because the quotient map $U_a \twoheadrightarrow U_a/U_{2a}$ only has a non-canonical T-equivariant section.

Proof If a is non-divisible then the k-group U_a can be taken to be $U_{(a)}$, and it admits the required unique linear structure. If $2a \in \Phi(G,T)$ then a is non-divisible and $2a$ is non-multipliable, and $U_{(a)}$ has a linear structure in which the $2a$-weight space U_{2a} admits the required unique linear structure. Thus, the only issue is to prove the uniqueness of U_a.

Let $T_a = (\ker a)^0_{\text{red}}$. The pseudo-reductive k-subgroup $Z_G(T_a)$ in G containing T has Lie algebra whose nontrivial T-weights are precisely the nonzero rational multiples of a that lie in $\Phi(G,T)$. If H is a T-stable smooth connected k-subgroup of G such that T_a acts trivially on $\mathrm{Lie}(H)$ then $H \subseteq Z_G(T_a)$ by Corollary A.8.11. Hence, we lose no generality by replacing G with $Z_G(T_a)$, so by Theorem 2.3.10 the set $\Phi(G,T)$ is equal to either $\{\pm a\}$, $\{\pm a/2, \pm a\}$, or $\{\pm a, \pm 2a\}$, with the latter two only possible when $\mathrm{char}(k) = 2$.

Since T_a acts trivially on $\mathrm{Lie}(G)$ in all cases, by Corollary A.8.11 we have that T_a is central in G whereas T is non-central. Thus, T-conjugation on G factors through T/T_a, so for any $\lambda : \mathrm{GL}_1 \to T$ such that $\langle a, \lambda \rangle > 0$ the map λ is non-central in G and the k-subgroups $P_G(\pm\lambda)$, $U_G(\pm\lambda)$, and $Z_G(\lambda)$ are independent of λ by Remark 2.1.7. The quotient $Z_G(\lambda) \simeq P_G(\lambda)/U_G(\lambda)$ is a pseudo-reductive k-group in which the maximal k-torus T acts with no nontrivial weights (due to Proposition 2.1.8(1)) and thus is central by Corollary A.8.11. We conclude that $Z_G(\lambda)$ is commutative, by Proposition 1.2.4. Note also that $U_{(a)} = U_G(\lambda)$ by definition.

For the proof of uniqueness of U_a, let $U \subseteq G$ be a smooth connected k-subgroup that is normalized by T and such that $\mathrm{Lie}(U)$ is the a-root space in $\mathrm{Lie}(G)$ when a is non-multipliable and is the span of the root spaces for a and $2a$ when a is multipliable. We first claim that U must be unipotent. More generally, if a smooth connected affine k-group H admits a T-action for which $\mathrm{Lie}(H)$ has only nontrivial T-weights then H is unipotent. Indeed, if H is not unipotent then it contains a nontrivial k-torus, so the smooth connected k-group $T \ltimes H$ contains a maximal k-torus T' strictly containing T. But then $T' = T \ltimes S$ for a nontrivial k-torus S of H, so $\mathrm{Lie}(S)$ is a nonzero subspace of $\mathrm{Lie}(H)$ with trivial T-action, contrary to our hypothesis on H. Thus, indeed U is unipotent. In particular, $T \cap U = 1$.

The smooth connected k-subgroup $TU = T \ltimes U$ in G has Lie algebra which admits as its set of T-weights either $\{0, a\}$ or $\{0, a, 2a\}$. These weights have nonnegative pairing against λ, so by Proposition 2.2.4 we therefore have $TU \subseteq P_G(\lambda)$.

We claim that $U \subseteq U_G(\lambda) = U_{(a)}$, or equivalently that U has trivial image in $P_G(\lambda)/U_G(\lambda)$. This quotient is the pseudo-reductive k-group $Z_G(\lambda)$ that we have seen is commutative, so it contains no nontrivial smooth connected unipotent k-subgroup. Hence, indeed $U \subseteq U_G(\lambda) = U_{(a)}$, and as long as a is non-divisible this inclusion is an equality because the Lie algebras coincide inside of $\mathrm{Lie}(G)$.

It remains to show that if $a/2 \in \Phi(G, T)$ then U_a is unique. Note that in such cases $\mathrm{char}(k) = 2$. The proof of Theorem 2.3.10 shows that $\Phi(G_{\overline{k}}^{\mathrm{red}}, T_{\overline{k}}) = \{\pm a\}$. The smooth connected unipotent k-subgroup $U_G(\lambda)$ has Lie algebra that is the span of the root spaces for a and $a/2$ in $\mathrm{Lie}(G)$, and $U_G(\lambda)$ is normalized by $T_a \cdot \lambda(\mathrm{GL}_1) = T$. The same method as above proves that any smooth connected unipotent k-subgroup U of G that is normalized by T and has Lie algebra whose T-weights are a subset of $\{a, a/2\}$ necessarily lies in $U_G(\lambda)$.

Recall that $U_G(\lambda)$ is a vector group that admits a T-equivariant linear structure, with weights a and $a/2$. The resulting weight space for a in $U_G(\lambda)$ can be constructed without reference to such a choice of linear structure on $U_G(\lambda)$: Since $a/2$ is primitive in $\mathrm{X}(T)$ by Theorem 2.3.10 (as a root in a reductive group cannot be divisible by more than 2 in the character lattice), we have $T_a = \ker(a/2)$, so the a-weight space in $U_G(\lambda)$ is the centralizer scheme $Z_{U_G(\lambda)}(\mu)$ for the action by the 2-torsion group scheme

$$\mu := \ker(a)/T_a = (T/T_a)[2] \simeq \mu_2$$

on $U_G(\lambda)$. (See Proposition A.8.10 for a general discussion of such a centralizer scheme, including that it is smooth with Lie algebra equal to the a-weight space in $\mathrm{Lie}(G)$.)

Consider any T-stable smooth connected unipotent k-subgroup $U \subseteq G$ such that $\mathrm{Lie}(U)$ is the a-root space of $\mathrm{Lie}(G)$. We have seen above that $U \subseteq U_{(a)} = U_G(\lambda)$, and by Corollary A.8.11 the μ-action on U is trivial. Thus, $U \subseteq Z_{U_G(\lambda)}(\mu)^0$. This inclusion is an equality on Lie algebras, so it is an equality of k-groups. Hence, the uniqueness of U_a is proved in all cases. □

Remark 2.3.12 If a is non-divisible in $\Phi(G,T)$ then $U_{(a)} = U_a$. If a is divisible then $U_{(a)} = U_{a/2}$.

Definition 2.3.13 With hypotheses and notation as in Proposition 2.3.11, for each $a \in \Phi(G,T)$ the k-subgroup $U_a \subseteq G$ is the *root group* associated to a.

Remark 2.3.14 If $a \in \Phi(G,T)$ is multipliable then $U_{2a} \subseteq U_a$ and $\mathrm{Lie}(U_a)$ is the span of the root spaces for a and $2a$ in $\mathrm{Lie}(G)$, and otherwise it is the root space for a. Since $Z_G(T)$ is smooth, the unique characterization of U_a in Proposition 2.3.11 implies that U_a is normalized by $Z_G(T)$; this refines Remark 2.3.5.

For isotropic connected semisimple groups over a field, if the relative root system is non-reduced then the root groups whose Lie algebra supports more than one weight can be non-commutative. In contrast, all root groups in a pseudo-split pseudo-reductive group are commutative, even when the root system is non-reduced (which happens over imperfect fields of characteristic 2, as we show in Chapter 9).

As in the theory of connected reductive groups, the formation of root groups is well behaved with respect to central quotients:

Proposition 2.3.15 *Let G be a smooth connected affine k-group and T a k-split torus in G. Let Z be a central closed k-subgroup of G, define $\overline{G} = G/Z$, and let $\overline{T}$ be the image of T in $\overline{G}$. The inclusion of* $\mathrm{X}(\overline{T})$ *into* $\mathrm{X}(T)$ *carries the set $\Phi' := \Phi(\overline{G},\overline{T})$ of nontrivial $\overline{T}$-weights on* $\mathrm{Lie}(\overline{G})$ *bijectively onto the set $\Phi := \Phi(G,T)$. Moreover, for any $a \in \Phi$ and the corresponding $a' \in \Phi'$, the quotient map $\pi : G \to \overline{G}$ carries $U^G_{(a)}$ isomorphically onto $U^{\overline{G}}_{(a')}$.*

Proof We identify $\mathrm{X}(\overline{T})$ with a subgroup of $\mathrm{X}(T)$ via $\pi|_T$. As Z is central, the adjoint action of G on $\mathrm{Lie}(G)$ factors through an action by $\overline{G}$ compatible via $\mathrm{Lie}(\pi)$ with the adjoint action of $\overline{G}$ on $\mathrm{Lie}(\overline{G})$. In particular, the action of T on $\mathrm{Lie}(G)$ is through $\overline{T}$, so $\Phi \subseteq \mathrm{X}(\overline{T})$ inside $\mathrm{X}(T)$. That is, every $a \in \Phi$

factors uniquely through a nontrivial character a' of $\overline{T}$. By Remark 2.3.6, every element of Φ' has a positive rational multiple contained in Φ and for every $a \in \Phi$ the map π carries $U^G_{(a)}$ onto $U^{\overline{G}}_{(a')}$. Since $U^G_{(a)} \subseteq U_{Z_G(T_a)}(\lambda)$ for a nontrivial cocharacter $\lambda \in X_*(T)$ and $Z \subseteq Z_{Z_G(T_a)}(\lambda)$ (as Z is central in G), clearly $U^G_{(a)} \cap Z = 1$. Hence, the quotient map $\pi_a : U^G_{(a)} \twoheadrightarrow U^{\overline{G}}_{(a')}$ is an isomorphism for all $a \in \Phi$, so $U^{\overline{G}}_{(a')} \neq 1$ for such a. Consideration of the isomorphism $\mathrm{Lie}(\pi_a)$ then shows that a' is a weight on $\mathrm{Lie}(\overline{G})$ (so $\Phi \subseteq \Phi'$) and moreover that $\mathbf{Q}_{>0}a' \cap \Phi' = \mathbf{Q}_{>0}a \cap \Phi$, so $\Phi = \Phi'$ as desired. □

Remark 2.3.16 The interaction of $U_{(a)}$ with purely inseparable Weil restriction is governed by Example 2.3.2 as follows. Let k'/k be a purely inseparable finite extension of fields, and G' a smooth connected affine k'-group. Let $G = \mathrm{R}_{k'/k}(G')$. As we saw in Example 2.3.2, G is pseudo-split over k if and only if G' is pseudo-split over k'. Assume G is pseudo-split, with T a split maximal k-torus in G, and let T' be the corresponding split maximal k'-torus in G', so we may identify $\Phi(G,T)$ and $\Phi(G',T')$ as in Example 2.3.2. For any $a \in \Phi(G,T)$ and the corresponding $a' \in \Phi(G',T')$, the quotient map $G_{k'} \to G'$ carries $(U^G_{(a)})_{k'}$ into $U^{G'}_{(a')}$ due to Corollary 2.1.9, so $U^G_{(a)} \subseteq \mathrm{R}_{k'/k}(U^{G'}_{(a')})$. Lie algebra considerations via Lemma 2.3.3 imply that this containment is an equality.

2.4 Representability of automorphism functors

In the discussion preceding Theorem 1.3.9, we introduced two subgroup functors of the automorphism functor $\underline{\mathrm{Aut}}(G)$ of a smooth affine group G over a field k: the functors $\underline{\mathrm{Aut}}_{G,T}$ and $\underline{\mathrm{Aut}}_{G,C}$ of automorphisms respectively restricting to the identity on a maximal torus T and a Cartan subgroup $C = Z_G(T)$. We now prove representability results for these functors, upon which Theorem 1.3.9 was conditionally proved.

Theorem 2.4.1 *Let G be a smooth connected affine group over a field k, and let T be a maximal k-torus in G. Let $C = Z_G(T)$. The functor $\underline{\mathrm{Aut}}_{G,C}$ is represented by an affine k-group scheme $\mathrm{Aut}_{G,C}$ of finite type. If G is pseudo-reductive then the maximal smooth closed k-subgroup scheme $Z_{G,C}$ (in the sense of Lemma C.4.1) of $\mathrm{Aut}_{G,C}$ is commutative and has no nontrivial smooth unipotent k-subgroups. In particular, the identity component $Z^0_{G,C}$ is pseudo-reductive.*

Proof By finite Galois descent (which is effective for affine schemes), it suffices to prove the result after a finite Galois extension of k. Thus, we can assume that T is k-split. Let Φ denote the set of nontrivial weights for the T-action on $\mathrm{Lie}(G)$. (If Φ is empty then $G = C$ by Corollary A.8.11, so $\underline{\mathrm{Aut}}_{G,C} = 1$ in such cases.)

Step 1. For each $a \in \Phi$ let $T_a = (\ker a)^0_{\mathrm{red}}$ so that, by definition, $U_{(a)} = U_{Z_G(T_a)}(\lambda_a)$ for any $\lambda_a \in \mathrm{X}_*(T)$ satisfying $\langle a, \lambda_a \rangle > 0$. The k-group $U_{(a)}$ has a natural T-action, so we can define the automorphism functor $\underline{\mathrm{Aut}}_{U_{(a)},T}$ classifying T-equivariant automorphisms of $U_{(a)}$. Since scheme-theoretic centralizers and the "$U_H(\lambda)$" construction are characterized by the functors they represent on arbitrary k-algebras (or even k-schemes), for any k-scheme S and S-group automorphism ϕ of G_S that restricts to the identity on T_S, we see that ϕ restricts to an S-group automorphism of each $Z_G(T_a)_S$ and hence of each $U_{Z_G(T_a)}(\lambda_a)_S = (U_{(a)})_S$. That is, we get a group functor map $\underline{\mathrm{Aut}}_{G,T} \to \underline{\mathrm{Aut}}_{U_{(a)},T}$ for each $a \in \Phi$.

Consider the smooth connected k-subgroup N of G generated by the k-subgroups $U_{(a)}$. This is normalized by every $U_{(a)}$ and by C (as C normalizes each $U_{(a)}$), which collectively generate G (Lemma 2.3.3), so N is normal in G. Clearly $G = C \cdot N$, which is to say that the map $C \ltimes N \to G$ defined by multiplication is surjective. By [Bo2, 2.2], there is an ordered finite sequence $\{a_i\}_{i \in I}$ in Φ (with repetitions allowed, and we can assume that every $a \in \Phi$ occurs at least once) such that the multiplication map of k-schemes $\prod_{i \in I} U_{(a_i)} \to N$ is surjective. Thus, we get a surjective map of smooth affine k-schemes

$$Y := C \times \prod_{i \in I} U_{(a_i)} \to G.$$

In particular, this identifies the coordinate ring $\mathscr{O}(G)$ with a k-subalgebra of the coordinate ring $\mathscr{O}(Y)$, and $\underline{\mathrm{Aut}}_{G,C}$ with a subfunctor of $\prod_i \underline{\mathrm{Aut}}_{U_{(a_i)},T}$.

Step 2. Now the idea is to prove that each $\underline{\mathrm{Aut}}_{U_{(a)},T}$ is represented by an affine k-scheme of finite type, to be denoted by $\mathrm{Aut}_{U_{(a)},T}$, and that the subfunctor $\underline{\mathrm{Aut}}_{G,C}$ of $\prod_i \underline{\mathrm{Aut}}_{U_{(a_i)},T}$ is represented by a closed subscheme of the affine product $\prod_i \mathrm{Aut}_{U_{(a_i)},T}$. The representability for each $\underline{\mathrm{Aut}}_{U_{(a)},T}$ follows from:

Lemma 2.4.2 *Let U be a smooth connected unipotent k-group equipped with an action by a split k-torus T such that the weights of T on* $\mathrm{Lie}(U)$ *are nontrivial and are $\mathbf{Q}_{>0}$-multiples of each other in $\mathrm{X}(T)_{\mathbf{Q}}$. The functor $\underline{\mathrm{Aut}}_{U,T}$ classifying T-equivariant automorphisms of U is represented by an affine k-group $\mathrm{Aut}_{U,T}$ of finite type.*

Proof We may assume $U \neq 1$ and can choose $a \in \mathrm{X}(T)$ such that all weights a' which occur in $\mathrm{Lie}(U)$ have the form $a' = n_{a'}a$ for integers $n_{a'} \geqslant 1$. The k-subgroup $\ker a$ acts trivially on $\mathrm{Lie}(U)$, so it acts trivially on U by Corollary A.8.11. Hence, we may replace T with $T/(\ker a)$ to reduce to the case when $a : T \to \mathrm{GL}_1$ is an isomorphism. The coordinate ring $k[U]$ acquires a k-linear grading $\bigoplus_{n \geqslant 0} k[U]_n$ via the T-action (Lemma A.8.8). We claim that $k[U]_0 = k$. It suffices to check the analogous property for the complete local ring $A = \widehat{\mathscr{O}}_{U,e}$ at the origin (which contains $k[U]$ equivariantly). If $\mathfrak{m}$ denotes the maximal ideal of A, then all weights on $\mathfrak{m}/\mathfrak{m}^2$ are positive multiples of a (as this space is equivariantly dual to $\mathrm{Lie}(U)$). Thus, the same holds for its higher tensor powers, and hence for the quotients $\mathfrak{m}^i/\mathfrak{m}^{i+1}$ for all $i \geqslant 1$. It follows that the space of T-invariant elements of A is k, as desired.

Since $k[U]$ is finitely generated as a k-algebra, we can choose a finite set of homogeneous generators, all with degree > 0. Thus, $k[U]_n$ has finite k-dimension for all $n \geqslant 1$. Pick $N > 0$ such that $k[U]$ is generated by the $k[U]_n$ for $1 \leqslant n \leqslant N$. Letting $V = \bigoplus_{1 \leqslant n \leqslant N} k[U]_n$ with its evident grading, the functor of graded linear automorphisms of V is represented by a product of general linear groups, and the symmetric algebra $\mathrm{Sym}(V)$ inherits a natural grading so that the quotient map $\pi : \mathrm{Sym}(V) \twoheadrightarrow k[U]$ respects the gradings. The condition on a graded linear automorphism f of V that $\mathrm{Sym}(f)$ and $\mathrm{Sym}(f^{-1})$ preserve $\ker \pi$ is a Zariski-closed condition, so the graded k-algebra automorphism functor of U is represented by an affine k-group of finite type. This is the functor $\underline{\mathrm{Aut}}_{U,T}$, so we are done. □

Returning to the setting of interest, let R be the coordinate ring of $\prod_i \mathrm{Aut}_{U_{(a_i)},T}$, so there is a canonical R-scheme automorphism f of Y_R that is a direct product of the identity on C_R and the pullback of the universal T-equivariant automorphism of each $U_{(a_i)}$. Our aim is to prove that the subfunctor $\underline{\mathrm{Aut}}_{G,C}$ of $\prod_i \underline{\mathrm{Aut}}_{U_{(a_i)},T}$ is represented by a quotient of R. It is easier to think about this in another way: for any k-algebra A and T_A-equivariant A-group automorphisms ϕ_i of $(U_{(a_i)})_A$ for each i we get a direct product automorphism ϕ of Y_A (using the identity on C_A), and we claim that the condition on A-algebras B that ϕ_B arises from a (necessarily unique) B-group automorphism of G_B is represented by a quotient of A. (Applying this with $A = R$ and the canonical automorphism of Y_R then gives a quotient of R that represents the subfunctor $\underline{\mathrm{Aut}}_{G,C}$.)

STEP 3. We now proceed in two stages: find a quotient of A representing the condition that ϕ_B extends to a B-scheme endomorphism of G_B, and then impose the conditions of being a B-group endomorphism and finally automorphism. Now we have a very concrete situation: for an inclusion of k-vector spaces $V \hookrightarrow W$ (such as $\mathscr{O}(G) \hookrightarrow \mathscr{O}(Y)$) and an A-linear endomorphism f of W_A, we consider the condition on A-algebras B that f_B carries V_B into itself. This says that the composite A-linear map

$$V_A \hookrightarrow W_A \xrightarrow{f} W_A \to (W/V)_A$$

vanishes after scalar extension to B. Using k-bases of V and W/V to express this composite A-linear map as a "matrix" (with at most finitely many nonzero entries in each column), the condition is represented by the quotient A/J where J is the ideal generated by the entries of this matrix. Back in the original setting, this proves the existence of a universal quotient of R representing the functor of scheme endomorphisms of G that restrict to T-equivariant automorphisms of the k-subgroups $U_{(a)}$ and restrict to the identity map on C.

The condition to be a group scheme endomorphism is a further Zariski-closed condition (defined by vanishing of entries of a huge matrix expressing the compatibility of a scheme endomorphism with the group scheme structure), so we have constructed an affine k-scheme of finite type which represents the functor of *endomorphisms* of the group G that restrict to the identity on C and to automorphisms on the $U_{(a)}$. But we can likewise construct a universal *pair* of such endomorphisms and impose the Zariski-closed condition that they are inverse to each other. This represents $\underline{\mathrm{Aut}}_{G,C}$. (By Remark 2.4.7 below, the final step using pairs of endomorphisms is actually not necessary.)

STEP 4. It remains to analyze the structure of the maximal smooth closed k-subgroup $Z_{G,C}$ of the representing group for $\underline{\mathrm{Aut}}_{G,C}$: we wish to prove that if G is pseudo-reductive then $Z_{G,C}$ is commutative and contains no nontrivial smooth unipotent k-subgroups. We may assume $k = k_s$, so for commutativity it suffices to prove that the group $Z_{G,C}(k)$ is commutative. By construction $Z_{G,C}(k) = \underline{\mathrm{Aut}}_{G,C}(k)$, so it suffices to show that the group of k-automorphisms of G restricting to the identity on C is commutative.

The pseudo-reductivity of G implies (by Proposition 1.2.2) that k-automorphisms of G are faithfully represented by their effect on $G_{\overline{k}}^{\mathrm{red}}$. Since $T_{\overline{k}}$ is a maximal torus in $G_{\overline{k}}^{\mathrm{red}}$, the problem is reduced to the commutativity of the group of automorphisms of $G_{\overline{k}}^{\mathrm{red}}$ that are trivial on $T_{\overline{k}}$. The Isogeny Theorem (Theorem A.4.10) identifies this with the group of points of the image of $T_{\overline{k}}$ in the maximal adjoint quotient. In particular, it is commutative.

To prove that $Z_{G,C}$ contains no nontrivial smooth unipotent k-subgroups (again assuming that G is pseudo-reductive), we separately treat the cases when $\mathrm{char}(k) = 0$ and $\mathrm{char}(k) = p > 0$. In characteristic 0, the group G is connected reductive, so by Corollary A.4.7(2) the action of T/Z_G on G is final among all smooth connected k-groups equipped with a left action on G fixing $C = T$ pointwise. In particular, the $Z_{G,C}$-action on G factors through this torus action which visibly restricts to the identity action on C, so by universality it follows that $Z_{G,C} = T/Z_G$. This is connected and reductive, so the case $\mathrm{char}(k) = 0$ is settled. If instead $\mathrm{char}(k) = p > 0$, we observe that a nontrivial smooth unipotent k-group has nontrivial p-torsion k-points (since $k = k_s$). We saw above that $Z_{G,C}(k)$ is a subgroup of the group of geometric points of the torus image of $T_{\overline{k}}$ in the maximal adjoint quotient of $G_{\overline{k}}^{\mathrm{red}}$, and in characteristic $p > 0$ there is no nontrivial p-torsion in the geometric points of a torus. □

The preceding proof used $\underline{\mathrm{Aut}}_{G,C}$ rather than $\underline{\mathrm{Aut}}_{G,T}$ only because the smooth connected normal k-subgroup N generated by the root groups satisfies $C \cdot N = G$, and in general $T \cdot N$ may be strictly smaller (e.g., G may contain a non-reductive smooth connected commutative direct factor). By Proposition A.2.11, the smooth connected k-subgroup G_t of G generated by the k-tori is normal and satisfies $(G_t)_K = (G_K)_t$ for any extension field K/k.

Definition 2.4.3 A smooth connected affine group G over a field k is *generated by tori* if $G = G_t$. (For example, if G is perfect then it is generated by tori, by Proposition A.2.11.)

When $G = G_t$, the proof of Theorem 2.4.1 works with T in place of C:

Corollary 2.4.4 *Let G be a smooth connected affine k-group such that $G = G_t$, and let T be a maximal k-torus of G. The functor $\underline{\mathrm{Aut}}_{G,T}$ is represented by a finite type affine k-group scheme* $\mathrm{Aut}_{G,T}$. *The natural map* $\mathrm{Aut}_{G,C} \rightarrow \mathrm{Aut}_{G,T}$ *is a closed immersion, and the maximal smooth closed k-subgroups of* $\mathrm{Aut}_{G,C}$ *and* $\mathrm{Aut}_{G,T}$ *coincide when G is pseudo-reductive.*

By Proposition A.2.11, the equality $G = G_t$ holds when G is perfect. Also, maximal smooth closed k-subgroups are constructed by Lemma C.4.1.

Proof Using notation from the proof of Theorem 2.4.1, we may again pass to a finite Galois extension of k so that T is k-split, and define N as before. To prove the representability claim we just have to check that $T \cdot N = G$ (and then the proof of Theorem 2.4.1 applies without change by using T instead of C everywhere). To verify this equality we may replace G with its smooth quotient G/N to get to the case in which T acts on $\mathrm{Lie}(G)$ with no nontrivial weights. Hence, T is central (Corollary A.8.11). But $G = G_t$, so centrality of T implies $G = T$.

Since the representability is now settled in general, we return to the initial setup and consider properties of $\mathrm{Aut}_{G,T}$. The map $\mathrm{Aut}_{G,C} \to \mathrm{Aut}_{G,T}$ between affine finite type k-groups has trivial kernel, so it is a closed immersion (Proposition A.2.1). To compare their maximal smooth closed k-subgroups when G is pseudo-reductive over k, we may extend scalars to k_s and then the claim amounts to the equality of sets of k-points. This equality is Example 1.3.8. □

Example 2.4.5 Over any imperfect field k it is easy to give examples of pairs (G,T) with G perfect pseudo-reductive such that $\mathrm{Aut}_{G,T}$ is not smooth. Consider k-groups $G = \mathrm{R}_{k'/k}(G')$ for a nontrivial purely inseparable finite extension k'/k and a nontrivial split connected semisimple k'-group G' that is simply connected (e.g., SL_n, $n \geqslant 2$). In these cases, Theorem 1.3.9 gives an explicit description of the common maximal smooth closed k-subgroup of $\mathrm{Aut}_{G,C}$ and $\mathrm{Aut}_{G,T}$: it is $\mathrm{R}_{k'/k}(T'/Z_{G'})$ for the maximal k'-torus T' of G' such that $C = \mathrm{R}_{k'/k}(T')$. Thus, we just need to exhibit a closed k-subgroup H of $\mathrm{Aut}_{G,T}$ with $\dim H > 0$ such that $Z_{G,C} \cap H = 1$.

Choose C corresponding to a T' that is k'-split, so $(G',T') \simeq (\mathscr{G}_{k'}, \mathscr{T}_{k'})$ for a k-split pair $(\mathscr{G}, \mathscr{T})$ over k. The algebraic group H of k-algebra automorphisms of k' acts naturally on $G = \mathrm{R}_{k'/k}(\mathscr{G}_{k'})$ since for any k-algebra A we have

$$G(A) = \mathscr{G}_{k'}(A') = \mathscr{G}(A')$$

for $A' := k' \otimes_k A$. The H-action clearly restricts to the identity action on $T = \mathscr{T}$. By computing $H(\overline{k})$ we see that $\dim H > 0$ since k'/k is purely inseparable and nontrivial. Finally, to prove $Z_{G,C} \cap H = 1$ we compute A-points for any k-algebra A:

$$(Z_{G,C} \cap H)(A) = (T'/Z_{G'})(A') \cap \mathrm{Aut}_A(A') \subseteq \mathrm{Aut}_A(\mathrm{R}_{A'/A}(\mathscr{G} \otimes_k A')).$$

This intersection is trivial since the representation of $\mathrm{Aut}_A(A')$ on

$$\mathrm{Lie}(\mathrm{R}_{A'/A}(\mathscr{G} \otimes_k A')) = \mathrm{Lie}(\mathscr{G}) \otimes_k A'$$

is faithful on $\mathrm{Lie}(\mathscr{T}) \otimes_k A'$ and the action of $(T'/Z_{G'})(A')$ restricts to the identity on this A'-submodule.

The method of proof of Theorem 2.4.1 and Corollary 2.4.4 gives an interesting result for endomorphisms:

Corollary 2.4.6 *Let G be a smooth connected affine group over a field k, T a maximal k-torus, and $C = Z_G(T)$. For any endomorphism h of G that restricts to the identity on C, the following conditions are equivalent:*

(i) *h is surjective,*
(ii) *h is an automorphism,*
(iii) Lie(h) *is an automorphism.*

The same holds when h merely restricts to the identity on T provided that $G = G_t$ (e.g., $G = \mathscr{D}(G)$).

If G is pseudo-reductive then all of these equivalent conditions hold for any endomorphism h such that h restricts to the identity on T.

Remark 2.4.7 By using flatness considerations, the equivalence among conditions (i), (ii), and (iii) holds for S-group endomorphisms of G_S for any k-scheme S.

Proof Since $C = Z_G(T)$ is pseudo-reductive when G is pseudo-reductive, for such G if $h|_T = \mathrm{id}_T$ then h not only induces an endomorphism of C but in fact this endomorphism is the identity (Proposition 1.2.2). Thus, when we later verify that all of the conditions hold in the pseudo-reductive case, it will suffice to verify just one of them. (We will verify (iii).)

It is clear that (ii) implies (iii) and that (iii) implies (i). To prove (i) implies (ii) we may and do assume $k = \overline{k}$. Now assume h is surjective. For each nontrivial T-weight a on Lie(G), consider the smooth connected unipotent k-subgroup $U_{(a)}$. Since h is surjective and restricts to the identity on T, by Corollary 2.1.9 it follows that h restricts to a surjective endomorphism of each $U_{(a)}$.

Consider the resulting injective T-equivariant pullback map h_a^* on $k[U_{(a)}]$. As we saw in the proof of Lemma 2.4.2, these coordinate rings have finite-dimensional graded pieces under the grading determined by the T-action. Since h_a^* respects the grading and is injective in each degree, it must be an isomorphism in each degree and hence an isomorphism. That is, h restricts to an automorphism of each $U_{(a)}$. Define Y as in the proof of Theorem 2.4.1, or if $G = G_t$ and we only assume h to be the identity on T then define Y similarly except with T in place of C in the definition of Y as a direct product scheme. In an evident manner, h^* induces an automorphism f of $\mathscr{O}(Y)$ that preserves the subspace $\mathscr{O}(G)$.

The finite-dimensionality of the graded parts of each $k[U_{(a)}]$ and the condition that $h|_C = \mathrm{id}_C$ or $h|_T = \mathrm{id}_T$ imply that $\mathscr{O}(Y)$ is the directed union of f-stable finite-dimensional subspaces V_α. Thus, if W is any k-linear subspace of $\mathscr{O}(Y)$ that is carried into itself under f (such as $\mathscr{O}(G)$) then f restricts to an injective linear endomorphism of every $W \cap V_\alpha$. But these are finite-dimensional, so f is an automorphism on every $W \cap V_\alpha$ and thus also on W. Taking $W = \mathscr{O}(G)$ then implies that h is an automorphism.

Finally, we check that (iii) holds (so (i) and (ii) hold) when G is pseudo-reductive. Equivalently, we will prove that $\mathrm{Lie}(h)$ is injective. We may and do assume $k = k_s$, so T is split. Since $\ker \mathrm{Lie}(h)$ has a weight space decomposition with respect to T, it suffices to check the injectivity of $\mathrm{Lie}(h)$ on each T-weight space in $\mathrm{Lie}(G)$. The trivial weight space is $\mathrm{Lie}(Z_G(T))$, on which $\mathrm{Lie}(h)$ is the identity (since an endomorphism of $Z_G(T)$ is uniquely determined by its restriction to T, by Proposition 1.2.2). For each $a \in \Phi(G,T)$, h restricts to an endomorphism of $Z_G(T_a)$ (with $T_a = (\ker a)^0_{\mathrm{red}}$) and hence also restricts to an endomorphism of $U_{Z_G(T_a)}(\lambda_a) = U_{(a)}$ (with $\lambda_a \in \mathrm{X}_*(T)$ satisfying $\langle a, \lambda_a\rangle > 0$). Consider the resulting T-equivariant endomorphism h_a of U_a. Since $\mathrm{Lie}(U_a)$ contains the a-root space of $\mathrm{Lie}(G)$, it suffices to check that h_a is an automorphism for all a.

If a is not multipliable then U_a has a unique T-equivariant linear structure (Lemma 2.3.8), and for such a an argument similar to the proof of uniqueness implies that h_a is a linear endomorphism. Thus, for such a the kernel $\ker h_a$ is smooth, connected, and unipotent. To deduce that h_a is an isomorphism for such a it suffices to show that $\ker h_a = 0$, or more generally that $\ker h$ contains no nontrivial smooth connected unipotent k-subgroup. If H is a smooth connected k-subgroup of $\ker h$ then its $G(k)$-conjugates generate a smooth connected k-subgroup $N \subseteq \ker h$ that is normal in G (as $k = k_s$). But G is pseudo-reductive, so N is pseudo-reductive. The intersection $T \cap N$ is a maximal k-torus of N (Corollary A.2.7), yet $T \cap N \subseteq T \cap \ker h = 1$ since h restricts to the identity on T, so N is unipotent. In view of the pseudo-reductivity of N it follows that $N = 1$, so $H = 1$ as desired. We conclude that h_a is a linear isomorphism for non-multipliable a.

If $a \in \Phi(G,T)$ is multipliable (so $\mathrm{char}(k) = 2$ and all multipliable a are primitive vectors in the lattice $\mathrm{X}(T)$) then h_a restricts to the automorphism h_{2a} of U_{2a}, so by the snake lemma for commutative k-group schemes of finite type we see that the kernel of h_a is identified with the kernel of the induced endomorphism of U_a/U_{2a}. But U_a/U_{2a} admits a unique T-equivariant linear structure, and similarly the induced endomorphism of this quotient is again linear so it has a smooth connected kernel. In other words, $\ker h_a$ is smooth and connected in the multipliable case as well, so once again it must be trivial (due to the pseudo-reductivity of G). In fact, this shows that the linear endomorphism induced on U_a/U_{2a} is an automorphism, so $h_a : U_a \to U_a$ is also an automorphism. □

3

Basic structure theory

3.1 Perfect normal subgroups of pseudo-reductive groups

In the general study of standardness for pseudo-reductive groups, the most important case turns out to be where G is perfect and the connected semisimple quotient $G_{\overline{k}}^{\mathrm{ss}} = G_{\overline{k}}/\mathscr{R}(G_{\overline{k}})$ is simple. It is therefore convenient to introduce the following definition of Tits [Ti3, II] and a related lemma.

Definition 3.1.1 A *pseudo-simple* group H over a field k is a smooth connected affine k-group that is non-commutative and has no nontrivial smooth connected proper normal k-subgroup. We say that H is *absolutely pseudo-simple* (over k) if H_{k_s} is pseudo-simple over k_s.

By a specialization argument (as in the proof of Proposition 1.1.9(1)), if K/k is a separable extension field then a k-group H is absolutely pseudo-simple if and only if H_K is absolutely pseudo-simple over K.

Lemma 3.1.2 *A group scheme H over a field k is pseudo-simple over k if and only if H is pseudo-reductive over k, $H = \mathscr{D}(H)$, and $H_{k_{\mathrm{p}}}^{\mathrm{ss}}$ is k_{p}-simple, where k_{p} is the perfect closure of k. In particular, H is absolutely pseudo-simple if and only if H is pseudo-reductive, $H = \mathscr{D}(H)$, and $H_{\overline{k}}^{\mathrm{ss}}$ is simple.*

This lemma will be used repeatedly, often without comment, when $k = k_s$.

Proof First assume that H is pseudo-simple over k, so $\mathscr{D}(H) = H$ since we require H to be non-commutative (in the definition of pseudo-simplicity). In particular, $\mathscr{R}(H_{k_{\mathrm{p}}}) = \mathscr{R}_u(H_{k_{\mathrm{p}}})$ is unipotent. Also, since H cannot be solvable (as $\mathscr{D}(H) = H \neq 1$), we cannot have $\mathscr{R}_{u,k}(H) = H$, so $\mathscr{R}_{u,k}(H) = 1$. That is, H is pseudo-reductive over k.

To prove that $H_{k_{\mathrm{p}}}^{\mathrm{ss}}$ is simple, assume to the contrary. Note that $H_{k_{\mathrm{p}}}^{\mathrm{ss}} \neq 1$ since H is not solvable. Pick a maximal k-torus $T \subseteq H$, so $T_{k_{\mathrm{p}}}$ is identified with a maximal torus in $H^{\mathrm{ss}}_{k_{\mathrm{p}}}$. Consider the nontrivial decomposition of $H^{\mathrm{ss}}_{k_{\mathrm{p}}}$ into an almost direct product of k_{p}-simple smooth connected normal k_{p}-subgroups; this is constructed via Galois descent from the analogous such result for $H^{\mathrm{ss}}_{\overline{k}}$ [Bo2, 14.10] (according to which each smooth connected normal $\overline{k}$-subgroup is an almost direct product among a uniquely determined set of the minimal nontrivial ones). This decomposition over k_{p} induces a nontrivial decomposition of $T_{k_{\mathrm{p}}}$ into an almost direct product of two nontrivial proper k_{p}-subtori S' and S'' that are respectively maximal k_{p}-tori in complementary almost direct products among the k_{p}-simple factors of $H^{\mathrm{ss}}_{k_{\mathrm{p}}}$.

Since k_{p}/k is purely inseparable, there is a unique k-subtorus $T' \subseteq T$ descending S'. The normal k-subgroup N of H obtained by Galois descent of the k_s-subgroup of H_{k_s} generated by the $H(k_s)$-conjugates of T'_{k_s} is a nontrivial smooth connected normal k-subgroup of H, and $N_{k_{\mathrm{p}}}$ has proper image in $H^{\mathrm{ss}}_{k_{\mathrm{p}}}$, so $N \neq H$. This contradicts the pseudo-simplicity of H over k, so $H^{\mathrm{ss}}_{k_{\mathrm{p}}}$ is simple.

Conversely, suppose that H is pseudo-reductive over k, $H = \mathscr{D}(H)$, and $H^{\mathrm{ss}}_{k_{\mathrm{p}}}$ is simple. Clearly H is non-commutative and $H^{\mathrm{red}}_{k_{\mathrm{p}}} = H^{\mathrm{ss}}_{k_{\mathrm{p}}}$, so $\mathscr{R}(H_{k_{\mathrm{p}}})$ is unipotent. Let $N \subseteq H$ be a nontrivial smooth connected normal k-subgroup. This is pseudo-reductive and hence not unipotent, so the image of $N_{k_{\mathrm{p}}}$ in $H^{\mathrm{ss}}_{k_{\mathrm{p}}}$ must be nontrivial and hence full. It follows that $(H/N)^{\mathrm{ss}}_{k_{\mathrm{p}}} = 1$, so H/N is solvable yet it is also equal to its own derived group. We conclude that $N = H$. □

Example 3.1.3 As an application of Lemma 3.1.2, we can construct absolutely pseudo-simple groups from certain pseudo-reductive groups, as follows. Let G be a non-commutative pseudo-reductive group over a field k, and assume that the nontrivial semisimple group $G^{\mathrm{ss}}_{\overline{k}}$ is simple. We claim that $\mathscr{D}(G)$ is absolutely pseudo-simple and that the natural map $\mathscr{D}(G)^{\mathrm{ss}}_{\overline{k}} \to G^{\mathrm{ss}}_{\overline{k}}$ is a central isogeny. This second claim follows from Proposition A.4.8 (since $\mathscr{D}(G)^{\mathrm{ss}}_{\overline{k}} \to G^{\mathrm{ss}}_{\overline{k}}$ is surjective), so it remains to prove the absolute pseudo-simplicity.

The k-group $\mathscr{D}(G)$ is certainly pseudo-reductive, and by Proposition 1.2.6 it is equal to its own derived group. Thus, by Lemma 3.1.2 the k-group $\mathscr{D}(G)$ is absolutely pseudo-simple provided that $\mathscr{D}(G)^{\mathrm{ss}}_{\overline{k}}$ is simple. The $\overline{k}$-group $\mathscr{D}(G)^{\mathrm{ss}}_{\overline{k}}$ has a central isogeny to the simple $G^{\mathrm{ss}}_{\overline{k}}$, so we are done.

The relationship between $\Phi(G^{\mathrm{red}}_{\overline{k}}, T_{\overline{k}})$ and $\Phi(G, T)$ when T is k-split (see Theorem 2.3.10) enables us to describe derived groups of smooth connected normal k-subgroups in non-commutative pseudo-split pseudo-reductive groups over any field k. This requires some preparations in the pseudo-split case (such as over a separably closed field), as follows.

Let G be a pseudo-split non-commutative pseudo-reductive group over a field k, and let T be a k-split maximal k-torus in G. The root system

$$\Phi := \Phi(G_{\overline{k}}^{\mathrm{red}}, T_{\overline{k}})$$

in $\mathrm{X}(T_{\overline{k}}) = \mathrm{X}(T)$ consists of the non-multipliable elements in $\Phi(G,T)$, and it is not empty since G is not commutative. The structure theory of connected reductive groups implies that under the decomposition $\Phi = \coprod_{i \in I} \Phi_i$ into irreducible root systems there is a natural bijection between the non-empty index set I and the set of (pairwise commuting) simple semisimple smooth connected normal subgroups of $G_{\overline{k}}^{\mathrm{red}}$.

Explicitly, for the k-split maximal k-torus

$$S := T \cap \mathscr{D}(G)$$

of $\mathscr{D}(G)$ (Corollary A.2.7), the decomposition of Φ corresponds to an isogeny decomposition of S into a product of subtori S_i, and the ith simple factor of $\mathscr{D}(G)_{\overline{k}}^{\mathrm{ss}} = \mathscr{D}(G_{\overline{k}}^{\mathrm{red}})$ (Example A.4.9) contains $(S_i)_{\overline{k}}$ as a maximal torus with associated root system Φ_i.

In general, the smooth connected normal subgroups of $\mathscr{D}(G)_{\overline{k}}^{\mathrm{ss}} = \mathscr{D}(G_{\overline{k}}^{\mathrm{red}})$ are exactly the almost direct products among the simple factors. Thus, we obtain a bijection between the set of such subgroups and the set of subsets of I. We wish to lift this to a description of the derived groups of smooth connected normal k-subgroups of G, and then get an analogous result when G is not assumed to be pseudo-split.

Continuing to assume that G is pseudo-split, for $i \in I$ let G_i denote the smooth connected k-subgroup of G generated by the $U_{(a)}$ for $a \in \Phi_i$. For any subset J of I, let $\Phi_J = \coprod_{j \in J} \Phi_j$ and let G_J denote the subgroup generated by the G_j for $j \in J$. Note that the formation of the G_i and the G_J is compatible with any separable extension of the ground field (given that G is pseudo-split). These subgroups will be the building blocks for the derived groups of the smooth connected normal k-subgroups of G, so we first prove that they are examples of such subgroups:

Lemma 3.1.4 *Each G_J is normal in G and is perfect. For every $i \in I$, G_i is absolutely pseudo-simple over k, and hence G_i is a minimal nontrivial smooth connected normal k-subgroup of G. If G is perfect then it is pseudo-simple over k if and only if $\Phi(G,T)$ is irreducible, in which case it is absolutely pseudo-simple over k.*

Proof By Proposition 2.3.9, $U_{(a)}$ commutes with $U_{(a')}$ for all $a \in \Phi_J$ and $a' \notin \Phi_J$. In particular, G_J commutes with $U_{(a')}$ for all $a' \notin \Phi_J$. By Remark 2.3.5, each G_J is normalized by $Z_G(T)$. Since G is generated by $Z_G(T)$ and the k-subgroups $U_{(a)}$ for $a \in \Phi$ (as their Lie algebras span Lie(G)), it follows that each G_J is normal in G. Hence, G_J is pseudo-reductive over k for all J.

To prove that every k-subgroup G_J is perfect, it suffices to prove the stronger claim that for every $a \in \Phi$ the smooth connected k-subgroup H_a generated by $U_{(a)}$ and $U_{(-a)}$ is perfect. The image of $(H_a)_{\overline{k}}$ in $G_{\overline{k}}^{\text{red}}$ is a simple semisimple group (of rank 1), so the maximal k-tori of H_a are not central.

Since H_a is normalized by T, the k-torus $S_a := T \cap H_a$ is a maximal k-torus of H_a (Corollary A.2.7), so S_a is 1-dimensional and has nontrivial conjugation action on $U_{(\pm a)}$. (In Proposition 3.2.3 we will give a more direct description of S_a.) The T-action on H_a makes the codimension-1 subtorus $T_a := (\ker a)^0_{\text{red}}$ act trivially (as can be checked against the generating subgroups $U_{(\pm a)}$), so T is an almost direct product of S_a and T_a, and $a|_{S_a} \neq 1$. Thus, commutators of S_a against $U_{(\pm a)}$ generate $U_{(\pm a)}$. This shows that H_a is perfect for all $a \in \Phi$, so G_J is perfect for all J.

To prove that each G_i is absolutely pseudo-simple over k (in which case it is clearly minimal as a nontrivial smooth connected normal k-subgroup of G), observe that the natural map $(G_i)_{\overline{k}} \to G_{\overline{k}}^{\text{red}}$ has unipotent kernel and simple semisimple image, so $(G_i)_{\overline{k}}^{\text{ss}}$ is simple. Hence, G_i is absolutely pseudo-simple over k, by Lemma 3.1.2.

Since each G_i is absolutely pseudo-simple over k and normal in G, and $G_{i'} \neq G_i$ if $i' \neq i$ (as the $(G_i)_{\overline{k}}$ have pairwise distinct images in $G_{\overline{k}}^{\text{red}}$), if G is pseudo-simple over k then there must be exactly one i and the corresponding G_i must coincide with G, so in such cases G is absolutely pseudo-simple over k and $\Phi = \Phi(G_{\overline{k}}^{\text{red}}, T_{\overline{k}})$ is irreducible. Likewise, if Φ is irreducible then $G_{\overline{k}}^{\text{ss}}$ is simple, so if also G is perfect then it is absolutely pseudo-simple over k (by Lemma 3.1.2). In general (for any pseudo-reductive G and k-split maximal k-torus T) the root system Φ is the set of non-multipliable elements in the root system $\Phi(G, T)$, by Theorem 2.3.10, so the sets of irreducible components of $\Phi(G, T)$ and Φ are in natural bijective correspondence. This establishes the final assertion of the lemma. □

By Proposition A.4.8 and Lemma 3.1.4, the image of the natural map $(G_J)_{\overline{k}} \to G_{\overline{k}}^{\text{red}}$ is identified with the quotient $(G_J)_{\overline{k}}^{\text{ss}}$ of $(G_J)_{\overline{k}}$. By construction of G_J, this image is the smooth connected normal subgroup of $\mathscr{D}(G_{\overline{k}}^{\text{red}}) = \mathscr{D}(G)_{\overline{k}}^{\text{ss}}$ generated by the simple factors indexed by elements of J.

Lemma 3.1.5 *For any $J \subseteq I$ and $J' := I - J$, the k-subgroups G_J and $G_{J'}$ commute with each other and $G_J \cdot G_{J'} = \mathscr{D}(G)$. In particular, $G_I = \mathscr{D}(G)$; i.e., $\mathscr{D}(G)$ is generated by the root groups with respect to any given split maximal torus.*

Proof Since $(U_{(a)}, U_{(a')}) = 1$ for all $a \in \Phi_J$ and $a' \in \Phi_{J'}$, the groups G_J and $G_{J'}$ commute inside of G. In particular, the product $G_J \cdot G_{J'}$ is a smooth connected k-subgroup of G. It is contained in $\mathscr{D}(G)$, so the image of $(G_J \cdot G_{J'})_{\overline{k}}$ under the projection $G_{\overline{k}} \twoheadrightarrow G_{\overline{k}}^{\mathrm{red}}$ is equal to $\mathscr{D}(G)_{\overline{k}}^{\mathrm{ss}}$ (as the image is contained in the connected semisimple $\mathscr{D}(G)_{\overline{k}}^{\mathrm{ss}}$ yet also contains all of its simple factors).

We conclude that the quotient $\mathscr{D}(G)/(G_J \cdot G_{J'})$ has trivial maximal semisimple quotient over $\overline{k}$. This says that $\mathscr{D}(G)/(G_J \cdot G_{J'})$ is solvable, yet it is its own derived group since $\mathscr{D}(G)$ is its own derived group (Proposition 1.2.6). Thus, this quotient is trivial; i.e., $G_J \cdot G_{J'} = \mathscr{D}(G)$. □

The following result uses the concept of a smooth connected affine k-group being "generated by tori" (see Definition 2.4.3); note that any perfect smooth connected affine k-group is generated by tori, by Proposition A.2.11.

Proposition 3.1.6 *Let G be a smooth connected affine k-group, and $G' = G_{k_{\mathrm{p}}}^{\mathrm{red}}$. For any smooth connected normal k-subgroup N in G, let N' denote the image of $N_{k_{\mathrm{p}}}$ in G'. Then $N \mapsto N'$ defines an inclusion-preserving (in both directions) bijection between the set of smooth connected normal k-subgroups of G generated by tori and the set of smooth connected normal k_{p}-subgroups of G'.*

Under this bijection, N is perfect if and only if N' is perfect.

Proof Let $N_t(G)$ denote the set of smooth connected normal k-subgroups of G generated by tori (in the sense of Definition 2.4.3). If k'/k is an extension field such that k is separably algebraically closed in k' (e.g., $k' = k_{\mathrm{p}}$) and $U' \subseteq G_{k'}$ is a normal unipotent closed k'-subgroup scheme with associated quotient map $\pi : G_{k'} \to G_{k'}/U'$ then we claim that the natural map $N_t(G) \to N_t(G_{k'}/U')$ defined by $N \mapsto \pi(N_{k'})$ is bijective, and that N is perfect if and only if $\pi(N_{k'})$ is perfect. This recovers the desired result by taking $k' = k_{\mathrm{p}}$ and $U' = \mathscr{R}_{u,k_{\mathrm{p}}}(G_{k_{\mathrm{p}}})$.

The factorization $N_t(G) \to N_t(G_{k'}) \to N_t(G_{k'}/U')$ reduces our task to two special cases: $U' = 1$ or $k' = k$. For the case $U' = 1$, the perfectness assertion is obvious and so we just have to show that $N_t(G) \to N_t(G_{k'})$ is bijective. Injectivity is obvious, and for surjectivity we choose $N' \in N_t(G_{k'})$ and seek $N \in N_t(G)$ such that $N' = N_{k'}$. Let $T \subseteq G$ be a maximal k-torus, so $T_{k'}$ is a maximal k'-torus in $G_{k'}$ and hence $S' := T_{k'} \cap N'$ is a maximal k'-torus in N' (by normality of N' in $G_{k'}$). Since $N' \in N_t(G_{k'})$, N' is normally generated by S' in G'. But $\mathrm{Gal}(k'_s/k') \to \mathrm{Gal}(k_s/k)$ is surjective since k is separably algebraically closed in k', so there is a (unique) k-subtorus $S \subseteq T$

such that $S_{k'} = S'$. The smooth connected normal k-subgroup N in G normally generated by S clearly satisfies $N_{k'} = N'$.

Now we may assume $k' = k$ and aim to show that if $U \subseteq G$ is a normal unipotent closed k-subgroup scheme, and $\pi : G \to G/U$ is the quotient map, then (i) every $\overline{N} \in N_t(G/U)$ has the form $\pi(N)$ for a unique $N \in N_t(G)$, (ii) $N \in N_t(G)$ is perfect if and only if $\pi(N)$ is perfect. In view of the case already established, we may replace k with its perfect closure so that k is perfect. For any $\overline{N}$, the k-scheme $\pi^{-1}(\overline{N})^0_{\rm red}$ is a smooth connected normal k-subgroup of G since k is perfect, and it visibly maps onto $\overline{N}$. Clearly $N := (\pi^{-1}(\overline{N})^0_{\rm red})_t \in N_t(G)$ with H_t defined as in Proposition A.2.11, and $\pi(N) = \overline{N}$ since $\overline{N}$ is generated by tori. Thus, to establish (i) it remains to show that if $M, N \in N_t(G)$ have the same image $\overline{N}$ under π then $M = N$ inside G. For any maximal k-torus T in G, by normality $T \cap M$ and $T \cap N$ are maximal k-tori in M and N respectively. These subtori of T are carried by π onto subtori of the k-torus $\pi(T) \cap \overline{N}$, yet their images must be maximal in the common quotient $\overline{N}$ of M and N, so $\pi(T \cap M) = \pi(T) \cap \overline{N} = \pi(T \cap N)$. Since $\pi : T \to \pi(T)$ is an isomorphism (as U is unipotent), it follows that $T \cap M = T \cap N$. As we vary T, M is generated by the tori $T \cap M$ and N is generated by the tori $T \cap N$, so $M = N$. This proves (i), and then (ii) follows by taking $M = \mathscr{D}(N)_t$ when $\overline{N}$ is perfect. Note that if $\overline{N}$ is perfect then it is generated by tori and so $\mathscr{D}(N)_t$ does project onto $\overline{N}$. □

Now we are ready to describe the possibilities for the derived group $\mathscr{D}(N)$ of a smooth connected normal k-subgroup N in any non-commutative pseudo-reductive group G over an arbitrary field k. The idea is to construct $\mathscr{D}(N)$ as a product from among a collection of "explicit" minimal possibilities, much as one does for smooth connected normal subgroups of a nontrivial connected semisimple group over a field. Observe that $\mathscr{D}(N)$ is perfect, by Proposition 1.2.6, so it is equivalent to describe those N that are perfect.

Remark 3.1.7 It is natural to wonder (based on analogy with the case of connected semisimple groups) if every smooth connected normal k-subgroup N of a perfect pseudo-reductive k-group is perfect. This is false over every imperfect field, as was shown in Example 1.6.4.

Proposition 3.1.8 *Let G be a smooth connected affine k-group. Consider smooth connected normal k-subgroups $N \subseteq G$ that are perfect. The set $\{N_i\}_{i \in I}$ of minimal such nontrivial k-subgroups of G is finite and the N_i are* *pseudo-simple over k and pairwise commute. The image* $\mathscr{N}$ *of the natural homomorphism*

$$\pi : \prod_{i \in I} N_i \to G \tag{3.1.1}$$

is the maximal perfect smooth connected normal k-subgroup of G. The kernel of π *is central and it does not contain any nontrivial smooth connected k-subgroups.*

For each subset $J \subseteq I$*, define* N_J *to be the subgroup of G generated by the* N_j *for* $j \in J$*. Every N has the form* N_J *for a unique J, and* $N_i \subseteq N$ *if and only if* $i \in J$*.*

If G is pseudo-reductive then $\mathscr{D}(G)$ is its maximal perfect smooth connected normal subgroup. If, moreover, G is a pseudo-split (e.g., $k = k_s$) then $G_{k_p}^{\mathrm{red}}$ is k_p-split, and the minimal nontrivial N are precisely the G_i as considered above. Hence, if G is pseudo-split pseudo-reductive then each N is equal to G_J for some (unique) J, so in particular the G_i are absolutely pseudo-simple over k.

Proof Let $G' = G_{k_p}^{\mathrm{red}}$. Under the correspondence in Proposition 3.1.6, the N_i correspond to the k_p-simple factors N_i' of $\mathscr{D}(G')$. The triviality of the commutator group (N_i, N_j) for $i \neq j$ is a consequence of the triviality of its image (N_i', N_j') in G' for all $i \neq j$. Likewise, the classification of all possible N in terms of subsets of I is a consequence of the analogous classification of perfect smooth connected normal k_p-subgroups of G' in terms of products among the k_p-simple factors of $\mathscr{D}(G')$. In particular, the maximal perfect smooth connected normal subgroup of G corresponds to $\mathscr{D}(G')$ and is the image of π.

The centrality of $\ker \pi$ in $\prod_{i \in I} N_i$ is an instance of the general fact that for a group H generated by a finite collection of pairwise commuting subgroups $\{N_i\}$, the multiplication map from the product $\prod N_i$ has central kernel. Indeed, if (n_i) lies in the kernel then $n_i = (\prod_{j \neq i} n_j)^{-1}$ centralizes N_i for all i, so $n_i \in Z_{N_i}$ for all i and hence (n_i) is indeed central. □

Example 3.1.9 Over any imperfect field k there are standard pseudo-reductive G for which the kernel of (3.1.1) has dimension > 0. To make such a G, let k'/k be a nontrivial purely inseparable finite extension and G' a connected semisimple k'-group that is absolutely simple and simply connected such that the center $Z' = Z_{G'}$ is not k'-étale. (If $\mathrm{char}(k) = p > 0$ then one such G' is SL_p.) Let T' be a maximal k'-torus in G' and define

$$C = \mathrm{R}_{k'/k}(T' \times T')/\mathrm{R}_{k'/k}(Z')$$

using the diagonal embedding of $\mathrm{R}_{k'/k}(Z')$. Let $j_1 : \mathrm{R}_{k'/k}(T') \to C$ be the map induced by the inclusion of T' into the first factor of $T' \times T'$.

Observe that C is pseudo-reductive because it fits into a short exact sequence of k-groups

$$1 \to \mathrm{R}_{k'/k}(T') \xrightarrow{j_1} C \to \mathrm{R}_{k'/k}(T')/\mathrm{R}_{k'/k}(Z') \to 1$$

in which the outer terms are pseudo-reductive. (For the third term we use that it is naturally a k-subgroup of the commutative pseudo-reductive k-group $\mathrm{R}_{k'/k}(T'/Z')$.) Since $\mathrm{R}_{k'/k}(Z')$ is central in $\mathrm{R}_{k'/k}(G')$, the k-group C has a natural left action on $\mathrm{R}_{k'/k}(G' \times G')$ via conjugation by $\mathrm{R}_{k'/k}(T' \times T')$. Thus, it makes sense to form the semidirect product

$$\mathrm{R}_{k'/k}(G' \times G') \rtimes C.$$

We embed $\mathrm{R}_{k'/k}(T' \times T')$ as a central k-subgroup of this semidirect product via the anti-diagonal embedding $\alpha : x \mapsto (x^{-1}, x \bmod \mathrm{R}_{k'/k}(Z'))$.

Consider the k-group

$$\begin{aligned} G = \operatorname{coker} \alpha &= (\mathrm{R}_{k'/k}(G' \times G') \rtimes C)/\mathrm{R}_{k'/k}(T' \times T') \\ &= \mathrm{R}_{k'/k}(G' \times G')/\mathrm{R}_{k'/k}(Z'). \end{aligned}$$

The k-group G fits into a pushout diagram

$$\begin{array}{ccc} \mathrm{R}_{k'/k}(T' \times T') & \longrightarrow & \mathrm{R}_{k'/k}(G' \times G') \\ \downarrow & & \downarrow \\ C & \longrightarrow & G \end{array}$$

of k-groups. Since C is pseudo-reductive over k, this is an instance of the construction of standard pseudo-reductive groups, and so it follows from Proposition 1.4.3 that G is pseudo-reductive over k.

The proof of Proposition 3.1.8 shows that the set $\{N_i\}$ consists of two copies of $\mathrm{R}_{k'/k}(G')$ with the evident embeddings into G, so $\ker \pi = \mathrm{R}_{k'/k}(Z')$. This has dimension > 0 since k'/k contains an intermediate extension of degree p over k and $Z'_{k'_s}$ contains μ_p, where $p = \mathrm{char}(k)$.

Remark 3.1.10 For the interested reader, we now show how to eliminate the perfectness hypothesis on N in Proposition 1.2.7. Using notation as in that result, but not assuming N to be perfect, since the smooth connected derived group $\mathscr{D}(N)$ is perfect (Proposition 1.2.6 applied to N) it follows from Proposition 1.2.7 that $\mathscr{D}(N)$ is normal in G. To prove that N is normal in G, we may assume $k = k_s$ (so G is pseudo-split) and then define $\{G_i\}_{i \in I}$ and G_J (for $J \subseteq I$) as above Lemma 3.1.4. It then follows from Proposition 3.1.8 in the pseudo-split case that $\mathscr{D}(N) = G_J$ for some subset $J \subseteq I$. Hence, by Lemma 1.2.5(i) the subgroups G_{I-J} and N of G commute with each other. Let C be a Cartan k-subgroup of G, so by two applications of Lemma 1.2.5(ii) (to the normal H in G and then the normal N in H) we see that $C_N := C \cap N$ is a Cartan k-subgroup of N. Thus, by Proposition 1.2.6 and Lemma 3.1.5, we have $G = C \cdot \mathscr{D}(G) = C \cdot G_I = C \cdot G_{I-J} \cdot G_J$ and $N = C_N \cdot \mathscr{D}(N) = C_N \cdot G_J$. Since G_{I-J} commutes with N, and the commutative C commutes with C_N and normalizes G_J, a direct computation gives

$$(G, N) = (C \cdot G_{I-J} \cdot G_J, C_N \cdot G_J) \subseteq G_J = \mathscr{D}(N) \subseteq N$$

(to verify the first inclusion, work modulo G_J and use that C_N commutes with both C and G_{I-J}), so N is normal in G.

3.2 Root datum for pseudo-reductive groups

To get more insight into the properties of the k-groups U_a in Proposition 2.3.11, we need to introduce a root datum so that we can speak of positive systems of roots. We begin with a preliminary lemma.

Lemma 3.2.1 *Let $f : H \to H'$ be a surjective map between smooth connected affine groups over a field k, and assume that $(\ker f_{\overline{k}})^0_{\mathrm{red}}$ is unipotent. Then for a k-torus T in H and its image T' in H', the induced map $N_H(T) \to N_{H'}(T')$ is surjective and the induced map $N_H(T)/Z_H(T) \to N_{H'}(T')/Z_{H'}(T')$ is an isomorphism. In particular, $Z_H(T) \to Z_{H'}(T')$ is surjective.*

See Remark C.4.6 for a generalization with weaker hypotheses on H, H', T, and $\ker f$.

Proof We may and do assume that k is algebraically closed. First we check surjectivity for the map between normalizers. Choose $h' \in N_{H'}(T')(k)$ and $h \in H(k)$ lifting h' via f. Then hTh^{-1} is a torus of H carried by f onto $h'T'h'^{-1} = T'$, so hTh^{-1} is a maximal torus of the smooth connected k-group $f^{-1}(T')^0_{\mathrm{red}}$. But $U := (\ker f)^0_{\mathrm{red}}$ is unipotent and $f^{-1}(T')^0_{\mathrm{red}}/U \to T'$ an isogeny, so $f^{-1}(T')^0_{\mathrm{red}}$ is solvable and contains T as a maximal torus.

Conjugacy of maximal tori in smooth connected solvable k-groups implies that $hTh^{-1} = uTu^{-1}$ for some $u \in U(k)$. Thus, replacing h with $u^{-1}h$ does the job.

Now consider the induced map $N_H(T)/Z_H(T) \to N_{H'}(T')/Z_{H'}(T')$. This is a surjection, and both source and target are finite étale (Lemma A.2.9), so to prove it is an isomorphism we just have to check injectivity on k-points. In other words, if $h \in H(k)$ normalizes T and if $f(h)$ centralizes T' then we want h to centralize T. Since $(\ker f)^0_{\rm red}$ is unipotent, the map $T \to T'$ induced by f is an isogeny. Thus, the desired result is clear. □

Now let G be a pseudo-split pseudo-reductive group over a field k and let T be a split maximal k-torus in G.

Definition 3.2.2 Let $R = \Phi(G,T), X = {\rm X}(T)$, and $X^\vee = {\rm X}_*(T)$. The elements of R are called the *roots* of G relative to T, and the subset $R^\vee \subseteq X^\vee$ of *coroots* of G relative to T is defined as follows. For any non-multipliable $a \in R$ (i.e., $a \in \Phi(G^{\rm red}_{\overline{k}}, T_{\overline{k}})$) the coroot $a^\vee \in X^\vee = {\rm X}_*(T_{\overline{k}})$ is the 1-parameter k-subgroup of T corresponding to a viewed as a root for $G^{\rm red}_{\overline{k}}$ relative to $T_{\overline{k}}$. For a multipliable $a \in R$, define $a^\vee = 2 \cdot (2a)^\vee$.

It is trivial to check that $a \mapsto a^\vee$ is a bijection from R to $R^\vee$, with the non-multipliable (resp. non-divisible) roots corresponding to the non-divisible (resp. non-multipliable) coroots. It is also straightforward to check that $\langle a, a^\vee \rangle = 2$ for all $a \in R$, and that the respective reflections

$$r_a : x \mapsto x - \langle x, a^\vee \rangle a, \quad r^\vee_a : \lambda \mapsto \lambda - \langle a, \lambda \rangle a^\vee$$

of X and $X^\vee$ are dual to each other and satisfy $r_a = r_{2a}$ and $r^\vee_a = r^\vee_{2a}$ for multipliable a.

We can describe the image of $a^\vee : {\rm GL}_1 \to T$ in concrete terms via root groups, just as in the connected reductive case, as follows. Let G_a denote the smooth connected k-subgroup of G generated by U_a and U_{-a}. This is normalized by T, so $T \cap G_a$ is a maximal k-torus of G_a (Corollary A.2.7). In Proposition 3.4.1 we will give a useful alternative description of G_a and its structure. The coroot $a^\vee$ is valued in G_a:

Proposition 3.2.3 *The k-subgroup $a^\vee({\rm GL}_1)$ in T is $T \cap G_a$.*

Proof Let $a' = a$ when a is non-multipliable and $a' = 2a$ when a is multipliable. The quotient map $G_{\overline{k}} \to G^{\rm red}_{\overline{k}}$ carries $(U_a)_{\overline{k}}$ into the 1-dimensional a'-root group of $(G^{\rm red}_{\overline{k}}, T_{\overline{k}})$ (Remark 2.3.6), and the image of $(U_a)_{\overline{k}}$ is nontrivial since G is pseudo-reductive (Lemma 1.2.1). Thus, $(G_a)_{\overline{k}}$ is carried onto the rank-1 semisimple subgroup $(G^{\rm red}_{\overline{k}})_{a'}$ with unipotent kernel. Hence, the maximal torus $(T \cap G_a)_{\overline{k}}$ maps isomorphically onto the maximal torus $T_{\overline{k}} \cap (G^{\rm red}_{\overline{k}})_{a'}$ that is precisely $a'^{\vee}_{\overline{k}}(\mathrm{GL}_1)$ by the definition of coroots in connected reductive groups over $\overline{k}$. It follows that the unique k-torus in T that descends the subtorus $a'^{\vee}_{\overline{k}}(\mathrm{GL}_1)$ in $T_{\overline{k}}$ is $T \cap G_a$, so $a'^{\vee}$ has image $T \cap G_a$. But $a^{\vee} \in \{a'^{\vee}, 2a'^{\vee}\}$ by definition, so $a^{\vee}$ has the same image as $a'^{\vee}$. □

We wish to naturally associate a root datum to any pseudo-split pair (G, T) as above. See Definition A.4.2 for the combinatorial notion of root datum (refining the classical notion of root system). A root datum $(X, R, X^{\vee}, R^{\vee})$ implicitly involves the specification of a bijection $a \mapsto a^{\vee}$ between R and $R^{\vee}$ in order to define the reflections r_a and $r_a^{\vee}$ in the axioms. This bijection can be omitted from the notation, due to:

Lemma 3.2.4 *The bijection $a \mapsto a^{\vee}$ in a root datum $(X, R, X^{\vee}, R^{\vee})$ is uniquely determined. That is, if $f : R \to R^{\vee}$ is a bijection with respect to which the axioms are satisfied then $f(a) = a^{\vee}$ for all $a \in R$. Moreover, for all $a, b \in R$ we have $r_a(b)^{\vee} = r_a^{\vee}(b^{\vee})$.*

Proof By [SGA3, XXI, 1.1.4], a coroot c^* is uniquely determined by the function $a \mapsto \langle a, c^* \rangle$ on R. Thus, to prove the uniqueness it suffices to show that the functional $\langle \cdot, c^* \rangle$ on the $\mathbf{Q}$-vector space V spanned by R is uniquely determined by the root c alone for any possible bijection $c \mapsto c^*$ from R to $R^{\vee}$ that satisfies the root datum axioms. Fixing such a bijection, we may define the reflection $r_c : x \mapsto x - \langle x, c^* \rangle c$ of V that carries R into itself and negates c. Such a reflection of V is uniquely determined by the nonzero c [Bou2, VI, 1.1, Lemma 1]. Thus, given c, the reflection r_c of V is independent of the choice of bijection $c \mapsto c^*$ satisfying the root datum axioms. From the formula for r_c we conclude that for each $c \in R$ the functional $\langle \cdot, c^* \rangle$ on V is independent of the choice of $c \mapsto c^*$, so this bijection is unique.

For a fixed $a \in R$, since the respective reflections r_a and $r_a^{\vee}$ on X and $X^{\vee}$ are dual to each other and preserve R and $R^{\vee}$, by the preceding uniqueness they must intertwine the bijection $b \mapsto b^{\vee}$ with itself. That is, $r_a(b)^{\vee} = r_a^{\vee}(b^{\vee})$ for all $b \in R$. □

Definition 3.2.5 The *root datum* associated to the pseudo-split pair (G, T) with pseudo-reductive G is the 4-tuple

$$R(G, T) = (\mathrm{X}(T), \Phi(G, T), \mathrm{X}_*(T), \Phi(G, T)^{\vee})$$

equipped with the bijection $a \mapsto a^\vee$ between $\Phi(G,T)$ and $\Phi(G,T)^\vee$ as defined above.

To justify this terminology, we must verify that the axioms for a root datum (as in Definition A.4.2) are satisfied. By Theorem 2.3.10, the root datum for $(G^{\mathrm{red}}_{\overline{k}}, T_{\overline{k}})$ is obtained by removing the multipliable roots and their associated (divisible) coroots for (G,T). Thus, since the roots and coroots for $(G^{\mathrm{red}}_{\overline{k}}, T_{\overline{k}})$ satisfy the axioms for a root datum, we only need to check that the reflections r_a carry multipliable elements of $\Phi(G,T)$ into $\Phi(G,T)$ and that the reflections $r_a^\vee$ carry divisible elements of $\Phi(G,T)^\vee$ into $\Phi(G,T)^\vee$ (for any $a \in \Phi(G,T)$).

In view of how the reflections were defined, this amounts to checking that if $a, b \in \Phi(G^{\mathrm{red}}_{\overline{k}}, T_{\overline{k}})$ with b divisible in $\Phi(G,T)$ (which can only occur when k is imperfect with characteristic 2) then $r_a(b/2) \in \Phi(G,T)$ (i.e., $r_a(b)$ is divisible) and $r_{a^\vee}(2b^\vee) \in \Phi(G,T)^\vee$. By Lemma 3.2.4 we have $r_{a^\vee}(b^\vee) = r_a(b)^\vee$ for $a, b \in \Phi(G^{\mathrm{red}}_{\overline{k}}, T_{\overline{k}})$, so since $c^\vee$ was defined to be $2(2c)^\vee$ for multipliable $c \in \Phi(G,T)$, it suffices to check that $r_a(b/2) \in \Phi(G,T)$ when b is divisible.

The reflections r_a of $\mathrm{X}(T_{\overline{k}})$ are (by definition) induced by the action of $N_{G^{\mathrm{red}}_{\overline{k}}}(T_{\overline{k}})$ on $T_{\overline{k}}$, so the problem is to show that this action on $\mathrm{X}(T_{\overline{k}}) = \mathrm{X}(T) = \mathrm{X}(T_{k_s})$ carries $\Phi(G,T) = \Phi(G_{k_s}, T_{k_s})$ into itself. This property is an immediate consequence of Lemma 3.2.1 (applied to $G_{\overline{k}} \twoheadrightarrow G^{\mathrm{red}}_{\overline{k}}$) because the quotient $N_G(T)/Z_G(T)$ is étale (Lemma A.2.9) and hence the set of its geometric points is exactly $N_G(T)(k_s)/Z_G(T)(k_s)$ (since $Z_G(T)$ is smooth).

If $(X, R, X^\vee, R^\vee)$ is any root datum then its *root lattice* Q is the $\mathbf{Z}$-span of R in X and its *coroot lattice* $Q^\vee \subseteq X^\vee$ is the $\mathbf{Z}$-span of $R^\vee$. We call X the *character lattice* and $X^\vee$ the *cocharacter lattice* of the root datum. The root datum is *semisimple* if the root lattice has finite index in the character lattice, in which case its *weight lattice* P is the $\mathbf{Z}$-dual of $Q^\vee$ inside of $Q_{\mathbf{R}}$. For a semisimple root datum we have $Q \subseteq X \subseteq P$, and the root datum is *simply connected* when $X = P$ (equivalently, $X^\vee = Q^\vee$) and is *adjoint* when $X = Q$.

Example 3.2.6 The root datum $R(G,T)$ is semisimple if and only if $G^{\mathrm{red}}_{\overline{k}}$ is semisimple, in which case the coroot lattices (and hence the weight lattices) for $R(G,T)$ and $R(G^{\mathrm{red}}_{\overline{k}}, T_{\overline{k}})$ coincide since removing some divisible coroots does not affect the coroot lattice. Hence, $R(G,T)$ is semisimple and simply connected precisely when $G^{\mathrm{red}}_{\overline{k}}$ is semisimple and simply connected.

The Weyl group of the root system $(R, Q_{\mathbf{R}})$ attached to a root datum $(X, R, X^\vee, R^\vee)$ is naturally identified with a subgroup of $\mathrm{Aut}(X)$ via the canonical decomposition $X_{\mathbf{Q}} = Q_{\mathbf{Q}} \oplus (Q_{\mathbf{Q}}^{\vee})^{\perp}$, with $(Q_{\mathbf{Q}}^{\vee})^{\perp}$ denoting the annihilator of $Q_{\mathbf{Q}}^{\vee}$ in $X_{\mathbf{Q}}$. In the case of the root datum arising from a pseudo-split pair (G, T) as above, the Weyl group of the associated root system is related to (G, T) as follows.

For any smooth connected affine k-group G and k-torus T in G, the *Weyl group* $W(G, T)$ for the pair (G, T) is the finite étale quotient k-group $N_G(T)/Z_G(T)$. Since $Z_G(T)$ is smooth, we have

$$N_G(T)(k_s)/Z_G(T)(k_s) = W(G, T)(k_s) = W(G, T)(\overline{k}).$$

When T is maximal and k-split and G is pseudo-reductive, we can do better:

Proposition 3.2.7 *Let G be a pseudo-split pseudo-reductive group over a field k, and let T be a split maximal k-torus in G. The finite étale k-group $W(G, T)$ is constant and the natural map $W(G, T)(k) \to \mathrm{Aut}(X)$ carries $W(G, T)(k)$ isomorphically onto the Weyl group of the root system associated to (G, T). Also, the natural map $W(G, T)_{\overline{k}} \to W(G_{\overline{k}}^{\mathrm{red}}, T_{\overline{k}})$ is an isomorphism.*

In Corollary 3.4.3(1) we will prove that $N_G(T)(k)/Z_G(T)(k) \hookrightarrow W(G, T)(k)$ is an equality. In Proposition C.2.10 we will prove this and the constancy of $W(G, T)$ without a pseudo-reductive hypothesis.

Proof Since $r_a = r_{2a}$ for multipliable $a \in \Phi(G, T)$, the Weyl group W of the root system of (G, T) coincides with that of the root system for $(G_{\overline{k}}^{\mathrm{red}}, T_{\overline{k}})$. The k-rational character group of T naturally coincides with the character groups of T_{k_s} and $T_{\overline{k}}$; denote them all as X, so the diagram

$$\begin{array}{ccc} W(G,T)(k_s) & \longrightarrow & W(G_{\overline{k}}^{\mathrm{red}}, T_{\overline{k}}) \\ & \searrow & \downarrow \\ & & \mathrm{Aut}(X) \end{array}$$

commutes, with the right side an isomorphism onto W. The horizontal arrow in the diagram is bijective, by Lemma 3.2.1, so the diagonal arrow is bijective onto W. Since this diagonal arrow is $\mathrm{Gal}(k_s/k)$-equivariant yet the Galois action on X is trivial, $W(G, T)(k_s)$ also has trivial Galois action. Thus, $W(G, T)$ is a constant k-group. □

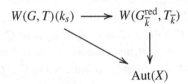

The set $\Phi(G, T)$ is empty if and only if $G_{\overline{k}}^{\mathrm{red}}$ is a torus, which is to say that G is solvable, or equivalently commutative. Since $\Phi(G, T)$ is a root system in $Q_{\mathbf{R}}$, where $Q = Q(G,T)$ is the root lattice of $R(G,T)$, we can speak of a *positive system of roots* in $\Phi(G,T)$, as for any root system.

Example 3.2.8 If the root system $\Phi(G,T)$ is irreducible and non-reduced (of which there is exactly one example, BC_r, in each rank $r \geqslant 1$) then removing the multipliable roots yields a root system of type C. Thus, if $R(G,T)$ is a non-reduced root datum then certain simple factors of type C in $G_{\overline{k}}^{\mathrm{red}}$ will play a distinguished role.

Remark 3.2.9 Let k be a separably closed field. We know that when $\mathrm{char}(k) \neq 2$, all pseudo-reductive k-groups have a reduced root system (by Theorem 2.3.10). In all characteristics we claim that standard pseudo-reductive k-groups have a reduced root system. To prove this, first observe (via Proposition 2.3.15) that reducedness of the root system is insensitive to passage to a pseudo-reductive central quotient. Thus, from the definition of standardness it suffices to prove that for any separably closed field k, nonzero finite reduced k-algebra k', and reductive k'-group G' with connected fibers, the Weil restriction $\mathrm{R}_{k'/k}(G')$ has a reduced root system. This follows from Example 2.3.2. In Corollary 4.1.6 we will obtain the reducedness of the root system in a more precise form that keeps track of the root spaces.

Let G be a nontrivial perfect pseudo-reductive group over a field k, and let $\{G_i\}_{i\in I}$ be the non-empty set of minimal nontrivial perfect smooth connected normal k-subgroups of G. By Proposition 3.1.8, the k-subgroups G_i pairwise commute and are pseudo-simple over k, and the natural homomorphism

$$\pi : \prod G_i \to G$$

is surjective with central kernel that contains no nontrivial smooth connected k-subgroup (in particular, it contains no nontrivial k-tori). The following refinement relates maximal k-tori of G to collections of maximal k-tori in the k-subgroups G_i, and in the pseudo-split case relates the corresponding root systems inside of a character lattice.

Proposition 3.2.10 *Using the preceding notation with a nontrivial perfect pseudo-reductive group G over a field k, there are mutually inverse bijective correspondences between the set of maximal k-tori T in G and the set of choices $\{T_i\}$ of maximal k-tori T_i in G_i via $T \mapsto \{T \cap G_i\}$ and $\{T_i\} \mapsto \pi(\prod T_i)$. The natural map $\pi : \prod T_i \to T := \pi(\prod T_i)$ is an isogeny, and T is k-split if and only if every T_i is k-split.*

If T is a k-split maximal torus in G, the resulting isomorphism $\mathrm{X}(T)_{\mathbf{Q}} = \prod \mathrm{X}(T_i)_{\mathbf{Q}}$ identifies the root system $\Phi(G,T)$ with the direct sum of the root systems $\Phi(G_i,T_i)$. In particular, if G is pseudo-split then it has a non-reduced root system if and only if some G_i has a non-reduced root system.

Proof Given a choice of the T_i, their product $\prod T_i$ is a maximal k-torus in $\prod G_i$, so it maps onto a maximal k-torus T in the quotient G. Since $\ker \pi$ contains no nontrivial k-tori, the map $\prod T_i \to T$ is an isogeny. Each $T \cap G_i$ is a maximal k-torus in G_i (Corollary A.2.7), and the containment $T_i \subseteq T \cap G_i$ is an equality by maximality of T_i. In the reverse direction, if T is a maximal k-torus in G and we define $T_i := T \cap G_i$ then every T_i is a maximal k-torus in G_i, so $\pi(\prod T_i)$ is a maximal k-torus in the image G of π. Thus, the containment $\pi(\prod T_i) \subseteq T$ is an equality. In particular, if every T_i is k-split then so is T, and the converse is obvious.

Since $\prod G_i \to G$ is a central quotient map between pseudo-reductive groups, and it carries $\prod T_i$ onto T, Proposition 2.3.15 reduces the desired relation among root systems in the pseudo-split case to the special case $G = \prod G_i$, and this case is trivial. □

See Proposition C.2.32 for the link between k-isotropic pseudo-simple normal k-subgroups and irreducible components of the relative root system (relative to a maximal k-split torus).

3.3 Unipotent groups associated to semigroups of weights

For a split connected reductive group G over a field k, and a k-split maximal k-torus T, it is useful to know when a set of root groups $\{U_a\}_{a\in\Psi}$ directly spans (in any order) a smooth connected unipotent k-subgroup U in the sense that multiplication defines a k-scheme isomorphism $\prod_i U_{a_i} \simeq U$ for any enumeration $\{a_i\}$ of Ψ. There is a sufficient criterion for this to hold [Bo2, 14.5] (expressed in terms of additivity and positivity properties of Ψ inside of $\Phi(G,T)$), and we seek an analogue in the pseudo-reductive case. This will be done by a method (inspired by [BoTi2, §2]) which is completely different from the reductive case, and moreover works for arbitrary smooth connected affine k-groups equipped with a left action by a split torus. Rather than proving direct spanning simultaneously with the construction of the ambient unipotent group (as in the reductive case), we will first build the unipotent group and only afterwards infer direct spanning properties.

We use the following conventions and notation with semigroups. We require semigroups to be associative and commutative, but do not require an identity element or non-emptiness. If Γ and Γ' are subsemigroups of a semigroup X then $\Gamma+\Gamma'$ denotes the subsemigroup of sums $\gamma+\gamma'$ with $\gamma \in \Gamma$ and $\gamma' \in \Gamma'$, and $\langle \Gamma, \Gamma' \rangle$ denotes the subsemigroup $\Gamma \cup \Gamma' \cup (\Gamma + \Gamma')$ generated by Γ and Γ'. Finally, if Σ is a subset of X then $\langle \Sigma \rangle$ denotes the subsemigroup of finite sums of elements of Σ; this is empty if Σ is empty.

Let G be a smooth affine group over a field k, S a k-split torus equipped with a left action on G, and $\Psi(G,S)$ the set of S-weights that occur in $\mathrm{Lie}(G)$. (If S is a split k-torus in $G^0 \neq 1$ acting by conjugation then $\Psi(G,S) = \Phi(G,S) \cup \{0\}$.) Let $\mathscr{O}(G)$ be the coordinate ring of G, and $I_G = \ker(e^* : \mathscr{O}(G) \to k)$ the augmentation ideal (so $\mathrm{Lie}(G) = (I_G/I_G^2)^*$). We make S act on the k-algebra $\mathscr{O}(G)$ by $(s.f)(g) = f(s.g)$, so $\Psi(G,S)$ coincides with the set of S-weights on the quotient I_G/I_G^2; this is consistent with the convention implicit in (2.1.2).

Remark 3.3.1 We will need to form centralizers for S-actions. In general, for any locally finite type k-group G and k-group M of multiplicative type acting on G, Proposition A.8.10(1) provides the existence of a closed subscheme $G^M \subseteq G$ representing the functor of M-fixed points in G, and this is smooth when G is smooth. In the special case that G is smooth and affine and M is a k-torus, this can be seen more directly by applying Proposition A.2.5 to $M \ltimes G^0$.

For a pair of smooth connected k-subgroups $H, H' \subseteq G$ that are normalized by S, we wish to understand the S-weights that appear in the Lie algebra of the k-subgroup they generate, as well as the Lie algebra of their commutator (H, H'). It turns out to be simpler to work with the set of S-weights appearing on the entire augmentation ideal of such groups, rather than focusing on just their Lie algebras, so we will begin with a study of this aspect in a more general setting.

(It is essential that we require S-stability for the k-subgroup and not merely for its Lie algebra. The reason for this is illustrated by the following example. For a field k of characteristic $p > 0$, let $G = \mathbf{G}_{\mathrm{a}} \times \mathbf{G}_{\mathrm{a}}$ be endowed with an action of a 2-dimensional k-split k-torus T via $t.(x,x') = (\chi(t)x, \chi'(t)x')$, where $\chi, \chi' \in \mathrm{X}(T)$ are linearly independent. Let $H \subseteq G$ be the first $\mathbf{G}_{\mathrm{a}}$-factor and let $H' \subseteq G$ be the graph of the relative Frobenius $F_{\mathbf{G}_{\mathrm{a}}/k}$, which is to say $H' = \{(x, x^p)\}$. Clearly H is T-stable but H' is not. The Lie algebras $\mathrm{Lie}(H)$ and $\mathrm{Lie}(H')$ coincide in $\mathrm{Lie}(G)$ with the weight space for χ, and the k-groups H and H' generate the k-group G whose Lie algebra supports the weight χ' that is linearly independent from χ.)

Let V be a geometrically integral S-stable closed subscheme of G that passes through the identity e. (A natural example is the closure of the image of the commutator map $H \times H' \to G$ for smooth connected S-stable k-subgroups H and H' of G.) Define $I_V = \ker(e^* : \mathscr{O}(V) \to k)$, and let $\mathrm{wt}(V)$ denote the set of S-weights which occur in I_V. Since $\mathscr{O}(V)$ is a domain, products of (nonzero) eigenfunctions for S in I_V are (nonzero) eigenfunctions, so $\mathrm{wt}(V)$ is a subsemigroup of $\mathrm{X}(S)$. (If $V = \{e\}$ then $\mathrm{wt}(V)$ is empty.) The formation of $\mathrm{wt}(V)$ commutes with any extension of the ground field.

Example 3.3.2 Let H be an S-stable smooth connected k-subgroup of G. We claim that $\mathrm{wt}(H) = \langle \Psi(H,S) \rangle$. To prove this, since $\Psi(H,S)$ is the set of S-weights occurring in I_H/I_H^2 we certainly have $\Psi(H,S) \subseteq \mathrm{wt}(H)$, so $\langle \Psi(H,S) \rangle \subseteq \mathrm{wt}(H)$. For the reverse inclusion, since $\mathscr{O}(H)$ is a noetherian domain it injects into the noetherian local ring $\mathscr{O}_{H,e} = \mathscr{O}(H)_{I_H}$. Thus, $\mathrm{wt}(H)$ is contained in the set of S-weights occurring on the maximal ideal $\mathfrak{m}$ of $\mathscr{O}_{H,e}$. Using the separated $\mathfrak{m}$-adic filtration, every S-weight on $\mathfrak{m}$ occurs in I_H^r/I_H^{r+1} for some $r \geqslant 1$. But this subquotient is an S-equivariant quotient of $(I_H/I_H^2)^{\otimes r}$, so the weights obtained all lie in $\langle \Psi(H,S) \rangle$.

Example 3.3.3 Assume $\mathrm{char}(k) = p > 0$, and consider the k-homomorphism $F_{G/k} : G \to G^{(p)}$ given by relative Frobenius. This is S-equivariant if we make S act on $G^{(p)}$ via the composition of the natural $S^{(p)}$-action (defined via base change on k by the p-power map) and the k-homomorphism $F_{S/k} : S \to S^{(p)}$. Since $F_{\mathrm{GL}_1/\mathbf{F}_p} : \mathrm{GL}_1 \to \mathrm{GL}_1$ is the p-power map and S is k-split, for any geometrically integral closed subscheme $V \subseteq G$ passing through e we have $\mathrm{wt}(V^{(p)}) = p\,\mathrm{wt}(V)$. In particular, if H is an S-stable smooth connected k-subgroup then $\mathrm{wt}(H^{(p)}) = p\,\mathrm{wt}(H)$. In the special case that H is commutative we have $\ker F_{H/k} \subseteq H[p]$, so the S-equivariant $p : H \to H$ factors through $F_{H/k}$ and hence $\mathrm{wt}(pH) \subseteq p\,\mathrm{wt}(H)$.

Lemma 3.3.4 *Let $V, V' \subseteq G$ be S-stable geometrically integral closed subschemes passing through the identity e. Let VV' denote the closure of the image of the multiplication map $V \times V' \to G$, and let (V, V') denote the closure of the image of the commutator map $V \times V' \to G$ defined by $(v, v') \mapsto vv'v^{-1}v'^{-1}$. Then* $\mathrm{wt}(VV') = \langle \mathrm{wt}(V), \mathrm{wt}(V') \rangle$ *and* $\mathrm{wt}((V, V')) \subseteq \mathrm{wt}(V) + \mathrm{wt}(V')$.

Proof The dominant map $V \times V' \to VV'$ induces an injection $\mathscr{O}(VV') \hookrightarrow \mathscr{O}(V) \otimes_k \mathscr{O}(V')$, so we get an S-equivariant embedding

$$I_{VV'} \subseteq I_V \oplus I_{V'} \oplus (I_V \otimes_k I_{V'}).$$

Thus, $\mathrm{wt}(VV') \subseteq \langle \mathrm{wt}(V), \mathrm{wt}(V') \rangle$. Since $\mathrm{wt}(VV')$ is a subsemigroup of $\mathrm{X}(S)$, to prove the reverse inclusion we just need to show that $\mathrm{wt}(V)$ and $\mathrm{wt}(V')$ are contained in $\mathrm{wt}(VV')$, and this follows from the fact that V and V' are contained in VV' (so I_V and $I_{V'}$ are S-equivariant quotients of $I_{VV'}$). The commutator map $V \times V' \to (V, V')$ restricts to the constant map e on the slices $V \times \{e\}$ and $\{e\} \times V'$, so the same method shows that $\mathrm{wt}((V, V')) \subseteq \mathrm{wt}(V) + \mathrm{wt}(V')$. □

By [Bo2, 2.2], for sufficiently large n the images of the multiplication maps $V^n \to G$ stabilize at a common geometrically integral image $\langle V \rangle$ that contains e. Hence, the multiplication map $\langle V \rangle \times \langle V \rangle \to G$ factors through $\langle V \rangle$, so $\langle V \rangle$ is a closed k-subsemigroup of G passing through the identity. But any closed k-subsemigroup scheme H in G passing through the identity must be a k-subgroup of G. To see this, observe that for any k-algebra R and point $h \in H(R)$, the left multiplication mapping $\ell_h : H_R \to H_R$ is a monomorphism from H_R into itself, so it must be an isomorphism [EGA, IV_4, 17.9.6]. In particular, since H passes through e it follows that H is stable under inversion, so it is indeed a k-subgroup of G. This proves that $\langle V \rangle$ is the smooth connected k-subgroup of G generated by V, and repeated application of Lemma 3.3.4 gives that $\mathrm{wt}(\langle V \rangle) = \mathrm{wt}(V)$. Since $\langle V \rangle$ is a group we also conclude that $\mathrm{wt}(V^{-1}) = \mathrm{wt}(V)$.

Proposition 3.3.5 *If $H', H'' \subseteq G$ are smooth connected k-subgroups normalized by the S-action and H denotes the k-subgroup they generate, then* $\mathrm{wt}(H) = \langle \mathrm{wt}(H'), \mathrm{wt}(H'') \rangle$. *Also,* $\mathrm{wt}((H', H'')) \subseteq \mathrm{wt}(H') + \mathrm{wt}(H'')$. *If H'' normalizes H', then any S-weight of* $\mathrm{Lie}(H'H'')$ *admits a positive rational multiple that is an S-weight of either* $\mathrm{Lie}(H')$ *or* $\mathrm{Lie}(H'')$.

Proof The first two assertions of the proposition are obtained by repeated application of Lemma 3.3.4. To prove the last assertion, assume H'' normalizes H' and consider the S-equivariant surjective homomorphism $H' \rtimes H'' \to H'H''$ defined by multiplication. By Proposition B.4.6, any S-weight a that occurs on $\mathrm{Lie}(H'H'')$ is a positive integral multiple of an S-weight that occurs on $\mathrm{Lie}(H' \rtimes H'') = \mathrm{Lie}(H') \oplus \mathrm{Lie}(H'')$. Thus, some positive rational multiple of a is an S-weight of either $\mathrm{Lie}(H')$ or $\mathrm{Lie}(H'')$. □

It is natural to ask what constraints are imposed on the subsets $\mathrm{wt}(H) \subseteq \mathrm{X}(S)$ beyond the subsemigroup property. There are no further constraints, and there is a canonical way to go in reverse:

Proposition 3.3.6 *Let G be a smooth affine group over a field k, and S a k-split k-torus equipped with an action on G. For every subsemigroup $A \subseteq \mathrm{X}(S)$, there is a unique smooth connected S-stable k-subgroup $H_A(G)$ in G such that* $\mathrm{Lie}(H_A(G))$ *is the span of the weight spaces in* $\mathrm{Lie}(G)$ *for all $a \in \Psi(G, S) \cap A$. The k-subgroups $H_A(G)$ and $H_A(G^0)$ coincide, and if $0 \notin A$ then $H_A(G)$ is unipotent.*

The k-subgroup $H_A(G)$ is maximal in the sense that a smooth connected S-stable k-subgroup H in G satisfies $\Psi(H,S) \subseteq A$ if and only if $H \subseteq H_A(G)$.

When $0 \notin A$ we will often write $U_A(G)$ instead of $H_A(G)$.

Proof If H' and H'' are k-subgroups satisfying the requirements for $H_A(G)$ then by Proposition 3.3.5 the k-subgroup H they generate is another, yet its Lie algebra cannot be any larger, so by smoothness and connectedness we have $H' = H = H''$. This proves uniqueness. In particular, once existence is proved in general then $H_A(G^0) = H_A(G)$. To prove existence of $H_A(G)$, the case $A = \emptyset$ is trivial ($H = 1$), so by Proposition 3.3.5 it suffices to treat the case where A is the cyclic subsemigroup $\langle a \rangle$ generated by a single element $a \in \mathrm{X}(S)$.

If $a = 0$ then $(G^S)^0$ works. If $a \neq 0$ then $M_a := \ker a$ is a (k-split) k-subgroup of S, so G^{M_a} makes sense and is smooth since G is smooth (Proposition A.8.10(2)). The 1-dimensional split torus $S_a = S/M_a$ naturally acts on G^{M_a} and is identified with GL_1 via a. By Proposition A.8.10(1), $\mathrm{Lie}(G^{M_a})$ is the span of the weight spaces for the characters in $\Psi(G,S)$ that factor through S_a, which is to say that they are the $\mathbf{Z}$-multiples of a which occur on $\mathrm{Lie}(G)$. Letting $\lambda : \mathrm{GL}_1 \simeq S_a$ be the isomorphism inverse to a, $U_{G^{M_a}}(\lambda)$ is a smooth *connected* k-subgroup of G by Proposition 2.1.8(3),(4), and its Lie algebra is the span of the weight spaces for the elements of $\Psi(G,S) \cap \langle a \rangle$. This completes the proof of existence.

For any A, consider an S-stable smooth connected k-subgroup H in G such that $\Psi(H,S) \subseteq A$. Example 3.3.2 and Proposition 3.3.5 imply that the smooth connected k-subgroup H' generated by H and $H_A(G)$ satisfies $\mathrm{wt}(H') \subseteq A$, so $\mathrm{Lie}(H')$ has all S-weights in A. But $\mathrm{Lie}(H_A(G))$ is the span of all such weight spaces, so the inclusion $H_A(G) \subseteq H'$ is an equality on Lie algebras and hence is an equality of k-groups. In particular, $H \subseteq H_A(G)$.

Now assume $0 \notin A$, so S acts on $H_A(G)$ with only nontrivial weights on $\mathrm{Lie}(H_A(G))$. To prove that $H_A(G)$ is unipotent, we use the elementary fact (explained in the proof of Proposition 2.3.11) that if H is a smooth connected affine k-group and a split k-torus S acts on H with only nontrivial weights on $\mathrm{Lie}(H)$ then H is unipotent. □

Example 3.3.7 Let G be a pseudo-split pseudo-reductive group over k and T a split maximal k-torus. For any $a \in \Phi(G,T)$ the k-subgroup $U_{\langle a \rangle}(G)$ is the root group U_a in Proposition 2.3.11 (in view of the constraints on multipliability among roots in Theorem 2.3.10).

It is immediate from Proposition 3.3.6 that if $A \subseteq A'$ then $H_A(G) \subseteq H_{A'}(G)$, so if A and A' are two subsemigroups in $\mathrm{X}(S)$ then the smooth connected

k-subgroup of G generated by $H_A(G)$ and $H_{A'}(G)$ is contained in $H_{\langle A,A'\rangle}(G)$. Likewise, it follows from Proposition 3.3.5 that $(H_A(G), H_{A'}(G)) \subseteq H_{A+A'}(G)$. Since torus centralizers and $U_G(\lambda)$ are both functorial in k-group homomorphisms, it follows from the construction that if $f : G' \to G$ is an S-equivariant k-homomorphism then f carries $H_A(G')$ into $H_A(G)$; the same even holds for S-equivariant R-group homomorphisms $G'_R \to G_R$ for any k-algebra R since the theory of $U_G(\lambda)$ works over any ring. In the special case of smooth S-equivariant surjective homomorphisms $f : G' \to G$, the induced map $f_A : H_A(G') \to H_A(G)$ is surjective on Lie algebras (as $\mathrm{Lie}(f)$ is an S-equivariant quotient map), so f_A is a smooth surjection (due to connectedness).

We now focus on the case $U_A(G)$ for cyclic $A = \langle a \rangle$ with a nonzero $a \in \mathrm{X}(S)$. There is a natural decreasing filtration on $A = \langle a \rangle$ given by the subsemigroups A^j consisting of the characters na for $n \geqslant j$ (with $j \geqslant 1$). This gives rise to a decreasing filtration of $H_A(G)$ by smooth connected k-subgroups $H_{A^j}(G)$. The following result generalizes the structure of root groups in the pseudo-split pseudo-reductive case (including the 2-step filtration $\{1\} \subseteq U_{2a} \subseteq U_a$ for multipliable roots a), as well as the structure of non-commutative root groups which arise in the theory of relative root systems for non-split connected reductive groups.

Lemma 3.3.8 *Using notation as in Proposition 3.3.6, if A is a subsemigroup of $\mathrm{X}(S)$ that does not contain 0 then the unipotent subgroup $U := H_A(G)$ is k-split. For $A = \langle a \rangle$ with $a \in \mathrm{X}(S) - \{0\}$, each $U_{A^j}(G)$ is normal in $U_A(G)$ and the quotient $V_j = U_{A^j}(G)/U_{A^{j+1}}(G)$ is a vector group such that $\Psi(V_j, S) = \{ja\}$ if $V_j \neq 0$. In particular, V_j has a unique S-equivariant linear structure.*

Proof The k-split smooth connected normal k-subgroup $U_{\mathrm{split}} \subseteq U$ as in Theorem B.3.4 is stable under the action of S (as we may check over k_s) and $\overline{U} := U/U_{\mathrm{split}}$ is k-wound. Thus, the S-action on $\overline{U}$ must be trivial (Proposition B.4.4). Since $\mathrm{Lie}(\overline{U})$ is an S-equivariant quotient of $\mathrm{Lie}(U)$, yet all S-weights on $\mathrm{Lie}(U)$ lie in A and $0 \notin A$, we see that $\mathrm{Lie}(\overline{U}) = 0$; i.e., $U = U_{\mathrm{split}}$. This gives that U is k-split. For the remainder of the argument, we consider the case $A = \langle a \rangle$ with $a \in \mathrm{X}(S) - \{0\}$.

The commutator $(U_A(G), U_{A^j}(G))$ is contained in $U_{A+A^j}(G) = U_{A^{j+1}}(G) \subseteq U_{A^j}(G)$, so $U_A(G)$ normalizes $U_{A^j}(G)$. Likewise,

$$(U_{A^j}(G), U_{A^j}(G)) \subseteq U_{A^j+A^j}(G) = U_{A^{2j}}(G) \subseteq U_{A^{j+1}}(G),$$

so each $V_j := U_{A^j}(G)/U_{A^{j+1}}(G)$ is commutative. By construction, V_j is also smooth, connected, and unipotent. Since $A^j - A^{j+1} = \{ja\}$, $\mathrm{Lie}(V_j)$ clearly has

ja as its unique S-weight if $V_j \neq 0$. If $\mathrm{char}(k) = 0$ then a smooth connected commutative unipotent k-group is a vector group and admits a unique linear structure, so we are done in such cases.

Now assume $\mathrm{char}(k) = p > 0$. By Proposition B.4.1, to show that V_j is a vector group and admits an S-equivariant linear structure, it suffices to show that V_j is p-torsion. The linear structure will then be unique, due to uniqueness of the S-weight on $\mathrm{Lie}(V_j)$ when $V_j \neq 0$, exactly as in the proof of Lemma 2.3.8. We may assume $V_j \neq 0$, so by Example 3.3.2, $\mathrm{wt}(V_j) = \langle ja \rangle$. Thus, Example 3.3.3 implies $\mathrm{wt}(pV_j) \subseteq p\langle ja \rangle$, so $\mathrm{Lie}(pV_j)$ has no occurrence of the weight ja. Since ja is the only S-weight on $\mathrm{Lie}(V_j)$, necessarily $pV_j = 0$. □

Returning to the consideration of general A (not necessarily cyclic), we now study the behavior of $H_A(G)$ with respect to S-stable smooth connected k-subgroups G' of G, generalizing Remark 2.3.6 (via Remark 2.1.11). We allow for the possibility $0 \in A$. By functoriality $H_A(G') \subseteq G' \cap H_A(G)$, and the Lie algebras coincide. But $H_A(G')$ is smooth and connected, so $G' \cap H_A(G)$ has dimension at least as large as that of its Lie algebra. This proves:

Lemma 3.3.9 *For any S-stable smooth connected k-subgroup G' of G, the intersection $G' \cap H_A(G)$ is smooth and $(G' \cap H_A(G))^0 = H_A(G')$.*

It is natural to ask when $G' \cap H_A(G)$ is connected. Connectedness holds if A is a saturated subgroup in $\mathrm{X}(S)$ (i.e., $\mathrm{X}(S)/A$ is torsion-free). Indeed, in such cases $S_A := \bigcap_{a \in A} \ker a$ is a k-subtorus of S (since A is saturated) and it centralizes $H_A(G)$ (by Corollary A.8.11), so there is an inclusion $G' \cap H_A(G) \subseteq G'^{S_A}$ which is an equality on Lie algebras (since A is saturated). This must then be an equality of k-groups, due to the connectedness of G'^{S_A} (which follows from the formula $Z_{G' \rtimes S_A}(S_A) = G'^{S_A} \times S_A$).

In general, $G' \cap H_A(G)$ can be disconnected. For example, consider the inclusion $G' = \mathrm{PGL}_2 \to \mathrm{GL}(\mathfrak{pgl}_2) \simeq \mathrm{GL}_3 =: G$ defined by the adjoint representation. Let S be a split maximal k-torus in PGL_2, and A the index-2 sublattice in $\mathrm{X}(S) \simeq \mathbf{Z}$, so $H_A(G) = Z_G(S[2])$ and hence $G' \cap H_A(G) = Z_{G'}(S[2])$. By inspection this is $N_{G'}(S)$, which is disconnected (with identity component S). Disconnectedness never occurs when $0 \notin A$, and more generally we have:

Proposition 3.3.10 *Let G be a smooth affine k-group, and S a k-split k-torus equipped with an action on G. Let G' be an S-stable smooth connected closed k-subgroup of G. We have $G' \cap H_A(G) = H_A(G')$ (equivalently, $G' \cap H_A(G)$ is connected) in any of the following cases:*

(a) $0 \notin A$,
(b) G' *is solvable,*
(c) G' *is normal in* G.

Proof Suppose $G' \cap H_A(G)$ is disconnected. We will show that $0 \in A$, so (a) follows. Consider the idempotent f on $G' \cap H_A(G)$ that vanishes on the identity component (and hence at e) but equals 1 elsewhere, so f is S-invariant and lies in the augmentation ideal of $\mathscr{O}(G' \cap H_A(G))$. The coordinate ring $\mathscr{O}(G' \cap H_A(G))$ is a quotient of $\mathscr{O}(H_A(G))$, so f lifts to an S-invariant element in $I_{H_A(G)}$. But the only S-weights in $I_{H_A(G)}$ are elements of A, so $0 \in A$.

To prove (b) and (c), we may assume $0 \in A$ and $k = \overline{k}$. We will prove (c) conditional on (b), and then prove (b). Assuming G' is normal in G, let $U' = \mathscr{R}_u(G')$ and $U = \mathscr{R}_u(G)$, so $U' = G' \cap U$ by Proposition A.4.8. (Equivalently, the natural map $G'^{\text{red}} \to G^{\text{red}}$ has trivial kernel.) We will reduce to the case where G' is reductive, and then to the case where G is also reductive.

Applying (b) to U' gives that $U' \cap H_A(G)$ is connected, so $G' \cap H_A(G)$ is connected if and only if the k-group

$$(G' \cap H_A(G))/(U' \cap H_A(G)) = ((G' \cap H_A(G))U')/U'$$

is connected. But $H_A(G)U'/U' = H_A(G/U')$, so to prove (c) we may replace G' with G'/U' and replace G with G/U' to arrange that G' is reductive. In particular, $G' \cap U = 1$, so G' is naturally a k-subgroup of G/U and we have the inclusions

$$H_A(G') \hookrightarrow G' \cap H_A(G) \hookrightarrow G' \cap H_A(G/U).$$

Thus, if the outer terms are equal then so are the first two, so we may replace G with G/U to arrange that G is also reductive. With G reductive, it follows from Corollary A.4.7(2) that the S-action on G is given by composing some k-homomorphism $S \to G/Z_G$ with the natural left action of G/Z_G on G.

The image of S in G/Z_G is contained in a maximal k-torus $\overline{T}$, and $\overline{T} = T/Z_G$ for a unique maximal k-torus T in G. Thus, by Lie algebra considerations we can describe $H_A(G)$ explicitly (using that $0 \in A$): it is generated by $T = Z_G(T)$ and the root groups U_a for $a \in \Phi(G,T) \subseteq \mathrm{X}(\overline{T})$ such that $a|_S \in A$ (via the restriction map $\mathrm{X}(\overline{T}) \to \mathrm{X}(S)$). An analogous procedure describes $H_A(G/Z_G)$, using $\overline{T}$ and root groups in G/Z_G. Comparing these similar descriptions of $H_A(G)$ and $H_A(G/Z_G)$, we see that $H_A(G/Z_G)$ is the image of $H_A(G)$ (even though $G \to G/Z_G$ may not be smooth). But $Z_G \subseteq T \subseteq H_A(G)$, so $H_A(G)$ is the full preimage of $H_A(G/Z_G)$ in G. Running through the same argument for the reductive G', we get that $H_A(G')$ is the full preimage of $H_A(G'/Z_{G'})$ in G'.

Our aim is to prove that the containment

$$H_A(G') \subseteq G' \cap H_A(G)$$

is an equality. Since G' is normal in G, we have $Z_{G'} \subseteq Z_G$ (Proposition A.1.13). Thus both sides contain $Z_{G'}$, so it is equivalent to check the equality of the images in $G'/Z_{G'}$. The induced map $G'/Z_{G'} \to G/Z_G$ between maximal adjoint quotients has trivial kernel (Proposition A.4.8), so since $Z_G \subseteq H_A(G)$ it follows that the image of $G' \cap H_A(G)$ in $G'/Z_{G'}$ is $(G'/Z_{G'}) \cap H_A(G/Z_G)$. Hence, we can replace G and G' with their respective maximal adjoint quotients, in which case G' is a direct factor of G and so the desired equality is clear. This proves (c), conditional on (b).

Finally, we prove (b). Since $H_A(G) = H_A(G^0)$, we may replace G with G^0 so that G is connected. In particular, G^S is connected (by Proposition A.2.5 applied to $S \ltimes G$). Let us fix a 1-parameter k-subgroup $\lambda : \mathrm{GL}_1 \to S$ such that $\langle \chi, \lambda \rangle \neq 0$ for all nonzero $\chi \in \Psi(G, S)$. (If $\Psi(G, S) = \{0\}$ then $G = G^S$ and there is nothing to do.) Thus, the containment $G^S \subseteq Z_G(\lambda)$ is an equality since it induces an equality of Lie algebras and $Z_G(\lambda)$ is smooth and connected (due to the smoothness and connectedness of G). This implies

$$\begin{aligned}(G' \cap H_A(G))^S = (G' \cap H_A(G)) \cap G^S &= (G' \cap H_A(G)) \cap Z_G(\lambda) \\ &= Z_{G' \cap H_A(G)}(\lambda).\end{aligned}$$

But we have reduced to the case $0 \in A$, so $(G' \cap H_A(G))^S = G'^S$. This is connected since G' is connected (Proposition A.2.5), so we conclude that $Z_{G' \cap H_A(G)}(\lambda)$ is connected. Applying Proposition 2.1.12(2) to the smooth connected solvable k-group G', its k-subgroup $G' \cap H_A(G)$ must therefore be connected. □

We can now reap the fruit of our labors.

Theorem 3.3.11 *Let G be a solvable smooth connected affine k-group equipped with an action by a split k-torus S. Let Ψ denote the set of S-weights on* Lie(G). *Suppose there is a disjoint union decomposition $\Psi = \coprod_{i=1}^{n} \Psi_i$ such that Ψ_i is disjoint from the subsemigroup $A_j := \langle \Psi_j \rangle$ whenever $i \neq j$. The k-scheme map*

$$H_{A_1}(G) \times \cdots \times H_{A_n}(G) \to G \tag{3.3.1}$$

defined by multiplication is an isomorphism.

Proof We may assume $k=\overline{k}$. Pick a maximal k-torus S' in $S \ltimes G$ containing S. This meets the smooth connected normal k-subgroup G in a maximal k-torus T (Corollary A.2.7), so the maximal k-torus T in G is centralized by the S-action on G. In particular, T normalizes $H_{A_i}(G)$ for all i.

The first step is to reduce to the problem with the solvable G replaced by its unipotent radical U. If $0 \notin \Psi$ then G is unipotent (as we have already observed at the end of the proof of Proposition 3.3.6), so there is nothing to do. If $0 \in \Psi$ then exactly one Ψ_i contains 0, and $0 \notin A_j$ for all $j \neq i$. Thus, by Proposition 3.3.6, $H_{A_j}(G)$ is unipotent for all $j \neq i$. Since $G = T \ltimes U$, we then have $H_{A_j}(G) = H_{A_j}(U)$ for all $j \neq i$. Clearly $T \subseteq H_{A_i}(G)$, and $U \cap H_{A_i}(G)$ is connected (by Proposition 3.3.10, either (b) or (c)), so inside of $G = T \ltimes U$ we have

$$H_{A_i}(G) = T \ltimes (U \cap H_{A_i}(G)) = T \ltimes H_{A_i}(U).$$

This factor of T can be passed across all subgroups $H_{A_j}(G)$ (as all are normalized by T), so (3.3.1) for G is reduced to the same problem for U in place of G. Thus, we may and do assume that G is unipotent.

Now we proceed by induction on $\dim G$. We may assume $G \neq 1$, so there exists a nontrivial smooth connected central k-subgroup $Z \subseteq G$ that is an S-stable vector group admitting an S-equivariant linear structure. (To construct Z, let N be the last nontrivial term in the descending central series of G, so N is commutative. Set $Z = N$ if $\mathrm{char}(k) = 0$, and set $Z = p^e N$ with maximal $e \geqslant 0$ such that $p^e N$ is nontrivial when $\mathrm{char}(k) = p > 0$, so N is also p-torsion in such cases. The vector group property and S-equivariant linear structure when $\mathrm{char}(k) > 0$ are obtained from Lemma B.1.10 and Theorem B.4.3.) The analogue of (3.3.1) for Z is immediate from the weight space decomposition, and

$$Z \cap H_{A_i}(G) = H_{A_i}(Z)$$

for all i, by Proposition 3.3.10 (either (b) or (c)). We can assume $G/Z \neq 1$ or else we would be done, and since $\dim(G/Z) < \dim G$ the analogue of (3.3.1) holds for G/Z by induction.

The image of $H_{A_i}(G)$ under the smooth surjection $G \to G/Z$ is $H_{A_i}(G/Z)$ for all i, so our problem is now reduced to a special case of the following general claim: if G is a smooth k-group scheme of finite type, Z is a central k-subgroup of G, and $H_1, \ldots, H_n$ are smooth k-subgroups such that the natural k-scheme maps

$$(Z \cap H_1) \times \cdots \times (Z \cap H_n) \to Z,$$
$$\mathrm{im}(H_1 \to G/Z) \times \cdots \times \mathrm{im}(H_n \to G/Z) \to G/Z$$

defined by multiplication are isomorphisms then the natural map

$$\mu : H_1 \times \cdots \times H_n \to G$$

defined by multiplication is an isomorphism. If these were ordinary groups then an elementary diagram chase yields the bijectivity (so μ is bijective on geometric points, which is insufficient). Likewise, for the analogous claim with sheaves of groups on a topological space, a straightforward variant of the same diagram chase establishes the isomorphism result. Since we can view the k-schemes $G, Z, G/Z, H_i$, and so on as sheaves of groups for the fppf topology on the category of k-schemes, the sheaf version of the diagram chase can be used to establish that μ is an isomorphism of k-schemes. □

Corollary 3.3.12 *Let G be a pseudo-split pseudo-reductive group over a field k, and let T be a split maximal k-torus in G. For any $\lambda \in \mathrm{X}_*(T)$, $U_G(\lambda)$ is directly spanned in any order by the k-groups U_a for non-divisible $a \in \Phi(G,T)$ such that $\langle a, \lambda \rangle > 0$. Moreover, a root $a \in \Phi(G,T)$ occurs as a weight for the T-action on* $\mathrm{Lie}(P_G(\lambda))$ *if and only if $U_{(a)} \subseteq P_G(\lambda)$.*

In particular, if Φ^+ is a positive system of roots in $\Phi(G,T)$ then the U_a for non-divisible $a \in \Phi^+$ directly span (in any order) the unique smooth connected unipotent k-subgroup U_{Φ^+} in G normalized by T whose Lie algebra is the span of the root spaces in $\mathrm{Lie}(G)$ *for roots in Φ^+.*

Proof The hypotheses in Theorem 3.3.11 are satisfied by taking the sets Ψ_i to be $\langle a_i \rangle \bigcap \Phi(G,T)$ for any enumeration $\{a_1, \dots\}$ of the set of non-divisible $a \in \Phi(G,T)$ such that $\langle a, \lambda \rangle > 0$. For any non-divisible $a \in \Phi$ we have $H_{\langle a \rangle}(G) = U_a$ by Example 3.3.7, so the asserted description of $U_G(\lambda)$ follows. By Proposition 2.1.8(1), $\mathrm{Lie}(P_G(\lambda))$ is the span of the weight spaces in $\mathrm{Lie}(G)$ for the weights $a \in \Phi(G,T) \cup \{0\}$ such that $\langle a, \lambda \rangle \geqslant 0$. In other words, $P_G(\lambda) = H_A(G)$ for the saturated subsemigroup $A = \mathrm{X}(T)_{\lambda \geqslant 0}$. For each $a \in \Phi(G,T)$ we have $U_{(a)} = H_{\langle a \rangle_{\mathrm{sat}}}(G)$ where $\langle a \rangle_{\mathrm{sat}} := \mathrm{X}(T) \cap \mathbf{Q} \cdot a$ is the saturation of $\langle a \rangle$ in $\mathrm{X}(T)$. But for every T-weight a on $\mathrm{Lie}(H_A(G))$ we have $\langle a \rangle_{\mathrm{sat}} \subseteq A$ since A is saturated, so $U_{(a)} \subseteq H_A(G) = P_G(\lambda)$.

The assertions concerning U_{Φ^+} (both existence and uniqueness) are obtained by choosing λ (as we may, see Proposition 2.2.8(3)) so that $\Phi^+ = \Phi(G,T)_{\lambda > 0}$. □

Corollary 3.3.13 *Let G be a pseudo-split pseudo-reductive group over a field k and let T be a split maximal k-torus in G. Let Φ^+ denote a positive system of roots in $\Phi = \Phi(G,T)$.*

(1) *Let $\Psi \subseteq \Phi^+$ be a closed set of roots, and let $\Psi_{\rm nd}$ be the set of elements of Ψ not in 2Ψ. The set $\{U_c\}_{c\in\Psi_{\rm nd}}$ directly spans, in any order, a T-stable smooth connected unipotent k-subgroup U_Ψ of G. This is the unique T-stable smooth connected k-subgroup of G whose Lie algebra is the span of the subspaces* $\mathrm{Lie}(U_c)$ *for all $c \in \Psi$.*

(2) *For any (possibly dependent) $a, b \in \Phi^+$, let (a, b) denote the closed set of roots*

$$(\langle a\rangle + \langle b\rangle) \cap \Phi^+ = \{c \in \Phi^+ \mid c = ma + nb \text{ for some integers } m, n \geqslant 1\}.$$

Then $(U_a, U_b) \subseteq U_{(a,b)}$.

It is understood that $U_\Psi = 1$ when Ψ is empty.

Proof Let $A \subseteq \mathrm{X}(T)$ denote the subsemigroup generated by Ψ, so by Proposition 2.2.7 we have $\Psi = \Phi \cap A$ (and hence $\Phi^+ \cap A = \Psi$ as well). Since $\Phi^+ = \Phi_{\lambda>0}$ for some $\lambda \in \mathrm{X}_*(T)$, we have $A \subseteq \mathrm{X}(T)_{\lambda>0}$, so $0 \notin A$. Hence, $H_A(G)$ is unipotent. We apply Theorem 3.3.11 to $H_A(G)$ and the set $\Psi_{\rm nd}$ decomposed into a disjoint union of singletons $\Psi_i = \{a_i\}$ for an enumeration $\{a_1, \ldots\}$ of $\Psi_{\rm nd}$. Since $\Psi_{\rm nd}$ generates the semigroup A and $A \cap \Phi = \Psi$, $H_A(G)$ satisfies the requirements to be U_Ψ. The unique characterization of U_Ψ follows from the uniqueness in Proposition 3.3.6 (since Ψ generates A as a semigroup). This proves (1).

To prove (2), since $U_a = H_{\langle a\rangle}(G)$ and $U_b = H_{\langle b\rangle}(G)$ we have $(U_a, U_b) \subseteq H_{\langle a\rangle+\langle b\rangle}(G) = U_{(a,b)}$. □

Example 3.3.14 Consider the set Φ' of non-multipliable roots in $\Phi := \Phi(G,T)$, where G is pseudo-reductive over a separably closed field k and T is a maximal k-torus of G. The root system Φ' is naturally identified with the root system $\Phi(G_{\overline{k}}^{\rm red}, T_{\overline{k}})$. The construction $\Phi^+ \mapsto \Phi^+ \cap \Phi'$ defines a bijection from the set of positive systems of roots of Φ to those of Φ'.

For a fixed choice of Φ^+ we claim that the associated positive system of roots Φ'^+ in Φ' is closed in Φ. This is only of interest when Φ is non-reduced (so k is imperfect of characteristic 2), as otherwise we are in the trivial case $\Phi' = \Phi$.

Choose distinct $c, c' \in \Phi'^+$, so c and c' are linearly independent, and assume $c+c' \in \Phi^+$. We have to show that $c+c'$ is not multipliable in Φ. Since

$c + c' \in \Phi$, the linearly independent roots c and c' lie in the same irreducible component of Φ. Using the natural bijection between the set of irreducible components of Φ and those of Φ', we may assume that Φ is irreducible. Let $r \geqslant 2$ denote its rank. We may also assume that $c + c'$ is multipliable, so Φ is non-reduced.

By [Bou2, VI, §4.14], there is a unique non-reduced irreducible root system of rank r. Relative to any positive system of roots, it has a basis of simple positive roots $\{a_1, \ldots, a_r\}$ such that the positive roots are of two types: the roots of the first type are $\sum_{j=i}^{i'} a_j$ with $1 \leqslant i \leqslant i' \leqslant r$ and the roots of the second type are

$$\sum_{j=i}^{i'-1} a_j + 2\sum_{j=i'}^{r} a_j$$

with $1 \leqslant i \leqslant i' \leqslant r$. The divisible positive roots are precisely the ones of the second type with $i' = i$; that is, the roots $2\sum_{j=i}^{r} a_j$ with $1 \leqslant i \leqslant r$. Hence, the multipliable positive roots are $\sum_{j=i}^{r} a_j$ with $1 \leqslant i \leqslant r$. We have to show that if $r \geqslant 2$ then there is no pair of distinct positive non-multipliable roots c and c' such that $c + c'$ is multipliable. (This is also true, and trivial, when $r = 1$.)

Suppose such c, c' exist. Since multipliable positive roots have all nonzero coefficients equal to 1 relative to the basis of simple positive roots, necessarily c and c' are positive roots of the first type. Also, since $c + c'$ is assumed to be multipliable, by swapping c and c' if necessary we must have

$$c = \sum_{j=i}^{i'-1} a_j, \quad c' = \sum_{j=i'}^{r} a_j$$

with $i < i' \leqslant r$. But then c' is multipliable, a contradiction.

Generalizing the relationship between unipotent radicals of Borel subgroups and positive systems of roots in the setting of split connected reductive groups over a field, we wish to express $\mathscr{R}_{u,k}(P)$ as a direct span (in any order) of root groups associated to the non-divisible elements in a positive system of roots, with P any minimal pseudo-parabolic k-subgroup of G such that $T \subseteq P$. As a first step, we now show that such P actually exist by proving the equivalence of two senses of minimality: for any pseudo-parabolic k-subgroup P of G containing T such that P is minimal with respect to the property of containing T, necessarily P is minimal as a pseudo-parabolic k-subgroup of G. Indeed, we pick a pseudo-parabolic k-subgroup Q of G contained in such a P and will prove that $Q = P$. Let S be a maximal k-torus in Q (so it is also maximal in G, as Q contains $Z_G(\mu)$ for some 1-parameter subgroup μ over k), so T and S are maximal k-tori in P. By Proposition A.2.10, some $g \in P(k_s)$ conjugates S_{k_s} into T_{k_s}, so it conjugates Q_{k_s} to a pseudo-parabolic k_s-subgroup Q' of G_{k_s} contained in P_{k_s} and containing T_{k_s}. By Proposition 2.2.11, $Q' = P_{G_{k_s}}(\mu')$ for some $\mu' \in \mathrm{X}_*(T_{k_s})$. But μ' is defined over k since T is k-split, so Q' descends to a pseudo-parabolic k-subgroup of G that is contained in P and contains T. The minimality hypothesis on P then forces this k-descent of Q' to equal P, so $Q' = P_{k_s}$ and hence $Q = P$ as desired.

Now we are in position to relate minimal pseudo-parabolic k-subgroups P containing T with positive systems of roots in $\Phi(G,T)$. In any root system, there are finitely many choices of a positive system of roots, and the finite Weyl group of the root system acts simply transitively on this finite set of positive systems, by [Bou2, VI, §1.5, Thm. 2(i)] and [Bou2, VI, §1.7, Cor. 1]. In a split connected reductive group over a field, the set of Borel subgroups containing a given split maximal torus T is in natural bijection with the set of positive systems of roots in the associated (reduced) root system (and correspondingly the Weyl group associated to T acts simply transitively on the finite set of Borel subgroups containing T). The analogous result for minimal pseudo-parabolic subgroups in pseudo-split pseudo-reductive groups over a field is given by:

Proposition 3.3.15 *Let k be a field, and let T be a split maximal k-torus in a non-commutative pseudo-split pseudo-reductive k-group G.*

(1) *The minimal pseudo-parabolic k-subgroups of G that contain T are precisely the subgroups $P = P_G(\lambda)$ for $\lambda \in \mathrm{X}_*(T)$ such that the set*

$$\Phi(P,T) = \Phi_{\lambda \geqslant 0} := \{a \in \Phi(G,T) \mid \langle a, \lambda \rangle \geqslant 0\}$$

of nontrivial T-weights occurring on $\mathrm{Lie}(P)$ *is a positive system of roots in $\Phi := \Phi(G,T)$.*

(2) *Assigning to each such P the set of roots in Φ appearing in* $\mathrm{Lie}(P)$ *defines a bijection between the set of such P and the set of positive systems of roots in Φ. The group $W(G,T)(k)$ acts simply transitively on the set of such P.*

(3) *For any minimal pseudo-parabolic k-subgroup P of G, the maximal pseudo-reductive quotient $P/\mathscr{R}_{u,k}(P)$ over k is commutative.*

Proof By Proposition 2.2.11, if P is a pseudo-parabolic k-subgroup of G containing T then $P = P_G(\lambda)$ for some k-rational cocharacter λ of T. For any such λ, the set of roots $\Phi_{\lambda \geqslant 0}$ is the set of weights for T on $P_G(\lambda)$ (Proposition 2.1.8(1)). By Proposition 2.2.8, it contains a positive system of roots Φ^+ of $\Phi(G, T)$.

STEP 1. Fix a positive system of roots Φ^+ contained in $\Phi_{\lambda \geqslant 0}$, so there exists $\mu \in X_*(T)$ such that

$$\Phi^+ = \{a \in \Phi \mid \langle a, \mu \rangle > 0\}.$$

All other elements of Φ are contained in $-\Phi^+$, so $\Phi^+ = \Phi_{\mu \geqslant 0}$. Consider $P_G(\mu)$. We claim that $P_G(\mu) \subseteq P_G(\lambda)$. It suffices to show that $Z_G(\mu) \subseteq Z_G(\lambda)$ and $U_G(\mu) \subseteq P_G(\lambda)$.

Since $Z_G(\mu)$ contains T, so it contains the image of λ, the containment $Z_G(\mu) \subseteq Z_G(\lambda)$ holds provided that $Z_G(\mu)$ is commutative. The k-group $Z_G(\mu)$ is pseudo-reductive, so by Proposition 1.2.3 it is commutative provided that T is a central subgroup. By Proposition 2.1.8(1), the T-weights in the Lie algebra of $U_G(\pm\mu)$ are $\pm\Phi^+$, so the only weight occurring for T in $\mathrm{Lie}(Z_G(\mu))$ is the trivial weight. Hence, by Corollary A.8.11, T is indeed central in $Z_G(\mu)$.

To prove $U_G(\mu) \subseteq P_G(\lambda)$, it suffices (by Corollary 3.3.12) to show that $U_a \subseteq P_G(\lambda)$ for any non-divisible $a \in \Phi(G, T)$ satisfying $\langle a, \mu \rangle > 0$ (i.e., $a \in \Phi^+$). Such a lie in $\Phi_{\lambda \geqslant 0}$, so by Corollary 3.3.12 we have $U_a = U_{(a)} \subseteq P_G(\lambda)$.

With the containment $P_G(\mu) \subseteq P_G(\lambda)$ proved, we see that if $P_G(\lambda)$ is a minimal pseudo-parabolic k-subgroup then this containment is an equality. In particular, $\Phi_{\lambda \geqslant 0} = \Phi^+$ when $P := P_G(\lambda)$ is minimal. In this case, $U_G(\lambda) = U_G(\mu)$ and

$$P_G(\lambda)/U_G(\lambda) = P_G(\mu)/U_G(\mu) \simeq Z_G(\mu)$$

is a commutative pseudo-reductive group. This proves (3).

STEP 2. Conversely, if $\Phi_{\lambda \geqslant 0}$ is a positive system of roots Φ^+ of $\Phi(G, T)$ then we claim that a pseudo-parabolic k-subgroup P of G contained in $P_G(\lambda)$ must equal $P_G(\lambda)$. To prove this we may and do assume $k = k_s$. The hypothesis says that the set of nontrivial T-weights that occurs in $\mathrm{Lie}(P_G(\lambda))$ is a positive system of roots in $\Phi(G, T)$, and this is preserved if we change λ and T by conjugation by a common k-point of $P_G(\lambda)$. Hence, we can assume that T is contained in P, since maximal k-tori in a pseudo-parabolic k-subgroup of G are maximal k-tori of G and all maximal k-tori in $P_G(\lambda)$ are $P_G(\lambda)(k)$-conjugate (as $k = k_s$, so Proposition A.2.10 applies). By Proposition 2.2.11 we may therefore write $P = P_G(\lambda')$ for some k-homomorphism $\lambda' : \mathrm{GL}_1 \to T$, so the containment $\mathrm{Lie}(P) \subseteq \mathrm{Lie}(P_G(\lambda))$ implies

$$\Phi_{\lambda' \geqslant 0} \subseteq \Phi_{\lambda \geqslant 0} = \Phi^+.$$

But $\Phi_{\lambda' \geqslant 0}$ is a parabolic set of roots, so it contains a positive system of roots. All positive systems of roots have the same size, so $\Phi_{\lambda' \geqslant 0} = \Phi^+$. Thus, $\mathrm{Lie}(P)$ and $\mathrm{Lie}(P_G(\lambda))$ are the spans in $\mathrm{Lie}(G)$ for the weight spaces for T associated to a common set of weights (namely $\Phi^+ \cup \{0\}$), so $\mathrm{Lie}(P) = \mathrm{Lie}(P_G(\lambda))$. Hence, the containment $P \subseteq P_G(\lambda)$ is an equality, so $P_G(\lambda)$ is indeed minimal.

Step 3. Finally, we show that $W(G,T)(k)$ acts simply transitively on the set of minimal parabolic k-subgroups containing T. The Weyl group of any root system acts simply transitively on the set of positive systems of roots, so by Proposition 3.2.7 the group $W(G,T)(k)$ acts simply transitively on the set of positive systems of roots in $\Phi(G,T)$. Since we have shown that every minimal pseudo-parabolic k-subgroup containing T has its Lie algebra equal to the span of the weight spaces for a positive system of roots together with the trivial weight, it remains to show that any minimal pseudo-parabolic k-subgroup P in G containing T is determined by $\mathrm{Lie}(P) \subseteq \mathrm{Lie}(G)$.

We saw at the outset (using that T is k-split) that we can choose $\lambda : \mathrm{GL}_1 \to T$ such that $P = P_G(\lambda)$. The centralizer $Z_G(\lambda) \simeq P/\mathscr{R}_{u,k}(P)$ is commutative (since (3) has already been proved), so it equals $Z_G(T)$. But

$$P = Z_G(\lambda) \ltimes U_G(\lambda) = Z_G(T) \ltimes \mathscr{R}_{u,k}(P),$$

so we just have to show that $U_G(\lambda)$ is determined by $\mathrm{Lie}(P)$. The k-group $U_G(\lambda)$ is generated by the root groups U_a for non-divisible $a \in \Phi_{\lambda > 0}$, and these a are exactly the non-divisible nontrivial T-weights on $\mathrm{Lie}(P)$ (since $Z_G(\lambda) = Z_G(T)$), so $U_G(\lambda)$ is determined by $\mathrm{Lie}(P)$ and T. □

Corollary 3.3.16 *Let G be a pseudo-split pseudo-reductive group over a field k, and let T be a split maximal k-torus in G. Let Φ^+ be a positive system of roots in $\Phi(G,T)$, and let $\{c_i\}$ be an enumeration of the non-divisible elements of Φ^+. Let P be the minimal pseudo-parabolic k-subgroup of G containing T that corresponds to Φ^+.*

The multiplication map of k-scheme $\prod_i U_{c_i} \to G$ is an isomorphism onto $\mathscr{R}_{u,k}(P)$, and the multiplication map of k-schemes

$$\prod_i U_{c_i} \times Z_G(T) \times \prod_i U_{-c_i} \to G \qquad (3.3.2)$$

is an open immersion.

Proof To prove that the k-groups U_{c_i} directly span $\mathscr{R}_{u,k}(P)$ in any order, we apply Corollary 3.3.12 with $\lambda \in \mathrm{X}_*(T)$ such that Φ^+ is the set of $c \in \Phi(G,T)$ satisfying $\langle c, \lambda \rangle > 0$. Such a λ is provided by Proposition 2.2.8(3). Combining this with Proposition 2.1.8(2),(3) yields the open immersion claim. □

We often refer to the open subscheme (3.3.2) in G as the *open Bruhat cell* of G (determined by T and Φ^+), though sometimes we may put the positive root groups on the right and the negative root groups on the left; the context will make this clear (if it matters). The open Bruhat cell depends on Φ^+, but we omit mention of this choice if it is clear from the context (or does not matter).

The following application of Proposition 3.3.15 will be useful later. For an analogous result over an arbitrary field see Theorem C.2.5.

Proposition 3.3.17 *Let G be a smooth connected affine group over a separably closed field k. All minimal pseudo-parabolic k-subgroups of G are $G(k)$-conjugate.*

Proof Since $k = k_s$, so smooth surjections between k-schemes induce surjections on k-points, by Proposition 2.2.10 we may replace G with $G/\mathscr{R}_{u,k}(G)$ to reduce to the case where G is pseudo-reductive over k. Let P and Q be minimal pseudo-parabolic k-subgroups of G. By definition of pseudo-parabolicity, a maximal k-torus in P is also maximal in G (and likewise for Q). The $G(k)$-conjugacy of maximal k-tori in G (Proposition A.2.10) thereby reduces us to the case where P and Q contain a common maximal k-torus T. Proposition 3.3.15(2) now completes the argument. □

3.4 Bruhat decomposition and Levi subgroups

If G is a smooth connected affine group over a field k, then a *Levi k-subgroup* of G is a smooth connected k-subgroup $L \subseteq G$ such that for the quotient map $\pi : G_{\overline{k}} \twoheadrightarrow G' := G_{\overline{k}}^{\mathrm{red}}$, the restriction $\pi|_{L_{\overline{k}}}$ is an isomorphism (i.e., $L_{\overline{k}} \ltimes \mathscr{R}_u(G_{\overline{k}}) = G_{\overline{k}}$). It is a well-known fact (due to G.D. Mostow) that if k has characteristic 0 then Levi k-subgroups always exist. However, over every algebraically closed field with nonzero characteristic there are smooth connected affine groups not admitting a Levi subgroup. For the convenience of the interested reader, a family of counterexamples is given in §A.6 (which will not be used in what follows).

The main results of this section are that for a pseudo-split pseudo-reductive group G over a field k, Levi k-subgroups always exist and there is a Bruhat decomposition for $G(k)$ relative to a choice of minimal pseudo-parabolic k-subgroup of G. The Bruhat decomposition of $G(k)$ for any smooth connected affine group G over an arbitrary field k will be given in Theorem C.2.8. The existence of Levi k-subgroups in the pseudo-reductive case over $k = k_s$ will be especially useful in our classification of pseudo-reductive groups in characteristic 2 (and will provide some motivation in our classification work in characteristic 3). The hypothesis that G is pseudo-split is necessary for the existence of Levi k-subgroups; in Example 7.2.2 we construct standard pseudo-reductive k-groups having no Levi k-subgroup.

The construction of Bruhat decompositions and Levi subgroups in the pseudo-split case rests on the following two results.

Proposition 3.4.1 *Let G be a pseudo-split pseudo-reductive group over a field k, and let T be a split maximal k-torus in G. Choose $a \in \Phi(G,T)$, define $T_a = (\ker a)^0_{\mathrm{red}}$, and let $U_{\pm a}$ be the root groups in G associated to $\pm a$ as in Definition 2.3.13.*

(1) *The smooth connected k-subgroup G_a generated by U_a and U_{-a} is pseudo-reductive and perfect with split maximal k-torus $S = T \cap G_a$, and $(G_a)^{\mathrm{red}}_{\overline{k}}$ is of type* A_1. *The roots $\pm a \in \Phi(G_a, S)$ are not divisible.*
(2) *If a is not divisible in $\Phi(G,T)$ then $G_a = \mathscr{D}(Z_G(T_a))$.*
(3) *Assume a is divisible in $\Phi(G,T)$ (so k is imperfect and* $\mathrm{char}(k) = 2$). *For $\mu := S[2]$, the identity component $H_a = Z_{Z_G(T_a)}(\mu)^0$ is pseudo-split pseudo-reductive with $\Phi(H_a, T) = \{\pm a\}$, and H_a is generated by $Z_G(T)$ and G_a. Moreover, $G_a = \mathscr{D}(H_a)$ for such a.*

Note that in (3), the k-group scheme H_a is pseudo-reductive; this follows from Proposition A.8.14(2) (taking G and H there to be $Z_G(T_a)$ and μ respectively in the above notation). By Remark A.8.15, in part (3) above the centralizer $Z_{Z_G(T_a)}(\mu)$ is actually connected (since the maximal reductive quotient of $Z_G(T_a)_{\overline{k}}$ has derived group SL_2 in the divisible case). We will not use this.

Proof Since $Z_G(T_a)$ is pseudo-reductive (Proposition 1.2.4), we can rename it as G to reduce to the case where T_a is central in G. By Corollary A.2.7 $S = T \cap \mathscr{D}(G)$ is a (split) maximal k-torus in the derived group $\mathscr{D}(G)$, and $\mathscr{D}(G)^{\mathrm{red}}_{\overline{k}} = \mathscr{D}(G^{\mathrm{red}}_{\overline{k}})$ (Proposition A.4.8). The set of roots for $(\mathscr{D}(G^{\mathrm{red}}_{\overline{k}}), S)$ coincides with the set of roots for $(G^{\mathrm{red}}_{\overline{k}}, T_{\overline{k}})$, which in turn is the set of non-multipliable roots in $\Phi(G,T)$ (Theorem 2.3.10). This is $\{\pm a\}$ when a is not multipliable, and $\{\pm 2a\}$ otherwise, so it has rank 1. Hence, $\dim S = 1$. (This also follows from Proposition 3.2.3.) It is clear that $a|_S \neq 1$, so $T = S \cdot T_a$. By Proposition 3.2.3, $S = a^\vee(\mathrm{GL}_1)$.

If a is not divisible in $\Phi(G,T)$ then $U_{\pm a} = U_{(\pm a)}$, so it follows from Lemma 3.1.5 that in such cases U_a and U_{-a} generate $\mathscr{D}(G)$. In the root system $\Phi(\mathscr{D}(G), S)$ of rank 1 the roots $\pm a$ are clearly not divisible when a is not divisible in $\Phi(G,T)$.

Now suppose that a is divisible in $\Phi(G,T)$, so k is imperfect with characteristic 2 and $\Phi(G,T) = \{\pm a, \pm a/2\}$. In this case, the non-multipliable roots $\pm a$ are the $T_{\overline{k}}$-roots for $G_{\overline{k}}^{\mathrm{red}}$, and hence the $S_{\overline{k}}$-roots for its derived group $\mathscr{D}(G)_{\overline{k}}^{\mathrm{red}}$ (see Proposition A.4.8). But $\mathscr{D}(G)_{\overline{k}}^{\mathrm{red}}$ is semisimple of type A_1, so the divisibility of the root a in its character lattice $\mathrm{X}(S_{\overline{k}})$ implies that either of $\pm a/2$ generates $\mathrm{X}(S)$. Hence, by Proposition A.8.10 and Proposition A.8.14(2),(3) it follows that the k-subgroup $H := Z_G(S[2])^0$ is pseudo-reductive with T-roots consisting of those for (G,T) that are trivial on $S[2]$. Thus, $\Phi(H,T) = \{\pm a\}$, so Lie algebra considerations imply that H is generated by $Z_G(T)$, U_a, and U_{-a}. Applying Lemma 3.1.5 to H gives that $\mathscr{D}(H)$ is generated by U_a and U_{-a}. In particular, $\Phi(\mathscr{D}(H), S) = \{\pm a\}$, so $\pm a$ are not divisible in $\Phi(\mathscr{D}(H), S)$. □

Proposition 3.4.2 (Tits) *Let k be a field, G a pseudo-split pseudo-reductive k-group, and T a split maximal k-torus of G. For $a \in \Phi := \Phi(G,T)$ let the root groups $U_{\pm a}$ be as in Definition 2.3.13.*

Fix a and $u \in U_a(k) - \{1\}$. There exist unique $u', u'' \in U_{-a}(k)$ such that $u'uu''$ normalizes T. The induced action of $m(u) := u'uu''$ on the character group $\mathrm{X}(T)$ is the reflection r_a associated to the root a, $u' = m(u)um(u)^{-1} = u''$, and $m(u)^2 = a^\vee(-1) \in T(k)$.

For $z \in Z_G(T)(\overline{k})$, if $zuz^{-1} \in U_a(k)$ then $zu'z^{-1} \in U_{-a}(k)$ and $m(zuz^{-1}) = zm(u)z^{-1}$.

Proof By arguing separately with Proposition 3.4.1 depending on whether or not a is divisible, the smooth connected k-subgroup of G generated by $C := Z_G(T)$, U_a, and U_{-a} is pseudo-reductive, and by replacing G with this k-subgroup (as we may) we gain the properties that $G_{\overline{k}}^{\mathrm{red}}$ has a root system of rank 1 with root groups $U_{\pm a} = U_{(\pm a)}$ (so $\pm a$ are now not divisible). Let $U = U_a$ and $U' = U_{-a}$, so Proposition 3.4.1(2) shows that $\mathscr{D}(G)$ is generated by U and U'. For the Cartan k-subgroup C, the multiplication map $U' \times C \times U \to G$ is an open immersion (by Proposition 2.1.8(2),(3)). By Proposition 3.2.7, the normalizer $N = N_G(T)$ has two connected components: C and $N - C$. Galois theory and uniqueness for u' and u'' allow us to assume $k = k_s$, so we can choose $n \in N(k) - C(k)$. Thus, $N - C = nC$.

Let Y be the preimage of $N - C$ under the multiplication map $U' \times U \times U' \to G$. Note that C acts on Y through conjugation on all factors. The condition $u'uu'' = nc$ can be rewritten as $n^{-1}u = (n^{-1}u'^{-1}n)cu''^{-1}$. Since $U \times C \times U'$ is open in G via multiplication and $U' = nUn^{-1}$, the C-equivariant projection $Y \to U$ is an isomorphism onto the open subscheme $\Omega \subseteq U$ consisting of points u of U such that $n^{-1}u$ lies in the open Bruhat cell $U \times C \times U'$ of $(G, T, \{a\})$. Thus, if $(v', v, v'') \in Y(\overline{k})$ and $v \in U(k)$ then v' and v'' are uniquely determined and lie in $U'(k)$.

Letting $R = \mathscr{R}_u(G_{\overline{k}})$, we claim that $\Omega_{\overline{k}} = U_{\overline{k}} \cap (G_{\overline{k}} - R)$. The pseudo-reductive derived group $H = \mathscr{D}(G)$ contains $U_{\pm a}$ and has maximal k-torus $S := T \cap H$ with dimension 1, so $a|_S \neq 1$. Since $T = S \cdot T_a$ for the central codimension-1 torus $T_a = (\ker a)^0_{\mathrm{red}}$, a point in $H(k)$ normalizes (resp. centralizes) T if and only if it normalizes (resp. centralizes) S. Thus, to justify the description of Ω and to prove all other claims we may replace G with H so that G is perfect with $G^{\mathrm{red}}_{\overline{k}}$ semisimple of type A_1 (i.e., SL_2 or PGL_2).

The equality $\Omega_{\overline{k}} = U_{\overline{k}} \cap (G_{\overline{k}} - R)$ is a trivial calculation when $k = \overline{k}$ for the groups SL_2 and PGL_2 (with $R = \{1\}$). For the general case, it suffices to check that the open cell $(UCU')_{\overline{k}}$ in $G_{\overline{k}}$ is the full preimage of its image under the projection $\pi : G_{\overline{k}} \to G_{\overline{k}}/R = G^{\mathrm{red}}_{\overline{k}}$. Equivalently, if we choose $\lambda : \mathrm{GL}_1 \to T$ such that $\langle a, \lambda \rangle < 0$ and define $P' := P_G(\lambda) = CU'$ then we want $U_{\overline{k}}P'_{\overline{k}}R = U_{\overline{k}}P'_{\overline{k}}$. Let $\mu : \mathrm{GL}_1 \times R \to R$ denote the left action induced by conjugation on R by $\lambda_{\overline{k}}$. By Remark 2.1.11 and Proposition 2.1.12(1), the multiplication map $U_R(-\mu) \times P_R(\mu) \to R$ is an isomorphism, with $U_R(-\mu) = U_G(-\lambda)_{\overline{k}} \cap R = U_{\overline{k}} \cap R$ and $P_R(\mu) = P_G(\lambda)_{\overline{k}} \cap R = P'_{\overline{k}} \cap R$ both smooth connected $\overline{k}$-subgroups of R. From this decomposition of R (and the normality of R in $G_{\overline{k}}$) we see that $U_{\overline{k}}P'_{\overline{k}}R = U_{\overline{k}}RP'_{\overline{k}} = U_{\overline{k}}P'_{\overline{k}}$, as desired.

Since $\ker(\pi|_{U(k)})$ and $\ker(\pi|_{U'(k)})$ are normalized by $T(k)$, either would be infinite if nontrivial, due to the existence of linear structures on U and U' compatible with the T-action. But $R(\overline{k}) \cap G(k)$ is finite since G is pseudo-reductive, so π is injective on $U(k)$ and $U'(k)$. Thus, $U(k) - \{1\} \subseteq \Omega(k)$ and so for each $u \in U(k) - \{1\}$ there exist unique $u', u'' \in U'(k)$ such that $m(u) := u'uu'' \in N(k) - C(k)$. To prove uniqueness only requiring $u'uu'' \in N(k)$, we need $u'uu'' \notin C(k)$ for any $u', u'' \in U'(k)$. This is trivial for $G^{\mathrm{red}}_{\overline{k}}$ (i.e., SL_2 or PGL_2), so it holds for G since $\pi(u) \neq 1$. Likewise, since $\pi|_{U'(k)}$ is injective, the identities $u' = u''$ and $u' = m(u)um(u)^{-1}$ in $U'(k)$ follow from the trivial case of $G^{\mathrm{red}}_{\overline{k}}$. This argument also shows $\pi(m(u)) = m(\pi(u))$, and the C-action on $Y \simeq \Omega \subseteq U$ implies that if $c \in C(\overline{k})$ and $cuc^{-1} \in U(k)$ then $cu'c^{-1} \in U'(k)$ and $m(cuc^{-1}) = cm(u)c^{-1}$.

Finally, we prove $m(u)^2 = a^\vee(-1)$. If $\mathrm{char}(k) = 2$ then u and $u' = u''$ have order 2, so clearly $m(u)^2 = 1$. Now assume $\mathrm{char}(k) \neq 2$, so $\Phi = \{\pm a\}$. We may and do assume $k = k_s$, so we can choose $t \in T(k)$ such that $a(t) = -1$; this uses that $\mathrm{char}(k) \neq 2$ and either a or $a/2$ generates $\mathrm{X}(T)$. Conjugation by t carries the points $u \in U(k)$ and $u' = u'' \in U'(k)$ to their inverses, so $tm(u)^{-1}t^{-1} = m(u)$ and hence $m(u)^2 = m(u)tm(u)^{-1}t^{-1}$. This lies in $T(k)$ since $m(u) \in N(k)$. The desired identity $m(u)^2 \stackrel{?}{=} a^\vee(-1)$ therefore compares two elements of $T(k)$. Since π carries $T_{\overline{k}}$ isomorphically onto a maximal torus in $G_{\overline{k}}^{\mathrm{red}}$, and π identifies Φ with $\Phi(G_{\overline{k}}^{\mathrm{red}}, T_{\overline{k}})$, it suffices to check the equality for $G_{\overline{k}}^{\mathrm{red}}$ (i.e., SL_2 or PGL_2), which is trivial. □

Corollary 3.4.3 *Let G be a pseudo-split pseudo-reductive group over a field k, and T a split maximal k-torus in G.*

(1) *The inclusion $N_G(T)(k)/Z_G(T)(k) \to W(G,T)(k)$ is bijective.*
(2) *If G is also absolutely pseudo-simple then for any two roots $a, a' \in \Phi(G,T)$ with the same length there exists $n \in N_G(T)(k)$ whose conjugation action on G carries a to a'.*

Proof The group $W(G,T)(k)$ coincides with the Weyl group of $\Phi(G,T)$ (Proposition 3.2.7), and so it is generated by the reflections r_a associated to the roots. But for each such a we can choose a nontrivial $u \in U_a(k)$ since U_a is a nonzero vector group over k, and so the element $m(u) \in N_G(T)(k)$ as in Proposition 3.4.2 maps to r_a. This proves (1).

Assertion (2) follows from (1) and the fact that $\Phi(G,T)$ is an irreducible root system when G is absolutely pseudo-simple since the Weyl group of an irreducible (possibly non-reduced) root system acts transitively on the set of roots of a given length. □

Example 3.4.4 To appreciate how remarkable it is that $N_G(T)(k) \to W(G,T)(k)$ is surjective when G is pseudo-split, we explain why naive Galois cohomology considerations cannot yield the result. The obstructions to surjectivity lie in $\mathrm{H}^1(k, Z_G(T))$, and this cohomology group can be nonzero, even when G is perfect and standard. To give some examples, let k be an imperfect field of characteristic $p > 0$, and choose purely inseparable nontrivial finite extensions $k''/k'/k$ so that $k' = k^{1/p} \cap k''$ and $[k'' : k'] = p$. Let $G = \mathrm{R}_{k''/k}(\mathrm{SL}_p)/\mathrm{R}_{k'/k}(\mu_p)$. Letting D denote the diagonal k''-torus of SL_p, the k-subgroup $C := \mathrm{R}_{k''/k}(D)/\mathrm{R}_{k'/k}(\mu_p)$ in G is a Cartan k-subgroup (by Proposition A.2.8) and it clearly contains a k-split maximal k-torus. The pseudo-reductivity of G is a special case of the self-contained argument in Example 5.3.7.

The explicit k''-isomorphism $D \simeq \mathrm{GL}_1^{p-1}$ as given in Example 4.2.6 implies that the k-group C admits as a direct factor

$$Q := \mathrm{R}_{k''/k}(\mathrm{GL}_1)/\mathrm{R}_{k'/k}(\mu_p) \simeq \mathrm{coker}(\mathrm{R}_{k'/k}(\mathrm{GL}_1) \xrightarrow{(i,\mathrm{N})} \mathrm{R}_{k''/k}(\mathrm{GL}_1) \times \mathrm{GL}_1),$$

where i is the natural map and $\mathrm{N} = \mathrm{N}_{k'/k}$ is the norm, so it suffices that $\mathrm{H}^1(k, Q) \neq 0$. The smooth presentation of Q yields an exact sequence in Galois cohomology

$$\mathrm{H}^1(k, Q) \to \mathrm{Br}(k') \to \mathrm{Br}(k'') \times \mathrm{Br}(k),$$

so if $\mathrm{Br}(k') \to \mathrm{Br}(k'') \times \mathrm{Br}(k)$ is not injective then we are done. This happens for the rational function field $k = k_0(x, y)$ over an algebraically closed field k_0 with $k' = k(x^{1/p})$ and $k'' = k(x^{1/p^2})$. (Setting $u = x^{1/p}$ and $v = y$, this is the natural map $\mathrm{Br}(K) \to \mathrm{Br}(K(u^{1/p})) \times \mathrm{Br}(K(v^{1/p}))$ for $K = k_0(u, v)$. Let $f(T) = T^p - T - v/u$. The degree-p cyclic splitting field K'/K of f has generator $\sigma \in \mathrm{Gal}(K'/K)$ given by $\sigma : r \mapsto r + 1$ for a root r of f, and r has norm v/u in K whereas u and v are not norms from K' (as r is algebraic over $k(v/u)$ but v/u is algebraically independent from u and v over k). The cyclic algebra in $\mathrm{Br}(K)$ defined by $\sigma^p = u$ is non-split but splits over $K(u^{1/p})$ and $K(v^{1/p})$.)

Here is the Bruhat decomposition in the pseudo-split case (e.g., $k = k_s$); see Theorem C.2.8 for a generalization to arbitrary smooth connected affine groups.

Theorem 3.4.5 (Borel–Tits) *Let G be a pseudo-split pseudo-reductive group over a field k, T a split maximal k-torus, and P a minimal pseudo-parabolic k-subgroup of G containing T. The natural map $N_G(T)(k)/Z_G(T)(k) \to P(k)\backslash G(k)/P(k)$ is bijective.*

Proof Let $N = N_G(T)$, $Z = Z_G(T)$, and $\Phi = \Phi(G, T)$. By Corollary 3.4.3(1), $W := N(k)/Z(k)$ is equal to $W(G, T)(k)$ via the natural map. For $a \in \Phi$, let the root group U_a be as in Proposition 2.3.11, and define M_a to be the set of elements of $N(k)$ whose image in the Weyl group $W(G, T)(k)$ is the reflection r_a in a. Let Φ^+ be the positive system of roots $\Phi(P, T)$ (see Proposition 3.3.15(2)), and $\Phi^- = -\Phi^+$. Let $\Delta \subseteq \Phi^+$ be the basis of Φ given by the simple positive roots. Choose a 1-parameter k-subgroup $\lambda : \mathrm{GL}_1 \to T$ such that $P = P_G(\lambda)$ (Proposition 2.2.11).

We claim that the datum $(Z(k), (U_a(k), M_a)_{a\in\Phi})$ is a generating root datum (*donnée radicielle génératrice*) of type Φ in $G(k)$ in the sense of [BrTi, (6.1.1)]. This requires verifying a list of six axioms (DR1)–(DR6) and a group generation claim, as follows. Axiom (DR1) is the obvious fact that $Z(k)$ is a subgroup of $G(k)$ and each $U_a(k)$ is a nontrivial subgroup of $G(k)$. Axiom (DR2) (which needs to be verified only when b is not a positive multiple of $-a$) is an immediate consequence of Corollary 3.3.13(2), and (DR3) is the obvious fact that $U_{2a}(k)$ is a proper subgroup of $U_a(k)$ when a and $2a$ lie in Φ. By Proposition 3.4.2, for all $a \in \Phi$ we have $U_{-a}(k) - \{1\} \subseteq U_a(k)M_aU_a(k)$ with M_a a right $Z(k)$-coset in $G(k)$; this is precisely (DR4). The definition of M_a implies that $nU_bn^{-1} = U_{r_a(b)}$ for all $a, b \in \Phi$ and $n \in M_a$, which is (DR5). Finally, by Corollary 3.3.12, the subgroup $U^+ \subseteq G(k)$ generated by the groups $U_a(k)$ for $a \in \Phi^+$ is $U_G(\lambda)(k)$ and the subgroup $U^- \subseteq G(k)$ generated by the groups $U_a(k)$ for $a \in \Phi^-$ is $U_G(-\lambda)(k)$. Hence, $Z(k)U^+ = P(k)$, so $Z(k)U^+ \cap U^- = \{1\}$ by Proposition 2.1.8(3). The triviality of this intersection is (DR6).

Having verified the axioms, the remaining condition to verify is that $Z(k)$ and the subgroups $U_a(k)$ generate $G(k)$. We certainly get the k-points of the open Bruhat cell $\Omega = U_G(-\lambda)ZU_G(\lambda)$ (with $\Omega(k) = U^-Z(k)U^+$). Thus, it suffices to show that the dense open $V := g\Omega \cap \Omega$ has a k-point for each $g \in G(k)$. If $k = k_s$ (the only case we need) this is obvious. In general, since V is stable under right translation by $P = P_G(\lambda) = ZU_G(\lambda)$, it is the pullback of the dense open V/P in G/P that is contained in Ω/P. By the self-contained Lemma C.2.1 (whose proof simplifies a bit in the pseudo-reductive case) we have $(G/P)(k) = G(k)/P(k)$, so $(V/P)(k) = V(k)/P(k)$. Hence, it suffices to show that $(V/P)(k)$ is non-empty. The coset space Ω/P is isomorphic to $U_G(-\lambda)$, which is an affine space over k, so V/P is a dense open subscheme of an affine space over k and hence has a k-point if k is infinite. This verifies that $(Z(k), (U_a(k), M_a)_{a\in\Phi})$ is a generating root datum when k is infinite. If k is finite then (by the perfectness of finite fields) the pseudo-split k-group G is (connected) reductive, so the desired Bruhat decomposition is classical [Bo2, 21.15] (which even formally implies the group generation claim in such cases).

It now follows from [BrTi, Cor. 6.1.11(ii), Prop. 6.1.12] that if we define $R = \{r_a\}_{a\in\Delta}$ to be the indexed set of images in W of the sets M_a for $a \in \Delta$ then $(G(k), P(k), N(k), R)$ is a (saturated) Tits system with Weyl group W in the sense of [Bou2, Ch. IV, §2.1, Def. 1]. There is a Bruhat decomposition for any group equipped with a Tits system [Bou2, Ch. IV, §2.3, Thm. 1]. For the

Tits system just constructed in $G(k)$, the Bruhat decomposition is precisely the desired bijectivity of the map $W \to P(k)\backslash G(k)/P(k)$. □

The rest of this section is devoted to the proof of the existence of Levi k-subgroups in any pseudo-split pseudo-reductive group G over a field. Our existence result will also include a uniqueness aspect, so we first set up some notation.

Let G be a pseudo-split pseudo-reductive group over a field k, T a split maximal k-torus of G, $\Phi = \Phi(G, T)$, and Φ' the set of non-multipliable roots in Φ. We will view $T' := T_{\overline{k}}$ as a maximal torus of $G' := G_{\overline{k}}^{\mathrm{red}}$, so the identification $\mathrm{X}(T) = \mathrm{X}(T')$ carries Φ' over to $\Phi(G', T')$. Let Φ'^{+} be a positive system of roots in Φ' (equivalently, the set of non-multipliable elements in a positive system of roots in Φ), and let Δ be the set of simple roots in Φ'^{+}. (Example 3.3.14 shows that the set Φ'^{+} of non-multipliable positive roots is closed.)

By Proposition 2.3.11, for each $a \in \Phi'$ the root group U_a in G is a vector group which admits a unique linear structure relative to which the conjugation action of T is linear, and this linear action is through the character a. A *k-linear subgroup* of U_a is a k-subgroup which corresponds to a k-vector subspace of $U_a(k)$. Since $a \neq 1$, the k-linear subgroups of U_a are precisely the smooth connected k-subgroups of U_a that are normalized by T. Our main result on Levi subgroups (also a consequence of Theorem C.2.30) is:

Theorem 3.4.6 *Let G be a pseudo-split pseudo-reductive group over a field k, T a split maximal torus, and Φ'^{+} a positive system of roots in the set Φ' of non-multipliable roots in $\Phi = \Phi(G, T)$. For each simple positive $a \in \Phi'^{+}$, let E_a be a smooth connected 1-dimensional k-subgroup of U_a that is normalized by T (i.e., E_a is a k-linear subgroup of U_a of dimension 1).*

There is a unique Levi k-subgroup L of G containing T and every E_a. Moreover, every Levi k-subgroup of G containing T arises in this way.

Proof If L is to be a Levi k-subgroup of G containing T then the isomorphism $L_{\overline{k}} \simeq G_{\overline{k}}^{\mathrm{red}}$ carrying $T_{\overline{k}}$ onto a maximal $\overline{k}$-torus implies that inside of $\mathrm{X}(T) = \mathrm{X}(T_{\overline{k}})$ we have $\Phi(L, T) = \Phi(G_{\overline{k}}^{\mathrm{red}}, T_{\overline{k}}) = \Phi'$. For each $a \in \Phi'$ the a-root group of (L, T) lies in $Z_G(T_a)$ for $T_a := (\ker a)^0_{\mathrm{red}}$, and in fact it is contained in $U_{Z_G(T_a)}(a^\vee) =: U_{(a)}$ (as we can compute the limiting behavior via the isomorphism $L_{\overline{k}} \simeq G_{\overline{k}}^{\mathrm{red}}$). But U_a is the subgroup of $U_{(a)}$ on which $\ker(a)$ acts trivially, so the a-root group of (L, T) for each simple positive $a \in \Phi'^{+}$ can be taken as a choice for E_a. Hence, the proposed recipe will have to yield all possible L containing T.

Let Δ be the set of simple positive roots in Φ'^{+}, and for each $a \in \Delta$ make a choice of E_a and a nontrivial $u_a \in E_a(k)$. (Such a k-point u_a exists since $E_a \simeq \mathbf{G}_{\mathrm{a}}$.) By Proposition 3.4.2, there is a unique $v_a \in U_{-a}(k)$ such that $m(u_a) := v_a u_a v_a$ normalizes T, and the induced action of $m(u_a)$ on $\mathrm{X}(T)$ is the reflection r_a associated to a. The Weyl group $W(G_{\overline{k}}^{\mathrm{red}}, T_{\overline{k}}) = W(\Phi')$ is generated by the r_a, and each Weyl group orbit in Φ' meets Δ, so the subgroups

$$(m(u_{a_1}) \cdots m(u_{a_q}))E_a(m(u_{a_1}) \cdots m(u_{a_q}))^{-1} \subseteq G$$

for varying $a \in \Delta$, $q \geqslant 0$, and $a_1, \ldots, a_q \in \Delta$ (allowing repetitions) must vary through precisely the root groups of (L, T) for any possible L. But the connected reductive k-group L is generated by its k-split maximal k-torus T and its root groups with respect to T, so this establishes the uniqueness of L.

The real work is to prove that such an L exists for any specification of the k-subgroups E_a. Let k'/k be the minimal field of definition for the geometric unipotent radical of G; k'/k is a finite purely inseparable extension, and $\mathscr{R}_{u,k'}(G_{k'})$ is a k'-descent of $\mathscr{R}_u(G_{\overline{k}})$. Thus, the quotient

$$\pi : G_{k'} \twoheadrightarrow G_{k'}/\mathscr{R}_{u,k'}(G_{k'}) =: G'$$

is a k'-descent of the maximal reductive quotient of $G_{\overline{k}}$. By definition, the k-homomorphism $G \to \mathrm{R}_{k'/k}(G')$ corresponding to π (via the universal property of Weil restriction) is i_G. The kernel of i_G cannot contain any nontrivial smooth connected k-subgroups (due to Lemma 1.2.1). Thus, i_G is nontrivial on each E_a, so π is nontrivial on each $(E_a)_{k'}$.

We identify $T_{k'}$ with a k'-split maximal k'-torus T' in G' in the evident manner. Consider the $T_{k'}$-equivariant composite map

$$(E_a)_{k'} \hookrightarrow G_{k'} \twoheadrightarrow G'.$$

This lands in the a-root group U'_a for (G', T'), and the resulting $T_{k'}$-equivariant k'-homomorphism $(E_a)_{k'} \to U'_a$ must be *linear*. It has already been shown that this map is nontrivial, so it is an isomorphism. In particular, $\ker(i_G|_{E_a}) = 1$. We will use the resulting k-structure E_a on the k'-group U'_a for each $a \in \Delta$, together with the k-structure T on T', to specify a k-structure (G_0, T_0) on (G', T'). This rests on the general concept of *pinning* for split reductive pairs, which is discussed in Definition A.4.12 and Theorem A.4.13, and the procedure goes as follows.

For each $a \in \Phi'$, let G'_a denote the k'-subgroup of G' generated by U'_a and U'_{-a}: this subgroup is connected semisimple of rank 1, with $T' \cap G'_a$ as a k'-split maximal k'-torus. By Example A.4.1, the pair $(G'_a, T' \cap G'_a)$ is isomorphic to either (SL_2, D) or $(\mathrm{PGL}_2, D/\mu_2)$ where D is the diagonal torus. Thus, we can choose a central k'-isogeny $\varphi_a : \mathrm{SL}_2 \to G'_a$ which carries the diagonal k'-torus to $T' \cap G'_a$ and the upper triangular unipotent subgroup U^+ of SL_2 onto U'_a. By the construction of coroots, on the diagonal torus this map must be $\varphi_a(\mathrm{diag}(y, 1/y)) = a^\vee_{k'}(y)$. The collection $\{\varphi_a\}_{a \in \Delta}$ is a pinning in the sense of Definition A.4.12.

Using the parameterization $x : \mathbf{G}_{\mathrm{a}} \simeq U^+$ defined by $x(u) = \left(\begin{smallmatrix} 1 & u \\ 0 & 1 \end{smallmatrix}\right)$, for $a \in \Delta$ consider the composite k'-isomorphism

$$\mathbf{G}_{\mathrm{a}} \simeq U^+ \simeq U'_a \simeq (E_a)_{k'}.$$

This might not be defined over k, but we can fix that problem as follows. The k'-automorphism of SL_2 defined by any $\mathrm{diag}(c', 1) \in \mathrm{PGL}_2(k')$ restricts to the identity on the diagonal k'-torus and induces multiplication by $c' \in k'^\times$ on $U^+ \simeq \mathbf{G}_{\mathrm{a}}$. Composing φ_a with such an automorphism for a suitable c' allows us to impose the additional requirement that $\varphi_a : \mathbf{G}_{\mathrm{a}} \simeq (E_a)_{k'}$ *is defined over* k for each $a \in \Delta$.

Applying Theorem A.4.13, the pinning $\{\varphi_a\}_{a \in \Delta}$ determines a canonical k-structure (G_0, T_0) on (G', T'), with T_0 a k-split maximal k-torus of G_0. The resulting isomorphism $(T_0)_{k'} \simeq T' = T_{k'}$ must (uniquely) descend to a k-isomorphism $\theta : T_0 \simeq T$ since these k-tori are k-split. Hence, the k-subgroup inclusion

$$j_0 : G_0 \hookrightarrow \mathrm{R}_{k'/k}((G_0)_{k'}) \simeq \mathrm{R}_{k'/k}(G')$$

carries T_0 isomorphically onto the maximal k-torus T in $\mathrm{R}_{k'/k}(T') = \mathrm{R}_{k'/k}(T_{k'})$ via θ.

Using the equality $\Phi(G_0, T_0) = \Phi(G', T') = \Phi'$ inside $\mathrm{X}(T_0) = \mathrm{X}(T') = \mathrm{X}(T)$, the characterization of the k-structure (G_0, T_0) in terms of the pinning (in Theorem A.4.13) implies that for each $a \in \Delta$, the composite isomorphism of root groups

$$(U_a^{G_0})_{k'} \simeq U'_a \simeq (E_a)_{k'}$$

is defined over k. The conceptual significance of this fact is explained by the following lemma.

Lemma 3.4.7 *Using notation as above, consider the diagram*

$$\begin{array}{ccc} & & G \\ & & \downarrow{\scriptstyle i_G} \\ G_0 & \xrightarrow[j_0]{} & \mathrm{R}_{k'/k}(G') \end{array}$$

The restriction of j_0 to T_0 and to each root group $U_a^{G_0}$ ($a \in \Phi'$) uniquely factors through i_G via a homomorphism to G, with T_0 carried isomorphically onto T via θ.

Proof Since $\ker i_G$ contains no nontrivial smooth connected k-subgroups, its only smooth connected closed subscheme through the identity is $\{1\}$. (This follows from Lemma C.4.1, or arguments using subgroups generated by connected subvarieties passing through the identity.) Hence, the desired factorizations are unique if they exist. The map i_G carries T isomorphically onto the maximal k-torus in $\mathrm{R}_{k'/k}(T') = \mathrm{R}_{k'/k}(T_{k'})$ via the canonical map (due to how i_G and T' are defined), so j_0 carries T_0 isomorphically onto $i_G(T)$. We have seen in the discussion preceding this proof that this latter isomorphism is $i_G \circ \theta$, so the factorization of $j_0|_{T_0}$ is settled.

Consider $a \in \Delta$. The restriction of j_0 to $U_a^{G_0}$ lands inside of $\mathrm{R}_{k'/k}(G'_a)$. We claim that its image is exactly $i_G(E_a)$, so since $i_G|_{E_a}$ is a k-subgroup inclusion we would get the desired lifting on $U_a^{G_0}$ for each $a \in \Delta$. More generally, consider a k'-group H' of finite type and a pair of k-subgroups $H_1, H_2 \hookrightarrow \mathrm{R}_{k'/k}(H')$ such that the induced k'-homomorphisms $(H_i)_{k'} \to H'$ are isomorphisms. These latter isomorphisms say exactly that each H_i is a k-structure on H'. (The equivalence follows from the explicit descriptions in Proposition A.5.7.) Thus, the images of the H_i coincide (i.e., the H_i are k-isomorphic respecting their inclusions into $\mathrm{R}_{k'/k}(H')$) if and only if the composite isomorphism

$$(H_1)_{k'} \simeq H' \simeq (H_2)_{k'}$$

is defined over k. Applying this with $H_1 = U_a^{G_0}$, $H_2 = E_a$, $H' = U'_a$, and the inclusions $j_0|_{U_a^{G_0}}$ and $i_G|_{E_a}$, we are done when $a \in \Delta$.

Now consider an arbitrary $a \in \Phi'$. Let $W = W(\Phi')$, so $a = w.c$ for some $w \in W$ and $c \in \Delta$. We have equalities

$$\begin{aligned} N_{G_0}(T_0)(k)/T_0(k) = W(G_0, T_0)(k) = W = W(G, T)(k) \\ = N_G(T)(k)/Z_G(T)(k) \end{aligned}$$

since the pairs (G_0, T_0) and (G, T) are pseudo-split over k (see Corollary 3.4.3(1)). This identification of Weyl groups of pseudo-split pairs matches their identification with the Weyl group $W(\mathrm{R}_{k'/k}(G'), T)(k)$. Thus, if we can find representatives $g_0 \in N_{G_0}(T_0)(k)$ and $g \in N_G(T)(k)$ of w whose images in $G'(k')$ coincide then by composing the lifted map $U_c^{G_0} \simeq E_c$ (for $c \in \Delta$) with conjugation by g_0^{-1} on the source and by g on the target we get a k-homomorphism $U_a^{G_0} \simeq E_a$ that is the desired lift.

To find such g_0 and g, by writing w as a word in reflections $r_{a'}$ for $a' \in \Delta$ it suffices to treat the case $w = r_{a'}$. In this case we choose a nontrivial $u_0 \in U_{a'}^{G_0}(k)$ and let $u \in E_{a'}(k)$ be the corresponding nontrivial element via the lift $U_{a'}^{G_0} \simeq E_{a'}$ constructed above (so $j_0(u_0) = i_G(u)$ as nontrivial elements in $U'_{a'}(k')$; denote it as u'). Proposition 3.4.2 provides unique elements $v_0 \in U_{-a'}^{G_0}(k)$ and $v \in U_{-a'}(k)$ such that $m(u_0) := v_0 u_0 v_0$ and $m(u) := vuv$ respectively normalize T_0 and T and each represent $r_{a'}$. It follows that if $v' \in U'_{-a}(k')$ is either of the elements $j_0(v_0)$ or $i_G(v)$ then $v'u'v'$ normalizes T' and represents $r_{a'}$. Thus, the uniqueness aspect of Proposition 3.4.2 (applied to (G', T')) implies that $j_0(v_0) = i_G(v)$, so the elements $g_0 = v_0 u_0 v_0$ and $g = vuv$ have the same image in $G'(k')$, as required. $\square$

Lemma 3.4.7 now yields the desired Levi k-subgroup, as follows. We fix an enumeration $\{a_i\}$ of the given positive system of roots Φ'^{+}. We then have the open cell

$$\Omega := \prod_i U_{-a_i}^{G_0} \times T_0 \times \prod_i U_{a_i}^{G_0} \hookrightarrow G_0$$

defined via multiplication, so the restriction of j_0 to this open cell factors through i_G by forming the corresponding product (in the same order) of the unique lifts of $j_0|_{T_0}$ and $j_0|_{U_a^{G_0}}$ ($a \in \Phi'$). The resulting factorization diagram of pointed k-schemes

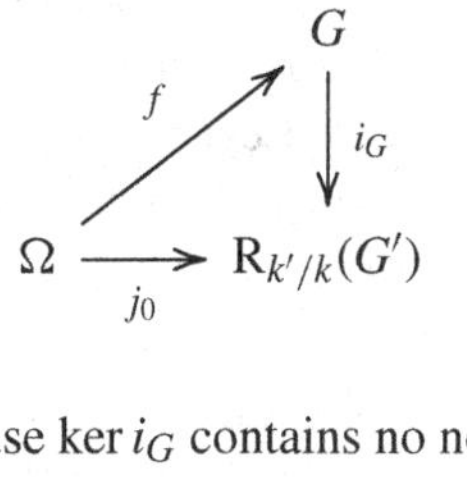

uniquely determines f because $\ker i_G$ contains no nontrivial smooth connected closed subscheme passing through the identity.

Applying the same uniqueness consideration to the product map $i_G \times i_G$, we deduce from the fact that j_0 and i_G are k-homomorphisms that f is a k-rational homomorphism. Hence, f uniquely extends to a k-homomorphism

$F : G_0 \to G$ (as we see via Galois descent by using the covering of $(G_0)_{k_s}$ by $\Omega(k_s)$-translates of Ω_{k_s}; cf. [SGA3, XVIII, Prop. 2.3]). This must be a k-subgroup inclusion since $i_G \circ F = j_0$ is a k-subgroup inclusion. By construction $F(G_0)$ contains T and every E_a $(a \in \Delta)$. It follows that F identifies G_0 with a connected reductive k-subgroup L of G having T as a maximal k-torus and Δ as a basis for a positive system of roots (with associated root groups E_a for $a \in \Delta$). Since $\pi \circ F_{k'} : (G_0)_{k'} \to G'$ induces $(T_0)_{k'} \simeq T'$ respecting the common basis Δ for the root systems, $\pi : L_{k'} \to G'$ induces an isomorphism between root data and hence is an isomorphism. Thus, L is the desired Levi k-subgroup of G. □

Corollary 3.4.8 *Let T be a maximal k-torus in a pseudo-reductive group G over a field k, and let $\Phi = \Phi(G_{k_s}, T_{k_s})$. Assume $G_{\overline{k}}^{\mathrm{red}}$ is semisimple of adjoint type, and for each $a \in \Phi$ let U_a denote the a-root group of (G_{k_s}, T_{k_s}).*

If the $(Z_G(T))(k_s)$-action on $\prod_{a\in\Delta}(U_a(k_s) - \{0\})$ makes it a principal homogeneous space then there is a cohomology class $c_{T,\Delta} \in \mathrm{H}^1(k, Z_G(T)/T)$ whose vanishing is necessary and sufficient for the existence of a Levi k-subgroup of G containing T. The formation of $c_{T,\Delta}$ commutes with separable extension on k.

This cohomological criterion will be used in Proposition 7.3.7 to prove the existence of Levi k-subgroups in certain "exotic" pseudo-reductive groups G over imperfect fields k of characteristic 3 such that $[k : k^3] < \infty$.

Proof The assumption that $G_{\overline{k}}^{\mathrm{red}}$ is semisimple of adjoint type has two useful consequences: it implies that there is no simple factor of simply connected type C, so Φ is reduced (even in characteristic 2), and Δ is a $\mathbf{Z}$-basis of $\mathrm{X}(T_{k_s})$. Thus, the principal homogeneous space hypothesis implies that the space of choices $\{\ell_a\}_{a\in\Delta}$ of k_s-lines $\ell_a \subseteq U_a(k_s)$ is a principal homogeneous space for $(Z_G(T)/T)(k_s)$.

Theorem 3.4.6 identifies this space of choices of such lines with the set of Levi k_s-subgroups L of G_{k_s} containing T_{k_s}. Concretely, the action of $(Z_G(T)/T)(k_s)$ on the set of such Levi k_s-subgroups is $g.L = \widetilde{g}L\widetilde{g}^{-1}$ for $\widetilde{g} \in Z_G(T)(k_s)$ representing g. Thus, if we choose some L over k_s then for each $\gamma \in \mathrm{Gal}(k_s/k)$ we have $\gamma(L) = g_\gamma.L$ for a unique $g_\gamma \in (Z_G(T)/T)(k_s)$. It is easy to check that $\gamma \mapsto g_\gamma$ is a continuous 1-cocycle whose cohomology class is independent of the initial choice of L, independent of the choice of k_s/k, and vanishes precisely when there is a $\mathrm{Gal}(k_s/k)$-stable choice of L (i.e., there exists a Levi k-subgroup of G containing T). This class is $c_{T,\Delta}$, and by the construction its formation is compatible with separable extension of k. □

As a consequence of the existence of Levi subgroups in pseudo-split pseudo-reductive groups, we get a better understanding of the relationship between pseudo-reductivity and reductivity, as well as between pseudo-parabolicity and parabolicity:

Proposition 3.4.9 *Let G be a perfect pseudo-reductive group over a field k.*

(1) *The root groups of G_{k_s} are all 1-dimensional if and only if G is reductive.*
(2) *Let P be a pseudo-parabolic k-subgroup of G that does not contain any nontrivial perfect smooth connected normal k-subgroup of G. The coset space G/P is complete (i.e., P is a parabolic subgroup) if and only if G is reductive.*

Proof The structure theory of connected reductive groups [Bo2, 13.18] implies that G_{k_s} has 1-dimensional root groups if it is reductive. Likewise, by Proposition 2.2.9, the pseudo-parabolic k-subgroups of G are precisely the parabolic k-subgroups when G is reductive. This settles the "if" direction in both parts.

For the converse implication in (1), we may assume $k = k_s$, $G \neq 1$, and that G has 1-dimensional root groups. Let T be a maximal k-torus of G. The root system $\Phi := \Phi(G, T)$ is reduced since any multipliable root a has root group $U_a = U_{(a)}$ with dimension at least 2.

Since $k = k_s$, by Theorem 3.4.6 there is a Levi k-subgroup L of G which contains T and has the same root system (with respect to T) as does G. But the root groups U_a of G are 1-dimensional, so L contains U_a for all $a \in \Phi$. It follows that L is the smooth connected k-subgroup of G generated by T and the k-groups U_a for $a \in \Phi$. Since G is generated by $Z_G(T)$ and the k-groups U_a for $a \in \Phi$ (Lemma 2.3.3), and $Z_G(T)$ normalizes each root group U_a, we conclude that L is a normal k-subgroup of G. Thus, the quotient G/L makes sense as a k-group and, being a quotient of the commutative group $Z_G(T)$, it is commutative. But we assumed that G is perfect, so commutativity of G/L forces triviality of G/L; i.e., $G = L$ is reductive. This proves (1).

Now consider the converse implication in (2). We may assume $G \neq 1$, so $P \neq G$. Fix a maximal k-torus T in P, so T is also maximal in G. It suffices to prove that all root groups for T_{k_s} in G_{k_s} are 1-dimensional. Since $P \neq G$, G is non-commutative. The k_s-subgroup $P_{k_s} \subseteq G_{k_s}$ is pseudo-parabolic, and it contains no nontrivial perfect smooth connected normal k_s-subgroup N of G_{k_s}. (Indeed, otherwise P_{k_s} would contain the nontrivial perfect smooth connected normal k_s-subgroup of G_{k_s} generated by the $\mathrm{Gal}(k_s/k)$-orbit of N. This descends to a nontrivial perfect smooth connected normal k-subgroup of

G contained in P, contradicting our hypothesis on P.) Thus, we can assume $k = k_s$.

Let $\{G_i\}_{i \in I}$ be the set of minimal nontrivial perfect smooth connected normal k-subgroups of G. By Proposition 3.1.8 this is a finite set of pairwise commuting k-subgroups, the G_i are pseudo-simple over k, and the natural multiplication homomorphism $\pi : \prod G_i \to G$ is surjective with kernel containing no nontrivial smooth connected k-subgroup. Also, by the construction of the G_i (and the condition $k = k_s$), the semisimple quotients $(G_i)^{\mathrm{ss}}_{\overline{k}}$ are naturally identified (without repetition) with the simple connected semisimple normal $\overline{k}$-subgroups of $G^{\mathrm{red}}_{\overline{k}} = G^{\mathrm{ss}}_{\overline{k}}$.

By normality of G_i in G, $T_i = T \cap G_i$ is a maximal k-torus in G_i, so $\prod T_i$ is carried by π onto a maximal k-torus in G. But π carries $\prod T_i$ into T, so π restricts to a surjection of k-tori $\prod T_i \to T$ with kernel containing no nontrivial k-tori. Thus, this surjection of k-tori is an isogeny.

By Proposition 2.2.11, we can choose a 1-parameter k-subgroup $\lambda : \mathrm{GL}_1 \to T$ such that $P = P_G(\lambda)$, Moreover, we may replace λ with λ^n for any $n \geqslant 1$ (Remark 2.1.7). Thus, taking n to be the degree of the isogeny $\prod T_i \to T$ allows us to arrange that there exist 1-parameter k-subgroups $\lambda_i : \mathrm{GL}_1 \to T_i$ such that $\lambda = \pi \circ \Lambda$ where $\Lambda : \mathrm{GL}_1 \to \prod T_i$ has ith component equal to λ_i for all i.

Consider the pseudo-parabolic k-subgroup $P_i = P_{G_i}(\lambda_i)$ in G_i containing T_i. Thus, $\prod P_i = P_{\prod G_i}(\Lambda)$ is carried by π onto $P_G(\pi \circ \Lambda) = P_G(\lambda) = P$ (see Corollary 2.1.9). In particular, $P_i \subseteq G_i \cap P$ for all i. The GL_1-action on G_i through λ-conjugation coincides with the action defined by λ_i-conjugation because $\lambda = \pi \circ \Lambda$ and $G_{i'}$ centralizes G_i for all $i' \neq i$. Hence, $P_i = G_i \cap P$ by functorial considerations. Since no G_i is contained in P by hypothesis, it follows that $P_i \neq G_i$ for all i and that $G_i/P_i = (G_iP)/P$. But $(G_iP)/P$ is closed in G/P since G_iP is closed in G (as it is a k-subgroup), so we conclude that G_i/P_i is complete. Thus, each triple (G_i, P_i, T_i) satisfies the same hypotheses as on (G, P, T). The surjectivity of π implies that G is reductive if each G_i is reductive, so we may now treat the triples (G_i, P_i, T_i) separately. That is, we can assume that G has no smooth connected proper normal k-subgroup. In particular, the semisimple quotient $G^{\mathrm{red}}_{\overline{k}} = G^{\mathrm{ss}}_{\overline{k}}$ is simple.

Let $G' = G^{\mathrm{red}}_{\overline{k}}$ (so G' is semisimple and simple), so the natural projection $\pi : G_{\overline{k}} \twoheadrightarrow G'$ carries $T_{\overline{k}}$ isomorphically onto a maximal $\overline{k}$-torus T' of G'. The set Φ' of non-multipliable roots in Φ is a root system and is naturally identified with the root system $\Phi(G', T')$. In particular, Φ' is reduced and irreducible, so Φ is irreducible.

Since G/P is complete, it follows from the fixed point theorem of Borel that $P_{\overline{k}}$ contains a conjugate of the smooth connected solvable subgroup $\mathscr{R}_u(G_{\overline{k}})$ in

$G_{\overline{k}}$. Hence, by normality of the unipotent radical, $\mathscr{R}_u(G_{\overline{k}}) \subseteq P_{\overline{k}}$. By hypothesis $P \neq G$, so Lie(G) is strictly larger than Lie(P). In other words, since $P = P_G(\lambda)$,

$$\Phi_{\lambda<0} := \{a \in \Phi \mid \langle a, \lambda \rangle < 0\}$$

is non-empty. Since $\mathscr{R}_u(G_{\overline{k}}) \subseteq P_{\overline{k}}$, it follows that no positive rational multiple of $a \in \Phi_{\lambda<0}$ occurs as a T'-weight on Lie($\mathscr{R}_u(G_{\overline{k}})$). But the exact sequence of smooth connected $\overline{k}$-groups

$$1 \to \mathscr{R}_u(G_{\overline{k}}) \to G_{\overline{k}} \to G' \to 1$$

induces a $T_{\overline{k}}$-equivariant exact sequence of Lie algebras, so the span of the root spaces in Lie(G) for the roots that are positive rational multiples of a is 1-dimensional over k since the same holds in Lie(G') over $\overline{k}$.

The k-subspace Lie($U_{(a)}$) $\subseteq$ Lie(G) is the span of the root spaces for the positive rational multiples of a in Φ, so we conclude that $U_{(a)}$ is 1-dimensional when $a \in \Phi'_{\lambda<0} = \Phi' \cap \Phi_{\lambda<0}$. In particular, such an a is also not divisible, and $U_a = U_{(a)}$. Since the Weyl group of an irreducible root system acts transitively on the set of roots of a common length [Bou2, VI, §1.3, Prop. 11], and $N_G(T)(k)$ maps onto the common Weyl group of Φ and Φ' (Proposition 3.2.7), conjugation by $N_G(T)(k)$ is transitive on the set of elements of Φ' with a common length. We are aiming to prove that all root groups for (G, T) are 1-dimensional, so it now suffices to show that if Φ' is any reduced and irreducible root system then $\Phi'_{\lambda<0}$ contains roots of all lengths for any nonzero $\lambda \in \mathrm{X}_*(T)$.

Let $W = W(\Phi)$ be the Weyl group of Φ. It suffices to show that the W-orbit of any $a \in \Phi$ meets the locus where λ does not vanish, as the orbit is stable under negation. Assuming to the contrary, the hyperplane $\ker \lambda_{\mathbf{Q}}$ in $\mathrm{X}_*(T)_{\mathbf{Q}}$ would contain a nonzero W-stable subspace, and this contradicts the irreducibility of the W-action on $\mathrm{X}_*(T)_{\mathbf{Q}}$ [Bou2, VI, §1.2, Cor.]. □

The perfectness hypothesis on G in Proposition 3.4.9 is reasonable because the criteria in (1) and (2) cannot distinguish between the pseudo-reductive G and its perfect derived group $\mathscr{D}(G)$, and over any imperfect field k there are non-reductive standard pseudo-reductive G such that $\mathscr{D}(G)$ is nontrivial and reductive.

To be precise, for (1) the root groups of $\mathscr{D}(G)_{k_s}$ and G_{k_s} coincide (using a maximal k-torus T of G and the maximal k-torus $T \cap \mathscr{D}(G)$ of $\mathscr{D}(G)$). For (2), the results we will prove in §3.5 imply that $P \mapsto \mathscr{D}(G) \cap P$ is a bijective correspondence between the sets of pseudo-parabolic k-subgroups in

G and $\mathscr{D}(G)$, with $\mathscr{D}(G)/(\mathscr{D}(G) \cap P) \to G/P$ an isomorphism. Moreover, the perfect smooth connected normal k-subgroups of G and $\mathscr{D}(G)$ coincide, by Proposition 3.1.8.

Here is an example of a non-reductive standard pseudo-reductive G over any imperfect field k of characteristic $p > 0$ such that $\mathscr{D}(G)$ is reductive. Consider a nontrivial finite extension k'/k with $k'^p \subseteq k$ and expand SL_p by expanding a GL_1-factor in a split maximal k-torus T to be $\mathrm{R}_{k'/k}(\mathrm{GL}_1)$. That is, we identify T with GL_1^{p-1} so that the center $\mu = Z_{\mathrm{SL}_p} \simeq \mu_p$ is in the first factor (see the displayed isomorphism in Example 4.2.6), and define $C = \mathrm{R}_{k'/k}(\mathrm{GL}_1) \times \mathrm{GL}_1^{p-2}$, so $T \to T/Z$ factors through the natural inclusion $T \hookrightarrow C$ by using $\mathrm{N}_{k'/k} : \mathrm{R}_{k'/k}(\mathrm{GL}_1) \to \mathrm{GL}_1$. Then $G = (\mathrm{SL}_p \rtimes C)/T$ is a standard pseudo-reductive k-group with Cartan k-subgroup C that is not a torus (so G is not reductive) but $\mathscr{D}(G) = \mathrm{SL}_p$.

In general there is a natural intermediate group between G and $\mathscr{D}(G)$ whose reductivity is equivalent to that of $\mathscr{D}(G)$ and whose derived group coincides with $\mathscr{D}(G)$: the k-subgroup G_t generated by the k-tori of G (see Proposition A.2.11). The perfect group $\mathscr{D}(G)$ is contained in G_t, by Proposition A.2.11, so $G_t/\mathscr{D}(G)$ is the maximal k-torus in $G/\mathscr{D}(G)$. Thus, in Proposition 3.4.9 it suffices to impose the slightly weaker hypothesis $G = G_t$.

3.5 Classification of pseudo-parabolic subgroups

Let G be a pseudo-split pseudo-reductive group over a field k, and T a split maximal k-torus in G. An interesting application of Corollary 3.3.13, to be discussed below, is to parameterize all pseudo-parabolic k-subgroups of G that contain T, generalizing Proposition 3.3.15(2). This involves the notion of a parabolic set Π of roots in a root system Φ, as discussed in Definition 2.2.6 and Proposition 2.2.8. In particular, for any such Π, the set Ψ of non-divisible $a \in \Phi$ with $-a \notin \Pi$ (so $a \in \Pi$) is a closed set of roots contained in a suitable positive system of roots $\Phi^+ \subseteq \Phi$ (so in the case $\Phi = \Phi(G,T)$, U_Ψ makes sense).

Proposition 3.5.1 *Let G be a pseudo-reductive group over a field k, and let T be a split maximal k-torus in G.*

(1) *For every pseudo-parabolic k-subgroup $P \subseteq G$ containing T,* $\mathrm{Lie}(P)$ *is the span of* $\mathrm{Lie}(T)$ *and some root spaces for T acting on* $\mathrm{Lie}(G)$*, and* $\mathrm{Lie}(P)$ *uniquely determines P.*
(2) *The possibilities for the set $\Phi(P,T)$ of nontrivial T-weights on* $\mathrm{Lie}(P)$ *are precisely the parabolic subsets of $\Phi(G,T)$, and for each P we have*

$\mathscr{R}_{u,k}(P) = U_\Psi$, where Ψ is the set of non-divisible roots in $\Phi(G,T)$ whose negative is not in $\Phi(P,T)$.

(3) *If P and P' are pseudo-parabolic k-subgroups containing T then the following are equivalent: $P \subseteq P'$, $\mathrm{Lie}(P) \subseteq \mathrm{Lie}(P')$, and $\Phi(P,T) \subseteq \Phi(P',T)$.*

(4) *A pseudo-parabolic k-subgroup P in G is minimal if and only if $P/\mathscr{R}_{u,k}(P)$ is commutative.*

Since $\Phi(P,T)$ in (2) is a parabolic subset of $\Phi(G,T)$, the set Ψ in (2) is contained in $\Phi(P,T)$.

Proof By definition, $P = P_G(\lambda)$ for some k-homomorphism $\lambda : \mathrm{GL}_1 \to G$. By Proposition 2.2.11, we can arrange that $\lambda(\mathrm{GL}_1) \subseteq T$ (which conversely forces $T \subseteq Z_G(\lambda) \subseteq P_G(\lambda)$). It then follows from Proposition 2.1.8(1) that

$$\Phi(P,T) = \Phi_{\lambda \geqslant 0} := \{a \in \Phi(G,T) \mid \langle a, \lambda \rangle \geqslant 0\}.$$

By Proposition 2.2.8, these are exactly the parabolic sets of roots (as we vary λ) and the set $\Psi = \Phi_{\lambda > 0}$ is a closed set of roots contained in a positive system of roots, so by Corollary 3.3.13(1) the k-group U_Ψ makes sense.

The quotient $P/U_G(\lambda) \simeq Z_G(\lambda)$ is pseudo-reductive over k, so $\mathscr{R}_{u,k}(P) = U_G(\lambda)$ has Lie algebra that is the span of the root spaces for the roots in $\Phi_{\lambda>0}$. By construction we have $U_\Psi \subseteq U_G(\lambda)$ (due to the limiting property that characterizes points of $U_G(\lambda)$), so the equality of their Lie algebras yields equality of groups. In other words, $\mathscr{R}_{u,k}(P) = U_\Psi$.

We shall now prove that if $\mathrm{Lie}(P) \subseteq \mathrm{Lie}(P')$ then $P \subseteq P'$, from which everything else remaining to be proved for (1), (2), and (3) will follow. Consideration of weight spaces gives $\Phi(P,T) \subseteq \Phi(P',T')$. Choose $\lambda' : \mathrm{GL}_1 \to T$ such that $P' = P_G(\lambda')$. Thus, every $a \in \Phi(P,T)$ is contained in $\Phi(P',T) = \Phi_{\lambda' \geqslant 0}$. That is, $\langle a, \lambda' \rangle \geqslant 0$. In particular, all points of $U_{(a)}$ satisfy the limiting property to belong to $P_G(\lambda') = P'$. (To do the limiting calculation, choose a T-equivariant linear structure on the vector group $U_{(a)}$, and work one root space at a time.) Clearly $Z_G(T) \subseteq P'$ for the same reason (as λ' is valued in T). Hence, it suffices to show that the k-group $P = P_G(\lambda)$ is generated by $Z_G(T)$ and the k-groups $U_{(a)}$ for all $a \in \Phi(P,T) = \Phi_{\lambda \geqslant 0}$. Certainly $Z_G(T)$ and all such k-groups $U_{(a)}$ are contained in P (by the limiting property that characterizes points of P), and the Lie algebras of $Z_G(T)$ and the k-groups $U_{(a)}$ span $\mathrm{Lie}(P)$, so we are done.

Finally, we prove (4). If P is minimal then $P/\mathscr{R}_{u,k}(P)$ is commutative by Proposition 3.3.15(3). Conversely, suppose $P = P_G(\lambda)$ is a pseudo-parabolic k-subgroup such that $P/\mathscr{R}_{u,k}(P) = Z_G(\lambda)$ is commutative. We wish to prove that P is minimal. Suppose P strictly contains a pseudo-parabolic k-subgroup Q, and let $T \subseteq Q$ be a maximal k-torus of G. We may and do arrange that λ takes values in T. In particular, $T \subseteq Z_G(\lambda)$. By parts (1), (2), and (3), $\Phi(P,T)$ must strictly contain a positive system of roots and $Z_G(\lambda)$ must contain a nontrivial T-weight on its Lie algebra. But this is impossible since $Z_G(\lambda)$ is commutative and contains T. □

Proposition 3.5.2 *Let G be a smooth connected affine group over a field k, and P a smooth connected k-subgroup containing a maximal k-torus T in G.*

(1) *Let K/k be a separable extension of fields. The k-subgroup P in G is pseudo-parabolic if and only if the K-subgroup P_K in G_K is pseudo-parabolic.*
(2) *The pseudo-parabolic k-subgroups of G containing T are in natural bijective correspondence with* $\mathrm{Gal}(k_s/k)$*-stable parabolic subsets of* $\Phi((G/\mathscr{R}_{u,k}(G))_{k_s}, T_{k_s})$.

The separability hypothesis in (1) can be dropped for the "if" direction; see Corollary 3.5.9.

Proof Once (1) is proved, (2) is immediate from Proposition 2.2.10 and Proposition 3.5.1. To prove (1), since the formation of $\mathscr{R}_{u,k}(G)$ commutes with separable extension on k (Proposition 1.1.9(1)), we may assume G is pseudo-reductive over k (see Proposition 2.2.10). The "converse" implication is trivial, so to prove the other direction it suffices to treat two cases: $K = k_s$, and k separably closed.

First we carry out the descent from $K = k_s$ down to k. We have to show that if $P_{k_s} = P_{G_{k_s}}(\lambda')$ for a 1-parameter k_s-subgroup $\lambda' : (\mathrm{GL}_1)_{k_s} \to G_{k_s}$ then we can change the choice of λ' if necessary so that it is defined over k. By Proposition 2.2.11 we may arrange that λ' takes values in T_{k_s}. We pick a finite Galois extension k'/k contained in k_s such that $T_{k'}$ is k'-split and λ' descends to a k'-homomorphism. In other words, we have $P_{k'} = P_{G_{k'}}(\lambda')$ for a 1-parameter k'-subgroup $\lambda' : (\mathrm{GL}_1)_{k'} \to T_{k'} \subseteq G_{k'}$. Exactly as in the proof of Proposition 2.2.9, the 1-parameter k-subgroup

$$\lambda = \sum_{\gamma \in \mathrm{Gal}(k'/k)} \gamma(\lambda') \in \mathrm{Hom}_k(\mathrm{GL}_1, T)$$

has the property that the pseudo-parabolic k'-subgroups $P_{k'} = P_{G_{k'}}(\lambda')$ and $P_G(\lambda)_{k'}$ share the same sets of $T_{k'}$-weights on their Lie algebras over k'. By Proposition 3.5.1 (applied over k_s), it follows that $P_{k_s} = P_G(\lambda)_{k_s}$, so $P = P_G(\lambda)$.

It remains to prove that P is pseudo-parabolic when P_K is pseudo-parabolic provided that $k = k_s$. We may assume K/k is finitely generated, so by separability the field K is the fraction field of a smooth k-algebra domain A. Since G_K is pseudo-reductive, $P_K = P_{G_K}(\lambda)$ for a K-homomorphism $\lambda : \mathrm{GL}_1 \to G_K$. By localizing A, we can "spread out" λ to an A-homomorphism $\mu : \mathrm{GL}_1 \to G_A$. The theory of 1-parameter limits in §2.1 was developed over any base ring, so $P_{G_A}(\mu)$ makes sense. The closed A-subgroups P_A and $P_{G_A}(\mu)$ in G_A coincide over K, so by localizing A we can arrange that $P_A = P_{G_A}(\mu)$ inside G_A. Specializing at a k-point of the k-smooth A then gives the required description of P as a pseudo-parabolic k-subgroup of G. □

Remark 3.5.3 Assertion (4) of Proposition 3.5.1 can be improved: for any separable extension K/k, P is minimal in G if and only if P_K is minimal in G_K, since the commutativity of $P/\mathscr{R}_{u,k}(P)$ is a necessary and sufficient minimality criterion that is insensitive to separable extension on k.

Proposition 3.5.4 *Let G be a pseudo-reductive group over a field k, and T a maximal k-torus of G. There is a natural inclusion-preserving bijection between the set of pseudo-parabolic k-subgroups P of G containing T and the set of parabolic $\overline{k}$-subgroups P' of $G' := G_{\overline{k}}^{\mathrm{red}}$ containing the (isomorphic) image T' of $T_{\overline{k}}$ in G'.*

Explicitly, P' is the image of $P_{\overline{k}}$ under the quotient map $G_{\overline{k}} \twoheadrightarrow G'$, and if T is k-split then $\Phi(P',T') = \Phi(P,T) \cap \Phi(G',T')$. For such T, the $\overline{k}$-group P' is a Borel subgroup of G' if and only if P is a minimal pseudo-parabolic k-subgroup of G.

For k-split T, an equivalent way to formulate the comparison of sets of nontrivial T-weights occurring on the Lie algebras of the pseudo-parabolic subgroups containing T is that $\Phi(P',T')$ is the set of non-multipliable roots of $\Phi(P,T)$. In Corollary 3.5.11 we will strengthen Proposition 3.5.4 by showing that $T \subseteq P$ if and only if $T' \subseteq P'$ (with P' denoting the image of $P_{\overline{k}}$ in G').

Proof By Galois descent and Remark 3.5.3, we may assume $k = k_s$. Let $\Phi = \Phi(G,T)$ and $\Phi' = \Phi(G',T')$, so Φ' is the set of non-multipliable elements of Φ (Theorem 2.3.10). By Proposition 2.2.11, the possibilities for P are precisely the k-subgroups $P_G(\lambda)$ for $\lambda \in \mathrm{X}_*(T)$. Hence, by Corollary 2.1.9, the image P' of $P_{\overline{k}} = P_{G_{\overline{k}}}(\lambda_{\overline{k}})$ in G' is a pseudo-parabolic subgroup of G' containing T'. Proposition 2.2.9 ensures that the pseudo-parabolic subgroups of G' are the same as its parabolic subgroups, so the proposed inclusion-preserving correspondence at least makes sense.

We have to prove the following:

(i) P' determines P,
(ii) every parabolic subgroup of G' containing T' has the form P' for some pseudo-parabolic k-subgroup P of G containing T,
(iii) $\Phi(P', T') = \Phi(P, T) \cap \Phi'$ for all such P (from which it follows via Proposition A.2.10, Proposition 3.3.17, and Proposition 3.5.1 that P is minimal in G if and only if P' is minimal in G', which is to say that P' is a Borel subgroup).

The proof of (ii) is straightforward, as follows. Every parabolic subgroup P' in G' containing T' has the form $P_{G'}(\lambda')$ for some $\lambda' \in \mathrm{X}_*(T')$. But k is separably closed, so $\lambda' = \lambda_{\overline{k}}$ for a unique $\lambda \in \mathrm{X}_*(T)$. Thus, $P := P_G(\lambda)$ is carried onto P', establishing (ii). In view of the characterization of parabolic subsets of a root system (Proposition 2.2.8), the correspondence $\Psi \mapsto \Psi \cap \Phi'$ is an inclusion-preserving bijection between the set of parabolic subsets of Φ and the set of parabolic subsets of Φ' (as Φ' is the set of non-multipliable roots of Φ). Hence, if we prove (iii) then (i) follows. We will now prove (iii).

Consider any P containing T, so $P = P_G(\lambda)$ for some $\lambda \in \mathrm{X}_*(T)$. By Corollary 3.3.12, P is generated by $Z_G(T)$ and the k-subgroups $U_{(a)}$ for all $a \in \Phi(P, T)$. The formation of $U_{(a)}$ is unaffected by replacing a with any positive rational multiple that is also a root, so P is also generated by $Z_G(T)$ and the k-subgroups $U_{(a)}$ for $a \in \Phi(P, T) \cap \Phi'$. The compatibility of quotient maps with the formation of torus centralizers in smooth connected affine groups (Proposition A.2.8) implies that the quotient map $\pi : G_{\overline{k}} \twoheadrightarrow G'$ carries $Z_G(T)_{\overline{k}} = Z_{G_{\overline{k}}}(T_{\overline{k}})$ onto T' (viewed as a maximal torus of G'). Hence, if U'_a denotes the root group in G' associated to any $a \in \Phi'$, we will be done if $(U_{(a)})_{\overline{k}}$ maps onto U'_a for all $a \in \Phi'$. Letting $T_a = (\ker a)^0_{\mathrm{red}}$ and T'_a be the image of $(T_a)_{\overline{k}}$ in G', the quotient map π carries $Z_G(T_a)_{\overline{k}}$ onto $Z_{G'}(T'_a)$ by Proposition A.2.8. Hence, it follows from the definition of $U_{(a)}$ and Corollary 2.1.9 that π carries $(U_{(a)})_{\overline{k}}$ onto $U'_{(a)}$. But Φ' is a reduced root system, so $U'_{(a)} = U'_a$ by Proposition 2.3.11. □

Let G be a smooth connected affine group over a field k. Due to the analogy with parabolic subgroups in connected reductive groups, it is natural to wonder if a smooth closed k-subgroup Q of G containing a pseudo-parabolic k-subgroup P of G is automatically pseudo-parabolic (hence connected), and to wonder if $P = N_G(P)$. To settle both questions (in the affirmative), we first handle the normalizer question, using the following lemma of independent interest.

Lemma 3.5.5 *Let G be a smooth connected affine group over a field k, T a maximal k-torus of G, and P a pseudo-parabolic k-subgroup of G containing T. Let $U = \mathscr{R}_{u,k}(P)$, and let $\overline{T}$ be the image of T in the maximal pseudo-reductive quotient $\overline{P} = P/U$.*

The formation of images and preimages under the quotient map $\pi : P \twoheadrightarrow \overline{P}$ defines a bijection between the set of pseudo-parabolic k-subgroups Q of G contained in P and the set of pseudo-parabolic k-subgroups of $\overline{P}$. Moreover, Q contains T if and only if $\pi(Q)$ contains $\overline{T}$.

Proof Observe that $\mathscr{R}_{u,k}(G) \subseteq P$ by definition of pseudo-parabolicity, so $\mathscr{R}_{u,k}(G) \subseteq U$. Hence, by Proposition 2.2.10 we may replace G with $G/\mathscr{R}_{u,k}(G)$ and replace T and P with their respective images in this quotient to reduce to the case where G is pseudo-reductive over k. The proposed bijectivity also allows us to assume $k = k_s$ (by Galois descent and Proposition 3.5.2(1)).

The pseudo-reductivity of G implies that any pseudo-parabolic k-subgroup Q of G containing T has the form $P_G(\mu)$ for a 1-parameter k-subgroup $\mu : \mathrm{GL}_1 \to T$ (see Proposition 2.2.11). Assume Q is contained in P, so $Q = P_G(\mu) \cap P = P_P(\mu)$. Thus, the image $\overline{Q}$ of Q in $\overline{P}$ is $P_{\overline{P}}(\pi \circ \mu)$ (Corollary 2.1.9). This is a pseudo-parabolic k-subgroup of the pseudo-reductive k-group $\overline{P}$, and it contains $\overline{T}$ since $\pi \circ \mu$ takes values in $\overline{T}$.

The key point is that Q is the preimage of $\overline{Q}$, which is to say that Q contains $U = \mathscr{R}_{u,k}(P)$. To prove this, recall from Proposition 3.5.1(2) that U is directly spanned by the T-root groups U_a for a in the set Ψ of non-divisible roots in $\Phi(G,T)$ such that $-a$ is not a T-weight on $\mathrm{Lie}(P)$. This applies just as well to similarly describe $\mathscr{R}_{u,k}(Q)$ in terms of T-weights on $\mathrm{Lie}(Q)$, and since $\mathrm{Lie}(Q) \subseteq \mathrm{Lie}(P)$ we see that $-a$ is not a T-weight on $\mathrm{Lie}(Q)$ for each $a \in \Psi$. Hence, such U_a are contained in $\mathscr{R}_{u,k}(Q)$, so $\mathscr{R}_{u,k}(P) = U_\Psi \subseteq \mathscr{R}_{u,k}(Q)$. This proves that $Q = \pi^{-1}(\pi(Q))$.

It remains to show that every pseudo-parabolic k-subgroup of $\overline{P}$ is the image of a pseudo-parabolic k-subgroup of G contained in P. Since every pseudo-parabolic k-subgroup of G contains a maximal k-torus of G (and likewise for $\overline{P}$) and every maximal k-torus of $\overline{P} = P/U$ is the image of a maximal k-torus of P (due to smoothness of U), by varying the choice of T, it suffices to restrict attention to a pseudo-parabolic k-subgroup $\overline{Q}$ in $\overline{P}$ containing $\overline{T}$. We will use root system considerations to construct a pseudo-parabolic k-subgroup Q' of G contained in P such that $T \ltimes U \subseteq Q'$ and the pseudo-parabolic image $\overline{Q}'$ of Q' in $\overline{P}$ satisfies $\Phi(\overline{Q}', \overline{T}) = \Phi(\overline{Q}, \overline{T})$. Proposition 3.5.1(3) applied to $(\overline{P}, \overline{T})$ will then force $\overline{Q}' = \overline{Q}$, so $\pi^{-1}(\overline{Q}) = Q'$ is a pseudo-parabolic k-subgroup of G containing T, as desired.

Let $\Phi = \Phi(G, T)$. Since P is pseudo-parabolic and contains T, $P = P_G(\lambda)$ for a 1-parameter k-subgroup $\lambda : \mathrm{GL}_1 \to T$ (Proposition 2.2.11). Hence, $P = Z_G(\lambda) \ltimes U$ with U directly spanned by the root groups U_a for $a \in \Phi_{\lambda>0}$, and $\mathrm{Lie}(Z_G(\lambda))$ is the span of the T-weight spaces in $\mathrm{Lie}(G)$ for the trivial weight and the roots in $\Phi_{\lambda=0}$. By Proposition 2.2.8, there is a positive system of roots Φ^+ in Φ such that the parabolic set $\Phi(P, T) = \Phi_{\lambda\geqslant 0}$ has the form $\Phi^+ \cup [I]$ for a (unique) subset I of the set Δ of simple positive roots in Φ^+ (with $[I]$ denoting the set of $\mathbf{Z}$-linear combinations of I that lie in Φ). In particular, $[I]$ is the set of elements in $\Phi(P, T)$ whose negative is also in $\Phi(P, T)$, so $[I]$ is equal to the set $\Phi_{\lambda=0}$ of T-weights on $\mathrm{Lie}(Z_G(\lambda))$.

The map $Z_G(\lambda) \to \overline{P} = P/U$ induced by π is an isomorphism, so $[I] = \Phi_{\lambda=0}$ is identified with the root system $\Phi(\overline{P}, \overline{T})$ in which $\Phi(\overline{Q}, \overline{T})$ is a parabolic set of roots. That is, I is identified with a basis of simple positive roots for $\Phi(\overline{P}, \overline{T})$ and there is a linear form μ on the $\mathbf{Q}$-linear span of $[I]$ such that the parabolic subset $\Phi(\overline{Q}, \overline{T}) \subseteq [I]$ is the locus in $[I]$ at which $\mu \geqslant 0$.

Now we note that λ vanishes on I and its values on $\Delta - I$ are > 0. Since Δ is a $\mathbf{Q}$-basis for $\mathrm{X}(T)_{\mathbf{Q}}$, we can define a unique linear form ℓ on $\mathrm{X}(T)_{\mathbf{Q}}$ by requiring it to agree with λ on $\Delta - I$ and with $\varepsilon\mu$ on I, where $\varepsilon \in \mathbf{Q}_{>0}$ is so small that $\Phi_{\lambda<0} \subseteq \Phi_{\ell<0}$. Then the parabolic set of roots $\Phi_{\ell\geqslant 0}$ is contained in $\Phi_{\lambda\geqslant 0} = \Phi(P, T)$. By Proposition 3.5.1 there is a unique pseudo-parabolic k-subgroup $Q' \subseteq G$ containing T with $\Phi(Q', T) = \Phi_{\ell\geqslant 0}$, and $Q' \subseteq P$ because $\Phi(Q', T) \subseteq \Phi(P, T)$. Hence, by the unique characterization of root groups in terms of the T-action on G, for each $a \in \Phi_{\lambda>0}$ we have $U_a \subseteq Q'$. The collection of such U_a directly spans $U = \mathscr{R}_{u,k}(P)$, so $U \subseteq Q'$. Hence, $T \ltimes U \subseteq Q'$.

It remains to prove that the image $\overline{Q}'$ of Q' in $\overline{P}$ satisfies $\Phi(\overline{Q}', \overline{T}) = \Phi(\overline{Q}, \overline{T})$. By definition, $\overline{Q}' = Q'/U$. Since U is directly spanned by root groups associated to certain non-divisible roots, $\mathrm{Lie}(U)$ is spanned by the root spaces in $\mathrm{Lie}(G)$ for all roots in $\Phi(U, T)$. Hence, $\Phi(\overline{Q}', \overline{T})$ consists of the roots in $\Phi(Q', T) = \Phi_{\ell\geqslant 0}$ not in $\Phi(U, T) = \Phi_{\lambda>0}$. By our choice of ℓ we have $\Phi_{\lambda<0} \subseteq \Phi_{\ell<0}$, so the complement of $\Phi(U, T)$ in $\Phi(Q', T)$ consists of the locus in $\Phi(Q', T) = \Phi_{\ell\geqslant 0}$ at which λ vanishes. But $\Phi_{\lambda=0} = [I]$, and ℓ coincides with $\varepsilon\mu$ on $[I]$. Hence, $\Phi(\overline{Q}', \overline{T})$ is the locus in $[I]$ at which $\mu \geqslant 0$, and this is $\Phi(\overline{Q}, \overline{T})$ due to how μ was initially chosen. □

Remark 3.5.6 Consider a smooth connected affine group G over a field k and a pseudo-parabolic k-subgroup P. Since every maximal k-torus in P is maximal in G, by applying Proposition 2.2.10 to P it follows from Lemma 3.5.5 that every pseudo-parabolic k-subgroup P' of P is a pseudo-parabolic k-subgroup of G. (Proposition 3.5.8 gives a converse.)

Proposition 3.5.7 *Let G be a smooth connected affine group over a field k. A pseudo-parabolic k-subgroup P in G is its own scheme-theoretic normalizer in G, as is $P_G(\lambda)$ for any 1-parameter k-subgroup λ of G.*

Proof We may assume $k = k_s$. The proof that $P_G(\lambda)$ is its own normalizer will be given at the end. As a first step in the study of $N_G(P)$, we show that $\mathrm{Lie}(N_G(P)) = \mathrm{Lie}(P)$, so $N_G(P)$ is smooth with identity component P. This will then reduce the problem for $N_G(P)$ to a comparison of k-points. Our argument will rest on a general description of Lie algebras of scheme-theoretic normalizers and centralizers of smooth k-subgroups H of G: $\mathrm{Lie}(Z_G(H)) = \bigcap_{h \in H(k)} \ker(\mathrm{Ad}(h) - 1)$ and

$$\mathrm{Lie}(N_G(H)) = \{X \in \mathrm{Lie}(G) \mid \mathrm{Ad}(h)(X) - X \in \mathrm{Lie}(H) \text{ for all } h \in H(k)\}.$$

(In nonzero characteristic these are generally strictly smaller than the corresponding Lie-theoretic centralizer and Lie-theoretic normalizer of $\mathrm{Lie}(H)$ in $\mathrm{Lie}(G)$.) For the convenience of the reader, let us now recall the proofs of these formulas. By Zariski-density of $H(k)$ in H we have $Z_G(H) = \bigcap_{h \in H(k)} Z_G(h)$, so $\mathrm{Lie}(Z_G(H))$ is the intersection of the Lie algebras $\mathrm{Lie}(Z_G(h))$. The centralizer $Z_G(h)$ is the fiber over the identity of the k-scheme map $g \mapsto (hgh^{-1})g^{-1}$, so passing to tangent spaces at the identity yields $\mathrm{Lie}(Z_G(h)) = \ker(\mathrm{Ad}(h) - 1)$. This establishes the formula for $\mathrm{Lie}(Z_G(H))$. To handle $N_G(H)$, for each $h \in H(k)$ let $f_h : G \to G$ denote $x \mapsto hxh^{-1}x^{-1}$ (so $f_h(1) = 1$). By the Zariski-density of $H(k)$ in H, we have $N_G(H) = \bigcap_{h \in H(k)} f_h^{-1}(H)$. Passing to tangent spaces at the identity yields the desired description of $\mathrm{Lie}(N_G(H))$.

By pseudo-parabolicity, $P = P_G(\lambda)\mathscr{R}_{u,k}(G)$ for a k-homomorphism $\lambda : \mathrm{GL}_1 \to G$. Let T be a maximal k-torus in G containing $\lambda(\mathrm{GL}_1)$, so $Z_G(T) \subseteq Z_G(\lambda) \subseteq P_G(\lambda) \subseteq P$. Suppose there exists $X \in \mathrm{Lie}(N_G(P))$ not contained in $\mathrm{Lie}(P)$. Both of these Lie subalgebras of $\mathfrak{g} := \mathrm{Lie}(G)$ are stable under the adjoint action of T (as T is contained in $N_G(P)$), so we can find such an X that is an eigenvector for the T-action on $\mathfrak{g}$. But $\mathrm{Lie}(Z_G(T)) \subseteq \mathrm{Lie}(P)$ (due to the corresponding group containment inside of G), so

$$X \notin \mathrm{Lie}(Z_G(T)) = \bigcap_{t \in T(k)} \ker(\mathrm{Ad}(t) - 1).$$

It follows that if $a : T \to \mathrm{GL}_1$ is the eigencharacter for X then $a \neq 1$. Choose $t_0 \in T(k)$ such that $a(t_0) \neq 1$, so $\mathrm{Ad}(t_0)(X) - X = (a(t_0) - 1)X \notin \mathrm{Lie}(P)$. But $X \in \mathrm{Lie}(N_G(P))$ by hypothesis, so since $t_0 \in P(k)$ we have $\mathrm{Ad}(t_0)(X) - X \in \mathrm{Lie}(P)$. This is a contradiction.

We have shown that Lie(P) = Lie($N_G(P)$), so $P = N_G(P)^0$ as subschemes of G. In particular, $N_G(P)$ is smooth. It remains to show that if Q is a smooth k-subgroup of G containing P and normalizing P then $Q = P$. (The point is that we do not assume Q is connected.) Since Q is smooth, its conjugation action on G preserves $\mathscr{R}_{u,k}(G)$. But $\mathscr{R}_{u,k}(G) \subseteq P$ by definition of pseudo-parabolicity, so we can pass to the quotients $G/\mathscr{R}_{u,k}(G)$ and $P/\mathscr{R}_{u,k}(G)$ to reduce to the case where G is pseudo-reductive.

By smoothness, to prove $Q = P$ it is equivalent to show that $Q(k) = P(k)$. For any $q \in Q(k)$, qTq^{-1} is a maximal k-torus in the smooth connected k-group P. Since all maximal k-tori in P are $P(k)$-conjugate (as $k = k_s$), there exists $h \in P(k)$ such that $qTq^{-1} = hTh^{-1}$. Hence, $h^{-1}q \in N_G(T)(k)$, so

$$Q(k) = P(k) \cdot (Q(k) \cap N_G(T)(k)).$$

Thus, it suffices to show that $Q(k) \cap N_G(T)(k) \subseteq P(k)$.

Let $U = \mathscr{R}_{u,k}(P)$, so $\overline{P} = P/U$ is a pseudo-reductive k-group and T maps isomorphically onto a maximal k-torus $\overline{T}$ in $\overline{P}$. Fix a positive system of roots Φ^+ in $\Phi(\overline{P}, \overline{T})$. For $q \in Q(k) \cap N_G(T)(k)$, conjugation by q on P induces an automorphism of $\overline{P}$ that preserves $\overline{T}$, so q likewise acts on $\Phi(\overline{P}, \overline{T})$. In particular, $q.\Phi^+$ is a positive system of roots in $\Phi(\overline{P}, \overline{T})$. Since $W(\overline{P}, \overline{T})(k)$ acts transitively on the set of positive systems of roots in $\Phi(\overline{P}, \overline{T})$, there exists $w \in W(\overline{P}, \overline{T})(k)$ such that $w.(q.\Phi^+) = \Phi^+$. Let $n_w \in N_{\overline{P}}(\overline{T})(k)$ be a representative of w. Using Lemma 3.2.1 we now choose a lift g of n_w in $P(k)$ which normalizes T.

We can replace q with gq, so $q.\Phi^+ = \Phi^+$. The set of roots Φ^+ corresponds to a unique minimal pseudo-parabolic k-subgroup $\overline{P}'$ in $\overline{P}$ containing $\overline{T}$ (Proposition 3.3.15), and since q normalizes T and $q.\Phi^+ = \Phi^+$ it follows that the q-action on $\overline{P}$ carries $\overline{P}'$ back onto itself. Hence, the preimage P' of $\overline{P}'$ in P is normalized by q. By Proposition 3.5.5, P' is a minimal pseudo-parabolic k-subgroup of G, so the set $\Phi(P', T)$ of nontrivial T-weights on Lie(P') is a positive system of roots in $\Phi(G, T)$. The action on $\Phi(G, T)$ by the element $q \in N_G(T)(k)$ preserves this positive system of roots (since it preserves P'). Thus, by simple transitivity of the action of $W(G, T)(k)$ on the set of positive systems of roots in $\Phi(G, T)$ (by Proposition 3.2.7), we conclude that q has trivial image in $W(G, T)(k)$. That is, $q \in Z_G(T)(k) \subseteq P(k)$, so we have proved that $N_G(P) = P$.

To prove that $\mathscr{P} := P_G(\lambda)$ is its own scheme-theoretic normalizer for any λ, we can assume $k = \overline{k}$. Let $\mathscr{U} = \mathscr{R}_u(G)$, so the image of $\mathscr{P}$ in $G^{\rm red} = G/\mathscr{U}$ is a parabolic subgroup. Parabolic subgroups in connected reductive groups are their own scheme-theoretic normalizers (either as a special case of what we just proved above, or by [SGA3, XXII, 5.8.5]), so we are reduced to showing that $\mathscr{U} \cap N_G(\mathscr{P}) = \mathscr{U} \cap \mathscr{P}$. Since λ defines a GL_1-action on the normal subgroup $\mathscr{U}$ of G, by applying Proposition 2.1.12 to $\mathscr{U}$ we have $U_{\mathscr{U}}(-\lambda) \times P_{\mathscr{U}}(\lambda) = \mathscr{U}$ via multiplication. But $P_{\mathscr{U}}(\lambda) = \mathscr{U} \cap P_G(\lambda) = \mathscr{U} \cap \mathscr{P}$, so the problem is to prove that $U_{\mathscr{U}}(-\lambda) \cap N_G(\mathscr{P}) = 1$. More generally, we shall prove that $U_G(-\lambda) \cap N_G(\mathscr{P}) = 1$. Let R be a k-algebra and choose $u \in U_G(-\lambda)(R)$ that normalizes $\mathscr{P}_R = P_{G_R}(\lambda_R)$. We will show that $u \in Z_G(\lambda)(R)$ and hence $u = 1$. Since λ normalizes $U_G(-\lambda)$, for any R-algebra R' and $t \in R'^{\times}$ the commutator $(u, \lambda(t))$ lies in $U_G(-\lambda)(R')$. But it also lies in $P_G(\lambda)(R')$ since u normalizes $P_{G_R}(\lambda_R)$ (in which λ_R takes its values). Hence, this commutator is trivial, so $u \in Z_G(\lambda)(R)$. □

Using Proposition 3.5.7, we can now analyze subgroups containing pseudo-parabolic subgroups:

Proposition 3.5.8 *Let G be a smooth connected affine group over a field k, and let P be a pseudo-parabolic k-subgroup of G. Any smooth closed k-subgroup Q of G containing P is pseudo-parabolic in G. (In particular, Q is connected.) Moreover, P is a pseudo-parabolic k-subgroup of Q.*

Proof By definition of pseudo-parabolicity, P contains $\mathscr{R}_{u,k}(G)$. Using Proposition 2.2.10, we may pass to images in $G/\mathscr{R}_{u,k}(G)$ to reduce to the case where G is pseudo-reductive over k. By Proposition 3.5.2, we can assume $k = k_s$. Connectedness of P implies that Q^0 contains P. If we can prove that Q^0 is pseudo-parabolic in G then Q^0 is its own normalizer in G by Proposition 3.5.7, yet Q normalizes Q^0 so then $Q = Q^0$ and we would be done. Hence, we may replace Q with Q^0 to reduce to the case where Q is connected. Moreover, we can assume that P is a minimal pseudo-parabolic k-subgroup of G.

Fix a maximal k-torus T of G contained in P, so $Z_G(T) \subseteq P$ (Proposition 2.2.11). We may and do identify $T_{\overline{k}}$ with a maximal torus T' in $G' := G_{\overline{k}}^{\mathrm{red}}$. Let $\Phi = \Phi(G,T)$, and $\Phi' = \Phi(G',T')$, so Φ' is the set of non-multipliable roots in Φ (Theorem 2.3.10). The set $\Phi(P,T)$ of roots in Φ occurring as T-weights on $\mathrm{Lie}(P)$ is a positive system of roots in Φ (Proposition 3.3.15), and by Proposition 3.5.4 the image P' of $P_{\overline{k}}$ under the quotient map $\pi : G_{\overline{k}} \twoheadrightarrow G'$ is the Borel subgroup of G' containing T' such that $\Phi(P',T') = \Phi(P,T) \cap \Phi'$ is the positive system of roots in Φ' corresponding to $\Phi(P,T)$.

Consider the image $Q' = \pi(Q_{\overline{k}})$ in G'. This is a smooth connected subgroup of G' containing the Borel subgroup P', so it is a parabolic subgroup containing T'. In particular, it is a pseudo-parabolic subgroup containing

T' (by Proposition 2.2.9), so $Q' = P_{G'}(\lambda')$ for some $\lambda' \in X_*(T')$ (by Proposition 2.2.11). Since $k = k_s$ we have $X_*(T) = X_*(T')$ via the isomorphism $T_{\overline{k}} \simeq T'$, so $\lambda' = \lambda_{\overline{k}}$ for a unique $\lambda \in X_*(T)$. We will prove that $P_G(\lambda) \subseteq Q$ and that the Lie algebras of these two groups coincide inside of $\mathrm{Lie}(G)$, whence the groups coincide inside of G (by smoothness and connectedness) and we will be done with the proof that Q is pseudo-parabolic in G.

By Corollary 3.3.12, the k-group $P_G(\lambda)$ is generated by $Z_G(T)$ and the k-groups $U_{(a)}$ for non-multipliable $a \in \Phi(P_G(\lambda), T)$; i.e., for $a \in \Phi'$ such that $\langle a, \lambda\rangle \geqslant 0$. But $\pi(P_G(\lambda)) = P_{G'}(\lambda') = Q'$, so by Proposition 3.5.4 we see that $U_{(a')} \subseteq P_G(\lambda)$ for $a' \in \Phi(Q', T')$ (viewed as a character of T in the natural way), so $P_G(\lambda)$ is generated by $Z_G(T)$ and the k-groups $U_{(a')}$ for $a' \in \Phi(Q', T')$. It follows from the discussion near the end of Remark 2.3.6 that for $a \in \Phi(Q, T)$ its unique positive rational multiple $a' \in \Phi'$ occurs in $\Phi(Q', T')$. Now since $Z_G(T) \subseteq P \subseteq Q$, it suffices to prove $U_{(a)} \subseteq Q$ for all $a \in \Phi(Q, T)$ (as then $P_G(\lambda) \subseteq Q$ with a reverse inclusion of Lie algebras inside $\mathrm{Lie}(G)$).

To prove this, choose $a \in \Phi(Q, T)$. If a lies in the positive system of roots $\Phi^+ := \Phi(P, T) \subseteq \Phi(Q, T)$ inside Φ then $U_{(a)} \subseteq P \subseteq Q$. Thus, if in general a is carried by $N_{Q(k)}(T)$ into Φ^+ then we are done. It is equivalent to say that the non-multipliable root $a' \in \Phi'$ is carried by $N_{Q(k)}(T)$ into Φ^+. To construct enough such elements of $N_{Q(k)}(T)$ requires some preliminary considerations with finite étale groups.

Since $Z_G(T) \subseteq P \subseteq Q$, we have $Z_Q(T) = Z_G(T)$. By Lemma A.2.9, $N_Q(T)$ is smooth and $W(Q, T) := N_Q(T)/Z_Q(T)$ is finite étale over k. Likewise, $Z_{Q'}(T') = Z_{G'}(T') = T'$, $N_{Q'}(T')$ is smooth, and $W(Q', T') := N_{Q'}(T')/T'$ is finite étale over $\overline{k}$. There is a natural map of finite étale $\overline{k}$-groups $W(Q, T)_{\overline{k}} \to W(Q', T')$ induced by the surjective map $Q_{\overline{k}} \twoheadrightarrow Q'$ that carries $T_{\overline{k}}$ isomorphically onto T'. This latter isomorphism of tori implies that a point of $N_Q(T)$ valued in any k-algebra R centralizes T_R inside of Q_R if and only if the corresponding $R_{\overline{k}}$-point of $N_{Q'}(T')$ centralizes $T'_{R_{\overline{k}}}$ inside of $Q'_{R_{\overline{k}}} = (Q_R)_{\overline{k}}$. Hence, the map $W(Q, T)_{\overline{k}} \to W(Q', T')$ between finite étale $\overline{k}$-groups has trivial kernel. By Lemma 3.2.1, the map $\pi : N_Q(T)_{\overline{k}} = N_{Q_{\overline{k}}}(T_{\overline{k}}) \to N_{Q'}(T')$ is surjective. In particular, $W(Q, T)_{\overline{k}} \to W(Q', T')$ is an isomorphism.

Since $W(Q, T)$ is étale and $k = k_s$, we have $W(Q, T)(k) = W(Q, T)(\overline{k})$. But it was just proved that $W(Q, T)_{\overline{k}} \to W(Q', T')$ is an isomorphism of $\overline{k}$-groups, so $W(Q, T)(k) \to W(Q', T')(\overline{k})$ is an isomorphism. The formation of coset spaces for smooth k-groups commutes with the formation of k-points since $k = k_s$, so the group $N_Q(T)(k) = N_{Q(k)}(T)$ maps onto $W(Q', T')(\overline{k})$. Hence, the natural map $N_{Q(k)}(T) \to W(Q', T')(\overline{k})$ is surjective. In particular, the $N_{Q(k)}(T)$-action on Φ restricts to the $W(Q', T')(\overline{k})$-action on Φ' via the surjective map $N_{Q(k)}(T) \twoheadrightarrow W(Q', T')(\overline{k})$. The classification of parabolic subgroups of connected reductive $\overline{k}$-groups (as "standard" parabolic subgroups containing a given Borel subgroup) shows that the $W(Q', T')(\overline{k})$-action on Φ' carries each element of $\Phi(Q', T')$ into the positive system of roots $\Phi(P', T') = \Phi' \cap \Phi^+$. Applying this to $a' \in \Phi(Q', T')$, we have proved that Q is pseudo-parabolic in G.

Finally, we prove that P is pseudo-parabolic in Q. Since G is pseudo-reductive, $P = P_G(\mu)$ for some $\mu : \mathrm{GL}_1 \to G$. But μ takes values in $Z_G(\mu) \subseteq P \subseteq Q$, so $P_Q(\mu)$ makes sense and $P_Q(\mu) = P_G(\mu) \cap Q = P \cap Q = P$. Hence, P is pseudo-parabolic in Q provided that P contains $\mathscr{R}_{u,k}(Q)$. The k-group $\mathscr{R}_{u,k}(Q)$ is directly spanned by the groups $U_{(a)}$ for those $a \in \Phi(Q, T)$ such that $-a \notin \Phi(Q, T)$. Since $\mathrm{Lie}(P) \subseteq \mathrm{Lie}(Q)$, we have $-a \notin \Phi(P, T)$. Also, by pseudo-parabolicity of P, some positive system of roots in $\Phi(G, T)$ is contained in $\Phi(P, T)$. Hence, $a \in \Phi(P, T)$, so $U_{(a)} \subseteq P$ by Corollary 3.3.12. Varying over all such a, indeed $\mathscr{R}_{u,k}(Q) \subseteq P$. □

Corollary 3.5.9 *Let G be a smooth connected affine group over a field k, P a k-subgroup of G, and K/k an extension field. If P_K is pseudo-parabolic in G_K then P is pseudo-parabolic in G.*

This is a refinement of Proposition 3.5.2(1).

Proof We may and do assume $k = k_s$. Since P_K is smooth and connected, so is P. Since $\mathscr{R}_{u,k}(G)_K \subseteq \mathscr{R}_{u,K}(G_K) \subseteq P_K$, necessarily $\mathscr{R}_{u,k}(G) \subseteq P$. Thus, by Proposition 2.2.10, we can pass to quotients by $\mathscr{R}_{u,k}(G)$ so that G is pseudo-reductive over k. Let $T \subseteq P$ be a maximal k-torus, so T_K is a maximal K-torus in P_K and hence in G_K. Let $\pi : G_K \to G' = G_K/\mathscr{R}_{u,K}(G_K)$ be the natural projection onto the maximal pseudo-reductive quotient over K.

The pseudo-parabolic K-subgroup P_K of G_K must contain $\mathscr{R}_{u,K}(G_K)$, so the quotient $P' = \pi(P_K)$ is a pseudo-parabolic K-subgroup of G' in which T_K is a maximal K-torus. Hence, by Proposition 2.2.11, there exists a 1-parameter K-subgroup $\lambda' : \mathrm{GL}_1 \to T_K$ such that $P' = P_{G'}(\lambda')$. Since $k = k_s$, λ' descends uniquely to a 1-parameter k-subgroup $\lambda : \mathrm{GL}_1 \to T$, so $P_G(\lambda)_K = P_{G_K}(\lambda')$. Thus, $\pi(P_G(\lambda)_K) = P_{G'}(\lambda') = P' = \pi(P_K)$, so $P_K = P_G(\lambda)_K \mathscr{R}_{u,K}(G_K)$. This implies that P contains the pseudo-parabolic k-subgroup $P_G(\lambda)$ of G, and so P is a pseudo-parabolic k-subgroup of G by Proposition 3.5.8. □

Corollary 3.5.10 *Let G be a smooth connected affine group over a field k, and let P be a pseudo-parabolic k-subgroup of G. For any separable field extension K/k, the normalizers in $G(K)$ of $\mathscr{R}_{u,k}(P)_K = \mathscr{R}_{u,K}(P_K)$ and $\mathscr{R}_{us,k}(P)_K = \mathscr{R}_{us,K}(P_K)$ both equal $P(K)$.*

Proof Since P_K is pseudo-parabolic in G_K (Proposition 3.5.2(1)), we can rename K as k. By Galois descent we can also assume $k = k_s$. The k-subgroup $\mathscr{R}_{us,k}(P)$ in $\mathscr{R}_{u,k}(P)$ is stable under k-automorphisms, so its normalizer in $G(k)$ contains the normalizer of $\mathscr{R}_{u,k}(P)$ (which in turn clearly contains $P(k)$). Thus, letting Q denote the Zariski closure of $N_{G(k)}(\mathscr{R}_{us,k}(P))$ in G, it suffices to prove that $Q = P$. By its definition, Q is a smooth k-subgroup of G containing P (since $k = k_s$), so it is pseudo-parabolic in G by Proposition 3.5.8. Choose a maximal k-torus T in P, so T is also maximal in G and lies in Q. Let $G' = G_{\overline{k}}^{\mathrm{red}}$ and let T' denote the (isomorphic) image of $T_{\overline{k}}$ in G'. By Proposition 3.5.4, to prove $P = Q$ it suffices to prove the equality of the corresponding parabolic image subgroups P' and Q' in G' that contain T'.

Since $Q(k)$ has dense image in Q', as $k = k_s$, it follows that Q' normalizes the image of $\mathscr{R}_{us,k}(P)_{\overline{k}}$ in G'. We claim that the image of $\mathscr{R}_{us,k}(P)_{\overline{k}}$ in G' is exactly $\mathscr{R}_u(P')$. By Corollary 2.2.5, this image is the same as that of $\mathscr{R}_{u,k}(P)_{\overline{k}}$. Hence, we may first replace G with $G/\mathscr{R}_{u,k}(G)$ so that G is pseudo-reductive over k. It follows that $P = P_G(\lambda)$ for some $\lambda : \mathrm{GL}_1 \to T$, so $\mathscr{R}_{u,k}(P) = U_G(\lambda)$. Let $\lambda' : \mathrm{GL}_1 \to T'$ correspond to $\lambda_{\overline{k}}$, so the quotient map $G_{\overline{k}} \twoheadrightarrow G'$ carries $U_G(\lambda)_{\overline{k}} = U_{G_{\overline{k}}}(\lambda_{\overline{k}})$ onto $U_{G'}(\lambda')$ and carries $P_{\overline{k}}$ onto $P_{G'}(\lambda')$ (Corollary 2.1.9). We conclude that $P' = P_{G'}(\lambda')$, so the k-group $\mathscr{R}_u(P') = U_{G'}(\lambda')$ is indeed the image of $\mathscr{R}_{us,k}(P)_{\overline{k}}$.

The containments $P' \subseteq Q'$ and $Q'(\overline{k}) \subseteq N_{G'(\overline{k})}(\mathscr{R}_u(P'))$ now allow us to replace G with G' so as to reduce to checking that in a connected reductive group G over an algebraically closed field, any parabolic subgroup P is the normalizer in G of $\mathscr{R}_u(P)$ at the level of geometric points. This is an immediate consequence of the explicit description [Bo2, 21.12] of the parabolic subgroups containing a fixed Borel subgroup. (The point is that if Q is any parabolic subgroup strictly containing P then relative to a maximal k-torus of P the Lie algebra of Q supports a pair of opposite roots one of which is supported on $\mathrm{Lie}(\mathscr{R}_u(P))$. A suitable Weyl element coming from $Q(k)$ switches these roots, so Q cannot normalize $\mathscr{R}_u(P)$.) □

Here is a strengthening of the correspondence in Proposition 3.5.4:

Corollary 3.5.11 *Let G be a smooth connected affine group over a field k, and let $G' = G_{\overline{k}}^{\mathrm{red}}$. Choose a maximal k-torus T in G, and let T' be the* *(isomorphic) image of* $T_{\overline{k}}$ *in* G'*. For a pseudo-parabolic* k*-subgroup* P *in* G *and the corresponding parabolic image* P' *of* $P_{\overline{k}}$ *in* G', $T \subseteq P$ *if and only if* $T' \subseteq P'$*. In particular,* P' *determines* P *without reference to a choice of* T*.*

Note that we do not claim that every parabolic $\overline{k}$-subgroup of G' has the form P' for some pseudo-parabolic k-subgroup P in G; this is generally not true.

Proof Choose P such that $T' \subseteq P'$. By Proposition 3.5.4, there is a unique pseudo-parabolic k-subgroup Q in G containing T such that $Q' = P'$ inside of G'. It is necessary and sufficient to prove $P = Q$ (so then $T \subseteq P$). We can replace Q with the smooth connected k-subgroup of G generated by P and Q, as this subgroup is pseudo-parabolic by Proposition 3.5.8, so we can assume $P \subseteq Q$.

Let S be a maximal k-torus in P, so it is maximal in G and contained in Q. The bijectivity in Proposition 3.5.4 for pseudo-parabolic subgroups containing S implies $P = Q$ since $P' = Q'$. □

As an application of our understanding of pseudo-parabolic subgroups in smooth connected affine groups G over a general field k, we can prove $G(k)$-conjugacy results due to Borel and Tits for maximal k-split tori and minimal pseudo-parabolic k-subgroups of G. We only use these facts in Chapter 9, so the development is given in Appendix C. The following result concerning intersections of pseudo-parabolic subgroups is used in §C.2 in the proof of the Bruhat decomposition for $G(k)$ with any (smooth connected affine) G and any field k.

Proposition 3.5.12 *Let* G *be a smooth connected affine group over a field* k*. Let* P *and* Q *be pseudo-parabolic* k*-subgroups of* G*. Then:*

(1) *The scheme-theoretic intersection* $P \cap Q$ *is smooth and connected, and it contains a maximal* k*-torus of* G*.*
(2) *The* k*-subgroup* $(P \cap Q)\mathscr{R}_{u,k}(P)$ *in* G *is pseudo-parabolic.*
(3) *If* P *is* $G(k)$*-conjugate to* Q *and if the smooth connected* k*-subgroup* $P \cap Q$ *in* G *is pseudo-parabolic then* $P = Q$*.*

Proof We first explain how to reduce the assertion over k to the assertion over k_s, so grant the entire result over separably closed ground fields. Note that P_{k_s} and Q_{k_s} are pseudo-parabolic k_s-subgroups of G_{k_s}, so (2) over k_s implies (2) over k, and similarly for (3). The smoothness and connectedness of $(P \cap Q)_{k_s} = P_{k_s} \cap Q_{k_s}$ imply the smoothness and connectedness of $P \cap Q$ (as for any finite type k-scheme in place of $P \cap Q$), so all maximal k-tori of $P \cap Q$ have the same dimension. Hence, $P \cap Q$ contains a maximal k-torus of G if and only if the common dimension of the maximal k-tori of $P \cap Q$ coincides with the common dimension of the maximal k-tori of G. These dimensions can be computed after scalar extension to k_s, so (1) over k_s implies (1) over k. We may and do now assume for the remainder of the proof that $k = k_s$.

By Proposition 2.2.10 we can replace G with $G/\mathscr{R}_{u,k}(G)$ and replace P and Q with their respective images in $G/\mathscr{R}_{u,k}(G)$ to assume that G is pseudo-reductive. To prove that $P \cap Q$ contains a maximal k-torus of G, we may assume that P and Q are minimal pseudo-parabolic k-subgroups. By conjugacy of such subgroups (Proposition 3.3.17), there exists $g \in G(k)$ such that $Q = gPg^{-1}$. Choose a maximal k-torus S of P, so S is maximal in G and hence the Bruhat decomposition of $G(k)$ in Theorem 3.4.5 yields $g = pnp'$ for some $p, p' \in P(k)$ and $n \in N_G(S)(k)$. Hence, $Q = gPg^{-1} = pnPn^{-1}p^{-1}$, and so $P \cap Q = p(P \cap nPn^{-1})p^{-1}$. Since n normalizes S, we conclude that $P \cap Q$ contains the maximal k-torus pSp^{-1} of G.

We now drop minimality hypotheses on P and Q, and we fix a maximal k-torus T of G contained in $P \cap Q$. Let $\Phi = \Phi(G, T)$, $Z = Z_G(T)$, and $\mathfrak{z} = \mathrm{Lie}(Z)$. For each $a \in \Phi$, let $U_{(a)}$ denote the k-subgroup as in Definition 2.3.4, and let $\mathfrak{u}_a$ be the a-root space in the Lie algebra $\mathfrak{g}$ of G. By Proposition 2.2.11 there exist 1-parameter k-subgroups $\lambda, \mu : \mathrm{GL}_1 \rightrightarrows T$ such that $P = P_G(\lambda)$ and $Q = P_G(\mu)$, so $Z \subseteq P \cap Q$. By Corollary 3.3.12, for $a \in \Phi$, $U_{(a)} \subseteq P$ if and only if $a \in \Phi(P, T)$, and likewise for Q. Hence, $U_{(a)} \subseteq P \cap Q$ for every $a \in \Phi(P, T) \cap \Phi(Q, T)$, so the smooth connected k-subgroup H of G generated by Z and the k-groups $U_{(a)}$ with $a \in \Phi(P, T) \cap \Phi(Q, T)$ is contained in $P \cap Q$.

The T-weight space decomposition of $\mathrm{Lie}(G)$ yields the T-equivariant decompositions

$$\mathrm{Lie}(P) = \mathfrak{z} \oplus \left(\bigoplus_{a \in \Phi(P,T)} \mathfrak{u}_a \right), \quad \mathrm{Lie}(Q) = \mathfrak{z} \oplus \left(\bigoplus_{a \in \Phi(Q,T)} \mathfrak{u}_a \right),$$

so

$$\begin{aligned} \mathrm{Lie}(H) \subseteq \mathrm{Lie}(P \cap Q) &= \mathrm{Lie}(P) \cap \mathrm{Lie}(Q) \\ &= \mathfrak{z} \oplus \left(\bigoplus_{a \in \Phi(P,T) \cap \Phi(Q,T)} \mathfrak{u}_a \right) \\ &\subseteq \mathrm{Lie}(H). \end{aligned}$$

Thus, the inclusion $H \subseteq P \cap Q$ of k-subgroup schemes of G induces an equality on Lie algebras. But H is smooth and connected so $H = (P \cap Q)^0$. In particular,

$(P \cap Q)^0$ is smooth and hence $P \cap Q$ is smooth. This concludes the proof of (1) apart from the connectedness claim.

According to Corollary 2.2.5, $\mathscr{R}_{us,k}(P) = U_G(\lambda) = \mathscr{R}_{u,k}(P)$. By Proposition 3.5.8, any smooth closed k-subgroup of G containing a pseudo-parabolic k-subgroup is pseudo-parabolic, and in particular is connected. Thus, to prove (2) it suffices to show that $(P \cap Q)^0 \mathscr{R}_{us,k}(P)$ is a pseudo-parabolic k-subgroup of G, as then

$$(P \cap Q)\mathscr{R}_{us,k}(P) = (P \cap Q)^0 \mathscr{R}_{us,k}(P). \tag{3.5.1}$$

We claim that the intersection $J := (P \cap Q)^0 \cap \mathscr{R}_{us,k}(P)$ is smooth. Its Lie algebra is the intersection of the Lie algebras of P, Q, and $\mathscr{R}_{us,k}(P)$, so $\mathrm{Lie}(J)$ is spanned by the $\mathfrak{u}_a$ for $a \in \Phi$ such that $U_{(a)} \subseteq J$ (see Proposition 3.5.1(2)). Hence, as in our proof of smoothness of $P \cap Q$, the smooth connected k-subgroup of J^0 generated by the $U_{(a)}$ that are contained in J must coincide with J^0. In particular, J^0 is smooth, so J is smooth. It follows that the surjective homomorphism

$$(P \cap Q)^0 \ltimes \mathscr{R}_{us,k}(P) \to (P \cap Q)^0 \mathscr{R}_{us,k}(P)$$

has smooth kernel, so

$$\begin{aligned} \mathrm{Lie}((P \cap Q)^0 \mathscr{R}_{us,k}(P)) &= \mathrm{Lie}((P \cap Q)^0) + \mathrm{Lie}(\mathscr{R}_{us,k}(P)) \\ &= \mathrm{Lie}(P) \cap \mathrm{Lie}(Q) + \mathrm{Lie}(U_G(\lambda)). \end{aligned}$$

For any $a \in \Phi$, we have $a \in \Phi(P, T)$ if and only if $\langle a, \lambda \rangle \geqslant 0$, and moreover a occurs as a T-weight on the Lie algebra of $\mathscr{R}_{us,k}(P) = U_G(\lambda)$ if and only if $\langle a, \lambda \rangle > 0$. Hence, if $a \in \Phi$ satisfies $\langle a, \lambda \rangle \neq 0$, then exactly one $b \in \{\pm a\}$ occurs as a T-weight in $\mathrm{Lie}(\mathscr{R}_{us,k}(P))$, with the corresponding $U_{(b)}$ contained in $\mathscr{R}_{us,k}(P)$, whereas if $\langle a, \lambda \rangle = 0$ then both a and $-a$ lie in $\Phi(P, T)$ and at least one of them lies in $\Phi(Q, T)$ (since $\Phi(Q, T)$ contains a positive system of roots in Φ). But $U_{(b)} \subseteq (P \cap Q)^0$ for each $b \in \Phi(P, T) \cap \Phi(Q, T)$. Thus, the pseudo-parabolicity of $(P \cap Q)^0 \mathscr{R}_{us,k}(P)$ (and so the proof of (2)) is a consequence of the following surprising general lemma.

Lemma 3.5.13 *Let G be a pseudo-reductive group over a field k, and T a maximal k-torus in G. Let H be a smooth k-subgroup of G such that:*

(i) $Z_G(T) \subseteq H$,
(ii) $\Phi(G_{k_s}, T_{k_s}) = \Phi(H_{k_s}, T_{k_s}) \cup -\Phi(H_{k_s}, T_{k_s})$, *and*
(iii) $a \in \Phi(H_{k_s}, T_{k_s})$ *implies* $U^G_{(a)} \subseteq H_{k_s}$.

Then H is pseudo-parabolic in G.

Proof We may and do assume $k = k_s$. Observe also that by (iii), if an element $a \in \Phi(G,T)$ lies in $\Phi(H,T)$ then so does every positive rational multiple of a in $\Phi(G,T)$.

Let H' denote the image of $H_{\overline{k}}$ in $G' = G_{\overline{k}}^{\text{red}}$ and let T' be the maximal torus image of $T_{\overline{k}}$ in G', so H' contains T'. The quotient map $H_{\overline{k}} \to H'$ cannot kill any $(U_{(a)}^G)_{\overline{k}}$ for $a \in \Phi(H,T)$ (due to the pseudo-reductivity of G over k), so $(U_{(a)}^G)_{\overline{k}}$ is carried T'-equivariantly onto a root group of G' contained in H'. Thus, this root group must correspond to a positive rational multiple of a, so it follows from Remark 2.3.6 and Theorem 2.3.10 that $\Phi(H',T')$ consists of the non-multipliable roots in $\Phi(H,T)$. This argument also shows that H' satisfies the analogues of (i), (ii), and (iii) relative to (G',T').

Assume H' is parabolic in G', so there is a unique pseudo-parabolic k-subgroup P in G such that $T \subseteq P$ and $P_{\overline{k}}$ maps onto H' (Proposition 3.5.4), and $\Phi(H',T')$ is the set of non-multipliable roots in $\Phi(P,T)$. Hence, $\Phi(P,T)$ and $\Phi(H,T)$ have the same non-multipliable elements, so $\Phi(P,T) = \Phi(H,T)$. By (i) and (iii), we would then have $H^0 = P$, so $H = P$ by Proposition 3.5.7.

It remains to show that H' is parabolic in G', which is to say that we can assume k is algebraically closed and G is reductive. Since H contains the maximal torus T of G, this is also a maximal torus of H and so there is a Borel subgroup B of H containing T. The containment $\mathscr{R}_u(H) \subseteq B$ and the reductivity of $H/\mathscr{R}_u(H)$ imply that $\Phi(H,T)$ is the union of $\Phi(B,T)$ and $-\Phi(B,T)$, so likewise for $\Phi(G,T)$ by (ii). In particular, the size of $\Phi(B,T)$ is at least as large as $(1/2)\#\Phi(G,T)$, so

$$\dim B = \dim T + \#\Phi(B,T) \geqslant \dim T + (1/2)\#\Phi(G,T).$$

This lower bound is the common dimension of the Borel subgroups of the connected reductive G, so for dimension reasons B is also a Borel subgroup of G. Hence, H is parabolic in G. □

Returning to the proof of Proposition 3.5.12, to complete the proof of (1) it remains to prove that $P \cap Q$ is connected. By (3.5.1), we have $P \cap Q \subseteq (P \cap Q)^0 \mathscr{R}_{us,k}(P)$, so by consideration of geometric points it suffices to show that the $\overline{k}$-points of the triple intersection $I := P \cap Q \cap \mathscr{R}_{us,k}(P)$ are contained in $(P \cap Q)^0$. In other words, we need to show that every $g \in I(\overline{k})$ is contained in a connected subset of $P \cap Q$ that contains the identity. Consider the $\overline{k}$-scheme map $\phi : \mathrm{GL}_1 \to G_{\overline{k}}$ defined by $t \mapsto \lambda(t) g \lambda(t)^{-1}$. This map lands inside of the closed subscheme $I_{\overline{k}}$ since T normalizes each of P, Q, and $\mathscr{R}_{us,k}(P)$. Moreover, by definition of $\mathscr{R}_{us,k}(P) = U_G(\lambda)$ it follows that ϕ extends to a k-scheme map $\widetilde{\phi} : \mathbf{A}^1_{\overline{k}} \to G_{\overline{k}}$ carrying 0 to 1. But the image of $\widetilde{\phi}$ must remain inside of $I_{\overline{k}}$, and it is connected and contains $\widetilde{\phi}(1) = g$ and $\widetilde{\phi}(0) = 1$.

It remains to prove (3), so we assume that $Q = gPg^{-1}$ for some $g \in G(k)$ and that the smooth connected k-subgroup $P \cap Q$ in G is pseudo-parabolic. In particular, $P \cap Q$ contains a minimal pseudo-parabolic k-subgroup P' of G. Thus, $g^{-1}P'g$ is a minimal pseudo-parabolic k-subgroup of G, yet it is also contained in $g^{-1}Qg = P$, so by Remark 3.5.6 and Proposition 3.5.8 both P' and $g^{-1}P'g$ are minimal pseudo-parabolic k-subgroups of P. Hence, by the $P(k)$-conjugacy of such k-subgroups (Proposition 3.3.17), there exists $p \in P(k)$ such that $p^{-1}g^{-1}P'gp = P'$. This says that $gp \in N_G(P')(k) = P'(k)$ (see Proposition 3.5.7), so $Q = gPg^{-1} = P$ as desired. □

One consequence of Corollary 3.5.10 is that in a smooth connected affine group G over a field k, any pseudo-parabolic k-subgroup is uniquely determined by its k-unipotent radical. The following result (which is also a significant refinement of Proposition 3.5.1(3)) shows something much stronger: containment relations among pseudo-parabolic k-subgroups are equivalent to containment relations (in the opposite direction) among their k-unipotent radicals. (For connected reductive G this is a well-known fact.)

Proposition 3.5.14 *Let G be a smooth connected affine group over a field k, and let P and Q be pseudo-parabolic k-subgroups of G. The following conditions are equivalent:*

(i) $P \subseteq Q$,
(ii) $\mathscr{R}_{u,k}(Q) \subseteq \mathscr{R}_{u,k}(P)$,
(iii) $\mathscr{R}_{us,k}(Q) \subseteq \mathscr{R}_{us,k}(P)$.

Proof We first prove the equivalence of (i) and (ii). We may and do assume $k = k_s$, and (by Proposition 2.2.10) that G is pseudo-reductive over k, as well as non-commutative (so the root system associated to G is non-empty). Assume (i), and let T be a maximal k-torus in P (so also $T \subseteq Q$). Let Ψ_P denote the set of non-divisible roots $a \in \Phi(G,T)$ such that $-a$ is not a T-weight on $\mathrm{Lie}(P)$, and define Ψ_Q similarly. The set $\Phi(P,T)$ of nontrivial T-weights on $\mathrm{Lie}(P)$ is contained in $\Phi(Q,T)$, so $\Psi_Q \subseteq \Psi_P$. This latter containment implies (ii), due to Proposition 3.5.1(2).

Conversely, assume (ii) holds. By Proposition 3.5.12(1), there is a maximal k-torus T contained in both P and Q. By Proposition 3.5.1(3), to prove (i) it is equivalent to prove that $\Phi(P,T) \subseteq \Phi(Q,T)$. Suppose to the contrary that some $a \in \Phi(P,T)$ does not lie in $\Phi(Q,T)$. Since

$\Phi(P,T)$ is a parabolic subset of $\Phi(G,T)$, we may arrange that a is non-divisible in $\Phi(G,T)$. It follows from Proposition 3.5.1(2) that $-a$ is a T-weight on $\mathrm{Lie}(\mathscr{R}_{u,k}(Q))$, so $-a$ is a T-weight on $\mathrm{Lie}(\mathscr{R}_{u,k}(P))$ due to (ii). Once again applying Proposition 3.5.1(2), we conclude that $a \notin \Phi(P,T)$, a contradiction. We have proved that (i) and (ii) are equivalent.

Next, assume (ii). Consider the natural k-homomorphism $\mathscr{R}_{us,k}(Q) \to \mathscr{R}_{u,k}(P)/\mathscr{R}_{us,k}(P)$. By Corollary B.3.5 the target of this map is k-wound, so the k-homomorphism is trivial. Thus, (iii) holds. Conversely, if we assume (iii) then Corollary 2.2.5 implies that $\mathscr{R}_{u,k}(Q) = \mathscr{R}_{us,k}(Q)\mathscr{R}_{u,k}(G) \subseteq \mathscr{R}_{us,k}(P)\mathscr{R}_{u,k}(G) = \mathscr{R}_{u,k}(P)$. □

PART II

Standard presentations and their applications

4

Variation of $(G', k'/k, T', C)$

4.1 Absolutely simple and simply connected fibers

Let k be a field and G a standard pseudo-reductive k-group in the sense of Definition 1.4.4. By definition of being standard, G arises from a 4-tuple $(G', k'/k, T', C)$ (and a factorization diagram involving C as in (1.4.1), which we suppress from the notation), but several such 4-tuples can give rise to the same standard k-group G. We now focus our efforts on finding a canonical way to choose the 4-tuple (if it exists), as that will make it easier to carry out Galois descent arguments.

Observe that the k-group G is solvable if and only if $G' = T'$, or equivalently (by Proposition 1.2.3) if and only if G is commutative. In such cases it is natural to use the choices $k' = k$, $G' = T' = \{1\}$, and $C = G$, so we shall now focus on the non-commutative case. It is proved below that a non-commutative standard pseudo-reductive k-group G can always be obtained from a 4-tuple $(G', k'/k, T', C)$ such that the fibers of G' over the factor fields k'_i of k' are *absolutely simple and simply connected*. Thus, there is no loss of generality in the definition of standardness for a non-commutative pseudo-reductive k-group if we require that $G' \to \operatorname{Spec} k'$ has absolutely simple and simply connected fibers. In §4.2 we will show that with these finer requirements, the pair $(G', k'/k)$ (equipped with certain additional structure) is uniquely determined up to unique k-isomorphism by the non-commutative k-group G.

There is complete flexibility in the choice of maximal k'-torus $T' \subseteq G'$ appearing in the standard construction of G, in the following sense. Such choices are generally not $G'(k')$-conjugate to each other (when k is not separably closed), but if we make another choice of maximal k'-torus $S' \subseteq G'$, then Proposition 4.1.4(2), (3) will show that the factorization diagram of k-groups

$$\mathrm{R}_{k'/k}(T') \xrightarrow{\phi} C \xrightarrow{\psi} \mathrm{R}_{k'/k}(\overline{T}') \tag{4.1.1}$$

used to construct G can be changed by replacing C with an étale twist (and replacing T' and $\overline{T}'$ with S' and $\overline{S}' = S'/Z_{G'}$ respectively) so that the resulting standard pseudo-reductive k-group is k-isomorphic to the given k-group G.

Theorem 4.1.1 *Let k be a field, and G a non-commutative standard pseudo-reductive k-group. The k-group G arises from a 4-tuple $(G', k'/k, T', C)$ as in Definition 1.4.4 such that G' has absolutely simple and simply connected fibers over* Spec k'.

Proof Pick a 4-tuple $(G', k'/k, T', C)$ as in the standard construction so that there is an isomorphism of k-groups

$$(\mathrm{R}_{k'/k}(G') \rtimes C)/\mathrm{R}_{k'/k}(T') \simeq G.$$

Here it is only assumed that the fibers of $G' \to \mathrm{Spec}\, k'$ are connected reductive. We fix such an isomorphism, and try to improve the choice of the 4-tuple.

Let G'_i and T'_i respectively denote the fiber of G' and fiber of the maximal k'-torus $T' \subseteq G'$ over the factor field k'_i of k'. If $G'_{i_0} = T'_{i_0}$ for some i_0 then we can drop the i_0-factor from k' and G' as well as from the flanking terms in (4.1.1) (leaving C unchanged) without affecting G. (Since G is non-commutative, we cannot have $G'_i = T'_i$ for all i.) Thus, we can assume that all G'_i are non-commutative, so the connected semisimple derived group $\mathscr{D}(G'_i)$ is nontrivial for all i.

Let $\widetilde{G}' \to \mathscr{D}(G')$ be the (fiberwise) simply connected central cover of $\mathscr{D}(G')$ over Spec k', and let $Z' \subseteq T'$ be the maximal central k'-torus in G'. Under the central k'-isogeny $\widetilde{G}' \times Z' \to G'$, the preimage of T' is $\widetilde{T}' \times Z'$ for a maximal k'-torus $\widetilde{T}' \subseteq \widetilde{G}'$. Thus, T' viewed as a quotient of $\widetilde{T}' \times Z'$ inherits a natural left action on $\widetilde{G}' \times Z'$ induced by conjugation. This yields a pushout presentation

$$G' \simeq ((\widetilde{G}' \times Z') \rtimes T')/(\widetilde{T}' \times Z')$$

involving the quotient by a smooth k'-group, so the formation of this presentation commutes with $\mathrm{R}_{k'/k}$ (Corollary A.5.4).

We therefore arrive at a commutative diagram

$$\begin{array}{ccccccc}
\mathrm{R}_{k'/k}(\widetilde{T}' \times Z') & \longrightarrow & \mathrm{R}_{k'/k}(T') & \stackrel{\phi}{\longrightarrow} & C & \stackrel{\psi}{\longrightarrow} & \mathrm{R}_{k'/k}(\overline{T}') \\
\downarrow & & \downarrow & & \downarrow & & \\
\mathrm{R}_{k'/k}(\widetilde{G}' \times Z') & \longrightarrow & \mathrm{R}_{k'/k}(G') & \longrightarrow & G & &
\end{array}$$

in which $\overline{T}' = T'/Z_{G'}$ and the two small rectangles are non-commutative pushouts in the sense of Remark 1.4.5, resting on the natural left action of $\overline{T}'$ on $\widetilde{G}'$ that is trivial on Z'. Hence, the larger rectangle is such a pushout.

The k'-torus $\overline{T}'$ is the quotient of $\widetilde{T}' \times Z'$ modulo the scheme-theoretic center of $\widetilde{G}' \times Z'$, so $(\widetilde{G}' \times Z', k'/k, \widetilde{T}' \times Z', C)$ gives rise to G via the standard construction in Definition 1.4.4. The 4-tuple $(\widetilde{G}', k'/k, \widetilde{T}', C)$ also makes sense as input into the standard construction (with $\widetilde{T}'/Z_{\widetilde{G}'} = \overline{T}'$), and it also yields G.

Remark 4.1.2 Beware that the map $\widetilde{\phi} : \mathrm{R}_{k'/k}(\widetilde{T}') \to C$ induced by ϕ can fail to be surjective, even if ϕ is surjective. For example, suppose $G = \mathrm{R}_{k'/k}(G')$ with $G' \to \mathrm{Spec}\, k'$ having connected semisimple fibers. If the central isogeny $\widetilde{G}' \to G'$ is non-étale on the fiber over some factor field k'_i that is not k-étale, then $\widetilde{\phi}$ is not surjective.

Continuing with the proof of Theorem 4.1.1, rename $\widetilde{G}'$ and $\widetilde{T}'$ as G' and T' respectively, so the fibers of G' over the factor fields of k' are semisimple and simply connected. By applying Proposition A.5.14 over the factor fields of k', there is a k'-group isomorphism $G' \simeq \mathrm{R}_{k''/k'}(G'')$ for a faithfully flat finite étale k'-algebra k'' and a semisimple k''-group G'' whose fibers are connected, absolutely simple, and simply connected. The behavior of maximal tori with respect to Weil restriction (Proposition A.5.15(2)) implies that T' is the maximal k'-torus in $\mathrm{R}_{k''/k'}(T'')$ for a unique maximal k''-torus T'' in G''. But since k'' is k'-étale, $\mathrm{R}_{k''/k'}(T'')$ is a k'-torus, so it is equal to T'. We can therefore replace k' and (G', T') with k'' and (G'', T'') respectively to recover the same output G without changing (1.4.1) but achieving absolute simplicity for the simply connected fibers of G' over $\mathrm{Spec}\, k'$. This proves Theorem 4.1.1. □

Definition 4.1.3 Let G be a non-commutative standard pseudo-reductive group over a field k. A *standard presentation* of G is a 4-tuple $(G', k'/k, T', C)$ as in Definition 1.4.4 such that G' is a semisimple k'-group whose fibers are connected, absolutely simple, and simply connected, along with a choice of isomorphism (1.4.3). We also call this a *standard presentation of G resting on* $(G', k'/k)$.

For later purposes we need to understand the effect on a standard presentation when we change the choice of maximal k'-torus T' in G'. This is addressed in part (3) of:

Proposition 4.1.4 *Let G be a non-commutative standard pseudo-reductive k-group, and fix a standard presentation $(G', k'/k, T', C)$ of G.*

(1) *Let $j : \mathrm{R}_{k'/k}(G') \to G$ be the natural map, and let $\mathscr{T}$ be the maximal k-torus in $\mathrm{R}_{k'/k}(T')$. The image of j is $\mathscr{D}(G)$, and if T is the maximal k-torus in G that is an almost direct product of the maximal central k-torus Z of G and $j(\mathscr{T})$ then $C = Z_G(T)$.*
(2) *The natural map $\mathscr{T} \to j(\mathscr{T})$ is an isogeny, and there is a bijective correspondence between the set of maximal k'-tori S' in G' and the set of maximal k-tori in G defined by $S' \mapsto j(\mathscr{S}) \cdot Z$, where $\mathscr{S}$ is the maximal k-torus in $\mathrm{R}_{k'/k}(S')$.*
(3) *Let S be a maximal k-torus in G, S' the associated maximal k'-torus in G', and $\mathscr{S}$ the maximal k-torus in $\mathrm{R}_{k'/k}(S')$. The conjugation action of $Z_G(S)$ on $\mathscr{D}(G) = j(\mathrm{R}_{k'/k}(G'))$ uniquely lifts to an action of $Z_G(S)$ on $\mathrm{R}_{k'/k}(G')$ fixing the maximal k-torus $\mathscr{S}$. For the standard pseudo-reductive k-group associated to $(G', k'/k, S', Z_G(S))$ and the resulting factorization diagram*

$$\mathrm{R}_{k'/k}(S') \longrightarrow Z_G(S) \longrightarrow \mathrm{R}_{k'/k}(S'/Z_{G'})$$

via Theorem 1.3.9 for $(G', k'/k, S')$, the natural k-homomorphism

$$\gamma : (\mathrm{R}_{k'/k}(G') \rtimes Z_G(S))/\mathrm{R}_{k'/k}(S') \to G$$

is an isomorphism. In particular, there is a natural bijective correspondence between the set of standard presentations of G resting on $(G', k'/k)$ and the set of maximal k-tori of G.

Proof Since the formation of the maximal central torus in a smooth connected affine group over a field commutes with any extension of the ground field, by Galois descent we may and do assume $k = k_s$. The k-group $\mathrm{R}_{k'/k}(G')$ is perfect (Corollary A.7.11), so the image of j is $\mathscr{D}(G)$ since C is commutative.

The image of $\mathrm{R}_{k'/k}(G')$ in G meets C in the image of the Cartan k-subgroup $\mathrm{R}_{k'/k}(T')$ of $\mathrm{R}_{k'/k}(G')$. Thus, C is the image under the quotient map

$$\mathrm{R}_{k'/k}(G') \rtimes C \twoheadrightarrow G$$

of the Cartan k-subgroup $\mathrm{R}_{k'/k}(T') \rtimes C$. Hence, C is a Cartan k-subgroup of G, so it is $Z_G(T)$ where T is the maximal k-torus in the commutative C. Since the composite map in the factorization diagram (1.4.1) underlying the standard presentation induces an isogeny between maximal k-tori (as we can check on maximal reductive quotients over $\overline{k}$), this diagram canonically expresses T as an almost direct product of the image $j(\mathscr{T})$ of the maximal k-torus $\mathscr{T}$ in $\mathrm{R}_{k'/k}(T')$ and the maximal k-torus in $\ker\psi$. This latter k-torus is the maximal central k-torus Z in G, as C is a Cartan k-subgroup, so (1) is proved.

The common kernel $\ker j = \ker\phi$ contains no nontrivial k-tori (it is even contained in the torsion group $\ker(\psi\circ\phi) = \mathrm{R}_{k'/k}(Z_{G'})$), so $\mathscr{T} \to j(\mathscr{T})$ is an isogeny. By rational conjugacy of maximal tori over a separably closed field and the identification of C with $Z_G(T)$, (2) now follows by applying $G(k)$-conjugacy throughout. The same rational conjugacy considerations prove (3), provided we check that the $Z_G(S)$-action on $\mathrm{R}_{k'/k}(G')$ is uniquely determined by the conjugation action on the quotient $\mathrm{R}_{k'/k}(G')/\ker\phi = \mathrm{im}(j) = \mathscr{D}(G)$.

By Theorem 1.3.9, the $Z_G(S)$-action on $\mathrm{R}_{k'/k}(G')$ is classified by a k-homomorphism

$$Z_G(S) \to \mathrm{R}_{k'/k}(S'/Z_{G'})$$

to a *commutative* target. Thus, we just have to check that if the action on $\mathrm{R}_{k'/k}(G')/\ker\phi$ is trivial then the action on $\mathrm{R}_{k'/k}(G')$ is trivial. But $\ker\phi \subseteq \mathrm{R}_{k'/k}(Z_{G'})$, so by considering k-points it suffices to prove that if f is a k-automorphism of $\mathrm{R}_{k'/k}(G')$ that induces the trivial automorphism on the quotient $\mathrm{R}_{k'/k}(G')/\mathrm{R}_{k'/k}(Z_{G'})$ then f is the identity. This condition says that $g \mapsto f(g)g^{-1}$ is a pointed k-scheme map from $\mathrm{R}_{k'/k}(G')$ into $\mathrm{R}_{k'/k}(Z_{G'})$, or equivalently a pointed k'-scheme map $\mathrm{R}_{k'/k}(G')_{k'} \to Z_{G'}$. Such a map is trivial since the source has smooth geometrically connected fibers over k' (by Proposition A.5.11(3)) whereas $Z_{G'}$ has fibers that are finite. □

Remark 4.1.5 By Proposition 4.1.4, it makes sense to say that a standard presentation of a non-commutative pseudo-reductive k-group G is *compatible* (or not) with a uniquely determined maximal k-torus T of G (which can be chosen arbitrarily) when the pair $(G', k'/k)$ is fixed and the fibers of G' over $\mathrm{Spec}\, k'$ are connected, absolutely simple, and simply connected. In §4.2 we will prove that G canonically determines $(G', k'/k)$, so choosing a standard presentation of G is "the same" as choosing a maximal k-torus of G.

Corollary 4.1.6 *Let G be a non-commutative standard pseudo-reductive group over a separably closed field k, and $(G', k'/k, T', C)$ a standard presentation. Let T be the associated maximal k-torus of G. Let $\mathfrak{g} = \mathrm{Lie}(G)$ and $\mathfrak{g}' = \mathrm{Lie}(G')$. Let $\{k'_i\}_{i \in I}$ be the set of factor fields of k', and (G'_i, T'_i) the k'_i-fiber of (G', T'). Let $\mathfrak{g}'_i = \mathrm{Lie}(G'_i)$, so $\mathfrak{g}' = \prod \mathfrak{g}'_i$.*

There is a canonical decomposition of root systems

$$\Phi(G, T) = \coprod \Phi(G'_i, T'_i)$$

(i.e., the root systems $\Phi(G'_i, T'_i)$ are the irreducible components of $\Phi(G, T)$), and if $a \in \Phi(G, T)$ goes over to $a'_i \in \Phi(G'_i, T'_i)$ for a unique $i \in I$ then the natural map

$$\mathfrak{g}' = \mathrm{Lie}(\mathrm{R}_{k'/k}(G')) \xrightarrow{j} \mathrm{Lie}(G) = \mathfrak{g}$$

carries $(\mathfrak{g}'_i)_{a'_i}$ isomorphically onto $\mathfrak{g}_a$. In particular, $\Phi(G, T)$ is always a reduced root system when G is standard.

Proof Let Z be the maximal k-torus in C that is central in G (i.e., the maximal k-torus in the kernel of $\psi : C \to \mathrm{R}_{k'/k}(T'/Z_{G'})$), and let $\mathscr{T}$ be the maximal k-torus in $\mathrm{R}_{k'/k}(T')$. Thus, $\mathscr{T} \to j(\mathscr{T})$ is an isogeny and T is the almost direct product of $j(\mathscr{T})$ and Z. In particular, all elements of $\Phi(G, T)$ factor through the quotient T/Z that is an isogenous quotient of $\mathscr{T}$.

Since $\ker j \subseteq \mathrm{R}_{k'/k}(Z_{G'})$ and $\mathrm{R}_{k'/k}(Z_{G'})$ is central in $\mathrm{R}_{k'/k}(G')$, the root spaces for $\mathscr{T}$ acting on $\mathrm{Lie}(\mathrm{R}_{k'/k}(G'))$ map isomorphically onto the root spaces for T acting on $\mathfrak{g}$, so $\Phi(G, T) = \Phi(\mathrm{R}_{k'/k}(G'), \mathscr{T})$. This reduces the problem to the special case $G = \mathrm{R}_{k'/k}(G')$ and T the maximal k-torus in $\mathrm{R}_{k'/k}(T')$. In particular, $\mathfrak{g} = \prod \mathfrak{g}'_i$.

Decomposing according to the factor fields of k' reduces us to the case where k' is a field. Under the canonical identification of $\mathrm{X}(T)$ and $\mathrm{X}(T')$ (as $k = k_s$), the distinct weight spaces for T' acting k'-linearly on $\mathfrak{g}'$ correspond to the distinct weight spaces for T acting k-linearly on $\mathfrak{g}'$. □

4.2 Uniqueness of $(G', k'/k)$

We have shown in §4.1 that for any non-commutative standard pseudo-reductive k-group G there is a nonzero finite reduced k-algebra k' such that

G can be obtained (via the standard construction in Definition 1.4.4) from data $(G', k'/k, T', C)$ with a k'-group G' whose fibers over $\mathrm{Spec}\, k'$ are connected, semisimple, absolutely simple, and *simply connected*.

We wish to reconstruct the pair $(G', k'/k)$ (up to k-isomorphism) from the non-commutative (hence non-solvable) standard pseudo-reductive k-group G by a procedure that will make sense for an *arbitrary* non-solvable smooth connected affine k-group (perhaps not even pseudo-reductive). This rests on a certain category that is associated to any smooth connected affine group G over a field k, as follows.

Consider pairs $(F/k, q: G_F \twoheadrightarrow H)$ with F a reduced artinian k-algebra (i.e., a product of finitely many extension fields of k, not necessarily of finite k-degree), H an adjoint semisimple F-group with absolutely simple and connected fibers, and q a quotient homomorphism over F that is a *maximal* simple adjoint semisimple quotient on geometric fibers. Define a *morphism*

$$(F/k, q : G_F \twoheadrightarrow H) \to (F'/k, q' : G_{F'} \twoheadrightarrow H')$$

to consist of a k-algebra map $\alpha: F \to F'$ and commutative diagram of F'-group maps

$$\begin{array}{ccc} G_{F'} & = & G_{F'} \\ {\scriptstyle \alpha^*(q)}\downarrow & & \downarrow{\scriptstyle q'} \\ \alpha^*(H) & \xrightarrow[\simeq]{} & H' \end{array} \tag{4.2.1}$$

Note that the bottom arrow in this diagram is uniquely determined by α (if it exists). Let $\mathscr{C}_{G/k}$ denote the category just defined. There is a final object in $\mathscr{C}_{G/k}$, corresponding to $F = 0$ and H the empty group scheme over $\mathrm{Spec}(0)$, and if G is solvable then this is the only object (up to isomorphism) in the category.

Let $\mathscr{C}_{G/k}^{\mathrm{fin}}$ be the full subcategory of objects $(F/k, q)$ in $\mathscr{C}_{G/k}$ for which F is finite-dimensional as a k-vector space. If k/k is a separable extension of fields then there is a natural "base change" functor $\mathscr{C}_{G/k}^{\mathrm{fin}} \to \mathscr{C}_{G_{\mathsf{k}}/\mathsf{k}}$ defined by $(F/k, q) \rightsquigarrow (F/k, q)_{\mathsf{k}} := (F_{\mathsf{k}}/\mathsf{k}, q \otimes_F F_{\mathsf{k}})$ where $F_{\mathsf{k}} = \mathsf{k} \otimes_k F$ (a finite product of fields since k/k is separable). This functor is left adjoint in an evident sense to the forgetful functor $\mathscr{C}_{G_{\mathsf{k}}/\mathsf{k}} \to \mathscr{C}_{G/k}$ when k/k is separable.

Proposition 4.2.1 *There exists an initial object $(K/k, \pi_G : G_K \twoheadrightarrow \overline{\mathscr{G}})$ in the category $\mathscr{C}_{G/k}$. The structure map* $\mathrm{Spec}\, K \to \mathrm{Spec}\, k$ *is finite, and the formation of $(K/k, \pi_G)$ commutes with any separable extension k/k in the sense that $(K/k, \pi_G)_{\mathsf{k}}$ is initial in $\mathscr{C}_{G_{\mathsf{k}}/\mathsf{k}}$.*

We call $(K/k, \pi_G)$ the *universal maximal absolutely simple adjoint quotient* of G. Before we prove it exists, let us briefly explain why a certain possible generalization can fail to exist. Consider the larger category $\mathfrak{C}_{G/k}$ in which we allow fiberwise (over Spec F) absolutely simple adjoint semisimple quotients $q : G_F \twoheadrightarrow H$ that may not be maximal on geometric fibers over Spec F (and we drop the requirement that the bottom arrow in (4.2.1) is an isomorphism). The initial object in Proposition 4.2.1 can fail to be initial in $\mathfrak{C}_{G/k}$ when k is imperfect because composing an adjoint semisimple quotient of $G_{\overline{k}}$ with an iterated Frobenius isogeny can decrease its field of definition over k.

Proof For any separable extension k/k, the adjointness between the forgetful functor $\mathscr{C}_{G_{\mathrm{k}}/\mathrm{k}} \to \mathscr{C}_{G/k}$ and the base change functor $(F/k, q) \rightsquigarrow (F/k, q)_{\mathrm{k}}$ for k-finite F implies that such base change carries an initial object of $\mathscr{C}_{G/k}$ to an initial object of $\mathscr{C}_{G_{\mathrm{k}}/\mathrm{k}}$ provided that such objects lie in $\mathscr{C}^{\mathrm{fin}}_{G/k}$ and $\mathscr{C}^{\mathrm{fin}}_{G_{\mathrm{k}}/\mathrm{k}}$ respectively.

Now we construct an object $(K/k, \pi_G : G_K \twoheadrightarrow \overline{\mathscr{G}})$ in $\mathscr{C}_{G/k}$ so that $[K : k] < \infty$ and for any $(F/k, q : G_F \twoheadrightarrow H)$ in $\mathscr{C}_{G/k}$ there exists a unique k-algebra map $\alpha : K \to F$ such that $\alpha^*(\pi_G) : G_F \twoheadrightarrow \alpha^*(\overline{\mathscr{G}})$ is isomorphic to H (necessarily uniquely) as quotients of G_F. The case of solvable G is trivial (as then $\mathscr{C}_{G/k}$ consists of only the empty object), so we now assume G is non-solvable.

As a first step, we show that if an object $(K/k, \pi_G)$ has been constructed so that there exists a unique α as above for any $(F/k, q)$ with F a separably closed field then the existence and uniqueness of α hold in general. Consider a general pair $(F/k, q)$, say with $F = \prod F_i$ for fields F_i/k and $q_i : G_{F_i} \twoheadrightarrow H_i$ the F_i-fiber of q. To give a k-algebra map $\alpha : K \to F = \prod F_i$ such that the quotient map $\alpha^*(\pi_G)$ over F is (necessarily uniquely) identified with H as a quotient of G_F is equivalent to giving k-algebra maps $\alpha_i : K \to F_i$ such that the quotient map $\alpha_i^*(\pi_G)$ over F_i is identified with q_i as a quotient of G_{F_i} for each i (the link is that α_i is the F_i-component of α). Hence, there is no loss of generality in restricting attention to the case when F is a field. Applying the assumption that existence and uniqueness are known for $(F_s/k, q_{F_s})$, there exists a unique k-algebra map $\alpha_s : K \to F_s$ such that there is a (necessarily unique) identification $\theta_s : \alpha_s^*(\pi_G) = q_{F_s}$ as quotients of G_{F_s}. But then for every $\sigma \in \mathrm{Gal}(F_s/F)$ the map $\sigma \circ \alpha_s$ works as well, using the composition of the identifications $\sigma^*(\theta_s)$ and $\sigma^*(q_{F_s}) = q_{F_s}$ of quotients of $\sigma^*(G_{F_s}) = G_{F_s}$. Hence, uniqueness for $(F_s/k, q)$ implies that $\sigma \circ \alpha_s = \alpha_s$ for all σ, which is to say that $\alpha_s(K) \subseteq F$. Letting $\alpha : K \to F$ be the k-algebra map induced by α_s, the identification θ_s of quotients $\alpha^*(\pi_G)_{F_s} = \alpha_s^*(\pi_G)$ and q_{F_s} of G_{F_s} is unique (as for identifications of quotients in general), so it is automatically $\mathrm{Gal}(F_s/F)$-equivariant and thus descends to an identification

$\theta : \alpha^*(\pi_G) = q_F$ of quotients of G_F. This proves existence of the desired α for $(F/k, q)$, and its uniqueness is immediate from the assumed uniqueness for $(F_s/k, q_{F_s})$.

Now we may and do only consider the mapping property to pairs $(F/k, q)$ for which F is a separably closed field. We next wish to reduce to the construction of the desired $(K/k, \pi_G)$ to the case $k = k_s$. Assume there is an initial object $(K_s/k_s, \pi_{G_{k_s}})$ in $\mathscr{C}_{G_{k_s}/k_s}$ with K_s a finite k_s-algebra. By expressing k_s as a direct limit of finite Galois extensions k/k, for a sufficiently large k/k we can write $K_s = K' \otimes_{\mathsf{k}} k_s$ for a finite k-algebra K' and descend $\pi_{G_{k_s}}$ to a quotient map $\pi' : G_{K'} \twoheadrightarrow \overline{\mathscr{G}}'$ in $\mathscr{C}_{G_{\mathsf{k}}/\mathsf{k}}$. For any pair $(F/\mathsf{k}, q)$ with F a separably closed field, we may choose a k-embedding of k_s into F, so there is a unique morphism $(K'/\mathsf{k}, \pi')_{k_s} = (K_s/k_s, \pi_{G_{k_s}}) \to (F/k_s, q)$ in $\mathscr{C}_{G_{k_s}/k_s}$. Thus, by Galois descent with respect to k_s/k, there is a unique morphism $(K'/\mathsf{k}, \pi') \to (F/\mathsf{k}, q)$ in $\mathscr{C}_{G_{\mathsf{k}}/\mathsf{k}}$. It follows that $(K'/\mathsf{k}, \pi')$ is initial in $\mathscr{C}_{G_{\mathsf{k}}/\mathsf{k}}$.

The universal property of $(K'/\mathsf{k}, \pi')$ provides a natural action by $\mathrm{Gal}(\mathsf{k}/k)$ on $(K'/\mathsf{k}, \pi')$ covering its natural action on k and respecting the k-structure on G_{k}. Since $\overline{\mathscr{G}}'$ is affine, this Galois descent datum is effective. That is, $(K'/\mathsf{k}, \pi')$ descends to an object $(K/k, \pi_G)$ in $\mathscr{C}_{G/k}$, and the same argument that descended from k_s to k (now using pairs $(F/k, q)$ with F a separably closed field) also shows that $(K/k, \pi_G)$ is initial.

Now we may assume $k = k_s$, and we will construct an object $(K/k, \pi_G)$ in $\mathscr{C}_{G/k}$ for which the desired existence and uniqueness of α hold relative to all objects $(F/k, q)$ with F a separably closed field. As we have seen above, such a $(K/k, \pi_G)$ would then be an initial object in $\mathscr{C}_{G/k}$.

Let $K = \prod k_i$ where $\{k_i\}$ is the set of fields of definition over k for the kernels of the projections $\pi'_i : G_{\overline{k}} \twoheadrightarrow \overline{G}_i$ onto the distinct maximal simple adjoint semisimple quotients. Let $\pi_G : G_K \twoheadrightarrow \overline{\mathscr{G}}$ have k_i-fiber $G_{k_i} \twoheadrightarrow \overline{\mathscr{G}}_i$ that is the unique descent of π'_i to its minimal field of definition k_i over k. Since $k = k_s$, $\overline{k}$ has a unique k-embedding into the algebraic closure $\overline{F}$ of $F = F_s$. The maximal simple adjoint semisimple quotient map $q : G_F \twoheadrightarrow H$ over $F = F_s$ identifies $H_{\overline{F}}$ with a unique simple factor of $G^{\mathrm{ss}}_{\overline{F}}/Z_{G^{\mathrm{ss}}_{\overline{F}}}$.

Since $\mathscr{R}(G_{\overline{F}}) = \mathscr{R}(G_{\overline{k}})_{\overline{F}}$ inside of $G_{\overline{F}}$, there is a unique isomorphism

$$G^{\mathrm{ss}}_{\overline{F}}/Z_{G^{\mathrm{ss}}_{\overline{F}}} \simeq (G^{\mathrm{ss}}_{\overline{k}}/Z_{G^{\mathrm{ss}}_{\overline{k}}})_{\overline{F}}$$

as quotients of $G_{\overline{F}}$. This isomorphism respects the formation of simple factors over $\overline{k}$ and $\overline{F}$, so we conclude that there is a unique i such that $q_{\overline{F}}$ is identified with $(\pi'_i) \otimes_{\overline{k}} \overline{F}$. In other words, the kernel of the projection from $G_{\overline{F}}$ onto its ith maximal simple adjoint semisimple quotient is defined over k_i/k and over

F/k. Hence, it is defined over $k_i \cap F$ (as an extension of k). But k_i/k is the minimal field of definition of this quotient viewed over $\overline{k}$, so in particular this quotient cannot be descended over a proper subfield of k_i/k. It follows that $k_i \cap F = k_i$, so $k_i \subseteq F$ inside of $\overline{F}$. This inclusion of fields over k is the required unique α. □

Proposition 4.2.2 *For any short exact sequence*

$$1 \to G' \xrightarrow{j} G \xrightarrow{p} G'' \to 1$$

of smooth connected affine groups over a field k, the universal maximal absolutely simple adjoint quotient $(K/k, \pi_G)$ for G is canonically identified with the "disjoint union" of the objects $(K'/k, \widetilde{\pi}_{G'})$ and $(K''/k, \pi_{G''} \circ p_{K''})$ for $\widetilde{\pi}_{G'} : G_{K'} \twoheadrightarrow \overline{\mathscr{G}}'$ uniquely determined by the condition $\widetilde{\pi}_{G'} \circ j_{K'} = \pi_{G'}$. That is, Spec K *is identified with* Spec $K' \coprod$ Spec K'' *over which π_G is the disjoint union of $\widetilde{\pi}_{G'}$ and $\pi_{G''} \circ p_{K''}$.*

Proof Consider the following two assertions:

(i) for any $(F''/k, q'')$ in $\mathscr{C}_{G''/k}$, the pair $(F''/k, q'' \circ p)$ is an object in $\mathscr{C}_{G/k}$,
(ii) for any $(F'/k, q')$ in $\mathscr{C}_{G'/k}$, there is a unique $(F'/k, \widetilde{q}')$ in $\mathscr{C}_{G/k}$ such that $(F'/k, \widetilde{q}' \circ j_{F'}) = (F'/k, q')$.

Once (i) and (ii) are proved, applying them to initial objects defines a canonical map $(K/k, \pi_G) \to ((K' \times K'')/k, \widetilde{\pi}_{G'} \coprod \pi_{G''} \circ p_{K''})$ in $\mathscr{C}_{G/k}$ whose formation commutes with any separable extension on k, and we will show that this canonical map is an isomorphism. In view of the uniqueness assertions, by Galois descent we can assume $k = k_s$.

To prove (i) and the existence in (ii), it is easy to reduce to the case of the initial objects attached to G' and G''. By the final assertion in Proposition A.4.8, there is an induced *short exact* sequence among the maximal adjoint semisimple quotients of $G'_{\overline{k}}$, $G_{\overline{k}}$, and $G''_{\overline{k}}$. Since connected semisimple groups of adjoint type over $\overline{k}$ are canonically a direct product of their simple factors, the maximal adjoint semisimple quotient of $G_{\overline{k}}$ is canonically the product of those for $G'_{\overline{k}}$ and $G''_{\overline{k}}$. Thus, (i) is clear. In view of how the universal maximal absolutely simple adjoint quotients (i.e., initial objects in $\mathscr{C}_{G'/k}$, $\mathscr{C}_{G/k}$, and $\mathscr{C}_{G''/k}$) are constructed, it remains to (a) prove the uniqueness in (ii), and (b) show that for each maximal simple adjoint semisimple quotient $q : G_{\overline{k}} \twoheadrightarrow H$ the field of definition F/k for the kernel is the same as for the corresponding such quotient for $G'_{\overline{k}}$ or $G''_{\overline{k}}$.

Consider (b). The case of q factoring through $G''_{\overline{k}}$ is trivial, so we just have to check that if q is nontrivial on $G'_{\overline{k}}$ then the field of definition F'/k for the kernel of the resulting maximal simple adjoint semisimple quotient map $q' = q \circ j_{\overline{k}} : G'_{\overline{k}} \twoheadrightarrow H$ is equal to F. Clearly $F' \subseteq F$ over $k = k_s$, and we need to prove this is an equality. We can rename F' as k, so it suffices to prove that any adjoint semisimple quotient $q' : G' \twoheadrightarrow H$ over k that is maximal as such over $\overline{k}$ lifts uniquely to a quotient map $q : G \twoheadrightarrow H$ over k. This would also settle (a) by letting k vary through the factor fields of F'.

The uniqueness of q follows from the structure of $G^{\rm ss}_{\overline{k}}/Z_{G^{\rm ss}_{\overline{k}}}$. For the existence, since G is smooth its conjugation action on G' induces an action of $G_{\overline{k}}$ on $G'^{\rm ss}_{\overline{k}}$, and hence likewise on the *maximal* adjoint semisimple quotient $G'^{\rm ss}_{\overline{k}}/Z_{G'^{\rm ss}_{\overline{k}}}$. By connectedness of $G_{\overline{k}}$, this latter action preserves each of the simple factors, so q' is a G-equivariant quotient (as that can be checked over $\overline{k}$). We can therefore form the "non-commutative pushout" of G along the quotient map q' as in Remark 1.4.5 to reduce to the case where G' is a simple adjoint semisimple k-group.

By Corollary A.4.7(2), the G-action on G' is conjugation through a k-homomorphism $f : G \to G'$. But this action restricts to the inner action of G' on itself, so $f(g')$-conjugation on G' coincides with g'-conjugation on G' for all points g' of G'. Since $Z_{G'} = 1$, it follows that $f|_{G'} = {\rm id}_{G'}$, so f expresses G' as a direct factor of G over k. □

Using the universal property of Weil restriction, the canonical map π_G in Proposition 4.2.1 corresponds to a map of k-groups

$$f_G : G \to {\rm R}_{K/k}(\overline{\mathscr{G}}) = \prod {\rm R}_{k_i/k}(\overline{\mathscr{G}}_i), \qquad (4.2.2)$$

where $\{k_i\}$ is the set of factor fields of K and $\overline{\mathscr{G}}_i$ is the k_i-fiber of $\overline{\mathscr{G}}$. By construction, the formation of f_G commutes with any separable extension on the ground field.

Definition 4.2.3 Let G be a non-solvable smooth connected affine group over a field k, and consider the initial object $(K/k, \pi_G : G_K \to \overline{\mathscr{G}})$ as in Proposition 4.2.1. The *simply connected datum* associated to G is the triple $(\widetilde{\mathscr{G}}, K/k, f_G)$ where f_G is as in (4.2.2) and $\widetilde{\mathscr{G}} \to \overline{\mathscr{G}}$ is the simply connected central cover of $\overline{\mathscr{G}}$ over K.

The formation of the simply connected datum $(\widetilde{\mathscr{G}}, K/k, f_G)$ is compatible with any separable extension on k. Our main interest in the preceding considerations is due to the fact that we can describe $(K/k, \pi_G)$, or equivalently

$(K/k, f_G)$, when G is a non-commutative (hence non-solvable) standard pseudo-reductive group over a field k:

Proposition 4.2.4 *Let k be a field and G a non-commutative standard pseudo-reductive k-group arising from a 4-tuple $(G', k'/k, T', C)$ such that $G' \to \operatorname{Spec} k'$ has absolutely simple and simply connected fibers. The simply connected datum $(\widetilde{\mathscr{G}}, K/k, f_G)$ associated to G is naturally isomorphic to $(G', k'/k, f)$, where $f : G \to \mathrm{R}_{k'/k}(\overline{G}')$ is the displayed map in Remark 1.4.8.*

Proof By two applications of Proposition 4.2.2, we are reduced to proving that if $G = \mathrm{R}_{k'/k}(G')$ then π_G is the natural quotient map $q : \mathrm{R}_{k'/k}(G')_{k'} \twoheadrightarrow G' \twoheadrightarrow \overline{G}'$. That is, we claim that q is a maximal absolutely simple adjoint quotient of $G_{k'}$ on geometric fibers over $\operatorname{Spec} k'$ and that the resulting natural map $(K/k, \pi_G) \to (k'/k, q)$ is an isomorphism.

We can assume $k = k_s$, and then another application of Proposition 4.2.2 (for split exact sequences arising from the direct product decomposition of G according to the factor fields of k') reduces us to the case where k' is a field. In this case q is a k'-descent of the maximal adjoint semisimple quotient of $\mathrm{R}_{k'/k}(G')_{\overline{k}}$, by Proposition A.5.11, so (due to how $(K/k, \pi_G)$ was constructed in the proof of Proposition 4.2.1) we just have to check that k'/k is the minimal field of definition over k for $\ker q$. This follows from Theorem 1.6.2(1). □

The identification of f_G with f in Proposition 4.2.4 implies that if G is a non-commutative standard pseudo-reductive group then $\ker f_G$ is central in G (Remark 1.4.8). The image of f_G is mysterious in general, but it is always pseudo-reductive. This is a special case of the following result that will be useful in our work in characteristics 2 and 3:

Proposition 4.2.5 *Let G be a smooth connected affine group over a field k, k' a nonzero finite reduced k-algebra, and $\pi : G_{k'} \twoheadrightarrow G'$ a quotient over k' such that the fibers of G' over k' are pseudo-reductive. The associated map of k-groups $f : G \to \mathrm{R}_{k'/k}(G')$ has pseudo-reductive image.*

The case of most interest to us is when G' has reductive fibers over k'.

Proof Let $H = f(G)$, so H is a smooth connected affine k-group. Let $i : H \to \mathrm{R}_{k'/k}(G')$ be the inclusion. We first wish to replace (G, f) with (H, i). By applying functoriality of the universal property of Weil restriction to the quotient map $G \to H$, π factors as

$$G_{k'} \twoheadrightarrow H_{k'} \overset{i_{k'}}{\hookrightarrow} \mathrm{R}_{k'/k}(G')_{k'} \overset{q}{\to} G'$$

where the final map is the natural one as in Proposition A.5.7. Moreover, the uniqueness aspect of the universal property of Weil restriction ensures that i corresponds to $q \circ i_{k'}$. Since surjectivity of π implies that $q \circ i_{k'}$ is surjective, we can replace (G, π, f) with $(H, q \circ i_{k'}, i)$ to reduce to the case $\ker f = 1$.

By surjectivity, the map π carries $\mathscr{R}_{u,k}(G)_{k'}$ onto a smooth connected unipotent normal k'-subgroup of G'. But G' is pseudo-reductive over k', so $\mathscr{R}_{u,k}(G)_{k'}$ has trivial image in G'. Equivalently, $\mathscr{R}_{u,k}(G)$ is killed by the k-homomorphism $f : G \to \mathrm{R}_{k'/k}(G')$ corresponding to π. Since $\ker f = 1$, we have $\mathscr{R}_{u,k}(G) = 1$. □

The concept of simply connected datum $(\widetilde{\mathscr{G}}, K/k, f_G)$ associated to a non-solvable smooth connected affine k-group G is closely related to the universal maximal absolutely simple adjoint quotient as in Proposition 4.2.1. In particular, by construction, the factor fields of K/k are identified with the fields of definition over k for representatives of the $\mathrm{Aut}(\overline{k}/k)$-orbits of maximal simple *adjoint* semisimple quotients of $G_{\overline{k}}$. It is essential that we work with the maximal simple *adjoint* semisimple quotients of $G_{\overline{k}}$ rather than with the maximal simple semisimple quotients of $G_{\overline{k}}$, as otherwise the corresponding fields of definition over k for the quotients could increase and so there would be no analogue of Proposition 4.2.4 that reconstructs k'/k from G in the non-commutative standard pseudo-reductive case. This is best understood with an example:

Example 4.2.6 Let k be a field, G' be a connected semisimple k-group, and $T' \subseteq G'$ a maximal k-torus. Let $T' \to C \to T'/Z_{G'}$ be a factorization of the canonical quotient map through a commutative pseudo-reductive k-group C such that T' maps onto the maximal k-torus in C. Assume that C is not a torus. (We will soon exhibit some examples satisfying these conditions.)

These hypotheses force $Z_{G'} \neq 1$. Indeed, if $Z_{G'} = 1$ then the hypotheses yield a direct product decomposition of the k-group C as $T' \times C'$, with T' the maximal k-torus of C and C' a smooth connected commutative k-group. But C is not a torus, so C' is nontrivial and unipotent, contrary to the pseudo-reductivity of C over k.

Under these assumptions, $G := (G' \rtimes C)/T'$ is a standard pseudo-reductive k-group whose maximal adjoint semisimple quotient $G'/Z_{G'}$ is defined over k (as a quotient of G). The initial object in $\mathscr{C}_{G/k}$ is $(k/k, G \twoheadrightarrow G'/Z_{G'})$. However, $\mathscr{R}(G_{\overline{k}})$ is the subgroup $\mathscr{R}_u(C_{\overline{k}})$ in $G_{\overline{k}}$ whose field of definition K over k cannot equal k (as C is pseudo-reductive over k but not a torus). Thus, the maximal semisimple quotient of $G_{\overline{k}}$ has a *strictly larger* field of definition over k than the maximal adjoint semisimple quotient, and these quotients are both simple if G' is absolutely simple over k. (In these examples, $\mathscr{R}(G_{\overline{k}})$ is unipotent but $\mathscr{D}(G) \neq G$ because G has a nontrivial commutative quotient given by the quotient of C modulo the image of T' in C. Failure of perfectness of G is a crucial feature of these examples, as we will see in Proposition 5.3.3.)

The conditions for this construction can be satisfied over any imperfect field k of characteristic $p > 0$, as follows. Take G' to be SL_p and T' to be the split diagonal torus. Let K/k be any purely inseparable extension of degree p and let $C = \mathrm{R}_{K/k}(\mathrm{GL}_1) \times \mathrm{GL}_1^{p-2}$. Consider the k-isomorphism $T' \simeq \mathrm{GL}_1^{p-1}$ defined by the map

$$\mathrm{diag}(t_1, \ldots, t_{p-1}, 1/(t_1 \cdots t_{p-1})) \mapsto (t_1, t_2/t_1, \ldots, t_{p-1}/t_{p-2}).$$

This carries the center $Z_{G'} = \mu_p$ over to the canonical inclusion of μ_p into the first factor of GL_1^{p-1}. Define the maps $T' \to C$ and $C \to T'/Z_{G'}$ to each be the identity on the last $p-2$ factors of T' and C, and to respectively be the inclusion $\mathrm{GL}_1 \to \mathrm{R}_{K/k}(\mathrm{GL}_1)$ and the norm map $\mathrm{R}_{K/k}(\mathrm{GL}_1) \to \mathrm{GL}_1 \simeq \mathrm{GL}_1/\mu_p$ on the first factor. This satisfies all of the requirements.

5

Ubiquity of the standard construction

5.1 Main theorem and central extensions

Although it is difficult to say much about the general structure of commutative pseudo-reductive groups, the commutative case is essentially the only mystery. This follows from the ubiquity of standard pseudo-reductive groups, as we now make precise over arbitrary ground fields k, assuming $[k:k^2] \leqslant 2$ when $\mathrm{char}(k) = 2$.

Theorem 5.1.1 *Let G be a pseudo-reductive group over a field k.*

(1) *The k-group G is standard over k in the sense of Definition 1.4.4 except possibly when* $\mathrm{char}(k) = p \in \{2, 3\}$*, k is imperfect, and the Dynkin diagram of the maximal semisimple quotient $G^{\mathrm{ss}}_{\overline{k}}$ contains an edge with multiplicity p or has an isolated vertex when $p = 2$.*
(2) *Assume that k is imperfect with* $\mathrm{char}(k) = p \in \{2, 3\}$*, and $[k : k^2] \leqslant 2$ if $p = 2$. The pseudo-reductive k-group G is standard if and only if the root system Φ of G_{k_s} is reduced and the root spaces in* $\mathrm{Lie}(G_{k_s})$ *for any two roots in the same irreducible component of Φ have the same dimension.*
(3) *If G_{k_s} has a non-reduced root system (so* $\mathrm{char}(k) = 2$*) and $[k : k^2] = 2$ then there is a unique decomposition $G = G_1 \times G_2$ where $(G_1)_{k_s}$ has a reduced root system and G_2 is totally non-reduced in the sense of Definition 10.1.1.*
(4) *If* $\mathrm{char}(k) \in \{2, 3\}$*, with $[k : k^2] \leqslant 2$ when* $\mathrm{char}(k) = 2$*, and G_{k_s} has a reduced root system then G is generalized standard in the sense of Definition 10.1.9.*

Remark 5.1.2 Every connected reductive k-group G over any field k is trivially a standard pseudo-reductive k-group by taking $k' = k$ and $G' = G$ in Definition 1.4.4. Also, by Proposition 1.2.3, if a pseudo-reductive k-group

G is solvable then it is commutative and hence is standard over k. Over a perfect field every pseudo-reductive group is reductive. Thus, Theorem 5.1.1 is only interesting when k is imperfect and G is non-solvable. We also note that the totally non-reduced factor in part (3) is classified by (the proof of) Theorem 9.9.3(1) and Proposition 10.1.4.

Before we take up the task of proving that standardness always holds, apart from some exceptional situations in characteristics 2 and 3, we wish to prove over an arbitrary ground field k that standardness is well behaved with respect to basic operations such as formation of torus centralizers, passage to derived groups or smooth connected normal k-subgroups, and scalar extension to k_s. We begin with an important criterion for the splitting of certain central extensions. It is the key to controlling connectedness properties of kernels and the size of Cartan subgroups in the classification of pseudo-reductive groups.

Proposition 5.1.3 *Let k be a field and G a nontrivial perfect pseudo-reductive k-group. Let T be a maximal k-torus in G, and assume that a dense open locus in $Z_G(T)_{k_s}$ can be expressed group-theoretically in terms of root groups: for some finite sequence of (not necessarily distinct) roots $a_1, \ldots, a_n \in \Phi(G_{k_s}, T_{k_s})$ there exist rational k_s-scheme maps $h_i : Z_G(T)_{k_s} \dashrightarrow U_{a_i}$ to the corresponding root groups such that their product $\prod h_i : Z_G(T)_{k_s} \dashrightarrow G_{k_s}$ is the natural inclusion.*

For any commutative k-group scheme Z of finite type that contains no nontrivial smooth connected k-subgroup, every central extension

$$1 \to Z \to E \to G \to 1 \tag{5.1.1}$$

over k is uniquely split.

In Proposition 5.1.3 we do not require E (or equivalently, Z) to be affine, though we reduce to this case in the proof. The result is only used with affine E, so the reader can restrict to that case.

There will be several classes of G for which the hypotheses of Proposition 5.1.3 will be verified: Weil restrictions of simply connected semisimple groups, and later exotic constructions in characteristics 2 and 3. Before we take up the proof of Proposition 5.1.3, we now verify its assumptions in the first of these classes of G.

Example 5.1.4 Consider $G = \mathrm{R}_{k'/k}(G')$ for a nonzero finite reduced k-algebra k' and a nontrivial semisimple k'-group G' whose fibers are connected and simply connected. To verify the assumptions in Proposition 5.1.3 for this nontrivial G, we can assume $k = k_s$ and that k' is a field.

By Corollary A.7.11, G is perfect. There is a unique maximal k'-torus T' in G' such that T is the maximal k-torus in $\mathrm{R}_{k'/k}(T')$ (Proposition A.5.15(2)). Explicitly, the natural map $G_{k'} \to G'$ is a k'-descent of the geometric maximal reductive quotient (Proposition A.5.11(2)) and this carries $T_{k'}$ isomorphically onto T', thereby identifying $\mathrm{X}(T)$ and $\mathrm{X}(T')$. Hence, it follows from the natural identification of $\mathrm{Lie}(G)$ with the Lie algebra over k underlying $\mathrm{Lie}(G')$ (Corollary A.7.6) that the root data $R(G,T)$ and $R(G',T')$ naturally coincide. Likewise, by the unique characterization of root groups in Proposition 2.3.11 we have $U_a = \mathrm{R}_{k'/k}(U'_{a'})$ for corresponding roots $a \in \Phi(G,T)$ and $a' \in \Phi(G',T')$. Since $Z_G(T)$ is equal to $\mathrm{R}_{k'/k}(Z_{G'}(T')) = \mathrm{R}_{k'/k}(T')$ (Proposition A.5.15(1)), to construct the required rational maps $h_i : Z_G(T) \dashrightarrow G$ over k it suffices (by Proposition A.5.9) to do so for (G',T') over k' and then apply Weil restriction, as any non-empty open subset of a smooth connected scheme over a field is Zariski-dense. Hence, we may and do assume $k' = k$ and $G' = G$.

Since G is simply connected, if we choose a positive system of roots in $\Phi(G,T)$ then the coroots associated to the simple positive roots form a basis for the cocharacter group of T, and for each positive root a the root groups U_a and U_{-a} generate a copy of SL_2 in G with $a^\vee(\mathrm{GL}_1)$ as a maximal k-torus. Thus, it suffices to treat the case $G = \mathrm{SL}_2$ with T equal to the diagonal k-torus D. In this case, the standard isomorphisms $x_{\pm a} : \mathbf{G}_\mathrm{a} \simeq U_{\pm a}$ and the isomorphism $a^\vee : \mathrm{GL}_1 \simeq D$ defined by $t \mapsto \mathrm{diag}(t, 1/t)$ satisfy

$$a^\vee(t) = x_a(t)x_{-a}(-1/t)x_a(t-1)x_{-a}(1)x_a(-1),$$

from which we see the evident choices for $h_1, \ldots, h_5$ as rational maps (that are even morphisms)

$$D \underset{\simeq}{\overset{a^\vee}{\longleftarrow}} \mathrm{GL}_1 \dashrightarrow U_{a_i}.$$

Remark 5.1.5 The hypothesis in Proposition 5.1.3 that Z contains no nontrivial smooth connected k-subgroup appears to be necessary. For example, suppose that k has characteristic $p > 0$, K/k is an extension with $[K:k] = p^2$, and $K^p \subseteq k$ (which forces K/k to be non-primitive); such a K can be found when k is finitely generated with transcendence degree > 1 over $\mathbf{F}_p$. A construction using a nonzero k-linear map $\phi : \Omega^2_{K/k} \to k$ appears to give rise to a non-split central extension of $R_{K/k}(SL_n)$ by $\mathbf{G}_a$ for any $n > 2$. (To make this extension, argue as follows. The universal central extension of $SL_n(K) = R_{K/k}(SL_n)(k)$ is by $K_2(K)$ since $n > 2$. Form the pushout by the map $K_2(K) \to k$ defined by $\{f, g\} \mapsto \phi((df/f) \wedge (dg/g))$. This is classified by a 2-cocycle given generically over $R_{K/k}(SL_n)^2$ by a rational function. Such a pushout seems likely to give rise to a non-split central extension of $R_{K/k}(SL_n)$ by $\mathbf{G}_a$ as k-groups.)

The imprimitivity of K/k in such examples is essential: it can be proved that if k is a field of arbitrary characteristic and K is a nonzero primitively generated finite k-algebra then for any simply connected semisimple K-group $\widetilde{G}$ (with connected fibers), any central extension of $R_{K/k}(\widetilde{G})$ by a k-group scheme of finite type is split.

Proof of Proposition 5.1.3 Since $G = \mathscr{D}(G)$ by hypothesis, the splitting of (5.1.1)) is unique if it exists. Hence, by Galois descent it suffices to treat the case where k is separably closed. (The scalar extension from k to k_s does not ruin the property of Z not containing a nontrivial smooth connected subgroup over the ground field.)

The commutator map of k-schemes $c_E : E \times E \to E$ uniquely factors through a k-scheme map

$$\widetilde{c}_E : G \times G \simeq (E \times E)/(Z \times Z) \to E$$

lifting the commutator map of $E/Z = G$ since (5.1.1) is a central extension, and clearly the element $\widetilde{c}_E(g_1, g_2) \in E(k)$ lifts $g_1g_2g_1^{-1}g_2^{-1}$ for $g_i \in G(k)$. Since $G = \mathscr{D}(G)$ and $k = k_s$, the commutator subgroup of $G(k)$ is Zariski-dense in G. We just showed that all commutators in $G(k)$ lift to $E(k)$, so it follows that $E(k)$ has image in $G(k)$ that is Zariski-dense in G. Hence, the Zariski closure $H \subseteq E$ of $E(k)$ maps onto G, and so does the identity component H^0 (as G is connected). Our goal is to find a section $G \to E$ as k-groups, so we can replace E with the k-subgroup H^0 and replace Z with $Z \cap H^0$ to reduce to the case where E is smooth and connected.

Obviously $\mathscr{D}(E)$ also maps onto G, so we can replace E with $\mathscr{D}(E)$. This reduces us to the case where E is affine (or equivalently, Z is affine), since the derived group of a smooth connected group over a field is always affine. (Indeed, it suffices to prove such affineness over the perfect closure, and by Theorem A.3.7 every smooth connected group over a perfect field is an extension of an abelian variety by a smooth connected affine group. Thus, the derived group is affine.)

Consider the surjective map $\pi : E \twoheadrightarrow G$ between smooth connected affine k-groups. If it admits a k-rational section over some dense open subset $\Omega \subseteq G$ then $\pi^{-1}(\Omega) \simeq \Omega \times Z$ as k-schemes. But $\pi^{-1}(\Omega)$ is smooth and connected since E is smooth and connected, so Z would have to be smooth and connected. Since Z has no nontrivial smooth connected k-subgroup, this would force $Z = 1$, so π would be an isomorphism and hence we would have the required splitting over k.

We have reduced the proof of Proposition 5.1.3 to showing that any central extension

$$1 \to Z \to E \overset{\pi}{\to} G \to 1 \tag{5.1.2}$$

of k-group schemes with E a smooth connected affine k-group admits a k-rational section over a dense open subscheme of G. We will construct such a section with E an arbitrary affine k-group scheme of finite type.

Let $T \subseteq G$ be a maximal k-torus, and $\Phi = \Phi(G, T)$. Fix a positive system of roots $\Phi^+ \subseteq \Phi$, and let U^+ denote the k-unipotent radical of the minimal pseudo-parabolic k-subgroup $P \subseteq G$ corresponding to Φ^+, so U^+ is directly spanned (in any order) by the root groups associated to the non-divisible positive roots. Likewise define U^- using $\Phi^- := -\Phi^+$, so the direct product scheme $\Omega := U^- \times Z_G(T) \times U^+$ is a dense open subscheme of G via multiplication (Corollary 3.3.16). We will construct a section to π over a dense open locus in Ω.

By hypothesis, there is a dense open subscheme $V \subseteq Z_G(T)$ and a finite set of k-morphisms $h_i : V \to U_{a_i}$ for a sequence of (not necessarily distinct) roots $a_1, \ldots, a_n$ such that $v = \prod h_i(v)$ for all $v \in V$. We can take the a_i to be non-divisible, since $U_a \subseteq U_{(a)} = U_{a/2}$ for each divisible root a. Thus, if we can find a section $\sigma_a : U_a \to \pi^{-1}(U_a)$ to π over the entirety of each U_a ($a \in \Phi$) then $\prod(\sigma_{a_i} \circ h_i) : V \to E$ is a section to π over the dense open V in $Z_G(T)$. Since $U^{\pm}$ is directly spanned by the U_a for non-divisible $a \in \Phi^{\pm}$, by suitably multiplying all of these sections using the group law on E we get a section to π over the dense open $U^- \cdot V \cdot U^+$ in G, as desired.

It remains to split the central extension

$$1 \to Z \to E_a = \pi^{-1}(U_a) \to U_a \to 1$$

obtained by restricting the given central extension to U_a for each $a \in \Phi$. The centrality of Z in E implies that the G-action on G via conjugation lifts to an action of $G = E/Z$ on E that is trivial on Z. In particular, E_a is a

T-equivariant central extension of U_a, and by Proposition 2.3.11 each U_a is a vector group admitting a linear structure with respect to which the weights of T are positive rational multiples of a. The following lemma therefore finishes the proof of Proposition 5.1.3. □

Lemma 5.1.6 *Let k be a field, C a commutative affine k-group scheme of finite type, and V a k-vector group equipped with an action by a k-split k-torus S such that the weights of S on* Lie(V) *are all nontrivial and any two such linearly dependent weights are $\mathbf{Q}_{>0}$-multiples of each other in* $\mathrm{X}(S)_{\mathbf{Q}}$. *Every S-equivariant central extension E of V by C (giving C the trivial S-action) admits a unique S-equivariant splitting over k.*

In the intended applications, C is a group scheme that cannot be assumed to be smooth.

Proof We may assume $V \neq 0$. By using uniqueness of the linear structure on V if $\mathrm{char}(k) = 0$ and using Theorem B.4.3 if $\mathrm{char}(k) > 0$, we may and do choose a linear structure on V with respect to which the S-action is linear, so the weights occurring in V match the ones in $\mathrm{Lie}(V)$. For any k-rational character $\chi \in \mathrm{X}(S)$ let V_χ denote the χ-weight space for V with its S-action and let $\mathbf{G}_{\mathrm{a}}(\chi)$ denote $\mathbf{G}_{\mathrm{a}}$ equipped with an S-action through χ. We have $V \simeq \prod V_{\chi_i}$ for some nontrivial and pairwise distinct $\chi_i \in \mathrm{X}(S)$.

For a nontrivial $\chi \in \mathrm{X}(S)$, the S-action on $\mathbf{G}_{\mathrm{a}}(\chi)$ is transitive on the complement of the origin. Thus, $\mathbf{G}_{\mathrm{a}}(\chi)$ has no nonzero S-equivariant quotient with a trivial S-action, so the group $\mathrm{Hom}_S(V, C)$ of S-equivariant k-homomorphisms vanishes. An S-equivariant central extension E of V by C over k therefore admits at most one S-equivariant splitting over k. To construct a splitting, by Galois descent we may and do now assume $k = k_s$. In particular, k is infinite.

As a first step, we show that E is commutative as a k-group. Since E is an S-equivariant central extension, the commutator map $E \times E \to E$ factors uniquely through an S-equivariant biadditive alternating map $V \times V \to C$. Thus, by biadditivity it suffices to show that for any pair of (possibly equal) weights χ and χ' for S acting on V, the induced S-equivariant map $f : V_\chi \times V_{\chi'} \to C$ of k-schemes is the constant map to $0 \in C(k)$. (Note that C is not necessarily smooth.)

The fiber $f^{-1}(0)$ is a closed subscheme of $V_\chi \times V_{\chi'}$, so to prove $f = 0$ it suffices (by k-smoothness of the weight spaces, and the condition $k = k_s$) to check that f vanishes on k-points. The S-equivariance implies that for any $(v, v') \in V_\chi(k) \times V_{\chi'}(k)$, the restriction of f to the Zariski closure of the

S-orbit $S.(v, v')$ is equal to the constant value $f(v, v') \in C(k)$. Hence, to show $f(v, v') = 0$ it suffices to find a k-point in the Zariski closure of $S.(v, v')$ at which f vanishes.

If χ and χ' are independent characters then there is a 1-parameter subgroup $a : \mathrm{GL}_1 \to S$ over k such that the composition $\chi' \circ a$ is trivial and $\chi \circ a \in \mathbf{Z}$ is equal to some $n \geqslant 1$. Thus, for $t \in k^\times$ we have $a(t).(v, v') = (t^n v, v')$. It follows from the infinitude of k that $(0, v')$ lies in the Zariski closure of $S.(v, v')$, so $f(v, v') = f(0, v') = 0$. If χ and χ' are dependent characters then (by hypothesis) they are $\mathbf{Q}_{>0}$-multiples of each other in $\mathrm{X}(S)_{\mathbf{Q}}$, so $\chi^m = \chi'^n$ for some $m, n \geqslant 1$. Thus, a variation on the argument for independent characters gives that the Zariski closure of $S.(v, v')$ contains $(0, 0)$, so $f(v, v') = f(0, 0) = 0$. This completes the proof that E is commutative.

Since k is infinite, there exists $s_0 \in S(k)$ such that $\chi(s_0) \in k^\times - \{1\}$ for all weights χ of S occurring on V. Consider the k-homomorphism $E \to E$ defined by $x \mapsto s_0.x - x$. This kills the central k-subgroup scheme C, so it factors through an S-equivariant k-homomorphism $h : V \to E$. The S-equivariant composite mapping $V \xrightarrow{h} E \twoheadrightarrow E/C = V$ is the map $v \mapsto s_0.v - v$ which is multiplication by $\chi(s_0) - 1 \in k^\times$ on V_χ for each weight χ of S on V. This is an automorphism of V, so $h(V)$ is an S-stable k-subgroup of E mapping isomorphically onto $E/C = V$. In other words, $h(V)$ provides an S-equivariant splitting of the given central extension structure on E. □

As an application of Proposition 5.1.3 (and Example 5.1.4), we have:

Proposition 5.1.7 *Let k be a field, k' a nonzero finite reduced k-algebra, and G' a semisimple k'-group whose fibers are connected, absolutely simple, and simply connected. Let $Z \subseteq \mathrm{R}_{k'/k}(Z_{G'}) = Z_{\mathrm{R}_{k'/k}(G')}$ be a central k-subgroup scheme, and $G := \mathrm{R}_{k'/k}(G')/Z$ the smooth connected affine quotient over k.*

(1) *Let $(G'_0, k'_0/k)$ be another such pair, $Z_0 \subseteq \mathrm{R}_{k'_0/k}(Z_{G'_0})$ a central k-subgroup scheme, and $G_0 := \mathrm{R}_{k'_0/k}(G'_0)/Z_0$. For any separable field extension F/k, every F-isomorphism $f : G_F \simeq (G_0)_F$ uniquely lifts to an F-homomorphism $\widetilde{f} : \mathrm{R}_{k'/k}(G')_F \to \mathrm{R}_{k'_0/k}(G'_0)_F$, and $\widetilde{f}$ is an isomorphism. Moreover, if $(G'_0, k'_0/k, Z_0) = (G', k'/k, Z)$ then $\widetilde{f}$ restricts to the identity on a smooth connected F-subgroup of $\mathrm{R}_{k'/k}(G')_F$ if f restricts to the identity on its image in G_F.*
(2) *If H is a smooth connected k-group equipped with a left action on G then this uniquely lifts to a left action on $\mathrm{R}_{k'/k}(G')$. This lifted action restricts to the trivial action on a smooth connected k-subgroup of $\mathrm{R}_{k'/k}(G')$ if H acts trivially on its image in G.*

An especially useful class of smooth connected subgroups in (1) is the maximal F-tori.

Proof We first reduce (2) to a special case of (1), as follows. Let $\mathscr{G} = \mathrm{R}_{k'/k}(G')$, and consider a smooth connected k-group H equipped with a left action on G. The action map

$$H \times G \to H \times G$$

defined by $(h, g) \mapsto (h, h.g)$ is an H-group automorphism, so it pulls back over the generic point $\eta : \operatorname{Spec} F \to H$ to an F-group automorphism of G_F. Since $F = k(H)$ is separable over k, by (1) this automorphism uniquely lifts to an F-group automorphism α_F of $\mathscr{G}_F$.

For a suitable dense open subscheme $U \subseteq H$, the lifted F-group automorphism α_F of $\mathscr{G}_F$ extends to a U-group automorphism of $U \times \mathscr{G}$. We claim that the resulting dominant composite map

$$\alpha : U \times \mathscr{G} \simeq U \times \mathscr{G} \stackrel{\mathrm{pr}_2}{\to} \mathscr{G}$$

uniquely extends to an action of H on $\mathscr{G}$, which would settle (2) conditional on (1). The uniqueness is clear, so we need to verify existence, for which we may assume $k = k_s$. For each $u \in U(k)$, restricting α to the u-slice defines a k-group automorphism θ_u of $\mathscr{G}$. Define a map $Uu \times \mathscr{G} \to \mathscr{G}$ by $(u'u, g) \mapsto \alpha(u', \theta_u(g))$. The associated Uu-map $Uu \times \mathscr{G} \to Uu \times \mathscr{G}$ has F-fiber equal to an F-group endomorphism of $\mathscr{G}_F$ lifting the generic fiber of the action of H on G. Thus, the uniqueness aspect of (1) implies that these maps $Uu \times \mathscr{G} \to \mathscr{G}$ agree on overlaps (as that can be checked on $\mathscr{G}_F$) and so glue to a k-map $H \times \mathscr{G} \to \mathscr{G}$.

The associated H-map $H \times \mathscr{G} \to H \times \mathscr{G}$ is an H-group map (since it is an F-group map over the generic point of H). Specializing at $1 \in H$ gives an endomorphism of $\mathscr{G}$ covering the identity of G, so it must be the identity (by (1)), and likewise for any $h, h' \in H(k)$ the endomorphism of $\mathscr{G}$ induced by hh' is the composition of the endomorphisms induced by h and h' separately (as this holds with G in place of $\mathscr{G}$). It follows that H acts on $\mathscr{G}$ in an associative manner, so its action is through automorphisms, thereby finishing the proof of (2).

Now we prove (1). By Galois descent and uniqueness we may assume $F = F_s$, and since F/k is separable we can extend scalars to F to reduce to the case $F = k$. To prove the uniqueness of the lifted homomorphism it suffices to observe that a map of pointed k-schemes $\mathrm{R}_{k'/k}(G') \to \mathrm{R}_{k'_0/k}(Z_{G'_0})$ is trivial since it corresponds to a map of pointed k'_0-schemes $\mathrm{R}_{k'/k}(G')_{k'_0} \to Z_{G'_0}$ in which the source has smooth connected fibers over $\operatorname{Spec} k'_0$ and the target has finite fibers over $\operatorname{Spec} k'_0$. The same argument handles the condition $\widetilde{f} = \mathrm{id}$ on smooth connected k-subgroups when $(G'_0, k'/k, Z_0) = (G', k'/k, Z)$.

It remains to show that for the central quotient maps $\pi : \mathrm{R}_{k'/k}(G') \twoheadrightarrow \mathrm{R}_{k'/k}(G')/Z = G$ and $\pi_0 : \mathrm{R}_{k'_0/k}(G'_0) \twoheadrightarrow \mathrm{R}_{k'_0/k}(G'_0)/Z_0 = G_0 \simeq G$ (using the k-isomorphism f), π lifts through π_0 to a k-isomorphism $\widetilde{f}$. Consider the pullback diagram

$$\begin{array}{ccccccccc} 1 & \longrightarrow & Z_0 & \longrightarrow & E & \longrightarrow & \mathrm{R}_{k'/k}(G') & \longrightarrow & 1 \\ & & \| & & \downarrow & & \downarrow{\scriptstyle \pi} & & \\ 1 & \longrightarrow & Z_0 & \longrightarrow & \mathrm{R}_{k'_0/k}(G'_0) & \underset{\pi_0}{\longrightarrow} & G & \longrightarrow & 1 \end{array} \qquad (5.1.3)$$

The top line is a central extension of $\mathrm{R}_{k'/k}(G')$ by the k-group Z_0. We claim that Z_0 contains no nontrivial smooth connected k-subgroup. It suffices to prove this claim for the ambient central k-subgroup $\mathrm{R}_{k'_0/k}(Z_{G'_0})$, and so by the universal property of Weil restriction it suffices to prove the analogous fact for each fiber of $Z_{G'_0}$ over $\operatorname{Spec} k'_0$. These fibers are finite, so the claim is clear.

Since Z_0 contains no nontrivial smooth connected k-subgroup, any central extension of $\mathscr{G} = \mathrm{R}_{k'/k}(G')$ by Z_0 over k (such as the top row of (5.1.3)) is uniquely split over k (Example 5.1.4). The unique splitting of the top row of (5.1.3) provides a canonical copy of $\mathscr{G}$ as a k-subgroup of E, and its map down to the middle term along the bottom row of (5.1.3) is a homomorphism $\widetilde{f} : \mathscr{G} \to \mathrm{R}_{k'_0/k}(G'_0)$ that carries Z into Z_0 and induces the isomorphism f modulo Z and Z_0. It remains only to prove that $\widetilde{f}$ is an isomorphism. But we can apply the same lifting result to f^{-1}, and this provides a homomorphism $\widetilde{f}'$ carrying Z_0 into Z such that the composites $\widetilde{f} \circ \widetilde{f}'$ and $\widetilde{f}' \circ \widetilde{f}$ respectively preserve Z_0 and Z and reduce to the identity modulo Z_0 and modulo Z. Hence, uniqueness forces $\widetilde{f}'$ to be an inverse to $\widetilde{f}$. □

5.2 Properties of standardness and standard presentations

In order to prove standardness for a pseudo-reductive group, it is convenient to record how standard presentations (when they exist!) behave with respect to operations on groups. We begin with passage to the derived group.

Proposition 5.2.1 *A non-commutative pseudo-reductive k-group G is standard if and only if $\mathscr{D}(G)$ is standard. If either admits a standard presentation resting on the pair $(G', k'/k)$ then so does the other.*

Proof Choose a maximal k-torus T in G, and let $S = T \cap \mathscr{D}(G)$ be the corresponding maximal k-torus in $\mathscr{D}(G)$. We have $G = Z_G(T) \cdot \mathscr{D}(G)$, by Proposition 1.2.6. By Proposition 4.1.4, we may and do restrict our attention to standard presentations that are compatible with T and S.

First assume that G is standard, so it arises from a 4-tuple $(G', k'/k, T', C)$ compatible with T. In particular, $j : \mathrm{R}_{k'/k}(G') \to G$ has image $\mathscr{D}(G)$ and carries $\mathrm{R}_{k'/k}(T')$ into $C = Z_G(T)$. Hence,

$$j(\mathrm{R}_{k'/k}(T')) \subseteq \mathscr{D}(G) \cap C \subseteq Z_{\mathscr{D}(G)}(S).$$

Thus, we can form the standard pseudo-reductive k-subgroup

$$(\mathrm{R}_{k'/k}(G') \rtimes Z_{\mathscr{D}(G)}(S))/\mathrm{R}_{k'/k}(T')$$

of the standard pseudo-reductive k-group

$$(\mathrm{R}_{k'/k}(G') \rtimes Z_G(T))/\mathrm{R}_{k'/k}(T') = G.$$

This subgroup of G is clearly contained in $\mathscr{D}(G)$, yet it contains $j(\mathrm{R}_{k'/k}(G')) = \mathscr{D}(G)$. Thus, we have constructed a standard presentation of $\mathscr{D}(G)$, and it rests on the same pair $(G', k'/k)$.

Conversely, suppose that $\mathscr{D}(G)$ is standard, and choose a 4-tuple $(G', k'/k, T', \mathscr{C})$ giving rise to a standard presentation of $\mathscr{D}(G)$ that is compatible with S. Hence, $\mathscr{C} = Z_{\mathscr{D}(G)}(S)$ and there is a factorization diagram

$$\mathrm{R}_{k'/k}(T') \stackrel{\phi}{\to} \mathscr{C} \stackrel{\psi}{\to} \mathrm{R}_{k'/k}(T'/Z_{G'})$$

and an isomorphism

$$\delta : (\mathrm{R}_{k'/k}(G') \rtimes \mathscr{C})/\mathrm{R}_{k'/k}(T') \simeq \mathscr{D}(G)$$

with the maximal k-torus $\mathscr{T}$ in $\mathrm{R}_{k'/k}(T')$ mapping into the maximal k-torus S in $\mathscr{D}(G)$. Since $\mathrm{R}_{k'/k}(G')$ is carried onto $\mathscr{D}^2(G) = \mathscr{D}(G)$, it carries $\mathscr{T}$ onto S (even by an isogeny).

The kernel $\ker j = \ker \phi$ is central (as for any standard presentation). Observe that

$$j(\mathrm{R}_{k'/k}(T')) \subseteq Z_{\mathscr{D}(G)}(S) \subseteq Z_G(T) =: C,$$

so to make a standard presentation of G (compatible with T, and resting on $(G', k'/k)$) we are led to try to lift the natural conjugation action of C on the central quotient $\mathscr{D}(G)$ of $\mathrm{R}_{k'/k}(G')$ to an action on $\mathrm{R}_{k'/k}(G')$ that restricts to the identity on the maximal k-torus $\mathscr{T}$. Such a lift is provided by Proposition 5.1.7, and it is even unique. Using this action and the universal property of $\mathrm{R}_{k'/k}(T'/Z_{G'})$ from Theorem 1.3.9, we get a factorization diagram

$$\mathrm{R}_{k'/k}(T') \to C \to \mathrm{R}_{k'/k}(T'/Z_{G'}) \tag{5.2.1}$$

and hence a standard pseudo-reductive k-group

$$(\mathrm{R}_{k'/k}(G') \rtimes C)/\mathrm{R}_{k'/k}(T').$$

There is an evident natural k-homomorphism

$$\gamma : (\mathrm{R}_{k'/k}(G') \rtimes C)/\mathrm{R}_{k'/k}(T') \to G$$

that restricts to the chosen standard presentation δ of $\mathscr{D}(G)$. In particular, γ is surjective since $G = Z_G(T) \cdot \mathscr{D}(G)$.

To prove that γ is an isomorphism, consider a point (g, z) of $\mathrm{R}_{k'/k}(G') \rtimes C$ (valued in some k-algebra) such that it maps to 1 in G. We have to show that this point lies in the central k-subgroup $\mathrm{R}_{k'/k}(T')$. Since g maps into $\mathscr{D}(G)$, it follows that z is a point of $C \cap \mathscr{D}(G) \subseteq Z_{\mathscr{D}(G)}(S) = \mathscr{C}$. Hence, the triviality of $\ker \delta$ does the job. □

A consequence of Proposition 5.2.1 is a description of all isomorphisms between standard pseudo-reductive k-groups. This will enable us to reduce the verification of standardness to a problem over separably closed fields.

Proposition 5.2.2 *Let $(G', k'/k, T', C_1)$ and $(G'', k''/k, T'', C_2)$ be two 4-tuples that give standard presentations of a non-commutative pseudo-reductive k-group G, and choose both to be compatible with a fixed maximal k-torus T in G.*

Under the resulting identifications of C_1 and C_2 with $C := Z_G(T)$, the composite isomorphism

$$(\mathrm{R}_{k'/k}(G') \rtimes C)/\mathrm{R}_{k'/k}(T') \simeq G \simeq (\mathrm{R}_{k''/k}(G'') \rtimes C)/\mathrm{R}_{k''/k}(T'')$$

is induced by a unique pair consisting of a k-algebra isomorphism $\alpha : k'' \simeq k'$ and an isomorphism of pairs $(G', T') \simeq (G'', T'')$ over Spec α *respecting the given factorization diagram* (5.2.1) *and its counterpart for $(G'', k''/k, T'', C)$.*

In particular, a standard presentation of G compatible with a specified maximal k-torus of G is unique up to unique isomorphism.

Proof The uniqueness reduces the problem to the case $k = k_s$ by Galois descent. The images of $\mathrm{R}_{k'/k}(G')$ and $\mathrm{R}_{k''/k}(G'')$ in G coincide with $\mathscr{D}(G)$, and give rise to standard presentations of $\mathscr{D}(G)$ compatible with the maximal k-torus $S = T \cap \mathscr{D}(G)$ (Proposition 5.2.1). Thus, we can replace G with $\mathscr{D}(G)$ to reduce to the case where G is perfect (as Proposition 5.1.7(2) ensures C-equivariance of the resulting k-isomorphism $\mathrm{R}_{k'/k}(G') \simeq \mathrm{R}_{k''/k}(G'')$). Letting $Z_1 = \ker\phi_1$ and $Z_2 = \ker\phi_2$ be the central kernels arising from the two standard presentations of G, the method of proof of Proposition 5.1.7(1) works without change to show that the k-isomorphism

$$f : \mathrm{R}_{k'/k}(G')/Z_1 \simeq G \simeq \mathrm{R}_{k''/k}(G'')/Z_2$$

uniquely lifts to a k-isomorphism $\widetilde{f} : \mathrm{R}_{k'/k}(G') \simeq \mathrm{R}_{k''/k}(G'')$ carrying Z_1 into Z_2 (via an isomorphism, by working with f^{-1} and its unique lift $\widetilde{f}^{-1}$). The bijective correspondence between sets of maximal k'-tori of G', maximal k-tori of G, and maximal k''-tori of G'' implies that $\widetilde{f}$ carries the maximal k-torus of $\mathrm{R}_{k'/k}(T')$ over to the maximal k-torus of $\mathrm{R}_{k''/k}(T'')$. Passing to centralizers of these maximal k-tori, the Cartan k-subgroup $\mathrm{R}_{k'/k}(T')$ is carried over to the Cartan k-subgroup $\mathrm{R}_{k''/k}(T'')$.

It remains to prove that any k-group isomorphism $f\colon \mathrm{R}_{k'/k}(G') \simeq \mathrm{R}_{k''/k}(G'')$ arises from a unique k-algebra isomorphism $\alpha\colon k'' \simeq k'$ and a group isomorphism $\phi : G' \simeq G''$ over $\operatorname{Spec}\alpha$, and that if f carries $\mathrm{R}_{k'/k}(C')$ over to $\mathrm{R}_{k''/k}(C'')$ for a Cartan k'-subgroup $C' \subseteq G'$ and a Cartan k''-subgroup $C'' \subseteq G''$ then $\phi(C') = C''$ (so ϕ carries the maximal k'-torus of C' over to the maximal k''-torus of C''). The compatibility with Cartans is automatic, due to the behavior of Weil restriction with respect to the formation of Cartan subgroups (Proposition A.5.15(3)), so the problem is to prove the existence and uniqueness of (α, ϕ).

First consider the case when k' and k'' are both fields. The simply connected datum associated to $\mathrm{R}_{k'/k}(G')$ has the form $(G', k'/k, \pi)$, and similarly for $\mathrm{R}_{k''/k}(G'')$, so functoriality of the simply connected datum with respect to isomorphisms implies that $k' = k''$ (as purely inseparable extensions of k) and that there is an isomorphism $f' : G' \simeq G''$ over $k' = k''$ which lifts the isomorphism induced by $f_{k'}$ on maximal adjoint quotients over k'. But then f' coincides with the map induced by $f_{k'}$ on maximal reductive quotients over k', so $f = \mathrm{R}_{k'/k}(f')$ by Proposition 1.2.2 and Proposition A.5.11.

To reduce the general case to the settled case of fields, let $\{k'_i\}$ and $\{k''_j\}$ be the respective factor fields of k' and k'', and let G'_i and G''_j respectively denote the k'_i-fiber of G' and the k''_j-fiber of G''. We claim that f arises from a collection of k-isomorphisms $\mathrm{R}_{k'_i/k}(G'_i) \simeq \mathrm{R}_{k''_{j(i)}/k}(G''_{j(i)})$ for a bijection $i \mapsto j(i)$ between the index sets, so applying the settled field case to these isomorphisms would then complete the proof in general. To get this structure for f, it suffices to prove an intrinsic characterization of the collection of nontrivial smooth connected k-subgroups $G_i := \mathrm{R}_{k'_i/k}(G'_i)$ in $\mathrm{R}_{k'/k}(G')$: they are the smooth connected normal k-subgroups that are pseudo-simple over k (see Definition 3.1.1). This consists of two parts, namely that the G_i are themselves pseudo-simple, and that there are no others.

To prove the pseudo-simplicity of the G_i, first note that each G_i is perfect, by Corollary A.7.11. By pseudo-reductivity of G_i and Lemma 3.1.2, it suffices to prove that $(G_i)^{\mathrm{ss}}_{\overline{k}}$ is simple. Since k'_i/k is purely inseparable, the natural quotient map $q_i : (G_i)_{k'_i} \twoheadrightarrow G'_i$ has smooth connected unipotent kernel (Proposition A.5.11), so $(G_i)^{\mathrm{red}}_{\overline{k}} = G'_i \otimes_{k'_i} \overline{k}$ is simple.

Finally, consider a nontrivial smooth connected normal k-subgroup N in $\mathrm{R}_{k'/k}(G')$. We seek some G_i inside N. Since $N \subseteq \prod G_i$, its smooth connected normal image in some G_{i_0} is nontrivial. This image is full by pseudo-simplicity of G_{i_0}, so (N, G_{i_0}) also maps onto G_{i_0} since G_{i_0} is perfect. But $(N, G_{i_0}) \subseteq G_{i_0}$, so $(N, G_{i_0}) = G_{i_0}$. Normality of N implies that $(N, G_{i_0}) \subseteq N$, so $G_{i_0} \subseteq N$. □

Corollary 5.2.3 *Let G be a pseudo-reductive k-group. If G_{k_s} is standard over k_s, then so is G over k.*

In Corollary 5.3.9 we will generalize this to allow any separable extension of fields in place of k_s/k. An interesting case is the extension k_v/k for a global function field k and a place v of k.

Proof We may and do assume that G is non-commutative, and choose a maximal k-torus T of G. By Proposition 4.1.4, we can consider the unique standard presentation of G_{k_s} compatible with T_{k_s}. Let $(G', k'/k)$ arise from the simply connected datum associated to G. The standard presentation of G_{k_s} compatible with T_{k_s} rests on the pair $(G'_{k'_s}, k'_s/k_s)$ for $k'_s := k_s \otimes_k k'$ (due to the compatibility of the simply connected datum with respect to separable extension of the ground field). Uniqueness of the standard presentation compatible with T_{k_s} allows us to apply Galois descent to the data in this presentation to make a k-homomorphism $j : \mathrm{R}_{k'/k}(G') \to G$ that descends the corresponding map in the standard presentation for G_{k_s} over k_s. By compatibility of the presentation with T_{k_s} over k_s, this map carries $\mathrm{R}_{k'/k}(T')$ into $C := Z_G(T)$. Moreover, the map $C_{k_s} \to \mathrm{R}_{k'/k}(T'/Z_{G'})_{k_s}$ is Galois-equivariant and so descends, thereby defining a standard k-group

$$(\mathrm{R}_{k'/k}(G') \rtimes C)/\mathrm{R}_{k'/k}(T'/Z_{G'})$$

equipped with an evident k-homomorphism γ to G. But over k_s this recovers the standard presentation for G_{k_s} compatible with T_{k_s}, so γ_{k_s} is an isomorphism and hence γ is an isomorphism. □

Now we can establish good behavior of standardness relative to torus centralizers and normal subgroups. We first take up the case of torus centralizers. Let G be a standard pseudo-reductive group over a field k. For any k-torus S in G, the centralizer $Z_G(S)$ is a pseudo-reductive k-group, by Proposition 1.2.4. Choose a maximal k-torus T in G containing S. Suppose that G is non-commutative, and let $(G', k'/k, T', C)$ be the standard presentation compatible with T. In particular, $C = Z_G(T) \subseteq Z_G(S)$.

The action of C on $\mathrm{R}_{k'/k}(G')$ centralizes $\mathrm{R}_{k'/k}(T')$, so it also centralizes $\mathrm{R}_{k'/k}(S')$ for any k'-torus S' in T'. Since $\mathrm{R}_{k'/k}(Z_{G'}(S'))$ is the centralizer of $\mathrm{R}_{k'/k}(S')$ in $\mathrm{R}_{k'/k}(G')$ (Proposition A.5.15(3)), it follows that C normalizes $\mathrm{R}_{k'/k}(Z_{G'}(S'))$ for any k-torus S' in T'. Thus, the semidirect product $\mathrm{R}_{k'/k}(Z_{G'}(S')) \rtimes C$ makes sense.

Proposition 5.2.4 *For a standard pseudo-reductive k-group G and a k-torus S in G, the centralizer $Z_G(S)$ is a standard pseudo-reductive k-group. More specifically, if G is non-commutative then with notation as above, there is a canonical k'-torus $S' \subseteq T'$ such that*

$$(\mathrm{R}_{k'/k}(Z_{G'}(S')) \rtimes C)/\mathrm{R}_{k'/k}(T') = Z_G(S) \tag{5.2.2}$$

inside $(\mathrm{R}_{k'/k}(G') \rtimes C)/\mathrm{R}_{k'/k}(T') = G$. The formation of S' respects separable extension on k.

The k'-torus S' is not uniquely determined, but our construction of it will be canonical. We have little control over $Z_{G'}(S')$ beyond that it has connected reductive fibers, so this result illustrates the usefulness of the generality of the standard construction in the non-commutative case (not requiring G' to have absolutely simple and simply connected fibers).

Proof If G is commutative then $Z_G(S) = G$ is clearly standard, so we may and do assume that G is non-commutative. For any k'-torus S' in T', the fibers of

$Z_{G'}(S')$ over $\mathrm{Spec}(k')$ are connected reductive but generally are not semisimple, let alone absolutely simple and simply connected.

Step 1. Let $\overline{G}' = G'/Z_{G'}$, and consider the map of k'-groups $G_{k'} \twoheadrightarrow \overline{G}'$ which corresponds to the natural k-map $f : G \to \mathrm{R}_{k'/k}(\overline{G}')$ (i.e., the map f in Remark 1.4.8) via the universal property of Weil restriction. This k'-group map is surjective, so since the formation of torus centralizers is compatible with the formation of images under maps between smooth connected affine groups over a field (Proposition A.2.8), by working on fibers over the factor fields of k' it follows that the image $\overline{M}'$ of $Z_G(S)_{k'}$ in $\overline{G}'$ is the centralizer of the image of $S_{k'}$ in $\overline{G}'$. In particular, $\overline{M}'$ is a reductive k'-group with connected fibers since $\overline{G}'$ is a semisimple k'-group with connected fibers.

Let $\overline{S}'$ be the maximal central torus in $\overline{M}'$ (formed fiberwise over $\mathrm{Spec}(k')$). The image of $S_{k'}$ in $\overline{G}'$ is central in $\overline{M}'$ and so is contained in $\overline{S}'$. Since the image of $T_{k'}$ in $\overline{G}'$ contains the maximal torus $\overline{T}' = T'/Z_{G'}$, it is equal to $\overline{T}'$. Hence, $C_{k'} = Z_G(T)_{k'}$ has image $\overline{T}'$ in $\overline{G}'$ as well. In particular, $\overline{T}'$ is a maximal torus of $\overline{M}'$.

Under a central isogeny $H_1 \to H_2$ between connected semisimple groups over a field, consideration of maximal tori shows that every (not necessarily maximal) torus T_2 in H_2 is the image of a unique torus T_1 in H_1 (so T_1 is the identity component of the underlying reduced scheme of the preimage of T_2 in H_1). Applying this to the central isogeny $G' \twoheadrightarrow \overline{G}'$ fiberwise over $\mathrm{Spec}(k')$, there is a unique k'-torus S' in G' mapping onto $\overline{S}'$ in $\overline{G}'$. We shall prove that S' satisfies (5.2.2). Observe that, by construction, the formation of the k'-torus S' in G' commutes with separable extension on k. In particular, to prove that (5.2.2) holds we may and do now assume that k is separably closed.

Step 2. First we check that the inclusion "$\subseteq$" holds in (5.2.2). To see that $j : \mathrm{R}_{k'/k}(G') \to G$ carries $\mathrm{R}_{k'/k}(Z_{G'}(S'))$ into $Z_G(S)$ it suffices to check on k-points since k is separably closed. That is, we must prove that j carries the subgroup $Z_{G'(k')}(S') \subseteq G'(k') = \mathrm{R}_{k'/k}(G')(k)$ into $(Z_G(S))(k) = Z_{G(k)}(S)$. In other words, if a point $g' \in G'(k')$ centralizes S' then we want $g := j(g') \in G(k)$ to centralize S. Since $\overline{S}'(k') = \mathrm{R}_{k'/k}(\overline{S}')(k)$ contains the image of $S'(k') = \mathrm{R}_{k'/k}(S')(k)$ under the natural map $\mathrm{R}_{k'/k}(T') \to \mathrm{R}_{k'/k}(\overline{T}')$ with $S'(k')$ having Zariski-dense image in $\overline{S}'$, the point $\overline{g}' = f'(g) \in \overline{G}'(k')$ centralizes $\overline{S}'(k')$. Thus, if $\{k_i'\}$ is the set of factor fields of k' and $\overline{G}_i'$ and $\overline{S}_i'$ denote the k_i'-fibers of $\overline{G}'$ and $\overline{S}'$ respectively, then when $\overline{g}'$ is viewed as a $\overline{k}$-point of $\prod(\overline{G}_i' \otimes_{k_i'} \overline{k})$ it centralizes $\prod(\overline{S}_i' \otimes_{k_i'} \overline{k})$.

Since k is separably closed, the finite extensions k_i'/k are purely inseparable. By construction of f', the natural map

$$G_{\overline{k}} \xrightarrow{f'_{\overline{k}}} \mathrm{R}_{k'/k}(\overline{G}')_{\overline{k}} \twoheadrightarrow \prod (\overline{G}'_i \otimes_{k'_i} \overline{k})$$

is the projection to the maximal adjoint semisimple quotient. Also, as we have noted already (using the definition of $\overline{S}'$), the image of $S_{\overline{k}}$ in the ith factor of this quotient is contained in $\overline{S}'_i \otimes_{k'_i} \overline{k}$. Hence, for the point $g \in G(k)$ and the k-torus S in G, the image of g as a $\overline{k}$-point of the maximal adjoint semisimple quotient of $G_{\overline{k}}$ centralizes the image of $S_{\overline{k}}$ in this quotient. Since the maximal semisimple quotient $G^{\mathrm{ss}}_{\overline{k}}$ maps via a central isogeny onto the maximal adjoint semisimple quotient of $G_{\overline{k}}$, it follows that the image of g in $G^{\mathrm{ss}}_{\overline{k}}$ centralizes the image of $S_{\overline{k}}$ in $G^{\mathrm{ss}}_{\overline{k}}$. By the following lemma, $g \in Z_G(S)(k)$ as desired (thereby completing the proof of the containment "$\subseteq$" in (5.2.2)).

Lemma 5.2.5 *Let H be a pseudo-reductive group over a field F, and T an F-torus in H. A point $h \in H(F)$ centralizes T if and only if its image in $H^{\mathrm{ss}}_{\overline{F}}(\overline{F})$ centralizes the image of $T_{\overline{F}}$ in $H^{\mathrm{ss}}_{\overline{F}}$.*

Proof The "only if" direction is trivial, so assume that the image of h centralizes the image of $T_{\overline{F}}$ in $H^{\mathrm{ss}}_{\overline{F}}$. We first check that h centralizes $T_{\overline{F}}$ viewed as a torus in the maximal reductive quotient $H^{\mathrm{red}}_{\overline{F}}$. If Z denotes the maximal central torus in $H^{\mathrm{red}}_{\overline{F}}$ then it is equivalent to check that the image of h in $H^{\mathrm{red}}_{\overline{F}}$ centralizes the torus $Z \cdot T_{\overline{F}}$ in $H^{\mathrm{red}}_{\overline{F}}$. But any torus in $H^{\mathrm{red}}_{\overline{F}}$ that contains Z is uniquely an almost direct product of Z and a torus in $\mathscr{D}(H^{\mathrm{red}}_{\overline{F}})$. Writing $Z \cdot T_{\overline{F}}$ as an almost direct product $Z \cdot S$ with a torus S in $\mathscr{D}(H^{\mathrm{red}}_{\overline{F}})$, we need to prove that the image of h centralizes S. Since $\mathscr{D}(H^{\mathrm{red}}_{\overline{F}}) \to H^{\mathrm{ss}}_{\overline{F}}$ is an isogeny, it is equivalent for h to centralize the image of S in $H^{\mathrm{ss}}_{\overline{F}}$, and this image of S is equal to the image of $T_{\overline{F}}$ that is assumed to be centralized by the image of h.

Since we have proved that h centralizes $T_{\overline{F}}$ viewed as a torus in $H^{\mathrm{red}}_{\overline{F}}$, it remains to show that a point $h \in H(F)$ that centralizes $T_{\overline{F}}$ in $H^{\mathrm{red}}_{\overline{F}}$ also centralizes T in H. Consider the natural map $T \to H$ defined by $t \mapsto (t, h) = tht^{-1}h^{-1}$. The closure X of its image has $\overline{F}$-fiber that is reduced and irreducible and contained in $\mathscr{R}_u(H_{\overline{F}})$, so by pseudo-reductivity of H over F we can apply Lemma 1.2.1 to deduce that $X = \{1\}$. In other words, h centralizes T in H. □

Step 3. It remains to prove that the inclusion "$\subseteq$" in (5.2.2) is an equality, and for this it suffices to check the reverse inclusion on k-points. Since $k = k_s$ we have

$$Z_G(S)(k) \subseteq G(k) = (G'(k') \rtimes C(k))/T'(k'),$$

so a point $g \in Z_G(S)(k)$ is represented by a pair $(g', c) \in G'(k') \rtimes C(k)$. The problem is to show that $g' \in Z_{G'(k')}(S')$.

Recall from Step 1 that under the quotient map $G_{k'} \twoheadrightarrow \overline{G}'$, $Z_G(S)_{k'}$ maps onto the connected reductive k'-subgroup $\overline{M}'$ in which $\overline{S}'$ is (by definition) the maximal central torus, and that $C_{k'}$ maps onto the maximal k'-torus $\overline{T}'$ of $\overline{M}'$. In particular, $f'((1, c) \bmod T')$ centralizes $\overline{S}'$, so $\overline{g}' := f'(j(g')) \in \overline{G}'(k')$ centralizes the k'-torus $\overline{S}'$ in $\overline{G}'$. Thus, under the isogeny $G' \to \overline{G}'$ between semisimple k'-groups, the point $g' \in G'(k')$ maps into $Z_{\overline{G}'}(\overline{S}')(k')$. Since S' is (by definition) the unique k'-torus in G' mapping onto $\overline{S}'$, and the map $S' \to \overline{S}'$ is an isogeny, it follows that $g' \in Z_{G'}(S')(k')$. This proves the equality (5.2.2). □

The next result records the preservation of standardness under passage to normal subgroups:

Proposition 5.2.6 *Any smooth connected normal k-subgroup N of a standard pseudo-reductive k-group G is a standard pseudo-reductive k-group.*

Proof Since $\mathscr{R}_{u,k}(N)$ is normal in the pseudo-reductive k-group G, it is trivial. Thus, N is pseudo-reductive over k. To prove N is standard, by Corollary 5.2.3 we may assume that k is separably closed. The case of commutative G is trivial, so we may also assume that G is non-commutative.

Let S be a maximal k-torus of N, and let T be a maximal k-torus of G containing S. Let $C = Z_G(T)$, so by Proposition 4.1.4 (also see Remark 4.1.5) we have an associated standard presentation $(\mathrm{R}_{k'/k}(G') \rtimes C)/\mathrm{R}_{k'/k}(T') = G$.

Let $\{k'_i\}_{i \in I}$ be the set of factor fields of k'. For each i the finite extension k'_i/k is purely inseparable and the quotient map $G_{k'_i} \twoheadrightarrow \overline{G}'_i$ carries $N_{k'_i}$ onto a smooth connected normal k'_i-subgroup of $\overline{G}'_i$. The connected semisimple k'_i-group $\overline{G}'_i$ is simple (even absolutely simple), so $N_{k'_i}$ has either trivial or full image in $\overline{G}'_i$. Suppose for some i_0 that $N_{k'_{i_0}}$ has trivial image in $\overline{G}'_{i_0}$. Since

$$
\begin{aligned}
N(k) \subseteq G(k) &= (G'(k') \rtimes C(k))/T'(k') \\
&= \left(\left(\prod G'_i(k'_i)\right) \rtimes C(k)\right) \Big/ \left(\prod T'_i(k'_i)\right),
\end{aligned}
$$

any $n \in N(k)$ is represented by a point $((g'_i), c)$ whose image in $\overline{G}'_{i_0}(k'_{i_0})$ is trivial. But $C(k)$ is carried into $\overline{T}'_{i_0}(k'_{i_0})$, so g'_{i_0} is also carried into this subgroup.

For each i, the map $G'_i \to \overline{G}'_i$ is a central isogeny between connected semisimple k'_i-groups, so T'_i is the preimage of $\overline{T}'_i$. In particular, $g'_{i_0} \in T'_{i_0}(k'_{i_0})$.

We can therefore modify our choice of representative for $n \in N(k)$ so that $g'_{i_0} = 1$ and all other g'_i remain unchanged.

Let J be the set of i for which $N_{k'_i}$ maps onto $\overline{G}'_i$, define $k'_J := \prod_{j \in J} k'_j$ to be the corresponding factor ring of k', and define $G'_J := \coprod_{j \in J} G'_j$ and $T'_J = \coprod_{j \in J} T'_j$ to be the respective k'_J-fibers of the k'-groups G' and T'. The preceding argument shows that N is contained in the standard pseudo-reductive k-subgroup

$$G_J := (\mathrm{R}_{k'_J/k}(G'_J) \rtimes C)/\mathrm{R}_{k'_J/k}(T'_J) \subseteq (\mathrm{R}_{k'/k}(G') \rtimes C)/\mathrm{R}_{k'/k}(T') = G.$$

Observe that J labels the simple factors of the maximal semisimple quotient of $(G_J)_{\overline{k}}$. If J is empty then N is contained in C, in which case N is commutative and therefore standard. By induction on the number of simple factors of $G^{\mathrm{ss}}_{\overline{k}}$, we may reduce to the case where J is non-empty and is equal to the set of all i, which is to say that $N_{k'_i}$ maps onto $\overline{G}'_i$ for all i. Consider the preimage H of N in $\mathrm{R}_{k'/k}(G') \rtimes C$. This is smooth and connected since it fits into the exact sequence

$$1 \to \mathrm{R}_{k'/k}(T') \to H \to N \to 1$$

and it is normal in $\mathrm{R}_{k'/k}(G') \rtimes C$. Provided that H contains $\mathrm{R}_{k'/k}(G')$ we will have that

$$N = (\mathrm{R}_{k'/k}(G') \rtimes C_N)/\mathrm{R}_{k'/k}(T')$$

with C_N denoting the image of H in C (so C_N contains the image of $\mathrm{R}_{k'/k}(T')$ in C, and it is pseudo-reductive since it is a smooth connected k-subgroup of the commutative pseudo-reductive k-group C). Such an expression for N is a standard presentation, so we would be done.

It remains to prove that a smooth connected normal k-subgroup H in $\mathrm{R}_{k'/k}(G') \rtimes C$ such that $H_{k'_i}$ maps onto $\overline{G}'_i$ for all i must contain $\mathrm{R}_{k'/k}(G')$. Equivalently, we claim that H contains $\mathrm{R}_{k'_i/k}(G'_i)$ for each i. For $i \in I$, consider the smooth connected commutator k-group $H_i = (H, \mathrm{R}_{k'_i/k}(G'_i))$. This is contained in both H and $\mathrm{R}_{k'_i/k}(G'_i)$ due to the normality of both the subgroups, it is a normal k-subgroup of $\mathrm{R}_{k'_i/k}(G'_i)$, and $(H_i)_{k'_i}$ maps onto $\mathscr{D}(\overline{G}'_i) = \overline{G}'_i$. In particular, $H_i \neq 1$ for all i. But the k-group $\mathrm{R}_{k'_i/k}(G'_i)$ has no nontrivial smooth connected normal proper k-subgroups (as we showed at the end of the proof of Proposition 5.2.2). This proves that $\mathrm{R}_{k'_i/k}(G_i) = H_i \subseteq H$ for all i, as required. □

5.3 A standardness criterion

The main result in this section is a criterion for standardness of absolutely pseudo-simple groups. Before discussing it, we reduce the verification of standardness to the (absolutely) pseudo-simple case provided that the ground field is separably closed:

Proposition 5.3.1 *Let G be a non-commutative pseudo-reductive group over $k = k_s$, and $\{G_i\}$ the collection of pseudo-simple normal k-subgroups of G. If each G_i is standard then so is G.*

Proof By Proposition 5.2.1 we may replace G with $\mathscr{D}(G)$, so G is perfect. Let T be a maximal k-torus of G, and let $T_i := T \cap G_i$ be the corresponding maximal k-torus of G_i (Corollary A.2.7). By Proposition 4.1.4, we may construct standard presentations

$$(\mathrm{R}_{k_i'/k}(G_i') \rtimes C_i)/\mathrm{R}_{k_i'/k}(Z_{G_i'}(T_i')) \simeq G_i$$

compatibly with the T_i, where $C_i = Z_{G_i}(T_i)$. The perfectness of G_i implies that the map $j_i : \mathrm{R}_{k_i'/k}(G_i') \to G_i$ is surjective, so $G_i = \mathrm{R}_{k_i'/k}(G_i')/Z_i$ for the central k-subgroup $Z_i = \ker \phi_i$ arising from the standard presentation of G_i, and j_i carries the maximal k-torus in $\mathrm{R}_{k_i'/k}(T_i')$ onto T_i. Moreover, since each G_i is absolutely pseudo-simple, each k_i' must be a field.

Let $k' = \prod k_i'$, $G' = \coprod G_i'$, and $T' = \coprod T_i'$, so T' is a maximal k'-torus of G'. Let T be the maximal k-torus $\prod T_i$ in $\mathrm{R}_{k'/k}(T')$. The G_i pairwise commute and generate the perfect G (Proposition 3.1.8), so the product map

$$j : \mathrm{R}_{k'/k}(G') = \prod \mathrm{R}_{k_i'/k}(G_i') \stackrel{\prod j_i}{\longrightarrow} \prod G_i \twoheadrightarrow G$$

is surjective. Since G is pseudo-reductive, and any *pseudo-reductive* central quotient of $\mathrm{R}_{k'/k}(G')$ is automatically standard (see Remark 1.4.6), it suffices to prove that $\ker j$ is central. To prove centrality of this normal subgroup scheme, it suffices to prove that conjugation by T is trivial on its Lie algebra (as then Corollary A.8.11 will imply centrality of $(\ker j)^0$ due to Proposition A.2.10 and Proposition A.2.11, so Lemma 5.3.2 below will yield centrality of $\ker j$ in the perfect $\mathrm{R}_{k'/k}(G')$). In other words, we just have to show that the induced map of Lie algebras is injective on the root spaces with respect to T. Each such root space occurs within the Lie algebra of a single factor $\mathrm{R}_{k_i'/k}(G_i')$, so it suffices to check that each quotient map $\mathrm{R}_{k_i'/k}(G_i') \to G_i$ is injective on the root spaces for T_i. But this is a *central* quotient, so even the maps between root groups are isomorphisms (due to the structure of the open Bruhat cell, and the fact that central subgroup schemes are contained in Cartan subgroups). □

The following lemma was used to bypass connectedness problems in the preceding argument, and it will be used later on for a similar reason.

Lemma 5.3.2 *Let G be a smooth connected group over a field k and assume that G/Z_G is perfect. If $N \subseteq G$ is a closed normal k-subgroup scheme then N is central in G if and only if N^0 is central in G.*

Proof We may assume that k is algebraically closed and that N^0 is central. The finite étale component group $\Gamma = N/N^0$ is normal in the connected G/N^0, so it is central. Thus, $N = \coprod_{\gamma \in \Gamma} N_\gamma$ where $N_1 = N^0$ is central in G and N_γ denotes the coset of N^0 corresponding to $\gamma \in \Gamma$.

The centrality of Γ in G/N^0 says that G-conjugation on N carries each N_γ back into itself. This action is N^0-invariant since N^0 is central in G by hypothesis, so since each N_γ is an N^0-torsor we see that the G-conjugation on each N_γ is given by $gn_\gamma g^{-1} = f_\gamma(g \bmod N^0) \cdot n_\gamma$ for a k-homomorphism $f_\gamma : G/N^0 \to N^0$. By construction it is clear that the composition of f_γ with $G \to G/N^0$ factors through the quotient G/Z_G of G. But G/Z_G is a smooth connected group equal to its own derived group and N^0 is a commutative group scheme, so all such homomorphisms f_γ are trivial. □

For absolutely pseudo-simple groups we will give a useful standardness criterion. This rests on an alternative interpretation, for absolutely pseudo-simple groups, of the k-algebra k' appearing in the universal maximal absolutely simple quotient:

Proposition 5.3.3 *Let G be a perfect smooth connected affine group over a field k. The maximal semisimple quotient of $G_{\overline{k}}$ and the maximal adjoint semisimple quotient of $G_{\overline{k}}$ have the same field of definition over k.*

That is, if $\pi : G_{\overline{k}} \twoheadrightarrow G_{\overline{k}}^{\rm ss}/Z_{G_{\overline{k}}^{\rm ss}}$ denotes the maximal adjoint semisimple quotient and the purely inseparable finite extension K/k is the field of definition of $\ker \pi \subseteq G_{\overline{k}}$ over k then the (unipotent) radical $\mathscr{R}(G_{\overline{k}}) \subseteq G_{\overline{k}}$ has field of definition over k equal to K.

Example 4.2.6 shows that the perfectness hypothesis on G cannot be dropped in this proposition. The proof of Proposition 5.3.3 will use faithfully flat descent with base rings that may not be reduced (such as $k' \otimes_k k'$ for an algebraic extension k'/k that may not be separable), so we first require a lemma for certain groups over a general ring:

Lemma 5.3.4 *Let k be a field, and G be a smooth k-group of finite type. For any k-algebra A and any A-homomorphism $f : G_A \to Z$ to a commutative A-group Z, $f|_{\mathscr{D}(G)_A} = 1$.*

Proof We may assume $k = k_s$ and $A \neq 0$. Since G is smooth, the commutator subgroup C of $G(k)$ is Zariski-dense in $G' := \mathscr{D}(G)$ (because $k = k_s$). Thus, C is a collection of sections of $G' \to \mathrm{Spec}\, k$ that is schematically dense (in the sense of [EGA, IV$_3$, 11.10.1]). By [EGA, IV$_3$, 11.10.6], for any extension field K/k, a K-morphism from G'_K to a separated K-scheme is therefore uniquely determined by its restriction to $C \subseteq G'(k) \subseteq G'_K(K)$. The locally closed $\ker f \subseteq G_A$ therefore has underlying topological space that contains G'_A since we can work on fibers over points $x \in \mathrm{Spec}\, A$ and use that the fiber Z_x is separated (as are any group schemes over a field). Thus, $G'_A \cap \ker f$ is a closed subscheme of G'_A containing C, so $G'_A \cap \ker f = G'_A$ (by schematic density). This says $f|_{G'_A} = 1$. □

Proof of Proposition 5.3.3 We may assume that k is separably closed (so K is too). Observe that $G_{\overline{k}}^{\mathrm{red}} = G_{\overline{k}}^{\mathrm{ss}}$ because G is perfect, so $\mathscr{R}(G_{\overline{k}})$ is unipotent.

First we address the minimality property for K as a field of definition over k for the geometric radical, assuming that a K-descent of the subgroup $\mathscr{R}(G_{\overline{k}}) \subseteq G_{\overline{k}}$ exists. Suppose that $\mathscr{R}(G_{\overline{k}})$ descends to a K'-subgroup of $G_{K'}$, with K' a subfield of K containing k. The quotient $G_{\overline{k}}^{\mathrm{ss}}$ of $G_{\overline{k}}$ descends to a quotient Q of $G_{K'}$, so likewise the maximal adjoint semisimple quotient $G_{\overline{k}}^{\mathrm{ss}}/Z_{G_{\overline{k}}^{\mathrm{ss}}}$ of $G_{\overline{k}}$ descends to a quotient Q/Z_Q of $G_{K'}$. But K is the minimal field of definition over k for the maximal adjoint semisimple quotient of $G_{\overline{k}}$, so K' must equal K.

It remains to descend the unipotent subgroup $\mathscr{R}(G_{\overline{k}}) \subseteq G_{\overline{k}}$ to a K-subgroup of G_K. We will do this via faithfully flat descent. Let $\Pi : G_K \twoheadrightarrow \overline{G}$ be the K-descent of π, and let $\widetilde{G} \twoheadrightarrow \overline{G}$ be the simply connected central cover over K. The maximal semisimple (even maximal reductive) quotient $G_{\overline{k}}^{\mathrm{ss}}$ of $G_{\overline{k}}$ is a central cover of $\overline{G} \otimes_K \overline{k}$ and so is uniquely isomorphic over $\overline{G} \otimes_K \overline{k}$ to $(\widetilde{G} \otimes_K \overline{k})/M$ for a unique $\overline{k}$-subgroup scheme M in the finite multiplicative center $Z_{\widetilde{G}_{\overline{k}}}$. The finite multiplicative center $Z_{\widetilde{G}}$ of $\widetilde{G}$ has constant Cartier dual over K since K is separably closed, so the subgroup $M \subseteq Z_{\widetilde{G}_{\overline{k}}}$ has the form $\mu_{\overline{k}}$ for a unique K-subgroup $\mu \subseteq Z_{\widetilde{G}}$. Thus, the quotient map $\Pi : G_K \twoheadrightarrow \overline{G}$ over K admits a factorization through the central cover $\widetilde{G}/\mu \to \overline{G}$ after we extend scalars to $\overline{k}$. Such a factorization is unique because there are no nontrivial homomorphisms from the smooth connected group $G_{\overline{k}}$ into the finite commutative group $(Z_{\widetilde{G}}/\mu)_{\overline{k}}$.

To descend $\mathscr{R}(G_{\overline{k}})$ to a K-subgroup of G_K it suffices to descend the unique factorization

$$G_{\overline{k}} \to G_{\overline{k}}^{\mathrm{ss}} = (\widetilde{G}/\mu)_{\overline{k}} \to \overline{G}_{\overline{k}} \tag{5.3.1}$$

over $\overline{k}$ to a factorization

$$G_K \stackrel{?}{\to} \widetilde{G}/\mu \to \overline{G} \tag{5.3.2}$$

over K using the canonical quotient map $\widetilde{G}/\mu \to \overline{G}$ over K. Consider the central extension E of $\overline{G}$ by $Z_{\widetilde{G}}/\mu$ over K given by

$$1 \to Z_{\widetilde{G}}/\mu \to \widetilde{G}/\mu \to \overline{G} \to 1.$$

Pulling this back along $\Pi : G_K \twoheadrightarrow \overline{G}$ gives a central extension $\Pi^*(E)$ of G_K by $Z_{\widetilde{G}}/\mu$ over K. This pullback central extension has at most one splitting over K, and likewise after any ground field extension (e.g., to $\overline{k}$). By (5.3.1), the central extension $\Pi^*(E)_{\overline{k}}$ splits, so to give a splitting of $\Pi^*(E)$ over K is exactly to descend (5.3.1) to a diagram (5.3.2) over K.

More generally, if $\mathscr{E}$ is a central extension of G_K by a commutative finite type K-group scheme Q and the extension $\mathscr{E}_{\overline{k}}$ is (necessarily uniquely) split then we claim that the extension $\mathscr{E}$ is split over K. To prove this via faithfully flat descent it suffices to show the two pullback splittings over $\overline{k} \otimes_K \overline{k}$ coincide. Since we are splitting a central extension, any two splittings over a K-algebra A (such as $\overline{k} \otimes_K \overline{k}$) are related via multiplication against a homomorphism from G_A to Q_A. By Lemma 5.3.4 all such homomorphisms are trivial (since $G = \mathscr{D}(G)$), so Proposition 5.3.3 is proved. □

Let k be an arbitrary field and G a perfect smooth connected affine k-group. (We are most interested in the case where G is also pseudo-reductive over k with $G_{\overline{k}}^{\mathrm{ss}}$ a simple $\overline{k}$-group, which is to say that G is an absolutely pseudo-simple k-group.) Let K be the field of definition over k for the kernel of the projection from $G_{\overline{k}}$ onto its maximal adjoint semisimple quotient. By Proposition 5.3.3, $\mathscr{R}(G_{\overline{k}})$ descends to a K-subgroup $\mathscr{R}(G_K)$ in G_K.

Let $G' = G_K/\mathscr{R}(G_K)$ (the K-descent of $G_{\overline{k}}^{\mathrm{ss}}$), so there is a canonical central K-isogeny $\alpha_G : G' \to \overline{G}'$ onto a connected semisimple K-group of adjoint type. Hence, the simply connected central cover $\widetilde{G}' \to \overline{G}'$ uniquely factors as $\widetilde{G}' \stackrel{\theta}{\to} G' \stackrel{\alpha_G}{\to} \overline{G}'$ with θ a central K-isogeny. In fact, by the preceding considerations the central isogeny θ arises from a unique $\overline{G}'$-isomorphism $\widetilde{G}'/\mu \simeq G'$ over K.

Associated to the canonical quotient map $\pi_G : G_K \twoheadrightarrow G' \simeq \widetilde{G}'/\mu$ of K-groups there is a canonical homomorphism $i_G : G \to \mathrm{R}_{K/k}(\widetilde{G}'/\mu)$ of k-groups that coincides with the canonical map in (1.6.1). Since G is perfect, the map i_G factors through the derived group of its target. Proposition 1.3.4 identifies this derived group. The resulting factorization is so important that it deserves a notation:

Definition 5.3.5 For an absolutely pseudo-simple k-group G and the field of definition K/k for its geometric radical,

$$\xi_G : G \to \mathrm{R}_{K/k}(\widetilde{G}')/\mathrm{R}_{K/k}(\mu) \tag{5.3.3}$$

is the unique k-homomorphism through which i_G factors.

Our definition of ξ_G makes sense more generally when G is pseudo-reductive and perfect, but it is the "wrong" notion except when G is absolutely pseudo-simple over k. Indeed, the absolutely pseudo-simple case is the only one for which the simply connected datum $(\widetilde{G}', k'/k, f)$ of G has k' a field that is moreover purely inseparable over k. Beyond this case, K/k as defined above does not encode useful information about the maximal étale subalgebra of k' or the individual factor fields of k'.

For any absolutely pseudo-simple k-group G we have $\xi_G(G) = i_G(G)$, so it follows from Proposition 4.2.5 that $\xi_G(G)$ is always pseudo-reductive. Also, since separable extension of the ground field preserves absolute pseudo-simplicity (Lemma 3.1.2) and is compatible with the formation of the field of definition K/k (Proposition 1.1.9(2)), it follows that the formation of ξ_G commutes with separable extension of the ground field.

Remark 5.3.6 Let us assume now that G is a standard pseudo-reductive k-group which is absolutely pseudo-simple. We now prove two important properties of ξ_G for such G: it is surjective and $\ker \xi_G$ is a central k-subgroup scheme in G.

It follows from Proposition 4.2.4 that G admits a standard presentation $(\widetilde{G}', K/k, \widetilde{T}', C)$ with $\widetilde{G}'$ and K as above, $\widetilde{T}'$ a maximal K-torus of $\widetilde{G}'$, and $C = Z_G(T)$, with T the maximal k-torus of G contained in the image of $\mathrm{R}_{K/k}(\widetilde{T}')$. The composite map

$$\mathrm{R}_{k'/k}(\widetilde{G}') \xrightarrow{j} G \xrightarrow{\xi_G} \mathrm{R}_{k'/k}(\widetilde{G}')/\mathrm{R}_{k'/k}(\mu)$$

is the canonical surjection, as we may check after composition with the inclusion into $\mathrm{R}_{k'/k}(\widetilde{G}'/\mu)$, so ξ_G is surjective when G is standard over k.

The map $j : \mathrm{R}_{k'/k}(\widetilde{G}') \to G$ is surjective since $G = \mathscr{D}(G)$, so $\ker \xi_G$ is a quotient of $\mathrm{R}_{k'/k}(\mu)$ and hence is a central k-subgroup scheme of the quotient G of $\mathrm{R}_{k'/k}(\widetilde{G}')$.

The aim of the rest of this section is to show (in Theorem 5.3.8) that the surjectivity of ξ_G and centrality of $\ker \xi_G$ are not only necessary for standardness of G when G is absolutely pseudo-simple over k, as we just saw, but are also sufficient.

Let k be an arbitrary field and G an absolutely pseudo-simple k-group. In particular, by Lemma 3.1.2 the k-group G is a non-commutative pseudo-reductive k-group, and by Proposition 5.3.3 the field of definition K/k of the kernel of the projection from $G_{\overline{k}}$ onto its maximal *adjoint* semisimple quotient coincides with the field of definition over k for the radical $\mathscr{R}(G_{\overline{k}}) \subseteq G_{\overline{k}}$.

Consider ξ_G as in (5.3.3). By functoriality, $\ker \xi_G$ is the maximal k-subgroup scheme $Z \subseteq G$ such that $Z_K \subseteq \mathscr{R}(G_K)$. Thus, $\ker \xi_G$ is unipotent; *a priori* it may be disconnected. Note that $\ker \xi_G$ cannot contain a nontrivial smooth connected k-subgroup U, for otherwise the unipotent k-group scheme $\ker \xi_G$ would contain the nontrivial smooth connected *normal* k-subgroup of G generated by the $G(k_s)$-conjugates of U_{k_s}; this normal k-subgroup would have to be unipotent, contradicting the pseudo-reductivity of G over k. In particular, $(\ker \xi_G)(k_s)$ is finite. As the following example shows, *it can happen, even when the absolutely pseudo-simple k-group G is a standard pseudo-reductive group, that* $\dim(\ker \xi_G) > 0$. (By Theorem 5.3.8, an essential feature of all such examples is that G_K^{ss} has fundamental group with order divisible by $\mathrm{char}(k)$.) Here is an example:

Example 5.3.7 Let k be an imperfect field with characteristic $p > 0$ and let k'/k be a nontrivial finite purely inseparable extension. Let $\mathsf{k} \subseteq k'$ be the intermediate field $k' \cap k^{1/p^n}$ with some $n > 0$ (so $\mathsf{k} \neq k$). We will be most interested in the case that k is a proper subfield of k' (so n is not too large, relative to k'/k), but at the outset we do not impose this condition.

Let G' be a connected semisimple k'-group that is absolutely simple and simply connected. Let $\mu' \subseteq Z_{G'}$ be a k'-subgroup with p-part that is killed by p^n but not by any strictly smaller power of p. (As an example, we can take $G' = \mathrm{SL}_{p^n}$ and $\mu' = Z_{G'} = \mu_{p^n}$.) By the classification of simply connected groups, the p-part of $\mu'_{k'_s}$ must be μ_{p^n} except if $p = 2$ and G' is of type D_{2m}, in which case another possibility is $\mu_2 \times \mu_2$ over k'_s.

Choose a maximal k'-torus $T' \subseteq G'$ and define T to be the maximal k-torus in $\mathrm{R}_{k'/k}(T')$. Let $\mu = T \cap \mathrm{R}_{k'/k}(\mu')$, so $\mu_{k'}$ is naturally identified with μ'. The quotient

$$G := \mathrm{R}_{k'/k}(G')/\mathrm{R}_{\mathsf{k}/k}(\mu_{\mathsf{k}})$$

is equal to its own derived group since $\mathrm{R}_{k'/k}(G')$ is perfect (as G' is simply connected, so Corollary A.7.11 applies). Moreover, $G_{\overline{k}}^{\mathrm{red}} = (G'/\mu') \otimes_{k'} \overline{k}$ and G is the standard pseudo-reductive k-group associated to the factorization

$$\mathrm{R}_{k'/k}(T') \to C := \mathrm{R}_{k'/k}(T')/\mathrm{R}_{\mathsf{k}/k}(\mu_{\mathsf{k}}) \to \mathrm{R}_{k'/k}(T'/Z_{G'})$$

provided that the smooth connected commutative k-group C is pseudo-reductive over k (a property to be verified below). The map ξ_G makes sense, and the common kernel $\ker \xi_G = \ker i_G$ is identified with

$$Q := \mathrm{R}_{k'/k}(\mu')/\mathrm{R}_{\mathsf{k}/k}(\mu_{\mathsf{k}}).$$

By consideration of $\overline{k}$-points we will compute $\dim Q$ and show that it is > 0 whenever $\mathsf{k} \neq k'$ (i.e., when n is not too large relative to k'/k).

To check that C is pseudo-reductive over k, note that the quotient $\mathrm{R}_{k'/k}(T')/\mathrm{R}_{k'/k}(\mu')$ of C is pseudo-reductive over k because it is a k-subgroup of $\mathrm{R}_{k'/k}(T'/\mu')$, so if $U \subseteq C$ is a smooth connected unipotent k-subgroup then

$$U \subseteq \ker(C \twoheadrightarrow \mathrm{R}_{k'/k}(T')/\mathrm{R}_{k'/k}(\mu')) = Q.$$

Thus, to prove $U = 1$ it suffices to prove $Q(k_s)$ is finite. By decomposing μ' into its primary parts and applying Proposition A.5.13 to the (étale) prime-to-p-parts, we can pass to the case where μ' has p-power order. Hence, $\mu'_{k'_s} = \mu_{p^n}$ except possibly when $p = 2$ and G' is of type D_{2m}, in which case $\mu'_{k'_s}$ may also be $\mu_2 \times \mu_2$. We shall prove $Q(k_s) = \{1\}$.

We may assume k is separably closed, and then we can assume $\mu' = \mu_{p^n}$ even in the D_{2m} case when $p = 2$. In the exact sequence

$$1 \to \mathrm{R}_{\mathsf{k}/k}(\mu_{p^n})/\mu_{p^n} \to \mathrm{R}_{k'/k}(\mu_{p^n})/\mu_{p^n} \to Q \to 1, \tag{5.3.4}$$

the term $\mathrm{R}_{k'/k}(\mu_{p^n})/\mu_{p^n} = (\mathrm{R}_{k'/k}(\mathrm{GL}_1)/\mathrm{GL}_1)[p^n]$ (see the snake lemma argument in Example 1.3.2 with $m = p^n$) has $(k'^{\times}/k^{\times})[p^n] = \mathsf{k}^{\times}/k^{\times}$ as its k-points due to the definition of k. In other words, in (5.3.4) the map on k-points between the left and middle terms is an isomorphism. Thus $Q(k) = \{1\}$ because the kernel $\mathrm{R}_{\mathsf{k}/k}(\mu_{p^n})/\mu_{p^n} = \mathrm{R}_{\mathsf{k}/k}(\mathrm{GL}_1)/\mathrm{GL}_1$ in (5.3.4) is k-smooth (with dimension $[\mathsf{k} : k] - 1$) and k is separably closed. This concludes the verification that G is a standard pseudo-reductive k-group.

A calculation with $\overline{k}$-points shows that $(\mathrm{R}_{k'/k}(\mathrm{GL}_1)/\mathrm{GL}_1)[p^n]$ has dimension $[k' : k] - [kk'^{p^n} : k]$ because the image of the p^n-power map on $k' \otimes_k \overline{k}$ is $kk'^{p^n} \otimes_k \overline{k}$, so

$$\dim Q = ([k' : k] - [kk'^{p^n} : k]) - ([\mathsf{k} : k] - 1). \tag{5.3.5}$$

The surjection $k \otimes_{\mathsf{k}^{p^n}} k'^{p^n} \twoheadrightarrow kk'^{p^n}$ implies $[kk'^{p^n} : k] \leqslant [k'^{p^n} : \mathsf{k}^{p^n}] = [k' : \mathsf{k}]$ with equality if and only if k and k'^{p^n} are linearly disjoint over $\mathsf{k}^{p^n} = k \cap k'^{p^n}$. Since $[\mathsf{k} : k] > 1$, we therefore have

$$\dim Q \geqslant [k' : k] + 1 - ([k' : \mathsf{k}] + [\mathsf{k} : k]) \geqslant [k' : k] + 1 - ([k' : k] + 1) = 0,$$

with equality between outer terms if and only if k and k'^{p^n} are linearly disjoint over $k \cap k'^{p^n}$ and $[k' : \mathsf{k}] = 1$ (i.e., $k'^{p^n} \subseteq k$). Thus, $\dim Q > 0$ provided that $k'^{p^n} \not\subseteq k$, which is to say that k is a proper subfield of k'. In the special case $k' = k(a^{1/p^{2n}})$ with $a \in k - k^p$ the formula (5.3.5) implies $\dim Q = (p^n - 1)^2$. Thus, we can make the common kernel Q of ξ_G and i_G have arbitrarily large dimension (with $Q(k_s) = 1$).

Theorem 5.3.8 *Let k be a field and let G be an absolutely pseudo-simple k-group. Let ξ_G be as in* (5.3.3). *If ξ_G is surjective then* $\ker \xi_G$ *is connected, and in general the pseudo-reductive k-group G is standard (over k) if and only if ξ_G is surjective and* $\ker \xi_G$ *is central in G.*

If G is standard then ξ_G is an isomorphism provided that $G^{\mathrm{ss}}_{\overline{k}}$ has fundamental group with order not divisible by $\mathrm{char}(k)$.

Proof This theorem is trivial if $\mathrm{char}(k) = 0$ because in such cases $K = k$ and ξ_G is an isomorphism whenever G is a connected semisimple k-group. We may and do now assume $\mathrm{char}(k) = p > 0$. In Remark 5.3.6 we proved that if the pseudo-reductive group G is standard then ξ_G is surjective and $\ker \xi_G$ is central. Thus, in the remainder of the proof we may assume that ξ_G is surjective. That is, we have an exact sequence

$$1 \to Z \to G \overset{\xi_G}{\to} \mathrm{R}_{K/k}(\widetilde{G}')/\mathrm{R}_{K/k}(\mu) \to 1, \tag{5.3.6}$$

where $Z = \ker \xi_G$ and $\mu := \ker(\widetilde{G}' \twoheadrightarrow G' = G^{\mathrm{ss}}_K)$.

We will now use Proposition 5.1.3 to prove the connectedness of $\ker \xi_G$ when ξ_G is surjective. Since Z is normal in the k-smooth G, so is its identity component Z^0. Thus, it makes sense to push forward (5.3.6) along $Z \to Z/Z^0$, yielding the exact sequence

$$1 \to Z/Z^0 \to G/Z^0 \to \mathrm{R}_{K/k}(\widetilde{G'})/\mathrm{R}_{K/k}(\mu) \to 1 \tag{5.3.7}$$

that is a central extension since Z/Z^0 is a finite étale normal subgroup of the connected k-group G/Z^0. Let $\pi : \mathrm{R}_{K/k}(\widetilde{G'}) \to \mathrm{R}_{K/k}(\widetilde{G'})/\mathrm{R}_{K/k}(\mu)$ be the quotient map. The π-pullback of (5.3.7) is therefore a central extension fitting into the top row of the diagram

$$\begin{array}{ccccccccc} 1 & \longrightarrow & Z/Z^0 & \longrightarrow & \mathcal{G} & \longrightarrow & \mathrm{R}_{K/k}(\widetilde{G'}) & \longrightarrow & 1 \\ & & \Vert & & \downarrow & & \downarrow{\scriptstyle \pi} & & \\ 1 & \longrightarrow & Z/Z^0 & \longrightarrow & G/Z^0 & \longrightarrow & \mathrm{R}_{K/k}(\widetilde{G'})/\mathrm{R}_{K/k}(\mu) & \longrightarrow & 1 \end{array} \tag{5.3.8}$$

$$h : \mathrm{R}_{K/k}(\widetilde{G'}) \to G/Z^0 \text{ (diagonal arrow)}$$

in which the squares commute and the arrow labeled h has not yet been constructed. Since Z/Z^0 is finite, Proposition 5.1.3 and Example 5.1.4 imply that the top row in (5.3.8) is split over k. Equivalently, we can define h to make the diagram (5.3.8) commutative.

An elementary diagram chase shows that h is surjective, so h is a quotient map between smooth affine k-groups. Also, the commutativity of the above diagram implies that $h^{-1}(Z/Z^0) = \mathrm{R}_{K/k}(\mu)$, so h induces a quotient map $\mathrm{R}_{K/k}(\mu) \to Z/Z^0$. We will show that this map is the trivial map, so $Z = Z^0$ (i.e., $\ker \xi_G$ is connected), as desired. To verify such triviality, first observe that the k-group $Z = \ker \xi_G$ is unipotent (since Z_K is a K-subgroup of the unipotent K-group $\mathscr{R}(G_K)$). Thus, its finite étale quotient Z/Z^0 is unipotent and hence is a p-group. Consider the primary decomposition $\mu = M \times M'$, with M the p-part and M' the prime-to-p-part, so

$$\mathrm{R}_{K/k}(\mu) = \mathrm{R}_{K/k}(M) \times \mathrm{R}_{K/k}(M').$$

The factor $\mathrm{R}_{K/k}(M')$ must therefore have trivial image in Z/Z^0 for torsion-order reasons. To prove that $\mathrm{R}_{K/k}(M)$ also has trivial image in Z/Z^0, it suffices to show that $\mathrm{R}_{K/k}(M)$ is connected.

By extending scalars to $L = \overline{k}$ and renaming $K \otimes_k L$ as A, it suffices to prove that $\mathrm{R}_{A/L}(\mu_{p^n})$ is connected for any algebraically closed field L of characteristic $p > 0$, any finite local L-algebra A, and any $n \geqslant 1$. Since A has residue field L, we have a canonical decomposition $\mathrm{R}_{A/L}(\mathrm{GL}_1) = \mathrm{GL}_1 \times U$ via the unipotent radical of 1-units of A. Using this decomposition, it suffices to show that the p^n-torsion subgroup scheme $U[p^n]$ is connected. Let $\underline{\mathfrak{m}}$ denote the affine space over L associated to the maximal ideal $\mathfrak{m}$ of A. Since $(1+x)^{p^n} = 1 + x^{p^n}$ in characteristic p, the underlying scheme of $U[p^n]$

is identified with the kernel of the endomorphism $t \mapsto t^{p^n}$ of the smooth connected additive L-group $\underline{\mathfrak{m}}$. But $c \cdot x^{p^n} = (c^{1/p^n}x)^{p^n}$ for any $c \in L$ and $x \in A$, so this kernel is a vector subspace of $\mathfrak{m}$ at the level of L-points. Hence, it is connected.

It remains to show that if the (connected) group scheme $Z = \ker \xi_G$ is central in G then G is standard, and that moreover $Z = 1$ (i.e., ξ_G is an isomorphism) when μ has order not divisible by p. By unipotence of Z and its normality in G, the pseudo-reductivity of G over k implies that Z has no nontrivial smooth connected k-subgroup (by Lemma 1.2.1). Thus, if we now form a pullback diagram of central extensions similar to (5.3.8) in which Z/Z^0 has been replaced with Z and G/Z^0 with G, we may again apply Proposition 5.1.3 to get a k-splitting after pullback. That is, we get a commutative diagram of k-groups

$$\begin{array}{ccccccccc} & & & & & & \mathrm{R}_{K/k}(\widetilde{G}') & & \\ & & & & & \overset{j}{\swarrow} & \downarrow \pi & & \\ 1 & \longrightarrow & Z & \longrightarrow & G & \xrightarrow[\xi_G]{} & \mathrm{R}_{K/k}(\widetilde{G}')/\mathrm{R}_{K/k}(\mu) & \longrightarrow & 1 \end{array}$$

such that the composition of j and the natural map $f' : G \to \mathrm{R}_{K/k}(\widetilde{G}'/Z_{\widetilde{G}'}) = \mathrm{R}_{K/k}(\overline{G}')$ is the canonical map. The image of j is normal since the target of ξ_G is a central quotient of G. Thus, this image is a nontrivial smooth connected normal k-subgroup of G, whence j is surjective since G is absolutely pseudo-simple. But $\ker j$ is central since $\ker \pi$ is central, so the pseudo-reductive G is a central quotient of $\mathrm{R}_{K/k}(\widetilde{G}')$. It then follows from Remark 1.4.6 that G must be standard.

Since j is a surjective homomorphism between smooth affine k-groups, it is a quotient map. Hence, it restricts to a quotient map $\mathrm{R}_{K/k}(\mu) \to Z$ between k-group schemes (i.e., it is faithfully flat). In particular, if p does not divide the order of μ then $\mathrm{R}_{K/k}(\mu)$ is finite étale of order prime to p (Proposition A.5.13), so its quotient Z inherits these properties in such cases. But Z is a unipotent k-group, so $Z = 1$ when μ has order prime to p. □

Corollary 5.3.9 *Let F/k be a separable extension of fields, and G a smooth connected affine k-group. Then G is a standard pseudo-reductive k-group if and only if G_F is a standard pseudo-reductive F-group.*

Proof By Proposition 1.1.9(1), pseudo-reductivity for G over k is equivalent to that of G_F over F, so we may assume that G, or equivalently G_F, is pseudo-reductive. Standardness is insensitive to passage to a separable closure of the ground field (Corollary 5.2.3), so we may assume k and F are separably closed. We can then replace G with $\mathscr{D}(G)$, by Proposition 5.2.1, so G is perfect. We may assume $G \neq 1$, so by Proposition 3.1.8 (over separably closed fields) and Proposition 5.3.1 we can even assume that G and G_F are pseudo-simple. Thus, G is standard if and only if ξ_G is surjective with a central kernel (Theorem 5.3.8), and similarly for G_F. Separability of F/k implies that $(\xi_G)_F = \xi_{G_F}$ by Proposition 1.1.9(2), so we are done. □

Proposition 5.3.10 *Let G be an absolutely pseudo-simple group over a field k. If the root system of G_{k_s} is reduced then* $\ker \xi_G$ *is central, and in such cases G is standard if and only if ξ_G is surjective.*

Proof We may and do assume $k = k_s$. Let T be a maximal k-torus in G, so G is generated by the $G(k)$-conjugates of T (since $G = \mathscr{D}(G)$). To prove centrality of $\ker \xi_G$, by Lemma 5.3.2 and Corollary A.8.11 it is enough to show that $\ker \xi_G$ has no nontrivial T-weights on its Lie algebra.

Let K/k be the field of definition over k for the kernel of the projection from $G_{\overline{k}}$ onto its maximal adjoint semisimple quotient. By Proposition 5.3.3 this is also the field of definition over k for $\mathscr{R}(G_{\overline{k}}) \subseteq G_{\overline{k}}$. By definition of ξ_G, it has the same kernel as $G \to \mathrm{R}_{K/k}(G')$. The reducedness hypothesis on $\Phi(G,T)$ implies that $\Phi(G,T) = \Phi(G', T_K)$ (by Theorem 2.3.10).

For each $a \in \Phi(G,T)$, consider the associated T-stable root group $U_a \subseteq G$ from Proposition 2.3.11, so U_a is a k-vector group admitting a unique linear structure with respect to which T acts linearly, and $\mathrm{Lie}(U_a)$ is the a-root space for T on $\mathrm{Lie}(G)$. Since $a \in \Phi(G', T_K)$, the restriction of ξ_G to U_a is a T-equivariant map between a-root groups. The 1-dimensional a_K-root group of G' has a unique linear structure, and the map induced by ξ_G between a-root groups respects the linear structures since GL_1-compatibility follows from the nontriviality of a. Thus, $\ker(\xi_G|_{U_a})$ is also a vector group, so it is a smooth connected k-subgroup of G whose geometric fiber lies in the unipotent radical. By pseudo-reductivity of G it follows that $\ker(\xi_G|_{U_a})$ is trivial, so passing to the Lie algebra gives that $\mathrm{Lie}(\xi_G)$ is injective on the a-root space in $\mathrm{Lie}(G)$ for all $a \in \Phi(G,T)$. This gives the centrality of $\ker \xi_G$. By Theorem 5.3.8, the equivalence of standardness and surjectivity of ξ_G follows. □

Remark 5.3.11 By Theorem 2.3.10, the reducedness hypothesis in Proposition 5.3.10 always holds except possibly when k is imperfect of characteristic 2 with $G_{\overline{k}}^{\mathrm{ss}}$ of simply connected type C_n for some $n \geqslant 1$ (i.e., Sp_{2n}), as this type is the only irreducible and reduced root datum for which some roots are divisible in the character lattice.

6

Classification results

Let G be an absolutely pseudo-simple group over a field k. We have seen in Proposition 5.3.3 that the field of definition K/k of the kernel of the projection of $G_{\overline{k}}$ onto its maximal *adjoint* semisimple quotient is also the field of definition over k for $\mathscr{R}(G_{\overline{k}}) \subseteq G_{\overline{k}}$. Let $\mu = \ker(\widetilde{G}' \twoheadrightarrow G_K^{\mathrm{ss}} := G_K/\mathscr{R}(G_K))$, with $\widetilde{G}'$ the simply connected central cover of G_K^{ss}, so we have a natural map $\xi_G : G \to \mathrm{R}_{K/k}(\widetilde{G}')/\mathrm{R}_{K/k}(\mu)$ as in (5.3.3). Observe that if μ is étale (e.g., $G_{\overline{k}}^{\mathrm{ss}}$ is of type A_1 and $\mathrm{char}(k) \neq 2$) then ξ_G is identified with the natural map $i_G : G \to \mathrm{R}_{K/k}(G')$ from (1.3.1).

In §5.3 it was shown that the pseudo-reductive G is standard if and only if ξ_G is surjective with central kernel, that $\ker \xi_G$ is connected if ξ_G is surjective, and that $\ker \xi_G$ is central if $\Phi(G_{k_s}, T_{k_s})$ is reduced (e.g., $\mathrm{char}(k) \neq 2$). The definition of ξ_G makes sense even if we replace K with a finite extension (i.e., any finite extension of k over which the subgroup $\mathscr{R}(G_{\overline{k}}) \subseteq G_{\overline{k}}$ is defined), so for the structure of $\ker \xi_G$ it is irrelevant if we choose K/k to be the (minimal) field of definition of $\mathscr{R}_u(G_{\overline{k}}) = \mathscr{R}(G_{\overline{k}}) \subseteq G_{\overline{k}}$ over k or merely to be some finite extension of k over which this $\overline{k}$-subgroup can be defined.

In contrast, the property of ξ_G being surjective or not depends very much on working with the minimal K/k. It appears to be quite difficult to directly analyze the image of ξ_G without using a more precise understanding of the structure of G. Our strategy is to first handle the case where $G_{\overline{k}}^{\mathrm{ss}}$ is of type A_1 by means of concrete calculations with SL_2 and $\mathfrak{sl}_2$, and to then deduce results in the general case from this case via consideration of torus centralizers. For later purposes in characteristic 2 we also include a separate analysis of types A_2 and G_2 away from characteristic 3 (but allowing characteristic 2). The study of type A_1 in characteristic 2 leads to completely new problems and examples, as we will see in Chapters 9 and 10.

6.1 The A_1 case away from characteristic 2

The goal of this section is to prove the following special case of Theorem 5.1.1.

Theorem 6.1.1 *Let k be a field with $\mathrm{char}(k) \neq 2$ and let G be a pseudo-reductive k-group such that $G^{\mathrm{ss}}_{\overline{k}}$ is isomorphic to SL_2 or PGL_2. The pseudo-reductive k-group G is standard.*

If G is absolutely pseudo-simple over k and if K is the field of definition over k for the $\overline{k}$-subgroup

$$\mathscr{R}_u(G_{\overline{k}}) = \mathscr{R}(G_{\overline{k}}) \subseteq G_{\overline{k}}$$

then the natural map $\xi_G = i_G : G \to \mathrm{R}_{K/k}(G^{\mathrm{ss}}_K)$ is an isomorphism.

To prove Theorem 6.1.1, by Corollary 5.2.3 we may assume $k = k_s$. Since $\mathscr{D}(G)^{\mathrm{ss}}_{\overline{k}} \to G^{\mathrm{ss}}_{\overline{k}}$ is a central isogeny (Proposition A.4.8), by Propositions 1.2.6 and 5.2.1 we can also assume G is perfect. Thus, by Lemma 3.1.2, G is absolutely pseudo-simple over k. To reduce the PGL_2 case to the SL_2 case, we apply the following lemma that will be useful in some later situations as well.

Lemma 6.1.2 *Let G be an absolutely pseudo-simple group over a field k, with K/k the field of definition of its geometric radical. Let $\widetilde{G}' \to G^{\mathrm{ss}}_K$ be the simply connected central cover, and μ the kernel of this central isogeny. Consider the cartesian diagram*

$$\begin{array}{ccc} \mathscr{G} & \xrightarrow{\xi} & \mathrm{R}_{K/k}(\widetilde{G}')/\mathrm{R}_{K/k}(\mu^0) \\ {\scriptstyle f}\downarrow & & \downarrow{\scriptstyle g} \\ G & \xrightarrow[\xi_G]{} & \mathrm{R}_{K/k}(\widetilde{G}')/\mathrm{R}_{K/k}(\mu) \end{array} \qquad (6.1.1)$$

in which g and ξ_G are the natural maps.

The map g is an étale central isogeny, the k-group $\mathscr{G}$ is absolutely pseudo-simple (in particular, connected) with K/k the field of definition of its geometric radical, $\mathscr{G}^{\mathrm{ss}}_K = \widetilde{G}'/\mu^0$, and ξ is identified with $\xi_{\mathscr{G}}$. Moreover, ξ_G is surjective if and only if the map $\xi_{\mathscr{G}}$ is surjective, and $\ker \xi_G$ is central if and only if $\ker \xi_{\mathscr{G}}$ is central.

Proof The connected-étale sequence $1 \to \mu^0 \to \mu \to \mu^{\text{ét}} \to 1$ is canonically split: if $\text{char}(k) = 0$ then $\mu^0 = 1$, and if $\text{char}(k) = p > 0$ then the (functorial) decomposition of the torsion group μ into a direct product of its p-part μ^0 and its prime-to-p-part gives the splitting. Thus, the inclusion $\mathrm{R}_{K/k}(\mu^0) \to \mathrm{R}_{K/k}(\mu)$ has cokernel $\mathrm{R}_{K/k}(\mu^{\text{ét}})$ that is finite étale (Proposition A.5.13), so the homomorphism g is an étale central isogeny. The map f is therefore also an étale central isogeny, so $\mathscr{G}^0$ is an absolutely pseudo-simple group. It is obvious that ξ_G is surjective if and only if ξ is surjective and that $\ker \xi$ is central if and only if $\ker \xi_G$ is central.

We now show that $\mathscr{G}$ is connected. It suffices to show that the kernel $\ker g$ is contained in the identity component of $\mathscr{G}$. Fix a maximal k-torus T of G; we will prove that $f^{-1}(T)$ is connected. Since T_K is a maximal K-torus of $G_K^{\text{ss}} = \widetilde{G}'/\mu$, its preimage $\widetilde{T}'$ in $\widetilde{G}'$ is a maximal K-torus in $\widetilde{G}'$. The map ξ_G carries T isomorphically onto the maximal k-torus in $\mathrm{R}_{K/k}(\widetilde{T}')/\mathrm{R}_{K/k}(\mu)$ (recall that $\ker \xi_G$ is unipotent), so in view of the short exact sequence

$$1 \to \ker g \to \mathrm{R}_{K/k}(\widetilde{T}')/\mathrm{R}_{K/k}(\mu^0) \to \mathrm{R}_{K/k}(\widetilde{T}')/\mathrm{R}_{K/k}(\mu) \to 1$$

we just have to show that $\ker g$ is contained in the maximal k-torus of $\mathrm{R}_{K/k}(\widetilde{T}')/\mathrm{R}_{K/k}(\mu^0)$ (since the quotient map g carries maximal tori onto maximal tori). But $\ker g$ is finite étale, and for n not divisible by $\text{char}(k)$, the n-torsion in $\mathrm{R}_{K/k}(\widetilde{T}')/\mathrm{R}_{K/k}(\mu^0)$ is identified with the n-torsion in $\mathrm{R}_{K/k}(\widetilde{T}')$, which in turn clearly lies in the maximal k-torus of $\mathrm{R}_{K/k}(\widetilde{T}')$. This completes the proof that $\mathscr{G}$ is connected. The field of definition over k for the geometric radical of $\mathscr{G}$ must contain the corresponding field of definition K/k for G. This containment is an equality since $f_K^{-1}(\mathscr{R}(G_K))^0$ is a K-descent of $\mathscr{R}(\mathscr{G}_{\overline{k}})$ (due to f_K-pullback preserving smoothness, since f is étale).

It remains to show that ξ is identified with $\xi_{\mathscr{G}}$. Applying extension of scalars by K/k to (6.1.1), and using the canonical map $q : \mathrm{R}_{K/k}(H)_K \twoheadrightarrow H$ for any affine K-group H of finite type, we get the following commutative diagram

$$\begin{array}{ccc} \mathscr{G}_K & \xrightarrow{h} & \widetilde{G}'/\mu^0 \\ {\scriptstyle f_K}\downarrow & & \downarrow{\scriptstyle \pi} \\ G_K & \longrightarrow & \widetilde{G}'/\mu \end{array}$$

in which the bottom and right maps are the canonical ones. We have to show that the top map is the maximal semisimple quotient of $\mathscr{G}_K$. By connectedness considerations, the map h is surjective since $\pi \circ h$ is surjective with π an étale isogeny. The identity component $(\ker h)^0$ must equal $f_K^{-1}(\mathscr{R}(G_K))^0 = \mathscr{R}(\mathscr{G}_K)$, so $(\ker h)^0$ is smooth and unipotent. Hence, h factors as

$$\mathscr{G}_K \to \mathscr{G}_K^{\mathrm{ss}} \to \widetilde{G}'/\mu^0$$

with the first map equal to the canonical quotient map and the second an étale isogeny. But the final term is the quotient of a simply connected semisimple group by an infinitesimal central subgroup, so it admits no nontrivial connected finite étale isogenous covers (as a k-group). In other words, $\mathscr{G}_K^{\mathrm{ss}} = \widetilde{G}'/\mu^0$, or more precisely h is the maximal semisimple quotient of $\mathscr{G}_K$. □

Returning to the proof of Theorem 6.1.1, when $G_K^{\mathrm{ss}} \simeq \mathrm{SL}_2$ we will prove that ξ_G is surjective. Once this is shown, standardness will follow from the hypothesis $\mathrm{char}(k) \neq 2$ (which ensures that G has a reduced root system, so Proposition 5.3.10 applies). Since G_K^{ss} has fundamental group of order 1 or 2, standardness will imply (by the final claim in Theorem 5.3.8) that ξ_G is an isomorphism.

We will now choose a convenient K-isomorphism $G_K^{\mathrm{ss}} \simeq \mathrm{SL}_2$. This requires some preparations. Fix a maximal k-torus $T \subseteq G$, so the K-torus T_K in G_K maps isomorphically onto a maximal K-torus in $G_K^{\mathrm{red}} = G_K^{\mathrm{ss}} \simeq \mathrm{SL}_2$. We abuse notation by identifying T_K with its image in G_K^{ss} since no confusion seems likely (i.e., the context will make clear whether we are viewing T_K as a K-torus in G_K or in G_K^{ss}). Since $\mathrm{char}(k) \neq 2$, the root system of (G, T) is reduced and $\Phi(G_K^{\mathrm{ss}}, T_K) = \Phi(G, T)$ inside $\mathrm{X}(T) = \mathrm{X}(T_K)$. This root system has the form $\{\pm a_0\}$, and Proposition 2.1.8 gives that the multiplication map

$$U_{a_0} \times Z_G(T) \times U_{-a_0} \to G \tag{6.1.2}$$

is an open immersion of k-schemes.

By Remark 2.3.6, the image of $(U_{\pm a_0})_K \subseteq G_K$ in the G_K-quotient $G_K^{\mathrm{ss}} \simeq \mathrm{SL}_2$ is the $\pm a_0$-root group for (G_K^{ss}, T_K). Since T_K maps isomorphically onto a maximal K-torus in SL_2, we can choose a K-isomorphism $G_K^{\mathrm{ss}} \simeq \mathrm{SL}_2$ which carries T_K into the diagonal K-torus D_K of SL_2 arising from the diagonal k-torus $D \subseteq \mathrm{SL}_2$. Applying conjugation by a representative of the nontrivial class in $W(\mathrm{SL}_2, D)(k)$ if necessary, we can arrange that $(U_{a_0})_K$ maps onto the upper triangular unipotent matrix group U^+, and $(U_{-a_0})_K$ maps onto the lower triangular unipotent matrix group U^-. In particular, $\xi_G(U_{\pm a_0}) \subseteq \mathrm{R}_{K/k}(U^{\pm})$ and these images under ξ_G are nontrivial. The isomorphism $T_K \simeq D_K$ of K-tori uniquely descends to a k-isomorphism $T \simeq D$ since k is separably closed. In particular, $D(k) = \xi_G(T(k))$.

Note that applying the K-automorphism of SL_2 arising from the action of an element of the diagonal torus in $\mathrm{PGL}_2 = \underline{\mathrm{Aut}}(\mathrm{SL}_2)$ will generally change our chosen K-isomorphism $G_K^{\mathrm{ss}} \simeq \mathrm{SL}_2$ but will preserve the properties that we have arranged in the preceding considerations. It now suffices to prove the following result.

Lemma 6.1.3 *With notation as above* (*so* $\mathrm{char}(k) \neq 2$), *the map* $G \stackrel{\xi_G}{\to} \mathrm{R}_{K/k}(G_K^{\mathrm{ss}}) \simeq \mathrm{R}_{K/k}(\mathrm{SL}_2)$ *is surjective.*

Proof We will prove that the image is $\mathrm{R}_{L/k}(\mathrm{SL}_2)$ for some intermediate field $L \subseteq K$ over k after applying a suitable K-automorphism of SL_2 arising from the diagonal torus in PGL_2. Granting this for a moment, we first claim that the L-map $\pi : G_L \to \mathrm{SL}_2$ corresponding to the k-map $G \to \mathrm{R}_{L/k}(\mathrm{SL}_2)$ induced by ξ_G must be a descent of the canonical quotient K-map $\Pi : G_K \twoheadrightarrow G_K^{\mathrm{ss}} = \mathrm{SL}_2$.

Rather generally, consider a tower $K/L/k$ of finite extensions of fields and affine (or just quasi-projective) schemes X and Y of finite type over k and L respectively such that there is a commutative diagram

$$\begin{array}{ccc} X & \xrightarrow{f} & \mathrm{R}_{K/k}(Y_K) \\ & {\scriptstyle h}\searrow & \uparrow{\scriptstyle \mathrm{R}_{L/k}(j_{Y,K/L})} \\ & & \mathrm{R}_{L/k}(Y) \end{array} \tag{6.1.3}$$

where $j_{Y,K/L} : Y \hookrightarrow \mathrm{R}_{K/L}(Y_K)$ is as in (A.5.3). We claim that the associated maps $X_K \to Y_K$ and $X_L \to Y$ obtained from the universal property of Weil restriction are related via the first being the scalar extension of the second. To verify this, we simply use the mechanism by which these associated maps are constructed: they are respectively the top and bottom composites in the following diagram:

$$\begin{array}{ccccc} X_K & \xrightarrow{f_K} & \mathrm{R}_{K/k}(Y_K)_K & \xrightarrow{q_{Y,K/k}} & Y_K \\ \| & & \uparrow{\scriptstyle \mathrm{R}_{L/k}(j_{Y,K/L})_K} & & \| \\ X_K & \xrightarrow[(h_L)_K]{} & (\mathrm{R}_{L/k}(Y)_L)_K & \xrightarrow[(q_{Y,L/k})_K]{} & Y_K \end{array}$$

It therefore suffices to prove that the diagram commutes. The commutativity of the left part is a tautology, and by Proposition A.5.7 the right part evaluated on A-points for any K-algebra A comes out to be the diagram

$$\begin{array}{ccc} Y(K \otimes_k A) & \longrightarrow & Y(A) \\ \uparrow & & \| \\ Y(L \otimes_k A) & \longrightarrow & Y(A) \end{array}$$

obtained by applying the functor Y to the evident commutative diagram of L-algebras.

Returning to our situation, the L-subgroup $\ker \pi \subseteq G_L$ must then be a descent of the K-subgroup $\ker \Pi = \mathscr{R}(G_K) \subseteq G_K$. Minimality of K/k as a field of definition of the radical of $G_{\overline{k}}$ therefore forces the inclusion $L \subseteq K$ over k to be an equality. Hence, granting the existence of L/k as above, ξ_G has image $\mathrm{R}_{K/k}(\mathrm{SL}_2)$, which is to say that ξ_G is surjective, as desired.

Now we turn to the task of constructing L/k within K as claimed. Let $\mathscr{G} \subseteq \mathrm{R}_{K/k}(\mathrm{SL}_2)$ denote the image of ξ_G. The arguments following (6.1.2) show that $\mathscr{G}$ is generated by the smooth connected k-subgroup $\xi_G(Z_G(T)) \subseteq \mathrm{R}_{K/k}(T_K)$ and some nontrivial smooth connected k-subgroups of $\mathrm{R}_{K/k}(U^+)$ and $\mathrm{R}_{K/k}(U^-)$, with $D \subseteq \xi_G(Z_G(T))$. Recall that $D(k) = \xi_G(T(k))$, so $D(k) \subseteq \xi_G(Z_G(T)(k))$.

We conclude that $\mathscr{G}$ is the Zariski closure in $\mathrm{R}_{K/k}(\mathrm{SL}_2)$ of the subgroup of

$$\mathrm{R}_{K/k}(\mathrm{SL}_2)(k) = \mathrm{SL}_2(K)$$

generated by the subgroup $\xi_G(Z_G(T)(k)) \subseteq D(K)$ that contains $D(k)$ and the subgroups

$$\mathscr{G}(k) \cap U^+(K) = \begin{pmatrix} 1 & V^+ \\ 0 & 1 \end{pmatrix}, \quad \mathscr{G}(k) \cap U^-(K) = \begin{pmatrix} 1 & 0 \\ V^- & 1 \end{pmatrix}$$

for some nonzero additive subgroups $V^\pm \subseteq K$.

Fix the K-isomorphisms $t : \mathrm{GL}_1 \simeq D_K$ and $u^\pm : \mathbf{G}_\mathrm{a} \simeq U^\pm$ via

$$t(x) = \begin{pmatrix} x & 0 \\ 0 & 1/x \end{pmatrix}, \quad u^+(x) = \begin{pmatrix} 1 & x \\ 0 & 1 \end{pmatrix}, \quad u^-(x) = \begin{pmatrix} 1 & 0 \\ x & 1 \end{pmatrix}.$$

Since $t(x)u^\pm(y)t(x)^{-1} = u^\pm(x^{\pm 2}y)$ and $D(k)$ is the image of $k^\times \subseteq K^\times$ under the K-isomorphism $t : \mathrm{GL}_1 \simeq D_K$, the additive subgroups $V^\pm \subseteq K$ are stable under scaling by squares in $k^\times$. Pick a nonzero $y \in V^-$, so $y \in K^\times$. Using the natural left action of PGL_2 on SL_2 over K, changing the isomorphism $G_K^{\mathrm{ss}} \simeq \mathrm{SL}_2$ by the action of the diagonal element $\begin{pmatrix} y & 0 \\ 0 & 1 \end{pmatrix} \in \mathrm{PGL}_2(K)$ is harmless and brings us to the case where $1 \in V^-$.

We shall now prove that for any nonzero additive subgroups $V^\pm \subseteq K$ such that $1 \in V^-$ and both V^+ and $V^-$are stable under scaling by $(k^\times)^2$, the Zariski closure in $\mathrm{R}_{K/k}(\mathrm{SL}_2)$ of the subgroup generated by

$$D(k), \begin{pmatrix} 1 & V^+ \\ 0 & 1 \end{pmatrix}, \begin{pmatrix} 1 & 0 \\ V^- & 1 \end{pmatrix} \tag{6.1.4}$$

is $\mathrm{R}_{L'/k}(\mathrm{SL}_2)$ where $L' \subseteq K$ is the subfield $k(V^+, V^-)$ generated over k by V^+ and V^-. In our preceding setup this will give $\mathrm{R}_{L'/k}(\mathrm{SL}_2) \subseteq \mathscr{G}$, and we will then have to address the contribution from the subgroup $\xi_G(Z_G(T)(k)) \subseteq D(K) = t(K^\times)$ that contains $t(k^\times) = D(k)$.

Since the subset $(k^\times)^2 \subseteq k$ is Zariski-dense in the closed k-subgroup $\mathbf{G}_\mathrm{a} \subseteq \mathrm{R}_{K/k}(\mathbf{G}_\mathrm{a})$ over k, we may replace V^+ and V^- with their k-linear spans within K without affecting the Zariski closure in $\mathrm{R}_{K/k}(\mathrm{SL}_2)$ of the subgroup of $(\mathrm{R}_{K/k}(\mathrm{SL}_2))(k) = \mathrm{SL}_2(K)$ generated by the groups in (6.1.4). That is, we may assume that the additive subgroups $V^\pm \subseteq K$ are nonzero k-linear subspaces (with $1 \in V^-$).

For $L' = k(V^+, V^-)$, the Zariski closure $\mathscr{H} \subseteq \mathrm{R}_{K/k}(\mathrm{SL}_2)$ of the subgroup of $\mathrm{SL}_2(K)$ generated by the groups in (6.1.4) is a smooth connected k-subgroup of the smooth connected k-subgroup $\mathrm{R}_{L'/k}(\mathrm{SL}_2)$. Hence, to prove $\mathscr{H} = \mathrm{R}_{L'/k}(\mathrm{SL}_2)$ it is sufficient to compare dimensions, and for that purpose it is enough to compare Lie algebras. More specifically, within the Lie algebra $\mathfrak{sl}_2(K)$ viewed over k, it suffices to show that the Lie subalgebra $\mathfrak{g}$ over k generated by the k-subalgebras

$$\left\{ \begin{pmatrix} x & 0 \\ 0 & -x \end{pmatrix} \middle| \, x \in k \right\}, \begin{pmatrix} 0 & V^+ \\ 0 & 0 \end{pmatrix}, \begin{pmatrix} 0 & 0 \\ V^- & 0 \end{pmatrix}$$

is equal to $\mathfrak{sl}_2(L')$.

For any $x \in K$, define

$$\mathrm{d}t(x) = \begin{pmatrix} x & 0 \\ 0 & -x \end{pmatrix}, \ \mathrm{d}u^+(x) = \begin{pmatrix} 0 & x \\ 0 & 0 \end{pmatrix}, \ \mathrm{d}u^-(x) = \begin{pmatrix} 0 & 0 \\ x & 0 \end{pmatrix}, \tag{6.1.5}$$

so $\mathrm{d}u^-(1) \in \mathfrak{g}$ since $1 \in V^-$. Also, by hypothesis $\mathrm{d}t(k) \subseteq \mathfrak{g}$. But we have more in $\mathfrak{g}$ on the diagonal: $\mathfrak{g}$ contains $[\mathrm{d}u^+(x), \mathrm{d}u^-(1)] = \mathrm{d}t(x)$ for all $x \in V^+$.

Hence, for any $x, x' \in V^+$, $\mathfrak{g}$ contains $[\mathrm{d}t(x), \mathrm{d}u^+(x')] = \mathrm{d}u^+(2xx')$. Since $\mathrm{char}(k) \neq 2$, it follows that $\mathfrak{g}$ contains the image under $\mathrm{d}u^+$ of the k-linear span A^+ of all (non-empty) products among elements in V^+.

For a nonzero $x_0 \in V^+ \subseteq K$, say with minimal polynomial

$$f(X) = X^n + c_{n-1}X^{n-1} + \cdots + c_0 \in k[X]$$

over k, we have $c_0 \in k^\times$ and $c_0 = -x^n - \sum_{j=1}^{n-1} c_j x^j \in A^+$, so $1 \in A^+$. Hence, A^+ is a k-subalgebra of K, which is to say it is an intermediate field. We may replace V^+ with A^+ for the purposes of carrying out the determination of $\mathfrak{g}$, so we can assume that V^+ is a subfield of K over k. In particular, $1 \in V^+$, so by running through the analogous calculations with focus on the lower triangular part we may replace V^- with the subfield it generates in K over k. Since $\mathfrak{g}$ contains $\mathrm{d}t(V^+)$ and $\mathrm{d}t(V^-)$, the computation of $[\mathrm{d}t(x), \mathrm{d}u^+(x')]$ shows that we can replace V^+ with the k-linear span of the set $V^- \cdot V^+$ of products, which is to say with L'. The same goes with the lower left corner, so $\mathfrak{g}$ contains $\mathrm{d}u^\pm(L')$. These generate $\mathfrak{sl}_2(L')$.

Our arguments show that $\mathscr{G} = \xi_G(G)$ is the Zariski closure in $\mathrm{R}_{K/k}(\mathrm{SL}_2)$ of the group Γ generated by $\mathrm{SL}_2(L') = \mathrm{R}_{L'/k}(\mathrm{SL}_2)(k)$ and the subgroup $Z := \xi_G(Z_G(T)(k)) \subseteq t(K^\times)$ containing $t(k^\times)$. For any $z \in K^\times$ and $x \in L'$, $t(z)u^\pm(x)t(z)^{-1} = u^\pm(z^{\pm 2}x)$. Hence, Γ contains $u^\pm(z^{\pm 2}L')$ for any $z \in K^\times$ such that $t(z) \in Z$. Since Z corresponds under t to a multiplicative subgroup of $K^\times$, it follows that Γ contains $u^\pm(L)$ where $L \subseteq K$ is the subfield generated by L' and squares of elements $z \in K^\times$ for which $t(z) \in Z$. Since the subfield $L \subseteq K$ is separably closed with $\mathrm{char}(L) \neq 2$, it is stable under the formation of square roots within K. Hence, $Z \subseteq t(L^\times)$. But $u^+(L)$ and $u^-(L)$ generate $\mathrm{SL}_2(L)$, whose diagonal is the group $t(L^\times)$ that contains Z. Thus, consideration of k-points gives $\mathscr{G} = \mathrm{R}_{L/k}(\mathrm{SL}_2)$, as required. □

6.2 Types A_2 and G_2 away from characteristic 3

Consider an absolutely pseudo-simple group G over a field k with $\mathrm{char}(k) \neq 3$. The main case of interest will be $\mathrm{char}(k) = 2$. Let K/k be the purely inseparable finite extension that is the field of definition over k for the (necessarily unipotent) radical $\mathscr{R}(G_{\overline{k}})$ as a subgroup of $G_{\overline{k}}$. This radical descends to a K-subgroup $\mathscr{R}(G_K) \subseteq G_K$ and by Proposition 5.3.3 the extension K/k is also the field of definition over k for the maximal adjoint semisimple quotient of $G_{\overline{k}}$. Define $G' = G_K^{\mathrm{ss}} = G_K/\mathscr{R}(G_K)$. Let $\widetilde{G}' \twoheadrightarrow G'$ be the simply connected central cover, and let μ be the kernel of this covering map.

Since $G = \mathscr{D}(G)$, the canonical map $i_G : G \to \mathrm{R}_{K/k}(G')$ factors through a canonical k-group map $\xi_G : G \to \mathrm{R}_{K/k}(\widetilde{G}')/\mathrm{R}_{K/k}(\mu)$ whose target is the derived subgroup of $\mathrm{R}_{K/k}(\widetilde{G}'/\mu) = \mathrm{R}_{K/k}(G')$ (Proposition 1.3.4). The formation of ξ_G commutes with separable algebraic extension on k, and the reason for our interest in ξ_G is that the absolutely pseudo-simple k-group G is standard if and only if ξ_G is surjective and $\ker \xi_G$ is central in G (Theorem 5.3.8), and that centrality is automatic when $G^{\mathrm{ss}}_{\overline{k}}$ is not of simply connected type C_n for some $n \geqslant 1$ (Proposition 5.3.10 and Theorem 2.3.10).

Assume that $G^{\mathrm{ss}}_{\overline{k}}$ is of type A_2 or G_2. Our aim is to prove that any such G is standard. Since $\mathrm{char}(k) \neq 3$ and $G^{\mathrm{ss}}_{\overline{k}}$ is isomorphic to SL_3, PGL_3, or G_2, the simply connected central covering map $\widetilde{G}' \to G' := G^{\mathrm{ss}}_K$ has étale kernel μ. Hence,

$$\xi_G = i_G : G \to \mathrm{R}_{K/k}(G') \tag{6.2.1}$$

since $\mathrm{R}_{K/k}$ commutes with the formation of quotients by smooth (e.g., étale) subgroups. To prove the standardness of such G, it suffices to prove the following analogue of Theorem 6.1.1 that is the first main goal of this section:

Proposition 6.2.1 *Let k be a field with* $\mathrm{char}(k) \neq 3$, *and G an absolutely pseudo-simple k-group such that $G^{\mathrm{ss}}_{\overline{k}}$ is of type* A_2 *or* G_2. *The associated natural map* (6.2.1) *is an isomorphism, and hence G is standard.*

To prove Proposition 6.2.1, we may and do assume $k = k_s$. Thus, by Theorem 5.3.8 and Lemma 6.1.2, we may also assume μ is infinitesimal. In our present setting, μ is infinitesimal precisely when $\mu = 1$, so G' is isomorphic to either SL_3 or G_2.

Now we fix notation related to roots. Choose a maximal k-torus T in G, so the quotient map

$$\pi : G_K \twoheadrightarrow G^{\mathrm{red}}_K = G^{\mathrm{ss}}_K = G'$$

carries T_K isomorphically onto a maximal K-torus T' of G'. Choose a positive system of roots Φ^+ in the rank-2 root system $\Phi(G', T') \subseteq \mathrm{X}(T') = \mathrm{X}(T)$ and let $\Delta = \{a, b\}$ be the basis of Φ^+, with a long and b short in the G_2 case. The image of the coroot $b^\vee$ normalizes the root group $U'_{\pm a} \simeq \mathbf{G}_{\mathrm{a}}$ in G', and relative to the unique linear structure on $U'_{\pm a}$ the conjugation action on it by the point $b^\vee(s) \in G'(K)$ for $s \in K^\times$ is multiplication by $s^{\langle \pm a, b^\vee \rangle} \in K^\times$ (i.e., $s^{\mp 1}$ for simply connected type A_2 and $s^{\mp 3}$ for type G_2). Note in particular that the exponent of s is not divisible by $\mathrm{char}(K) = \mathrm{char}(k)$.

Since G' is simple but not of type C_n ($n \geqslant 1$), the elements of the reduced and irreducible root system $\Phi(G', T')$ are not divisible by 2 in $X(T')$. Thus, by Theorem 2.3.10, $\Phi(G, T) = \Phi(G', T')$. The proof of Proposition 6.2.1 will proceed by showing that (6.2.1) is surjective, in which case standardness follows from Proposition 5.3.10 and the isomorphism property for ξ_G follows from the final claim in Theorem 5.3.8 (as $\mathrm{char}(k) \neq 3$). The proof of surjectivity requires some input from subgroups of G that we can understand in concrete terms and which have a nontrivial image under (6.2.1). As in our work in §6.1 away from characteristic 2, to do this we begin by analyzing root groups for (G', T').

For each $c \in \Phi(G, T)$, consider the root group $U_c \subseteq G$ as in Proposition 2.3.11. Since the K-subgroup $(U_c)_K \subseteq G_K$ has Lie algebra $K \otimes_k \mathrm{Lie}(U_c)$ that is equal to the c_K-root space in $\mathrm{Lie}(G_K)$, this Lie algebra maps onto the c_K-root space in $\mathrm{Lie}(G')$ under $\mathrm{Lie}(\pi)$. In particular, the map $U_c \to \mathrm{R}_{K/k}(G')$ induced by ξ_G is nonzero on Lie algebras. Hence, we deduce (from T-equivariance of ξ_G) the nontriviality of the k-subgroup $\underline{V}_c := \xi_G(U_c) \subseteq \mathrm{R}_{K/k}(G')$ that is contained in $\mathrm{R}_{K/k}(U'_c)$, where U'_c is the c_K-root group of G'. By T-equivariance, the map ξ_G also carries $Z_G(T)$ into $\mathrm{R}_{K/k}(Z_{G'}(T')) = \mathrm{R}_{K/k}(T')$.

Using the basis Δ, choose a pinning of the pair (G', T') (see Definition A.4.12). In particular, for $c \in \Delta$, the root groups $U'_{\pm c}$ of (G', T') are thereby equipped with a specified identification $u_{\pm c} : \mathbf{G}_a \simeq U'_{\pm c}$ over K. (We will make the choice of pinning more precise below.) Hence, $\underline{V}_{\pm c} = \xi_G(U_{\pm c})$ is identified with a smooth connected nontrivial k-subgroup of $\mathrm{R}_{K/k}(\mathbf{G}_a)$, and it is stable under the natural GL_1-action, so it is a vector subgroup. In other words, it corresponds to a nonzero k-subspace $V_{\pm c} \subseteq K$. We wish to get to the situation in which the k-subspaces $V_{-a} \subseteq U'_{-a}(K) = K$ and $V_{-b} \subseteq U'_{-b}(K) = K$ both contain 1 (with the equalities $U'_c(K) = K$ obtained via our choice of pinning of (G', T')). Choose $c \in \{-a, -b\}$. For $s, s' \in K^\times$ and coroots λ and λ', conjugation by $\lambda(s)\lambda'(s')$ on G' acts on the c_K-root group $U'_c \simeq \mathbf{G}_a$ of G' by multiplication by $s^{\langle c,\lambda\rangle} s'^{\langle c,\lambda'\rangle} \in K^\times$. Thus, by taking $\lambda = a^\vee$ and $\lambda' = b^\vee$, it suffices to find $s, s' \in K^\times$ such that the elements

$$s^{\langle a,a^\vee\rangle} s'^{\langle a,b^\vee\rangle}, \; s^{\langle b,a^\vee\rangle} s'^{\langle b,b^\vee\rangle} \in K^\times$$

coincide respectively with any given pair of elements of $K^\times$ (such as some nonzero elements $v_{-a} \in V_{-a}$ and $v_{-b} \in V_{-b}$ with respect to the initial choice of pinning). Since K is separably closed, such $s, s' \in K^\times$ can be found since the determinant of the Cartan matrix of (G', T') is not divisible by $\mathrm{char}(K)$ (as $\mathrm{char}(k) \neq 3$). Applying conjugation by $a^\vee(s)b^\vee(s')$ for such s, s', adapted to the pair (v_{-a}, v_{-b}), gives a new pinning such that both V_{-a} and V_{-b} contain 1.

Fix $c \in \{-a, -b\}$. The pinning on the *simply connected* pair (G', T') over K provides a canonical identification of SL_2 with the K-subgroup $G'_c \subseteq G'$ generated by U'_c and U'_{-c} and this identification carries U'_c over to the standard upper triangular unipotent K-subgroup of SL_2. We likewise get a corresponding copy of $\mathfrak{sl}_2(K)$ as a Lie K-subalgebra of $\mathrm{Lie}(G')$. In terms of the resulting identification of $\mathfrak{sl}_2(K)$ as a Lie k-subalgebra of $\mathrm{Lie}(\mathrm{R}_{K/k}(G')) = \mathrm{Lie}(G')$, the Lie k-subalgebra $\mathfrak{g} := \mathrm{Lie}(\xi_G(G))$ must contain

$$\begin{pmatrix} 0 & x \\ 0 & 0 \end{pmatrix}, \begin{pmatrix} 0 & 0 \\ x' & 0 \end{pmatrix}$$

for all $x \in V_c$ and $x' \in V_{-c}$. But

$$\left[\begin{pmatrix} 0 & x \\ 0 & 0 \end{pmatrix}, \begin{pmatrix} 0 & 0 \\ x' & 0 \end{pmatrix}\right] = \begin{pmatrix} xx' & 0 \\ 0 & -xx' \end{pmatrix},$$

so

$$\mathrm{d}a^\vee(V_a \cdot V_{-a}),\ \mathrm{d}b^\vee(V_b \cdot V_{-b}) \subseteq \mathfrak{g}, \tag{6.2.2}$$

where $V_c \cdot V_{-c}$ denotes the k-subspace of K generated by the products xx' for $x \in V_c$ and $x' \in V_{-c}$.

We claim that inside of K,

$$V_a \subseteq V_a \cdot V_{-a} \subseteq V_{-b}. \tag{6.2.3}$$

The first inclusion holds because $1 \in V_{-a}$. Since $1 \in V_{-b}$, to prove the second inclusion it suffices to prove $V_a \cdot V_{-a} \cdot V_{-b} \subseteq V_{-b}$ (where the triple product is defined analogously to the products $V_c \cdot V_{-c}$). The adjoint representation of any smooth affine k-group induces the adjoint representation of its Lie algebra, so for any $x \in V_a \cdot V_{-a}$ and $y \in V_{-b}$ it follows from (6.2.2) that the Lie k-subalgebra $\mathfrak{g}$ in $\mathrm{Lie}(G')$ contains $[\mathrm{d}a^\vee(x), \mathrm{d}u_{-b}(y)]$. This equals $\mathrm{d}u_{-b}(\langle -b, a^\vee\rangle xy)$ because $a^\vee(s)u_{-b}(t)a^\vee(s)^{-1} = u_{-b}(s^{\langle -b, a^\vee\rangle}t)$ for all $s \in K^\times$ and $t \in K$. Passing to tangent space at the identity yields that $\langle -b, a^\vee\rangle \cdot xy \in V_{-b}$ for all $x \in V_a \cdot V_{-a}$ and $y \in V_{-b}$. But $\langle -b, a^\vee\rangle$ has nonzero image in the field K, so (6.2.3) is proved.

Lemma 6.2.2 *The k-subspaces $V_{\pm a}, V_{\pm b} \subseteq K$ coincide with a common subfield L containing k.*

Proof By Corollary 3.4.3(2), $\dim_k V_c = \dim_k V_{-c}$ for all roots c. Thus, the inclusion $V_a \subseteq V_{-b}$ implies

$$\dim_k V_{-a} = \dim_k V_a \leqslant \dim_k V_{-b} = \dim_k V_b.$$

But we can interchange the roles of a and b in our proof of (6.2.3), so also $V_b \subseteq V_{-a}$. Hence, all four k-dimensions coincide, so in fact the containments $V_a \subseteq V_{-b}$ and $V_b \subseteq V_{-a}$ are equalities, so $1 \in V_a$ and $1 \in V_b$. Also, the containments in (6.2.3) are forced to be equalities, and likewise for the analogous containments with a and b interchanged. This shows that $V_{\pm a}$ and $V_{\pm b}$ are all equal to a common k-subspace $V \subseteq K$ that contains 1 and is stable under multiplication against itself. Hence, it is a subfield $L \subseteq K$ containing k, as desired. □

Our fixed pinning on (G', T') via the basis $\{a, b\}$ of Φ specifies a K-isomorphism of G' with either SL_3 or G_2. Thus, by a refinement of the Isomorphism Theorem (see Theorem A.4.13 for this refinement), there is a unique descent of (G', T') to a pair $(\mathscr{G}, \mathscr{T})$ over k such that the pinning is defined over k. In particular, $\mathscr{T}$ defines the same k-structure on $T' = T_K$ that T does, so the subsets $\mathscr{T}(k), T(k) \subseteq T'(K)$ coincide.

By Lemma 6.2.2, $\underline{V}_{\pm a}$ and $\underline{V}_{\pm b}$ generate $R_{L/k}(\mathscr{G}_L)$, as we may check using k-rational points and the open Bruhat cell of $(\mathscr{G}_L, \mathscr{T}_L)$. Hence, we have proved the containment $R_{L/k}(\mathscr{G}_L) \subseteq \xi_G(G)$ inside of $R_{K/k}(G') = R_{K/k}(\mathscr{G}_K)$. We can do better: for every $c \in \Phi(G, T)$, necessarily $\underline{V}_c \subseteq R_{L/k}(\mathscr{G}_L)$ inside of $\xi_G(G)$. To prove this, observe that $W(\mathscr{G}, \mathscr{T})(k) = W(G', T')(K) = W(G, T)(K)$ (the final equality by Proposition 3.2.7). Thus, we can find $n \in N_{\mathscr{G}}(\mathscr{T})(k)$ that conjugates an arbitrary but fixed $c \in \Phi(G, T)$ to one of a or b. This moves $\underline{V}_c$ to $\underline{V}_a$ or $\underline{V}_b$, each of which lies in $R_{L/k}(\mathscr{G}_L)$, so we get the desired containment for c.

We have done enough analysis of root groups to finally prove the following result, which completes the proof of Proposition 6.2.1.

Lemma 6.2.3 *The map ξ_G is surjective.*

Proof Lemma 3.1.5 implies that the root groups U_c for $c \in \Phi(G, T)$ generate G, and hence the $\underline{V}_c = \xi_G(U_c)$ generate $\xi_G(G)$. But for each $c \in \Phi(G, T)$, $\underline{V}_c$ is contained in $R_{L/k}(\mathscr{G}_L)$, and therefore $\xi_G(G) \subseteq R_{L/k}(\mathscr{G}_L)$. As we observed above, $R_{L/k}(\mathscr{G}_L) \subseteq \xi_G(G)$. We conclude that $\xi_G(G) = R_{L/k}(\mathscr{G}_L)$, so the map $\xi_G : G \to R_{K/k}(G') = R_{K/k}(\mathscr{G}_K)$ factors through $R_{L/k}(\mathscr{G}_L)$. Hence, as in the proof of Lemma 6.1.3 (using (6.1.3)) it follows from the functorial definition of Weil restriction that the natural quotient map

$$\pi : G_K \twoheadrightarrow G' \simeq \mathscr{G}_K$$

descends to an L-group map $G_L \twoheadrightarrow \mathscr{G}_L$. But the K-subgroup $\ker \pi = \mathscr{R}(G_K)$ in G_K has (minimal) field of definition over k equal to K (by definition of K/k), so $L = K$; i.e., $\xi_G(G) = \mathrm{R}_{K/k}(G')$. □

Our study of types A_2 and G_2 will be most useful in characteristic 2, and will be used to prove a general standardness result away from characteristic 3 in Proposition 6.3.6 when the Dynkin diagram of $G^{\mathrm{ss}}_{\overline{k}}$ has no edges of multiplicity 2 and no isolated vertices. We end this section with an application in a special case when there is an edge with multiplicity 2, type F_4:

Proposition 6.2.4 *Let G be an absolutely pseudo-simple group over a separably closed field k, and assume that $G^{\mathrm{ss}}_{\overline{k}}$ is of type* F_4. *Let T be a maximal k-torus of G, and $\Phi = \Phi(G,T)$. For each $c \in \Phi$, let G_c be the k-subgroup generated by the $\pm c$-root groups. Then ξ_{G_c} is an isomorphism.*

Proof The case $\mathrm{char}(k) \neq 2$ is settled by Theorem 6.1.1, so now assume $\mathrm{char}(k) = 2$. Choose a positive system of roots Φ^+ such that c lies in the set Δ of simple positive roots. The vertex corresponding to c is adjacent to that corresponding to $c' \in \Delta$ of the same length. Let S be the codimension-2 k-torus $(\ker c \cap \ker c')^0_{\mathrm{red}}$ in T, and consider $H = \mathscr{D}(Z_G(S))$. The k-group H is pseudo-reductive and $H^{\mathrm{red}}_{\overline{k}}$ is identified with $\mathscr{D}(Z_{G^{\mathrm{ss}}_{\overline{k}}}(S_{\overline{k}}))$ by Proposition A.4.8. This is simply connected of type A_2 since $G^{\mathrm{ss}}_{\overline{k}}$ is the (simply connected) group of type F_4. In particular, H is absolutely pseudo-simple.

By Proposition 6.2.1, the pseudo-reductive k-group H is standard. Since $H^{\mathrm{ss}}_{\overline{k}}$ is simply connected, the final part of Theorem 5.3.8 implies that ξ_H is an isomorphism. That is, $H \simeq \mathrm{R}_{K/k}(H^{\mathrm{ss}}_K)$ for the field of definition K/k of $\mathscr{R}(H_{\overline{k}})$, with $H^{\mathrm{ss}}_K \simeq \mathrm{SL}_3$. This latter isomorphism can be arranged to carry the maximal K-torus $(T \cap H)_K$ to the diagonal K-torus D. Hence, for each $a \in \Phi(H, T \cap H) = \Phi(\mathrm{SL}_3, D)$ we have that H_a is carried isomorphically onto $\mathrm{R}_{K/k}(\mathrm{SL}_2)$ by ξ_H. Choose $a = c$, and note that $H_c = G_c$ by construction of H, so $G_c \simeq \mathrm{R}_{K/k}(\mathrm{SL}_2)$ and hence ξ_{G_c} is an isomorphism. □

6.3 General cases away from characteristics 2 and 3

We shall now prove Theorem 1.5.1. This result is obtained by combining Corollary 6.3.5 and Proposition 6.3.6. It will be convenient to first carry out some general considerations with torus centralizers, refining Proposition 3.4.1. *We initially allow the ground field to have any characteristic* so that some of our conclusions can be applied later in our work in characteristics 2 and 3.

Let G be a perfect pseudo-reductive group over a separably closed field k, T a maximal k-torus in G, and K/k the field of definition for the $\overline{k}$-subgroup $\mathscr{R}(G_{\overline{k}}) \subseteq G_{\overline{k}}$. For each $a \in \Phi(G,T)$, let $G_a \subseteq G$ be the smooth connected k-subgroup generated by the root groups U_a and U_{-a}, so by Proposition 3.4.1 the k-group G_a is absolutely pseudo-simple of rank 1 and it has the following alternative description: if a is not divisible in $\Phi(G,T)$ (as always happens except possibly when k is imperfect of characteristic 2) then $G_a = \mathscr{D}(Z_G(T_a))$ where $T_a = (\ker a)^0_{\mathrm{red}}$, and if a is divisible in $\Phi(G,T)$ (so k is imperfect of characteristic 2) then $G_a = \mathscr{D}(Z_{Z_G(T_a)}(\mu)^0)$ where $\mu = S[2]$ for the rank-1 maximal k-torus $S = T \cap G_{a/2}$ in $G_{a/2}$.

Proposition 6.3.1 *For each $a \in \Phi(G,T)$, let K_a/k be the field of definition of $\mathscr{R}((G_a)_{\overline{k}}) \subseteq (G_a)_{\overline{k}}$.*

(1) *The fields K_a are all contained in K over k, and they generate K over k.*

(2) *Let $\pi_a : (G_a)_{K_a} \twoheadrightarrow \mathscr{G}'_a$ be the maximal semisimple quotient over K_a. The natural map $\pi : (G_a)_K \to G' := G_K^{\mathrm{ss}}$ uniquely factors through $(\pi_a)_K$ via a K-subgroup inclusion of $(\mathscr{G}'_a)_K$ into G'; equivalently, $\mathscr{R}((G_a)_{\overline{k}}) = (G_a)_{\overline{k}} \cap \mathscr{R}(G_{\overline{k}})$ for all a. The image of $(\mathscr{G}'_a)_K$ in G' is the subgroup $G'_{a'}$ generated by the $\pm a'_K$-root groups of (G',T'), where T' is the torus T_K of G_K viewed as a maximal K-torus of G' and $a' = 2a$ or $a' = a$ according as a is multipliable or not.*

(3) *Letting $i_G : G \to \mathrm{R}_{K/k}(G_K) \to \mathrm{R}_{K/k}(G')$ be the natural composite map, the diagram*

$$\begin{array}{ccccccc} G_a & \xrightarrow{\xi_{G_a}} & \mathscr{D}(\mathrm{R}_{K_a/k}(\mathscr{G}'_a)) & \longrightarrow & \mathrm{R}_{K_a/k}(\mathscr{G}'_a) & \longrightarrow & \mathrm{R}_{K/k}(G'_{a'}) \\ \downarrow & & & & & & \downarrow \\ G & & & \xrightarrow{\quad i_G \quad} & & & \mathrm{R}_{K/k}(G') \end{array} \qquad (6.3.1)$$

of k-homomorphisms (with all unlabeled maps equal to the natural inclusions) commutes.

Note that in (1) there is no ambiguity concerning how each K_a embeds into K over k since they are all purely inseparable over k.

Proof By Proposition A.4.8 applied initially to $Z_G(T_a)$ in G and then to $\mathscr{D}(Z_G(T_a))$ in $Z_G(T_a)$, it follows that the geometric radical of $\mathscr{D}(Z_G(T_a))$ is defined over K. Hence, if a is not divisible in $\Phi(G,T)$ (in which case $G_a = \mathscr{D}(Z_G(T_a)))$ then $K_a \subseteq K$ over k. To handle the divisible case, we first use Proposition A.8.14(2) to see that $H_a := Z_{Z_G(T_a)}(\mu)^0$ (with μ the 2-torsion in $T \cap G_{a/2}$) is a pseudo-reductive k-group whose geometric unipotent radical $\mathscr{R}_u((H_a)_{\overline{k}})$ equals $\mathscr{R}_u(Z_G(T_a)_{\overline{k}}) \cap (H_a)_{\overline{k}} = \mathscr{R}_u(G_{\overline{k}}) \cap (H_a)_{\overline{k}}$ (with connectedness due to the fact that the μ-action arises from the action on $Z_G(T_a)$ by the torus $T \cap G_{a/2}$, as explained in the proof of Proposition A.8.14(2)). In particular, this geometric unipotent radical has field of definition over k that is contained in the field of definition over k for the geometric unipotent radical of $Z_G(T_a)$. But this latter field is contained in K by Proposition A.4.8 (applied over $\overline{k}$), and so a further application of Proposition A.4.8 to the derived subgroup G_a of H_a gives that $K_a \subseteq K$ in the divisible case as well. This proves the first assertion in (1), so the formulation of (2) now makes sense.

Before we can prove the second assertion in (1), we need to prove (2). We treat the formulation in terms of geometric radicals, or equivalently (by perfectness of G and G_a) geometric unipotent radicals. By applying Proposition A.4.8 twice over $\overline{k}$ (for passage to torus centralizers and to derived subgroups thereof) we can handle the case when a is not divisible in $\Phi(G,T)$, and the same method works when a is divisible by using Proposition A.8.14(2) over $\overline{k}$ to handle μ-centralizers. Thus, (2) is proved apart from identifying the image of $(\mathscr{G}'_a)_K$ in G'. By definition G_a is generated by the root groups $U_{\pm a}$ in G, so in view of Theorem 2.3.10 and Remark 2.3.6, the quotient map $G_K \twoheadrightarrow G'$ carries $(U_{\pm a})_K$ onto the $\pm a'_K$-root groups in G'. Thus, the image of $(G_a)_K$ in G' is generated by the $\pm a'_K$-root groups for (G', T'). This coincides with the image of the inclusion $(\mathscr{G}'_a)_K \hookrightarrow G'$.

Returning to the second claim in (1), we let $L \subseteq K$ be a subfield that contains K_a for all $a \in \Phi(G,T)$ (so $k \subseteq L$) and we aim to prove that $L = K$. By Proposition 5.3.3, K/k is equal to the field of definition of the kernel of the projection from $G_{\overline{k}}$ onto its maximal adjoint semisimple quotient. Hence, it suffices to construct an L-descent of this latter quotient of $G_{\overline{k}}$. Let R be the smooth connected normal L-subgroup of G_L generated by the $G(k)$-conjugates of the radical $\mathscr{R}((G_a)_L)$ as we vary $a \in \Phi(G,T)$. (This radical over L makes sense since L contains K_a for all a, and R is normal in G_L since $G(k)$ is Zariski-dense in G_L due to the fact that $k = k_s$.) We shall use R to descend the maximal adjoint semisimple quotient of $G_{\overline{k}}$ to a quotient of G_L.

Since $\mathscr{R}((G_a)_{\overline{k}}) = (G_a)_{\overline{k}} \cap \mathscr{R}(G_{\overline{k}})$ by (2) we see that $R_{\overline{k}}$ is contained in $\mathscr{R}(G_{\overline{k}})$. Let us show that it suffices to prove that $\mathscr{R}(G_{\overline{k}})/R_{\overline{k}}$ is central in

$(G_L/R)_{\overline{k}}$. If this centrality holds then the central quotient $(G_L/R)/Z_{(G_L/R)}$ is semisimple and so over $\overline{k}$ it dominates the *maximal* adjoint semisimple quotient of $G_{\overline{k}}$. That is, the quotient of $(G_L/R)/Z_{(G_L/R)}$ modulo its finite scheme-theoretic center (over L) is an adjoint semisimple quotient of G_L that descends the maximal adjoint semisimple quotient of $G_{\overline{k}}$, as desired.

To prove the centrality of a smooth connected normal $\overline{k}$-subgroup H of $(G_L/R)_{\overline{k}}$ (such as the radical) it suffices to prove that H is centralized by the maximal $\overline{k}$-torus image of $T_{\overline{k}}$, since the conjugates of a maximal torus of $(G_L/R)_{\overline{k}}$ generate $(G_L/R)_{\overline{k}}$ (as G_L/R is perfect, since G is). By Corollary A.8.11, a torus action on a smooth connected group is trivial if and only if the induced action on the Lie algebra is trivial. Thus, it remains to prove that $T_{\overline{k}}$ acts trivially on the Lie algebra of $\mathscr{R}(G_{\overline{k}})/R_{\overline{k}}$. Equivalently, for nontrivial weights a of T on $\mathrm{Lie}(G)$ we claim $\mathrm{Lie}(G_{\overline{k}})_a \cap \mathrm{Lie}(\mathscr{R}(G_{\overline{k}})) \subseteq \mathrm{Lie}(R_{\overline{k}})$. Since the a-root space of $\mathrm{Lie}(G)$ is contained in $\mathrm{Lie}(G_a)$ (as G_a contains the a-root group U_a of G), we conclude that

$$(G_a)_{\overline{k}} \cap \mathscr{R}(G_{\overline{k}}) = \mathscr{R}((G_a)_{\overline{k}}) = \mathscr{R}((G_a)_L) \otimes_L \overline{k} \subseteq R \otimes_L \overline{k} = R_{\overline{k}}$$

and our claim follows from this since the formation of Lie subalgebras of subgroup schemes is compatible with (scheme-theoretic) intersections.

Finally, we prove (3). The composite map across the top in (6.3.1) corresponds (under the universal property of Weil restriction) to the natural quotient map $(G_a)_K \twoheadrightarrow G'_{a'}$, and i_G likewise corresponds to the natural quotient map $G_K \twoheadrightarrow G' = G_K^{\mathrm{ss}}$. It therefore suffices to prove that the diagram of K-groups

$$\begin{array}{ccc} (G_a)_K & \longrightarrow & G'_{a'} \\ \downarrow & & \downarrow \\ G_K & \longrightarrow & G' \end{array}$$

commutes, and this follows from the definition of $G'_{a'}$. □

Let G be absolutely pseudo-simple over a field k. By Theorem 5.3.8, the pseudo-reductive k-group G is standard over k if and only if ξ_G is surjective and $\ker \xi_G$ is central in G. Moreover, by Proposition 5.3.10, if $\mathrm{char}(k) \neq 2$ then $\ker \xi_G$ is always central in G and hence G is standard if and only if ξ_G is surjective. We therefore wish to prove the surjectivity of ξ_G when $\mathrm{char}(k) \neq 2$, subject to an extra hypothesis in characteristic 3. A key ingredient is the following proposition:

Proposition 6.3.2 *Let k be a separably closed field, and G an absolutely pseudo-simple k-group whose root system is reduced. Let K/k be the field of definition over k for the subgroup $\mathscr{R}(G_{\overline{k}}) = \mathscr{R}_u(G_{\overline{k}})$ in $G_{\overline{k}}$. Fix a maximal k-torus $T \subseteq G$ and for each $a \in \Phi := \Phi(G,T)$ let $K_a \subseteq K$ be the field of definition over k for the kernel of the projection from $(G_a)_{\overline{k}}$ onto its maximal semisimple quotient, where G_a is generated by the root groups U_a and U_{-a}.*

(1) *If all root lengths in Φ coincide then $K_a = K$ for all $a \in \Phi$.*
(2) *Assume there are two distinct root lengths and ξ_{G_a} is surjective for all a. Then $K_a = K$ for all a except possibly when k is imperfect with* $\mathrm{char}(k) =$ *$p \in \{2,3\}$ and $G^{\mathrm{ss}}_{\overline{k}}$ has an edge of multiplicity p in its Dynkin diagram. In such exceptional cases, $K_b = K$ for all short roots b and the fields K_a for long a are all equal to a common subfield $K_> \subseteq K$ satisfying $kK^p \subseteq K_> \subseteq K$.*

Before we prove Proposition 6.3.2, we make some remarks. Since all choices for T are $G(k)$-conjugate, in case (2) the subfield $K_> \subseteq K$ is independent of T, so it is an intrinsic invariant of G. In particular, if we consider an arbitrary field k and an absolutely pseudo-simple k-group G such that G_{k_s} satisfies the requirements for case (2) then (with K/k denoting the field of definition of the geometric radical) we can define the subfield $K_> \subseteq K$ over k by Galois descent of the subfield $(K_s)_> \subseteq K_s$. Also, Theorem 6.1.1 ensures that the surjectivity hypothesis in (2) holds when $\mathrm{char}(k) \neq 2$ (in fact ξ_{G_a} is always an isomorphism in such cases) and it will turn out that in characteristic 2 it is satisfied whenever $[k : k^2] = 2$ (by Proposition 9.2.4).

Proof Let $G' = G^{\mathrm{ss}}_K$ and let T' be T_K viewed naturally as a maximal K-torus of G'. Since Φ is assumed to be reduced, by Theorem 2.3.10 we have $R(G,T) = R(G',T')$ as root data, compatibly with the identification of character lattices and cocharacter lattices for T and T'. The reducedness of the root system also implies that $\mathscr{D}(Z_G(T_a))$ coincides with the k-subgroup G_a considered in Proposition 6.3.1.

By Corollary 3.4.3(2), if a pair of roots $a, b \in \Phi$ have the same length then some $n \in N_G(T)(k)$ moves a to b and thus conjugates G_a into G_b. Hence, $K_a = K_b$ inside of K in such cases. It follows that the subfields K_a/k are all equal when there is only one root length, whereas in the case of two root lengths we get a pair of subfields $K_<$ and $K_>$ in K over k (with $K_< = K_a$ for short a and $K_> = K_a$ for long a). Since the fields K_a generate K over k (Proposition 6.3.1(1)), it follows that the fields K_a are all equal to K in case all roots are of the same length and that $K_>$ and $K_<$ generate K over k in case there are roots of unequal lengths. This proves (1). Of course, if k is perfect then we always have $K_< = K_> = K$ since $K = k$ in such cases.

To prove (2) we can now assume that there are two distinct root lengths, that ξ_{G_a} is surjective for all a, and that k is imperfect. It suffices to prove that $K_> \subseteq K_<$, and that the reverse inclusion holds except possibly when $\mathrm{char}(k) = p \in \{2,3\}$ and there is an edge of multiplicity p in the Dynkin diagram of Φ, in which case $kK_<^p \subseteq K_> \subseteq K_<$.

The key point is to analyze some root groups for (G,T). Choose any $a \in \Phi$. By Proposition 6.3.1, whose notation we now freely use, the field of definition K_a/k for $\mathscr{R}((G_a)_{\overline{k}}) \subseteq (G_a)_{\overline{k}}$ is contained in K and $(G_a)_K^{\mathrm{ss}}$ is identified with the K-subgroup $G'_a \subseteq G'$ of rank 1 generated by the $\pm a_K$-root groups of G'.

Recall from Corollary 2.1.9 that for any surjective homomorphism between pseudo-reductive k-groups equipped with a compatible choice of maximal k-tori, Cartan k-subgroups are carried onto Cartan k-subgroups and root groups are carried onto root groups when the root systems are reduced. Thus, since the map ξ_{G_a} in (6.3.1) is surjective by hypothesis, it carries the Cartan k-subgroup $Z_{G_a}(T \cap G_a)$ onto the centralizer of $\mathrm{R}_{K_a/k}(\mathscr{T}'_a)$ in the k-subgroup $\mathscr{D}(\mathrm{R}_{K_a/k}(\mathscr{G}'_a))$ of $\mathrm{R}_{K/k}(G'_a)$ for the K_a-descent $\mathscr{G}'_a = (G_a)_{K_a}^{\mathrm{ss}}$ of $G'_a = (G_a)_K^{\mathrm{ss}}$ in which $(T \cap G_a)_{K_a}$ is identified with the K_a-descent $\mathscr{T}'_a$ of the maximal K-torus $T' \cap G'_a$ of G'_a (centralizing $\mathrm{R}_{K_a/k}(\mathscr{T}'_a)$ is the same as centralizing its maximal k-torus, due to Proposition 1.2.2).

We claim that $\xi_{G_a}(Z_{G_a}(T \cap G_a))$ is the image of

$$\mathrm{R}_{K_a/k}(a^\vee_{K_a}) : \mathrm{R}_{K_a/k}(\mathrm{GL}_1) \to \mathrm{R}_{K_a/k}(T_{K_a}) \hookrightarrow \mathrm{R}_{K/k}(T_K) = \mathrm{R}_{K/k}(T').$$

To prove this, we will separately treat the cases where $\mathscr{G}'_a$ is simply connected or adjoint (i.e., isomorphic to SL_2 or PGL_2). First consider the simply connected case, so $\mathrm{R}_{K_a/k}(\mathscr{G}'_a)$ is its own derived group (since $\mathscr{G}'_a \simeq \mathrm{SL}_2$). Thus, $\xi_{G_a}(Z_{G_a}(T \cap G_a)) = \mathrm{R}_{K_a/k}(\mathscr{T}'_a)$ inside $\mathrm{R}_{K/k}(T' \cap G'_a)$, so we prove $\mathrm{R}_{K_a/k}(\mathscr{T}'_a)$ is the image of $\mathrm{R}_{K_a/k}(a^\vee_{K_a})$ inside $\mathrm{R}_{K/k}(T')$. Since we are in the simply connected case, the 1-parameter subgroup $a^\vee_K : \mathrm{GL}_1 \to T' \cap G'_a \subseteq T' = T_K$ has trivial kernel and so is a direct factor. Thus, likewise $a^\vee_{K_a} : \mathrm{GL}_1 \to T_{K_a}$ is a direct factor, so by left-exactness of Weil restriction, the image of the map $\mathrm{R}_{K_a/k}(a^\vee_{K_a})$ is obtained by applying $\mathrm{R}_{K_a/k}$ to the image of $a^\vee_{K_a}$. This image is $a^\vee(\mathrm{GL}_1)_{K_a} \subseteq T_{K_a}$, and since $a^\vee(\mathrm{GL}_1) = T \cap G_a$ we get that $a^\vee(\mathrm{GL}_1)_{K_a}$ is the subtorus $(T \cap G_a)_{K_a}$ which (by definition) goes over to $\mathscr{T}'_a$ under the quotient map $(G_a)_{K_a} \twoheadrightarrow \mathscr{G}'_a$. Hence, we get the desired result when $\mathscr{G}'_a$ is simply connected.

Suppose instead that $\mathscr{G}'_a$ is adjoint. In this case $\mathscr{G}'_a \simeq \mathrm{PGL}_2 = \mathrm{SL}_2/\mu_2$, so if $q_a : \widetilde{\mathscr{G}}'_a \to \mathscr{G}'_a$ denotes the simply connected central cover and $\mu \simeq \mu_2$ denotes the kernel of the covering map then Proposition 1.3.4 provides an identification

$$\mathrm{R}_{K_a/k}(\widetilde{\mathscr{G}}'_a)/\mathrm{R}_{K_a/k}(\mu) = \mathscr{D}(\mathrm{R}_{K_a/k}(\mathscr{G}'_a)).$$

Letting $\widetilde{\mathscr{T}}'_a \subseteq \widetilde{\mathscr{G}}'_a$ denote the q_a-preimage of the maximal K_a-torus $\mathscr{T}'_a = (T \cap G_a)_{K_a} \subseteq \mathscr{G}'_a$, we have

$$\xi_{G_a}(Z_{G_a}(T \cap G_a)) = \mathrm{R}_{K_a/k}(\widetilde{\mathscr{T}}'_a)/\mathrm{R}_{K_a/k}(\mu).$$

In this adjoint case the coroot $a^\vee : \mathrm{GL}_1 \to T \cap G_a$ is an isogeny between 1-dimensional tori and has kernel of order 2, so over K_a there is a unique factorization

$$\begin{array}{ccc} \mathrm{GL}_1 & \longrightarrow & \widetilde{\mathscr{T}}'_a \\ & {}_{a^\vee_{K_a}}\searrow & \downarrow{\scriptstyle q_a} \\ & & \mathscr{T}'_a \end{array}$$

in which the horizontal map is an isomorphism for degree reasons. Thus, once again $\xi_{G_a}(Z_{G_a}(T \cap G_a))$ is the image of $\mathrm{R}_{K_a/k}(a^\vee_{K_a})$.

Now consider the action of $Z_{G_a}(T \cap G_a)$ on the b-root group of (G, T) for any $b \in \Phi$. This b-root group coincides with that of $(G_b, T \cap G_b)$, which the surjective homomorphism $\xi_{G_b} : G_b \twoheadrightarrow \mathscr{D}(\mathrm{R}_{K_b/k}(\mathscr{G}'_b))$ carries onto $\mathrm{R}_{K_b/k}(\mathscr{U}'_b)$ where $\mathscr{U}'_b$ is the b-root group in $\mathscr{G}'_b$ that is a K_b-descent of the b_K-root group U'_b of (G', T'). (The intervention of the derived group may shrink Cartan subgroups, but preserves root groups.) Applying i_G and using (6.3.1), we conclude that under the natural action of $\mathrm{R}_{K/k}(\mathrm{GL}_1)$ on $\mathrm{R}_{K/k}(U'_b)$ induced by $a^\vee_K : \mathrm{GL}_1 \to T'$ and the action of $\mathrm{R}_{K/k}(T')$ on $\mathrm{R}_{K/k}(G')$ (i.e., $t.x = t^{\langle b, a^\vee \rangle}x$), the k-subgroup $\mathrm{R}_{K_a/k}(\mathrm{GL}_1)$ preserves $\mathrm{R}_{K_b/k}(\mathscr{U}'_b)$.

Upon fixing a K_b-group isomorphism $\mathscr{U}'_b \simeq \mathbf{G}_\mathrm{a}$ we recover on $\mathscr{U}'_b(K) = U'_b(K)$ the natural $K^\times$-action arising from the natural 1-dimensional K-vector space structure on $U'_b(K)$, so the restriction to $\mathrm{R}_{K_a/k}(\mathrm{GL}_1)(k) = K_a^\times$ of the $\langle b, a^\vee \rangle$-power map on $K^\times$ preserves the subgroup $\mathrm{R}_{K_b/k}(\mathscr{U}'_b)(k) = \mathscr{U}'_b(K_b)$ inside of $\mathscr{U}'_b(K) = U'_b(K)$. But this subgroup is a K_b-line, and the only $K^\times$-scalings on a 1-dimensional K-vector space which preserve a K_b-line are scalings by $K_b^\times$. Hence, the $\langle b, a^\vee \rangle$-power map on $K_a^\times$ takes values in $K_b^\times$. Likewise, the $\langle a, b^\vee \rangle$-power map on $K^\times$ carries $K_b^\times$ into $K_a^\times$. Since there are distinct root lengths, there is a non-orthogonal pair $\{a, b\}$ of simple positive roots (with respect to a fixed ordering on Φ) with a long and b short. Then $\langle b, a^\vee \rangle = -1$, so $K_> \subseteq K_<$ and hence $K_< = K$.

The edge joining the vertices corresponding to a and b in the Dynkin diagram has multiplicity 2 or 3, which is to say $\langle a, b^\vee \rangle$ is equal to -2 or -3 respectively. Hence, $K^2 \subseteq K_>$ or $K^3 \subseteq K_>$ respectively. Since k is separably closed we therefore get $K_> = K$ except possibly when $p = \mathrm{char}(k) \in \{2, 3\}$ and the Dynkin diagram has an edge with multiplicity p, in which case $kK^p \subseteq K_> \subseteq K$. □

Definition 6.3.3 Let k be a field. A pseudo-reductive k-group G is *exceptional* if it is absolutely pseudo-simple and the following properties hold:

(i) the root system Φ of G_{k_s} is reduced,
(ii) k is imperfect with $\mathrm{char}(k) = p \in \{2, 3\}$, and the Dynkin diagram of Φ has an edge of multiplicity p,
(iii) for every $a \in \Phi$ the map $\xi_{(G_{k_s})_a}$ is surjective,
(iv) $K_> \neq K$, where K/k is the field of definition of the geometric radical of G.

Note that when properties (i), (ii), and (iii) hold, Proposition 6.3.2 (and the remarks immediately following it) provides the existence of the field $K_>$ used in (iv). There is some redundancy in this definition: properties (i) and (iii) always hold when $p = 3$ (by Theorem 2.3.10 and Theorem 6.1.1). Later in Proposition 9.2.4 we will prove that (iii) is redundant when $p = 2$ if $[k : k^2] = 2$ since we assume reducedness of Φ in (i). If Φ has distinct root lengths, then Proposition 6.3.2 implies that (iv) holds, i.e., $K_> \neq K$, only if (ii) holds. The definition is formulated so that we can use it now for both values of p without logical dependence on results not yet proved when $p = 2$.

Under hypotheses that always hold away from characteristic 2 and that will later be proved to hold in characteristic 2 when $[k : k^2] = 2$, the exceptional pseudo-reductive groups account for all absolutely pseudo-simple k-groups that are not standard and have a reduced root system:

Theorem 6.3.4 *Let G be an absolutely pseudo-simple group over a field k, and let T be a maximal k-torus in G. Assume that $\Phi(G_{k_s}, T_{k_s})$ is reduced and that $\xi_{(G_{k_s})_a}$ is surjective for all $a \in \Phi(G_{k_s}, T_{k_s})$, where $(G_{k_s})_a \subseteq G_{k_s}$ is the pseudo-simple k_s-subgroup of rank 1 generated by the $\pm a$-root groups.*

The pseudo-reductive k-group G is standard if and only if G is not exceptional.

Proof We may assume that k is separably closed. By reducedness of the root system and Proposition 5.3.10, standardness of G is equivalent to the surjectivity of ξ_G.

Assuming that G is standard, we will prove that the inclusion $K_a \subseteq K$ is an equality for all $a \in \Phi(G,T)$ (and hence G is not exceptional). Since G is pseudo-simple, standardness implies (by Remark 5.3.6) that $G \simeq \mathrm{R}_{K/k}(\widetilde{G}')/Z$ for a connected semisimple K-group $\widetilde{G}'$ that is (absolutely) simple and simply connected, with Z a k-subgroup of $\mathrm{R}_{K/k}(Z_{\widetilde{G}'})$ such that $C := \mathrm{R}_{K/k}(\widetilde{T}')/Z$ is pseudo-reductive for some (equivalently, any) maximal K-torus $\widetilde{T}'$ of $\widetilde{G}'$. Note that the pseudo-reductivity of C is also a consequence of the pseudo-reductivity of the quotient $G = \mathrm{R}_{K/k}(\widetilde{G}')/Z$. Indeed, $\mathrm{R}_{K/k}(\widetilde{T}')$ is a Cartan k-subgroup of $\mathrm{R}_{K/k}(\widetilde{G}')$ (Proposition A.5.15(3)), so its image C is a Cartan k-subgroup of G (Proposition A.2.8), and Cartan k-subgroups of pseudo-reductive groups are always pseudo-reductive (Proposition 1.2.4). The idea for proving $K_a = K$ for all a in the standard case is to reduce to the case of rank 1.

By Proposition 4.1.4(2), the choice of $\widetilde{T}'$ corresponds canonically to a choice of maximal k-torus T in G, and by Corollary 4.1.6 we naturally have $\Phi(G,T) = \Phi(\widetilde{G}',\widetilde{T}')$. Beware that the canonical map $\widetilde{G}' \to G'$ carries $\widetilde{T}'$ onto T' with nontrivial kernel if G' is not simply connected, so $\mathrm{X}(T)$ is just a finite-index subgroup of $\mathrm{X}(\widetilde{T}')$. The precise relationship is that the maximal k-torus $\mathscr{T}$ in $\mathrm{R}_{K/k}(\widetilde{T}')$ maps isogenously onto T in the quotient G of $\mathrm{R}_{K/k}(\widetilde{G}')$, since this quotient map is j as in Remark 5.3.6. We also have $\Phi(\widetilde{G}',\widetilde{T}') = \Phi(\mathrm{R}_{K/k}(\widetilde{G}'),\mathscr{T})$ via the equality $\mathrm{X}(\widetilde{T}') = \mathrm{X}(\mathscr{T})$ arising from the natural identification of $\mathscr{T}_K$ with $\widetilde{T}'$.

For each root a in the common root system Φ of (G,T), $(\widetilde{G}',\widetilde{T}')$, and $(\mathrm{R}_{K/k}(\widetilde{G}'),\mathscr{T})$, define T_a, $\widetilde{T}'_a$, and $\mathscr{T}'_a$ to be the codimension-1 tori contained in the kernel of a in the respective groups. Since Φ is reduced, the root groups $U_{\pm a}$ in G are the only ones with trivial T_a-action, and similarly for the root groups $\widetilde{U}'_{\pm a}$ in $\widetilde{G}'$ (relative to $\widetilde{T}'_a$) and the root groups $\mathrm{R}_{K/k}(\widetilde{U}'_{\pm a})$ in $\mathrm{R}_{K/k}(\widetilde{G}')$ (relative to $\mathscr{T}'_a$). Thus, the quotient map $j : \mathrm{R}_{K/k}(\widetilde{G}') \to G$ carries $\mathrm{R}_{K/k}(\widetilde{U}'_{\pm a})$ onto $U_{\pm a}$ (Remark 2.3.6) for all a. But $\widetilde{G}'$ is simply connected, so $\widetilde{G}'_a \simeq \mathrm{SL}_2$. Hence, $\widetilde{G}'_a(K)$ is generated by its subgroups $\widetilde{U}'_{\pm a}(K)$, so $\mathrm{R}_{K/k}(\widetilde{G}')_a = \mathrm{R}_{K/k}(\widetilde{G}'_a)$, and this group is mapped onto G_a by j. This says that the pseudo-reductive subgroup G_a is standard with the same associated field K/k in a standard presentation. Thus, the link between the parameters in standard presentations and the simply connected datum (Proposition 4.2.4) implies that K/k is the field of definition for the maximal adjoint semisimple quotient of $(G_a)_{\overline{k}}$. By perfectness of G_a, this field coincides with the field of definition K_a/k of the geometric radical of G_a (Proposition 5.3.3). Thus we have proved that $K_a = K$ for all a when G is standard.

Now we assume G is not exceptional (so $K_a = K$ for all a by Proposition 6.3.2) and prove that G is standard. Since we assume that all maps ξ_{G_a} are surjective, it follows from the commutativity of (6.3.1) that for every $a \in \Phi(G,T)$ the image group $\xi_G(G_a)$, inside of $\mathrm{R}_{K/k}(G')$, is equal to $\mathrm{R}_{K/k}(G'_a)$. The full image of ξ_G is therefore a smooth connected k-subgroup of $\mathrm{R}_{K/k}(G')$ that contains the k-subgroups $\mathrm{R}_{K/k}(G'_a)$ as we vary through all $a \in \Phi(G,T) = \Phi(G',T')$. Thus, it is enough to show that these k-subgroups of $\mathrm{R}_{K/k}(G')$ generate the target $\mathrm{R}_{K/k}(\widetilde{G}')/\mathrm{R}_{K/k}(\mu) \subseteq \mathrm{R}_{K/k}(G')$ of ξ_G, with μ denoting the kernel of the simply connected central covering map $\pi : \widetilde{G}' \to G' = G_K^{\mathrm{ss}}$.

The preimage $\widetilde{T}' = \pi^{-1}(T')$ is a maximal K-torus of $\widetilde{G}'$. For each $a \in \Phi(G',T')$ there is a unique root $\widetilde{a} \in \Phi(\widetilde{G}',\widetilde{T}')$ such that the root group $\widetilde{U}'_{\widetilde{a}}$ in $\widetilde{G}'$ relative to $\widetilde{T}'$ lies in $\pi^{-1}(U'_a)$, and π restricts to an isomorphism $\widetilde{U}'_{\widetilde{a}} \simeq U'_a$. Thus, likewise $\mathrm{R}_{K/k}(\pi)$ restricts to an isomorphism $\mathrm{R}_{K/k}(\widetilde{U}'_{\widetilde{a}}) \simeq \mathrm{R}_{K/k}(U'_a)$. The map $a \mapsto \widetilde{a}$ is a bijection between the root systems $\Phi(G',T')$ and $\Phi(\widetilde{G}',\widetilde{T}')$, so it suffices to observe that since $\mathrm{R}_{K/k}(\widetilde{G}')$ is a perfect pseudo-reductive group (Corollary A.7.11), by Lemma 3.1.5 it is generated by the root groups $\mathrm{R}_{K/k}(\widetilde{U}'_{\widetilde{a}})$ for $\widetilde{a} \in \Phi(\widetilde{G}',\widetilde{T}')$. □

Corollary 6.3.5 *Let k be a field with* char$(k) \neq 2$, *and G a pseudo-reductive k-group. If k is imperfect with* char$(k) = 3$, *assume $G_{\overline{k}}^{\mathrm{ss}}$ has no factor of type* G_2. *Then G is standard over k.*

Proof By Corollary 5.2.3 and Proposition 5.3.1 we may assume that $k = k_s$ and that G is (absolutely) pseudo-simple over k. Theorem 2.3.10 implies reducedness of the root system $\Phi = \Phi(G,T)$ (for any maximal k-torus T in G), and by Theorem 6.1.1 all maps ξ_{G_a} ($a \in \Phi$) are isomorphisms. Thus, by Theorem 6.3.4, a necessary condition for G to be non-standard is that k is imperfect and the Dynkin diagram of Φ has an edge with multiplicity $p = \mathrm{char}(k) \in \{2,3\}$. But we assume $\mathrm{char}(k) \neq 2$, and that if k is imperfect with characteristic 3 then the irreducible Dynkin diagram of Φ is not G_2 (the unique one with an edge of multiplicity 3). □

We end with a standardness result that is only of interest over imperfect fields of characteristic 2, where it gives the best analogue of Corollary 6.3.5 (and, together with Corollary 6.3.5, finishes the proof of Theorem 5.1.1(1)).

Proposition 6.3.6 *Let k be a field with* char$(k) \neq 3$, *and G a pseudo-reductive k-group such that the Dynkin diagram of $G_{\overline{k}}^{\mathrm{ss}}$ has no isolated vertices and no edges with multiplicity* 2. *The pseudo-reductive k-group G is standard.*

Proof We may assume that k is separably closed (Corollary 5.2.3), and we may also assume that G is not commutative and hence is non-solvable. By Proposition 5.3.1 we can assume that G is pseudo-simple (so $\mathscr{D}(G) = G$ and $G^{ss}_{\overline{k}}$ is simple). Finally, by Proposition 6.2.1 we can assume that the simple group $G^{ss}_{\overline{k}}$ is not of type A_2 or G_2, and by hypothesis it is not of type $C_1 = A_1$ nor of type $C_2 = B_2$. Hence, it has rank at least 3.

Lemma 6.1.2 further reduces our problem to the case where the covering group μ for the simply connected central cover $\widetilde{G}'$ of $G' := G^{red}_K = G^{ss}_K$ is infinitesimal (hence of degree not divisible by 3, since $\mathrm{char}(k) \neq 3$). Let T be a maximal k-torus in G, so $T' = T_K$ may be viewed as a maximal K-torus of G'. The scheme-theoretic preimage $\widetilde{T}'$ of T' in $\widetilde{G}'$ is a maximal K-torus of $\widetilde{G}'$ with $\widetilde{T}'/\mu = T'$ in $\widetilde{G}'/\mu = G'$. Let $\Phi = \Phi(G', T') \subseteq \mathrm{X}(T') = \mathrm{X}(T)$, so Φ has rank at least 3.

The list of options for the irreducible Dynkin diagram ruled out from consideration implies that Φ is not of type B_n, C_n ($n \geqslant 1$), or F_4, so $\Phi = \Phi(G, T)$ is reduced and all roots are of the same length. In particular, the Weyl group $W(G, T)(k) = W(\Phi)$ (see Proposition 3.2.7) acts transitively on Φ and for all $a \in \Phi$ the field of definition over k for the geometric radical of G_a is K (Proposition 6.3.2(1)). Moreover, reducedness of the root system implies that to prove standardness it is enough to show that ξ_G is surjective (Proposition 5.3.10).

We fix a positive system of roots $\Phi^+ \subseteq \Phi$, and choose a simple positive root a, as well as another simple positive root b that is adjacent to a in the Dynkin diagram associated to (Φ, Φ^+). Define the k-subtori

$$T_a = (\ker a)^0_{\mathrm{red}}, \quad T_{a,b} = ((\ker a) \cap (\ker b))^0_{\mathrm{red}}$$

of T, and let G_a and $G_{a,b}$ respectively denote the pseudo-simple derived groups of the centralizers $Z_G(T_a)$ and $Z_G(T_{a,b})$. The root systems of (G_a, T_a) and $(G_{a,b}, T_{a,b})$ are of type A_1 and A_2 respectively (as a and b have the same length).

Let $K_{a,b}/k$ be the finite extension that is the field of definition over k for the (unipotent) radical $\mathscr{R}((G_{a,b})_{\overline{k}})$. Since $\Phi(G, T)$ is reduced, G_a can also be described as the k-subgroup of G generated by the root groups U_a and U_{-a} (Proposition 3.4.1(2)). We can therefore determine G_a in exactly the same way using $(G_{a,b}, T \cap G_{a,b})$ in place of (G, T), so $K_a = K_{a,b}$. Hence, $K_{a,b} = K$.

The absolutely pseudo-simple group $G_{a,b}$ is of type A_2 so it is standard with $\xi_{G_{a,b}}$ an isomorphism by Proposition 6.2.1. Since $\mathrm{char}(k) \neq 3$, the maps $i_{G_{a,b}}$ and $\xi_{G_{a,b}}$ coincide (see Definition 5.3.5). The isomorphism property of $i_{G_{a,b}}$

implies that the root groups in $G_{a,b}$ have the form $\mathrm{R}_{K/k}(\mathbf{G}_{\mathrm{a}})$. These root groups have dimension $[K : k]$, and the $\pm a$-root groups of (G, T) coincide with those of both $(G_a, T \cap G_a)$ and $(G_{a,b}, T \cap G_{a,b})$. Since these root groups of $(G_a, T \cap G_a)$ are its only ones, the root groups of $(G_a, T \cap G_a)$ have dimension $[K : k]$.

Let

$$G'_a := (G_a)^{\mathrm{ss}}_{K} = (G_a)_K / \mathscr{R}((G_a)_K), \quad G'_{a,b} := (G_{a,b})^{\mathrm{ss}}_{K} = (G_{a,b})_K / \mathscr{R}((G_{a,b})_K)$$

denote the maximal semisimple quotients (over their minimal fields of definition over k) of $(G_a)_K$ and $(G_{a,b})_K$. Using Proposition A.4.8, we view G'_a and $G'_{a,b}$ as subgroups of G' with $G'_a \subseteq G'_{a,b}$. The map $i_{G_a} : G_a \to \mathrm{R}_{K/k}(G'_a)$ carries the $\pm a$-root groups of G_a into those of $\mathrm{R}_{K/k}(G'_a)$ (apply Remark 2.3.6 to $(G_a)_K \twoheadrightarrow G'_a$), yet $\ker i_{G_a} = 1$ due to the commutative diagram of natural maps

$$\begin{array}{ccc} G_a & \longrightarrow & G_{a,b} \\ {\scriptstyle i_{G_a}}\downarrow & & \simeq \downarrow {\scriptstyle i_{G_{a,b}} = \xi_{G_{a,b}}} \\ \mathrm{R}_{K/k}(G'_a) & \longrightarrow & \mathrm{R}_{K/k}(G'_{a,b}) \end{array}$$

But we have seen that the root groups of $(G_a, T \cap G_a)$ have dimension $[K : k]$, as do those of the target $\mathrm{R}_{K/k}(G'_a)$, so the image of i_{G_a} contains the root groups of $\mathrm{R}_{K/k}(G'_a)$.

Consider the commutative diagram

$$\begin{array}{ccc} G_a & \longrightarrow & G \\ {\scriptstyle i_{G_a}}\downarrow & & \downarrow {\scriptstyle i_G} \\ \mathrm{R}_{K/k}(G'_a) & \longrightarrow & \mathrm{R}_{K/k}(G') \end{array}$$

We just proved that the image of the map along the left contains $\mathrm{R}_{K/k}(U'_a)$ for the root group U'_a of (G', T'). Any $a \in \Phi$ is simple positive for some Φ^+, and the natural map $\widetilde{G}' \to G'$ restricts to an isomorphism between root groups (relative to $\widetilde{T}'$ and T' respectively), so it follows from the factorization

$$i_G : G \stackrel{\xi_G}{\to} \mathrm{R}_{K/k}(\widetilde{G}')/\mathrm{R}_{K/k}(\mu) \hookrightarrow \mathrm{R}_{K/k}(\widetilde{G}'/\mu) = \mathrm{R}_{K/k}(G')$$

that the image of ξ_G contains the images of the Weil restrictions of all root groups of $(\widetilde{G}', \widetilde{T}')$. Thus, to deduce the surjectivity of ξ_G it suffices to prove that

$R_{K/k}(\widetilde{G}')$ is generated by the Weil restrictions of the root groups of $(\widetilde{G}', \widetilde{T}')$. As in the proof of Theorem 6.3.4, this follows from Corollary A.7.11 and Lemma 3.1.5. □

Remark 6.3.7 Note that Corollary 6.3.5 and Proposition 6.3.6 together imply Theorem 1.5.1 (which is Theorem 5.1.1(1)).

PART III

General classification and applications

7

The exotic constructions

7.1 Calculations in characteristics 2 and 3

Over imperfect fields k of characteristics 2 and 3 there are pseudo-reductive groups that are not standard. Interesting classes of such groups (to be called *exotic*) were discovered by Tits, resting on special properties of the group G_2 in characteristic 3 and the group F_4 in characteristic 2, as well as groups in characteristic 2 of type B_n and C_n with $n \geqslant 2$. A basic common feature of these types is that the unique pair of non-orthogonal positive simple roots $\{a, b\}$ with distinct root lengths, say with a long and b short, satisfies $\langle a, b^\vee \rangle = -p$. In other words, the Dynkin diagram has an edge with multiplicity p.

The definition of exotic pseudo-reductive k-groups requires a lot of preparation (essentially all of §§7.1 and 7.2), due to the intervention of subtleties related to the field of definition over k of a Levi k_s-subgroup (see Example 7.2.2). The case of G_2 provides everything we need for a complete result in characteristic 3.

Tits' method for constructing exotic examples in characteristics 2 and 3 uses general axiomatic arguments with root systems that he sketched in [Ti2, §5] over a separably closed field (and applied to the root systems of a split G_2 in characteristic 3 and a split F_4 in characteristic 2, as well as variants for types B and C in rank $\geqslant 2$ in characteristic 2). We will use an alternative approach that is well suited to working with arbitrary k-forms of type G_2 (resp. types F_4 and B_n and C_n with $n \geqslant 2$) for any imperfect k of characteristic 3 (resp. 2). This section largely focuses on the calculations with (forms of) these types that underlie the constructions to be given in §7.2.

Let k be an arbitrary field of characteristic $p \in \{2, 3\}$, and let G be a connected semisimple k-group that is absolutely simple and simply connected with Dynkin diagram having an edge with multiplicity p. We do not assume that G is k-split.

Fix a maximal k-torus T in G. Let $\mathfrak{g} = \mathrm{Lie}(G)$ and $\mathfrak{t} = \mathrm{Lie}(T)$, and let k' be a separable algebraic extension of k over which T splits. Let $T' = T_{k'}$ and $G' = G_{k'}$. For each root $c \in \Phi := \Phi(G', T')$, let U'_c be the associated root group in G', which is to say the unique T'-stable smooth connected unipotent k'-subgroup of G' whose Lie algebra is the 1-dimensional c-root space $\mathfrak{g}'_c$ in $\mathfrak{g}' = \mathrm{Lie}(G') = \mathfrak{g}_{k'}$.

There are two lengths for the roots in Φ, and we call a root group U'_c *short* (resp. *long*) when c is a short root (resp. long root). We also write $\Phi_<$ (resp. $\Phi_>$) to denote the set of short (resp. long) roots in Φ. If Φ^+ is a positive system of roots then we define $\Phi^+_< = \Phi^+ \cap \Phi_<$ and $\Phi^+_> = \Phi^+ \cap \Phi_>$. We write Δ for the set of simple positive roots in Φ^+, and $\Delta_<$ (resp. $\Delta_>$) to denote the set of short roots (resp. long roots) in Δ.

Since G is assumed to be simply connected, the coroots $c^\vee$ for $c \in \Delta$ form a basis of the cocharacter group $\mathrm{X}_*(T')$. In particular, in such cases k'-subgroups $c^\vee : \mathrm{GL}_1 \hookrightarrow T'$ for $c \in \Delta$ have tangent lines $\mathrm{Lie}(c^\vee(\mathrm{GL}_1))$ that form a k'-basis of $\mathfrak{t}' := \mathrm{Lie}(T') = \mathfrak{t}_{k'}$. (Recall that for any $c \in \Phi$, $c^\vee(\mathrm{GL}_1)$ is the unique 1-dimensional k'-torus that normalizes U'_c and U'_{-c} and lies in the SL_2 generated by U'_c and U'_{-c}. This is essentially how the coroots $c^\vee$ are defined.) Likewise, the roots $c \in \Delta$ form a basis of the character group $\mathrm{X}(T'/Z_{G'})$ of the k'-split maximal k'-torus $T'/Z_{G'}$ in the adjoint quotient $G'/Z_{G'}$. In the specified characteristics something remarkable happens:

Lemma 7.1.1 *Let Φ^+ be a positive system of roots in Φ, and Δ the basis of simple positive roots.*

(1) *The k'-lines* $\mathrm{Lie}(c^\vee(\mathrm{GL}_1)) \subseteq \mathfrak{t}_{k'}$ *for $c \in \Delta_<$ directly span a subspace that coincides with the span of the k'-lines* $\mathrm{Lie}(c^\vee(\mathrm{GL}_1))$ *for all $c \in \Phi_<$. In particular, this subspace is independent of the ordering on Φ.*
(2) *For all $a \in \Phi_>$ and $b \in \Phi_<$, $\langle a, b^\vee \rangle \in p\mathbf{Z}$.*
(3) *Under the isomorphism $T'/Z_{G'} \simeq \prod_{c\in\Delta} \mathrm{GL}_1$ defined by $\bar{t} \mapsto (c(\bar{t}))$, the subgroup $\prod_{c\in\Delta_<} \mu_p$ goes over to the intersection in $(T'/Z_{G'})[p]$ of the kernels of the long roots in Φ when these roots are viewed as characters on $T'/Z_{G'}$, and it is not annihilated by any short root. In particular, this k'-subgroup is independent of the ordering on Φ.*

By Galois descent, the k'-subspace in (1) and the k'-subgroup in (3) respectively descend to a canonical k-subspace in $\mathfrak{t}$ and a k-subgroup in $(T/Z_G)[p]$ (independent of k'/k); we respectively denote them as $\mathfrak{t}_<$ and $\mu_<$.

Proof We may assume $k' = k$. Write

$$\Delta_< = \{b_j\}, \quad \Delta_> = \{a_i\}.$$

By [Bou2, VI, §1.2, Prop. 7] and the irreducibility of Φ, up to scaling there is a unique $W(\Phi)$-invariant inner product $(\cdot,\cdot)$ on the vector space $\mathrm{X}(T')_{\mathbf{R}}$. Fix a choice of such an inner product, and let $a_0 \in \Delta_>$ and $b_0 \in \Delta_<$ be the unique long and short roots in Δ that are not orthogonal under this inner product (equivalently, they correspond to adjacent vertices in the Dynkin diagram). By inspection, in each case we have $\langle a_0, b_0^\vee\rangle = -p$ and $\langle b_0, a_0^\vee\rangle = -1$, so $(a_0, a_0) = p(b_0, b_0)$. It follows that $(a,a) = p(b,b)$ for any $a \in \Phi_>$ and $b \in \Phi_<$. We use the inner product to identify the dual spaces $\mathrm{X}_*(T')_{\mathbf{R}}$ and $\mathrm{X}(T')_{\mathbf{R}}$, and this carries $c^\vee$ to $2c/(c,c)$ for all $c \in \Phi$. In other words, it identifies the *inverse root system* $(\mathrm{X}_*(T')_{\mathbf{R}}, \Phi^\vee)$ (terminology from [Bou2, VI, §1.1]) with the *dual root system* $(\mathrm{X}(T')_{\mathbf{R}}, \{2c/(c,c)\}_{c\in\Phi})$ (terminology from [Hum1, 4.2]).

The positive system of coroots $\{c^\vee\}_{c\in\Phi^+}$ has basis $\Delta^\vee = \{c^\vee\}_{c\in\Delta}$, so the sets of simple positive short and long coroots are respectively given by

$$\Delta^\vee_< = (\Delta_>)^\vee := \{a_i^\vee\}, \quad \Delta^\vee_> = (\Delta_<)^\vee := \{b_j^\vee\}.$$

Thus, if $a \in \Phi_>$ with unique expression $a = \sum m_i a_i + \sum n_j b_j$ for integers $m_i, n_j \in \mathbf{Z}$ then

$$a^\vee = \frac{2a}{(a,a)} = \sum_i m_i \cdot \frac{2a_i}{(a,a)} + \sum_j n_j \cdot \frac{2b_j}{(a,a)} = \sum_i m_i a_i^\vee + \sum_j (n_j/p) b_j^\vee.$$

But these coefficients must be integral, so $p|n_j$ for all j; i.e., each long root has its $\Delta_<$ coefficients in $p\mathbf{Z}$. The same argument applied to short roots gives that if $b \in \Phi_<$ with unique expression $b = \sum m_i a_i + \sum n_j b_j$ then $b^\vee = \sum p m_i a_i^\vee + \sum n_j b_j^\vee$. Hence, the element $b^\vee \in \Phi^\vee_>$ has $\Delta^\vee_<$ coefficients in $p\mathbf{Z}$, so $\mathrm{Lie}(b^\vee(\mathrm{GL}_1))$ in characteristic p has no contribution from $\Delta^\vee_< = (\Delta_>)^\vee$. This proves (1).

The assertion in (2) is a well-known property of root systems that are reduced and irreducible (in view of our hypotheses on the edge multiplicities in the Dynkin diagram). To prove (3), by construction it is clear that $\mu_<$ contains the intersection of the kernels in $(T'/Z_{G'})[p]$ for the long roots. To prove that this containment is an equality, we just have to show that $\mu_<$ is killed by all long roots. But we saw above that every $a \in \Phi_>$ has $\Delta_<$ coefficients in $p\mathbf{Z}$, so $a(\mu_<) = 1$ by definition of $\mu_<$.

It remains to show that if $c \in \Phi_<$ then $c(\mu_<)$ is nontrivial. Since $N_G(T)(k)$ acts transitively on the set of positive systems of roots in Φ, and for some choice the root c is a simple positive root, it suffices to check for any $g \in N_G(T)(k)$ that the expansion of $\mu_<$ in $(T'/Z_{G'})[p]$ relative to the μ_p-decomposition indexed by the basis $g(\Delta)$ has vanishing components along

$g(\Delta_>)$ (as then by consideration of order of the group scheme it must be $\prod_{c \in g(\Delta_<)} \mu_p$). But this vanishing is a special case of the fact that $\mu_<$ is killed by all long roots in Φ. □

With the maximal k-torus T in G fixed, we let $\mathfrak{g}_< \subseteq \mathfrak{g}$ denote the k-descent of the k'-span of the root spaces in $\mathfrak{g}'$ for roots in $\Phi_<$ and we likewise let $\mathfrak{g}_> \subseteq \mathfrak{g}$ denote the k-descent of the k'-span of the root spaces in $\mathfrak{g}'$ for the roots in $\Phi_>$. Both $\mathfrak{g}_<$ and $\mathfrak{g}_>$ are independent of k'/k, and neither is a Lie subalgebra of $\mathfrak{g}$ (since any pair of opposite root groups $U_{\pm c}$ generate an SL_2 in the usual manner with diagonal $c^\vee(\mathrm{GL}_1)$, as G is simply connected, and the Lie bracket between opposite root spaces in $\mathfrak{sl}_2$ is non-vanishing and valued in the diagonal subalgebra).

There is a remarkable nontrivial k-isogeny $\pi : G \to \overline{G}$ onto a simply connected absolutely simple k-group whose root system is isomorphic to the dual of the root system Φ of G. For the convenience of the reader and to set some notation to be used later, we recall how this is constructed. The key point is to give a nontrivial factorization over k of the Frobenius isogeny $F_{G/k} : G \to G^{(p)}$. This rests on the following lemma.

Lemma 7.1.2 *Every proper G-submodule of $\mathfrak{g}$ under the adjoint action of G is contained in $\mathfrak{t} \oplus \mathfrak{g}_<$, and the subspace $\mathfrak{n} = \mathfrak{t}_< \oplus \mathfrak{g}_< \subseteq \mathfrak{g}$ is a G-stable p-Lie subalgebra of $\mathfrak{g}$. Every nonzero G-submodule of $\mathfrak{g}$ distinct from $\mathfrak{z} :=$ Lie(Z_G) contains $\mathfrak{n}$. In particular, except for types B_n $(n \geqslant 2)$ and C_{2m} $(m \geqslant 2)$, $\mathfrak{n}$ is the unique irreducible G-submodule of $\mathfrak{g}$.*

If G is of type B_n (with $n \geqslant 2$) or of type C_n (with n even), the center $Z_G = \mu_2$ is contained in $\prod_{c \in \Delta_<} c^\vee(\mathrm{GL}_1)$, so $\mathfrak{z} \subseteq \mathfrak{t}_< \subseteq \mathfrak{n}$ in these cases. Also, for types F_4 and G_2 the subspace $\mathfrak{n}$ is the unique nonzero proper G-submodule of $\mathfrak{g}$.

Proof We may assume that $k = k_s$, and we first address the uniqueness aspect. Since a G-stable subspace in $\mathfrak{g}$ is a Lie ideal (as $\mathrm{ad}_{\mathfrak{g}}$ is obtained by differentiating the adjoint representation of G; see Proposition A.7.5), we are aiming to find nonzero proper G-stable Lie ideals $\mathfrak{h} \subseteq \mathfrak{g}$. The Weyl group $W(G,T)(k)$ maps isomorphically onto the Weyl group of the irreducible and reduced root system Φ, so $N_G(T)(k)$ acts transitively on all roots of a given length. Hence, by consideration of T-weight spaces it follows that a G-stable subspace of $\mathfrak{g}$ must be a direct sum among $\mathfrak{g}_<$, $\mathfrak{g}_>$, and a subspace of $\mathfrak{t}$. Each pair of opposite root groups for (G,T) generates an SL_2 since G is simply connected, so each pair of root spaces in $\mathfrak{g}$ for opposite roots generates an $\mathfrak{sl}_2$. We conclude that if a G-stable subspace $\mathfrak{h}$ contains a root space then it contains the $\mathfrak{sl}_2$ arising from such a root. In particular, if $\mathfrak{h} \neq \mathfrak{g}$ then it cannot contain both $\mathfrak{g}_<$ and $\mathfrak{g}_>$.

Choose a positive system of roots and consider the unique pair $\{a, b\}$ of non-orthogonal simple positive roots, with a long and b short. For the codimension-2 subtorus

$$S = (\ker a \cap \ker b)^0_{\mathrm{red}},$$

the connected semisimple group $\mathscr{D}(Z_G(S))$ is of rank 2 with root datum of type B_2 when $p = 2$ and of type G_2 when $p = 3$ (as $\{a, b\}$ is a basis of the root system for $(Z_G(S), T)$ and $\{a^\vee, b^\vee\}$ is a basis of the corresponding coroot system). The Chevalley commutation relations among the root groups in these two cases [Hum2, §33.4(b), §33.5(b)] (where α is short and β is long!) give that $[\mathfrak{g}_a, \mathfrak{g}_b] = \mathfrak{g}_{a+b}$, and $a+b$ is a short root. Thus, if a G-stable subspace of $\mathfrak{g}$ contains $\mathfrak{g}_>$ then it contains $\mathfrak{g}_<$. We conclude that a proper G-stable subspace $\mathfrak{h}$ must be contained in $\mathfrak{t} \oplus \mathfrak{g}_<$.

If $\mathfrak{h} \subseteq \mathfrak{t} = \mathrm{Lie}(T)$ then by $G(k)$-conjugacy of all maximal k-tori we have $\mathfrak{h} \subseteq \bigcap_T \mathrm{Lie}(T) = \mathrm{Lie}(\bigcap_T T)$. But G is generated by its maximal k-tori (as they generate a smooth connected normal k-subgroup), so $\bigcap_T T = \bigcap_T Z_G(T) = Z_G$. In other words, in such cases $\mathfrak{h} \subseteq \mathfrak{z}$. This vanishes for F_4 and G_2, and it is 1-dimensional for types B and C. Thus, all other $\mathfrak{h}$ (if any exist) contain $\mathfrak{t}_< \oplus \mathfrak{g}_<$.

It remains to prove that $\mathfrak{n} := \mathfrak{t}_< \oplus \mathfrak{g}_<$ is a G-stable p-Lie subalgebra. Since $\mathfrak{n}$ is spanned by the Lie algebras of smooth k-subgroups of G, by functoriality of the p-operation and the universal Lie-theoretic formula (see [Bo2, 3.1(iii)]) for its interaction with addition we see that $\mathfrak{n}$ is stable under the p-operation provided that $\mathfrak{n}$ is a Lie subalgebra of $\mathfrak{g}$. This will even be a Lie ideal if it is G-stable, so it remains to check G-stability of $\mathfrak{n}$.

It suffices to check stability under the actions of T and the root groups U_c for $c \in \Phi$. Stability under the T-action is obvious. Choose $c \in \Phi$ and $c' \in \Delta_<$, and fix an isomorphism $x_c : \mathbf{G}_\mathrm{a} \simeq U_c$. Consider the basis vectors $X_c = dx_c(\partial_u|_{u=0}) \in \mathrm{Lie}(U_c)$ and $H_{c'} = dc'^\vee(\partial_t|_{t=1}) \in \mathrm{Lie}(c'^\vee(\mathrm{GL}_1))$. For $u \in \mathbf{G}_\mathrm{a}$, $\mathrm{Ad}(x_c(u))(H_{c'}) = -u\langle c, c'^\vee\rangle X_c + H_{c'}$. Only long c need to be considered (due to how $\mathfrak{n}$ is defined), and then for long c and short c' we have $\langle c, c'^\vee\rangle \in p\mathbf{Z}$ (Lemma 7.1.1).

To handle the behavior of $\mathfrak{g}_<$ under the adjoint action of U_c on $\mathfrak{g}$ for $c \in \Phi$, we have to study U_c-conjugation on $\mathrm{Lie}(U_{c'})$ for $c' \in \Phi_<$. If $c' = \pm c$ there is nothing to do since U_c and U_{-c} generate an SL_2 whose diagonal is $c^\vee(\mathrm{GL}_1)$ and hence has Lie algebra contained in $\mathfrak{n}$. Thus, we can assume c and c' are independent, and hence we can choose Φ^+ to contain c and c'.

Fix a k-isomorphism $x_b : \mathbf{G}_a \simeq U_b$ for all $b \in \Phi^+$. Letting (c, c') denote the set of roots in Φ of the form $ic + jc'$ with $i, j \geqslant 1$, equipped with a fixed ordering, there is an identity

$$x_c(s)x_{c'}(t)x_c(s)^{-1} = \prod_{ic+jc' \in (c,c')} x_{ic+jc'}(r_{i,j}s^i t^j) \cdot x_{c'}(t)$$

with $r_{i,j} \in k$. Passing to the t-derivative at $t = 0$ reduces us to the problem of proving that $r_{i,1} = 0$ if $ic + c'$ is a long root (with c' short). But such c and c' do not exist, by inspection of the root system they generate (with rank 2). □

By Proposition A.7.14, there is a unique k-subgroup scheme $N \subseteq \ker F_{G/k}$ whose Lie algebra is $\mathfrak{n}$ inside of $\mathrm{Lie}(\ker F_{G/k}) = \mathfrak{g}$, and by Example A.7.16 it is normal in G (due to Lemma 7.1.2). Another characterization of N (due to Lemma 7.1.2) is that it is the unique minimal non-central normal k-subgroup scheme of G with trivial Frobenius. Concretely, in case $k = k_s$ (or more generally, if T is k-split) the k-group N is directly spanned by the kernels of the Frobenius isogenies $U_c \to U_c^{(p)}$ for all short roots c and the Frobenius kernels $\ker F_{c^\vee(\mathrm{GL}_1)/k}$ for the short roots c contained in a basis of the root system.

Definition 7.1.3 With notation as above, the canonical quotient map $\pi : G \to G/N$ is the *very special isogeny* associated to the absolutely simple and simply connected k-group G. More generally, if k is a nonzero finite product of fields and G is a semisimple k-group whose fibers are connected, absolutely simple, and simply connected with Dynkin diagram having an edge of multiplicity p then we define the *very special isogeny* for G by performing this construction on fibers over $\mathrm{Spec}\, k$.

(We will generally write $\overline{G}$ to denote the quotient G/N, making sure to avoid confusion with the similar notation used to denote adjoint semisimple quotients of connected semisimple groups. We call $\overline{G}$ the *very special quotient* of G.)

The Frobenius isogeny $F_{G/k} : G \to G^{(p)}$ over the field k has kernel that contains N (or equivalently, the Lie algebra $\mathrm{Lie}(\ker F_{G/k}) = \mathfrak{g}$ contains $\mathfrak{n}$), so we get a unique factorization

$$G \xrightarrow{\pi} \overline{G} \xrightarrow{\overline{\pi}} G^{(p)} \tag{7.1.1}$$

of $F_{G/k}$ through π.

Remark 7.1.4 In general, if G is a connected semisimple absolutely simple group over a field k, an isogeny $G \to G'$ is called *special* (in the sense of

Borel and Tits) if the induced map $\mathfrak{g} \to \mathfrak{g}'$ on Lie algebras does not factor through the G-coinvariants of $\mathfrak{g}$. (Equivalently, if we fix a maximal k-torus T, the Lie algebra map is nonzero on the root space in $\mathfrak{g}_{\overline{k}}$ for some element of $\Phi(G_{\overline{k}}, T_{\overline{k}})$.) For example, every central isogeny is special and the Frobenius isogeny in nonzero characteristic is never special.

By construction, the map π in Definition 7.1.3 is special in the sense of Borel and Tits, thereby explaining our terminology. Borel and Tits classified the special isogenies between absolutely simple semisimple groups, and non-central ones only exist in characteristics 2 and 3.

It is important that we determine the root datum for $\overline{G}$. More precisely, if T is a k-split maximal k-torus of G and $\overline{T} := \pi(T)$ denotes the associated k-split maximal k-torus of $\overline{G}$ under the very special isogeny then we want to relate $\overline{\Phi} := \Phi(\overline{G}, \overline{T})$ to $\Phi := \Phi(G,T)$ and determine where $\overline{G}$ sits within its central isogeny class (e.g., whether it is simply connected or adjoint). For each $c \in \Phi$ (resp. $\overline{c} \in \overline{\Phi}$) we denote by U_c (resp. $\overline{U}_{\overline{c}}$) the corresponding root group.

Proposition 7.1.5 *The root system $\overline{\Phi}$ is dual to that of Φ, and $\overline{G}$ is simply connected. More precisely:*

(1) *For each long root $c \in \Phi_>$, the map π carries U_c isomorphically onto the root group $\overline{U}_{\overline{c}}$ for a short root $\overline{c} \in \overline{\Phi}_<$ satisfying $\overline{c} \circ \pi = c$ in $\mathrm{X}(T)$ and $\overline{c}^\vee = \pi \circ c^\vee$ in $\mathrm{X}_*(\overline{T})$.*
(2) *For each short root $c \in \Phi_<$ the map π restricts to a purely inseparable isogeny $U_c \to \overline{U}_{\overline{c}}$ of degree p (i.e., it induces an isomorphism $U_c/\ker(F_{U_c/k}) \simeq \overline{U}_{\overline{c}}$) for a long root $\overline{c} \in \overline{\Phi}_>$ satisfying $\overline{c} \circ \pi = pc$ in $\mathrm{X}(T)$ and $\overline{c}^\vee = (1/p)(\pi \circ c^\vee)$ in $\mathrm{X}_*(\overline{T})$.*
(3) *In the factorization (7.1.1), $\overline{\pi}$ is the very special isogeny of $\overline{G}$.*
(4) *The bijection $\Phi \to \overline{\Phi}$ defined by $c \mapsto \overline{c}$ carries the basis of a positive system of roots to the basis of the corresponding positive system of roots.*

For a reduced root system R spanning a finite-dimensional $\mathbf{R}$-vector space V and a choice of $W(R)$-invariant inner product $(\cdot,\cdot)$ on V (which is unique up to scaling factors associated to the irreducible components of R [Bou2, VI, §1.2, Prop. 7]), the *dual root system* is the subset $\{2a/(a,a)\}_{a\in R}$ in V. In the setting of the proposition, if we choose the inner product uniquely so that $(a,a) = 2$ for the long roots then (1) and (2) imply that $\overline{c} \mapsto \overline{c} \circ \pi$ identifies $\overline{\Phi}$ with the dual of Φ.

Proof Since π is bijective on geometric points and carries T onto $\overline{T}$, by Lemma 3.2.1 it carries $N_G(T)$ onto $N_{\overline{G}}(\overline{T})$ and induces an isomorphism $W(G,T) \to W(\overline{G}, \overline{T})$.

Root groups relative to a split maximal torus are characterized as being the minimal nontrivial smooth connected unipotent subgroups normalized by the torus. Hence, for each $c \in \Phi$ there is a unique $\overline{c} \in \overline{\Phi}$ such that $\pi(U_c) = \overline{U}_{\overline{c}}$. Thus, $c \mapsto \overline{c}$ defines a bijection $\Phi \to \overline{\Phi}$. This bijection is defined in terms of root groups, not in terms of the finite-index inclusion $\mathrm{X}(\overline{T}) \hookrightarrow \mathrm{X}(T)$ of character lattices, so we are not claiming that this correspondence is well behaved with respect to the additive structure (and it generally is not). However, since π carries Borel subgroups onto Borel subgroups, the bijection $\Phi \to \overline{\Phi}$ carries positive systems of roots to positive systems of roots.

The line $\mathrm{Lie}(U_c) \subseteq \mathrm{Lie}(G)$ is the c-root space, and $\mathrm{Lie}(\ker \pi) = \ker(\mathrm{Lie}(\pi)) = \mathfrak{t}_< \oplus \mathfrak{g}_<$ by construction of π. Hence, if c is long then $\pi : U_c \to \overline{U}_{\overline{c}}$ is an isomorphism. Likewise, if c is short then $\ker(\pi|_{U_c})$ is nontrivial yet is infinitesimal of height $\leqslant 1$ (as this is true for $\ker \pi = N$ by construction), so it is the unique such k-subgroup of $U_c \simeq \mathbf{G}_\mathrm{a}$, namely α_p. This has order p by inspection (or because U_c is smooth of dimension 1). Since the roots of a given length constitute a single Weyl orbit of roots, it follows that the map $c \mapsto \overline{c}$ either takes all short (resp. long) roots to short (resp. long) roots, or it maps short roots onto long roots and long roots onto short roots.

Now we choose $c \in \Phi$ and explicitly compute $\overline{c} \circ \pi$ in terms of c and compute $\pi \circ c^\vee$ in terms of $\overline{c}^\vee$. Fixing isomorphisms $x_c : \mathbf{G}_\mathrm{a} \simeq U_c$ and $x_{\overline{c}} : \mathbf{G}_\mathrm{a} \simeq \overline{U}_{\overline{c}}$, $\pi : U_c \to \overline{U}_{\overline{c}}$ becomes $u \mapsto au$ for some $a \in k^\times$ when c is long and $u \mapsto au^p$ for some $a \in k^\times$ when c is short. The T-action on U_c is through c and the $\overline{T}$-action on $\overline{U}_{\overline{c}}$ is through $\overline{c}$, so by equivariance of $\pi : U_c \to \overline{U}_{\overline{c}}$ with respect to $\pi : T \to \overline{T}$ we get $a \cdot c(t)u = \overline{c}(\pi(t)) \cdot au$ for $u \in \mathbf{G}_\mathrm{a}$ and $t \in T$ when c is long, and $a \cdot (c(t)u)^p = \overline{c}(\pi(t)) \cdot au^p$ for $u \in \mathbf{G}_\mathrm{a}$ and $t \in T$ when c is short. Hence, $\overline{c} \circ \pi$ is equal to c when c is long and to pc when c is short. In particular, $\overline{-c} = -\overline{c}$.

To relate $\overline{c}^\vee$ and $\pi \circ c^\vee$, we first observe that in the subgroup G_c generated by U_c and U_{-c}, $c^\vee(\mathrm{GL}_1)$ is the unique maximal k-torus normalizing $U_{\pm c}$. Since $\pi(G_c) = \overline{G}_{\overline{c}}$, it follows that $\pi(c^\vee(\mathrm{GL}_1)) = \overline{c}^\vee(\mathrm{GL}_1)$. Thus, $\pi \circ c^\vee = q_c \cdot \overline{c}^\vee$ for some $q_c \in \mathbf{Q}^\times$. For long c we get

$$2 = c \circ c^\vee = \overline{c} \circ \pi \circ c^\vee = q_c \cdot \overline{c} \circ \overline{c}^\vee = 2q_c,$$

so $\pi \circ c^\vee = \overline{c}^\vee$ for long c. Likewise, if c is short then

$$2 = c \circ c^\vee = (1/p) \cdot (\overline{c} \circ \pi \circ c^\vee) = (q_c/p) \cdot \overline{c} \circ \overline{c}^\vee = 2q_c/p,$$

so $\pi \circ c^\vee = p \cdot \overline{c}^\vee$ for short c. It follows from these formulas that $\overline{\Phi}$ is the dual root system to Φ via the isomorphism $\mathrm{X}(\overline{T})_\mathbf{R} \simeq \mathrm{X}(T)_\mathbf{R}$ induced by π that is equivariant with respect to the actions of the Weyl groups $W(\Phi) = W(G,T)(k)$ and $W(\overline{\Phi}) = W(\overline{G},\overline{T})(k)$. Since the ratio of the squared root lengths in Φ is p, it even follows from the explicit formula for $\overline{c} \circ \pi$ in terms of c that $c \mapsto \overline{c}$ carries long roots to short roots and carries short roots to long roots.

The duality of the root systems implies that the Dynkin diagrams for $\overline{\Phi}$ and Φ are dual to each other, so if G has type G_2 (resp. B_n, resp. C_n, resp. F_4) then $\overline{G}$ has type G_2 (resp. C_n, resp. B_n, resp. F_4). In the case of types F_4 and G_2 it is automatic that $\overline{G}$ is simply connected. If G is simply connected of type C_n ($n \geqslant 2$), to show that the quotient $\overline{G}$ of type B_n is simply connected we fix a short root $\overline{c}$ in $\overline{\Phi}$ and let $\overline{G}_{\overline{c}}$ be the subgroup of $\overline{G}$ generated by $\overline{U}_{\pm\overline{c}}$. Then $\overline{G}$ is simply connected if and only if the rank-1 group $\overline{G}_{\overline{c}}$ is simply connected (i.e., it is isomorphic to SL_2). If $c \in \Phi_>$ corresponds to $\overline{c}$ then π carries G_c onto $\overline{G}_{\overline{c}}$ and $\mathrm{Lie}(\ker \pi)$ meets the Lie algebra $\mathfrak{g}_{-c} \oplus \mathrm{Lie}(c^\vee(\mathrm{GL}_1)) \oplus \mathfrak{g}_c$ of G_c trivially (by weight space considerations with respect to T, note that $\mathrm{Lie}(c^\vee(\mathrm{GL}_1)) \cap \mathfrak{t}_<$ is trivial by Lemma 7.1.1(1)). Hence, $G_c \to \overline{G}_{\overline{c}}$ is an isomorphism, and $G_c \simeq \mathrm{SL}_2$ since G is simply connected. Thus, $\overline{G}$ is simply connected of type B_n when G is simply connected of type C_n.

Now consider the factorization (7.1.1) with G simply connected of type C_n ($n \geqslant 2$), or more generally any example for which the very special quotient $\overline{G}$ is known to be simply connected (i.e., all examples so far except possibly G of type B_n, $n \geqslant 3$). We have shown that π carries long root groups isomorphically onto the short root groups with degree equal to $p^{\#\Phi_< + \#\Delta_<}$, and the composite map in the diagram (7.1.1) has kernel equal to the k-group $\ker F_{G/k}$ of order $p^{\dim G}$ whose own Frobenius morphism vanishes. Thus, the other isogeny $\overline{\pi} : \overline{G} \to G^{(p)}$ in (7.1.1) has degree $p^{\#\Phi_> + \#\Delta_>} = p^{\#\overline{\Phi}_< + \#\overline{\Delta}_<}$ and kernel of height $\leqslant 1$ that contains the root spaces for the short roots. The Lie subalgebra generated by root spaces for short roots contains $\overline{\mathfrak{t}}_<$ in $\overline{\mathfrak{g}}$, so $\overline{\mathfrak{t}}_< \oplus \overline{\mathfrak{g}}_< \subseteq \ker \overline{\pi}$ and this must be an equality for dimension reasons. Hence, $\overline{\pi}$ is the very special isogeny of $\overline{G}$, so the very special quotient of $\overline{G}$ is $G^{(p)}$, which is simply connected.

Up to isomorphism there is only one connected semisimple absolutely simple group of type B_n that is split and simply connected, and we just realized it as a very special quotient $\overline{G}$ above. But in that argument we also determined the very special isogeny of $\overline{G}$ and observed that its own very special quotient is simply connected. Thus, we can run the preceding argument with G simply connected of type B_n to deduce that (7.1.1) exhibits $\overline{\pi}$ as the very special isogeny of $\overline{G}$ in the remaining case where G is of type B_n. □

Remark 7.1.6 The very special isogeny from simply connected type B_n to simply connected type C_n in characteristic 2 is a classical construction. Let (V, q) be a quadratic space of odd dimension $2n + 1$, $n \geqslant 2$, that is non-degenerate (in the sense of [EKM, 7.A]): if $V^\perp \subseteq V$ is the 1-dimensional radical of the alternating bilinear form $B_q(v,w) = q(v+w) - q(v) - q(w)$ then B_q induces a non-degenerate alternating form $\overline{B}_q$ on the $2n$-dimensional space $\overline{V} = V/V^{\perp}$ and the very special isogeny is the map $\mathrm{Spin}(q) \to \mathrm{SO}(q) \to \mathrm{Sp}(\overline{B}_q)$.

To use the above results to construct some new pseudo-reductive groups over imperfect fields of characteristics 2 or 3, we need refined information concerning the group-theoretic properties of the root groups relative to the long roots and short roots in $\Phi(G,T)$ when T is k-split.

Proposition 7.1.7 *Let k be a field of characteristic $p \in \{2,3\}$ and let G be a connected semisimple group that is absolutely simple and simply connected with Dynkin diagram having an edge with multiplicity p. Let T be a maximal k-torus in G.*

(1) *Assume that T is k-split. The long root groups of (G,T) generate a k-subgroup $G_> \subseteq G$ containing T that is semisimple and simply connected, and the root groups of $(G_>,T)$ are precisely the long root groups of (G,T) with $\Phi(G_>,T) = \Phi_>$. The short root groups of (G,T) generate a k-subgroup $G_< \subseteq G$ containing T that is of the same type as $\overline{G}_>$, and its center equals Z_G. The root groups of $(G_<,T)$ are precisely the short root groups of (G,T) with $\Phi(G_<,T) = \Phi_<$.*
(2) *The pair $(\Phi_>,\Phi)$ is in the list $(\mathrm{D}_4,\mathrm{F}_4)$, $(\mathrm{D}_n,\mathrm{B}_n)$ for $n \geqslant 2$, $(\mathrm{A}_1^n,\mathrm{C}_n)$ for $n \geqslant 2$, and $(\mathrm{A}_2,\mathrm{G}_2)$.*
(3) *In general, the k_s-subgroups $(G_{k_s})_>$ and $(G_{k_s})_<$ of G_{k_s} arising from the k_s-split torus T_{k_s} uniquely descend to respective k-subgroups $G_>$ and $G_<$ of G containing T.*

The Lie algebras are given by $\mathrm{Lie}(G_>) = \mathfrak{t} \oplus \mathfrak{g}_>$ *and* $\mathrm{Lie}(G_<) = \mathfrak{t} \oplus \mathfrak{g}_<$.

It is understood that D_3 means A_3 and D_2 means $\mathrm{A}_1 \times \mathrm{A}_1$. We introduce $G_<$ because it is a useful device for the study of $G_>$ in certain cases, but in later applications we will only use $G_>$. The fact that $G_>$ is always simply connected (in contrast with $G_<$ for G of certain types in characteristic 2) is useful.

Proof Assertion (3) follows via Galois descent from k_s to k once (1) is proved, so for the proof of (3) it suffices to focus on assertion (1). In particular, T is k-split (so G is k-split). The basic idea for proving that $G_>$ is no larger than expected (i.e., it contains no short root groups) is to indirectly construct $G_>$ by another procedure (resting on a technique of Borel and de Siebenthal).

Let $\mu_< \subseteq (T/Z_G)[p]$ be the canonical k-subgroup as in Lemma 7.1.1(3). Using the natural left action of G/Z_G on G, we get an action of $\mu_<$ on G.

Let $H = Z_G(\mu_<)^0$ denote the identity component of the centralizer of the $\mu_<$-action on G. By Proposition A.8.12, H is a connected reductive group. The k-split T is a maximal k-torus in H, so the connected reductive k-group H is k-split and its Lie algebra is the span of $\mathfrak{t}$ and the root spaces $\mathfrak{g}_c$ for which $c|_{\mu_<} = 1$ (Proposition A.8.10(3)). By Lemma 7.1.1(3), such c are precisely the long roots. We conclude that the root groups of H are the long root groups of G, so $\Phi(H,T)$ is the root system $\Phi_>$ formed by the long roots in $\Phi(G,T)$.

It is a classical fact that when Φ is one of the types B_n ($n \geqslant 2$), C_n ($n \geqslant 3$), F_4, or G_2 then $\Phi_>$ is respectively of type D_n, A_1^n, D_4, and A_2 (with D_n interpreted as explained above when $n = 2,3$). Since the root system Φ and the root subsystem $\Phi_>$ have the same rank, the semisimple derived group $\mathscr{D}(H)$ contains the maximal torus T, so $H = \mathscr{D}(H)$ and hence $H = G_>$. In particular, $G_>$ contains T.

In all cases, the preimage M of $\mu_<$ under $T \to T/Z_G$ is a central subgroup scheme in H. Its order is $\#\mu_< \cdot \#Z_G = p^{\#\Delta_<} \cdot \#Z_G$. In case $p = 3$ the k-group $G_>$ is k-isomorphic to either the simply connected SL_3 or the adjoint PGL_3, but the central subgroup M has order 3, so $G_>$ must be simply connected. In each case it is easy to check that M has order equal to the order of the center of the simply connected group of the same type as $G_>$, so $G_>$ is always simply connected (with M equal to its center).

Consider the very special isogeny $\pi : G \to \overline{G}$ that carries T onto $\overline{T}$. The map π carries long root groups onto short root groups and carries short root groups onto long root groups. The map π therefore induces a purely inseparable isogeny from $G_<$ onto $\overline{G}_>$ and from $G_>$ onto $\overline{G}_<$. So we conclude that $G_<$ is a smooth connected semisimple k-subgroup of G containing T and its root system $\Phi(G_<, T)$ equals $\Phi_<$. To prove that the center of $G_<$ equals the center Z_G of G, it suffices to observe that the former certainly contains the latter, and to recall that the short roots in Φ generate the root lattice so the center of $G_<$ cannot be bigger. $\square$

7.2 Basic exotic pseudo-reductive groups

Building on the results in §7.1, we are now in position to present a twisted version of Tits' examples of non-standard pseudo-reductive groups over imperfect fields with characteristics 2 and 3. Let k be a field of characteristic $p \in \{2,3\}$ and let k'/k be a finite extension with $k'^p \subseteq k$. Let G' be a connected semisimple k'-group that is absolutely simple and simply connected with Dynkin diagram having an edge with multiplicity p, and let $\pi' : G' \to \overline{G}'$ be the very special k'-isogeny. Consider the k-map

$$f = \mathrm{R}_{k'/k}(\pi') : \mathrm{R}_{k'/k}(G') \to \mathrm{R}_{k'/k}(\overline{G}').$$

We wish to consider f-preimages of certain Levi k-subgroups of $\mathrm{R}_{k'/k}(\overline{G}')$.

As a preliminary step, it is convenient to parameterize Levi k-subgroups of $\mathrm{R}_{k'/k}(\overline{G}')$. These are related to k-isomorphism classes of k-structures on $\overline{G}'$ in the sense of the following general lemma.

Lemma 7.2.1 *Let k'/k be a finite and purely inseparable extension of fields and let G' be a connected reductive k'-group. There is a natural bijection between the set of Levi k-subgroups of $\mathrm{R}_{k'/k}(G')$ and the set of k-isomorphism classes of pairs (G, i) where G is a k-group and $i : G_{k'} \simeq G'$ is a k'-isomorphism. In particular, $\mathrm{R}_{k'/k}(G')$ admits a Levi k-subgroup if and only if the k'-group G' admits a descent to a k-group.*

Each abstract k-group isomorphism between a pair of Levi k-subgroups of $\mathrm{R}_{k'/k}(G')$ is induced by the action of a unique element of $\mathrm{Aut}_{k'}(G')$. In particular, if G' is k'-split then $\mathrm{Aut}_{k'}(G')$ acts transitively on the set of k-split Levi k-subgroups of $\mathrm{R}_{k'/k}(G')$.

This lemma becomes more concrete when G' is semisimple and the Dynkin diagram for $G'_{k'_s}$ has no nontrivial automorphisms, since then $\mathrm{Aut}_{k'}(G') = (G'/Z_{G'})(k') = \mathrm{R}_{k'/k}(G'/Z_{G'})(k)$ (by Corollary A.4.7(1)).

Proof By Theorem 1.6.2(2), the unipotent radical of $\mathrm{R}_{k'/k}(G')_{\overline{k}}$ has field of definition over k equal to k'. The maximal reductive quotient of $\mathrm{R}_{k'/k}(G')_{k'}$ is naturally identified with G', so if $j : G \hookrightarrow \mathrm{R}_{k'/k}(G')$ is a Levi k-subgroup then the natural map $i : G_{k'} \to G'$ corresponding (via the universal property of Weil restriction) to j is an isomorphism.

Conversely, if G is an abstract k-group equipped with a k'-isomorphism $i : G_{k'} \simeq G'$ then the k-map $G \to \mathrm{R}_{k'/k}(G')$ corresponding to i has trivial kernel (as it is the composition of $G \hookrightarrow \mathrm{R}_{k'/k}(G_{k'})$ with $\mathrm{R}_{k'/k}(i)$) and it is a Levi k-subgroup since k'/k is purely inseparable. This establishes the desired bijection.

It remains to check that any k-group isomorphism α between two Levi k-subgroups G and H of $\mathrm{R}_{k'/k}(G')$ has the form $\mathrm{R}_{k'/k}(\alpha')$ for a unique $\alpha' \in \mathrm{Aut}_{k'}(G')$. The inclusions of G and H as Levi k-subgroups induce k'-isomorphisms $G_{k'} \simeq G'$ and $H_{k'} \simeq G'$, so a k-isomorphism $\alpha : G \simeq H$ induces a k'-automorphism α' of G'. The induced k-automorphism of $\mathrm{R}_{k'/k}(G')$ carries G into H via α, so it suffices to show that the k-automorphism $\mathrm{R}_{k'/k}(\alpha')$ of $\mathrm{R}_{k'/k}(G')$ uniquely determines α'. Since the natural map $q : \mathrm{R}_{k'/k}(G')_{k'} \twoheadrightarrow G'$ is the maximal reductive quotient and intertwines $\mathrm{R}_{k'/k}(\alpha')$ and α', the desired uniqueness follows from Proposition 1.2.2. □

Example 7.2.2 Let k be an imperfect field of characteristic $p > 0$, and k'/k a finite purely inseparable extension. Let G' be a k'-form of SL_p, GL_p, or PGL_p. We seek examples of such G' that do not admit a descent to a k-group, since then Lemma 7.2.1 implies that the (standard) pseudo-reductive k-group $\mathrm{R}_{k'/k}(G')$ has no Levi k-subgroup. Such examples demonstrate the need for the pseudo-split hypothesis in Theorem 3.4.6.

For any $n > 1$, the automorphism groups of SL_n, GL_n, and PGL_n over a field are given as follows: $\mathrm{Aut}(\mathrm{SL}_2) = \mathrm{Aut}(\mathrm{PGL}_2) = \mathrm{PGL}_2$ and in all other cases it is the semidirect product $(\mathbf{Z}/2\mathbf{Z}) \ltimes \mathrm{PGL}_n$ in which $\mathbf{Z}/2\mathbf{Z}$ is generated by transpose-inverse. We now use the classical identification of $\mathrm{H}^1(k, \mathrm{PGL}_n)$ with the subset $\mathrm{Br}(k)_{\mathrm{ind}|n} \subseteq \mathrm{Br}(k)[n]$ consisting of elements with index dividing n (where the *index* of a Brauer class is d when the central division algebra representative has dimension d^2). This carries the effect of transpose-inverse on $\mathrm{H}^1(k, \mathrm{PGL}_n)$ over to inversion on $\mathrm{Br}(k)_{\mathrm{ind}|n}$. Thus, the fibers of the map

$$\mathrm{Br}(k)_{\mathrm{ind}|n} = \mathrm{H}^1(k, \mathrm{PGL}_n) \to \mathrm{H}^1(k, (\mathbf{Z}/2\mathbf{Z}) \ltimes \mathrm{PGL}_n)$$

are Brauer classes up to inversion (using [Ser3, I, §5.5, Prop. 39(ii)]). Since the restriction map $\mathrm{H}^1(k, \mathbf{Z}/2\mathbf{Z}) \to \mathrm{H}^1(k', \mathbf{Z}/2\mathbf{Z})$ is an isomorphism, by taking $n = p$ we conclude that the existence of the desired examples would follow from the non-surjectivity of the restriction map $\mathrm{Br}(k)_{\mathrm{ind}|p} \to \mathrm{Br}(k')_{\mathrm{ind}|p}$. (Note, in contrast, that $\mathrm{Br}(k) \to \mathrm{Br}(k')$ is surjective since it is the map in degree-2 Galois cohomology induced by the maximal k-torus inclusion $\mathrm{GL}_1 \to \mathrm{R}_{k'/k}(\mathrm{GL}_1)$ whose cokernel is unipotent and hence has vanishing H^2 [Oes, IV, 2.2].)

Consider the case $k' = k^{1/p}$ (assumed to be of finite degree over k). In this case the isomorphism $k' \simeq k$ defined by $t \mapsto t^p$ carries the inclusion $k \hookrightarrow k'$ over to the p-power map on k. Thus, the resulting isomorphism $\mathrm{Br}(k') \simeq \mathrm{Br}(k)$ carries the restriction map $\mathrm{Br}(k) \to \mathrm{Br}(k')$ over to the p-power map on $\mathrm{Br}(k)$. In particular, $\mathrm{Br}(k)[p]$ is killed by restriction to $\mathrm{Br}(k')$, so any central division algebra over k' of rank p^2 cannot arise from a k-algebra. For example, if k is a local or global function field (so $\mathrm{Br}(k')_{\mathrm{ind}|n} = \mathrm{Br}(k')[n]$ for any $n \geqslant 1$, by local and global class field theory) and $k' = k^{1/p}$ then local and global class field theory for k' ensure the existence of a central division algebra over k' of rank p^2, so we obtain the desired G' over k' for which $\mathrm{R}_{k'/k}(G')$ has no Levi k-subgroup.

The following is another example. Let k be a local function field of characteristic 2 and k' be a quadratic extension of k. Let q' be a non-degenerate quadratic form over k' in $2n+1$ variables and of Witt index $n-1$. (See [EKM, 7.A, 8.A] for the notions of non-degeneracy and Witt index of finite-dimensional quadratic spaces over any field F. In the non-degenerate case, the

Witt index equals the F-rank of the associated special orthogonal group.) Then the k'-group $G' := \mathrm{SO}(q')$ does not admit a descent to a k-group. To prove this, we first note that the existence of such a descent is equivalent to the existence of a quadratic form q over k such that q' is a nonzero scalar multiple of q by extension of scalars to k'. Since k' is a local function field, if a non-degenerate quadratic form over k' in $2n+1$ variables is obtained by extension of scalars from a quadratic form over k then its Witt index is necessarily n. Thus, no such q exists, so G' cannot be descended to a k-group.

Returning to the very special k'-isogeny $\pi' : G' \to \overline{G}'$, assume there is given a k-descent G of G', or equivalently (by Lemma 7.2.1) a Levi k-subgroup $G \subseteq \mathrm{R}_{k'/k}(G')$. We want to relate k-structures on G' and k-structures on $\overline{G}'$, or equivalently Levi k-subgroups of $\mathrm{R}_{k'/k}(G')$ and Levi k-subgroups of $\mathrm{R}_{k'/k}(\overline{G}')$.

This can go in two directions: given a k-structure on G' can we make one on $\overline{G}'$ with respect to which it is compatible via π', and given a k-structure on $\overline{G}'$ is there one on G' with respect to which it is compatible via π'? The easy (and less useful) direction is going from G' down to $\overline{G}'$. Namely, if G is a k-structure on G' then the very special k-isogeny $\pi : G \to \overline{G}$ provides a unique k-structure $\overline{G}$ on $\overline{G}'$ with respect to which the very special k'-isogeny π' descends to the very special k-isogeny π. In particular, $\overline{G}$ is naturally identified with a Levi k-subgroup of $\mathrm{R}_{k'/k}(\overline{G}')$.

To go in the reverse direction, we need to work with preimages under the k-homomorphism

$$f = \mathrm{R}_{k'/k}(\pi') : \mathrm{R}_{k'/k}(G') \to \mathrm{R}_{k'/k}(\overline{G}').$$

It is the preimage $f^{-1}(\overline{G})$ that will be Tits' construction of exotic pseudo-reductive groups, at least when there is a G compatible with $\overline{G}$ via π'. The point is that if there is a compatible k-descent pair $(G, \overline{G})$ as above then (by functoriality) f carries the Levi k-subgroup $G \subseteq \mathrm{R}_{k'/k}(G')$ into the Levi k-subgroup $\overline{G} \subseteq \mathrm{R}_{k'/k}(\overline{G}')$ via the very special k-isogeny π, so if we are given $\overline{G}$ and seek to find G then a natural idea is to look for a Levi k-subgroup of $f^{-1}(\overline{G})$ (assuming $f^{-1}(\overline{G})$ is smooth).

But there is a serious difficulty with this preimage procedure: f is not surjective when $k' \neq k$, so for some choices of k-descent $\overline{G}$ of $\overline{G}'$ the preimage $f^{-1}(\overline{G})$ may not map onto $\overline{G}$ (and may not even be smooth). The reason that f is not surjective when $k' \neq k$ is that

$$\ker f = \mathrm{R}_{k'/k}(\ker \pi')$$

has dimension > 0 in such cases, since it contains a k-subgroup $\mathrm{R}_{k'/k}(\mu')$ whose k_s-fiber contains the positive-dimensional $\mathrm{R}_{k'_s/k_s}(\mu_p) = \mathrm{R}_{k'_s/k_s}(\mathrm{GL}_1)[p]$.

In Proposition 7.3.1 it is proved that failure of $f^{-1}(\overline{G})$ to map onto $\overline{G}$ is actually equivalent to non-smoothness of $f^{-1}(\overline{G})$.

Here is the main result concerning properties of $f^{-1}(\overline{G})$ when there does exist a k-structure G on G' compatible via π' with the chosen k-structure $\overline{G}$ on $\overline{G}'$.

Theorem 7.2.3 *Let k be a field of characteristic $p \in \{2, 3\}$ and let k'/k be a finite extension with $k'^p \subseteq k$. Let G be a connected semisimple k-group that is absolutely simple and simply connected with Dynkin diagram having an edge with multiplicity p. Let $\pi : G \to \overline{G}$ be the very special isogeny, T a maximal k-torus of G, $\Phi = \Phi(G_{k_s}, T_{k_s})$, and Δ the basis of a positive system of roots in Φ.*

Let $f = \mathrm{R}_{k'/k}(\pi_{k'})$ and let $\mathscr{G} = f^{-1}(\overline{G})$.

(1) *The k-group $\mathscr{G}$ is absolutely pseudo-simple, and*

$$\dim \mathscr{G} = (\#\Delta_< + \#\Phi_<)[k' : k] + (\#\Delta_> + \#\Phi_>).$$

(2) *Let $T' = T_{k'}$ be the associated maximal k'-torus in $G' := G_{k'}$, and $\mathfrak{n} \subseteq \mathfrak{g}$ as in Lemma 7.1.2. We have*

$$\mathrm{Lie}(\mathscr{G}) = \mathrm{Lie}(G_>) + \mathfrak{n}_{k'} \tag{7.2.1}$$

in $\mathrm{Lie}(\mathrm{R}_{k'/k}(G')) = \mathrm{Lie}(G') = \mathfrak{g}_{k'}$.

If T is k-split then $\mathscr{G}$ is the k-subgroup of $\mathrm{R}_{k'/k}(G')$ generated by $G_>$ and the Weil restrictions $\mathrm{R}_{k'/k}(U'_c)$ of the short root groups U'_c of (G', T'). Also, still assuming T is k-split, the Cartan k-subgroup $Z_{\mathscr{G}}(T)$ is the smooth connected k-subgroup of $\mathrm{R}_{k'/k}(T')$ generated by T and the k-subgroups $\mathrm{R}_{k'/k}(c^\vee_{k'}(\mathrm{GL}_1))$ for c varying through short roots in $\Phi(G, T) = \Phi(G', T')$. More precisely, if Δ is a basis of $\Phi(G, T)$ then

$$Z_{\mathscr{G}}(T) = \prod_{a \in \Delta_>} a^\vee(\mathrm{GL}_1) \times \prod_{b \in \Delta_<} \mathrm{R}_{k'/k}(b^\vee_{k'}(\mathrm{GL}_1)).$$

(3) *The natural map $q : \mathscr{G}_{k'} \to G'$ corresponding to the inclusion $\mathscr{G} \hookrightarrow \mathrm{R}_{k'/k}(G')$ is a smooth surjection with connected unipotent kernel.*

Roughly speaking, $\mathscr{G}$ is obtained from the k-group G by thickening the short root groups from k to k' and thickening part of a maximal k-torus from $k^\times$ to $k'^\times$ (using the k-subtorus arising from the coroots associated to short simple positive roots). In particular, if $k' = k$ then $\mathscr{G} = G$, so this construction is only interesting when $k' \neq k$.

Proof Observe that (2) over k_s implies that $\mathscr{G}$ is smooth, connected, and perfect with the expected dimension as in (1) (using (7.2.1)), so we may and do assume $k = k_s$. We first prove (2), then address the absolute pseudo-simplicity claim in (1).

STEP 1. To prove (2), let $\mathscr{H}$ be the smooth connected k-subgroup of $\mathrm{R}_{k'/k}(G')$ generated by $G_>$ and the k-subgroups $\mathrm{R}_{k'/k}(U'_c)$ for short $c \in \Phi(G,T)$. Our first aim is to show $\mathscr{H} = \mathscr{G}$. By functoriality of the natural map

$$X \to \mathrm{R}_{k'/k}(X_{k'})$$

for affine k-schemes X, f carries $G_>$ into $\pi(G_>) = \overline{G}_< \subseteq \overline{G}$. For each short root $c \in \Phi(G,T)$, say with $\overline{c} \in \Phi(\overline{G},\overline{T})$ the associated long root such that π carries U_c onto $\overline{U}_{\overline{c}}$, we claim that the k-map $\mathrm{R}_{k'/k}(U'_c) \to \mathrm{R}_{k'/k}(\overline{U}'_{\overline{c}})$ induced by f has image contained in $\overline{U}_{\overline{c}}$.

It suffices to check this on k-points, where it becomes the assertion that the map $U_c(k') \to \overline{U}_{\overline{c}}(k')$ induced by π has image contained in $\overline{U}_{\overline{c}}(k)$. Under suitable k-isomorphisms $\mathbf{G}_\mathrm{a} \simeq U_c$ and $\mathbf{G}_\mathrm{a} \simeq \overline{U}_{\overline{c}}$, the k-map $\pi : U_c \to \overline{U}_{\overline{c}}$ becomes $x \mapsto x^p$. Since $k'^p \subseteq k$ by hypothesis, we get the desired containment. Thus, $f(\mathscr{H}) \subseteq \overline{G}$, so we have an inclusion $\mathscr{H} \subseteq \mathscr{G}$ as subschemes of $\mathrm{R}_{k'/k}(G')$.

For any long root $c \in \Phi$, π carries U_c isomorphically onto $\overline{U}_{\overline{c}}$ for some short root $\overline{c} \in \overline{\Phi}$. Hence, f carries U_c isomorphically onto $\overline{U}_{\overline{c}}$ as well. We have just seen that if $c \in \Phi$ is short and $\overline{c} \in \overline{\Phi}$ is the long root associated to c via π then f carries the smooth connected $\mathrm{R}_{k'/k}(U'_c)$ into $\overline{U}_{\overline{c}} \simeq \mathbf{G}_\mathrm{a}$ by a nonzero map, so $f(\mathrm{R}_{k'/k}(U'_c)) = \overline{U}_{\overline{c}}$ for dimension reasons. Since f carries T onto $\overline{T}$, we conclude that f carries $\mathscr{H}$ onto $\overline{G}$. To prove $\mathscr{H} = \mathscr{G}$, it remains to show that the k-subgroup

$$\ker f = \mathrm{R}_{k'/k}(\ker \pi_{k'})$$

is contained in $\mathscr{H}$.

STEP 2. The kernel of f is directly spanned (as a k-group scheme) by the k-groups

$$\mathrm{R}_{k'/k}(\ker F_{U'_c/k'}) \subseteq \mathscr{H},$$

for $c \in \Phi_<$, and the k-group $\mathrm{R}_{k'/k}(\mu_{k'})$ where μ is directly spanned by the Frobenius kernels of the k-tori $c^\vee(\mathrm{GL}_1)$ for short roots c in a basis of a positive system of roots over k. Hence, we just have to check that $\mathrm{R}_{k'/k}(\mu_{k'}) \subseteq \mathscr{H}$.

Due to the direct spanning description of μ, by treating the short simple positive roots c separately, this becomes a problem for SL_2 over k: if D is the diagonal k-torus and $U^{\pm}$ are the two nontrivial D-stable smooth connected unipotent k-subgroups, then we claim that $\mathrm{R}_{k'/k}(U^+_{k'})$ and $\mathrm{R}_{k'/k}(U^-_{k'})$ generate a k-subgroup of $\mathrm{R}_{k'/k}(\mathrm{SL}_2)$ containing the k-group scheme $\mathrm{R}_{k'/k}(D[p]_{k'})$. In fact, they generate the entirety of $\mathrm{R}_{k'/k}(\mathrm{SL}_2)$ because on k-points of the Weil restrictions we get the k-subgroups $U^+(k')$ and $U^-(k')$ in $\mathrm{SL}_2(k')$ that generate $\mathrm{SL}_2(k')$ (a classical fact for any field k'). This completes the proof that $\mathscr{H} = \mathscr{G}$.

Step 3. To prove the rest of (2), let $\overline{G}' = \overline{G}_{k'}$ and $\overline{T}' = \overline{T}_{k'}$, with $\overline{T} = \pi(T) \subseteq \overline{G}$. By Proposition A.5.15(1), $Z_{\mathrm{R}_{k'/k}(G')}(T) = \mathrm{R}_{k'/k}(T')$. Hence, the smooth k-group $Z_{\mathscr{G}}(T)$ is the scheme-theoretic intersection $\mathscr{G} \cap \mathrm{R}_{k'/k}(T')$. This is the scheme-theoretic preimage of $\overline{G} \cap \mathrm{R}_{k'/k}(\overline{T}') = \overline{T}$ under the map

$$f_T : \mathrm{R}_{k'/k}(T') \to \mathrm{R}_{k'/k}(\overline{T}').$$

By decomposing T and $\overline{T}$ compatibly according to the coroots associated to a basis Δ of a positive system of roots in Φ (as we may do since G and $\overline{G}$ are simply connected), f_T decomposes into a direct product of two kinds of nontrivial maps: the identity on $\mathrm{R}_{k'/k}(\mathrm{GL}_1)$ for the coroots $c^\vee$ associated to long $c \in \Delta$ and the p-power map on $\mathrm{R}_{k'/k}(\mathrm{GL}_1)$ for the coroots $c^\vee$ associated to short $c \in \Delta$. The second of these two maps lands in the k-subgroup $\mathrm{GL}_1 \subseteq \mathrm{R}_{k'/k}(\mathrm{GL}_1)$ (since $k'^p \subseteq k$) and so it maps onto GL_1 for dimension reasons. This gives the desired description of $Z_{\mathscr{G}}(T) = f_T^{-1}(\overline{T})$ for part (2).

To complete the proof of (2), it only remains to determine the Lie algebra of $\mathscr{G} = f^{-1}(\overline{G})$. This Lie algebra is the preimage of the k-subspace

$$\overline{\mathfrak{g}} := \mathrm{Lie}(\overline{G}) \subseteq \overline{\mathfrak{g}}'$$

under the k'-linear map $\mathrm{Lie}(\pi') : \mathfrak{g}' \to \overline{\mathfrak{g}}'$ arising from $\pi' = \pi_{k'}$. To determine $\mathrm{Lie}(\pi')^{-1}(\overline{\mathfrak{g}})$ we first observe that it is T-stable, so it admits a k-linear weight space decomposition with respect to T. Hence, we can determine $\mathrm{Lie}(\pi')^{-1}(\overline{\mathfrak{g}})$ by working one weight at a time, as follows.

For the trivial weight, under the map

$$\mathfrak{t}_{k'} = \mathfrak{t}' \to \overline{\mathfrak{t}}' = \overline{\mathfrak{t}}_{k'}$$

(with $\mathfrak{t} = \mathrm{Lie}(T)$ and $\overline{\mathfrak{t}} = \mathrm{Lie}(\overline{T})$) the preimage of $\overline{\mathfrak{t}}$ is $\mathfrak{t} + \mathfrak{t}'_<$ (as a Lie-theoretic consequence of our determination that $f_T^{-1}(\overline{T}) = Z_{\mathscr{G}}(T)$). For each $c \in \Phi_>$ the map $\mathrm{Lie}(\pi')$ carries $\mathfrak{g}'_c$ isomorphically onto $\overline{\mathfrak{g}}'_{\overline{c}}$ for a suitable $\overline{c} \in \overline{\Phi}_<$.

Likewise, for each $c \in \Phi_<$ the map $\mathrm{Lie}(\pi')$ kills the entire weight space $\mathfrak{g}'_c$ since π carries U_c onto $\overline{U}_{\overline{c}}$ by a map that is given in suitable coordinates by $x \mapsto x^p$ and hence vanishes on Lie algebras over k. This gives the expected components for $\mathrm{Lie}(\pi')^{-1}(\overline{\mathfrak{g}})$ in the weight spaces for the nontrivial weights of the T-action on the k-Lie algebra $\mathfrak{g}'$, namely $\mathfrak{g}_c$ for $c \in \Phi_>$ and $\mathfrak{g}'_c$ for $c \in \Phi_<$. This completes the proof of (2).

STEP 4. To prove (1), since $\mathscr{G}$ is now known to be smooth, connected, and perfect of the expected dimension it remains to prove that it is absolutely pseudo-simple over k. Observe that $G \subseteq \mathscr{G} \subseteq \mathrm{R}_{k'/k}(G_{k'})$, with G a Levi k-subgroup of $\mathrm{R}_{k'/k}(G_{k'})$ (since k'/k is purely inseparable). Also, $\mathrm{R}_{k'/k}(G_{k'})$ is a pseudo-reductive k-group, by Proposition 1.1.10. Hence, the pseudo-reductivity of $\mathscr{G}$ and (3) follow from the following useful general result.

Lemma 7.2.4 *Let H be a pseudo-reductive group over any field k, and $L \subseteq H$ a Levi k-subgroup. Let K/k be the field of definition for $\mathscr{R}_u(H_{\overline{k}}) \subseteq H_{\overline{k}}$. Any smooth connected k-subgroup $M \subseteq H$ containing L is also pseudo-reductive over k, with L a Levi k-subgroup of M. The geometric unipotent radical $\mathscr{R}_u(M_{\overline{k}}) \subseteq M_{\overline{k}}$ is defined over K and $\mathscr{R}_{u,K}(M_K) = M_K \cap \mathscr{R}_{u,K}(H_K)$, so the natural map $M_K \to H' := H_K/\mathscr{R}_{u,K}(H_K)$ is a smooth surjection with connected unipotent kernel $\mathscr{R}_{u,K}(M_K)$.*

In particular, $M_K^{\mathrm{red}} = H'$.

Proof Since L is a Levi k-subgroup of H, the natural map $L_K \ltimes \mathscr{R}_{u,K}(H_K) \to H_K$ is an isomorphism by descent from $\overline{k}$. But $L_K \subseteq M_K$, so

$$M_K = L_K \ltimes (M_K \cap \mathscr{R}_{u,K}(H_K)).$$

Hence, $M_K \cap \mathscr{R}_{u,K}(H_K)$ is smooth and connected, so it must be $\mathscr{R}_{u,K}(M_K)$. This shows that L is a Levi k-subgroup of M, and that $\mathscr{R}_{u,K}(M_K) \subseteq \mathscr{R}_{u,K}(H_K)$. Now we can apply Lemma 1.2.1 to $X = \mathscr{R}_{u,k}(M)$ to deduce that M is pseudo-reductive over k. □

STEP 5. It remains to show that $\mathscr{G}$ is absolutely pseudo-simple. It is a perfect pseudo-reductive group and $\mathscr{G}^{\mathrm{ss}}_{\overline{k}} \simeq G'_{\overline{k}}$ is simple, so Lemma 3.1.2 applies to $\mathscr{G}$, completing the proof of Theorem 7.2.3. □

Corollary 7.2.5 *Let k be a field of characteristic $p \in \{2,3\}$, and let $(G, k'/k)$ be as in Theorem 7.2.3. Let $\mathscr{G} \subseteq \mathrm{R}_{k'/k}(G_{k'})$ be the associated pseudo-reductive k-subgroup as in Theorem 7.2.3.*

(1) *The scheme-theoretic center $Z_{\mathscr{G}}$ is equal to $\mathscr{G} \cap \mathrm{R}_{k'/k}((Z_G)_{k'})$; in particular, $\mathscr{G}/Z_{\mathscr{G}}$ is naturally a k-subgroup of $\mathrm{R}_{k'/k}((G/Z_G)_{k'})$.*

(2) *The center* $Z_{\mathscr{G}}$ *is trivial if* G *is of type* F_4 *or* G_2. *If* G *is of type* B_n *with* $n \geqslant 2$ *or type* C_n *with even* $n \geqslant 2$ *then* $Z_{\mathscr{G}} = \mathrm{R}_{k'/k}((Z_G)_{k'}) \simeq \mathrm{R}_{k'/k}(\mu_2)$ *inside of* $\mathrm{R}_{k'/k}(G_{k'})$. *If* G *is of type* C_n *with odd* $n \geqslant 3$ *then* $Z_{\mathscr{G}} = Z_G \simeq \mu_2$ *inside of* $\mathrm{R}_{k'/k}((Z_G)_{k'})$.

(3) *Assume there exists a k-split maximal k-torus* T *of* G. *For each long root* $c \in \Phi(G,T)$, *with associated root group* $U_c \subseteq G$ *and* 1*-parameter subgroup* $c^\vee : \mathrm{GL}_1 \to T$, *we have*

$$\mathscr{G} \cap \mathrm{R}_{k'/k}((U_c)_{k'}) = U_c, \quad \mathscr{G} \cap \mathrm{R}_{k'/k}(c^\vee(\mathrm{GL}_1)_{k'}) = c^\vee(\mathrm{GL}_1).$$

Proof To establish the description of $Z_{\mathscr{G}}$, we may and do assume $k = k_s$. Letting T vary through the maximal k-tori of G, we get (by Theorem 7.2.3(2))

$$\begin{aligned} Z_{\mathscr{G}} \subseteq \bigcap\nolimits_T Z_{\mathscr{G}}(T) \subseteq \bigcap\nolimits_T \mathrm{R}_{k'/k}(T_{k'}) &\\ = \mathrm{R}_{k'/k}\left(\bigcap\nolimits_T T_{k'}\right) &\\ = \mathrm{R}_{k'/k}\left(\left(\bigcap\nolimits_T T\right)_{k'}\right) &\\ = \mathrm{R}_{k'/k}((Z_G)_{k'}). &\end{aligned}$$

This implies (1), so $Z_{\mathscr{G}} = 1$ when $Z_G = 1$, settling (2) for types F_4 and G_2. For type B_n with $n \geqslant 2$ or type C_n with even $n \geqslant 2$, the center Z_G is a copy of μ_2 diagonally embedded into the product of the k-groups $c^\vee(\mu_2)$ for certain short simple positive roots c relative to a fixed choice of positive system of roots (the unique such c for type B_n, and $n/2$ of them for type C_n). Since $\mathrm{R}_{k'/k}(c^\vee(\mathrm{GL}_1)_{k'}) \subseteq \mathscr{G}$ for short c, it follows that $\mathrm{R}_{k'/k}((Z_G)_{k'}) \subseteq \mathscr{G}$ in these cases and it is clearly central in $\mathscr{G}$ (as it is even central in $\mathrm{R}_{k'/k}(G_{k'})$). In contrast, for type C_n with odd $n \geqslant 3$ the center $Z_G \simeq \mu_2$ has nontrivial projection to the direct factor $c^\vee(\mathrm{GL}_1)$ of T for the unique long positive simple root c. Thus, once we establish that $\mathscr{G} \cap \mathrm{R}_{k'/k}(c^\vee(\mathrm{GL}_1)_{k'}) = c^\vee(\mathrm{GL}_1)$ then we will get the containment $Z_{\mathscr{G}} \subseteq Z_G = \mu_2$, with equality then being immediate.

Now let T be a maximal k-torus of G and assume that T is k-split. Let $\overline{T} = \pi(T)$ be the associated k-split maximal k-torus in $\overline{G}$ and for $c \in \Phi(G,T)_>$ let $\overline{c} \in \Phi(\overline{G}, \overline{T})$ be the short root such that $\pi : G \to \overline{G}$ carries U_c isomorphically onto $\overline{U}_{\overline{c}}$, so $\pi^{-1}(\overline{U}_{\overline{c}}) = U_c \times \ker \pi$. Hence,

$$\begin{aligned} \mathscr{G} \cap (\mathrm{R}_{k'/k}((U_c)_{k'}) \times \ker f) &= f^{-1}(\overline{G}) \cap f^{-1}(\mathrm{R}_{k'/k}((\overline{U}_{\overline{c}})_{k'})) \\ &= f^{-1}(\overline{G} \cap \mathrm{R}_{k'/k}((\overline{U}_{\overline{c}})_{k'})) \\ &= f^{-1}(\overline{U}_{\overline{c}}). \end{aligned}$$

But the map f restricts to π on $U_c \subseteq G \subseteq \mathrm{R}_{k'/k}(G_{k'})$, so $f^{-1}(\overline{U}_{\overline{c}}) = U_c \times \ker f$. Hence,

$$\mathscr{G} \cap \mathrm{R}_{k'/k}((U_c)_{k'}) = U_c.$$

The analogous claim concerning $c^\vee(\mathrm{GL}_1)$ with a long root c was settled in the determination of $f_T^{-1}(\overline{T}) = Z_{\mathscr{G}}(T)$ in the proof of Theorem 7.2.3 since $\mathrm{R}_{k'/k}(c^\vee(\mathrm{GL}_1)_{k'}) \cap \ker f = 1$ (as we can check via Lie algebras and the triviality of étale multiplicative subgroups of $\ker \pi$). □

By Lemma 7.2.4, the pseudo-simple k-group $\mathscr{G}$ contains G as a Levi k-subgroup. In Proposition 8.1.1 we will show that such $\mathscr{G}$ are *never* standard pseudo-reductive k-groups, and so over imperfect fields of characteristics 2 and 3 we will need to incorporate these k-groups into our classification of pseudo-reductive k-groups. Observe that these k-groups $\mathscr{G}$ (and their Weil restrictions) contain Levi k-subgroups (and not merely Levi k_s-subgroups).

To circumvent the fact that (in the non-pseudo-split case) pseudo-reductive groups can fail to have Levi k-subgroups (Example 7.2.2), we now introduce a class of non-standard pseudo-reductive k-groups in characteristic $p \in \{2, 3\}$, including the above k-groups $\mathscr{G}$ as well as variants, whose analysis only requires working with Levi subgroups over k_s. The definition of this class of k-groups (Definition 7.2.6) will involve a condition that amounts to the existence of a Levi subgroup over k_s. In view of the existence of Levi k_s-subgroups in pseudo-reductive groups over k_s (Theorem 3.4.6), imposing the requirement of the existence of a Levi subgroup over k_s is not unreasonable. However, it is also not especially elegant. In Proposition 7.3.1 we establish an equivalence between this hypothesis and several other properties, some of which look more natural than the existence of a Levi k_s-subgroup. The proof of this equivalence is somewhat technical, and its statement is not used in the rest of §7.2.

Definition 7.2.6 Let k be an imperfect field with characteristic $p \in \{2, 3\}$. A *basic exotic pseudo-reductive* k-group is a k-group $\mathscr{G}$ that is k-isomorphic to a k-group scheme of the form $f^{-1}(\overline{G})$ where:

(i) $f = \mathrm{R}_{k'/k}(\pi')$ for a nontrivial finite extension k'/k satisfying $k'^p \subseteq k$ and π' denoting the very special k'-isogeny $G' \to \overline{G}'$ for a connected semisimple k'-group G' that is absolutely simple and simply connected with an edge of multiplicity p in its Dynkin diagram,
(ii) the k-subgroup $\overline{G} \subseteq \mathrm{R}_{k'/k}(\overline{G}')$ is a Levi k-subgroup,
(iii) $f^{-1}(\overline{G})_{k_s}$ contains a Levi k_s-subgroup of $\mathrm{R}_{k'/k}(G')_{k_s} = \mathrm{R}_{k'_s/k_s}(G'_{k'_s})$.

We make a similar definition when k is a nonzero finite product of fields, treating the fibers over $\mathrm{Spec}\, k$ separately.

The Levi k_s-subgroup in (iii) allows us (via Lemma 7.2.1) to apply Theorem 7.2.3 and Corollary 7.2.5 over k_s to see that a k-group $\mathscr{G}$ as in Definition 7.2.6 is absolutely pseudo-simple (and in particular, smooth and connected) over k with center $Z_{\mathscr{G}} = \mathscr{G} \cap \mathrm{R}_{k'/k}(Z_{G'})$ (which is trivial precisely when G' is of type F_4 or G_2), so $\mathscr{G}/Z_{\mathscr{G}} \subseteq \mathrm{R}_{k'/k}(G'/Z_{G'})$.

Note that in Definition 7.2.6 we require $k' \neq k$. This will be needed to prove several interesting properties of basic exotic pseudo-reductive k-groups, including that they are non-standard (and in particular, not reductive). By using Weil restriction one can give a more general construction of non-standard absolutely pseudo-simple pseudo-reductive k-groups, to be called *exotic*, as we will see in Definition 8.2.2.

It is a very useful fact that a basic exotic pseudo-reductive k-group $\mathscr{G}$ as in Definition 7.2.6 uniquely determines both the pair $(G', k'/k)$ and the Levi k-subgroup $\overline{G} \subseteq \mathrm{R}_{k'/k}(\overline{G}')$ that give rise to it, up to unique k-isomorphism. This is part of the following proposition.

Proposition 7.2.7 *Let k be an imperfect field of characteristic $p \in \{2, 3\}$. Let k'/k be a nontrivial finite extension with $k'^p \subseteq k$, and let G' be a connected semisimple k'-group that is simply connected and absolutely simple with an edge of multiplicity p in its Dynkin diagram. Denote its very special k'-isogeny as $\pi' : G' \to \overline{G}'$.*

Let $\overline{G} \subseteq \mathrm{R}_{k'/k}(\overline{G}')$ be a Levi k-subgroup such that for $f = \mathrm{R}_{k'/k}(\pi')$ the preimage $f^{-1}(\overline{G})_{k_s} = f_{k_s}^{-1}(\overline{G}_{k_s})$ contains a Levi k_s-subgroup of $\mathrm{R}_{k'/k}(G')_{k_s} = \mathrm{R}_{k'_s/k_s}(G'_{k'_s})$. Let $\mathscr{G} = f^{-1}(\overline{G}) \subseteq \mathrm{R}_{k'/k}(G')$ be the associated basic exotic pseudo-reductive k-group.

1. *The field of definition over k for the unipotent radical of $\mathscr{G}_{\overline{k}}$ is equal to k' (so $\mathscr{G}$ is not reductive), and if T is a maximal k-torus of $\mathscr{G}$ then $\Phi(\mathscr{G}_{k_s}, T_{k_s}) = \Phi(G'_{k'_s}, T_{k'_s})$ inside of $\mathrm{X}(T_{k_s}) = \mathrm{X}(T_{k'_s})$. In particular, $\Phi(\mathscr{G}_{k_s}, T_{k_s})$ is reduced.*
2. *The k'-descent of the maximal reductive quotient of $\mathscr{G}_{\overline{k}}$ is uniquely k'-isomorphic to the canonical k'-map $\mathscr{G}_{k'} \to G'$ corresponding to the inclusion $\mathscr{G} \hookrightarrow \mathrm{R}_{k'/k}(G')$, and the image of $\mathscr{G}$ under the Weil restriction $\mathrm{R}_{k'/k}(\pi')$ of the very special k'-isogeny $\pi' : G' \to \overline{G}'$ is the Levi k-subgroup $\overline{G} \subseteq \mathrm{R}_{k'/k}(\overline{G}')$.*
3. *Let K/k be any separable extension field. An arbitrary k-group $\mathscr{H}$ is a basic exotic pseudo-reductive group if and only if the K-group $\mathscr{H}_K$ is a basic exotic pseudo-reductive group.*

Proof By Proposition 1.1.9(2), the field of definition E_k over k of the $\overline{k}$-unipotent radical of a smooth affine k-group is a finite purely inseparable extension of k whose formation commutes with any separable extension on k. Hence, (1) and (2) together imply (3) via Galois descent and specialization.

To prove (1) and (2) we may assume that k is separably closed. In particular, by the hypotheses, the k-group $\mathscr{G} = f^{-1}(\overline{G})$ contains a Levi k-subgroup G of $\mathrm{R}_{k'/k}(G')$. The identification of root systems in (1) is immediate from (7.2.1). The natural map $G \to \overline{G}$ induced by f must be the very special k-isogeny of G because over k' it is identified with the very special k'-isogeny $G' \to \overline{G}'$ (due to Lemma 7.2.1 and the fact that G and $\overline{G}$ are Levi k-subgroups of $\mathrm{R}_{k'/k}(G')$ and $\mathrm{R}_{k'/k}(\overline{G}')$ respectively).

By Theorem 7.2.3(3), the natural map $q : \mathscr{G}_{k'} \to G'$ adjoint to the inclusion $\mathscr{G} \hookrightarrow \mathrm{R}_{k'/k}(G')$ is a k'-descent of the maximal reductive quotient of $\mathscr{G}_{\overline{k}}$. Thus, our problem is to prove that the k'-subgroup $\ker q \subseteq \mathscr{G}_{k'}$ has field of definition over k equal to k' (and then the rest is straightforward). It suffices to show that the k'-subspace $\mathrm{Lie}(\ker q) = \ker(\mathrm{Lie}(q))$ in $\mathrm{Lie}(\mathscr{G}_{k'}) = k' \otimes_k \mathrm{Lie}(\mathscr{G})$ cannot be defined over a proper subfield of k' containing k.

The k'-linear map $k' \otimes_k \mathrm{Lie}(\mathscr{G}) \to \mathrm{Lie}(G')$ is induced by the k-linear inclusion

$$\mathrm{Lie}(\mathscr{G}) \hookrightarrow \mathrm{Lie}(G') = k' \otimes_k \mathrm{Lie}(G)$$

whose image is the k-subspace described by (7.2.1) in terms of the weight spaces on $\mathfrak{g} = \mathrm{Lie}(G)$ relative to a choice of maximal k-torus T in G. Since a descent of the k'-subspace $\ker(\mathrm{Lie}(q))$ to a K-subspace of $K \otimes_k \mathrm{Lie}(\mathscr{G})$ for an intermediate field $k \subseteq K \subseteq k'$ must be equivariant for the action by T_K, it suffices to study the descent problem separately on the different weight spaces in $k' \otimes_k \mathrm{Lie}(\mathscr{G})$. In particular, it suffices to show that the canonical k'-linear map $k' \otimes_k \mathfrak{g}'_c \to \mathfrak{g}'_c$ on $\mathfrak{g}'_c = k' \otimes_k \mathfrak{g}_c$ for a short root $c \in \Phi(G,T)$ has kernel whose (minimal) field of definition over k is k'. This holds by Lemma A.7.7. □

By Proposition 7.2.7(2), for $\mathscr{G}$ as in Definition 7.2.6 (so $\mathscr{G}$ is k-smooth and connected) the inclusion $\mathscr{G} \hookrightarrow \mathrm{R}_{k'/k}(G')$ induces an isomorphism on maximal reductive quotients over $\overline{k}$. Hence, a Levi k_s-subgroup of $\mathscr{G}_{k_s}$ is the same as one of $\mathrm{R}_{k'/k}(G')_{k_s} = \mathrm{R}_{k'_s/k_s}(G'_{k'_s})$ that lies in $\mathscr{G}_{k_s}$ (and likewise over k). In particular, in condition (iii) in Definition 7.2.6 it is equivalent to require that $\mathscr{G}$ is a smooth connected k-group such that $\mathscr{G}_{k_s}$ admits a Levi k_s-subgroup, which is a natural condition in view of Theorem 3.4.6.

We prefer to think in terms of Levi k_s-subgroups of $\mathrm{R}_{k'_s/k_s}(G'_{k'_s})$ rather than of $\mathscr{G}_{k_s}$ because the former have a useful interpretation as explained in

Lemma 7.2.1. Also, in Proposition 7.2.7(2) we saw that the Levi k-subgroup $\overline{G}$ in (ii) is canonically determined by the k-group $\mathscr{G}$.

Remark 7.2.8 Let $\mathscr{G}$ be a basic exotic pseudo-reductive group over an imperfect field k of characteristic $p \in \{2, 3\}$, with $(G', k'/k, \overline{G})$ the corresponding triple of data underlying its construction. Assume $k = k_s$, so the k-group $\overline{G}$ is k-split. The very special k'-isogeny $G' \to \overline{G}' = \overline{G}_{k'}$ descends to the very special k-isogeny $G \to \overline{G}$ using any Levi k-subgroup G in $\mathscr{G}$. Let T be a maximal k-torus in G, so $T' := T_{k'}$ is a maximal k'-torus in $G_{k'} = G'$. Let U_c and U'_c denote the corresponding root groups in G and G' for any $c \in \Phi(G, T) = \Phi(G', T')$.

Pick a positive system of roots Φ^+ in $\Phi(\mathscr{G}, T) = \Phi(G, T)$ and let Δ be the set of positive simple roots. The corresponding open Bruhat cell (as in Corollary 3.3.16) can be written down explicitly: by Theorem 7.2.3(2) and Corollary 7.2.5, it is the product scheme

$$\prod_{c \in \Phi^-_<} \mathrm{R}_{k'/k}(U'_c) \times \prod_{c \in \Phi^-_>} U_c \times Z_{\mathscr{G}}(T) \times \prod_{c \in \Phi^+_<} \mathrm{R}_{k'/k}(U'_c) \times \prod_{c \in \Phi^+_>} U_c \qquad (7.2.2)$$

(embedded in $\mathscr{G}$ via multiplication) where

$$Z_{\mathscr{G}}(T) = \prod_{b \in \Delta_<} \mathrm{R}_{k'/k}(b^\vee(\mathrm{GL}_1)) \times \prod_{a \in \Delta_>} a^\vee(\mathrm{GL}_1).$$

7.3 Algebraic and arithmetic aspects of basic exotic pseudo-reductive groups

In our work with basic exotic pseudo-reductive groups over imperfect fields of characteristics 2 and 3, there are several specialized results that are not logically relevant in the proof of the classification theorems but are nonetheless of interest, so we record them here. Some are very useful in arithmetic applications (such as in [Con2]).

Let k be an imperfect field of characteristic $p \in \{2, 3\}$. In the definition of a basic exotic pseudo-reductive k-group (Definition 7.2.6), we imposed a hypothesis (hypothesis (iii)) requiring the existence of a Levi k_s-subgroup (for an auxiliary k-group). That definition was entirely sufficient for our needs, but the hypothesis (iii) over k_s may appear to be *ad hoc*. The following proposition establishes an equivalence between that hypothesis in Definition 7.2.6 and several alternative conditions that may seem more natural.

Proposition 7.3.1 *Let k be an imperfect field with characteristic $p \in \{2,3\}$, k'/k a finite extension satisfying $k'^p \subseteq k$, and G' a connected semisimple k'-group that is absolutely simple and simply connected with an edge of multiplicity p in its Dynkin diagram. Let*

$$\pi' : G' \to \overline{G}'$$

be the associated very special k'-isogeny, and let $f = \mathrm{R}_{k'/k}(\pi')$. Let $\overline{G}$ be a Levi k-subgroup of $\mathrm{R}_{k'/k}(\overline{G}')$.

The following conditions on the k-subgroup scheme $H = f^{-1}(\overline{G}) \subseteq \mathrm{R}_{k'/k}(G')$ are equivalent:

(a) *H is k-smooth,*
(b) *H_{k_s} has maximal dimension among the $f_{k_s}^{-1}(L)$ for Levi k_s-subgroups L of $\mathrm{R}_{k'_s/k_s}(\overline{G}'_{k'_s})$,*
(c) *the natural map $f : H \to \overline{G}$ is surjective,*
(d) *$\overline{G}$ is contained in the image of f,*
(e) *H_{k_s} contains a Levi k_s-subgroup of $\mathrm{R}_{k'/k}(G')_{k_s}$.*

When these five equivalent conditions hold, $H = f^{-1}(\overline{G})$ is connected.

Proof We may and do assume that k is separably closed. The key to the proof is to show that $\mathrm{im}(\mathrm{Lie}(f)) \cap \mathrm{Lie}(\overline{G})$ has dimension independent of $\overline{G}$. To show this, we will first check:

Lemma 7.3.2 *The Lie ideal* $\mathrm{im}(\mathrm{Lie}(f))$ *in* $\mathrm{Lie}(\mathrm{R}_{k'/k}(\overline{G}'))$ *is stable under the natural action of* $\mathrm{R}_{k'/k}(\overline{G}'/Z_{\overline{G}'})$.

Proof Consider the factorization in k'-isogenies

$$G' \stackrel{\pi'}{\to} \overline{G}' \stackrel{\overline{\pi}'}{\to} G'^{(p)}$$

of the Frobenius isogeny $F_{G'/k'}$ of G'. By Proposition 7.1.5(3), $\overline{\pi}'$ is the very special isogeny of $\overline{G}'$. Moreover, the proof of that result showed that the dimension of $\ker(\mathrm{Lie}(\overline{\pi}')) = \mathrm{Lie}(\ker \overline{\pi}')$ is equal to the dimension of $\mathrm{im}(\mathrm{Lie}(\pi'))$ so the containment $\mathrm{im}(\mathrm{Lie}(\pi')) \subseteq \ker \mathrm{Lie}(\overline{\pi}')$ inside of $\mathrm{Lie}(\overline{G}')$ is an equality. Viewing these k'-vector spaces as k-vector spaces, this says that $\mathrm{im}(\mathrm{Lie}(f))$ is the Lie algebra of the kernel of the map $\mathrm{R}_{k'/k}(\overline{\pi}')$, so the Lie ideal $\mathrm{im}(\mathrm{Lie}(f))$ in $\mathrm{Lie}(\mathrm{R}_{k'/k}(\overline{G}'))$ is stable under the natural action of $\mathrm{R}_{k'/k}(\overline{G}'/Z_{\overline{G}'})$, as desired. □

To deduce that the k-dimension of $\mathrm{im}(\mathrm{Lie}(f)) \cap \mathrm{Lie}(\overline{G})$ is independent of the choice of $\overline{G}$, it now suffices to show that all choices of $\overline{G}$ are conjugate to each other under the natural action of $\mathrm{R}_{k'/k}(\overline{G}'/Z_{\overline{G}'})(k)$ on $\mathrm{R}_{k'/k}(\overline{G}')$. Such conjugacy follows from Lemma 7.2.1 (since $\overline{G}'$ is a k'-split form of B_n ($n \geqslant 2$), C_n ($n \geqslant 2$), F_4, or G_2, as k' is separably closed).

In general $\dim H \leqslant \dim \mathrm{Lie}(H)$ with equality if and only if H is smooth, and

$$\begin{aligned}\dim \mathrm{Lie}(H) &= \dim \mathrm{Lie}(f)^{-1}(\mathrm{Lie}(\overline{G}))\\ &= \dim \mathrm{Lie}(\ker f) + \dim(\mathrm{im}(\mathrm{Lie}(f)) \cap \mathrm{Lie}(\overline{G})),\end{aligned}$$

where the first equality uses that the formation of Lie algebras commutes with fiber products. Thus, $\dim \mathrm{Lie}(H)$ is independent of $\overline{G}$ due to the result proved above. Since k' is separably closed, G' is k'-split, so it admits a k-descent G. Hence, by Theorem 7.2.3, there exists a Levi k-subgroup $\overline{G}$ in the image of f such that $H = f^{-1}(\overline{G})$ is smooth. Thus, for this H we have $\dim H = \dim \mathrm{Lie}(H)$, so the equivalence of (a) and (b) follows.

Since we have just shown that, when varying the Levi k-subgroup $\overline{G}$ in $\mathrm{R}_{k'/k}(\overline{G}')$, the containment $\overline{G} \subseteq \mathrm{im}(f)$ does sometimes occur, the maximal value attained by

$$\dim H = \dim \ker f + \dim(\mathrm{im}(f) \cap \overline{G})$$

as we vary $\overline{G}$ is obtained precisely when $\overline{G} \subseteq \mathrm{im}(f)$. In other words, (b) and (d) are equivalent. Since $H = f^{-1}(\overline{G})$, we see that (c) and (d) are equivalent. The Levi k-subgroup $G \subseteq H$ in condition (e) (with $k = k_s$!) is a k-descent of G' such that the map $\pi : G \to \overline{G}$ descends π', so it identifies the k-group $H = f^{-1}(\overline{G})$ with an instance of the k-group $\mathscr{G}$ as in Theorem 7.2.3. Hence, Theorem 7.2.3(1) gives that (e) implies (a), and also that H is connected when (e) holds.

To conclude the proof, we need to show that (d) implies (e). We have put ourselves in the case where k is separably closed, so we have to show that if (d) holds with k separably closed then the k-group $H = f^{-1}(\overline{G})$ contains a Levi k-subgroup of $\mathrm{R}_{k'/k}(G')$. Since (d) implies (a), H is smooth. Since (d) implies (c), the natural map $H \to \overline{G}$ is surjective and hence the natural map $H_{k'} \to G'$ is also surjective. Fix a maximal k-torus T in H, so T is also a maximal torus of $\mathrm{R}_{k'/k}(G')$. It follows that the image $\overline{T}$ of T in $\overline{G}$ is a maximal k-torus and $T' := T_{k'}$ is a maximal torus of the maximal reductive quotient G' of $\mathrm{R}_{k'/k}(G')_{k'}$. The identity component H^0 of H is pseudo-reductive over k because H^0 maps onto $\overline{G}$ with kernel containing no nontrivial k-points.

By Corollary A.7.6 we have $\Phi(\mathrm{R}_{k'/k}(G'), T) = \Phi(G', T')$ (via the equality $\mathrm{X}(T) = \mathrm{X}(T')$). The surjectivity of the natural map $H_{k'} \to G'$ implies that the containment of root systems $\Phi(H^0, T) \subseteq \Phi(G', T')$ is an equality (by Remark 2.3.6 and the reducedness of $\Phi(G', T')$). Using the equality of root systems $\Phi(H^0, T) = \Phi(\mathrm{R}_{k'/k}(G'), T)$, we can choose a common positive system of roots and let Δ be the resulting basis of simple positive roots. For each $a \in \Delta$, let E_a be a 1-dimensional k-linear subgroup of the a-root group U_a of (H^0, T). By Theorem 3.4.6, there is a unique Levi k-subgroup L of H^0 containing T and E_a for each $a \in \Delta$. Explicitly, if we choose $u_a \in E_a(k) - \{1\}$ for each $a \in \Delta$ then there is a unique $v_a \in U_{-a}(k)$ such that $n_a := v_a u_a v_a$ normalizes T and induces the reflection r_a on $\mathrm{X}(T)$, and L is generated by T and the k-subgroups $\{E_a, n_a E_a n_a^{-1}\}_{a \in \Delta}$. The containment of $U_{\pm a}$ in the $\pm a$-root group for $(\mathrm{R}_{k'/k}(G'), T)$ for each $a \in \Delta$ implies that L coincides with the unique Levi k-subgroup of $\mathrm{R}_{k'/k}(G')$ containing T and E_a for each $a \in \Delta$. Thus, H contains a Levi k-subgroup of $\mathrm{R}_{k'/k}(G')$. □

In arithmetic applications, a very useful feature of standard pseudo-reductive groups is that a standard presentation can be given in terms of a simply connected semisimple group. It was proved by Harder (resp. Bruhat and Tits) that connected semisimple simply connected groups over global function fields (resp. local function fields) have vanishing degree-1 Galois cohomology. For basic exotic pseudo-reductive groups, there is a related vanishing theorem that plays an equally essential role when proving general theorems for algebraic groups over local or global function fields of characteristics 2 or 3. The key point is a consequence of Proposition 7.2.7 that allows us, in certain contexts, to replace basic exotic pseudo-reductive groups over local or global function fields of characteristics 2 and 3 with forms of simply connected semisimple groups over the same ground field:

Proposition 7.3.3 *Let k be an imperfect field of characteristic $p \in \{2, 3\}$ such that $[k : k^p]$ is finite. Let $\mathscr{G}$ be the basic exotic pseudo-reductive k-group arising from a triple $(G', k'/k, \overline{G})$ with $k' = k^{1/p}$.*

(1) *The natural map $\mathscr{G}(k) \to \overline{G}(k)$ is an isomorphism of groups and the natural map $\mathrm{H}^1(k, \mathscr{G}) \to \mathrm{H}^1(k, \overline{G})$ is bijective.*
(2) *If k is complete with respect to an absolute value then the natural map $\mathscr{G}(k) \to \overline{G}(k)$ is a topological isomorphism.*
(3) *If k is a global function field then the natural map $\mathscr{G}(\mathbf{A}_k) \to \overline{G}(\mathbf{A}_k)$ on adelic points is a topological isomorphism.*

Note that if $[k : k^p] = p$ then every basic exotic pseudo-reductive k-group arises as in the above proposition. This applies when k is a local or global function field of characteristic p.

Before we take up the proof of Proposition 7.3.3, it is convenient to first review the description of open Bruhat cells (Corollary 3.3.16) in basic exotic pseudo-reductive groups over imperfect fields k of characteristic p. Let k be such a field, and k'/k a nontrivial finite extension such that $k'^p \subseteq k$. Let G' be a connected semisimple k'-group that is absolutely simple and simply connected of type B_n or C_n (with $n \geqslant 2$) or F_4 when $p = 2$ and of type G_2 when $p = 3$, with associated very special k'-isogeny $\pi' : G' \to \overline{G}'$. Let $f = R_{k'/k}(\pi')$ and let $\overline{G} \subseteq R_{k'/k}(\overline{G}')$ be a Levi k-subgroup such that $f^{-1}(\overline{G})_{k_s}$ contains a Levi k_s-subgroup of $R_{k'/k}(G')_{k_s}$, so $\mathscr{G} := f^{-1}(\overline{G})$ is a basic exotic pseudo-reductive k-group.

Assume that $\mathscr{G}$ contains a Levi k-subgroup G. This is a property that holds after some finite separable extension on k (since it holds over k_s). Under this assumption, the map $\pi : G \to \overline{G}$ induced by f must be the very special k-isogeny of G and more specifically over k' it is identified with π'.

Impose the further hypothesis that G is k-split, so we may choose a k-split maximal k-torus T in G. Let $\overline{T} = \pi(T)$ be the corresponding k-split maximal k-torus in $\overline{G}$. Let Φ^+ be a positive system of roots in $\Phi = \Phi(G, T)$, and let $\overline{\Phi}^+$ be the corresponding positive system of roots in $\overline{\Phi} = \Phi(\overline{G}, \overline{T})$. Let $V \subseteq G$ be the open Bruhat cell of G associated to T and the choice of Φ^+.

By calculating as in the determination of where $\mathscr{G}$ meets thickened root groups and tori in the proof of Corollary 7.2.5, the open subscheme $R_{k'/k}(V_{k'}) \subseteq R_{k'/k}(G_{k'})$ meets the closed k-subgroup $\mathscr{G} \subseteq R_{k'/k}(G_{k'})$ in the open subscheme $\mathscr{V} \subseteq \mathscr{G}$ given by (7.2.2) in Remark 7.2.8 (where we assumed $k = k_s$, which is not necessary under the above hypotheses). There is an analogous open subscheme $\overline{V}$ of $\overline{G}$ given by the open Bruhat cell

$$\prod_{\overline{c} \in \overline{\Phi}^-} U_{\overline{c}} \times \overline{T} \times \prod_{\overline{c} \in \overline{\Phi}^+} U_{\overline{c}}.$$

The k-map $f = R_{k'/k}(\pi_{k'})$ carries $\mathscr{V}$ into the open Bruhat cell $\overline{V}$ of $\overline{G}$ compatibly with the product decompositions (with long roots of Φ going over to short roots of $\overline{\Phi}$ and short roots of Φ going over to long roots of $\overline{\Phi}$). It is also compatible with product decompositions of $Z_{\mathscr{G}}(T)$ and $\overline{T}$ according to the simple positive coroots since the bijection $\Phi \to \overline{\Phi}$ carries the basis of Φ^+ to the basis of $\overline{\Phi}^+$ (Proposition 7.1.5(4)).

The open locus $\mathscr{V}$ is the open Bruhat cell in the pseudo-reductive k-group $\mathscr{G}$ equipped with the k-split maximal k-torus $T \subseteq G \subseteq \mathscr{G}$ and the choice of Φ^+.

Proof of Proposition 7.3.3 The hypotheses and the formation of $\mathscr{G}$ and $(G', k'/k, \overline{G})$ are compatible with separable algebraic extension on k. Thus, once it is proved that $\mathscr{G}(k) \to \overline{G}(k)$ is bijective in general, the natural $\mathrm{Gal}(k_s/k)$-equivariant map $\mathscr{G}(k_s) \to \overline{G}(k_s)$ is bijective. Hence, $\mathrm{H}^1(k, \mathscr{G}) \to \mathrm{H}^1(k, \overline{G})$ is bijective. For part (1) we can therefore focus on the group-theoretic aspects of the assertion.

By Galois equivariance, to prove that the map $\mathscr{G}(k) \to \overline{G}(k)$ is an isomorphism of abstract groups or (in the case of fields complete with respect to an absolute value) a topological isomorphism, it is harmless to replace k with a finite separable extension. Thus, we can assume that the k-group $\mathscr{G} = f^{-1}(\overline{G})$ contains a Levi k-subgroup G (and hence G' is canonically identified with $G_{k'}$). By making a further finite separable extension on k we can assume that G is k-split.

Fix a k-split maximal k-torus T in G and a positive system of roots Φ^+ in $\Phi(G,T)$, so we get an associated k-split maximal k-torus $\overline{T}$ in the quotient $\overline{G}$ and an associated positive system of roots $\overline{\Phi}^+ \subseteq \Phi(\overline{G}, \overline{T})$. In terms of this data, construct both the open Bruhat cells $\mathscr{V} \subseteq \mathscr{G}$ and $\overline{V} \subseteq \overline{G}$ as above. We shall now prove that the map $\mathscr{V}(k) \to \overline{V}(k)$ is bijective in general and a homeomorphism when k is complete with respect to an absolute value.

We may work factor by factor in the product decompositions of $\mathscr{V}$ and $\overline{V}$ and we may treat long roots in $\Phi(G,T)$ separately from short roots. On some factors the map $\mathscr{V}(k) \to \overline{V}(k)$ induces the identity on k or $k^\times$, and on other factors it is the p-power map from k' to k or from $k'^\times$ to $k^\times$. These p-power maps are bijective since $k' = k^{1/p}$, and they are trivially homeomorphisms when k is topologized by an absolute value. In such cases, $\mathrm{R}_{k'/k}(\mathbf{G}_\mathrm{a})(k) = k'$ is endowed with its topology as a finite-dimensional k-vector space and this coincides with the valuation topology on k' when k is complete.

For an arbitrary imperfect field k of characteristic p such that $k' = k^{1/p}$ has finite degree over k we now prove that the map of groups $\mathscr{G}(k) \to \overline{G}(k)$ is bijective. It is injective because the kernel is a subgroup of $(\ker f)(k) \subseteq (\ker \pi')(k')$ and π' is a purely inseparable isogeny. To prove surjectivity, first note that the bijectivity of the map $\mathscr{V}(k) \to \overline{V}(k)$ proved above implies that for the positive Borel k-subgroup $\overline{B} \subseteq \overline{G}$ the abstract subgroup $\overline{B}(k) \subseteq \overline{G}(k)$ lies in the image of $\mathscr{G}(k)$. The map $W(G,T) \to W(\overline{G}, \overline{T})$ between Weyl groups induced by π is an isomorphism, so each $\overline{w} \in W(\overline{G}, \overline{T})(k) = N_{\overline{G}}(\overline{T})(k)/\overline{T}(k)$ is represented by the image of some point of $N_G(T)(k) \subseteq G(k) \subseteq \mathrm{R}_{k'/k}(G_{k'})(k)$. But $G(k) \subseteq \mathscr{G}(k)$, so $f(\mathscr{G}(k))$ contains a set of representatives for $W(\overline{G}, \overline{T})(k)$. Such representatives together with $\overline{B}(k)$ generate the group $\overline{G}(k)$, by the Bruhat decomposition for the k-split $\overline{G}$, so we obtain the desired surjectivity on k-points. When k is complete with respect to an absolute value we have shown that the continuous bijective map on k-points restricts to a topological isomorphism between open neighborhoods of the identity points, so it is a topological isomorphism globally.

Finally, we assume that k is a global function field of characteristic p and we check that the continuous map $\mathscr{G}(\mathbf{A}_k) \to \overline{G}(\mathbf{A}_k)$ on adelic points is a topological isomorphism. For any place v of k, $k^{1/p} \otimes_k k_v = k_v^{1/p}$. Thus, the natural map $\mathscr{G}(k_v) \to \overline{G}(k_v)$ is a topological isomorphism. Hence, the continuous map $\mathscr{G}(\mathbf{A}_k) \to \overline{G}(\mathbf{A}_k)$ between topological groups is injective. These groups are locally compact, Hausdorff, and separable, so the map is a topological quotient map (and hence a homeomorphism) if it is surjective. Our problem is thereby reduced to showing that the map $\mathscr{G} \to \overline{G}$ that is surjective on k-points and k_v-points for all v is also surjective on $\mathbf{A}_k$-points. By Proposition A.2.13 (and Example A.2.14), such surjectivity follows from the fact that $\overline{G}$ is a connected semisimple k-group which is absolutely simple and simply connected. (It is also enough to consider k-split $\overline{G}$, since the bijectivity on $\mathbf{A}_k$-points follows by Galois theory from the same on $\mathbf{A}_{k'}$-points for a finite Galois extension k'/k that splits $\overline{G}$.) □

A basic exotic pseudo-reductive k-group $\mathscr{G}$ uniquely determines, up to unique k-isomorphism, the triple $(G', k'/k, \overline{G})$ of data that gives rise to it (Proposition 7.2.7), so it is natural to try to relate properties of $\mathscr{G}$ and G'. Here is one such relationship, assuming $k' = k^{1/p}$.

Corollary 7.3.4 *Let k be an imperfect field of characteristic $p \in \{2, 3\}$ such that $[k : k^p]$ is finite. Let $\mathscr{G}$ and $(G', k'/k, \overline{G})$ be as in Definition 7.2.6 with $k' = k^{1/p}$, and let $f : \mathscr{G} \to \overline{G}$ denote the canonical surjection. If $\overline{T} \subseteq \overline{G}$ is a maximal k-torus (resp. maximal k-split k-torus) then there exists a unique maximal k-torus (resp. maximal k-split k-torus) $\mathscr{T} \subseteq \mathscr{G}$ such that $\mathscr{T} \subseteq f^{-1}(\overline{T})$, and $\mathscr{T} \to \overline{T}$ is an isogeny.*

In particular, there is a natural bijection between the set of maximal k-tori (resp. maximal k-split k-tori) in $\mathscr{G}$ and $\overline{G}$, and $\mathscr{G}$ is k-isotropic (resp. pseudo-split over k) if and only if $\overline{G}$ is k-isotropic (resp. k-split).

Proof By Galois descent, we may and do assume that $k = k_s$. Since $f : \mathscr{G} \to \overline{G}$ is surjective, it carries maximal k-tori onto maximal k-tori. By Proposition 7.3.3(1) we have $\mathscr{G}(k) \simeq \overline{G}(k)$ via f since $k' = k^{1/p}$, so by the conjugacy of maximal k-tori in smooth affine k-groups (Proposition A.2.10) each maximal k-torus $\overline{T}$ of $\overline{G}$ is the image of one in $\mathscr{G}$. It remains to show that $f^{-1}(\overline{T})$ contains a unique maximal k-torus of $\mathscr{G}$. Since $(f^{-1}(\overline{T}))(k)$ is commutative (as it is isomorphic to a subgroup of $\overline{T}(k)$), the desired uniqueness follows. □

A variant on the construction in Corollary 7.3.4, using maximal tori in G' rather than in $\overline{G}$, works for any imperfect field k of characteristic $p \in \{2, 3\}$ and nontrivial finite extension k'/k with $k'^p \subseteq k$; see Proposition 11.1.3.

A defect of Corollary 7.3.4 is its restrictions on $[k : k^p]$ and k'. However, these hypotheses are always satisfied in arithmetic cases (local and global function fields) and under such hypotheses there are no further constraints involving descent of fields of definition of tori as in Proposition 11.1.3. It is for this reason that Corollary 7.3.4 is easier to use than Proposition 11.1.3, at least when its hypotheses are satisfied.

The following result relates k-forms of $\mathscr{G}$ to k-forms of $\overline{G}$ when $k' = k^{1/p}$.

Proposition 7.3.5 *Let k be an imperfect field of characteristic $p \in \{2, 3\}$ such that $[k : k^p]$ is finite. Let $\mathscr{G}_1$ and $\mathscr{G}_2$ be basic exotic pseudo-reductive k-groups such that the associated triples $(G'_1, k'_1/k, \overline{G}_1)$ and $(G'_2, k'_2/k, \overline{G}_2)$ satisfy $k'_1 = k'_2 = k^{1/p}$. Let $f_i : \mathscr{G}_i \to \overline{G}_i$ be the canonical surjection.*

(1) *For every k-isomorphism $\alpha : \mathscr{G}_1 \simeq \mathscr{G}_2$ there is a unique k-isomorphism $\overline{\alpha} : \overline{G}_1 \simeq \overline{G}_2$ such that $f_2 \circ \alpha = \overline{\alpha} \circ f_1$. The resulting map $\mathrm{Isom}_k(\mathscr{G}_1, \mathscr{G}_2) \to \mathrm{Isom}_k(\overline{G}_1, \overline{G}_2)$ is bijective.*
(2) *Assume $[k : k^p] = p$, and let Φ denote one of the following types: G_2 if $p = 3$, and B_n $(n \geqslant 2)$, C_n $(n \geqslant 2)$, or F_4 if $p = 2$. The assignment $\mathscr{G} \mapsto \overline{G}$ is a bijection from the set of k-isomorphism classes of basic exotic pseudo-reductive k-groups of type Φ to the set of k-isomorphism classes of absolutely simple simply connected k-groups of type $\Phi^\vee$.*

Notable cases for which the hypothesis in (2) is satisfied are local and global function fields. Thus, for such k the basic exotic pseudo-reductive k-groups of type B_n (resp. C_n) for $n \geqslant 2$ are classified up to k-isomorphism by the classification of k-forms of simply connected absolutely simple groups of type C_n (resp. B_n) [Spr, 17.2].

Proof To prove (1), by Galois descent we may and do assume $k = k_s$, so each $\mathscr{G}_i$ contains a Levi k-subgroup G_i and the map $G_i \to \overline{G}_i$ induced by f_i is the very special isogeny for G_i. Since G_1 and G_2 are absolutely simple and simply connected semisimple k-groups of the same type and $k = k_s$, there exists a k-isomorphism $G_1 \simeq G_2$ over a k-isomorphism $\overline{G}_1 \simeq \overline{G}_2$ and hence a k-isomorphism $\mathscr{G}_1 \simeq \mathscr{G}_2$ over a k-isomorphism $\overline{G}_1 \simeq \overline{G}_2$. For the proof of (1) it therefore suffices to work with a single $\mathscr{G}$ (corresponding to a triple $(G', k'/k, \overline{G})$) and study k-automorphisms of $\mathscr{G}$ and $\overline{G}$; this makes the notation a bit simpler. Let $G \subseteq \mathscr{G}$ be a Levi k-subgroup, so $f|_G : G \to \overline{G}$ is the very special isogeny for G. By our hypotheses, $k' = k^{1/p}$.

We claim that the reductive quotient map $f : \mathscr{G} \to \overline{G}$ is maximal over k in the sense that any surjective k-homomorphism $\mathscr{G} \to H$ onto a reductive k-group uniquely factors through f. To prove this, first observe by a noetherian argument that a maximal such quotient exists (it is $\mathscr{G}/\mathscr{N}$ for the minimal normal k-subgroup scheme $\mathscr{N}$ such that $\mathscr{R}_u(\mathscr{G}_{\overline{k}}) \subseteq \mathscr{N}_{\overline{k}}$), and it is clearly unique. If $\Pi : \mathscr{G} \twoheadrightarrow \overline{H}$ denotes this quotient then Π dominates f via a unique map $\phi : \overline{H} \to \overline{G}$. We assume that ϕ is not an isomorphism and seek a contradiction. By Theorem 7.2.3(3), the maximal reductive quotient of $\mathscr{G}_{k'}$ is the natural map $q : \mathscr{G}_{k'} \to G'$ which restricts to an isomorphism $G_{k'} \simeq G'$ on the Levi k'-subgroup $G_{k'}$. The reductive quotient $\Pi_{k'}$ over k' induces a quotient map $\pi : G' \to \overline{H}_{k'}$, and the composite map

$$G' \stackrel{\pi}{\longrightarrow} \overline{H}_{k'} \stackrel{\phi_{k'}}{\longrightarrow} \overline{G}_{k'}$$

is the very special isogeny of G' (due to the isomorphism $G_{k'} \simeq G'$, which carries $\phi_{k'} \circ \pi$ over to $(f|_G)_{k'}$). This is a factorization of the very special isogeny as a composition of surjections with $\phi_{k'}$ not an isomorphism, so by Lemma 7.1.2 either π is an isomorphism (i.e., $\Pi_{k'} = q$) or else $\mathrm{char}(k) = 2$ with G' of type B or C and $\ker \pi = Z_{G'} = \mu_2$. Hence, the k-subgroup $\ker \Pi \subseteq \mathscr{G}$ is a k-descent of either $\ker q$ or $q^{-1}(Z_{G'})$ respectively, so to complete the proof that ϕ is an isomorphism it suffices to prove that neither of these descends to a k-subgroup of $\mathscr{G}$. Proposition 7.2.7(1) gives minimality of k'/k as a field of definition for $\ker q$ (a descent of $\mathscr{R}_u(\mathscr{G}_{\overline{k}})$), and then Proposition 5.3.3 gives the same for $q^{-1}(Z_{G'})$ since $\mathscr{G}$ is perfect.

In view of the uniqueness of the maximal reductive quotient map f over k, it follows that for every k-automorphism α of $\mathscr{G}$ there is a unique k-automorphism $\overline{\alpha}$ of $\overline{G}$ such that $f \circ \alpha = \overline{\alpha} \circ f$. The resulting map $\mathrm{Aut}_k(\mathscr{G}) \to \mathrm{Aut}_k(\overline{G})$ is injective by Proposition 1.2.2 (since the maximal reductive quotient $G'_{\overline{k}}$ of $\mathscr{G}_{\overline{k}}$ maps onto $\overline{G}_{\overline{k}}$ via an isogeny; see Proposition 7.2.7(2)). To prove surjectivity, consider a Levi k-subgroup G in $\mathscr{G}$ (which exists since $k = k_s$). As we noted earlier, f restricts to a map $\pi : G \to \overline{G}$ that is the very special isogeny. In the factorization

$$G \stackrel{\pi}{\to} \overline{G} \to G^{(p)}$$

of $F_{G/k}$, the second map is the very special isogeny for $\overline{G}$ (as follows from Proposition 7.1.5(1),(2) applied to G and Lemma 7.1.2 applied to $\overline{G}$, together with degree considerations).

We conclude that for any k-automorphism $\overline{\alpha}$ of $\overline{G}$, there is an induced k-automorphism α' of the very special quotient $G^{(p)}$, which in turn lies over the k-automorphism $\overline{\alpha}^{(p)}$ of $\overline{G}^{(p)}$ via $\pi^{(p)}$. The p-power map of k that is used to define $G^{(p)}$ is identified with the inclusion of k into $k' := k^{1/p}$, so α' may be viewed as a k'-automorphism of $G_{k'}$ that lies over the k'-automorphism $\overline{\alpha}_{k'}$ of

$\overline{G}_{k'}$ via $\pi_{k'}$. Thus, the map

$$R_{k'/k}(\pi_{k'}) : R_{k'/k}(G_{k'}) \to R_{k'/k}(\overline{G}_{k'})$$

is compatible with $R_{k'/k}(\alpha')$ and $R_{k'/k}(\overline{\alpha}_{k'})$. Since $R_{k'/k}(\overline{\alpha}_{k'})$ restricts to $\overline{\alpha}$ on the Levi k-subgroup $\overline{G}$, we conclude that $R_{k'/k}(\alpha')$ restricts to a k-automorphism of $R_{k'/k}(\pi)^{-1}(\overline{G}) = \mathscr{G}$ that is carried to $\alpha' \in \mathrm{Aut}_k(\overline{G})$ via $f : \mathscr{G} \twoheadrightarrow \overline{G}$.

Now we deduce (2) from (1). Under the assumption in (2), the condition $k'_1 = k'_2 = k^{1/p}$ required in (1) is always satisfied since k'_i/k is a nontrivial finite extension satisfying $k'_i \subseteq k^{1/p}$. In particular, since all possibilities for $\overline{G}$ become isomorphic over k_s, it follows that all possibilities for $\mathscr{G}$ become isomorphic over k_s. Moreover, at least one $\mathscr{G}$ exists, namely the one associated to a triple $(G_{k'}, k'/k, \overline{G})$ with k-split G. Thus, upon fixing one such $\mathscr{G}$, the set of all k-forms is in natural bijection with $H^1(k_s/k, \mathrm{Aut}_{k_s}(\mathscr{G}_{k_s}))$. Moreover, (1) provides a natural $\mathrm{Gal}(k_s/k)$-equivariant isomorphism $\mathrm{Aut}_{k_s}(\mathscr{G}_{k_s}) \to \mathrm{Aut}_{k_s}(\overline{G}_{k_s})$, and the induced bijection in degree-1 Galois cohomology corresponds exactly to the assignment considered in (2). Thus, that assignment is bijective. □

An interesting consequence of Proposition 7.3.5 is the existence of basic exotic pseudo-reductive k-groups without a Levi k-subgroup, over any local or global function field k of characteristic 2. In Example 7.2.2 we saw standard pseudo-reductive k-groups without Levi k-subgroups over any local or global function field, but examples in the basic exotic case are especially interesting because they demonstrate that in the condition in Definition 7.2.6(iii) on existence of a Levi subgroup it really is necessary to extend scalars to k_s.

We recall that the *k-rank* of a connected smooth affine k-group is the dimension of any maximal k-split torus it contains (see Theorem C.2.3).

Proposition 7.3.6 *Let k be a local or global function field of characteristic $p = 2$, and $\mathscr{G}$ a basic exotic pseudo-reductive k-group with $\mathscr{G}^{\mathrm{ss}}_{\overline{k}}$ of rank $n \geqslant 2$. Assume $\mathscr{G}$ is not pseudo-split over k.*

The k-group $\mathscr{G}$ does not contain a Levi k-subgroup if the root system of $\mathscr{G}^{\mathrm{ss}}_{\overline{k}}$ is either of type B_n with $n \geqslant 2$ or of type C_n with n even. If the root system of $\mathscr{G}^{\mathrm{ss}}_{\overline{k}}$ is of type C_n with n odd, then $\mathscr{G}$ contains a Levi k-subgroup.

If $n \geqslant 3$ and if k'/k denotes the field of definition of $\mathscr{R}_u(\mathscr{G}_{\overline{k}})$ then the k'-rank of $\mathscr{G}^{\mathrm{ss}}_{k'} = \mathscr{G}^{\mathrm{red}}_{k'}$ is larger than the k-rank of $\mathscr{G}$.

By Proposition 7.3.5, every $\mathscr{G}$ of type F_4 or G_2 over a local or global function field (of characteristic 2 or 3 respectively) is pseudo-split since absolutely simple semisimple groups of these types are split over such fields (due to vanishing results for H^1 of simply connected groups, such as the automorphism groups for types F_4 and G_2). However, there do exist non-pseudo-split $\mathscr{G}$ with $\mathscr{G}^{\mathrm{ss}}_{\overline{k}}$ of any type B_n or C_n with $n \geqslant 2$ since non-split absolutely simple and semisimple k-groups of each of these types exist (due to the classification of such groups). The hypothesis that $\mathscr{G}$ is not pseudo-split is necessary, since by Theorem 3.4.6 a pseudo-split pseudo-reductive group over any field k admits a Levi k-subgroup. The inequality $\mathrm{rank}_k(\mathscr{G}) < \mathrm{rank}_{k'}(\mathscr{G}^{\mathrm{ss}}_{k'})$ is interesting because this never occurs for absolutely pseudo-simple standard pseudo-reductive k-groups (due to Proposition 4.1.4(2)).

Proof In all cases, $\mathscr{G}$ corresponds to a triple $(G', k'/k, \overline{G})$ giving rise to $\mathscr{G}$ with $k' = k^{1/p}$ equal to the field of definition over k for $\mathscr{R}_u(\mathscr{G}_{\overline{k}})$. There is a surjective k-homomorphism $f : \mathscr{G} \twoheadrightarrow \overline{G}$. For $n \geqslant 2$, a non-split absolutely simple and semisimple group of type B_n (resp. type C_n) over a local or global function field has rank $n-1$ (resp. rank $n/2$ or $(n-1)/2$, depending on the parity of n). The k-group $\overline{G}$ has type dual to that of $\mathscr{G}^{\mathrm{ss}}_{\overline{k}}$ with the same absolute rank, so by Corollary 7.3.4 it follows that $\mathscr{G}$ has k-rank equal to $n/2$ or $(n-1)/2$ if $\mathscr{G}^{\mathrm{ss}}_{\overline{k}}$ is of type B_n and k-rank equal to $n-1$ if it is of type C_n.

Suppose there is a Levi k-subgroup G in $\mathscr{G}$. If $\mathscr{G}^{\mathrm{ss}}_{\overline{k}}$ is of type B_n then G is non-split of type B_n and hence of k-rank $n-1$. But $\mathscr{G}$ has k-rank $n/2$ or $(n-1)/2$ (depending on the parity of $n \geqslant 3$) which is strictly smaller than $n-1$, so no such G can exist.

Now let us assume that $\mathscr{G}^{\mathrm{ss}}_{\overline{k}}$ is of type C_n with n even. Then G is a non-split absolutely simple simply connected k-group of type C_n, and hence its k-rank is $n/2$. Let S be a maximal k-split torus of G and let $A = \mathscr{D}(Z_G(S))$. In the Tits index of G (see [Ti0, Table II]) there are $n/2$ non-distinguished vertices, and they are all isolated. Thus, there is a central quaternion division algebra D over k such that A is the direct product of $n/2$ copies of the anisotropic k-group $\mathrm{SL}_{1,D}$. Since the root corresponding to each non-distinguished vertex in the Tits index of G is short, the very special isogeny $\pi : G \to \overline{G}$ restricts to a quotient map $A \to \overline{A} := \pi(A) = \mathscr{D}(Z_{\overline{G}}(\pi(S)))$ whose kernel is the Frobenius kernel of A, so this latter map is identified with the Frobenius morphism for A. Hence, $\overline{A}$ is isomorphic to the direct product of $n/2$ copies of SL_2. We conclude that the k-rank of $\overline{G}$ is equal to $\dim \pi(S) + n/2 = n$. Corollary 7.3.4 implies then that the k-rank of $\mathscr{G}$ is also n, and hence $\mathscr{G}$ is pseudo-split, contrary to our hypothesis. Hence, $\mathscr{G}$ cannot contain a Levi k-subgroup.

To prove the existence of Levi k-subgroups when $\mathscr{G}^{\mathrm{ss}}_{\overline{k}}$ is of type C_n with odd $n > 1$, we carry out a Galois-twisting argument beginning in the pseudo-split case. To do this, let G be an absolutely simple and simply connected k-split k-group of type C_n with odd $n > 1$ and let $\pi : G \to \overline{G}$ be the very special isogeny. Let $\mathscr{G}$ be the pseudo-split basic exotic pseudo-reductive k-group corresponding to the triple $(G', k'/k, \overline{G})$, where $k' = k^{1/2}$ and $G' = G_{k'}$. Then $\mathscr{G} \subseteq \mathrm{R}_{k'/k}(G')$ and the inclusion $G \hookrightarrow \mathrm{R}_{k'/k}(G')$ is a Levi k-subgroup of $\mathscr{G}$. It suffices to show that all k-forms of $\mathscr{G}$ admit a Levi k-subgroup (since over k_s there is only one isomorphism class of basic exotic pseudo-reductive group of each semisimple type, by the end of Proposition 7.2.7(2)).

As n is odd, the center Z of $\mathscr{G}$ is isomorphic to μ_2 and is contained in G, and π maps Z isomorphically onto the center $\overline{Z}$ of $\overline{G}$. Therefore, π induces a homomorphism $\mathrm{Aut}_{k_s}(G_{k_s}) = (G/Z)(k_s) \to (\overline{G}/\overline{Z})(k_s) = \mathrm{Aut}_{k_s}(\overline{G}_{k_s})$. By Proposition 7.3.5(1), we have

$$\mathrm{H}^1(k_s/k, \mathrm{Aut}_{k_s}(\mathscr{G}_{k_s})) = \mathrm{H}^1(k_s/k, \mathrm{Aut}_{k_s}(\overline{G}_{k_s})) = \mathrm{H}^1(k_s/k, (\overline{G}/\overline{Z})(k_s)).$$

The well-known triviality of degree-1 Galois cohomology of simply connected groups over local and global function fields implies that in the commutative diagram of natural maps

$$\begin{array}{ccc} \mathrm{H}^1(k_s/k, (G/Z)(k_s)) & \xrightarrow{\delta} & \mathrm{H}^2(k, Z) \\ \downarrow & & \downarrow \simeq \\ \mathrm{H}^1(k_s/k, (\overline{G}/\overline{Z})(k_s)) & \xrightarrow[\overline{\delta}]{} & \mathrm{H}^2(k, \overline{Z}) \end{array}$$

(in which the right side is the isomorphism $Z \simeq \overline{Z}$ induced by π), the horizontal connecting maps are injective. The terms on the right side are identified with $\mathrm{H}^2(k, \mu_2) = \mathrm{Br}(k)[2]$, and for local and global function fields this is the set of isomorphism classes of central simple quaternion algebras over k. The explicit construction of all k-forms of groups of types B_n and C_n in terms of quaternion division algebras implies that the connecting maps are bijective. Thus, the k-forms of both $\mathscr{G}$ and $\overline{G}$ are compatibly parameterized by $\mathrm{H}^2(k, \overline{Z})$.

For any continuous 1-cocycle $c : \mathrm{Gal}(k_s/k) \to (G/Z)(k_s)$ the Galois-twists ${}_c\mathscr{G}$ and ${}_cG$ make sense, with ${}_cG$ a Levi k-subgroup in ${}_c\mathscr{G}$. Thus, it suffices to show that as c varies, we obtain all k-forms of $\mathscr{G}$; that is, for each $x \in \mathrm{H}^2(k, \overline{Z})$ we seek a continuous 1-cocycle $c = c(x)$ on $\mathrm{Gal}(k_s/k)$ with values in $(G/Z)(k_s) \subseteq \mathrm{Aut}_{k_s}(\mathscr{G}_{k_s})$ such that the image of the cohomology class of $c(x)$ in $\mathrm{H}^2(k, \overline{Z})$ is equal to x. The existence of such a $c(x)$ is immediate from the above commutative diagram of natural maps.

Now we turn to a comparison of the k-rank of $\mathscr{G}$ and the k'-rank of $\mathscr{G}^{\mathrm{ss}}_{k'}$ when $\mathscr{G}$ is not pseudo-split over k and $\mathscr{G}^{\mathrm{ss}}_{k'}$ is either of type B_n or of type C_n

with $n \geqslant 3$. The natural quotient map $q : \mathscr{G}_{k'} \twoheadrightarrow G'$ is a k'-descent of the maximal reductive quotient of $\mathscr{G}_{\overline{k}}$ (Proposition 7.2.7(2)), so $\mathscr{G}_{k'}^{\rm ss} = G'$.

It follows that if $\mathscr{G}_{\overline{k}}^{\rm ss}$ is of type B_n then the k'-rank of G' is at least $n-1$. We saw above that in such cases the non-pseudo-split $\mathscr{G}$ has k-rank $n/2$ or $(n-1)/2$, which is strictly smaller than the k'-rank of $G' = \mathscr{G}_{k'}^{\rm ss}$ if $n \geqslant 3$. Suppose instead that $\mathscr{G}_{\overline{k}}^{\rm ss}$ is of type C_n, so the k-rank of $\mathscr{G}$ is $n-1$. The quotient $G' = \mathscr{G}_{k'}^{\rm ss}$ of $\mathscr{G}_{k'}$ therefore has k'-rank at least $n-1$. But G' is of type C_n, so if G' were not k'-split then its k'-rank would be $n/2$ or $(n-1)/2$, which is strictly smaller than $n-1$. Hence, G' is k'-split, so its k'-rank is n, which strictly exceeds the k-rank of $\mathscr{G}$. □

Let k be an imperfect field of characteristic $p \in \{2,3\}$ and let $\mathscr{G}$ be a basic exotic pseudo-reductive k-group with associated unique triple (up to unique k-isomorphism) $(G', k'/k, \overline{G})$ as in Definition 7.2.6 and Proposition 7.2.7, so $\mathscr{G}_{\overline{k}}^{\rm red} \simeq G'_{\overline{k}}$. Assume this is of type F_4 when $p=2$ and of type G_2 when $p=3$, so it is adjoint and thus we can try to use Corollary 3.4.8 to construct Levi k-subgroups in $\mathscr{G}$. The case of most interest is $k' = k^{1/p}$ (so $[k : k^p]$ is finite). If $[k : k^p] = p$ then this always holds.

Proposition 7.3.7 *Let $\mathscr{G}$ be a basic exotic pseudo-reductive k-group such that $\mathscr{G}_{\overline{k}}^{\rm red}$ is of type F_4 or G_2. Let $\mathscr{T}$ be a maximal k-torus in $\mathscr{G}$. If there is a finite extension K/k of degree not divisible by p such that $\mathscr{T}_K$ is K-split then $\mathscr{G}$ admits a Levi k-subgroup containing $\mathscr{T}$.*

If $k' = k^{1/p}$ and $p=3$ then there exists a Levi k-subgroup of $\mathscr{G}$.

We do not know any examples without a Levi k-subgroup when $p=2$ for type F_4, nor when $p=3$ for type G_2.

Proof First assume that such a K/k exists. The explicit structure of $Z_{\mathscr{G}}(\mathscr{T})_{k_s}$ and root groups for $(\mathscr{G}_{k_s}, \mathscr{T}_{k_s})$ shows that the principal homogeneous space hypothesis in Corollary 3.4.8 is satisfied. Thus, a necessary and sufficient condition for the existence of a Levi k-subgroup containing $\mathscr{T}$ is the vanishing of a certain class $c \in \mathrm{H}^1(k, Z_{\mathscr{G}}(\mathscr{T})/\mathscr{T})$ whose formation commutes with separable extension on k. By hypothesis, $c|_K$ therefore vanishes for a separable extension K/k of degree not divisible by p. But the smooth connected commutative k-group $Z_{\mathscr{G}}(\mathscr{T})/\mathscr{T}$ is unipotent, so its group of k_s-points is p-power torsion (in fact it is p-torsion by inspection since $k'^p \subseteq k$). Hence, the splitting of a class c by a finite separable extension of degree not divisible by p implies that $c=1$.

Now assume that $p=3$ and $k' = k^{1/p}$. In this case it suffices to prove that there exists a $\mathscr{T}$ split over a separable quadratic extension of k. The hypothesis $k' = k^{1/p}$ is preserved under separable algebraic extension on k, so by Corollary 7.3.4 the problem of constructing a maximal k-torus in $\mathscr{G}$ that is split by a separable quadratic extension of k is reduced to the analogous problem for the k-form $\overline{G}$ of G_2. For any field F with characteristic different from 2, the set of isomorphism classes of F-forms of G_2 is naturally identified (functorially in F) with the subset of $H^3(F, \mathbf{Z}/2\mathbf{Z})$ consisting of triple cup products from $H^1(F, \mathbf{Z}/2\mathbf{Z})$ [Spr, 17.4.6]. In particular, any such form is split by some quadratic extension of F. Thus, Lemma 7.3.8 below completes the proof. □

Lemma 7.3.8 *Let F be a field and G a connected reductive F-group that is split by a separable quadratic extension F'/F. There exists a maximal F-torus in G that is F'-split.*

Proof Let $B' \subseteq G_{F'}$ be a Borel subgroup defined over F' and let $\overline{B}'$ denote its conjugate by the nontrivial element of $\mathrm{Gal}(F'/F)$. By [Bo2, 14.22(i), 20.7(i)], $B' \cap \overline{B}'$ is a smooth connected F'-subgroup of $G_{F'}$ and it contains a maximal F'-torus of $G_{F'}$.

Since B' is an F'-Borel subgroup of the F'-split connected reductive F'-group $G_{F'}$, its F'-unipotent radical $\mathscr{R}_{u,F'}(B')$ descends $\mathscr{R}_u(B'_{\overline{F}'})$, so every maximal F'-torus in B' maps isomorphically onto the quotient $B'/\mathscr{R}_{u,F'}(B')$ that is an F'-split torus. Thus, all maximal F'-tori in B' are F'-split. The smooth connected $\mathrm{Gal}(F'/F)$-stable F'-subgroup $B' \cap \overline{B}'$ in $G_{F'}$ descends to a smooth connected F-subgroup H in G, and the maximal F-tori in H are also maximal in G (as may be checked over F'). Choosing such an F-torus T, we conclude that $T_{F'}$ is F'-split since it is a maximal F'-torus in B'. □

8

Preparations for classification in characteristics 2 and 3

To complete the proof of Theorem 5.1.1 we need to understand the significance of the constructions in Chapter 7, especially the notion of basic exotic pseudo-reductive group that was introduced in Definition 7.2.6 over any imperfect field k of characteristic 2 or 3.

8.1 Further properties of basic exotic pseudo-reductive groups

This section lays the groundwork for the definition (in §8.2) of the concept of an *exotic pseudo-reductive group* over an imperfect field k of characteristic 2 or 3, generalizing the concept of a basic exotic pseudo-reductive group. The real work will be to show that it accounts for all exceptional absolutely pseudo-simple groups (in the sense of Definition 6.3.3), at least when $G^{ss}_{\overline{k}}$ is simply connected.

Proposition 8.1.1 *Let k be an imperfect field of characteristic* 2 *or* 3*. A basic exotic pseudo-reductive k-group $\mathscr{G}$ is never standard over k.*

Proof By Corollary 5.2.3 we may assume $k = k_s$. Let $(G', k'/k, \overline{G})$ be the unique triple giving rise to $\mathscr{G}$ as in Proposition 7.2.7. By Corollary 7.2.5(3) the long root groups of $\mathscr{G}$ have dimension 1, and by construction the short root groups have dimension > 1 (in fact, $[k' : k]$). The pseudo-reductive k-group $\mathscr{G}$ is absolutely pseudo-simple (Theorem 7.2.3 and Definition 7.2.6), and an absolutely pseudo-simple standard pseudo-reductive group has all root groups of the same dimension. Hence, $\mathscr{G}$ is not standard. □

We wish to apply Proposition 5.1.3 in the context of basic exotic pseudo-reductive groups, as well as their Weil restrictions:

Proposition 8.1.2 *Let k be an imperfect field of characteristic $p \in \{2,3\}$, K a nonzero finite reduced k-algebra, and $\mathscr{G}$ a K-group whose fibers are basic exotic pseudo-reductive groups in the sense of Definition 7.2.6.*

The k-group $\mathrm{R}_{K/k}(\mathscr{G})$ is pseudo-reductive, perfect, and not reductive. Moreover, if Z is a commutative k-group scheme of finite type with no nontrivial smooth connected k-subgroup then any central extension of $\mathrm{R}_{K/k}(\mathscr{G})$ by Z over k is split over k.

Proof By Proposition 1.1.10, $\mathrm{R}_{K/k}(\mathscr{G})$ is pseudo-reductive over k. It is not reductive, by Proposition 7.2.7(1) applied to the (non-reductive) quotient $\mathscr{G}$ of $\mathrm{R}_{K/k}(\mathscr{G})_K$ (also see Lemma A.5.10). To prove that $\mathrm{R}_{K/k}(\mathscr{G})$ is its own derived group, we can assume k is separably closed and K is a field. By Proposition A.7.10, it suffices to prove that $\mathscr{G}$ is perfect and that $\mathrm{Lie}(\mathscr{G})$ has a vanishing space of coinvariants with respect to the adjoint action by $\mathscr{G}$. The key point is to explicitly describe the open Bruhat cell in $\mathscr{G}$ relative to a maximal K-torus $\mathscr{T}$ and a positive system of roots Φ^+ in $\Phi(\mathscr{G},\mathscr{T})$. By Theorem 7.2.3(2) and Corollary 7.2.5, for each root c the opposite root groups U_c and U_{-c} in $\mathscr{G}$ generate a copy of SL_2 (with diagonal $c^\vee(\mathrm{GL}_1)$) if c is long and a copy of $\mathrm{R}_{K/k}(\mathrm{SL}_2)$ (with diagonal $\mathrm{R}_{K/k}(c^\vee(\mathrm{GL}_1))$) if c is short. Moreover, if $\Delta \subseteq \Phi^+$ is the set of simple positive roots then

$$Z_{\mathscr{G}}(\mathscr{T}) = \prod_{a\in\Delta_>} a^\vee(\mathrm{GL}_1) \times \prod_{b\in\Delta_<} \mathrm{R}_{K/k}(b^\vee(\mathrm{GL}_1)). \tag{8.1.1}$$

Thus, the vanishing of the $\mathscr{G}$-coinvariants of $\mathrm{Lie}(\mathscr{G})$ is reduced to the analogous vanishing property for SL_2 and for $\mathrm{R}_{K/k}(\mathrm{SL}_2)$. Since $\mathrm{Lie}(\mathrm{R}_{K/k}(\mathrm{SL}_2)) = \mathfrak{sl}_2(K)$ as a Lie algebra over k (Corollary A.7.6), it has vanishing space of coinvariants over k with respect to the adjoint action of $\mathrm{R}_{K/k}(\mathrm{SL}_2)(k) = \mathrm{SL}_2(K)$. The subgroup of $\mathscr{G}(k)$ generated by copies of $\mathrm{SL}_2(k)$ and $\mathrm{SL}_2(K)$ corresponding to the long and short roots is Zariski-dense in $\mathscr{G}$, so $\mathscr{G}$ is perfect. This completes the proof that $\mathrm{R}_{K/k}(\mathscr{G})$ is perfect.

It remains to split central extensions of $\mathrm{R}_{K/k}(\mathscr{G})$ by commutative k-groups Z of finite type that contain no nontrivial smooth connected k-subgroups. Since $\mathrm{R}_{K/k}(\mathscr{G})$ is perfect for any k (perhaps not separably closed), such a splitting is provided by Proposition 5.1.3 when $Z_{\mathscr{G}}(\mathscr{T})_{k_s}$ is rationally generated by root groups, which is to say that there exist rational maps h_i on $Z_{\mathscr{G}}(\mathscr{T})$ as in Proposition 5.1.3 when $k = k_s$. By (8.1.1), it suffices to check the analogous property for Cartan k-subgroups of SL_2 and $\mathrm{R}_{K/k}(\mathrm{SL}_2)$. This in turn is settled by Example 5.1.4. □

Remark 8.1.3 Let K/k be a purely inseparable finite extension of fields. In view of Corollary A.7.11, it is natural to ask if $\mathrm{R}_{K/k}(H)$ is perfect for absolutely pseudo-simple K-groups H such that $H^{\mathrm{ss}}_{\overline{K}}$ is simply connected. Basic exotic H constitute a nontrivial class of examples beyond the reductive case for which the answer is affirmative, but for imperfect k of characteristic 2 such that $[k : k^2] > 2$ there is an abundant supply of counterexamples (e.g., see the discussion in §9.1 from 9.1.8 onwards). This phenomenon will be explored at length in [CP].

Corollary 8.1.4 *Let $(\mathscr{G}, K/k)$ be as in Proposition 8.1.2, $\{K_i\}$ the set of factor fields of K, and $\mathscr{G}_i$ the fiber of $\mathscr{G}$ over K_i.*

Every smooth connected normal k-subgroup of the pseudo-reductive k-group $G := \mathrm{R}_{K/k}(\mathscr{G})$ is a product of some of the k-subgroups $\mathrm{R}_{K_i/k}(\mathscr{G}_i)$. In particular, G is absolutely pseudo-simple if and only if K/k is a purely inseparable field extension.

Proof By Galois descent we may reduce to the case of separably closed k. We first check that each factor $G_i = \mathrm{R}_{K_i/k}(\mathscr{G}_i)$ is pseudo-simple. The natural map $(G_i)_{K_i} \twoheadrightarrow \mathscr{G}_i$ has smooth connected unipotent kernel since K_i/k is purely inseparable (Proposition A.5.11). By Theorem 7.2.3(1), $\mathscr{G}_i$ is absolutely pseudo-simple over k_i. Since G_i is perfect (by Proposition 8.1.2), pseudo-simplicity of G_i follows from Lemma 3.1.2.

Now let N be a smooth connected normal k-subgroup of G. To prove that N has the asserted form, the pseudo-simplicity of the factors G_i reduces us to the case in which N maps onto each factor. Thus, the smooth connected k-subgroup (N, G_i) is normal and also maps onto each G_i (as G_i is perfect). But (N, G_i) is contained in the pseudo-simple G_i, so $(N, G_i) = G_i$ for all i. Since (N, G_i) is contained in N, we get $N = G$. □

We next wish to make precise the sense in which the pseudo-reductive k-group $G = \mathrm{R}_{K/k}(\mathscr{G})$ for a basic exotic pseudo-reductive K-group $\mathscr{G}$ over a finite extension K/k in characteristic $p \in \{2, 3\}$ uniquely determines the pair $(\mathscr{G}, K/k)$ up to canonical k-isomorphism (much like the uniqueness of the pair $(G', k'/k)$ in the non-commutative standard case via Proposition 4.2.4). We also want to reconstruct the k-group identification $G \simeq \mathrm{R}_{K/k}(\mathscr{G})$ from the k-group G. This rests on the determination of the universal maximal absolutely simple adjoint quotient of G in the sense defined after Proposition 4.2.1, as we now explain.

Proposition 8.1.5 *Let k be imperfect of characteristic $p \in \{2, 3\}$, K/k a nonzero finite reduced k-algebra, and $\mathscr{G}$ a K-group with basic exotic* *fibers arising from a triple* $(G', K'/K, \overline{G})$. *Let* $G = \mathrm{R}_{K/k}(\mathscr{G}) \subseteq \mathrm{R}_{K/k}(\mathrm{R}_{K'/K}(G')) = \mathrm{R}_{K'/k}(G')$.

(1) *The universal maximal absolutely simple adjoint quotient of G is* $(K'/k, f)$, *where*

$$f : G_{K'} \to \mathrm{R}_{K'/k}(G')_{K'} \twoheadrightarrow G' \to G'/Z_{G'}$$

is the canonical composite map. In particular, the universal maximal absolutely simple adjoint quotient map f uniquely factors through the simply connected central cover $G' \to G'/Z_{G'}$ *of this quotient.*

(2) *The k-subalgebra* $K \subseteq K'$ *is the maximal one such that the k-subspace*

$$\mathrm{Lie}(G) \subseteq \mathrm{Lie}(\mathrm{R}_{K'/k}(G')) = \mathrm{Lie}(G')$$

of the K'*-module* $\mathrm{Lie}(G')$ *is a K-submodule. Also, the K-map* $G_K \to \mathrm{R}_{K'/K}(G')$ *corresponding to* $G \to \mathrm{R}_{K'/k}(G') = \mathrm{R}_{K/k}(\mathrm{R}_{K'/K}(G'))$ *has image* $\mathscr{G}$.

The case $K = k$ (equivalently $G = \mathscr{G}$) is the most important one, but for the purpose of obtaining a classification theorem for all pseudo-reductive groups in characteristics 2 and 3 we need to permit the possibility that $K \neq k$ and that K may not be étale over k.

Proof If $\{K'_i\}$ and $\{K_i\}$ are the compatible respective collections of factors fields of K' and K then every intermediate k-algebra F between K' and K has the form $F = \prod F_i$ for some intermediate fields $K_i \subseteq F_i \subseteq K'_i$. This applies to the maximal k-subalgebra as defined for (2), so we may use Galois descent to reduce to the case $k = k_s$ and pass to factor fields of K to reduce to the case where K and K' are fields.

We first prove (2), and then (1). The k-subspace $\mathrm{Lie}(G) \subseteq \mathrm{Lie}(G')$ is the K-subspace $\mathrm{Lie}(\mathscr{G})$, so to prove the maximality claim concerning K in (2) it suffices to treat the case $k = K$ (i.e., $G = \mathscr{G}$). Since K is separably closed, $\mathscr{G}$ contains a Levi K-subgroup L of $\mathrm{R}_{K'/K}(G')$. The K-subspace $\mathrm{Lie}(\mathscr{G}) \subseteq \mathrm{Lie}(G') = K' \otimes_K \mathrm{Lie}(L)$ is therefore described explicitly in (7.2.1). Inspection of the contribution from long roots therefore gives the maximality claim for K.

For the second claim in (2) we use a functoriality argument as follows. For any affine schemes X and Y of finite type over k and K respectively, a k-map $\phi : X \to \mathrm{R}_{K/k}(Y)$ corresponds to the K-map

$$X_K \xrightarrow{\phi_K} \mathrm{R}_{K/k}(Y)_K \to Y,$$

where the second map is given on R-valued points for a K-algebra R by the map $Y(K \otimes_k R) \to Y(R)$ arising from the K-algebra multiplication map $K \otimes_k R \to R$ (see Proposition A.5.7).

Thus, if $j : \mathscr{G} \hookrightarrow \mathrm{R}_{K'/K}(G')$ is the inclusion then $\alpha := \mathrm{R}_{K/k}(j) : G \hookrightarrow \mathrm{R}_{K'/k}(G')$ is the inclusion and the map $G_K \to \mathrm{R}_{K'/K}(G')$ of interest in (2) is the composition of α_K with the natural projection

$$q' : \mathrm{R}_{K'/k}(G')_K = \mathrm{R}_{K/k}(\mathrm{R}_{K'/K}(G'))_K \twoheadrightarrow \mathrm{R}_{K'/K}(G').$$

Consider $q : G_K \to \mathscr{G}$ that corresponds to the equality $G = \mathrm{R}_{K/k}(\mathscr{G})$ defining G. The diagram of K-groups

$$\begin{array}{ccc} G_K & \xrightarrow{\alpha_K} & \mathrm{R}_{K'/k}(G')_K \\ {\scriptstyle q}\downarrow & & \downarrow{\scriptstyle q'} \\ \mathscr{G} & \xrightarrow[j]{} & \mathrm{R}_{K'/K}(G') \end{array}$$

commutes and the left side is a smooth surjection with connected unipotent kernel (Proposition A.5.11), so the final assertion of (2) is proved.

Turning to the proof of (1), the natural map

$$G_K = \mathrm{R}_{K/k}(\mathscr{G})_K \to \mathscr{G}$$

is a smooth surjection with a connected unipotent kernel as observed above. Hence, by extending scalars from K to K' we obtain a smooth surjection

$$G_{K'} \to \mathscr{G}_{K'} \tag{8.1.2}$$

over K' with a smooth connected unipotent kernel. By Theorem 7.2.3(3), $\mathscr{G}_{K'}$ admits a canonical smooth surjection onto G' with a smooth connected unipotent kernel. Composing this with (8.1.2), we get a quotient map

$$\Pi : G_{K'} \twoheadrightarrow G'$$

over K' with a smooth connected unipotent kernel. Thus, Π is a K'-descent of the maximal simple semisimple quotient of $G_{\overline{k}}$. In particular, the maximal simple adjoint quotient of $G_{\overline{k}}$ is defined over K' (as a quotient) and as such is naturally identified with the quotient $G'/Z_{G'}$. In view of how the universal maximal absolutely simple adjoint quotient was constructed over a separably closed ground field in the proof of Proposition 4.2.1, our problem is to show that K'/k is the minimal extension over which the maximal adjoint semisimple quotient of $G_{\overline{k}}$ is defined (since this quotient is simple).

The k-group G is perfect (Proposition 8.1.2), so the minimal field of definition over k for the maximal adjoint semisimple quotient of $G_{\overline{k}}$ is the same as the minimal field of definition over k for the maximal semisimple quotient of $G_{\overline{k}}$ (Proposition 5.3.3). Hence, it suffices to prove that $\ker \Pi$ as a K'-subgroup of $G_{K'}$ cannot be defined over a proper subfield containing k. As in the proof of (2), let $L \subseteq \mathscr{G}$ be a Levi K-subgroup of $\mathrm{R}_{K'/K}(G')$. Thus, we have a k-subgroup inclusion $j : \mathrm{R}_{K/k}(L) \hookrightarrow \mathrm{R}_{K/k}(\mathscr{G}) = G$.

Let $h = q_L : \mathrm{R}_{K/k}(L)_K \twoheadrightarrow L$ denote the canonical K-descent of the maximal reductive quotient of $\mathrm{R}_{K/k}(L)_{\overline{k}}$. Since K/k is purely inseparable, by Theorem 1.6.2(2) the K-subgroup $\ker h \subseteq \mathrm{R}_{K/k}(L)_K$ has minimal field of definition over k equal to K. But $h_{K'}$ factors as

$$\mathrm{R}_{K/k}(L)_{K'} \overset{j_{K'}}{\hookrightarrow} G_{K'} \overset{\Pi}{\twoheadrightarrow} G' = L_{K'},$$

so

$$(\ker h)_{K'} = j_{K'}^{-1}(\ker(G_{K'} \overset{\Pi}{\twoheadrightarrow} G')).$$

Thus, if the field of definition over k for the K'-subgroup $\ker \Pi \subseteq G_{K'}$ is $F \subseteq K'$ then $(\ker h)_{K'}$ also descends to F (as j is a k-map). Since $(\ker h)_{K'}$ descends to K, its minimal field of definition over k is therefore contained in $F \cap K$. Hence, the K-subgroup $\ker h \subseteq \mathrm{R}_{K/k}(L)_K$ descends to $F \cap K$. But we observed above that K is the minimal field of definition over k for the K-subgroup $\ker h$ is K, so $F \cap K = K$, which is to say $K \subseteq F$.

Our problem is therefore reduced to showing that K' is the field of definition over K (rather than over k) for $\ker \Pi$. But Π factors as the composition

$$G_{K'} \twoheadrightarrow \mathscr{G}_{K'} \twoheadrightarrow G' \tag{8.1.3}$$

in which the first step is a scalar extension of the quotient map $q : G_K \twoheadrightarrow \mathscr{G}$ over K and the second step is the canonical quotient map. Since we are now studying fields of definition over K, it is equivalent to prove that the kernel of the second step in (8.1.3) (viewed as a K'-subgroup of $\mathscr{G}_{K'}$) has minimal field of definition over K equal to K'. This is exactly Proposition 7.2.7(1). □

8.2 Exceptional and exotic pseudo-reductive groups

Let k be an imperfect field of characteristic 2 or 3. Propositions 7.2.7 and 8.1.5 show that for any finite extension of fields K/k and basic exotic pseudo-reductive K-group $\mathscr{G}$ arising from a triple $(G', K'/K, \overline{G})$, if G is a k-group that is k-isomorphic to $\mathrm{R}_{K/k}(\mathscr{G})$ then G canonically determines the 5-tuple $(K/k, \mathscr{G}, G', K'/K, \overline{G})$ (note that $\mathrm{Aut}(K'/K) = \{1\}$ since K'/K is purely inseparable) as well as a canonical isomorphism $G \simeq \mathrm{R}_{K/k}(\mathscr{G})$. Since G is absolutely pseudo-simple precisely when K/k is purely inseparable (Corollary 8.1.4), such pseudo-reductive groups provide an explicit description of all absolutely pseudo-simple k-groups G such that $G_{\overline{k}}^{\mathrm{ss}}$ is simply connected and G is exceptional in the sense of Definition 6.3.3:

Theorem 8.2.1 *Let k be an imperfect field of characteristic $p \in \{2, 3\}$ and let G be an absolutely pseudo-simple k-group. Then G is exceptional in the sense of Definition 6.3.3 with $G_{\overline{k}}^{\mathrm{ss}}$ simply connected (so $\xi_G = i_G$) if and only if G is the Weil restriction of a basic exotic pseudo-reductive group over a finite purely inseparable extension of k.*

When these two equivalent conditions hold, G is explicitly described as follows. Let K/k be the field of definition for the geometric radical of G, and $G_K^{\mathrm{ss}} \to \overline{G_K^{\mathrm{ss}}}$ the very special K-isogeny. There is a unique proper subfield $\mathsf{k} \subseteq K$ so that $kK^p \subseteq \mathsf{k}$ and the k-map $\xi_G : G \to \mathrm{R}_{K/k}(G_K^{\mathrm{ss}})$ is an isomorphism onto $\mathrm{R}_{\mathsf{k}/k}(\mathscr{G})$, where $\mathscr{G} \subseteq \mathrm{R}_{K/\mathsf{k}}(G_K^{\mathrm{ss}})$ is the basic exotic pseudo-reductive k-subgroup associated to $(G_K^{\mathrm{ss}}, K/\mathsf{k}, L)$ for a Levi k-subgroup L of $\mathrm{R}_{K/\mathsf{k}}(\overline{G_K^{\mathrm{ss}}})$ whose preimage in $\mathrm{R}_{K/\mathsf{k}}(G_K^{\mathrm{ss}})$ is k-smooth.

Proof Since K/k is purely inseparable, the formation of K is compatible with separable algebraic extension on k. Thus, by Galois descent and Proposition 7.2.7(3) we may and do assume $k = k_s$. Fix a maximal k-torus T in G and let $\Phi = \Phi(G, T)$.

Step 1. First we check that if k/k is a finite extension and G is a basic exotic pseudo-reductive k-group such that $G = \mathrm{R}_{\mathsf{k}/k}(\mathsf{G})$ then G is exceptional with $G_{\overline{k}}^{\mathrm{ss}}$ simply connected. (Note that all k-groups of the form $\mathrm{R}_{\mathsf{k}/k}(\mathsf{G})$ are absolutely pseudo-simple, by Proposition 8.1.2 and Corollary 8.1.4 since k/k is purely inseparable, but we are assuming G to be absolutely pseudo-simple at the outset.) By Proposition A.5.15(2) there is a unique maximal k-torus T in G such that T is the maximal k-torus in $\mathrm{R}_{\mathsf{k}/k}(\mathsf{T})$, and since k/k is purely inseparable it follows from Example 2.3.2 that naturally $\mathsf{T} = T_{\mathsf{k}}$ and moreover $\Phi(\mathsf{G}, \mathsf{T}) = \Phi$ via the equality $\mathrm{X}(\mathsf{T}) = \mathrm{X}(T)$. By Proposition 7.2.7(1), $\Phi(\mathsf{G}, \mathsf{T})$ is reduced with connected Dynkin diagram having an edge with multiplicity p, so these properties hold for Φ. The natural map $G_{\mathsf{k}} \to \mathsf{G}$ is a smooth surjection with connected unipotent kernel (Proposition A.5.11), so $G_{\overline{k}}^{\mathrm{ss}} \simeq \mathsf{G}_{\overline{k}}^{\mathrm{ss}}$; this latter group is simply connected (Proposition 7.2.7(2)).

To complete the proof that G is exceptional, it remains to analyze properties of the maps ξ_{G_c} and the fields K_c/k, where c varies through Φ and K_c/k is the field of definition for the geometric radical of the k-subgroup G_c generated by the root groups $U_{\pm c}$. The structure of the open Bruhat cell for (G, T) is given in Remark 7.2.8, from which we get the analogous description for (G, T) by applying $\mathrm{R}_{\mathsf{k}/k}$. This explicit description gives a nontrivial finite extension k'/k satisfying $\mathsf{k}'^p \subseteq \mathsf{k}$ such that $G_c \simeq \mathrm{R}_{\mathsf{k}/k}(\mathrm{SL}_2)$ for long c and $G_c \simeq \mathrm{R}_{\mathsf{k}'/k}(\mathrm{SL}_2)$ for short c. In particular, all maps ξ_{G_c} are isomorphisms and K_c is equal to k when c is long and to k' when c is short. But the fields K_c generate K over k (Proposition 6.3.1(1)), so $K = \mathsf{k}'$ and $K_> = \mathsf{k}$, where $K_>$ is the common subfield K_c for all long roots c. Thus, $K_> \neq K$.

Now we may and do assume that G is exceptional with $G_{\overline{k}}^{\mathrm{ss}}$ simply connected. We need to prove that G is a Weil restriction of the type considered in Step 1, and that more specifically this description is obtained from G by the intrinsic process given in the second part of the statement of the theorem.

Step 2. As in Step 1, for each $c \in \Phi$ let G_c denote the absolutely pseudo-simple k-subgroup of rank 1 in G generated by the root groups $U_{\pm c}$. By Proposition 3.4.1(2), for all $c \in \Phi$ we have $G_c = \mathscr{D}(Z_G(T_c))$ for the codimension-1 k-torus $T_c = (\ker c)^0_{\mathrm{red}}$ in T. Let K_c/k be the field of definition for the geometric radical of G_c. By hypothesis we are in the exceptional case of Proposition 6.3.2, so $K_c = K$ when the root $c \in \Phi$ is short, and for long $c \in \Phi$ the fields K_c are all equal to a common proper subfield $\mathsf{k} \subsetneq K$ over k satisfying $kK^p \subseteq \mathsf{k}$.

As a first step towards obtaining the desired description of G, we prove that the maps ξ_{G_c} are isomorphisms. (This is automatic when $p = 3$, by Theorem 6.1.1, but we will handle both p in a uniform manner.) Since Φ is reduced (by definition of G being exceptional), the root system of G_c is also reduced. But ξ_{G_c} is surjective (again, by definition of G being exceptional), so G_c is standard and $\ker \xi_{G_c}$ is central by Proposition 5.3.10. The K-group $(G_c)_K^{\mathrm{ss}}$ is identified with the subgroup of G_K^{ss} generated by the root groups for $\pm c_K$ (Proposition 6.3.1(2)), and this subgroup is SL_2 (rather than PGL_2) since $G_{\overline{k}}^{\mathrm{ss}}$ is assumed to be simply connected. Therefore, $\mathscr{G}_c' := (G_c)_{K_c}^{\mathrm{ss}}$ is isomorphic to SL_2, and ξ_{G_c} expresses G_c as a central extension of $\mathrm{R}_{K_c/k}(\mathscr{G}_c') = \mathrm{R}_{K_c/k}(\mathrm{SL}_2)$ by the unipotent k-group scheme $\ker \xi_{G_c}$ that contains no nontrivial smooth connected k-subgroup schemes. Since SL_2 is simply connected, this extension uniquely splits (Example 5.1.4) and consequently $\ker \xi_{G_c} = 1$ since G_c is smooth and connected. Thus, $\xi_{G_c} : G_c \to \mathrm{R}_{K_c/k}(\mathscr{G}_c')$ is indeed an isomorphism for all c.

Step 3. Next, we will use a Levi k-subgroup of G to construct a basic exotic pseudo-reductive k-group $\mathscr{G}$ such that $G \simeq \mathrm{R}_{\mathsf{k}/k}(\mathscr{G})$.

Let $T' := T_K$ viewed as a maximal K-torus in $G' := G_K^{\mathrm{ss}}$ (so $\Phi(G', T') = \Phi$ via the identification $\mathrm{X}(T') = \mathrm{X}(T)$, since Φ is reduced and G is perfect). By Theorem 3.4.6, there exists a Levi k-subgroup L in G containing T. Since the map $\xi_G : G \to \mathrm{R}_{K/k}(G')$ induces the composite map

$$G_K \to \mathrm{R}_{K/k}(G')_K \to G' = G_K^{\mathrm{ss}}$$

that is the canonical quotient map (by the definition of ξ_G), it follows that $(\xi_G)_{\overline{k}}$ respects unipotent radicals and induces an isomorphism between maximal reductive quotients. In particular, $\xi_G(L)$ is a Levi k-subgroup of $\mathrm{R}_{K/k}(G')$ and $\xi_G|_L : L \to \xi_G(L)$ is an isomorphism. But $\xi_G(L) \subseteq \xi_G(G)$, so $\xi_G(L)$ is also a Levi k-subgroup of $\xi_G(G)$. We may therefore rename $\xi_G(L)$ as L so that L is a common Levi k-subgroup of $\xi_G(G)$ and $\mathrm{R}_{K/k}(G')$ that contains T.

The composite map $L_K \hookrightarrow \mathrm{R}_{K/k}(G')_K \twoheadrightarrow G'$ is an isomorphism; we will use it to identify G' with L_K. This identification recovers that of T_K with the maximal K-torus T' of G', so (L, T) is a k-descent of (G', T'). In particular, $(L_{\mathsf{k}}, T_{\mathsf{k}})$ is a k-descent of (G', T'), so L_{k} is a Levi k-subgroup of $\mathrm{R}_{K/\mathsf{k}}(G')$ via the canonical map $L_{\mathsf{k}} \hookrightarrow \mathrm{R}_{K/\mathsf{k}}(L_K) = \mathrm{R}_{K/\mathsf{k}}(G')$, so it gives rise to a basic exotic pseudo-reductive k-subgroup $\mathscr{G} \subseteq \mathrm{R}_{K/\mathsf{k}}(G')$. Explicitly, using Theorem 7.2.3(2) (with k in place of k and K in place of k', and the very special isogeny $L_{\mathsf{k}} \twoheadrightarrow \overline{L}_{\mathsf{k}}$), $\mathscr{G}$ is generated by the long root groups of L_{k} relative to T_{k} and the Weil restrictions from K to k of the short root groups of L_K $(= G')$ relative to $T_K = T'$. We shall prove that $\xi_G(G) = \mathrm{R}_{\mathsf{k}/k}(\mathscr{G})$ inside of $\mathrm{R}_{K/k}(G')$ and that $\xi_G : G \twoheadrightarrow \xi_G(G)$ is an isomorphism.

Step 4. For $c \in \Phi(G, T) = \Phi(L, T)$, let L_c be the subgroup of L generated by the root groups in L corresponding to the roots $\pm c$. In Step 2 we saw that for each $c \in \Phi$, ξ_G carries G_c isomorphically onto its image in $\mathrm{R}_{K/k}(G'_c) = \mathrm{R}_{K/k}((L_c)_K)$, and that this image is k-isomorphic to $\mathrm{R}_{K_c/k}((L_c)_{K_c})$. (Note that $L_c \simeq \mathrm{SL}_2$ since $L_{\overline{k}} \simeq G_{\overline{k}}^{\mathrm{ss}}$ is simply connected.) Since L is the image of a Levi k-subgroup of G under ξ_G, we have the containments

$$L_c \subseteq \xi_G(G_c) \subseteq \mathrm{R}_{K/k}((L_c)_K).$$

Rather generally, we claim that if $F'/F/k$ is a tower of finite extensions of $k = k_s$ and if H is any connected reductive k-group (such as $L_c = \mathrm{SL}_2$) then the only intermediate group M such that

$$H \subseteq M \subseteq \mathrm{R}_{F'/k}(H_{F'})$$

and which is isomorphic to $\mathrm{R}_{F/k}(H_F)$ is the canonical copy of $\mathrm{R}_{F/k}(H_F)$ inside of $\mathrm{R}_{F'/k}(H_{F'})$. This claim would imply that $\xi_G(G_c)$ equals $\mathrm{R}_{K_c/k}((L_c)_{K_c})$ inside of $\mathrm{R}_{K/k}((L_c)_K)$ ($\subseteq \mathrm{R}_{K/k}(L_K)$) for all $c \in \Phi$. (That is, if $c \in \Phi_<$ then $\xi_G(G_c) = \mathrm{R}_{K/k}((L_c)_K)$, and if $c \in \Phi_>$ then $\xi_G(G_c) = \mathrm{R}_{\mathrm{k}/k}((L_c)_{\mathrm{k}})$ inside of $\mathrm{R}_{K/k}(L_K)$.)

To prove the claim, first observe that H is embedded as a Levi k-subgroup of M (as it is visibly a Levi k-subgroup of the ambient k-group $\mathrm{R}_{F'/k}(H_{F'})$). Choose a k-isomorphism $\theta : M \simeq \mathrm{R}_{F/k}(H_F)$. The composite inclusion ι

$$H \hookrightarrow M \overset{\theta}{\simeq} \mathrm{R}_{F/k}(H_F)$$

is a Levi k-subgroup. In particular, the induced composite F-map

$$\phi : H_F \xrightarrow{\iota_F} \mathrm{R}_{F/k}(H_F)_F \overset{q_{H_F}}{\twoheadrightarrow} H_F$$

is an isomorphism. Composing θ with $\mathrm{R}_{F/k}(\phi)^{-1}$ then brings us to the case where θ extends the *identity map* between the given copy of H inside M and the canonical copy inside of $\mathrm{R}_{F/k}(H_F)$. Our problem is therefore reduced to verifying that the *only* k-homomorphism $f : \mathrm{R}_{F/k}(H_F) \to \mathrm{R}_{F'/k}(H_{F'})$ extending the identity on the k-subgroup H is the canonical inclusion. The map $f_{\overline{k}}$ respects the unipotent radicals since the composite map

$$\mathrm{R}_{F/k}(H_F)_{\overline{k}} \xrightarrow{f_{\overline{k}}} \mathrm{R}_{F'/k}(H_{F'})_{\overline{k}} \twoheadrightarrow H_{\overline{k}}$$

is surjective (as the Levi $\overline{k}$-subgroup $H_{\overline{k}}$ provides a section). It follows that f induces a homomorphism $f_{\overline{k}}^{\mathrm{red}}$ between the maximal geometric reductive quotients, and the canonical identification of these quotients over $\overline{k}$ with $H_{\overline{k}}$ (via the canonical Levi k-subgroups H) identifies $f_{\overline{k}}^{\mathrm{red}}$ with the identity map. But the canonical inclusion of $\mathrm{R}_{F/k}(H_F)$ into $\mathrm{R}_{F'/k}(H_{F'})$ has the same property, so by Proposition 1.2.2 it must equal f.

Step 5. As G is perfect, it is generated by the k-subgroups G_c for all $c \in \Phi(G,T) = \Phi(G',T')$ (Lemma 3.1.5). Thus, by (6.3.1) and Step 4, the k-subgroup

$$\xi_G(G) \subseteq \mathrm{R}_{K/k}(G') = \mathrm{R}_{K/k}(L_K)$$

is generated by the k-subgroups $\xi_G(G_c) = \mathrm{R}_{K_c/k}((L_c)_{K_c})$ for all c. These latter subgroups generate $\mathrm{R}_{\mathrm{k}/k}(\mathscr{G})$ due to the structure of an open Bruhat cell in $\mathscr{G}$ with respect to $T' = T_K$ via (7.2.2) (with $(G, k'/k)$ there replaced with $(L_{\mathrm{k}}, K/\mathrm{k})$). Hence, $\xi_G(G) = \mathrm{R}_{\mathrm{k}/k}(\mathscr{G})$.

The same considerations with an open Bruhat cell for $\mathscr{G}$ and the isomorphism property of the maps ξ_{G_c} (with both long and short c) imply that the image of $\mathrm{Lie}(\xi_G)$ equals the Lie algebra of $\xi_G(G) = \mathrm{R}_{\mathsf{k}/k}(\mathscr{G})$. We conclude that $\ker \xi_G$ is a smooth unipotent normal k-subgroup of G. But G is pseudo-reductive over k, so $\ker \xi_G$ must be étale and hence central in G since G is connected. Applying Proposition 8.1.2 to $\mathscr{G}$ and $Z = \ker \xi_G$ then gives that $\ker \xi_G = 1$ since G is connected. Thus, the map $\xi_G : G \to \xi_G(G) = \mathrm{R}_{\mathsf{k}/k}(\mathscr{G})$ is an isomorphism as desired.

The uniqueness claims in Theorem 8.2.1 (concerning k and a Levi k-subgroup of $\mathrm{R}_{K/\mathsf{k}}(\overline{G_K^{\mathrm{ss}}})$) follow from Proposition 8.1.5(2). This completes the proof of Theorem 8.2.1. □

Theorem 8.2.1 justifies the interest in the following definition.

Definition 8.2.2 Let k be an imperfect field of characteristic $p \in \{2,3\}$. A pseudo-reductive k-group G is *exotic* (over k) if $G \simeq \mathrm{R}_{K/k}(\mathscr{G})$ for a nonzero finite reduced k-algebra K and a K-group $\mathscr{G}$ whose fibers are basic exotic pseudo-reductive groups (in the sense of Definition 7.2.6).

Exotic pseudo-reductive groups are perfect and non-reductive, by Proposition 8.1.2. By Corollary 8.1.4, an exotic pseudo-reductive k-group G is uniquely a direct product of pseudo-simple exotic pseudo-reductive k-groups, and the set of such pseudo-simple factors is in natural bijective correspondence with the set of factor fields of K. In particular,

$$G_{\overline{k}}^{\mathrm{ss}} \simeq \prod_{\sigma : K \to \overline{k}} (\mathscr{G} \otimes_{K,\sigma} \overline{k})^{\mathrm{ss}}$$

(product over k-algebra maps σ) is simply connected and its $\mathrm{Aut}(\overline{k}/k)$-orbits of simple factors are in bijective correspondence with the set of factor fields of K. Thus, G is pseudo-simple over k if and only if K is a field, it is absolutely pseudo-simple if and only if K/k is purely inseparable, and in the pseudo-simple case the connected components of the Dynkin diagram of $G_{\overline{k}}^{\mathrm{ss}}$ are all of the same type. We call this the *type* of a pseudo-simple exotic pseudo-reductive G (as it is the type which underlies the construction of $\mathscr{G}$ over the field K).

Remark 8.2.3 The root system (over k_s) of an exotic pseudo-reductive k-group is always reduced (even in characteristic 2): in the basic exotic case this is part of Proposition 7.2.7(1), and in general it follows from the behavior of root systems with respect to Weil restriction (Example 2.3.2). The description of the root spaces shows that in the absolutely pseudo-simple case over $k = k_s$ the root spaces do not all have the same dimension. Hence, such absolutely pseudo-simple groups are never standard.

We wish to describe isomorphisms between exotic pseudo-reductive groups in terms of isomorphisms among the data used to construct such groups. Proposition 5.2.2 motivates our description of all such isomorphisms in Proposition 8.2.4 below, but we require some preparations before we state the description. We will consider 5-tuples $(K, K', G', \mathscr{G}, \overline{G})$ satisfying:

(a) K is a nonzero finite reduced algebra over an imperfect field k with characteristic $p \in \{2,3\}$; let $\{K_i\}$ be its set of factor fields.
(b) $\mathscr{G}$ is a K-group whose fiber $\mathscr{G}_i$ over K_i is a basic exotic pseudo-reductive K_i-group for every i.
(c) For each i, $(G'_i, K'_i/K_i)$ is the pair that is uniquely associated to $\mathscr{G}_i$ (up to unique K_i-isomorphism) as in Definition 7.2.6, with G'_i absolutely simple and simply connected with Dynkin diagram having an edge with multiplicity p.
(d) $K' = \prod K'_i$, $G' = \coprod G'_i$, $\pi' : G' \to \overline{G}'$ is the very special K-isogeny, and

$$\mathscr{G} = \mathrm{R}_{K'/K}(\pi')^{-1}(\overline{G})$$

for a Levi K-subgroup $\overline{G}$ of $\mathrm{R}_{K'/K}(\overline{G}')$. (The K-group $\mathscr{G}$ determines $\overline{G}$ because $\overline{G}$ is the image of $\mathscr{G}$ under $\mathrm{R}_{K'/K}(\pi')$, by Proposition 7.2.7(2). Proposition 7.3.1 gives necessary and sufficient conditions on $\overline{G}$ so that its preimage under $\mathrm{R}_{K'/K}(\pi')$ is K-smooth.)

By Definition 8.2.2, the pseudo-reductive k-group $\mathrm{R}_{K/k}(\mathscr{G})$ is exotic, and by Proposition 8.1.5 every exotic pseudo-reductive k-group is k-isomorphic to $\mathrm{R}_{K/k}(\mathscr{G})$ for such a pair $(\mathscr{G}, K/k)$ that is unique up to canonical k-isomorphism. The description of isomorphisms among k-groups of the form $\mathrm{R}_{K/k}(\mathscr{G})$ is provided by the following result.

Proposition 8.2.4 *Let* $(K, K', G', \mathscr{G}, \overline{G})$ *and* $(L, L', H', \mathscr{H}, \overline{H})$ *be 5-tuples over* k *satisfying the above properties* (a)–(d)*. Consider a pair* (α, φ) *where* α *is a* k*-algebra isomorphism* $L \simeq K$ *that lifts (necessarily uniquely) to a* k*-algebra isomorphism* $\alpha' : L' \simeq K'$ *and* φ *is a group isomorphism* $G' \simeq H'$ *over* $\mathrm{Spec}(\alpha')$.

(1) *The induced* k*-isomorphism*

$$\mathrm{R}_{\alpha'/k}(\varphi) : \mathrm{R}_{K'/k}(G') = \mathrm{R}_{K/k}(\mathrm{R}_{K'/K}(G')) \to \mathrm{R}_{L'/k}(H')$$

restricts to a k-isomorphism

$$\sigma : \mathrm{R}_{K/k}(\mathscr{G}) \simeq \mathrm{R}_{L/k}(\mathscr{H}) \tag{8.2.1}$$

between the associated exotic pseudo-reductive k-groups if and only if the isomorphism $\overline{\varphi} : \overline{G}' \simeq \overline{H}'$ *over* α' *induced by* φ *satisfies* $\mathrm{R}_{\alpha'/\alpha}(\overline{\varphi})(\overline{G}) = \overline{H}$ *inside of* $\mathrm{R}_{K'/K}(\overline{G}') \simeq \mathrm{R}_{L'/L}(\overline{H}')$*, and also if and only if* $\mathrm{R}_{\alpha'/\alpha}(\varphi)(\mathscr{G}) = \mathscr{H}$*.*

Conversely, any k-isomorphism $\sigma : \mathrm{R}_{K/k}(\mathscr{G}) \simeq \mathrm{R}_{L/k}(\mathscr{H})$ *between exotic pseudo-reductive k-groups arises in this way from a unique such pair* (α, φ)*.*

(2) *Let S and T be maximal k-tori in* $\mathrm{R}_{K/k}(\mathscr{G})$*, and let* $\mathscr{S}$ *and* $\mathscr{T}$ *be the corresponding maximal K-tori in* $\mathscr{G}$ *via Proposition A.5.15(2). For a k-automorphism* σ *of* $\mathrm{R}_{K/k}(\mathscr{G})$ *arising from a pair* (α, φ)*, the conditions* $\sigma(S) = T$ *and* $\mathrm{R}_{\alpha'/\alpha}(\varphi)(\mathscr{S}) = \mathscr{T}$ *are equivalent. The same holds for maximal split tori.*

Proof Part (2) is a consequence of the bijectivity in Proposition A.5.15(2), so we focus our attention on (1). In particular, by Galois descent we may assume that k is separably closed.

Since Weil restriction is compatible with the formation of preimages, if $\mathrm{R}_{\alpha'/\alpha}(\overline{\varphi})(\overline{G}) = \overline{H}$ then the map $\mathrm{R}_{\alpha'/k}(\varphi)$ does induce a k-isomorphism σ as in (8.2.1). Also, the equivalence involving $\mathrm{R}_{\alpha'/k}(\varphi)$ and $\mathrm{R}_{\alpha'/\alpha}(\overline{\varphi})$ formally implies the one involving $\mathrm{R}_{\alpha'/k}(\varphi)$ and $\mathrm{R}_{\alpha'/\alpha}(\varphi)$ (using the end of Proposition 7.2.7(2)). Finally, the necessity in the second equivalence in (1) is seen by passing to maximal pseudo-reductive quotients of fibers over $L \simeq K$, so the equivalences in (1) are now all proved.

It remains to show that every k-isomorphism $\sigma : \mathrm{R}_{K/k}(\mathscr{G}) \simeq \mathrm{R}_{L/k}(\mathscr{H})$ arises from a unique pair (α, φ). By Corollary 8.1.4, in the k-group $\mathrm{R}_{K/k}(\mathscr{G}) = \prod \mathrm{R}_{K_i/k}(\mathscr{G}_i)$ the factors $\mathrm{R}_{K_i/k}(\mathscr{G}_i)$ are intrinsically determined: they are the (absolutely) pseudo-simple smooth connected normal k-subgroups. Thus, k-isomorphisms between two such k-groups must arise from a collection of isomorphisms between the pseudo-simple factors. This reduces our problem to the case where K and L are fields, in which case if the k-isomorphism σ exists then Proposition 8.1.5 shows that there is a pair (α, φ) giving rise to σ. For the uniqueness of (α, φ), since $k = k_s$ and K is a field we just have to show for $\varphi \in \mathrm{Aut}_{K'}(G')$ that if $\mathrm{R}_{K'/k}(\varphi)$ restricts to the identity on $G = \mathrm{R}_{K/k}(\mathscr{G})$ then $\varphi = \mathrm{id}_{G'}$. By Proposition 8.1.5(1), φ induces the identity on $G'/Z_{G'}$, so indeed φ is the identity. □

Corollary 8.2.5 *Let k be an imperfect field of characteristic $p \in \{2,3\}$ and G an exotic absolutely pseudo-simple k-group of type F_4 when $p = 2$ and type G_2 when $p = 3$. Every k-automorphism of G is inner.*

Proof Consider the unique triple $(G', K'/K, \overline{G})$ that gives rise to G (see Proposition 8.1.5), so $G = \mathrm{R}_{K/k}(\mathscr{G})$ for the K-subgroup $\mathscr{G} \subseteq \mathrm{R}_{K'/K}(G')$ that is the preimage of the Levi K-subgroup $\overline{G} \subseteq \mathrm{R}_{K'/K}(\overline{G}')$. The condition of G being absolutely pseudo-simple says precisely that K is a field purely inseparable over k. In particular, K has no nontrivial k-automorphisms. Thus, by Proposition 8.2.4, every k-automorphism of G can be written as $\mathrm{R}_{K/k}(\sigma)$ for the K-automorphism σ of $\mathscr{G}$ induced by $\mathrm{R}_{K'/K}(\varphi)$ for a unique K'-automorphism φ of G' such that the induced automorphism $\overline{\varphi}$ of the very special quotient $\overline{G}' = \overline{G}_{K'}$ of G' is defined over K. It therefore suffices to prove that any such φ is conjugation by a point of the subgroup $\mathscr{G}(K) \subseteq \mathrm{R}_{K'/K}(G')(K) = G'(K')$.

Since G' is absolutely simple of adjoint type and has no nontrivial diagram automorphisms, by Corollary A.4.7(1) its K'-automorphism φ is conjugation by a unique point $g' \in G'(K')$. Thus, $\overline{\varphi}$ is conjugation on $\overline{G}' = \overline{G}_{K'}$ by the image point $\overline{g}' \in \overline{G}'(K')$. But recall that $\overline{\varphi}$ descends to a K-automorphism of $\overline{G}$. This descent must be conjugation by a point $\overline{g} \in \overline{G}(K)$ (by again applying Corollary A.4.7(1)). The K'-points $\overline{g}, \overline{g}' \in \overline{G}'(K')$ induce the same conjugation automorphism $\overline{\varphi}$ of $\overline{G}'$, so these points are equal (as $Z_{\overline{G}'} = 1$). This says that the pair $(\overline{g}, g') \in \overline{G}(K) \times G'(K')$ lies in the subset

$$\overline{G}(K) \times_{\overline{G}'(K')} G'(K') = (\overline{G} \times_{\mathrm{R}_{K'/K}(\overline{G}')} \mathrm{R}_{K'/K}(G'))(K) = \mathscr{G}(K).$$

By construction, conjugation on $\mathscr{G}$ by this point in $\mathscr{G}(K)$ induces φ. □

Here is an interesting application to the structure of the automorphism schemes that were studied in Theorem 2.4.1.

Proposition 8.2.6 *Consider a 5-tuple $(K, K', G', \mathscr{G}, \overline{G})$ as above, a Cartan K-subgroup $\mathscr{C}$ of $\mathscr{G}$, and the associated Cartan k-subgroup $C = \mathrm{R}_{K/k}(\mathscr{C})$ of $G = \mathrm{R}_{K/k}(\mathscr{G})$.*

Let $Z_{\mathscr{G},\mathscr{C}}$ denote the commutative fiberwise maximal smooth K-subgroup of the automorphism scheme $\mathrm{Aut}_{\mathscr{G},\mathscr{C}}$, and define $Z_{G,C} \subseteq \mathrm{Aut}_{G,C}$ similarly. The k-homomorphism

$$\mathrm{R}_{K/k}(\mathrm{Aut}_{\mathscr{G},\mathscr{C}}) \to \mathrm{Aut}_{G,C},$$

defined by applying $\mathrm{R}_{K/k}$ to automorphisms, carries $\mathrm{R}_{K/k}(Z_{\mathscr{G},\mathscr{C}})$ isomorphically onto $Z_{G,C}$.

Strictly speaking, we have only defined $\mathrm{Aut}_{\mathscr{G},\mathscr{C}}$ when the base is a field. For a base such as Spec K we define it fiberwise.

Proof We may and do assume $k = k_s$, so the factor fields K_i of K are purely inseparable over k. In particular, a k-algebra automorphism of K is uniquely determined by its effect on the primitive idempotents of K. But according to Corollary 8.1.4, these idempotents are in natural bijection with the pseudo-simple factors G_i of G. Since $C = \prod C_i$ where $C_i = C \cap G_i$ is a Cartan k-subgroup of G_i, with each $C_i \neq 1$, the description of k-automorphisms of G in Proposition 8.2.4(1) implies that those automorphisms that restrict to the identity on C are precisely $\mathrm{R}_{K/k}(\varphi)$ for a K-automorphism φ of $\mathscr{G}$ such that $\mathrm{R}_{K/k}(\varphi)$ restricts to the identity on $C = \mathrm{R}_{K/k}(\mathscr{C})$. Passing to k-points shows that it is equivalent to require that φ restricts to the identity on $\mathscr{C}$. This proves that the natural map $\mathrm{Aut}_{\mathscr{G},\mathscr{C}}(K) \to \mathrm{Aut}_{G,C}(k)$ is bijective.

By construction $Z_{\mathscr{G},\mathscr{C}}(K) = \mathrm{Aut}_{\mathscr{G},\mathscr{C}}(K)$, and similarly $Z_{G,C}(k) = \mathrm{Aut}_{G,C}(k)$. Thus, the map on k-points induced by the natural map $\mathrm{R}_{K/k}(Z_{\mathscr{G},\mathscr{C}}) \to Z_{G,C}$ is $\mathrm{Aut}_{\mathscr{G},\mathscr{C}}(K) \to \mathrm{Aut}_{G,C}(k)$.

In general, if $f : H' \to H$ is a k-homomorphism between (possibly disconnected) smooth k-groups of finite type then we claim f is an isomorphism if and only if $H'(F) \to H(F)$ is bijective for all separable extension fields F/k. In case H' and H are connected, hence integral, this was shown near the beginning of the proof of Theorem 1.3.9. In general, we can assume $k = k_s$, so each connected component is geometrically connected and contains a k-point. Thus, the function field at any generic point coincides with the function field of the identity component. In particular, each generic point of H is the image of a k-morphism $h : \mathrm{Spec}\, k(H^0) \to H$, so taking $F = k(H^0)$ provides a k-morphism $h' : \mathrm{Spec}\, k(H^0) \to H'$ lifting h. Taking h to be the canonical map h_0 hitting the generic point of H^0, the associated map h'_0 lands in some connected component H'_i of H'. The resulting k-map $f : H'_i \to H^0$ between integral k-schemes is injective on $k(H'_i)$-valued points and admits a lift h'_0 of the canonical k-map h_0, so the method of proof in the integral case can be applied to deduce that $H'_i \to H^0$ is birational. It follows via translation by the inverse of any k-point of H'_i that the k-homomorphism $f^0 : H'^0 \to H^0$ is birational, hence an isomorphism, so $\ker f$ is finite étale. But f is injective on k-points with $k = k_s$, so f is an open immersion. Bijectivity on k-valued points then forces f to be an isomorphism, as desired.

We conclude that $\mathrm{R}_{K/k}(Z_{\mathscr{G},\mathscr{C}}) \to Z_{G,C}$ is an isomorphism provided that it is bijective on F-points for all separable extension fields F/k. It even suffices to check bijectivity on F_s-points, as then passing to $\mathrm{Gal}(F_s/F)$-invariants would give the bijectivity for F-points. Since $K \otimes_k F$ is reduced (as F/k is separable), we can therefore rename F as k to reduce to the case $F = k$ which we have already settled. □

As an application of Proposition 7.2.7, we can refine Proposition 8.2.6 as follows. Let $\mathscr{G}$ be the basic exotic pseudo-reductive k-group arising from a triple $(G', k'/k, \overline{G})$. Let $\mathscr{C}$ be a Cartan k-subgroup of $\mathscr{G}$, and let $Z_{\mathscr{G},\mathscr{C}}$ be the maximal smooth k-subgroup of the k-group $\mathrm{Aut}_{\mathscr{G},\mathscr{C}}$. Using the maximal reductive quotient map $q : \mathscr{G}_{k'} \twoheadrightarrow G'$, let $T' = q(\mathscr{C}_{k'})$ be the associated maximal k'-torus of G'. The natural action of $(Z_{\mathscr{G},\mathscr{C}})_{k'}$ on $\mathscr{G}_{k'}$ restricting to the identity on $\mathscr{C}_{k'}$ factors through q because $Z_{\mathscr{G},\mathscr{C}}$ is k-smooth (so we can check the preservation of $\ker q = \mathscr{R}_{u,k'}(\mathscr{G}_{k'})$). In this way we get an action of $(Z_{\mathscr{G},\mathscr{C}})_{k'}$ on G' that restricts to the identity on T', and the resulting k'-homomorphism

$$(Z_{\mathscr{G},\mathscr{C}})_{k'} \to \mathrm{Aut}_{G',T'}$$

factors through the maximal smooth k'-subgroup $T'/Z_{G'}$ (Theorem 1.3.9). Thus, by the universal property of Weil restriction, we thereby obtain a k-homomorphism

$$Z_{\mathscr{G},\mathscr{C}} \to \mathrm{R}_{k'/k}(T'/Z_{G'}). \tag{8.2.2}$$

Corollary 8.2.7 *The map in* (8.2.2) *is a closed immersion, and the image of*

$$Z_{\mathscr{G},\mathscr{C}}(k) \hookrightarrow \mathrm{R}_{k'/k}(T'/Z_{G'})(k) = (T'/Z_{G'})(k')$$

consists of the points whose action on $\mathrm{R}_{k'/k}(G')$ *preserves each root group of* $(\mathscr{G}, \mathscr{T})$*, where* $\mathscr{T}$ *is the maximal k-torus in* $\mathscr{C}$*. An equivalent requirement on such points when* $k = k_s$ *is that they preserve the root groups corresponding to long roots in a basis of the root system.*

Proof To prove that (8.2.2) is a closed immersion, it suffices to prove that its scheme-theoretic kernel is trivial (Proposition A.2.1). By Proposition 7.2.7(2), the natural inclusion $\mathscr{G} \to \mathrm{R}_{k'/k}(G')$ corresponds to the maximal reductive quotient map $q : \mathscr{G}_{k'} \to G'$ over k'. Thus, if a point of $Z_{\mathscr{G},\mathscr{C}}$ has trivial image in $\mathrm{R}_{k'/k}(T'/Z_{G'})$ then its action on $\mathrm{R}_{k'/k}(G')$ is trivial on $\mathscr{G}$. But $Z_{\mathscr{G},\mathscr{C}}$ acts through its subgroup inclusion into $\mathrm{Aut}_{\mathscr{G},\mathscr{C}}$, so a point of $Z_{\mathscr{G},\mathscr{C}}$ acting trivially on $\mathscr{G}$ is itself trivial. This proves that (8.2.2) has trivial kernel, as desired.

To identify the image of (8.2.2) on k-points, we first observe that for any k-automorphism σ of $\mathscr{G}$, $\sigma_{k'}$ induces a unique k'-automorphism φ of the maximal reductive quotient G' over k'. Since the quotient map q over k' corresponds to the inclusion of $\mathscr{G}$ into $\mathrm{R}_{k'/k}(G')$, it follows that the k-automorphisms of $\mathscr{G}$ are in natural bijective correspondence with the k'-automorphisms φ of G' such that $\mathrm{R}_{k'/k}(\varphi)$ preserves $\mathscr{G}$ inside $\mathrm{R}_{k'/k}(G')$ (as any endomorphism with trivial scheme-theoretic kernel must be an automorphism). Moreover, if $\sigma \in \mathrm{Aut}_k(\mathscr{G})$ restricts to the identity on $\mathscr{C}$ then the corresponding $\varphi \in \mathrm{Aut}_{k'}(G')$ restricts to the identity on $T' = q(\mathscr{C}_{k'})$, and the converse holds as well since $\mathscr{C} \subseteq \mathrm{R}_{k'/k}(T')$. Hence, $\mathrm{Aut}_{\mathscr{G},\mathscr{C}}(k)$ corresponds (via Weil restriction) to the points in

$$\mathrm{Aut}_{G',T'}(k') = (T'/Z_{G'})(k') = \mathrm{R}_{k'/k}(T'/Z_{G'})(k)$$

whose natural action on $\mathrm{R}_{k'/k}(G')$ restricts to an automorphism of $\mathscr{G}$.

But the $T'/Z_{G'}$-action on G' respects all root groups and is trivial on the Cartan k'-subgroup T', so it restricts to an automorphism of any open Bruhat cell attached to (G', T'). Since open Bruhat cells for $(\mathscr{G}, \mathscr{T})$ are direct products (via multiplication) of $\mathscr{C}$ and the root groups, it follows that a point in $(T'/Z_{G'})(k')$ comes from $Z_{\mathscr{G},\mathscr{T}}(k)$ if and only if its action on $\mathrm{R}_{k'/k}(G')$ preserves the root groups of $(\mathscr{G}, \mathscr{T})$.

Finally, for $k = k_s$ we check that preservation of the root groups corresponding to long roots in just one basis of the root system is sufficient. Using the natural identification of the k-groups $W(\mathscr{G}, \mathscr{T})$ and $\mathrm{R}_{k'/k}(W(G', T'))$ via Proposition 3.2.7 and the identification of these Weyl groups with the Weyl group of the common reduced root system, we infer that the root groups of $(\mathscr{G}, \mathscr{T})$ corresponding to every long root are preserved. By Theorem 7.2.3(2), the short root groups coincide with those of $(\mathrm{R}_{k'/k}(G'), T)$ (where T is the maximal k-torus in $\mathrm{R}_{k'/k}(T')$), so there is nothing to check for the short roots. □

Over imperfect fields k of characteristic 2 or 3, certain exotic pseudo-reductive groups will uniquely arise as direct factors in the classification. A related result will hold in characteristic 2 (assuming $[k : k^2] = 2$) for non-reduced components of the root system. The direct factor result in the exotic case will be treated now, and the method will be applicable when we later study how non-reduced root systems arise.

As a first step, we check that the property of a k-group being exotic (in the sense of Definition 8.2.2) is unaffected by extending the ground field to k_s. More generally:

Proposition 8.2.8 *Let k be an imperfect field of characteristic $p \in \{2, 3\}$, and F/k a separable extension. A k-group G is exotic over k if and only if G_F is exotic over F.*

Proof If G_F is exotic then it is non-commutative and pseudo-reductive, so the same holds for G over k. Thus, throughout this proof we may and do assume that G is non-commutative and pseudo-reductive over k (and hence it is non-solvable). We now use Proposition 8.1.5 to give a necessary and sufficient criterion for G to be exotic, and the criterion will be visibly insensitive to separable extension on k.

Consider the simply connected datum $(G', K'/k, f)$ for the non-solvable smooth connected affine k-group G (see Definition 4.2.3), with a nonzero finite reduced k-algebra K' and a quotient map $f : G_{K'} \to G'/Z_{G'}$ over K' with adjoint semisimple target having absolutely simple fibers. There is at most one factorization

$$\begin{array}{ccc} G_{K'} & \xrightarrow{\widetilde{f}} & G' \\ & \underset{f}{\searrow} & \downarrow \\ & & G'/Z_{G'} \end{array} \tag{8.2.3}$$

Suppose such a lift $\widetilde{f}$ exists and that each fiber of G' over Spec K' has an edge of multiplicity p in its Dynkin diagram; by Proposition 8.1.5(1) this is a necessary condition for G to be exotic. In particular, there is a (fiberwise) very special isogeny $\pi : G' \to \overline{G}'$ over Spec K'.

By the universal property of Weil restriction, $\widetilde{f}$ corresponds to a k-group homomorphism $j : G \to \mathrm{R}_{K'/k}(G')$, and again using Proposition 8.1.5(1) we see that if G is exotic then $\ker j = 1$. We now assume $\ker j = 1$, so Lie(G) is identified with a k-subspace of $\mathrm{Lie}(\mathrm{R}_{K'/k}(G')) = \mathrm{Lie}(G')$. Let $K \subseteq K'$ be the maximal k-subalgebra of K' such that Lie(G) is a K-submodule of the K-module Lie(G'). Note that K is a nonzero finite reduced k-algebra.

Expressing the target of j as $\mathrm{R}_{K/k}(\mathrm{R}_{K'/K}(G'))$, the map j corresponds to a K-group map $G_K \to \mathrm{R}_{K'/K}(G')$. Let $\mathscr{G}$ be the image of this map, and consider the image $\overline{G}$ of $\mathscr{G}$ under the map

$$h : \mathrm{R}_{K'/K}(\pi) : \mathrm{R}_{K'/K}(G') \to \mathrm{R}_{K'/K}(\overline{G}'),$$

so $\mathscr{G} \subseteq h^{-1}(\overline{G})$. By Proposition 8.1.5(2), if G is exotic then the following properties must hold: $K'^p \subseteq K$, the fiber-degrees of the map Spec $K' \to$ Spec K are all > 1, $\overline{G}$ is a Levi K-subgroup of $\mathrm{R}_{K'/K}(\overline{G}')$, $\mathscr{G} = h^{-1}(\overline{G})$, and the map $G \to \mathrm{R}_{K/k}(\mathscr{G})$ is an isomorphism. Assuming all of these hold, the K-smoothness of $h^{-1}(\overline{G})$ forces $\mathscr{G}$ to be the fiberwise basic exotic pseudo-reductive K-group associated to $(G', K'/K, \overline{G})$ by Proposition 7.3.1. We can avoid Proposition 7.3.1 by observing the alternative necessary condition that for $K_s := K \otimes_k k_s$, the K_s-group $\mathscr{G}_{K_s}$ contains a Levi K_s-subgroup of $\mathrm{R}_{K'/K}(G')_{K_s}$, which again implies the same conclusion relating $\mathscr{G}$ to $(G', K'/K, \overline{G})$. But then the isomorphism $G \simeq \mathrm{R}_{K/k}(\mathscr{G})$ implies that G is exotic.

We can summarize the preceding considerations as an intrinsic characterization of when a non-commutative pseudo-reductive k-group G is exotic. It is necessary and sufficient that the following hold. First, we form the simply connected datum $(G', K'/k, f)$ associated to G. It is necessary that each fiber of G' over K' (all of which are absolutely simple and simply connected) has an edge of multiplicity p in its Dynkin diagram, and that we can lift f as in (8.2.3). Assuming these hold, we can form the very special K'-isogeny $\pi : G' \to \overline{G}'$, and $\widetilde{f}$ provides us $j : G \to \mathrm{R}_{K'/k}(G')$. Necessarily ker $j = 1$, and then we can define the k-subalgebra $K \subseteq K'$, the K-group $\mathscr{G}$, and the K-subgroup $\overline{G} \subseteq \mathrm{R}_{K'/K}(\overline{G}')$ as above. It is necessary that $K'^p \subseteq K$ (which forces K and K' to have compatible decompositions into factor fields: $K' = \prod K'_i, K = \prod K_i$), each fiber of Spec $K' \to$ Spec K has degree > 1, $\overline{G}$ is a Levi K-subgroup of $\mathrm{R}_{K'/K}(\overline{G}')$, $\mathscr{G} = \mathrm{R}_{K'/K}(\pi)^{-1}(\overline{G})$, and $\mathscr{G}_{K_s}$ contains a Levi K_s-subgroup of $\mathrm{R}_{K'/K}(G')_{K_s}$ or equivalently a Levi K_s-subgroup of itself. (This final condition can be omitted if we are willing to use Proposition 7.3.1, or to introduce the condition that $\mathscr{G}$ is pseudo-reductive and invoke the existence of Levi subgroups in pseudo-reductive groups over a separably closed field via Theorem 3.4.6.) Moreover, when all of these conditions do hold the k-group G is exotic. We just have to check that these conditions interact well with separable extension of the ground field.

To be precise, since the formation of the simply connected datum commutes with any separable extension of the ground field (by Proposition 4.2.1) and any tensor product of fields $F \otimes_k K$ is a finite product of fields when F/k is separable and K/k is finite, it suffices to verify three further compatibilities:

(i) the existence or not of $\widetilde{f}$ is insensitive to separable extension of the ground field,
(ii) for a k-algebra A and k-subspace W of an A-module V, the formation of the maximal k-subalgebra $F \subseteq A$ such that W is an F-submodule commutes with any extension on k,
(iii) the existence or not of a Levi subgroup in a smooth connected affine group over a separably closed field is insensitive to separable extension to another separably closed field.

In (ii), F is the kernel of the k-linear map $A \to \mathrm{Hom}_k(W, V/W)$ defined by

$$a \mapsto (w \mapsto aw \bmod W).$$

Thus, the compatibility in (ii) follows from the injectivity of the natural map

$$L \otimes_k \mathrm{Hom}_k(W, W') \to \mathrm{Hom}_L(W_L, W'_L)$$

for any field L/k. Assertion (iii) is an immediate consequence of specialization arguments as in the proof of Proposition 1.1.9(1). For (i), we first note that the existence of $\widetilde{f}$ is equivalent to the splitting of the f-pullback central extension

$$1 \to Z_{G'} \to E \to G_{K'} \to 1$$

with $E = G_{K'} \times_{f,G'/Z_{G'}} G'$. Since $G_{K'}$ is K'-smooth with connected fibers and $Z_{G'}$ is K'-finite, a necessary and sufficient condition for the existence of a splitting is that the fiberwise connected component E^0_{red} of the identity section in the underlying reduced scheme E_{red} maps isomorphically onto $G_{K'}$ as a K'-scheme. But the formation of the underlying reduced scheme of a K'-scheme commutes with separable extension on the factor fields of K', so we are done. □

To complete the theory of exotic pseudo-reductive groups what remains is to determine their automorphisms that pointwise fix a Cartan subgroup:

Example 8.2.9 Let k be an imperfect field of characteristic $p \in \{2, 3\}$, K a nonzero finite reduced k-algebra, and $G = \mathrm{R}_{K/k}(\mathscr{G})$ for a K-group $\mathscr{G}$ whose fibers are basic exotic pseudo-reductive groups all of *the same type* (which we call the *type* of G). Let $(G', K'/K, \overline{G})$ be the triple giving rise to $\mathscr{G}$ over $\mathrm{Spec}\, K$. Let $T \subseteq G$ be a maximal k-torus, $C = Z_G(T)$, and $\mathscr{T} \subseteq \mathscr{G}$ the unique maximal K-torus such that T is the maximal k-torus in $\mathrm{R}_{K/k}(\mathscr{T})$ (see Proposition A.5.15(2)). Let $\mathscr{C} = Z_{\mathscr{G}}(\mathscr{T})$ be the corresponding Cartan K-subgroup of $\mathscr{G}$. The identification of G' with the maximal reductive quotient of $G_{K'}$ carries $T_{K'}$ isomorphically onto a maximal K'-torus $T' \subseteq G_{K'}$.

Let $\underline{\mathrm{Aut}}_{G,C}$ be the functor considered in Theorem 2.4.1 (where it was shown to be represented by an affine k-group of finite type). The action of $\mathrm{R}_{K/k}(\mathscr{G}/Z_{\mathscr{G}})$ on $G = \mathrm{R}_{K/k}(\mathscr{G})$ is faithful (use the canonical quotient map $\mathrm{R}_{K/k}(\mathscr{G})_K \twoheadrightarrow \mathscr{G}$; see Proposition A.5.11(1)), so the group scheme $\underline{\mathrm{Aut}}_{G,C}$ naturally contains $\mathrm{R}_{K/k}(\mathscr{C}/Z_{\mathscr{G}})$ as a subgroup. This latter k-group is smooth and connected, so it must lie in the maximal smooth connected k-subgroup $Z_{G,C}$. Explicitly, Proposition 8.2.6 identifies $Z_{G,C}$ with $\mathrm{R}_{K/k}(Z_{\mathscr{G},\mathscr{C}})$, and Corollary 8.2.7 determines $Z_{\mathscr{G},\mathscr{C}}$ as a K-subgroup of $\mathrm{R}_{K'/K}(T'/Z_{G'})$. This computes $Z_{G,C}$ as a k-subgroup of $\mathrm{R}_{K'/k}(T'/Z_{G'})$.

At this point something wonderful happens: in characteristic 3, pseudo-simple exotic pseudo-reductive normal k-subgroups always uniquely split off as a direct factor. Our proof of this splitting behavior also works over imperfect fields of characteristic 2 for pseudo-simple exotic pseudo-reductive normal k-subgroups of type F_4, as we show in Theorem 8.2.10 below. Such splitting behavior rests on two properties that exotic absolutely pseudo-simple groups G of types F_4 and G_2 share in common with $G_{\overline{K}}^{\mathrm{ss}}$: trivial center and only inner automorphisms.

Theorem 8.2.10 *Let k be an imperfect field with characteristic $p \in \{2, 3\}$. Let G be a non-standard pseudo-reductive k-group.*

(1) *If $p = 3$ then there is a unique decomposition $G = \mathscr{G}_1 \times \mathscr{G}_2$ into commuting k-subgroups such that $\mathscr{G}_1$ is a (possibly trivial) standard pseudo-reductive k-group and $\mathscr{G}_2$ is an exotic pseudo-reductive k-group in the sense of Definition 8.2.2.*

(2) *If $p = 2$ and G_{k_s} contains a pseudo-simple normal k-subgroup that is exotic of type F_4 then there is a unique decomposition $G = \mathscr{G}_1 \times \mathscr{G}_2$ into commuting k-subgroups such that $(\mathscr{G}_1)_{k_s}$ has no pseudo-simple normal k_s-subgroups that are exotic with type F_4 and $\mathscr{G}_2$ is an exotic pseudo-reductive k-group whose pseudo-simple factors are all of type F_4.*

Theorem 8.2.10(1) and Corollary 6.3.5 yield the classification of pseudo-reductive groups in characteristic 3. In Chapter 10 we will handle characteristics 2 and 3 by a uniform method when the root system over k_s is reduced (assuming $[k : k^2] \leqslant 2$ when $\mathrm{char}(k) = 2$).

Proof Since we are claiming uniqueness, by Galois descent and both Corollary 5.2.3 and Proposition 8.2.8 (and Corollary 8.1.4) we may and do assume that k is separably closed. The non-standardness hypothesis on G implies that the pseudo-reductive k-group $G' := \mathscr{D}(G)$ is not standard (Proposition 5.2.1). In particular, G' is non-commutative. By Proposition 1.2.6, G' is its own derived group. Thus, by Proposition 3.1.8, the collection $\{G'_i\}_{i \in I}$ of pseudo-simple normal k-subgroups of G' is pairwise commuting and in natural bijection with the set of simple factors of the maximal semisimple quotient of $G'_{\overline{k}}$, and the G'_i generate G'. By Proposition 1.2.7 each G'_i is normal in G.

Step 1. The first part of the argument is devoted to proving that the smooth connected k-subgroup generated by the G'_i that are exotic with type F_4 or G_2 is actually a direct product of such subgroups. Consider the disjoint union decomposition $I = I_1 \coprod I_2$ in which I_2 is the set of $i \in I$ such that G'_i is exotic of type F_4 when $p = 2$ and type G_2 when $p = 3$. Note that if $p = 3$ then every pseudo-simple k-group is either standard or exotic (Corollary 6.3.5 and Theorem 8.2.1), so when $p = 3$ we can equivalently say that I_1 is the set of $i \in I$ such that G'_i is standard. (This description is not generally true when $p = 2$.) By Proposition 5.3.1 non-standardness of G' implies that of some G'_i, so if $p = 3$ then the hypothesis on G says $I_2 \neq \emptyset$. When $p = 2$, it is assumed in (2) that $I_2 \neq \emptyset$.

Let H_1 (resp. $\mathscr{G}_2$) denote the smooth connected normal k-subgroup of G' generated by the G'_i for $i \in I_1$ (resp. $i \in I_2$). Note that $\mathscr{G}_2 \neq 1$ since I_2 is non-empty, and H_1 commutes with $\mathscr{G}_2$. By Proposition 3.1.8 applied to G', the G'_i for $i \in I_1$ (resp. $i \in I_2$) are the pseudo-simple normal k-subgroups of H_1 (resp. of $\mathscr{G}_2$). Also, observe that H_1 is non-commutative if it is nontrivial (i.e., if I_1 is non-empty). Hence, by Proposition 5.3.1, the pseudo-reductive k-group H_1 is standard if $p = 3$.

For any $i \in I_2$, the k-group G'_i has trivial intersection with the subgroup generated by the G'_j for all $j \in I - \{i\}$ since such an intersection is central in G'_i yet $Z_{G'_i} = 1$ (Corollary 7.2.5). Thus, the natural map $\prod_{i \in I_2} G'_i \to \mathscr{G}_2$ is an isomorphism (so $\mathscr{G}_2$ is exotic) and the natural map $H_1 \times \mathscr{G}_2 \to G$ has trivial kernel. By Proposition 3.1.8, this proves that a perfect pseudo-reductive k-group whose pseudo-simple normal k-subgroups are all exotic of type F_4 or G_2 is a direct product of such k-subgroups.

STEP 2. We next carry out the construction of $\mathscr{G}_1$ as a subgroup generated by H_1 and a piece of a Cartan k-subgroup of G (the idea being to fill out what should be the Cartan k-subgroup of $\mathscr{G}_1$, in case G is not perfect). The quotient $G/(H_1 \times \mathscr{G}_2)$ is commutative since $H_1 \times \mathscr{G}_2 = G'$.

Let T be a maximal k-torus in G. Since the formation of torus centralizers is compatible with quotients of smooth connected affine groups, it follows from the commutativity of $G/(H_1 \times \mathscr{G}_2)$ that $Z_G(T)$ maps onto $G/(H_1 \times \mathscr{G}_2)$. Equivalently, $G = Z_G(T) \cdot (H_1 \times \mathscr{G}_2)$. Note that $Z_G(T)$ is commutative by Proposition 1.2.4, but $Z_G(T) \cap \mathscr{G}_2 \neq 1$ (e.g., this intersection contains the k-subtorus of T that is maximal in the non-solvable normal k-subgroup $\mathscr{G}_2$ of G). We seek a smooth connected k-subgroup Z_1 of $Z_G(T)$ which centralizes $\mathscr{G}_2$ and such that $G = Z_1 \cdot (H_1 \times \mathscr{G}_2)$. For a suitable such Z_1 we will be able to take $\mathscr{G}_1 = Z_1 \cdot H_1$.

The k-subtorus $T_2 = T \cap \mathscr{G}_2$ of T is maximal in $\mathscr{G}_2$. The image of $(\mathscr{G}_2)_{\overline{k}}$ in $G^{\rm red}_{\overline{k}}$ is a product among simple factors of the derived group of $G^{\rm red}_{\overline{k}}$, so $Z_{\mathscr{G}_2}(T_2)_{\overline{k}}$ and $(T_2)_{\overline{k}}$ have the same image in $G^{\rm red}_{\overline{k}}$. In particular, the commutator $(Z_{\mathscr{G}_2}(T_2), Z_G(T))_{\overline{k}}$ has trivial image in $G^{\rm red}_{\overline{k}}$, so $(Z_{\mathscr{G}_2}(T_2), Z_G(T)) = 1$ by Lemma 1.2.1. Equivalently (in view of the commutativity of $Z_G(T)$), we have $Z_{\mathscr{G}_2}(T_2) \subseteq Z_G(T)$.

Let $Z = Z_G(T)$ and $Z_2 = Z_{\mathscr{G}_2}(T_2)$. Every $z \in Z(k)$ centralizes T_2 and so the conjugation action of z on the normal subgroup $\mathscr{G}_2 \subseteq G$ is an automorphism of $\mathscr{G}_2$ which is trivial on the maximal k-torus T_2. By Corollary 8.2.5 every such automorphism of $\mathscr{G}_2$ is inner, so there exists $z_2 \in \mathscr{G}_2(k)$ such that zz_2^{-1}

centralizes $\mathscr{G}_2$. In particular, $z_2 \in Z_{\mathscr{G}_2}(T_2)(k) = Z_2(k)$, so $Z(k) = C \cdot Z_2(k)$ with C denoting the subgroup of elements of $Z(k)$ that centralize $\mathscr{G}_2$.

The identity component Z_1 of the Zariski closure of C in $Z = Z_G(T)$ centralizes $\mathscr{G}_2$, and $Z_1(k)Z_2(k)$ has finite index in $Z(k) = C \cdot Z_2(k)$ (since $C/Z_1(k)$ is finite, due to finiteness of the component group of the Zariski closure of C in Z). Hence, $(Z/Z_1Z_2)(k) = Z(k)/(Z_1Z_2)(k)$ is finite, yet $k = k_s$ and $Z/(Z_1Z_2)$ is smooth and connected, so $Z = Z_1Z_2$. But Z_1 centralizes $\mathscr{G}_2$ and Z_2 is contained in $\mathscr{G}_2$, so $G = Z \cdot (H_1 \times \mathscr{G}_2) = (Z_1 \cdot H_1) \cdot \mathscr{G}_2$ with $Z_1 \cdot H_1$ centralizing $\mathscr{G}_2$.

Let $\mathscr{G}_1 = Z_1 \cdot H_1$, so $\mathscr{G}_1$ is a smooth connected k-subgroup of G that centralizes the normal k-subgroup $\mathscr{G}_2$ and satisfies $G = \mathscr{G}_1 \cdot \mathscr{G}_2$. Thus, $\mathscr{G}_1$ is normal in G and so is pseudo-reductive over k. The intersection $\mathscr{G}_1 \cap \mathscr{G}_2$ is central in $\mathscr{G}_2$, so $\mathscr{G}_1 \cap \mathscr{G}_2 = 1$ since $Z_{\mathscr{G}_2} = 1$. Hence, the natural map $\mathscr{G}_1 \times \mathscr{G}_2 \to G$ is an isomorphism.

Step 3. Finally, we prove that the pair $(\mathscr{G}_1, \mathscr{G}_2)$ really works, and that it is unique. Among the pseudo-simple normal k-subgroups G'_i of G, the ones that lie in $\mathscr{G}_1$ are precisely the G'_i for $i \in I_1$. It follows once again via Proposition 1.2.7 and the normality of $\mathscr{G}_1$ in G that every pseudo-simple normal k-subgroup of $\mathscr{G}_1$ is normal in G and hence (by Proposition 3.1.8) is equal to some G'_i. If $p = 3$ then these are standard by definition of I_1, so $\mathscr{G}_1$ is standard by Proposition 5.3.1. This concludes the proof of the existence of the desired decomposition of G.

The uniqueness of the decomposition of G is a consequence of the construction of $\mathscr{G}_1$ and $\mathscr{G}_2$, as follows. Let $\mathscr{G}'_1 \times \mathscr{G}'_2$ be a decomposition of G with $\mathscr{G}'_1$ having no exotic pseudo-simple normal k-subgroups of type F_4 or G_2 and $\mathscr{G}'_2$ an exotic pseudo-reductive k-group whose pseudo-simple factors are all of type F_4 or G_2. Since the scheme-theoretic centers $Z_{\mathscr{G}'_2}$ and $Z_{\mathscr{G}_2}$ are trivial, the decompositions $G = \mathscr{G}_1 \times \mathscr{G}_2$ and $G = \mathscr{G}'_1 \times \mathscr{G}'_2$ imply $\mathscr{G}_1 = Z_G(\mathscr{G}_2)$ and $\mathscr{G}'_1 = Z_G(\mathscr{G}'_2)$. Hence, it suffices to prove that $\mathscr{G}'_2 = \mathscr{G}_2$. By Proposition 3.1.8 the pseudo-simple normal k-subgroups of $\mathscr{G}'_2$ are among the G'_i for $i \in I_2$, so by definition of $\mathscr{G}'_2$ being exotic we have $\mathscr{G}'_2 = \prod_{j \in J} G'_j$ for some subset $J \subseteq I_2$. We just have to rule out the possibility $J \neq I_2$.

If there exists $i \in I_2 - J$ then G'_i centralizes $\mathscr{G}'_2$, so G'_i is contained in $Z_G(\mathscr{G}'_2) = \mathscr{G}'_1$. But $\mathscr{G}'_1$ has no exotic pseudo-simple normal k-subgroups of type F_4 or G_2 and G'_i is an exotic pseudo-simple k-group of type F_4 or G_2, so we have a contradiction. □

Corollary 8.2.11 *Let G be a pseudo-reductive group over an imperfect field k with* $\mathrm{char}(k) = 3$, *and let T be a maximal k-torus in G. The k-group G is a standard pseudo-reductive k-group if and only if for every simple smooth* *connected normal subgroup* $H \subseteq G^{\mathrm{ss}}_{\overline{k}}$ *of type* G_2 *and any two roots for* $T_{\overline{k}}$ *arising in* $\mathrm{Lie}(H)$*, the* $\overline{k}$*-dimensions of the corresponding root spaces in* $\mathrm{Lie}(G_{\overline{k}})$ *are equal.*

Proof The necessity is clear, and by Theorem 8.2.10 the sufficiency is immediate from Proposition A.2.10 and the description (7.2.1) of the Lie algebras of exotic pseudo-reductive groups. □

9

Absolutely pseudo-simple groups in characteristic 2

Let G be an absolutely pseudo-simple group over a field k of characteristic 2 and let K/k be the field of definition of $\mathscr{R}(G_{\overline{k}}) \subseteq G_{\overline{k}}$. The classification of possibilities for such G (subject to the hypothesis that G is pseudo-split, unless $[k : k^2] \leqslant 2$) will proceed in several stages, inspired by our work in Chapter 6, as follows. Recall that in §6.2 we proved standardness when $G^{\rm ss}_{\overline{k}}$ is either of the rank-2 types A_2 or G_2 whose Dynkin diagram does not have an edge with multiplicity 2 (in contrast with the other rank-2 type, namely $B_2 = C_2$). In Proposition 6.3.6 these rank-2 cases were used to prove standardness whenever the simple semisimple $\overline{k}$-group $G^{\rm ss}_{\overline{k}}$ is of any type with rank at least 2 that does not have an edge of multiplicity 2 (i.e., we allow all types of rank $\geqslant 2$ aside from B_n ($n \geqslant 2$), C_n ($n \geqslant 2$), and F_4). In Definition 8.2.2 we introduced a class of non-standard pseudo-reductive k-groups with reduced root systems of rank $\geqslant 2$.

To go further, we need to understand rank-1 cases: type $C_1 = A_1$, where most of the essential new difficulties in characteristic 2 arise (e.g., non-reduced root systems). This rests on structural results from §3.4: the existence of Levi subgroups and Bruhat decomposition for *pseudo-split* pseudo-reductive groups. In §9.2 we determine all possibilities for the image of ξ_G when $G^{\rm ss}_{\overline{k}}$ is of type A_1 and G is pseudo-split. Consider such a G, and let K/k be the field of definition for the subgroup $\mathscr{R}_u(G_{\overline{k}}) \subseteq G_{\overline{k}}$. Upon fixing the K-isomorphism class of $G^{\rm ss}_{\overline{k}}$ (i.e., SL_2 or PGL_2), the existence of a Levi k-subgroup in G and the Bruhat decomposition for $G(k)$ will enable us to show that the possibilities for the image of ξ_G are parameterized by the $K^\times$-homothety classes of nonzero kK^2-subspaces $V \subseteq K$ such that K is generated as a k-algebra by the ratios v'/v for $v, v' \in V - \{0\}$ (see Definition 9.1.1 and Proposition 9.1.7).

The results in Theorem 9.8.1, Proposition 9.8.4, Theorem 9.8.6, and Proposition 9.8.9 provide a classification of all *pseudo-split* absolutely pseudo-simple groups with a non-reduced root system over any imperfect field of characteristic 2, subject to an additional minimality condition introduced in Definition 9.4.4. If $[k : k^2] = 2$ then the only possibility for V as above is K. We use this to prove in §9.2 that if $[k : k^2] = 2$ and G has rank 1 and a reduced root system over k_s then G is standard (Proposition 9.2.4). (If $[k : k^2] \geqslant 4$ then any finite extension K/k contained in $k^{1/2}$ with $[K : k] \geqslant 4$ admits a $V \neq K$, such as the k-span of 1 and a 2-basis of K/k. The k-groups associated to such V are not covered by the constructions in our main classification in Theorem 5.1.1.) This underlies the proof that the pseudo-split and minimality hypotheses automatically hold when $[k : k^2] = 2$ (see Theorem 9.9.3(1)), yielding a better classification result over such k.

Building on preliminary work in §9.3, in §9.4 we introduce and study the notion of pseudo-reductive groups of minimal type over fields of arbitrary characteristic. This is crucial in [CP] to go beyond the case $[k : k^2] = 2$ when $\mathrm{char}(k) = 2$.

9.1 Subgroups of $\mathrm{R}_{K/k}(\mathrm{SL}_2)$ and $\mathrm{R}_{K/k}(\mathrm{PGL}_2)$

Let K/k be a finite purely inseparable extension of an arbitrary *infinite* field k of characteristic 2. For any non-empty subset $S \subseteq K^\times$, we define $k\langle S\rangle \subseteq K$ to be the k-subalgebra generated by the ratios s'/s for $s, s' \in S$, so $k\langle \lambda S\rangle = k\langle S\rangle$ for any $\lambda \in K^\times$. When S is a nontrivial additive subgroup of K we write $k\langle S\rangle$ to denote $k\langle S - \{0\}\rangle$. Note that $k\langle S\rangle$ and $k[S]$ are fields since K is algebraic over k, so $k\langle S\rangle \subseteq k[S]$ with equality when $1 \in S$. If $F := k\langle S\rangle$ then $kF^2 = k\langle S^2\rangle$ (where S^2 denotes the set of elements s^2 for $s \in S$). In the special case that $1 \in S$, if S is a $k\langle S^2\rangle$-subspace of K then $S - \{0\}$ is stable under inversion since $1/s = (1^2/s^2)s \in S$ for $s \in S - \{0\}$. Likewise, if V is a k-subspace of K containing 1 then V is a $k\langle V^2\rangle$-subspace of K if and only if $V^2 \cdot V \subseteq V$. Indeed, the "if" direction is obvious, and the converse follows from the equality $k\langle S\rangle = k[S]$ for any subset $S \subseteq K^\times$ containing 1.

Consider a nonzero k-subspace V of K such that V is a $k\langle V^2\rangle$-subspace of K. For any $v_1, v_2 \in V - \{0\}$, the inclusion $(v_1/v_2)^2 V \subseteq V$ is an equality for k-dimension reasons and so $(1/v_1)^2 V = (1/v_2)^2 V$ inside K. In other words, for any $v \in V - \{0\}$ the nonzero k-subspace $V^- := (1/v)^2 V \subseteq K$ is independent of v, and it is a $k\langle (V^-)^2\rangle$-subspace since $k\langle (V^-)^2\rangle = k\langle V^2\rangle$. Thus, V^- satisfies the same hypotheses as V, and $k\langle (V^-)^2\rangle = k\langle V^2\rangle$ inside K. Also, for any $\lambda \in K^\times$ the k-subspace λV satisfies the same hypotheses as V does, with the same associated field $k\langle (\lambda V)^2\rangle = k\langle V^2\rangle$, and $(\lambda V)^- = (1/\lambda)V^-$. Note that $V = V^-$ if $1 \in V$ or if V is a kK^2-subspace of K. Iterating $V \mapsto V^-$ twice recovers V; i.e., $(V^-)^- = V$. We will use V and V^- to define interesting k-subgroups of $\mathrm{R}_{K/k}(\mathrm{SL}_2)$ and $\mathrm{R}_{K/k}(\mathrm{PGL}_2)$.

For reasons of efficiency of exposition, we wish to treat subgroups of $\mathrm{R}_{K/k}(\mathrm{SL}_2)$ and $\mathrm{R}_{K/k}(\mathrm{PGL}_2)$ at the same time, so we now introduce some notation. Let L be one of the two k-groups SL_2 or PGL_2, and let U^+ and U^- be the respective upper and lower triangular unipotent k-subgroups of L equipped with their canonical k-isomorphism with $\mathbf{G}_\mathrm{a}$. These are the root groups of L relative to the diagonal k-torus D of L.

Definition 9.1.1 Let V be a nonzero k-subspace of K. The Zariski closure in $\mathrm{R}_{K/k}(D_K)$ of the subgroup of $D(K)$ generated by the matrices $\begin{pmatrix} v'/v & 0 \\ 0 & v/v' \end{pmatrix} \in \mathrm{SL}_2(K)$ for $v, v' \in V - \{0\}$ will be denoted $V^*_{K/k}$. (Note that $\mathrm{SL}_2(K) \subseteq \mathrm{PGL}_2(K)$ since $\mathrm{char}(k) = 2$. Although the notation $V^*_{K/k}$ does not specify whether it is being computed inside $\mathrm{R}_{K/k}(\mathrm{SL}_2)$ or $\mathrm{R}_{K/k}(\mathrm{PGL}_2)$, the context will always make the intended meaning clear. Also, for any $\lambda \in K^\times$ clearly $(\lambda V)^*_{K/k} = V^*_{K/k}$.) In particular, if F is a subfield of K containing k then $F^*_{K/k}$ denotes the group $V^*_{K/k}$ for $V = F$ (and it depends on L).

Define $\underline{V} \subseteq \mathrm{R}_{K/k}(\mathbf{G}_\mathrm{a})$ to be the k-subgroup corresponding to V. If V is a $k\langle V^2\rangle$-subspace of K then define H_V to be the k-subgroup of $\mathrm{R}_{K/k}(L_K)$ generated by the k-subgroups $U^+_{\underline{V}} \subseteq \mathrm{R}_{K/k}(U^+_K)$ and $U^-_{\underline{V}^-} \subseteq \mathrm{R}_{K/k}(U^-_K)$ respectively corresponding to $\underline{V}$ and $\underline{V^-}$. (If we need to keep track of K/k then we will write $H_{V,K/k}$.)

For a nonzero k-subspace V of K it appears to be very difficult to explicitly describe the subset $V^*_{K/k}(k) \subseteq K^\times$ in general (via the map $\mathrm{diag}(x, 1/x) \mapsto x$ when $L = \mathrm{SL}_2$ and $\mathrm{diag}(x, 1) \mapsto x$ when $L = \mathrm{PGL}_2$), but when $L = \mathrm{SL}_2$ we can compute the k-algebra it generates: this is exactly $k\langle V\rangle$. Indeed, the inclusion $k\langle V\rangle \subseteq k[V^*_{K/k}(k)]$ is obvious, and for the reverse inclusion we simply observe that by the definition of $V^*_{K/k}$ as a Zariski closure inside $\mathrm{R}_{K/k}(\mathrm{GL}_1)$ (with $L = \mathrm{SL}_2$) we have $V^*_{K/k} \subseteq \mathrm{R}_{k\langle V\rangle/k}(\mathrm{GL}_1)$, so $V^*_{K/k}(k) \subseteq k\langle V\rangle^\times$, yielding the reverse inclusion at the level of k-subalgebras of K.

If F is an extension of k contained in K such that $F^2 \cdot V = V$ then the k-subgroup $F^*_{K/k}$ of $\mathrm{R}_{K/k}(L_K)$ normalizes H_V. Also, if V is a $k\langle V^2\rangle$-subspace of K and $1 \in V$ then $V^*_{K/k}$ is the Zariski closure of the subgroup of $D(K)$ generated by the matrices $\begin{pmatrix} v & 0 \\ 0 & 1/v \end{pmatrix}$ because $V - \{0\}$ is stable under inversion in such cases (as $v^{-1} = v^{-2} \cdot v$ for $v \in V - \{0\}$, and $k\langle V^2\rangle = k[V^2]$ since $1 \in V^2$).

Remark 9.1.2 For later computations with k_s-points, it is useful to observe that for any nonzero k-subspace V of K, the formation of several associated structures over k naturally commutes with separable extension on k: $V^*_{K/k}$, $k\langle V\rangle$ when $L = \mathrm{SL}_2$, and H_V when V is also a $k\langle V^2\rangle$-subspace of K. To be precise, consider a separable extension field k' of k and the field $K' = k' \otimes_k K$ and

k'-subspace $V' = k' \otimes_k V \subseteq K'$. (Clearly V' is a $k'\langle V'^2\rangle$-subspace of K' when V is a $k\langle V^2\rangle$-subspace of K.) Since $V - \{0\}$ is Zariski-dense in $(\underline{V} - \{0\})_{k'}$ (as k is infinite), it is easy to see that $(V^*_{K/k})_{k'} = (V')^*_{K'/k'}$. Thus, if $L = \mathrm{SL}_2$ then the inclusion $k' \otimes_k k\langle V\rangle \subseteq k'\langle V'\rangle$ of subfields of K is an equality because $(V')^*_{K'/k'} = (V^*_{K/k})_{k'} \subseteq \mathrm{R}_{k\langle V\rangle/k}(\mathrm{GL}_1)_{k'}$ and $k'\langle V'\rangle$ is generated as a k'-algebra by the subgroup $(V')^*_{K'/k'}(k') \subseteq K'^{\times}$ when $L = \mathrm{SL}_2$ (with $V^*_{K/k}(k)$ Zariski-dense in $(V^*_{K/k})_{k'} = (V')^*_{K'/k'}$). Finally, if V is a $k\langle V^2\rangle$-subspace of K then inside $\mathrm{R}_{K/k}(L_K)_{k'} = \mathrm{R}_{K'/k'}(L_{K'})$ we have $(U^+_V)_{k'} = U^+_{V'}$ and $(U^-_{V^-})_{k'} = U^-_{(V')^-}$, so $(H_V)_{k'} = H_{V'}$.

There is a variant that will be useful later: if W is a nonzero K^2-subspace of K and $W' := W \otimes_{K^2} K'^2 \subseteq K'$ then we claim that the inclusion $k' \otimes_k k\langle W\rangle \subseteq k'\langle W'\rangle$ inside K' is an equality. Since $w_1/w_2 = (1/w_2)^2 w_1 \cdot w_2$ and $w_1 w_2 = (w_1^2 w_2)/w_1$ for $w_i \in W - \{0\}$, clearly $k\langle W\rangle = k[W \cdot W]$ where $W \cdot W$ denotes the set of products of pairs of elements of W. Similarly, $k'\langle W'\rangle = k'[W' \cdot W']$, so the ratios are replaced with products throughout and hence the desired equality is clear.

The smooth commutative closed k-subgroup $V^*_{K/k}$ of the commutative pseudo-reductive k-group $\mathrm{R}_{K/k}(D_K)$ is clearly connected, so it is pseudo-reductive. Also, the maximal k-torus D of $\mathrm{R}_{K/k}(D_K)$ lies in $V^*_{K/k}$ because $c = cv/v$ for any $c \in k^{\times}$ and $v \in V - \{0\}$. Assuming that V is a $k\langle V^2\rangle$-subspace of K, for any $\lambda \in K^{\times}$ and $g = \mathrm{diag}(\lambda, 1) \in \mathrm{PGL}_2(K)$ we have $gH_Vg^{-1} = H_{\lambda V}$. Taking $\lambda = 1/v$ for $v \in V - \{0\}$, the conjugate of H_V by $\mathrm{diag}(1/v, 1)$ is $H_{(1/v)V}$ with $1 \in (1/v)V = ((1/v)V)^-$. The k-group L is generated by the k-points of its upper and lower triangular unipotent subgroups (as k is infinite), so $L \subseteq H_{(1/v)V}$ inside $\mathrm{R}_{K/k}(L_K)$. Thus, H_V contains the conjugate of L by $\mathrm{diag}(v, 1)$ for any $v \in V - \{0\}$, so H_V contains D.

Remark 9.1.3 The group $H_{K,K/k}$ (i.e., $H_{V,K/k}$ for $V = K$) is the subgroup of $\mathrm{R}_{K/k}(L_K)$ (where L is either SL_2 or PGL_2) generated by the upper and lower triangular unipotent k-subgroups. If $L = \mathrm{SL}_2$ then $H_{K,K/k} = \mathrm{R}_{K/k}(\mathrm{SL}_2)$, and if $L = \mathrm{PGL}_2$ then $H_{K,K/k} = \mathscr{D}(\mathrm{R}_{K/k}(\mathrm{PGL}_2)) = \mathrm{R}_{K/k}(\mathrm{SL}_2)/\mathrm{R}_{K/k}(\mu_2)$.

Proposition 9.1.4 *Let L be the k-group SL_2 or PGL_2. Let V be a nonzero k-subspace of K, and assume that V is a $k\langle V^2\rangle$-subspace of K. The smooth connected k-subgroup $H := H_V \subseteq \mathrm{R}_{K/k}(L_K)$ is absolutely pseudo-simple. Its root groups with respect to the torus $D\,(\subseteq \mathrm{R}_{K/k}(D_K))$ are U^+_V and $U^-_{V^-}$, and the Cartan subgroup $Z_H(D)$ is equal to $V^*_{K/k}$. In particular, the k-subgroup H_V inside $\mathrm{R}_{K/k}(L_K)$ determines the k-subspace $V \subseteq K$.*

Moreover, the natural map $\pi : H_K \to L_K$ is a K-descent of the maximal reductive quotient of $H_{\overline{k}}$, and the field of definition over k for the *geometric unipotent radical of H is $k\langle V\rangle$. (In particular, if V' is a nonzero k-subspace of K that is a $k\langle V'^2\rangle$-subspace of K and $H_V \simeq H_{V'}$ as k-groups then $k\langle V\rangle = k\langle V'\rangle$ as purely inseparable extensions of k.)*

Proof Conjugation by $\mathrm{diag}(\lambda, 1)$ for $\lambda \in K^\times$ shows that the problems for V and λV are equivalent. Thus, we may and do assume $1 \in V$. In particular, the following properties hold: $k\langle V\rangle = k[V]$, $k\langle V^2\rangle = k[V^2]$ (so V is a $k[V^2]$-subspace of K), $V^- = V$, $L \subseteq H$, $V^2 \subseteq V$, and $V^*_{K/k}$ contains a Zariski-dense subset generated by the k-points $\mathrm{diag}(v, 1/v)$ for $v \in V - \{0\}$.

By Corollary A.5.16, the k-subgroup L in $\mathrm{R}_{K/k}(L_K)$ is a Levi k-subgroup. As L is contained in H, Lemma 7.2.4 implies that H is pseudo-reductive and L is a Levi k-subgroup of H such that π is a K-descent of the maximal reductive quotient of $H_{\overline{k}}$. Hence, the field of definition over k of $\mathscr{R}_u(H_{\overline{k}})$ is contained in K and $\mathscr{D}(H)$ is absolutely pseudo-simple of type A_1.

The root groups of H with respect to D are $U_H^\pm := H \cap \mathrm{R}_{K/k}(U_K^\pm)$. We will show that $U_H^\pm = U_V^\pm$ and $Z_H(D) = V^*_{K/k}$. We first check that $V^*_{K/k} \subseteq Z_H(D)$, or equivalently that $V^*_{K/k} \subseteq H$. The subset $V - \{0\} \subseteq K^\times$ is stable under inversion (since $1 \in V$ and V is a subspace of K over $k\langle V^2\rangle = k[V^2]$), so it suffices to apply the identity

$$\begin{pmatrix} t & 0 \\ 0 & t^{-1} \end{pmatrix} = \begin{pmatrix} 1 & 1 \\ 0 & 1 \end{pmatrix} \begin{pmatrix} 1 & 0 \\ t-1 & 1 \end{pmatrix} \begin{pmatrix} 1 & -1/t \\ 0 & 1 \end{pmatrix} \begin{pmatrix} 1 & 0 \\ t(1-t) & 1 \end{pmatrix} \tag{9.1.1}$$

that expresses a diagonal matrix in SL_2 as a product of upper and lower triangular unipotent matrices. (Note that for $t \in V$, $t(1-t) = t - t^2 \in V$ because V contains V^2.) Since $V^2 \cdot V \subseteq V$, it also follows from the description of $V^*_{K/k}$ as a Zariski closure that it normalizes the smooth connected k-subgroup $U_V^\pm \subseteq \mathrm{R}_{K/k}(U_K^\pm)$. Hence, it makes sense to define the following subgroups of $H(k)$:

$$\mathscr{Z} = V^*_{K/k}(k), \;\; \mathscr{U} = U_V^+(k), \;\; \mathscr{P} = \mathscr{Z} \ltimes \mathscr{U}.$$

Note that $\mathscr{Z}$ is Zariski-dense in $V^*_{K/k}$ due to the definition of $V^*_{K/k}$ as a Zariski closure.

The Bruhat decomposition of $\mathrm{R}_{K/k}(L_K)(k) = L(K)$ (i.e., $\mathrm{SL}_2(K)$ or $\mathrm{PGL}_2(K)$) gives that for

$$n := \begin{pmatrix} 0 & 1 \\ -1 & 0 \end{pmatrix}, \tag{9.1.2}$$

the map $\mathscr{U} \times \mathscr{P} \to H(k)$ defined by $(u, p) \mapsto unp$ is injective and we have the disjoint union containment

$$\mathscr{U} n \mathscr{P} \bigcup \mathscr{P} \subseteq H(k).$$

It is easy to check that $\mathscr{H} := \mathscr{U} n \mathscr{P} \bigcup \mathscr{P}$ is stable under inversion and multiplication, so it is a subgroup of $H(k)$. As it contains the subgroups $\mathscr{U}$ and $n\mathscr{U}n^{-1}$, which are Zariski-dense in U_V^+ and U_V^- respectively, and H is generated by U_V^+ and U_V^-, we see that $\mathscr{H}$ is *Zariski-dense* in H.

By Proposition 2.1.8(2),(3) applied to the standard parameterization λ : $\mathrm{GL}_1 \simeq D$, the multiplication map

$$n^{-1}U_H^+ n \times Z_H(D) \times U_H^+ \to H$$

is an open immersion, and clearly its left-translate by n meets the Zariski-dense $\mathscr{H}$ in $\mathscr{U} n \mathscr{P}$. Hence, $n^{-1}\mathscr{U} n \mathscr{P}$ is Zariski-dense in H. Therefore, its closure $n^{-1}U_V^+ n \cdot V_{K/k}^* \cdot U_V^+$ in $n^{-1}U_H^+ n \cdot Z_H(D) \cdot U_H^+$ must be full, so $Z_H(D) = V_{K/k}^*$ and $U_H^+ = U_V^+$; hence $U_H^- = nU_H^+ n^{-1} = nU_V^+ n^{-1} = U_V^-$ and H is generated by the vector groups $U_V^{\pm}$ on which D acts linearly through nontrivial characters, so H is perfect. Now it follows from Lemma 3.1.2 that H is absolutely pseudo-simple (with root groups $U_V^{\pm} = U_H^{\pm}$ relative to D).

Our remaining task is to show that the field of definition over k for $\ker \pi \subseteq H_K$ is equal to $k[V]$. For this, we may and do assume, after replacing K by $k[V]$, that $K = k[V]$. We will now show that if $F \subseteq K$ is a subfield containing k such that $\ker \pi$ descends to an F-subgroup $R \subseteq H_F$ then $F = K$.

Since the Levi k-subgroup L in H yields a Levi F-subgroup L_F in H_F, this F-subgroup maps isomorphically onto the quotient H_F/R. Hence, the K-map π descends to an F-map $\pi_0 : H_F \twoheadrightarrow L_F$. (The point is that the descent π_0 of π has target that is the *canonical* F-descent of the target L_K of π over K.) Passing to Lie algebras of root groups relative to both the maximal F-torus D_F of H_F and the diagonal F-torus of the target of π_0, we get an abstract F-linear map $F \otimes_k V \to F$ that descends the canonical K-linear map $K \otimes_k V \to K$ induced by $\mathrm{Lie}(\pi)$.

The map π is induced by the canonical quotient map

$$\mathrm{R}_{K/k}(L_K)_K \to L_K$$

that on Lie algebras is the natural multiplication map

$$K \otimes_k \mathfrak{l}_K \to \mathfrak{l}_K$$

(with $\mathfrak{l}_K = \mathrm{Lie}(L_K)$), so the map $K \otimes_k V \to K$ induced by $\mathrm{Lie}(\pi)$ is $c \otimes v \mapsto cv$. Hence, the existence of π_0 implies that $F \cdot V \subseteq F$ inside of K, and in particular $V \subseteq F$. But F is a subfield of K containing k, so F contains $k[V] = K$. □

In the following proposition (and only in this proposition in §9.1) we let k be an infinite field of *arbitrary* characteristic and K a purely inseparable finite extension field of k. Let L be either of the k-groups SL_2 or PGL_2, and D the diagonal k-torus of L. We will now prove the following converse of Proposition 9.1.4; it can be used to give an alternative proof of Theorem 6.1.1.

Proposition 9.1.5 *Let G be a pseudo-semisimple k-subgroup of $\mathrm{R}_{K/k}(L_K)$ that contains the diagonal k-torus D $(\subseteq \mathrm{R}_{K/k}(D_K))$.*

(1) *If k is of characteristic 2 then there exist a unique nonzero k-subspace $V \subseteq K$ which is a $k\langle V^2\rangle$-subspace of K such that $G = H_V$. In particular, $1 \in V$ if and only if the element $\left(\begin{smallmatrix}1&1\\0&1\end{smallmatrix}\right) \in L(K)$ lies in $G(k)$.*
(2) *If $\mathrm{char}(k) \neq 2$ then there exist a subfield F of K that contains k and $\lambda \in K^\times$ such that the conjugate of G under $\mathrm{diag}(\lambda, 1)$ is the k-subgroup $\mathrm{R}_{F/k}(L_F)$ of $\mathrm{R}_{K/k}(L_K)$.*

Proof The root groups of (G, D) are the intersections $U_G^\pm$ of the root groups $\mathrm{R}_{K/k}(U_K^\pm)$ of $\mathrm{R}_{K/k}(L_K)$ with G. These subgroups are stable under the natural conjugation action of D and are k-subgroups of the upper and lower triangular unipotent subgroups of $\mathrm{R}_{K/k}(L_K)$. Thus, by the Zariski-density of $(k^\times)^2$ in GL_1, the k-groups $U_G^\pm$ correspond to k-subspaces of K. Hence, $V^\pm := U_G^\pm(k)$ are k-subspaces of $\mathrm{R}_{K/k}(\mathbf{G}_\mathrm{a})(k) = K$. They are nonzero since the split unipotent k-groups $U_G^\pm$ are nontrivial (due to the absolute pseudo-simplicity of G). In assertion (1), if $G = H_V$ for some V then necessarily $V = V^+$, so uniqueness of V is clear. (The problem is to prove existence.)

Applying conjugation by $\mathrm{diag}(\lambda, 1)$ for $\lambda \in K^\times$ has no effect on D nor (when $\mathrm{char}(k) = 2$) on whether or not G has the form H_V for a nonzero k-subspace V of K that is a $k\langle V^2\rangle$-subspace, and it replaces $V^\pm$ with $\lambda^{\pm 1}V^\pm$. Thus, by choosing $\lambda = 1/v$ for $v \in V^+ - \{0\}$ we may reduce to the case $1 \in V^+$, which is to say $\left(\begin{smallmatrix}1&1\\0&1\end{smallmatrix}\right) \in G(k)$.

By Proposition 3.4.2, there is a unique element in $N_G(D)(k) - Z_G(D)(k)$ of the form $\nu = u'uu'$, with $u = \left(\begin{smallmatrix}1&1\\0&1\end{smallmatrix}\right)$ and $u' \in U_G^-(k)$. We see by explicit computation (in $\mathrm{SL}_2(K)$ or $\mathrm{PGL}_2(K)$) that $u' = \left(\begin{smallmatrix}1&0\\-1&1\end{smallmatrix}\right)$ and $\nu = \left(\begin{smallmatrix}0&1\\-1&0\end{smallmatrix}\right)$. Thus, the standard Weyl element $n := \left(\begin{smallmatrix}0&1\\-1&0\end{smallmatrix}\right)$ lies in $G(k)$. Using conjugation by n, we see that $V^+ = V^-$. We will denote this common k-subspace of K by V, so $1 \in V$.

Next, we prove that the subset $V - \{0\} \subseteq K^\times$ is closed under inversion and that $V^2 \cdot V \subseteq V$ (i.e., $v^2 V = V$ for any $v \in V - \{0\}$). Let $Z = Z_G(D)$. The Bruhat decomposition of $G(k)$ relative to the minimal pseudo-parabolic

k-subgroup $Z \ltimes U_V^-$ is provided by Theorem 3.4.5, and it says

$$G(k) = U_V^-(k)nZ(k)U_V^-(k) \coprod Z(k)U_V^-(k). \tag{9.1.3}$$

Since $U_V^+ \cap (Z \ltimes U_V^-) = \{1\}$, we have $U_V^+(k) - \{1\} \subseteq U_V^-(k)nZ(k)U_V^-(k)$.

Multiplication defines an open immersion $U_V^+ \times Z \times U_V^- \to G$ (Proposition 2.1.8), so the equality $U_V^+ = n^{-1}U_V^- n$ and left translation by n imply that the map $U_V^- \times Z \times U_V^- \to G$ defined by $(u', z, u'') \mapsto u'nzu''$ is an open immersion. In particular, for $u', u'' \in U_V^-(k)$ and $z \in Z(k)$ the product $u'nzu'' \in G(k)$ uniquely determines u', u'', and z. Thus, for each *nonzero* $x \in V$ there exist unique $y_1, y_2 \in V$ and $z \in Z(k)$ such that

$$\begin{pmatrix} 1 & x \\ 0 & 1 \end{pmatrix} = \begin{pmatrix} 1 & 0 \\ y_1 & 1 \end{pmatrix} \cdot \begin{pmatrix} 0 & 1 \\ -1 & 0 \end{pmatrix} \cdot z \cdot \begin{pmatrix} 1 & 0 \\ y_2 & 1 \end{pmatrix} \tag{9.1.4}$$

in $\mathrm{SL}_2(K)$ or $\mathrm{PGL}_2(K)$. All terms in (9.1.4) aside from z come from $\mathrm{SL}_2(K)$, so even when $L = \mathrm{PGL}_2$ we have that z uniquely arises from $\mathrm{SL}_2(K)$. Hence, (9.1.4) holds in $\mathrm{SL}_2(K)$ upon replacing z with a unique matrix $\left(\begin{smallmatrix} t^{-1} & 0 \\ 0 & t \end{smallmatrix}\right)$ with $t \in K^\times$. We wish to compute y_1, y_2, and t in terms of x.

Multiplying out the right side of (9.1.4) gives

$$\begin{pmatrix} 1 & x \\ 0 & 1 \end{pmatrix} = \begin{pmatrix} ty_2 & t \\ ty_1y_2 - t^{-1} & ty_1 \end{pmatrix}$$

in $\mathrm{SL}_2(K)$, so $t = x$ and $y_1 = 1/t = y_2$. Hence, $1/x = y_1 \in V$. This shows that the subset $V - \{0\} \subseteq K^\times$ is stable under inversion and moreover that if $x \in V - \{0\}$ then

$$\begin{pmatrix} x & 0 \\ 0 & x^{-1} \end{pmatrix} \in G(k) \tag{9.1.5}$$

inside $\mathrm{SL}_2(K)$ or $\mathrm{PGL}_2(K)$. Thus, for any $x' \in V$ the product

$$\begin{pmatrix} x & 0 \\ 0 & x^{-1} \end{pmatrix} \cdot \begin{pmatrix} 1 & x' \\ 0 & 1 \end{pmatrix} \cdot \begin{pmatrix} x^{-1} & 0 \\ 0 & x \end{pmatrix} = \begin{pmatrix} 1 & x^2x' \\ 0 & 1 \end{pmatrix}$$

lies in $G(k)$. In other words, $x^2x' \in V$ for all $x, x' \in V$ (as the case $x = 0$ is trivial). This says exactly that $V^2 \cdot V \subseteq V$.

Assume $\mathrm{char}(k) \neq 2$. For $x, x' \in V$, both $(1 + x)^2x'$ and x^2x' lie in V, so $xx' \in V$. Thus the k-subspace V of K is a k-subalgebra, hence a subfield F. Clearly $G = \mathrm{R}_{F/k}(L_F)$.

Suppose instead that $\mathrm{char}(k) = 2$. Since G is pseudo-semisimple, by Lemma 3.1.5 it is generated by the root groups $U_G^{\pm} = U_V^{\pm}$ (see Definition 9.1.1), but by definition H_V is the subgroup of $\mathrm{R}_{K/k}(L_K)$ generated by these two subgroups since $(1/v^2)V = V$ for any $v \in V - \{0\}$. Hence, $G = H_V$. □

The following simple result is the reason why the classification of pseudo-simple groups over fields k of characteristic 2 such that $[k : k^2] = 2$ is simpler than the classification over an arbitrary field of characteristic 2. Note that if $[k : k^2] = 2$ then the nontrivial purely inseparable finite extensions K of k are the fields $k^{1/2^m}$ with $m \geqslant 1$, so $kK^2 = K^2$ and hence $[K : kK^2] = [K : K^2] = 2$.

Lemma 9.1.6 *Let k be a field of characteristic 2, K a purely inseparable finite extension of k such that $[K : kK^2] \leqslant 2$, and V a nonzero k-subspace of K that is also a $k\langle V^2\rangle$-subspace. If $k\langle V\rangle = K$ then $V = K$.*

Note that if $K = kK^2$ then $\Omega^1_{K/k} = 0$, so K/k is separable and hence $K = k$.

Proof It is harmless to replace V with a $K^{\times}$-multiple, so we may and do assume $1 \in V$. Hence, $k[V] = k\langle V\rangle = K$, so $k\langle V^2\rangle = kK^2$. Thus, V is a nonzero kK^2-subspace of K. The case $K = kK^2$ is trivial, so we may assume $[K : kK^2] = 2$. If $\dim_{kK^2} V = 1$ then $k\langle V\rangle = kK^2 \neq K$, a contradiction. Hence, by kK^2-dimension reasons, V must be equal to K. □

Proposition 9.1.7 *Let V and V' be nonzero k-subspaces of K such that $k\langle V\rangle = K = k\langle V'\rangle$ (so $k\langle V^2\rangle = kK^2 = k\langle V'^2\rangle$). Assume that V and V' are kK^2-subspaces of K. The k-groups $H := H_V$ and $H' := H_{V'}$ are k-isomorphic if and only if $V' = \lambda V$ for some $\lambda \in K^{\times}$, and necessarily $\lambda \in V \cap V' - \{0\}$ if $1 \in V \cap V'$. Conversely, for any $\lambda \in K^{\times}$ we have $H_{\lambda V} \simeq H_V$ as k-groups, and if $1 \in V$ and $\lambda \in V - \{0\}$ then $1 \in \lambda V$.*

Proof For any $\lambda \in K^{\times}$ we have already noted that $H_{\lambda V}$ is the conjugate of H_V by the element $\mathrm{diag}(\lambda, 1) \in \mathrm{PGL}_2(K) = \mathrm{R}_{K/k}(\mathrm{PGL}_2)(k)$, and we also know that if $1 \in V$ then $V - \{0\}$ is stable under inversion. Thus, if $1 \in V$ and $\lambda \in V - \{0\}$ then clearly $1 \in \lambda V$. We may now restrict attention to the cases in which $1 \in V \cap V'$, so $k[V] = k\langle V\rangle = K = k\langle V'\rangle = k[V']$. Our problem is to show that if H and H' are k-isomorphic then $V' = \lambda V$ for some $\lambda \in K^{\times}$. (Note that then necessarily $\lambda \in V \cap V' - \{0\}$ since $1 \in V \cap V'$ and $V - \{0\}$ is stable under inversion.)

Consider an arbitrary isomorphism $f : H \simeq H'$ as k-groups. The isomorphism f_K induces an isomorphism between the maximal reductive quotients $H_K/\mathscr{R}_{u,K}(H_K)$ and $H'_K/\mathscr{R}_{u,K}(H'_K)$ over K. These quotients are each identified with L_K, so we obtain a K-automorphism of L_K. Such an automorphism is given by the "conjugation" action of a unique element $g \in \mathrm{PGL}_2(K)$, and it is clear that the g-action on $\mathrm{R}_{K/k}(L_K)$ carries H over to H' as k-subgroups.

Let D be the diagonal k-torus in L, so D is a split maximal k-torus in each of H and H'. By the $H'(k)$-conjugacy of maximal split k-tori in H' (Theorem C.2.3), we can modify g so that it preserves D. Modifying g by a representative of the nontrivial element of $N_L(D)(k)/D(k) = N_H(D)(k)/Z_H(D)(k)$ if necessary brings us to the case that g centralizes D. But $D(k)$ has Zariski-dense image in the diagonal maximal K-torus of PGL_2, so $g = \left(\begin{smallmatrix} \lambda & 0 \\ 0 & 1 \end{smallmatrix}\right)$ for some $\lambda \in K^\times$. The action of g on $\mathrm{R}_{K/k}(U^+) = \mathrm{R}_{K/k}(\mathbf{G}_\mathrm{a})$ is multiplication by λ, so intersecting these root groups with H and H' gives that $V' = \lambda V$. □

9.1.8 We finish our discussion of the basic properties of $H_{V,K/k}$ by addressing its behavior with respect to inseparable Weil restriction. Consider a nonzero kK^2-subspace V of K such that $k\langle V\rangle = K$. For a subfield k_0 over which k is finite and purely inseparable, we have $k_0\langle V\rangle = K$ because if $v_0 \in V - \{0\}$ and $c \in k^\times$ then $c = (cv_0)/v_0 \in k_0\langle V\rangle$. Thus, the perfect pseudo-reductive k_0-group $H_{V,K/k_0} \subseteq \mathrm{R}_{K/k_0}(L_K) = \mathrm{R}_{k/k_0}(\mathrm{R}_{K/k}(L_K))$ makes sense and inspection of root groups shows that it is contained inside $\mathrm{R}_{k/k_0}(H_{V,K/k})$. The resulting inclusion $H_{V,K/k_0} \subseteq \mathscr{D}(\mathrm{R}_{k/k_0}(H_{V,K/k}))$ is an equality, due to root group considerations and Proposition 9.1.5 (applied to $G = \mathscr{D}(\mathrm{R}_{k/k_0}(H_{V,K/k}))$).

Suppose $L = \mathrm{SL}_2$, so the k_0-group $\mathrm{R}_{k/k_0}(H_{V,K/k})$ is equal to its own derived subgroup when $V = K$. This covers all cases when $[k : k^2] = 2$ (Lemma 9.1.6), but for more general k (and $L = \mathrm{SL}_2$) $\mathrm{R}_{k/k_0}(H_{V,K/k})$ can fail to be perfect, in contrast with the preservation of perfectness of simply connected semisimple groups under inseparable Weil restriction (Corollary A.7.11). This complicates the formulation of structural results for pseudo-reductive groups in characteristic 2 when $[k : k^2] > 2$ (in contrast with the case $\mathrm{char}(k) \neq 2$).

To understand the cause of such non-perfectness, first note that (by consideration of Cartan subgroup factors in open cells) $\mathrm{R}_{k/k_0}(H_{V,K/k})$ is perfect if and only if the inclusion $V^*_{K/k_0} \subseteq \mathrm{R}_{k/k_0}(V^*_{K/k})$ is an equality. Thus, the loss of perfectness can be attributed to the failure of the formation of the k-group $V^*_{K/k}$ to commute with purely inseparable Weil restriction. More specifically, the k-group $H_{V,K/k}$ can admit central extensions

$$1 \to Z \to G \to H_{V,K/k} \to 1$$

with G smooth and Z a nontrivial infinitesimal commutative k-group (so the hypothesis in Proposition 5.1.3 given in terms of a Cartan subgroup (of $H_{V,K/k}$) can fail). The relevance of such central extensions is that if $k \neq k_0$ then $\mathrm{R}_{k/k_0}(Z)$ has positive dimension (as we may check over $(k_0)_s$) whereas smoothness of G implies that $\mathrm{R}_{k/k_0}(G)$ and $\mathrm{R}_{k/k_0}(H_{V,K/k})$ have the same dimension. Since R_{k/k_0} is pushforward through the finite flat covering $f : \mathrm{Spec}(k) \to \mathrm{Spec}(k_0)$, the left-exact sequence of fppf group sheaves

$$1 \to \mathrm{R}_{k/k_0}(G)/\mathrm{R}_{k/k_0}(Z) \to \mathrm{R}_{k/k_0}(H_{V,K/k}) \stackrel{\delta}{\to} \mathrm{R}^1 f_*(Z)$$

with *commutative* right term obstructs perfectness of $\mathrm{R}_{k/k_0}(H_{V,K/k})$.

We shall make examples (with $[k : k^2] > 2$) for which $H_{V,K/k}$ admits a smooth central extension by α_2, using $(K/k, V)$ as above with $K^2 \subseteq k$ and $1 \in V$ (so $k[V] = K$ and $V - \{0\}$ is stable under inversion). Our main tool will be the Clifford algebra C over k associated to the (degenerate) quadratic form $q : V \to k$ defined by $q(v) = v^2$ that satisfies $q(V - \{0\}) \subseteq k^\times$. Let $i_q : V \to C$ denote the canonical k-linear injection into the odd part (so $1 \notin i_q(V)$ inside C).

This Clifford algebra is *commutative* since the associated symmetric bilinear form B_q vanishes and $\mathrm{char}(k) = 2$, and all squares of C lie in k, so C is artin local with residue field contained in $k^{1/2}$. Explicitly, C is the quotient of $\mathrm{Sym}(V)$ modulo the relations $v \cdot v - q(v)$ for $v \in V$. Since the unique k-algebra map $r : C \to K = k[V]$ which carries $i_q(v)$ to v is surjective, K is the residue field of C and $\ker r$ is nilpotent. For n as in (9.1.2), consider the birational group structure on the dense open subscheme $U_V^+ n \times V_{K/k}^* \times U_V^+$ of the k-subgroup $H_{V,K/k} \subseteq \mathrm{R}_{K/k}(\mathrm{SL}_2)$ (with $K^2 \subseteq k$). The explicit description of this birational group structure ultimately rests on the fact that K is a commutative k-algebra containing V in which squaring restricts to q on V (forcing $V - \{0\}$ to consist of units).

Define $H_{V,C/k} \subseteq \mathrm{R}_{C/k}(\mathrm{SL}_2)$ to be the smooth connected k-subgroup generated by the two copies of V viewed as the k-subspace $i_q(V)$ inside the standard "root groups" $\mathrm{R}_{C/k}(U_C^\pm)$. The quotient map $\mathrm{R}_{C/k}(\mathrm{SL}_2) \to \mathrm{R}_{K/k}(\mathrm{SL}_2)$ carries $H_{V,C/k}$ onto $H_{V,K/k}$. In general $H_{V,C/k}$ can fail to be pseudo-reductive (see Remark 9.1.11 below), but the examples of interest that we eventually construct will be absolutely pseudo-simple.

Proposition 9.1.9 *The kernel* $N = \ker(\pi : H_{V,C/k} \twoheadrightarrow H_{V,K/k})$ *is central and the Cartan k-subgroup* $Z_{H_{V,C/k}}(D)$ *has dimension* $1 + d(d-1)/2$ *with* $d := \dim_k V$ *(so* $\dim V_{K/k}^* \leqslant 1 + d(d-1)/2$*, with equality if and only if N is finite).*

Proof We begin by establishing some basic properties of $H := H_{V,C/k}$, analogous to known properties of the absolutely pseudo-simple $H_{V,K/k}$. For any

$v \in V - \{0\}$, the involution

$$n(v) := \begin{pmatrix} 0 & i_q(v) \\ -1/i_q(v) & 0 \end{pmatrix} \in \mathrm{SL}_2(C) = \mathrm{R}_{C/k}(\mathrm{SL}_2)(k)$$

lies in $H(k)$ since conjugation by $\mathrm{diag}(i_q(v), 1)$ on $\mathrm{R}_{C/k}(\mathrm{SL}_2)$ brings the standard involution in (9.1.2) to $n(v)$ and brings the vector subgroup $\mathbf{G}_{\mathrm{a}} \subseteq \mathrm{R}_{C/k}(U_C^{\pm})$ corresponding to the line $k \subseteq C$ over to the vector subgroup corresponding to the line $ki_q(v) = ki_q(v)^{-1}$ (so this conjugation moves the standard Levi k-subgroup SL_2 to H). Moreover, for $v, v' \in V - \{0\}$ and $t := i_q(v')/i_q(v) \in C^{\times}$, the identity

$$\begin{aligned} &n(v)\begin{pmatrix} 1 & -i_q(v') \\ 0 & 1 \end{pmatrix} n(v) \\ &= \begin{pmatrix} 1 & q(v)i_q(1/v') \\ 0 & 1 \end{pmatrix} n(v) \begin{pmatrix} t & 0 \\ 0 & t^{-1} \end{pmatrix} \begin{pmatrix} 1 & q(v)i_q(1/v') \\ 0 & 1 \end{pmatrix} \end{aligned}$$

in $\mathrm{SL}_2(C)$ (following from the observation that $i_q(v')^2 = q(v') \in k^{\times}$, so $i_q(1/v') = i_q(q(v')^{-1}v') = i_q(v')/q(v') = i_q(v')^{-1}$) forces $Z_H(D)$ to contain the smooth Zariski closure $V_{C/k}^{*} \subseteq \mathrm{R}_{C/k}(\mathrm{GL}_1)$ of the subgroup of $C^{\times}$ generated by ratios of elements of $i_q(V) - \{0\}$. Note that $V_{C/k}^{*}$ is connected.

Inside $\mathrm{R}_{C/k}(\mathrm{SL}_2)$, the strictly upper and lower triangular unipotent k-subgroups $U_C^{\pm} = \mathrm{R}_{C/k}(\mathbf{G}_{\mathrm{a}})$ contain unique smooth connected k-subgroups $U_{i_q(V)}^{\pm}$ that recover the k-subspace $i_q(V) \subseteq C$ on k-points. These do *not* contain the respective root groups $U^{\pm}$ for the diagonal k-torus in the standard Levi k-subgroup SL_2 because $1 \notin i_q(V)$ inside C. Nonetheless, since $i_q(V) - \{0\} \subseteq C^{\times}$ and all squares of C lie in k, the role of Proposition 2.1.8 in the proof of Proposition 9.1.4 can be adapted almost without change by using C instead of K and using the involution $n(v_0)$ instead of (9.1.2) (for a fixed $v_0 \in V - \{0\}$) to prove the following: $H \cap \mathrm{R}_{C/k}(U_C^{\pm}) = U_{i_q(V)}^{\pm}$, $Z_H(D) = V_{C/k}^{*}$, and H is generated by the vector subgroups $U_{i_q(V)}^{\pm}$ (so H is perfect). Note that H is generally *not* pseudo-reductive, and it admits no analogue of (9.1.1) since $1 \notin i_q(V)$ inside C.

The D-weight spaces in $\mathrm{Lie}(H) \subseteq \mathfrak{sl}_2(C)$ for the nontrivial weights that occur are $\mathrm{Lie}(U_{i_q(V)}^{\pm})$, and these map isomorphically onto their images in $\mathfrak{sl}_2(K)$ since the quotient map $C \twoheadrightarrow K$ carries $i_q(V)$ isomorphically onto V. Hence, $\mathrm{Lie}(N)$ has no nontrivial D-weight, so by Corollary A.8.11 the conjugation action of the maximal k-torus D on the identity component N^0 (which inherits normality in H from that of N) is trivial. By Proposition A.2.10 $N_{k_s}^0$ is therefore centralized by all maximal k_s-tori of H_{k_s}, yet any perfect smooth connected affine k_s-group G coincides with its smooth connected normal k_s-subgroup G_t generated by the maximal k_s-tori (since G/G_t has no nontrivial k_s-tori, so it is unipotent and hence trivial due to perfectness). Thus, $N^0_{k_s}$ is a central k_s-subgroup scheme in H_{k_s}, so N^0 is a central k-subgroup scheme in H. It follows from Lemma 5.3.2 and the perfectness of H that N is central in H.

Next, we show that the k-subgroup $Z_H(D) = V^*_{C/k} \subseteq R_{C/k}(GL_1)$ has dimension $1 + d(d-1)/2$. The ratios $c = i_q(cv)/i_q(v)$ for $c \in k^\times$ and a fixed $v \in V - \{0\}$ account for the 1-dimensional k-torus D inside $Z_H(D)$, so our problem is to show that the geometric unipotent radical of $V^*_{C/k}$ has dimension $d(d-1)/2$. In the remainder of this proof, to simplify the notation we will identify $i_q(V)$ with V and so view V as a k-subspace of C. The overlap $\underline{V}^\times$ of $\underline{V}$ with the dense open subscheme $R_{C/k}(GL_1)$ inside $\underline{C} := R_{C/k}(\mathbf{G}_a)$ is dense open inside $\underline{V}$ (containing $V - \{0\}$), so it is clear that ratios of elements of $\underline{V}^\times(\overline{k})$ lie in $V^*_{C/k}(\overline{k})$. Observe that the units in $C_{\overline{k}}$ are the elements x whose image x_0 under the unique $\overline{k}$-algebra map $C_{\overline{k}} \to \overline{k}$ lies in $\overline{k}^\times$ (or equivalently such that the element $x^2 \in \overline{k} \subseteq C_{\overline{k}}$ is nonzero). We conclude that $(V^*_{C/k})_{\overline{k}}$ is the Zariski closure of the subgroup of $C^\times_{\overline{k}}$ generated by ratios of elements $x \in V_{\overline{k}}$ such that $x_0 \in \overline{k}^\times$.

Let $\widetilde{V}$ be the $\overline{k}$-subspace of (nilpotent) elements of $C_{\overline{k}}$ of the form $x - x_0$ for $x \in V_{\overline{k}}$ (with $x_0 \in \overline{k}$); this is d-dimensional over $\overline{k}$ since $V_{\overline{k}} \cap \overline{k} = 0$ inside $C_{\overline{k}}$. For $x, y \in V_{\overline{k}}$ such that $x_0, y_0 \in \overline{k}^\times$, we have

$$\frac{x}{y} = \frac{x_0}{y_0} \cdot \frac{1 + (x/x_0 - 1)}{1 + (y/y_0 - 1)},$$

so $\mathscr{R}_u((V^*_{C/k})_{\overline{k}})$ is the Zariski closure of the subgroup of $C^\times_{\overline{k}}$ generated by ratios $(1 + \widetilde{v})/(1 + \widetilde{w})$ for $\widetilde{v}, \widetilde{w} \in \widetilde{V}$ each having the form $z - 1$ for $z \in V_{\overline{k}}$ such that $z_0 = 1$. Let $\mathbf{1} \in V_{\overline{k}}$ be $1 \in V$, so inside $C_{\overline{k}}$ we have $\mathbf{1}^2 = 1$, $\mathbf{1}_0 = 1$, and $z - 1 = (z - \mathbf{1}) + (\mathbf{1} - 1)$. As we vary z, such elements $z - 1$ exhaust the $(\mathbf{1} - 1)$-translate of the hyperplane kernel V' of the nonzero $\overline{k}$-linear reduction map $V_{\overline{k}} \to \overline{k}$. In other words, $1 + \widetilde{v}$ and $1 + \widetilde{w}$ vary precisely through $\mathbf{1} + V' = \mathbf{1}(1 + \mathbf{1} \cdot V')$. Since $0 \in \mathbf{1} \cdot V'$ and all nilpotents in $C_{\overline{k}}$ have vanishing square (so $1/(1+c) = 1 + c$ for all nilpotent $c \in C_{\overline{k}}$), it follows that $\mathscr{R}_u((V^*_{C/k})_{\overline{k}})$ is the Zariski closure of the group generated by units in $1 + \mathbf{1} \cdot V'$.

We may write $C_{\overline{k}} = \overline{k}[T_1, \dots, T_d]/(T_j^2)$ with $T_j = v_j - \sqrt{q(v_j)}$ for any basis $\{v_1, \dots, v_d\}$ of $V_{\overline{k}}$. In particular, we may and do choose $v_1, \dots, v_{d-1}$ to be a basis of the hyperplane V', so $T_j = v_j$ for $j \neq d$. For any unit u of $C_{\overline{k}}$ we get the presentation $C_{\overline{k}} = \overline{k}[z_1, \dots, z_d]/(z_j^2)$ for $z_j := uT_j$, and by choosing $u = \mathbf{1}$ we have that $\mathbf{1} \cdot V'$ is the $\overline{k}$-span of $z_1, \dots, z_{d-1}$. Hence, for $m := d - 1 \geqslant 0$

we are reduced to showing that $m(m+1)/2$ is the dimension of the algebraic group $\mathscr{G}_m$ of 1-units in $\overline{k}[z_1,\dots,z_m]/(z_j^2)$ generated by elements $1+\ell$ for linear forms ℓ in the z_j (with ℓ understood to have vanishing constant term).

The identity $1+\alpha\beta = (1+\alpha)(1+\beta)(1+\alpha+\beta)$ for square-zero elements α and β in an $\mathbf{F}_2$-algebra implies that $\mathscr{G}_m$ contains all 1-units of the type $1+\ell\ell'$ for linear forms ℓ and ℓ' in $z_1,\dots,z_m$. Setting $\ell' = z_m$, we see that $\mathscr{G}_m$ contains all 1-units of the type $(1+az_m)(1+\ell z_m) = 1+(a+\ell)z_m$ for constant a and linear forms ℓ that we may assume do not involve z_m (since $z_m^2 = 0$). Letting a and $\ell(z_1,\dots,z_{m-1})$ vary, we conclude that $\mathscr{G}_m$ contains the m-dimensional vector group of 1-units

$$V_m = \{1+(a+\ell(z_1,\dots,z_{m-1}))z_m\}.$$

Obviously $\mathscr{G}_{m-1} \cap V_m = 1$, so $\mathscr{G}_{m-1} \times V_m \subseteq \mathscr{G}_m$ as $\overline{k}$-groups. We will prove that this inclusion is an equality, yielding $\dim \mathscr{G}_m = m(m+1)/2$ by induction on m.

The problem is to show equality on $\overline{k}$-points, so for any linear form $L = \sum a_i z_i$ we want to show that $1+L \in \mathscr{G}_{m-1} \times V_m$ inside $\mathscr{G}_m$. In other words, we seek a point $1+(a+\ell)z_m \in V_m$ such that $1+L \in \mathscr{G}_{m-1}\cdot(1+(a+\ell)z_m)$. It is equivalent to require that $(1+L)(1+(a+\ell)z_m)$ does not involve z_m. Setting $a = a_m$ and $\ell(z_1,\dots,z_{m-1}) = a_m \sum_{i<m} a_i z_i$ does the job. □

Since $K = k[V]$, $\Omega^1_{K/k}$ is spanned over K by the elements dv for $v \in V$. Hence, $\Omega^1_{K/k}$ has a K-basis of elements da_j for some $a_j \in V$; $\{a_j\}$ is a 2-basis of K/k: $a_j^2 = c_j \in k - k^2$ and $K = k(a_1,\dots,a_n) = k[x_1,\dots,x_n]/(x_j^2 - c_j)$ (with $[K:k] = 2^n$), so $d := \dim_k V \geqslant n+1$ (since $1 \in V$). Using notation from the preceding proof, when viewing the Cartan k-subgroup $V^*_{C/k}$ inside the algebraic group $\mathrm{R}_{C/k}(\mathrm{GL}_1)$ of units of $C = \mathrm{Sym}(V)/(v\cdot v - q(v))$ we see that the composite map $V^*_{C/k} \twoheadrightarrow V^*_{K/k} \hookrightarrow \mathrm{R}_{K/k}(\mathrm{GL}_1)$ is induced by the map of unit groups arising from the map of k-algebras $C \twoheadrightarrow K$. Extending scalars to $\overline{k}$ and defining $V'' = \mathbf{1}\cdot V'$, the subalgebra $\mathrm{Sym}(V'')/(v''^2) \subseteq C_{\overline{k}}$ maps onto $\overline{k}\otimes_k K$ since $\overline{k}\otimes_k K$ is generated as a $\overline{k}$-algebra by the image of $V_{\overline{k}} = \overline{k}\cdot\mathbf{1} + V'$ (hence by the image of V') and $\mathbf{1} \mapsto 1$. The description of $\mathscr{R}_u((V^*_{C/k})_{\overline{k}})$ in the preceding proof shows that it lies inside the unit group of $\mathrm{Sym}(V'')/(v''^2)$, and this geometric unipotent radical contains $N_{\overline{k}}$, with N as in Proposition 9.1.9 (since the geometric fibers of $V^*_{C/k}$ and $V^*_{K/k}$ are functorially the direct product of their unipotent radicals and $D_{\overline{k}}$).

We conclude that $N_{\overline{k}}$ is a subgroup scheme of the kernel of the map of unit groups induced by the quotent map $\mathrm{Sym}(V'')/(v''^2) \twoheadrightarrow \overline{k}\otimes_k K$. Since the map

$V' \simeq V''$ defined by $v' \mapsto \mathbf{1} \cdot v'$ induces an isomorphism $\mathrm{Sym}(V')/(v'^2) \simeq \mathrm{Sym}(V'')/(v''^2)$ commuting with the quotient map from each onto $\overline{k} \otimes_k K$, by passing to unit groups we see that $N_{\overline{k}}$ is isomorphic to the intersection of two subgroups of the algebraic unit group of $\mathrm{Sym}(V')/(v'^2)$: (i) the Zariski closure $\mathscr{G}_{d-1}$ of the subgroup of $(\mathrm{Sym}(V')/(v'^2))^\times$ generated by the 1-units $1 + v'$ ($v' \in V'$), and (ii) the kernel of the map of unit groups induced by the surjection of $\overline{k}$-algebras $\mathrm{Sym}(V')/(v'^2) \twoheadrightarrow \overline{k} \otimes_k K$. If $d = n + 1$ then this latter surjective $\overline{k}$-algebra map is an isomorphism for dimension reasons (as $\dim V' = d - 1$), forcing $N = 1$ and hence $H_{V,C/k} \simeq H_{V,K/k}$. Thus, the only interesting cases are when $d \geqslant n + 2$. By Proposition 9.1.9, to find finite N necessarily $1 + d(d-1)/2 = \dim V^*_{K/k} \leqslant [K : k] = 2^n$. In the case $d = n+2$ the inequality $1 + d(d-1)/2 \leqslant 2^n$ holds only for $n \geqslant 4$, and it is an equality precisely for $n = 4$. This motivates focusing on the case $[K : k] = 16$ and $\dim V = 6$, as we do in the following example.

Example 9.1.10 Consider k satisfying $[k : k^2] \geqslant 16$, so we may and do choose $K \subseteq k^{1/2}$ satisfying $[K : k] = 16$. Let $\{e_1, \dots, e_4\}$ be a 2-basis of K/k and choose a nonzero quadratic form $Q \in k[x_1, \dots, x_4]$ such that $Q_{\overline{k}}$ is not a square. The element $v = Q(e_1, \dots, e_4) \in K$ is k-linearly independent from $\{1, e_1, \dots, e_4\}$, so we may define

$$V := k \oplus ke_1 \oplus ke_2 \oplus ke_3 \oplus ke_4 \oplus kv \subseteq K.$$

If Q is non-degenerate then we claim $N \simeq \alpha_2$ (so $V^*_{K/k} = \mathrm{R}_{K/k}(\mathrm{GL}_1)$, for dimension reasons, and $H_{V,C/k}$ is absolutely pseudo-simple).

Since a k-group is isomorphic to α_2 if and only if its $\overline{k}$-fiber is, it suffices to work over $\overline{k}$. We shall use notation from the calculations over $\overline{k}$ in the proof of Proposition 9.1.9, such as V', $\mathbf{1}$, and $\mathscr{G}_m$. Let $c_j = e_j^2 \in k^\times$, so the 5-dimensional $\overline{k}$-subspace V' inside $C_{\overline{k}}$ has as a basis the vectors $v'_j := i_q(e_j) - \sqrt{c_j}\, i_q(\mathbf{1})$ for $1 \leqslant j \leqslant 4$ and $v'_5 := i_q(v) - Q(\sqrt{c_1}, \dots, \sqrt{c_4})\, i_q(\mathbf{1})$. In particular, the natural map

$$f : \overline{k}[v'_1, \dots, v'_5]/(v_j'^2) = \mathrm{Sym}(V')/(v'^2) \twoheadrightarrow \overline{k} \otimes_k K$$

is injective on $\overline{k}[v'_1, \dots, v'_4]/(v_j'^2)$ and so its kernel is the ideal J generated by

$$\begin{aligned} v'_5 - Q(v'_1 + \sqrt{c_1}, \dots, v'_4 + \sqrt{c_4}) + Q(\sqrt{c_1}, \dots, \sqrt{c_4}) \\ = v'_5 - Q(v'_1, \dots, v'_4) - L(\sqrt{c_1}, \dots, \sqrt{c_4}) \end{aligned}$$

with the linear $L = B_Q((v'_1, \ldots, v'_4), \cdot)$, where B_Q is the symmetric bilinear form attached to Q. The map f is unaffected by precomposition with the $\overline{k}$-algebra automorphism defined by $v'_j \mapsto \mathbf{1} \cdot v'_j$ since $f(\mathbf{1}) = 1$. Thus, we are reduced to showing $\mathscr{G}_5 \cap (1+J) \simeq \alpha_2$ as subgroup schemes of the scheme of units of $\mathrm{Sym}(V')/(v'^2)$.

It is harmless to translate v'_5 by a $\overline{k}$-linear combination of $v'_1, \ldots, v'_4$, so we replace v'_5 with $v'_5 + L(\sqrt{c_1}, \ldots, \sqrt{c_4})$ to arrive at an algebraic problem that has nothing to do with K or the c_j: if I denotes the ideal in $A := \overline{k}[x_1, \ldots, x_5]/(x_j^2)$ generated by $x_5 - Q(x_1, \ldots, x_4)$ and $\mathscr{G}_5$ is the group scheme of units in A generated by $1 + \ell$ for linear forms ℓ then $\mathscr{G}_5 \cap (1+I) \simeq \alpha_2$ when Q is non-degenerate. Applying a $\overline{k}$-linear change of coordinates in $x_1, \ldots, x_4$, we may assume $Q = x_1x_2 + x_3x_4$.

For any $\overline{k}$-algebra R and element $a \in R$ and linear forms ℓ, ℓ' in $x_1, \ldots, x_5$ over R,

$$1 + (a+\ell)\ell' = (1 + a\ell')(1+\ell)(1+\ell')(1+\ell+\ell') \in \mathscr{G}_5(R).$$

Thus, points of $\mathscr{G}_5(R)$ are precisely $(1+h)(1+(a+\ell)x_5)$ for a unique $a \in R$, linear form $\ell = \sum_{i \leqslant 4} t_i x_i$ over R, and $h \in R[x_1, \ldots, x_4]/(x_j^2)$ with vanishing constant term such that $1 + h \in \mathscr{G}_4(R)$. It is easy to see (ignoring the condition $1 + h \in \mathscr{G}_4(R)$ and the specific Q) that $(1+h)(1+(a+\ell)x_5) \in 1 + R \otimes_{\overline{k}} I$ if and only if $h = (a+\ell)Q$, in which case

$$(1+h)(1+(a+\ell)x_5) = 1 + (x_5 + Q)((a+\ell) + a^2x_5) \in 1 + R \otimes_{\overline{k}} I.$$

The condition $1 + h = 1 + (a+\ell)Q \in \mathscr{G}_4(R)$ says that there exist (necessarily unique) $a' \in R$ and a linear form $\ell'(x_1, x_2, x_3)$ over R such that

$$(1 + (a+\ell)Q)(1 + (a' + \ell')x_4) \in \mathscr{G}_3(R),$$

so this product does not involve x_4. Using that $Q = x_1x_2 + x_3x_4$, the condition of not involving x_4 says exactly $a' = 0$, $\ell' = ax_3$, $\ell = t_3x_3$, and $a \in \alpha_2(R)$, in which case

$$(1 + (a+\ell)Q)(1 + (a' + \ell')x_4) = 1 + ax_1x_2 + t_3x_1x_2x_3.$$

Since $1 + ax_1x_2 = (1 + ax_1)(1 + x_2)(1 + ax_1 + x_2)$, the right side is easily checked to lie in $\mathscr{G}_3(R)$ if and only if $t_3 = 0$, so the necessary and sufficient conditions on the initial h, a, and ℓ are: $h = aQ$, $\ell = 0$, and $a \in \alpha_2(R)$. This yields the 1-units

$$(1 + aQ)(1 + ax_5) = 1 + a(x_5 + x_1x_2 + x_3x_4)$$

that constitute an α_2 as R varies.

In cases with non-degenerate Q we claim that the map $f : G := H_{V,C/k} \twoheadrightarrow H_{V,K/k} \subseteq \mathrm{R}_{K/k}(\mathrm{SL}_2)$ is i_G, so $\ker i_G$ is a central α_2. To compute i_G, first note that the central unipotent $(\ker f)_{\overline{k}}$ must be contained inside $\mathscr{R}_u(G_{\overline{k}})$ (as the connected reductive group $G_{\overline{k}}/\mathscr{R}_u(G_{\overline{k}})$ has no nontrivial unipotent central subgroup scheme, since its scheme-theoretic center is contained in any maximal torus). Hence, the geometric unipotent radical of G is the scheme-theoretic preimage of that of $H_{V,K/k}$, so the minimal field of definition over k for $\mathscr{R}_u(G_{\overline{k}}) \subseteq G_{\overline{k}}$ coincides with the one for the geometric unipotent radical of $H_{V,K/k}$, which is to say it is K (by Proposition 9.1.4). Thus, i_G is the composition of $H_{V,C/k} \twoheadrightarrow H_{V,C/k}/(\ker f) = H_{V,K/k}$ and $i_{H_{V,K/k}}$. This composition is precisely f, due to the description of $i_{H_{V,K/k}}$ provided by Proposition 9.1.4.

Remark 9.1.11 In the preceding example, if Q is degenerate with only an ordinary double point singularity, which is to say that $Q_{\overline{k}}$ is a product of independent linear forms, then N is smooth and connected of dimension 1 (so $H_{V,C/k}$ is not pseudo-reductive and $V^*_{K/k}$ has codimension 1 in $\mathrm{R}_{K/k}(\mathrm{GL}_1)$). Indeed, we may make a $\overline{k}$-linear change of coordinates so $Q_{\overline{k}} = x_1x_2$, and then computations similar to the above show that $N_{\overline{k}}$ is the vector group of 1-units of the form $1 + aQ$.

If we instead choose $v = h(e)$ where $h \in k[x_1, \ldots, x_4]$ is homogeneous of degree 4 with $h_{\overline{k}}$ a product of independent linear forms then $N = \mathbf{Z}/2\mathbf{Z}$ (so $V^*_{K/k} = \mathrm{R}_{K/k}(\mathrm{GL}_1)$ and $H_{V,C/k}$ is absolutely pseudo-simple with $\ker i_{H_{V,C/k}} = N = \mathbf{Z}/2\mathbf{Z}$). Indeed, it suffices to show that $N_{\overline{k}} = \mathbf{Z}/2\mathbf{Z}$, and we can modify the preceding calculations after applying a $\overline{k}$-linear change of variables to bring $h_{\overline{k}}$ to $x_1x_2x_3x_4$ and replace x_j with $x_j/\sqrt{c_j}$ for $j \leqslant 4$: for $Q := h(x+1) - h(1)$, we want the group scheme of pairs (a, ℓ) for which $1 + (a + \ell)Q \in \mathscr{G}_4$ to be $\mathbf{Z}/2\mathbf{Z}$. The preliminary condition that for some $(a', \ell'(x_1, x_2, x_3))$ the product $(1 + (a + \ell)Q)(1 + (a' + \ell'(x_1, x_2, x_3)x_4))$ does not involve x_4 says exactly that $\ell = (a + a^2)x_4$, in which case

$$\begin{aligned} 1 + (a + \ell)Q = (1 + ax_4)(1 + a(x_1 + x_2 + x_3) \\ + a(x_1x_2 + x_1x_3 + x_2x_3) + ax_1x_2x_3). \end{aligned}$$

The left side lies in $\mathscr{G}_4$ if and only if the second factor on the right lies in $\mathscr{G}_3$, which in turn says exactly that $a^2 = a$, yielding that $N = \mathbf{Z}/2\mathbf{Z}$ and $1 + (a + \ell)Q = \prod_{j \leqslant 4}(1 + ax_j)$.

9.2 Pseudo-split pseudo-simple groups of type A_1

We begin this section with a result that will be used frequently in this chapter:

Lemma 9.2.1 *Let G be a smooth connected affine k-group and K the minimal field of definition over k for the geometric unipotent radical $\mathscr{R}_u(G_{\overline{k}})$. Let $G' = G_K^{\text{red}} = G_K/\mathscr{R}_{u,K}(G_K)$, and let $i_G : G \to \mathrm{R}_{K/k}(G')$ be the natural map. Let N be a closed k-subgroup scheme of G that is normal in G and contained in $\ker i_G$ (e.g., $N = \ker i_G$).*

1. *The extension K/k is the minimal field of definition for the geometric unipotent radical of G/N and the natural map $G' \to (G/N)' := (G/N)_K^{\text{red}}$ is an isomorphism. In particular, the composition of $G \to G/N$ with $i_{G/N} : G/N \to \mathrm{R}_{K/k}((G/N)') = \mathrm{R}_{K/k}(G')$ is equal to i_G.*
2. *The extension K/k is the minimal field of definition for the geometric unipotent radical of $i_G(G)$ and the natural map $G' \to i_G(G)' := i_G(G)_K^{\text{red}}$ is an isomorphism. If G is pseudo-reductive then the same holds for $i_G(G)$, and likewise for the properties of being pseudo-semisimple or pseudo-simple over k.*

Proof The quotient map $G_K \to G'$ is the composite of the natural maps $G_K \to \mathrm{R}_{K/k}(G')_K \twoheadrightarrow G'$, so $N_{\overline{k}} \subseteq (\ker i_G)_{\overline{k}} \subseteq \mathscr{R}_u(G_{\overline{k}})$. Hence,

$$\mathscr{R}_u((G/N)_{\overline{k}}) = \mathscr{R}_u(G_{\overline{k}})/N_{\overline{k}}, \quad G_{\overline{k}}^{\text{red}} = (G/N)_{\overline{k}}^{\text{red}},$$

so K/k is the (minimal) field of definition for the geometric unipotent radical $\mathscr{R}_u(G_{\overline{k}})/N_{\overline{k}}$ of G/N and the natural map $G' \to (G/N)' := (G/N)_K^{\text{red}}$ is an isomorphism.

As $G/\ker i_G \simeq i_G(G)$, applying the above for $N = \ker i_G$ gives that the extension K/k is the minimal field of definition for the geometric unipotent radical of $i_G(G)$ and the natural map $G' \to i_G(G)' := i_G(G)_K^{\text{red}}$ is an isomorphism. Moreover, it follows from Proposition 4.2.5 that $i_G(G)$ is pseudo-reductive if G is pseudo-reductive, so likewise for pseudo-semisimplicity, and in such cases Lemma 3.1.2 implies that $i_G(G)$ is pseudo-simple over k if G is. □

9.2.2 Let k be an infinite field of characteristic 2 and let G be a *pseudo-split* absolutely pseudo-simple k-group. Let K/k denote the field of definition of $\mathscr{R}_u(G_{\overline{k}}) \subseteq G_{\overline{k}}$. Assume G_K^{ss} is of type A_1, which is to say that it is K-isomorphic to either SL_2 or PGL_2. Our complete description of all such G requires the construction of pseudo-split pseudo-semisimple groups with non-reduced root system BC_1 of rank 1, so for now we restrict ourselves to a more limited goal: we seek to describe the image of the homomorphism $i_G : G \to R_{K/k}(G_K^{ss})$, at least after fixing a suitable K-isomorphism of G_K^{ss} with SL_2 or PGL_2, and to prove that G is standard when its root system is reduced and $[k : k^2] \leqslant 2$.

In the rest of §9.2, L will denote the k-group SL_2 or PGL_2 according as G_K^{ss} is simply connected or adjoint, so G_K^{ss} is isomorphic to L_K. We shall denote the diagonal k-torus of L by D. We fix a split maximal k-torus T in G (so $T \simeq GL_1$) and fix a K-group isomorphism from G_K^{ss} to L_K such that T_K is carried to the diagonal K-torus D_K. This isomorphism is unique up to composition with the action of an element of $PGL_2(K)$ normalizing the diagonal torus, so any construction we make that is invariant by such an action is intrinsic to (G, T). We identify G_K^{ss} with L_K in terms of the chosen isomorphism, and will view i_G as a homomorphism from G into $R_{K/k}(L_K)$.

By Proposition 3.2.7, $W(G, T)(k)$ maps isomorphically onto the group $W(L_K, D_K)(K)$ of order 2. Thus, the elements of $N_G(T)(k) - Z_G(T)(k)$ represent the unique nontrivial class in $W(G, T)(k)$. Fix a nontrivial k-point u in a root group for (G, T). By Proposition 3.4.2, there is a unique element in $N_G(T)(k) - Z_G(T)(k)$ of the form $v = u'uu'$, where u' is a k-point in the opposite root group (and u' is then uniquely determined by u), and $v^2 = 1$. The image of u under the map $\pi : G(k) \to L(K)$ lies in one of the two root groups for (L_K, D_K). Change the K-isomorphism of G_K^{ss} with L_K by conjugation by the Weyl element $\left(\begin{smallmatrix} 0 & 1 \\ -1 & 0 \end{smallmatrix}\right)$ if necessary so that $\pi(u)$ lies in the upper triangular unipotent subgroup: $\pi(u) = \left(\begin{smallmatrix} 1 & t \\ 0 & 1 \end{smallmatrix}\right)$ for some $t \in K$. Finiteness of $(\ker i_G)(k_s)$ and non-centrality of u imply that $\pi(u) \neq 1$, so $t \in K^{\times}$. Thus, composing the K-isomorphism from G_K^{ss} to L_K with conjugation by $\left(\begin{smallmatrix} t^{-1} & 0 \\ 0 & 1 \end{smallmatrix}\right) \in PGL_2(K)$ arranges that $t = 1$. But then we see by explicit calculation that $\pi(u') = \left(\begin{smallmatrix} 1 & 0 \\ -1 & 1 \end{smallmatrix}\right)$, so for the resulting k-map

$$i_G : G \to R_{K/k}(L_K)$$

we have

$$n := \begin{pmatrix} 0 & 1 \\ -1 & 0 \end{pmatrix} = \pi(v) \in i_G(N_G(T)(k)[2]). \tag{9.2.1}$$

By pseudo-reductivity of G, $\ker i_G$ does not contain a nontrivial smooth connected unipotent k-subgroup.

The root system $\Phi(L_K, D_K) \subseteq X(D_K) = X(D) \simeq \mathbf{Z}$ is equal to $\{\pm a\}$ for a unique character a of D (or equivalently of D_K) such that the a-root group is the upper triangular unipotent subgroup U^+ of L. We write U^- for the opposite root group; i.e., the lower triangular unipotent subgroup of L. We shall use the standard isomorphisms $U^{\pm} \simeq \mathbf{G}_a$.

By our choice of the isomorphism $G_K^{ss} \simeq L_K$, the image $H := i_G(G)$ contains the diagonal torus D of $\mathrm{R}_{K/k}(L_K)$, and moreover the image of $G(k)$ contains the unipotent element $\left(\begin{smallmatrix}1 & 1\\ 0 & 1\end{smallmatrix}\right)$ and the Weyl element $n = \left(\begin{smallmatrix}0 & 1\\ -1 & 0\end{smallmatrix}\right)$. Therefore, H contains the k-subgroup L of $\mathrm{R}_{K/k}(L_K)$. We note here that by Lemma 9.2.1 the field of definition over k for the unipotent radical of $H_{\overline{k}}$ is K. Propositions 9.1.5 and 9.1.4 now imply that $i_G(G) =: H = H_V$ for a kK^2-subspace V of K containing 1 such that $k[V] = K$. Thus, we have:

Proposition 9.2.3 *Let G, K, and L be as above (so L is either the k-group* SL_2 *or* PGL_2, *and $G_K^{ss} \simeq L_K$). There is a K-isomorphism $G_K^{ss} \to L_K$ so that under the resulting identification of i_G with a k-homomorphism from G to $\mathrm{R}_{K/k}(L_K)$, $i_G(G) = H_V$ for a kK^2-subspace V of K such that $1 \in V$ and $k[V] = K$.*

If $V = K$ then $\xi_G(G) = i_G(G) = \mathscr{D}(\mathrm{R}_{K/k}(G_K^{ss}))$ (cf. Remark 9.1.3). *In particular, ξ_G is surjective if $V = K$.*

The condition $V = K$ holds whenever $[K : kK^2] \leqslant 2$ (by Lemma 9.1.6). Proposition 9.2.3 enables us to give a complete classification result in the A_1 case over a field k of characteristic 2 when the root system is reduced and $[K : kK^2] \leqslant 2$:

Proposition 9.2.4 *Let G be an absolutely pseudo-simple group over a field k of characteristic* 2, *and let K/k be the field of definition over k for the subgroup $\mathscr{R}(G_{\overline{k}}) \subseteq G_{\overline{k}}$. Assume $[K : kK^2] \leqslant 2$, G_K^{ss} is of type* A_1, *and G_{k_s} has a reduced root system. Then ξ_G is surjective and G is standard. If $G_{\overline{k}}^{ss} \simeq$* SL_2 *then the natural map $\xi_G : G \to \mathrm{R}_{K/k}(G_K^{ss})$ is an isomorphism.*

The hypothesis in Proposition 9.2.4 that the root system is reduced is satisfied whenever $G_{\overline{k}}^{ss} \simeq \mathrm{PGL}_2$ (by Theorem 2.3.10, since $\Phi(\mathrm{PGL}_2, \mathrm{GL}_1)$ consists of non-divisible elements in $\mathrm{X}(\mathrm{GL}_1)$). In §§9.6–9.8, for any imperfect field k of characteristic 2 we will construct absolutely pseudo-simple k-groups G with $G_{\overline{k}}^{ss} \simeq \mathrm{SL}_2$ such that G_{k_s} has a non-reduced root system (so G is not standard).

Proof By Corollary 5.3.9, we may and do assume $k = k_s$. Let T be a maximal k-torus in G, so G is generated by the $G(k)$-conjugates of T (since $G = \mathscr{D}(G)$).

It follows from Proposition 9.2.3 that ξ_G is surjective, so the hypothesis that the root system $\Phi(G, T)$ is reduced implies that G is standard and $\ker \xi_G$

is central in G (Proposition 5.3.10). In case G_K^{ss} is SL_2, the surjective map $\xi_G : G \to R_{K/k}(G_K^{ss})$ with central unipotent kernel Z is a (uniquely) split extension by Z due to Proposition 5.1.3 and Example 5.1.4 (since G_K^{ss} is simply connected). This forces Z to be a direct factor of G, so Z is smooth and connected. Thus, $Z = 1$ by the pseudo-reductivity of G. □

We conclude this section with a result that gives some more instances of standardness in the pseudo-split absolutely pseudo-simple case when $\Phi(G,T)$ is reduced. This will be used as a step towards the complete classification in characteristic 2 when $[k : k^2] = 2$. Since standardness has been proved over k in the rank-1 case when $[k : k^2] = 2$ and the root system is reduced (Proposition 9.2.4), as well as away from the types B_n ($n \geqslant 2$), C_n ($n \geqslant 2$), and F_4 in higher rank without any hypotheses on $[k : k^2]$, we may as well assume G_K^{ss} is one of these remaining higher-rank types. In other words, it has an edge with multiplicity 2 in its Dynkin diagram.

Let G be a *pseudo-split* absolutely pseudo-simple group over a field k of characteristic 2 such that $[k : k^2] = 2$, and let K/k be the field of definition of the geometric radical of G. Let T be a split maximal k-torus in G, and let $\Phi := \Phi(G,T)$. For each $a \in \Phi$, let $G_a \subseteq G$ be the smooth connected k-subgroup generated by U_a and U_{-a} as in Proposition 3.4.1, so G_a is absolutely pseudo-simple of rank 1 containing $S_a := T \cap G_a$ as a maximal k-torus. In particular, $\Phi(G_a, S_a) \subseteq \Phi$. Let K_a/k be the field of definition over k for the geometric radical of G_a, so $K_a \subseteq K$ by Proposition 6.3.1(1).

Proposition 9.2.5 *Using notation and hypotheses as above (so $[k : k^2] = 2$), assume that Φ is reduced and G_K^{ss} has an edge in its Dynkin diagram with multiplicity 2.*

(1) *For all long roots $a \in \Phi$, the subfield $K_a \subseteq K$ is equal to a common subfield $K_> \subseteq K$ independent of a. The subfield is either K or K^2, and the k-group G is standard if and only if $K_> = K$.*
(2) *If G is not standard and $G_{\overline{k}}^{ss}$ is simply connected then G is an exotic pseudo-reductive group.*

See Definition 8.2.2 for the definition of an exotic pseudo-reductive group.

Proof By Corollary 5.3.9 (resp. Proposition 8.2.8), G is standard (resp. exotic) over k if and only if G_{k_s} is standard (resp. exotic) over k_s. Thus, since the formation of each of the purely inseparable extensions K and K_a of k is compatible with scalar extension to k_s, it is harmless to extend the ground field to k_s. Hence, we may and do assume $k = k_s$. We may also assume $K \neq k$.

Choose a root $a \in \Phi$. Since Φ is reduced, the K-group $(G_a)_K^{\rm ss}$ is identified with the K-subgroup G'_a in $G' = G_K^{\rm ss}$ generated by the $\pm a_K$-root groups (Proposition 6.3.1). Since $\Phi(G_a, S_a) \subseteq \Phi$, it is a reduced root system. Thus, by Proposition 9.2.4, $\xi_{G_a} : G_a \to {\rm R}_{K_a/k}(G'_a)$ is surjective, so we apply Proposition 6.3.2(2) to see that $K_a = K$ for short a and the fields K_a are equal to a common $K_>$ containing kK^2 for long a. Since we assume $[k : k^2] = 2$ and $K \neq k$, we have $kK^2 = K^2$ and $[K : K^2] = 2$, so in the chain of inclusions $K^2 \subseteq K_> \subseteq K$ one must be an equality.

Standardness of G_a from Proposition 9.2.4 implies that $[K_a : k]$ is the dimension of the a-root group for each long root a of Φ. But in the absolutely pseudo-simple standard case, all root groups have the same dimension. Hence, $K_> = K$ when G is standard (as $K = K_a$ for short $a \in \Phi$). Conversely, assume $K_> = K$. Since ξ_{G_a} is surjective for all a (by Proposition 9.2.4), the standardness of G follows from Theorem 6.3.4 (as G is not exceptional in the sense of Definition 6.3.3 when $K_> = K$).

To prove part (2), recall from the outset of this proof that $(G_a)_{\overline{k}}^{\rm ss} \simeq (G_{\overline{k}}^{\rm ss})_a$. If $G_{\overline{k}}^{\rm ss}$ is simply connected then this rank-1 group is ${\rm SL}_2$ (rather than ${\rm PGL}_2$). Hence, by Proposition 9.2.4 the map ξ_{G_a} is an isomorphism for all a, so the hypotheses in Theorem 8.2.1 are satisfied (we have $K_> \neq K$ due to the established standardness criterion in part (1)), from which it follows that G is exotic. □

9.3 Root groups and module schemes

Let k be a field and let G be a pseudo-reductive k-group. Let K/k be the (minimal) field of definition over k for the geometric unipotent radical of $G_{\overline{k}}$, and let $i_G : G \to {\rm R}_{K/k}(G')$ be the canonical map with $G' := G_K/\mathscr{R}_{u,K}(G_K)$. The connected reductive quotient G' is absolutely simple when G is absolutely pseudo-simple.

Let $T \subseteq G$ be a maximal k-torus, so T_K is identified with a maximal K-torus T' in G'. By Theorem 2.3.10, we have

$$\Phi(G'_{K_s}, T'_{K_s}) \subseteq \Phi(G_{k_s}, T_{k_s}) \subseteq \Phi(G'_{K_s}, T'_{K_s}) \coprod \frac{1}{2} \cdot \Phi(G'_{K_s}, T'_{K_s}),$$

with $\Phi(G'_{K_s}, T'_{K_s})$ equal to the set of non-multipliable roots in $\Phi(G_{k_s}, T_{k_s})$. Assume G is absolutely pseudo-simple and the irreducible $\Phi(G_{k_s}, T_{k_s})$ is non-reduced, so the following properties hold: ${\rm char}(k) = 2$, G is not reductive, and $K \neq k$. The irreducible $\Phi(G_{k_s}, T_{k_s})$ is the unique non-reduced irreducible root system ${\rm BC}_n$ of rank $n = \dim T$, so the root datum $R(G'_{K_s}, T'_{K_s})$ consisting of the non-multipliable roots (and the non-divisible coroots) is the reduced root datum of a simply connected group of type C_n; i.e., $G'_{K_s} \simeq \mathrm{Sp}_{2n}$.

Our primary aim is the classification and construction of all absolutely pseudo-simple and pseudo-split k-groups (e.g., $k = k_s$). In §§9.6–9.8 we will use Weil's method of birational group laws to construct pseudo-split absolutely pseudo-simple k-groups with root system BC_n for any $n \geqslant 1$ over any imperfect field k of characteristic 2. The general determination of étale k-forms of such groups appears to be very hard, but miraculously when $[k : k^2] = 2$ (e.g., local or global function fields of characteristic 2, the cases of most immediate interest for number theory in characteristic 2) we will show in Theorem 9.9.3(1) that *such G have no nontrivial k-forms* (for the étale topology over k)! The case of general imperfect fields k of characteristic 2 exhibits many features and possibilities that do not occur when $[k : k^2] = 2$.

In this section we undertake a preliminary study of root groups in *any* pseudo-split pseudo-reductive G over any field k, especially the relationship between root groups for a pair of roots of the form $\{a, 2a\}$. This is an essential ingredient in our general classification in characteristic 2, and it also provides useful clues for how to build examples.

In the rest of §9.3 we assume T is k-split and G is pseudo-reductive. The natural map

$$i_G : G \to \mathrm{R}_{K/k}(G')$$

carries the commutative Cartan k-subgroup $Z_G(T)$ into $\mathrm{R}_{K/k}(T')$. For $a \in \Phi(G,T)$, let $U_{\pm a}$ be the $\pm a$-root groups and let $G_a = \langle U_a, U_{-a}\rangle$. By Proposition 3.4.1(1), G_a is absolutely pseudo-simple with $T \cap G_a$ a k-split maximal k-torus; this torus is $a^\vee(\mathrm{GL}_1)$. The k-homomorphism

$$\chi_a : Z_G(T) \xrightarrow{i_G} \mathrm{R}_{K/k}(T') \xrightarrow{\mathrm{R}_{K/k}(a_K)} \mathrm{R}_{K/k}(\mathrm{GL}_1) \qquad (9.3.1)$$

will be useful in the study of the structure of U_a when $\mathrm{char}(k) = 2$ and G is absolutely pseudo-simple since there are many possibilities for U_a in characteristic 2 when $[k : k^2] > 2$ (as we saw in Propositions 9.1.4 and 9.1.5 even for G_a with a reduced root system).

Definition 9.3.1 Let $K'_a \subseteq K$ be the subfield over k such that $K'_a \otimes_k k_s$ is the subfield of $K_s = K \otimes_k k_s$ generated over k_s by the $\mathrm{Gal}(k_s/k)$-stable subgroup $\chi_a(Z_G(T)(k_s)) \subseteq K_s^\times$.

Clearly χ_a factors through a k-homomorphism $\chi'_a : Z_G(T) \to \mathrm{R}_{K'_a/k}(\mathrm{GL}_1)$, as it suffices to check this over k_s.

Remark 9.3.2 The subfield $K'_a \subseteq K$ over k can be characterized as the smallest k-subalgebra $F \subseteq K$ such that $\chi_a : Z_G(T) \to \mathrm{R}_{K/k}(\mathrm{GL}_1)$ factors through $\mathrm{R}_{F/k}(\mathrm{GL}_1)$. Consequently, in the definition of K'_a we can replace $Z_G(T)(k_s)$ with any subgroup $\Gamma \subseteq Z_G(T)(k_s)$ that is Zariski-dense in $Z_G(T)_{k_s}$. In particular, if $h : \mathscr{C} \to Z_G(T)$ is a surjective k-homomorphism from a smooth commutative affine k-group $\mathscr{C}$ then the subgroup $h(\mathscr{C}(k_s)) \subseteq Z_G(T)(k_s)$ also generates $K'_a \otimes_k k_s$ as a k_s-algebra.

Remark 9.3.3 Assume that G is absolutely pseudo-simple. Let us explain why K'_a and χ'_a are only of interest for such G in the non-standard case in characteristic 2. First suppose that the pseudo-split absolutely pseudo-simple G is standard, with char(k) arbitrary. We claim that χ'_a is always surjective and that $K'_a = K$ except when char$(k) = 2$ and $G^{\mathrm{ss}}_{\overline{k}}$ is of type C_n ($n \geqslant 1$) with a long, in which case $K'_a = kK^2$. (We make the convention that for type $\mathrm{C}_1 = \mathrm{A}_1$, the roots are long because they are divisible in the weight lattice.) By Theorem 4.1.1, Proposition 4.1.4(3), and Proposition 4.2.4 there is a split connected semisimple K-group $(\mathscr{G}, \mathscr{T})$ that is absolutely simple and simply connected such that $G \simeq \mathrm{R}_{K/k}(\mathscr{G})/Z$ for a closed k-subgroup $Z \subseteq \mathrm{R}_{K/k}(Z_{\mathscr{G}})$ and $Z_G(T) = \mathrm{R}_{K/k}(\mathscr{T})/Z$. Thus, the image of χ_a is the same as that of $\mathrm{R}_{K/k}(a_K) : \mathrm{R}_{K/k}(\mathscr{T}) \to \mathrm{R}_{K/k}(\mathrm{GL}_1)$, and $\mathrm{X}_*(\mathscr{T})$ has a basis given by the coroots associated to a basis Δ of $\Phi(G, T) = \Phi(\mathscr{G}, \mathscr{T})$. For $b \in \Delta$, the restriction of $\mathrm{R}_{K/k}(a_K)$ to $\mathrm{R}_{K/k}(b^\vee(\mathrm{GL}_1))$ is the $\langle a, b^\vee \rangle$-power endomorphism of $\mathrm{R}_{K/k}(\mathrm{GL}_1)$, so by taking $b = a$ we settle the case char$(k) \neq 2$ (as squaring is a surjective endomorphism of $\mathrm{R}_{K/k}(\mathrm{GL}_1)$ away from characteristic 2).

Now suppose char$(k) = 2$ (and G is standard as above). We may choose Δ to contain a, so if there exists $b \in \Delta$ such that $\langle a, b^\vee \rangle$ is odd (i.e., ± 1 or ± 3) then we again see that $K'_a = K$ and χ'_a is surjective. This settles the cases when $\Phi(G, T)$ has rank $\geqslant 2$ with no double bond, as well as any cases in which a has an adjacent vertex b in the Dynkin diagram with the same length (as then $\langle a, b^\vee \rangle = -1$). Thus, it remains to treat the unique short root $a \in \Delta$ for type B_n with $n \geqslant 2$ as well as the unique long root $a \in \Delta$ for type C_n with $n \geqslant 1$. For the short $a \in \Delta$ in type B_n with $n \geqslant 2$ there is a unique long $b \in \Delta$ adjacent to a in the Dynkin diagram, and $\langle a, b^\vee \rangle = -1$. If instead $a \in \Delta$ is the unique long root for type C_n with $n \geqslant 2$ then $\langle a, b^\vee \rangle = -2$ for the unique short root $b \in \Delta$ adjacent to a in the Dynkin diagram. Hence, it remains to check that the squaring endomorphism of $\mathrm{R}_{K/k}(\mathrm{GL}_1)$ has image $\mathrm{R}_{kK^2/k}(\mathrm{GL}_1)$, or in other words that $(K^\times)^2 = K^2 - \{0\}$ is Zariski-dense in $\mathrm{R}_{kK^2/k}(\mathrm{GL}_1)$ if $k = k_s$. Since the open subscheme $R_{kK^2/k}(GL_1)$ is dense in $R_{kK^2/k}(\mathbf{G}_a)$, if $k = k_s$ it is equivalent that K^2 is Zariski-dense in $R_{kK^2/k}(\mathbf{G}_a)$. Since K^2 spans kK^2 over k and contains k^2, the Zariski-density of k^2 in $\mathbf{G}_a$ if $k = k_s$ yields the desired result.

Away from characteristic 2, the only non-standard absolutely pseudo-simple groups are the exotic ones in characteristic 3 (by Proposition 8.2.10). Consider a pseudo-split exotic absolutely pseudo-simple G of any type; i.e., type G_2 in characteristic 3, or types B_n ($n \geqslant 2$), C_n ($n \geqslant 2$), or F_4 in characteristic 2. We claim that χ'_a is surjective for all roots a, $K'_a = K$ for short a for types G_2, F_4, and C_n ($n \geqslant 3$), and $K'_a = K_>$ for short a for type B_n ($n \geqslant 2$). Moreover, if a is a long root then we claim that $K'_a = K_>$ provided that G is not of type C_n with $n \geqslant 2$ whereas $K'_a = kK^2$ when G is of type C_n with $n \geqslant 2$. These assertions are proved via arguments similar to the ones given in the standard case above by using Proposition 7.1.7(2), the properties of basic exotic groups given in Proposition 7.2.7, and the description of their open Bruhat cell in Remark 7.2.8.

Our main interest in the fields K'_a is in the case when a is a multipliable root (so char$(k) = 2$ and k is imperfect, hence infinite). More specifically, if a root $c \in \Phi(G,T)$ is multipliable then it is obvious (by computing over k_s) that $K'_{2c} = kK'^2_c$ but it is rather less evident yet more important that if G is absolutely pseudo-simple then $K'_c = K$ for such c (see Proposition 9.5.2). Applying this latter general equality to the pseudo-split and absolutely pseudo-simple rank-1 group G_c implies that the field analogous to K'_c that is defined by using $Z_{G_c}(T) = Z_{G_c}(c^\vee(GL_1))$ in place of $Z_G(T)$ in Definition 9.3.1 for c is equal to the field of definition K_c/k for the geometric unipotent radical of G_c. Unlike K'_c, the subfield K_c of K over k (see Proposition 6.3.1) can be a *proper* subfield of K when G has rank $n \geqslant 2$. Absolutely pseudo-simple examples with $K_c \neq K$ for any rank $\geqslant 2$ and any K/k with $[K : kK^2] > 2$ are obtained by applying Theorem 9.8.1 to the linear algebra data built at the end of Remark 9.6.10 (see 9.6.8 and 9.6.12). Hence, the general equality $K'_c = K$ for multipliable c in the absolutely pseudo-simple case is a "global" property of (G,T) in the sense that it cannot be proved by studying G_c in isolation when G has rank $\geqslant 2$.

Let $c \in \Phi(G,T)$ be a multipliable root (with pseudo-reductive G and k-split T). The commutative root groups U_{2c} and U_c are nontrivial vector groups admitting compatible T-equivariant linear structures (by Proposition 2.3.11), so the same holds for U_c/U_{2c}. By Lemma 2.3.8, the T-equivariant linear structures on U_{2c} and U_c/U_{2c} are unique. Our interest in χ'_c is due to its role in the characterization of "module scheme" structures on the k-groups U_c/U_{2c} and U_{2c} enhancing the unique T-equivariant linear structures on these vector groups. This is most conveniently expressed with the following terminology.

Definition 9.3.4 Let F be a field, and A a finite-dimensional commutative F-algebra. Let $\underline{A} = \mathrm{R}_{A/F}(\mathbf{G}_{\mathrm{a}})$, viewed as a ring scheme over F in the evident way; i.e., its functor of points assigns to every commutative F-algebra B the commutative ring $A \otimes_F B$. (This is the natural ring scheme structure on the affine space associated to the F-vector space A.)

An $\underline{A}$*-module scheme* over F is a commutative F-group scheme M of finite type such that for every commutative F-algebra B the group $M(B)$ is equipped with a module structure over $\underline{A}(B) = A \otimes_F B$ functorially in B and the associated action map $A \to \mathrm{End}(\mathrm{Lie}(M))$ is an F-algebra homomorphism.

For a commutative F-group M of finite type, clearly an $\underline{A}$-module structure on M is the "same" as an $\underline{F}$-module structure on M equipped with an F-algebra homomorphism $A \to \mathrm{End}_{\underline{F}}(M)$. Also, the Lie algebra condition in the definition of an $\underline{A}$-module scheme is precisely the condition that the induced action of $\mathrm{GL}_1 = \underline{F}^{\times}$ on $\mathrm{Lie}(M)$ is the canonical scaling action on this Lie algebra (so the vector space structure on $\mathrm{Lie}(M)$ over $\underline{F}(F) = F$ arising from the $\underline{F}$-action on M is the usual F-vector space structure).

The F-groups associated to finitely generated A-modules are smooth affine $\underline{A}$-modules. This construction is reversed via the formation of F-points in the following strong sense:

Lemma 9.3.5 *The functor $M \rightsquigarrow M(F)$ is an equivalence from the category of smooth affine $\underline{A}$-module schemes over F to the category of finitely generated A-modules. In particular, any $\underline{A}$-linear homomorphism $f : M' \to M$ has smooth connected kernel, and f is a closed immersion if and only if f is injective on F-points.*

Proof It clearly suffices to treat the case $A = F$. By Galois descent, we may assume $F = F_s$. Our aim is to show that $M(F)$ is finite-dimensional as an F-vector space and that the natural $\underline{F}$-linear F-homomorphism $\underline{F} \otimes_F M(F) \to M$ is an isomorphism. (For any finite-dimensional F-vector space V, $\underline{F} \otimes_F V$ is the $\underline{F}$-module scheme representing the functor $B \rightsquigarrow \underline{F}(B) \otimes_F V = B \otimes_F V$; i.e., it is the affine space over F covariantly associated to V.)

For any finite ordered subset $\{m_1, \dots, m_n\}$ in $M(F)$ we obtain an $\underline{F}$-linear homomorphism $f : \underline{F}^n \to M$ via

$$(a_1, \dots, a_n) \mapsto \sum a_i m_i.$$

If $m \in M(F)$ is not in the smooth connected closed image of f then we can use the unique $\underline{F}$-linear map $\underline{F} \to M$ carrying 1 to m to get an $\underline{F}$-linear homomorphism $f' : \underline{F}^{n+1} \to M$ such that $\dim f'(\underline{F}^{n+1}) > \dim f(\underline{F}^n)$. Hence, by choosing f with image of maximal dimension we see that the closed image of f contains the Zariski-dense $M(F)$, so f is surjective. In particular, M is connected. The group $(\ker f)(F)$ is an F-subspace of F^n, so by using a basis of F^n extending a basis of $(\ker f)(F)$ we can pass to the quotient of $\underline{F}^n$ by the Zariski closure of $(\ker f)(F)$ to reduce to the case $(\ker f)(F) = 0$ with $M \neq 0$.

The case $\mathrm{char}(F) = 0$ is now settled, so we may assume $\mathrm{char}(F) = p > 0$. In this case M is p-torsion since $M(F)$ is p-torsion and $F = F_s$, so the smooth connected M is unipotent. The Lie algebra condition in the definition of an $\underline{F}$-module implies that the action on M by the unit group GL_1 of the ring scheme $\underline{F}$ induces a GL_1-action on $\mathrm{Lie}(M)$ that is the usual unit-scaling action on this F-vector space. In particular, there is no occurrence of the trivial weight on $\mathrm{Lie}(M)$ under the action by $\underline{F}^\times = \mathrm{GL}_1$, so by Theorem B.4.3 there is an F-group isomorphism $M \simeq \mathbf{G}_\mathrm{a}^n$ such that the GL_1-action on M via the given $\underline{F}$-module structure is carried to an action

$$t.(x_1, \ldots, x_n) = (t^{e_1}x_1, \ldots, t^{e_n}x_n)$$

for nonzero integers $e_1, \ldots, e_n$. Consideration of the limiting behavior as $t \to 0$ forces all $e_i > 0$, and by additivity of the action we have $e_i = p^{r_i}$ for integers $r_i \geqslant 0$. Since the only weight for the induced GL_1-action on $\mathrm{Lie}(M)$ is 1, necessarily $r_i = 0$ for all i. □

For a vector group V over k, Lemma 9.3.5 implies that a "linear structure" on V in the sense of Definition A.1.1 is the same as a $\underline{k}$-module structure on V.

Proposition 9.3.6 *Let $c \in \Phi(G,T)$ be a multipliable root.*

(i) *There exists a unique $Z_G(T)$-equivariant $\underline{K'_c}$-module structure on U_c/U_{2c} so that the $Z_G(T)$-action on U_c/U_{2c} is the composition of χ'_c with the action of the group of units $\mathrm{R}_{K'_c/k}(\mathrm{GL}_1)$ of the k-ring scheme $\underline{K'_c}$.*

(ii) *The k-subgroup $i_G(U_{2c})$ is a $\underline{kK'^2_c}$-submodule of the $\underline{K}$-module $\mathrm{R}_{K/k}(U'_{2c})$ on which the $Z_G(T)$-action is through χ'^2_c. The map $i_G : U_{2c} \to i_G(U_{2c})$ is an isomorphism that defines a $Z_G(T)$-equivariant $\underline{kK'^2_c}$-module structure on U_{2c} extending its unique T-equivariant linear structure. The k-subgroup $i_G(U_c)$ is also a $\underline{kK'^2_c}$-submodule of the $\underline{K}$-module $\mathrm{R}_{K/k}(U'_{2c})$, and the induced map $i_G : U_c/U_{2c} \to \mathrm{R}_{K/k}(U'_{2c})/i_G(U_{2c})$ is linear over squaring $K'_c \to kK'^2_c$.*

Proof By uniqueness and Galois descent, we may and do assume $k = k_s$ (so the content in (i) is existence). As before, let $G' = G_K^{\rm red}$. The $2c$-root group $U'_{2c} \subseteq G'$ for (G', T') is the image of $(U_c)_K$ under the natural quotient map $G_K \twoheadrightarrow G'$ (as this quotient map carries T_K isomorphically onto T'), so i_G carries U_c into $\mathrm{R}_{K/k}(U'_{2c}) \subseteq \mathrm{R}_{K/k}(G')$.

We first claim that the map $i_G : U_c \to \mathrm{R}_{K/k}(U'_{2c})$ is injective on k-points, or equivalently the finite group $(\ker i_G)(k)$ meets $U_c(k)$ trivially. (In Remark 9.3.8 we will see that $U_c \cap \ker i_G$ is nontrivial, hence non-smooth.) Since this map is equivariant with respect to $T \to \mathrm{R}_{K/k}(T')$, the finite subgroup $U_c(k) \cap (\ker i_G)(k)$ in $U_c(k)$ is $T(k)$-stable. But all $T(k)$-orbits in $U_c(k) - \{0\}$ are infinite since k is infinite and U_c admits a T-equivariant linear structure having weights c and $2c$ where $c \neq 0$. Thus, $U_c(k) \cap (\ker i_G)(k) = 0$ as claimed.

The restriction of i_G to U_{2c} is a linear homomorphism $\ell_{2c} : U_{2c} \to \mathrm{R}_{K/k}(U'_{2c})$ relative to the canonical linear structures (determined by the respective actions of T and $\mathrm{R}_{K/k}(T')$), so by injectivity on k-points it is a k-subgroup inclusion. Define

$$V_c := U_c/U_{2c}, \quad V' := i_G(U_{2c}) \subseteq \mathrm{R}_{K/k}(U'_{2c}), \quad L' = \mathrm{R}_{K/k}(U'_{2c})/V'.$$

These vector groups are equipped with canonical T-equivariant linear structures. We have shown that i_G identifies $U_c(k)$ with a subgroup of the K-vector space $\mathrm{R}_{K/k}(U'_{2c})(k) = U'_{2c}(K)$ such that $U_{2c}(k)$ is carried k-linearly isomorphically onto the k-subspace $V'(k)$. Hence, the group $V_c(k) = U_c(k)/U_{2c}(k)$ is naturally a subgroup of the k-vector space $L'(k) = U'_{2c}(K)/V'(k)$.

Denote the k-homomorphism $V_c \to L'$ induced by i_G as j; it is injective on k-points. (The kernel $\ker j$ is nontrivial, hence non-smooth, since $U_c \cap \ker i_G$ is non-smooth; see Remark 9.3.8.) For $t \in T$, j is equivariant with respect to the actions of t on U_c and of $i_G(t) \in \mathrm{R}_{K/k}(T')$ on $\mathrm{R}_{K/k}(U'_{2c})$. Hence, $j(c(t)v) = (2c)(t)j(v)$ for all $v \in V_c$ and $t \in T$ relative to the $\underline{k}$-linear structure on V_c and $\underline{K}$-linear structure on U'_{2c}. But $(2c)(t) = c(t)^2$ and the k-scheme $c(T) = \mathrm{GL}_1$ is Zariski-dense in $\underline{k}$, so $j(\lambda v) = \lambda^2 j(v)$ for all $v \in V_c$ and $\lambda \in \underline{k} = \mathbf{G}_{\rm a}$. Therefore, $j(V_c(k))$ is a k^2-linear subspace of the k-vector space $L'(k)$.

The natural action of $Z_G(T)$ on V_c must respect the *unique* T-equivariant linear structure on V_c, so the $\underline{k}$-module V_c is a linear representation of $Z_G(T)$. Likewise, $Z_G(T)$ acts linearly on the $\underline{k}$-module $L' = \mathrm{R}_{K/k}(U'_{2c})/V'$ since its action is defined via the k-homomorphism $i_G : Z_G(T) \to \mathrm{R}_{K/k}(T')$ and the T'-action on $U'_{2c} \simeq \mathbf{G}_{\rm a}$ (with V' a $\underline{k}$-submodule of $\mathrm{R}_{K/k}(U'_{2c})$). We will construct a $\underline{K'_c}$-module structure on V_c extending its T-equivariant $\underline{k}$-module structure by using the additive inclusion of $V_c(k)$ into $L'(k) = \mathrm{R}_{K/k}(U'_{2c})(k)/V'(k)$ and the $\underline{K}$-module structure on $\mathrm{R}_{K/k}(U'_{2c})$.

Let $A \subseteq \mathrm{End}_{\underline{k}}(V_c) = \mathrm{End}_k(V_c(k))$ denote the k-subalgebra generated by the image of $Z_G(T)(k)$ in $\mathrm{GL}(V_c(k))$ (so $Z_G(T)(k) \subseteq A^\times$), and let $A' \subseteq \mathrm{End}_k(L'(k))$ denote the k^2-subalgebra generated by the image of $Z_G(T)(k)$ in $\mathrm{GL}(L'(k))$ (so $Z_G(T)(k)$ acts through $A'^\times$). The $Z_G(T)(k)$-equivariance of the additive inclusion $j : V_c(k) \hookrightarrow L'(k)$ and its Frobenius semilinearity relative to the k-linear structures imply that the subset $j(V_c(k)) \subseteq L'(k)$ is A'-stable with the A'-action on $j(V_c(k))$ defining a surjective ring map $A' \twoheadrightarrow A$ that (i) carries λ^2 to λ for $\lambda \in k$, and (ii) is compatible with the natural homomorphisms from $Z_G(T)(k)$ into $A'^\times$ and $A^\times$.

Since T' acts on U'_{2c} through the K-homomorphism $(2c)_K : T' \to \mathrm{GL}_1$, every element of $Z_G(T)(k)$ acts on $U'_{2c}(K)$ through an element of $K^\times$ that preserves the k-subspace $V'(k)$. The subring

$$F = \{\lambda \in K \mid \lambda \cdot V'(k) \subseteq V'(k) \text{ inside } U'_{2c}(K)\}$$

of K contains k, so it is a field since K/k is an algebraic extension. Thus, the natural k-algebra map $F \to \mathrm{End}_k(L'(k))$ is *injective*, and visibly its image contains A'. Hence, A' is identified with a k^2-subalgebra of F, so A' is a field since F is algebraic over k^2. It follows that the ring map $A' \twoheadrightarrow A$ is an isomorphism, so A is a field of finite degree over k.

The subfield K'_c of K contains k, so K'_c is a finite purely inseparable extension of k. Thus, there is at most one k-embedding of K'_c into A. We claim that there is a natural k-embedding of K'_c into A, with A equal to the embedded K'_c, and that the resulting map $Z_G(T) \to K'^\times_c$ recovers χ'_c on k-points. This would complete the proof of part (i) of the proposition, since V_c is visibly an $\underline{A}$-module scheme compatibly with its initial $\underline{k}$-linear structure and the claim would imply that A is generated as a k-algebra by the subset $\chi'_c(Z_G(T)(k)) \subseteq A^\times$.

Since A' is a subextension over k^2 of the purely inseparable extension K of k^2, it is purely inseparable over k^2. Moreover, by the definition of K'_c we see that the subfield $A' \subseteq K$ generated over k^2 by the image of $Z_G(T)(k) \longrightarrow T'(K) \xrightarrow{(2c)_K} K^\times$ contains K'^2_c. In view of the isomorphism of abstract fields $A' \simeq A$ over the square root isomorphism $k^2 \simeq k$, A is uniquely identified with $A'^{1/2}$ as extensions of k. Thus, A contains K'_c over k. The homomorphism $Z_G(T)(k) \to A^\times$ lands inside $K'^\times_c \subseteq K^\times$ via χ_c on k-points due to two facts: the isomorphism $A' \simeq A$ via square roots is induced by restriction to the $Z_G(T)(k)$-stable subset $V_c(k) \subseteq L'(k)$, and the $Z_G(T)(k)$-action on $L'(k) = U'_{2c}(K)/V'(k)$ lands inside K'^2_c via the composite map

$$\chi_c^2 : Z_G(T) \xrightarrow{i_G} \mathrm{R}_{K/k}(T') \xrightarrow{\mathrm{R}_{K/k}((2c)_K)} \mathrm{R}_{K/k}(\mathrm{GL}_1)$$

(since T' acts on U'_{2c} through $(2c)_K$). The extension field A of K'_c is generated as a k-algebra by the image of $Z_G(T)(k)$ in $A^\times$, but we just saw that this image is contained in $K'^{\times}_c$, so the inclusion of k-algebras $K'_c \hookrightarrow A$ is an equality. Hence, (i) is proved and we also see that $A' = A^2 = K'^2_c$.

We will now show that the k-subgroup $i_G(U_{2c}) = V'$ is a $\underline{kK'^2_c}$-submodule of the $\underline{K}$-module $\mathrm{R}_{K/k}(U'_{2c})$. The action of $Z_G(T)$ on $\mathrm{R}_{K/k}(U'_{2c})$ is through the $\mathrm{R}_{K/k}(T')$-action via the $\underline{K}^\times$-valued $\mathrm{R}_{K/k}((2c)_K)$ (i.e., through $\chi'_{2c} = \chi'^2_c$) and vector subgroup $V' = i_G(U_{2c})$ is stable under this action. As the k-linear subspace $V'(k) \subseteq \mathrm{R}_{K/k}(U'_{2c})(k) = U'_{2c}(K)$ is by definition stable under the action of the subring $A' \subseteq F \subseteq K$, and $A' = K'^2_c$ inside K, V' is a $\underline{kK'^2_c}$-submodule of $\mathrm{R}_{K/k}(U'_{2c})$ and the action of $Z_G(T)$ on it is through χ'^2_c. Therefore, the quotient $L' = \mathrm{R}_{K/k}(U'_{2c})/V'$ admits a natural $\underline{kK'^2_c}$-module structure. The $\underline{k}$-linear isomorphism $i_G : U_{2c} \simeq V'$ transfers the $\underline{kK'^2_c}$-module structure on V' to a $\underline{kK'^2_c}$-module structure on U_{2c} that is $Z_G(T)$-equivariant and extends the unique T-equivariant $\underline{k}$-module structure on U_{2c}. The final 2-linearity assertion in (ii) now makes sense and is immediate because: the $Z_G(T)$-actions on U_c/U_{2c} and $\mathrm{R}_{K/k}(U'_{2c})$ are respectively through χ'_c (by (i)) and χ'^2_c, and $\chi'_c(Z_G(T)(k))$ generates K'_c as a k-algebra (by definition, since $k_s = k$).

Finally, we show that $i_G(U_c)$ is a $\underline{kK'^2_c}$-submodule of the $\underline{K}$-module $\mathrm{R}_{K/k}(U'_{2c})$. For this purpose, we first prove that $j(V_c)$ is a $\underline{kK'^2_c}$-submodule of L'. We have shown above that with respect to the squaring isomorphism $K'_c \simeq K'^2_c$, j is equivariant for the action of K'_c on $V_c(k)$ and the action of the subfield $K'^2_c \subseteq K$ on $L'(k)$ induced by the K-action on $\mathrm{R}_{K/k}(U'_{2c})$. Thus, the smooth closed k-subgroup $j(V_c) \subseteq L'$ is stable under the action of K'^2_c on L' since $j(V_c(k))$ is Zariski-dense in $j(V_c)$ (as $k = k_s$). In particular, $j(V_c)$ is stable under the k^2-action on L'. A k^2-stable smooth closed subscheme of L' is $\underline{k}$-stable since k^2 is Zariski-dense in $\underline{k}$, so $j(V_c)$ is stable under the actions of k and K'^2_c on L'. This proves that $j(V_c)$ is a $\underline{kK'^2_c}$-submodule of L'.

Now consider the natural k-homomorphism $h : i_G(U_c) \to j(V_c)$ between smooth affine k-groups. This is obviously surjective, so it is faithfully flat and hence identifies $j(V_c)$ with the quotient of $i_G(U_c)$ modulo the scheme-theoretic kernel of h. By working with algebraic groups as fppf group sheaves, we see that $\ker h = i_G(U_{2c})$ as k-subgroup schemes of $i_G(U_c)$, so $i_G(U_c)/i_G(U_{2c}) \to j(V_c)$ is an isomorphism. (The reason we argue using the fppf topology and not more classical methods is because the scheme $\ker(i_G|_{U_c}) = U_c \cap \ker i_G$ is not smooth; see Remark 9.3.8.) So, the fact that $i_G(U_{2c})$ and $j(V_c)$ are $\underline{kK'^2_c}$-submodules of $\mathrm{R}_{K/k}(U'_{2c})$ and L' respectively implies that $i_G(U_c)$ is a $\underline{kK'^2_c}$-submodule of the $\underline{K}$-module $\mathrm{R}_{K/k}(U'_{2c})$. □

In what follows, we continue to use the notation $G' := G_K^{\mathrm{red}}$ and $V' := i_G(U_{2c})$ as in the preceding proof; V' is a $\underline{kK_c'^2}$-submodule of $\mathrm{R}_{K/k}(U'_{2c})$ and $i_G : U_{2c} \to V'$ is a $\underline{k}$-linear isomorphism, to be denoted by ℓ_{2c}, by means of which we define the $Z_G(T)$-equivariant $\underline{kK_c'^2}$-module structure on U_{2c} extending its unique T-equivariant $\underline{k}$-module structure. For the vector groups $V_c = U_c/U_{2c}$ and $L' = \mathrm{R}_{K/k}(U'_{2c})/V'$ over k equipped with their canonical $Z_G(T)$-equivariant module scheme structures over $\underline{K_c'}$ and $\underline{kK_c'^2}$ respectively, the natural k-homomorphism $j : V_c \to L'$ induced by i_G is 2-linear in the sense that $j(\lambda v) = \lambda^2 j(v)$ for $v \in V_c$ and $\lambda \in \underline{K_c'}$ by Proposition 9.3.6(ii). Since the additive j is a quadratic map, we denote it as q_c.

We will now show that the map $U_c \twoheadrightarrow U_c/U_{2c} =: V_c$ admits a $Z_G(T)$-equivariant k-homomorphic section satisfying an additional linearity property.

Proposition 9.3.7 *The quotient map $U_c \to V_c$ admits a homomorphic section $s : V_c \to U_c$ that is equivariant for the natural actions of $C := Z_G(T)$, and the C-action on U_c factors through an action by the quotient $i_G(C) \subseteq \mathrm{R}_{K/k}(T')$. In particular, $C \cap \ker i_G$ acts trivially on U_c. Moreover, s can be chosen so that the k-homomorphism $i_G \circ s : V_c \to \mathrm{R}_{K/k}(U'_{2c})$ is linear over the squaring map $\underline{K_c'} \to \underline{K}$.*

Proof Consider the commutative diagram of short exact sequences of k-groups

$$\begin{array}{ccccccccc} 0 & \longrightarrow & U_{2c} & \longrightarrow & U_c & \longrightarrow & V_c & \longrightarrow & 0 \\ & & {\scriptstyle \ell_{2c}}\downarrow{\scriptstyle \simeq} & & {\scriptstyle i_G}\downarrow & & \downarrow{\scriptstyle q_c} & & \\ 0 & \longrightarrow & V' & \longrightarrow & \mathrm{R}_{K/k}(U'_{2c}) & \longrightarrow & L' & \longrightarrow & 0 \end{array} \tag{9.3.2}$$

in which the outer vertical maps are induced by i_G. The diagram is equivariant for the natural C-actions on the middle terms of each row (and the induced actions on the outer terms of each row). By Proposition 9.3.6(ii), the $\underline{K_c'}$-module structure on V_c makes the right vertical map linear over the squaring map $\underline{K_c'} \to \underline{kK_c'^2}$, where the $\underline{kK_c'^2}$-module structure on L' arises from the $\underline{K}$-module structure on $\mathrm{R}_{K/k}(U'_{2c})$.

Since ℓ_{2c} is an isomorphism, diagram (9.3.2) identifies the top row as a C-equivariant pullback of the bottom row along q_c. Thus, to make a C-equivariant splitting of the top row with the desired linearity property it suffices to split the bottom row equivariantly with respect to the action on

$\mathrm{R}_{K/k}(U'_{2c})$ by a k-subgroup of $\mathrm{R}_{K/k}(\mathrm{GL}_1)$ that preserves V' and through which the actions of C and $\mathrm{R}_{kK_c'^2/k}(\mathrm{GL}_1)$ factor.

The construction of the $\underline{kK_c'^2}$-linear structure on V' in Proposition 9.3.6 showed that V' is a $\underline{kK_c'^2}$-submodule of the $\underline{K}$-module $\mathrm{R}_{K/k}(U'_{2c})$. Let $F \subseteq K$ be the maximal subfield containing $kK_c'^2$ such that $V'(k)$ is an F-subspace of the 1-dimensional K-vector space $U'_{2c}(K)$, so the bottom row of (9.3.2) is an application of $\mathrm{R}_{F/k}$ to a short exact sequence of finite-dimensional F-vector spaces. The formation of F commutes with separable algebraic extension on k (by Galois descent), so every element of $C(k_s)$ acts on $\mathrm{R}_{K/k}(U'_{2c})_{k_s}$ through an element of $(K \otimes_k k_s)^\times$ that lies in $(F \otimes_k k_s)^\times = \mathrm{R}_{F/k}(\mathrm{GL}_1)(k_s)$. Thus, the map $C \to \mathrm{R}_{K/k}(\mathrm{GL}_1)$ which gives the action of C on $\mathrm{R}_{K/k}(U'_{2c})$ factors through the k-subgroup $\mathrm{R}_{F/k}(\mathrm{GL}_1)$, so the desired C-equivariant splitting of the bottom row of (9.3.2) (and hence of the top row via q_c-pullback) follows from the fact that any short exact sequence of F-vector spaces splits. This provides s with the desired linearity property.

Since (9.3.2) is a C-equivariant pullback diagram, to show that the action of C on U_c factors through the quotient $i_G(C)$ it suffices to show that the C-actions on $\mathrm{R}_{K/k}(U'_{2c})$ and V_c factor through $i_G(C)$. The case of $\mathrm{R}_{K/k}(U'_{2c})$ is clear since the middle vertical arrow in (9.3.2) is equivariant with respect to $i_G : C \to \mathrm{R}_{K/k}(T')$. For V_c we apply Proposition 9.3.6, which gives that the action of C on V_c factors through χ'_c, whose kernel contains $C \cap \ker i_G$. □

Remark 9.3.8 The C-equivariant splitting identifies $U_c \cap \ker i_G$ with the kernel of the 2-linear map $q_c : V_c \to L'$ that is injective on k-points, so with respect to a $\underline{k}$-module isomorphism $V_c \simeq \mathbf{G}_\mathrm{a}^n$ it is defined by an equation $\sum \alpha_i x_i^2 = 0$ for nonzero $\alpha_i \in k$ (with $1 \leqslant i \leqslant \dim V_c$). By inspection, this is not smooth.

9.4 Central quotients

The classification of pseudo-split absolutely pseudo-simple groups in characteristic 2 involves preliminary work with central quotients. A central quotient of an absolutely pseudo-simple group can fail to be pseudo-reductive (e.g., $\mathrm{R}_{k'/k}(\mathrm{SL}_p)/\mu_p$ for any nontrivial finite purely inseparable extension k'/k with $\mathrm{char}(k) = p > 0$; see Example 1.3.5), so we now give a pseudo-reductivity criterion for central quotients.

Lemma 9.4.1 *Let G be a pseudo-reductive group over a field k, $Z \subseteq G$ a central closed k-subgroup scheme, and T a maximal k-torus. The quotient*

G/Z is pseudo-reductive if and only if the smooth connected commutative k-group $Z_G(T)/Z$ is pseudo-reductive.

Proof Let $H = G/Z$. The image S of T in H is a maximal k-torus, and the centralizer of S in H is the image $Z_G(T)/Z$ of $Z_G(T)$ in H (see Proposition A.2.8), so $Z_G(T)/Z$ coincides with the Cartan k-subgroup $Z_H(S)$ of H. Hence, if H is pseudo-reductive then $Z_G(T)/Z$ is pseudo-reductive by Proposition 1.2.4.

Now assuming that the commutative k-group $Z_G(T)/Z = Z_H(S)$ is pseudo-reductive, we aim to prove that H is pseudo-reductive. We may and do assume $k = k_s$, so all k-tori are split. The commutative pseudo-reductive k-group $Z_H(S)$ contains no nontrivial smooth connected unipotent k-subgroup. Let $U = \mathscr{R}_{u,k}(H)$, so $Z_H(S) \cap U$ is smooth and connected (since S normalizes U). This intersection is also unipotent, so it is trivial.

Pick a cocharacter $\lambda \in \mathrm{X}_*(T)$ that does not annihilate any elements of $\Phi(G,T)$, so $Z_G(T) = Z_G(\lambda)$. The 1-parameter subgroup λ operates on $H = G/Z$ through its image in S and $U_G(\pm\lambda) \cap Z \subseteq U_G(\pm\lambda) \cap Z_G(\lambda) = 1$, so the natural k-homomorphism $U_G(\pm\lambda) \to U_H(\pm\lambda)$ is an isomorphism. Moreover, since the k-group $Z_H(\lambda) \cap U = Z_H(S) \cap U$ is trivial, by Proposition 2.1.12(1) the multiplication map of schemes

$$U_U(-\lambda) \times U_U(\lambda) \to U$$

is an isomorphism. Let U^- and U^+ be the smooth connected k-subgroups of $U_G(-\lambda)$ and $U_G(\lambda)$ respectively corresponding to $U_U(-\lambda)$ and $U_U(\lambda)$ under the isomorphisms $U_G(-\lambda) \to U_H(-\lambda)$ and $U_G(\lambda) \to U_H(\lambda)$. To prove that $U = 1$ it is equivalent to show that the subgroups U^- and U^+ of G are trivial. Observe that $U^{\pm}_{\overline{k}} \subseteq \mathscr{R}_u(G_{\overline{k}})$. Indeed, the central quotient map $G^{\mathrm{red}}_{\overline{k}} \to H^{\mathrm{red}}_{\overline{k}}$ must kill the smooth connected unipotent image of $U^{\pm}_{\overline{k}}$ in $G^{\mathrm{red}}_{\overline{k}}$, and the center of $G^{\mathrm{red}}_{\overline{k}}$ contains no nontrivial smooth connected unipotent subgroups. Hence, $U^{\pm}_{\overline{k}}$ has trivial image in $G^{\mathrm{red}}_{\overline{k}}$, as claimed. By the pseudo-reductivity of G over k, it follows (via Lemma 1.2.1) that U^- and U^+ are trivial, so $U = 1$. This proves that the k-group $H = G/Z$ is pseudo-reductive. □

Let k be a field, K/k a purely inseparable finite extension, G an absolutely pseudo-simple k-group, and T a maximal k-torus of G. Let $G' := G^{\mathrm{ss}}_K = G_K/\mathscr{R}_K(G_K)$, and $i_G : G \to \mathrm{R}_{K/k}(G')$ the natural k-homomorphism. The cases with $\ker i_G = 1$ and simply connected G' correspond to certain "large" absolutely pseudo-simple k-subgroups of the k-group $\mathrm{R}_{K/k}(G')$ that is itself absolutely pseudo-simple (since G' is simply connected). Many such cases with G a *proper* k-subgroup of $\mathrm{R}_{K/k}(G')$ arose in the study of basic exotic pseudo-simple groups in characteristics 2 and 3 (see Proposition 7.2.7(2) for the non-surjectivity of i_G in those cases). Going beyond the basic exotic cases, if G is an absolutely pseudo-simple group over a field k then the image $i_G(G)$ is always absolutely pseudo-simple, without any hypotheses on $\ker i_G$ or $G^{\mathrm{ss}}_{\overline{k}}$ (see Proposition 4.2.5 and Lemma 3.1.2).

In the following proposition we will show that every absolutely pseudo-simple group G over a field k admits a canonical absolutely pseudo-simple *central* quotient $\mathscr{G}$ such that (i) the normal k-subgroup scheme $\ker i_{\mathscr{G}}$ in $\mathscr{G}$ has trivial intersection with a Cartan k-subgroup $\mathscr{C}$ of $\mathscr{G}$ (so the same holds for all Cartan k-subgroups of $\mathscr{G}$) and (ii) $\mathscr{G}$ shares some common invariants with G, such as the same K/k and the same root datum. In particular, $G^{\mathrm{ss}}_K \to \mathscr{G}^{\mathrm{ss}}_K$ is an isomorphism, so $i_{\mathscr{G}}(\mathscr{G}) = i_G(G)$. An absolutely pseudo-simple k-group $\mathscr{G}$ is called *of minimal type* if it satisfies (i).

If G is of minimal type with associated invariants $(G', K/k)$ and if $C = Z_G(T)$ is a Cartan k-subgroup of G then for the maximal K-torus $T' = T_K \subseteq G^{\mathrm{ss}}_K = G'$ we see that i_G carries C isomorphically onto a smooth connected k-subgroup of $\mathrm{R}_{K/k}(T')$ containing the maximal k-torus $T \subseteq \mathrm{R}_{K/k}(T')$. Hence, an advantage of the minimal type case for classification purposes with a given $(G', K/k)$ is that the structure of Cartan k-subgroups is more accessible, due to such C occurring among the smooth connected k-subgroups of $\mathrm{R}_{K/k}(T')$ that contain its maximal k-torus (with T' varying through maximal K-tori of G'). The existence of a canonical minimal type central quotient in general is an important step towards the classification of absolutely pseudo-simple k-groups G. Note that such G of minimal type with $\ker i_G \neq 1$ can only exist when k is imperfect of characteristic 2 and G_{k_s} has a non-reduced root system, as otherwise the kernel $\ker i_G = \ker \xi_G$ is central in G by Proposition 5.3.10.

For technical reasons, it is convenient to define the "minimal type" condition beyond the absolutely pseudo-simple case. The following result records the properties that will underlie this notion when it is defined for arbitrary pseudo-reductive groups.

Proposition 9.4.2 *Let G be a pseudo-reductive group over a field k, and let K/k denote the (minimal) field of definition over k for the unipotent radical of $G_{\overline{k}}$. Let $T \subseteq G$ be a maximal k-torus.*

(i) *The unipotent group scheme $\mathscr{C}_G = Z_G(T) \cap \ker i_G$ is central in G and independent of T, and the quotient $\mathscr{G} := G/\mathscr{C}_G$ is pseudo-reductive with the image of T (to be identified with T) a maximal k-torus. If G is absolutely pseudo-simple then so is $\mathscr{G}$.*

(ii) *The k-subgroup $\mathscr{C}_{\mathscr{G}} \subseteq \mathscr{G}$ is trivial.*

(iii) *If T is k-split then $R(G,T) = R(\mathscr{G},T)$ as root data for $X(T)$ and the induced map between root groups $U_a \to \mathscr{U}_a$ is an isomorphism for all $a \in \Phi(G,T)$. In particular, the root system $\Phi(\mathscr{G}_{k_s}, T_{k_s})$ is non-reduced if and only if $\Phi(G_{k_s}, T_{k_s})$ is non-reduced.*

(iv) *For every maximal k-torus $\mathscr{T} \subseteq \mathscr{G}$ there is a unique maximal k-torus $T \subseteq G$ mapping onto $\mathscr{T}$, and $T \to \mathscr{T}$ is an isomorphism. This defines a bijection between the sets of maximal k-tori of G and $\mathscr{G}$, so G is pseudo-split over k if and only if $\mathscr{G}$ is pseudo-split over k.*

Obviously the formation of $\mathscr{C}_G$ is compatible with separable extension on k.

Proof We may assume $k = k_s$ (e.g., for (iv) it suffices to work over k_s due to Galois descent). Since we do not yet know that $\mathscr{C}_G$ is intrinsic to G, we shall temporarily denote it as $\mathscr{C}$. The unipotence of the group scheme $\mathscr{C}$ implies that $T \cap \mathscr{C} = 1$, so T can be viewed as a maximal k-torus of $\mathscr{G}$ once it is known that $\mathscr{C}$ is normal in G. Thus, (i) implies (iii) by Proposition 2.3.15. Also, the implication (i) $\Rightarrow$ (ii) is immediate via Lemma 9.2.1. Thus, we just need to prove (i) and (iv). Since all maximal k-tori in G are $G(k)$-conjugate (as $k = k_s$), once centrality of $\mathscr{C}$ is proved it will follow that $\mathscr{C}$ is independent of T. The centrality of $\mathscr{C}$ and pseudo-reductivity of $\mathscr{G}$ in (i), as well as the assertions in (iv), are all that remain to be proved (since then absolute pseudo-simplicity of G implies the same for $\mathscr{G}$ due to the natural isomorphism $G_{\overline{k}}^{\rm red} \simeq \mathscr{G}_{\overline{k}}^{\rm red}$ and the perfectness of G).

The centrality of $\mathscr{C}$ in G is reduced to the fact that the k-subgroup $\mathscr{C}$ of the commutative Cartan k-subgroup $Z_G(T)$ centralizes all root groups U_a for $a \in \Phi(G,T)$, since these root groups and $Z_G(T)$ generate G. The case of multipliable roots (necessarily in characteristic 2) was handled in Proposition 9.3.7, so we only need to consider non-multipliable a. The map $i_G : G \to \mathrm{R}_{K/k}(G')$ identifies T with the maximal k-torus of the Cartan k-subgroup $\mathrm{R}_{K/k}(T')$ of $\mathrm{R}_{K/k}(G')$, where $T' = T_K$. Letting

$$\Phi(G',T') = \Phi(\mathrm{R}_{K/k}(G'),T)$$

(viewed inside $X(T')$ via the canonical identification $X(T) = X(T_K)$), by Theorem 2.3.10 we see that $\Phi(G,T)$ is the (disjoint) union of $\Phi(G',T')$ and the set of elements $c \in \Phi(G,T)$ such that $2c \in \Phi(G',T')$. In particular, $\Phi(G',T')$ is the set of non-multipliable elements in $\Phi(G,T)$. If $a \in \Phi(G,T)$ is non-multipliable then the map i_G restricts to a k-homomorphism $i_{G,a} : U_a \to \mathrm{R}_{K/k}(U'_a)$ that is equivariant with respect to $Z_G(T) \to \mathrm{R}_{K/k}(T')$.

For any $\chi \in X(T)$ and the associated character $\chi_K \in X(T')$ via the identification of T' with T_K, the diagram

$$\begin{array}{ccc} T & \longrightarrow & R_{K/k}(T') \\ {\scriptstyle\chi}\downarrow & & \downarrow{\scriptstyle R_{K/k}(\chi_K)} \\ GL_1 & \longrightarrow & R_{K/k}(GL_1) \end{array}$$

with canonical horizontal maps visibly commutes. Since the identification of $X(T)$ with $X(T')$ underlies the identification of $\Phi(G', T')$ with a subset of $\Phi(G, T)$, we see that for non-multipliable a the homomorphism $i_{G,a} : U_a \to R_{K/k}(U'_a)$ induced by i_G is linear over $\underline{k} \hookrightarrow \underline{K}$ with respect to the unique T-equivariant linear structure on the vector group U_a and the canonical $R_{K/k}(T')$-equivariant $\underline{K}$-module structure on $R_{K/k}(U'_a)$. Thus, for such a the map $i_{G,a}$ corresponds to a linear homomorphism between finite-dimensional k-vector spaces, so the group scheme $\ker(i_{G,a})$ is smooth and connected. But $\ker i_G$ contains no nontrivial smooth connected k-subgroups, so $\ker(i_{G,a}) = 1$. Hence, $\mathscr{C}$ centralizes U_a for non-multipliable a since the k-group $i_G(\mathscr{C}) = 1$ centralizes $R_{K/k}(U'_a)$. The centrality of $\mathscr{C}$ in G is now established, so $\mathscr{C}$ is independent of T and thus we may and do denote it as $\mathscr{C}_G$.

Now we prove that $\mathscr{G}$ is pseudo-reductive. Identify T as a maximal k-torus in the central quotient $\mathscr{G}$ of G modulo the unipotent group scheme $\mathscr{C}_G$, so $Z_{\mathscr{G}}(T)$ is a Cartan k-subgroup of $\mathscr{G}$. In view of the definition of $\mathscr{G}$, it follows that $Z_{\mathscr{G}}(T)$ maps isomorphically onto its image in $i_G(G)$, and this image must be a Cartan k-subgroup since $i_G(G)$ is compatibly a quotient of G and $\mathscr{G}$. But $i_G(G)$ is pseudo-reductive by Proposition 4.2.5, and Cartan k-subgroups of pseudo-reductive k-groups are pseudo-reductive, so $Z_{\mathscr{G}}(T)$ is pseudo-reductive. Thus, $\mathscr{G}$ is pseudo-reductive by Lemma 9.4.1.

Finally, we prove (iv). It suffices to prove the following general assertion concerning central extensions of tori. Let T be a torus over a field F, and let U be a commutative unipotent F-group scheme. We claim that for any central extension

$$1 \to U \to E \to T \to 1$$

with affine F-group schemes of finite type, E is commutative and the extension is uniquely split. The uniqueness of the splitting is clear, since $\mathrm{Hom}(T, U) = 1$, so by Galois descent we may assume $F = F_s$. The commutator pairing $E \times E \to E$ factors through a bi-additive pairing $B : T \times T \to U$ of F-group schemes (since $T = E/U$ in the sense of fppf group sheaves with U central in

E), and for any $t \in T(F)$ the homomorphism $B(t, \cdot) : T \to U$ must be trivial. The Zariski-density of $T(F)$ in the smooth T then implies that $B = 0$, so E is commutative. If $\mathrm{char}(F) = 0$ then U is smooth and connected, so E is also smooth and connected, and hence the general structure of smooth connected commutative affine F-groups provides the splitting.

Now assume $\mathrm{char}(F) = p > 0$. For any $n \geqslant 1$ not divisible by p, we may form the F-subgroup scheme $E[n]$ in E. By the snake lemma, $E[n] \to T[n]$ is an isomorphism (since $n : U \to U$ is an isomorphism). The Zariski closure in E of the union of the finite constant F-subgroups $E[n]$ is a smooth commutative F-subgroup Z such that the p-power endomorphism of Z is surjective (as it is an automorphism on the members of the Zariski-dense collection of finite constant F-subgroups $E[n]$). This forces $\mathscr{R}_u(Z_{\overline{F}}) = 1$ and the finite étale Z/Z^0 to have order not divisible by p, so Z^0 is a torus and $Z \cap U = 1$. Thus, $Z \to T$ has trivial scheme-theoretic kernel. But by design, the map $Z \to T$ is surjective (its image contains $T[n]$ for all n not divisible by p), so it is an isomorphism. □

The passage between G and its central quotient $\mathscr{G}$ is rather mild, due to:

Corollary 9.4.3 *The field of definition over k for the geometric unipotent radical of $\mathscr{G}$ is K, and the induced quotient map $G_K/\mathscr{R}_{u,K}(G_K) \to \mathscr{G}_K/\mathscr{R}_{u,K}(\mathscr{G}_K)$ is an isomorphism. In particular, i_G factors as the composition of the central quotient map $G \twoheadrightarrow \mathscr{G}$ and the map $i_{\mathscr{G}}$.*

Proof Let $N = \mathscr{C}_G$ in Lemma 9.2.1. □

Definition 9.4.4 A pseudo-reductive group G over a field k is of *minimal type* when $\mathscr{C}_G = 1$; i.e., i_G carries some (equivalently, any) Cartan k-subgroup of G isomorphically onto its image in $i_G(G)$.

It is clear that if k' is a separable extension of k then G is of minimal type if and only if $G_{k'}$ is of minimal type. Over any imperfect field k, there are pseudo-reductive k-groups that are *not* of minimal type (see Examples 1.6.3 and 5.3.7 for commutative and standard absolutely pseudo-simple examples respectively, and Example 9.1.10 for non-standard absolutely pseudo-simple examples in characteristic 2, with $\ker i_G$ central in each case). Such groups are rare when $\mathrm{char}(k) \neq 2$.

Proposition 9.4.5 *Let G be a pseudo-split pseudo-reductive group over a field k, with K the field of definition over k for the geometric unipotent radical, and let $T \subseteq G$ be a split maximal k-torus. Let $a \in \Phi(G, T)$ be a root. Let N be a smooth connected normal k-subgroup of G, and μ a closed k-subgroup scheme* *of T. Letting H denote either N or $Z_G(\mu)^0$, we have $\mathscr{C}_H = H \cap \mathscr{C}_G$. In particular, if G is of minimal type then so is any such H, so G_a is of minimal type.*

Note that $Z_G(\mu)^0$ is pseudo-reductive by Proposition A.8.14(2) (and $T \cap N$ is a maximal k-torus in N by Corollary A.2.7, visibly split since T is).

Proof Proposition 3.4.1 describes G_a as the derived group of the identity component of the centralizer of a closed k-subgroup scheme of T, so the final assertion concerning G_a is a consequence of the others. To prove that $\mathscr{C}_H = H \cap \mathscr{C}_G$ for any H as in the statement of the proposition, we first check that $\mathscr{R}_u(H_{\overline{k}}) = H_{\overline{k}} \cap \mathscr{R}_u(G_{\overline{k}})$, and then use this to prove $\ker i_H = H \cap \ker i_G$ (and then compare suitable Cartan k-subgroups).

The equality $\mathscr{R}_u(H_{\overline{k}}) = H_{\overline{k}} \cap \mathscr{R}_u(G_{\overline{k}})$ for $H = N$ is an application of Proposition A.4.8 over $\overline{k}$, and for $H = Z_G(\mu)^0$ it is an application of Proposition A.8.14(2) over the perfect field $\overline{k}$ (since $(Z_G(\mu)^0)_{\overline{k}} = Z_{G_{\overline{k}}}(\mu_{\overline{k}})^0$). The k-group $\ker i_H$ contains every closed k-subgroup of H whose $\overline{k}$-fiber is contained in $\mathscr{R}_u(H_{\overline{k}}) = H_{\overline{k}} \cap \mathscr{R}_u(G_{\overline{k}})$, so it contains $H \cap \ker i_G$. On the other hand, the $\overline{k}$-fiber of $\ker i_H$ is contained in $\mathscr{R}_u(H_{\overline{k}}) \subseteq \mathscr{R}_u(G_{\overline{k}})$, so $\ker i_H \subseteq \ker i_G$. Thus, $\ker i_H = H \cap \ker i_G$. It remains to show that $T \cap H$ is a maximal k-torus of H and $H \cap Z_G(T) = Z_H(T \cap H)$. The case $H = Z_G(\mu)^0$ is obvious since $Z_G(T) \subseteq Z_G(\mu)^0$. On the other hand, if $H = N$ then Corollary A.2.7 implies that $T \cap N$ is a maximal torus of N, and the equality $N \cap Z_G(T) = Z_N(T \cap N)$ follows from Lemma 1.2.5(ii), according to which T is an almost direct product of $T \cap \mathscr{D}(N)$ and the maximal k-subtorus of T that centralizes N. □

By Proposition 9.4.2 and Corollary 9.4.3, any pseudo-reductive group G over a field k is canonically a central extension of a pseudo-reductive $\mathscr{G}$ of minimal type that has the same root datum as G over k_s (with $\mathscr{G}$ absolutely pseudo-simple when G is). Moreover, by Proposition 2.3.15 the structure of the root groups of G is encoded in that of its central quotient $\mathscr{G}$. Hence, the tasks of classifying absolutely pseudo-simple G over separably closed fields (or over general fields under the pseudo-split hypothesis) and understanding its internal structure fall into two parts: classifying those G of minimal type, and classifying smooth connected central extensions of the pseudo-split minimal type absolutely pseudo-simple k-groups by commutative unipotent k-group schemes containing no nontrivial smooth connected k-subgroups (subject to perfectness of the central extension). This second part seems to be difficult in general, but often it is straightforward to reduce general problems to the study of "mild" central quotients, and so to pass to the minimal type case. The proof of the following result illustrates this principle.

Proposition 9.4.6 *Let G be a pseudo-split absolutely pseudo-simple group over a field k, with K the field of definition over k for the geometric unipotent radical, and let $T \subseteq G$ be a split maximal k-torus. Let $a \in \Phi(G,T)$ be a root. Assume that either* $\mathrm{char}(k) \neq 2$ *or a is multipliable. The field of definition K_a/k for the geometric unipotent radical of G_a is contained in K'_a. If G has rank 1 then $K'_a = K_a = K$.*

The definition of K'_a is given in terms of Galois descent through k_s/k in Definition 9.3.1.

Proof To prove a containment among subextensions of the purely inseparable extension K/k it is harmless to check after applying $k_s \otimes_k (\cdot)$, so we may and do assume $k = k_s$. In the initial part of the arguments below, a may be any root.

Consider the central quotient $\mathscr{G} = G/\mathscr{C}_G$ of minimal type. The quotient map $\pi : G \to \mathscr{G}$ carries T isomorphically onto a split maximal k-torus $\mathscr{T}$ of $\mathscr{G}$ since $\mathscr{C}_G$ is unipotent, and the field of definition over k for the geometric unipotent radical of $\mathscr{G}$ is K by Corollary 9.4.3. Under the equality of root data in Proposition 9.4.2(iii), the subfield K'_a of K is the same for (G,T) and $(\mathscr{G},\mathscr{T})$ due to Corollary 9.4.3 and the definition of this field for each pair (see the discussion of equivalent variations on this definition in Remark 9.3.2).

Applying Corollary 9.4.3 to G_a, the purely inseparable finite extension K_a of k is unaffected by replacing (G,T) with $(\mathscr{G},\mathscr{T})$ since $\mathscr{C}_{G_a} = G_a \cap \mathscr{C}_G$ by Proposition 9.4.5. Thus, we may assume G is of minimal type.

It remains to treat G of minimal type. By Proposition 9.4.5, G_a is of minimal type. By Proposition 3.4.1(1), the k-group G_a is absolutely pseudo-simple and $T \cap G_a$ is a 1-dimensional split maximal k-torus of G_a. Using the explicit description of G_a provided by Proposition 3.4.1(2), an application of the formula "$Z_H(T \cap H) = H \cap Z_G(T)$" from the proof of Proposition 9.4.5 gives that the centralizer $Z_{G_a}(T) = G_a \cap Z_G(T)$ is equal to $Z_{G_a}(T \cap G_a)$ with $T \cap G_a$ a 1-dimensional split maximal k-torus in G_a. Since the subfield $K'_a := k[\chi_a(Z_G(T)(k))] \subseteq K$ contains the subfield

$$k[\chi_a(Z_{G_a}(T)(k))] = k[\chi_a(Z_{G_a}(T \cap G_a)(k))] \subseteq K_a \subseteq K,$$

by using Proposition 6.3.1 we may replace G with G_a (thereby possibly shrinking K'_a) to reduce to the case $G = G_a$. The root system $\Phi(G,T)$ equals either $\{\pm a\}$ or $\{\pm a, \pm 2a\}$ and obviously $K_a = K$. Moreover, $G' := G_K^{\mathrm{ss}}$ is isomorphic to PGL_2 or SL_2 (if a is multipliable then G' is isomorphic to SL_2) with T_K corresponding to the diagonal K-torus D and the root group $U'_a \subseteq G'$ corresponding to the subgroup of upper triangular unipotent matrices.

It suffices in such cases to prove that $K'_a = K$, which is to say that $\chi_a(Z_G(T)(k))$ generates K over k.

We first treat the case when a is multipliable (so $\mathrm{char}(k) = 2$ and $G' = \mathrm{SL}_2$). The image $H = i_G(G) \subseteq \mathrm{R}_{K/k}(G') = \mathrm{R}_{K/k}(\mathrm{SL}_2)$ is an absolutely pseudo-simple k-subgroup containing the diagonal k-torus $D_0 \subseteq \mathrm{R}_{K/k}(D)$, and its geometric unipotent radical has field of definition over k equal to K (by Lemma 9.2.1). Thus, by Proposition 9.1.5, if the isomorphism $G' \simeq \mathrm{SL}_2$ is chosen appropriately then $H = H_{V,K/k}$ for a kK^2-subspace $V \subseteq K$ that contains 1 and generates K as a k-algebra. Hence, it suffices to show that $V - \{0\} \subseteq \chi_a(Z_G(T)(k))$ inside K.

The Cartan k-subgroup $Z_H(D_0)$ is the image of $Z_G(T)$ under i_G, and the surjective homomorphism $Z_G(T) \twoheadrightarrow Z_H(D_0)$ via i_G is an isomorphism since G is of minimal type. In view of the definition of χ_a and the fact that a is multipliable, it is obvious that $\chi_a(Z_G(T)(k)) = Z_H(D_0)(k)$ as a subgroup of $D_0(K) = D(K) = K^\times$, where the final equality is the K-isomorphism $a_K : D_K \simeq \mathrm{GL}_1$ defined by $\mathrm{diag}(x, 1/x) \mapsto x$. Hence, we are reduced to proving that $V - \{0\}$ lies in the group of first diagonal entries of elements of $Z_H(D_0)(k)$. By Proposition 9.1.4, $Z_H(D_0)$ is identified via a_K with the Zariski closure in $\mathrm{R}_{K/k}(\mathrm{GL}_1)$ of the subgroup generated by $V - \{0\} \subseteq K^\times = \mathrm{R}_{K/k}(\mathrm{GL}_1)(k)$, so we are done in the case of a multipliable root.

Now we may suppose $\mathrm{char}(k) \neq 2$. Since G' is K-isomorphic to SL_2 or PGL_2, the map $i_G : G \to \mathrm{R}_{K/k}(G')$ is an isomorphism by Theorem 6.1.1. Thus, if $G' \simeq \mathrm{PGL}_2$ then $\chi_a(Z_G(T)(k)) = K^\times$ and if $G' \simeq \mathrm{SL}_2$ then $\chi_a(Z_G(T)(k)) = (K^\times)^2$. In either case, $\chi_a(Z_G(T)(k))$ generates K over k (using in the SL_2 case that $k[K^2] = K$ since $2x = (x+1)^2 - x^2 - 1$ and $\mathrm{char}(k) \neq 2$). □

The proof of the following important general result occupies much of the rest of §9.4:

Theorem 9.4.7 *Let G be a pseudo-reductive group of minimal type. The k-group* $\ker i_G$ *is connected and commutative,* $\ker i_G = 1$ *if and only if G_{k_s} has a reduced root system, and if* $\ker i_G \neq 1$ *then it is non-central. If G contains a split maximal k-torus T then* $\mathrm{Lie}(\ker i_G)$ *is the span of the weight spaces for the multipliable roots in $\Phi(G,T)$ and* $\ker i_G$ *is the direct product of its intersections with the root groups for the multipliable roots. In particular,* $\ker i_G \subseteq \mathscr{D}(G)$.

Keep in mind that if a is a multipliable root then $\mathrm{Lie}(U_a)$ is the span of the root spaces for a and $2a$. For any $n \geqslant 1$ and imperfect field k of characteristic 2, Theorem 9.8.1 will provide a minimal type pseudo-split absolutely pseudo-simple k-group G of rank n with $\ker i_G \neq 1$.

Proof We may and do assume $k = k_s$ and $G \neq 1$. Let T be a maximal k-torus of G (so $T \neq 1$ and T is k-split). The containment $\ker i_G \subseteq \mathscr{D}(G)$ is immediate once the direct product description of $\ker i_G$ is proved because any root group U_a is contained in $\mathscr{D}(G)$. (Indeed, U_a is a T-equivariant direct product of copies $\mathbf{G}_{\rm a}(\chi_j)$ of $\mathbf{G}_{\rm a}$ on which T acts through nontrivial characters χ_j of T, and $tut^{-1}u^{-1} = \chi_j(t)u - u$ for $t \in T$ and $u \in \mathbf{G}_{\rm a}(\chi_j)$.)

Choose a cocharacter λ of T such that $\langle \chi, \lambda \rangle \neq 0$ for all $\chi \in \Phi(G,T)$, so the inclusion $Z_G(T) \subseteq Z_G(\lambda)$ is an equality due to connectedness and Lie algebra considerations. As G is of minimal type, so $(\ker i_G) \cap Z_G(T) = 1$, the GL_1-action on the T-stable $\ker i_G$ via λ satisfies $Z_{\ker i_G}(\lambda) = 1$ (notation as in Remark 2.1.11 applied to $\ker i_G$ with its GL_1-action). By applying Proposition 2.1.12(2) to the unipotent smooth *connected* $\overline{k}$-group $\mathscr{R}_u(G_{\overline{k}})$ and its closed subgroup $H = (\ker i_G)_{\overline{k}}$, we deduce from the triviality of $Z_H(\lambda_{\overline{k}})$ and the resulting equality $P_H(\lambda_{\overline{k}}) = U_H(\lambda_{\overline{k}})$ that the multiplication map

$$U_{\ker i_G}(-\lambda) \times U_{\ker i_G}(\lambda) \to \ker i_G \tag{9.4.1}$$

is an isomorphism of pointed schemes. Thus, by Proposition 2.1.8(4), $\ker i_G$ is connected. Hence $\ker i_G \neq 1$ if and only if $\mathrm{Lie}(\ker i_G) \neq 0$, and in such cases $\ker i_G$ is non-central because T does not centralize it, as

$$\mathrm{Lie}(\ker i_G) \cap \mathfrak{g}^T = \mathrm{Lie}((\ker i_G)^T) = 0,$$

due to G being of minimal type.

Enumerate the set of non-divisible roots in $\Phi(G,T)_{\lambda>0}$ as $\{a_1, \dots, a_n\}$, so the multiplication map $\prod U_{a_j} \to U_G(\lambda)$ is an isomorphism. Letting K/k be the minimal field of definition over k for the geometric unipotent radical of G, $T' := T_K$ is identified with a maximal K-torus in $G' := G_K^{\rm red}$ and the reduced root system $\Phi(G', T')$ is identified with the set of non-multipliable roots in $\Phi(G,T)$. For each non-divisible root $a \in \Phi(G,T)$ we define the root $a' \in \Phi(G', T')$ to be a_K when a is non-multipliable and to be $2a_K$ when a is multipliable, so $\{a'_1, \dots, a'_n\}$ is an enumeration of the positive system of roots $\Phi(G', T')_{\lambda_K>0}$ and the map $i_G : G \to \mathrm{R}_{K/k}(G')$ carries U_a into $\mathrm{R}_{K/k}(U'_{a'})$ for all $a \in \Phi(G,T)$. In particular, if a is multipliable then the entire a-root space in $\mathrm{Lie}(U_a)$ is contained in $\ker(\mathrm{Lie}(i_G)) = \mathrm{Lie}(\ker i_G)$ due to the equivariance of $\mathrm{Lie}(i_G) : \mathrm{Lie}(U_a) \to \mathrm{Lie}(U'_{2a_K})$ with respect to $T \to \mathrm{R}_{K/k}(T')$ and $k \to K$, and in general i_G carries $U_G(\lambda) = \prod U_{a_j}$ into $\mathrm{R}_{K/k}(U_{G'}(\lambda_K)) = \prod \mathrm{R}_{K/k}(U'_{a'_j})$ (and similarly for $-\lambda$ using $\{-a'_j\}$). Thus,

$$U_{\ker i_G}(\pm\lambda) = (\ker i_G) \cap U_G(\pm\lambda) = \prod_j ((\ker i_G) \cap U_{\pm a_j}).$$

We conclude via (9.4.1) that $\ker i_G$ is directly spanned by its k-subgroup schemes $(\ker i_G) \cap U_a$ for non-divisible $a \in \Phi(G,T)$ by putting the negative roots ahead of all positive roots (with whatever enumeration we wish among the negative roots and among the positive roots).

Consider non-multipliable $a \in \Phi(G,T)$ (so $a_K \in \Phi(G',T')$). The quotient map $G_K \to G'$ must carry $(U_a)_K$ into the 1-dimensional a-root group via a map that is linear with respect to the unique T'-equivariant linear structures on these vector groups. Thus, the map $i_G : G \to \mathrm{R}_{K/k}(G')$ restricts to a *linear* map $U_a \to \mathrm{R}_{K/k}(U_a)$. By linearity, the kernel $\ker(i_G|_{U_a}) = (\ker i_G) \cap U_a$ is smooth and connected. Since $\ker i_G$ contains no nontrivial smooth connected k-subgroup, it follows that $(\ker i_G) \cap U_a = 1$ for non-multipliable a.

We conclude that $\ker i_G = 1$ if $\Phi(G,T)$ is reduced, and that if $\Phi(G,T)$ is non-reduced (so $\mathrm{char}(k) = 2$ and k is imperfect) then $\ker i_G$ is directly spanned by its closed k-subgroups $(\ker i_G) \cap U_a$ for multipliable $a \in \Phi(G,T)$ (with negative roots put ahead of all positive roots). For multipliable a we have $(\ker i_G) \cap U_{2a} = 1$ (since $2a$ is non-multipliable) and the entire a-root space has already been seen to lie inside $\mathrm{Lie}(\ker i_G)$, so $\mathrm{Lie}(\ker i_G)$ is spanned by the root spaces for the multipliable roots. In particular, $\ker i_G \neq 1$ whenever $\Phi(G,T)$ is non-reduced. Since each U_a is commutative (Proposition 2.3.11), our remaining problem is to show that the k-subgroups $(\ker i_G) \cap U_a$ for multipliable a pairwise commute with each other.

For multipliable a, the quotient map $G_K \to G'$ carries U_a onto the 1-dimensional a'-root group, where $a' = 2a_K$ is a long root in an irreducible factor of $\Phi(G',T')$ of type C. If b is a multipliable root distinct from $\pm a$ (so b and a are linearly independent) then the root groups of (G',T') for the roots a' and b' commute with each other (this is obvious if a' and b' lie in distinct irreducible components of the root system, and otherwise follows from the fact that in an irreducible root system of type C the sum of two linearly independent long roots is not a root). Hence, for such a and b the smooth connected commutator k-subgroup (U_a, U_b) is trivial by Lemma 1.2.1.

It remains to show that $(\ker i_G) \cap U_a$ commutes with $(\ker i_G) \cap U_{-a}$ for multipliable $a \in \Phi(G,T)$. Since $U_{\pm a}$ is also the $\pm a$-root group of $G_a = \langle U_a, U_{-a}\rangle$ relative to its maximal torus $T \cap G_a$, and $(\ker i_G) \cap G_a = \ker i_{G_a}$ with G_a of minimal type (see Proposition 9.4.5 and its proof), we may replace (G,T) with $(G_a, T \cap G_a)$. Now G has rank 1 and its root system is non-reduced (i.e., BC_1) with a as a multipliable root, and $G' \simeq \mathrm{Sp}_2 = \mathrm{SL}_2$. Since $i_G(N_G(T)(k))$ contains an element representing the nontrivial class in $W(G',T')$ (see (9.2.1)), this latter isomorphism can be chosen so that T' goes over to the diagonal K-torus D, the $\pm a'$-root group of $G' = \mathrm{SL}_2$ is the upper/lower triangular unipotent subgroup $U^{\pm}$, and a chosen element $\nu \in$

$N_G(T)(k) - Z_G(T)(k)$ goes over to the standard Weyl element $\left(\begin{smallmatrix}0 & 1\\ -1 & 0\end{smallmatrix}\right)$ (and the element $\nu^2 \in Z_G(T)(k)$ is trivial since $\left(\begin{smallmatrix}0 & 1\\ -1 & 0\end{smallmatrix}\right)^2 = 1$, as $\mathrm{char}(k) = 2$). The resulting isomorphism $T' \simeq D = \mathrm{GL}_1$ (using the standard identification of GL_1 with D via $\lambda \mapsto \mathrm{diag}(\lambda, 1/\lambda)$) induces an *inclusion* of $Z_G(T)$ into $\mathrm{R}_{K/k}(\mathrm{GL}_1)$ since G is of minimal type.

The map $i_G : G \to \mathrm{R}_{K/k}(G') = \mathrm{R}_{K/k}(\mathrm{SL}_2)$ carries $Z_G(T)$ into $\mathrm{R}_{K/k}(D)$ and carries $U_{\pm a}$ into $\mathrm{R}_{K/k}(U^{\pm})$. By Propositions 9.3.6, 9.3.7, and 9.4.6, we may construct a $Z_G(T)$-equivariant identification $u_+ : \underline{V} \times \underline{V'} \simeq U_a$, where V and V' are finite-dimensional nonzero vector spaces over K and kK^2 respectively, such that $Z_G(T)$ acts on $\underline{V}$ through

$$\chi_a : Z_G(T) \xrightarrow{i_G} \mathrm{R}_{K/k}(T_K) \xrightarrow{\mathrm{R}_{K/k}(a_K)} \mathrm{R}_{K/k}(\mathrm{GL}_1) \tag{9.4.2}$$

and acts on $\underline{V'}$ through χ_a^2, with $K = k[\chi_a(Z_G(T)(k))]$. The effect of ν-conjugation on $Z_G(T)$ is inversion (as it suffices to check this on the maximal k-torus T since $Z_G(T)$ is commutative and pseudo-reductive), so it swaps U_{-a} and U_a. Composing this isomorphism $U_a \simeq U_{-a}$ with u_+, we get an identification $u_- : \underline{V} \times \underline{V'} \simeq U_{-a}$ under which the $Z_G(T)$-action on U_{-a} becomes scaling by $\chi_{-a} = 1/\chi_a$ on $\underline{V}$ and scaling by χ_{-a}^2 on $\underline{V'}$.

Identifying U_{-a} and U_a via ν-conjugation is i_G-compatible with identifying U^- and U^+ via conjugation by the standard Weyl element. This latter identification swaps the standard parameterizations $x_{\pm} : \mathbf{G}_\mathrm{a} \simeq U^{\pm}$ (i.e., no sign intervention) since $\mathrm{char}(k) = 2$. The subset $\chi_a^2(Z_G(T)(k)) \subseteq kK^2$ contains a Zariski-dense set of k-points in $\mathbf{G}_\mathrm{a}$ and generates kK^2 as a k-algebra, so the role of i_G in the definition of $\chi_{\pm a}^2$ implies that the maps $i_G : \underline{V'} = U_{\pm 2a} \to \mathrm{R}_{K/k}(U^{\pm}) = \mathrm{R}_{K/k}(\mathbf{G}_\mathrm{a}) =: \underline{K}$ correspond to a common *linear* inclusion of kK^2-vector spaces $V' \hookrightarrow K$. In this way we identify V' with a kK^2-subspace of K. Similarly, the maps $i_G : \underline{V} \hookrightarrow U_{\pm a} \to \mathrm{R}_{K/k}(U^{\pm}) = \underline{K}$ correspond to a common additive injection $q : V \to K$, and by Proposition 9.3.7 we may arrange that this map is 2-linear over K in the sense that $q(\lambda v) = \lambda^2 q(v)$ for $\lambda \in K$ and $v \in V$. Define the k-homomorphism $\underline{q} : \underline{V} \to \underline{K}$ on points in a k-algebra R by $\underline{q}(r \otimes v) = q(v) \otimes r^2 \in K \otimes_k R$, so with respect to a K-basis $\{e_1, \ldots, e_r\}$ of V we have $\underline{q}(x_1, \ldots, x_r) = \sum c_i x_i^2$ for $x_i \in \underline{K}$ and some $c_i \in K$.

Let $h_{\pm} : U_{\pm a} \to \mathrm{R}_{K/k}(U^{\pm}) = \mathrm{R}_{K/k}(\mathbf{G}_\mathrm{a})$ be the natural maps, so the composite maps $h_{\pm} \circ u_{\pm}$ are each identified with the map $\underline{V} \times \underline{V'} \to \mathrm{R}_{K/k}(\mathbf{G}_\mathrm{a})$ defined by $(v, v') \mapsto \underline{q}(v) + v'$.

Lemma 9.4.8 *For points (v, v') and (w, w') of $\underline{V} \times \underline{V'}$ valued in a common k-algebra, define the points $x = \underline{q}(v) + v'$ and $y = \underline{q}(w) + w'$ of $\mathrm{R}_{K/k}(\mathbf{G}_\mathrm{a})$.*

On the dense open locus in $(\underline{V} \times \underline{V}')^2$ *defined by the condition that* $1 + xy$ *is a unit, the commutator*

$$u_-(v, v')u_+(w, w')u_-(v, v')^{-1}u_+(w, w')^{-1}$$

is valued in the open cell $\Omega = U_{-a} \times Z_G(T) \times U_a$ *with respective components*

$$u_-\left(\frac{xy}{1+xy}v + \frac{x}{1+xy}w, \left(\frac{xy}{1+xy}\right)^2 v' + \left(\frac{x}{1+xy}\right)^2 w'\right) \in U_{-a},$$

$$1 + xy \in Z_G(T) \subseteq \mathrm{R}_{K/k}(\mathrm{GL}_1),$$

$$u_+\left(\frac{y}{1+xy}v + \frac{xy}{1+xy}w, \left(\frac{y}{1+xy}\right)^2 v' + \left(\frac{xy}{1+xy}\right)^2 w'\right) \in U_a.$$

Note that the formulas for the components in $U_{\pm a}$ make sense since $\underline{V}'(A)$ is a $kK^2 \otimes_k A$-submodule of $\mathrm{R}_{K/k}(\mathbf{G}_\mathrm{a})(A) = K \otimes_k A$ for any k-algebra A.

Proof The image Ω' of Ω_K in $G' = \mathrm{SL}_2$ is the open cell $U^- \times D \times U^+$. It is shown in the proof of Proposition 3.4.2 that $\Omega_{\overline{k}}$ is the preimage of $\Omega'_{\overline{k}}$ under $G_{\overline{k}} \to G_{\overline{k}}^{\mathrm{red}}$, so Ω_K is the preimage of Ω' and hence $\mathrm{R}_{K/k}(\Omega_K)$ is the preimage of $\mathrm{R}_{K/k}(\Omega')$ in $\mathrm{R}_{K/k}(G_K)$. Since $G \cap \mathrm{R}_{K/k}(\Omega_K) = \Omega$, we conclude that $\Omega = i_G^{-1}(\mathrm{R}_{K/k}(\Omega'))$.

We are proposing an equality of two maps

$$\{((v, v'), (w, w')) \in (\underline{V} \times \underline{V}')^2 \mid 1 + x(v, v')y(w, w') \in \mathrm{R}_{K/k}(\mathrm{GL}_1)\} \rightrightarrows G$$

between smooth k-schemes, and that on this open domain the function $1 + xy$ is valued in the closed subscheme $Z_G(T)$ of $\mathrm{R}_{K/k}(\mathrm{GL}_1)$, so it suffices to check equality on k-points (since $k = k_s$). Since i_G is injective on k-points (the finite étale normal k-subgroup $(\ker i_G)(k) \subseteq G$ is central and hence trivial, as G is of *minimal type*) and $\Omega = i_G^{-1}(\mathrm{R}_{K/k}(\Omega'))$, it suffices to compose with $i_G : G \to \mathrm{R}_{K/k}(\mathrm{SL}_2)$ and check the corresponding properties in $\mathrm{SL}_2(K)$.

The 2-linearity of the map $q : V \to K$ between K-vector spaces reduces our problem to the elementary general assertion that for $x, y \in K$ satisfying $1 - xy \in K^\times$, the commutator

$$\begin{pmatrix} 1 & 0 \\ x & 1 \end{pmatrix}\begin{pmatrix} 1 & y \\ 0 & 1 \end{pmatrix}\begin{pmatrix} 1 & 0 \\ -x & 1 \end{pmatrix}\begin{pmatrix} 1 & -y \\ 0 & 1 \end{pmatrix}$$

in $\mathrm{SL}_2(K)$ lies in the open cell $U^- \times D \times U^+$ and is equal to

$$\begin{pmatrix} 1 & 0 \\ -a(x,y) & 1 \end{pmatrix} \begin{pmatrix} 1-xy & 0 \\ 0 & 1/(1-xy) \end{pmatrix} \begin{pmatrix} 1 & a(y,x) \\ 0 & 1 \end{pmatrix}$$

with

$$a(x,y) = \frac{x^2 y}{1-xy} = -\left(\frac{xy}{1-xy}\right)^2 x + \left(\frac{x}{1-xy}\right)^2 y. \qquad \square$$

Continuing with the proof of Theorem 9.4.7, consider points $u_-(v, v') \in \ker h_-$ and $u_+(w, w') \in \ker h_+$ valued in a common k-algebra A. The associated elements $x = \underline{q}(v) + v' = h_-(u_-(v, v'))$ and $y = \underline{q}(w) + w' = h_+(u_+(w, w'))$ in $K \otimes_k A$ both vanish, so $1 + xy = 1 \in (K \otimes_k A)^\times$. Hence, we may apply the lemma to compute the commutator of $u_-(v, v')$ and $u_+(w, w')$ in $G(A)$, and obviously the formula in this case collapses to the identity element. This shows that the k-subgroup schemes $\ker h_\pm = (\ker i_G) \cap U_{\pm a}$ in G commute with each other, as desired. $\square$

We conclude our initial discussion of minimal type k-groups with a result concerning the center when the root system over k_s is non-reduced.

Proposition 9.4.9 *Let G be an absolutely pseudo-simple k-group of minimal type. If its root system over k_s is non-reduced then the (scheme-theoretic) center Z_G of G is trivial.*

Proof We may and do assume $k = k_s$. Let T be a (split) maximal k-torus of G, so $\Phi(G,T) = \mathrm{BC}_n$ for some $n \geqslant 1$. Let K/k be the field of definition for $\mathscr{R}_u(G_{\overline{k}}) \subseteq G_{\overline{k}}$ and define $G' = G_K^{\mathrm{ss}}$ with $T' := T_K \hookrightarrow G'$ a maximal K-torus. We have $G' \simeq \mathrm{Sp}_{2n}$ by Theorem 2.3.10, and the minimal type hypothesis implies that $i_G : G \to \mathrm{R}_{K/k}(G')$ restricts to an inclusion $Z_G(T) \hookrightarrow \mathrm{R}_{K/k}(T')$. Let Δ be a basis of $\Phi(G,T)$, so Δ contains a unique multipliable root a.

The quotient map $G_K \twoheadrightarrow G'$ must carry $(Z_G)_K$ into $Z_{G'}$, so i_G defines an inclusion of Z_G into $\mathrm{R}_{K/k}(Z_{G'})$. By (9.4.2), for any k-algebra R and $z \in Z_G(T)(R)$, the adjoint action of z on the a-root space is multiplication by $a(i_G(z)) \in R_K^\times$, with $R_K := K \otimes_k R$ and $i_G(z) \in T'(R_K)$. Assume $z \in Z_G$, so $a(i_G(z)) = 1$ and $i_G(z) \in \mathrm{R}_{K/k}(Z_{G'})(R) = Z_{G'}(R_K)$. Thus, $b(i_G(z)) = 1$ for every $b \in \Phi(G', T')$. But Δ is a $\mathbf{Z}$-basis of $\mathrm{X}(T')$ and $\Delta - \{a\} \subseteq \Phi(G', T')$, so $i_G(z) = 1$. This implies that $z = 1$, so $Z_G = 1$. $\square$

9.5 Computations with a multipliable root

Consider an absolutely pseudo-simple and pseudo-split group G over a field k of characteristic 2, and assume G has a non-reduced root system. (We do not yet impose a minimal type hypothesis on G.) Let K be the field of definition over k for the geometric unipotent radical of G, so $G' := G_K/\mathscr{R}_K(G_K)$ is K-isomorphic to Sp_{2n}. Let T be a split maximal k-torus of G. We fix a basis Δ of the root system $\Phi(G,T)$ of type BC_n (with $n = \dim T$). Let $c \in \Delta$ be the unique short root, so c is multipliable and Δ is a basis for the reduced root system (of type B_n) consisting of non-divisible elements of $\Phi(G,T)$ whereas

$$\Delta' := (\Delta - \{c\}) \cup \{2c\}$$

is a basis for the reduced root system $\Phi(G', T_K)$ (of type C_n) consisting of the non-multipliable elements of $\Phi(G,T)$. For each root $a \in \Phi(G,T)$, define $a' = a$ if a is not multipliable and define $a' = 2a$ if a is multipliable, so the quotient map $G_K \twoheadrightarrow G'$ carries $(U_a)_K$ onto $U'_{a'}$. There are K_a-isomorphisms $j_a : (G_a)^{\mathrm{ss}}_{K_a} \simeq \mathrm{SL}_2$ carrying $a^\vee(\mathrm{GL}_1)_{K_a}$ onto the diagonal torus since $G^{\mathrm{ss}}_{\overline{k}} \simeq \mathrm{Sp}_{2n}$ is simply connected. It is essential for later computations that these isomorphisms j_a can all be obtained from a single K-isomorphism in the following sense:

Lemma 9.5.1 *There exists a K-isomorphism $G^{\mathrm{ss}}_K \simeq \mathrm{Sp}_{2n}$ carrying T_K to the diagonal K-torus D of Sp_{2n} whose restriction to a K-isomorphism*

$$(G_a)^{\mathrm{ss}}_K = (G_K)^{\mathrm{ss}}_{a'} \simeq (\mathrm{Sp}_{2n})_{a'} = \mathrm{SL}_2$$

respects the natural K_a-structure on both sides for all $a \in \Phi(G,T)$.

The value of this compatibility is that computations inside $\mathrm{Sp}_{2n}(K)$ (governed by the numerology of the C_n root system) can keep track of rationality properties over *subfields* of K containing k such as the K_a.

Proof We shall use a Levi k-subgroup $L \subseteq G$ containing T; such a subgroup exists by Theorem 3.4.6. Use the natural isomorphism $L_K \simeq G^{\mathrm{ss}}_K$ to identify (L,T) as a k-descent of (G^{ss}_K, T_K). The identification of $\Phi(L,T)$ with the set of non-multipliable roots in $\Phi(G,T)$ thereby identifies $L_{a'}$ as a k-structure on $(G_a)^{\mathrm{ss}}_{K_a}$ for all $a \in \Phi(G,T)$. Thus, for every $a \in \Phi(G,T)$ we obtain a K_a-isomorphism $(L_{a'})_{K_a} \simeq (G_a)^{\mathrm{ss}}_{K_a}$ that is compatible with the K-isomorphism $L_K \simeq G^{\mathrm{ss}}_K$.

Choose a k-isomorphism $L \simeq \mathrm{Sp}_{2n}$ carrying T onto the diagonal k-torus. The resulting K-isomorphism $G^{\mathrm{ss}}_K \simeq L_K \simeq \mathrm{Sp}_{2n}$ has the desired properties. □

In what follows we use $(G, T, \Delta, c, \Delta')$ as at the start of §9.5 and an isomorphism built as in the proof of Lemma 9.5.1, resting on a fixed choice of Levi k-subgroup L containing T.

If $n \geqslant 2$, let b be the simple positive root adjacent to $2c$ in the Dynkin diagram of $\Phi(G_K^{\rm ss}, T_K)$. All non-multipliable roots in Δ are conjugate to b under $N_G(T)(k)$, so they have the same (purely inseparable) field of definition K_b over k for the geometric unipotent radical of the associated rank-1 pseudo-simple group. Thus, by Proposition 6.3.1(1) and the inclusion $K_{2c} \subseteq K_c$, the field K is generated over k by K_c and K_b.

Proposition 9.5.2 *If G has rank 1 then $K'_c = K_c = K$. If G has rank > 1 then $K'_c = K_b = K$, $kK^2 \subseteq K_c \subseteq K$, and the k-subgroup $U_b \hookrightarrow \mathrm{R}_{K/k}(U'_b) = \mathrm{R}_{K/k}(\mathbf{G}_{\rm a})$ is a $\underline{K_c}$-submodule.*

In particular, $K'_{2c} = kK^2$, U_c/U_{2c} is naturally a $\underline{K}$-module scheme, and U_{2c} is naturally a $\underline{kK^2}$-module scheme (see Proposition 9.3.6).

Example 9.8.18 provides G (over suitable k) such that $K_c \neq K$.

Proof The obvious equality $K'_{2c} = kK'^2_c$ reduces the determination of K'_{2c} to that of K'_c. The case of rank 1 is settled by Proposition 9.4.6 (with $a = c$) and Proposition 9.3.6, so we now assume that G has rank $n \geqslant 2$. The first step is to show that $K_b^2 \subseteq K_c \subseteq K_b = K'_c$ over k. We may and do assume $k = k_s$, so $K'_a = k[\chi_a(Z_G(T)(k))]$ for all $a \in \Phi(G, T)$.

As we saw in the proof of Proposition 9.4.6, we can pass to the minimal type central quotient $G/\mathscr{C}_G$ of G without changing the fields K_a or K'_a for any $a \in \Phi(G, T)$, and also without affecting the formation of the root groups (in view of the product structure of open cells), so we now assume that G is *of minimal type*. Thus, each G_a is also of minimal type (by Proposition 9.4.5), so i_G carries the Cartan k-subgroup $Z_{G_a}(T \cap G_a) = G_a \cap Z_G(T)$ of G_a isomorphically onto its image $C_a \subseteq i_G(G)$ that is a Cartan k-subgroup of $i_G(G_a) = i_{G_a}(G_a) \subseteq i_G(G) \subseteq \mathrm{R}_{K/k}(G_K^{\rm ss})$. We identify $Z_G(T)$ with its image $C := i_G(Z_G(T))$, and for any $a \in \Phi(G, T)$ we identify $G_a \cap Z_G(T)$ with C_a.

For a in the basis Δ of $\Phi(G, T)$, let $a' = a$ if $a \neq c$ and let $c' = 2c$, so $\Delta' = \{a' \mid a \in \Delta\}$. As we saw in the proof of Lemma 9.5.1, the image $i_{G_a}(G_a) \subseteq \mathrm{R}_{K/k}(G_K^{\rm ss}) = \mathrm{R}_{K/k}(L_K)$ is contained in the pseudo-simple k-subgroup $\mathrm{R}_{K_a/k}((L_{a'})_{K_a})$, where $L_{a'}$ is the k-subgroup of L generated by the root groups for $\pm a'$. Fix an isomorphism $L_{a'} \simeq \mathrm{SL}_2$ carrying $T \cap L_{a'}$ to the diagonal torus and carrying the a'-root group to the standard positive root group; in other words, we are choosing a pinning of $(L_{a'}, T \cap L_{a'}, \{a'\})$, or equivalently a k-isomorphism of the a'-root group of (L, T) with $\mathbf{G}_{\rm a}$. This identifies $i_{G_a}(U_a)$ with a k-subgroup of $\mathrm{R}_{K_a/k}(\mathbf{G}_a)$ that contains the canonical k-subgroup $\mathbf{G}_a$ as the a'-root group of (L, T). In particular, $1 \in (i_{G_a}(U_a))(k)$ inside $\mathrm{R}_{K_a/k}(\mathbf{G}_a)(k) = K_a$ via the chosen identification $L_{a'} = \mathrm{SL}_2$. Hence, by Proposition 9.1.4, Proposition 9.1.5, and Lemma 9.2.1, $i_{G_a}(G_a)$ coincides with the pseudo-simple k-subgroup $H_{V_a} \subseteq \mathrm{R}_{K_a/k}(\mathrm{SL}_2) = \mathrm{R}_{K_a/k}((L_{a'})_{K_a})$ associated to a kK_a^2-subspace $V_a \subseteq K_a$ containing 1 that generates K_a as a k-algebra. In particular, under the isomorphism $a'^\vee : \mathrm{GL}_1 \simeq T \cap L_{a'} \subseteq L_{a'} = \mathrm{SL}_2$ given by $x \mapsto \mathrm{diag}(x, 1/x)$, the Cartan k-subgroup $C_a = i_G(Z_{G_a}(T \cap G_a)) \subseteq H_{V_a}$ is the Zariski closure $(V_a)^*_{K_a/k} \subseteq \mathrm{R}_{K_a/k}(\mathrm{GL}_1)$ of the subgroup of $K_a^\times$ generated by $V_a - \{0\}$.

For any $a \in \Delta$, inside $T(K) \subseteq L(K)$ the elements of $C_a(k)$ are precisely the points $a'^\vee_K(t)$ for $t \in (V_a)^*_{K_a/k}(k) \subseteq K_a^\times \subseteq K^\times$. For any distinct $a_1, a_2 \in \Delta$ we conclude that the conjugation action by $C_{a_1}(k)$ on the k-points of the k-subgroup $V_{a_2} \subseteq i_{G_{a_2}}(G_{a_2})$ inside the a'_2-root group $\mathrm{R}_{K/k}(\mathbf{G}_a)$ of $(\mathrm{R}_{K/k}(L_K), T)$ (with (L, T, Δ') pinned as above) is given by the formula

$$(a'_1)^\vee(t).v_2 = t^{\langle a'_2, (a'_1)^\vee \rangle} v_2 \in K \tag{9.5.1}$$

for $t \in (V_{a_1})^*_{K_{a_1}/k}(k) \subseteq K_{a_1}^\times \subseteq K^\times$ and $v_2 \in V_{a_2} \subseteq K_{a_2} \subseteq K$. In particular, $t^{\langle a'_2, (a'_1)^\vee \rangle} V_{a_2} = V_{a_2}$ inside K for all $t \in (V_{a_1})^*_{K_{a_1}/k}(k)$. Since $1 \in V_{a_2}$, it follows that the $\langle a'_2, (a'_1)^\vee \rangle$-power map carries $V_{a_1} - \{0\}$ into $V_{a_2} \subseteq K_{a_2}$.

Since $\langle b, (2c)^\vee \rangle = -1$, we conclude that inversion carries $V_c - \{0\}$ into K_b, but $V_c - \{0\}$ is stable under inversion since $1/v = (1/v)^2 v$ and V_c is a kK_c^2-subspace of K_c. But V_c generates K_c as a k-algebra, so $K_c \subseteq K_b$ and hence $K_b = K$ (since K is generated by its subfields K_c and K_b over k). Likewise, since $\langle 2c, b^\vee \rangle = -2$ we see that $K^2 = K_b^2 \subseteq K_c$, so $kK^2 \subseteq K_c$.

To prove that $K'_c = K$ it suffices to show that $V_b - \{0\} \subseteq \chi_c(C_b(k))$ inside $K^\times$ (since V_b generates $K_b = K$ as a k-algebra and $\chi_c(Z_G(T)(k)) \subseteq K'_c \subseteq K$). More generally, we claim that the restriction of $\chi_c : Z_G(T)(k) \to K^\times$ to the subset $C_b(k) = (V_b)^*_{K_b/k}(k) \subseteq K^\times$ (inclusion via the inverse of the isomorphism $b^\vee : \mathrm{GL}_1 \simeq T \cap G_b$) is $z \mapsto 1/z$. By definition $\chi_c = \mathrm{R}_{K/k}(c_K) \circ i_G$, and i_G identifies C_b with a k-subgroup of $\mathrm{R}_{K/k}((T \cap G_b)_K)$, so since $b^\vee : \mathrm{GL}_1 \to T \cap G_b$ is a k-isomorphism we see that $\chi_c : C_b(k) \to K^\times$ is $z \mapsto z^{\langle c, b^\vee \rangle} = 1/z$.

Finally, we show that V_b is a K_c-subspace of K. Applying (9.5.1) with $a_1 = c$ and $a_2 = b$ gives that V_b is stable under scaling by $t^{\langle b, (2c)^\vee \rangle} = t^{-1}$ for all $t \in (V_c)^*_{K_c/k}(k)$. In particular, it is stable under scaling by the multiplicative inverses of elements of $V_c - \{0\}$. The identity $v = v^2 \cdot (1/v)$ therefore implies that the kK^2-subspace V_b inside K is a subspace over $k[V_c] = K_c$. □

Fix a $Z_G(T)$-equivariant splitting $U_c = V_c \times U_{2c}$ provided by Proposition 9.3.7. By Proposition 9.3.6 and the identification of K'_c in Proposition 9.5.2, the k-group $V_c = U_c/U_{2c}$ is naturally a $\underline{K}$-module scheme and U_{2c} and $i_G(U_c)$ are naturally $\underline{kK^2}$-module schemes making the natural map $U_{2c} \to i_G(U_{2c})$ linear over $\underline{kK^2}$ and the natural map $q_c : U_c/U_{2c} \to \mathrm{R}_{K/k}(U'_{2c})/i_G(U_{2c})$ linear over the squaring map $\underline{K} \to \underline{kK^2}$. The splitting provided by Proposition 9.3.7 has the property that i_G carries $U_c = V_c \times U_{2c}$ into $\mathrm{R}_{K/k}(U'_{2c}) = \mathrm{R}_{K/k}(\mathbf{G}_a)$ by a $Z_G(T)$-equivariant additive map

$$h : (v, u) \mapsto q_c(v) + \ell_{2c}(u) \tag{9.5.2}$$

where ℓ_{2c} is a $\underline{kK^2}$-linear inclusion and q_c is 2-linear over $\underline{K}$ (i.e., it is additive and $q_c(\lambda v) = \lambda^2 q_c(v)$ for all $\lambda \in \underline{K}$). Since $(\ker i_G)(k_s)$ is finite and all $T(k_s)$-orbits in $U_c(k_s) - \{0\}$ are infinite, it follows via T-equivariance that (9.5.2) is injective on k_s-points. We shall view $h(U_c(k))$ as an additive subgroup of $K = \mathrm{R}_{K/k}(\mathbf{G}_a)(k)$. This subgroup is a nonzero K^2-subspace of K.

Proposition 9.5.3 *Let G, K/k, and c be as above.*

(i) *If $[k : k^2] < \infty$, then*

$$\dim_K V_c(k) + \dim_{K^2} U_{2c}(k) \leqslant \dim_{K^2}(i_G(U_c)(k)) \leqslant [K : K^2] < \infty.$$

(ii) *The subfields K_c and $k\langle h(U_c(k))\rangle$ of K coincide.*

(iii) *If $[K : kK^2] = 2$ then $K_c = K$, $h(U_{2c}(k))$ is 1-dimensional as a kK^2-vector space, and $i_G(G_c) \simeq \mathrm{R}_{K/k}(\mathrm{SL}_2)$. If, moreover, $[k : k^2] = 2$ then $V_c(k)$ is 1-dimensional as a K-vector space.*

The hypothesis in (iii) holds if $[k : k^2] = 2$ (as then $K = k^{1/2^m}$ for some $m \geqslant 1$).

Proof First we prove (i). Finiteness of $[k : k^2]$ implies finiteness of $[F : F^2]$ for any finite extension F of k, so $[K : K^2]$ is finite. Hence, all dimensions in (i) are finite (since $q_c(V_c(k))$ is a K^2-subspace of the 1-dimensional K-vector space $U'_{2c}(K)$ and the quotient space $i_G(U_c)(k)/i_G(U_{2c})(k)$ is a K^2-subspace of $U'_{2c}(K)/i_G(U_{2c})(k)$).

Choose a K-basis $\{v_1, \ldots, v_d\}$ of $V_c(k)$ and let $\alpha_i = q_c(v_i) \in K$, so $q_c(\sum x_i v_i) = \sum \alpha_i x_i^2$ for $x_i \in K$. The map $i_G : U_c = V_c \times U_{2c} \to i_G(U_c) \subseteq \mathrm{R}_{K/k}(U'_{2c})$ is injective on k-points (since (9.5.2) is injective on k_s-points), so for any $x_1, \ldots, x_d \in K$ and $u \in U_{2c}(k)$ the element

$$\sum \alpha_i x_i^2 + \ell_{2c}(u) = i_G\left(\sum x_i v_i, u\right) \in i_G(U_c)(k) \subseteq \mathrm{R}_{K/k}(U'_{2c})(k) = U'_{2c}(K)$$

vanishes if and only if all x_i and u vanish. In other words, the union of $\{\alpha_1, \ldots, \alpha_d\}$ and a K^2-basis of $U_{2c}(k)$ is a K^2-linearly independent set in the kK^2-subspace $i_G(U_c)(k)$ of the 1-dimensional K-vector space $U'_{2c}(K)$. This proves (i).

To prove assertion (ii) we note that by Lemma 9.2.1, K_c is also the field of definition over k for the geometric unipotent radical $\mathscr{R}_u(i_G(G_c)_{\overline{k}}) \subseteq i_G(G)_{\overline{k}}$. We may choose the isomorphism $(G_K^{\mathrm{ss}}, T) \simeq (\mathrm{Sp}_{2n}, D)$ so that $1 \in \ell_{2c}(U_c(k))$. Hence, $1 \in h(U_c(k)) = i_G(U_c(k))$ inside $(G_c)_K^{\mathrm{ss}}(K) = \mathrm{SL}_2(K)$. Thus, by Proposition 9.1.5, $i_G(G_c)$ equals H_V where V is the k-linear span of the subgroup $h(U_c(k))$ of K. Proposition 9.1.4 implies now that $K_c = k[V] = k[h(U_c(k))]$. This proves (ii).

As $h(U_c(k))$ is a K^2-subspace of K containing 1, K_c is a field extension of kK^2 contained in K. Now let us assume that $[K : kK^2] = 2$, so $[K : K_c] \leqslant 2$. The 1-dimensional K-vector space $U'_{2c}(K)$ is 2-dimensional over kK^2, so its nonzero proper kK^2-subspace $h(U_{2c}(k))$ must be 1-dimensional, and then the k-subspace V spanned by $h(U_c(k))$ must be all of K. This, in particular, implies that $i_G(G_c) = \mathrm{R}_{K/k}(\mathrm{SL}_2)$ and $K_c = K$.

Finally, we assume $[k : k^2] = 2$ (so $[K : kK^2] = 2$ and hence $K_c = K$), and we show that $V_c(k)$ is 1-dimensional as a K-vector space. The k-group $V_c \simeq U_c/U_{2c}$ is nonzero and $kK^2 = K^2$, so we conclude from assertion (i) that $\dim_K V_c(k) + 1 \leqslant 2$. Hence, $V_c(k)$ is 1-dimensional as a K-vector space. □

Proposition 9.5.4 *If* $[K : kK^2] = 2$ *then the map* $\xi_G = i_G : G \to \mathrm{R}_{K/k}(G')$ *is surjective.*

The hypothesis holds when $[k : k^2] = 2$.

Proof Let $\Phi' = \Phi(G', T')$; this is the set of non-multipliable roots in $\Phi(G, T)$. Let Δ be a basis of $\Phi(G, T)$, with c the multipliable root in Δ, so $\Delta' := (\Delta - \{c\}) \cup \{2c\}$ is a basis of Φ'. The group $\mathrm{R}_{K/k}(G')$ is generated by its k-subgroups $\mathrm{R}_{K/k}(G'_{a'})$ for $a' \in \Delta'$. By Propositions 9.5.2 and 9.5.3(iii), $K_a = K$ for *all* $a \in \Delta$ (using that $[K : kK^2] = 2$ to handle $a = c$). Also, $(G_a)_K^{\mathrm{ss}} = \mathrm{SL}_2$ for all $a \in \Delta$ since $G' = \mathrm{Sp}_{2n}$ is simply connected. Thus, $\xi_{G_a} = i_{G_a}$, so the compatibility between $\xi_G|_{G_a}$ and ξ_{G_a} as in Proposition 6.3.1 reduces our problem to proving the surjectivity of ξ_{G_a} for each $a \in \Delta$.

If $a \neq c$ then G_a has rank 1 and a reduced root system, so the surjectivity of ξ_{G_a} is part of Proposition 9.2.4. Now suppose $a = c$, which is to say that we are in the rank-1 case of the original problem (with a non-reduced root system). Since $\mathrm{R}_{K/k}(G'_{2c})$ is generated by $\mathrm{R}_{K/k}(U'_{\pm 2c})$ (as $G'_{2c} = \mathrm{SL}_2$), it suffices to prove that the map

$$i_G : U_c = V_c \times U_{2c} \to \mathrm{R}_{K/k}(U'_{2c})$$

and its analogue for $-c$ are both surjective. By symmetry it suffices to treat c. So we aim to prove that the 2-linear map $V_c \to \mathrm{R}_{K/k}(U'_{2c})/i_G(U_{2c})$ from a $\underline{K}$-module to a $\underline{kK^2}$-module is surjective. The smooth image is necessarily a nonzero $\underline{kK^2}$-submodule since K^2 is Zariski-dense in $\underline{kK^2}$, and $U'_{2c}(K)/i_G(U_{2c})(k)$ has kK^2-dimension $[K : kK^2] - 1 = 1$ by Proposition 9.5.3(iii), so it remains to note that a smooth affine $\underline{kK^2}$-module corresponding to a 1-dimensional kK^2-vector space has no nontrivial smooth proper $\underline{kK^2}$-submodules (due to Lemma 9.3.5). □

9.6 Birational group law preparations

Our method of construction of basic exotic pseudo-reductive groups in Chapter 7 rests on using the degeneration of Chevalley commutation relations in special characteristics to make new groups by modifying root groups and Cartan subgroups in purely inseparable Weil restrictions of certain connected semisimple groups. This idea succeeds due to the existence of very special isogenies in characteristics 2 and 3. But groups built in this way always have a reduced root system.

The construction of pseudo-reductive groups with a non-reduced root system will rest on an entirely different phenomenon: divisibility of roots in the weight lattice. The only type for which this happens is simply connected type C_n ($n \geqslant 1$), or in other words (forms of) Sp_{2n}, in which case the long roots are precisely the roots that are divisible in the weight lattice (with the convention for $n = 1$ that both roots are viewed as long, as they are divisible by 2 in the weight lattice). A surprising feature is that these groups with a non-reduced root system are not constructed directly as affine schemes, in contrast with our construction of exotic pseudo-reductive groups as Weil restrictions of fiber products of affine schemes. To construct absolutely pseudo-simple groups with a non-reduced root system of rank $n \geqslant 1$, we will need to use special properties of Sp_{2n} in characteristic 2 and Weil's technique of (strict) birational group laws. Thus, now we briefly digress to review some basic definitions and results concerning birational group laws, and refer the reader to [BLR, Ch. 5] for a complete treatment.

Definition 9.6.1 For any field k, a *birational group law* on a smooth separated k-scheme X of finite type is a rational map

$$m : X \times X \dashrightarrow X$$

such that the universal translation rational maps $X \times X \dashrightarrow X \times X$ defined by

$$\lambda : (x, y) \mapsto (x, m(x, y)), \quad \rho : (x, y) \mapsto (m(x, y), y)$$

are birational (so m is dominant) and the rational composition law m is associative in the sense of rational maps (i.e., $m(x, m(y, z)) = m(m(x, y), z)$ as rational maps from X^3 to X; such composite rational maps make sense since m is dominant).

Note that there is no axiom concerning inversion or an identity element. (The case $X = \emptyset$ admits a birational group law.) To give an example, let G be a smooth (possibly disconnected) k-group of finite type and $X \subseteq G$ a dense open subscheme. The k-scheme X is smooth and separated (since G is), and the group law m_G on G defines a birational group law on X. These birational group laws satisfy some additional properties. To formulate such properties, we require a new concept. For any smooth k-scheme Y of finite type, an open subscheme $U \subseteq Y \times Y$ is *Y-dense* if it has dense intersection with all fibers of both projection maps $Y \times Y \rightrightarrows Y$. (In case Y is geometrically connected over k, and hence is geometrically irreducible, it is equivalent to say that U meets all fibers of both projection maps $Y \times Y \rightrightarrows Y$.) The property of Y-density is insensitive to ground field extension, and when k is algebraically closed it suffices to check fibers over $Y(k)$.

If (X', m') is a birational group law on a smooth separated k-scheme X' of finite type and $X \subseteq X'$ is a dense open subscheme then for the rational map $m : X \times X \dashrightarrow X$ induced by m' the pair (X, m) is a birational group law. Consider a birational group law (X, m) which arises as a dense open subscheme of a smooth k-group G of finite type. The open subscheme $U := m_G^{-1}(X) \cap (X \times X)$ in $X \times X$ is the domain of definition of the rational map $m : X \times X \dashrightarrow X$, and it is X-dense since for any $x \in X(\overline{k})$ the open overlaps $(x^{-1} \cdot X_{\overline{k}}) \cap X_{\overline{k}}$ and $(X_{\overline{k}} \cdot x^{-1}) \cap X_{\overline{k}}$ inside $G_{\overline{k}}$ are dense in $X_{\overline{k}}$. Moreover, the universal translation morphisms $\lambda : U \to X \times X$ and $\rho : U \to X \times X$ are open immersions (induced by the universal translation isomorphisms $G \times G \simeq G \times G$ defined by $(g, g') \mapsto (g, gg'), (gg', g')$) and each has X-dense image. This motivates the following refinement of the concept of a birational group law.

Definition 9.6.2 A *strict birational group law* on a smooth separated k-scheme X of finite type is a birational group law m on X such that there is an open subscheme $U \subseteq \mathrm{dom}(m)$ which is X-dense and for which the k-morphisms $\lambda : U \to X \times X$ and $\rho : U \to X \times X$ respectively defined by $u \mapsto (p_1(u), m(u))$ and $u \mapsto (m(u), p_2(u))$ are open immersions with X-dense images in $X \times X$.

Remark 9.6.3 Since U is required to be X-dense and ρ and λ are assumed to be open immersions, if X is geometrically connected then the X-density conditions on the open images of ρ and λ amount to the condition $m(U) = X$ (because a smooth connected scheme of finite type over a field is irreducible, so its non-empty open subsets are dense). Geometrically connected X are sufficient for our needs.

By [BLR, 5.2/2], for any birational group law (X', m') over a field there is a dense open subscheme $X \subseteq X'$ such that the induced birational group law (X, m) on X is strict. The main result of Weil concerns strict birational group laws:

Theorem 9.6.4 (Weil) *If (X, m) is a strict birational group law over a field k and $X \neq \emptyset$ then up to unique k-isomorphism there is a dense open immersion of X into a smooth k-group G of finite type such that the group law $G \times G \to G$ restricts to the rational map m on $X \times X$.*

The formation of G is functorial in X in the sense that if $f : X \dashrightarrow G'$ is a rational map to a smooth k-group of finite type and the two rational maps from $X \times X$ to G' defined by $f \circ m$ and $m_{G'} \circ (f \times f)$ coincide then f uniquely extends to a k-homomorphism $G \to G'$.

The composition $f \circ m$ makes sense as a rational map because $m : X \times X \dashrightarrow X$ is dominant.

Proof The existence and uniqueness up to unique isomorphism of G are given in [BLR, 5.2/3] when k is separably closed. By uniqueness up to unique isomorphism, the general case follows from the separably closed case by Galois descent. To ensure the effectivity of the descent, we need to use the fact that a smooth k-group of finite type is necessarily quasi-projective. This quasi-projectivity follows from [BLR, 6.4/2], which is formulated over a Dedekind base but whose proof is applicable in a simpler form over a field. (In the cases we need, it will be shown after some work that the groups of interest over k_s are affine. Thus, in such cases the effectivity of the descent can be seen without reliance on general quasi-projectivity results for abstract group varieties.)

To prove the functoriality assertion, by Galois descent we may assume $k = k_s$ and we identify X as a dense open locus in G (so $X(k)$ is dense in G, and hence is also dense in $G_{\overline{k}}$). We may replace X with a dense open subset so that f is a morphism on X. Our problem is to extend $f : X \to G'$ to a k-scheme morphism $\widetilde{f} : G \to G'$, since then $\widetilde{f} \circ m_G = m_{G'} \circ (\widetilde{f} \times \widetilde{f})$ due to the equality of the rational maps $f \circ m_G$ and $m_{G'} \circ (f \times f)$. The "rational homomorphism" hypothesis on f implies that the morphisms $m_G^{-1}(X) \cap (X \times X) \rightrightarrows G'$ defined by $(x_1, x_2) \mapsto f(x_1 x_2)$ and $(x_1, x_2) \mapsto f(x_1)f(x_2)$ coincide as rational maps and thus as morphisms. Hence, for $x_0 \in X(k)$, the map $f_{x_0} : X \cdot x_0 \to G'$ defined by $xx_0 \mapsto f(x)f(x_0)$ agrees with f on the dense open overlap $X \cap (X \cdot x_0)$. For $x_0, x_1 \in X(k)$, the morphisms f_{x_0} and f_{x_1} therefore coincide on the open locus $\Omega = (X \cdot x_0) \cap (X \cdot x_1)$ inside G since they agree (with f) on the dense open overlap of Ω with X.

We glue the maps f_{x_0} to a k-morphism $\widetilde{f} : \bigcup_{x_0 \in X(k)} X \cdot x_0 \to G'$ extending f. The open subsets $\{X \cdot x\}_{x \in X(k)}$ cover G because for any $g \in G(\overline{k})$ the non-empty open subset $X_{\overline{k}}^{-1} \cdot g$ meets the dense subset $X(k) \subseteq G_{\overline{k}}$. Thus, $\widetilde{f}$ is a morphism from G to G' extending f. □

9.6.5 We now turn to the task of using birational group laws to construct pseudo-split absolutely pseudo-simple groups with non-reduced root system BC_n, over any imperfect field k of characteristic 2. The construction provides smooth connected group varieties that are not known at the outset to be affine, as Weil's method involves abstract gluing procedures that cannot be carried out without leaving the affine framework. (To prove that the smooth connected k-groups G constructed in this way are indeed pseudo-reductive, and in particular are *affine*, we will need to separately construct what should be $i_G(G)$ prior to knowing that G is pseudo-reductive or even affine.)

Fix an integer $n \geqslant 1$, and use the following notation when working with Sp_{2n} over an arbitrary base ring (or over $\mathbf{F}_2$-algebras, which suffices for our needs). View Sp_{2n} as a subgroup of GL_{2n} and denote its points as $\left(\begin{smallmatrix} a & b \\ c & d \end{smallmatrix}\right)$, where $a, b, c, d \in \mathrm{Mat}_n$ are $n \times n$ matrices. For any positive integer N and $N \times N$ matrix g, we denote by 1_N the identity in GL_N and we denote the transpose of g as ${}^{\mathrm{t}}g$. We define Sp_{2n} to be the automorphism group of the symplectic form $\left(\begin{smallmatrix} 0 & 1_n \\ -1_n & 0 \end{smallmatrix}\right)$, which is to say

$$\begin{aligned}\mathrm{Sp}_{2n} &= \left\{ g \in \mathrm{GL}_{2n} \,\middle|\, {}^{\mathrm{t}}g \begin{pmatrix} 0 & 1_n \\ -1_n & 0 \end{pmatrix} g = \begin{pmatrix} 0 & 1_n \\ -1_n & 0 \end{pmatrix} \right\} \\ &= \left\{ \begin{pmatrix} a & b \\ c & d \end{pmatrix} \in \mathrm{Mat}_{2n} \,\middle|\, {}^{\mathrm{t}}a \cdot c = {}^{\mathrm{t}}c \cdot a,\ {}^{\mathrm{t}}b \cdot d = {}^{\mathrm{t}}d \cdot b,\ {}^{\mathrm{t}}a \cdot d - {}^{\mathrm{t}}c \cdot b = 1_n \right\}.\end{aligned}$$

Let $D_n \subseteq \mathrm{GL}_n$ be the diagonal torus and let $U_n^{\pm} \subseteq \mathrm{GL}_n$ denote the upper and lower triangular unipotent subgroups. The n-dimensional split torus in Sp_{2n} given by

$$D := \left\{ \begin{pmatrix} t^{-1} & 0 \\ 0 & t \end{pmatrix} \middle| \, t \in D_n \right\}$$

is its own centralizer, so it is maximal in Sp_{2n}. Likewise, letting $\mathrm{Sym}_n \subseteq \mathrm{Mat}_n$ denote the space of symmetric $n \times n$ matrices, the closed subscheme

$$U := \left\{ \begin{pmatrix} {}^{\mathrm{t}}u^{-1} & mu \\ 0 & u \end{pmatrix} \middle| \, m \in \mathrm{Sym}_n, \ u \in U_n^+ \right\} \subseteq \mathrm{Sp}_{2n}$$

is a subgroup because

$$\begin{pmatrix} {}^{\mathrm{t}}u_1^{-1} & m_1 u_1 \\ 0 & u_1 \end{pmatrix} \begin{pmatrix} {}^{\mathrm{t}}u_2^{-1} & m_2 u_2 \\ 0 & u_2 \end{pmatrix} = \begin{pmatrix} {}^{\mathrm{t}}(u_1 u_2)^{-1} & ({}^{\mathrm{t}}u_1^{-1} m_2 u_1^{-1} + m_1)(u_1 u_2) \\ 0 & u_1 u_2 \end{pmatrix}$$

and

$$\begin{pmatrix} {}^{\mathrm{t}}u^{-1} & mu \\ 0 & u \end{pmatrix}^{-1} = \begin{pmatrix} {}^{\mathrm{t}}u & (-{}^{\mathrm{t}}umu)u^{-1} \\ 0 & u^{-1} \end{pmatrix}.$$

Since

$$\begin{pmatrix} {}^{\mathrm{t}}u^{-1} & mu \\ 0 & u \end{pmatrix} = \begin{pmatrix} 1_n & m \\ 0 & 1_n \end{pmatrix} \begin{pmatrix} {}^{\mathrm{t}}u^{-1} & 0 \\ 0 & u \end{pmatrix},$$

clearly U is smooth with connected unipotent fibers of dimension $n(n-1)/2 + n(n+1)/2 = n^2$. It is also easy to check that U is normalized by D.

Let $B_n^{\pm} := D_n \ltimes U_n^{\pm} \subseteq \mathrm{GL}_n$ denote the upper and lower triangular Borel subgroups, so

$$B := D \ltimes U = \left\{ \begin{pmatrix} {}^{\mathrm{t}}b^{-1} & mb \\ 0 & b \end{pmatrix} \middle| \, m \in \mathrm{Sym}_n, \ b \in B_n^+ \right\} \qquad (9.6.1)$$

is a Borel subgroup of Sp_{2n} with unipotent radical U. This corresponds to the "standard" positive system of roots Φ^+ in $\Phi := \Phi(\mathrm{Sp}_{2n}, D)$. If it is necessary to specify n, we will denote this data as Φ_n and Φ_n^+.

The matrix entries of $u \in U_n^+$ and $m \in \mathrm{Sym}_n$ are the root groups for (Sp_{2n}, D) contained in B (i.e., root groups for the roots in Φ^+). More explicitly, for $1 \leqslant i < j \leqslant n$ we define $u_{ij} : \mathbf{G}_{\mathrm{a}} \to \mathrm{GL}_n$ to carry x to the matrix in U_n^+ whose ij entry is x and whose other off-diagonal entries vanish (these are the root groups for (GL_n, D_n) contained in U_n^+), so the closed subgroup $[u]_{ij} : \mathbf{G}_{\mathrm{a}} \hookrightarrow U$ defined by

$$[u]_{ij} : x \mapsto \begin{pmatrix} {}^{\mathrm{t}}u_{ij}(x)^{-1} & 0 \\ 0 & u_{ij}(x) \end{pmatrix} \tag{9.6.2}$$

is normalized by D since for

$$t := \mathrm{diag}(t_1^{-1}, \ldots, t_n^{-1}, t_1, \ldots, t_n) \in D$$

we have $t[u]_{ij}(x)t^{-1} = [u]_{ij}(t_i t_j^{-1} x)$. Likewise, for $1 \leqslant i < j \leqslant n$ we define $m_{ij} : \mathbf{G}_{\mathrm{a}} \to \mathrm{Sym}_n$ to carry x to the symmetric matrix whose ij and ji entries are x and all other entries vanish, and for $1 \leqslant i \leqslant n$ we define $m_{ii} : \mathbf{G}_{\mathrm{a}} \to \mathrm{Sym}_n$ to carry x to the symmetric matrix whose ii entry is x and whose other entries vanish. The closed subgroup $[m]_{ij} : \mathbf{G}_{\mathrm{a}} \hookrightarrow \mathrm{Sp}_{2n}$ defined by

$$[m]_{ij} : x \mapsto \begin{pmatrix} 1_n & m_{ij}(x) \\ 0 & 1_n \end{pmatrix} \tag{9.6.3}$$

for $1 \leqslant i \leqslant j \leqslant n$ is normalized by D with t as above satisfying $t[m]_{ij}(x)t^{-1} = [m]_{ij}((t_i t_j)^{-1} x)$.

The standard representative for the longest Weyl element in $W(\mathrm{Sp}_{2n}, D)$ is $w := \left(\begin{smallmatrix} 0 & 1_n \\ -1_n & 0 \end{smallmatrix}\right)$; this matrix normalizes D via inversion. Explicitly, we have the conjugation formula

$$w \begin{pmatrix} {}^{\mathrm{t}}u^{-1} & mu \\ 0 & u \end{pmatrix} w^{-1} = \begin{pmatrix} u & 0 \\ -mu & {}^{\mathrm{t}}u^{-1} \end{pmatrix}$$

(so one can check directly that w acts on the root groups for D in Sp_{2n} in the required manner).

Remark 9.6.6 Inspection of the preceding descriptions of root groups shows that the long roots in Φ^+ (equivalently, the roots that are divisible in $\mathrm{X}(D)$) correspond precisely to the root groups $[m]_{ii}$. In particular, these root groups *pairwise commute* with each other (as is also easily deduced from the structure of the root system: a sum of two positive long roots is not a root).

We end our general discussion of Sp_{2n} by recording a special feature in characteristic 2:

Proposition 9.6.7 *Let (H, S) be a split connected semisimple group over a field k of characteristic 2, and assume $\Phi = \Phi(H, S)$ is irreducible of type C_n with $n \geqslant 2$. Let Φ^+ be a positive system of roots in Φ. The subset $\Phi_<$ of short roots in Φ is a root system of type D_n, and the root groups U_a for the short positive roots $a \in \Phi^+$ directly span, in any order, a smooth connected unipotent k-subgroup of H.*

As usual, D_2 means $A_1 \times A_1$, so SO_{2n} has type D_n for all $n \geqslant 2$.

Proof This result can be proved entirely by arguments with root systems (using degeneration of the Chevalley commutation relations in characteristic 2), but we give a more direct argument in terms of algebraic groups.

The assertions concern the subgroup structure of U_{Φ^+}, so it is harmless to replace H with its simply connected central cover. That is, we can assume $H = \mathrm{Sp}_{2n}$. In characteristic 2, since $n \geqslant 2$ there is a very special isogeny

$$\mathrm{Spin}_{2n+1} \to \mathrm{SO}_{2n+1} \xrightarrow{\pi} \mathrm{Sp}_{2n}$$

as in Remark 7.1.6. Explicitly, for the quadratic form

$$q_{2n+1} = x_0^2 + x_1 x_{n+1} + \cdots + x_n x_{2n}$$

on $V = k^{2n+1}$ (with standard basis denoted $\{e_0, \ldots, e_{2n}\}$), the defect line $V^\perp$ is ke_0 and the induced symplectic form on the quotient $V/V^\perp = k^{2n}$ is the one used in our conventions for computing with Sp_{2n} (noting that $-1 = 1$ in k). The maximal torus T in $\mathrm{SO}_{2n+1} = \mathrm{SO}(q_{2n+1})$ defined by

$$\mathrm{diag}(1, 1/t_1, 1/t_2, \ldots, 1/t_n, t_1, \ldots, t_n)$$

is carried isomorphically onto the diagonal maximal torus D with which we are doing computations in Sp_{2n}. (Conceptually, the isomorphism between tori is due to the unipotence of the scheme-theoretic kernel of $\pi : \mathrm{SO}_{2n+1} \twoheadrightarrow \mathrm{Sp}_{2n}$.)

Since the central isogeny $\mathrm{Spin}_{2n+1} \to \mathrm{SO}_{2n+1}$ induces an isomorphism between root groups and root systems (though not root data), as a special case of the general properties of very special isogenies in Proposition 7.1.5 we deduce that under the isogeny $\pi : \mathrm{SO}_{2n+1} \to \mathrm{Sp}_{2n}$ and the induced isomorphism between diagonal maximal tori the following hold:

(a) the long roots of the root system $\Phi(\mathrm{SO}_{2n+1}, T)$ of type B_n are identified with the short roots of the root system $\Phi(\mathrm{Sp}_{2n}, D)$ of type C_n,

(b) π induces an isomorphism between the corresponding root groups (for long root groups in SO_{2n+1} and short root groups in Sp_{2n}).

Thus, our problem for short roots of Sp_{2n} reduces to a problem for long roots of SO_{2n+1}.

It suffices to prove that for $n \geqslant 2$ the set of long roots in $\Psi = \Phi(\mathrm{SO}_{2n+1}, T)$ is the simply laced root system of type D_n and for a choice of positive system of roots Ψ^+ the root groups for the associated set $\Psi^+_>$ of positive long roots directly span, in any order, a smooth connected unipotent k-subgroup of SO_{2n+1}. The elements of $\Psi^+_>$ are a positive system of roots for the standard subgroup $\mathrm{SO}_{2n} \subseteq \mathrm{SO}_{2n+1}$ of type D_n containing the diagonal maximal torus (where $\mathrm{SO}_{2n} = \mathrm{SO}(q_{2n})$ for the restriction $q_{2n} = \sum_{1 \leqslant i \leqslant n} x_i x_{n+i}$ of q_{2n+1} to $k^{2n} = \bigoplus_{1 \leqslant i \leqslant 2n} k e_i$). Thus, the associated root groups directly span, in any order, the unipotent radical of the corresponding Borel subgroup of SO_{2n} containing T. □

9.6.8 Now fix an imperfect field k of characteristic 2 and a nontrivial purely inseparable finite extension K/k. Choose a pair of *nonzero* K^2-subspaces $V^{(2)}, V' \subseteq K$ such that the following conditions hold:

(i) V' is a kK^2-subspace of K,
(ii) $V^{(2)} \cap V' = 0$,
(iii) $\dim_{K^2} V^{(2)} < \infty$.

(Condition (iii) is automatic if $[k : k^2]$ is finite.) For any K/k as above, it is always possible to find a pair $(V^{(2)}, V')$ satisfying these conditions. Indeed, it suffices to check that $K \neq kK^2$ (as then we can choose $V' = kK^2$ and $V^{(2)}$ to be the K^2-line spanned by an element of $K - kK^2$). If $K = kK^2$ then $\Omega^1_{K/k} = 0$, forcing the nontrivial finite extension K/k to be separable, contradicting the hypothesis that it is purely inseparable.

Define the K-vector space $V := K \otimes_{\iota, K^2} V^{(2)}$ where $\iota : K^2 \to K$ is the square root isomorphism, and define the additive bijection $q : V \to V^{(2)}$ to be $\lambda \otimes v^{(2)} \mapsto \lambda^2 v^{(2)}$ (so $q(\lambda v) = \lambda^2 q(v)$ for $\lambda \in K$ and $v \in V$). Also define the kK^2-subalgebra

$$K_0 := k\langle V^{(2)} + V' \rangle \subseteq K$$

generated by ratios of nonzero elements of $V^{(2)} + V'$, and let V_0 denote the k-span of $V^{(2)} = q(V)$ and V' inside K. (If $[k : k^2] = 2$, so $[K : K^2] = 2$, the inclusion $V^{(2)} \oplus V' \to K$ is an equality for K^2-dimension reasons. In particular, $V_0 = K_0 = K$ in such cases.) For any proper nonzero kK^2-subspace V' of K

we can always choose the nonzero K^2-subspace $V^{(2)}$ of K so that $K_0 = K$. Indeed, $V^{(2)}$ can be taken to be the K^2-span of a kK^2-linearly independent set that, together with a kK^2-basis of V', is a kK^2-basis of the finite-dimensional kK^2-vector space K.

Remark 9.6.9 By construction, the additive map $V \times V' \to K$ defined by $(v, v') \mapsto q(v) + v'$ has trivial kernel (since $V^{(2)} \cap V' = 0$). However, the associated k-homomorphism $h : \underline{V} \times \underline{V'} \to \underline{K}$ *always* has nontrivial kernel (though without nontrivial k-points) since $\underline{q} : \underline{V} \to \underline{K}$ has nontrivial kernel (a hypersurface of the form $\sum \alpha_i x_i^2 = 0$ with nonzero $\{\alpha_1, \dots, \alpha_d\}$ for $d = \dim_K V$). If $k \subseteq K^2$ then $\underline{q}(\underline{V}) = \underline{q(V)}$ (since $q(V) = V^{(2)}$ is a subspace of K over $K^2 = kK^2$), so $\ker h = \ker \underline{q}$ in these cases. However, if $k \not\subseteq K^2$ then for any proper nonzero kK^2-subspace $V' \subseteq K$ we can choose $V^{(2)}$ such that $\underline{q}(\underline{V}) \cap \underline{V'} \neq 0$ (e.g., if V'' is a K^2-linear complement of V' in K that is *not* a kK^2-linear complement then $kV'' \cap V' \neq 0$, so for a suitable finite-dimensional K^2-subspace $V^{(2)}$ of V'' we have $kV^{(2)} \cap V' \neq 0$). Observe that $\ker h$ is strictly larger than $\ker \underline{q}$ for such $(V^{(2)}, V')$.

Remark 9.6.10 The data $V^{(2)}, V' \subseteq K$ axiomatize aspects of our study of general pseudo-split absolutely pseudo-simple groups with a non-reduced root system and a multipliable positive simple root c in Proposition 9.3.6, Proposition 9.4.6, and Proposition 9.5.2. To be precise, the inclusion $V' \hookrightarrow K$ corresponds to the $\underline{kK^2}$-linear inclusion $\ell_{2c} : U_{2c} \to \mathrm{R}_{K/k}(U'_{2c})$ on k-points, $V^{(2)}$ corresponds to the K^2-subspace $q_c(V_c(k)) \subseteq U'_{2c}(K) = K$, and the bijective $q : V \to V^{(2)}$ corresponds to the map q_c from $V_c(k)$ onto its image inside $\mathrm{R}_{K/k}(U'_{2c})(k)$. Also, the k-subgroup $i_{G_c}(G_c) \subseteq \mathrm{R}_{K/k}(\mathrm{SL}_2)$ has root group $\underline{V_0}$ inside $\mathrm{R}_{K/k}(U^+) = \mathrm{R}_{K/k}(\mathbf{G}_a)$ (relative to $i_{G_c}(c^\vee(\mathrm{GL}_1))$), so $i_{G_c}(G_c) = H_{V_0}$. Hence, K_0 corresponds to K_c (since $1 \in U_{2c}(k)$ inside $U'_{2c}(K) = K$).

Suppose $[k : k^2] = 2$, so the purely inseparable nontrivial finite extensions F of k are precisely $k^{1/2^m}$ for $m > 0$, and hence $kF^2 = F^2 \neq F$ for any such F. Since $kK^2 = K^2$ and $[K : K^2] = 2$, necessarily V' is a K^2-line in K (as it is a K^2-subspace and $V^{(2)} \neq 0$), so $V' = K^2$ if $1 \in V'$. Thus, if $1 \in V'$ then the choices for $V^{(2)}$ are precisely the K^2-lines complementary to K^2 in the 2-dimensional K^2-vector space K, and $K_0 = K$. (Writing $V^{(2)} = K^2\alpha$ for $\alpha \in K - K^2$, the map $q : V \to V^{(2)}$ is $\lambda \otimes \alpha \mapsto \lambda^2\alpha$.)

Similarly, if $[K : kK^2] = 2$ (as happens whenever $[k : k^2] = 2$, but can happen in other cases; e.g., $K = \kappa(x, y)$ for a perfect field κ of characteristic 2 and $k = \kappa(x, y^2)$) then V' is a kK^2-line for kK^2-dimension reasons and $V^{(2)}$ must contain an element of $K - V'$, so again $K_0 = K$. However, if $[K : kK^2] \geqslant 4$

then we can always choose $V^{(2)}$ and V' such that $1 \in V'$ and $K_0 \neq K$. One such choice is $V' = kK^2$ and $V^{(2)} = K^2\alpha$ for any $\alpha \in K - kK^2$.

Remark 9.6.11 Let us explain the behavior of $\{V', V^{(2)}, V, K, q, V_0, K_0\}$ with respect to a *separable* (possibly non-algebraic) extension of the ground field $k \to k'$. The k'-algebra $K' := k' \otimes_k K$ is a field (purely inseparable over k' and separable over K) since K/k is purely inseparable, and $k' \otimes_k kK^2 \to k'K'^2$ is clearly an isomorphism. Likewise, $K'^2 \otimes_{K^2} K$ is a field since K'^2/K^2 is separable and K/K^2 is purely inseparable, and $K'^2 \otimes_{K^2} K \to K'$ is an equality if the separable extension k'/k is algebraic (as then K'/K is separable algebraic) but it is generally not an equality if k'/k is not algebraic (consider $k' = k(X)$). Note that $K'^2 = k'^2 \otimes_{k^2} K^2$.

Clearly $K'^2 \otimes_{K^2} V^{(2)} = k'^2 \otimes_{k^2} V^{(2)}$ and $k'K'^2 \otimes_{kK^2} V' = k' \otimes_k V'$, with these respectively subspaces of $K'^2 \otimes_{K^2} K$ and K' over K'^2 and $k'K'^2$, so via the inclusion $K'^2 \otimes_{K^2} K \hookrightarrow K'$ we view both inside K'. The K'-vector space $K' \otimes_K V = k' \otimes_k V$ is naturally identified with $K' \otimes_{\iota', K'^2} (K'^2 \otimes_{K^2} V^{(2)})$ (where $\iota' : K'^2 \to K'$ is the square root isomorphism). Also, the 2-linear map $q' : K' \otimes_K V \to K'^2 \otimes_{K^2} V^{(2)}$ defined by $\lambda' \otimes v \mapsto \lambda'^2 \otimes q(v)$ is clearly the (bijective) 2-linear map attached to the K'^2-subspace $K'^2 \otimes_{K^2} V^{(2)}$ inside K', and the subfield $k' \otimes_k K_0 \subseteq K'$ over $k' \otimes_k kK^2 = k'K'^2$ coincides with

$$k'\langle K'^2 \otimes_{K^2} V^{(2)} + k' \otimes_k V' \rangle$$

by Remark 9.1.2.

Let $kV^{(2)}$ denote the kK^2-span of $V^{(2)}$ inside K, so the k-subgroup $\underline{kV^{(2)}}$ of $\underline{K} = \mathrm{R}_{K/k}(\mathbf{G}_\mathrm{a})$ is the Zariski closure of the subgroup $V^{(2)} \subseteq K = \underline{K}(k)$ and q gives rise to a k-homomorphism $\underline{q} : \underline{V} \to \underline{kV^{(2)}}$. Clearly $k' \otimes_k kV^{(2)}$ is the $k'K'^2$-span of $K'^2 \otimes_{K^2} V^{(2)}$ inside K', and the k'-homomorphism $\underline{q}_{k'}$ defined by base change $k \to k'$ is thereby identified with the k'-homomorphism attached to q' (via the natural isomorphism $k' \otimes_k V = K' \otimes_K V$). Likewise, $k' \otimes_k V_0 = V'_0$ and $k' \otimes_k K_0 = K'_0$ inside $k' \otimes_k K = K'$.

Finally, we have to check that conditions (i), (ii), and (iii) in 9.6.8 hold for K'/k', $k' \otimes_k V'$, and $K'^2 \otimes_{K^2} V^{(2)}$. This is clear for (i) and (iii). By the equality $\underline{q}_{k'} = \underline{q'}$, (ii) means that $h : \underline{V} \times \underline{V'} \to \underline{K}$ as in Remark 9.6.9 has no nontrivial k'-points in its kernel. More generally we claim that the group scheme $\ker(h_{k'}) = (\ker h)_{k'}$ has trivial maximal smooth closed k'-subgroup. Since k'/k is separable, by Lemma C.4.1 it is equivalent to show that $(\ker h)(k_s) = 1$. In other words, we may assume $k' = k_s$, so now k' is *algebraic* over k. Thus, $kk'^2 = k'$, so the natural map of fields $K'^2 \otimes_{K^2} kK^2 \to k'K'^2$ is an equality. It follows that $k'K'^2 \otimes_{kK^2} V' = K'^2 \otimes_{K^2} V'$, so inside K' its intersection with $K'^2 \otimes_{K^2} V^{(2)}$ equal to $K'^2 \otimes_{K^2} (V^{(2)} \cap V') = 0$ as desired.

9.6.12 Fix $n \geqslant 1$ and write the long roots in $\Phi := \Phi(\mathrm{Sp}_{2n}, D)$ in the form $2c$ with $c \in \mathrm{X}(D)$. We shall use $(K/k, V^{(2)}, V', n)$ to construct a pseudo-split absolutely pseudo-simple k-group (G, T) with a non-reduced root system of rank n such that:

(a) G is of minimal type,
(b) K is the field of definition over k for the geometric radical (provided that we also require $K_0 = K$ when $n = 1$),
(c) for a suitable K-isomorphism

$$(G_K^{\mathrm{ss}}, T_K) \simeq (\mathrm{Sp}_{2n}, D) \tag{9.6.4}$$

(identifying the set of non-multipliable elements in $\Phi(G, T)$ with $\Phi := \Phi(\mathrm{Sp}_{2n}, D)$) the resulting inclusion $i_G(G) \hookrightarrow \mathrm{R}_{K/k}(\mathrm{Sp}_{2n})$ identifies the map $U_{2c} \simeq i_G(U_{2c}) \hookrightarrow \mathrm{R}_{K/k}(U'_{2c}) = \mathrm{R}_{K/k}(\mathbf{G}_{\mathrm{a}})$ with the inclusion $\underline{V'} \hookrightarrow \underline{K}$ and identifies the map $i_G : U_c \to \mathrm{R}_{K/k}(U'_{2c})$ with the k-homomorphism $\underline{V} \times \underline{V'} \to \mathrm{R}_{K/k}(\mathbf{G}_{\mathrm{a}})$ associated to $(v, v') \mapsto \underline{q}(v) + v'$.

The isomorphism (9.6.4) will be chosen so that the map $\xi_G : G \to \mathrm{R}_{K/k}(G_K^{\mathrm{ss}}) = \mathrm{R}_{K/k}(\mathrm{Sp}_{2n})$ carries T isomorphically onto the maximal k-torus in $\mathrm{R}_{K/k}(D)$ and carries a Levi k-subgroup $L \subseteq G$ containing T isomorphically onto the canonical Levi k-subgroup $\mathrm{Sp}_{2n} \subseteq \mathrm{R}_{K/k}(\mathrm{Sp}_{2n})$. For all $a \in \Phi(G, T)$, we have $i_G(G_a) = i_{G_a}(G_a) \subseteq \mathrm{R}_{K/k}(\mathrm{SL}_2)$ (using Proposition 6.3.1 for the equality), and this k-subgroup contains the canonical k-subgroup SL_2. Thus, since $i_{G_a}(G_a)$ is absolutely pseudo-simple by Lemma 9.2.1, it follows from Proposition 9.1.5 that

$$i_G(G_{2c}) = H_{V'}, \quad i_G(G_c) = H_{kV^{(2)}+V'}$$

inside $\mathrm{R}_{K/k}(\mathrm{SL}_2)$. Hence, $K_{2c} = k\langle V' \rangle$ and $K_c = K_0$ inside K by Proposition 9.1.4. (If $[K : kK^2] > 2$ and $n = 2$ then there are more pseudo-split absolutely pseudo-simple k-groups with root system BC_n and K/k as the field of definition of the geometric radical; see 9.8.3.)

The idea for constructing G is to define a birational group law on a variant of the open Bruhat cell of $\mathrm{R}_{K/k}(\mathrm{Sp}_{2n})$ in which, for each long root $a \in \Phi(\mathrm{Sp}_{2n}, D)$ (so $a \in 2\mathrm{X}(D)$), the a-root group $\mathrm{R}_{K/k}(\mathbf{G}_{\mathrm{a}})$ is "replaced" with $\underline{V} \times \underline{V'}$ via the k-homomorphism $\underline{V} \times \underline{V'} \to \mathrm{R}_{K/k}(\mathbf{G}_{\mathrm{a}})$ defined by $(v, v') \mapsto \underline{q}(v) + v'$ (with V and q as in 9.6.8). This k-homomorphism is generally not surjective (unless $[K : kK^2] = 2$), its kernel is not smooth, and it is $\mathrm{R}_{K/k}(D)$-equivariant when letting $\mathrm{R}_{K/k}(D)$ act on $\underline{V} \times \underline{V'}$ through the $\mathrm{R}_{kK^2/k}(\mathrm{GL}_1)$-valued $\mathrm{R}_{K/k}(a)$ on the $\underline{kK^2}$-module $\underline{V'}$ and via the $\mathrm{R}_{K/k}(\mathrm{GL}_1)$-valued $\mathrm{R}_{K/k}(a/2)$ on the $\underline{K}$-module $\underline{V}$.

For the K-group Sp_{2n}, choose the positive system of roots $\Phi^+ \subseteq \Phi(\mathrm{Sp}_{2n}, D)$ corresponding to the Borel K-subgroup $B = D \ltimes U$ as in (9.6.1). Define $\Phi^+_>$ and $\Phi^+_<$ to be respectively the sets of long roots and short roots in Φ^+, with the convention for $n = 1$ that we declare all roots to be long, which is to say $\Phi^+_< = \emptyset$. This ensures that in all cases the long roots are the ones that lie in $2\mathrm{X}(D)$. Fix enumerations of $\Phi^+_>$ and $\Phi^+_<$, so we have the direct spanning expression

$$\prod_{2c \in \Phi^+_>} U_{2c} \times \prod_{b \in \Phi^+_<} U_b \simeq U \tag{9.6.5}$$

as K-schemes via multiplication, where U_a denotes the root group for $a \in \Phi(\mathrm{Sp}_{2n}, D)$. The U_a for $a \in \Phi^+$ are identified with $\mathbf{G}_{\mathrm{a}}$ over K via the root group parameterizations $[u]_{ij}$ $(1 \leqslant i < j \leqslant n)$ and $[m]_{ij}$ $(1 \leqslant i \leqslant j \leqslant n)$ defined in (9.6.2) and (9.6.3). The enumeration among the long root groups is unimportant in (9.6.5) since these root groups pairwise commute (Remark 9.6.6).

Using the chosen enumerations of $\Phi^+_>$ and $\Phi^+_<$, define the product scheme

$$\mathscr{U} = \prod_{2c \in \Phi^+_>} \mathscr{U}_c \times \prod_{b \in \Phi^+_<} \mathscr{U}_b$$

over k where $\mathscr{U}_b := \mathrm{R}_{K/k}(\mathbf{G}_{\mathrm{a}})$ for $b \in \Phi^+_<$ and $\mathscr{U}_c := \underline{V} \times \underline{V'}$ for $2c \in \Phi^+_>$. Finally, define the map of pointed k-schemes $f : \mathscr{U} \to \mathrm{R}_{K/k}(U)$ to be the direct product of k-maps f_a (respecting the origins) as follows: if a is short then $f_a : \mathscr{U}_a \to \mathrm{R}_{K/k}(U_a)$ is the identity map of $\mathrm{R}_{K/k}(\mathbf{G}_{\mathrm{a}})$, and if $a = 2c$ is long then f_a is the k-homomorphism

$$\mathscr{U}_c := \underline{V} \times \underline{V'} \to \mathrm{R}_{K/k}(\mathbf{G}_{\mathrm{a}}) = \mathrm{R}_{K/k}(U_{2c})$$

given by $(v, v') \mapsto \underline{q}(v) + v'$ that is injective on k-points since $V^{(2)} \cap V' = 0$. Note that $\mathscr{U}$ generally depends on the pair $(V^{(2)}, V')$, as does f, though we omit this dependence from the notation. Also, the formation of $\mathscr{U}$ and f is compatible with any separable extension k'/k (see Remark 9.6.11), and f is injective on k-points (hence also on k'-points for any *separable* extension k'/k).

Remark 9.6.13 We will see in Remark 9.6.15 that up to unique k-isomorphism f is independent of the choice of enumeration of Φ^+.

The first serious construction towards a birational group law is the following result that provides a group law on the pointed smooth affine k-scheme $\mathscr{U}$.

Theorem 9.6.14 *There is a unique group law on the pointed k-scheme $\mathscr{U}$ making the k-morphism $f : \mathscr{U} \to \mathrm{R}_{K/k}(U)$ a homomorphism, and on $\mathscr{U}_a$ for each a this group law restricts to the usual additive law. There is also a unique action of $\mathrm{R}_{K/k}(D)$ on the k-group $\mathscr{U}$ making f equivariant with respect to the natural action of $\mathrm{R}_{K/k}(D)$ on $\mathrm{R}_{K/k}(U)$.*

Proof Since k is an infinite field and $\mathscr{U}$ and $\mathrm{R}_{K/k}(D)$ are rational varieties, the locus of k-points is Zariski-dense in each variety. But f is injective on k-points, so uniqueness of the group law follows. Likewise, if there is to be such a group law m on $\mathscr{U}$ then the distinguished point in $\mathscr{U}(k)$ must be the identity in $\mathrm{R}_{K/k}(U)(k) = U(K)$ and the restriction of m to $\mathscr{U}_a \times \mathscr{U}_a$ (via the distinguished point in all other factors) will automatically be the usual additive group law on $\mathscr{U}_a$ (because f_a is a k-homomorphism). Since f is injective on k-points and carries the "origin" of its source to the identity point of its target, to make the group law it suffices to fill in the top arrows to make commutative diagrams of k-schemes

$$\begin{array}{ccc} \mathscr{U} \times \mathscr{U} & \xrightarrow{?} & \mathscr{U} \\ {\scriptstyle f\times f}\downarrow & & \downarrow{\scriptstyle f} \\ \mathrm{R}_{K/k}(U) \times \mathrm{R}_{K/k}(U) & \xrightarrow[\mu]{} & \mathrm{R}_{K/k}(U) \end{array} \qquad \begin{array}{ccc} \mathscr{U} & \xrightarrow{?} & \mathscr{U} \\ {\scriptstyle f}\downarrow & & \downarrow{\scriptstyle f} \\ \mathrm{R}_{K/k}(U) & \xrightarrow[i]{} & \mathrm{R}_{K/k}(U) \end{array}$$

where μ and i are multiplication and inversion in the k-group law of $\mathrm{R}_{K/k}(U)$. The case $n = 1$ is trivial, by equipping $\mathscr{U} = \underline{V} \times \underline{V}'$ with the evident k-group law. Thus, in what follows we assume $n \geqslant 2$ (so $\Phi^+_<$ is non-empty).

Consider the composite maps

$$h = \mu \circ (f \times f) : \mathscr{U}^2 = \mathscr{U} \times \mathscr{U} \to \mathrm{R}_{K/k}(U), \quad g = i \circ f : \mathscr{U} \to \mathrm{R}_{K/k}(U)$$

along the left and bottom sides. Let h_a and g_a respectively denote the composition of h and g with the projection $\mathrm{R}_{K/k}(U) \to \mathrm{R}_{K/k}(U_a)$ for each $a \in \Phi^+$. To factor h and g through f means that h_a and g_a factor through f_a for all a. For short a the map f_a is an isomorphism, so there is nothing to do. For long a, the problem is to show that the maps

$$h_a : \mathscr{U}^2 \to \mathrm{R}_{K/k}(\mathbf{G}_\mathrm{a}), \ \ g_a : \mathscr{U} \to \mathrm{R}_{K/k}(\mathbf{G}_\mathrm{a})$$

have the respective forms $\underline{q} \circ h_a^{(1)} + h_a^{(2)}$ and $\underline{q} \circ g_a^{(1)} + g_a^{(2)}$ for some k-scheme morphisms

$$h_a^{(1)} : \mathscr{U}^2 \to \underline{V}, \ h_a^{(2)} : \mathscr{U}^2 \to \underline{V}', \ g_a^{(1)} : \mathscr{U} \to \underline{V}, \ g_a^{(2)} : \mathscr{U} \to \underline{V}'$$

(viewing $\underline{V}'$ as a k-subgroup of $\underline{K} = \mathrm{R}_{K/k}(\mathbf{G}_\mathrm{a})$).

First consider the problem for μ, keeping in mind two facts: our enumeration of Φ^+ implicit in (9.6.5) puts the long positive root groups first, and these root groups pairwise commute (Remark 9.6.6) whereas the short positive root groups directly span a smooth connected K-subgroup $U_< \subseteq \mathrm{Sp}_{2n}$ (Proposition 9.6.7). Let $U_>$ be the commutative k-subgroup $\prod_{a\in\Phi_>^+} U_a$ of U, so multiplication in U defines a K-scheme isomorphism $U_> \times U_< \to U$ and hence a k-scheme isomorphism

$$\mathrm{R}_{K/k}(U_>) \times \mathrm{R}_{K/k}(U_<) \simeq \mathrm{R}_{K/k}(U).$$

Correspondingly, we define direct factors

$$\mathscr{U}_> = \prod_{a\in\Phi_>^+} (\underline{V} \times \underline{V}'), \ \ \mathscr{U}_< = \mathrm{R}_{K/k}(U_<)$$

of $\mathscr{U}$ (using our choice of enumeration of Φ^+). The K-subgroup properties for $U_<$ and $U_>$ reduce our problem for μ to studying the projections to the long root factors in the target of the composite map

$$\mathscr{U}_< \times \mathscr{U}_> \stackrel{1\times f}{\to} \mathrm{R}_{K/k}(U_<) \times \mathrm{R}_{K/k}(U_>) \stackrel{\mu}{\to} \mathrm{R}_{K/k}(U) = \mathrm{R}_{K/k}(U_>) \times \mathrm{R}_{K/k}(U_<).$$

To be precise, for $(u_<, u_>) \in \mathrm{R}_{K/k}(U_<) \times \mathscr{U}_>$ we claim that when the point $u_< \cdot f(u_>) \in \mathrm{R}_{K/k}(U)$ is expanded via the product decomposition $U = U_> \times U_<$ and the enumeration used in (9.6.5), the projection to $\mathrm{R}_{K/k}(U_a) = \mathrm{R}_{K/k}(\mathbf{G}_\mathrm{a})$ for each $a \in \Phi_>^+$ has the form $\underline{q}(h_a^{(1)}(u_<, u_>)) + h_a^{(2)}(u_<, u_>)$ for k-scheme morphisms

$$h_a^{(1)} : \mathscr{U}_< \times \mathscr{U}_> \to \underline{V}, \ h_a^{(2)} : \mathscr{U}_< \times \mathscr{U}_> \to \underline{V}'.$$

This is essentially a problem concerning how long positive root groups "pass through" short positive root groups via conjugation in the K-group law on U.

Since $U_>$ is *commutative* and the map $\underline{V} \times \underline{V}' \to \mathrm{R}_{K/k}(\mathbf{G}_\mathrm{a})$ defined by $(v, v') \mapsto \underline{q}(v) + v'$ is a k-homomorphism, it suffices to establish the following property of each long positive root.

PROPERTY P(c): Let $2c \in \Phi^+_>$ be a long positive root, and let $u_{2c} : \mathbf{G}_\mathrm{a} \simeq U_{2c}$ be the associated parameterization over K (as in (9.6.3) with $i = j$). For points $v \in \underline{V}$, $v' \in \underline{V}'$, and $u_< \in \mathrm{R}_{K/k}(U_<)$, the projection into $\mathrm{R}_{K/k}(U_>)$ of the point

$$u_< \cdot u_{2c}(\underline{q}(v) + v') \in \mathrm{R}_{K/k}(U) = \mathrm{R}_{K/k}(U_>) \times \mathrm{R}_{K/k}(U_<)$$

has component along each long root factor $\mathrm{R}_{K/k}(U_{2a}) = \mathrm{R}_{K/k}(\mathbf{G}_\mathrm{a})$ of the form $\underline{q}(h^{(1)}_{a,c}(u_<, v, v')) + h^{(2)}_{a,c}(u_<, v, v')$ for k-scheme morphisms

$$h^{(1)}_{a,c} : \mathrm{R}_{K/k}(U_<) \times \underline{V} \times \underline{V}' \to \underline{V}, \;\; h^{(2)}_{a,c} : \mathrm{R}_{K/k}(U_<) \times \underline{V} \times \underline{V}' \to \underline{V}'.$$

We write $U_<$ as the direct span of the short root groups U_b according to the choice of enumeration of $\Phi^+_<$ used in (9.6.5), and so are led to consider the commutation relations between the root groups U_b and U_{2c} for a short positive root b and a long positive root $2c$. If $b + 2c$ is not a root then U_b commutes with U_{2c}. If $b + 2c$ is a root then the roots of the form $mb + 2m'c$ with $m, m' \geqslant 1$ are the long $2b + 2c$ and the short $b + 2c$ (as we see either by inspection of the root system C_n for $n \geqslant 2$, or by verification in the rank-2 case since such b and $2c$ generate a root system of type $\mathrm{B}_2 = \mathrm{C}_2$ when $b + 2c$ is a root).

There is a unique positive root of maximal height (the height relative to the basis of Φ associated to Φ^+). This root is long, and therefore lies in $2\mathrm{X}(D)$. We shall denote it by $2\tilde{c}$, so $a + 2\tilde{c}$ is not a root for any positive root a. Hence, $U_{2\tilde{c}}$ commutes with *all* positive root groups, so P($\tilde{c}$) holds by defining $h^{(1)}_{a,\tilde{c}}$ and $h^{(2)}_{a,\tilde{c}}$ to vanish for $a \neq \tilde{c}$ and defining

$$h^{(1)}_{\tilde{c},\tilde{c}}(u_<, v, v') = \underline{q}(v), \;\; h^{(2)}_{\tilde{c},\tilde{c}}(u_<, v, v') = v'.$$

We shall now prove P(c) for all long positive roots $2c$ by *descending* induction on the height of the root $2c$; the base case of maximal height has just been settled.

Let $2c$ be a long positive root and assume that P(c') holds for all long positive roots $2c'$ whose height is strictly larger than that of $2c$. Consider a short positive root b such that $b + 2c$ is a root. Then U_{2c}, U_{b+2c}, and U_{2b+2c} commute with each other since none of $b + 4c$, $2b + 4c$, and $3b + 4c$ is a root. For $a \in \Phi^+$, use (9.6.2) and (9.6.3) to define an isomorphism $u_a : \mathbf{G}_\mathrm{a} \simeq U_a$ over K (or even over $\mathbf{F}_2$). The Chevalley commutation relations (whose sign ambiguities are invisible in characteristic 2!) yield

$$\begin{aligned} u_b(x)u_{2c}(y) &= u_{2b+2c}(x^2y)u_{b+2c}(xy)u_{2c}(y)u_b(x) \\ &= u_{2c}(y)u_{2b+2c}(x^2y)u_{b+2c}(xy)u_b(x), \end{aligned}$$

with the last two factors on the right in $U_<$ and the first two in $U_>$. Thus, for any k-algebra A and the associated quadratic form $q_A : V \otimes_k A \to K \otimes_k A$ over $K \otimes_k A$ (using the K-vector space structure on V), any elements $x \in K \otimes_k A$, $v \in V \otimes_k A$, and $v' \in V' \otimes_k A \subseteq K \otimes_k A$ satisfy

$$\begin{aligned} u_b(x)u_{2c}(q_A(v)+v') &\in u_{2c}(q_A(v)+v')u_{2b+2c}(q_A(xv)+x^2v')\cdot \mathrm{R}_{K/k}(U_<)(A) \\ &\subseteq \mathrm{R}_{K/k}(U)(A) \end{aligned} \tag{9.6.6}$$

with $x^2v' \in V' \otimes_k A$ inside $K \otimes_k A$ (since V' is a kK^2-subspace of K). Note that the long positive root $2b+2c$ appearing in (9.6.6) has strictly larger height than $2c$. This will enable us to prove P(c) by descending induction on the height of $2c$, but to do this we need to introduce another inductive assertion depending on c.

For $1 \leqslant r \leqslant n^2-n$ and the fixed enumeration $\{b_1,\ldots,b_{n^2-n}\}$ of $\Phi^+_<$, consider the product

$$\begin{aligned} u_{b_1}(x_1)\cdots u_{b_r}(x_r)u_{2c}(q_A(v)+v') &\in \mathrm{R}_{K/k}(U)(A) \\ &= \mathrm{R}_{K/k}(U_>)(A) \times \mathrm{R}_{K/k}(U_<)(A) \end{aligned}$$

for $x_i \in K \otimes_k A = \mathrm{R}_{K/k}(U_{a_i})(A)$, $v \in V \otimes_k A = \underline{V}(A)$, $v' \in V' \otimes_k A = \underline{V'}(A)$ with a varying k-algebra A. We shall study the following property depending on the fixed c and varying r.

Property P$'(c,r)$: For every long positive root $2a$, the $\mathrm{R}_{K/k}(U_>)$-component of

$$u_{b_1}(x_1)\cdots u_{b_r}(x_r)u_{2c}(\underline{q}(v)+v') \in \mathrm{R}_{K/k}(U_>) \times \mathrm{R}_{K/k}(U_<)$$

has its $\mathrm{R}_{K/k}(U_{2a})$-factor of the form

$$u_{2a}(\underline{q}(h^{(1)}_{r,a,c}(x_1,\ldots,x_r,v,v')) + h^{(2)}_{r,a,c}(x_1,\ldots,x_r,v,v'))$$

for some k-scheme morphisms

$$h^{(1)}_{r,a,c} : \mathrm{R}_{K/k}(\mathbf{G}_a)^r \times \underline{V} \times \underline{V'} \to \underline{V},\ h^{(2)}_{r,a,c} : \mathrm{R}_{K/k}(\mathbf{G}_a)^r \times \underline{V} \times \underline{V'} \to \underline{V'}.$$

Clearly $\mathrm{P}'(c, n^2 - n)$ is the same as $\mathrm{P}(c)$, so we shall now prove $\mathrm{P}'(c,r)$ via induction on $r \geqslant 1$ (with c fixed) since the base case $r = 1$ is settled by (9.6.6) and its trivial analogue if $b + 2c$ is not a root. In general, if $1 < r \leqslant n^2 - n$ and $b_r + 2c$ is not a root, then on A-valued points

$$\begin{aligned} &u_{b_1}(x_1)\cdots u_{b_r}(x_r)u_{2c}(q_A(v) + v') \\ &\quad = u_{b_1}(x_1)\cdots u_{b_{r-1}}(x_{r-1})u_{2c}(q_A(v) + v')u_{b_r}(x_r) \end{aligned}$$

lies in $u_{b_1}(x_1)\cdots u_{b_{r-1}}(x_{r-1})u_{2c}(q_A(v) + v') \cdot \mathrm{R}_{K/k}(U_<)(A)$, and in this case $\mathrm{P}'(c,r)$ follows at once from $\mathrm{P}'(c, r-1)$. On the other hand, if $b_r + 2c$ is a root (so $2b_r + 2c$ is also a root) then (9.6.6) enables us to rewrite $u_{b_1}(x_1)\cdots u_{b_r}(x_r)u_{2c}(q_A(v) + v')$ as an element in

$$(u_{b_1}(x_1)\cdots u_{b_{r-1}}(x_{r-1})u_{2c}(q_A(v)+v'))u_{2b_r+2c}(q_A(x_r v) + x_r^2 v') \cdot \mathrm{R}_{K/k}(U_<)(A).$$

Now we apply $\mathrm{P}'(c, r-1)$ to rewrite

$$u_{b_1}(x_1)\cdots u_{b_{r-1}}(x_{r-1})u_{2c}(q_A(v) + v'),$$

and then we conclude by applying $\mathrm{P}(2b_r+2c)$ for the long positive root $2b_r+2c$ with larger height than $2c$.

This nested induction completes our construction of the desired morphism $\mathscr{U} \times \mathscr{U} \to \mathscr{U}$ lifting multiplication μ on $\mathrm{R}_{K/k}(U)$ through $f : \mathscr{U} \to \mathrm{R}_{K/k}(U)$. The analogous problem for lifting inversion i on $\mathrm{R}_{K/k}(U)$ through f to an endomorphism of $\mathscr{U}$ proceeds in exactly the same way, using the additional elementary fact that $u_a(x)^{-1} = u_a(-x)$ for any root a.

To finish the proof of Theorem 9.6.14, it remains to define an action of $\mathrm{R}_{K/k}(D)$ on the k-scheme $\mathscr{U}$ making f an $\mathrm{R}_{K/k}(D)$-equivariant map. Indeed, f is a k-homomorphism that is injective on k-points, so it would then be automatic that the action of $\mathrm{R}_{K/k}(D)$ on $\mathscr{U}$ respects the group law just constructed on $\mathscr{U}$ since the action of $\mathrm{R}_{K/k}(D)$ on $\mathrm{R}_{K/k}(U)$ respects the group law on $\mathrm{R}_{K/k}(U)$. For each $b \in \Phi_<^+$ we have $\mathscr{U}_b = \mathrm{R}_{K/k}(U_b)$ via f, so we make $\mathrm{R}_{K/k}(D)$ act on $\mathscr{U}_b$ through its usual action on $\mathrm{R}_{K/k}(U_b)$ (via $\mathrm{R}_{K/k}(b)$). For $a = 2c \in \Phi_>^+$ the action of $\mathrm{R}_{K/k}(D)$ on $\mathrm{R}_{K/k}(U_a)$ is through the $\mathrm{R}_{K/k}(\mathrm{GL}_1)$-valued Weil restriction of $a \in \mathrm{X}(D)$. But $\mathrm{R}_{K/k}(a)$ is valued in $[2]\mathrm{R}_{K/k}(\mathrm{GL}_1) = \mathrm{R}_{kK^2/k}(\mathrm{GL}_1)$, so $\mathrm{R}_{K/k}(a)$ defines a scaling action of $\mathrm{R}_{K/k}(D)$ on the $\underline{kK^2}$-submodule $\underline{V'} \subseteq \mathrm{R}_{K/k}(\mathbf{G}_\mathrm{a})$. Likewise, $\mathrm{R}_{K/k}(c)$ defines a scaling action of $\mathrm{R}_{K/k}(D)$ on the $\underline{K}$-module $\underline{V}$. Using these to make $\mathrm{R}_{K/k}(D)$ act on $\mathscr{U}_c = \underline{V} \times \underline{V'}$, it is easy to check that $f|_{\mathscr{U}_c} : (v, v') \mapsto \underline{q}(v) + v'$ is $\mathrm{R}_{K/k}(D)$-equivariant, so we are done (as f is a direct product of $\mathrm{R}_{K/k}(D)$-equivariant maps of pointed k-schemes). □

Remark 9.6.15 For later purposes, we now address the precise sense in which the k-group $\mathscr{U}$ is independent of the choice of enumeration of Φ^+ (putting long roots before short roots), up to a *unique* k-isomorphism. Consider an arbitrary enumeration $\tau : \{1, \ldots, n^2\} \simeq \Phi^+$, without regard to the lengths of roots. Since the root groups $\{U_a\}_{a \in \Phi^+}$ directly span U in any order, the multiplication map $\prod_{i=1}^{n^2} U_{\tau(i)} \to U$ is a k-scheme isomorphism. For each $a \in \Phi^+$ we define $a^\natural = a$ when a is short and define $a^\natural = a/2$ when a is long, so we get a canonical k-scheme morphism

$$F_\tau : \prod_{i=1}^{n^2} \mathscr{U}_{\tau(i)^\natural} \to \mathrm{R}_{K/k}(U)$$

induced by multiplication in $\mathrm{R}_{K/k}(U)$ and the natural maps $f_a : \mathscr{U}_{a^\natural} \to \mathrm{R}_{K/k}(U_a)$ for all $a \in \Phi^+$ (i.e., via (9.6.2) and (9.6.3) the map f_a is the identity map of $\mathrm{R}_{K/k}(\mathbf{G}_\mathrm{a})$ for all $a \in \Phi^+_<$ and the map $f_{2c} : \underline{V} \times \underline{V}' \to \mathrm{R}_{K/k}(\mathbf{G}_\mathrm{a})$ is $(v, v') \mapsto \underline{q}(v) + v'$ for all $2c \in \Phi^+_>$). Clearly F_τ is injective on k-points. We claim that $\prod_{i \geqslant 1} \mathscr{U}_{\tau(i)^\natural}$ admits a unique k-group structure such that F_τ is a k-homomorphism, and that there is a unique k-group isomorphism filling in the top of a commutative diagram

$$\begin{array}{ccc} \prod_{i=1}^{n^2} \mathscr{U}_{\tau(i)^\natural} & \xrightarrow[\simeq]{?} & \mathscr{U} \\ & \searrow^{F_\tau} & \downarrow f \\ & & \mathrm{R}_{K/k}(U) \end{array}$$

(where for definiteness $\mathscr{U}$ is constructed using the initial choice of enumeration of Φ^+; the specific choice does not matter once this claim is proved).

The uniqueness is clear since f and F_τ are injective on k-points. To prove existence, it suffices to show that the unipotent smooth connected k-group $\mathscr{U}$ is directly spanned in any order by its k-subgroups $\mathscr{U}_{a^\natural}$ for $a \in \Phi^+$. The set of weights in $\mathrm{Lie}(\mathscr{U})$ for the action of the split maximal k-torus inside $\mathrm{R}_{K/k}(D)$ on $\mathscr{U}$ is clearly $\Phi^+ \cup \{a^\natural \mid a \in \Phi^+_>\}$, and the $\mathscr{U}_{a^\natural}$ are the "root groups" for non-divisible positive roots, so an application of Theorem 3.3.11 yields the desired direct spanning result.

9.7 Construction of birational group laws

By Theorem 9.6.14 we can define a k-group $\mathscr{B} := \mathrm{R}_{K/k}(D) \ltimes \mathscr{U}$, and naturally extend $f : \mathscr{U} \to \mathrm{R}_{K/k}(U)$ to a homomorphism $\mathscr{B} \to B$ (injective on

k-points) that we denote again by f. In this section we will use $\mathscr{B}$ and the k-group law on $\mathrm{R}_{K/k}(\mathrm{Sp}_{2n})$ to build an abstract k-group via the method of birational group laws, and in §9.8 we show that our construction provides all pseudo-split absolutely pseudo-simple k-groups of minimal type with root system BC_n.

To formulate the main existence result (from which all others will be derived), let $\mathbf{w}$ denote a formal symbol which we imagine as a nontrivial 2-torsion k-point normalizing $\mathrm{R}_{K/k}(D)$ via inversion (much as the element $w = \begin{pmatrix} 0 & 1_n \\ -1_n & 0 \end{pmatrix} \in \mathrm{Sp}_{2n}(K)$ has order 2 in characteristic 2 and acts on D via inversion). The right w-translate $\mathrm{R}_{K/k}(U)w\mathrm{R}_{K/k}(B)$ of the standard open cell in $\mathrm{R}_{K/k}(\mathrm{Sp}_{2n})$ will guide our construction. Denoting the product $\mathscr{U} \times \mathscr{B}$ in the suggestive form $\mathscr{U}\mathbf{w} \times \mathscr{B}$, it admits a *strict* birational group law in a canonical manner:

Theorem 9.7.1 *There exists a unique strict birational group law $\widetilde{\mu}_{V^{(2)},V'}$ on $\mathscr{U}\mathbf{w} \times \mathscr{B}$ making $f : \mathscr{U}\mathbf{w} \times \mathscr{B} \to \mathrm{R}_{K/k}(UwB)$ defined by $(u, b) \mapsto f(u)wf(b)$ compatible with the birational group laws.*

Remark 9.6.15 provides a precise sense in which this result is independent of the initial choice of enumeration of Φ^+ (though for definiteness, in the proof we use a fixed such choice putting the long roots before the short roots).

Proof As in the proof of Theorem 9.6.14, the uniqueness in the theorem follows from the injectivity of f on k-points. Since $\mathscr{B}$ and $\mathscr{U}$ have been endowed with k-group structures making f a k-homomorphism, and we aim to make $\mathbf{w}$ have order 2 and normalize $\mathrm{R}_{K/k}(D)$ via inversion, the main issue in the construction is to decide which points g of $\mathscr{U}$ should satisfy $\mathbf{w}g\mathbf{w}^{-1} \in \mathscr{U}\mathbf{w} \times \mathscr{B}$. To figure this out, we first examine the birational group law on the product scheme

$$\mathrm{R}_{K/k}(UwB) = \mathrm{R}_{K/k}(U)w \times \mathrm{R}_{K/k}(D) \times \mathrm{R}_{K/k}(U)$$

in which w has order 2 and normalizes $\mathrm{R}_{K/k}(D)$ via inversion. The *domain of definition* Ω for the birational group law

$$\mathrm{R}_{K/k}(UwB) \times \mathrm{R}_{K/k}(UwB) \dashrightarrow \mathrm{R}_{K/k}(UwB)$$

consists of the points $(u_1wt_1u_1', u_2wt_2u_2')$ such that $wt_1u_1'u_2w \in \mathrm{R}_{K/k}(UwB)$, or equivalently

$$wt_1u_1'u_2t_1^{-1}w \in \mathrm{R}_{K/k}(UwB).$$

(In particular, (w, w) is *not* contained in $\Omega(k)$.) This is a strict birational group law since it is an open locus in a smooth connected k-group.

Define "**w**-conjugation" on $\mathrm{R}_{K/k}(D)$ to be inversion (as for w-conjugation). Let $\widetilde{\Omega} = (f \times f)^{-1}(\Omega)$ be the preimage of Ω in $(\mathscr{U}\mathbf{w} \times \mathscr{B})^2$. Since f can fail to be surjective or even fail to be dominant, the $(\mathscr{U}\mathbf{w} \times \mathscr{B})$-density of $\widetilde{\Omega}$ does not follow formally from the fact that Ω is $\mathrm{R}_{K/k}(UwB)$-dense. Nonetheless, it will be important to establish that $\widetilde{\Omega}$ is $\mathscr{U}\mathbf{w} \times \mathscr{B}$-dense, so let us now show that this property follows formally if we can prove that $\widetilde{\Omega} \neq \emptyset$. The problem is to show that both projections

$$\mathrm{pr}_1, \mathrm{pr}_2 : \widetilde{\Omega} \rightrightarrows \mathscr{U}\mathbf{w} \times \mathscr{B}$$

have all geometric fibers non-empty. For pr_2, the non-emptiness of its geometric fibers is the condition that for all $\widetilde{u}_2, \widetilde{u}'_2 \in \mathscr{U}(\overline{k})$ and $t_2 \in \mathrm{R}_{K/k}(D)(\overline{k})$ there exist $\widetilde{u}_1, \widetilde{u}'_1 \in \mathscr{U}(\overline{k})$ and $t_1 \in \mathrm{R}_{K/k}(D)(\overline{k})$ such that $wt_1f(\widetilde{u}'_1\widetilde{u}_2)w \in \mathrm{R}_{K/k}(UwB)(\overline{k})$. Since $\mathscr{U}(\overline{k})\widetilde{u}_2 = \mathscr{U}(\overline{k})$, this is equivalent to the condition that $w\mathrm{R}_{K/k}(D)(\overline{k})f(\mathscr{U}(\overline{k}))$ meets $\mathrm{R}_{K/k}(UB^-)(\overline{k})$. This latter formulation has nothing to do with the initial triple $(\widetilde{u}_2, \widetilde{u}'_2, t_2)$, so it is equivalent to say that pr_2 has *some* non-empty geometric fiber, which is to say that $\widetilde{\Omega}$ is non-empty. The case of pr_1 goes in the same way.

Using the group laws of $\mathscr{B}$ and B and the **w**-conjugation and w-conjugation on $\mathrm{R}_{K/k}(D)$ via inversion, we have a commutative diagram

$$\begin{array}{ccc} (\mathscr{U}\mathbf{w} \times \mathscr{B}) \times (\mathscr{U}\mathbf{w} \times \mathscr{B}) & \longrightarrow & \mathscr{U}\mathbf{w} \times \mathscr{U}\mathbf{w} \times \mathscr{B} \\ {\scriptstyle f\times f}\downarrow & & \downarrow{\scriptstyle f} \\ \mathrm{R}_{K/k}(UwB) \times \mathrm{R}_{K/k}(UwB) & \longrightarrow & \mathrm{R}_{K/k}(Uw \times Uw \times B) \end{array} \tag{9.7.1}$$

in which the horizontal maps are surjective. Explicitly, the bottom row is

$$(u_1wt_1u'_1, u_2wt_2u'_2) \mapsto (u_1w, t_1(u'_1u_2)t_1^{-1}w, t_1^{-1}t_2u'_2)$$

and the top row is similar except with **w** replacing w.

Letting $U' \subseteq U$ be the open locus defined by

$$U' = \{u \in U \mid wuw^{-1} \in UwB\},$$

Ω is the full preimage of $\mathrm{R}_{K/k}(Uw \times U'w \times B)$ in $\mathrm{R}_{K/k}(UwB)^2$ along the bottom row of (9.7.1). Define the open subspace $\mathscr{U}' := f^{-1}(\mathrm{R}_{K/k}(U')) \subseteq \mathscr{U}$, so $\widetilde{\Omega}$ is the full preimage of $\mathscr{U}\mathbf{w} \times \mathscr{U}'\mathbf{w} \times \mathscr{B}$ along the top row of the commutative diagram (9.7.1). Thus, if $\widetilde{\Omega} \neq \emptyset$ then $\mathscr{U}' \neq \emptyset$.

The key point is to show that $\widetilde{\Omega}, \mathscr{U}' \neq \emptyset$ and that the diagram of k-schemes

$$\begin{array}{ccc} \mathscr{U}' & \xrightarrow{\ ?\ } & \mathscr{U}\mathbf{w} \times \mathscr{B} \\ {\scriptstyle f}\big\downarrow & & \big\downarrow{\scriptstyle f} \\ \mathrm{R}_{K/k}(U') & \longrightarrow & \mathrm{R}_{K/k}(UwB) \end{array} \qquad (9.7.2)$$

(using $u \mapsto wuw^{-1}$ on the bottom) can be uniquely filled in to be commutative. The uniqueness is clear since f is injective on k-points. Let us temporarily grant the non-emptiness of $\widetilde{\Omega}$ (and hence of $\mathscr{U}'$), which will later be proved by exhibiting a point in $U'(k)$, and also admit the existence in (9.7.2). We can use these properties along with the commutativity of (9.7.1) to fill in the top of a commutative diagram of k-schemes

$$\begin{array}{ccc} \widetilde{\Omega} & \xrightarrow{\widetilde{\mu}_{V^{(2)},V'}} & \mathscr{U}\mathbf{w} \times \mathscr{B} \\ {\scriptstyle f\times f}\big\downarrow & & \big\downarrow{\scriptstyle f} \\ \Omega & \xrightarrow[\mu]{} & \mathrm{R}_{K/k}(UwB) \end{array} \qquad (9.7.3)$$

where μ is the birational group law on $\mathrm{R}_{K/k}(UwB)$. (The map $\widetilde{\mu}_{V^{(2)},V'}$ depends on $V^{(2)}$ and V' through the dependence of f on this data.) By equivariance for left $\mathscr{U}$-translation and right $\mathscr{B}$-translation, $\widetilde{\mu}_{V^{(2)},V'}$ is *surjective*. Consideration of k-points (using that f is injective on k-points and μ is surjective and associative in the sense of dominant rational maps) implies that $\widetilde{\mu}_{V^{(2)},V'}$ is automatically associative in the sense of dominant rational maps. Thus, by Remark 9.6.3, $\widetilde{\mu}_{V^{(2)},V'}$ would be a *strict* birational group law provided that the universal left and right translations $\rho_{V^{(2)},V'}, \lambda_{V^{(2)},V'} : \widetilde{\Omega} \rightrightarrows (\mathscr{U}\mathbf{w} \times \mathscr{B})^2$ are open immersions. The proof of these open immersion properties requires Zariski's Main Theorem, so we settle it at the end, after we prove $\widetilde{\Omega} \neq \emptyset$ and fill in (9.7.2) as a commutative square.

To analyze (9.7.2) and prove $\widetilde{\Omega}$ is not empty (and hence is $\mathscr{U}\mathbf{w} \times \mathscr{B}$-dense), we need to describe $U' \subseteq U$ and the components of wgw^{-1} in UwB for any point g of U' (i.e., any $g \in U$ such that $wgw^{-1} \in UwB$). This will be expressed in terms of the Bruhat decomposition for GL_n associated to the choice of D_n and U_n^+ (equivalently, the choice of D_n and B_n^+), so we define the *standard open Bruhat cell of* GL_n to be $U_n^+ \times B_n^- = U_n^+ \times D_n \times U_n^-$.

Lemma 9.7.2 *For*

$$g = \begin{pmatrix} {}^{\mathrm{t}}u_0^{-1} & m_0 u_0 \\ 0 & u_0 \end{pmatrix} \in U,$$

wgw^{-1} lies in UwB (i.e., $g \in U'$) if and only if m_0u_0 lies in the standard open Bruhat cell of GL_n (so $m_0 \in \mathrm{GL}_n$, and in particular $m_0 \neq 0$), in which case

$$wgw^{-1} = \begin{pmatrix} {}^{\mathrm{t}}u^{-1} & -m_0^{-1}\,u \\ 0 & u \end{pmatrix} w \begin{pmatrix} {}^{\mathrm{t}}b^{-1} & -u^{-1}\,m_0\,{}^{\mathrm{t}}u^{-1}\,b \\ 0 & b \end{pmatrix},$$

where $u\,{}^{\mathrm{t}}b^{-1}$ is the decomposition of m_0u_0 in $U_n^+B_n^-$ (with $u \in U_n^+$, $b \in B_n^+$).

In particular, for any $x \in K^\times$ the point

$$\gamma_0(x) = \begin{pmatrix} 1_n & x \cdot 1_n \\ 0 & 1_n \end{pmatrix} \in U(K)$$

corresponding to $(m_0, u_0) = (x \cdot 1_n, 1_n)$ lies in $U'(K)$.

Proof We compute $wgw^{-1} = \begin{pmatrix} u_0 & 0 \\ -m_0u_0 & {}^{\mathrm{t}}u_0^{-1} \end{pmatrix}$, and an arbitrary point of UwB is

$$\begin{pmatrix} {}^{\mathrm{t}}u^{-1} & mu \\ 0 & u \end{pmatrix} w \begin{pmatrix} {}^{\mathrm{t}}b^{-1} & m'b \\ 0 & b \end{pmatrix} = \begin{pmatrix} -mu\,{}^{\mathrm{t}}b^{-1} & -mum'b + {}^{\mathrm{t}}u^{-1}\,b \\ -u\,{}^{\mathrm{t}}b^{-1} & -um'b \end{pmatrix}.$$

Hence, necessary and sufficient conditions on $u \in U_n^+$, $b \in B_n^+$, and $m, m' \in \mathrm{Sym}_n$ are:

(i) $mu\,{}^{\mathrm{t}}b^{-1} = -u_0$,
(ii) $mum'b = {}^{\mathrm{t}}u^{-1}\,b$,
(iii) $u\,{}^{\mathrm{t}}b^{-1} = m_0u_0$,
(iv) $um'b = -{}^{\mathrm{t}}u_0^{-1}$.

Condition (iii) encodes precisely the Bruhat constraint on $m_0u_0 \in \mathrm{GL}_n$, and when this holds (i) says $m = -m_0^{-1}$ and (iv) says $m' = -\,u^{-1}m_0\,{}^{\mathrm{t}}u^{-1}$. When these all hold, one checks by direct calculation that (ii) holds. □

Choose $x \in V' - \{0\}$, so we can define a point $u \in \mathscr{U}(k)$ with trivial component along $\mathscr{U}_b$ for all $b \in \Phi_<^+$ and component $(0, x) \in (\underline{V} \times \underline{V'})(k) = \mathscr{U}_c(k)$ for all $2c \in \Phi_>^+$. The image of u under $\mathscr{U} \to \mathrm{R}_{K/k}(U)$ is the point $\gamma_0(x) \in U'(K) = \mathrm{R}_{K/k}(U')(k) \subseteq \mathrm{R}_{K/k}(U)(k)$, so $u \in \mathscr{U}'(k)$. Clearly $(\mathbf{w}u, \mathbf{w}) \in \widetilde{\Omega}(k)$. Hence, $\widetilde{\Omega}, \mathscr{U}' \neq \emptyset$.

Now we prepare to fill in (9.7.2) as a commutative diagram. Consider any symmetric $m_0 \in \mathrm{R}_{K/k}(\mathrm{GL}_n)$ and any $u_0 \in \mathrm{R}_{K/k}(U_n^+)$ such that m_0u_0 lies in the Weil restriction of the standard open Bruhat cell of GL_n. By Lemma 9.7.2, this says exactly that the point

$$g := \begin{pmatrix} {}^{\mathrm{t}}u_0^{-1} & m_0u_0 \\ 0 & u_0 \end{pmatrix} \in \mathrm{R}_{K/k}(U)$$

lies in $\mathrm{R}_{K/k}(U')$, so

$$wgw^{-1} \in \mathrm{R}_{K/k}(UwB) = \mathrm{R}_{K/k}(U)w \times \mathrm{R}_{K/k}(D) \times \mathrm{R}_{K/k}(U).$$

By Remark 9.6.6, our interest is in the diagonal entries of the $\mathrm{R}_{K/k}(\mathrm{Sym}_n)$-components of both $\mathrm{R}_{K/k}(U)$-components of wgw^{-1}. In the proof of Lemma 9.7.2, we saw that these two components, as functorial points of $\mathrm{R}_{K/k}(\mathrm{Sym}_n)$, are $m = -m_0^{-1}$ and $m' = -u^{-1}m_0\,{}^{\mathrm{t}}u^{-1}$, where ub' is the decomposition of m_0u_0 in $\mathrm{R}_{K/k}(U_n^+B_n^-)$, with $u \in \mathrm{R}_{K/k}(U_n^+)$ and $b' \in \mathrm{R}_{K/k}(B_n^-)$. To construct (9.7.2), it remains to prove:

Proposition 9.7.3 *Choose $g \in \mathrm{R}_{K/k}(U')$. Writing as usual*

$$g = \begin{pmatrix} {}^{\mathrm{t}}u_0^{-1} & m_0u_0 \\ 0 & u_0 \end{pmatrix},$$

let ub' be the "Bruhat decomposition" of $m_0u_0 \in \mathrm{R}_{K/k}(U_n^+)\mathrm{R}_{K/k}(B_n^-)$ via Lemma 9.7.2. Identify the pointed k-scheme $\mathrm{R}_{K/k}(U)$ with

$$\prod_{a\in\Phi^+} \mathrm{R}_{K/k}(U_a) = \mathrm{R}_{K/k}(\mathbf{G}_{\mathrm{a}})^{n^2}$$

according to our initial choice of enumeration of Φ^+ (putting the long roots before the short roots) and the parameterizations $u_a : \mathbf{G}_{\mathrm{a}} \simeq U_a$ via the matrix entries of $m_0 \in \mathrm{R}_{K/k}(\mathrm{Sym}_n)$ and $u_0 \in \mathrm{R}_{K/k}(U_n^+)$; this identifies $\mathrm{R}_{K/k}(U')$ with a dense open subscheme $\mathscr{W} \subseteq \mathrm{R}_{K/k}(\mathbf{G}_{\mathrm{a}})^{n^2}$.

Choose $1 \leqslant i \leqslant n$. Viewing the ii entry of $-m_0^{-1}$ (resp. $-u^{-1}m_0\,{}^{\mathrm{t}}u^{-1}$) as an $\mathrm{R}_{K/k}(\mathbf{G}_{\mathrm{a}})$-valued function on $\mathscr{W}$ over k (taking g to be the universal point), its composition with $f : \mathscr{U}' \to \mathrm{R}_{K/k}(U') \simeq \mathscr{W}$ has the form $\underline{q} \circ \varphi_i + \varphi_i'$ (resp. $\underline{q} \circ \psi_i + \psi_i'$) for k-scheme morphisms $\varphi_i : \mathscr{U}' \to \underline{V}$ and $\varphi_i' : \mathscr{U}' \to \underline{V'}$ (resp. $\psi_i : \mathscr{U}' \to \underline{V}$ and $\psi_i' : \mathscr{U}' \to \underline{V'}$).

Proof We first settle the case $n = 1$ by a direct calculation, and then treat the case $n \geqslant 2$ by general considerations with the root system C_n for $n \geqslant 2$. Assuming $n = 1$, we have $U_n^+ = 1$, $\mathscr{W} = R_{K/k}(GL_1) \subseteq R_{K/k}(\mathbf{G}_a)$, and

$$\mathscr{U}' = \{(v, v') \mid \underline{q}(v) + v' \in R_{K/k}(GL_1)\} \subseteq \underline{V} \times \underline{V}'.$$

Thus, in this case the assertion for $-u^{-1}m_0{}^{t}u^{-1} = m_0$ is proved by defining ψ_1 and ψ_1' to be the respective restrictions of the projections $\underline{V} \times \underline{V}' \rightrightarrows \underline{V}, \underline{V}'$ to the dense open subscheme $\mathscr{U}'$ as just described. The assertion for $-m_0^{-1} = m_0^{-1}$ with $n = 1$ rests on the following computation. Consider $(v, v') \in \underline{V} \times \underline{V}'$ such that the point $m_0 := \underline{q}(v) + v' \in \underline{K}$ is valued in $\underline{K}^\times = R_{K/k}(GL_1)$ (i.e., $(v, v') \in \mathscr{U}'$). Clearly

$$m_0^{-1} = m_0^{-2} \cdot m_0 = \underline{q}(m_0^{-1}v) + m_0^{-2}v',$$

with $m_0^{-1}v$ defined as a point of $\underline{V}$ via the $\underline{K}$-module structure on $\underline{V}$ and with $m_0^{-2}v'$ defined as a point of $\underline{V}'$ inside $\underline{K}$ via the fact that V' is a kK^2-subspace of K. Hence, we can define φ_1 and φ_1' at any point (v, v') in the open subscheme $\mathscr{W} \subseteq \underline{V} \times \underline{V}'$ by the following formulas:

$$\varphi_1(v, v') = (\underline{q}(v) + v')^{-1} \cdot v, \quad \varphi_1'(v, v') = (\underline{q}(v) + v')^{-2} \cdot v'.$$

Now we assume $n \geqslant 2$, so $\Phi_<^+$ is non-empty. Perhaps surprisingly, it is easier to treat the case of the diagonal entries of $u^{-1}m_0{}^{t}u^{-1}$ (rather than of $-m_0^{-1}$) even though u is a mysterious function of $(m_0, u_0) \in \mathscr{W}$, so let us first consider such matrix entries. The following lemma will provide what we need.

Lemma 9.7.4 *Let $M = (m_{ij})$ be a symmetric $n \times n$ matrix over an $\mathbf{F}_2$-algebra R (with $n \geqslant 1$), and let $g = (g_{ij}) \in \mathrm{Mat}_n(R')$ for an R-algebra R'. The ii entry of ${}^{t}gMg$ is $\sum_j g_{ji}^2 m_{jj}$.*

Proof Let $\langle\cdot,\cdot\rangle_M$ denote the symmetric bilinear form on affine n-space over R associated to M (i.e., $\langle v, w\rangle_M = \sum m_{ij}v_iw_j$). Since R is an $\mathbf{F}_2$-algebra, the associated quadratic form q_M encodes precisely the diagonal entries m_{ii}: for any point $v = (v_1, \ldots, v_n)$ in affine n-space,

$$q_M(v) := \langle v, v\rangle_M = \sum m_{jj}v_j^2.$$

This involves each v_i only through its square. Now from the computation

$$q_{{}^t gMg}(v) = q_M(gv) = \sum_j m_{jj}\Big(\sum_i g_{ji}v_i\Big)^2 = \sum_i\Big(\sum_j g_{ji}^2 m_{jj}\Big)v_i^2$$

our assertion is obvious. □

For any k-algebra A, any finite sequence of A-valued points (v_α, v'_α) of $\underline{V}\times\underline{V}'$ (i.e., $v_\alpha \in V\otimes_k A$ and $v'_\alpha \in V'\otimes_k A \subseteq K\otimes_k A$), and any $f_\alpha \in \mathrm{R}_{K/k}(\mathbf{G}_\mathrm{a})(A) = K\otimes_k A$, clearly

$$\sum f_\alpha^2\cdot(q_A(v_\alpha)+v'_\alpha) = q_A\Big(\sum f_\alpha\cdot v_\alpha\Big)+\sum f_\alpha^2\cdot v'_\alpha \in (\underline{V}\times\underline{V}')(A) \tag{9.7.4}$$

via the $\underline{K}$-module structure on $\underline{V}$ and the $\underline{kK^2}$-module structure on $\underline{V}'$. Thus, Lemma 9.7.4 implies the existence of the required ψ_i and ψ'_i whenever $n \geqslant 2$.

The proof of Proposition 9.7.3 is reduced to showing that if $m_0 \in \mathrm{R}_{K/k}(\mathrm{GL}_n)$ is *symmetric* and has jj entry of the form $\underline{q}(v_j)+v'_j$ with $(v_j, v'_j) \in \underline{V}\times\underline{V}'$ for every $1 \leqslant j \leqslant n$ then for each $1 \leqslant i \leqslant n$ the ii entry of $m_0^{-1} \in \mathrm{R}_{K/k}(\mathrm{Sym}_n)$ has the form $\underline{q}(h_i)+h'_i$ where h_i is a $\underline{V}$-valued algebraic function (resp. h'_i is a $\underline{V}'$-valued algebraic function) of the pairs (v_j, v'_j) and of the $\underline{K}$-valued off-diagonal entries of m_0. To be precise, h_i and h'_i are required to be k-morphisms from the open subscheme $\mathscr{U}'\subseteq\mathscr{U}$ into $\underline{V}$ and $\underline{V}'$ respectively. By using (9.7.4), the existence of such h_i and h'_i follows from the general lemma below that concerns symmetric matrices over $\mathbf{F}_2$-algebras (such as the coordinate ring of the affine scheme $\mathrm{R}_{K/k}(\mathrm{Sym}_n\cap\mathrm{GL}_n)$). □

Lemma 9.7.5 *Let R be an $\mathbf{F}_2$-algebra and let $m = (r_{ij}) \in \mathrm{Sym}_n(R)\cap\mathrm{GL}_n(R)$ be an invertible symmetric $n\times n$ matrix over R, with $n \geqslant 1$. For $1 \leqslant i \leqslant n$, the ii entry in m^{-1} has the form $\sum_j f_{ij}^2\cdot r_{jj}$ for some $f_{ij} \in R$.*

By applying the lemma to the "universal m" (over the coordinate ring of the affine variety $\mathrm{Sym}_n\cap\mathrm{GL}_n$ over $\mathbf{F}_2$), we obtain universal formulas $f_{ij} = h_{ij}/\det(x_{ij})^{e_{ij}}$ for polynomials h_{ij} over $\mathbf{F}_2$ in the symmetric matrix entries. For example, if $n = 1$ then $f_{11} = 1/x_{11}$, and if $n = 2$ then it is easy to verify that

$$f_{11} = \frac{x_{22}}{\det(x_{ij})},\quad f_{12} = \frac{x_{12}}{\det(x_{ij})},\quad f_{21} = \frac{x_{12}}{\det(x_{ij})},\quad f_{22} = \frac{x_{11}}{\det(x_{ij})}$$

(where $x_{12} = x_{21}$).

Proof Let $\mathbf{F}_2[\underline{x}] = \mathbf{F}_2[x_{ij}]_{1 \leqslant i,j \leqslant n}$, so $A := \mathbf{F}_2[\underline{x}]/(x_{ij} - x_{ji})_{1 \leqslant i < j \leqslant n}$ is the coordinate ring of Sym_n over $\mathbf{F}_2$. The domain $A[1/\det]$ is the coordinate ring of the open subscheme $\mathrm{Sym}_n \cap \mathrm{GL}_n$, and $F := \mathrm{Frac}(A)$ is the function field of Sym_n. Let $m = (x_{ij})$ be the universal symmetric matrix over A, so m is invertible over $A[1/\det]$. It suffices to show that m^{-1} has ii entry of the form $\sum_j f_{ij}^2 \cdot x_{jj}$ for some elements $f_{ij} \in A[1/\det]$.

We wish to exploit the fact that F has an F^2-basis given by the products among $\{x_{ij}\}_{i \leqslant j}$ (without repetition), so we first show that it suffices to construct the f_{ij} as elements of F. More specifically, if k is a field of characteristic $p > 0$ and A is a polynomial ring over k in some variables $y_1, \ldots, y_N$ (such as the x_{ij} for $1 \leqslant i \leqslant j \leqslant n$ as above), then for $f_1, \ldots, f_N \in \mathrm{Frac}(A)$ such that $f := \sum f_i^p \cdot y_i \in A[1/a]$ for some nonzero $a \in A$ we claim that $f_i \in A[1/a]$ for all i. To prove this, we may replace f with $a^{Np} f$ for $N \geqslant 0$ to arrange that $f \in A$. It suffices to show that $f_i \in A$ for all i, so we may assume $a = 1$. Suppose that some f_i has denominator divisible by an irreducible $\pi \in A$. By replacing f with a suitable A-multiple we can arrange that $f_i = g_i/\pi$ with $g_i \in A$ for all i and $\pi \nmid g_{i_0}$ for some i_0. Thus, $g := \sum g_i^p y_i = \pi^p f \in \pi^p A$, so $\partial g/\partial y_{i_0} \in \pi^p A$ as well. But this derivative is clearly equal to $g_{i_0}^p$, so $\pi^p | g_{i_0}^p$, contradicting that the irreducible π does not divide g_{i_0}. Now we may and do work field-theoretically.

For each subset I of $\{1, \ldots, n\} \times \{1, \ldots, n\}$ consisting of ordered pairs (j, j') with $j \leqslant j'$, define $x_I = \prod_{(j,j') \in I} x_{jj'}$ (so $x_\emptyset = 1$). These x_I are an F^2-basis of F. Choose $1 \leqslant i \leqslant n$, and let $h \in A[1/\det]$ denote the (visibly nonzero) ii entry of m^{-1}. This is homogeneous of degree -1 (i.e., $h = \phi_1/\phi_2$ for homogeneous polynomials ϕ_1 and ϕ_2 over $\mathbf{F}_2$ such that $\deg \phi_1 = \deg \phi_2 - 1$), and there is a unique expansion

$$h = \sum_I f_I^2 \cdot x_I$$

with $f_I \in F$. Clearly if f_I is nonzero then it is homogeneous of degree d_I that satisfies $2d_I + \#I = -1$, so $\#I$ is odd. In particular, $f_\emptyset = 0$. Hence, it suffices to prove that $f_I = 0$ for all *non-empty* I that are not singletons of the form $\{(j, j)\}$.

For any $1 \leqslant j \leqslant j' \leqslant n$,

$$\frac{\partial h}{\partial x_{jj'}} = \sum_I f_I^2 \cdot \frac{\partial x_I}{\partial x_{jj'}}.$$

The derivative $\partial x_I/\partial x_{jj'}$ vanishes when $(j, j') \notin I$ and it is equal to $x_{I - \{(j,j')\}}$ when $(j, j') \in I$, so if $\partial h/\partial x_{jj'} = 0$ then $f_I = 0$ whenever $(j, j') \in I$. Hence, if we can show that $\partial h/\partial x_{jj'} = 0$ whenever $1 \leqslant j < j' \leqslant n$ then $f_I = 0$

except possibly when I is a subset of $\{(1,1),\ldots,(n,n)\}$. The vanishing of these partial derivatives (as we let h vary through the diagonal entries of m^{-1}) says exactly that the first-order partial derivatives of $m \mapsto m^{-1}$ in the off-diagonal directions have vanishing diagonal. To prove this vanishing, we shall carry out a computation using a "generic" *symmetric* $n \times n$ matrix $\varepsilon = (\varepsilon_{ij})$ with independent and orthogonal first-order infinitesimals and vanishing diagonal. (The vanishing of the diagonal of ε corresponds to the fact that we only seek to compute the first-order partial derivatives in off-diagonal directions.) That is, we apply the scalar extension functor

$$(\cdot) \otimes_{\mathbf{F}_2} \mathbf{F}_2[\varepsilon_{ij}]_{i<j}/(\varepsilon_{ij}\varepsilon_{i'j'})_{i<j;i'<j'}$$

and aim to show that $(m+\varepsilon)^{-1} + m^{-1}$ has vanishing diagonal.

Writing $\varepsilon = mm^{-1}\varepsilon$, we find that

$$(m+\varepsilon)^{-1} - m^{-1} = m^{-1}\varepsilon m^{-1}$$

(due to the orthogonality of distinct $\varepsilon_{jj'}$). Thus, our problem is to prove that the matrix $m^{-1}\varepsilon m^{-1}$ has vanishing diagonal. This matrix is symmetric (since m and ε are symmetric), and a symmetric matrix T over an $\mathbf{F}_2$-algebra has vanishing diagonal if and only if the associated quadratic form q_T as in the proof of Lemma 9.7.4 vanishes. The symmetry of m^{-1} implies that $q_{m^{-1}\varepsilon m^{-1}}$ is obtained from q_ε by an invertible linear change of variables, so we just need to show that $q_\varepsilon = 0$. This vanishing is clear since ε is symmetric and has vanishing diagonal.

Since $f_\emptyset = 0$ and we have shown that $f_I = 0$ whenever I contains an element (j,j') with $j < j'$, to show that each h has the desired form it suffices to prove that $\partial^2 h/\partial x_{jj}\partial x_{j'j'} = 0$ for any $1 \leqslant j < j' \leqslant n$. This must be proved for h varying through the diagonal entries of m^{-1}, so now we consider the *diagonal* matrix $\varepsilon = \mathrm{diag}(\varepsilon_1,\ldots,\varepsilon_n)$ over the ring

$$\mathbf{F}_2[\varepsilon_1,\ldots,\varepsilon_n]/(\varepsilon_1^2,\ldots,\varepsilon_n^2);$$

we do not impose the condition $\varepsilon_j\varepsilon_{j'} = 0$ for $j < j'$, since such a product will be used to compute the second-order partial derivative in the "directions" of x_{jj} and $x_{j'j'}$. The desired vanishing for specific second-order partial derivatives of the diagonal entries of m^{-1} means that for $1 \leqslant j < j' \leqslant n$ the coefficient matrix of $\varepsilon_j\varepsilon_{j'}$ in $(m+\varepsilon)^{-1} + m^{-1}$ has vanishing diagonal. Varying over all such pairs (j,j'), this says precisely that the degree-2 homogeneous part in the ε_i for $(m+\varepsilon)^{-1} + m^{-1}$ has vanishing diagonal. Clearly

$$\begin{aligned}(m+\varepsilon)^{-1} - m^{-1} &= (1+m^{-1}\varepsilon)^{-1}m^{-1} - m^{-1} \\ &\equiv m^{-1}(-\varepsilon + \varepsilon m^{-1}\varepsilon)m^{-1} \bmod (\varepsilon_1, \ldots, \varepsilon_n)^3,\end{aligned}$$

so our problem is to prove the vanishing of the diagonal of $(m^{-1}\varepsilon)m^{-1}(\varepsilon m^{-1})$. But m^{-1} is *symmetric*, as is ε, so by Lemma 9.7.4 with $g = \varepsilon m^{-1}$ we see that the diagonal of interest involves the matrix entries of $g = \varepsilon m^{-1}$ through their *squares*. These squares vanish since ε is diagonal and $\varepsilon_i^2 = 0$ for all i, so we are done. □

Our analysis of (9.7.2) is complete, so we obtain $\widetilde{\mu}_{V(2),V'}$ as in (9.7.3). This is independent of the enumeration of Φ^+ (in the sense made precise by Remark 9.6.15). To finish the proof of Theorem 9.7.1, we have to show that for $\widetilde{\Omega} = (f \times f)^{-1}(\Omega) \subseteq (\mathscr{U}\mathbf{w} \times \mathscr{B})^2$, the universal translation morphisms

$$\rho_{V(2),V'}, \lambda_{V(2),V'} : \widetilde{\Omega} \rightrightarrows (\mathscr{U}\mathbf{w} \times \mathscr{B})^2$$

are open immersions. We treat the case of the universal left translation, and the case of right translation goes the same way. We will show that $\lambda_{V(2),V'}$ is injective on R-points for any k-algebra R. That is, if $x, y \in (\mathscr{U}\mathbf{w} \times \mathscr{B})(R)$ and $(x, y) \in \widetilde{\Omega}(R)$ then we claim that the ordered pair $(x, \widetilde{\mu}_{V(2),V'}(x, y))$ of points in $(\mathscr{U}\mathbf{w} \times \mathscr{B})(R)$ determines y. Granting this for a moment, let us deduce that $\lambda_{V(2),V'}$ is an open immersion. The map $\lambda_{V(2),V'}$ has finite fibers (as follows just from the assumed injectivity of $\lambda_{V(2),V'}$ on geometric points), so since $\widetilde{\Omega}$ and $(\mathscr{U}\mathbf{w} \times \mathscr{B})^2$ are geometrically irreducible (over k) of the same dimension it follows that $\lambda_{V(2),V'}$ is dominant and generically finite. Hence, the induced map

$$\operatorname{Spec} k(\widetilde{\Omega}) \to \operatorname{Spec}(k((\mathscr{U}\mathbf{w} \times \mathscr{B})^2))$$

at generic points is finite and a functorial monomorphism. But if F'/F is a finite extension of fields such that $\operatorname{Spec} F' \to \operatorname{Spec} F$ is a monomorphism of schemes (equivalently, F' has at most one F-algebra map to any F-algebra) then the projections $\operatorname{Spec}(F' \otimes_F F') \rightrightarrows \operatorname{Spec} F'$ coincide, so $F' = F$. Thus, $\lambda_{V(2),V'}$ is birational. By Zariski's Main Theorem [EGA, IV$_3$, 8.12.10], any birational morphism of finite type between separated integral normal noetherian schemes is an open immersion if it has finite fibers. This implies that $\lambda_{V(2),V'}$ is an open immersion, conditional on $\lambda_{V(2),V'}$ being a functorial injection.

To prove the injectivity of $\lambda_{V(2),V'}$ on R-points for every k-algebra R, first observe that inversion on Sp_{2n} preserves $UwB = UwDU$ (since w normalizes D), so inversion on $\mathrm{R}_{K/k}(\mathrm{Sp}_{2n})$ preserves $\mathrm{R}_{K/k}(UwB)$. Letting i denote the resulting involution of $\mathrm{R}_{K/k}(UwB)$, we seek an involution I of $\mathscr{U}\mathbf{w} \times \mathscr{B}$ lifting

i with enough properties to undo $\widetilde{\mu}_{V^{(2)},V'}$ on the entirety of $\widetilde{\Omega}$. This is made precise by Lemma 9.7.6 below, from which the functorial injectivity of $\lambda_{V^{(2)},V'}$ follows. □

Lemma 9.7.6 *There is a unique endomorphism I of $\mathscr{U}\mathbf{w} \times \mathscr{B}$ over k fitting into a commutative diagram of k-schemes*

$$\begin{array}{ccc} \mathscr{U}\mathbf{w} \times \mathscr{B} & \xrightarrow{I} & \mathscr{U}\mathbf{w} \times \mathscr{B} \\ {\scriptstyle f}\downarrow & & \downarrow{\scriptstyle f} \\ \mathrm{R}_{K/k}(UwB) & \underset{i}{\xrightarrow{\simeq}} & \mathrm{R}_{K/k}(UwB) \end{array} \tag{9.7.5}$$

and I is an involution fixing $(\mathbf{w}, 1)$.

Moreover, for $(x, y) \in \widetilde{\Omega} \subseteq (\mathscr{U}\mathbf{w} \times \mathscr{B})^2$, *the point* $(I(x), \widetilde{\mu}_{V^{(2)},V'}(x, y))$ *also lies in* $\widetilde{\Omega}$ *and* $\widetilde{\mu}_{V^{(2)},V'}(I(x), \widetilde{\mu}_{V^{(2)},V'}(x, y)) = y$.

Proof We use the same technique as in the proof of Theorem 9.7.1, except that matters will be much simpler because we do not have to contend with domains of definition and we can write things down explicitly. Uniqueness follows as usual from the injectivity of f on k-points, and for the same reason I is automatically an involution fixing $(\mathbf{w}, 1)$ if it exists since i is an involution fixing w (i.e., $w^2 = 1$, as $\mathrm{char}(k) = 2$). Since w normalizes D via inversion, in $UwB = UwDU$ the formula for inversion is $i(uwtu') = u'^{-1}wtu^{-1}$. Hence, we define I on $\mathscr{U}\mathbf{w} \times \mathscr{B} = \mathscr{U}\mathbf{w} \times (\mathrm{R}_{K/k}(D) \ltimes \mathscr{U})$ by $I(u\mathbf{w}, tu') := (u'^{-1}\mathbf{w}, tu^{-1})$. This clearly makes (9.7.5) commutative.

By definition, $\widetilde{\Omega}$ is the full preimage of the open locus Ω of points in $\mathrm{R}_{K/k}(UwB)^2$ whose product in $\mathrm{R}_{K/k}(\mathrm{Sp}_{2n})$ is contained in $\mathrm{R}_{K/k}(UwB)$. Thus, to check if $(I(x), \widetilde{\mu}_{V^{(2)},V'}(x, y)) \in \widetilde{\Omega}$ for any $(x, y) \in \widetilde{\Omega}$ it suffices to apply $f \times f$ and check that $(i(x'), \mu(x', y')) \in \Omega$ for any $(x', y') \in \Omega$. But this is clear since $\mu(i(x'), \mu(x', y')) = y' \in \mathrm{R}_{K/k}(UwB)$. We conclude that the expression $\widetilde{\mu}_{V^{(2)},V'}(I(x), \widetilde{\mu}_{V^{(2)},V'}(x, y))$ makes sense as a map from $\widetilde{\Omega}$ to $\mathscr{U}\mathbf{w} \times \mathscr{B}$. To check that it agrees with $\mathrm{pr}_2 : \widetilde{\Omega} \to \mathscr{U}\mathbf{w} \times \mathscr{B}$, we are faced with the task of comparing two k-morphisms $\widetilde{\Omega} \rightrightarrows \mathscr{U}\mathbf{w} \times \mathscr{B}$. Since $\widetilde{\Omega}$ is a rational variety, it suffices to compare on k-points. Thus, by applying f we are reduced to the analogous agreement for two maps $\Omega \rightrightarrows \mathrm{R}_{K/k}(UwB)$, in which case the agreement is obvious by working inside the k-group $\mathrm{R}_{K/k}(\mathrm{Sp}_{2n})$. □

Using the strict birational group law $\widetilde{\mu}_{V^{(2)},V'}$ on $\mathscr{U}\mathbf{w} \times \mathscr{B}$ that is built above, Weil's Theorem on the unique extension of strict birational group laws to actual groups (Theorem 9.6.4) allows us to make the following general definition.

Definition 9.7.7 Let k be an imperfect field of characteristic 2, K/k a nontrivial purely inseparable finite extension, and $V^{(2)}, V' \subseteq K$ a pair of nonzero K^2-subspaces as in 9.6.8. For each $n \geqslant 1$, define $G_{K/k,V^{(2)},V',n}$ to be the smooth connected k-group containing $(\mathscr{U}\mathbf{w} \times \mathscr{B}, \widetilde{\mu}_{V^{(2)},V'})$ as a dense open locus.

By construction, $G_{K/k,V^{(2)},V',n}$ is equipped with a canonical k-homomorphism to $\mathrm{R}_{K/k}(\mathrm{Sp}_{2n})$ (the unique one extending f). Remark 9.6.11 ensures that the formation of $G_{K/k,V^{(2)},V',n}$ and its k-homomorphism to $\mathrm{R}_{K/k}(\mathrm{Sp}_{2n})$ are compatible with separable extension on k. To simplify notation, we generally write $G_{V^{(2)},V',n}$ rather than $G_{K/k,V^{(2)},V',n}$ when K/k is understood from context. Note that up to unique k-isomorphism, $G_{V^{(2)},V',n}$ is independent of any choice of enumeration of Φ_n^+ (see Remark 9.6.15).

9.7.8 The group $\mathscr{B} = \mathrm{R}_{K/k}(D) \ltimes \mathscr{U}$ used in the construction of $G_{V^{(2)},V',n}$ is usually too large if we want $G_{V^{(2)},V',n}$ to be perfect. To explain the difficulty, let Δ_n be the basis of Φ_n^+, and let $2c$ be its unique long element, so $\Delta_n^\vee$ is a $\mathbf{Z}$-basis of $\mathrm{X}_*(D)$ (since Sp_{2n} is simply connected). Under the product decomposition $\mathrm{R}_{K/k}(D) = \prod_{a\in\Delta_n} \mathrm{R}_{K/k}(a^\vee(\mathrm{GL}_1))$, the $2c$-coroot factor needs to be made smaller because for V and q as in 9.6.8 the image of $\mathscr{U}_{\pm c} = \underline{V} \times \underline{V'}$ in $\mathrm{R}_{K/k}(\mathbf{G}_\mathrm{a})$ under $(v, v') \mapsto \underline{q}(v) + v'$ is usually not full. To be precise, let $V_0 \subseteq K$ be the kK^2-subspace that is the k-span of the K^2-subspace $V^{(2)}$ and the kK^2-subspace V', let $K_0 = k\langle V_0\rangle$, and define $(V_0)^*_{K_0/k} \subseteq \mathrm{R}_{K_0/k}(\mathrm{GL}_1)$ to be the Zariski closure of the subgroup of $K_0^\times$ generated by ratios of elements of $V_0 - \{0\}$. We wish to replace the $2c$-coroot factor of $\mathrm{R}_{K/k}(D)$ with $(V_0)^*_{K_0/k}$ in the definition of $\mathscr{B}$; i.e., we want to use the k-groups

$$C_{V^{(2)},V',n} = (V_0)^*_{K_0/k} \times \prod_{a\in\Delta_n-\{2c\}} \mathrm{R}_{K/k}(\mathrm{GL}_1), \qquad (9.7.6)$$
$$\mathscr{B}_{V^{(2)},V',n} = C_{V^{(2)},V',n} \ltimes \mathscr{U}$$

rather than $\mathrm{R}_{K/k}(D)$ and $\mathscr{B}$ respectively. This is motivated by the description of Cartan k-subgroups in Proposition 9.1.4.

Note that $C_{V^{(2)},V',n}$ is commutative and pseudo-reductive, since it is a smooth connected k-subgroup of $\mathrm{R}_{K/k}(D)$, and its formation is compatible with separable extension on k (due to Remark 9.6.11). It is unpleasant to carry out the constructions from 9.6.12 up to Definition 9.7.7 using the smaller groups in (9.7.6) because that entails keeping track of membership in the k-subgroup $C_{V^{(2)},V',n} \subseteq \mathrm{R}_{K/k}(D)$ when carrying out conjugation computations. Thus, we prefer to construct the k-group $G = G_{V^{(2)},V',n}$ as above via a strict birational group law using $\mathrm{R}_{K/k}(D)$ and $\mathscr{B}$, and then pass to $\mathscr{D}(G)$ (which we will prove is absolutely pseudo-simple with the "expected" open cell; e.g., $C_{V^{(2)},V',n}$ is a Cartan k-subgroup of $\mathscr{D}(G)$).

It is not obvious from the definition that $G_{V^{(2)},V',n}$ is *affine*, let alone pseudo-reductive over k, nor that $\mathscr{D}(G_{V^{(2)},V',n})$ is pseudo-split and absolutely pseudo-simple over k with a non-reduced root system of type BC_n and the "expected" root groups (and with K/k as the field of definition of its geometric unipotent radical, provided that for $n = 1$ we impose the additional requirement $K_0 = K$ that is also necessary due to Proposition 9.1.4).

9.7.9 For later purposes, it is convenient to record the behavior of $G_{V^{(2)},V',n}$ under $K^\times$-scaling on $(V^{(2)}, V')$ inside K. This involves K-automorphisms of Sp_{2n} extending the identity on the diagonal torus and inducing specific scaling operations on certain root groups, so we now define these automorphisms.

Write points of the K-group Sp_{2n} in the form $g = \left(\begin{smallmatrix} a & b \\ c & d \end{smallmatrix}\right)$ with $a, b, c, d \in \mathrm{Mat}_n$ subject to the conditions ${}^{\mathrm{t}}a \cdot c = {}^{\mathrm{t}}c \cdot a$, ${}^{\mathrm{t}}b \cdot d = {}^{\mathrm{t}}d \cdot b$, ${}^{\mathrm{t}}a \cdot d - {}^{\mathrm{t}}c \cdot b = 1_n$. (These conditions say exactly that the point $g \in \mathrm{Mat}_{2n}$ preserves the standard symplectic form in $2n$ dimensions, so they force g to be invertible.) For $\lambda \in K^\times$ it is easy to check that $\left(\begin{smallmatrix} a & \lambda b \\ \lambda^{-1} c & d \end{smallmatrix}\right)$ is also a point in Sp_{2n} and that the map

$$i_\lambda : \begin{pmatrix} a & b \\ c & d \end{pmatrix} \mapsto \begin{pmatrix} a & \lambda b \\ \lambda^{-1} c & d \end{pmatrix}$$

is a K-group automorphism of Sp_{2n} that is the identity on D and induces multiplication by λ on *every* long positive root group (i.e., on the $[m]_{ii}$ from (9.6.3)). We can describe i_λ in terms of coroots as follows. Let $\mu = \mu_2$ denote the center of Sp_{2n} and let $\Phi_n^+ = \Phi(\mathrm{Sp}_{2n}, D)^+$ be the positive system of roots corresponding to (9.6.1), so the basis for Φ_n^+ has the form $\{a_1, \ldots, a_n\}$ with a_n long and a_i adjacent to a_{i+1} in the Dynkin diagram for $1 \leqslant i \leqslant n - 1$. Clearly $\{a_1, a_2, \ldots, a_n\}$ is a $\mathbf{Z}$-basis for $\mathrm{X}(D/\mu) \subseteq \mathrm{X}(D)_{\mathbf{Q}}$, and the automorphism i_λ of Sp_{2n} centralizing D is the action by the unique $t \in (D/\mu)(K)$ such that $a_n(t) = 1/\lambda$ and $a_j(t) = 1$ for all $j < n$. Clearly i_{λ^2} is conjugation by $\left(\begin{smallmatrix} \lambda \cdot 1_n & 0 \\ 0 & \lambda^{-1} \cdot 1_n \end{smallmatrix}\right)$, so i_λ coincides with conjugation by an element of $\mathrm{Sp}_{2n}(K)$, or equivalently by an element of $D(K)$, if and only if $\lambda \in (K^\times)^2$.

(The Isomorphism Theorem for pinned split connected reductive groups and the root group parameterizations in (9.6.2) and (9.6.3) provide a conceptual characterization of i_λ: it is the unique automorphism that is the identity on

D and on the simple short positive root groups $[u]_{i,i+1}$ $(1 \leqslant i < n)$, and is multiplication by λ on the simple long positive root group $[m]_{11}$.)

Proposition 9.7.10 *Let $(K/k, V^{(2)}, V')$ be as in 9.6.8, and define K_0 to be $k\langle V^{(2)} + V' \rangle \subseteq K$. Fix an integer $n \geqslant 1$, and if $n = 1$ then assume $K_0 = K$. Let D_0 be the diagonal k-torus in Sp_{2n}.*

For any $\lambda \in K^\times$ there is a unique k-isomorphism θ fitting into a commutative diagram

$$\begin{array}{ccc} G_{K/k,V^{(2)},V',n} & \xrightarrow[\simeq]{\theta} & G_{K/k,\lambda V^{(2)},\lambda V',n} \\ \downarrow & & \downarrow \\ \mathrm{R}_{K/k}(\mathrm{Sp}_{2n}) & \xrightarrow[\mathrm{R}_{K/k}(i_\lambda)]{} & \mathrm{R}_{K/k}(\mathrm{Sp}_{2n}) \end{array}$$

where the vertical maps are the canonical ones.

Proof The uniqueness follows from the injectivity on k_s-points in a dense open subset of the source for the vertical maps in the diagram. For similar reasons, it suffices to construct θ instead as a k-scheme isomorphism between the dense open subschemes "$\mathscr{U}\mathbf{w}\mathscr{B}$" in the two k-groups along the top, as then θ respects the birational group laws (since it suffices to check this by computing on k_s-points and using that $\mathrm{R}_{K/k}(i_\lambda)$ is a k-group isomorphism). Indeed, the functorial aspect of Theorem 9.6.4 extends such a rational map θ to a k-group isomorphism fitting into the desired commutative diagram. Note that both vertical maps carry $(\mathbf{w}, 1)$ to w, and $\mathrm{R}_{K/k}(i_\lambda)(w) = w\,d(\lambda)$ where $d(\lambda) := \begin{pmatrix} \lambda^{-1}.1_n & 0 \\ 0 & \lambda\cdot 1_n \end{pmatrix} \in D(K) = \mathrm{R}_{K/k}(D)(k)$.

Let $\mathscr{U}_{V^{(2)},V'}$ and $\mathscr{U}_{\lambda V^{(2)},\lambda V'}$ be the unipotent smooth connected k-subgroups that serve in the role of "$\mathscr{U}$" in the construction of $G_{K/k,V^{(2)},V',n}$ and $G_{K/k,\lambda V^{(2)},\lambda V',n}$. We claim that there is a k-scheme isomorphism θ_0 fitting into the top of a commutative diagram of k-schemes

$$\begin{array}{ccc} \mathscr{U}_{V^{(2)},V'} & \xrightarrow[\simeq]{\theta_0} & \mathscr{U}_{\lambda V^{(2)},\lambda V'} \\ \downarrow & & \downarrow \\ \mathrm{R}_{K/k}(U) & \xrightarrow[\mathrm{R}_{K/k}(i_\lambda)]{} & \mathrm{R}_{K/k}(U) \end{array}$$

where U is the unipotent radical of the upper triangular Borel subgroup of Sp_{2n} and the vertical maps are the natural ones as in Theorem 9.6.14. Since the vertical maps are defined as direct products according to factors indexed by Φ_n^+, and i_λ is likewise such a product map (either by inspection, or because it commutes with D-conjugation), to construct θ_0 it suffices to work separately with each root in Φ_n^+. At the factors for each short positive root in Φ_n there is nothing to do since the vertical maps are isomorphisms between such factors, so we focus on the factors at the long positive roots in Φ_n.

Let $\widetilde{V}$ be the K-vector space obtained from the K^2-vector space $\lambda V^{(2)}$ via scalar extension through the square root isomorphism $K^2 \simeq K$, so the 2-linear bijection $\widetilde{q} : \widetilde{V} \to \lambda V^{(2)}$ carries $1 \otimes \lambda v$ to λv. Since i_λ restricts to multiplication by λ on all long positive root groups, we can define the product map θ_0 : $\mathscr{U}_{V^{(2)},V'} \simeq \mathscr{U}_{\lambda V^{(2)},\lambda V'}$ on the factors indexed by long positive roots to be the product map $\underline{V} \times \underline{V'} \simeq \underline{\widetilde{V}} \times \underline{\lambda V'}$ induced by the kK^2-linear λ-multiplication $V' \simeq \lambda V'$ and the scalar extension along $K^2 \simeq K$ of the K^2-linear λ-multiplication $V^{(2)} \simeq \lambda V^{(2)}$. It is easy to see that this direct product map yields the desired commutative diagram. The isomorphism θ between dense open subschemes "$\mathscr{U}\mathbf{w}\mathscr{B}$" in $G_{K/k,V^{(2)},V',n}$ and $G_{K/k,\lambda V^{(2)},\lambda V',n}$ may now be defined by $u\mathbf{w}tu' \mapsto \theta_0(u)\mathbf{w}d(\lambda)t\theta_0(u')$. □

9.8 Properties of the groups $G_{V^{(2)},V',n}$

To establish that the groups we have built via Weil's birational method satisfy the expected properties, we construct a special class of absolutely pseudo-simple k-subgroups of $\mathrm{R}_{K/k}(\mathrm{Sp}_{2n})$ for any $n \geqslant 1$ and any nontrivial finite purely inseparable extension K of k. This construction rests on Theorem C.2.29, which gives a technique to build pseudo-split pseudo-reductive subgroups from a given split torus S and suitable subgroups of root groups for a linearly independent subset of $\mathrm{X}(S)$. As a first step, we now use Theorem C.2.29 to show that $G_{V^{(2)},V',n}$ is a pseudo-split pseudo-reductive k-group (in particular, affine) of rank n, with a non-reduced root system. In fact, we prove more:

Theorem 9.8.1 *Let k be an imperfect field of characteristic 2, K/k a purely inseparable nontrivial finite extension, and $V^{(2)}, V' \subseteq K$ a pair of nonzero K^2-subspaces as in 9.6.8. Define the kK^2-subalgebra $K_0 = k\langle V^{(2)} + V'\rangle \subseteq K$ and the kK^2-subspace $V_0 = kV^{(2)} + V' \subseteq K$, where $kV^{(2)}$ denotes the k-span of $V^{(2)}$ inside K. Define $q : V \to K$ as in 9.6.8.*

Fix $n \geqslant 1$ and let $\Phi = \Phi_n := \Phi(\mathrm{Sp}_{2n}, D)$. If $n = 1$, assume $K_0 = K$. Let $G = G_{V^{(2)},V',n}$, $\Omega = \mathscr{U}\mathbf{w} \times \mathscr{B}$, $C = C_{V^{(2)},V',n}$, and $\Omega' = \mathscr{U}\mathbf{w} \times \mathscr{B}_{V^{(2)},V',n}$, with $\mathscr{U} = \prod_{a\in\Phi^+} \mathscr{U}_a$ as in 9.6.12 and $C_{V^{(2)},V',n}$ and $\mathscr{B}_{V^{(2)},V',n}$ as in (9.7.6).

Let $\phi : G \to \mathrm{R}_{K/k}(\mathrm{Sp}_{2n})$ be the unique k-homomorphism extending the "rational homomorphism" $f : \Omega \to \mathrm{R}_{K/k}(UwB)$ via the functorial aspect of Theorem 9.6.4.

(1) *The smooth connected k-group G is pseudo-reductive of minimal type, $\mathscr{D}(G)$ is absolutely pseudo-simple and of minimal type, and if $1 \in V'$ then $(\mathbf{w}, 1) \in \mathscr{D}(G)(k)$.*

(2) *There is a unique Cartan k-subgroup of G carried isomorphically by ϕ onto $\mathrm{R}_{K/k}(D) \subseteq \mathrm{R}_{K/k}(\mathrm{Sp}_{2n})$, and under this identification the k-subgroup C of $\mathrm{R}_{K/k}(D)$ with a Cartan k-subgroup of $\mathscr{D}(G)$. If $D_0 \subseteq \mathrm{R}_{K/k}(D)$ denotes the k-split maximal k-torus then the subset $\Phi(\mathscr{D}(G), D_0) = \Phi(G, D_0) \subseteq \mathrm{X}(D_0) = \mathrm{X}(D)$ is the non-reduced irreducible root system obtained from Φ by including $a/2$ for every $a \in \Phi_>$. There exists a Levi k-subgroup of G containing D_0 that maps isomorphically onto a Levi k-subgroup of $\mathrm{R}_{K/k}(\mathrm{Sp}_{2n})$.*

Assume $1 \in V'$. Inversion on the commutative k-subgroup $\mathrm{R}_{K/k}(D) \subseteq \mathrm{R}_{K/k}(\mathrm{Sp}_{2n})$ uniquely lifts to an inclusion $\mathrm{R}_{K/k}(D) \hookrightarrow G$ via $t \mapsto (\mathbf{w}, t)(\mathbf{w}, 1)^{-1}$, and the canonical k-subgroup $\mathrm{Sp}_{2n} \subseteq \mathrm{R}_{K/k}(\mathrm{Sp}_{2n})$ uniquely lifts to a k-subgroup L of G. This is a common Levi k-subgroup of G and $\mathscr{D}(G)$ containing D_0 and $(\mathbf{w}, 1)$, and the root groups of (L, D_0) are the following canonical k-subgroups: $\mathbf{G}_{\mathrm{a}} \subseteq \mathrm{R}_{K/k}(\mathbf{G}_{\mathrm{a}}) = \mathscr{U}_a$ for $a \in \Phi^+_<$, $\mathbf{G}_{\mathrm{a}} \subseteq \underline{V}' \subseteq \underline{V} \times \underline{V}' = \mathscr{U}_{a/2}$ for $a \in \Phi^+_>$, and their $(\mathbf{w}, 1)$-conjugates (for $a \in -\Phi^+$).

The open cell in G associated to the positive system of roots in $\Phi(G, D_0)$ corresponding to Φ^+ is $\Omega \cdot (\mathbf{w}, 1)^{-1}$, with $\mathscr{U}$ the smooth connected unipotent k-subgroup generated by the root groups for positive roots. The analogous open cell in $\mathscr{D}(G)$ is $\Omega' \cdot (\mathbf{w}, 1)^{-1}$ (so $\Omega' = \Omega \cap \mathscr{D}(G)$).

(3) *The fields of definition over k for the $\overline{k}$-subgroups $\mathscr{R}_u(G_{\overline{k}}) \subseteq G_{\overline{k}}$ and $\mathscr{R}(\mathscr{D}(G)_{\overline{k}}) \subseteq \mathscr{D}(G)_{\overline{k}}$ both coincide with K, the k-homomorphisms ϕ and $\phi|_{\mathscr{D}(G)}$ respectively extending $f : \Omega \to \mathrm{R}_{K/k}(UwB)$ and $f' = f|_{\Omega'}$ coincide with i_G and $\xi_{\mathscr{D}(G)}$, and i_G is injective on k_s-points (so $\xi_{\mathscr{D}(G)}$ is also injective on k_s-points). For multipliable roots c, the common field of definition over k for the geometric unipotent radical of $\mathscr{D}(G)_c$ is K_0.*

(4) *Let $h : \underline{V} \times \underline{V}' \to \underline{K}$ be the k-homomorphism $(v, v') \mapsto \underline{q}(v) + v'$. The map $\xi_{\mathscr{D}(G)}$ is surjective if and only if h is surjective, in which case $G = \mathscr{D}(G)$. If $1 \in V'$ then the k-group $\ker i_G$ is identified with $\prod_{a \in \Phi_>} \ker h$ via multiplication inside G (using the definitions $\mathscr{U}_c = \underline{V} \times \underline{V}'$ and $\mathscr{U}_{-c} = (\mathbf{w}, 1)\mathscr{U}_c(\mathbf{w}, 1)^{-1}$ for all $2c \in \Phi^+_>$).*

Remark 9.8.2 Since $h(\underline{V})$ is a $\underline{kK^2}$-submodule of $\underline{K}$ (due to the Zariski-density of k^2 in $\underline{k} = \mathbf{G}_{\mathrm{a}}$) and the subset $V^{(2)} = h(\underline{V}(k))$ of K is not contained in the nonzero proper kK^2-subspace $\underline{V}'(k) = V' \subseteq K$, the surjectivity of h in Theorem 9.8.1(4) holds whenever $[K : kK^2] = 2$ (so in particular whenever $[k : k^2] = 2$). Such surjectivity clearly implies the equality $K_0 = K$ that is imposed as an assumption when $n = 1$. If $[K : kK^2] > 2$ then we can always find $V^{(2)}$ and V' such that $K_0 \neq K$ (so h is not surjective). See Remark 9.6.9 for a discussion of when the inclusion $\ker \underline{q} \subseteq \ker h$ is an equality.

Proof of Theorem 9.8.1 We may and do assume $k = k_s$ (see Remark 9.6.11). Let us show that it suffices to treat the cases with $1 \in V'$ (for which we exhibit an explicit Levi k-subgroup in part (2)). Choose $v' \in V' - \{0\}$ and let $\lambda = 1/v'$, so $1 \in \lambda V'$. Consider the isomorphism

$$\theta : G = G_{V^{(2)},V',n} \simeq G_{\lambda V^{(2)},\lambda V',n} =: G_\lambda$$

provided by Proposition 9.7.10. This lies over the automorphism $\mathrm{R}_{K/k}(i_\lambda)$ of $\mathrm{R}_{K/k}(\mathrm{Sp}_{2n})$ that is the identity on $\mathrm{R}_{K/k}(D)$. The theorem is assumed to be known for G_λ since $1 \in \lambda V'$. By construction, θ intertwines the maps f and f' in part (3) for G with the analogues for G_λ, so the results for G_λ imply everything for G. For the rest of the proof we shall assume $1 \in V'$, so $K_0 = k[V^{(2)} + V']$.

Since $\phi|_{\Omega(k)}$ recovers f on k-points and so is injective, if $g \in (\ker \phi)(k)$ then by choosing a k-point h in the dense open subset $\Omega \cap \Omega g^{-1}$ of G the points $h, hg \in \Omega(k)$ are carried to the same point by ϕ, forcing $h = hg$ and hence $g = 1$. The resulting triviality of $(\ker \phi)(k)$ implies that $\ker \phi$ contains no nontrivial smooth closed k-subgroup (as $k = k_s$) and that $(\mathbf{w}, 1)$ is the *unique* element of $G(k)$ carried to $w \in \mathrm{Sp}_{2n}(K) = \mathrm{R}_{K/k}(\mathrm{Sp}_{2n})(k)$.

Any homomorphism between smooth k-groups of finite type has closed image, so $\phi(G)$ is a smooth connected closed k-subgroup of $\mathrm{R}_{K/k}(\mathrm{Sp}_{2n})$. The key to proving that G is pseudo-reductive of rank n is to directly construct a *pseudo-reductive* k-subgroup $H \subseteq \mathrm{R}_{K/k}(\mathrm{Sp}_{2n})$ containing the diagonal k-torus D_0 inside $\mathscr{D}(H)$ such that $\phi(G) = H$ and $\mathscr{D}(H)$ is absolutely pseudo-simple with a known open cell relative to D_0.

STEP 1. We begin by constructing H and establishing its basic properties, as an application of Theorem C.2.29 to the pseudo-reductive k-group $\mathscr{G} := \mathrm{R}_{K/k}(\mathrm{Sp}_{2n})$ and its split maximal k-torus $D_0 \subseteq \mathrm{R}_{K/k}(D)$. The root system $\Phi(\mathscr{G}, D_0)$ is naturally identified with $\Phi_n := \Phi(\mathrm{Sp}_{2n}, D)$ via the natural identification of $\mathrm{X}(D)$ with $\mathrm{X}(D_0)$ (using the natural isomorphism $(D_0)_K \simeq D$). Let Δ_n be the basis of Φ_n associated to our conventional choice Φ_n^+. This contains $n - 1$ short roots and a single long root $2c$ with $c \in \mathrm{X}(D_0)$ (so $\Delta_n = \{2c\}$ if $n = 1$).

For each $a \in \Phi_n$, let $U'_a \subseteq \mathrm{Sp}_{2n}$ denote the associated root group over K. (In Step 5, after we have proved that G is pseudo-reductive with split maximal k-torus D_0, we will write U_a to denote the root group in G over k associated to each $a \in \Phi(G, D_0)$.) For each short $a \in \Delta_n \subseteq \mathrm{X}(D_0) = \mathrm{X}(D)$, define F_a to be the smooth connected k-subgroup of $\mathscr{G}$ generated by $\mathrm{R}_{K/k}(D)$ and $\mathrm{R}_{K/k}((\mathrm{Sp}_{2n})_a)$ (with $(\mathrm{Sp}_{2n})_a := \langle U'_a, U'_{-a}\rangle \simeq \mathrm{SL}_2$ since Sp_{2n} is simply connected). Note that $\mathrm{R}_{K/k}(D)$ normalizes $\mathrm{R}_{K/k}((\mathrm{Sp}_{2n})_a)$. The pseudo-reductive $\mathrm{R}_{K/k}((\mathrm{Sp}_{2n})_a)$ is normal in F_a, and the quotient

$$F_a/\mathrm{R}_{K/k}((\mathrm{Sp}_{2n})_a) = \prod_{\Delta_n^\vee - \{a^\vee\}} \mathrm{R}_{K/k}(\mathrm{GL}_1)$$

is visibly pseudo-reductive. Thus, F_a is pseudo-reductive and this computation makes it clear that $F_a \cap Z_{\mathscr{G}}(D_0) = \mathrm{R}_{K/k}(D)$ and $\mathscr{D}(F_a) = \mathrm{R}_{K/k}((\mathrm{Sp}_{2n})_a)$, so $\Phi(F_a, D_0) = \{a, -a\}$.

Define F_{2c} to be the smooth connected k-subgroup of $\mathscr{G}$ generated by $\mathrm{R}_{K/k}(D)$ and the absolutely pseudo-simple k-subgroup

$$H_{V_0} := H_{V_0, K_0/k} \subseteq \mathrm{R}_{K_0/k}((\mathrm{Sp}_{2n})_{2c}) \subseteq \mathrm{R}_{K/k}((\mathrm{Sp}_{2n})_{2c})$$

(implicitly using the evident K_0-structure on the K-group $(\mathrm{Sp}_{2n})_{2c}$). By Proposition 9.1.4,

$$\mathrm{R}_{K/k}(D) \cap H_{V_0} = (V_0)^*_{K_0/k}.$$

Assuming $\mathrm{R}_{K/k}(D)$ normalizes H_{V_0}, the preceding method establishes pseudo-reductivity of the k-group $\mathrm{R}_{K/k}(D) \cdot \mathrm{R}_{K/k}((\mathrm{Sp}_{2n})_{2c})$. This k-group has the Levi k-subgroup $D_0 \cdot (\mathrm{Sp}_{2n})_{2c}$ contained in F_{2c}, so F_{2c} would be pseudo-reductive by Lemma 7.2.4. We could then also verify the following additional properties for $2c$ as we did for the short roots:

$$F_{2c} \cap Z_{\mathscr{G}}(D_0) = \mathrm{R}_{K/k}(D), \quad \mathscr{D}(F_{2c}) = H_{V_0}, \quad \Phi(F_{2c}, D_0) = \{\pm 2c\}.$$

The assumed normalizing property is trivial to verify when $n = 1$ since V_0 is a K^2-subspace of K, so we just have to prove it when $n \geqslant 2$.

Let $b \in \Delta_n$ be the short root adjacent to $2c$ in the Dynkin diagram of (Sp_{2n}, D_0) (i.e., $\langle 2c, b^\vee\rangle \neq 0$ whereas $\langle 2c, a^\vee\rangle = 0$ for all $a \in \Delta_n - \{2c, b\}$). For every $a \in \Delta_n - \{2c, b\}$ the coroot group $a^\vee(\mathrm{GL}_1)$ centralizes $(\mathrm{Sp}_{2n})_{2c}$, so $\mathrm{R}_{K/k}(a^\vee(\mathrm{GL}_1))$ centralizes $\mathrm{R}_{K/k}((\mathrm{Sp}_{2n})_{2c}) \supseteq H_{V_0}$. Thus, we just have to check that $\mathrm{R}_{K/k}(b^\vee(\mathrm{GL}_1))$ normalizes H_{V_0}. It is enough to check this for conjugation by k-points on the standard open cell of H_{V_0}. (This open cell is the overlap of H_{V_0} with the standard open cell of $\mathrm{R}_{K/k}((\mathrm{Sp}_{2n})_{2c})$.)

The points of $R_{K/k}(b^\vee(GL_1))(k)$ are precisely $b^\vee(t)$ for $t \in K^\times$, and $b^\vee(t)$-conjugation on the open cell of $R_{K/k}((Sp_{2n})_{2c})$ is trivial on the direct factor $R_{K/k}((2c)^\vee(GL_1)) \supseteq (V_0)^*_{K_0/k}$ of $R_{K/k}(D)$ and is multiplication by $t^{\langle \pm 2c, b^\vee \rangle} = t^{\mp 2} \in (K^\times)^2$ on $R_{K/k}(U'_{\pm 2c}) = R_{K/k}(\mathbf{G}_a)$. This latter scaling action preserves the k-subgroup $\underline{V_0}$ because V_0 is a K^2-subspace of K. Since the k-subgroup $\underline{V_0}$ of $R_{K/k}(U'_{\pm 2c})$ is the $\pm 2c$-root group in H_{V_0}, the desired normalizing property is established.

For distinct $a_1, a_2 \in \Delta_n$ (not necessarily short), the a_1-root group for (F_{a_1}, D_0) and the $-a_2$-root group for (F_{a_2}, D_0) commute because even the a_1-root group and $-a_2$-root group for $(\mathscr{G}, D_0)$ commute (as this holds for the analogous root groups of (Sp_{2n}, D) over K, since $(a_1)_K$ and $-(a_2)_K$ are linearly independent and their sum $(a_1)_K - (a_2)_K$ is not in $\Phi(Sp_{2n}, D)$). We conclude via Theorem C.2.29 that the k-subgroups $\{F_a\}_{a \in \Delta_n}$ in $\mathscr{G}$ generate a pseudo-reductive k-subgroup $H \subseteq \mathscr{G}$ such that $D_0 \subseteq H$, $Z_H(D_0) = R_{K/k}(D)$, $\Delta_n \subseteq \Phi(H, D_0)$, the $\pm a$-root groups of (H, D_0) are $R_{K/k}(U'_{\pm a})$ for $a \in \Delta_n - \{2c\}$, and for $a = 2c$ the $\pm a$-root groups coincide with the k-subgroup $\underline{V_0}$ of $R_{K/k}(U'_{\pm 2c}) = R_{K/k}(\mathbf{G}_a)$ that contains the standard $\mathbf{G}_a$ (as $1 \in V'$). Thus, $H_a = R_{K/k}((Sp_{2n})_a)$ for $a \in \Delta_n - \{2c\}$ and $H_{2c} = H_{V_0, K_0/k}$ (see Proposition 9.1.4). In particular, for all $a \in \Delta_n$ we have the inclusion of k-groups

$$SL_2 = (Sp_{2n})_a \subseteq H_a \subseteq R_{K/k}((Sp_{2n})_a) \subseteq R_{K/k}(Sp_{2n}),$$

so the k-subgroup H inside $R_{K/k}(Sp_{2n})$ contains the canonical Levi k-subgroup Sp_{2n}.

In the same way, replacing $R_{K/k}(D)$ with $C := C_{V^{(2)},V',n}$ from (9.7.6) in the preceding definitions gives pseudo-reductive k-subgroups $F'_a \subseteq F_a$ with $Z_{F'_a}(D_0) = C$ for all $a \in \Delta_n$, and these generate a pseudo-reductive k-subgroup H' of H containing D_0 such that $Z_{H'}(D_0) = C$ and $H'_a = H_a$ for all $a \in \Delta_n$. Thus, by inspection of the definition of C, we see that H' is generated by its perfect k-subgroups $\{H'_a = H_a\}_{a \in \Delta_n}$, so $H' = \mathscr{D}(H)$. In particular, H' is pseudo-semisimple and $Sp_{2n} \subseteq H'$ inside $R_{K/k}(Sp_{2n}) = \mathscr{G}$, so

$$\Phi(Sp_{2n}, D_0) \subseteq \Phi(H, D_0) \subseteq \Phi(\mathscr{G}, D_0) = \Phi(Sp_{2n}, D)$$

inside $X(D_0) = X(D)$. It follows that H and H' have the same rank n, $\Phi(H', D_0) = \Phi(H, D_0) = \Phi(\mathscr{G}, D_0)$, and the natural maps $H^{red}_{\overline{k}} \to \mathscr{G}^{red}_{\overline{k}} = Sp_{2n}$ and $(H'_{\overline{k}})^{red} \to \mathscr{G}^{red}_{\overline{k}}$ are isomorphisms. By Proposition 1.2.6 and Lemma 3.1.2 we conclude that H' is absolutely pseudo-simple over k.

Step 2. Now we use H and $H' = \mathscr{D}(H)$ to analyze properties of G. The main point is to prove that $H = \phi(G)$ (so then $H' = \phi(\mathscr{D}(G))$). Since $\phi|_\Omega = f$, it is enough to prove that H meets the open subscheme $\mathrm{R}_{K/k}(UwB) \subseteq \mathrm{R}_{K/k}(\mathrm{Sp}_{2n})$ in precisely the smooth closed subscheme $f(\Omega) \subseteq \mathrm{R}_{K/k}(UwB)$. We have seen in Step 1 that H contains the canonical k-subgroup $\mathrm{Sp}_{2n} \subseteq \mathrm{R}_{K/k}(\mathrm{Sp}_{2n})$, so H contains the subgroup $\mathrm{Sp}_{2n}(k) \subseteq \mathrm{Sp}_{2n}(K)$. In particular, $H(k)$ contains $w = f(\mathbf{w}, 1)$.

Since $H \cap \mathrm{R}_{K/k}(U'_a)$ is equal to the a-root group of (H, D_0) (due to Lemma 2.3.3, Proposition 2.3.11, Remark 2.3.12, and Proposition 2.1.8(3)), relative to the direct spanning decomposition $\prod_{a\in\Phi_n^+} U'_a$ of U via multiplication (using any enumeration of Φ_n^+) we have

$$\prod_{a\in\Phi_n^+} (H \cap \mathrm{R}_{K/k}(U'_a)) = H \cap \mathrm{R}_{K/k}(U)$$

via multiplication due to Proposition 2.1.8(3). As H contains $\mathrm{R}_{K/k}(D)$,

$$H \cap \mathrm{R}_{K/k}(B) = \mathrm{R}_{K/k}(D) \ltimes (H \cap \mathrm{R}_{K/k}(U)).$$

The element $w \in H(k)$ swaps Φ_n^+ and $-\Phi_n^+$, so left w-translation and consideration of open cells show via Proposition 2.1.8(3) that

$$H \cap \mathrm{R}_{K/k}(UwB) = (H \cap \mathrm{R}_{K/k}(U))w(H \cap \mathrm{R}_{K/k}(B)).$$

For short $b \in \Delta_n$, the b-root group of (H, D_0) coincides with the b-root group of $(\mathrm{R}_{K/k}(\mathrm{Sp}_{2n}), D_0)$, so by Weyl conjugacy and dimension considerations the same holds for all short $b \in \Phi_n$. For the unique long root $2c \in \Delta_n$, the $2c$-root group of (H, D_0) is the k-subgroup $\underline{V_0}$ of $\mathrm{R}_{K/k}(\mathbf{G}_{\mathrm{a}})$ that is the image of the k-homomorphism

$$f_c : \mathscr{U}_c = \underline{V} \times \underline{V'} \to \mathrm{R}_{K/k}(\mathbf{G}_{\mathrm{a}}) = \mathrm{R}_{K/k}(U'_{2c})$$

defined by $(v, v') \mapsto \underline{q}(v) + v'$. We claim that for *every* long positive root $a \in \Phi_n^+$, the parameterization $u_a : \mathbf{G}_{\mathrm{a}} \simeq U'_a$ via "matrix entries" (using (9.6.3) with $i = j$) identifies the k-subgroup $H \cap \mathrm{R}_{K/k}(U'_a) \subseteq \mathrm{R}_{K/k}(\mathbf{G}_{\mathrm{a}})$ with $\underline{V_0}$. Since the subgroup $\mathrm{Sp}_{2n}(k) \subseteq H(k)$ contains representatives for $W(H, D_0) = W(\mathrm{Sp}_{2n}, D)$, it suffices to check that for any $g \in N_{\mathrm{Sp}_{2n}}(D_0)(k)$ and long positive roots $a_1, a_2 \in \Phi(\mathrm{Sp}_{2n}, D_0)$ satisfying $a_2 = g.a_1$, the induced isomorphism $\mathrm{R}_{K/k}(U'_{a_1}) \simeq \mathrm{R}_{K/k}(U'_{a_2})$ via conjugation by g intertwines the k-isomorphisms $\mathrm{R}_{K/k}(U'_{a_1}) \simeq \mathrm{R}_{K/k}(\mathbf{G}_{\mathrm{a}})$ and $\mathrm{R}_{K/k}(U'_{a_2}) \simeq \mathrm{R}_{K/k}(\mathbf{G}_{\mathrm{a}})$ up to the action of an element of $k^\times$ rather than just $K^\times$ (so it carries $\underline{V_0}$ onto itself, as V_0 is a k-subspace of K). Since the root group parameterizations u_{a_1} and u_{a_2} respect the k-structure (Sp_{2n}, D_0) on the pair (Sp_{2n}, D) over K and any two

$\mathbf{G}_a$-parameterizations of a root group over k are related through $k^\times$-scaling, we are done. This completes the proof that $H \cap \mathrm{R}_{K/k}(UwB) = f(\Omega)$, so indeed $H = \phi(G)$.

Now we can prove that G is pseudo-reductive of rank n. First, we prove G is affine. By Theorem A.3.9, there is a central extension

$$1 \to Z \to G \to G/Z \to 1$$

with G/Z affine and Z a semi-abelian variety. Since the k-group $H' = \mathscr{D}(\phi(G))$ in $\mathrm{R}_{K/k}(\mathrm{Sp}_{2n})$ is pseudo-semisimple and contains the maximal torus D_0, the scheme-theoretic center of $H = \phi(G)$ cannot contain any nontrivial k-torus (since $(H'_{\overline{k}})^{\mathrm{red}}$ is the derived group of $H^{\mathrm{red}}_{\overline{k}}$). But ϕ must carry the central k-subgroup Z into the center of $\phi(G)$, so we conclude that the maximal k-torus in Z is killed by ϕ. We have already seen that $(\ker \phi)(k) = 1$, so the maximal k-torus in Z must be trivial; i.e., Z is an abelian variety. The map $\phi|_Z$ must then be trivial (as ϕ has affine target), so likewise $Z = 1$. Thus, G is affine (so $\mathscr{R}_{u,k}(G)$ makes sense).

Similar arguments (now using the pseudo-reductivity of $\phi(G) = H$) show that $\mathscr{R}_{u,k}(G) = 1$. Hence, G is pseudo-reductive. Since $\ker \phi$ contains no nontrivial k-tori, we also conclude that G has the same rank as its quotient $\phi(G)$, namely n. (The absolute pseudo-simplicity of $\mathscr{D}(G)$ will be proved later in the next step.)

STEP 3. We compute the maximal reductive quotient of $G_{\overline{k}}$. Consider the composite map

$$\psi : G_K \stackrel{\phi_K}{\to} \mathrm{R}_{K/k}(\mathrm{Sp}_{2n})_K \twoheadrightarrow \mathrm{Sp}_{2n}\,.$$

This is surjective, since the k-subgroup inclusion $\phi(G) = H \hookrightarrow \mathrm{R}_{K/k}(\mathrm{Sp}_{2n})$ induces an isomorphism between maximal reductive quotients over $\overline{k}$. We claim that ψ is a descent to K of the maximal reductive quotient of $G_{\overline{k}}$ (so K contains the minimal field of definition over k for $\mathscr{R}_u(G_{\overline{k}})$ as a $\overline{k}$-subgroup of $G_{\overline{k}}$).

Consider the exact sequence

$$1 \to (\ker \psi)/(\ker \psi)^0 \to G_K/(\ker \psi)^0 \to \mathrm{Sp}_{2n} \to 1.$$

The term on the left is an étale normal k-subgroup of the connected middle term, so this exact sequence is a central extension and the middle term is therefore connected semisimple with an étale isogeny onto Sp_{2n}. Since Sp_{2n}

is simply connected, it follows that $G_K/(\ker \psi)^0 \to \mathrm{Sp}_{2n}$ is an isomorphism, so $\ker \psi$ is connected. The connected K-group $\ker \psi$ cannot contain any nontrivial K-tori since the rank of G_K coincides with the rank of its quotient Sp_{2n} by $\ker \psi$. Thus, to show that ψ descends the maximal reductive quotient of $\overline{k}$ we just need to check that $\ker \psi$ is smooth, or equivalently that ψ is smooth. It suffices to check smoothness on some dense open subscheme of G_K, so we restrict ψ to Ω_K.

The restriction $\psi|_{\Omega_K}$ is a direct product of copies of the following maps: the two natural K-homomorphisms

$$\pi' : (\mathrm{R}_{K/k}(\mathrm{GL}_1))_K \to \mathrm{GL}_1, \ \ \pi : (\mathrm{R}_{K/k}(\mathbf{G}_\mathrm{a}))_K \to \mathbf{G}_\mathrm{a}$$

(respectively corresponding to the coroots for the simple positive roots and to the short positive roots) and the K-homomorphism

$$(\underline{V} \times \underline{V}')_K \stackrel{h}{\to} (\mathrm{R}_{K/k}(\mathbf{G}_\mathrm{a}))_K \stackrel{\pi}{\twoheadrightarrow} \mathbf{G}_\mathrm{a} \tag{9.8.1}$$

where $h(v, v') = \underline{q}(v) + v'$ (corresponding to the long positive roots). All of these maps admit sections via the base change to K of the canonical inclusions $\mathrm{GL}_1 \to \mathrm{R}_{K/k}(\mathrm{GL}_1)$ and $\mathbf{G}_\mathrm{a} \to \mathrm{R}_{K/k}(\mathbf{G}_\mathrm{a})$ over k. (This makes sense for the composite map in (9.8.1) because the condition $1 \in V'$ ensures $\mathbf{G}_\mathrm{a} \subseteq \underline{V} \times \underline{V}'$ inside $\mathrm{R}_{K/k}(\mathbf{G}_\mathrm{a})$.) Thus, smoothness follows from the observation that if a surjective homomorphism $\varphi : \mathscr{H}' \twoheadrightarrow \mathscr{H}$ between finite type group schemes over a field F admits a homomorphic section $s : \mathscr{H} \to \mathscr{H}'$ then $\mathscr{H}' = \mathscr{H} \ltimes \ker(\varphi)$ as F-schemes, so if $\mathscr{H}'$ is F-smooth and connected then $\ker(\varphi)$ is F-smooth and connected, and hence φ is smooth.

Since ψ descends the maximal reductive quotient of $G_{\overline{k}}$, the semi-simplicity of this quotient and Proposition A.4.8 imply that $\psi|_{\mathscr{D}(G)_K}$ is a descent of the maximal reductive quotient of $\mathscr{D}(G)_{\overline{k}}$. Hence, $\mathscr{D}(G)$ is absolutely pseudo-simple (due to Proposition 1.2.6 and Lemma 3.1.2) and the minimal field of definition K' over k of $\mathscr{R}_u(\mathscr{D}(G)_{\overline{k}})$ inside $\mathscr{D}(G)_{\overline{k}}$ is a subfield of K (over k), with $\phi : \mathscr{D}(G) \to \mathrm{R}_{K/k}(\mathrm{Sp}_{2n})$ equal to the composition of $\xi_{\mathscr{D}(G)}$ and the canonical inclusion $\mathrm{R}_{K'/k}(\mathrm{Sp}_{2n}) \to \mathrm{R}_{K/k}(\mathrm{Sp}_{2n})$. In particular, $i_{\mathscr{D}(G)}(\mathscr{D}(G)) \simeq H'$ as k-groups, so we can determine K'/k by using the k-group H', due to Lemma 9.2.1; in Step 5 we will prove that $K' = K$ over k, so even the subgroup $\mathscr{R}_u(G_{\overline{k}}) \subseteq G_{\overline{k}}$ has K as its minimal field of definition over k and hence $\phi = i_G$.

Step 4. An open Bruhat cell of G (in the sense of Corollary 3.3.16) will now be related to Ω. Consider the translate $\Omega \cdot (\mathbf{w}, 1)^{-1} \subseteq G$. This is carried by ϕ

into the open cell

$$\mathrm{R}_{K/k}(U)\cdot\mathrm{R}_{K/k}(D)\cdot\mathrm{R}_{K/k}(U^-)$$

of $(\mathrm{R}_{K/k}(\mathrm{Sp}_{2n}), D_0, \Phi_n^+)$, where $U^- := wUw^{-1}$, and this restriction of ϕ is injective on k-points. The locally closed subschemes

$$(\mathscr{U}\mathbf{w},1)\cdot(\mathbf{w},1)^{-1},\quad (\mathbf{w},\mathrm{R}_{K/k}(D))\cdot(\mathbf{w},1)^{-1}$$

are carried respectively into $\mathrm{R}_{K/k}(U)$ and $\mathrm{R}_{K/k}(D)$ by ϕ. The restriction of ϕ to the second of these is identified with inversion on $\mathrm{R}_{K/k}(D)$, and the restriction of ϕ to the first of these is the canonical homomorphism $\mathscr{U}\to\mathrm{R}_{K/k}(U)$. Indeed, for the verification of these identities it suffices to check on k-points, where everything is clear. Since $\mathscr{U}(k)$ is Zariski-dense in $\mathscr{U}$, and $\mathrm{R}_{K/k}(D)(k)$ is Zariski-dense in $\mathrm{R}_{K/k}(D)$, the triviality of $(\ker\phi)(k)$ and the smoothness of the k-groups $\mathscr{U}$ and $\mathrm{R}_{K/k}(D)$ imply that the immersions $\mathscr{U}\to G$ and $\mathrm{R}_{K/k}(D)\to G$ respectively defined by $u\mapsto(u\mathbf{w},1)\cdot(\mathbf{w},1)^{-1}$ and $t\mapsto(\mathbf{w},t)\cdot(\mathbf{w},1)^{-1}$ are homomorphisms. Thus, these are closed k-subgroups of G (as for any smooth locally closed k-subgroup of a smooth k-group). In particular, $\mathrm{R}_{K/k}(D)$ viewed inside G via $t\mapsto(\mathbf{w},t)\cdot(\mathbf{w},1)^{-1}$ acquires its usual commutative k-group structure that is normalized by $(\mathbf{w},1)$ via inversion, so the maximal k-torus D_0 of $\mathrm{R}_{K/k}(D)$ is thereby identified as a maximal k-torus of G. Likewise, for each $a\in\Phi_n^+$ that is short (resp. long), the map $u\mapsto(u\mathbf{w},1)\cdot(\mathbf{w},1)^{-1}$ identifies $\mathscr{U}_a$ (resp. $\mathscr{U}_{a/2}$) with a closed k-subgroup of G.

We can run through this argument using $\mathscr{B}$ in place of $\mathscr{U}$, and again exploit the injectivity of $\phi|_{\Omega(\mathbf{w},1)^{-1}}$ on k-points to infer that for $a\in\Phi_n^+$ that is short (resp. long) the k-subgroup $\mathscr{U}_a$ (resp. $\mathscr{U}_{a/2}$) of G is normalized by the k-subgroup $\mathrm{R}_{K/k}(D)$ with the expected action induced by the group law of G (depending on whether a is short or long). In a similar manner, it follows via Remark 9.6.15 that these k-subgroups of G directly span $\mathscr{U}=(\mathscr{U}\mathbf{w},1)(\mathbf{w},1)^{-1}$ in any order via the group law of G. Since the birational group law is defined so that $(\mathbf{w},1)$ normalizes $\mathrm{R}_{K/k}(D)$ via inversion, we see via weight space considerations in the Lie algebras that:

(i) $\mathrm{R}_{K/k}(D)=Z_G(D_0)$ with $\phi: G\to\mathrm{R}_{K/k}(\mathrm{Sp}_{2n})$ restricting to the canonical inclusion on $\mathrm{R}_{K/k}(D)$,

(ii) the subset $\Phi(G,D_0)\subseteq\mathrm{X}(D_0)$ is the expected non-reduced and irreducible root system of rank n containing Φ_n (so $\Delta:=(\Delta_n-\{2c\})\cup\{c\}$ is a basis of $\Phi(G,D_0)$),

(iii) for each root a in the positive system of roots in $\Phi(G,D_0)$ corresponding to Φ_n^+, the corresponding root group for (G,D_0) (in the sense of

Definition 2.3.13) as a subgroup of $\mathscr{U}$ is $\mathscr{U}_a$ when a is a non-multipliable and non-divisible positive root, is $\mathscr{U}_a = \underline{V} \times \underline{V}'$ when a is a multipliable positive root, and is $\mathscr{U}_a = \underline{V}'$ inside $\mathscr{U}_{a/2}$ for divisible positive roots a.

We conclude that $\Omega(\mathbf{w},1)^{-1}$ is the open Bruhat cell of G corresponding to D_0 and Φ_n^+, and that $\phi|_{Z_G(D_0)}$ has trivial kernel (so G is of *minimal type* once we prove $\phi = i_G$, as will be done in Step 5). Recall also that $(\mathbf{w},1)$ is the *unique* element of $G(k)$ carried to $w \in \mathrm{Sp}_{2n}(K) = \mathrm{R}_{K/k}(\mathrm{Sp}_{2n})(k)$.

Let us push this computation a bit further. For all non-divisible $a \in \Phi(G,D_0)^+$, the natural k-homomorphism $\mathscr{U}_a \to \mathrm{R}_{K/k}(\mathbf{G}_\mathrm{a}) \subseteq \mathrm{R}_{K/k}(\mathrm{Sp}_{2n})$ and its composition with conjugation by $(\mathbf{w},1)$ on the source and w on the target both admit an evident k-homomorphic section over the canonical k-subgroup $\mathbf{G}_\mathrm{a} \subseteq \mathrm{R}_{K/k}(\mathbf{G}_\mathrm{a})$ (using that $1 \in V'$ when a is multipliable). For the smooth connected closed k-subgroup $L \subseteq \mathscr{D}(G)$ generated by these $\mathbf{G}_\mathrm{a}$ we see that $L(k)$ maps onto $\mathrm{Sp}_{2n}(k)$ (since the group of rational points of each k-group $(\mathrm{Sp}_{2n})_a = \mathrm{SL}_2$ is generated by the sets of rational points of its upper and lower triangular k-subgroups). In particular, $L(k)$ has image in $\mathrm{R}_{K/k}(\mathrm{Sp}_{2n})(k)$ that contains $w \in \mathrm{Sp}_{2n}(k) \subseteq \mathrm{Sp}_{2n}(K)$. But we have shown that $(\mathbf{w},1)$ is the only element of $G(k)$ carried onto w, so we conclude that $(\mathbf{w},1) \in L(k) \subseteq \mathscr{D}(G)(k)$.

An important consequence is that we can describe the open cell for $(\mathscr{D}(G), D_0)$ corresponding to the one just determined for (G, D_0). These share the same root groups (as for any pseudo-split pseudo-reductive group), so the only issue is to determine the Cartan k-subgroup $Z_{\mathscr{D}(G)}(D_0) \subseteq Z_G(D_0) = \mathrm{R}_{K/k}(D)$. We claim that this subgroup is C, so $\Omega' \cdot (\mathbf{w},1)^{-1}$ is the open cell. Since we have seen that the map $G \twoheadrightarrow H$ induces an isomorphism between Cartan k-subgroups relative to D_0, the map $\mathscr{D}(G) \twoheadrightarrow H'$ between derived groups induces an isomorphism between Cartan k-subgroups relative to D_0. This proves the desired equality $Z_{\mathscr{D}(G)}(D_0) = C$ inside $Z_G(D_0) = \mathrm{R}_{K/k}(D)$.

To summarize, the absolutely pseudo-simple $\mathscr{D}(G)$ has a known split maximal torus D_0, a known non-reduced root system and root groups, a known Cartan subgroup C, maximal geometric reductive quotient Sp_{2n} via $\phi_{\overline{k}}$, and it contains Ω' (since $(\mathbf{w},1) \in \mathscr{D}(G)(k)$). The pseudo-split (G, D_0) and $(\mathscr{D}(G), D_0)$ share the same root system and hence the same Weyl group, and the open Bruhat cell $\Omega(\mathbf{w},1)^{-1}$ of G meets $\mathscr{D}(G)$ in an open subscheme of $\mathscr{D}(G)$ that contains the open Bruhat cell $\Omega'(\mathbf{w},1)^{-1}$. The resulting containment

$$\Omega'(\mathbf{w},1)^{-1} \subseteq (\Omega(\mathbf{w},1)^{-1}) \cap \mathscr{D}(G)$$

between open subschemes of $\mathscr{D}(G)$ is an equality by Proposition 2.1.8(3). Thus, right translation by $(\mathbf{w},1) \in \mathscr{D}(G)(k)$ implies that $\Omega \cap \mathscr{D}(G) = \Omega'$.

STEP 5. As we explained at the end of Step 3, to prove that the minimal field of definition K'/k for the geometric unipotent radical of $\mathscr{D}(G)$ is equal to K (so $\phi = i_G$ and $\phi|_{\mathscr{D}(G)} = \xi_{\mathscr{D}(G)}$) it suffices to study the analogous such field for H'. In the rank-1 case we saw in Step 1 that the k-subgroup $H' = H_{2c}$ equals $H_{V_0,K_0/k}$, and $K_0 = K$ by hypothesis when $n = 1$, so the equality $K' = K$ follows from Proposition 9.1.4 when $n = 1$. Now let us assume that $n \geqslant 2$. Then by Proposition 9.5.2, K' is the minimal field of definition over k for the geometric unipotent radical of G_b for any non-multipliable non-divisible $b \in \Phi(G, D_0)$. But $\phi(G_b) = \phi(\langle U_b, U_{-b}\rangle)$ and $\phi(U_{\pm b}) = \mathrm{R}_{K/k}(U'_{\pm b})$. That is, $\phi(G_b) = \mathrm{R}_{K/k}((\mathrm{Sp}_{2n})_b) = \mathrm{R}_{K/k}(\mathrm{SL}_2)$, from which we easily read off that the inclusion $K' \subseteq K$ over k is an equality in general. This also completes the proof that $\phi = i_G$ and $\phi|_{\mathscr{D}(G)} = \xi_{\mathscr{D}(G)}$. These equalities and the triviality of the kernel of the restriction of $\phi : G \to H$ between Cartan k-subgroups imply that G and $\mathscr{D}(G)$ are *of minimal type.*

Since $\phi = i_G$, so $\ker\phi$ is a unipotent group scheme, we conclude from Theorem 9.4.7 that $\ker\phi$ is commutative and connected and is the direct product of its intersections with the root groups of (G, D_0) for the multipliable roots. Let $L \subseteq G$ be as defined in Step 4. As an application of the unipotence of $\ker\phi$, we will now show that the surjective k-homomorphism $\phi|_L : L \to \mathrm{Sp}_{2n}$ is an isomorphism, so L is a Levi k-subgroup of G and $\mathscr{D}(G)$.

Any maximal k-torus of L maps onto one of Sp_{2n}, and by construction $L(k)$ maps onto $\mathrm{Sp}_{2n}(k)$. Thus, the k-rational conjugacy of all maximal k-tori of Sp_{2n} implies that via a suitable $L(k)$-conjugation we can find a maximal k-torus S of L that maps onto $D_0 \subseteq \mathrm{Sp}_{2n}$. The unipotence of the group scheme $\ker\phi$ forces the surjection $\phi : S \twoheadrightarrow D_0$ between group schemes of multiplicative type to have vanishing kernel and hence be an isomorphism. By uniqueness of the lift of D_0 through $\phi : G \to \mathrm{R}_{K/k}(\mathrm{Sp}_{2n})$, we conclude that $S = D_0$ inside G; i.e., $D_0 \subseteq L$ inside G. Since $Z_L(D_0) \to Z_{\mathrm{Sp}_{2n}}(D_0) = D_0$ is visibly surjective yet has trivial kernel (as ϕ even carries $Z_G(D_0)$ isomorphically onto its image $\mathrm{R}_{K/k}(D)$ inside $\mathrm{R}_{K/k}(\mathrm{Sp}_{2n})$), it follows that $Z_L(D_0) = D_0$. The triviality of $(\ker\phi)(k)$ and the surjectivity of $\phi : L \to \mathrm{Sp}_{2n}$ force L to be pseudo-reductive, so (L, D_0) has a good theory of root groups and an open cell as a direct product of root groups and the Cartan k-subgroup $Z_L(D_0) = D_0$.

By construction of L,

$$\Phi_n \subseteq \Phi(L, D_0) \subseteq \Phi(G, D_0)$$

where Φ_n is the set of non-multipliable roots in $\Phi(G, D_0)$ and the quotient (Sp_{2n}, D_0) over k has 1-dimensional root groups that lift homomorphically into the corresponding root groups of (L, D_0). Thus, for any root group U_0 of

(L, D_0), the restriction of $\phi|_L : L \twoheadrightarrow \mathrm{Sp}_{2n}$ to U_0 is a map $U_0 \twoheadrightarrow \phi(U_0) = \mathbf{G}_\mathrm{a}$ with a homomorphic section, so the unipotent $\ker(\phi|_{U_0})$ is a direct factor of U_0 and therefore is smooth and connected. Since $\phi = i_G$, this forces $\ker(\phi|_{U_0}) = 1$, so each root group of (L, D_0) maps isomorphically onto its image in (Sp_{2n}, D_0), and hence $\ker(\phi|_L)$ has trivial intersection with an open cell of (L, D_0), so this kernel is étale and thus central in L. That is, $\phi : L \to \mathrm{Sp}_{2n}$ is a central (étale) isogeny. Since Sp_{2n} is simply connected, it follows that $\phi : L \to \mathrm{Sp}_{2n}$ is an isomorphism, and the root groups of (L, D_0) must be as expected (namely, the $\mathbf{G}_\mathrm{a}$ used in the construction of L). This completes the proof of part (3).

Step 6. To conclude, we study surjectivity of $\xi_{\mathscr{D}(G)} = \phi|_{\mathscr{D}(G)}$ and other aspects of part (4). Since $\xi_{\mathscr{D}(G)}$ restricts to the identity map on the maximal k-torus D_0, surjectivity of $\xi_{\mathscr{D}(G)} : \mathscr{D}(G) \to \mathrm{R}_{K/k}(\mathrm{Sp}_{2n})$ is equivalent to its surjectivity between associated root groups and between the Cartan k-subgroups (C and $\mathrm{R}_{K/k}(D)$ respectively) associated to D_0. By inspection of the construction of ϕ from the map $f : \mathscr{U}\mathbf{w} \times \mathscr{B} \to \mathrm{R}_{K/k}(UwB)$ and the description of the open cell for $\mathscr{D}(G)$ in terms of that for G via shrinking the Cartan k-subgroup from $\mathrm{R}_{K/k}(D)$ to C, it follows that $\xi_{\mathscr{D}(G)}$ is surjective if and only if the morphism $h : (v, v') \mapsto \underline{q}(v) + v'$ into $\mathrm{R}_{K/k}(\mathbf{G}_\mathrm{a})$ is surjective and $(V_0)^*_{K_0/k} = \mathrm{R}_{K/k}(\mathrm{GL}_1)$. The surjectivity of h amounts to the image $V^{(2)} + V'$ of h on k-points being Zariski-dense in K, or equivalently to the k-span V_0 of $V^{(2)} + V'$ being equal to K, in which case the field $K_0 = k[V^{(2)} + V']$ is necessarily equal to K. But in such cases obviously $(V_0)^*_{K_0/k} = \mathrm{R}_{K/k}(\mathrm{GL}_1)$, so the initial equivalence claim in part (4) is proved. If such surjectivity holds then necessarily $V_0 = K$, so $C = \mathrm{R}_{K/k}(D)$ and hence $G = \mathscr{D}(G)$.

In general (i.e., without any hypothesis on $\xi_{\mathscr{D}(G)}$), Theorem 9.4.7 says that $\ker i_G$ is the commuting direct product of its intersections with the root groups of (G, D_0) for the multipliable roots. (The Lie algebras of these root groups are spanned by the weight spaces for multipliable and divisible roots, and the divisible roots are precisely the longest roots in $\Phi(G, D_0)$.) Since $i_G = \phi$ and the restriction $\mathscr{U}_c \to \mathrm{R}_{K/k}(U'_{2c})$ of ϕ to the root group of each multipliable root c is identified with h (in Step 2), $\ker i_G = \mathscr{K}' \times \mathscr{K}$ where $\mathscr{K} := \prod_{a \in \Phi^+_>} \ker h$ and $\mathscr{K}'$ is the $(\mathbf{w}, 1)$-conjugate of $\mathscr{K}$. □

9.8.3 In rank 2 we can construct another class of pseudo-split absolutely pseudo-simple k-groups with a non-reduced root system (and with field of definition K/k for the geometric unipotent radical) by a variant on $\mathscr{D}(G_{K/k, V^{(2)}, V', 2})$, using smaller root groups for short roots in Φ_2. This exploits a special feature when $n = 2$: in type C_2 the positive short root groups pairwise commute in characteristic 2 (either by direct computation in $\mathrm{Sp}_4 \subseteq \mathrm{SL}_4$ over

$\mathbf{F}_2$, or by applying Proposition 9.6.7 with $n = 2$ and noting that $D_2 = A_1 \times A_1$). Now we explain how this works.

Let $(K/k, V^{(2)}, V')$, $V_0 := kV^{(2)} + V'$, and $K_0 := k\langle V^{(2)} + V' \rangle \subseteq K$ be as in 9.6.8. Note that $kK^2 \subseteq K_0$. Choose a nonzero K_0-subspace $V'' \subseteq K$ such that $k\langle V'' \rangle = K$ (equivalently $k[V''] = K$ when $1 \in V''$); we are primarily interested in such data with $V'' \neq K$. If k'/k is a separable extension then $V''_{k'} := k' \otimes_k V''$ is clearly a K'_0-subspace of K' satisfying $k'\langle V''_{k'} \rangle = K'$. For a given K/k, a triple $(V^{(2)}, V', V'')$ exists with $V'' \neq K$ if and only if $[K : kK^2] > 4$. (For the implication "$\Rightarrow$", first note that $V^{(2)} + V'$ strictly contains the nonzero kK^2-subspace V', so $[K_0 : kK^2] \geqslant 2$. We also have $[K : K_0] \geqslant 4$, since $\dim_{K_0} V''$ must be at least 2 because V'' is a proper K_0-subspace of K and $k\langle V'' \rangle = K$. To prove "$\Leftarrow$", suppose $[K : kK^2] = 2^r$ with $r > 2$. Let $\{\alpha_1, \ldots, \alpha_r\}$ be a 2-basis for K over kK^2; equivalently, $\{d\alpha_i\}$ is a K-basis of $\Omega^1_{K/k} = \Omega^1_{K/kK^2}$. We choose $V' = kK^2$, $V^{(2)} = \alpha_1 K^2$, and $V'' = kK^2 + (\bigoplus \alpha_i kK^2) + (\bigoplus_{i \geqslant 2} \alpha_1\alpha_i kK^2)$, so $V'' \neq K$ since $r > 2$.)

The set Δ_2 of simple roots in Φ_2^+ is $\{b, 2c\}$ with b short and $2c$ long. For short $a \in \Phi_2^+$ define $\mathscr{U}''_a := \underline{V''} \subseteq \mathrm{R}_{K/k}(\mathbf{G}_a) = \mathscr{U}_a$, and for long $a \in \Phi_2^+$ define $\mathscr{U}''_{a/2} := \mathscr{U}_{a/2} = \underline{V} \times \underline{V'}$. The proof of Theorem 9.6.14 adapts to this modified input because the two positive short root groups of Sp_4 *commute* in characteristic 2. This step only requires that V'' is a nonzero kK^2-subspace of K, and it yields a smooth connected k-subgroup $\mathscr{U}'' \subseteq \mathscr{U}$ directly spanned *in any order* by the k-subgroups $\mathscr{U}''_a$ for short $a \in \Phi_2^+$ and $\mathscr{U}''_{a/2}$ for long $a \in \Phi_2^+$.

We cannot define an analogue of $\mathscr{B}$ as a semidirect product of $\mathscr{U}''$ with $\mathrm{R}_{K/k}(D)$ because the action on $\mathscr{U}$ by

$$\mathrm{R}_{K/k}(D) = \mathrm{R}_{K/k}((2c)^\vee(\mathrm{GL}_1)) \times \mathrm{R}_{K/k}(b^\vee(\mathrm{GL}_1))$$

does not preserve the k-subgroup $\mathscr{U}'' \subseteq \mathscr{U}$ when $V'' \neq K$. The problem is that the action of $\mathrm{R}_{K/k}((2c)^\vee(\mathrm{GL}_1))(k) = (2c)^\vee(K)$ on $\mathscr{U}_b(k) = K$ is given by

$$(2c)^\vee(t).v = t^{\langle b,(2c)^\vee \rangle} v = t^{-1} v,$$

so if $\mathrm{R}_{K/k}(D)$ normalizes $\mathscr{U}''$ then the nonzero V'' is a K-subspace of K (i.e., $V'' = K$). However, if we define the k-subgroup

$$C'' := (V_0)^*_{K_0/k} \times (V'')^*_{K/k} \subseteq C_{V^{(2)},V',2} \subseteq \mathrm{R}_{K/k}(D) \tag{9.8.2}$$

for the kK^2-subspace $V_0 = kV^{(2)} + V' \subseteq K_0$ then we claim that C'' normalizes $\mathscr{U}''$. To prove this property we will use the requirement that V'' is a K_0-subspace of K (not just a kK^2-subspace).

For long $a \in \Phi_2^+$ the k-subgroup $\mathscr{U}''_{a/2} = \mathscr{U}_{a/2}$ is normalized by $\mathrm{R}_{K/k}(D)$, so we just have to show that for short $a \in \Phi_2^+$ the k-subgroup $\mathscr{U}''_a = \underline{V''} \subseteq \mathrm{R}_{K/k}(\mathbf{G}_\mathrm{a}) = \mathscr{U}_a$ is normalized by C''. This holds provided that the Zariski-dense set V'' of k-points in $\underline{V''}$ is normalized by the action on $\mathscr{U}_a(k)$ by the subset of points $(x_0/y_0, x''_0/y''_0) \in C''(k)$ for $x_0, y_0 \in V_0 - \{0\}$ and $x''_0, y''_0 \in V'' - \{0\}$ (since this subset of $C''(k)$ generates a subgroup of $C''(k)$ that is Zariski-dense in C'' by definition of the factors of C'' as Zariski closures). Any point

$$((2c)^\vee(t), b^\vee(t')) \in (2c)^\vee(K) \times b^\vee(K) = D(K) = \mathrm{R}_{K/k}(D)(k)$$

acts on $v \in \mathscr{U}_a(k) = K$ via $v \mapsto t^{\langle a,(2c)^\vee\rangle} t'^{\langle a,b^\vee\rangle} v$, so the problem is to show that if $t_0 = x_0/y_0$ for $x_0, y_0 \in V_0 - \{0\}$ and if $t'_0 = x''_0/y''_0$ for $x''_0, y''_0 \in V'' - \{0\}$ then V'' is stable under multiplication inside K against $t_0^{\langle a,(2c)^\vee\rangle}$ and $t_0'^{\langle a,b^\vee\rangle}$. Since $\langle a, b^\vee\rangle$ is even for short $a \in \Phi_2^+$ (either 0 or -2) and V'' is a K_0-subspace of K, the desired stability is now clear (in view of how K_0 is defined) and we can define $\mathscr{B}''$ to be $C'' \ltimes \mathscr{U}''$. The complexity of C'' makes it cumbersome to adapt the construction of $G_{K/k,V^{(2)},V',2}$ (in terms of Weil birational group law) to make new pseudo-split absolutely pseudo-simple groups of rank 2 with non-reduced root system and Cartan k-subgroup C'', so instead we will use Theorem C.2.29 to construct the desired new groups inside $G_{K/k,V^{(2)},V',2}$.

We shall denote the group $G_{K/k,V^{(2)},V',2}$ by G here and in the next proposition, and as before, D_0 will denote the maximal k-torus of $\mathrm{R}_{K/k}(D)$. Let $\Psi = \Phi(G, D_0) \subseteq \mathrm{X}(D_0)$, and let Ψ^+ be the positive system of roots in Ψ corresponding to Φ_2^+. The associated basis Δ of Ψ has the form $\{b, c\}$ with multipliable c (so $\{b, 2c\}$ is the basis of Φ_2 associated to Φ_2^+). We identify $(G_K^{\mathrm{ss}})_b = (\mathrm{Sp}_4)_b$ with SL_2 in the standard way, using the coroot $b^\vee$ and the parameterization of the b-root group of (Sp_4, D) via (9.6.2) and (9.6.3). This recovers the identification of $\mathscr{U}_b$ with $\mathrm{R}_{K/k}(\mathbf{G}_\mathrm{a})$ in the original definition of $\mathscr{U}_b$ because the map $\phi : G \to \mathrm{R}_{K/k}(\mathrm{Sp}_4)$ arising from the construction of G via the birational group method is identified with i_G (see Theorem 9.8.1(3)) and underlies how G_K^{ss} is identified with Sp_4; the map ϕ also identifies G_b with $\mathrm{R}_{K/k}(\mathrm{SL}_2)$.

For the given K_0-subspace V'' of K, let G''_b be the pseudo-split absolutely pseudo-simple k-subgroup of G_b corresponding to the k-subgroup $H_{V''}$ of $\mathrm{R}_{K/k}(\mathrm{SL}_2) = G_b$. Note that G''_b might *not* contain the standard Levi k-subgroup SL_2 inside G_b since possibly $1 \notin V''$. We claim that C'' normalizes both G''_b and G_c, with $F_b := C'' \cdot G''_b$ and $F_c := C'' \cdot G_c$ pseudo-reductive. To prove this claim it is harmless to apply a preliminary K-automorphism of Sp_4 that acts as the identity on the diagonal K-torus and acts as scaling by arbitrary $\lambda \in K^\times$ on the $2c$-root space and by arbitrary $\mu \in K^\times$ on the b-root space (so it has the effect of replacing V' with $\lambda V'$ and replacing V'' with $\mu V''$).

In 9.7.9 we introduced a class of K-automorphisms i_λ of Sp_{2n} (for $\lambda \in K^\times$) extending the identity on the diagonal torus and inducing specific scaling operations on certain root groups. For $n = 2$ there is another class of such automorphisms, as follows. We define an automorphism of Sp_4 by

$$I_\lambda : \begin{pmatrix} a & b \\ c & d \end{pmatrix} \mapsto \begin{pmatrix} j_{\lambda^{-1}}(a) & F_\lambda(b) \\ F_{\lambda^{-1}}(c) & j_\lambda(d) \end{pmatrix}$$

with j_λ denoting conjugation by $\mathrm{diag}(\lambda, 1)$ (i.e., the 2×2 analogue of i_λ) and

$$F_\lambda \begin{pmatrix} x & y \\ z & w \end{pmatrix} := \begin{pmatrix} x & \lambda y \\ \lambda z & \lambda^2 w \end{pmatrix}.$$

To verify that I_λ is valued in Sp_4 and is a homomorphism, one uses the following identities on Mat_2:

$$F_\lambda({}^t m) = {}^t F_\lambda(m), \quad j_\lambda({}^t m) = {}^t j_{\lambda^{-1}}(m), \quad j_\lambda(m) F_{\lambda^{-1}}(m') = F_{\lambda^{-1}}(mm'),$$
$$F_\lambda(m) j_\lambda(m') = F_\lambda(mm'), \quad F_{\lambda^{-1}}(m) F_\lambda(m') = j_\lambda(mm').$$

Also, by inspection, for any $\lambda, \mu \in K^\times$ the K-automorphisms I_λ and i_μ of Sp_4 commute. (By the Isomorphism Theorem, I_λ is characterized by the following properties: it is the identity on both D and the unique simple long positive root group, and it is multiplication by λ on the unique simple short positive root group.) Applying $i_\lambda \circ I_\mu$ with $\lambda = 1/v'$ and $\mu = 1/v''$ for nonzero $v' \in V'$ and $v'' \in V''$ thereby brings us to the case that $1 \in V'$ and $1 \in V''$.

Since now $1 \in V'$ and $1 \in V''$, we have $(\mathbf{w}, 1) \in G(k)$ and $(\mathrm{Sp}_4)_b \subset G''_b$. Note that $(\mathbf{w}, 1)$-conjugation negates all roots via the *identity map* on opposite root spaces relative to the standard parameterizations of these root spaces in characteristic 2, so such conjugation preserves G''_b via swapping its $\pm b$-root groups because $(V'')^- = V''$ (as V'' is a kK^2-subspace of K). The same holds for G_c and its $\pm c$-root groups. Since $(\mathbf{w}, 1)$ normalizes C'' via inversion, and C'' normalizes the unipotent k-subgroups $\mathscr{U}''_b$ and $\mathscr{U}_c$, it follows that both G''_b and G_c are normalized by C''.

The definition of C'' as a direct product implies that C'' is the centralizer of D_0 in F_b and F_c because $Z_{G''_b}(b^\vee(\mathrm{GL}_1)) = (V'')^*_{K/k}$ by Proposition 9.1.4 and $Z_{G_c}(c^\vee(\mathrm{GL}_1)) = (V_0)^*_{K_0/k}$ by Theorem 9.8.1(2) (applied to G_c with $n = 1$). The intersections of the pseudo-reductive G''_b and G_c with the direct product C'' are respectively equal to the direct factors indexed by the roots b and c, so

F_b/G''_b equals the c-factor of C'' and F_c/G_c equals the b-factor of C''. Hence, F_b and F_c are pseudo-reductive.

Let G'' be the smooth connected k-subgroup of G generated by F_b and F_c. (We denote G'' as $G_{K/k,V^{(2)},V',V''}$ or $G_{V^{(2)},V',V''}$ when we need to specify the linear algebra data upon which it depends.)

Proposition 9.8.4 *Let $(K/k, V^{(2)}, V', V'')$ and $G'' = G_{K/k,V^{(2)},V',V''}$ be as in* 9.8.3.

(1) *The k-group G'' is absolutely pseudo-simple and pseudo-split of rank* 2 *with root system* BC_2, *it is of minimal type, and $Z_{G''}(D_0)$ coincides with C'' in* (9.8.2). *If $1 \in V'$ and $1 \in V''$ then* $(\mathbf{w}, 1) \in G''(k)$.
(2) *The field of definition over k for $\mathscr{R}_u(G''_{\overline{k}}) \subseteq G''_{\overline{k}}$ is K, and $\xi_{G''} = i_G|_{G''}$ for $G = G_{K/k,V^{(2)},V',2}$.*
(3) *If $V'' \neq K$(so $[K : kK^2] > 4$) then G'' is not k-isomorphic to any of the rank-2 absolutely pseudo-simple k-groups $\mathscr{D}(G_{K/k,\widetilde{V}^{(2)},\widetilde{V}',2})$.*
(4) *If $V'' = K$ then the inclusion $G'' \subseteq \mathscr{D}(G_{K/k,V^{(2)},V',2})$ is an equality.*

Proof Applying $i_\lambda \circ I_\mu$ for suitable $\lambda, \mu \in K^\times$ as above brings us to the case that $1 \in V'$ and $1 \in V''$. By Remark 9.6.11 and the compatibility of V'' with separable extension on k, the formation of G'' is compatible with separable extension on k. By Theorem C.2.29, G'' is pseudo-reductive with k-split maximal torus D_0 having centralizer C'' in G'' and the root group for (G'', D_0) corresponding to b (resp. c) is $\mathscr{U}''_b$ (resp. $\mathscr{U}_c$). Also, G'' contains the Levi k-subgroup Sp_4 of G (since G''_b and G_c contain $(\mathrm{Sp}_4)_b$ and $(\mathrm{Sp}_4)_{2c}$ respectively, as $1 \in V'$ and $1 \in V''$), so the root system $\Phi(G'', D_0)$ coincides with $\Phi(G, D_0)$, which is the non-reduced root system of type BC_2. By Lemma 7.2.4, this Sp_4 is a Levi k-subgroup of G'' too.

Since the factors $(V_0)^*_{K_0/k}$ and $(V'')^*_{K/k}$ of the Cartan k-subgroup $C'' = Z_{G''}(D_0)$ are contained in G_c and G''_b respectively, with G_c and G''_b perfect groups, C'' is contained in the derived group $\mathscr{D}(G'')$. Thus, by Proposition 1.2.6 we conclude that $G'' = \mathscr{D}(G'')$, so G'' is perfect. Now since G'' contains the absolutely simple group Sp_4 as a Levi k-subgroup, its maximal semi-simple quotient over $\overline{k}$ is $(\mathrm{Sp}_4)_{\overline{k}}$, so Lemma 3.1.2 implies that G'' is absolutely pseudo-simple. The groups $\xi_{G''}(G''_b)$, $\xi_{G''}(G_c)$, and $\xi_{G''}(G_{2c})$ are respectively identified with $\xi_{G''_b}(G''_b)$, $\xi_{G_c}(G_c)$, and $\xi_{G_{2c}}(G_{2c})$ (by Proposition 6.3.1), and these are k-isomorphic to $H_{V''}$, H_{V_0}, and $H_{V'}$ respectively, so by Proposition 9.1.4 their geometric unipotent radicals have field of definition over k equal to $k\langle V''\rangle = K$, $k\langle V_0\rangle = K_0$, and $k\langle V'\rangle \subseteq K_0$ respectively. Thus, the geometric unipotent radicals of G''_b, G_c, and G_{2c} have fields of definition over k respectively equal to K, K_0, and $K_{2c} = k\langle V'\rangle \subseteq K_0$ as well (due to Lemma 9.2.1).

All elements of $\Phi(G'', D_0) = \mathrm{BC}_2$ are $W(\Phi(G'', D_0))$-conjugate to either b, c, or $2c$, so since $N_{G''}(D_0)(k)$ maps onto $W(\Phi(G'', D_0))$ (Proposition 3.2.7 and Corollary 3.4.3(1)) it follows that for all $a \in \Phi(G'', D_0)$ the geometric unipotent radical of G''_a has field of definition over k equal to either K, K_0, or $K_{2c} \subseteq K_0$. Hence, Proposition 6.3.1(1) and the inclusion $K_0 \subseteq K$ imply that the field of definition over k of the geometric unipotent radical of G'' is K. We conclude that $i_G|_{G''} = \xi_{G''}$, so $\xi_{G''}$ has trivial kernel on the Cartan subgroup C'' of G'' since even i_G has trivial kernel on $\mathrm{R}_{K/k}(D)$. Thus, G'' is of minimal type.

Since G''_b is isomorphic to $H_{V''}$, which is a proper k-subgroup of $\mathrm{R}_{K/k}(\mathrm{SL}_2)$ unless $V'' = K$, it is obvious that G'' is not k-isomorphic to any of the rank-2 absolutely pseudo-simple groups $\mathscr{D}(G_{K/k,\widetilde{V}^{(2)},\widetilde{V}',2})$ if $V'' \neq K$. When $V'' = K$ then $C'' = (V_0)^*_{K_0/k} \times \mathrm{R}_{K/k}(\mathrm{GL}_1)$, so the inclusion $G'' \subseteq \mathscr{D}(G)$ between absolutely pseudo-simple groups containing the common split maximal k-torus D_0 is an equality on D_0-centralizers (see (9.7.6) with $n = 2$) and on root groups for the simple non-divisible roots b and c. Thus, comparing dimensions of root groups (via Weyl conjugacy with simple non-divisible roots) and open Bruhat cells implies that $G'' = \mathscr{D}(G)$ when $V'' = K$. □

Remark 9.8.5 There is no analogue of 9.8.3 (or Proposition 9.8.4) with $V'' \neq K$ for rank $n > 2$. The conceptual reason is that if $n > 2$ then for any pair of adjacent roots in the A_{n-1} diagram of short roots in $\Delta_n \subseteq \Phi_n^+$ we can apply the standardness results for type A_2 away from characteristic 3 in Proposition 6.2.1 to deduce that the root spaces for such roots must coincide with K rather than be a more general nonzero K_0-subspace $\mathscr{V}$ of K such that $k\langle\mathscr{V}\rangle = K$. (The necessity of considering such K_0-subspaces is due to Proposition 9.5.2 and the Weyl conjugacy of roots of a common length in any irreducible root system.)

At a more practical level, let us try to carry out an analogous construction for $n > 2$ and indicate why the method can only work when the root groups for short elements of Φ_n have dimension $[K : k]$ (so no "new" group is obtained using $n > 2$). For each short $b \in \Delta_n$, pick a nonzero K_0-subspace $V_b \subseteq K$ such that $k\langle V_b\rangle = K$. In view of the desired minimal type condition, there is only one possibility for the Cartan k-subgroup containing T: by Proposition 9.1.5, Lemma 9.2.1 (in rank-1 cases), Proposition 9.4.5, Proposition 9.5.2, Theorem C.2.29, and the structure of Cartan k-subgroups in Proposition 9.1.4 we must use

$$C := (V_0)^*_{K_0/k} \times \prod_{b \in \Delta_n - \{2c\}} (V_b)^*_{K/k} \subseteq \mathrm{R}_{K/k}(D).$$

Since $n \geqslant 3$, such a b is adjacent to another short $b' \in \Delta_n$. As C normalizes each of the root spaces, computing with the b'-coroot factor of C forces the nonzero kK^2-subspace V_b of K to be a subspace over $k\langle V_{b'}\rangle = K$. Hence, $V_b = K$.

We now show that our constructions of minimal type pseudo-split absolutely pseudo-simple k-groups with a non-reduced root system (including the additional rank-2 constructions in 9.8.3 and Proposition 9.8.4) are exhaustive. The proof is long, but essentially involves nothing more than running in reverse the reasoning that motivated our constructions.

Theorem 9.8.6 *Let k be an imperfect field of characteristic 2, K/k a nontrivial purely inseparable finite extension, and $n \geqslant 1$ an integer.*

Let (G, T) be a pseudo-split absolutely pseudo-simple k-group of minimal type with rank $n \geqslant 1$. Assume $\Phi(G, T)$ is non-reduced and K/k is the field of definition over k for the geometric radical. Let $L \subseteq G$ be a Levi k-subgroup containing T (which exists by Theorem 3.4.6), so naturally $L_K \simeq G_K^{\mathrm{ss}}$. Choose a k-isomorphism $\iota : (L, T) \simeq (\mathrm{Sp}_{2n}, D_0)$, and use this to identify (G_K^{ss}, T_K) with (Sp_{2n}, D), where $D = (D_0)_K$.

(i) *If $n \neq 2$ then there exists a pair $(V^{(2)}, V')$ as in 9.6.8 satisfying $1 \in V'$ (with $k\langle V^{(2)} + V'\rangle = K$ when $n = 1$) and a unique k-isomorphism θ that fits into the top row of a commutative diagram*

$$\begin{array}{ccc} G & \xrightarrow[\simeq]{\theta} & \mathscr{D}(G_{K/k,V^{(2)},V',n}) \\ {\scriptstyle i_G}\downarrow & & \downarrow{\scriptstyle \xi} \\ \mathrm{R}_{K/k}(G_K^{\mathrm{ss}}) & \xrightarrow[\mathrm{R}_{K/k}(\iota_K)]{\simeq} & \mathrm{R}_{K/k}(\mathrm{Sp}_{2n}) \end{array}$$

and carries L over to Sp_{2n} via ι, where $\xi := \xi_{\mathscr{D}(G_{K/k,V^{(2)},V',n})}$.

(ii) *If $n = 2$ then there exists a triple $(V^{(2)}, V', V'')$ as in 9.8.3 satisfying $1 \in V' \cap V''$ and a unique k-isomorphism $G \simeq G_{K/k,V^{(2)},V',V''}$ fitting into an analogous commutative diagram and carrying L over to Sp_{2n} via ι.*

Proof The uniqueness of θ in (i) is immediate via the injectivity of ξ on k_s-points (see Theorem 9.8.1(3),(4)). The same holds for (ii) because

$$\xi_{\mathscr{D}(G_{K/k,V^{(2)},V',2})}|_{G_{K/k,V^{(2)},V',V''}} = \xi_{G_{K/k,V^{(2)},V',V''}}$$

(by Proposition 9.8.4(2)).

The condition that G is of minimal type implies that i_G carries $Z_G(T)$ isomorphically onto a closed k-subgroup of $\mathrm{R}_{K/k}(T_K) = \mathrm{R}_{K/k}(D)$. As we noted at the start of §9.3, $\Phi(G,T)$ is identified with the root system in $\mathrm{X}(D_0)$ consisting of Φ_n and the set of characters $a/2 \in \mathrm{X}(D_0)$ for the long roots a in Φ_n. In particular, Φ_n is the set of non-multipliable elements in $\Phi(G,T)$ and if Δ_n denotes the set of simple roots in Φ_n^+ with $2c$ its unique long element then $\Delta := (\Delta_n - \{2c\}) \cup \{c\}$ is a basis for the unique positive system of roots $\Phi(G,T)^+$ of $\Phi(G,T)$ meeting Φ_n in Φ_n^+. The set of coroots $a^\vee$ for $a \in \Delta_n$ is a basis for $\mathrm{X}_*(D_0)$, so we identify $Z_G(T)$ with a k-subgroup of

$$\prod_{a\in\Delta_n} \mathrm{R}_{K/k}(\mathrm{GL}_1) \simeq \mathrm{R}_{K/k}(D)$$

using the K-isomorphism $\prod_{a\in\Delta_n} \mathrm{GL}_1 \simeq D = (D_0)_K$ defined by $(t_a) \mapsto \prod a_K^\vee(t_a)$. Also, by Proposition 3.2.7 and Corollary 3.4.3(1), the natural map

$$N_G(T)(k)/Z_G(T)(k) \to W(\Phi) = W(\Phi_n) = N_{\mathrm{Sp}_{2n}}(D_0)(k)/D_0(k) \qquad (9.8.3)$$

is an isomorphism whose inverse is induced by the inclusion of $L = \mathrm{Sp}_{2n}$ as a Levi k-subgroup of G containing T.

Since $G = \mathscr{D}(G)$, by Lemma 3.1.5 and the Bruhat decomposition for pseudo-split pseudo-reductive groups (Theorem 3.4.5) we see that G is generated by its k-subgroups $G_a = \langle U_a, U_{-a}\rangle$ for $a \in \Delta$. (Note that $U_{\pm c}$ contains $U_{\pm 2c}$.) By Proposition 3.4.1, for all $a \in \Delta$ the k-group G_a is pseudo-split and absolutely pseudo-simple with the split maximal k-torus $a^\vee(\mathrm{GL}_1) = T \cap G_a$, and $\Phi(G_a, T \cap G_a) = \{\pm a\}$ when $a \neq c$ whereas $\Phi(G_c, T \cap G_c) = \{\pm c, \pm 2c\}$. (Here we use that $a^\vee(\mathrm{GL}_1)$ is an isogeny complement to the codimension-1 torus $(\ker a)^0_{\mathrm{red}}$ in T.)

By Proposition 9.4.5 the k-group G_a is of minimal type, and by Proposition 9.5.2 the field of definition K_a/k for the geometric unipotent radical of G_a is given as follows for $a \in \Delta$: if $n = 1$ then $K_a = K$, if $n > 1$ and $a \in \Delta - \{c\} = \Delta_n - \{2c\}$ then $K_a = K$, and if $n > 1$ then $kK^2 \subseteq K_c \subseteq K$. Since Sp_{2n} is simply connected, pairs of opposite root groups for (Sp_{2n}, D_0) generate SL_2 rather than PGL_2. Thus, for $a' \in \Delta_n$ there is a natural identification of the k-subgroup $L_{a'} = (\mathrm{Sp}_{2n})_{a'}$ with SL_2 that carries $a'^\vee(\mathrm{GL}_1)$ over to the diagonal k-torus and a' over to the standard "positive" root, so the a'-root group inside $L_{a'}$ goes over to the upper triangular unipotent subgroup of SL_2.

Lemma 9.8.7 *If $n > 2$ then $\xi_{G_b} : G_b \to \mathrm{R}_{K/k}((L_b)_K) = \mathrm{R}_{K/k}(\mathrm{SL}_2)$ is an isomorphism for all $b \in \Delta - \{c\}$. If $n = 2$ and b is the unique element in $\Delta - \{c\}$ then ξ_{G_b} carries G_b isomorphically onto $H_{V''}$ for a unique K_0-subspace V'' of K containing 1 such that $k[V''] = K$.*

In this lemma, when $n = 2$ we allow for the possibility that $V'' = K$.

Proof Consider any $b \in \Delta - \{c\} = \Delta_n - \{2c\}$ (with any $n \geqslant 2$), so $K_b = K$ and $(G_b, T \cap G_b)$ has a reduced root system. Such reducedness implies that $\ker \xi_{G_b}$ is central, by Proposition 5.3.10, so the fact that G_b is of minimal type forces this kernel to be trivial. In other words, the map

$$\xi_{G_b} : G_b \to \mathrm{R}_{K/k}((G_b)_K^{\mathrm{ss}}) = \mathrm{R}_{K/k}(\mathrm{SL}_2)$$

is a closed immersion. Under the identification of $(L_b)_K$ with $(G_b)_K^{\mathrm{ss}}$ (compatible with the isomorphism $L_K \simeq G_K^{\mathrm{ss}}$) we see that the inclusion $\xi_{G_b} : G_b \hookrightarrow \mathrm{R}_{K/k}((L_b)_K)$ is a k-subgroup containing L_b, so L_b is a Levi k-subgroup of G_b due to Lemma 7.2.4. Thus, by Proposition 9.1.5, G_b is identified with H_{V_b} for a unique kK^2-subspace $V_b \subseteq K$ containing 1 such that $k[V_b] = K$. In particular, the b-root group is $\underline{V_b} \subseteq \mathrm{R}_{K/k}(\mathbf{G}_{\mathrm{a}})$, and ξ_{G_b} is an isomorphism if and only if $V_b = K$. If $n = 2$ then V_b is a K_0-subspace of K due to Proposition 9.5.2.

It remains to show that if $n > 2$ then necessarily $V_b = K$. This was essentially shown in Remark 9.8.5 in a slightly different context, so now we present the same argument in terms of the k-group G whose existence is given to us. Since $b \neq c$, in the set $\Delta - \{c\} = \Delta_n - \{2c\}$ of short simple positive roots there is a root $b' \neq b$ that is adjacent to b in the Dynkin diagram. Thus, if we consider the codimension-2 torus $T_{b,b'} := ((\ker b) \cap (\ker b))^0_{\mathrm{red}}$ in T then $H := \mathscr{D}(Z_G(T_{b,b'}))$ is a pseudo-split absolutely pseudo-simple k-group with root system $\Phi(G,T) \cap (\mathbf{Z}b + \mathbf{Z}b')$ of type A_2 that admits $\{b, b'\}$ as a basis of simple roots. Explicitly, H is generated by G_b and $G_{b'}$, and in particular $H_b = G_b$ and $H_{b'} = G_{b'}$. Since $K_b = K_{b'} = K$ and the Weyl group of $(H, H \cap T)$ acts transitively on the set of roots (as A_2 is simply laced), it follows from Proposition 6.3.1(1) applied to H that K/k is the field of definition of the geometric unipotent radical of H. Thus, since $\mathrm{char}(k) \neq 3$, Proposition 6.2.1 implies that the natural map $i_H : H \to \mathrm{R}_{K/k}(H_K^{\mathrm{ss}})$ is an isomorphism, with H_K^{ss} isomorphic to SL_3 or PGL_3. In particular, the root groups of H relative to any split maximal k-torus have dimension $[K : k]$, so the inclusion $V_b \subseteq K$ is an equality. □

Now we turn our attention to the root groups of (G, T) for c and $2c$, which we can study by working with the minimal type pseudo-split absolutely pseudo-simple group $(G_c, T \cap G_c)$ of rank 1 with root system $\{\pm c, \pm 2c\}$. Keep in mind that the field of definition K_c/k for the geometric unipotent radical of G_c sits between kK^2 and K but need not equal K (except when $n = 1$). However, $K'_c = K$ in all cases by Proposition 9.5.2. As in the proof of the preceding lemma, L_{2c} is a Levi k-subgroup of G_c (with split maximal k-torus $c^\vee(\mathrm{GL}_1)$), and this is naturally identified with $\mathrm{Sp}_2 = \mathrm{SL}_2$ carrying $c^\vee(\mathrm{GL}_1)$ onto the diagonal k-torus and carrying the $2c$-root group onto the upper triangular unipotent subgroup.

By Proposition 9.3.6(ii), the map

$$\xi_{G_c} : G_c \to \mathrm{R}_{K_c/k}((G_c)^{\mathrm{ss}}_{K_c}) = \mathrm{R}_{K_c/k}((L_{2c})_{K_c}) = \mathrm{R}_{K_c/k}(\mathrm{SL}_2)$$

carries U_{2c} isomorphically onto a nonzero $\underline{kK^2}$-submodule of the upper triangular k-subgroup $\mathrm{R}_{K_c/k}(\mathbf{G}_\mathrm{a})$ inside $\mathrm{R}_{K_c/k}(\mathrm{SL}_2)$. (This makes sense since $kK^2 \subseteq K_c$.) Let V' denote the corresponding nonzero kK^2-subspace of K. Since the $2c$-root group in the Levi k-subgroup $L_{2c} \subseteq G_c$ is carried by ξ_{G_c} onto the natural k-subgroup $\mathbf{G}_\mathrm{a} \subseteq \mathrm{R}_{K_c/k}(\mathbf{G}_\mathrm{a})$, we deduce the important property that $1 \in V'$ inside K.

The identification of U_{2c} with $\underline{V'}$ carries the $Z_G(T)$-action on U_{2c} over to the scaling action on $\underline{V'}$ through the square of the character

$$\chi_c : Z_G(T) \xrightarrow{i_G} \mathrm{R}_{K/k}(D) \xrightarrow{\mathrm{R}_{K/k}(c_K)} \mathrm{R}_{K/k}(\mathrm{GL}_1)$$

as in (9.3.1). The root group U_c is a vector group (Proposition 2.3.11), and more specifically by Propositions 9.3.6, 9.3.7, and 9.5.2 (identifying K'_c as K) there is a $Z_G(T)$-equivariant decomposition of k-groups

$$U_c = \underline{V} \times \underline{V'}$$

where V is a nonzero finite-dimensional K-vector space such that the $Z_G(T)$-action on $\underline{V}$ is through the scaling action by χ_c relative to the $\underline{K}$-module structure on $\underline{V}$. As we saw in the discussion immediately preceding Proposition 9.5.3, the restriction of i_G to U_c is thereby identified with a k-homomorphism

$$\underline{V} \times \underline{V'} \to \mathrm{R}_{K/k}(\mathbf{G}_\mathrm{a}) \tag{9.8.4}$$

that on k-points recovers the canonical inclusion $V' \hookrightarrow K$ and defines an injective additive map $q_c : V \to K$ satisfying $q_c(\lambda v) = \lambda^2 q_c(v)$ for all $v \in V$ and $\lambda \in K$.

The injectivity of q_c identifies the K^2-subspace $q_c(V) \subseteq K$ with the scalar extension $V^{(2)}$ of V through the squaring isomorphism $K \to K^2$. Also, $V^{(2)} \cap V' = 0$ provided that $i_G|_{U_c}$ is injective on k-points. In fact, such injectivity even holds on k_s-points. Indeed, by pseudo-reductivity of G and the unipotence of $\ker(i_G)$, certainly $(\ker i_G)(k_s)$ is a finite subgroup of G_{k_s}. But this finite constant subgroup is visibly normal (as G is smooth), so it is central due to the connectedness of G. By design, this finite central k-subgroup lies in $\ker i_G$, so it must be trivial since G is of minimal type. Thus, $(\ker i_G)(k_s) = 1$ as claimed, so $V^{(2)} \cap V' = 0$. Proceeding in reverse, we may identify the K-vector space V with the scalar extension of the K^2-subspace $V^{(2)} \subseteq K$ via the square root isomorphism $K^2 \simeq K$, and this identifies q_c with the composition

$$V \xrightarrow{q} V^{(2)} \hookrightarrow K$$

in which q is the canonical bijection $v \mapsto 1 \otimes v$. In other words, we have constructed linear algebra data exactly as in 9.6.8 with $1 \in V'$.

The next step is to propagate the description of $i_G|_{U_a}$ for roots $a \in \Delta$ (in Lemma 9.8.7 and (9.8.4)) to all non-divisible *positive* roots in $\Phi(G, T)$. We need to do this in a manner that is compatible with the explicit parameterizations of the root groups for (Sp_{2n}, D_0) defined as in (9.6.2) and (9.6.3). For this purpose we will exploit the triviality of $\mathbf{F}_2^\times$ to avoid the intervention of unknown $k^\times$-scaling factors over k. (Though $k^\times$-scaling preserves k-subspaces of a k-vector space, it can change explicit formulas relative to specified trivializations of root groups.)

Let $W = N_G(T)(k)/Z_G(T)(k)$ be the Weyl group of the root system $\Phi(G, T)$. Using the evident split $\mathbf{F}_2$-descent (Sp_{2n}, S) of (Sp_{2n}, D_0), the natural maps

$$N_{\mathrm{Sp}_{2n}}(S)(\mathbf{F}_2) \to N_{\mathrm{Sp}_{2n}}(D_0)(k)/D_0(k) \to W$$

are isomorphisms (the first one since $S(\mathbf{F}_2) = 1$, as $\mathbf{F}_2^\times = 1$). In other words, the *subgroup* $N_{\mathrm{Sp}_{2n}}(S)(\mathbf{F}_2) \subseteq \mathrm{Sp}_{2n}(k) = L(k)$ is a set of representatives for W inside $L(k)$.

Let a be a non-divisible positive root in $\Phi(G, T)$. Then there exists $\nu \in N_{\mathrm{Sp}_{2n}}(S)(\mathbf{F}_2)$ such that $a \in \nu.\Delta$. That is, if a is multipliable then $a = \nu.c$ and if a is not multipliable then $a = \nu.b$ for some $b \in \Delta - \{c\} = \Delta_n - \{2c\}$.

Lemma 9.8.8 *Let $a \in \Phi(G, T)^+$ be non-divisible. Choose $\nu \in N_{\mathrm{Sp}_{2n}}(S)(\mathbf{F}_2) \subseteq \mathrm{Sp}_{2n}(k) = L(k)$ such that $a \in \nu.\Delta$.*

(i) *Assume a is multipliable (so $a = \nu.c$), as is necessary when $n = 1$. Using ν-conjugation to identify U_c with U_a, the composite homomorphism*

$$\underline{V} \times \underline{V}' = U_c \simeq U_a \stackrel{i_G}{\to} \mathrm{R}_{K/k}(U'_{2a}) = \mathrm{R}_{K/k}((L_{2a})_K) = \mathrm{R}_{K/k}(\mathbf{G}_\mathrm{a})$$

is equal to $(v, v') \mapsto \underline{q}(v) + v'$. *In particular, it is independent of the choice of* ν.

(ii) *Assume a is not multipliable and* $n = 2$. *Let* $b \in \Delta - \{c\}$ *be the unique element, so* $a = \nu.b$. *Using* ν*-conjugation to identify* U_b *with* U_a*, the composite homomorphism*

$$\underline{V}'' = U_b \simeq U_a \stackrel{i_G}{\to} \mathrm{R}_{K/k}(U'_a) = \mathrm{R}_{K/k}((L_a)_K) = \mathrm{R}_{K/k}(\mathbf{G}_\mathrm{a})$$

corresponds to the canonical inclusion of V'' *into* K. *In particular, it is independent of the choice of* ν.

(iii) *Assume a is not multipliable and* $n > 2$. *For any element* $b \in \Delta - \{c\}$ *such that* $a = \nu.b$*, using* ν*-conjugation to identify* U_b *with* U_a *yields a composite isomorphism*

$$\mathrm{R}_{K/k}(\mathbf{G}_\mathrm{a}) = U_b \simeq U_a \stackrel{i_G}{\to} \mathrm{R}_{K/k}(U'_a) = \mathrm{R}_{K/k}((L_a)_K) = \mathrm{R}_{K/k}(\mathbf{G}_\mathrm{a})$$

that is the identity map. In particular, it is independent of the choice of b and ν.

Proof Since $i_G : G \to \mathrm{R}_{K/k}(\mathrm{Sp}_{2n})$ restricts to the initial choice of k-isomorphism $\iota : L \simeq \mathrm{Sp}_{2n}$ (due to how we *defined* the K-isomorphism $G_K^{\mathrm{ss}} \simeq \mathrm{Sp}_{2n}$ via the isomorphism $L_K \simeq G_K^{\mathrm{ss}}$), i_G is equivariant with respect to conjugation on G by $\nu \in L(k)$ and conjugation on $\mathrm{R}_{K/k}(\mathrm{Sp}_{2n})$ by $\nu \in \mathrm{Sp}_{2n}(\mathbf{F}_2) \subseteq \mathrm{Sp}_{2n}(K)$. Thus, it suffices to show that for any $a_1, a_2 \in \Phi(\mathrm{Sp}_{2n}, S)$ and any $\nu \in N_{\mathrm{Sp}_{2n}}(S)(\mathbf{F}_2)$ such that $\nu(a_1) = a_2$, the ν-conjugation isomorphism between the a_i-root groups is carried to the identity automorphism of $\mathbf{G}_\mathrm{a}$ via the root group trivializations over $\mathbf{F}_2$ defined by (9.6.2) and (9.6.3). But this is obvious because the automorphism group of $\mathbf{G}_\mathrm{a}$ over $\mathbf{F}_2$ is $\mathbf{F}_2^\times = 1$. □

We will now complete the proof of Theorem 9.8.6. Denote $\Phi(G, T)$ by Ψ. Let Ψ^+ be the positive system of roots in Ψ determined by the "standard" positive system of roots in $\Phi(\mathrm{Sp}_{2n}, D)$. Choose an enumeration of the set Ψ^+_{nd} of non-divisible roots in Ψ^+. Multiplication defines a pointed k-scheme isomorphism

$$\prod_{a \in \Psi^+_{\mathrm{nd}}} U_a \simeq U_{\Psi^+}.$$

For $a \in \Psi^+_{\mathrm{nd}}$, the isomorphisms in Lemma 9.8.8 unambiguously identify the k-group U_a with $\underline{V} \times \underline{V}'$ for multipliable a and with $\underline{V}''$ (resp. $\mathbf{G}_\mathrm{a}$) for non-multipliable a when $n = 2$ (resp. $n > 2$). This identifies U_{Ψ^+} with the product scheme $\mathscr{U}$ used in Theorem 9.6.14 when $n \neq 2$, and with a closed subscheme of $\mathscr{U}$ (corresponding to the inclusion $\underline{V''} \hookrightarrow \mathrm{R}_{K/k}(\mathbf{G}_{\mathrm{a}})$ at positive roots of intermediate length in Ψ^+) when $n = 2$. To treat $n = 2$ on an equal footing with $n \neq 2$, let $j : U_{\Psi^+} \hookrightarrow \mathscr{U}$ be the closed immersion of pointed k-schemes just defined (so it is an isomorphism when $n \neq 2$ or when $n = 2$ with $V'' = K$).

By construction, the diagram of pointed schemes

$$\begin{array}{ccc} U_{\Psi^+} & \xrightarrow{\ j\ } & \mathscr{U} \\ & \searrow^{i_G} & \downarrow{\scriptstyle f} \\ & & \mathrm{R}_{K/k}(U) \end{array}$$

commutes, and i_G and f are k-homomorphisms with f injective on k-points. Hence, by Zariski-density of $U_{\Psi^+}(k)$ in U_{Ψ^+}, it follows that the horizontal map j is also a k-homomorphism. Using the inclusion of k-groups $Z_G(T) \hookrightarrow \mathrm{R}_{K/k}(D)$ defined by i_G, we thereby get a commutative diagram of pointed k-schemes

$$\begin{array}{ccc} Z_G(T) \ltimes U_{\Psi^+} & \longrightarrow & \mathscr{B} \\ & \searrow^{i_G} & \downarrow{\scriptstyle f} \\ & & \mathrm{R}_{K/k}(B) \end{array}$$

in which the diagonal and vertical maps are k-homomorphisms and f is injective on k-points. Thus, we may again conclude that the horizontal closed immersion is a k-homomorphism provided that $Z_G(T)(k)$ is Zariski-dense in $Z_G(T)$. Such density holds when $k = k_s$, and in general it is harmless to extend scalars to k_s to check the homomorphism property because everything we have done is compatible with such a scalar extension on k.

Let $w \in \mathrm{Sp}_{2n}(\mathbf{F}_2) \subseteq L(k)$ be the unique representative over $\mathbf{F}_2$ of the longest Weyl element. Consider the commutative diagram of pointed k-schemes

$$\begin{array}{ccc} U_{\Psi^+}w \times (Z_G(T) \ltimes U_{\Psi^+}) & \xrightarrow{\ J\ } & \mathscr{U}\mathbf{w} \times \mathscr{B} \\ & \searrow^{i_G} & \downarrow{\scriptstyle f} \\ & & \mathrm{R}_{K/k}(UwB) \end{array} \qquad (9.8.5)$$

in which the vertical map f is the restriction of $i_{G_{K/k,V^{(2)},V',n}}$ to the right $\mathbf{w}$-translate of the open cell relative to D_0 and Φ_n^+ (see Theorem 9.8.1(2)). The closed immersion J in (9.8.5) respects the strict birational group law on each side because i_G and f extend to k-homomorphisms and the extension ϕ of f is injective on k_s-points, so by Theorem 9.6.4 we can uniquely extend J to a k-homomorphism $\theta : G \to G_{K/k,V^{(2)},V',n}$ that has the following properties: it respects the evident maps from each side to $\mathrm{R}_{K/k}(\mathrm{Sp}_{2n})$, it extends the initial choice of k-isomorphism $\iota : (L,T) \simeq (\mathrm{Sp}_{2n}, D_0)$, and it restricts to a closed immersion between neighborhoods of 1 (as we may verify by applying a preliminary right translation by the inverse of the long Weyl element in $N_{\mathrm{Sp}_{2n}}(S)(\mathbf{F}_2) \subseteq \mathrm{Sp}_{2n}(k) = L(k)$). Hence, the normal k-subgroup scheme $\ker\theta$ in the smooth connected k-group G contains the identity section as an open subscheme, so $\ker\theta$ is étale. This forces $\ker\theta$ to be central (since G is connected), but $Z_G \subseteq Z_G(T)$ and the restriction of θ to $Z_G(T)$ is the closed immersion $Z_G(T) \hookrightarrow \mathrm{R}_{K/k}(D)$ arising from the hypothesis that G is of minimal type. We conclude that $\ker\theta = 1$, so θ is a closed immersion.

Since $G = \mathscr{D}(G)$, θ is a closed immersion into $\mathscr{D}(G_{V^{(2)},V',n})$. If $n \neq 2$ then comparison of dimensions of root groups for corresponding roots shows that this map between absolutely pseudo-simple groups is surjective between root groups, so it is surjective between k-groups and hence is an isomorphism. Thus, part (i) of Theorem 9.8.6 is proved. When $n = 2$ we can similarly settle part (ii) due to the realization of $G_{V^{(2)},V',V''}$ as an absolutely pseudo-simple closed k-subgroup of $G_{V^{(2)},V',2}$ with known root groups relative to D_0 (see 9.8.3) and a Cartan k-subgroup whose structure is as small as possible given the simple positive root groups (as we explained in Remark 9.8.5). □

Next, we characterize when our constructions yield isomorphic k-groups. This involves two aspects: showing that certain operations on the linear algebra data do not change the isomorphism class of the associated k-group, and showing that these operations account for all members of a common k-isomorphism class.

Proposition 9.8.9 *Let $(K/k, V^{(2)}, V')$ be as in 9.6.8, and define K_0 to be $k\langle V^{(2)} + V'\rangle \subseteq K$. Let $V'' \subseteq K$ be a nonzero K_0-subspace such that $k\langle V''\rangle = K$. Fix an integer $n \geqslant 1$, and if $n = 1$ then assume $K_0 = K$. For the diagonal k-torus D_0 in Sp_{2n}, let $D = (D_0)_K$.*

(i) *For $\lambda \in K^\times$ there is a unique k-isomorphism fitting into the top of a commutative diagram*

$$\begin{array}{ccc} G_{K/k,V^{(2)},V',n} & \xrightarrow[\simeq]{\theta} & G_{K/k,\lambda V^{(2)},\lambda V',n} \\ \downarrow & & \downarrow \\ \mathrm{R}_{K/k}(\mathrm{Sp}_{2n}) & \xrightarrow[\mathrm{R}_{K/k}(i_\lambda)]{} & \mathrm{R}_{K/k}(\mathrm{Sp}_{2n}) \end{array}$$

where the vertical maps are the canonical ones (so θ respects the natural inclusion of D_0 into each of $G_{K/k,V^{(2)},V',n}$ and $G_{K/k,\lambda V^{(2)},\lambda V',n}$). If $n = 2$ then there is an analogous unique k-isomorphism $G_{K/k,V^{(2)},V',V''} \simeq G_{K/k,\lambda V^{(2)},\lambda V',\mu V''}$ over $\mathrm{R}_{K/k}(i_\lambda \circ I_\mu)$ for any $\lambda, \mu \in K^\times$.

(ii) *Let $\widetilde{V}^{(2)}$ be a K^2-subspace of K such that $(K/k, \widetilde{V}^{(2)}, V')$ satisfies the conditions in 9.6.8 and $V^{(2)} \oplus V' = \widetilde{V}^{(2)} \oplus V'$ inside K. There exists a unique k-isomorphism $G_{K/k,V^{(2)},V',n} \simeq G_{K/k,\widetilde{V}^{(2)},V',n}$ respecting the maps from each to $\mathrm{R}_{K/k}(\mathrm{Sp}_{2n})$ (hence respecting the natural inclusion of D_0 into each of these k-groups). If $n = 2$ then the same holds for the pair of absolutely pseudo-simple k-groups $G_{K/k,V^{(2)},V',V''}$ and $G_{K/k,\widetilde{V}^{(2)},V',V''}$ attached to any nonzero K_0-subspace $V'' \subseteq K$ such that $k\langle V''\rangle = K$.*

(iii) *Let $\widetilde{V}^{(2)}$ and $\widetilde{V}'$ be K^2-subspaces of K that satisfy the conditions enumerated in 9.6.8, and let $\widetilde{K}_0 = k\langle \widetilde{V}^{(2)} + \widetilde{V}'\rangle$. If $n = 1$ then assume $\widetilde{K}_0 = K$. If $n = 2$ then let $\widetilde{V}'' \subseteq K$ be a nonzero $\widetilde{K}_0$-subspace such that $k\langle \widetilde{V}''\rangle = K$.*

If $\mathscr{D}(G_{K/k,V^{(2)},V',n}) \simeq \mathscr{D}(G_{K/k,\widetilde{V}^{(2)},\widetilde{V}',n})$ then $\widetilde{K}_0 = K_0$ and there exists $\lambda \in K^\times$ such that $\widetilde{V}' = \lambda V'$ and $\widetilde{V}^{(2)} + \widetilde{V}' = \lambda V^{(2)} + \lambda V'$. Likewise, if $n = 2$ and $G_{K/k,V^{(2)},V',V''} \simeq G_{K/k,\widetilde{V}^{(2)},\widetilde{V}',\widetilde{V}''}$ then there exist $\lambda \in K^\times$ and $\mu \in K^\times$ such that $\widetilde{V}'' = \mu V''$, $\widetilde{V}' = \lambda V'$, and $\widetilde{V}^{(2)} + \widetilde{V}' = \lambda V^{(2)} + \lambda V'$.

Part (i) largely restates Proposition 9.7.10, except we have incorporated additional constructions when $n = 2$. The main point is that (i) and (ii) give sufficient conditions on linear algebra data that ensure the associated pseudo-split pseudo-reductive k-groups are k-isomorphic, and (iii) says that these conditions *characterize* the k-isomorphism classes of such k-groups.

In Proposition 9.8.12 we will show that $Z_G = 1$ and in Theorem 9.9.3(2) we will show that if $[k:k^2]=2$ then all k-automorphisms of the k-groups G in case (i) are inner (i.e., arise from $G(k)$-conjugations). Whenever $[k:k^2]>2$, this conclusion is false: part (i) provides many k-groups $G := G_{K/k,V^{(2)},V',n}$ that admit a non-inner k-automorphism restricting to the identity on the Cartan k-subgroup $Z_G(D_0)$. To make such G, choose a subfield $K \subseteq k^{1/2}$ containing k with $[K : k] > 2$ and pick $\lambda \in K - k$.

Define $V' = K' := k[\lambda]$ (so $[K' : k] = 2$ and $K' \neq K$) and let $V^{(2)}$ be a K'-subspace of K complementary to V' (so $K_0 = K$ and $V^{(2)} \neq 0$). For the automorphism θ of G as in (i) arising from λ, we see by 9.7.9 that the induced K-automorphism $\theta_K^{\rm red} = i_\lambda$ of $G_K^{\rm red} = \mathrm{Sp}_{2n}$ is not an $\mathrm{Sp}_{2n}(K)$-conjugation because $\lambda \notin (K^\times)^2$. Thus, θ is non-inner.

Proof The assertion in (i) away from the exceptional rank-2 constructions was proved in Proposition 9.7.10. To handle the additional rank-2 constructions we need to avoid the additional complexity in the structure of the D_0-centralizers. First note that

$$(i_\lambda \circ I_\mu)(w) = w \cdot \delta(\lambda, \lambda\mu^2)$$

where $\delta(c_1, c_2) := \mathrm{diag}(1/c_1, 1/c_2, c_1, c_2) \in \mathrm{Sp}_4$. We thereby easily adapt the argument for $n \neq 2$ to build an isomorphism

$$G_{K/k,V^{(2)},V',2} \simeq G_{K/k,\lambda V^{(2)},\lambda V',2}$$

over $\mathrm{R}_{K/k}(i_\lambda \circ I_\mu)$, and passing to derived groups then settles the case of the exceptional rank-2 construction for $V'' = K$. The effect of I_μ on all short positive root groups is multiplication by μ, and its effect on long positive root groups is scaling by 1 on $[m]_{11}$ and by $\mu^2 \in (K^2)^\times \subseteq K_0^\times$ on $[m]_{22}$. Thus, we can compute on simple *positive* root groups (and keep track of opposite root groups via Proposition 9.1.5) to see that the isomorphism built for the case $V'' = K$ (and fixed λ, μ) restricts to the desired isomorphism for general V''. This settles (i) for $n = 2$.

To prove (ii) we treat $(G_{V^{(2)},V',n}, D_0)$ for any $n \geqslant 1$; the case of $(G_{V^{(2)},V',V''}, D_0)$ with $n = 2$ will proceed similarly via adjustments as in the proof of (i) that are left to the reader. To simplify the notation, we shall write G to denote $G_{V^{(2)},V',n}$ and $\widetilde{G}$ to denote $G_{\widetilde{V}^{(2)},V',n}$. Let V and $\widetilde{V}$ denote the K-vector spaces obtained from $V^{(2)}$ and $\widetilde{V}^{(2)}$ respectively via scalar extension along the square root isomorphism $K^2 \simeq K$.

Since the maps $i_G : G \to \mathrm{R}_{K/k}(\mathrm{Sp}_{2n})$ and $i_{\widetilde{G}} : \widetilde{G} \to \mathrm{R}_{K/k}(\mathrm{Sp}_{2n})$ are injective on k_s-points, there is at most one k-isomorphism $G \simeq \widetilde{G}$ intertwining i_G and $i_{\widetilde{G}}$. Thus, by Galois descent, to construct such a k-isomorphism we may assume $k = k_s$. It suffices to construct an isomorphism θ between their dense open Bruhat cells relative to (D_0, Φ_n^+) such that $i_{\widetilde{G}} \circ \theta = i_G$ on the open cell of G. Indeed, since i_G and $i_{\widetilde{G}}$ are k-homomorphisms that are injective on k-points, necessarily such a θ respects the birational group laws on the open Bruhat cells, so θ extends to a k-group isomorphism due to the functorial aspect of Theorem 9.6.4. We would then be done with the proof of (ii).

The construction of θ between open Bruhat cells shall use the explicit description of these open loci provided by Theorem 9.8.1(2). The Cartan k-subgroups are each identified with the Cartan k-subgroup $R_{K/k}(D)$ of $R_{K/k}(\mathrm{Sp}_{2n})$ via i_G and $i_{\widetilde{G}}$, so we just need to identify root groups (compatibly with i_G and $i_{\widetilde{G}}$) for corresponding non-divisible roots in $\Phi(G,T)$. That is, it suffices to build such compatible isomorphisms $\mathscr{U}_b \simeq \widetilde{\mathscr{U}}_b$ for short $b \in \Phi_n$ and $\mathscr{U}_c \simeq \widetilde{\mathscr{U}}_c$ for long $2c \in \Phi_n$. Using conjugation by $(\mathbf{w}, 1)$ reduces the case of negative roots to the case of positive roots (since i_G and $i_{\widetilde{G}}$ each carry $(\mathbf{w}, 1)$ to w), and the case of $b \in \Phi_<^+$ is obvious. For $2c \in \Phi_>^+$ the problem is reduced to one of linear algebra: we seek a k-group isomorphism

$$\varphi : \underline{V} \times \underline{V'} \simeq \underline{\widetilde{V}} \times \underline{V'}$$

that on k-points intertwines $(v, v') \mapsto \underline{q}(v) + v'$ and $(\widetilde{v}, v') \mapsto \underline{\widetilde{q}}(\widetilde{v}) + v'$ via the assumed equality $V^{(2)} + V' = \widetilde{V}^{(2)} + V'$ inside K.

Consider the K^2-linear composite maps

$$V^{(2)} \hookrightarrow V^{(2)} + V' = \widetilde{V}^{(2)} + V' = \widetilde{V}^{(2)} \oplus V' \rightrightarrows \widetilde{V}^{(2)}, V'.$$

The first component $V^{(2)} \to \widetilde{V}^{(2)}$ is a K^2-linear isomorphism, so it yields a K-linear isomorphism $\varphi_1 : V \simeq \widetilde{V}$ via scalar extension along the square root isomorphism $K^2 \simeq K$. The second component $V^{(2)} \to V'$ is K^2-linear, so its composition with the canonical bijection $q : V \to V^{(2)}$ is an additive map $\varphi_2 : V \to V'$ satisfying $\varphi_2(\lambda v) = \lambda^2 \varphi_2(v)$ for all $\lambda \in K$ and $v \in V$. By construction, the resulting k-homomorphism

$$\varphi : (v, v') \mapsto (\underline{\varphi}_1(v), \underline{\varphi}_2(v) + v')$$

is an isomorphism satisfying the desired properties. This completes the proof of (ii).

Finally, we prove (iii). By suitable $K^\times$-scaling via part (i), we may and do assume that 1 lies in each of V' and $\widetilde{V}'$, as well as in each of V'' and $\widetilde{V}''$ when $n = 2$. Thus, now $K_0 = k[V^{(2)} + V']$, $\widetilde{K}_0 = k[\widetilde{V}^{(2)} + \widetilde{V}']$, $k[V''] = k\langle V'' \rangle = K$, and $k[\widetilde{V}''] = k\langle \widetilde{V}'' \rangle = K$. In particular, $V^{(2)}, V' \subseteq K_0$ and $\widetilde{V}^{(2)}, \widetilde{V}' \subseteq \widetilde{K}_0$. Let $G := G_{K/k, V^{(2)}, V', n}$ and $\widetilde{G} := G_{K/k, \widetilde{V}^{(2)}, \widetilde{V}', n}$. Suppose there is a k-isomorphism $h : \mathscr{D}(G) \simeq \mathscr{D}(\widetilde{G})$. By k-rational conjugacy of maximal split k-tori (Theorem C.2.3) and the computation of the Weyl group for the root system in terms of k-points of a torus normalizer (see Proposition 3.2.7 and Corollary 3.4.3(1)), we can adjust the k-isomorphism so that it carries D_0 inside $\mathscr{D}(G)$ onto D_0 inside $\mathscr{D}(\widetilde{G})$ and preserves the standard positive system of roots associated to Φ_n^+. Thus, h must induce the identity map on the automorphism-free Dynkin diagram, and more specifically if c is the unique multipliable simple positive root in the common non-reduced root system then h must respect the c-root groups and the $-c$-root groups, so h restricts to an isomorphism between $\mathscr{D}(G)_c = \mathscr{D}(G_{K_0/k,V^{(2)},V',1})_c$ and $\mathscr{D}(\widetilde{G})_c = \mathscr{D}(G_{\widetilde{K}_0/k,\widetilde{V}^{(2)},\widetilde{V}',1})_c$. Theorem 9.8.1(3) implies that the field of definition over k of the geometric unipotent radicals of these derived groups are respectively K_0 and $\widetilde{K}_0$, so $\widetilde{K}_0 = K_0$ over k. This reduces the problem to the case $n = 1$ at the cost of replacing K with $K_0 = \widetilde{K}_0$ (as we may certainly do). Similarly, for the additional rank-2 constructions we use that Φ_2^+ has basis Δ_2 with a unique short element to obtain two induced isomorphisms: $\mathscr{D}(G_{K_0/k,V^{(2)},V',1})_c \simeq \mathscr{D}(G_{\widetilde{K}_0/k,\widetilde{V}^{(2)},\widetilde{V}',1})_c$ and $H_{V''} \simeq H_{\widetilde{V}''}$. The required μ is thereby obtained from Proposition 9.1.7, and once again $\widetilde{K}_0 = K_0$, so it remains to treat the case $n = 1$ (upon replacing K with K_0).

Now we may and do assume $n = 1$. As above, we may arrange that the isomorphism $h : \mathscr{D}(G) \simeq \mathscr{D}(\widetilde{G})$ restricts to the identity on D_0. The induced isomorphism $\mathscr{D}(G)_K^{\mathrm{ss}} \simeq \mathscr{D}(\widetilde{G})_K^{\mathrm{ss}}$ extending the identity on D is identified with a K-automorphism of SL_2 extending the identity on the diagonal K-torus, so it is induced by the action of a unique element $g = \mathrm{diag}(\lambda, 1) \in \mathrm{PGL}_2(K)$. Thus, we obtain the commutative diagram

$$\begin{array}{ccc} \mathscr{D}(G) & \xrightarrow[\simeq]{h} & \mathscr{D}(\widetilde{G}) \\ {\scriptstyle \xi_{\mathscr{D}(G)}}\downarrow & & \downarrow{\scriptstyle \xi_{\mathscr{D}(\widetilde{G})}} \\ \mathrm{R}_{K/k}(\mathrm{SL}_2) & \xrightarrow[g]{\simeq} & \mathrm{R}_{K/k}(\mathrm{SL}_2) \end{array}$$

in which the bottom row is the action of g. Since h must respect the $2c$-root groups, by evaluating on k-points and computing inside the upper triangular unipotent K-subgroup of SL_2 we see that multiplication by λ carries V' onto $\widetilde{V}'$ inside K. In other words, $\widetilde{V}' = \lambda V'$. Computing the image of k-points of c-root groups likewise gives that $\lambda(V^{(2)} + V') = \widetilde{V}^{(2)} + \widetilde{V}'$ inside K. □

Corollary 9.8.10 *Let G and $\mathscr{G}$ be absolutely pseudo-simple groups of minimal type over an imperfect field k of characteristic 2, and assume that their root systems over k_s are non-reduced. Let k'/k be a separable extension field.*

(1) *If G and $\mathscr{G}$ are pseudo-split over k and $G_{k'} \simeq \mathscr{G}_{k'}$ then $G \simeq \mathscr{G}$.*
(2) *The k-group G admits at most one pseudo-split k-form for the étale topology on k.*

In Theorem 9.9.3(1) we will prove that if $[k : k^2] = 2$ then all such G are pseudo-split. When $[k : k^2] > 2$, the existence of a pseudo-split k-form for the étale topology is a more difficult problem and we will address it in [CP, App. C].

Proof Applying (1) with $k' = k_s$ yields (2), so we can focus our efforts on (1). By using spreading out and specialization arguments (as in the proof of Proposition 1.1.9(1)), we may arrange that k' is a finite extension of k, and then also Galois. In particular, by Galois descent, the purely inseparable fields of definition over k for the geometric unipotent radicals of G and $\mathscr{G}$ are uniquely k-isomorphic; let K/k be this common extension. Let $n \geqslant 1$ be the common dimension of the maximal tori of G and $\mathscr{G}$.

For any two subfields F_1 and F_2 of K containing k, clearly $F_1 = F_2$ inside K if and only if $F_1 \otimes_k k' = F_2 \otimes_k k'$ inside $K \otimes_k k'$. Thus, by Theorem 9.8.6 and Remark 9.6.11, if $n \neq 2$ then there exist pairs $(V^{(2)}, V')$ and $(\mathscr{V}^{(2)}, \mathscr{V}')$ relative to $(K/k, n)$ as in 9.6.8 such that $k\langle V^{(2)} + V'\rangle$ and $k\langle \mathscr{V}^{(2)} + \mathscr{V}'\rangle$ coincide with a common subfield $K_0 \subseteq K$ over k (equal to K when $n = 1$) and

$$G \simeq \mathscr{D}(G_{K/k,V^{(2)},V',n}), \quad \mathscr{G} \simeq \mathscr{D}(G_{K/k,\mathscr{V}^{(2)},\mathscr{V}',n}).$$

Likewise, if $n = 2$ then there exist pairs $(V^{(2)}, V')$ and $(\mathscr{V}^{(2)}, \mathscr{V}')$ satisfying these linear algebra conditions and there exist nonzero K_0-subspaces $V'', \mathscr{V}'' \subseteq K$ satisfying $k\langle V''\rangle = K = k\langle \mathscr{V}''\rangle$ so that

$$G \simeq G_{K/k,V^{(2)},V',V''}, \quad \mathscr{G} \simeq G_{K/k,\mathscr{V}^{(2)},\mathscr{V}',\mathscr{V}''}.$$

By Proposition 9.8.9(iii) and the compatibility of our pseudo-split constructions with respect to separable extension of the ground field, if $n \neq 2$ then for $K' := K \otimes_k k'$ there exists $\lambda \in K'^{\times}$ such that $\lambda V'_{k'} = \mathscr{V}'_{k'}$ (equivalently, $\lambda V'_{K'^2} = \mathscr{V}'_{K'^2}$) and $\lambda(V^{(2)}_{K'^2} + V'_{K'^2}) = \mathscr{V}^{(2)}_{K'^2} + \mathscr{V}'_{K'^2}$ inside $K' = K \otimes_{K^2} K'^2$. If instead $n = 2$ then we get such a λ as well as an element $\mu \in K'^{\times}$ such that $\mu V''_{k'} = \mathscr{V}''_{k'}$ inside K'. Our problem is to show that we can find such scaling factors inside $K^{\times}$, as then these equalities inside K' descend to analogous equalities inside K, from which we deduce the existence of the desired k-isomorphisms (using Proposition 9.8.9(i),(ii)).

We treat the case $n \neq 2$ (as then we do not need to contend with V'' and $\mathscr{V}''$), and a straightforward variant of the same method will take care of the case $n = 2$. Identify the Galois groups $\mathrm{Gal}(k'/k)$, $\mathrm{Gal}(k'^2/k^2)$, $\mathrm{Gal}(K'/K)$, and $\mathrm{Gal}(K'^2/K^2)$ in the evident manner. For any γ in this common Galois group Γ, clearly $\gamma(\lambda)$ works in place of λ. Hence, the ratio $\gamma(\lambda)/\lambda$ lies in the k'^2-subalgebra $F' \subseteq K'$ of elements of K' whose scaling action preserves $V'_{k'} = V'_{K'^2}$ and $V^{(2)}_{K'^2} + V'_{K'^2}$. This subalgebra is a field since K' is algebraic over k'^2. By Galois descent, $F' = F \otimes_{k^2} k'^2$ for the analogous k^2-subalgebra F of K defined in terms of V' and $V^{(2)} + V'$. Thus, naturally $\Gamma = \mathrm{Gal}(F'/F)$, and $\gamma \mapsto \gamma(\lambda)/\lambda$ is a 1-cocycle on Γ valued in $F'^{\times}$. By Hilbert's Theorem 90, there exists $\alpha \in F'^{\times}$ such that $\gamma(\lambda)/\lambda = \gamma(\alpha)/\alpha$ for all $\gamma \in \Gamma$, so $\lambda/\alpha \in K^{\times}$. It is harmless to scale λ by $F'^{\times}$ in view of the definition of F', so we can replace λ with λ/α in order to arrange that $\lambda \in K^{\times}$ as desired. □

The following immediate consequence of Remark 9.8.2, Theorem 9.8.6, and Proposition 9.8.9 (using the discussion early in 9.8.3 concerning when there exists $V'' \neq K$) is of most interest to us when $[k : k^2] = 2$.

Corollary 9.8.11 *Assume $[K : kK^2] = 2$. The only V'' as in* 9.8.3 *for $n = 2$ is $V'' = K$, and if $1 \in V'$ then necessarily $V' = kK^2$ for any $n \geqslant 1$. In particular, the minimal type pseudo-split absolutely pseudo-simple k-groups of rank n with K as the field of definition over k for the geometric radical are precisely the k-groups $G_{V^{(2)},kK^2,n}$ for the nonzero K^2-subspaces $V^{(2)} \subseteq K$ meeting kK^2 trivially.*

The k-isomorphism classes of such groups are parameterized by the $(kK^2)^{\times}$-homothety class of $V^{(2)} \oplus kK^2$. If $[k : k^2] = 2$ then $V^{(2)} \oplus kK^2 = K$, so there is exactly one k-isomorphism class for each pair $(K/k, n)$.

Let $V^{(2)}$, V', V_0, and K_0 be as in 9.6.8. Let $(V_0)^*_{K_0/k}$ be the Zariski closure in $\mathrm{R}_{K_0/k}(\mathrm{GL}_1)$ of the k-subgroup generated by the ratios $x/y \in \mathrm{R}_{K_0/k}(\mathrm{GL}_1)(k)$ for $x, y \in V_0 - \{0\}$. Let $G = G_{K/k,V^{(2)},V',n}$ be the pseudo-reductive group constructed in §9.7 via a birational group law (assuming $K_0 = K$ when $n = 1$). By construction, there is a split maximal k-torus T in G and a K-isomorphism $G_K^{\mathrm{ss}} \simeq \mathrm{Sp}_{2n}$ such that the homomorphism $i_G : G \to \mathrm{R}_{K/k}(\mathrm{Sp}_{2n})$ carries T isomorphically onto the diagonal k-torus D_0 of $\mathrm{R}_{K/k}(\mathrm{Sp}_{2n})$ and maps $Z_G(T)$ isomorphically onto the Cartan subgroup $\mathrm{R}_{K/k}(D)$ of $\mathrm{R}_{K/k}(\mathrm{Sp}_{2n})$. From now on we will identify $Z_G(T)$ with $\mathrm{R}_{K/k}(D)$ and identify T with the maximal k-torus $D_0 \hookrightarrow \mathrm{R}_{K/k}(D)$ via i_G.

Proposition 9.8.12 *The (scheme-theoretic) centers of G and $\mathscr{D}(G)$ are trivial, and likewise for the rank-2 variants using $V'' \subseteq K$ as in* 9.8.3.

Since $\mathscr{D}(G)$ is absolutely pseudo-simple, the triviality of $Z_{\mathscr{D}(G)}$ (with any rank) is also a consequence of Proposition 9.4.9.

Proof In all cases (including the rank-2 examples from 9.8.3 and Proposition 9.8.4), the center Z (of either G or $\mathscr{D}(G)$) is contained in

$Z_G(T) = \mathrm{R}_{K/k}(D)$. The triviality of Z-conjugation on the a-root group for a in the basis Δ of BC_n associated to Φ_n^+ implies that the restriction to Z of the $Z_G(T)$-action on $\mathrm{R}_{K/k}(\mathbf{G}_a)$ through $\mathrm{R}_{K/k}(a_K)$-scaling is trivial on a nonzero $\underline{k}$-submodule (e.g., on $\underline{V}$ when a is multipliable). Thus, Z is contained in $\bigcap_{a\in\Delta} \ker \mathrm{R}_{K/k}(a_K)$. But Δ is a $\mathbf{Z}$-basis of the character group of $D_0 = T$, so this (scheme-theoretic) intersection is trivial. □

Observe that if we replace k with any extension field $k' \subseteq kK^2$ then the k'-group $G_{K/k',V^{(2)},V',n}$ makes sense since $k'K^2 = kK^2$ and the pair $(V^{(2)}, V')$ only "knows" about K and its subfields kK^2 and K^2, and likewise for $G_{K/k',V^{(2)},V',V''}$ as in Proposition 9.8.4 when $n = 2$. (The requirement for $n = 1$ that $k\langle V^{(2)} + V'\rangle = K$ is insensitive to replacing k with such an extension k' with $k'\langle V^{(2)} + V'\rangle = k\langle V^{(2)} + V'\rangle$ due to V' being a nonzero kK^2-subspace of K; similarly $k'\langle V''\rangle = k\langle V''\rangle$ if $n = 2$.) Likewise, if $k_0 \subseteq k$ is a subfield over which k is purely inseparable of finite degree then $k_0\langle V^{(2)}+V'\rangle = k\langle V^{(2)}+V'\rangle$ (since the ratios $cv'/v' = c$ for nonzero $v' \in V'$ and nonzero $c \in kK^2$ recover every element of $k^\times$) and likewise $k_0\langle V''\rangle = k\langle V''\rangle = K$ when $n = 2$. Thus, $G_{K/k_0,V^{(2)},V',n}$ makes sense, as does $G_{K/k_0,V^{(2)},V',V''}$.

The relationships between the constructions relative to K/k, K/k', and K/k_0 are recorded in the following result which implies that the most important case is $k = kK^2$ (i.e., $K^2 \subseteq k$).

Proposition 9.8.13 *Let $k' \subseteq kK^2$ be a subfield containing k and let $k_0 \subseteq k$ be a subfield over which k is purely inseparable of finite degree. Choose $V^{(2)}$ and V' as in 9.6.8. Denote $G_{K/k,V^{(2)},V',n}$, $G_{K/k',V^{(2)},V',n}$, and $G_{K/k_0,V^{(2)},V',n}$ by $G_{K/k}$, $G_{K/k'}$, and G_{K/k_0} respectively. There are unique isomorphisms of algebraic groups fitting into the top row of the commutative diagrams (with canonical bottom row):*

$$\begin{array}{ccc} G_{K/k} & \xrightarrow[\simeq]{\theta} & \mathrm{R}_{k'/k}(G_{K/k'}) \\ \downarrow{\scriptstyle i_{G_{K/k}}} & & \downarrow{\scriptstyle \mathrm{R}_{k'/k}(i_{G_{K/k'}})} \\ \mathrm{R}_{K/k}(\mathrm{Sp}_{2n}) & \xrightarrow[\simeq]{} & \mathrm{R}_{k'/k}(\mathrm{R}_{K/k'}(\mathrm{Sp}_{2n})) \end{array}$$

$$\begin{array}{ccc} G_{K/k_0} & \xrightarrow[\simeq]{\theta_0} & \mathrm{R}_{k/k_0}(G_{K/k}) \\ \downarrow{\scriptstyle i_{G_{K/k_0}}} & & \downarrow{\scriptstyle \mathrm{R}_{k/k_0}(i_{G_{K/k}})} \\ \mathrm{R}_{K/k_0}(\mathrm{Sp}_{2n}) & \xrightarrow[\simeq]{} & \mathrm{R}_{k/k_0}(\mathrm{R}_{K/k}(\mathrm{Sp}_{2n})) \end{array}$$

In particular, the maximal pseudo-reductive quotient $(G_{K/k})_{k'}^{\text{pred}}$ *of* $(G_{K/k})_{k'}$ *is identified with* $G_{K/k'}$, *so this has a* non-reduced *root system (of rank n).*

The analogue between $\mathscr{D}(G_{K/k})$ *and* $\mathscr{D}(G_{K/k'})$ *holds when using* $\mathscr{D} \circ \mathrm{R}_{k'/k}$, *as do its variants for* k/k_0 *and for the rank-2 constructions associated to proper* K_0*-subspaces* $V'' \subseteq K$ *as in* 9.8.3.

Proof We shall explain the existence and uniqueness of θ; the case of θ_0 is handled similarly and the analogues for derived groups and the additional rank-2 constructions follow immediately upon replacing a purely inseparable Weil restriction of $\mathscr{D}(G_{K/k})$ with the derived group of such a Weil restriction. (The Weil restrictions of these derived groups can fail to be perfect because of difficulties with the behavior of $V_{K/k}^{*}$ relative to purely inseparable Weil restriction, as we explained in 9.1.8 via obstructions that are illustrated in Example 9.1.10 whenever $[k : k^2] \geqslant 16$.)

The uniqueness is immediate since $(\ker i_{G_{K/k}})(k_s) = 1$. To prove existence, by the transitivity of Weil restriction with respect to field extensions we may and do restrict attention to the case $k' = kK^2$. Observe that it suffices to find an isomorphism θ between dense open subsets of $G_{K/k}$ and $\mathrm{R}_{k'/k}(G_{K/kK^2})$ that yields a commutative diagram of k-schemes. The reason is that the vertical maps are k-homomorphisms that are injective on k_s-points, so θ would automatically respect the birational group laws over k (as such compatibility can be checked using k_s-points in a dense open subset) and hence the functoriality aspect of Theorem 9.6.4 would extend θ to a k-group isomorphism making the desired diagram commute. By Theorem 9.8.1(2), the open Bruhat cell of $(G_{K/k}, D_0)$ relative to the positive system of roots determined by Φ_n^+ is naturally identified as a k-scheme with $\mathrm{R}_{kK^2/k}$ applied to the open Bruhat cell of $(G_{K/kK^2}, (D_0)_{kK^2})$ relative to the positive system of roots determined by Φ^+. This isomorphism θ between dense open subsets yields a commutative diagram since we can check such commutativity on k_s-points, so we are done. □

Remark 9.8.14 An interesting application of Proposition 9.8.13 occurs when $[k : k^2] = 2$, so the field $kK^2 = K^2$ contains all proper subfields of K over k. In such cases k' can be taken to be *any* proper subfield of K containing k, so the maximal pseudo-reductive quotient $G_{k'}^{\text{pred}}$ over k' has a non-reduced root system for all such k'. In 9.8.17 and Example 9.8.18 we will see that for more general k, if we consider certain proper subfields F of K containing k that do not lie inside kK^2 then it can happen that the non-reductive G_F^{pred} has a *reduced* root system.

Let G be an absolutely pseudo-simple k-group of minimal type whose irreducible root system over k_s is non-reduced, and let T be a maximal k-torus in G. For the Cartan k-subgroup $C = Z_G(T)$, Theorem 2.4.1 provides an affine k-group scheme $\mathrm{Aut}_{G,C}$ of finite type that classifies automorphisms of G restricting to the identity on C, and its maximal smooth closed k-subgroup $Z_{G,C}$ is commutative with pseudo-reductive identity component. Using our explicit knowledge of (G_{k_s}, C_{k_s}) via Theorem 9.8.6, we can determine $Z_{G,C}$ as follows.

Let K/k be the field of definition for the geometric unipotent radical of G. Define $G' = G_K^{\mathrm{ss}}$ and $T' = T_K \subseteq G'$. Define $\pi : T \to T^{\mathrm{ad}}$ to be the quotient corresponding to $T_K = T' \to T'/Z_{G'}$ (recall that since K/k is purely inseparable, base change to K defines an equivalence from the category of k-tori to the category of K-tori). The k-group $Z_{G,C}$ acts on G extending the identity on C, so $(Z_{G,C})_K$ acts on G_K extending the identity on C_K. This action must preserve $\mathscr{R}_{u,K}(G_K)$ because $(Z_{G,C})_K$ is K-smooth, so it induces an action on G' restricting to the identity on the image T' of C_K. Hence, we get a K-homomorphism $(Z_{G,C})_K \to \mathrm{Aut}_{G',T'}$. By Corollary A.4.7(1) applied to K_s-points, this factors through the centralizer of T' in $G'/Z_{G'}$, which is to say through $T'^{\mathrm{ad}} := T'/Z_{G'}$. The resulting K-homomorphism $(Z_{G,C})_K \to T'^{\mathrm{ad}}$ corresponds to a k-homomorphism

$$f : Z_{G,C} \to \mathrm{R}_{K/k}(T'^{\mathrm{ad}}) = \mathrm{R}_{K/k}(T_K^{\mathrm{ad}}).$$

The formation of f is compatible with separable extension on k, and f is injective on k-points because any k-automorphism of G is uniquely determined by the associated K-automorphism of G' (Proposition 1.2.2). Hence, f is injective on k'-valued points for any separable extension k'/k.

Proposition 9.8.15 *The minimal field $F_< \subseteq K$ over k such that f factors through $\mathrm{R}_{F_</k}(T_{F_<}^{\mathrm{ad}})$ is equal to K if $n \neq 2$ and contains kK^2 if $n = 2$. Moreover, there is a unique minimal subfield $F_\natural \subseteq K^{1/2}$ of finite degree over K such that f fits into a fiber product diagram*

$$\begin{array}{ccc} Z_{G,C} & \xrightarrow{\ \theta\ } & \mathrm{R}_{F_\natural/k}(T_{F_\natural}) \\ {\scriptstyle f}\big\downarrow & & \big\downarrow{\scriptstyle \mathrm{R}_{F_\natural/k}(\pi_{F_\natural})} \\ \mathrm{R}_{F_</k}(T_{F_<}^{\mathrm{ad}}) & \xrightarrow[\ j\ ]{} & \mathrm{R}_{F_\natural/k}(T_{F_\natural}^{\mathrm{ad}}), \end{array} \qquad (9.8.6)$$

where j is the canonical inclusion, and such a map θ is unique.

Assume T is split and Δ' is a basis of $\Phi' := \Phi(G',T')$ with respective subsets of short and long roots $\Delta'_<$ and $\Delta'_> = \{a\}$. Let $\Delta = \{a/2\} \cup \Delta'_<$, a basis of $\mathrm{X}(T)$ as well as of $\Phi := \Phi(G,T)$. Via the isomorphisms $T \simeq \mathrm{GL}_1^{\Delta}$ and $T^{\mathrm{ad}} \simeq \mathrm{GL}_1^{\Delta'}$ induced by evaluation of characters in the bases Δ and Δ' respectively, there is a unique identification

$$Z_{G,C} = \mathrm{R}_{F_\natural/k}(\mathrm{GL}_1) \times \prod_{b\in\Delta'_<} \mathrm{R}_{F_</k}(\mathrm{GL}_1) \subseteq \mathrm{R}_{F_\natural/k}(T_{F_\natural}) \tag{9.8.7}$$

over $\mathrm{R}_{F_\natural/k}(T^{\mathrm{ad}}_{F_\natural})$. In particular, $Z_{G,C}$ is connected.

The proof below shows that the formation of $F_< \subseteq K$ and $F_\natural \subseteq K^{1/2}$ commute with separable extension k'/k in the sense that the analogous subfields for $(G_{k'}, T_{k'})$ are $k' \otimes_k F_< \subseteq k' \otimes_k K$ and $k' \otimes_k F_\natural \subseteq (k' \otimes_k K)^{1/2}$ respectively.

Proof The minimality property of $F_<$ implies that its formation commutes with scalar extension to k_s. The uniqueness assertions concerning $F_\natural$ and (9.8.6) (including that of θ) allow us to use Galois descent relative to k_s/k to reduce the proof of existence of $F_\natural$ and (9.8.6) to the case of the ground field k_s. Thus, now we may and do assume $k = k_s$. In particular, T is k-split.

By Theorem 9.8.6, if $n \neq 2$ then there exists a pair $(V^{(2)}, V')$ as in 9.6.8 such that

$$(G,T) \simeq (\mathscr{D}(G_{K/k,V^{(2)},V',n}), D_0),$$

and if $n = 2$ then there is a triple $(V^{(2)}, V', V'')$ as in 9.8.3 such that

$$(G,T) \simeq (G_{K/k,V^{(2)},V',V''}, D_0),$$

where V'' is a nonzero K_0-subspace of K satisfying $k\langle V''\rangle = K$, with $K_0 := k\langle V^{(2)} + V'\rangle$ (so $K_0 = K$ if $n = 1$). Define $V'' = K$ if $n \neq 2$.

Let $F_<$ be the maximal subfield of K over kK^2 such that V'' is an $F_<$-subspace of K (so $F_< = K$ if $n \neq 2$). Since V'' is a nonzero K_0-subspace of K, clearly $K_0 \subseteq F_<$. Define F to be the maximal subfield of K over kK^2 such that the kK^2-subspace $V' \subseteq K$ is an F-subspace of K, so $F \subseteq k\langle V'\rangle \subseteq K_0 \subseteq F_<$.

Consider the nonzero finite-dimensional K-vector space $V = K \otimes_{\iota,K^2} V^{(2)}$, where $\iota : K^2 \simeq K$ is the square root isomorphism. Letting $q : V \simeq V^{(2)} \to K/V'$ be the 2-linear injection $\lambda \otimes x \mapsto \lambda^2 x \bmod V'$, define $F_\natural \subseteq \mathrm{End}_K(V)$ to be the K-algebra of K-linear endomorphisms $L : V \to V$ such that $q \circ L = \zeta_L \cdot q$ for some (necessarily unique) $\zeta_L \in F$. Since q is injective,

$L \mapsto \zeta_L$ defines an additive injection of $F_\natural$ into $F \subseteq K$ over squaring $K \simeq K^2$. Composing with the square root isomorphism $F \simeq F^{1/2}$ thereby identifies $F_\natural$ with a subfield of $F^{1/2}$ ($\subseteq F_<^{1/2} \subseteq K^{1/2}$) as K-algebras. Concretely, the subfield $F_\natural^2$ of F is the stabilizer in F of the nonzero K^2-subspace $V^{(2)} \subseteq K/V'$. The formation of F, $F_<$, and $F_\natural$ is naturally compatible with arbitrary (not necessarily algebraic) separable extension on k (using Remark 9.6.11 and the description of $F_\natural^2$ inside F).

Let a be the unique element in $\Delta'_>$. The isomorphisms $T \simeq \mathrm{GL}_1^\Delta$ and $T^{\mathrm{ad}} \simeq \mathrm{GL}_1^{\Delta'}$ defined by evaluation of characters in the bases Δ and Δ' identify $T \to T^{\mathrm{ad}}$ with the direct product of the squaring map $\mathrm{GL}_1 \to \mathrm{GL}_1$ along the components indexed by $a/2$ and a and the identity map between the remaining GL_1-factors (indexed by $\Delta'_<$). Since T is split, the right side of (9.8.7) makes sense using $F_\natural$ and $F_<$ as defined above. It is clear that this product agrees with the fiber product H formed by the bottom and right maps in (9.8.6), and a k-homomorphism $\theta : Z_{G,C} \to \mathrm{R}_{F_\natural/k}(T_{F_\natural})$ making a fiber square as in (9.8.6) is *unique* if it exists because the k-group scheme $\ker \mathrm{R}_{F_\natural/k}(\pi_{F_\natural}) = \mathrm{R}_{F_\natural/k}(\ker \pi) = \mathrm{R}_{F_\natural/k}(\mu_2)$ receives no nontrivial k-homomorphism from a smooth k-group. For similar reasons, the proposed identification over $\mathrm{R}_{F_\natural/k}(T^{\mathrm{ad}}_{F_\natural})$ in (9.8.7) is unique if it exists. Hence, the fiber product H made by (9.8.6) is smooth and if (9.8.7) holds then: H computes $Z_{G,C}$, $F_<$ is the minimal subfield of K such that f factors through $\mathrm{R}_{F_</k}(T^{\mathrm{ad}}_{F_<})$, and $F_\natural$ satisfies the desired uniqueness property.

Before we take up the proof of (9.8.7), let us interpret it on k-valued points. This equality amounts to the assertion that the subgroup

$$Z_{G,C}(k) \overset{f}{\hookrightarrow} T^{\mathrm{ad}}(K) \hookrightarrow T^{\mathrm{ad}}(K^{1/2})$$

is equal to the intersection of the subgroups $T^{\mathrm{ad}}(F_<)$ and $\pi(T(F_\natural)) \hookrightarrow T^{\mathrm{ad}}(F_\natural)$. If we can prove this equality and its analogue over every separably closed separable extension k'/k (via the inclusion $k' \otimes_k K^{1/2} \hookrightarrow (k' \otimes_k K)^{1/2}$) then we will be done. To see this, first note that once the equality is established on k'-points for *separably closed* separable extensions k'/k, it must then hold on k'-points for all separable extensions k'/k due to Galois theory. By taking k' to be the function field of the identity component of $Z_{G,C}$, we would thereby get a rational map θ^0 from $Z^0_{G,C}$ to $\mathrm{R}_{F_\natural/k}(T_{F_\natural})$ whose composition with $\mathrm{R}_{F_\natural/k}(\pi_{F_\natural})$ is the k-homomorphism $j \circ f|_{Z^0_{G,C}}$. The injectivity of the k-homomorphism $\mathrm{R}_{F_\natural/k}(\pi_{F_\natural})$ on separable points then would force θ^0 to be a rational homomorphism, and hence a k-homomorphism. Using k-point translations (recall we arranged that $k = k_s$) would similarly provide a k-homomorphism $\theta : Z_{G,C} \to \mathrm{R}_{F_\natural/k}(T_{F_\natural})$ such that $\mathrm{R}_{F_\natural/k}(\pi_{F_\natural}) \circ \theta = j \circ f$

and imply that θ is unique. This provides the diagram (9.8.6) as a commutative square that is cartesian on separable points over k, so the map from $Z_{G,C}$ to the *smooth* fiber product is a homomorphism between smooth affine k-groups that is bijective on k'-valued points for all separable extension fields k'/k. By the argument in the proof of Proposition 8.2.6, such a k-homomorphism must be an isomorphism, so we would indeed be done.

Finally, we now prove the desired bijectivity on k'-valued points for all separably closed separable extensions k'/k. We may rename k' as k to reduce to checking (9.8.7) on k-points. Note also that i_G is injective on k-points since G is of minimal type. Our problem is to show that for $\overline{t}' \in T'^{\mathrm{ad}}(K) = T^{\mathrm{ad}}(K)$, the $\overline{t}'$-action on $\mathrm{R}_{K/k}(G')$ lifts through i_G to an action on G restricting to the identity on C if and only if (i) $a(\overline{t}') \in (F_{\natural}^{\times})^2 \subseteq K^{\times}$ for the unique long $a \in \Delta'$, and (ii) $b(\overline{t}') \in F_{<}^{\times}$ for every short $b \in \Delta'$.

For $c \in \Phi = \Phi(G,T)$, i_G carries the c-root group of (G,T) into $\mathrm{R}_{K/k}(U'_c)$ if c is not multipliable and into $\mathrm{R}_{K/k}(U'_{2c})$ if $2c \in \Phi'$. Any C-invariant lift through $i_G : G \to \mathrm{R}_{K/k}(G')$ of the $\overline{t}'$-action on $\mathrm{R}_{K/k}(G')$ must certainly preserve each root group of (G,T). Conversely, suppose the $\overline{t}'$-action on $\mathrm{R}_{K/k}(U'_c)$ for $c \in \Phi'_{<}$ and on $\mathrm{R}_{K/k}(U'_{2c})$ if $2c \in \Phi'$ lifts to an action on the c-root group of (G,T) that preserves the $2c$-root group in the latter case. Then the direct product of these lifts defines a k-scheme automorphism of an open cell $\Omega \subseteq G$ (using the identity map on the direct factor C), and Ω is stable under conjugation by C. Hence, by injectivity of i_G on k-points, this lifted $\overline{t}'$-action on Ω is a C-equivariant rational homomorphism, so via k-point translations it extends to a genuine k-homomorphism that is birational and hence a k-group automorphism that does the job.

In other words, it is necessary and sufficient to prove that for all $\overline{t}' \in T'^{\mathrm{ad}}(K) = T^{\mathrm{ad}}(K)$, the $\overline{t}'$-action on $\mathrm{R}_{K/k}(G')$ lifts through i_G to an action on each root group of (G,T) if and only if

$$a(\overline{t}') \in (F_{\natural}^{\times})^2 \text{ for the unique } a \in \Delta'_{>}, \quad b(\overline{t}') \in F_{<}^{\times} \text{ for all } b \in \Delta'_{<}. \qquad (9.8.8)$$

Since $(\mathbf{w},1)$-conjugation on G intertwines with w-conjugation on $\mathrm{R}_{K/k}(G')$ via i_G, and both conjugations negate all roots, it suffices to study the lifting problem for positive roots. By definition we have $(F_{\natural})^2 \subseteq F_{<} \subseteq K \subseteq F_{\natural}$, and by inspection the long positive roots of Φ' (which are precisely the divisible positive roots in Φ) differ from a by twice a sum of short simple positive roots, so (9.8.8) is equivalent to the condition that $b(\overline{t}') \in F_{<}^{\times}$ for all $b \in \Phi'_{<}$ and $(2c)(\overline{t}') \in (F_{\natural}^{\times})^2$ for all multipliable roots $c \in \Phi$. (Note that for a multipliable root $c \in \Phi$, the sublattice $\mathrm{X}(T'^{\mathrm{ad}})$ inside $\mathrm{X}(T)$ contains $2c$ but does not contain

c, so it does not make sense to evaluate c on $\overline{t}'$; in particular, the value $(2c)(\overline{t}') \in K^\times$ need not lie in $(K^\times)^2$.)

For positive $b \in \Phi'_<$, the explicit description of (G,T) identifies $i_G : \mathscr{U}_b \to \mathrm{R}_{K/k}(U'_b)$ with the inclusion $\underline{V}'' \hookrightarrow \underline{K}$ (using $V'' = K$ if $n \neq 2$). In this way the $\overline{t}'$-action on $\mathrm{R}_{K/k}(U'_b)$ is multiplication by $b(\overline{t}')$ on $\underline{K}$. Thus, the liftability condition for such roots is precisely the condition $b(\overline{t}') \in F^\times_<$.

For a multipliable root $c \in \Phi^+$ we identify $i_G : \mathscr{U}_c \to \mathrm{R}_{K/k}(U'_{2c})$ with the map $\underline{V} \times \underline{V}' \to \underline{K}$ given by $(v, v') \mapsto \underline{q}(v) + v'$. In this way, the $\overline{t}'$-action on $\mathrm{R}_{K/k}(U'_{2c})$ is multiplication by $\lambda := (2c)(\overline{t}')$ on K. We want to show that the $\overline{t}'$-action has a lift through $i_G : \mathscr{U}_c \to \mathrm{R}_{K/k}(U'_{2c})$ preserving $\mathscr{U}_{2c}$ if and only if $\lambda \in (F^\times_\natural)^2$. Suppose such a lift exists. By working with k-points and using that $V \times V' \to K$ has image $V^{(2)} \oplus V'$, existence of the lift implies three conditions: $\lambda \in F^\times$, multiplication by λ on the F-vector space K/V' preserves $(V^{(2)} \oplus V')/V' = V^{(2)}$, and this action on $V^{(2)}$ intertwines through q with a C-equivariant k-automorphism L of $\underline{V} = \mathscr{U}_c/\mathscr{U}_{2c}$. To show that $\lambda \in (F^\times_\natural)^2$, by the definition of $F_\natural$ as a subfield of $K^{1/2}$ over K it suffices to show that L is $\underline{K}$-linear.

The k-vector space structure on $V = \underline{V}(k)$ underlies the *unique* T-equivariant linear structure on the vector group $\underline{V}$, so the C-equivariant k-automorphism L of $\underline{V}$ respects this linear structure since $T \subseteq C$. Hence, L is linear over the k-subalgebra $k[C(k)] \subseteq \mathrm{End}_k(V)$. The K-vector space structure on V is k-linear, so K-linearity of L is reduced to verifying that $K = k[C(k)]$ inside $\mathrm{End}_k(V)$. This equality follows from the direct product description of C in (9.7.6) or (9.8.2) provided by Theorem 9.8.1(2) or Proposition 9.8.4(1) respectively (since $k\langle V''\rangle = K$ and $\langle c, b^\vee\rangle = -1$ for some $b \in \Phi'_<$ if $n > 1$, and $K_0 = K$ if $n = 1$).

Conversely, assume that $\lambda \in (F^\times_\natural)^2$ (so $\lambda \in F^\times$). Multiplication by λ preserves V' inside K by definition of V' and λ, and by definition of $F_\natural$ this multiplier carries $V^{(2)}$ into $V^{(2)} \oplus V'$ via $v \mapsto L_1(v) + L_2(v)$ where L_1 is a K^2-linear automorphism of the finite-dimensional K^2-vector space $V^{(2)}$ and $L_2 : V^{(2)} \to V'$ is K^2-linear. The definition $V := K \otimes_{\iota, K^2} V^{(2)}$ motivates defining the K-linear automorphism $L : V \simeq V$ via $\mu \otimes x \mapsto \mu \otimes L_1(x)$ and the 2-linear map $Q : V \to V'$ via $\mu \otimes x \mapsto \mu^2 L_2(x)$. The k-group automorphism of $\mathscr{U}_c = \underline{V} \times \underline{V}'$ defined by

$$(v, v') \mapsto (\underline{L}(v), \underline{Q}(v) + \lambda v')$$

preserves $\underline{V}'$ and lifts the $\overline{t}'$-action on $\mathrm{R}_{K/k}(U'_{2c})$ since $\lambda = (2c)(\overline{t}')$. □

Example 9.8.16 In Theorem 9.9.3(2) we will show that if $[k : k^2] = 2$ then the map $G(k) \to \mathrm{Aut}_k(G)$ is bijective. (Recall that $Z_G = 1$ without restriction on $[k : k^2]$, by Proposition 9.8.12, and $G = \mathscr{D}(G)$ if $[k : k^2] = 2$ due to Theorem 9.8.1(4) and Remark 9.8.2.) In contrast, whenever $[k : k^2] > 2$ we claim that there exist pseudo-split minimal type absolutely pseudo-simple G over k with root system BC_n (any $n \geqslant 1$) such that the image of $Z_{G,C}(k) \to T'^{\mathrm{ad}}(K)$ is not contained in the image of $T'(K) = T(K)$, so this image is larger than the image of $C(k)$ and hence G admits k-automorphisms which are *not* inner.

Due to the natural inclusion $C \hookrightarrow \mathrm{R}_{K/k}(T') = \mathrm{R}_{K/k}(T_K)$, it suffices to show that the map $T(F_\natural) \hookrightarrow T^{\mathrm{ad}}(F_\natural)$ carries the set of k-points of the direct product in (9.8.7) onto a subset not contained in the image of $T(K)$. In other words, it suffices to show that as a subset of $T(F_\natural)$, the set of k-points of the direct product in (9.8.7) is not contained in $T(K)$. This amounts to $F_\natural$ strictly containing K, or equivalently $(F_\natural)^2$ strictly containing K^2. Equivalently, we seek $\alpha \in K$ not in K^2 such that α-scaling on K preserves V' and $V^{(2)} \oplus V'$.

Since $[k : k^2] > 2$, clearly $[K : K^2] > 2$. If kK^2 is strictly larger than K^2 then we can choose V' to be a nonzero proper kK^2-subspace of K and $V^{(2)}$ to be the $K^2(\alpha)$-span of a kK^2-basis of a kK^2-subspace of K complementary to V' for any $\alpha \in kK^2 - K^2$. If instead $kK^2 = K^2$ then $[K : K^2]$ is finite yet larger than 2, so for $\alpha \in K - K^2$ we can choose $V^{(2)}$ and V' to be nonzero complementary $K^2(\alpha)$-subspaces of K (since $K^2(\alpha)$ is a proper subfield of K, as $[K : K^2] > 2$).

We now review our general construction of pseudo-split absolutely pseudo-simple k-groups with a non-reduced root system, and indicate ways in which the case $[K : kK^2] = 2$ (which includes all cases with $[k : k^2] = 2$) is special relative to the possibilities that arise for more general K/k. (If $[k : k^2] > 2$ then we can always choose K over k such that $[K : kK^2] > 2$. Indeed, if $\{z_i\}$ is a 2-basis for k over $\mathbf{F}_2$ then products among the z_i without repetition are a k^2-basis for k, so there are at least two such z_i. For $K = k(z_1^{1/2}, z_2^{1/2})$ we have $kK^2 = k$, and $[K : k] = 4$ since a k-basis of $k^{1/2}$ is given by products among the $z_i^{1/2}$ without repetition, with the empty product understood to be 1.)

Fix an imperfect field k of characteristic 2, and a purely inseparable nontrivial finite extension K/k. Let V' be a nonzero kK^2-subspace of K and let $V^{(2)}$ be a nonzero finite-dimensional K^2-subspace of K such that $V^{(2)} \cap V' = 0$. Define $K_0 := k\langle V^{(2)} + V'\rangle$, define $V = K \otimes_{\iota,K^2} V^{(2)}$ where $\iota : K^2 \to K$ is the square root isomorphism, and define the 2-linear map $q : V \to K$ by $q(\lambda \otimes v_0) = \lambda^2 v_0$ (i.e., q is additive and $q(\lambda v) = \lambda^2 q(v)$ for all $v \in V$ and $\lambda \in K$). Note that $q(V) = V^{(2)}$ as a K^2-subspace of K.

The birational group law method applied to the map

$$\underline{V} \times \underline{V'} \to \mathrm{R}_{K/k}(\mathbf{G}_{\mathrm{a}}) \tag{9.8.9}$$

defined by $(v, v') \mapsto \underline{q}(v) + v'$ yields a pseudo-split and absolutely pseudo-simple k-group (G, T) of minimal type with a non-reduced root system of any desired rank $n > 1$ (and also of rank 1 if we assume that $K_0 = K$) such that the field of definition over k for the geometric unipotent radical of G is K. If G is of rank 1 then $i_G(G)$ is naturally identified with $H_{V_0} \subseteq \mathrm{R}_{K/k}(\mathrm{SL}_2)$, where $V_0 = kV^{(2)} + V'$. In particular, if G is of rank 1 then $Z_G(T)$ is identified with the Zariski closure $(V_0)^*_{K/k} \subseteq \mathrm{R}_{K/k}(\mathrm{GL}_1)$ of the subgroup of $K^\times$ generated by ratios x/y for $x, y \in V_0 - \{0\}$. Whenever $[K : kK^2] = 2$ we must have $V_0 = K$ since V_0 is a kK^2-subspace of K which properly contains the nonzero kK^2-subspace V' inside the 2-dimensional kK^2-vector space K. If $[K : kK^2] > 2$ then we can arrange $V_0 \neq K$. (Pick $\alpha \in K - kK^2$, so $[kK^2(\alpha) : kK^2] = 2$ and hence $[K : kK^2(\alpha)] > 1$. Choosing $V' = kK^2$ and $V^{(2)}$ to be the K^2-span of $\{\alpha v_i\}$ for a $kK^2(\alpha)$-basis $\{v_i\}$ of K containing 1, we have $V_0 \neq K$.) If $V_0 \neq K$ then the k-group $(V_0)^*_{K/k}$ can be a proper k-subgroup of $\mathrm{R}_{K/k}(\mathrm{GL}_1)$ (e.g., Remark 9.1.11 gives examples with codimension 1).

Let c be a multipliable root in $\Phi(G, T)$. The k-subgroup G_{2c} is of minimal type (by Proposition 9.4.5), and by 9.6.12 its geometric unipotent radical has field of definition K_{2c} over k equal to $k\langle V'\rangle$, whereas the analogous field K_c for G_c is $K_0 := k\langle V_0\rangle \subseteq K = K'_c$. Clearly $K'_{2c} = kK'^2_c = kK^2 \subseteq k\langle V'\rangle = K_{2c}$. Whenever $[K : kK^2] > 2$ the field $K_{2c} = k\langle V'\rangle$ can be strictly larger than K'_{2c} (even equal to K), by taking $V^{(2)} = K^2\alpha_1\alpha_2$ and $V' = kK^2 \oplus (\bigoplus_i kK^2\alpha_i)$ for a 2-basis $\{\alpha_i\}$ of K/k. In Example 9.8.18 we give examples for which $K'_{2c} \neq K_{2c}$ and $K_c \neq K$.

9.8.17 Assume $1 \in V'$, so $kK^2 \subseteq V'$. Over the field $F := k[V^{(2)}]$ between K and k generated by $V^{(2)}$ as a k-algebra, we claim that the maximal pseudo-reductive quotient G_F^{pred} (which is visibly absolutely pseudo-simple of rank n) has a *reduced* root system. To prove this, first recall that in general $U_{2c} \to \mathrm{R}_{K/k}(U'_{2c})$ is a $\underline{kK^2}$-linear inclusion, so since $kK^2 \subseteq F$ it suffices to prove that c is not a root of $(G_F^{\mathrm{pred}}, T_F)$ (as then $2c \in \Phi(G_F^{\mathrm{pred}}, T_F)$ and $2c$ is non-divisible in this root system). Applying scalar extension along $k \to F$ to (9.8.9) and composing with the canonical quotient map $\mathrm{R}_{K/k}(\mathbf{G}_{\mathrm{a}})_F \twoheadrightarrow \mathrm{R}_{K/F}(\mathbf{G}_{\mathrm{a}})$ (corresponding to the multiplication map $K \otimes_k F \to K$) yields an F-homomorphism

$$h_F : \underline{F \otimes_k V} \times \underline{F \otimes_k V'} \to \mathrm{R}_{K/F}(\mathbf{G}_\mathrm{a}).$$

Its kernel contains the *smooth connected* graph of the F-homomorphism $\underline{F \otimes_k V} \to \underline{F \otimes_k V'}$ corresponding to the 2-linear map of finite-dimensional F-vector spaces $F \otimes_k V \to F \otimes_k V'$ defined by $z \otimes v \mapsto z^2 q(v) \otimes 1$; this map makes sense since $q(V) = V^{(2)} \subseteq k[V^{(2)}] = F$. Any smooth connected F-subgroup of $\ker h_F$ has trivial image in the *pseudo-reductive* quotient G_F^{pred} (since any such subgroup is contained in the unipotent group $(\ker \xi_G)_F$), so the F-subgroup $(U_{2c})_F = \underline{V'}_F$ of G_F maps onto the $2c$-root group of $(G_F^{\mathrm{pred}}, T_F)$ and $c \notin \Phi(G_F^{\mathrm{pred}}, T_F)$.

We have $k[V^{(2)}] = K$ when $[K : kK^2] = 2$, because $k[V^{(2)}]$ is a kK^2-subalgebra of K containing an element outside $V' \supseteq kK^2$. If $[K : kK^2] > 2$ then we can always find V' and $V^{(2)}$ such that $k[V^{(2)}] \neq K$. Explicitly, when $[K : kK^2] = 2^r$ with $r \geqslant 2$ then for a 2-basis $\{z_1, \dots, z_r\}$ of K over k we can take $V' = kK^2[z_1]$ and $V^{(2)} = K^2[z_2, \dots, z_r]$. Cases with $k[V^{(2)}] \neq K$ are interesting because as we vary through the maximal pseudo-reductive quotients G_F^{pred} over purely inseparable finite extensions F of k we acquire reducedness of the root system over a strictly smaller extension of k than the minimal extension K that is required to gain reductivity. This is illustrated by the following example.

Example 9.8.18 Here is an explicit class of examples over a rational function field $k = \kappa(t_1, \dots, t_r)$ over any field κ of characteristic 2, with any $r \geqslant 2$. Define

$$K = k(\sqrt{t_1}, \dots, \sqrt{t_r}), \quad V' = k(\sqrt{t_{m+1}}, \dots, \sqrt{t_s}),$$
$$V^{(2)} = K^2\sqrt{t_1} \oplus \cdots \oplus K^2\sqrt{t_m}$$

with $1 \leqslant m < s \leqslant r$, so $kK^2 = k$ and hence $[K : kK^2] = 2^r$. Thus, using the square root isomorphism $\iota : K^2 \simeq K$,

$$V := K \otimes_{\iota, K^2} V^{(2)} = \bigoplus_{i \leqslant m} K(1 \otimes \sqrt{t_i})$$

and the 2-linear $q : V \to K$ is $\sum_{i \leqslant m} x_i(1 \otimes \sqrt{t_i}) \mapsto \sum \sqrt{t_i} \cdot x_i^2$ (for $x_i \in K$). Since $K^2 \subseteq k$, this data satisfies the requirements. Let G be the pseudo-split absolutely pseudo-simple k-group of rank $n \geqslant 2$ (with root system BC_n) built via this data. The field of definition K_{2c} over k for the geometric unipotent radical of G_{2c} is $k[V'] = k(\sqrt{t_{m+1}}, \dots, \sqrt{t_s}) \neq K$ and $K'_{2c} = kK^2 = k$ is a proper subfield of $K_{2c} \neq K$. Moreover, we see that

$$K_c = k[V^{(2)} + V'] = k(\sqrt{t_1}, \dots, \sqrt{t_s}),$$

so $K_c \neq K$ if $s < r$.

Note that the k-group $U_{2c} = \underline{V'}$ has dimension 2^{s-m}, which can be quite small in comparison with the dimension $[K : k]^m = 2^{rm}$ of the k-group $\underline{V} = U_c/U_{2c}$. Since $k[V^{(2)}] = k(\sqrt{t_1}, \dots, \sqrt{t_m}) \neq K$, this example exhibits the phenomenon (as explained in 9.8.17) that there can be a proper subfield F of K over k such that the maximal pseudo-reductive quotient G_F^{pred} of G_F has a reduced root system but is non-reductive (as $F \neq K$ over k).

9.9 Classification over fields k with $[k : k^2] = 2$

In the rest of this chapter k will be an imperfect field of characteristic 2 such that $[k : k^2] = 2$. Let K be a nontrivial purely inseparable finite extension of k. The fields $k^{1/2^m}$ for $m \geqslant 0$ are the only purely inseparable finite extensions of k, so $k \subseteq K^2$ and $[K : K^2] = 2$. As we noted in Remark 9.6.10, any $(V^{(2)}, V')$ as in 9.6.8 (over such a k) with $1 \in V'$ equals $(\alpha K^2, K^2)$ for some $\alpha \in K - K^2$ (so $K = k(\alpha)$), and $q : V \to V^{(2)}$ is $\lambda \otimes \alpha \mapsto \alpha\lambda^2$. We will denote the group $G_{K/k, V^{(2)}, V', n}$ as $G_{\alpha,n}$ in this section. Remark 9.8.2 and Theorem 9.8.1(4) ensure that $G_{\alpha,n}$ is its own derived group. By Corollary 9.8.11, up to k-isomorphism $G_{\alpha,n}$ is the unique pseudo-split absolutely pseudo-simple k-group of minimal type with rank n such that K is the field of definition over k for its geometric unipotent radical and the root system is BC_n. This implies, in particular, that up to k-isomorphism $G_{\alpha,n}$ is independent of the choice of $\alpha \in K - K^2$.

We fix a positive integer n and let Φ denote the root system of type BC_n. Choose a positive system of roots Φ^+ in Φ, and let Δ be the corresponding basis of Φ. Let c be the unique multipliable simple positive root. We also fix $\alpha \in K - K^2$ and will denote $G_{\alpha,n}$ simply by G. Let $T \subseteq G$ be a split maximal k-torus. Fix a Levi k-subgroup L of G containing T (which exists by Theorem 3.4.6, or by the construction of G). The natural map $L_K \to G_K^{\mathrm{ss}} \simeq \mathrm{Sp}_{2n}$ is an isomorphism and $\Phi(L, T)$ is identified with the root system of non-multipliable roots in Φ, so Φ^+ determines a positive system of roots in $\Phi(L, T)$ with

$$\Delta' := (\Delta - \{c\}) \cup \{2c\} = \Delta_n$$

a basis of $\Phi(L,T)$. Let D be the diagonal K-torus of Sp_{2n} and D_0 be the diagonal k-torus of Sp_{2n}. Choose a k-isomorphism $(L,T) \simeq (\mathrm{Sp}_{2n}, D_0)$ respecting the identification of based root systems. This identifies (G_K^{ss}, T_K) with (Sp_{2n}, D) respecting the k-structures (L,T) and (Sp_{2n}, D_0) on each. The induced map $\xi_G : G \to \mathrm{R}_{K/k}(\mathrm{Sp}_{2n})$ then identifies the Cartan subgroup $Z_G(T)$ of G with the Cartan subgroup $\mathrm{R}_{K/k}(Z_{\mathrm{Sp}_{2n}}(D))$.

Clearly D_0 is the maximal k-torus of $\mathrm{R}_{K/k}(D)$. Identify the root groups of $(\mathrm{R}_{K/k}(\mathrm{Sp}_{2n}), D_0)$ with $\mathrm{R}_{K/k}(\mathbf{G}_\mathrm{a})$ in the natural way. For every non-divisible non-multipliable root $a \in \Phi(G,T)$, ξ_G maps the root group U_a of G isomorphically onto the a-root group of $\mathrm{R}_{K/k}(\mathrm{Sp}_{2n})$. For any multipliable root a, the root group U_a of G is identified with $\mathrm{R}_{K/k}(\mathbf{G}_\mathrm{a}) \times \mathrm{R}_{K^2/k}(\mathbf{G}_\mathrm{a})$ so that $\xi_G|_{U_a}$ is identified with the map

$$\mathrm{R}_{K/k}(\mathbf{G}_\mathrm{a}) \times \mathrm{R}_{K^2/k}(\mathbf{G}_\mathrm{a}) \to \mathrm{R}_{K/k}(\mathbf{G}_\mathrm{a})$$

given by $(x,y) \mapsto \alpha x^2 + y$.

We now turn our attention to proving finer properties of the k-group G (assuming $[k : k^2] = 2$), and then use such properties to prove that up to k-isomorphism there exists exactly one absolutely pseudo-simple k-group of each absolute rank $n \geqslant 1$ with K/k the field of definition for its geometric radical. The main point is to show that the pseudo-split and minimal type hypotheses in Corollary 9.8.11 are unnecessary when $[k : k^2] = 2$. Our study of the properties of G begins by applying Proposition 5.1.3 to G:

Proposition 9.9.1 *For any commutative k-group scheme Z of finite type that does not contain a nontrivial smooth connected k-subgroup, every central extension*

$$1 \to Z \to E \to G \to 1$$

is uniquely split.

Proof By Proposition 5.1.3 and the perfectness of G, we just have to find a dense open $C \subseteq Z_G(T)$ and a finite ordered set of k-scheme maps h_i from C into root groups of (G,T) such that $\prod h_i : C \to G$ is the natural inclusion. The equality $G = G_{K/k,\alpha K^2,K^2,n}$ implies that ξ_G carries $Z_G(T)$ isomorphically onto $\mathrm{R}_{K/k}(D)$ (see Theorem 9.8.1(2), with ϕ there equal to ξ_G by Theorem 9.8.1(3)). In particular, this is naturally a direct product of copies of $\mathrm{R}_{K/k}(\mathrm{GL}_1)$ indexed by the elements of Δ', with the coroots $a'^\vee$ for $a' \in \Delta'$ defining a K-isomorphism $\prod_{a' \in \Delta'} \mathrm{GL}_1 \simeq D$ via $(t_{a'}) \mapsto \prod a'^\vee_K(t_{a'})$. For each $a' \in \Delta' - \{2c\}$ (i.e., a non-multipliable root in Δ), the "a'-factor" $\mathrm{R}_{K/k}(a'^\vee_K(\mathrm{GL}_1))$ of $\mathrm{R}_{K/k}(D)$

is the intersection of $Z_G(T)$ with $G_{a'}$. The evident identification of $G_{a'}$ with $\mathrm{R}_{K/k}(\mathrm{SL}_2)$ for such a' thereby reduces the construction of the desired generic formula for the "a'-factor" of $Z_G(T)$ to the formula given at the end of Example 5.1.4.

It remains to treat the $2c$-factor (corresponding to the unique multipliable simple root c). For this purpose, we consider the subgroup G_c (generated by the root groups $U_{\pm c}$). This is a pseudo-split absolutely pseudo-simple group of rank 1 (with split maximal torus $c^\vee(\mathrm{GL}_1) = T \cap G_c$) that inherits the "minimal type" property from G due to Proposition 9.4.5. Its root system is $\{\pm c, \pm 2c\}$, so by replacing G with G_c we may assume $n = 1$. (This has no effect on K since $K_c = K$ when $[k : k^2] = 2$, by Proposition 9.5.3(iii), and the role of "$\mathscr{U}\mathbf{w}\mathscr{B}$" in the construction of $G_{K/k,\alpha K^2,K^2,n}$ via birational group laws shows that $G_c \simeq G_{K/k,\alpha K^2,K^2,1}$.)

We now have $\Phi^+ = \{c, 2c\}$, and $U_{\pm c} = \mathrm{R}_{K/k}(\mathbf{G}_\mathrm{a}) \times \mathrm{R}_{K^2/k}(\mathbf{G}_\mathrm{a})$. We identify the upper triangular unipotent subgroup of $\mathrm{R}_{K/k}(\mathrm{SL}_2)$ with $\mathrm{R}_{K/k}(\mathbf{G}_\mathrm{a})$ in the canonical way. With this identification, the restriction of i_G to $U_c = \mathrm{R}_{K/k}(\mathbf{G}_\mathrm{a}) \times \mathrm{R}_{K^2/k}(\mathbf{G}_\mathrm{a})$ is $(v, v') \mapsto \underline{q}(v) + v'$ for the quadratic map $\underline{q} : \mathrm{R}_{K/k}(\mathbf{G}_\mathrm{a}) \to \mathrm{R}_{K/k}(\mathbf{G}_\mathrm{a})$ defined by $v \mapsto \alpha v^2$ with a fixed $\alpha \in K - K^2$. Dimension considerations with root groups show that the image of G under i_G is $\mathrm{R}_{K/k}(\mathrm{SL}_2)$ (i.e., i_G is surjective), and the Cartan subgroup $Z_G(T)$ of G is mapped isomorphically onto the diagonal Cartan subgroup of $\mathrm{R}_{K/k}(\mathrm{SL}_2)$ (since G is of minimal type).

The open subvariety $\mathrm{R}_{K/k}(\mathrm{GL}_1) \subseteq \mathrm{R}_{K/k}(\mathbf{G}_\mathrm{a})$ represents the functor of points $\alpha u + v$ (with $u, v \in \mathrm{R}_{K^2/k}(\mathbf{G}_\mathrm{a})$) such that $\alpha^2 u^2 + v^2 \in \mathrm{R}_{K^2/k}(\mathrm{GL}_1)$. We define the open locus $\Omega \subseteq \mathrm{R}_{K/k}(\mathrm{GL}_1)$ by imposing the additional condition that $u \in \mathrm{R}_{K^2/k}(\mathrm{GL}_1)$. Thus, points in Ω can be expressed as $\alpha u + v = u(\alpha + z)$, with $z = v/u$, so Ω is the scheme of pairs $(u, z) \in \mathrm{R}_{K^2/k}(\mathrm{GL}_1) \times \mathrm{R}_{K^2/k}(\mathbf{G}_\mathrm{a})$ such that $\alpha + z \in \mathrm{R}_{K/k}(\mathrm{GL}_1)$. We would like to write a generic element of Ω as a product of elements of U_c and U_{-c}. For this, we will use the following familiar identity in SL_2 written over a field of characteristic 2:

$$y(t^{-1})x(t)y(t^{-1}) = \begin{pmatrix} 1 & 0 \\ 1/t & 1 \end{pmatrix} \begin{pmatrix} 1 & t \\ 0 & 1 \end{pmatrix} \begin{pmatrix} 1 & 0 \\ 1/t & 1 \end{pmatrix} = \begin{pmatrix} 0 & t \\ 1/t & 0 \end{pmatrix}.$$

Taking $t = u$, we obtain

$$y(u^{-1})x(u)y(u^{-1}) = \begin{pmatrix} 0 & u \\ 1/u & 0 \end{pmatrix}. \tag{9.9.1}$$

Now consider the value

$$t = \frac{1}{\alpha + z} = \frac{\alpha + z}{\alpha^2 + z^2} = \frac{\alpha}{(\alpha + z)^2} + \frac{z}{\alpha^2 + z^2} = \alpha f(z)^2 + g(z),$$

for $z \in \mathrm{R}_{K^2/k}(\mathbf{G}_{\mathrm{a}})$ such that $\alpha + z \in \mathrm{R}_{K/k}(\mathrm{GL}_1)$, where $f(z) = (\alpha + z)^{-1}$ and $g(z) = z/(\alpha^2 + z^2) = zf(z)^2$. This gives the formula

$$y(\alpha + z)x(\alpha f(z)^2 + g(z))y(\alpha + z) = \begin{pmatrix} 0 & 1/(\alpha + z) \\ \alpha + z & 0 \end{pmatrix}. \tag{9.9.2}$$

Multiplying (9.9.1) and (9.9.2) yields that $\mathrm{diag}(u(\alpha+z), 1/u(\alpha+z))$ is equal to

$$y(u^{-1})x(u)y(u^{-1})y(\alpha + z)x(\alpha f(z)^2 + g(z))y(\alpha + z).$$

Using the identifications $U_{\pm c} = \mathrm{R}_{K/k}(\mathbf{G}_{\mathrm{a}}) \times \mathrm{R}_{K^2/k}(\mathbf{G}_{\mathrm{a}})$, the decomposition $K = \alpha K^2 \oplus K^2$ as k-vector spaces motivates us to consider the maps $\Omega \to U_{\pm c}$ defined by

$$h_1(u, z) = h_3(u, z) = (0, u^{-1}) \in U_{-c}, \;\; h_2(u, z) = (0, u) \in U_c,$$
$$h_4(u, z) = h_6(u, z) = (1, z) \in U_{-c}, \;\; h_5(u, z) = (f(z), g(z)) \in U_c.$$

To check that $\prod h_i : \Omega \to G$ is the natural inclusion on the open subscheme $\Omega \subseteq \mathrm{R}_{K/k}(\mathrm{GL}_1)$, we can assume $k = k_s$ and it suffices to work with k-points. Applying the map ξ_G that is injective on k-points, we can use the identities (9.9.1) and (9.9.2) to conclude the argument. □

Proposition 9.9.2 *The map $\xi_G : G \to \mathrm{R}_{K/k}(\mathrm{Sp}_{2n})$ is bijective on k-points if $[k : k^2] = 2$.*

Proof The vanishing of $(\ker \xi_G)(k)$ is part of Theorem 9.8.1(3). Thus, ξ_G carries $G(k)$ isomorphically onto a subgroup of $G_K^{\mathrm{ss}}(K) = \mathrm{Sp}_{2n}(K)$ that we want to show is the entire group.

Recall that $G = G_{\alpha,n}$ with $\alpha \in K - K^2$. By the proof of Theorem 9.8.1, the restriction of $\xi_G : G \to \mathrm{R}_{K/k}(\mathrm{Sp}_{2n})$ to the open cell $(\mathscr{U}\mathbf{w} \times \mathscr{B})(\mathbf{w}, 1)^{-1}$ is a direct product of k-scheme morphisms, each of which is one of the following three types: the identity map between Cartan k-subgroups, an isomorphism between root groups for non-divisible non-multipliable roots of G, or the homomorphism

$$\mathrm{R}_{K/k}(\mathbf{G}_{\mathrm{a}}) \times \mathrm{R}_{K^2/k}(\mathbf{G}_{\mathrm{a}}) \to \mathrm{R}_{K/k}(\mathbf{G}_{\mathrm{a}})$$

defined by $(x, y) \mapsto \alpha x^2 + y$. But this latter map is bijective on k-points since $[K : K^2] = 2$. Hence, the image of ξ_G on k-points contains the K-points of an open Bruhat cell of Sp_{2n}. Since K is infinite, for each $g \in \mathrm{Sp}_{2n}(K)$ the two subsets $g\Omega(K)$ and $\Omega(K)$ meet for any non-empty open $\Omega \subseteq \mathrm{Sp}_{2n}$ (such as an open Bruhat cell). Thus, $\mathrm{Sp}_{2n}(K)$ is the image of ξ_G on k-points. □

Proposition 9.9.2 identifies $G(k)$ with $\mathrm{Sp}_{2n}(K)$ via ξ_G when $[k : k^2] = 2$, but $\ker \xi_G$ is non-smooth and even has positive dimension when k is a proper subfield of K^2. (By Theorem 9.8.1(4) and Remark 9.6.9, this kernel is a product of $2n$ copies of $\mathrm{R}_{K^2/k}(\alpha_2)$, and the hypothesis $[k : k^2] = 2$ implies $\mathrm{R}_{K^2/k}(\alpha_2)$ has dimension $[K^2 : k]/2$ when k is a proper subfield of K^2.) Thus, although one may think of $G(k)$ as a "k-structure" on $\mathrm{Sp}_{2n}(K)$ when $[k : k^2] = 2$, the k-group G is quite different from $\mathrm{R}_{K/k}(\mathrm{Sp}_{2n})$.

Theorem 9.9.3 *Let k be a field of characteristic 2 such that $[k : k^2] = 2$ and let K be a purely inseparable nontrivial finite extension of k. Fix $n \geqslant 1$.*

(1) *Up to k-isomorphism there is a unique absolutely pseudo-simple k-group G with a non-reduced root system of rank n over k_s such that K is the field of definition over k for its geometric radical, and $G_K^{\mathrm{ss}} \simeq \mathrm{Sp}_{2n}$. These groups are pseudo-split and of minimal type over k.*
(2) *The center Z_G is trivial, the map $\xi_G : G(k) \to G_K^{\mathrm{ss}}(K) = \mathrm{Sp}_{2n}(K)$ is bijective, and all k-automorphisms of G are inner.*
(3) *For any commutative k-group scheme Z of finite type that does not contain a nontrivial smooth connected k-subgroup, any central extension of G by Z is uniquely split over k.*

Proof By Theorem 9.8.1 and Propositions 9.8.12, 9.9.1, and 9.9.2, there exists a pseudo-split absolutely pseudo-simple k-group satisfying all of the asserted properties except possibly the property of having only inner automorphisms. In fact, we constructed such a group $G_{K/k,\alpha K^2,K^2,n} =: G_{\alpha,n}$ using a birational group law resting on the k-homomorphism $f : \mathrm{R}_{K/k}(\mathbf{G}_{\mathrm{a}}) \times \mathrm{R}_{K^2/k}(\mathbf{G}_{\mathrm{a}}) \to \mathrm{R}_{K/k}(\mathbf{G}_{\mathrm{a}})$ defined by $(x, y) \mapsto \alpha x^2 + y$ for a choice of $\alpha \in K - K^2$. To prove the uniqueness in (1) (where there are no minimal type or pseudo-split hypotheses), we first treat the case $k = k_s$ and then use a Galois cohomology argument to deduce the general case from this.

Let G be a k-group as in (1) with $k = k_s$. We seek to prove that $G \simeq G_{\alpha,n}$. Since $k = k_s$, G is pseudo-split. Since $[k : k^2] = 2$, Corollary 9.8.11 implies that G is uniquely determined up to k-isomorphism by K/k and n *provided that* G is also of minimal type. By Proposition 9.4.2, G admits a minimal type pseudo-simple quotient $\mathscr{G}$ by a central unipotent k-subgroup scheme such that $\mathscr{G}$ has the same root datum and the same K/k. Thus, $\mathscr{G} \simeq G_{\alpha,n}$. This shows that G is a central extension of $G_{\alpha,n}$ by a commutative unipotent k-group scheme Z, and by centrality and unipotence of Z we see that Z contains no nontrivial smooth connected k-subgroup schemes (due to the pseudo-reductivity of G). We may therefore apply Proposition 9.9.1 to the exact sequence

$$1 \to Z \to G \to G_{\alpha,n} \to 1$$

to deduce that $G = Z \times G_{\alpha,n}$ as k-group schemes. But Z then inherits smoothness and connectedness from G, so the condition that Z contains no nontrivial smooth connected k-subgroups implies that $Z = 1$, so $G \simeq G_{\alpha,n}$!

Having proved (1) when $k = k_s$, before we carry out the proof for general k we prove one further property when $k = k_s$: all k-automorphisms ϕ of G are inner. In view of the uniqueness of G up to isomorphism over $k = k_s$, to prove this it suffices to analyze a single specific case; we will work with $G = G_{\alpha,n}$ for some α since in these cases we have described the structure of an open Bruhat cell explicitly. Composing with an inner automorphism, we can assume that ϕ preserves some maximal k-torus T (since $k = k_s$, so Proposition A.2.10 applies). Composing with conjugation by a suitable element of $N_G(T)(k)$ allows us to arrange in addition that the action of ϕ on $\Phi(G, T)$ preserves a choice of positive system of roots Φ^+ (Proposition 3.2.7). This corresponds to a positive system of roots in the set $\Phi(G_K^{\mathrm{ss}}, T_K)$ of non-multipliable roots of $\Phi(G, T)$. By Proposition 1.2.2, if two k-automorphisms of a pseudo-reductive group coincide on the maximal geometric reductive quotient over $\overline{k}$ then the automorphisms are equal. Hence, it suffices to find some $g \in G(k)$ such that g-conjugation and ϕ induce the same K-automorphism of $G' := G_K^{\mathrm{ss}}$.

Let $\Delta' = \{a_1, \ldots, a_{n-1}, a_n\} \subseteq \Phi(G', T_K)$ be the simple positive roots for G' with a_n the unique long simple root, so $a_n/2 \in \mathrm{X}(T_K) = \mathrm{X}(T)$ and $\Delta := \{a_1, \ldots, a_{n-1}, a_n/2\}$ is a basis of $\mathrm{X}(T)$. Since ϕ preserves Φ^+, it preserves the subset Δ'. It must then fix the elements of Δ' pointwise because the Dynkin diagram admits no nontrivial automorphisms. Hence, for each $1 \leqslant j \leqslant n$ the K-automorphism ϕ' induced by ϕ on G' acts on each root group $U'_{a_j} \simeq \mathbf{G}_{\mathrm{a}}$ via scaling by some $\lambda_j \in K^\times$. For $j = n$ something special happens: the map $\xi_G : G \to \mathrm{R}_{K/k}(G')$ carries the long root group U_{a_n} into $\mathrm{R}_{K/k}(U'_{a_n})$ by a $\underline{K}^2$-linear inclusion and $U_{a_n} \simeq \mathrm{R}_{K^2/k}(\mathbf{G}_{\mathrm{a}})$ as a $\underline{K}^2$-module (due to the explicit construction of $G = G_{\alpha,n}$ with $[k : k^2] = 2$). Since ξ_G intertwines ϕ and $\mathrm{R}_{K/k}(\phi')$, it follows that the scaling action by $\lambda_n \in K^\times$ on the 1-dimensional K-vector space $U'_{a_n}(K)$ preserves the 1-dimensional K^2-subspace $U_{a_n}(k)$. Hence, $\lambda_n \in (K^\times)^2$.

In view of the above explicit basis Δ of $\mathrm{X}(T)$, there exists a unique $g \in T(K) = \mathrm{R}_{K/k}(T_K)(k)$ such that $a_j(g) = \lambda_j \in K^\times$ for $j < n$ and $(a_n/2)(g) = \lambda_n^{1/2} \in K^\times$. The description of Cartan k-subgroups in $G_{\alpha,n}$ in Theorem 9.8.1(2) then identifies such a g with an element of $Z_G(T)(k)$ such that the conjugation by g on G' and the action of ϕ' have the same effect on all simple positive root groups of (G', T_K). Thus, composing ϕ with conjugation by g^{-1} on G brings us to the case where ϕ' is an automorphism of G' that preserves T_K, fixes Φ^+ pointwise inside of $\mathrm{X}(T)$, and is the identity on all simple positive root groups of (G', T_K).

In general, for any split connected semisimple group H over a field F and any F-automorphism f of H, if f preserves an F-split maximal F-torus S, fixes a positive system of roots $\Phi^+ \subseteq \Phi(H, S)$ pointwise, and is the identity on all simple positive root groups then $f = \mathrm{id}$. Indeed, by passing to the simply connected central cover of H we can assume H is simply connected, and for rank 1 this is a simple calculation (since $\mathrm{Aut}_F(\mathrm{SL}_2) = \mathrm{PGL}_2(F)$), so on the rank-1 connected semisimple subgroups of H generated by $U_{\pm a}$ for simple roots a we see that f induces the identity. In particular, f is the identity on the maximal F-torus S. It follows from the isomorphism case of the Isogeny Theorem (Theorem A.4.10) that f is "conjugation" by a unique F-point t of the image $\overline{S}$ of S in the adjoint quotient. However, the simple positive roots in Φ^+ are a *basis* of the character lattice of the maximal torus $\overline{S}$ in the adjoint quotient of H, so $t = 1$ (due to the hypothesis that f acts as the identity on the simple positive root groups).

Finally, we return to (1) in general (i.e., allow $k \neq k_s$) and also prove that all k-automorphisms of G are inner. Note that $[k_s : k_s^2] = 2$ (since $k_s^2 \otimes_{k^2} k \to k_s$ is a map of fields that is both separable and purely inseparable, hence an equality). The settled case over k_s therefore implies that $Z_G = 1$ for general k (or use Proposition 9.4.9). Hence, we can use Galois descent to prove that a k-automorphism is inner as a formal consequence of its validity in the settled case of a separably closed ground field.

To verify the uniqueness of G up to k-isomorphism it is equivalent to fix *one* example over k and show that it has no nontrivial k-forms for the étale topology over k (as any two examples become isomorphic over k_s, by what we proved above). To handle this analysis of k-forms, we may and do work with the example $G := G_{\alpha,n}$ for some $\alpha \in K - K^2$ because then the k-group G has a k-split maximal k-torus.

As every automorphism of G_{k_s} is inner, the triviality of the center implies that $\mathrm{Aut}_{k_s}(G_{k_s}) = G(k_s)$ as groups with $\mathrm{Gal}(k_s/k)$-action. But the natural map $G(k_s) \to G_K^{\mathrm{ss}}(K_s)$ is bijective (Proposition 9.9.2), so the set of k-forms of G for the étale topology over k is in bijection with the Galois cohomology set

$$\mathrm{H}^1(k_s/k, \mathrm{Aut}_{k_s}(G_{k_s})) = \mathrm{H}^1(k_s/k, G(k_s)) = \mathrm{H}^1(K_s/K, G_K^{\mathrm{ss}}(K_s)).$$

Since G has a k-split maximal k-torus, G_K^{ss} has a K-split maximal K-torus. Thus, $G_K^{\mathrm{ss}} \simeq \mathrm{Sp}_{2n}$. But $\mathrm{H}^1(K_s/K, \mathrm{Sp}_{2n}) = 1$ since up to K-isomorphism there is a unique non-degenerate symplectic space of dimension $2n$ over K. □

We conclude with analogues of Proposition 7.3.3 and Corollary 7.3.4.

Proposition 9.9.4 *Let k be imperfect of characteristic 2 such that $[k : k^2] = 2$. Let G be an absolutely pseudo-simple k-group with rank $n \geqslant 1$ and a non-reduced root system. Let K/k be the minimal field of definition of the geometric radical of G, $G' := G_K^{\mathrm{ss}} \simeq \mathrm{Sp}_{2n}$, and $\xi_G : G \to \mathrm{R}_{K/k}(G')$ the natural surjective map.*

(1) *The natural map $G(k) \to G'(K)$ is an isomorphism of groups and $\mathrm{H}^1(k, G)$ vanishes.*
(2) *If k is complete with respect to an absolute value then the natural map $G(k) \to G'(K)$ is a topological isomorphism.*
(3) *If k is a global function field then the natural map $G(\mathbf{A}_k) \to G'(\mathbf{A}_K)$ on adelic points is a topological isomorphism.*

Proof By Theorem 9.9.3(2) we get the group isomorphism claim in (1). Applying the result over k_s, we get $\mathrm{H}^1(k, G) \simeq \mathrm{H}^1(K, G')$. But $G' \simeq \mathrm{Sp}_{2n}$, whose degree-1 cohomology vanishes over any field whatsoever (as non-degenerate symplectic spaces over any field are determined up to isomorphism by their even rank). This settles (1). The completeness hypothesis in (2) ensures that the valuation topology on K is its topology as a finite-dimensional k-vector space. We shall prove (2) by using an explicit description of G, namely $G \simeq G_{\alpha,n}$ for a choice of $\alpha \in K - K^2$.

Since ξ_G is surjective with unipotent kernel, for a split maximal k-torus T in G the image $\xi_G(T)$ is a split maximal k-torus in $\mathrm{R}_{K/k}(G')$. This image is the split maximal k-torus in $\mathrm{R}_{K/k}(T')$ for a unique split maximal K-torus T' in G' (Proposition A.5.15(2)), and explicitly T' is the image of T_K under the quotient map $G_K \twoheadrightarrow G'$ corresponding to ξ_G via the universal property of Weil restriction. The natural isomorphism of $\mathrm{X}(T')$ with $\mathrm{X}(T)$ identifies the reduced root system $\Phi(G', T')$ with the set of non-multipliable roots in

$\Phi(G,T)$ (Theorem 2.3.10), so there is a natural bijective correspondence between positive systems of roots in the two root systems. Note that when T is identified with a maximal k-torus in $\mathrm{R}_{K/k}(G')$, the root system $\Phi(\mathrm{R}_{K/k}(G'),T)$ is identified with $\Phi(G',T')$ (Example 2.3.2).

For $G = G_{\alpha,n}$ there is an isomorphism of G' with Sp_{2n} (depending on a k-split maximal k-torus T in G and a positive system of roots Φ^+ in $\Phi(G,T)$) so that for the corresponding positive system of roots Φ'^+ in $\Phi(\mathrm{R}_{K/k}(G'),T) = \Phi(G',T')$ the restriction of ξ_G between open Bruhat cells for (G,T,Φ^+) and $(\mathrm{R}_{K/k}(G'),T,\Phi'^+)$ can be described explicitly. To be precise, by Theorem 9.8.1(3), the map ξ_G between translated open Bruhat cells is identified with the map

$$f: \mathscr{U}\mathbf{w} \times \mathscr{B} \to \mathrm{R}_{K/k}(UwB)$$

introduced in Theorem 9.7.1. This map is constructed as a direct product of several maps: the identity map on $\mathrm{R}_{K/k}(D)$ (for the standard maximal diagonal K-torus D in Sp_{2n}), copies of the identity map on $\mathrm{R}_{K/k}(\mathbf{G}_\mathrm{a})$ (for short roots of (Sp_{2n},D)), and copies of the map

$$\mathrm{R}_{K/k}(\mathbf{G}_\mathrm{a}) \times \mathrm{R}_{K^2/k}(\mathbf{G}_\mathrm{a}) \to \mathrm{R}_{K/k}(\mathbf{G}_\mathrm{a})$$

defined by $(x,y) \mapsto \alpha x^2 + y$. Thus, to establish (2) we just have to check that the bijective continuous map $K \times K^2 \to K$ defined by $(x,y) \mapsto \alpha x^2 + y$ has continuous inverse. Since $K = \alpha K^2 \oplus K^2$ and the squaring map $K \to K^2$ is a topological isomorphism (as we are in characteristic 2 and K is complete), the required continuity for (2) is clear.

To prove assertion (3), the established bijectivity on k-points and k_v-points for all v (in assertions (1) and (2)) allows us to use the same argument as in the proof of Proposition 7.3.3(3). In particular, we apply the surjectivity criterion in Proposition A.2.13 to the target group $\mathrm{R}_{K/k}(G') = \mathrm{R}_{K/k}(\mathrm{Sp}_{2n})$ (see Example A.2.14 and Remark A.2.15). □

Corollary 9.9.5 *Let $(G,K/k,G')$ be as in Proposition* 9.9.4.

(1) *If $T' \subseteq G'$ is a maximal K-torus then there exists a unique maximal k-torus $T \subseteq G$ such that $T \subseteq \xi_G^{-1}(\mathrm{R}_{K/k}(T'))$, and the map $T \to \mathrm{R}_{K/k}(T')$ is an isomorphism onto the maximal k-torus of $\mathrm{R}_{K/k}(T')$.*
(2) *The correspondence in* (1) *defines a bijection between sets of maximal tori, also characterized by the property $T(k) \subseteq T'(K)$ via the equality $G(k) = G'(K)$, and T is k-split if and only if T' is K-split.*
(3) *All maximal k-split k-tori in G are $G(k)$-conjugate.*

The assertion (3) is a special case of a conjugacy result for all smooth connected affine groups over any field (Theorem C.2.3), but we give a direct proof in this case.

Proof Since k is infinite and tori are unirational, the condition $T(k) \subseteq T'(K)$ is equivalent to the condition $T \subseteq \xi_G^{-1}(\mathrm{R}_{K/k}(T'))$. Also, the maximal k-torus S in $\mathrm{R}_{K/k}(T')$ is k-split if and only if T' is K-split, since the natural map $S_K \to T'$ is an isomorphism and K/k is purely inseparable. Thus, (2) is a consequence of (1).

To prove (1), since $\xi_G : G \to \mathrm{R}_{K/k}(G')$ is surjective with unipotent kernel, the method of proof of Corollary 7.3.4 carries over essentially unchanged (replacing Proposition 7.3.3(1) with Proposition 9.9.4(1)). The only novelty is the intervention of $\mathrm{R}_{K/k}$, so we use the good behavior of the formation of maximal tori with respect to Weil restriction (Proposition A.5.15(2)).

For (3), first observe that the connected reductive K-group $G' \simeq \mathrm{Sp}_{2n}$ is K-split, so its maximal K-split K-tori are all maximal as K-tori. Thus, if S is a maximal k-split k-torus in G then the K-split K-torus S_K of G' is contained in a K-split maximal K-torus T', so $\xi_G(S) \subseteq \mathrm{R}_{K/k}(T')$. The maximal k-torus T of G corresponding to T' is k-split, and it contains S. By maximality we must then have $S = T$, so S is maximal as a k-torus of G. In other words, the maximal k-split k-tori of G are precisely the maximal k-tori of G that are k-split. By (2), these naturally correspond to the K-split maximal K-tori of G'. The latter are $G'(K)$-conjugate to each other, so the definition of the bijective correspondence in (2) and the bijectivity of the map $\xi_G : G(k) \to G'(K)$ imply the $G(k)$-conjugacy of maximal k-split k-tori of G. □

10
General case

In characteristic 2, the notion of standard pseudo-reductive group has to be replaced with something that incorporates exotic constructions right from the beginning. The resulting "generalized standard" construction will classify all pseudo-reductive groups with a reduced root system (over k_s) in any characteristic, assuming $[k : k^2] \leqslant 2$ when $\mathrm{char}(k) = 2$. This only has novelty in characteristics 2 and 3, since in all other characteristics the generalized standard construction will literally be the standard construction. In [CP], many additional constructions are given in characteristic 2, yielding a classification theorem without restriction on $[k : k^2]$.

10.1 Factors with non-reduced root system and the generalized standard construction

Let k be a field. The first fact we wish to establish is that for any pseudo-reductive k-group G, if $\Phi(G_{k_s}, T_{k_s})$ is non-reduced (so k is imperfect with characteristic 2) then its non-reduced irreducible components can be uniquely split off as the root system arising from a k-group direct factor of G, provided $[k : k^2] = 2$. We will often use without comment the elementary observation that if k has characteristic $p > 0$ and $[k : k^p] = p$ then every finite extension of k satisfies the same degree condition.

Definition 10.1.1 A pseudo-reductive k-group G is *totally non-reduced* if $G = \mathscr{D}(G) \neq 1$ and every irreducible component of $\Phi(G_{k_s}, T_{k_s})$ is non-reduced, where T is a maximal k-torus of G.

Since the formation of the root system of a pseudo-simple group over a separably closed field is unaffected by a separable extension of the base field to another separably closed field, if k'/k is a separable extension then a pseudo-reductive k-group G is totally non-reduced if and only if $G_{k'}$ is totally non-reduced. By Proposition 3.2.10, G is totally non-reduced precisely when it is perfect and nontrivial and every pseudo-simple normal k_s-subgroup of G_{k_s} has a non-reduced root system.

Definition 10.1.2 A *basic non-reduced pseudo-simple k-group* is an absolutely pseudo-simple k-group G with a non-reduced root system such that the field of definition K/k for the geometric radical of G is quadratic over k.

For example, if k is imperfect with characteristic 2 and K/k is any purely inseparable quadratic extension then for any $\alpha \in K - k$ and $n \geqslant 1$, the k-group $G_{\alpha,n}$ introduced at the beginning of §9.9 is a basic non-reduced pseudo-reductive k-group. If $\mathrm{char}(k) = 2$ and $[k : k^2] = 2$ then up to k-isomorphism there is exactly one basic non-reduced pseudo-simple k-group of each rank $n \geqslant 1$ (Theorem 9.9.3(1)). If we allow the field of definition of the geometric radical to have larger degree, or we relax absolute pseudo-simplicity to perfectness, we can make more examples of totally non-reduced pseudo-reductive k-groups:

Example 10.1.3 Let k be an imperfect field of characteristic 2, k' a nonzero finite reduced k-algebra, and G' a k'-group such that each fiber over $\mathrm{Spec}\, k'$ is absolutely pseudo-simple. Also assume, for $k'_s := k' \otimes_k k_s$, that every fiber of $G'_{k'_s}$ over $\mathrm{Spec}\, k'_s$ has a non-reduced root system. Then we claim that $G := \mathrm{R}_{k'/k}(G')$ is totally non-reduced, and that it is perfect if $[k : k^2] = 2$.

We may assume $k = k_s$ and pass to factor fields of k' so that k' is a field. For a maximal k'-torus T' in G', the maximal k-torus T in $\mathrm{R}_{k'/k}(T')$ is a maximal k-torus in G (Proposition A.5.15(2)) and $T' \simeq T_{k'}$ since k'/k is purely inseparable. We then have $\Phi(G,T) = \Phi(G',T')$ via the identification of $\mathrm{X}(T)$ and $\mathrm{X}(T')$, by Example 2.3.2. Hence, $\Phi(G,T)$ is non-reduced and irreducible. To prove that G is perfect when $[k : k^2] = 2$ it suffices to prove that the group $G(k) = G'(k')$ is perfect. Since $[k' : k'^2] = 2$, by Theorem 9.9.3(2) we have $G'(k') \simeq \mathrm{Sp}_{2n}(k'')$ for some finite extension k''/k'. But $\mathrm{Sp}_{2n}(F)$ is perfect for any infinite field F, so we are done.

The preceding class of examples is exhaustive when $\mathrm{char}(k) = 2$ and $[k : k^2] = 2$:

Proposition 10.1.4 *Let k be an imperfect field of characteristic 2 such that $[k : k^2] = 2$, and let G be a totally non-reduced pseudo-reductive k-group.*

(1) *There exist a nonzero finite reduced k-algebra k' and a k'-group G' such that the fibers of $G' \to \operatorname{Spec} k'$ are basic non-reduced pseudo-simple k-groups and $G \simeq \mathrm{R}_{k'/k}(G')$.*
(2) *If $(G'', k''/k)$ is another such pair, then every k-group isomorphism*

$$\mathrm{R}_{k'/k}(G') \simeq \mathrm{R}_{k''/k}(G'')$$

is induced by a unique k-algebra isomorphism $\alpha : k'' \simeq k'$ and a unique group isomorphism $G' \simeq G''$ over $\operatorname{Spec}(\alpha)$. *In particular, the pair $(G', k'/k)$ in* (1) *is uniquely determined by G up to unique k-isomorphism.*
(3) *If $\{k'_i\}$ is the set of factor fields of k' then every smooth connected normal k-subgroup of G is a product among the factor groups $\mathrm{R}_{k'_i/k}(G'_i)$. In particular, G is pseudo-simple over k if and only if k' is a field, and it is absolutely pseudo-simple if and only if k'/k is purely inseparable.*

Proof By Galois descent, it suffices to treat the case where $k = k_s$. Let $\{G_i\}_{i\in I}$ be the set of pseudo-simple normal k-subgroups of G. By hypothesis G is totally non-reduced, so it follows from Proposition 3.2.10 that every G_i is totally non-reduced. But each G_i is (absolutely) pseudo-simple, so since $[k : k^2] = 2$ we may use Theorem 9.9.3 and Proposition 9.8.13 to infer that $Z_{G_i} = 1$ and $G_i \simeq \mathrm{R}_{k'_i/k}(G'_i)$ for a finite extension k'_i/k and a basic non-reduced pseudo-simple k'_i-group G'_i. The triviality of Z_{G_i} for all i implies that the surjective map $\prod G_i \to G$ with central kernel has trivial kernel and hence is an isomorphism. Thus, letting $k' = \prod k'_i$ and $G' = \coprod G'_i$ yields an isomorphism

$$G \simeq \prod G_i \simeq \mathrm{R}_{k'/k}(G').$$

This proves (1).

To prove (2), consider a second pair $(G'', k''/k)$, say with $\{k''_j\}_{j\in J}$ the factor fields of k'' and G''_j the k''_j-fiber of G'' (which we assume to be a basic non-reduced pseudo-simple k''_j-group). By Corollary 9.8.11 and Proposition 9.8.13, $\mathrm{R}_{k''_j/k}(G''_j)$ is absolutely pseudo-simple with a non-reduced root system. Consider any isomorphism

$$f : \prod_i \mathrm{R}_{k'_i/k}(G'_i) = \mathrm{R}_{k'/k}(G') \simeq \mathrm{R}_{k''/k}(G'') = \prod_j \mathrm{R}_{k''_j/k}(G''_j).$$

The k-subgroups $G_i \simeq \mathrm{R}_{k'_i/k}(G'_i)$ were constructed as the pseudo-simple normal k-subgroups of G, so by the proof of Corollary 8.1.4 every smooth connected normal k-subgroup of G is a product among the k-subgroups G_i. Applying this to the collection of subgroups $\mathrm{R}_{k''_j/k}(G''_j)$, we conclude that f identifies each $\mathrm{R}_{k''_j/k}(G''_j)$ with $\mathrm{R}_{k'_{i(j)}/k}(G'_{i(j)})$ for a unique $i(j)$, with $j \mapsto i(j)$ a bijection from J to I.

This reduces the proof of (2) to the case where k' and k'' are both fields, in which case the problem is to show that necessarily $k' = k''$ (as purely inseparable finite extensions of k) and that the isomorphism f is the Weil restriction of a unique isomorphism of groups $G' \simeq G''$ over the common ground field $k' = k''$. By Corollary 9.8.11 and Proposition 9.8.13, the k-group $\mathrm{R}_{k'/k}(G')$ intrinsically determines k': the field of definition over k for the geometric radical is $k'^{1/2}$. Hence, indeed $k' = k''$.

It remains, for the proof of (2), to show that if G' and H' are two basic non-reduced pseudo-simple k'-groups then the natural map

$$\mathrm{Isom}_{k'}(G', H') \to \mathrm{Isom}_k(\mathrm{R}_{k'/k}(G'), \mathrm{R}_{k'/k}(H'))$$

is bijective. More generally, for an arbitrary field k we claim that this map of Isom-sets is bijective for any finite purely inseparable extension of fields k'/k and any pseudo-reductive k'-groups G' and H'. By Proposition A.5.11 (with $B = k$, $B' = k'$, and $X' = G'$), the natural k'-homomorphism $q_{G'} : \mathrm{R}_{k'/k}(G')_{k'} \to G'$ is smooth and surjective with connected unipotent kernel, so it is the maximal pseudo-reductive quotient over k'. The same applies to $q_{H'}$, so any k-isomorphism $f : \mathrm{R}_{k'/k}(G') \simeq \mathrm{R}_{k'/k}(H')$ induces a k'-isomorphism $f' : G' \simeq H'$ between the maximal pseudo-reductive quotients over k'. Functoriality of Weil restriction implies that $\mathrm{R}_{k'/k}(\varphi)' = \varphi$ for any k'-isomorphism $\varphi : G' \simeq H'$, and by Proposition 1.2.2 any k-isomorphism f is determined by f'. Thus, $f \mapsto f'$ is an inverse to $\varphi \mapsto \mathrm{R}_{k'/k}(\varphi)$.

The asserted description of the smooth connected normal k-subgroups in (3) was established in the proof of assertion (2). □

Proposition 10.1.5 *Let k be an imperfect field with* $\mathrm{char}(k) = 2$ *such that* $[k : k^2] = 2$. *If G is a totally non-reduced pseudo-reductive k-group then* $Z_G = 1$ *and if G is absolutely pseudo-simple then all k-automorphisms of G are inner. Moreover, for any commutative k-group scheme Z of finite type not containing any nontrivial smooth connected k-subgroup, every central extension*

$$1 \to Z \to E \to G \to 1$$

over k is uniquely split over k.

Proof Once it is shown that the center is trivial, the assertion about inner automorphisms says exactly that every k-automorphism of G is conjugation by a unique $g \in G(k)$. This uniqueness and the asserted uniqueness for the splitting claim allow us to use Galois descent to reduce to the case $k = k_s$. Proposition 10.1.4 provides a decomposition $G \simeq \prod G_i$ with $G_i = \mathrm{R}_{k'_i/k}(G'_i)$ for finite extensions k'_i/k and a basic non-reduced pseudo-simple k'_i-group G'_i for each i. By Theorem 9.9.3(2) we have $Z_{G_i} = 1$ for all i and all automorphisms of G_i are inner. Thus, $Z_G = 1$ and in the absolutely pseudo-simple case (i.e., when there is only one i) every automorphism of G is inner.

Since k'_i/k is purely inseparable, the splitting result for each G_i is Theorem 9.9.3(3). The splitting result for the product $G = \prod G_i$ can be proved by exactly the same method, using the compatibility of open Bruhat cells with respect to Weil restriction and direct products. □

When $\mathrm{char}(k) = 2$ and $[k : k^2] = 2$, we have the following decomposition of a pseudo-reductive group:

Proposition 10.1.6 *Let k be a field of characteristic 2 such that $[k : k^2] = 2$. Let G be a pseudo-reductive k-group whose root system (over k_s) is non-reduced. There is a unique decomposition $G = G_1 \times G_2$ where G_2 is totally non-reduced and G_1 has a reduced root system (over k_s).*

Proof By the uniqueness and Galois descent we may assume $k = k_s$. Thus, if $\{G_i\}_{i \in I}$ is the set of pseudo-simple normal k-subgroups of G then by Proposition 3.2.10 some G_i has a non-reduced root system. Since absolutely pseudo-simple totally non-reduced pseudo-reductive k-groups have trivial center and only inner automorphisms over k, and Proposition 3.2.10 relates irreducible components of the root system to the root systems of the pseudo-simple normal k-subgroups (relative to compatible choices of maximal k-tori, as in Proposition 3.2.10), the proof of the splitting result in Theorem 8.2.10 applies verbatim (e.g., take G_2 to be the k-subgroup generated by the pseudo-simple normal k-subgroups of G whose root system is non-reduced). The reducedness of the root system of G_1 follows from the construction of G_1 and Proposition 3.2.10. □

In view of Propositions 10.1.6 and 10.1.4, to classify pseudo-reductive k-groups when $\mathrm{char}(k) = 2$ and $[k : k^2] = 2$, we can focus our attention on the case where the root system (over k_s) is reduced. There is no analogue of the splitting result in Theorem 8.2.10 for exotic pseudo-simple normal k_s-subgroups of types B_n or C_n with $n \geqslant 2$, due to the nontriviality of their center, so we need to revisit the definition of the standard construction. The idea is to use not only Weil restrictions of absolutely simple and simply connected groups, but also Weil restrictions of basic exotic pseudo-reductive groups. This will allow us to state and prove a general classification result in characteristic 2 (assuming that $[k : k^2] = 2$) in a manner that does not require treating specific types (with reduced root systems) in a special way. In fact, the methods will be largely characteristic-free.

Here is a generalization of the standard construction. Let k be an arbitrary field, k' a nonzero finite reduced k-algebra, and G' a k'-group whose fibers are absolutely pseudo-simple. Assume moreover that each fiber is either simply connected semisimple or (if $\mathrm{char}(k) = 2, 3$) basic exotic. Let $\mathscr{G} = \mathrm{R}_{k'/k}(G')$, and let T' be a maximal k'-torus in G', so the maximal k-torus $\mathscr{T}$ in $\mathrm{R}_{k'/k}(T')$ is a maximal k-torus in $\mathscr{G}$ (and all maximal k-tori of $\mathscr{G}$ uniquely arise in this way; see Proposition A.5.15(2)). Let $C' = Z_{G'}(T')$, so $\mathscr{C} := Z_{\mathscr{G}}(\mathscr{T}) = \mathrm{R}_{k'/k}(C')$ by Proposition A.5.15(1). By Theorem 2.4.1, $\underline{\mathrm{Aut}}_{\mathscr{G},\mathscr{C}}$ is represented by an affine k-group of finite type; let $Z_{\mathscr{G},\mathscr{C}}$ denote its (commutative) maximal smooth k-subgroup. We can similarly define the functor $\underline{\mathrm{Aut}}_{G',C'}$ on k'-schemes, and if $\{k'_i\}$ is the set of factor fields of k' then for any k'-algebra $R' = \prod R'_i$ (with $\mathrm{Spec}\, R'_i$ the k'_i-fiber of $\mathrm{Spec}\, R'$) we have $\underline{\mathrm{Aut}}_{G',C'}(R') = \prod \underline{\mathrm{Aut}}_{G'_i,C'_i}(R'_i)$ where (G'_i, C'_i) is the k'_i-fiber of (G', C'). Thus, by treating the fibers of G' over $\mathrm{Spec}\, k'$ separately we see that $\underline{\mathrm{Aut}}_{G',C'}$ is represented by the k'-group $\mathrm{Aut}_{G',C'}$ whose k'_i-fiber represents $\underline{\mathrm{Aut}}_{G'_i,C'_i}$. In particular, the maximal smooth k'-subgroup $Z_{G',C'}$ in $\underline{\mathrm{Aut}}_{G',C'}$ is $\coprod Z_{G'_i,C'_i}$ over $\mathrm{Spec}\, k' = \coprod \mathrm{Spec}\, k'_i$ and

$$\mathrm{R}_{k'/k}(\underline{\mathrm{Aut}}_{G',C'}) = \prod \mathrm{R}_{k'_i/k}(\underline{\mathrm{Aut}}_{G'_i,C'_i}).$$

Lemma 10.1.7 *The natural map* $\mathrm{R}_{k'/k}(\underline{\mathrm{Aut}}_{G',C'}) \to \underline{\mathrm{Aut}}_{\mathscr{G},\mathscr{C}}$ *defined by applying* $\mathrm{R}_{k'/k}$ *to automorphisms carries* $\mathrm{R}_{k'/k}(Z_{G',C'})$ *isomorphically onto* $Z_{\mathscr{G},\mathscr{C}}$.

Proof Since $\mathrm{R}_{k'/k}(Z_{G',C'}) = \prod \mathrm{R}_{k'_i/k}(Z_{G'_i,C'_i})$ is smooth, it lands in $Z_{\mathscr{G},\mathscr{C}}$. To check that this is an isomorphism, we can assume $k = k_s$ (so k'_i/k is purely inseparable for all i). If all fibers of G' are semisimple then the result is Theorem 1.3.9, and if all fibers of G' are basic exotic then it is Proposition 8.2.6. The general case goes exactly as in the proof of Proposition 8.2.6, since each k-group $\mathscr{G}_i = \mathrm{R}_{k'_i/k}(G'_i)$ is absolutely pseudo-simple (use the criterion in Lemma 3.1.2, along with Proposition A.5.11 and Corollary A.7.11 in the semisimple case and Corollary 8.1.4 in the exotic case) with $\mathscr{C}_i = \mathrm{R}_{k'_i/k}(C'_i)$ a Cartan k-subgroup. □

The preceding lemma provides an identification of $Z_{\mathscr{G},\mathscr{C}}$ with $R_{k'/k}(Z_{G',C'})$. The conjugation action of $\mathscr{C} = Z_{\mathscr{G}}(\mathscr{T})$ on $\mathscr{G}$ defines a homomorphism $\mathscr{C} \to \underline{\mathrm{Aut}}_{\mathscr{G},\mathscr{C}}$, and this factors through a unique k-homomorphism $\mathscr{C} \to Z_{\mathscr{G},\mathscr{C}}$. Consider any factorization

$$\mathscr{C} \xrightarrow{\phi} C \xrightarrow{\psi} Z_{\mathscr{G},\mathscr{C}}$$

where C is a commutative pseudo-reductive k-group. It is not really necessary to mention ψ here: we could instead just say that C is equipped with a left action on $\mathscr{G}$ restricting to the identity on $\mathscr{C}$, and that ϕ is a k-homomorphism respecting the actions on $\mathscr{G}$. However, it is more concrete and very useful in practice that we can encode the action in terms of a k-homomorphism ψ valued in $Z_{\mathscr{G},\mathscr{C}}$. For example:

Lemma 10.1.8 *The natural map $\mathscr{C} \to Z^0_{\mathscr{G},\mathscr{C}}$ between smooth connected commutative affine k-groups restricts to an isogeny between maximal k-tori.*

Proof We may assume $k = k_s$ and then pass to factor fields of k' to reduce to the case where k' is a field. Since $Z_{\mathscr{G},\mathscr{C}} = R_{k'/k}(Z_{G',C'})$, the compatibility of maximal tori with respect to Weil restriction (Proposition A.5.15(2)) reduces us to considering (G', T', k') instead of $(\mathscr{G}, \mathscr{T}, k)$. The simply connected semisimple case is obvious (with $Z_{G',C'} = T'/Z_{G'}$), and the basic exotic case follows from Corollary 8.2.7 (as we may compute by using Theorem 7.2.3(3) and a Levi k'-subgroup of G' containing T', which exists since k' is separably closed). □

Consider the semidirect product $\mathscr{G} \rtimes C$. The homomorphism $\alpha : \mathscr{C} \to \mathscr{G} \rtimes C$ defined by $c \mapsto (c^{-1}, \phi(c))$ is an embedding of $\mathscr{C}$ into $\mathscr{G} \rtimes C$ as a central subgroup. We will view $\mathscr{C}$ as a subgroup of $\mathscr{G} \rtimes C$ via α. Let $G = (\mathscr{G} \rtimes C)/\mathscr{C}$. By Proposition 1.4.3, G is pseudo-reductive over k. If all fibers of G' are absolutely simple and simply connected then $\mathscr{C} = R_{k'/k}(T')$ (Proposition A.5.15(3)), so this recovers the standard construction of non-commutative pseudo-reductive k-groups (in the precise form given in Theorem 4.1.1). More generally:

Definition 10.1.9 Let k be a field, and assume $[k : k^2] \leqslant 2$ if $\mathrm{char}(k) = 2$. A pseudo-reductive k-group G is *generalized standard* if it is commutative or if there exists a 4-tuple $(G', k'/k, T', C)$ consisting of a nonzero finite reduced k-algebra k', a k'-group G' whose fibers are absolutely pseudo-simple and either simply connected semisimple or (if $\mathrm{char}(k) = 2, 3$) basic exotic, a maximal k'-torus T' of G', a commutative pseudo-reductive k-group C, and a factorization diagram

$$\mathscr{C} \xrightarrow{\phi} C \xrightarrow{\psi} Z_{\mathscr{G},\mathscr{C}} = \mathrm{R}_{k'/k}(Z_{G',C'}) \tag{10.1.1}$$

with $\mathscr{G} = \mathrm{R}_{k'/k}(G')$, $C' = Z_{G'}(T')$, and $\mathscr{C} = \mathrm{R}_{k'/k}(C')$ such that there is an isomorphism of k-groups

$$G \simeq (\mathscr{G} \rtimes C)/\mathscr{C}. \tag{10.1.2}$$

The 4-tuple $(G', k'/k, T', C)$ equipped with the factorization (10.1.1) and isomorphism (10.1.2) is a *generalized standard presentation* of G.

Although Definition 10.1.9 makes perfectly good sense when $\mathrm{char}(k) = 2$ if $[k : k^2] > 2$, it is the wrong notion in all such cases. In forthcoming work [CP] an appropriate definition is made over arbitrary imperfect fields of characteristic 2, and Definition 10.1.9 is a special case when $[k : k^2] = 2$. It is clear that separable extension of the ground field preserves the property of being generalized standard (though descent of this property through such a ground field extension is not obvious; see Corollary 10.2.5 for separable algebraic extensions). In particular, exactly as in the standard case it follows that the root system (over k_s) of any generalized standard group is reduced, as this is true for the absolutely pseudo-simple groups that are either semisimple or basic exotic.

Remark 10.1.10 In the diagram (10.1.1), the kernel $\ker\phi =: Z$ is contained in the center $\mathrm{R}_{k'/k}(Z_{G'})$ of $\mathrm{R}_{k'/k}(G')$ (see Proposition A.5.15(1) for this determination of the center). Indeed, we just have to check that the composite map $\psi \circ \phi$ has kernel $\mathrm{R}_{k'/k}(Z_{G'})$. But this composite map is the canonical map (arising from the conjugation action of C' on G'), so by left-exactness of Weil restriction it is equivalent to prove that the conjugation action of C' on G' has functorial kernel $Z_{G'}$. By definition of the scheme-theoretic kernel, this is clear (since $Z_{G'}$ is certainly contained in C').

Also, in the diagram (10.1.1), the final term $\mathrm{R}_{k'/k}(Z_{G',C'})$ can be replaced with its identity component $\mathrm{R}_{k'/k}(Z^0_{G',C'})$. In fact, $Z_{G',C'}$ is always connected (see [CP]), but we do not need this fact.

As in the standard case, it is difficult to directly characterize all central k-subgroups $Z \subseteq \mathrm{R}_{k'/k}(Z_{G'})$ such that $\mathrm{R}_{k'/k}(G')/Z$ is pseudo-reductive. Note that the pseudo-reductivity of $\mathrm{R}_{k'/k}(G')/Z$ is equivalent to pseudo-reductivity of the Cartan k-subgroup $C := \mathrm{R}_{k'/k}(C')/Z$, due to the isomorphism

$$(\mathrm{R}_{k'/k}(G') \rtimes C)/\mathrm{R}_{k'/k}(C') \simeq \mathrm{R}_{k'/k}(G')/Z$$

and the pseudo-reductivity of Cartan k-subgroups of pseudo-reductive k-groups. The pseudo-reductivity of such a quotient C is generally not easy to determine (given an arbitrary Z).

Remark 10.1.11 In the setting of Definition 10.1.9, since the derived group of $\mathscr{G} \rtimes C$ is $\mathscr{G}$ it follows that the derived group $\mathscr{D}(G)$ of the generalized standard pseudo-reductive group G is naturally isomorphic to $\mathscr{G}/\ker\phi$. As $\mathscr{G} = \mathrm{R}_{k'/k}(G')$, and $\ker\phi$ is contained in $\mathrm{R}_{k'/k}(Z_{G'})$, there is a natural surjective homomorphism $\mathscr{D}(G)_{k'} \twoheadrightarrow G'/Z_{G'}$.

There is one property of central quotients of $\mathrm{R}_{k'/k}(G')$ that is very useful for applications to classification results:

Proposition 10.1.12 *Let $(G', k'/k)$ be as in Definition* 10.1.9, $Z \subseteq \mathrm{R}_{k'/k}(Z_{G'})$ *a k-subgroup scheme, and $G := \mathrm{R}_{k'/k}(G')/Z$ the smooth connected affine central quotient over k.*

(1) *Let $(G'_0, k'_0/k)$ be another such pair, $Z_0 \subseteq \mathrm{R}_{k'_0/k}(Z_{G'_0})$ a central k-subgroup scheme, and $G_0 := \mathrm{R}_{k'_0/k}(G'_0)/Z_0$. For any separable field extension F/k, every F-isomorphism $f : G_F \simeq (G_0)_F$ uniquely lifts to an F-homomorphism $\widetilde{f} : \mathrm{R}_{k'/k}(G')_F \to \mathrm{R}_{k'_0/k}(G'_0)_F$, and $\widetilde{f}$ is an isomorphism. Moreover, if $(G'_0, k'_0/k) = (G', k'/k)$ then $\widetilde{f}$ restricts to the identity on a smooth connected F-subgroup of $\mathrm{R}_{k'/k}(G')_F$ if f restricts to the identity on its image in G_F.*
(2) *If H is a smooth connected k-group equipped with a left action on G then this uniquely lifts to a left action on $\mathrm{R}_{k'/k}(G')$. This lifted action restricts to the trivial action on a smooth connected k-subgroup of $\mathrm{R}_{k'/k}(G')$ if H acts trivially on its image in G.*

Proof In the standard case this is Proposition 5.1.7, and the proof in general is identical, apart from two modifications. First, we replace references to Example 5.1.4 in the standard case with references to Proposition 8.1.2 in the exotic case. Second, the fibers of $Z_{G'}$ may not be finite: they could be Weil restrictions of μ_2 from an extension field. Such group schemes receive no nontrivial homomorphisms from a smooth connected group (due to the universal property of Weil restriction), so we can argue exactly as in the proof of Proposition 5.1.7. □

10.2 Classification via generalized standard groups

Our main theorem incorporating characteristic 2 is:

Theorem 10.2.1 *Let k be a field, and assume $[k : k^2] \leqslant 2$ if $\mathrm{char}(k) = 2$. Let G be a pseudo-reductive k-group.*

(1) *If G_{k_s} has a non-reduced root system then G has a unique decomposition $G_1 \times G_2$ for which $(G_1)_{k_s}$ has a reduced root system and G_2 is totally non-reduced.*
(2) *Assume G_{k_s} has a reduced root system, and let T be a maximal k-torus of G. Then G is generalized standard, and it is standard if and only if for all roots in a component of $\Phi(G_{k_s}, T_{k_s})$ the root spaces in $\mathrm{Lie}(G_{k_s})$ are of the same dimension.*

Since pseudo-split pseudo-simple groups with a non-reduced root system only exist over imperfect fields of characteristic 2, the assertion in (1) is settled by Proposition 10.1.6. Thus, to prove Theorem 10.2.1 it remains to prove (2), for which we can assume $\mathscr{D}(G) \neq 1$. Our proof of this special case will be modeled on the strategy from the standard case, namely to reduce to the case of perfect G with $k = k_s$, and then to the (absolutely) pseudo-simple case. In this latter case we have essentially already classified all possibilities as generalized standard groups aside from exceptional cases of types B_n and C_n ($n \geqslant 2$) for imperfect k of characteristic 2 when $G^{\mathrm{ss}}_{\overline{k}}$ is adjoint rather than simply connected (see Theorem 6.3.4, Theorem 8.2.1, and Proposition 9.2.4). We will need to do a bit more work to establish the generalized standard property in these remaining cases. Curiously, the reduction to the absolutely pseudo-simple case over $k = k_s$ will never use any hypotheses on $[k : k^2]$ when $\mathrm{char}(k) = 2$. However, since Definition 10.1.9 is only appropriate when $[k : k^2] \leqslant 2$ (and we want to be consistent with [CP]), in what follows we take k to be an *arbitrary* field subject to the requirement that $[k : k^2] \leqslant 2$ *when* $\mathrm{char}(k) = 2$.

As a first step, we give intrinsic interpretations of some ingredients in the generalized standard construction, much like in the standard case.

Proposition 10.2.2 *Let G be a non-commutative generalized standard group over a field k, arising from a 4-tuple $(G', k'/k, T', C)$ via a fixed choice of isomorphism* (10.1.2).

(1) *Let $j : \mathrm{R}_{k'/k}(G') \to G$ be the natural map, and let $\mathscr{T}$ be the maximal k-torus in $\mathrm{R}_{k'/k}(T')$. The image of j is $\mathscr{D}(G)$, and if T is the maximal*

k-torus of G that is an almost direct product of the maximal central k-torus Z and $j(\mathscr{T})$ then $C = Z_G(T)$.

(2) *The natural map $\mathscr{T} \to j(\mathscr{T})$ is an isogeny, and there is a bijective correspondence between the set of maximal k'-tori S' in G' and the set of maximal k-tori in G via $S' \mapsto j(\mathscr{S}) \cdot Z$ where $\mathscr{S}$ is the maximal k-torus in $\mathrm{R}_{k'/k}(S')$.*

(3) *Let S be a maximal k-torus in G, S' the associated maximal k'-torus in G', $C' = Z_{G'}(S')$, and $\mathscr{S}$ the maximal k-torus in $\mathrm{R}_{k'/k}(S')$. The conjugation action of $Z_G(S)$ on $\mathscr{D}(G) = j(\mathrm{R}_{k'/k}(G'))$ uniquely lifts to an action of $Z_G(S)$ on $\mathrm{R}_{k'/k}(G')$ restricting to the identity on the maximal torus $\mathscr{S}$.*

For the generalized standard pseudo-reductive k-group associated to $(G', k'/k, S', Z_G(S))$ and the resulting factorization diagram

$$\mathrm{R}_{k'/k}(Z_{G'}(S')) \longrightarrow Z_G(S) \longrightarrow \mathrm{R}_{k'/k}(Z_{G',C'}),$$

the natural k-homomorphism

$$\gamma : (\mathrm{R}_{k'/k}(G') \rtimes Z_G(S))/\mathrm{R}_{k'/k}(Z_{G'}(S')) \to G$$

is an isomorphism.

This result allows us to speak of a generalized standard presentation of a non-commutative pseudo-reductive k-group G being *compatible* with a fixed choice of maximal k-torus T of G, and part (3) shows that if there is such a presentation compatible with one choice of T then there is one for any choice of T.

Proof In the standard case this is Proposition 4.1.4, and the proof carries over to the general case with the following minor changes. We need to use Proposition 8.1.2 to get perfectness of purely inseparable Weil restrictions of exotic pseudo-reductive groups, and use Lemma 10.1.8 to see that the factorization diagram (10.1.1) canonically identifies the maximal k-torus S in C up to k-isogeny with an almost direct product of the maximal k-torus in $\mathrm{R}_{k'/k}(T')$ and the maximal k-torus in $\ker \psi$. Since $\ker \phi$ is central in $\mathrm{R}_{k'/k}(G')$ (as in the standard case) and contains no nontrivial k-tori (Lemma 10.1.8), up to writing $Z_{G'}(T')$ in place of T' and writing $Z_{G',T'}$ in place of $T'/Z_{G'}$ the earlier argument carries over unchanged. (We also need to make the same observation concerning the structure of $Z_{G'}$ as we did at the end of the proof of Proposition 10.1.12.) $\square$

Before we can pass to the case $k = k_s$, it is convenient to reduce to the consideration of the perfect pseudo-reductive k-group $\mathscr{D}(G)$:

Proposition 10.2.3 *A pseudo-reductive k-group G is generalized standard if and only if $\mathscr{D}(G)$ is generalized standard. If either admits a generalized standard presentation resting on the pair $(G', k'/k)$ then so does the other.*

Proof The proof of Proposition 5.2.1 carries over with just minor notational and cross-reference changes, as in the proof of Proposition 10.2.2. (For example, we use Proposition 10.1.12(2) instead of Proposition 5.1.7(2).) □

To pass to the case $k = k_s$, we have to show that the pair $(G', k'/k)$ is canonically determined by G. We cannot appeal to the simply connected datum as in the standard case, but the following analogue of Proposition 5.2.2 carries over:

Proposition 10.2.4 *Let $(G', k'/k, T', C_1)$ and $(G'', k''/k, T'', C_2)$ be two 4-tuples that give generalized standard presentations of a non-commutative pseudo-reductive k-group G, and choose both to be compatible with a fixed maximal k-torus T in G.*

Under the resulting identifications of C_1 and C_2 with $C := Z_G(T)$ via Proposition 10.2.2(1), *the composite isomorphism*

$$(\mathrm{R}_{k'/k}(G') \rtimes C)/\mathrm{R}_{k'/k}(Z_{G'}(T')) \simeq G \simeq (\mathrm{R}_{k''/k}(G'') \rtimes C)/\mathrm{R}_{k''/k}(Z_{G''}(T''))$$

is induced by a unique pair consisting of a k-algebra isomorphism $\alpha : k'' \simeq k'$ and an isomorphism of pairs $(G', T') \simeq (G'', T'')$ over Spec α *respecting the given factorization diagram* (10.1.1) *and its counterpart for $(G'', k''/k, T'', C)$.*

In particular, for each maximal k-torus T of G there is a unique generalized standard presentation of G compatible with T.

Proof In the standard case this is Proposition 5.2.2, and the proof works in general with the following minor changes. We use Proposition 10.1.12 instead of Proposition 5.1.7, and to prove that any k-group isomorphism $f : \mathrm{R}_{k'/k}(G') \simeq \mathrm{R}_{k''/k}(G'')$ arises from a unique k-algebra isomorphism $\alpha : k'' \simeq k'$ and a group isomorphism $\phi : G' \simeq G''$ over Spec α, we first have to check that the factor groups arising from the factor fields of k' and k'' are precisely the pseudo-simple smooth connected normal k-subgroups. An elementary argument with commutator subgroups (as in the proof of Corollary 8.1.4) shows that if $N_1, \dots, N_r$ are pseudo-simple k-groups then in $\prod N_i$ the smooth connected normal k-subgroups are the products of the N_i. The pseudo-simplicity of the factors arising from the exotic fibers of G' and G'' follows from Proposition 8.1.2 and Corollary 8.1.4. For absolutely simple and simply connected fibers the perfectness follows from Corollary A.7.11, and so for the pseudo-simplicity we can use the criterion in Lemma 3.1.2 when $k = k_s$. Considerations with Galois orbits of pseudo-simple factors of G_{k_s} then settles the pseudo-simplicity assertion in the standard case for general k.

Now we may and do assume that k' and k'' are fields. In this case, if G' and G'' are both absolutely simple then the existence and uniqueness of (α, ϕ) are provided by Proposition 5.2.2. If both are exotic then the result instead is provided by Proposition 8.2.4. The other possibility, that one is absolutely simple and the other is exotic, cannot happen since Weil restriction preserves standardness but the Weil restriction (through a ground field extension) of a basic exotic pseudo-reductive group is never standard (Remark 8.2.3). □

Corollary 10.2.5 *Let G be a non-commutative pseudo-reductive k-group. If G_{k_s} is generalized standard over k_s, then so is G over k.*

Proof In the standard case this is Corollary 5.2.3, and the proof works in general if we use Proposition 10.2.2 instead of Proposition 4.1.4 and write $Z_{G'}(T')$ instead of T' in several places. □

The following result has the same proof as its analogue (Corollary 4.1.6 and Proposition 5.3.1) in the standard case:

Proposition 10.2.6 *Let G be a non-commutative pseudo-reductive group over $k = k_s$, and $\{G_i\}$ the non-empty collection of pseudo-simple normal k-subgroups of G.*

(1) *If each G_i is generalized standard then so is G.*
(2) *Let T be a maximal k-torus of G and assume all G_i are generalized standard. Consider the generalized standard presentation $(G'_i, k'_i/k, T'_i, C_i)$ of G_i compatible with the maximal k-torus $T_i = T \cap G_i$. The identification of $\prod \mathrm{X}(T'_i)_{\mathbf{Q}} = \prod \mathrm{X}(T_i)_{\mathbf{Q}}$ with $\mathrm{X}(T \cap \mathscr{D}(G))_{\mathbf{Q}}$ via Proposition 3.2.10 also identifies the root systems $\Phi(G'_i, T'_i)$ with the irreducible components of $\Phi(G, T)$, and for each $a \in \Phi(G'_i, T'_i)$ the root space $\mathrm{Lie}(G)_a$ is identified with $\mathrm{Lie}(G'_i)_a$ as k-vector spaces.*

The preceding results reduce the proof of Theorem 10.2.1(2) to a special case: $k = k_s$ and G is absolutely pseudo-simple with a reduced root system. It must be proved that G is generalized standard, and that it is standard precisely when the dimension criterion on root spaces as in Theorem 10.2.1(2) is satisfied. Applying Proposition 10.2.6(2), the dimension criterion follows because it holds in the simply connected semisimple case (with all root groups of dimension 1) and it fails in the basic exotic cases (with long root groups of dimension 1 and short root groups of strictly larger dimension).

We may therefore focus our efforts on establishing the generalized standard property.

Let T be a maximal k-torus of G, $\Phi = \Phi(G, T)$, and for $a \in \Phi$ let G_a denote the rank-1 absolutely pseudo-simple subgroup generated by the root groups $U_{\pm a}$. Each G_a has a reduced root system, and Theorem 6.1.1 implies that away from characteristic 2 it is standard; Proposition 9.2.4 leads to the same conclusion when $\mathrm{char}(k) = 2$ (finally using that $[k : k^2] = 2$!). In particular, ξ_{G_a} is surjective for all $a \in \Phi$. Theorem 6.3.4 now yields standardness in all higher rank cases except when G is exceptional in the sense of Definition 6.3.3. So we may now assume that k is imperfect with characteristic $p \in \{2, 3\}$ and that $G^{\mathrm{ss}}_{\overline{k}}$ has type G_2 when $p = 3$, and type B_n $(n \geqslant 2)$, C_n $(n \geqslant 2)$, or F_4 when $p = 2$. In all of these cases, if $G^{\mathrm{ss}}_{\overline{k}}$ is simply connected then G was proved to be exotic (and hence generalized standard) in Theorem 8.2.1 (using the reducedness hypothesis on the root system in case $G^{\mathrm{ss}}_{\overline{k}}$ is simply connected of type C_n). Thus it only remains to address the cases with $p = 2$ and $G^{\mathrm{ss}}_{\overline{k}}$ of adjoint type B_n or C_n for $n \geqslant 2$ (so the root system for G is automatically reduced). Hence, *we now assume* $\mathrm{char}(k) = 2$, $[k : k^2] = 2$, *and that* $G^{\mathrm{ss}}_{\overline{k}}$ *is of adjoint type* B_n *or* C_n *with* $n \geqslant 2$.

Let K/k be the field of definition of the geometric radical of G, and let K_a/k be the corresponding field for G_a. We can assume that G is not standard, so $K \neq k$ and by Proposition 6.3.2 we get a proper subfield $\mathsf{k} \subseteq K$ containing kK^2 such that $K_a = K$ for all short $a \in \Phi$ and $K_a = \mathsf{k}$ for all long $a \in \Phi$. Let $G' = G^{\mathrm{ss}}_K$ and $T' = T_K$ viewed as a maximal K-torus of G'. For $a \in \Phi$, let G'_a denote the connected absolutely simple subgroup of rank 1 generated by the root groups in G' corresponding to the roots $\pm a$. By hypothesis G' is of adjoint type B_n or C_n, so its simply connected central cover $\widetilde{G}' \to G'$ is a double cover, with kernel $\mu \simeq \mu_2$. For each long $a \in \Phi$, G'_a is simply connected, hence it is isomorphic to SL_2 and $\xi_{G_a} : G_a \to \mathrm{R}_{\mathsf{k}/k}(G'_a)$ is an isomorphism (Proposition 9.2.4). In particular, G_a is isomorphic to $\mathrm{R}_{\mathsf{k}/k}(\mathrm{SL}_2)$ for long a.

Since G is perfect, the natural map $i_G : G \to \mathrm{R}_{K/k}(G')$ factors through the derived group of its target. This derived group is computed by Proposition 1.3.4, so we obtain a canonical k-homomorphism

$$\xi_G : G \to \mathscr{D}(\mathrm{R}_{K/k}(G')) = \mathrm{R}_{K/k}(\widetilde{G}')/\mathrm{R}_{K/k}(\mu).$$

A key point is to prove:

Proposition 10.2.7 *The k-group $\xi_G(G)$ is the image of $\mathrm{R}_{\mathsf{k}/k}(\widetilde{\mathscr{G}})$, where $\widetilde{\mathscr{G}} \subseteq \mathrm{R}_{K/\mathsf{k}}(\widetilde{G}')$ is a basic exotic pseudo-reductive k-group.*

Proof Let L be a Levi k-subgroup of G containing T. Then, as in Step 3 of the proof of Theorem 8.2.1, we can see that $\xi_G(L)$ is a Levi subgroup of both

$\xi_G(G)$ and $\mathrm{R}_{K/k}(G')$, and $\xi_G|_L : L \to \xi_G(L)$ is an isomorphism. We rename $\xi_G(L)$ as L. Then $T_K \subseteq L_K$ recovers the maximal K-torus T' of G', and hence (L, T) is a k-descent of (G', T').

Let $\pi : \widetilde{L} \to L$ be the simply connected central cover of L, and let $\widetilde{T}$ (resp. $\widetilde{T}'$) be the maximal torus of $\widetilde{L}$ (resp. $\widetilde{G}'$) which maps onto T (resp. T'). The pair $(\widetilde{L}, \widetilde{T})$ is a k-descent of the pair $(\widetilde{G}', \widetilde{T}')$, so $(\widetilde{L}_{\mathsf{k}}, \widetilde{T}_{\mathsf{k}})$ is a k-descent of $(\widetilde{G}', \widetilde{T}')$. Moreover, $\widetilde{L}_{\mathsf{k}}$ is a Levi k-subgroup of $\mathrm{R}_{K/\mathsf{k}}(\widetilde{G}')$ via the inclusion $\widetilde{L}_{\mathsf{k}} \hookrightarrow \mathrm{R}_{K/\mathsf{k}}(\widetilde{L}_K) = \mathrm{R}_{K/\mathsf{k}}(\widetilde{G}')$, so it gives rise to a basic exotic pseudo-reductive k-group $\mathscr{G} \subseteq \mathrm{R}_{K/\mathsf{k}}(\widetilde{L}_K)$ $(= \mathrm{R}_{K/\mathsf{k}}(\widetilde{G}'))$ containing $\widetilde{T}_{\mathsf{k}}$. Explicitly, $\mathscr{G}$ is the subgroup generated by the long root groups of $\widetilde{L}_{\mathsf{k}}$ $(\subseteq \mathrm{R}_{K/\mathsf{k}}(\widetilde{L}_K))$ relative to $\widetilde{T}_{\mathsf{k}}$ and the Weil restriction from K to k of the short root groups of $\widetilde{L}_K = \widetilde{G}'$ relative to $\widetilde{T}_K = \widetilde{T}'$, and these k-subgroups of $\mathscr{G}$ are its root groups with respect to $\widetilde{T}_{\mathsf{k}}$.

The k-group $\mathrm{R}_{\mathsf{k}/k}(\mathscr{G})$ is perfect, by Proposition 8.1.2, and any perfect pseudo-reductive k-group is generated by the root groups with respect to a fixed maximal k-torus (Lemma 3.1.5). Therefore, the image of $\mathrm{R}_{\mathsf{k}/k}(\mathscr{G})$ in $\mathrm{R}_{K/k}(L_K) = \mathrm{R}_{K/k}(G')$ is the subgroup generated by the Weil restriction from k to k of the long root groups of L_{k} $(\subseteq \mathrm{R}_{K/\mathsf{k}}(L_K))$ relative to T_{k} and the Weil restriction from K to k of the short root groups of $L_K = G'$ relative to $T_K = T'$.

We claim that these Weil restrictions are precisely the root groups of $\xi_G(G)$ with respect to T. Proposition 9.2.4 and (6.3.1) imply that $\xi_G(G_c) = \xi_{G_c}(G_c) = \mathscr{D}(\mathrm{R}_{K/k}((L_c)_K))$ for every short root c. Proposition 1.3.4 describes this derived group. If c is a long root then it follows from the discussion preceding this proposition that $\xi_G(G_c)$ $(= \xi_{G_c}(G_c))$ is a subgroup of $\mathrm{R}_{K/k}((L_c)_K)$ $(\subseteq \mathrm{R}_{K/k}(L_K))$ which contains L_c and is isomorphic to $\mathrm{R}_{\mathsf{k}/k}((L_c)_{\mathsf{k}})$, so by an observation in Step 4 of the proof of Theorem 8.2.1, $\xi_G(G_c)$ equals $\mathrm{R}_{\mathsf{k}/k}((L_c)_{\mathsf{k}})$ inside $\mathrm{R}_{K/k}((L_c)_K)$. Since the perfect pseudo-reductive k-group $\xi_G(G)$ is generated by its root groups $\xi_G(U_c)$ (Remark 2.3.6 and Lemma 3.1.5), we conclude that the image of $\mathrm{R}_{\mathsf{k}/k}(\mathscr{G})$ coincides with $\xi_G(G)$. □

Continuing with the proof of Theorem 10.2.1(2), now we will use Proposition 10.2.7 to deduce the generalized standard property for G. Let $\widetilde{T}'$ be the maximal K-torus of $\widetilde{G}'$ for which $T' := \widetilde{T}'/\mu$ is T_K viewed as a maximal K-torus of G'. The reducedness of Φ implies that $\Phi = \Phi(G', T') = \Phi(\widetilde{G}', \widetilde{T}')$ via the embedding $\mathrm{X}(T') \hookrightarrow \mathrm{X}(\widetilde{T}')$. Choose $c \in \Phi$. The proof of Proposition 10.2.7 implies that ξ_G carries G_c onto $\mathscr{D}(\mathrm{R}_{K_c/k}(\mathscr{G}_c'))$ via ξ_{G_c} for a K_c-structure $\mathscr{G}_c' = (G_c)_{K_c}^{\mathrm{ss}}$ on the K-subgroup G_c' of G' generated by the root groups $U_{\pm c}'$. By Proposition 9.2.4, the k-group G_c is standard, so by Remark 5.3.6 the map $\xi_{G_c} : G_c \to \mathscr{D}(\mathrm{R}_{K_c/k}(\mathscr{G}_c'))$ is a central quotient that carries the root groups for $(G_c, T \cap G_c)$ over k *isomorphically* onto the Weil restrictions (from K_c down to k) of the root groups for $(\mathscr{G}_c', (T \cap G_c)_{K_c})$.

Therefore, ξ_G carries the c-root group of (G,T) isomorphically onto a copy of $\mathrm{R}_{K_c/k}(\mathbf{G}_a)$ inside of the c-root group $\mathrm{R}_{K/k}(U'_c)$ of $(\mathrm{R}_{K/k}(G'),T)$ (with U'_c the c_K-root group of (G',T')).

The smooth connected affine k-group $\xi_G(G)$ is perfect, and it is also pseudo-reductive (Proposition 4.2.5). Consider the short exact sequence

$$1 \to \ker \xi_G \longrightarrow G \xrightarrow{\xi_G} \xi_G(G) \to 1.$$

The preceding analysis of root groups implies that the action of T on the Lie algebra of $\ker \xi_G$ is trivial. It then follows (by Lemma 5.3.2 and Corollary A.8.11) that $\ker \xi_G$ is *central* in G. As G is pseudo-simple, $\ker \xi_G$ does not contain any nontrivial smooth connected k-subgroup.

Let $\widetilde{\mathscr{G}}$ be as in Proposition 10.2.7 and $\pi : \mathrm{R}_{\mathsf{k}/k}(\widetilde{\mathscr{G}}) \to \xi_G(G)$ be a surjective homomorphism. We can form a pullback central extension

$$\begin{array}{ccccccccc} 1 & \longrightarrow & \ker \xi_G & \longrightarrow & E & \longrightarrow & \mathrm{R}_{\mathsf{k}/k}(\widetilde{\mathscr{G}}) & \longrightarrow & 1 \\ & & \| & & \downarrow & & \downarrow \pi & & \\ 1 & \longrightarrow & \ker \xi_G & \longrightarrow & G & \xrightarrow[\xi_G]{} & \xi_G(G) & \longrightarrow & 1 \end{array} \tag{10.2.1}$$

of $\mathrm{R}_{\mathsf{k}/k}(\widetilde{\mathscr{G}})$ which is uniquely split over k (by Proposition 8.1.2), so we obtain a factorization

$$\mathrm{R}_{\mathsf{k}/k}(\widetilde{\mathscr{G}}) \xrightarrow{j} G \overset{\xi_G}{\twoheadrightarrow} \xi_G(G)$$

of the surjective π. The kernel of π is central (as it is contained in the central subgroup $\mathrm{R}_{K/k}(\mu)$ of $\mathrm{R}_{K/k}(\widetilde{G}')$), and $\ker \xi_G$ is unipotent. Hence, $\ker j$ is central and j carries maximal tori onto maximal tori, so upon choosing compatible maximal tori it also carries root groups into root groups with trivial kernel. But π and ξ_G each restrict to *isomorphisms* between root groups, so it follows that j does too. In particular, since G is generated by its root groups relative to a maximal k-torus (due to perfectness of G; see Lemma 3.1.5), we conclude that j is *surjective*.

In other words, the pseudo-reductive G is a central quotient of $\mathrm{R}_{\mathsf{k}/k}(\widetilde{\mathscr{G}})$. Such a central quotient presentation can be promoted to a generalized standard presentation, as was explained immediately following Remark 10.1.10. This completes the proof of Theorem 10.2.1(2).

11
Applications

This chapter gives some applications over general fields. The reader is referred to [Con2] for arithmetic applications of Theorem 5.1.1 (and Proposition 7.3.3) to general finiteness theorems for all affine group schemes of finite type over global function fields (over finite fields).

11.1 Maximal tori in pseudo-reductive groups

Our concrete description of pseudo-reductive groups enables us to analyze the structure of such groups by means of maximal tori, as well as to analyze problems related to the structure of maximal tori. We begin with a toric criterion for a pseudo-reductive group to be reductive (away from characteristic 2).

Theorem 11.1.1 *Let k be a field with* $\mathrm{char}(k) \neq 2$. *A pseudo-reductive k-group G is reductive if and only if some (equivalently, every) Cartan k-subgroup is a torus.*

An interesting application of this result is given in Theorem C.1.9 (which is never used elsewhere in this monograph).

Proof It is a well-known fact that maximal tori in connected reductive groups over any field are their own centralizers [Bo2, 13.17, Cor. 2(c)], which is to say that Cartan subgroups are tori in such groups. For the converse, we will use that $\mathrm{char}(k) \neq 2$.

Let T be a maximal k-torus of G and $C = Z_G(T)$ the associated Cartan subgroup. Assume that $C = T$; we will prove that G is reductive. We may replace k with k_s so that k is separably closed. Let $\Phi = \Phi(G, T)$ be the root system of G with respect to T. For $a \in \Phi$, let G_a be the subgroup generated by the root groups $U_{\pm a}$. This is an absolutely pseudo-simple k-subgroup of rank 1 normalized by T (see Proposition 3.4.1 for an alternative description of G_a), and $S_a := T \cap G_a$ is the 1-dimensional maximal torus of G_a contained in T. The codimension-1 subtorus $T_a = (\ker a)^0_{\mathrm{red}}$ in T centralizes G_a, so T is an almost direct product of S_a and T_a. In particular, the Cartan k-subgroup $Z_{G_a}(S_a)$ is equal to $G_a \cap Z_G(T) = G_a \cap C = G_a \cap T = S_a$.

We now compute $Z_{G_a}(S_a)$ in another way, using that $\mathrm{char}(k) \neq 2$. By Theorem 6.1.1, G_a is isomorphic to $\mathrm{R}_{K/k}(\mathscr{G})$, where $\mathscr{G}$ is SL_2 or PGL_2 and K/k is a finite extension of fields. From this description over $k = k_s$, the Cartan k-subgroup $Z_{G_a}(S_a)$ is $\mathrm{R}_{K/k}(\mathrm{GL}_1)$. But we saw that $Z_{G_a}(S_a) = S_a \simeq \mathrm{GL}_1$, so comparison of dimensions forces $K = k$. Hence, G_a is reductive, so the root groups $U_{\pm a}$ are 1-dimensional. Thus, the equality $\Phi = \Phi(G^{\mathrm{red}}_{\overline{k}}, T_{\overline{k}})$ (as $\mathrm{char}(k) \neq 2$) implies that the smooth quotient map $G_{\overline{k}} \twoheadrightarrow G^{\mathrm{red}}_{\overline{k}}$ restricts to an isomorphism between Lie algebras (as we can see on separate $T_{\overline{k}}$-weight spaces, using 1-dimensionality for the nontrivial weights and the equality $C = T$ for the trivial weight). Hence, this quotient map is étale, so $G_{\overline{k}}$ is reductive and therefore so is G. □

Example 11.1.2 The reductivity criterion in Theorem 11.1.1 is false over every imperfect field k with characteristic 2, even for standard pseudo-reductive groups. Indeed, a counterexample is given by the non-reductive standard pseudo-reductive k-group $G = \mathrm{R}_{k'/k}(\mathrm{SL}_2)/\mathrm{R}_{k'/k}(\mu_2)$ for any purely inseparable quadratic extension k'/k with $\mathrm{char}(k) = 2$: G has the Cartan k-subgroup $C := \mathrm{R}_{k'/k}(\mathrm{GL}_1)/\mathrm{R}_{k'/k}(\mu_2)$ for the diagonal k'-torus $\mathrm{GL}_1 \subseteq \mathrm{SL}_2$, yet the natural inclusion $\mathrm{GL}_1/\mu_2 \subseteq C$ over k is an equality for dimension reasons since $[k' : k] = 2$.

For an exotic pseudo-reductive k-group $\mathrm{R}_{K/k}(\mathscr{G})$ over an imperfect field k of characteristic $p \in \{2, 3\}$ (with $\mathscr{G} \to \mathrm{Spec}\, K$ having fibers that are basic exotic pseudo-reductive groups), the following result parameterizes the maximal k-tori in $\mathrm{R}_{K/k}(\mathscr{G})$ in terms of the data $(G', K'/K, \overline{G})$ used to construct $\mathscr{G}$. The difficulty in the proof is due to the problem of lifting tori through very special isogenies.

Proposition 11.1.3 *With notation as above, there is a natural bijection between the set of maximal k-tori T in the k-group $\mathrm{R}_{K/k}(\mathscr{G}) \subseteq \mathrm{R}_{K'/k}(G')$ and the set of maximal K'-tori $T' \subseteq G'$ whose image $\overline{T}' \subseteq \overline{G}'$ under the very special isogeny $\pi' : G' \to \overline{G}' = \overline{G}_{K'}$ is defined over K relative to the Levi K-subgroup $\overline{G} \subseteq \mathrm{R}_{K'/K}(\overline{G}')$.*

The bijection is defined by the condition that T is the maximal k-torus in $\mathrm{R}_{K'/k}(T')$, and T' is the image of $T_{K'} \to G'$. The analogous such bijection holds between the set of maximal k-split tori in $\mathrm{R}_{K/k}(\mathscr{G})$ and the set of maximal K'-split tori S' in G' whose image $\overline{S}' \subseteq \overline{G}'$ under π' is defined over K relative to the Levi K-subgroup $\overline{G} \subseteq \mathrm{R}_{K'/K}(\overline{G}')$.

When $k = k_s$, $\mathrm{R}_{K'/k}(G')$ contains maximal k-tori which are not contained in the proper k-subgroup $\mathrm{R}_{K/k}(\mathscr{G})$. Thus, it is not possible to eliminate the constraint on T' related to the K-structure $\overline{G}$.

Proof The assertion for maximal split tori is immediate from the assertion for maximal tori, so we may assume that k is separably closed. By passing to factor fields we can also arrange that K' is a field; it is purely inseparable over k.

Let T be a maximal k-torus in $\mathrm{R}_{K/k}(\mathscr{G})$. By Proposition A.5.15(2), T is the maximal k-torus in $\mathrm{R}_{K/k}(\mathscr{T})$ for a unique maximal K-torus $\mathscr{T}$ of $\mathscr{G}$. Let G be a Levi K-subgroup of $\mathscr{G}$ containing $\mathscr{T}$ (Theorem 3.4.6). The image of T_K under the maximal pseudo-reductive quotient map $\mathrm{R}_{K/k}(\mathscr{G})_K \twoheadrightarrow \mathscr{G}$ is $\mathscr{T}$. Also, the image T' of $\mathscr{T}_{K'}$ under the maximal reductive quotient map $\mathscr{G}_{K'} \twoheadrightarrow G'$ is a maximal K'-torus in G'. Hence, T' is the image of $T_{K'}$ under the maximal reductive quotient map $\mathrm{R}_{K/k}(\mathscr{G})_{K'} \twoheadrightarrow G'$. It follows that T is a maximal k-torus of $\mathrm{R}_{K'/k}(T')$. Since the very special isogeny $G \to \overline{G}$ is a K-descent of the very special isogeny $G' \to \overline{G}'$, the image $\overline{T}'$ of T' in $\overline{G}'$ descends to the image $\overline{\mathscr{T}}$ of $\mathscr{T}$ in $\overline{G}$. In particular, this shows that the maximal K'-torus $T' \subseteq G'$ arising from the maximal k-torus $T \subseteq \mathrm{R}_{K/k}(\mathscr{G})$ must satisfy the asserted compatibility requirement with respect to the K-structure $\overline{G}$ on $\overline{G}'$.

The only remaining problem is to show that every maximal K'-torus T' in G' for which $\overline{T}' = \overline{T}_{K'}$ for a maximal K-torus $\overline{T}$ in $\overline{G}$ does arise from a maximal k-torus T of $\mathrm{R}_{K/k}(\mathscr{G})$. Since K and K' are separably closed, the K'-split K'-torus T' uniquely descends to a K-split K-torus $\mathscr{T}$, and the isogeny $T' \to \overline{T}'$ between K'-split K'-tori uniquely descends to an isogeny $\mathscr{T} \to \overline{T}$ between K-split K-tori. Thus, we get a commutative diagram

$$\begin{array}{ccc} \mathscr{T} & \longrightarrow & \mathrm{R}_{K'/K}(T') \\ \downarrow & & \downarrow \\ \overline{T} & \longrightarrow & \mathrm{R}_{K'/K}(\overline{T}') \end{array}$$

in which the bottom and right sides are induced by the bottom and right sides in the *cartesian* diagram

$$\begin{array}{ccc} \mathscr{G} & \longrightarrow & \mathrm{R}_{K'/K}(G') \\ \downarrow & & \downarrow \\ \overline{G} & \longrightarrow & \mathrm{R}_{K'/K}(\overline{G}') \end{array}$$

Thus, we get a K-homomorphism $h : \mathscr{T} \to \mathscr{G}$ whose composition with $\mathscr{G} \to \overline{G}$ is the K-isogeny $\mathscr{T} \to \overline{T}$ (so $h(\mathscr{T})$ is a maximal K-torus in $\mathscr{G}$) and whose composition with $\mathscr{G} \hookrightarrow \mathrm{R}_{K'/K}(G')$ is the inclusion of $\mathscr{T}$ into $\mathrm{R}_{K'/K}(\mathscr{T}_{K'}) = \mathrm{R}_{K'/K}(T')$ (so $\ker h = 1$). We conclude that $\mathscr{T}$ is a maximal K-torus in $\mathscr{G}$.

Since the inclusion $\mathscr{G} \hookrightarrow \mathrm{R}_{K'/K}(G')$ corresponds to a K'-group map $\mathscr{G}_{K'} \to G'$ that is a K'-descent of the maximal geometric reductive quotient (e.g., it is surjective) this latter map carries $\mathscr{T}_{K'}$ onto T'. Thus, if T is the maximal k-torus in $\mathrm{R}_{K/k}(\mathscr{T}) \subseteq \mathrm{R}_{K/k}(\mathscr{G})$ then $T_{K'} \to G'$ has image T'. □

At the end of [Ti3, II] Tits raised the question, for an imperfect field k, of characterizing those non-reductive absolutely pseudo-simple k-groups G for which the map i_G in (1.6.1) is an isomorphism. He proved this isomorphism property whenever $G^{\mathrm{ss}}_{\overline{k}}$ is unique in its central isogeny class (i.e., it is simply connected and adjoint), which is to say of type E_8, F_4, or G_2, subject to some restrictions on the characteristic (avoiding characteristic 2 for type F_4, avoiding characteristic 3 for type G_2). We can now prove a vast generalization in any characteristic, subject to a degree restriction in characteristic 2, as follows.

Let k be an imperfect field with characteristic $p > 0$, and let G be an absolutely pseudo-simple k-group that is not reductive. Let the nontrivial finite purely inseparable extension K/k be the field of definition over k of the $\overline{k}$-subgroup $\mathscr{R}_u(G_{\overline{k}}) = \mathscr{R}(G_{\overline{k}})$ in $G_{\overline{k}}$, and let $G' = G_K/\mathscr{R}(G_K)$ be the K-descent of the maximal semisimple quotient of $G_{\overline{k}}$. We seek necessary and sufficient conditions that characterize when the canonical morphism $i_G : G \to \mathrm{R}_{K/k}(G')$ is an isomorphism. Note that for the absolutely pseudo-simple k-group $G = \mathscr{D}(\mathrm{R}_{K/k}(\mathrm{PGL}_p))$ with K/k a nontrivial purely inseparable finite extension, $i_G : G \to \mathrm{R}_{K/k}(\mathrm{PGL}_p)$ is not an isomorphism since $\mathrm{R}_{K/k}(\mathrm{PGL}_p)$ is not perfect (Example 1.3.5). The source of this phenomenon is that PGL_p has fundamental group whose order is divisible by p. It turns out that if $p > 3$ then for any G as above, i_G is an isomorphism if and only if the fundamental group of $G^{\mathrm{ss}}_{\overline{k}}$ has order not divisible by p. More generally:

Proposition 11.1.4 *With notation as above, if i_G is an isomorphism then G_{k_s} has a reduced root system and $G^{\mathrm{ss}}_{\overline{k}}$ has fundamental group with order not divisible by p.*

Assume conversely that G_{k_s} has a reduced root system and $G^{\mathrm{ss}}_{\overline{k}}$ has fundamental group with order not divisible by p, and if $p = 2$ then assume $[k : k^2] = 2$.

(1) *The map i_G is an isomorphism except possibly when $p \in \{2,3\}$ and $G^{\mathrm{ss}}_{\overline{k}}$ is simply connected with an edge of multiplicity p in its Dynkin diagram.*
(2) *Assume $p \in \{2,3\}$ and that $G^{\mathrm{ss}}_{\overline{k}}$ is simply connected with an edge of multiplicity p in its Dynkin diagram. The map i_G is not an isomorphism onto $\mathrm{R}_{K/k}(G')$ precisely when it is an isomorphism onto $\mathrm{R}_{\mathsf{k}/k}(\mathscr{G})$, with $\mathsf{k} \subsetneq K$ a proper subfield over k containing K^p and $\mathscr{G} \subseteq \mathrm{R}_{K/\mathsf{k}}(G')$ a basic exotic pseudo-reductive k-group associated to the simply connected K-group G' and a choice of Levi k-subgroup of $\mathrm{R}_{K/\mathsf{k}}(\overline{G}')$, where $G' \to \overline{G}'$ is the very special K-isogeny.*

Proof If i_G is an isomorphism then G_{k_s} has a reduced root system (Example 2.3.2). Thus, for the rest of the argument we may and do assume that this root system is reduced.

Let $\widetilde{G}' \twoheadrightarrow G'$ be the simply connected central cover of G', and let μ denote the kernel of this covering map. Since G is absolutely pseudo-simple over k, the canonical map ξ_G as in (5.3.3) makes sense. By Proposition 1.3.4, the target $\mathrm{R}_{K/k}(\widetilde{G}')/\mathrm{R}_{K/k}(\mu)$ of ξ_G is the derived group of $\mathrm{R}_{K/k}(G')$, and by definition the composition of ξ_G with the canonical inclusion

$$\mathrm{R}_{K/k}(\widetilde{G}')/\mathrm{R}_{K/k}(\mu) \hookrightarrow \mathrm{R}_{K/k}(G')$$

onto the derived group is i_G. Hence, if i_G is an isomorphism then this inclusion is an equality. By dimension and connectedness considerations, equality happens if and only if $\dim \mathrm{R}_{K/k}(\mu) = 0$.

The finite extension K/k is nontrivial and purely inseparable, so $\mathrm{R}_{K/k}(\mu)$ has dimension > 0 whenever μ has nontrivial p-part (as we can check by working over k_s and replacing K with a subfield of degree p over k). Hence, i_G can be an isomorphism only if μ has order not divisible by p. The Cartier dual of $\mu_{\overline{k}}$ is the fundamental group of $G'_{\overline{k}}$, and absolutely simple groups of types E_8, F_4, and G_2 are both simply connected and adjoint whereas groups of types B_n and C_n in characteristic $p = 2$ have fundamental group of order not divisible by p precisely in the simply connected case.

Thus, for the remainder of the argument we may restrict attention to the case where $G^{\mathrm{ss}}_{\overline{k}}$ has fundamental group with order not divisible by p. In particular, the finite multiplicative K-group μ is étale, so

$$\mathrm{R}_{K/k}(\widetilde{G}')/\mathrm{R}_{K/k}(\mu) = \mathrm{R}_{K/k}(\widetilde{G}'/\mu) = \mathrm{R}_{K/k}(G').$$

Hence, now i_G is identified with ξ_G.

Since G_{k_s} has a reduced root system and $[k:k^2]=2$ when $p=2$, by Theorem 6.1.1, Theorem 8.2.1, and Proposition 9.2.4 it follows from Theorem 6.3.4 that G is standard when G is not exotic. Thus, the final claim in Theorem 5.3.8 ensures that the map $i_G = \xi_G$ is an isomorphism in such cases. In the exotic case, Theorem 8.2.1 gives that $i_G = \xi_G$ is an isomorphism onto the Weil restriction of a basic exotic pseudo-reductive group in the desired manner. □

Remark 11.1.5 It is natural to ask when Weil restrictions of the sort arising in Proposition 11.1.4 actually satisfy the initial hypotheses on the k-group G there. (That is, when are such Weil restrictions non-reductive and absolutely pseudo-simple?) This corresponds to a kind of converse to Proposition 11.1.4. To make it precise, fix an imperfect field k with characteristic $p > 0$ and consider a k-group $G := \mathrm{R}_{K/k}(G')$ with K/k a nontrivial purely inseparable finite extension and G' a connected semisimple K-group that is absolutely simple. The k-group G is pseudo-reductive over k but not reductive (by Example 1.6.1), and G_{k_s} has a reduced root system (Example 2.3.2). By Theorem 1.6.2(2), $G_{\overline{k}}^{\mathrm{ss}} \simeq G' \otimes_K \overline{k}$ and the field of definition of $\mathscr{R}(G_{\overline{k}}) = \mathscr{R}_u(G_{\overline{k}}) \subseteq G_{\overline{k}}$ over k is equal to K.

We claim that if the fundamental group of G' has order not divisible by p then G necessarily is absolutely pseudo-simple over k. To see this, note that the simply connected central cover $\widetilde{G}' \to G'$ is a finite étale cover, so the natural k-homomorphism

$$\mathrm{R}_{K/k}(\widetilde{G}') \to \mathrm{R}_{K/k}(G') = G$$

is a finite étale cover (Corollary A.5.4(2), Proposition A.5.13). Since $\mathrm{R}_{K/k}(\widetilde{G}')$ is its own derived group (Corollary A.7.11), it follows that G is perfect and hence G is absolutely pseudo-simple over k by Lemma 3.1.2.

In the non-standard direction, suppose $p \in \{2,3\}$ and consider a purely inseparable finite extension k/k, a basic exotic pseudo-reductive k-group $\mathscr{G}$, and the k-group $G := \mathrm{R}_{\mathsf{k}/k}(\mathscr{G})$. Once again, G_{k_s} has a reduced root system (Example 2.3.2). We claim that G is always absolutely pseudo-simple over k: this follows from Proposition 8.1.2 and Proposition 8.1.5 (due to the absolute pseudo-simplicity criterion in Lemma 3.1.2). Note that by Proposition 8.1.2, such k-groups G are not reductive.

11.2 Pseudo-semisimplicity

It is natural to seek a theory of pseudo-semisimplicity, as a special case of pseudo-reductivity. Intuition derived from experience with connected semisimple groups can be misleading, and even finding an appropriate definition leads to some surprises.

Consider a connected reductive group G over a field k. The derived group $\mathscr{D}(G)$ is semisimple, and in particular $\mathscr{D}(G)$ is its own derived group. If Z denotes the maximal central k-torus in G (i.e., maximal central connected reductive k-subgroup of G) then the natural map $\mathscr{D}(G) \times Z \to G$ is an isogeny. For pseudo-reductive G it is generally not true that G is generated by its derived group and its maximal central pseudo-reductive k-subgroup. Here are some counterexamples that are standard pseudo-reductive groups:

Example 11.2.1 For a field k, consider a pseudo-reductive k-group G containing no nontrivial central k-tori. We claim that the k-radical $\mathscr{R}_k(G)$ is necessarily trivial (exactly as in the connected reductive case), but that over any imperfect field k such G exist that are standard and have the additional property $G \neq \mathscr{D}(G)$ (which never happens in the connected reductive case). In particular, such G are not generated by $\mathscr{D}(G)$ and the maximal central pseudo-reductive k-subgroup.

To prove $\mathscr{R}_k(G) = 1$, first observe that since $\mathscr{R}_k(G)$ is solvable and pseudo-reductive (by normality in G), it is commutative. In particular, its maximal k-torus S is unique and thus is normal in G. But the automorphism functor of a k-torus is represented by an étale k-group, so the action on S by the connected G must be trivial. That is, the k-torus S is central in G, so by hypothesis $S = 1$. This forces $\mathscr{R}_k(G) = \mathscr{R}_{u,k}(G) = 1$.

To make such G for which $G \neq \mathscr{D}(G)$, let k be an imperfect field of characteristic $p > 0$ and let k'/k be a degree-p purely inseparable extension. Let $G = \mathrm{R}_{k'/k}(\mathrm{PGL}_p)$. This is pseudo-reductive and has trivial (scheme-theoretic) center since PGL_p has trivial center and the formation of the scheme-theoretic center commutes with Weil restriction through a finite extension of fields (Proposition A.5.15(1)). By Example 1.3.5, $G \neq \mathscr{D}(G)$. Here is an explicit nontrivial commutative quotient of the group $G(\overline{k}) = \mathrm{PGL}_p(A)$, where $A = \overline{k} \otimes_k k' \simeq \overline{k}[x]/(x^p)$. Since the pth powers in A are the elements of $\overline{k}$, the determinant on GL_p induces a surjective homomorphism from $\mathrm{PGL}_p(A)$ onto $A^\times/\overline{k}^\times \neq 1$.

There are (at least) two ways to characterize when a connected reductive k-group G is semisimple: the triviality of the k-radical $\mathscr{R}_k(G)$ and the perfectness of G. In the pseudo-reductive case the smooth connected solvable normal k-subgroup $\mathscr{R}_k(G)$ is always commutative (by Proposition 1.2.3), but Example 11.2.1 shows that requiring $\mathscr{R}_k(G) = 1$ does not generally imply that $G = \mathscr{D}(G)$ when k is imperfect. Thus, we are led to make the following definition.

Definition 11.2.2 Let k be a field. A k-group G is *pseudo-semisimple* if it is perfect and pseudo-reductive over k.

By Proposition 3.4.9, a pseudo-semisimple k-group G is reductive if and only if all root groups of G_{k_s} are 1-dimensional. (There is no analogue in the general pseudo-reductive case since commutative pseudo-reductive groups do not admit root groups.) As is the case for connected semisimple groups, any nontrivial pseudo-semisimple group is non-solvable. Likewise, pseudo-semisimplicity is insensitive to separable extension of the base field. For any pseudo-semisimple k-group G we have $\mathscr{R}_k(G) = 1$ because the perfectness of G forces the radical $\mathscr{R}(G_{\overline{k}})$ to be unipotent (i.e., $G_{\overline{k}}^{\text{red}}$ is semisimple) and hence $\mathscr{R}_k(G) = 1$ by Lemma 1.2.1.

For generalized standard pseudo-reductive groups over an arbitrary field k, there is a concrete characterization of pseudo-semisimplicity in terms of generalized standard presentations, as follows. By Definition 3.1.1, any pseudo-simple k-group is pseudo-semisimple, so if G is a non-commutative pseudo-reductive k-group admitting a generalized standard presentation $(G', k'/k, T', C)$, then by Proposition 10.2.2(1) the k-group G is pseudo-semisimple if and only if the given map $\phi : \mathrm{R}_{k'/k}(Z_{G'}(T')) \to C$ is surjective. (For the "only if" implication, use that G maps onto the commutative coker ϕ.)

By Proposition 10.1.4, if k is imperfect of characteristic 2 and $[k:k^2] = 2$ then all totally non-reduced pseudo-reductive k-groups are perfect. Since a product of k-groups is pseudo-semisimple if and only if each factor is pseudo-semisimple, we have therefore described the pseudo-semisimple groups within the classification of pseudo-reductive groups in Theorem 5.1.1.

Remark 11.2.3 We may restate Proposition 1.2.6 as follows: if G is a pseudo-reductive group over a field k then $\mathscr{D}(G)$ is pseudo-semisimple.

We now use Theorem 5.1.1 to make some of our earlier work with pseudo-semisimple groups more explicit. Proposition 3.1.8 describes the pseudo-semisimple normal k-groups N of any non-commutative pseudo-reductive k-group G as follows: if $\{N_i\}$ is the collection of such minimal nontrivial N then each N is generated by a collection of some of the N_i. It is natural to ask for an explicit description of each N_i (and hence of every N) when G is given in an explicit form as in our structure theorem (Theorem 5.1.1) for pseudo-reductive groups. The following discussion addresses this matter.

Let G be a non-commutative pseudo-reductive group over a field k, and assume $[k : k^2] = 2$ if k is imperfect with characteristic 2. In view of the splitting off of the non-reduced part of the root system in characteristic 2, and the general structure of totally non-reduced pseudo-reductive groups in such cases (see Propositions 10.1.4 and 10.1.6), any N_i that is totally non-reduced uniquely splits off as a direct factor of any N that meets it nontrivially. Thus, we may and do now assume that G_{k_s} has a reduced root system. By Theorem 5.1.1, G is generalized standard.

For a generalized standard group G over any field k whatsoever (i.e., no restriction on $[k : k^2]$ when $\mathrm{char}(k) = 2$), we shall determine the minimal such nontrivial N in terms of a generalized standard presentation of G, and then directly compute N and G/N in all cases. In particular, we will gain some insight into the phenomenon already seen in Example 1.6.4 whereby G/N can fail to be pseudo-reductive even when G and N are pseudo-semisimple.

Consider a generalized standard presentation

$$G = (\mathrm{R}_{k'/k}(G') \rtimes C)/\mathrm{R}_{k'/k}(Z_{G'}(T'))$$

using a suitable k-homomorphism $\phi : \mathrm{R}_{k'/k}(Z_{G'}(T')) \to C$. Here, C is a Cartan k-subgroup of G. Consider the decomposition $k' = \prod_{i \in I} k'_i$ into a product of fields and the associated decompositions

$$G' = \coprod G'_i, \ T' = \coprod T'_i$$

into a disjoint union of fibers over $\mathrm{Spec}\, k'$. For each subset $J \subseteq I$ define

$$k'_J = \prod_{j \in J} k'_j, \ G'_J = \coprod_{j \in J} G'_j, \ T'_J = \coprod_{j \in J} T'_j.$$

The Weil restriction $\mathrm{R}_{k'_J/k}(Z_{G'_J}(T'_J))$ is a direct factor of $\mathrm{R}_{k'/k}(Z_{G'}(T'))$; let C_J denote its image in C under ϕ. Note that C_J is pseudo-reductive since it is a smooth connected k-subgroup of the commutative pseudo-reductive k-group C. We therefore get a generalized standard pseudo-reductive normal k-subgroup

$$G_J := (\mathrm{R}_{k'_J/k}(G'_J) \rtimes C_J)/\mathrm{R}_{k'_J/k}(Z_{G'_J}(T'_J)) \subseteq G$$

that is pseudo-semisimple (due to the definition of C_J). Clearly the $G_{\{i\}}$ pairwise commute and the $G_{\{j\}}$ for $j \in J$ generate G_J.

Proposition 11.2.4 *With notation and hypotheses as above, every perfect smooth connected normal k-subgroup of G has the form $N = G_J$ for a unique subset $J \subseteq I$.*

Proof By Proposition 3.1.8, it suffices to prove that the $G_{\{i\}}$ are the minimal nontrivial perfect smooth connected normal k-subgroups of G (in which case J consists of exactly those i such that $G_{\{i\}} \subseteq N$). If we can prove this minimality property in the case of a separably closed base field then it follows in general by Galois descent considerations. Hence, to prove $N = G_J$ for some J we now may and do assume $k = k_s$.

By Proposition 3.1.6, it suffices to show that as we vary J, the image G'_J of $(G_J)_{\overline{k}}$ in $G^{\rm red}_{\overline{k}}$ varies without repetition through the perfect smooth connected normal subgroups of $G^{\rm red}_{\overline{k}}$. It is equivalent to verify the analogous assertion using $G^{\rm ss}_{\overline{k}}$ in place of $G^{\rm red}_{\overline{k}}$, since a connected reductive $\overline{k}$-group and its maximal semisimple quotient admit an evident natural correspondence between their sets of perfect smooth connected normal subgroups. For the same reason we can even pass to the maximal adjoint semisimple quotient $G^{\rm ss}_{\overline{k}}/Z_{G^{\rm ss}_{\overline{k}}}$. This adjoint group is the direct product of its simple factors, and (by using the generalized standard presentation of G) the simple factors are clearly the maximal adjoint semisimple quotients of the $(G'_i)_{\overline{k}}$. □

For $N = G_J$, the quotient $G/N = G/G_J$ is k-isomorphic to the k-group

$$(\mathrm{R}_{k'_{I-J}/k}(G'_{I-J}) \rtimes (C/C_J))/\mathrm{R}_{k'_{I-J}/k}(Z_{G'_{I-J}}(T'_{I-J})), \tag{11.2.1}$$

so this is pseudo-reductive if C/C_J is pseudo-reductive. More importantly, the converse is true: if G/G_J is pseudo-reductive then its k-subgroup C/C_J must be pseudo-reductive. The reason for this is that C/C_J is the image of the Cartan k-subgroup $C \subseteq G$, so it is a Cartan k-subgroup of G/G_J, and Cartan k-subgroups of pseudo-reductive k-groups are pseudo-reductive (Proposition 1.2.4).

Consider the special case that G is pseudo-semisimple (i.e., G is perfect), which is to say that the natural map $\phi : \mathrm{R}_{k'/k}(Z_{G'}(T')) \to C$ is surjective. Let $Z := \ker \phi$ and define the projection map

$$\pi_{I-J} : \mathrm{R}_{k'/k}(Z_{G'}(T')) \to \mathrm{R}_{k'_{I-J}/k}(Z_{G'_{I-J}}(T'_{I-J})).$$

We are given that the Cartan k-subgroup $\mathrm{R}_{k'/k}(Z_{G'}(T'))/Z$ of G is pseudo-reductive over k, and we have just seen that the quotient G/G_J is pseudo-reductive if and only if its Cartan k-subgroup

$$\mathrm{R}_{k'_{I-J}/k}(Z_{G'_{I-J}}(T'_{I-J}))/\pi_{I-J}(Z) \tag{11.2.2}$$

is pseudo-reductive.

It can happen for suitable triples $(G', k'/k, T')$ and k-subgroups $Z \subseteq \mathrm{R}_{k'/k}(Z_{G'}(T'))$ contained in $\mathrm{R}_{k'/k}(Z_{G'})$ that the group in (11.2.2) is not pseudo-reductive over k for some non-empty J even though $\mathrm{R}_{k'/k}(Z_{G'}(T'))/Z$ is pseudo-reductive over k! This is the essential content of the standard examples given in Example 1.6.4 (corresponding to the case $I = \{1,2\}$, $k'_1 = k'_2$, $G'_1 = G'_2$, $T'_1 = T'_2$, and $J = \{1\}$).

11.3 Unirationality

It is a well-known and very useful fact [Bo2, 18.2(ii)] that a connected reductive group over a field is unirational, and in particular has a Zariski-dense locus of rational points when the ground field is infinite. This ultimately rests on the unirationality of tori over any field, a result which has no analogue for general commutative pseudo-reductive groups:

Example 11.3.1 Over every imperfect field k there are commutative pseudo-reductive k-groups that are not unirational over k, and in case k is a rational function field $k_0(v)$ with $\mathrm{char}(k_0) > 0$ we can always find such examples whose locus of k-rational points is not Zariski-dense.

To make such examples, let k be imperfect with $\mathrm{char}(k) = p > 0$ and choose a p-power $q = p^r > 1$ and any $t \in k$ not in k^p. (The case $q > p$ will be of most interest for $p = 2$.) Consider the smooth k-group $U = \{y^q = x - tx^p\}$ in $\mathbf{G}_\mathrm{a} \times \mathbf{G}_\mathrm{a}$. This becomes isomorphic to $\mathbf{G}_\mathrm{a}$ over $k' = k(t')$ with $t' = t^{1/q}$. Indeed, if we define $x_0 = x$ and

$$x_j = y^{p^{r-j}} + t'^{p^{r-j}} x_{j-1}$$

for $1 \leqslant j \leqslant r$ then by induction on j we find $y^{p^{r-j}} = x_j - t'^{p^{r-j}} x_j^p$. Thus, for $z = x_r$ we have $y = z - t' z^p \in k'[z]$.

Since $x_{j-1} = t'^{-p^{r-j}}(x_j - y^{p^{r-j}})$ for $1 \leqslant j \leqslant r$, by descending induction on j we also have $x \in k'[z]$. It follows that $U_{k'} \simeq \mathbf{G}_\mathrm{a}$ as k'-groups via the coordinate z. By [Ru, Cor. 2.3.1], for an extension K/k we have $U_K \simeq \mathbf{G}_\mathrm{a}$ as K-groups if and only if K contains k' over k.

We claim that if $q > 2$ then the smooth geometrically connected k-scheme U of dimension 1 is not unirational over k. Suppose there were a dominant k-morphism $P \to U$ with P open in an affine space over k. By choosing P of minimal dimension we see via consideration of affine k-lines meeting P that P must be 1-dimensional, so $k(U)$ is k-rational by Lüroth's Theorem over k. But by [KMT, 6.9.2] such k-rationality can only happen if $p = 2$ and

$U \simeq \{w^2 = v - cv^2\}$ as k-schemes for some non-square c in k, in which case $U_K \simeq \mathbf{A}^1_K$ over $K = k(\sqrt{c})$ via the coordinate $z = w + \sqrt{c} \cdot v$ ($z^2 = v$, $z - \sqrt{c} \cdot z^2 = w$). Since $\mathbf{G}_{\mathrm{a}}$ is the unique group scheme structure on the affine line over a field with identity at the origin, in such a situation $U_K \simeq \mathbf{G}_{\mathrm{a}}$ as K-groups and hence K contains k' over k, so $q = [k' : k] \leqslant [K : k] = 2$.

Before we use such U to make non-unirational pseudo-reductive groups over k, we briefly digress to consider a special case in which $U(k)$ is not Zariski-dense in U.

Example 11.3.2 Suppose $k = k_0(t)$ for a subfield k_0 over which t is transcendental, and $q > 2$. Then $U(k) = \{(0,0)\}$ if $p \neq 2$ and $U(k) = \{(0,0), (1/t, 0)\}$ if $p = 2$, so $U(k)$ is not Zariski-dense in U. Indeed, we can assume k_0 is algebraically closed, and if $(x_0, y_0) \in U(k)$ then t-differentiation gives $x_0^p = x_0' := \mathrm{d}x_0/\mathrm{d}t$. Thus, if $x_0' = 0$ then $(x_0, y_0) = (0,0)$, so we can assume $x_0' \neq 0$. Hence, $e_a := \mathrm{ord}_a(x_0)$ is not divisible by p for some $a \in k_0$, so the equality $\mathrm{ord}_a(x_0') = \mathrm{ord}_a(x_0^p)$ gives $e_a - 1 = pe_a$. Thus, $p = 2$ and $e_a = -1$, so $x_0 = h_0^2/f_0$ for some separable monic non-constant $f_0 \in k_0[t]$ and some $h_0 \in k_0(t)^\times$ with divisor on Spec $k_0[t]$ disjoint from $\mathrm{div}(f_0)$. The identity

$$y_0^q = x_0 - tx_0^2 = \frac{h_0^2(f_0 - th_0^2)}{f_0^2}$$

then gives that y_0^q has a pole of order 2 at the zeros of f_0 away from $t = 0$. Since $q > 2$ this forces $f_0 = t$, so $h_0 = 1$ any hence $(x_0, y_0) = (1/t, 0)$.

Continuing with Example 11.3.1, consider a k-group C representing a class $\xi \in \mathrm{Ext}^1_k(U, \mathrm{GL}_1)$ (where we compute the Ext-group in the category of commutative affine k-group schemes of finite type). Obviously such a C is smooth and connected. If $q > 2$ then such a C is not unirational over k since its quotient U is not (and likewise if $k = k_0(t)$ with t transcendental over a subfield k_0 then the set $C(k)$ is not Zariski-dense in C when $q > 2$).

The induced map $h : \mathscr{R}_{u,k}(C) \to U$ has kernel $\mathscr{R}_{u,k}(C) \cap \mathrm{GL}_1$ that is both unipotent and of multiplicative type, hence trivial (Example A.1.7). Thus, if $\mathscr{R}_{u,k}(C) \neq 1$ then h is an isomorphism for dimension reasons (forcing $\xi = 0$). Hence, C is pseudo-reductive if $\xi \neq 0$, so to give examples of non-unirational commutative pseudo-reductive k-groups it suffices to prove $\mathrm{Ext}^1_k(U, \mathrm{GL}_1) \neq 0$ for all $q > 1$ (allowing $q = 2$). This will work out nicely for $p = 2$, but we will have to choose $t \in k$ carefully if $p > 2$.

Let $U' = \{y^p = x - tx^p\}$, so the map $(x, y) \mapsto (x, y^{q/p})$ defines an exact sequence of k-groups

$$0 \to \alpha_{q/p} \to U \to U' \to 0.$$

The pullback map

$$\mathrm{Ext}^1_k(U', \mathrm{GL}_1) \to \mathrm{Ext}^1_k(U, \mathrm{GL}_1)$$

is injective because $\mathrm{Hom}_k(\alpha_{q/p}, \mathrm{GL}_1) = 1$. Thus, it suffices to prove $\mathrm{Ext}^1_k(U', \mathrm{GL}_1) \neq 0$. According to [KMT, 6.13.3] (whose proof is explicit and works even if $p = 2$ and k is not separably closed, contrary to what is suggested by the statement there), this Ext-group is isomorphic (functorially in $k/\mathbf{F}_p(t)$) to the abelian group of $(p-1)$-tuples $(c_0, \ldots, c_{p-2}) \in k^{p-1}$ such that

$$c_{p-2} = \sum_{0 \leqslant j \leqslant p-2} c_j^p t^j. \tag{11.3.1}$$

We seek a nonzero solution to this equation in k for some choice of t. (Note that if $p = 3$ then this equation is $c_1 = c_0^3 + tc_1^3$, which we have seen in Example 11.3.2 has no nonzero solution if $k = k_0(t)$ with t transcendental over a subfield k_0. Hence, in general we will not be able to use an arbitrary $t \in k$.)

Set $c_j = 0$ for all $j < p-2$, so (11.3.1) with $c_{p-2} \neq 0$ collapses to $c_{p-2}^{p-1} = 1/t^{p-2}$. (For $p = 2$ the original equation is $c_0 = c_0^2$, so there is exactly one nonzero solution for each t. By [SGA3, XVII, 6.4(a)], this distinguished nontrivial extension corresponds to the k-group $\mathrm{R}_{k(\sqrt{t})/k}(\mathrm{GL}_1)$.) Thus, if at the start we choose $v \in k^\times$ that is not in k^p and define $t = v^{p-1}$ then t is not in k^p and we can set $c_{p-2} = 1/v^{p-2} \in k^\times$.

To make examples in which the locus of k-points is not Zariski-dense, consider $k = k_0(v)$ a rational function field and choose $t = v^{p-1}$. It suffices to show that if $q > 2$ then $U(k)$ is $\{(0,0)\} \cup \{(\zeta/v, 0) \mid \zeta^{p-1} = 1\}$. For $p = 2$ this is the settled case $t = v$ from Example 11.3.2, so we can assume $p > 2$. For $(x_0, y_0) \in U(k)$ differentiating the equation

$$y_0^q = x_0 - v^{p-1} x_0^p$$

gives $x_0' = -v^{p-2} x_0^p$, so if $x_0' = 0$ then $(x_0, y_0) = (0, 0)$.

We may assume k_0 is algebraically closed and $x_0' \neq 0$, so there exists some $a \in k_0$ such that $e_a := \mathrm{ord}_a(x_0)$ is not divisible by p. If $a \neq 0$ then

$$e_a - 1 = \mathrm{ord}_a(x_0') = \mathrm{ord}_a(-v^{p-2} x_0^p) = pe_a,$$

which is impossible since $p > 2$. Hence, necessarily $a = 0$ and $e_0 - 1 = p - 2 + pe_0$, so $e_0 = -1$. That is, $x_0 = x_1^p/v$ where $x_1 \in k_0(v)^\times$ has neither a zero nor a pole at 0. Then

$$y_0^q = x_0 - v^{p-1}x_0^p = \frac{x_1^p(1 - x_1^{p(p-1)})}{v} = \frac{(x_1(1 - x_1^{p-1}))^p}{v}$$

in $k_0(v)$, so $y_0 = 0$ and $x_1 = \zeta \in \mu_{p-1}(k_0)$. Hence, $x_0 = \zeta/v$.

We can push Example 11.3.1 a bit further:

Example 11.3.3 Let k be a (possibly archimedean) local field. It is a well-known fact [Ser3, III, §4.3] that $\mathrm{H}^1(k, G)$ is finite for any connected reductive k-group G. It is therefore natural to wonder if such finiteness holds for pseudo-reductive groups. This is false in the commutative case over every local function field k.

Indeed, choose an isomorphism $k \simeq \mathbf{F}((v))$ for a finite field $\mathbf{F}$ with characteristic $p > 0$, and let $U \subseteq \mathbf{G}_a \times \mathbf{G}_a$ be the k-group $\{y^q = x - v^{p-1}x^p\}$ with a p-power $q = p^r > 2$. We constructed above a non-split commutative extension C of U by GL_1 over k and showed that C is pseudo-reductive (and non-unirational) over k. We claim that $\mathrm{H}^1(k, C)$ is infinite.

Since U is p-torsion, the long exact Galois cohomology sequence associated to the extension structure on C provides an exact sequence

$$1 \to \mathrm{H}^1(k, C) \to \mathrm{H}^1(k, U) \to \mathrm{Br}(k)[p].$$

But $\mathrm{Br}(k) \simeq \mathbf{Q}/\mathbf{Z}$, so it suffices to prove that $\mathrm{H}^1(k, U)$ is infinite. The exact sequence

$$1 \to U \to \mathbf{G}_a \times \mathbf{G}_a \to \mathbf{G}_a \to 1$$

(with quotient map defined by $(x, y) \mapsto y^q - (x - v^{p-1}x^p)$) yields an isomorphism

$$\mathrm{H}^1(k, U) \simeq k/\{y^q - (x - v^{p-1}x^p) \mid x, y \in k\},$$

so we have to show that this quotient of k is infinite. Consideration of pole orders gives this when $p > 2$ since expressions of the form $y^q - (x - v^{p-1}x^p)$ with negative order only have orders that are 0 or -1 mod p. When $p = 2$ we have $4|q$ (since $q > 2$ by hypothesis), so a simple variant of this argument works by considering elements of k with negative order that is 2 mod 4.

By Example 11.3.1, a unirationality result for pseudo-reductive groups must avoid the commutative non-toric case. Likewise, by Example 11.3.3, a finiteness result for $\mathrm{H}^1(k, G)$ for pseudo-reductive G over local function fields k must avoid the commutative non-toric case. By Remark 11.2.3, it is natural to consider the pseudo-semisimple case for both problems. Since pseudo-semisimple groups are perfect, their unirationality is a special case of Proposition A.2.11. For pseudo-semisimple groups G over local function fields k the finiteness of $\mathrm{H}^1(k, G)$ requires arguments that are specific to the arithmetic case, so we refer the reader to [Con2] for the details on this result.

11.4 Structure of root groups and pseudo-parabolic subgroups

We saw in Proposition 2.2.9 that a pseudo-parabolic k-subgroup of a connected reductive k-group G is the same thing as a parabolic k-subgroup of G, and hence for any such subgroup P, G/P is proper over k . If G is a general smooth connected affine k-group then it is natural to inquire about the structure of G/P for pseudo-parabolic k-subgroups P of G. There is no loss of generality in restricting attention to the case of pseudo-reductive G, due to Proposition 2.2.10, and this is the case of most interest to us anyway.

Now let G be a pseudo-reductive group over a field k, and let P be a pseudo-parabolic k-subgroup. In general G/P can fail to be proper, as we saw explicitly in Example 2.1.3. It seems difficult to understand this failure of properness directly from the definitions. As an application of the theory of root groups in pseudo-reductive groups over a separably closed field, we saw in Proposition 3.4.9 that if G is perfect then it fails to be reductive precisely when G_{k_s} admits a root group with dimension larger than 1 and that in such cases G/P fails to be k-proper whenever P does not contain any nontrivial perfect smooth connected normal k-subgroups of G (i.e., P is not "too big").

In this section we shall use our main structure theorem (Theorem 5.1.1) to give a more explicit link from root groups and pseudo-parabolic k-subgroups in G to root groups and parabolic k-subgroups in connected reductive groups over finite extensions of k, assuming $[k : k^2] \leqslant 2$ when $\mathrm{char}(k) = 2$. This will clarify the structure of coset spaces G/P for pseudo-reductive G and pseudo-parabolic subgroups P over an arbitrary field k (assuming $[k : k^2] \leqslant 2$ when $\mathrm{char}(k) = 2$).

We begin with a result that provides a useful substitute in the pseudo-reductive case for the 1-dimensionality of root groups in the connected reductive case.

Proposition 11.4.1 *Let G be a pseudo-split pseudo-reductive group over a field k, and T a split maximal k-torus in G. If* $\mathrm{char}(k) = 2$ *then assume*

$[k:k^2] \leqslant 2$. *For each non-multipliable $a \in \Phi(G,T)$ there is a T-equivariant isomorphism $U_a \simeq \mathrm{R}_{k_a/k}(\mathbf{G}_\mathrm{a})$ for a purely inseparable finite extension field k_a/k, with T_{k_a} acting on $\mathbf{G}_\mathrm{a}$ through a_{k_a} and the quotient map $(U_a)_{k_a} \simeq (\mathrm{R}_{k_a/k}(\mathbf{G}_\mathrm{a}))_{k_a} \twoheadrightarrow \mathbf{G}_\mathrm{a}$ that is a k_a-descent of the quotient map from $(U_a)_{k_\mathrm{p}}$ onto the a-root group of $G^{\mathrm{red}}_{k_\mathrm{p}}$.*

If $\mathrm{char}(k) \neq 2$ *then the $Z_G(T)$-action on U_a is irreducible in the sense that it preserves no proper nontrivial smooth k-subgroup of U_a.*

We remind the reader that U_a is normalized by $Z_G(T)$ due to the intrinsic characterization of U_a in terms of the T-action on G; see Remark 2.3.5 (with $U_{(a)} = U_a$ when a is not divisible). Also, the avoidance of characteristic 2 for the irreducibility claim is essential. For example, if K/k is a purely inseparable quadratic extension in characteristic 2 then for $G = \mathrm{R}_{K/k}(\mathrm{SL}_2)$ the action of $Z_G(T)$ on U_a is the action of $\mathrm{R}_{K/k}(\mathrm{GL}_1)$ on $\mathrm{R}_{K/k}(\mathbf{G}_\mathrm{a})$ through scaling by squaring. But the squaring map on $\mathrm{R}_{K/k}(\mathrm{GL}_1)$ is valued in the k-torus GL_1, and $\mathrm{R}_{K/k}(\mathbf{G}_\mathrm{a}) \simeq \mathbf{G}_\mathrm{a}^2$ is not irreducible for its usual GL_1-action.

Proof By Theorem 2.3.10, since a is not multipliable it lies in the subset $\Phi(G^{\mathrm{red}}_{\overline{k}}, T_{\overline{k}})$. The structure of the k-subgroup G_a generated by U_a and U_{-a} is given in Proposition 3.4.1, and T is generated by $(\ker a)^0_{\mathrm{red}}$ and the 1-dimensional torus $G_a \cap T$, so replacing G with G_a does not affect U_a and brings us to the case where G is absolutely pseudo-simple of rank 1 with a reduced root system $\{\pm a\}$ (since we assumed at the outset that a is not multipliable). Thus, we may apply Theorem 6.1.1 away from characteristic 2 and Proposition 9.2.4 in characteristic 2 to conclude that $G \simeq \mathrm{R}_{K/k}(\mathrm{SL}_2)/Z$ for a purely inseparable finite extension K/k and a k-subgroup Z of the center $\mathrm{R}_{K/k}(\mu_2)$ such that $\mathrm{R}_{K/k}(\mathrm{GL}_1)/Z$ is pseudo-reductive.

Passage to central quotients does not affect the formation of root groups (Proposition 2.3.15), and any quotient map carries Cartan subgroups onto Cartan subgroups (Proposition A.2.8), so it suffices to treat the case $G = \mathrm{R}_{K/k}(\mathrm{SL}_2)$. This establishes the desired description of root groups, and reduces the irreducibility of the $Z_G(T)$-action on U_a (for $\mathrm{char}(k) \neq 2$) to the irreducibility of the action of $\mathrm{R}_{K/k}(\mathrm{GL}_1)$ on $\mathrm{R}_{K/k}(\mathbf{G}_\mathrm{a})$ through $(t,x) \mapsto t^2x$.

The squaring map on $\mathrm{R}_{K/k}(\mathrm{GL}_1)$ is surjective (hence faithfully flat) when $\mathrm{char}(k) \neq 2$, so it suffices to consider the action $(t,x) \mapsto tx$ instead. Suppose $H \subseteq \mathrm{R}_{K/k}(\mathbf{G}_\mathrm{a})$ is a smooth k-subgroup stable under the action of $\mathrm{R}_{K/k}(\mathrm{GL}_1)$. Then, $\mathrm{Lie}(H)$ is an $\mathrm{R}_{K/k}(\mathrm{GL}_1)$-stable k-linear subspace of K. Equivalently, it is a $K^\times$-stable k-linear subspace of K, so it is either 0 or K. If $\mathrm{Lie}(H) = 0$ then H is étale, and so it cannot be stable under the action of $\mathrm{R}_{K/k}(\mathrm{GL}_1)$ unless it is trivial. If $\mathrm{Lie}(H) = K$ then the inclusion $H \subseteq \mathrm{R}_{K/k}(\mathbf{G}_\mathrm{a})$ is an equality because H is smooth and $\mathrm{R}_{K/k}(\mathbf{G}_{\mathrm{a}})$ is smooth and connected. This completes the verification of $Z_G(T)$-irreducibility for root groups. □

Remark 11.4.2 The preceding method of proof also makes root groups explicit in terms of the parameters of a generalized standard presentation, assuming $[k : k^2] = 2$ in characteristic 2, as follows. Consider a generalized standard presentation

$$G \simeq (\mathrm{R}_{k'/k}(G') \rtimes C)/\mathrm{R}_{k'/k}(Z_{G'}(T'))$$

compatible with a k-split maximal k-torus T in G, so T' is k'-split. The behavior of root groups under central quotients (see Proposition 2.3.15) identifies each root group for (G, T) with a root group for $(\mathrm{R}_{k'_i/k}(G'_i), T'_i)$ for a unique i, where $\{k'_i\}$ is the set of factor fields of k' and (G'_i, T'_i) is the k'_i-fiber of (G, T). Moreover, the root systems $\Phi(G'_i, T'_i)$ are identified with the irreducible components of $\Phi(G, T)$. The corresponding root groups are $\mathrm{R}_{k'_i/k}(U'_{a_i})$ for $a'_i \in \Phi(G'_i, T'_i)$, where U'_{a_i} ranges through the root groups of (G'_i, T'_i).

When G'_i is semisimple, U'_{a_i} is k'_i-isomorphic to $\mathbf{G}_{\mathrm{a}}$. In the basic exotic case, so k is imperfect of characteristic $p \in \{2, 3\}$, we have $U'_{a_i} \simeq \mathbf{G}_{\mathrm{a}}$ over k'_i for long a_i, whereas $U'_{a_i} \simeq \mathrm{R}_{k''_i/k'_i}(\mathbf{G}_{\mathrm{a}})$ for a nontrivial finite extension k''_i/k'_i satisfying $k''_i \subseteq (k'_i)^{1/p}$ when a_i is short.

In the case of totally non-reduced pseudo-reductive groups with $[k : k^2] = 2$ we can again describe the root groups explicitly, and it comes out rather similarly to the exotic case. This is most easily seen by inspection of the open Bruhat cell for such k-groups, and we leave it to the interested reader. (For multipliable roots, the root group is not the Weil restriction of $\mathbf{G}_{\mathrm{a}}$ from a single extension field, but rather from a product of two extension fields.)

Now we turn our attention to the structure of pseudo-parabolic k-subgroups P of G and the coset spaces G/P. In contrast with the preceding considerations for root groups, we no longer assume G to be pseudo-split. We fix a maximal k-torus T and seek to describe the pseudo-parabolic k-subgroups P of G containing T in concrete terms so that we can understand the structure of G/P, assuming $[k : k^2] \leqslant 2$ when $\mathrm{char}(k) = 2$. Recall from Lemma 1.2.5(iii) that there is a bijective correspondence between the sets of maximal k-tori in G and $\mathscr{D}(G)$ via $T \mapsto T \cap \mathscr{D}(G)$ and $S \mapsto Z \cdot S$ with Z the maximal central k-torus in G (and $Z \cdot S$ an almost direct product). We claim that an analogous procedure works for pseudo-parabolic k-subgroups: it is $P \mapsto P \cap \mathscr{D}(G)$ and $Q \mapsto N_G(Q)$, with the natural map $\mathscr{D}(G)/Q \to G/P$ an isomorphism.

In one direction, choose a pseudo-parabolic k-subgroup P of G and write $P = P_G(\lambda)$. For a maximal k-torus T of P containing $\lambda(\mathrm{GL}_1)$, by Remark 2.1.7

we may replace λ with λ^n for a sufficiently divisible $n > 0$ depending on the degree of the isogeny $Z \times S \to T$ ($S = T \cap \mathscr{D}(G)$) so that we obtain the alternative description $P = P_G(\mu)$ for some $\mu : \mathrm{GL}_1 \to S$. Hence, $Q := P \cap \mathscr{D}(G) = P_{\mathscr{D}(G)}(\mu)$ is indeed a pseudo-parabolic k-subgroup of $\mathscr{D}(G)$. Since $G = Z_G(T) \cdot \mathscr{D}(G)$ by Proposition 1.2.6 and $Z_G(T) \subseteq Z_G(\mu) \subseteq P_G(\mu) = P$, the natural map $\mathscr{D}(G)/Q \to G/P$ is an isomorphism. Clearly P, and hence $Z_G(T)$, normalizes Q. Thus, since $G = Z_G(T) \cdot \mathscr{D}(G)$, we see that

$$P \subseteq N_G(Q) = Z_G(T) \cdot N_{\mathscr{D}(G)}(Q) = Z_G(T) \cdot Q \subseteq P,$$

so $P = N_G(Q)$.

In the reverse direction, if Q is a pseudo-parabolic k-subgroup of $\mathscr{D}(G)$ then we claim that $P := N_G(Q)$ is a pseudo-parabolic k-subgroup of G and $Q = P \cap \mathscr{D}(G)$. Writing $Q = P_{\mathscr{D}(G)}(\mu)$ for a 1-parameter k-subgroup μ of $\mathscr{D}(G)$, since $P_G(\mu) \cap \mathscr{D}(G) = Q$ we have $P_G(\mu) \subseteq P$. For a maximal k-torus S of Q containing $\mu(\mathrm{GL}_1)$, $T := Z \cdot S$ is a maximal k-torus of G. Clearly $Z_G(T) \subseteq Z_G(\mu)$, so

$$P = N_G(Q) = Z_G(T) \cdot N_{\mathscr{D}(G)}(Q) = Z_G(T) \cdot Q \subseteq P_G(\mu) \subseteq P,$$

so $P = P_G(\mu)$.

The preceding proves that $P \mapsto P \cap \mathscr{D}(G)$ and $Q \mapsto N_G(Q)$ are inverse bijective correspondences between the set of pseudo-parabolic k-subgroups P of G and the set of pseudo-parabolic k-subgroups Q of $\mathscr{D}(G)$. Moreover, it shows that if T is a maximal k-torus of G and $S = T \cap \mathscr{D}(G)$ is the corresponding maximal k-torus of $\mathscr{D}(G)$ then P contains T if and only if Q contains S. Thus, for the purpose of concretely describing P and G/P for all P containing T we lose no generality by replacing G with $\mathscr{D}(G)$; i.e., we can assume G is perfect (and $G \neq 1$).

We will address two cases: the nontrivial perfect generalized standard case over an arbitrary ground field k, and the totally non-reduced case when $\mathrm{char}(k) = 2$ with $[k : k^2] = 2$. By our main result (Theorem 5.1.1), these two cases cover all nontrivial perfect pseudo-reductive groups over any field k, provided we require $[k : k^2] \leqslant 2$ when $\mathrm{char}(k) = 2$. (We have used the elementary fact that a pseudo-parabolic k-subgroup in a direct product $G_1 \times G_2$ is precisely $P_1 \times P_2$ for pseudo-parabolic k-subgroups $P_i \subseteq G_i$.)

Remark 11.4.3 Let G be a nontrivial perfect generalized standard pseudo-reductive group over a field k, and let G' and k'/k be as in Definition 10.1.9. Since G is perfect, the natural map $j : \mathrm{R}_{k'/k}(G') \to G$ is a quotient map with central kernel. By Propositions 2.2.12 and 2.2.13, the set of pseudo-parabolic k-subgroups P of G is in bijective correspondence with the set of fiberwise pseudo-parabolic k'-subgroups P' of G' via the relations $P = j(\mathrm{R}_{k'/k}(P'))$ and $P' = \mathrm{im}(j^{-1}(P)_{k'} \hookrightarrow \mathrm{R}_{k'/k}(G')_{k'} \twoheadrightarrow G')$. In particular,

$$P'/Z_{G'} = \mathrm{im}(P_{k'} \to G'/Z_{G'}), \quad G/P = \mathrm{R}_{k'/k}(G'/P').$$

It remains to describe pseudo-parabolic subgroups containing a given maximal k-torus in a basic exotic pseudo-reductive k-group (with k imperfect of characteristic $p \in \{2, 3\}$) and in a basic non-reduced pseudo-simple group over an imperfect field k of characteristic 2 (assuming $[k : k^2] = 2$). The basic non-reduced case is easier (in both statement and proof), so we begin with that case.

Proposition 11.4.4 *Let k be an imperfect field of characteristic 2 such that $[k : k^2] = 2$, and let $K = k^{1/2}$. Choose $\alpha \in K - k$ and $n \geqslant 1$, and let $G = G_{\alpha,n}$ be the associated basic non-reduced pseudo-simple k-group. Let $G' = G_K^{\mathrm{ss}} \simeq \mathrm{Sp}_{2n}$.*

Under the non-central quotient map $\xi_G : G \to \mathrm{R}_{K/k}(G')$, there is a unique bijection between the set of pseudo-parabolic k-subgroups P of G and the set of parabolic K-subgroups P' of G' via either of the properties

$$\xi_G(P) = \mathrm{R}_{K/k}(P'), \quad P' = \mathrm{im}(P_K \to G').$$

Moreover, $P(k) = G(k) \cap P'(K)$ via $\xi_G : G(k) \simeq G'(K)$, and $P(k)$ is Zariski-dense in P.

By Corollary 3.5.11, if T is a maximal k-torus of G and T' is the corresponding maximal K-torus of G' as in Corollary 9.9.5 then $T \subseteq P$ if and only if $T' \subseteq P'$.

Proof By Proposition 3.5.4 and Corollary 3.5.11, for each P the image P' of P_K in G' is a pseudo-parabolic K-subgroup that determines P uniquely. The K-group P' is a parabolic K-subgroup, since G' is reductive (Proposition 2.2.9). Since every maximal K-torus in G' arises in a similar manner from a unique maximal k-torus in G (Corollary 9.9.5), it likewise follows from Proposition 3.5.4 that every parabolic K-subgroup of G' arises from a pseudo-parabolic k-subgroup P of G.

It now remains to prove that for any P and the corresponding parabolic K-subgroup $P' := \mathrm{im}(P_K \to G')$ we have $P(k) = G(k) \cap P'(K)$, $\xi_G(P) = \mathrm{R}_{K/k}(P')$, and $P(k)$ is Zariski-dense in P. Since $P = P_G(\lambda)$ for some

k-homomorphism $\lambda : \mathrm{GL}_1 \to G$, we see that $Z_G(\lambda)$ contains a maximal k-split k-torus of G (choose one containing $\lambda(\mathrm{GL}_1)$). But these are all maximal as k-tori in G and are $G(k)$-conjugate, by Corollary 9.9.5, so by applying a suitable $G(k)$-conjugation it suffices to treat the case of P containing a fixed k-split maximal k-torus of G. We will work with T corresponding to D_0 as in the construction of $G_{\alpha,n} \simeq G$.

The parabolic set of roots $\Phi(P,T)$ determines where P meets an open Bruhat cell Ω for (G,T) (relative to a choice of positive system of roots contained in $\Phi(P,T)$), and $\Phi(P',T')$ is the set of non-multipliable roots in $\Phi(P,T)$. Explicitly, $P \cap \Omega$ is a direct product of $Z_G(T)$ and some root groups for (G,T). But each root group is either $\mathrm{R}_{K/k}(\mathbf{G}_\mathrm{a})$ or $\mathrm{R}_{K/k}(\mathbf{G}_\mathrm{a}) \times \mathbf{G}_\mathrm{a}$, and $Z_G(T) = \mathrm{R}_{K/k}(\mathrm{GL}_1)^n$, and by inspection the k-points are Zariski-dense in each of these k-groups. Thus, $P(k)$ is Zariski-dense in P.

For each non-divisible root $a \in \Phi(G,T)$ and the corresponding non-multipliable root $a' \in \Phi(G',T')$, the explicit description of ξ_G on root groups gives that $\xi_G(U_a) = \mathrm{R}_{K/k}(U'_{a'})$. Likewise, ξ_G restricts to an isomorphism between Cartan k-subgroups, so $\xi_G(P)$ is open in $\mathrm{R}_{K/k}(P')$ (as they meet an open Bruhat cell in the same locus), and hence equality holds, due to connectedness considerations.

To prove that the containment $P(k) \subseteq G(k) \cap P'(K)$ is an equality, we work with the Zariski closure Q in G of the subgroup $G(k) \cap P'(K)$. This contains P since $P(k)$ is Zariski-dense in P, so Q is pseudo-parabolic (Proposition 3.5.8). But then $P_K \subseteq Q_K$, and (by construction of Q) a Zariski-dense locus in Q_K maps into P' in $G' = G_K^{\mathrm{ss}}$. Hence, $Q' = P'$, so the containment $P \subseteq Q$ is an equality. Thus, $G(k) \cap P'(K) \subseteq Q(k) = P(k)$, as required. □

Remark 11.4.5 Since G as in Proposition 11.4.4 is absolutely pseudo-simple and not reductive, it follows from Proposition 3.4.9(2) that G/P is non-proper for any pseudo-parabolic k-subgroup P of G distinct from G. This can be seen rather easily in the present situation: the natural map

$$G/P \to \mathrm{R}_{K/k}(G')/\mathrm{R}_{K/k}(P') = \mathrm{R}_{K/k}(G'/P')$$

is surjective with fiber over 1 isomorphic to $(\ker \xi_G)/(P \cap \ker \xi_G)$, and the explicit description of $\ker \xi_G$ in Theorem 9.8.1(4) implies (by root group considerations) that $P \cap \ker \xi_G$ has strictly smaller dimension than $\ker \xi_G$ when $P \neq G$. Since a positive-dimensional coset space for a unipotent group scheme is never proper, we get a direct proof of the non-properness of G/P when $P \neq G$.

Now we address the description of pseudo-parabolic k-subgroups in the basic exotic case over a field k with $\mathrm{char}(k) = p \in \{2,3\}$. Let $(G', k'/k, \overline{G})$ be the corresponding triple from the basic exotic construction, where $k'^p \subseteq k \subseteq k'$ with $\operatorname{Spec} k' \to \operatorname{Spec} k$ having all fiber-degrees > 1, $\pi' : G' \to \overline{G}' = \overline{G}_{k'}$ the (fiberwise) very special isogeny, and $\overline{G}$ a Levi k-subgroup of $\mathrm{R}_{k'/k}(\overline{G}')$ such that $G = \mathrm{R}_{k'/k}(\pi')^{-1}(\overline{G})$ is smooth. Then G is a basic exotic pseudo-reductive k-group and it canonically determines $(G', k'/k, \overline{G})$, by Proposition 8.1.5.

Proposition 11.4.6 *Let G be the basic exotic pseudo-reductive k-group corresponding to a triple $(G', k'/k, \overline{G})$.*

(1) *There is a natural bijection between the set of pseudo-parabolic k-subgroups $P \subseteq G$ and the set of parabolic k'-subgroups $P' \subseteq G'$ whose image $\overline{P}'$ under the very special isogeny $\pi' : G' \to \overline{G}' = \overline{G}_{k'}$ descends to a k-subgroup $\overline{P}$ in $\overline{G}$. Explicitly,*

$$P = G \cap \mathrm{R}_{k'/k}(P'), \quad P' = \mathrm{im}(P_{k'} \to G'),$$

and the coset space G/P is given by the fiber product

$$G/P = \overline{G}/\overline{P} \times_{\mathrm{R}_{k'/k}(\overline{G}'/\overline{P}')} \mathrm{R}_{k'/k}(G'/P'). \qquad (11.4.1)$$

In particular, the natural map $G/P \to \mathrm{R}_{k'/k}(G'/P')$ is a closed immersion.

(2) *Let $T \subseteq G$ be a maximal k-torus, and $T' := \mathrm{im}(T_{k'} \to G')$ the image maximal k'-torus in G'. Then P contains T if and only if P' contains T', and every pseudo-parabolic k'-subgroup of G' containing T' arises as such a P'.*

This proposition provides a concrete procedure reversing the one in Proposition 3.5.4 in the exotic case.

Proof We can pass to factor fields of k' to reduce to the case where k' is a field. We may and do also assume $k = k_s$ by Galois descent.

Recall from Proposition 7.2.3(3) that the natural map $G_{k'} \to G'$ is a k'-descent of the maximal geometric reductive quotient.

Let T be a maximal torus of G and T' be the image of $T_{k'}$ in G'. We naturally have $\Phi(G,T) = \Phi(G',T')$ via the identification $\mathrm{X}(T) = \mathrm{X}(T')$. The following sets are compatibly in natural bijective correspondence with the set of subsets of a fixed basis of the root system $\Phi(G,T)$ and therefore are in natural bijection with each other: the $G(k)$-conjugacy classes of pseudo-parabolic k-subgroups of G, the $G'(k')$-conjugacy classes of parabolic k'-subgroups of G', the $\overline{G}'(k')$-conjugacy classes of parabolic k'-subgroups of $\overline{G}'$, and the $\overline{G}(k)$-conjugacy classes of parabolic k-subgroups of $\overline{G}$.

Under this bijection, the $G(k)$-conjugacy class of a given pseudo-parabolic k-subgroup P of G corresponds to: the $G'(k')$-conjugacy class of the parabolic k'-subgroup P' obtained as the image of $P_{k'}$ under the quotient map $G_{k'} \twoheadrightarrow G'$, the $\overline{G}'(k')$-conjugacy class of the image $\overline{P}'$ of P' in $\overline{G}'$ under the very special isogeny π', and the $\overline{G}(k)$-conjugacy class of the image $\overline{P} = f(P)$ of P under the quotient map $G \twoheadrightarrow \overline{G}$ induced by $f := \mathrm{R}_{k'/k}(\pi')$. These conjugacy classes are naturally parameterized by $G(k)/P(k)$, $G'(k')/P'(k')$, $\overline{G}'(k')/\overline{P}'(k')$, and $\overline{G}(k)/\overline{P}(k)$ respectively.

There is a natural map $G(k)/P(k) \rightarrow G'(k')/P'(k')$ and we claim it is injective, which is to say $G(k) \cap P'(k') = P(k)$ inside $G'(k')$ (so P is uniquely determined inside G by the k'-subgroup P' of G'). We shall prove the more precise fact that $G \cap \mathrm{R}_{k'/k}(P') = P$ as closed subschemes of G. By writing $P = P_G(\lambda)$ we have $P' = P_{G'}(\lambda')$ for the composition λ' of $\lambda_{k'} : \mathrm{GL}_1 \rightarrow G_{k'}$ and the quotient map $G_{k'} \twoheadrightarrow G'$, and we know that $P_G(\lambda) = G \cap P_{\mathrm{R}_{k'/k}(G')}(\lambda)$ since G is a k-subgroup of $\mathrm{R}_{k'/k}(G')$. Moreover, from Proposition 2.1.13 we also know that $P_{\mathrm{R}_{k'/k}(G')}(\lambda) = \mathrm{R}_{k'/k}(P_{G'}(\lambda')) = \mathrm{R}_{k'/k}(P')$. Thus, $P = G \cap \mathrm{R}_{k'/k}(P')$ as claimed.

There are natural maps $G'(k')/P'(k') \rightarrow \overline{G}'(k')/\overline{P}'(k')$ and $G(k)/P(k) \rightarrow \overline{G}(k)/\overline{P}(k)$ induced by π' and f. Thus, to prove the first assertion of part (1) it suffices to show that the preimage of the $\overline{P}'(k')$-coset of a point in $\overline{G}(k)$ under the map

$$G'(k')/P'(k') \rightarrow \overline{G}'(k')/\overline{P}'(k')$$

is represented by a point of $G(k)$. Once this is proved, the explicit formulas for P in terms of P' and for P' in terms of P yield the first equivalence in part (2), and the second assertion in (2) is immediate from the identification of $\Phi(G,T)$ with $\Phi(G',T')$ and the bijective correspondence in Proposition 3.5.1 between the set of pseudo-parabolic subgroups containing a fixed split maximal torus (in a pseudo-reductive group) and the set of parabolic sets of roots in the associated root system.

To show that the preimage under $G'(k')/P'(k') \rightarrow \overline{G}'(k')/\overline{P}'(k')$ of the $\overline{P}'(k')$-coset of a point in $\overline{G}(k)$ is represented by a point in $G(k)$, we fix $T \subset P$ and work with the root system $\Phi := \Phi(G,T) = \Phi(G',T')$ and a positive system of roots Φ^+ contained in the parabolic subset $\Pi := \Phi(P,T) = \Phi(P',T')$. Observe that Π is the union of Φ^+ and the set of negative roots that are $\mathbf{Z}_{\leqslant 0}$-linear combinations of a unique set J of simple positive roots.

Let W denote the Weyl group of the root system Φ. We define W_J to be the subgroup generated by the reflections in the roots belonging to J (which is also naturally identified with the Weyl group of the maximal reductive quotients of the parabolic subgroups P' and $\overline{P}'$, relative to the maximal tori T' and $\overline{T}'$ of G' and $\overline{G}'$ respectively). By the proof of [Bo2, 21.24], each coset in W/W_J has a unique representative with shortest length; we let W^J denote the set of such representatives. By (the proof of) [Bo2, 21.29], there is a disjoint union decomposition

$$G'(k') = \coprod_{w \in W^J} U'_w(k') \cdot n_w \cdot P'(k')$$

where $n_w \in N_G(T)(k)$ represents w and each of the product sets is a direct product. (Here, U'_w denotes the k'-subgroup of G' directly spanned, in any order, by the root groups associated to the roots $a \in \Phi^+$ such that $w^{-1}(a)$ is negative.) We also have compatible decompositions

$$\overline{G}'(k') = \coprod_{w \in W^J} \overline{U}'_w(k') \cdot \overline{n}_w \cdot \overline{P}'(k'), \quad \overline{G}(k)\overline{P}'(k') = \coprod_{w \in W^J} \overline{\mathscr{U}}_w \cdot \overline{n}_w \cdot \overline{P}'(k')$$

of $\overline{G}'(k')$ and $\overline{G}(k)\overline{P}'(k')$ ($\subseteq \overline{G}'(k')$), with

$$\overline{U}'_w = \pi'(U'_w), \; \overline{\mathscr{U}}_w = \overline{G}(k) \cap \overline{U}'_w(k'),$$

and $\overline{n}_w = f(n_w)$. These compatible coset decompositions imply that under the natural map

$$G'(k') = \coprod_{w \in W^J} U'_w(k') \cdot n_w \cdot P'(k') \longrightarrow \coprod_{w \in W^J} \overline{U}'_w(k') \cdot \overline{n}_w \cdot \overline{P}'(k') = \overline{G}'(k'),$$

the preimage of $\overline{G}(k)\overline{P}'(k')$ consists of classes in $G'(k')/P'(k')$ whose unique representative $u'n_w$ ($w \in W^J$, $u' \in U'_w(k')$) has the property that it is carried by the map $U'_w(k') \to \overline{U}'_w(k')$ into an element of $\overline{\mathscr{U}}_w$ ($\subseteq \overline{G}(k)$). In particular, $\pi'(u') \in \overline{G}(k)$, so by the definition of $G = f^{-1}(\overline{G})$ it follows that $u' \in G(k)$ inside of $G'(k')$. Since the point n_w was chosen inside $G(k)$, we have proved that $u'n_w \in G(k)$ inside of $G'(k')$.

With assertion (2) and the necessary and sufficient criterion in (1) now proved, to finish the proof of (1) we have to establish that the natural map

$$j : G/P \to \overline{G}/\overline{P} \times_{\mathrm{R}_{k'/k}(\overline{G}'/\overline{P}')} \mathrm{R}_{k'/k}(G'/P')$$

is an isomorphism of k-schemes and $G/P \to \mathrm{R}_{k'/k}(G'/P')$ is a closed immersion. Since $P = G \cap \mathrm{R}_{k'/k}(P')$, j is a subfunctor inclusion because

$\mathrm{R}_{k'/k}(G')/\mathrm{R}_{k'/k}(P') = \mathrm{R}_{k'/k}(G'/P')$ (due to the good behavior of Weil restriction for torsors under smooth groups, by Corollary A.5.4(3)). Choose $\lambda \in \mathrm{X}_*(T)$ such that $P = P_G(\lambda)$ and define $U = U_G(-\lambda)$. Similarly define $\overline{U} \subseteq \overline{G}$, $\overline{U}' \subseteq \overline{G}'$, and $U' \subseteq G'$, so we have open subschemes $U \times P \subseteq G$, $\overline{U} \times \overline{P} \subseteq \overline{G}$, $\overline{U}' \times \overline{P}' \subseteq \overline{G}'$, and $U' \times P' \subseteq G'$ via multiplication (Proposition 2.1.8(3)). In this way we identify U, $\overline{U}$, $\overline{U}'$, and U' as respective open subschemes of G/P, $\overline{G}/\overline{P}$, $\overline{G}'/\overline{P}'$, and G'/P'. The map j restricts to the natural map

$$U \to \overline{U} \times_{\mathrm{R}_{k'/k}(\overline{U}')} \mathrm{R}_{k'/k}(U') =: \Omega_0$$

induced by the equality

$$G = f^{-1}(\overline{G}) = \overline{G} \times_{\mathrm{R}_{k'/k}(\overline{G}')} \mathrm{R}_{k'/k}(G').$$

The dynamic descriptions of U, $\overline{U}$, $\overline{U}'$, and U' imply that the restriction $j : U \to \Omega_0$ between open subschemes is an isomorphism. Thus, translation considerations over $\overline{k}$ imply that the monomorphism j is an open immersion. We will show that the target of j is covered by the translates $n_w\Omega_0$ for $w \in W := W(\Phi)$ (with $n_w \in N_G(T)(k)$), so then j is surjective and hence an isomorphism.

Let $\overline{\Omega}$ denote the open cell $(\overline{G}, \overline{T})$ relative to Φ^+. We claim $\overline{G}$ is covered by the translates $n_w\overline{\Omega}$ for $w \in W$. This is a general fact about split connected reductive groups: the scheme $\overline{G}$ is covered by the Bruhat cells $\overline{B}n_w\overline{B} = \overline{U}_w n_w \overline{B}$ (where the Borel subgroup $\overline{B}$ containing $\overline{T}$ corresponds to Φ^+), and $n_w^{-1}\overline{U}_w n_w \overline{B} \subseteq \overline{\Omega}$, so $\overline{U}_w n_w \overline{B} \subseteq n_w\overline{\Omega}$, establishing the claim. Since $\overline{U} \times \overline{P} = \overline{\Omega} \cdot \overline{P}$ as open subschemes of $\overline{G}$, we conclude that the quotient $\overline{G}/\overline{P}$ is covered by the translates $n_w\overline{U}$. Hence, the target of j is covered by the left n_w-translates of the open subscheme

$$\overline{U} \times_{\mathrm{R}_{k'/k}(\overline{G}'/\overline{P}')} \mathrm{R}_{k'/k}(G'/P') = \overline{U} \times_{\mathrm{R}_{k'/k}(\overline{U}')} \mathrm{R}_{k'/k}(\mathscr{U}')$$

where $\mathscr{U}' \subseteq G'/P'$ is the preimage of $\overline{U}'$ under $G'/P' \to \overline{G}'/\overline{P}'$. (Here we have used that Weil restriction is compatible with the formation of preimages of open subschemes.)

We need to show that the evident containment $U' \subseteq \mathscr{U}'$ of open subschemes of G'/P' is an equality, or equivalently that under the very special isogeny $q : G' \to \overline{G}'$ (which carries P' onto $\overline{P}'$) the preimage of the open subscheme $\overline{U}' \times \overline{P}'$ is the open subscheme $U' \times P'$. But q is purely inseparable and

$q(U' \times P') = \overline{U}' \times \overline{P}'$, so $q^{-1}(\overline{U}' \times \overline{P}') = U' \times P'$ as required. This completes the proof that j is an isomorphism of k-schemes.

The natural map $h : G/P \to \mathrm{R}_{k'/k}(G'/P')$ is thereby identified with a base change of the natural map $\overline{h} : \overline{G}/\overline{P} \to \mathrm{R}_{k'/k}(\overline{G}'/\overline{P}')$, so to show that h is a closed immersion it suffices to prove that $\overline{h}$ is. But $\overline{G}'/\overline{P}' = (\overline{G}/\overline{P})_{k'}$, so $\overline{h}$ is an instance of the general map $j_X : X \to \mathrm{R}_{k'/k}(X_{k'})$ for quasi-projective k-schemes X. This map is always a closed immersion, by Proposition A.5.7. □

Example 11.4.7 We now show that the constraint on $\overline{P}'$ in relation to $\overline{G}$ in Proposition 11.4.6 cannot be dropped (though part (2) of the proposition implies that this constraint is satisfied when P' contains a maximal k'-torus T' arising from a maximal k-torus T in G).

Let k be an imperfect field with $\mathrm{char}(k) \in \{2, 3\}$ and $\mathscr{G}$ a basic exotic pseudo-reductive k-group associated to a triple $(G', k'/k, \overline{G})$ with k' a field. Assume $k = k_s$, pick a Levi k-subgroup L in $\mathscr{G}$, and choose a Borel k-subgroup B in L and a maximal k-torus T in B. Since B is pseudo-parabolic in L, there is a 1-parameter k-subgroup $\lambda : \mathrm{GL}_1 \to T$ such that $B = P_L(\lambda)$. Define $\mathscr{B} = P_{\mathscr{G}}(\lambda)$.

By Proposition 2.3.10, the groups L, $\mathscr{G}$, and G' have the same root system (with respect to compatible maximal tori), so the characterization of minimality in terms of the root system in the pseudo-split case (Proposition 3.3.15(1)) implies that $\mathscr{B}$ is a minimal pseudo-parabolic k-subgroup of $\mathscr{G}$ and $B' := \mathrm{im}(\mathscr{B}_{k'} \to G')$ is a Borel k'-subgroup of G' with maximal k'-torus $T' := \mathrm{im}(T_{k'} \to G')$. Let $\overline{B}'$ be the image Borel subgroup of B' in the very special quotient $\overline{G}'$, and let $\overline{B}$ be the common image of B and $\mathscr{B}$ under the quotient map $\mathscr{G} \twoheadrightarrow \overline{G}$, so $\overline{B}$ is a Borel k-subgroup of $\overline{G}$ descending $\overline{B}'$.

Choose $g' \in G'(k')$ not in $\mathscr{G}(k)B'(k')$. Such a g' exists, as can be seen by working in an open Bruhat cell of $G' = L_{k'}$ relative to (B', T') (which arises by scalar extension from the open Bruhat cell of L relative to (B, T)). Consider $P' = g'B'g'^{-1}$. This is a Borel k'-subgroup of G', and we claim that it cannot arise from a pseudo-parabolic k-subgroup P in $\mathscr{G}$.

Suppose P' arises from some such P, which is to say that $P' = \mathrm{im}(P_{k'} \to G')$. Consideration of the root systems (relative to a maximal k-torus in P and the associated maximal k'-torus in P') shows that P must be a minimal pseudo-parabolic k-subgroup of $\mathscr{G}$. Thus, by rational conjugacy of minimal pseudo-parabolic subgroups over a separably closed field (Proposition 3.3.17), $P = g\mathscr{B}g^{-1}$ for some $g \in \mathscr{G}(k)$. Passing to images under the maximal reductive quotient map $q : \mathscr{G}_{k'} \twoheadrightarrow G'$, we get $P' = q(g)B'q(g)^{-1}$ in G'. But $P' = g'B'g'^{-1}$ by definition, so $q(g)^{-1}g' \in N_{G'}(B')(k') = B'(k')$. Thus, $g' \in \mathscr{G}(k)B'(k')$, contrary to how g' was chosen.

PART IV

Appendices

Appendix A
Background in linear algebraic groups

We assume the reader is familiar with the general theory of linear algebraic groups over a field, as well as with the special features of the connected reductive case over a field. More specifically, the book of Borel [Bo2] covers much of this background. However, we need to use non-smooth groups, and some definitions of common notions (such as unipotence, quotient, etc.) in [Bo2] in the smooth case need to be modified to work more generally. To avoid any confusion, in §A.1 we review a number of important definitions from the scheme-theoretic point of view that is necessary in our work; in general we follow [SGA3]. As an additional convenience to the reader, in §A.2 we record some facts from [Bo2] that we frequently apply, incorporating generalizations without smoothness hypotheses (since we need such generality).

In §A.3 we review the relative Frobenius morphism and discuss some results for non-affine groups (mainly for justifying a few examples); non-affine groups can be ignored for the development of the theory of pseudo-reductive groups. In §A.4 we record versions of the Existence, Isomorphism, and Isogeny Theorems for split connected reductive groups over an arbitrary field (since the literature on this topic aside from [SGA3] appears to always assume the ground field is algebraically closed). We also show how to deduce these results from the more well-known case of an algebraically closed ground field. The case of an arbitrary separably closed ground field arises often in our work.

Within the theory of linear algebraic groups, Weil restriction through a finite separable field extension is an important and widely used operation. Less familiar, but ubiquitous in our work, is the case of an inseparable finite extension of fields. In §A.5 we review the Weil restriction functor in a general setting, and an interesting example concerning Levi subgroups (and a variant on Weil restriction due to Greenberg) is the focus of attention in §A.6. In §A.7 we address the interaction of Weil restriction and Lie algebras.

Finally, in §A.8 we study group schemes M of multiplicative type and their actions on group schemes G of finite type over a field. This is a familiar situation when M is a torus and G is smooth and affine, but we need to allow for the possibility that M or G is not smooth.

A.1 Review of definitions

The *additive group* is denoted $\mathbf{G}_{\mathrm{a}}$ and the *multiplicative group* is denoted GL_1, always with the base ring understood from context. A *maximal k-torus T* in a group scheme G locally of finite type over a field k is a k-torus not strictly contained in another k-torus in G; see Corollary A.2.6 for the (well-known) invariance of this property with respect to any ground field extension when G is smooth and affine, and Lemma C.4.4 for the case of general smooth G (which we do not need).

Definition A.1.1 A *vector group* over a field k is a smooth commutative k-group scheme V that admits an isomorphism to $\mathbf{G}_{\mathrm{a}}^n$ for some $n \geqslant 0$. The GL_1-scaling action arising from such an isomorphism is a *linear structure* on V.

Observe that the GL_1-action on V arising from a linear structure induces the canonical $k^\times$-action on $\mathrm{Lie}(V)$ (e.g., if $\mathrm{char}(k) = p > 0$ then the composition of such a GL_1-action on V with the p-power map on GL_1 does not arise from a linear structure on V when $V \neq 0$).

Example A.1.2 If W is a finite-dimensional k-vector space then the *associated vector group* $\underline{W}$ represents the functor $R \rightsquigarrow R \otimes_k W$ on k-algebras and its formation commutes with any extension of the ground field. Explicitly, $\underline{W} = \mathrm{Spec}(\mathrm{Sym}(W^*))$ and it has an evident linear structure relative to which $\underline{W}(k)$ is naturally identified with W as a k-vector space.

When linear structures are specified on a pair of vector groups, a homomorphism respecting them is called *linear*. Over a field of characteristic 0 there is a unique linear structure and all homomorphisms are linear. Over a field with characteristic $p > 0$ the linear structure is not unique in dimension larger than 1 (e.g., $a.(x, y) := (ax + (a - a^p)y^p, ay)$ is a linear structure on $\mathbf{G}_{\mathrm{a}}^2$, obtained from the usual one via the non-linear k-group automorphism $(x, y) \mapsto (x + y^p, y)$ of $\mathbf{G}_{\mathrm{a}}^2$).

A smooth connected solvable group G over a field k is *k-split* if it admits a composition series over k whose successive quotients are k-isomorphic to GL_1 or $\mathbf{G}_{\mathrm{a}}$. Any quotient of a k-split smooth connected solvable k-group is k-split [Bo2, 15.4(i)]. A general discussion of k-split groups is in [Bo2, 15.1ff.].

The notion of unipotence for affine groups of finite type over a field is of most interest to us in the familiar smooth case, where it can be defined in terms of Jordan decomposition of geometric points. It is sometimes convenient (though not essential) to allow the non-smooth case, so we now record the definition in general and address alternative formulations.

Definition A.1.3 A group scheme U over a field k is *unipotent* if U is affine of finite type and $U_{\overline{k}}$ admits a finite composition series over $\overline{k}$ with each successive quotient isomorphic to a $\overline{k}$-subgroup of $\mathbf{G}_{\mathrm{a}}$.

The notions of composition series and successive quotients in Definition A.1.3 are meant in terms of the notions of normality and quotients for group schemes that will be reviewed in Definitions A.1.8 and A.1.11 and following pages. It follows from [SGA3, XVII, Prop. 2.2] that the quotient of a unipotent group scheme modulo a normal closed subgroup scheme is unipotent.

The definition of unipotence given in [SGA3, XVII, 1.3] does not assume U to be affine, but it is equivalent to Definition A.1.3 by [SGA3, XVII, 2.1].

Proposition A.1.4 *A group scheme U over a field k is unipotent if and only if U is k-isomorphic to a closed k-subgroup of the upper triangular unipotent subgroup of some* GL_n. *If U is smooth, connected, and unipotent then it is k-split if k is perfect, and in general if such a U is k-split then so is any quotient of U.*

Proof The equivalence with an upper triangular presentation over k is [SGA3, XVII, 3.5(i),(v)]. By [Bo2, I, 4.8], when U is a smooth affine k-group such an upper triangular form is equivalent to all elements of $U(\overline{k})$ being unipotent. (This latter condition is how unipotence for U is defined in [Bo2] and other classical references on linear algebraic groups, so for smooth affine k-groups the classical notion of unipotence is equivalent to Definition A.1.3.) The k-split property in the smooth connected case over a perfect field k is [Bo2, 15.5(ii)], and its preservation under passage to a quotient over any k is [Bo2, 15.4(i)]. □

Definition A.1.5 A group scheme T over a field k is a *k-torus* if $T_{\overline{k}} \simeq \mathrm{GL}_1^n$ for some $n \geqslant 0$. A k-group scheme M is *of multiplicative type* if $M_{\overline{k}}$ is isomorphic to a closed $\overline{k}$-subgroup scheme of a $\overline{k}$-torus. The *character group* of M is defined to be $\mathrm{X}(M) = \mathrm{Hom}_{\overline{k}}(M_{\overline{k}}, \mathrm{GL}_1)$ equipped with its natural action by $\mathrm{Gal}(k_s/k) = \mathrm{Aut}(\overline{k}/k)$.

By [Bo2, 8.11], if T is a k-torus then $T_{k_s} \simeq \mathrm{GL}_1^n$ over k_s for some $n \geqslant 0$. In [SGA3, IX, 1.1] a more general definition of *multiplicative type* is given, allowing k-group schemes that may not be of finite type. The distinction is that we require the geometric character group to be finitely generated as a $\mathbf{Z}$-module.

Example A.1.6 If $M = \mathrm{GL}_1$ then $\mathrm{X}(M) = \mathbf{Z}$ with trivial Galois action, and if $M = \mu_n$ then $\mathrm{X}(M)$ is the $\mathbf{Z}$-module $\mathbf{Z}/n\mathbf{Z}$ equipped with trivial Galois action.

Example A.1.7 A k-group scheme that is both unipotent and of multiplicative type is trivial. Indeed, we may assume k is algebraically closed, so it remains to show that a nontrivial k-subgroup scheme M of GL_1 cannot be a k-subgroup of $\mathbf{G}_{\mathrm{a}}$. Every nontrivial k-subgroup scheme of GL_1 contains μ_ℓ for some prime ℓ, so we just have to show that μ_ℓ has no nontrivial k-homomorphism to $\mathbf{G}_{\mathrm{a}}$. The case $\mathrm{char}(k) \neq \ell$ is clear, so we may assume $\ell = p = \mathrm{char}(k) > 0$. A k-homomorphism from μ_p to $\mathbf{G}_{\mathrm{a}}$ is an element $f \in k[t]/(t^p - 1)$ such that $f(xy) = f(x) + f(y)$ in $k[x, y]/(x^p - 1, y^p - 1)$. Direct comparison of monomial terms on both sides forces $f = 0$.

The following definitions for group schemes over a field k are scheme-theoretic versions of some familiar notions from the smooth case as in [Bo2].

Definition A.1.8 Let k be a ring. The *kernel* of a k-homomorphism $f : G \to G'$ between k-group schemes is $f^{-1}(1)$; it is denoted $\ker f$. We call G (or really (G, f)) a *k-subgroup* of G' when $\ker f = 1$, and a *closed k-subgroup* when f is a closed immersion. A k-subgroup G in G' is *normal* if $G(S)$ is a normal subgroup of $G'(S)$ for all k-schemes S (or equivalently, all affine k-schemes S).

Now assume k is a field. The kernel in Definition A.1.8 is the unique closed subscheme of G representing the functor on k-schemes

$$S \rightsquigarrow \ker(G(S) \to G'(S)).$$

The reason it is closed is that every group scheme over a field is separated (as the diagonal of a group scheme is a base change of the identity section, and any rational point in a scheme over a field is closed). In Proposition A.2.1 we will see that k-subgroups are always closed when working with k-groups of finite type.

Definition A.1.9 Let G be a group scheme over a field k. Let X be a locally finite type k-scheme equipped with a left G-action, and let V be a geometrically reduced (hence generically smooth) closed subscheme of X.

If V is separated then the *scheme-theoretic centralizer* $Z_G(V)$ of V in G is the unique closed k-subgroup of G such that, for each k-algebra A, the group $Z_G(V)(A)$ is the subgroup of $g \in G(A)$ such that the g-action on X_A is trivial on V_A.

If V is of finite type (e.g., X is of finite type) then the *scheme-theoretic normalizer* $N_G(V)$ of V in G is the unique closed subscheme of G such that, for each k-algebra A, $N_G(V)(A)$ is the set of g in $G(A)$ such that the g-action on X_A carries V_A back into itself. A k-subgroup H of G *normalizes* V if $H \subseteq N_G(V)$.

The functor $N_G(V)$ does in fact act on V via automorphisms since its A-points act via monomorphisms $V_A \to V_A$ and any monic endomorphism of a finitely presented scheme is an automorphism [EGA, IV_4, 17.9.6]. Thus, $N_G(V)$ really is a subgroup functor of G.

To see that $Z_G(V)$ and $N_G(V)$ exist as closed subschemes of G, since $V(k_s)$ is Zariski-dense in V_{k_s} (due to geometric reducedness) it follows from considerations with relative schematic density (cf. [EGA, IV_3, 11.10.6]) that $Z_G(V)$ (resp. $N_G(V)$) is represented by the $\text{Gal}(k_s/k)$-descent of the scheme-theoretic intersection of the centralizer schemes in G_{k_s} of the points $v \in V(k_s)$ (resp. intersection of the preimage of V_{k_s} under the maps $G_{k_s} \to X_{k_s}$ defined by $g \mapsto g.v$ for $v \in V(k_s)$). To make sense of such (generally infinite) intersections we need to check that the centralizers and preimages being used are closed subschemes. For the case of $N_G(V)$ it follows from the fact that V_{k_s} is closed in X_{k_s}. For the case of $Z_G(V)$, we use that the diagonal of V is closed in $V \times V$ (by the separatedness hypothesis) and that $V \times V$ is closed in $X \times X$.

It is an elementary consequence of the definitions that the formation of $Z_G(V)$ and $N_G(V)$ commutes with any extension of the ground field, but they need not be smooth, even when G and V are smooth. For example, if $\text{char}(k) = p > 0$ and $G = \text{SL}_p$ then $Z_G(G) = \mu_p$.

Definition A.1.10 The *identity component* H^0 of a locally finite type group scheme H over a field k is the unique connected open and closed subscheme containing the identity point; it is a geometrically irreducible normal subgroup scheme of finite type over k [SGA3, VI_A, 2.3.1, 2.4] (so its formation commutes with any extension of the ground field). The *scheme-theoretic center* Z_G of a smooth k-group scheme G is $Z_G(G)$; i.e., for any k-algebra A, $Z_G(A)$ is the group of $g \in G(A)$ whose conjugation action on G_A is trivial. A k-subgroup Z of a k-group scheme G is *central* if the conjugation action of Z on G is trivial.

Note that we only define the notion of scheme-theoretic center for smooth G. There is another group-theoretic construction which we need without smoothness restrictions: the formation of coset spaces, and especially quotients by normal subgroup schemes. In [Bo2], quotient maps between smooth affine groups over a field are generally required to be separable (equivalently, smooth). This restriction is due to the avoidance of non-smooth groups in [Bo2]. In our work it is crucial to go beyond this classical setting and use a more general theory of quotients as developed by Grothendieck:

Definition A.1.11 If G is a group scheme locally of finite type over a field k and H is a closed k-subgroup scheme of G then the *quotient* scheme G/H is taken in the sense of Grothendieck [SGA3, $\mathrm{VI_A}$, Thm. 3.2(iv)]. This is a locally finite type k-scheme equipped with faithfully flat map of k-schemes $G \to G/H$ that is initial for being invariant under right translation by H on G (and G/H is separated).

If H is normal in G then G/H admits a unique k-group structure making $G \to G/H$ a k-homomorphism, and its kernel is H. This is proved via descent theory. The k-group G/G^0 is étale [SGA3, $\mathrm{VI_A}$, 5.5.1]. An *isogeny* $f : G \to G'$ between locally finite type k-group schemes is a k-homomorphism that is flat and surjective with $H = \ker f$ a finite k-group. We call such an f a *central isogeny* when its kernel is contained in Z_G. Note that the k-finiteness of H is equivalent to the finiteness of $H(\overline{k})$, and by descent theory an isogeny is necessarily a finite morphism. Thus, the following example implies that a k-homomorphism $G \to G'$ between smooth k-groups of finite type is an isogeny if and only if $G(\overline{k}) \to G'(\overline{k})$ is surjective with finite kernel (thereby recovering the classical notion of isogeny for smooth connected groups).

Example A.1.12 Let $f : G \twoheadrightarrow G'$ be a surjective homomorphism between smooth affine groups over a field k, and let $H = \ker f$. We claim that the induced map $G/H \to G'$ is always an isomorphism. In other words, f has the universal property of a quotient map modulo H regardless of the structure of H (e.g., this works even when H is not smooth). The reason is that f has all fibers equidimensional of dimension $\dim G' - \dim G$, so regularity of G and G' implies that f is flat [Mat, 23.1] and hence the quotient property for f follows from descent theory and the isomorphism $H \times G \simeq G \times_{G'} G$. We often use this quotient property for such f when H is non-smooth. (The same argument shows in general that a surjective homomorphism between arbitrary affine groups of finite type over k is a quotient map if and only if it is flat.)

The quotient G/H is smooth when G is smooth (as it inherits geometric reducedness from G, and then homogeneity via the G-action propagates the generic smoothness globally). If G is smooth and affine and H is smooth then there is another notion of (smooth separated) quotient G/H defined in [Bo2, II, §6], and these quotients coincide because they have the same universal mapping property in the category of smooth separated k-schemes of finite type. In general, G/H is affine when G is affine and H is normal in G: for smooth H and G this follows from the construction of G/H in [Bo2, II, §§5 and 6], and the general case is deduced from the smooth case in the proof of [SGA3, $\mathrm{VI_B}$, 11.17].

Proposition A.1.13 *Let k be a field, G a smooth connected k-group, and N a smooth normal closed k-subgroup of G. If the center Z_N is of multiplicative type then Z_N is central in G. In particular, if G is reductive then $Z_N \subseteq Z_G$.*

In Definition A.1.15 we recall the definition of reductivity without connectedness hypotheses. It is visibly inherited by smooth normal closed k-subgroups.

Proof We may and do assume $k = \overline{k}$. The normality of N implies that Z_N is carried into itself under $G(k)$-conjugations. Thus, by smoothness of G, it follows that Z_N is normal in G. The conjugation action of G on Z_N corresponds to a homomorphism of group functors $f : G \to \underline{\mathrm{Aut}}(Z_N)$. Since Z_N is of multiplicative type, $\underline{\mathrm{Aut}}(Z_N)$ is represented by an étale k-scheme. Hence, connectedness of G forces f to be trivial. This implies that Z_N is central in G.

It remains to show that if G is reductive (with $k = \overline{k}$) then Z_N is of multiplicative type (without connectedness hypotheses on N). More specifically, we will show that Z_N lies in a k-torus of G. Any maximal torus T in a connected reductive group is its own scheme-theoretic centralizer (as this centralizer is smooth by [Bo2, Cor. 9.2], and has the same $\overline{k}$-points as T by [Bo2, 13.17]), so if S is a maximal k-torus in N^0 then $Z_G(S) \cap N^0 = Z_{N^0}(S) = S$. Since $Z_N \subseteq Z_G(S)$, Z_N is a commutative extension of a finite étale group by a closed k-subgroup of S. By the established normality of Z_N in G we can replace G with $Z_G(S)/S$ and replace N with the finite étale image of Z_N in $Z_G(S)/S$ to reduce to the case that N is finite étale. In such cases the normality of N and connectedness of G force N to be central in G, so $N \subseteq Z_G(T) = T$ for any maximal k-torus T of G. □

Let k be a field with $\mathrm{char}(k) = p > 0$, and $F_k : k \to k$ the p-power map. For a k-group G of finite type, the *relative Frobenius morphism* $F_{G/k} : G \to G^{(p)} := G \otimes_{k, F_k} k$ is very useful. Loosely speaking, it is defined by the p-power map in local coordinates over k. The definition and a systematic discussion of this map, including a proof that it is an isogeny of degree $p^{\dim G}$ when G is smooth, are given in §A.3. (This map is never flat in the non-smooth case.) In general the k-map $F_{G/k}$ is a k-homomorphism and its formation commutes with any scalar extension on k. For smooth G we call $F_{G/k}$ the *Frobenius isogeny*; the case of non-smooth G does arise in our work, such as infinitesimal G. The map $F_{G/k}$ is made explicit for affine G in Example A.3.4 (with $n = 1$ there).

When $\mathrm{char}(k) = p > 0$, there is yet another kind of quotient for smooth affine k-groups G defined in [Bo2, V, §17], the concept of quotient $G/\mathfrak{h}$ for any Lie subalgebra $\mathfrak{h} \subseteq \mathrm{Lie}(G)$ that is stable under the adjoint action of G and the p-operation on $\mathrm{Lie}(G)$; see §A.7 for a systematic discussion of Lie algebras for group schemes and the p-operation on them in characteristic $p > 0$.

In Proposition A.7.14 we review the sense in which any such $\mathfrak{h}$ uniquely "exponentiates" to an infinitesimal normal closed k-subgroup $H \subseteq G$ such that the relative Frobenius morphism $F_{H/k}$ is trivial and $\mathrm{Lie}(H) = \mathfrak{h}$ inside of $\mathrm{Lie}(G)$. In Example A.7.16 we explain why the quotient $G/\mathfrak{h}$ coincides with the quotient G/H.

If G is a smooth affine group over a field k, and H and H' are smooth closed k-subgroups, at least one of which is connected, then (H, H') denotes the commutator k-subgroup [Bo2, I, §2.3]. This is the unique smooth closed k-subgroup of G such that for any algebraically closed extension K/k the subgroup $(H, H')(K)$ in $G(K)$ is generated by the commutators $hh'h^{-1}h'^{-1}$ for $h \in H(K)$ and $h' \in H'(K)$. Beware that if H and H' are disconnected then no such k-subgroup may exist (e.g., $G = \mathrm{SL}_2$ over $\mathbf{Q}$ and suitable finite H and H' in $G(\mathbf{Q})$). Note that (H, H') is normalized by both H and H', due to commutator identities such as $h'_2(h, h'_1)h'^{-1}_2 = (h, h'_2)^{-1}(h, h'_2h'_1)$.

Definition A.1.14 The *derived group* $\mathscr{D}(G)$ of a smooth group G of finite type over a field k is the unique smooth closed k-subgroup such that $(\mathscr{D}(G))(K)$ is the commutator subgroup of $G(K)$ for any algebraically closed extension K/k.

Note that the derived group exists without connectedness hypotheses on G; see [Bo2, I, 2.4] for the affine case (which is all we need) and [SGA3, $\mathrm{VI_B}$, 7.2(vii), 7.10] for the general case. The formation of $\mathscr{D}(G)$ commutes with any extension of the base field, and the quotient map $G \to G/\mathscr{D}(G)$ is initial among all k-homomorphisms from G to a commutative k-group scheme (see Lemma 5.3.4 for a generalization). We never consider derived groups for non-smooth groups. A smooth k-group G of finite type is called *perfect* if $G = \mathscr{D}(G)$.

Definition A.1.15 For a smooth affine group G over a field k, the *geometric unipotent radical* (resp. *geometric radical*) of G is the maximal smooth connected normal closed $\overline{k}$-subgroup of $G_{\overline{k}}$ that is unipotent (resp. solvable). It is denoted $\mathscr{R}_u(G_{\overline{k}})$ (resp. $\mathscr{R}(G_{\overline{k}})$). We say G is *reductive* (resp. *semisimple*) if its geometric unipotent radical (resp. geometric radical) is trivial.

The *k-unipotent radical* $\mathscr{R}_{u,k}(G)$ (resp. *k-radical* $\mathscr{R}_k(G)$) is the maximal smooth connected unipotent (resp. solvable) normal k-subgroup of G.

A smooth affine k-group G is reductive (resp. semisimple) if and only if the *identity component* G^0 is reductive (resp. semisimple), so if G is reductive then the scheme-theoretic centralizer in G of a semisimple element of $\mathrm{Lie}(G)$ is reductive [Bo2, 13.19]. Such centralizers can be disconnected even when G is connected, but we only use their identity component (e.g., Proposition A.8.10(2) and its proof). The reader may safely follow the convention in [SGA3, XIX, 2.7] according to which reductive k-groups are connected.

Definition A.1.16 Let G be a smooth connected affine group over a field k. The *maximal geometric reductive* (resp. *semisimple*) quotient of $G_{\overline{k}}$ is the quotient $G_{\overline{k}}^{\text{red}} := G_{\overline{k}}/\mathscr{R}_u(G_{\overline{k}})$ (resp. $G_{\overline{k}}^{\text{ss}} := G_{\overline{k}}/\mathscr{R}(G_{\overline{k}})$) of $G_{\overline{k}}$; it uniquely dominates all other reductive (resp. semisimple) quotients of $G_{\overline{k}}$.

If K/k is the (minimal) field of definition (in the sense of Definition 1.1.6) for the closed subscheme $\mathscr{R}_u(G_{\overline{k}}) \subseteq G_{\overline{k}}$ then $G_K^{\text{red}} := G_K/\mathscr{R}_{u,K}(G_K)$ and $G_K^{\text{ss}} := G_K/\mathscr{R}_K(G_K)$. (Note that $\mathscr{R}_K(G_K)$ is a K-descent of $\mathscr{R}(G_{\overline{k}}) \subseteq G_{\overline{k}}$ since the schematic preimage in G_K of the maximal central K-torus in the reductive G_K^{red} is a K-descent of $\mathscr{R}(G_{\overline{k}})$.)

Since we are most interested in the case $\text{char}(k) \neq 0$, the possible existence of non-central finite normal subgroup schemes in G (e.g., the kernel of the Frobenius isogeny of G) forces us to distinguish between general isogenies and central ones. The quotient $G_{\overline{k}}^{\text{ss}}/Z_{G_{\overline{k}}^{\text{ss}}}$ of $G_{\overline{k}}^{\text{ss}}$ modulo its finite multiplicative scheme-theoretic center is called the *maximal (geometric) adjoint semisimple quotient* of $G_{\overline{k}}$.

A simple semisimple (resp. simple adjoint semisimple) quotient of G is called *maximal* if it is not dominated by any others except through isomorphisms. Every simple semisimple (resp. simple adjoint semisimple) quotient of $G_{\overline{k}}$ is uniquely dominated by a maximal one.

These notions of maximal quotient are sometimes considered as quotients of G rather than of $G_{\overline{k}}$, but only when the $\overline{k}$-subgroup $\mathscr{R}_u(G_{\overline{k}}) \subseteq G_{\overline{k}}$ is defined over k.

Remark A.1.17 Note that the maximal adjoint semisimple quotient of a connected reductive group G is G/Z_G, and up to unique isomorphism this is the only adjoint semisimple quotient of the form G/Z with $Z \subseteq G$ a central subgroup scheme. (For example, over any field of nonzero characteristic the central isogeny $\text{SL}_n \to \text{PGL}_n$ for $n > 1$ is the maximal adjoint semisimple quotient of SL_n, and its composite with the Frobenius isogeny of PGL_n is an adjoint semisimple quotient of SL_n modulo a non-central finite subgroup.)

Definition A.1.18 A *Levi k-subgroup* of a smooth connected affine group G over a field k is a smooth connected k-subgroup $L \subseteq G$ such that the natural map $L_{\overline{k}} \to G_{\overline{k}}/\mathscr{R}_u(G_{\overline{k}})$ is an isomorphism (equivalently, $\mathscr{R}_u(G_{\overline{k}}) \rtimes L_{\overline{k}} \to G_{\overline{k}}$ is an isomorphism). We use a similar definition when k is a nonzero finite product of fields and all fibers of $G \to \text{Spec}\, k$ are connected (treating the fibers over $\text{Spec}\, k$ separately).

We do not consider the notion of Levi subgroup in an affine k-group in the absence of the smoothness and connectedness requirements. This agrees with Borel's convention in [Bo2, 11.22]. Levi subgroups need not exist in nonzero characteristic, even over an algebraically closed field; see §A.6.

A.2 Some results from the general theory

If a homomorphism between smooth affine groups over a field has trivial kernel in the scheme-theoretic sense then it is injective on geometric points and on Lie algebras, so it is a closed immersion. The smoothness of the groups is not necessary:

Proposition A.2.1 *Let k be an artin local ring, G a k-group scheme of finite type, and G' a k-group scheme locally of finite type. If $f : G \to G'$ is a k-homomorphism and* $\ker f = 1$ *then f is a closed immersion.*

Proof This is [SGA3, $\mathrm{VI_B}$, 1.4.2]. □

Example A.2.2 Let G be a group scheme of finite type over an artin local ring k such that G admits a faithful linear representation in the sense that there is a subgroup functor inclusion $G \to \mathrm{GL}_n$. This is automatically a closed immersion, by Proposition A.2.1 (so G is affine).

For a field k and an affine k-scheme V equipped with a left action by a k-group scheme G, a finite-dimensional k-subspace $W \subseteq k[V]$ is *G-stable* if, for any k-algebra R, the R-submodule $R \otimes_k W \subseteq R[V]$ is $G(R)$-stable. In particular, we get a map of group functors $G \to \mathrm{GL}(W)$, so by Yoneda's Lemma it is a k-group scheme homomorphism (i.e., the G-action on W is "algebraic").

Proposition A.2.3 *Let k be a field and G an affine k-group scheme of finite type. Consider an affine k-scheme V endowed with a left action by G over k. The coordinate ring $k[V]$ is the directed union of G-stable finite-dimensional k-linear subspaces. In particular, taking $V = G$ with its natural left action, G is a closed k-subgroup of* GL_n *for some $n \geqslant 1$.*

Proof The existence of the directed union is proved in [SGA3, $\mathrm{VI_B}$, Lemme 11.8]. Next, we construct a closed k-subgroup inclusion of G into some GL_n. By taking $V = G$ with G-action by left translation, there is a G-stable k-linear subspace $W \subseteq k[G]$ of finite dimension that contains a finite generating set of $k[G]$ as a k-algebra. We claim that for any k-algebra R, the $G(R)$-action on $R \otimes_k W$ is faithful. Since this action is induced by the $G(R)$-action on the R-algebra $R[G]$, and $R \otimes_k W$ generates $R[G]$ as an R-algebra, it suffices to show that $G(R)$ acts faithfully on $R[G]$. But G is affine, so any $g \in G(R)$ is determined by the evaluations $f(g) = (g^{-1}.f)(1)$ for all $f \in R[G]$. This implies faithfulness of the action, so we get a monomorphism of group functors $G \hookrightarrow \mathrm{GL}(W)$. By Example A.2.2, this is a closed immersion. □

Proposition A.2.4 *Let k be a field and G an affine k-group scheme of finite type. Let H be a closed k-subgroup scheme. There is a nonzero* *finite-dimensional k-vector space V, a closed k-subgroup inclusion $\rho : G \hookrightarrow$* GL(V)*, and a line L in V such that H is the scheme-theoretic G-stabilizer of L viewed as a point in the projective space of lines in V.*

Proof This is proved in [Bo2, 5.1] when H and G are smooth, using computations with field-valued points. The same method of proof (ignoring the considerations with Lie algebras and convolutions there) works without smoothness by using points valued in k-algebras. Another reference, with a scheme-theoretic viewpoint, is [DG, II, §2, 3.5]. □

There are two very useful results concerning torus centralizers that are proved in [Bo2] and which we state here for ease of reference.

Proposition A.2.5 *Let G be a smooth connected affine group over a field k, and S a k-torus in G. The scheme-theoretic centralizer $Z_G(S)$ is smooth and connected, with Lie algebra* $\mathrm{Lie}(Z_G(S))$ *equal to the S-fixed part of* $\mathrm{Lie}(G)$.

Proof The smoothness and determination of the Lie algebra are [Bo2, Cor. 9.2], and the connectedness is [Bo2, 11.12]. □

Given a smooth affine group G defined over a field k, $G_{\overline{k}}$ contains a maximal $\overline{k}$-torus that is defined over k [Bo2, 18.2(i)]. Thus, if G is smooth, connected, and affine and does not contain a nontrivial k-torus then it is unipotent [Bo2, 11.5(2)].

Corollary A.2.6 *Let G be a smooth affine group over a field k, T a k-torus in G, and K/k an extension field. The k-torus T is maximal as such in G if and only if T_K is maximal as a K-torus in G_K. Likewise, if Z is the maximal central k-torus in G then Z_K is the maximal central K-torus in G_K.*

For the centrality, note that we do not assume G is connected.

Proof To prove that T_K is maximal in G_K when T is maximal in G, by Proposition A.2.5 we can replace G with $Z_G(T)$ so that T is central in G. Passing to G/T lets us assume $T = 1$, so G has no nontrivial k-tori. Thus, G^0 is unipotent, so G_K has no nontrivial K-tori.

Now consider the assertion concerning maximal central tori. The case $K = k_s$ is a consequence of Galois descent, so we may assume $k = k_s$ and $K = K_s$. We have to show that if a central K-torus S in G_K contains Z_K then $S = Z_K$. Let T be a maximal k-torus of G containing Z, so $T_K \cdot S$ is a K-torus in G_K yet T_K is maximal in G_K. Thus, S is an intermediate torus between T_K and Z_K that is central. But T is k-split, as $k = k_s$, so every K-subtorus of T_K descends to a k-subtorus of T. Thus, the k-descent of S is a central k-torus in G containing Z, so it equals Z by maximality of Z. That is, $S = Z_K$. □

Corollary A.2.7 *Let G be a smooth affine group over a field k, and T a maximal k-torus in G. For any smooth connected k-subgroup N in G that is normalized by T, the scheme-theoretic intersection $T \cap N$ is a maximal k-torus in N; in particular, it is smooth and connected.*

Note that this corollary is applicable whenever N is normal in G.

Proof By Corollary A.2.6, we can assume $k = \overline{k}$ (so all maximal k-tori in G are $G(k)$-conjugate). The affine k-group $H = N \rtimes T$ is smooth and connected with $Z_H(T) = (N \cap Z_G(T)) \rtimes T$, so $N \cap Z_G(T)$ is smooth and connected. Hence, to prove that $T \cap N$ is a maximal k-torus in N we can replace G with $Z_G(T)$ and replace N with $N \cap Z_G(T)$ so that T is central in G. Thus, T contains all k-tori in G, so $S := (T \cap N)^0_{\mathrm{red}}$ is the unique maximal k-torus in N. The quotient N/S is therefore unipotent, so every k-subgroup scheme of N of multiplicative type must be contained in S. But $T \cap N$ is such a k-subgroup, so the containment $S \subseteq T \cap N$ is an equality. □

Proposition A.2.8 *Let $f : G \twoheadrightarrow G'$ be a surjective homomorphism between smooth connected affine groups over a field k. Let S be a k-torus in G, and $S' = f(S)$ the image k-torus in G'. If S is maximal in G then S' is maximal in G', and in general $f(Z_G(S)) = Z_{G'}(S')$.*

Proof That S' is a maximal torus of G' is [Bo2, Prop. 11.14], and that $f(Z_G(S)) = Z_{G'}(S')$ is [Bo2, Cor. 2, 11.14]. □

Next, we wish to record a special case of a fundamental result of Grothendieck [SGA3, XII, 7.1(b)] concerning conjugacy of maximal tori for smooth group schemes over any base scheme. We focus on the case of a separably closed base field. (In Theorem C.2.3 we prove a finer result over an arbitrary field.) First we establish a lemma.

Lemma A.2.9 *Let G be a smooth affine group over a field k, and S a k-torus in G. Then $W(G,S) := N_G(S)/Z_G(S)$ is finite étale and $N_G(S)$ is smooth.*

Proof By Proposition A.2.5 (applied to G^0), $Z_G(S)$ is smooth. Thus, smoothness of $N_G(S)$ will follow once we prove that $W(G,S)$ is étale. Since $W(G,S)$ is finite type, we just have to prove that it is étale. The automorphism functor $\underline{\mathrm{Aut}}(S)$ of the k-torus S is an étale k-group (locally of finite type), and the conjugation action of $N_G(S)$ on S defines a k-group homomorphism $W(G,S) \to \underline{\mathrm{Aut}}(S)$ with trivial kernel. Thus, by Proposition A.2.1, $W(G,S)$ is a closed k-subgroup of $\underline{\mathrm{Aut}}(S)$. A closed subscheme of an étale scheme over a field is étale, so $W(G,S)$ is étale. □

Proposition A.2.10 (Grothendieck) *Let G be a smooth affine group over a separably closed field k. Any two maximal k-tori in G are $G(k)$-conjugate.*

Proof We may and do assume that G is connected. Let T and T' be two maximal k-tori in G. Consider the functor X on k-algebras defined as follows: $X(A)$ is the group of elements $g \in G(A)$ such that $gT_Ag^{-1} = T'_A$ in G_A. We claim that X is represented by a smooth non-empty closed subscheme of G, so $X(k)$ is non-empty since k is separably closed.

To see that X is represented by a closed subscheme of G, for each $n \geqslant 1$ that is not divisible by char(k) let X_n denote the subfunctor of G with $X_n(A)$ consisting of those $g \in G(A)$ such that $gT[n]_Ag^{-1} = T'[n]_A$ inside of G_A. Since $T[n]$ and $T'[n]$ are finite étale subschemes of G, they are constant and hence each X_n is represented by a closed subscheme of G. By Zariski-density of the collection of k-subgroups $S[n]$ for a k-torus S as n varies through all positive integers not divisible by char(k), we have $X = \bigcap_n X_n$ as subfunctors of G, so X is represented by a closed subscheme.

To prove that X is smooth and non-empty we may assume that k is algebraically closed. Then X is a translate of the normalizer $N_G(T)$ if X is non-empty, and $N_G(T)$ is smooth by Lemma A.2.9. Hence, it suffices to prove conjugacy when k is algebraically closed. This case is well known [Bo2, Cor. 11.3]. □

Proposition A.2.11 *Let G be a smooth connected affine group over a field k, and let G_t be the smooth connected k-subgroup generated by the k-tori of G. The k-group G_t is unirational and normal in G, with G/G_t unipotent (so $G = G_t$ if G is perfect or reductive).*

The formation of G_t commutes with any extension of the ground field. In particular, $(G_t)_{\overline{k}}$ contains all tori of $G_{\overline{k}}$, and $G(k)$ is Zariski-dense in G if k is infinite and G is perfect.

Proof Since G_t is generated by k-tori, by [Bo2, 2.2] there is a finite set of k-tori $T_1, \dots, T_n$ in G such that the multiplication map $T_1 \times \cdots \times T_n \to G_t$ is surjective. Hence, the unirationality of G_t follows from the unirationality of k-tori [Bo2, 8.13(2)].

Grant for a moment that the formation of G_t commutes with extension of the ground field, so its normality in G follows from working over $\overline{k}$. The smooth connected quotient G/G_t then makes sense as a k-group, and it has no nontrivial k-tori (due to the definition of G_t and Proposition A.2.8 applied to the smooth connected preimage in G of a maximal k-torus of G/G_t), so it is unipotent and hence solvable. A smooth connected solvable k-group is trivial if it is perfect, so $G = G_t$ when G is perfect.

It remains to prove that the formation of G_t commutes with extension of the ground field. Let Tor_G be the "variety of maximal tori" of G. Concretely this is $G/N_G(T_0)$ for a maximal k-torus T_0 in G, and there is a fiberwise maximal closed Tor_G-torus $\mathscr{T} \subseteq G \times \mathrm{Tor}_G$ with the property that for any field L/k the fibers $\mathscr{T}_x \subseteq G_L$ for $x \in \mathrm{Tor}_G(L)$ vary without repetition through precisely the set of maximal L-tori in G_L. To construct $\mathscr{T}$, we form the twisted embedding $T_0 \times (G/Z_G(T_0)) \to G \times (G/Z_G(T_0))$ defined via $(t, g) \mapsto (gtg^{-1}, g)$ and pass to the quotient by the free right action of the finite étale k-group $N_G(T_0)/Z_G(T_0)$ by right translation on $G/Z_G(T_0)$ and by conjugation on T_0. (The fibers over L-valued points for any L/k are as expected, since Proposition A.2.10 applies over L_s and then Galois descent does the rest.) The pair $(\mathrm{Tor}_G, \mathscr{T})$ satisfies a stronger universal property among maximal tori in G parameterized by arbitrary k-schemes [SGA3, XII, Cor. 1.10], but we do not need this fact.

What matters for our purposes is the important theorem of Chevalley–Grothendieck that Tor_G is a unirational (even rational) variety over k [SGA3, XIV, Thm. 6.1] (or see [BoSp, 7.1(i), 7.3, 7.9] for a more elementary proof). In particular, $\mathrm{Tor}_G(k)$ is Zariski-dense in Tor_G when k is infinite. Consider the k-map $f : \mathscr{T} \to G$ defined by restriction to $\mathscr{T}$ of the projection map $G \times \mathrm{Tor}_G \to G$. We shall prove that f lands in G_t when k is infinite, so the formation of G_t commutes with any extension of the ground field when k is infinite.

Consider the closed subset $Z = f^{-1}(G_t)$ in $\mathscr{T}$. To prove $Z = \mathscr{T}$ when k is infinite, it suffices to prove that the open subset $V := \mathscr{T} - Z$ in $\mathscr{T}$ is empty. Assume $V \neq \emptyset$, so V contains the unique generic point of $\mathscr{T}$, which the map $\pi : \mathscr{T} \to \mathrm{Tor}_G$ sends to the generic point of Tor_G. Thus, the constructible set $\pi(V)$ in Tor_G contains the generic point and so contains a dense open subset. But that must meet $\mathrm{Tor}_G(k)$ since k is infinite, which is a contradiction since V has empty fibers over $\mathrm{Tor}_G(k)$ (by definition of Z).

Now we address the case where k is finite. As a convenient notation, for any smooth connected k-group G we will denote by G_τ the Galois descent of the normal k_s-subgroup $(G_{k_s})_t$ in G_{k_s}, so G_τ is a smooth connected normal k-subgroup of G that contains all k-tori of G. We have to prove that G_τ is generated by the k-tori of G (i.e., $G_\tau = G_t$). We will treat the solvable case by an argument that works over any perfect field with nonzero characteristic, and then deduce the general case from the solvable case (using that any connected semisimple k-group contains a Borel k-subgroup, since k is finite).

So we now prove that if G is a smooth connected solvable group over a perfect field k of characteristic $p > 0$ then $G_\tau = G_t$. Writing $G = T \ltimes U$ for a maximal k-torus T in G and $U := \mathscr{R}_{u,k}(G)$, we first consider the case where U is commutative and p-torsion. Let U' be the maximal smooth connected T-stable k-subgroup of U such that U'^T is trivial; see Theorem B.4.3 for the existence of U'. We will use the following lemma.

Lemma A.2.12 *The $U'(k)$-conjugates of T generate $T \ltimes U'$.*

Proof We may replace U with U', and may assume $U' \neq 1$. By Theorem B.4.3 there is a T-equivariant isomorphism of U' onto the vector group

$\underline{V}$ associated to a finite-dimensional k-linear representation V of T with $V^T = 0$. It is well known that all k-linear representations of a k-torus are completely reducible over k. (A deduction from the case of split k-tori in Lemma A.8.8 is that if K/k is an extension of fields and $r : K \to k$ is a k-linear retraction then for any exact sequence $0 \to W' \to W \to W'' \to 0$ of k-linear representations of a k-group scheme G and any G_K-splitting $W''_K \to W_K$ of the associated K-linear sequence, a G-splitting of the original sequence is provided by $W'' \to W''_K \to W_K \to W$, using r at the final step.) By expressing V as a direct sum of T-irreducibles we may reduce to the case where V is irreducible over k.

Now choose $x \in V - \{0\} = U'(k) - \{1\}$, so x does not commute with the T-action (as $V^T = 0$) and hence $xTx^{-1} \neq T$. Thus, T and xTx^{-1} generate a smooth connected k-subgroup H in $T \ltimes U'$ that strictly contains T, so $H = T \ltimes (H \cap U')$ with $H \cap U'$ a nontrivial smooth connected k-group (as H is smooth and connected). Then $\mathrm{Lie}(H \cap U') \subseteq \mathrm{Lie}(U') = V$ is a nonzero T-stable subspace, so the irreducibility of the T-action on V implies $H \cap U' = U'$. □

By Lemma A.2.12 we have $(T \ltimes U')_t = T \ltimes U'$. Thus, to settle the case where U is commutative and p-torsion it suffices to prove that the inclusion $T \ltimes U' \subseteq (T \ltimes U)_\tau$ is an equality. This follows from the normality of $T \ltimes U'$ in $T \ltimes U$ and the unipotence of $(T \ltimes U)/(T \ltimes U')$.

Consider now the general solvable case. To prove that the containment $G_t \subseteq G_\tau$ is an equality we induct on $\dim G$ (the case $\dim G = 0$ is clear) and may replace G with G_τ so that $G_\tau = G$. Let $U = \mathscr{R}_{u,k}(G)$, which we may assume is nontrivial. Let $C(U)$ be the last nontrivial term in the descending central series for U (so $C(U)$ is central in U and normal in G), and let $n \geqslant 0$ be maximal such that $V := p^n C(U) \neq 1$. Hence, V is normal in G, central in U, and p-torsion. Clearly $(G/V)_\tau = G/V$ since $G_\tau = G$, so by induction on dimension it follows that $G/V = (G/V)_t = G_t V/V$. Thus $G_t V = G$. From this we see that if V is a central subgroup of G then G_t is a normal subgroup, and as G/G_t is a unipotent group, it follows that $G_\tau = G_t$.

Let us now assume that V is not a central subgroup of G, so (by centrality of V in U) the natural action of the torus $T := G/U$ on V is nontrivial. Let V' be the maximal smooth connected k-subgroup of V such that V'^T is trivial (cf. Theorem B.4.3), so V' is nontrivial and normal in G. Let $\pi : G \to G/V'$ be the quotient map. By induction on dimension, $(G/V')_t = G/V'$, so to prove that $G_t = G$ it suffices to show that for any maximal k-torus S in G/V', $\pi^{-1}(S)_t = \pi^{-1}(S)$. Since V' is commutative and p-torsion and V'^T is trivial, we can apply Lemma A.2.12 to conclude that $\pi^{-1}(S)_t = \pi^{-1}(S)$, and thus $G_t = G$. This completes our treatment of the case of solvable G.

Turning to the general case, consider an arbitrary k-split torus S in G, and choose a cocharacter $\lambda \in \mathrm{X}_*(S)$ such that $Z_G(\lambda) = Z_G(S)$. (The construction of such a λ is explained in a self-contained manner near the end of the proof of

Proposition 3.3.10, for part (b).) The multiplication map

$$U_G(\lambda) \times Z_G(S) \times U_G(-\lambda) \to G$$

is an open immersion (Proposition 2.1.8(3)), so G is generated by the k-subgroups $U_G(\pm\lambda)$ and the k-group $Z_G(S) = Z_G(\lambda)$ that normalizes each $U_G(\pm\lambda)$. Hence, the smooth connected k-subgroup H_S of G generated by S, $U_G(\lambda)$, and $U_G(-\lambda)$ is normal in G.

The k-subgroup of G generated by S and the image of the commutator map

$$c_\pm : S \times U_G(\pm\lambda) \to U_G(\pm\lambda)$$

is clearly inside $(S \ltimes U_G(\pm\lambda))_\tau$. Thus, provided that $c_\pm$ has dense image, $(S \ltimes U_G(\pm\lambda))_\tau = S \ltimes U_G(\pm\lambda)$ and hence $S \ltimes U_G(\pm\lambda)$ is generated by k-tori (by the solvable case). To prove that the image of $c_\pm$ is dense, it suffices to show that $\mathrm{Lie}(U_G(\pm\lambda))_{k_s}$ is spanned by the images of the derivative of $c_\pm$ at k_s-points in $c_\pm^{-1}(1)$. The image of this derivative at $(\lambda(t), 1)$ is the image of the endomorphism $v \mapsto \mathrm{Ad}_G(\lambda(t))(v) - v$ of the underlying k_s-vector space of $\mathrm{Lie}(U_G(\pm\lambda))_{k_s} = (\mathfrak{g}_{k_s})_{\pm\lambda > 0}$ (where $\mathfrak{g} = \mathrm{Lie}(G)$), so the desired spanning property holds.

We conclude that H_S is generated by k-tori. The k-subgroups H_S (with S varying) generate a smooth connected normal k-subgroup H of G that contains all k-split tori and is contained in G_t. The quotient G/H contains no nontrivial k-split k-tori (apply Proposition A.2.8 to the pullback of $G \to G/H$ along the inclusion into G/H of a maximal k-split k-torus), and hence likewise the same holds for its maximal semisimple quotient $(G/H)^{\mathrm{ss}}$. But $(G/H)^{\mathrm{ss}}$ contains a Borel k-subgroup (as k is finite) and hence contains a nontrivial split k-torus if $(G/H)^{\mathrm{ss}} \neq 1$. This forces $(G/H)^{\mathrm{ss}} = 1$, so G/H is solvable. By the settled solvable case, $(G/H)_\tau = (G/H)_t$. But $H \subseteq G_t$, so $(G/H)_t = G_t/H$ and $(G/H)_\tau = G_\tau/H$. Hence, $G_\tau = G_t$ as desired. □

We now turn to an arithmetic result concerning adelic points of linear algebraic groups. For any topological ring R and affine R-scheme X of finite type, the set $X(R)$ acquires a natural topological structure functorially in X by using any closed immersion of X into an affine n-space over R (and equipping $X(R)$ with the resulting subspace topology from R^n); the choice of closed immersion does not matter. In particular, if X is an affine R-group of finite type then $X(R)$ is naturally a topological group, and it is locally compact and Hausdorff when R is.

The cases of most interest to us are when R is a topological field (such as a local field), the adele ring $\mathbf{A}_k$ attached to a global field k, or the factor ring $\mathbf{A}_k^S = \mathbf{A}_k/(\prod_{v \in S} k_v)$ for a finite set S of places of k containing the archimedean places. In the adelic cases we generally consider X arising by scalar extension from an affine k-scheme of finite type (and then $X(\mathbf{A}_k) = X(\mathbf{A}_k^S) \times \prod_{v \in S} X(k_v)$ as topological spaces, and also as topological groups when X is a k-group).

Proposition A.2.13 *Let $f : H' \to H$ be a homomorphism between affine groups of finite type over a global field k. Assume f is surjective on k-points and also on k_v-points for all places v of k.*

Assume that there is a finitely generated subgroup Γ of $H(k)$ such that for some finite set S of places of k containing the archimedean places, the closure K of Γ in $H(\mathbf{A}_k^S)$ is open. Then $f : H'(\mathbf{A}_k) \to H(\mathbf{A}_k)$ is surjective.

Proof The subgroup $f(H'(\mathbf{A}_k)) \subseteq H(\mathbf{A}_k)$ is dense since f is surjective on k_v-points for all places v of k, so it is enough to show that $f(H'(\mathbf{A}_k))$ is open in $H(\mathbf{A}_k)$.

Since f is surjective on k-points, we can find a finitely generated subgroup $\Gamma' \subseteq H'(k)$ carried onto Γ under f. Pick a finite Σ containing S so that H' extends to an affine finite type $\mathscr{O}_{k,\Sigma}$-group H'_Σ. Any finite subset of $H'(k) = H'_\Sigma(k)$ lies in $H'_\Sigma(\mathscr{O}_{k,S'})$ for some finite S' containing Σ. Thus, we can find a finite S' so that the compact open subgroup $K' := H'_\Sigma(\prod_{v \notin S'} \mathscr{O}_v)$ in $H'(\mathbf{A}_k^{S'})$ contains the finitely generated Γ'. The image $f(K') \subseteq H(\mathbf{A}_k^{S'})$ is a compact subset containing $f(\Gamma') = \Gamma$. Thus, $f(K')$ contains the closure of Γ in $H(\mathbf{A}_k^{S'})$, which is the projection of the open subgroup K of $H(\mathbf{A}_k^S)$. Hence, $f(K')$ is open in $H(\mathbf{A}_k^{S'})$. Since $H(\mathbf{A}_k) = H(\mathbf{A}_k^{S'}) \times \prod_{v \in S'} H(k_v)$ and $H(k_v) = f(H'(k_v))$ for all v, we see that $f(H'(\mathbf{A}_k))$ is open in $H(\mathbf{A}_k)$. □

Example A.2.14 To illustrate Proposition A.2.13, consider a connected semisimple k-group H that is absolutely simple and simply connected (and we will be most interested in smooth H' that is not reductive). Let $\Gamma = H(k) \cap K$ for a compact open subgroup K of $H(\mathbf{A}_k^S)$ for a suitable finite set S of places of k containing the archimedean places. The hypothesis of Proposition A.2.13 is often satisfied: Γ is an "S-arithmetic subgroup" of $H(k)$, and if k is a number field then S-arithmetic subgroups of a reductive group are known to be finitely generated (in fact, finitely presented); see [PR, Thm. 5.11]. Finite generation (resp. finite presentation) of S-arithmetic subgroups of $G(k)$ for an absolutely simple semisimple group G over a global function field k holds provided that $\sum_{v \in S} k_v\text{-rank}\, G \geqslant 2$ (resp. $\sum_{v \in S} k_v\text{-rank}\, G > 2$); see [Be]. Moreover, the density of the S-arithmetic subgroup Γ in K is a consequence of the strong approximation property of H provided that $\sum_{v \in S} k_v\text{-rank}\, H > 0$.

For our work with exceptional constructions in characteristic 2, it is useful to record a class of smooth and possibly non-reductive affine k-groups H for which the hypothesis in Proposition A.2.13 is satisfied: any Weil restriction $H = \mathrm{R}_{k'/k}(G')$ for a finite extension field k'/k (possibly not separable) and connected semisimple k'-group G' that is absolutely simple and simply connected. (See §A.5 for a general discussion of Weil restriction.) Indeed, if S' is the set of places of k' above a choice of S for k then $\mathbf{A}_{k'}^{S'} = k' \otimes_k \mathbf{A}_k^S$ and the topology on $\mathbf{A}_{k'}^{S'}$ coincides with the one arising from its structure as a finite free $\mathbf{A}_k^S$-module, so the identification $H(\mathbf{A}_k^S) = G'(\mathbf{A}_{k'}^{S'})$ is topological. Since $H(k) = G'(k')$ as well, the subgroup $\Gamma = G'(k') \cap K'$, where K' is a compact open subgroup of $G'(\mathbf{A}_{k'}^{S'})$, is finitely generated and its closure in $H(\mathbf{A}_k^S)$ is the compact open subgroup K corresponding to the subgroup $K' \subseteq G'(\mathbf{A}_{k'}^{S'})$.

Remark A.2.15 When $H = \mathrm{R}_{k'/k}(G')$ for a k'-split simply connected semisimple group G' over a finite extension k'/k, it is not necessary to use the finite generation of S-arithmetic subgroups or the strong approximation property (over finite extensions of k) to verify the hypothesis in Proposition A.2.13. We now give an elementary construction of a finitely generated subgroup Γ_S of $H(k)$ whose closure in $H(\mathbf{A}_k^S)$ is open whenever $\#S > 1$. Since $H(k) = G'(k')$ and $H(\mathbf{A}_k^S) = G'(\mathbf{A}_{k'}^{S'})$ topologically for the finite set S' of places of k' over S, by replacing (H, k, S) with (G', k', S') we may and do assume that H is a k-split connected semisimple k-group that is simply connected.

Choose a maximal k-split torus $T \subseteq H$ and a positive system of roots Φ^+ in $\Phi(H, T)$ to get an open cell in H. The product structure of the open cell (with root groups and coroot groups in $T = \mathbf{G}_m^{\Delta^\vee}$ for the base Δ of Φ^+) reduces the problem to the case of SL_2 because $\mathrm{SL}_2(\mathscr{O}_v)$ $(v \notin S)$ is generated as a group by the $\mathscr{O}_v$-points of the standard open cell. Since $\mathscr{O}_{k,S}$ has finite class group and $\#S > 1$, separate arguments in the number field and function field cases imply that it is finitely generated as a module over the group ring $\mathbf{Z}[\mathscr{O}_{k,S}^\times]$ when using the $\mathscr{O}_{k,S}^\times$-action via ordinary multiplication and hence likewise for the action $u.x = u^{\pm 2}x$ since $\mathscr{O}_{k,S}^\times$ is finitely generated. Thus, the subgroup $\Gamma_S \subseteq \mathrm{SL}_2(\mathscr{O}_{k,S})$ generated by all upper triangular and lower triangular matrices is finitely generated. Each of the two standard "root groups" $\mathscr{O}_{k,S}$ has closure $\mathscr{O}_{k,S}^\wedge := \prod_{v \notin S} \mathscr{O}_v$ in $\mathbf{A}_k^S$. Combining this with the formula

$$\mathrm{diag}(t, 1/t) = u^+(t)u^-(-1/t)u^+(t-1)u^-(1)u^+(-1)$$

for $u^+(x) = \left(\begin{smallmatrix} 1 & x \\ 0 & 1 \end{smallmatrix}\right)$ and $u^-(x) = \left(\begin{smallmatrix} 1 & 0 \\ x & 1 \end{smallmatrix}\right)$, we see that Γ_S has open closure $\mathrm{SL}_2(\mathscr{O}_{k,S}^\wedge)$ inside $\mathrm{SL}_2(\mathbf{A}_k^S)$.

A.3 Frobenius morphisms and non-affine groups

We now turn to a pair of more technical topics. A useful but somewhat specialized construction in the study of algebraic groups in nonzero characteristic is the notion of a relative Frobenius morphism. We shall generally need this in the smooth case (in which case it is an isogeny), but it will also be used for groups that may be non-smooth of any dimension. Hence, as a convenient reference for the reader, we discuss Frobenius morphisms in general for schemes and address special properties in the case of group schemes. We also record some useful general structural results for (possibly non-affine) group schemes of finite type over a field, as is needed to justify some later examples.

For a fixed prime p, let S be an $\mathbf{F}_p$-scheme. The *absolute Frobenius morphism* $F_S : S \to S$ is given on affine open subsets by the p-power map on the coordinate ring. Equivalently, F_S is the identity on topological spaces and the p-power map on $\mathscr{O}_S$. For any S-scheme X and $n \geqslant 1$, the map $F_X^n : X \to X$ is generally not an S-map (aside from special cases, such as when S is the spectrum of a finite extension of $\mathbf{F}_p$ of degree dividing n). To promote this to an S-map, we are led to consider the S-scheme

$$X^{(p^n)} := S \times_{F_S^n, S} X$$

for any $n \geqslant 0$. Loosely speaking, $X^{(p^n)}$ is given by the same equations over S that define X except that the coefficients in the equations are replaced with their p^n-powers. The case of most interest to us will be when $S = \operatorname{Spec} k$ for a field k of characteristic p and X is an affine k-group of finite type (generally smooth, but not always).

The formation of F_S is functorial in S since all maps of $\mathbf{F}_p$-algebras are functorial with respect to the p-power map. This functoriality has some useful consequences. First, the formation of $X^{(p^n)}$ is compatible with base change on S in the sense that for any map of schemes $S' \to S$ and the associated S'-scheme $X' = X \times_S S'$, the natural S'-map

$$X'^{(p^n)} \to X^{(p^n)} \times_S S' \tag{A.3.1}$$

is an isomorphism. (This justifies suppressing the mention of S in the notation $X^{(p^n)}$.) Second, the compatibility of F_X^n and F_S^n yields the S-map

$$F_{X/S,n} : X \to X^{(p^n)} = S \times_{F_S^n, S} X \tag{A.3.2}$$

whose composite with the projection $\mathrm{pr}_2 : X^{(p^n)} \to X$ is F_X^n. Loosely speaking, the S-map $F_{X/S,n}$ is given by the p^n-power map on local coordinates over S. Via (A.3.1), it is straightforward to check that the formation of $F_{X/S,n}$ is compatible with base change on S. For example, if $X = \mathbf{A}_S^d$ is an affine space over S then via base change from $\operatorname{Spec} \mathbf{F}_p$ we may identify $F_{X/S,n}$ with the endomorphism of $\mathbf{A}_S^d$ defined by the p^n-power map on the standard coordinates. A third application of the functoriality of absolute Frobenius morphisms is that the formation of $F_{X/S,n}$ is functorial in X over S.

When $n = 1$ we write $F_{X/S}$ to denote $F_{X/S,1}$, so for $n > 1$ we have

$$F_{X/S,n} = F_{X^{(p^{n-1})}/S} \circ \cdots \circ F_{X^{(p)}/S} \circ F_{X/S}.$$

For example, if $S = \operatorname{Spec} k$ for an extension field $k/\mathbf{F}_p$ then when $[k : \mathbf{F}_p]$ is finite and divides n we have $F_{X/S,n} = F_X^n$ but otherwise $F_{X/S,n}$ and F_X^n generally have different targets.

Proposition A.3.1 *Let Z be a scheme that is smooth of pure relative dimension d over an $\mathbf{F}_p$-scheme S. The map $F_{Z/S,n} : Z \to Z^{(p^n)}$ is a finite locally free surjection with degree p^{nd}.*

This proposition will be of interest to us when $S = \operatorname{Spec} k$ for a field $k/\mathbf{F}_p$.

Proof The result is clear by inspection in the special case that Z is an affine space over S. To handle the general case we wish to work locally for the étale topology on Z to reduce to the case of an affine space. Thus, we now consider how $F_{Z/S,n}$ interacts with étale maps over S. Rather generally, for any étale map $f : X \to Y$ of S-schemes, consider the commutative diagram of S-maps

$$\begin{array}{ccc} X & \xrightarrow{\ f\ } & Y \\ {\scriptstyle F_{X/S,n}}\downarrow & & \downarrow{\scriptstyle F_{Y/S,n}} \\ X^{(p^n)} & \xrightarrow[f^{(p^n)}]{} & Y^{(p^n)} \end{array}$$

in which the horizontal maps are étale. We claim that this diagram is cartesian; that is, the natural map

$$h : X \to X^{(p^n)} \times_{Y^{(p^n)}} Y$$

is an isomorphism.

By inspection of the definition of relative Frobenius morphisms, the vertical maps in the diagram are finite, surjective, and radicial (i.e., injective on topological spaces and purely inseparable on residue field extensions). Hence, the map h over $X^{(p^n)}$ is finite, surjective, and radicial. But h is also a Y-map between étale Y-schemes, so h is étale. Since étale radicial maps are open immersions [EGA, IV_4, 17.9.1], it follows that h is an open immersion, so by surjectivity it is an isomorphism.

Applying the above cartesian property to the case of a Zariski-open inclusion $U \hookrightarrow Z$ we see that $F_{Z/S,n}^{-1}(U^{(p^n)}) = U$, so we may work Zariski-locally on Z. Hence, by smoothness, we can arrange (by [EGA, IV_4, 17.11.4]) that Z is étale over an affine space over S. Then the above cartesian property identifies

$F_{Z/S,n}$ with a base change of the n-fold relative Frobenius morphism of an affine space, so we are reduced to the settled case of an affine space. □

Example A.3.2 Let k be a field of characteristic $p > 0$ and let G be a connected reductive k-group that is k-split. Hence, by the Existence and Isomorphism Theorems (which we review in §A.4 over any field) there exists a split connected reductive $\mathbf{F}_p$-group G_0 such that $G \simeq (G_0)_k$, and G_0 is unique up to (generally non-unique) isomorphism. Under this identification, $F_{G/k,n}$ is carried to $F_{(G_0)_k,n} = (F_{G_0})_k^n$ via the natural identification of $(G_0)_k^{(p^n)} = (G_0^{(p^n)})_k$ with $(G_0)_k = G$. However, G_0 is not functorial in G, so unless G is defined to be $(G_0)_k$ (in particular, an $\mathbf{F}_p$-structure on G is specified, such as the natural one for $G = \mathrm{GL}_N$ over k) it is best not to view $F_{G/k,n}$ as an endomorphism of G except when $[k : \mathbf{F}_p]$ is finite and divides n.

The functoriality of the absolute Frobenius morphism implies, via the compatibility of the formation of $X^{(p^n)}$ with respect to fiber products in X over S, that $F_{(X \times_S Y)/S,n} = F_{X/S,n} \times F_{Y/S,n}$ for any S-schemes X and Y. Hence, if X is an S-group then $X^{(p^n)}$ is an S-group and $F_{X/S,n}$ is an S-group homomorphism.

Definition A.3.3 Let $X \to S$ be a map of $\mathbf{F}_p$-schemes. For any $n \geqslant 0$, the S-map $F_{X/S,n}$ in (A.3.2) is the *n-fold relative Frobenius morphism* of X over S, and for $n = 1$ it is the *relative Frobenius morphism* of X over S. If $S = \mathrm{Spec}\, k$ for a field k we write $F_{X/k,n}$ and $F_{X/k}$, and when G is a smooth k-group we call $F_{G/k} : G \to G^{(p)}$ the *Frobenius isogeny* of G over k.

In view of the preceding discussion, for any k-group G the map $F_{G/k,n}$ is a k-group homomorphism whose formation commutes with any extension of scalars on k, and it is functorial in G. In the smooth case it is an isogeny of degree $p^{n \dim G}$. By computing over $\overline{k}$ (where non-smoothness is the same as non-reducedness) we see that for G locally of finite type over k the map $F_{G/k}$ is never flat in the non-smooth case (e.g., $F_{G/k}$ vanishes for $G = \alpha_p$ and $G = \mu_p$).

Example A.3.4 Let G be an affine group scheme of finite type over a field $k/\mathbf{F}_p$. Choose a k-subgroup inclusion $j : G \hookrightarrow \mathrm{GL}_N$, so $G^{(p^n)}$ is likewise identified with a k-subgroup of GL_N via $j^{(p^n)}$. In this way, $F_{G/k,n}$ is induced by the endomorphism $(F^n_{\mathrm{GL}_N/\mathbf{F}_p})_k$ of GL_N that is defined by the p^n-power map on matrix entries.

In some general considerations with group schemes of finite type over a field it is not necessary to restrict attention to the affine case, and for such purposes the following general result is useful.

Proposition A.3.5 *Any group scheme G of finite type over a field k is quasi-projective.*

Proof The case of smooth G is originally due to Chow [Chow], and a modern proof is given in [Con1, Cor. 1.2] as an application of descent theory and Theorem A.3.7 below (which does not rely on the result we are presently proving). To handle the non-smooth case we may assume $\mathrm{char}(k) = p > 0$. For $n \geqslant 0$, consider the n-fold relative Frobenius morphism $F_{G/k,n} : G \to G^{(p^n)}$. The kernel is a finite (infinitesimal) normal k-subgroup scheme of G, so the quotient map $G \to G/\ker(F_{G/k,n})$ is a finite map. Hence, it suffices to prove quasi-projectivity of $G/\ker(F_{G/k,n})$ for some $n \geqslant 0$. By [SGA3, $\mathrm{VII_A}$, 8.3], if n is sufficiently large then this quotient of G is smooth, so we are done. □

Remark A.3.6 Although quasi-projectivity is a sufficient criterion for the existence of certain kinds of quotients, in view of the use of quotients G/H in the proof of Proposition A.3.5 (even in the smooth case) we note that the construction of such quotients in [SGA3, $\mathrm{VI_A}$, Thm. 3.2] does not use quasi-projectivity of G.

To conclude this section, we state an important structure theorem of Chevalley over perfect fields and a less widely known variation on it over any field.

Theorem A.3.7 (*Chevalley*) *Let G be a smooth connected group over a perfect field k. There is a unique short exact sequence of smooth connected k-groups*

$$1 \to H \to G \to A \to 1$$

with A an abelian variety and H affine.

Proof This is proved in [Chev]; see [Con1] for a modern exposition. □

Theorem A.3.7 is not applicable over imperfect fields if we insist on working with smooth H. Here are counterexamples, resting on our discussion of Weil restriction in §A.5.

Example A.3.8 Let k be an imperfect field, and k'/k a nontrivial purely inseparable finite extension. For any nonzero abelian variety A' over k', the Weil restriction $G = \mathrm{R}_{k'/k}(A')$ makes sense as a k-scheme since A' is quasi-projective over k'. By Proposition A.5.11, G is a smooth connected commutative k-group. There is no nontrivial smooth connected affine k-subgroup H in G. Indeed the inclusion map $j : H \hookrightarrow G$ corresponds to a k'-group map $H_{k'} \to A'$ that must be trivial (since $H_{k'}$ is a smooth connected affine k'-group and A' is an abelian variety over k'). Hence, j is also trivial, so $H = 1$ as claimed.

It follows that if Theorem A.3.7 were applicable to G then G would have to be an abelian variety. However, G is not an abelian variety. One reason is that the natural map $G_{k'} \to A'$ has nontrivial smooth connected unipotent kernel, by Proposition A.5.11; another reason is that $\mathrm{R}_{k'/k}$ applied to a smooth proper k'-scheme of positive dimension is never proper (Example A.5.6). Thus, G is a counterexample to the conclusion of Theorem A.3.7 over k when the perfectness hypothesis on k is dropped.

In [BLR, 9.2/1] there is given a version of Theorem A.3.7 over general fields by allowing the "affine part" to be non-smooth. For Galois cohomological and other purposes it is useful to avoid working with non-smooth groups. Thus, the following variant on Theorem A.3.7 (in which, surprisingly, the affine part goes on the right instead of on the left) is extremely useful. This result applies over arbitrary fields and only uses smooth groups, so it deserves to be better known.

Theorem A.3.9 (*Anti-Chevalley Theorem*) *Let k be a field, and G a smooth connected k-group. Define*

$$G^{\mathrm{aff}} = \mathrm{Spec}(\mathscr{O}(G))$$

equipped with its natural k-group scheme structure. This affine k-group is smooth (in particular, of finite type) and the natural map $G \to G^{\mathrm{aff}}$ fits into a short exact sequence of smooth connected k-groups

$$1 \to Z \to G \to G^{\mathrm{aff}} \to 1$$

with $\mathscr{O}(Z) = k$. The k-group Z is central in G and if $\mathrm{char}(k) > 0$ *then Z is semi-abelian.*

Proof This is [DG, III, §3, 8.2, 8.3] apart from the fact that Z is semi-abelian when $\mathrm{char}(k) > 0$. It remains to show that if $\mathrm{char}(k) = p > 0$ then any smooth connected commutative k-group H satisfying $\mathscr{O}(H) = k$ is semi-abelian (i.e., an extension of an abelian variety by a torus). This is proved in [Bri, Prop. 2.2] and [SS]; for the convenience of the reader, we now give another proof.

It suffices to check the semi-abelian property over an algebraic closure of the base field, so we may assume that k is algebraically closed. Hence, Theorem A.3.7 is applicable to H, and we have to show that the maximal smooth connected affine subgroup of H is a torus. Since H is commutative, it contains a unique maximal torus T. By Theorem A.3.7, H/T is an extension of an abelian variety A by a smooth connected unipotent group U. The k-group U is killed by p^N for sufficiently large N, so p^N-multiplication on the commutative H gives an isogeny-section s for the quotient $H/T \twoheadrightarrow A$. Hence, $(H/T)/s(A)$ is a unipotent quotient of H. But $\mathscr{O}(H) = k$, so H has no non-constant maps to an affine k-scheme. Thus, $(H/T)/s(A) = 1$. This forces $U = 1$ for dimension reasons. □

A.4 Split reductive groups: Existence, Isomorphism, and Isogeny Theorems

A connected reductive group G over a field k is called *split* if it contains a k-split maximal k-torus T. We will call such (G, T) a *split reductive pair* over k. If (G, T) is a split reductive pair over k then every root group for $T_{\overline{k}}$ in $G_{\overline{k}}$ is defined over k, and so every Borel subgroup of $G_{\overline{k}}$ containing $T_{\overline{k}}$ is defined over k [Bo2, 18.7].

Example A.4.1 Let G be a k-split connected semisimple group of rank 1, say with T a k-split maximal k-torus (of dimension 1). We claim that (G, T) is k-isomorphic to either SL_2 or PGL_2 equipped with its diagonal k-torus. Since SL_2 has μ_2 as its only finite normal subgroup scheme of order 2, it is equivalent to construct an isogeny $\mathrm{SL}_2 \to G$ of degree dividing 2 such that the diagonal k-torus is carried onto T. Over an algebraically closed field, this is [Spr, 7.2.4].

In general we have the result for $(G_{\overline{k}}, T_{\overline{k}})$, so $\Phi(G, T) = \{\pm a\}$. To handle the case of a general ground field, we claim that the proof in [Spr, 7.2.4] can be applied in such generality. This rests on several observations. First, for $N = N_G(T)$ there exists $n \in N(k)$ representing the nontrivial class in $N/T = \mathbf{Z}/2\mathbf{Z}$. This follows from the vanishing of $\mathrm{H}^1(k, T)$ (with $T \simeq \mathrm{GL}_1$); note that $n^{-1}U_a n = U_{-a}$. Consider the map

$$U_{-a} \times T \times U_a \to G$$

defined by multiplication. We claim that this is an isomorphism onto $G - n^{-1}B$, where $B = T \times U_a$ is $P_G(\lambda)$ for any $\lambda \in \mathrm{X}_*(T)$ such that $\langle a, \lambda \rangle > 0$. It suffices to check this property over $\overline{k}$ as well, in which case the verification is explained in the proof of [Spr, 7.2.4].

Finally, the k-groups $U_{\pm a}$ are smooth connected unipotent of dimension 1 (as may be checked over $\overline{k}$), and by Theorem B.4.3 each is k-split due to the nontrivial action by T. Thus, we can choose a k-group isomorphism $u : \mathbf{G}_\mathrm{a} \simeq U_a$. The proof of [Spr, 7.2.4] therefore applies over k to yield a k-rational homomorphism from SL_2 to G. By Galois descent from k_s (over which there is a Zariski-dense locus of rational points), this rational homomorphism uniquely extends to a k-group homomorphism, and the proof of [Spr, 7.2.4] shows that this is an isogeny with degree dividing 2 that carries the diagonal torus onto T, as desired. (Note that these properties can be proved by working over $\overline{k}$.)

Every G becomes split over a finite separable extension of the ground field (since each maximal torus over the ground field is split by such an extension). In particular, if $k = k_s$ then every connected reductive k-group is k-split. The split reductive pairs over a field are classified in terms of their root data (which are reduced). This is made precise by the Existence and Isomorphism Theorems for split connected reductive groups over a field (which we will need over a general separably closed field), and there is a stronger formulation that records information about isogenies. The literature on these topics (aside from [SGA3, XXV]) generally treats only an algebraically closed field, so below we explain how to deduce the Isomorphism and Isogeny Theorems over a general field from the case of an algebraically closed field.

Before we address the main theorems in this direction, we review the definitions of the notions of root data and isogenies between them (see [Spr, 7.4] and [SGA3, XXI] for further discussion, including the finiteness of their Weyl groups and the relationship with root systems and reductive groups; also see Lemma 3.2.4).

Definition A.4.2 A *root datum* is a 4-tuple $(X, R, X^\vee, R^\vee)$ where X and $X^\vee$ are finite free $\mathbf{Z}$-modules equipped with a perfect duality $\langle\cdot,\cdot\rangle : X \times X^\vee \to \mathbf{Z}$ and R and $R^\vee$ are finite subsets of X and $X^\vee$ respectively equipped with a bijection $R \to R^\vee$ (denoted $a \mapsto a^\vee$) such that the following two axioms hold: (RD1) $\langle a, a^\vee\rangle = 2$ for all $a \in R$, and (RD2) the (dual) reflections $r_a : x \mapsto x - \langle x, a^\vee\rangle a$ of X and $r_a^\vee : \lambda \mapsto \lambda - \langle a, \lambda\rangle a^\vee$ of $X^\vee$ satisfy $r_a(R) = R$ and $r_a^\vee(R^\vee) = R^\vee$ (so $-a = r_a(a) \in R$).

The *Weyl group* of the root datum is the subgroup $W(R) \subseteq \mathrm{Aut}(X)$ generated by the involutions r_a for $a \in R$. A root datum $(X, R, X^\vee, R^\vee)$ is *reduced* if $X \cap \mathbf{Q}a = \{\pm a\}$ inside $X_{\mathbf{Q}}$ for all $a \in R$.

Let $p \geqslant 1$ be a prime number or 1. A *p-morphism* of reduced root data

$$(X', R', X'^\vee, R'^\vee) \to (X, R, X^\vee, R^\vee)$$

is a triple (f, ι, q) with $f : X' \to X$ an injective $\mathbf{Z}$-linear map having finite cokernel (i.e., $f_{\mathbf{Q}}$ is an isomorphism), $\iota : a \mapsto a'$ a bijection from R to R' (the other direction!), and $q : R \to \{p^n\}_{n \geqslant 0}$ a map of sets such that

$$f(a') = q(a)a, \;\; f^\vee(a^\vee) = q(a)a'^\vee$$

for all $a \in R$. A *composition* of p-morphisms is defined by composition in f and ι and multiplication in q.

If $q(a) = 1$ for all a then (f, ι, q) is called *central*. If $p > 1, f(X') = pX$, and $q(a) = p$ for all $a \in R$ then (f, ι, q) is called a *Frobenius isogeny*.

Remark A.4.3 The injectivity of f implies that $(-a)'$ is proportional to a', so $(-a)' = -a'$ since ι is injective and the root data are assumed to be reduced. It then also follows that $q(-a) = q(a)$ in such cases.

The terminology in Definition A.4.2 follows [Spr]; in [SGA3, XXI, 6.8.1] the notion of p-morphism does not require $f_{\mathbf{Q}}$ to be an isomorphism. (This extra generality corresponds to considering maps $(G, T) \to (G', T')$ that may only be an isogeny between derived groups.)

Example A.4.4 Consider split reductive pairs (G, T) and (G', T') over a field k with characteristic exponent $p \geqslant 1$, and let $(X, R, X^\vee, R^\vee) = R(G, T)$ and $(X', R', X'^\vee, R'^\vee) = R(G', T')$ be the corresponding root data. Let $\phi : G \to G'$ be a k-isogeny such that $\phi(T) = T'$. We now explain how to naturally associate to ϕ a p-morphism (f, ι, q) between the corresponding root data.

The map

$$f := \mathrm{X}(\phi) : X' \to X$$

is injective with finite cokernel since $\phi : T \to T'$ is an isogeny of k-tori. For each $a \in R$ let U_a be the corresponding root group in G, and likewise let $U'_{a'}$ be the root group in G' associated to any $a' \in R'$. Root groups are uniquely characterized as nontrivial smooth connected unipotent k-subgroups normalized by the split maximal torus and on which a codimension-1 torus acts trivially, so for every $a \in R$ there is a unique $a' \in R'$ such that $\phi(U_a) = U'_{a'}$. Since ϕ is an isogeny, it follows that the map $\iota : R \to R'$ defined by $a \mapsto a'$ is injective, so it is bijective since the number of roots is determined by the dimension of the group and the dimension of the maximal tori.

The isogeny ϕ must respect the unique codimension-1 subtori $T_a = (\ker a)^0_{\mathrm{red}}$ and $T'_{a'} = (\ker a')^0_{\mathrm{red}}$ that act trivially on the respective root groups U_a and $U'_{a'}$, which is to say $\phi(T_a) = T'_{a'}$. In particular, $(-a)' = -a'$ since $T_{-a} = T_a$ and $U'_{\pm a'}$ are the only two root groups on which $T'_{a'}$ acts trivially. By using conjugation against suitable Weyl elements, the isogenies $U_a \to U'_{a'}$ and $U_{-a} \to U'_{-a'}$ have the same degree, which must be a p-power $q(a) = q(-a)$.

Upon choosing k-group isomorphisms $x_a : \mathbf{G}_\mathrm{a} \simeq U_a$ and $x_{a'} : \mathbf{G}_\mathrm{a} \simeq U'_{a'}$, we have

$$\phi(x_a(t)) = x_{a'}(c_a t^{q(a)})$$

for a unique $c_a \in k^\times$. Thus, $f(a') = q(a)a$ for all a. Let G_a be the semisimple rank-1 group generated by U_a and U_{-a}, and define $G'_{a'}$ similarly in terms of $U'_{a'}$ and $U'_{-a'} = U'_{(-a)'}$. The map f induces an isogeny $G_a \to G'_{a'}$ which restricts to an isogeny $T \cap G_a \to T' \cap G'_{a'}$ between their maximal tori. In view of the construction of coroots, it follows that $f^\vee(a^\vee) := \phi \circ a^\vee$ is a rational multiple of $a'^\vee$. We claim that the multiplier is $q(a)$:

$$f^\vee(a^\vee) = q(a)a'^\vee. \tag{A.4.1}$$

Indeed, writing $f^\vee(a^\vee) = ca'^\vee$ with $c \in \mathbf{Q}$, we have

$$2c = c\langle a', a'^\vee\rangle = \langle a', f^\vee(a^\vee)\rangle = \langle f(a'), a^\vee\rangle = q(a)\langle a, a^\vee\rangle = 2q(a).$$

We have associated to the isogeny $\phi : (G, T) \to (G', T')$ a p-morphism $f(\phi) = (f, \iota, q)$.

As a prelude for understanding the proofs of the Isomorphism and Isogeny Theorems, we now make a good choice of normalization of the parameterizations x_a and $x_{a'}$ so that we can relate c_a and c_{-a}. Since we are working with split reductive pairs, so Example A.4.1 applies to the pairs $(G_a, T \cap G_a)$ and $(G'_{a'}, T' \cap G'_{a'})$, we can choose the root group parameterizations x_a and $x_{a'}$ so that the elements

$$\begin{aligned} n_a &:= x_a(1)x_{-a}(-1)x_a(1) \in G(k), \\ n_{a'} &:= x_{a'}(1)x_{-a'}(-1)x_{a'}(1) \in G'(k) \end{aligned} \tag{A.4.2}$$

respectively lie in $N_G(T)(k)$ and $N_{G'}(T')(k)$ and represent the reflections r_a and $r_{a'}$ in $W(G, T)(k)$ and $W(G', T')(k)$ respectively. By [Spr, 8.1.4(iv)] (whose proof is valid for a split reductive pair over any ground field, due to Example A.4.1, and amounts to calculations with SL_2), these requirements determine the collections $\{x_a\}$ and $\{x_{a'}\}$ up to multiplication on $\mathbf{G}_\mathrm{a}$ by unique constants $u_a, u_{a'} \in k^\times$ satisfying $u_a u_{-a} = 1$ and $u_{a'}u_{-a'} = 1$ for all a and a'. Thus, if $q(a) = 1$ for all a then $c_a c_{-a} = 1$ for all a. We shall now prove that $c_a c_{-a} = 1$ for all $a \in R$, regardless of the value of each $q(a)$.

To prove $c_a c_{-a} = 1$ for all a, we first recall that by direct calculations with SL_2 (as in [Spr, 8.1.4] and its proof), we have

$$x_a(t)x_{-a}(-t^{-1})x_a(t) = a^\vee(t)n_a \in N_G(T) \tag{A.4.3}$$

for all geometric points t of GL_1. Applying ϕ then yields

$$\phi(x_a(t))\phi(x_{-a}(-t^{-1}))\phi(x_a(t)) \in N_{G'}(T')$$

with $\phi(x_a(t)) = x_{a'}(c_a t^{q(a)})$ a non-vanishing geometric point of $U'_{a'}$. In particular,

$$x_{a'}(c_a t^{q(a)})\phi(x_{-a}(-t^{-1}))x_{a'}(c_a t^{q(a)}) \in N_{G'}(T') \tag{A.4.4}$$

with $\phi(x_{-a}(-t^{-1})) \in U'_{(-a)'} = U'_{-a'}$ a nontrivial element. But we also have

$$x_{a'}(c_a t^{q(a)}) x_{-a'}(-(c_a t^{q(a)})^{-1}) x_{a'}(c_a t^{q(a)}) \in N_{G'}(T'). \tag{A.4.5}$$

The properties (A.4.4) and (A.4.5) with middle terms nontrivial in $U'_{-a'}$ force the middle terms to be equal [Spr, 8.1.4(iii)]. That is, $\phi(x_{-a}(-t^{-1})) = x_{-a'}(-(c_a t^{q(a)})^{-1})$. But there is another formula for this element:

$$\phi(x_{-a}(-t^{-1})) = x_{(-a)'}(c_{-a}(-t^{-1})^{q(-a)}) = x_{-a'}(-c_{-a}(t^{q(a)})^{-1}),$$

since $(-1)^{q(a)} = -1$ in k. Comparing the two formulas for $\phi(x_{-a}(-t^{-1}))$, we get $c_{-a} = c_a^{-1}$ as desired.

Before we state the Existence and Isomorphism Theorems over a general base field, it is convenient to record a preliminary lemma that will reduce certain difficulties with roots to the case of simple positive roots.

Lemma A.4.5 *Let G be a split connected reductive group over a field k, and T a k-split maximal k-torus. For each $a \in \Phi(G,T)$, let U_a be the corresponding root group and let G_a be the corresponding k-subgroup of type A_1 generated by U_a and U_{-a}.*

For any choice of positive system of roots in $\Phi(G,T)$, the k-subgroups G_a for simple positive a generate $\mathscr{D}(G)$.

Proof We may and do assume that k is algebraically closed (since the structure theory may be more familiar to the reader in that case). Since all G_a are contained in $\mathscr{D}(G)$, we may replace G with $\mathscr{D}(G)$ and replace T with the k-split subtorus $T \cap \mathscr{D}(G)$ to reduce to the case when G is semisimple. Let H be the smooth connected closed k-subgroup of G generated by the k-subgroups G_a. We have $T \subseteq H$ since T is generated by the tori $a^\vee(\mathrm{GL}_1)$ for simple positive a (as the corresponding coroots $a^\vee$ generate a lattice of finite index in $\mathrm{X}_*(T)$, due to the semisimplicity of G).

In view of the structure of an open Bruhat cell of (G,T), it suffices to prove that $G_b \subseteq H$ for all b. Since every root is carried to a simple positive root by some element of $W(G,T)$, it suffices for H to contain a set of representatives for $W(G,T)$. The Weyl group $W(G,T)$ is generated by the reflections r_a for simple positive a, so it is enough to find a representative in H for each such a. For every root a, the subgroup G_a contains a representative n_a for r_a. Thus, $n_a \in G_a \subseteq H$ for simple positive a. □

Theorem A.4.6 (*Existence and Isomorphism Theorems*) *Let k be a field. A split reductive pair (G,T) over k is determined up to isomorphism by its* *reduced root datum, and the isomorphism is unique up to the natural action α_t of a unique $t \in (T/Z_G)(k)$. Conversely, every reduced root datum arises from a split reductive pair (G,T) over k.*

Proof We first address the uniqueness of $t \in (T/Z_G)(k)$. That is, if the t-action on G is trivial then we claim $t = 1$. It suffices to check this when $k = \overline{k}$, in which case t lifts to $T(k)$, so the result is clear.

Turning to the other assertions, the case of algebraically closed k is proved in [Spr, 9.6.2, 10.1.1]. To establish the result over a general field k, first consider the Existence Theorem. Every separably closed field contains an algebraic closure of its prime field, so the Existence Theorem over separably closed fields is clear. This will be sufficient for our needs: the Existence Theorem and its consequences are used in this monograph only over a separably closed field.

For the reader who is interested in the Existence Theorem directly over an arbitrary field, we just make some general remarks and give a reference for a complete proof. Consideration with central isogenies reduces the problem to the case of irreducible and reduced root data that are "simply connected" in the sense that $R^\vee$ generates $X^\vee$. This reduces the problem to the construction of absolutely simple and simply connected split groups over every field. For the classical types one can write down the groups explicitly, and in the other cases it gets messy to argue by hand. For a uniform approach, a lot of work is required to overcome subtleties with "structure constants". We refer the reader to [SGA3, XXV, §§2–3] for a proof of the Existence Theorem over **Z** (and in particular, over prime fields), which takes as input the Existence Theorem over **C**.

We now give a complete proof of the Isomorphism Theorem over a general field k as a consequence of the result over $\overline{k}$. This will be an application of descent theory. Choose split reductive pairs (G,T) and (G',T') over k and an isomorphism

$$f : (X', R', X'^\vee, R'^\vee) \simeq (X, R, X^\vee, R^\vee)$$

between their root data. We seek a k-isomorphism $\phi : (G,T) \simeq (G',T')$, unique up to $(T'/Z_{G'})(k)$-conjugation, such that $f = f(\phi)$.

Let us first address uniqueness. In view of the result over $K := \overline{k}$, we have to show that if $t' \in (T'/Z_{G'})(K)$ is such that the t'-action $\alpha_{t'}$ on G'_K is defined over k then $t' \in (T'/Z_{G'})(k)$. Since $Z_{G'}$ is the scheme-theoretic center of G', $T'/Z_{G'}$ is a subfunctor of the automorphism functor $\underline{\mathrm{Aut}}(G')$ of G'. That is, for any k-algebra R, an element of $(T'/Z_{G'})(R)$ is uniquely determined by its action on G'_R. Hence, the two pullback maps

$$(T'/Z_{G'})(K) \rightrightarrows (T'/Z_{G'})(K \otimes_k K)$$

carry t' to the same point since the analogue holds for the pullback maps

$$\mathrm{Aut}(G'_K) \rightrightarrows \mathrm{Aut}(G'_{K\otimes_k K})$$

applied to the automorphism $\alpha_{t'}$ (which we assume is defined over k). By descent theory, it follows that $t' \in (T'/Z_{G'})(k)$, as desired.

For the existence of the k-isomorphism ϕ inducing f, if we can settle the problem over k_s then the uniqueness result implies that the obstruction to further descent down to k is a class in $\mathrm{H}^1(k_s/k, T'/Z_{G'})$. But $T'/Z_{G'}$ is a k-split torus, so the obstruction vanishes. Thus, it suffices to consider the case of a separably closed base field. We will only use that k is infinite. By the result over K we get a K-isomorphism ϕ such that $f = f(\phi)$. We seek $t' \in (T'/Z_{G'})(K)$ such that $\alpha_{t'} \circ \phi$ is defined over k.

The isomorphism $T_K \simeq T'_K$ induced by ϕ must descend to a k-isomorphism $T \simeq T'$ since the k-tori are k-split. For any $a \in \Phi(G,T)$, let $T_a = (\ker a)^0_{\mathrm{red}}$. The root group $(U_a)_K$ is the unique nontrivial smooth connected unipotent K-subgroup of G_K that is normalized by T_K, centralized by $(T_a)_K$, and has a single weight χ on its Lie algebra which moreover satisfies $\langle \chi, a^\vee \rangle > 0$. We have a similar description for $(U'_{a'})_K$ in terms of T'_K, where $f(a') = a$. Since $f^\vee(a^\vee) = a'^\vee$, it follows that ϕ carries $(U_a)_K$ isomorphically onto $(U'_{a'})_K$.

Now choose normalized k-isomorphisms $x_a : \mathbf{G}_\mathrm{a} \simeq U_a$ and $x_{a'} : \mathbf{G}_\mathrm{a} \simeq U'_{a'}$ for all $a \in R$ and $a' \in R'$ in accordance with the requirements as in (A.4.2). The parameterizations x_a and $x_{a'}$ convert $\phi : (U_a)_{K'} \simeq (U'_{a'})_{K'}$ into multiplication by some unit $c_a \in K^\times$. That is, $(x_{a'})_K(c_a t) = \phi((x_a)_K(t))$ for $t \in \mathbf{G}_\mathrm{a}$. Since $f^\vee(a^\vee) = a'^\vee$, it follows from the uniqueness argument in Example A.4.4 (resting on [Spr, 8.1.4(iv)]) that $c_{-a}c_a = 1$ for all a. Fixing a positive system of roots Φ^+ in $R' = \Phi(G', T')$, the simple positive roots form a basis of the character group of $T'/Z_{G'}$. It follows that there is a unique $t' \in (T'/Z_{G'})(K)$ such that $a'(t') = c_a^{-1}$ for all simple positive a'. By replacing ϕ with $\alpha_{t'} \circ \phi$ we get to the case when $c_a = 1$ for all simple positive a, so likewise $c_{-a} = 1$ for such a.

We have reached the situation in which $\phi \circ (x_a)_K = (x_{a'})_K$ for all simple positive a and simple negative a. Letting $G_a = \langle U_a, U_{-a} \rangle$ be the type-A_1 subgroup associated to a root $a \in R$, the K-group isomorphism $\phi : (G_a)_K \simeq (G'_{a'})_K$ is defined over k between open Bruhat cells of G_a and $G'_{a'}$, so by Zariski-density of these open Bruhat cells (and descent theory) the restriction of ϕ to $(G_a)_K$ is defined over k for all simple positive a.

By Lemma A.4.5, the k-subgroups G_a for simple positive a generate $\mathscr{D}(G)$. Thus, the groups $G_a(k)$ for such a generate a subgroup $\Gamma \subseteq \mathscr{D}(G)(k)$ that is Zariski-dense (as k is infinite and the G_a are unirational). The map ϕ must carry Γ into $\mathscr{D}(G')(k)$, so by descent theory and Zariski-density considerations it follows that $\phi|_{\mathscr{D}(G)_K}$ descends to a k-homomorphism $\mathscr{D}(G) \to G'$. If $Z \subseteq T$ is the maximal central k-torus then we have also seen that $\phi|_{Z_K}$ descends to a k-homomorphism $Z \to T' \hookrightarrow G'$.

Let $\pi : Z \times \mathscr{D}(G) \to G$ denote the canonical central isogeny over k. We have shown that $\phi \circ \pi_K$ is defined over k. Since π is faithfully flat, it follows by descent theory that ϕ is also defined over k. This settles the Isomorphism Theorem for split reductive pairs over any field. □

As an application of the uniqueness aspect of the Isomorphism Theorem, we get the following corollary concerning automorphisms of connected reductive groups. For the convenience of the reader we give direct proofs over a field, but a much more general version can be deduced from the structure of the automorphism group scheme of a connected reductive group [SGA3, XXIV, 1.7].

Corollary A.4.7 *Let G be a connected reductive group over a field k, and consider the natural left action of G/Z_G on G induced by conjugation.*

(1) *Assume G is absolutely simple and semisimple, and assume that the Dynkin diagram of G_{k_s} has no nontrivial automorphisms. Every k-automorphism of G is induced by the action of a unique element of $(G/Z_G)(k)$.*
(2) *Let H be a smooth connected k-group equipped with a left action $H \times G \to G$ on G as a k-group scheme (i.e., a homomorphism of group functors $H \to \underline{\mathrm{Aut}}(G)$ on the category of k-schemes or k-algebras). This action uniquely factors as a k-homomorphism $H \to G/Z_G$ followed by the natural action of G/Z_G on G.*

Part (2) of this corollary implies that the "natural" algebraic structure imposed on the automorphism group of a connected reductive group G over a field (as an extension of an étale group by G/Z_G) has the expected universal property among smooth groups acting on G.

Proof First we prove (1). In view of the uniqueness claim, we may assume $k = k_s$. The scheme-theoretic center Z_G of G represents the kernel of the homomorphism of group functors $G \to \underline{\mathrm{Aut}}(G)$, so G/Z_G is a subgroup functor of $\underline{\mathrm{Aut}}(G)$. In particular, any $g \in (G/Z_G)(k)$ is uniquely determined by its associated automorphism of G. We have to prove that all k-automorphisms of G are obtained from such g.

Fix a maximal k-torus T in G. By $G(k)$-conjugacy of maximal k-tori (Proposition A.2.10), we can restrict attention to automorphisms f of G that satisfy $f(T) = T$. Hence, f induces an automorphism of the root system of (G, T). Fix a positive system of roots Φ^+ in $\Phi(G, T)$. Since the group $W(G, T)(k) = N_G(T)(k)/T(k)$ acts transitively on the set of positive systems of roots in $\Phi(G, T)$, by composing f with conjugation by a suitable element of $N_G(T)(k)$ we can assume that $f(\Phi^+) = \Phi^+$.

The hypothesis concerning automorphisms of the Dynkin diagram implies that the f-action on Φ^+ must fix the simple positive roots pointwise, and so f acts as the identity on $\Phi(G,T)$. Since $\Phi(G,T)$ generates a finite-index subgroup of $X(T)$, f acts as the identity on $X(T)$. Hence, f acts as the identity on the root datum, so by the uniqueness aspect of the Isomorphism Theorem (applied in the case of automorphisms) we see that f is the action by a point in $(T/Z_G)(k)$. This proves (1).

To prove (2), let Z be the maximal central torus in G, so the H-action on $G = Z \cdot \mathscr{D}(G)$ restricts to actions on Z and $\mathscr{D}(G)$. The automorphism scheme of a torus is étale, so the H-action on Z is trivial due to the connectedness of H. Since G and $\mathscr{D}(G)$ have the same maximal adjoint quotient, we can replace G with $\mathscr{D}(G)$ to reduce to the case when G is semisimple. Also, since G has no nontrivial automorphisms that induce the identity on its maximal adjoint quotient, it suffices to treat the case when G is adjoint.

We now reduce to the case when k is algebraically closed (so that we can use general theorems about Borel subgroups in smooth connected affine groups). Suppose we can construct a $\overline{k}$-homomorphism $f : H_{\overline{k}} \to (G/Z_G)_{\overline{k}}$ such that when the given natural transformation of group functors $H \to \underline{\mathrm{Aut}}(G)$ on the category of k-schemes (or k-algebras) is viewed on the category of $\overline{k}$-schemes (or $\overline{k}$-algebras), it factors as

$$H_{\overline{k}} \xrightarrow{f} (G/Z_G)_{\overline{k}} \hookrightarrow \underline{\mathrm{Aut}}(G_{\overline{k}}) = \underline{\mathrm{Aut}}(G)_{\overline{k}}. \qquad \text{(A.4.6)}$$

Since the second map in (A.4.6) is a subfunctor inclusion, such a map f is unique and remains unique after extension of the base ring from $\overline{k}$ to any $\overline{k}$-algebra. In particular, over $\overline{k} \otimes_k \overline{k}$ we get the equality of pullback maps $p_1^*(f) = p_2^*(f)$ along the two projections $\mathrm{Spec}(\overline{k} \otimes_k \overline{k}) \rightrightarrows \mathrm{Spec}\,\overline{k}$ since the composite natural transformation (A.4.6) is defined over k. Thus, by descent theory, such an f must descend to a k-homomorphism and such a descent does the job. For the remainder of the argument, we now may and do assume k is algebraically closed.

Let us next reduce the problem to a group-theoretic one on geometric points. Assume we can prove, over any algebraically closed field F/k, that the map of groups $H(F) \to \mathrm{Aut}(G_F)$ factors through the subgroup $(G/Z_G)(F)$ of automorphisms arising from the faithful action of $(G/Z_G)(F)$. Taking $F = k$ yields a group homomorphism $H(k) \to (G/Z_G)(k)$, and the problem is to prove that it is algebraic (i.e., arises from a k-homomorphism $H \to G/Z_G$).

Consider the generic point $\eta \in H(k(H))$. Applying our group-theoretic hypothesis to a geometric point over η valued in an algebraically closed extension K of $k(H)$, we get a point $g \in (G/Z_G)(K)$ whose image in $\mathrm{Aut}(G_K) = \underline{\mathrm{Aut}}(G)(K)$ coincides with the image of η. Since η lands in $\underline{\mathrm{Aut}}(G)(k(H))$, so does g. But G/Z_G is a subfunctor of $\underline{\mathrm{Aut}}(G)$, so $g \in (G/Z_G)(k(H))$. (More specifically, the two pullbacks of g into $(G/Z_G)(K \otimes_{k(H)} K)$ coincide, so

$g \in (G/Z_G)(k(H))$ by descent theory.) In other words, g corresponds to a rational map of varieties from H to G/Z_G over k. Let $U \subseteq H$ be a dense open subscheme on which this rational map is defined, so the composite map

$$U \times G \xrightarrow{g \times 1} (G/Z_G) \times G \longrightarrow G$$

(the second step being the action map) coincides with the restriction of the given action map $H \times G \to G$ since such equality holds after pullback along $\eta : \operatorname{Spec} k(H) \to U$ (due to how g was chosen). Hence, the group-theoretic map $H(k) \to (G/Z_G)(k)$ is "algebraic" on the dense open $U \subseteq H$, so by translations it is everywhere algebraic as required.

It remains to prove, over any algebraically closed field k, that for each $h \in H(k)$ the h-action on G arises from the action of some element of $(G/Z_G)(k)$. Since $H(k)$ is generated by the subgroups $B(k)$ for Borel subgroups $B \subseteq H$ (as H is a smooth connected affine k-group), it suffices to treat such B in place of H. That is, we can assume H is solvable.

Consider the semidirect product $G' = G \rtimes H$ using the action of H on G. Since H is solvable, we can choose a Borel subgroup B' of G' containing H. Hence, $B' = B \rtimes H$ for a smooth connected solvable k-subgroup B of G which must be a Borel subgroup. Explicitly, $B = B' \cap G$ inside of G', with H normalizing B. Since H is smooth, the H-action on B induces an H-action on the maximal reductive quotient $B/\mathscr{R}_u(B)$ that is a torus. By connectedness of H, its action on this torus quotient must be trivial.

Fix a maximal torus T in B (so it is also maximal in G). For any $h \in H(k)$, the automorphism of B induced by h carries T to another maximal torus $h.T$ of B, so there exists $b(h) \in B(k)$ such that $b(h)Tb(h)^{-1} = h.T$. The composition ϕ of the h-action on G with conjugation by $b(h)^{-1}$ is an automorphism of G that preserves the pair (B, T). The induced action by ϕ on T is trivial, as we can check by using equivariance with respect to the isomorphism $T \simeq B/\mathscr{R}_u(B)$. Thus, ϕ induces the identity on the root datum of (G, T), so by the uniqueness aspect of the Isomorphism Theorem it must arise from the action of a point in $(T/Z_G)(k)$. □

Before we turn to the Isogeny Theorem over a general field, we address the behavior of maximal reductive quotients under passage to smooth connected normal subgroups or torus centralizers. Consider a smooth connected affine group H over a field k such that the subgroup $\mathscr{R}_u(H_{\overline{k}}) \subseteq H_{\overline{k}}$ is defined over k, or equivalently $\mathscr{R}_{u,k}(H)_{\overline{k}} = \mathscr{R}_u(H_{\overline{k}})$ (as happens when $k = \overline{k}$). Let H' be a smooth connected k-subgroup of H that is either normal or is the centralizer of a k-torus in H. (By Proposition A.2.5, every torus centralizer subgroup in H is smooth and connected.)

The image of H' in $H^{\text{red}} := H/\mathscr{R}_{u,k}(H)$ is normal when H' is normal in H, and likewise the image of H' in H^{red} is a torus centralizer when H' is a torus centralizer in H (Proposition A.2.8). But a smooth connected subgroup of a connected reductive group is reductive when it is normal or a torus centralizer [Bo2, 8.18, 13.19], so in both cases $\mathscr{R}_{u,k}(H')$ has vanishing image in H^{red}; i.e., $\mathscr{R}_{u,k}(H') \subseteq H' \cap \mathscr{R}_{u,k}(H)$. A finer relation holds:

Proposition A.4.8 *Let H be a smooth connected affine group over a field k such that $\mathscr{R}_{u,k}(H)_{\overline{k}} = \mathscr{R}_u(H_{\overline{k}})$. Let $H' \subseteq H$ be a smooth connected k-subgroup. Assume either that H' is normal in H or that it is the centralizer of a k-torus in H.*

The inclusion $\mathscr{R}_{u,k}(H') \subseteq H' \cap \mathscr{R}_{u,k}(H)$ is an equality inside of H' and is a k-descent of the subgroup $\mathscr{R}_u(H'_{\overline{k}}) \subseteq H'_{\overline{k}}$. In particular, the natural homomorphism $\phi : H'^{\mathrm{red}} \to H^{\mathrm{red}}$ between maximal reductive quotients over k has trivial kernel.

Moreover, if H' is normal in H then the induced map $H'^{\mathrm{ss}} \to H^{\mathrm{ss}}$ has finite central kernel. In particular, if k is perfect then the formation of the maximal adjoint semisimple quotient on the category of smooth connected affine k-groups carries short exact sequences to short exact sequences.

Proof Since the first claim says exactly that $H' \cap \mathscr{R}_{u,k}(H)$ is smooth and connected, we may and do assume that k is algebraically closed. First we treat the case when H' is normal in H, so $\mathscr{R}_u(H')$ is normal in H. Clearly

$$\ker \phi = (H' \cap \mathscr{R}_u(H))/\mathscr{R}_u(H'),$$

so passing to $H'/\mathscr{R}_u(H')$ and $H/\mathscr{R}_u(H')$ lets us assume that H' is reductive. Our aim is to prove that $H' \cap \mathscr{R}_u(H) = 1$. Consider the commutator subgroup $(\mathscr{R}_u(H), H')$. This is a smooth connected normal subgroup of H', so it inherits reductivity from H', but it is contained in the unipotent group $\mathscr{R}_u(H)$ and so is unipotent. Hence, this commutator subgroup is trivial, so $\mathscr{R}_u(H)$ commutes with H'. It follows that the unipotent group scheme $H' \cap \mathscr{R}_u(H)$ is central in H'. The reductivity of H' forces all central closed subgroup schemes of H' to be of multiplicative type, so $H' \cap \mathscr{R}_u(H) = 1$.

Now suppose $H' = Z_H(S)$ for a torus S in H. We have to prove that the containment $\mathscr{R}_u(H') \subseteq H' \cap \mathscr{R}_u(H)$ of unipotent normal subgroup schemes of H' is an equality. It suffices to show that the unipotent group scheme $H' \cap \mathscr{R}_u(H)$ is smooth and connected. This intersection is the centralizer in the smooth connected unipotent group $\mathscr{R}_u(H)$ under the natural conjugation action by the torus S. For any torus action on a smooth connected affine group, the centralizer of the action is smooth and connected (by applying Proposition A.2.5 to a suitable semidirect product).

Finally, we need to show that if H' is normal in H then $H'^{\mathrm{ss}} \to H^{\mathrm{ss}}$ has finite central kernel. By the parts already proved, we can replace $H' \to H$ with $H'^{\mathrm{red}} \to H^{\mathrm{red}}$, so H' and H are reductive. The maps $\mathscr{D}(H') \to H'^{\mathrm{ss}}$ and

$\mathscr{D}(H) \to H^{\mathrm{ss}}$ are therefore central isogenies, so the kernel of $H'^{\mathrm{ss}} \to H^{\mathrm{ss}}$ is the image of the part of $\mathscr{D}(H')$ in the finite central subgroup $\ker(\mathscr{D}(H) \to H^{\mathrm{ss}})$ of $\mathscr{D}(H)$. Hence, we get the required finiteness and centrality. □

Example A.4.9 To illustrate Proposition A.4.8, let H be a smooth connected affine group over a field k and let $H' = \mathscr{D}(H)$. Since $H'_{\overline{k}} = \mathscr{D}(H_{\overline{k}})$, applying the proposition to $H_{\overline{k}}$ and $H'_{\overline{k}}$ gives that $\mathscr{D}(H)^{\mathrm{ss}}_{\overline{k}} = \mathscr{D}(H^{\mathrm{red}}_{\overline{k}})$ inside of $H^{\mathrm{red}}_{\overline{k}}$.

Here is the Isogeny Theorem over a general base field; it is a "converse" to Example A.4.4.

Theorem A.4.10 *(Isogeny Theorem) Let k be a field with characteristic exponent $p \geqslant 1$, and let (G, T) and (G', T') be two split reductive pairs over k. Every p-morphism $R(G', T') \to R(G, T)$ between the root data has the form $f(\phi)$ for a k-isogeny $\phi : G \to G'$ carrying T onto T', with ϕ unique up to the $(T'/Z_{G'})(k)$-action on G'.*

Moreover, ϕ is central if and only if $f(\phi)$ is central, ϕ is an isomorphism if and only if $f(\phi)$ is an isomorphism, and ϕ is a Frobenius isogeny (i.e., $\ker \phi = \ker F_{G/k}$) if and only if $f(\phi)$ is a Frobenius isogeny.

Proof The existence and uniqueness aspects for the isogeny over an algebraically closed base field are proved in [Spr, 9.6.5]. Also, the characterization of isomorphisms is an immediate consequence of the characterization of central isogenies. To prove the characterization of central and Frobenius isogenies over a general base field it suffices to consider the case of an algebraically closed field. If $p > 1$ then it is clear that the Frobenius isogeny of G induces the Frobenius p-morphism. Thus, since an isogeny of groups over an algebraically closed field k is determined by the isogeny of root data up to an automorphism of the target, the case of Frobenius isogenies is settled over such k.

Continuing to assume k is algebraically closed, $f(\phi)$ is central if and only if ϕ restricts to an isomorphism between root groups; i.e., $\mathrm{Lie}(\ker\phi)$ has trivial T-action (equivalently, no nontrivial T-weights). Thus, if ϕ is a central isogeny then $f(\phi)$ is central. Conversely, assume $f(\phi)$ is central. The infinitesimal $(\ker \phi)^0$ has the form $\mathrm{Spec}\, R$ for a finite local k-algebra $(R, \mathfrak{m})$ such that the T-action on the k-algebra R is trivial on $\mathfrak{m}/\mathfrak{m}^2$. The T-action is therefore trivial on $\mathfrak{m}^i/\mathfrak{m}^{i+1}$ for all $i \geqslant 0$, so the T-action on R is unipotent. But T is a torus, so the T-action on R is trivial. All tori of G are conjugate, so they all centralize the normal $(\ker \phi)^0$, and hence so does G (as G is generated by its maximal tori). Since G is a central quotient of $Z \times \mathscr{D}(G)$ for a central k-torus Z of G and the connected semisimple $\mathscr{D}(G)$ that is its own derived group, $\ker \phi$ is central in G (cf. Lemma 5.3.2). That is, ϕ is central when $f(\phi)$ is central.

It remains to address the existence and uniqueness (up to suitable conjugation) for ϕ in the case of a general base field. That is, given a p-morphism

$$f : (X', R', X'^{\vee}, R'^{\vee}) \to (X, R, X^{\vee}, R^{\vee})$$

we seek to construct a k-isogeny $\phi : (G,T) \to (G',T')$, unique up to the $(T'/Z_{G'})(k)$-action, such that $f(\phi) = f$. By the same cohomological vanishing argument with split tori as in the proof of Theorem A.4.6, it suffices to treat the case of separably closed k.

Let $K = \overline{k}$, so the settled case over K provides an isogeny $\phi : G_K \to G'_K$ carrying T_K onto T'_K and satisfying $f(\phi) = f$. We seek $t' \in (T'/Z_{G'})(K)$ such that the composition of ϕ with the action of t' is defined over k. Since T and T' are k-split tori, the induced K-isogeny $\phi : T_K \to T'_K$ descends to a k-isogeny $T \to T'$. By definition of f being a p-morphism, we have $f(a') = q(a)a$ and $f^{\vee}(a^{\vee}) = q(a)a'^{\vee}$ for all $a \in R$, and by Remark A.4.3 we have $q(a) = q(-a)$ and $(-a)' = -a'$. Thus, the calculations in Example A.4.4 (especially the proof there that $c_a c_{-a} = 1$ for all a) allow us to use the method of proof of the Isomorphism Theorem over a general base field (Theorem A.4.6) to prove the Isogeny Theorem over a general base field; essentially no changes are required in the argument, apart from some intervention of $q(a)$-powers in some formulas. □

Recall that a connected semisimple group G over a field k is called *simply connected* if it admits no nontrivial central cover (i.e., any central k-isogeny $G' \to G$ from a connected semisimple k-group is an isomorphism).

Corollary A.4.11 *Let k be a field.*

(1) *Let G be a connected semisimple k-group. There is a central isogeny $\pi : \widetilde{G} \to G$ from a connected semisimple k-group $\widetilde{G}$ that is simply connected. The pair $(\widetilde{G}, \pi)$ is unique up to unique isomorphism, and its formation commutes with any extension of the base field.*
(2) *If k'/k is an extension of fields then any split reductive pair (G', T') over k' descends to a split reductive pair (G, T) over k.*

The pair $(\widetilde{G}, \pi)$ is called the *simply connected central cover* of G.

Proof Part (2) is immediate from the Existence and Isomorphism Theorems, since a split reductive pair is classified up to isomorphism by its root datum. Also, in view of the claimed uniqueness up to unique isomorphism for the simply connected central cover, it suffices (by Galois descent) to prove part (1) over separably closed fields.

By Theorem A.4.10, a connected semisimple group is simply connected if and only if the coroots generate $X^{\vee}$. By taking the character lattice to be the

$\mathbf{Z}$-dual of the coroot lattice, we deduce the existence of the simply connected central cover up to an isomorphism. The isomorphism between any two simply connected central covers is unique because for any isogeny $G' \to G$ between smooth connected groups, there is no nontrivial G-automorphism of G'. □

We also need a refined version of the Isomorphism Theorem, resting on the notion of a *pinning*. Let (G, T) be a split reductive pair over a field k, and for each $a \in \Phi(G, T)$ let G_a denote the type-A_1 subgroup of G generated by the root groups $U_{\pm a}$. Let Φ^+ be a positive system of roots in $\Phi(G, T)$, and let Δ be the set of simple positive roots.

Definition A.4.12 A *pinning* of (G, T, Φ^+) consists of a collection of central k-isogenies

$$\{\varphi_a : \mathrm{SL}_2 \to G_a\}_{a \in \Delta}$$

such that φ_a satisfies two properties:

$$\varphi_a(\mathrm{diag}(y, 1/y)) = a^\vee(y)$$

for all $y \in \mathrm{GL}_1$, and φ_a carries the upper triangular unipotent subgroup U^+ onto the root group $U_a \subseteq G_a$ (so $\varphi_a : U^+ \simeq U_a$).

Observe that if $\{\varphi_a\}_{a \in \Delta}$ is a pinning then, for each $a \in \Delta$, the kernel of φ_a is either trivial or μ_2. Our definition of a pinning is not literally the same as the definition in [SGA3, XXIII, 1.1], but they encode the same information. (In one direction, given a pinning in the above sense, the elements $X_a = \mathrm{Lie}(\varphi_a)\left(\begin{smallmatrix} 0 & 1 \\ 0 & 0 \end{smallmatrix}\right) \in \mathrm{Lie}(U_a)$ for $a \in \Delta$ are nonzero. In the other direction, the definition of a pinning in [SGA3, XXIII, 1.1] uniquely recovers maps φ_a as above because $\mathrm{Isom}(\mathbf{G}_\mathrm{a}, U_a) \simeq \mathrm{Isom}(k, \mathrm{Lie}(U_a))$ is an identification of $k^\times$-torsors and φ_a is unique up to precisely the natural faithful action on SL_2 by $\overline{D}(k) = k^\times$, where $\overline{D} = \mathrm{GL}_1$ is the diagonal torus in PGL_2. We do *not* claim to canonically determine φ_c for all $c \in \Phi(G, T)$ in terms of the choice of the maps φ_a for $a \in \Delta$, as there are $k^\times$-scaling ambiguities on the root groups U_c; we will return to this in Remark A.4.14. For our purposes, Definition A.4.12 is sufficient.) We are interested in pinnings due to the following result, which is a refinement of the Isomorphism Theorem.

Theorem A.4.13 *Let (G, T) be a split reductive pair over a field k. Let $\Phi^+ \subseteq \Phi(G, T)$ be a positive system of roots, and Δ the set of simple positive roots. Let $\{\varphi_a\}_{a \in \Delta}$ be a pinning.*

For any subfield $k_0 \subseteq k$ there is a descent (G_0, T_0) of (G, T) to a split reductive pair over k_0 such that the pinning maps $\varphi_a : \mathrm{SL}_2 \to G = (G_0)_k$ are *defined over k_0 for all $a \in \Delta$. That is, $(T_0)_k = T$ inside of $(G_0)_k = G$ and for each $a \in \Delta$ the k_0-structure imposed on G_a from SL_2 or $\mathrm{PGL}_2 = \mathrm{SL}_2/\mu_2$ via φ_a coincides with the k_0-structure arising from the k_0-subgroup $(G_0)_a \subseteq G_0$. Such a reductive pair (G_0, T_0) is unique up to unique isomorphism.*

If G is semisimple and the maps φ_a are defined over k_0 relative to a k_0-structure G_0 on G then T automatically descends to a k_0-split maximal k_0-torus T_0 in G_0: it is generated in G_0 by the images of the restrictions of the φ_a to the diagonal torus in SL_2 (as this same procedure in G over k recovers T, since G is semisimple). The same holds without semisimplicity if we assume the maps φ_a are defined over k_0 with respect to G_0 and that the maximal central k_0-torus in G_0 is k_0-split (as happens when k_0 is separably closed).

Proof By part (2) of Corollary A.4.11, there exists a split reductive pair (G_0, T_0) over k_0 equipped with an isomorphism

$$\phi : (G, T) \simeq ((G_0)_k, (T_0)_k).$$

Note that we are free to precompose ϕ with the action on G by any point in $(T/Z_G)(k)$. The problem is to find such a ϕ so that each $\phi \circ \varphi_a : \mathrm{SL}_2 \to (G_0)_k$ is defined over k_0. Via our initial choice of ϕ we may identify the character and cocharacter groups of T with those of T_0.

For each $a \in \Delta$, $(G_0)_a$ is k_0-isomorphic to SL_2 or PGL_2. Fix such a k_0-isomorphism h_a so that it carries the k_0-split torus $a^\vee(\mathrm{GL}_1) \subseteq T_0$ over to the diagonal torus in SL_2 or PGL_2. (We can do this since all rank-1 split tori in SL_2 or PGL_2 over a field are conjugate via a rational point.) Consider the restriction of $h_a \circ \phi \circ \varphi_a$ between the rank-1 split diagonal tori. Identify these tori with GL_1 via $y \mapsto \mathrm{diag}(y, 1/y)$ in SL_2 and $y \mapsto \mathrm{diag}(y, 1)$ in PGL_2, so the map $h_a \circ \phi \circ \varphi_a$ between diagonal tori is either the standard map (i.e., the identity map of GL_1 in the SL_2 case or the squaring map in the PGL_2 case) or its composite with inversion. Replacing h_a by precomposition with a representative of the nontrivial element of the Weyl group of $(G_0, a^\vee(\mathrm{GL}_1))$ if necessary brings us to the case when $h_a \circ \phi \circ \varphi_a$ is the standard map between diagonal tori for all $a \in \Delta$.

The map $h_a \circ \phi \circ \varphi_a$ restricts to a k-isomorphism between upper triangular unipotent subgroups. Using the standard coordinate on these copies of $\mathbf{G}_a$, the k-isomorphism between them must be given by $t \mapsto c_a t$ for some $c_a \in k^\times$. The torus T/Z_G in the k-split adjoint quotient G/Z_G has character group for which Δ is a basis. Hence, there is a unique $t \in (T/Z_G)(k)$ such that $a(t) = c_a$ for all $a \in \Delta$. Composing ϕ with the action of t^{-1} leaves $h_a|_{a^\vee(\mathrm{GL}_1)}$ unaffected and brings us to the case $c_a = 1$ for all $a \in \Delta$.

We are now in the case that each $h_a \circ \phi \circ \varphi_a$ is a central k-isogeny from SL_2 to either SL_2 or PGL_2 such that it is the standard map between diagonal tori and the identity map on upper triangular unipotent subgroups (with their standard coordinate). It suffices to check that the only such map is the standard map (which is defined over the prime field, let alone over k_0). Passing to the quotient by the center in the PGL_2 case if necessary, this comes down to the well-known fact that a k-automorphism of SL_2 or PGL_2 that is the identity on the upper triangular Borel subgroup is the identity map.

It remains to prove the uniqueness of (G_0, T_0) up to unique isomorphism. Let (G'_0, T'_0) be another such k_0-descent of (G, T). We have to prove that the composite k-isomorphism $\phi : (G'_0)_k \simeq G \simeq (G_0)_k$ is defined over k_0. Since ϕ carries $(T'_0)_k$ over to $(T_0)_k$, and T'_0 and T_0 are k_0-split, this isomorphism between maximal k-tori descends to a k_0-isomorphism $T'_0 \simeq T_0$. This identifies $\mathrm{X}(T'_0)$ with $\mathrm{X}(T_0)$, carrying $\Phi(G'_0, T'_0)$ to $\Phi(G_0, T_0)$. Since the pinning of (G, T) is assumed to be defined over k_0 relative to both k_0-structures on (G, T), for each root $a \in \Phi(G_0, T_0) = \Phi(G'_0, T'_0)$ the restriction of ϕ between the a-root groups viewed over k is also defined over k_0. Thus, the restriction of ϕ to the open Bruhat cells C' and C associated to a common choice of positive system of roots is defined over k_0. We can conclude by applying the general fact that if Y' and Y are separated and geometrically integral schemes of finite type over k_0 (such as G'_0 and G_0) and $U \subseteq Y$ and $U' \subseteq Y'$ are dense open subschemes (such as C' and C) then a k-morphism $h : Y'_k \to Y_k$ is defined over k_0 if it restricts to a k-morphism $h_\eta : U'_k \to U_k$ that is defined over k_0. (Indeed, the graph of h in $Y' \times Y$ is the closure of the graph of h_η in $U'_k \times U_k$. Thus, the graph in $U' \times U$ of the k_0-descent of h_η has closure in $Y' \times Y$ that projects isomorphically onto Y' since this can be checked over k. This closure in $Y' \times Y$ is therefore the graph of a k_0-descent of h.) □

Remark A.4.14 The method of proof of Theorem A.4.13 shows that assigning to each pinned k-split triple $(G, T, \{\varphi_a\}_{a\in\Delta})$ the based root datum $(R(G, T), \Phi^+, \Phi^{\vee +})$ is an equivalence of categories when we use isomorphisms on both sides. For a generalization allowing isogenies and more general base rings, see [SGA3, XXIII, 4.1]. An analogous theory over $\mathbf{Z}$ allows one to define φ_c for each root c by using the Weyl group action to move c into Δ, but this involves some ambiguities in scaling by $\mathbf{Z}^\times = \{\pm 1\}$ on root groups.

A.5 Weil restriction generalities

Let $B \to B'$ be a finite flat map of noetherian rings, and X' a quasi-projective B'-scheme. The *Weil restriction* $\mathrm{R}_{B'/B}(X')$ is a finite type B-scheme satisfying the universal property

$$\mathrm{R}_{B'/B}(X')(A) = X'(B' \otimes_B A)$$

for B-algebras A. For example, if X' is the final object Spec B' then $\mathrm{R}_{B'/B}(X')$ is the final object Spec B (even if $B' = 0$ and $B \neq 0$!). The existence and basic properties of Weil restriction are developed in [BLR, 7.6], to which we refer the reader for a proof of the existence of Weil restriction. One should think of the functor $\mathrm{R}_{B'/B}$ as analogous to the pushforward operation on sheaves. We will develop many properties of this functor in general, and then focus on special properties for Weil restriction relative to a finite extension of fields and for smooth affine groups over a field. (Other references for these special cases are [Pink] and [Oes, A.3] respectively.)

The role of the quasi-projective (and not merely finite type) hypothesis on X' is to ensure that any finite set of points in the underlying topological space of X' is contained in an affine open subscheme of X'. Under this quasi-projectivity hypothesis, $\mathrm{R}_{B'/B}(X')$ is constructed as a gluing of B-schemes $\mathrm{R}_{B'/B}(U'_i)$ for a suitable finite affine open cover $\{U'_i\}$ of X'. To be precise, the gluing is initially done (as a locally finite type B-scheme) using all open subschemes of all open affines in X'. To find finitely many open affines in X' whose Weil restrictions cover $\mathrm{R}_{B'/B}(X')$ (so $\mathrm{R}_{B'/B}(X')$ is finite type over B) we may assume $B \to B'$ has constant fiber-degree $n > 0$. Consider the quasi-compact *topological* n-fold power $P := X'^n$ (with the product topology). By quasi-projectivity of X' over B', any n-tuple in X' (allowing repetitions) is contained in an open affine U', so the corresponding open subsets U'^n cover$P$. By quasi-compactness of P, there are finitely many $\{U'_i\}$ such that the powers $U_i'^n$ cover P. We claim that the open affines $\mathrm{R}_{B'/B}(U'_i)$ cover $\mathrm{R}_{B'/B}(X')$. Consider a point $x \in \mathrm{R}_{B'/B}(X')$, say with residue field k. The composite map $\mathrm{Spec}(k) \stackrel{x}{\to} \mathrm{R}_{B'/B}(X') \to \mathrm{Spec}(B)$ arising from x makes the natural map $x : \mathrm{Spec}(k) \to \mathrm{R}_{B'/B}(X')$ into a B-morphism, which in turn corresponds to a B'-morphism $x' : \mathrm{Spec}(k \otimes_B B') \to X'$. But $\mathrm{Spec}(k' \otimes_B B')$ has at most n points, so the image of the B'-map x' consists of at most n points of X'. Thus, this B'-map factors through one of the open subschemes U'_{i_0} (as it suffices to check this topologically), so the k-point x of $\mathrm{R}_{B'/B}(X')$ over B lies in the open subscheme $\mathrm{R}_{B'/B}(U'_{i_0})$. This proves that $\mathrm{R}_{B'/B}(X')$ is covered by the finitely many open affine $\mathrm{R}_{B'/B}(U'_i)$. Beware that $\mathrm{R}_{B'/B}$ generally fails to commute with the formation of disjoint unions and Zariski-open covers; this is illustrated in Example A.5.3. Also, although $\mathrm{R}_{B'/B}$ does preserve quasi-projectivity, this is not seen from the construction; we address it in Proposition A.5.8.

Weil restriction is right adjoint to base change along $\mathrm{Spec}(B') \to \mathrm{Spec}(B)$ in the sense that there is a bijection

$$\mathrm{Hom}_B(Y, \mathrm{R}_{B'/B}(X')) \simeq \mathrm{Hom}_{B'}(Y_{B'}, X') \tag{A.5.1}$$

natural in X' and any B-scheme Y (as we see by working Zariski-locally on open affines of Y). In Proposition A.5.7 below we give an explicit description of the adjunction morphisms $Y \to \mathrm{R}_{B'/B}(Y_{B'})$ for quasi-projective B-schemes

Y and $\mathrm{R}_{B'/B}(X')_{B'} \to X'$ for quasi-projective B'-schemes X. The case of affine X' and Y is the one which is most essential for our purposes. However, non-affine cases do naturally arise, at least in examples (such as Weil restriction of a quasi-projective coset space G/H).

If $f' : X' \to Y'$ is a map between quasi-projective B'-schemes then there is a naturally induced map $\mathrm{R}_{B'/B}(f')$ between their Weil restrictions. If f' is a monomorphism then so is $\mathrm{R}_{B'/B}(f')$ (as for pushforwards of sheaves), but there are many interesting examples of surjective f' for which $\mathrm{R}_{B'/B}(f')$ is not surjective. Such failure of surjectivity arises often in the study of pseudo-reductive groups.

The usefulness of Weil restriction in the study of groups over a field rests on:

Proposition A.5.1 *Let k be a field and k' a nonzero finite reduced k-algebra (i.e., $k' = \prod k'_i$ for a non-empty finite collection of finite extension fields k'_i/k). The Weil restriction $G = \mathrm{R}_{k'/k}(G')$ exists as a k-scheme of finite type for any k'-group scheme G' of finite type, and it has a natural k-group structure.*

Proof It suffices to prove that G' is quasi-projective over $\mathrm{Spec}(k')$. Since

$$\mathrm{Spec}(k') = \coprod \mathrm{Spec}(k'_i)$$

with $\{k'_i\}$ the set of factor fields of k', it is equivalent to prove that $G' \otimes_{k'} k'_i$ is quasi-projective over k'_i for all i. Apply Proposition A.3.5 to conclude. □

We are most interested in applying Weil restriction through finite extension of the ground field in the setting of smooth affine groups, but applying it to non-smooth schemes is useful too (e.g., scheme-theoretic kernels). Weil restriction will be shown to preserve smoothness in general, but it can destroy reducedness when working with affine group schemes over imperfect ground fields. To give examples, let k be an imperfect field of characteristic $p > 0$ and $k' = k(a')$ where $a'^p = a \in k - k^p$. The domain $A' = k'[x,y]/(x^p + a'y^{p^2})$ is the coordinate ring of a reduced (non-smooth) k'-subgroup of $\mathbf{G}_{\mathrm{a}}^2$ and

$$\mathrm{R}_{k'/k}(\mathrm{Spec}(A')) = \mathrm{Spec}(D[y_0,\ldots,y_{p-1}]/(f(y_0,\ldots,y_{p-1})^p)),$$

where

$$f(t_0,\ldots,t_{p-1}) = t_0^p + at_1^p + \cdots + a^{p-1}t_{p-1}^p$$

and $D = k[x_0,\ldots,x_{p-1}]/(f(x_0,\ldots,x_{p-1}))$ is a domain. Loss of reducedness can also occur for Weil restriction through a *separable* extension of (imperfect) fields, since nontrivial tensor powers of non-smooth reduced algebras can be non-reduced. More specifically, for a nontrivial finite separable extension K and the coordinate ring $A = k[x,y]/(x^p + ay^p)$ of a reduced (non-smooth) k-group, $\mathrm{R}_{K/k}(\mathrm{Spec}(A_K))$ is non-reduced since its scalar extension to k_s has coordinate ring equal to the $[K:k]$-fold tensor power of the domain A_{k_s} over k_s and even the tensor square is non-reduced (as the fraction field of A_{k_s} is $k_s'(y)$). Immediately after the statement of Proposition A.5.9 we give an example of a geometrically reduced and geometrically irreducible plane curve that becomes disconnected with a non-reduced connected component after applying a purely inseparable Weil restriction.

Returning to the case of a general noetherian base ring, some of the useful properties of Weil restriction are recorded in the following result.

Proposition A.5.2 *Let $B \to B'$ be a finite flat map of noetherian rings, and X' a quasi-projective B'-scheme.*

(1) *The formation of $\mathrm{R}_{B'/B}(X')$ is naturally compatible with extension of scalars on B.*
(2) *The B-scheme $\mathrm{R}_{B'/B}(X')$ is affine when X' is affine.*
(3) *The formation of $\mathrm{R}_{B'/B}(X')$ is compatible with fiber products in X'; i.e., for any pair of B'-maps $X_1', X_2' \rightrightarrows X_3'$ between quasi-projective B'-schemes, the natural map*

$$\mathrm{R}_{B'/B}(X_1' \times_{X_3'} X_2') \to \mathrm{R}_{B'/B}(X_1') \times_{\mathrm{R}_{B'/B}(X_3')} \mathrm{R}_{B'/B}(X_2')$$

is an isomorphism. In particular, if X' is a B'-group then $\mathrm{R}_{B'/B}(X')$ is naturally a B-group.
(4) *If $f: X' \to Y'$ is a smooth map between quasi-projective B'-schemes then the induced map $\mathrm{R}_{B'/B}(f)$ is smooth, and similarly for the property of being étale or an open immersion. In particular, if X' is B'-smooth then $\mathrm{R}_{B'/B}(X')$ is B-smooth.*

Proof The assertion in (1) follows from the functorial characterization (A.5.1), since for any B-algebra A and B-scheme Y the base change Y_A represents the restriction to the category of A-schemes of the functor represented by Y on the category of B-schemes (and similarly over B' using B'-algebras A'). Part (2) is a consequence of the construction of Weil restriction.

The compatibility with fiber products in (3) is proved by the functorial characterization (A.5.1). The assertion in (3) concerning B'-groups and B-groups follows from the categorical characterization of group scheme structures in terms of diagrams with fiber products and the fact that Weil restriction carries final objects to final objects.

The verification of smoothness and étaleness is done via the infinitesimal criterion: if A is a B-algebra and J is an ideal in A with $J^2 = 0$ then in the B'-algebra $A' := B' \otimes_A A'$ the ideal $J' = JA'$ satisfies $J'^2 = 0$, so the map $R_{B'/B}(X')(A) \to R_{B'/B}(X')(A/J)$ is identified with the map $X'(A') \to X'(A'/J')$. This is surjective (resp. bijective) when X' is smooth (resp. étale). Since an open immersion is the same thing as an étale monomorphism [EGA, IV$_4$, 17.9.1], the property of being an open immersion is also preserved. $\square$

One checks via the functorial characterization that $R_{B'/B}$ is a left-exact functor from quasi-projective B'-groups to finite type B-groups. That is, if $f' : G' \to H'$ is a homomorphism between quasi-projective B'-groups then $\ker(R_{B'/B}(f'))$ is naturally identified with $R_{B'/B}(\ker f')$. The behavior under quotient maps is more delicate, due to subtleties in the behavior of Weil restriction with respect to surjectivity. Before we address criteria for $R_{B'/B}$ to interact well with certain kinds of coverings, we give an instructive example.

Example A.5.3 Let k be a field and k'/k a finite separable extension of degree $d > 1$. Let $X' = \mathbf{A}^1_{k'}$, and consider the open covering $\{U'_0, U'_1\}$ where $U'_0 = \{t \neq 0\}$ and $U'_1 = \{t \neq 1\}$. Then $X = R_{k'/k}(\mathbf{A}^1_{k'})$ contains the Weil restrictions $U_i = R_{k'/k}(U'_i)$ as open subschemes, but these do not cover X. Indeed, a choice of k_s-isomorphism $k_s \otimes_k k' \simeq k_s^d$ identifies X_{k_s} with $\mathbf{A}^d_{k_s}$, in which $(U_0)_{k_s}$ is the locus of points $(t_1, \ldots, t_d)$ such that $t_j \neq 0$ for all j and $(U_1)_{k_s}$ is the locus defined by $t_j \neq 1$ for all j. Thus, points such as $(1, 0, \ldots, 0)$ in $X(k_s)$ are not in $U_0(k_s) \cup U_1(k_s)$.

In the opposite extreme, consider $k' = k^d$ with $d > 1$. For any quasi-projective k'-scheme X' with fibers $X_1, \ldots, X_d$ over the points of $\operatorname{Spec} k'$, we have $R_{k'/k}(X') = \prod X_i$, so this is non-empty if and only if all X_i are non-empty. In particular, if X'_i denotes the open locus in X' with fiber X_i over the ith point of $\operatorname{Spec} k'$ and empty fiber over the other points then $\{X'_i\}$ is an open cover of X' but each $R_{k'/k}(X'_i)$ is *empty* and so these do not cover $R_{k'/k}(X')$ if every X'_i is non-empty.

The bad behavior of Weil restriction with respect to surjectivity and covering properties can be avoided in some cases:

Corollary A.5.4 *Let $B \to B'$ be a finite flat map of noetherian rings, and $f' : X' \to Y'$ a smooth surjective map between quasi-projective B'-schemes. Let $X = R_{B'/B}(X')$ and $Y = R_{B'/B}(Y')$.*

(1) *The induced smooth map*

$$f = R_{B'/B}(f') : X \to Y$$

is surjective.

(2) *Assume that* $\operatorname{Spec} B' \to \operatorname{Spec} B$ *is surjective and radicial. Then $R_{B'/B}$ commutes with the formation of finite disjoint unions, and if $\{U'_i\}$ is an open (resp. étale) cover of Y' then $\{R_{B'/B}(U'_i)\}$ is such a cover of Y.*

(3) *If X' is a G'-torsor over Y' for a smooth quasi-projective B'-group G' then X equipped with its natural action by $\mathrm{R}_{B'/B}(G')$ is a torsor over Y. In particular, if $f' : G' \to H'$ is a smooth surjection between quasi-projective B'-groups and $K' = \ker f'$ then the induced map*

$$\mathrm{R}_{B'/B}(G')/\mathrm{R}_{B'/B}(K') \to \mathrm{R}_{B'/B}(H') \tag{A.5.2}$$

is an isomorphism (i.e., $\mathrm{R}_{B'/B}(H')$ is a quotient of $\mathrm{R}_{B'/B}(G')$ modulo the smooth normal B-subgroup $\mathrm{R}_{B'/B}(K')$).

Proof To check surjectivity of f in (1) we can assume $B = k$ for an algebraically closed field k. Then B' is a finite product of finite local k-algebras, so by passing to factor rings we can assume that B' is local. In particular, its residue field is k.

The map $X(k) \to Y(k)$ is the map $X'(B') \to Y'(B')$, and to prove surjectivity of this latter map amounts to showing that for all $y' \in Y'(B')$ the pullback B'-scheme $X'_{y'}$ admits a B'-point. But $X'_{y'}$ is smooth and surjective onto Spec B' since f' is a smooth surjection, so by the functorial criterion for smoothness we just need to find a rational point over the residue field of B'. This residue field is the algebraically closed field k, so the required rational point exists.

For (2), first assume $Y' = \coprod Y'_i$, so each Y'_i is open in Y'. Thus, each $Y_i := \mathrm{R}_{B'/B}(Y'_i)$ is open in $Y := \mathrm{R}_{B'/B}(Y')$ by Proposition A.5.2(4), and we want to prove that they are disjoint and cover Y. We may again reduce to the case when $B = k$ is an algebraically closed field, so the radicial hypothesis implies that B' is a finite local k-algebra (with residue field k). The overlap $Y_i \cap Y_j$ is $Y_i \times_Y Y_j$, but this is $\mathrm{R}_{B'/B}(Y'_i \times_{Y'} Y'_j) = \mathrm{R}_{B'/k}(\emptyset)$, which is empty because the only k-scheme Y for which $Y_{B'}$ is empty is the empty scheme (as $B' \neq 0$). To see the covering property we compute with k-points: $Y(k) = Y'(B')$ and $Y_i(k) = Y'_i(B')$, so we just need to observe that every map Spec $B' \to Y' = \coprod Y'_i$ factors through some Y'_i because B' is local.

Now consider an open cover $\{U'_i\}$ of Y', so $\{\mathrm{R}_{B'/B}(U'_i)\}$ is a collection of open subschemes of Y. Passing to a finite subcover, $U' := \coprod U'_i$ is quasi-projective and $\mathrm{R}_{B'/B}(U')$ is the disjoint union of the $\mathrm{R}_{B'/B}(U'_i)$ by what we just showed. Since $U' \to Y'$ is a smooth surjective map, so is its Weil restriction. Hence, the $\mathrm{R}_{B'/B}(U'_i)$ are an open cover of Y. The same argument works for étale covers.

To prove (3), let $G = \mathrm{R}_{B'/B}(G')$. Assuming that X' is a G'-torsor over Y', we have to show that the G-action on X makes it a torsor over Y. Since G' is smooth, the map $X' \to Y'$ is a smooth surjection. Hence, $X \to Y$ is a smooth surjection by (1), so the desired torsor property holds provided that the natural map $G \times X \to X \times_Y X$ is an isomorphism. By Proposition A.5.2(3) this is the Weil restriction of the natural map $G' \times X' \to X' \times_{Y'} X'$ which is an isomorphism (since X' is a G'-torsor over Y'). □

Proposition A.5.5 *Let $B \to B'$ be a finite flat map of noetherian rings, and $f' : X' \to Y'$ a closed immersion between quasi-projective B'-schemes. The map $f = \mathrm{R}_{B'/B}(f')$ is a closed immersion.*

In particular, $\mathrm{R}_{B'/B}(X')$ is separated.

Proof The separatedness assertion is reduced to the closed immersion claim applied to $\Delta_{X'/Y'}$ (due to the compatibility of $\mathrm{R}_{B'/B}$ with respect to fiber products). To verify the closed immersion assertion, we first reduce to the affine case, as follows. Let $U' \to Y'$ be an affine étale cover (such as a finite disjoint union of the constituents of an affine open covering of Y'). Thus, $V' = X' \times_{Y'} U'$ is a closed subscheme of U' and is an affine étale cover of X'. The resulting commutative diagram

$$\begin{array}{ccc} \mathrm{R}_{B'/B}(V') & \longrightarrow & \mathrm{R}_{B'/B}(X') \\ \downarrow & & \downarrow f \\ \mathrm{R}_{B'/B}(U') & \longrightarrow & \mathrm{R}_{B'/B}(Y') \end{array}$$

is cartesian with étale surjections in the horizontal direction (by Corollary A.5.4(1)), so to prove that the right side is a closed immersion it suffices to do so for the left side. This reduces us to the case when Y' and X' are affine. By choosing a closed immersion of Y' into an affine space over B', we can even reduce to the special case when Y' is an affine space over B'. In this special case, the closed immersion property for f is the key ingredient in the construction of Weil restrictions in the affine case. □

In contrast with the special case of closed immersions, Weil restriction through a finite flat map generally does *not* preserve properness unless the finite flat map is étale. Conceptually, this can be predicted via the valuative criterion (non-étale finite flat extensions of a discrete valuation ring are generally not Dedekind). More concretely, we can build counterexamples as follows. Let k'/k be a nontrivial purely inseparable finite extension of fields and G' a connected reductive k'-group with B' a Borel k'-subgroup. For

$$G = \mathrm{R}_{k'/k}(G'), \quad B = \mathrm{R}_{k'/k}(B'),$$

it follows from Corollary A.5.4 that $G/B = \mathrm{R}_{k'/k}(G'/B')$. This *always* fails to be proper when G' is not a torus (i.e., $\dim G'/B' > 0$), due to:

Example A.5.6 Let $k \to k'$ be a ring map such that k' is locally free of finite constant rank $r \geqslant 1$ as a k-module, and let X' be a smooth quasi-projective k'-scheme with all fibers non-empty and of pure relative dimension d'. Thus, X'

admits an étale affine cover U' that is étale over $\mathbf{A}^{d'}_{k'}$ (e.g., a disjoint union of a suitable finite open affine cover). It follows from Corollary A.5.4 that the Weil restriction $X = \mathrm{R}_{k'/k}(X')$ admits an étale cover that is étale over $\mathrm{R}_{k'/k}(\mathbf{A}^{d'}_{k'})$. But Zariski-locally over Spec k this latter Weil restriction is isomorphic to $\mathbf{A}^{rd'}_k$, so X is k-smooth with pure relative dimension $d = rd'$. We claim that if $d' > 0$ and k' is not étale over k (e.g., a nontrivial purely inseparable finite extension of fields, or $k' = k[t]/(t^2)$ with nonzero k) then X is *never* k-proper, even if X' is k'-proper.

A concrete example with imperfect k of characteristic $p > 0$ is to take $X' = Y_{k'}$ for a k-smooth Y of pure dimension $d' > 0$ and $k' = k(a^{1/p})$ for some a in k that is not a p-power in k. In this case $k' \otimes_k \overline{k} = \overline{k}[t]/(t^p)$, so

$$X_{\overline{k}} = \mathrm{R}_{(\overline{k}[t]/(t^p))/\overline{k}}(Y_{\overline{k}} \otimes_{\overline{k}} (\overline{k}[t]/(t^p))),$$

which is the bundle of p-jets to $Y_{\overline{k}}$. For example, when $p = 2$ this is the tangent bundle to $Y_{\overline{k}}$. This is a vector bundle over $Y_{\overline{k}}$ with nonzero rank, so it cannot be proper when $Y_{\overline{k}}$ is proper.

To establish failure of properness in all cases with $d' > 0$ and k' not k-étale we may pass to fibers over a suitable geometric point of Spec(k) to reduce to the case that k is an algebraically closed field. Passing to factor rings of k' allows us to assume that k' is a finite local k-algebra distinct from k. Since $d' > 0$, by smoothness Spec$(k'[\epsilon])$ occurs as a closed subscheme of X' over k'. Thus, $\mathrm{R}_{k'/k}(X')$ contains the affine $Z := \mathrm{R}_{k'/k}(\mathrm{Spec}(k'[\epsilon]))$ as a closed subscheme. To show non-properness of $\mathrm{R}_{k'/k}(X')$ it suffices to prove that Z is not k-finite. But $Z(k)$ is infinite since k' is non-reduced.

To do calculations with Weil restriction, it is extremely useful to make explicit the pair of adjunction morphisms relating Weil restriction and base change:

Proposition A.5.7 *Let $B \to B'$ be a finite flat map between noetherian rings, X' a quasi-projective B'-scheme, and Y a quasi-projective B-scheme. Using the bifunctorial identification* (A.5.1), *let*

$$q_{X'} = q_{X',B'/B} : \mathrm{R}_{B'/B}(X')_{B'} \to X', \;\; j_Y = j_{Y,B'/B} : Y \to \mathrm{R}_{B'/B}(Y_{B'}) \quad \text{(A.5.3)}$$

respectively be the canonical B'-morphism corresponding to the identity map on $\mathrm{R}_{B'/B}(X')$ and the canonical B-morphism corresponding to the identity map on $Y_{B'}$.

For any B'-algebra A', the map $q_{X'}$ on A'-points is $X'(B' \otimes_B A') \to X'(A')$ induced by the B'-algebra homomorphism $B' \otimes_B A' \to A'$ defined by $b' \otimes a' \mapsto b'a'$. For any B-algebra A, the map j_Y on A-points is given by the map

$Y(A) \to Y_{B'}(B' \otimes_B A) = Y(B' \otimes_B A)$ induced by the B-algebra homomorphism $A \to B' \otimes_B A$ defined by $a \mapsto 1 \otimes a$.

Moreover, j_Y is a closed immersion if $\mathrm{Spec}\, B' \to \mathrm{Spec}\, B$ *is surjective and j_Y is an isomorphism if* $Y = \mathrm{Spec}\, B$.

Proof It will be convenient to work with contravariant functors rather than with schemes so that we can bypass representability issues. (In the affine case the following abstract proof using functors can be expressed in simpler terms.) Let $\mathscr{C}$ be the category of B-schemes and $\mathscr{C}'$ be the category of B'-schemes. For any contravariant set-valued functor F on $\mathscr{C}$, we write $F_{B'}$ to denote the restriction of F to $\mathscr{C}'$, by which we mean the composition of F with the forgetful map $\mathscr{C}' \to \mathscr{C}$. We call $F_{B'}$ the *base change* of F. (For example, if F is represented by a B-scheme Y then $F_{B'}$ is represented by the base change $Y_{B'}$.) Likewise, for any contravariant set-valued functor F' on $\mathscr{C}'$ we write $\mathrm{R}_{B'/B}(F')$ to denote the functor on $\mathscr{C}$ defined by $Z \rightsquigarrow F'(Z_{B'})$.

If F' is represented by a B-scheme X' then representability of $\mathrm{R}_{B'/B}(F')$ is equivalent to the functor $A \rightsquigarrow X'(B' \otimes_B A)$ on B-algebras being represented by a B'-scheme (which is known to occur when X' is quasi-projective over B'). The bijection (A.5.1) expresses the fact that if F' is the functor represented by a quasi-projective B'-scheme X' then its Weil restriction $\mathrm{R}_{B'/B}(F')$ is the functor represented by the B-scheme $\mathrm{R}_{B'/B}(X')$. This will allow us to reduce our problem to an assertion about functors, for which it is easier to see what is going on.

There is a natural transformation

$$j_F : F \to \mathrm{R}_{B'/B}(F_{B'})$$

defined by the map $F(Z) \to F_{B'}(Z_{B'}) = F(Z_{B'})$ associated to the canonical B-morphism $Z_{B'} \to Z$ for any B-scheme Z, and there is a natural transformation

$$q_{F'} : \mathrm{R}_{B'/B}(F')_{B'} \to F'$$

defined by the map $F'(Z'_{B'}) \to F'(Z')$ associated to the canonical B'-morphism

$$Z' \to Z'_{B'} = Z' \times_{\mathrm{Spec}(B)} \mathrm{Spec}(B')$$

for any B'-scheme Z'.

By considering affine Z and Z', when F is represented by a quasi-projective B-scheme Y (resp. when F' is represented by a quasi-projective B'-scheme X') then j_F corresponds to j_Y (resp. $q_{F'}$ corresponds to $q_{X'}$). Thus, apart from the closed immersion and isomorphism claims for j_Y at the end, it suffices to show that in general j_F and $q_{F'}$ define inverse correspondences

$$\begin{aligned}&\mathrm{Hom}_{\mathscr{C}}(F, \mathrm{R}_{B'/B}(F')) \to \mathrm{Hom}_{\mathscr{C}'}(F_{B'}, F'),\\ &\mathrm{Hom}_{\mathscr{C}'}(F_{B'}, F') \to \mathrm{Hom}_{\mathscr{C}}(F, \mathrm{R}_{B'/B}(F'))\end{aligned} \tag{A.5.4}$$

via $T \mapsto q_F \circ T_{B'}$ and $T' \mapsto \mathrm{R}_{B'/B}(T') \circ j_F$ respectively, and that if F is represented by a B-scheme Y then the resulting composite bijection

$$\begin{aligned}\mathrm{R}_{B'/B}(F')(Y) &= \mathrm{Hom}_{\mathscr{C}}(\mathrm{Hom}_B(\cdot, Y), \mathrm{R}_{B'/B}(F'))\\ &\simeq \mathrm{Hom}_{\mathscr{C}'}(\mathrm{Hom}(\cdot, Y_{B'}), F') = F'(Y_{B'})\end{aligned} \tag{A.5.5}$$

is the canonical identification $\beta_{F',Y}$ that occurs in the definition of the functor $\mathrm{R}_{B'/B}(F')$.

The fact that j_F and $q_{F'}$ define inverse correspondences goes exactly like the proof of adjointness of the pullback and pushforward functors between categories of sheaves on a topological space. The problem is therefore to compute (A.5.5): if $F = \mathrm{Hom}_B(\cdot, Y)$ for a B-scheme Y then we claim that the bijective correspondence (A.5.4) is $\beta_{F',Y}$. In other words, if $f' \in F'(Y_{B'})$ and

$$[f'] : F_{B'} = \mathrm{Hom}_B(\cdot, Y)|_{\mathscr{C}'} = \mathrm{Hom}_{B'}(\cdot, Y_{B'}) \to F'$$

denotes the associated natural transformation then we claim that the composite natural transformation

$$F \stackrel{j_F}{\to} \mathrm{R}_{B'/B}(F_{B'}) \stackrel{[f']}{\to} \mathrm{R}_{B'/B}(F') \tag{A.5.6}$$

is the point in $\mathrm{R}_{B'/B}(F')(Y)$ corresponding to f' via $\beta_{F',Y}$.

For $F = \mathrm{Hom}_B(\cdot, Y)$, it follows from the definition of j_F that the diagram

$$\begin{array}{ccc} F(Z) & \xrightarrow{j_F} & \mathrm{R}_{B'/B}(F_{B'})(Z) \\ & \searrow & \simeq \Big\downarrow \beta_{F_{B'},Z} \\ & & F_{B'}(Z_{B'}) \end{array}$$

commutes for any B-scheme Z. In particular, for any $a \in F(Z)$ the point $\beta_{F_{B'},Z}(j_F(a)) \in F_{B'}(Z_{B'})$ is equal to $a_{B'}$. (Note also that if $\pi : Z_{B'} \to Z$ is the natural map then the identification $F_{B'}(Z_{B'}) \simeq F(Z_{B'})$ carries $a_{B'}$ to $a \circ \pi$.) Hence, if $f' \in F'(Y_{B'})$ and $[f'] : F'_{B'} \to F'$ is the associated natural transformation then in the commutative diagram of sets

$$\begin{array}{ccccc} F(Z) & \xrightarrow{j_F} & \mathrm{R}_{B'/B}(F_{B'})(Z) & \xrightarrow{[f']} & \mathrm{R}_{B'/B}(F')(Z) \\ & \searrow & \simeq \Big\downarrow \beta_{F_{B'},Z} & & \beta_{F',Z} \Big\downarrow \simeq \\ & & F_{B'}(Z_{B'}) & \xrightarrow[{[f']}]{} & F'(Z_{B'}) \end{array}$$

an element $a \in F(Z)$ is carried to $a_{B'}$ in $F_{B'}(Z_{B'})$, and hence to $[f'](a_{B'}) = F'(a_{B'})(f')$ in $F'(Z_{B'})$. Setting $Z = Y$ and $a = \mathrm{id}_Y$, it follows that any $f' = F'(\mathrm{id}_{Y_{B'}})(f') \in F'(Y_{B'})$ is carried by $\beta_{F',Y}$ to the element of $\mathrm{R}_{B'/B}(F')(Y)$ associated to the composite transformation (A.5.6), as desired.

Finally, we check that j_Y is a closed immersion when $\mathrm{Spec}\, B' \to \mathrm{Spec}\, B$ is surjective, and that it is an isomorphism when $Y = \mathrm{Spec}\, B$. First consider the closed immersion claim under a surjectivity hypothesis. From the above functorial description of j_Y and consideration of points valued in any B-algebra A, if Z is a closed or open subscheme of Y then $j_Y^{-1}(\mathrm{R}_{B'/B}(Z_{B'})) = Z$ (using that $\mathrm{Spec}(B' \otimes_B A) \to \mathrm{Spec}\, A$ is faithfully flat). As U varies through the set of all affine open subschemes of Y, $\mathrm{R}_{B'/B}(U_{B'})$ varies through an affine open cover of $\mathrm{R}_{B'/B}(Y_{B'})$ (since any finite subset of $Y_{B'}$ lies over an open affine in Y). Hence, we may assume that Y is affine, and then even that $Y = \mathbf{A}^n_B$. Zariski-localization on B lets us arrange that B' admits a B-basis $\{e_1 = 1, \dots, e_d\}$. Thus, via the description of j_Y it follows that $\mathrm{R}_{B'/B}(Y_{B'})$ is identified with $\mathbf{A}^{nd}_B$ such that j_Y is the inclusion along the first n coordinates. This is a closed immersion by inspection. The functor F represented by $\mathrm{Spec}(B)$ assigns a one-point set to every B-scheme, so the identification of $j_{\mathrm{Spec}(B)}$ with j_F settles the isomorphism claim when $Y = \mathrm{Spec}\, B$. □

Proposition A.5.8 *Let $B \to B'$ be a finite flat map of noetherian rings, and X' a quasi-projective B'-scheme. The B-scheme $\mathrm{R}_{B'/B}(X')$ is quasi-projective.*

Proof For a line bundle $\mathscr{L}$ on a scheme Z, and any $s \in \mathscr{L}(Z)$, define Z_s to be the open locus where s generates $\mathscr{L}$. Following [EGA, II, 4.5.3], if Z is quasi-compact and separated then $\mathscr{L}$ is *ample* if there are sections s_i of tensor powers $\mathscr{L}^{\otimes n_i}$ ($n_i \geqslant 1$) such that the Z_{s_i} are affine and cover Z. Assuming Z is separated of finite type over a ring A, it is quasi-projective over A (in the sense of admitting a locally closed immersion into a projective space over A) if and only if it admits an ample line bundle [EGA, II, 5.3.1, 5.3.2], in which case every finite subset of Z is contained in some affine Z_s [EGA, II, 4.5.4]. It therefore suffices to construct an ample line bundle on $X = \mathrm{R}_{B'/B}(X')$.

Choose an ample line bundle $\mathscr{N}'$ on X', and consider the natural map $q : X_{B'} \to X'$ and the finite flat map $\pi : X_{B'} \to X$. The pullback $\mathscr{L}' = q^*(\mathscr{N}')$ is a line bundle on $X_{B'}$, and we claim that its norm $\mathscr{L} = \mathrm{N}_{B'/B}(\mathscr{L}')$

is ample on X. Observe that $\mathscr{L}^{\otimes n}$ is similarly constructed from $\mathscr{N}'^{\otimes n}$ for any $n \in \mathbf{Z}$. For any $s' \in \mathscr{N}'(X')$, define $\mathrm{N}(s') \in \mathscr{L}(X)$ to be the norm of $q^*(s') \in \mathscr{L}'(X_{B'})$. Since any line bundle on a semi-local ring is trivial, the locus $X_{\mathrm{N}(s')}$ equals $X - \pi(q^{-1}(X' - X'_{s'}))$. But it is immediate from functorial considerations that for any open subscheme U' in X', $U := X - \pi(q^{-1}(X' - U'))$ is $\mathrm{R}_{B'/B}(U')$, so $X_{\mathrm{N}(s')} = \mathrm{R}_{B'/B}(X'_{s'})$. The same works for tensor powers. Also, if $\{U'_i\}$ is an open cover of X' then the corresponding opens $U_i = X - \pi(q^{-1}(X' - U'_i))$ constitute an open cover of X provided that any finite subset of $X_{B'}$ (e.g., the fibers of π) is disjoint from some $q^{-1}(X' - U'_i)$. For example, if every finite subset of X' is contained in some U'_i then the U_i cover X.

By [EGA, II, 4.5.4], there is a collection $\{s'_i\}$ of global sections of powers $\mathscr{N}'^{\otimes n_i}$ with $n_i \geqslant 1$ such that the $X'_{s'_i}$ are an open affine cover of X' and every finite subset of X' is contained in some $X'_{s'_i}$. Thus, the corresponding $X_{\mathrm{N}(s'_i)}$ are an open affine cover of X, so $\mathscr{L}$ is ample on X. □

Now we specialize to the case of Weil restriction through a finite extension of fields and for group schemes.

Proposition A.5.9 *Let k be a field, k' a nonzero finite k-algebra, and $X' \to \operatorname{Spec} k'$ a smooth surjective quasi-projective map with geometrically connected fibers. The smooth k-scheme $X := \mathrm{R}_{k'/k}(X')$ is non-empty and geometrically connected.*

Before we prove the proposition, we explain why the preservation of connectedness without further restrictions on the k-algebra k' (e.g., it may not be étale) depends on smoothness. Let k be an imperfect field of characteristic $p > 0$ and k' a nontrivial purely inseparable finite extension of k. Choose $a' \in k' - k$, a power q of p such that $k'^q \subseteq k$, and an integer $m > 2q$ not divisible by p. Let X' be the geometrically integral plane curve over k' defined by $y^q = a'x^q + x^m$. The open subscheme $U' = X' - \{(0,0)\} = X' \cap \{x \neq 0\}$ is k'-smooth, so $U = \mathrm{R}_{k'/k}(U')$ is smooth and geometrically connected (by Proposition A.5.9) and it is an open subscheme of $X := \mathrm{R}_{k'/k}(X')$. We claim that X is the disjoint union of U and an open subscheme V such that $V_{\overline{k}}$ is nowhere reduced, and that if we replace q with qp then V is nowhere reduced.

Indeed, if we choose a k-basis $\{a'_i\}$ of k' beginning with $a'_0 = 1$ and $a'_1 = a'$, and substitute $\sum a'_i x_i$ for x and $\sum a'_i y_i$ for y in the equation of X' then for $c_i := a'^q_i \in k^\times$ we have

$$\sum c_i y_i^q = a'_1 \left(\sum c_i x_i^q\right) + \left(\sum a'_i x_i\right)^{m-2q} \left(\sum c_i x_i^q\right)^2.$$

Expanding $(\sum a'_i x_i)^{m-2q}$ as $\sum a'_j f_j$ with $f_j(x) \in k[x_0, x_1, \dots]$ and comparing a'_i terms on both sides shows that X is defined by the system of equations

$$\sum c_i y_i^q = f_0(x)\left(\sum c_i x_i^q\right)^2, \ \sum c_i x_i^q + f_1(x)\left(\sum c_i x_i^q\right)^2 = 0,$$
$$f_j(x)\left(\sum c_i x_i^q\right)^2 = 0 \quad (j \geqslant 2).$$

Letting $h := \sum c_i x_i^q$, for $P \in X(\overline{k})$ such that $h(P) = 0$ the relation $h(1 + f_1 h) = 0$ forces $h = 0$ in $\mathscr{O}_{X_{\overline{k}},P}$ since $(1 + f_1 h)(P) = 1$. Thus, Zariski-locally on X, h is 0 or a unit, so "$h = 0$" and "h is a unit" defines a separation of X. For $P \in X(\overline{k})$ the associated $P' : \operatorname{Spec}(k' \otimes_k \overline{k}) \to X'$ corresponds to $(x, y) \in (k' \otimes_k \overline{k})^2$ satisfying $y^q = (a' \otimes 1)x^q + x^m$ with $x = \sum a'_i \otimes x_i$ and $y = \sum a'_i \otimes y_i$ for $x_i, y_i \in \overline{k}$, so $h(P) = \sum c_i x_i^q$ is the image of x^q in the residue field $\overline{k}$ of the artin local ring $k' \otimes_k \overline{k}$. Hence, $h(P) = 0$ precisely when x is nilpotent, forcing y to be nilpotent and thus P' to be supported at $(0, 0) \in X'(\overline{k})$. In contrast, $h(P) \in \overline{k}^\times$ precisely when x is a unit; i.e., P' is supported in $U(\overline{k})$.

Thus, $V := X - U$ is an open subscheme of X with $h = 0$ in $\mathscr{O}_X(V)$, so $\sum c_i y_i^q = f_0 h^2 = 0$ on V as well. Hence, V is defined as a scheme by the vanishing of $\sum c_i x_i^q$ and $\sum c_i y_i^q$, so $V_{\overline{k}}$ is defined by the qth powers of a pair of independent linear forms (forcing $V_{\overline{k}}$ to be nowhere reduced). If we replace q with qp then we see similarly that V is nowhere reduced. (If there exists $b \in k$ not in k'^p, as happens for suitable k' when $[k : k^p] > p$, we can replace y^q with $y^q - b$ in the equation of X' so that X' is regular at its unique non-smooth point.)

Proof of Proposition A.5.9 By extension of scalars we can assume $k = \overline{k}$. Passing to local factor rings of k' reduces the problem to the case when k' is local. Let $X'_0 := X'_{\text{red}}$; this is the fiber of the k'-smooth X' over the residue field $k'_0 = k$ of k'. We have $X(k) = X'(k')$, and $X'(k') \to X'_0(k'_0) = X'_0(k)$ is surjective since X' is k'-smooth. Since $X'_0(k)$ is non-empty, we conclude that X is non-empty.

The functor $\mathrm{R}_{k'/k}$ carries Zariski-open covers to Zariski-open covers, by Corollary A.5.4(2). The smooth and geometrically connected X'_0 must be geometrically integral, so every non-empty open in X' is dense. In particular, any two non-empty open affines have a non-empty separated overlap. Since $\mathrm{R}_{k'/k}$ is compatible with open immersions and open covers, to prove that X is connected we can reduce to the case when X' is affine. By further Zariski-localization we can arrange that the k'-smooth X' is étale over an affine space over k'. That is, X' is an étale cover of a dense open $U' \subseteq \mathbf{A}^n_{k'}$ for some $n \geqslant 1$. Via the natural homeomorphism $\mathbf{A}^n_{k'} \twoheadrightarrow \mathbf{A}^n_k$ we see that $U' = U_{k'}$ for a dense open $U \subseteq \mathbf{A}^n_k$. By the topological invariance of the étale site [SGA1, IX, Thm. 4.10], the functor $E \rightsquigarrow E_{k'}$ is an equivalence between the category of étale U-schemes and the category of étale U'-schemes, so $X' = Y_{k'}$ for an étale U-scheme Y. Thus, we have reduced to the case when $X' = Y_{k'}$ for a smooth connected affine k-scheme Y, say $Y = \operatorname{Spec} B$.

Now consider the natural map $j : Y \to \mathrm{R}_{k'/k}(Y_{k'}) = X$ that is a closed immersion (Proposition A.5.7). Since Y is connected, it suffices to retract X onto Y by a homotopy. To be precise, for each $y' \in X(k) = Y(k')$ there is an associated specialization $y \in Y(k'_0) = Y(k)$, and we will deform y' to y. The point y' corresponds to a map $y'^* : B \to k'$ carrying $\mathfrak{m}_y$ into the nilpotent maximal ideal of k'. This latter ideal has vanishing dth power for some $d \geqslant 1$, so y'^* corresponds to a k-algebra map

$$k[u_1, \ldots, u_n]/(u_1, \ldots, u_n)^d = B/\mathfrak{m}_y^d \to k'$$

where $\{u_1, \ldots, u_n\}$ is a regular system of parameters of the local ring B_y. Precomposing with the scaling map $(u_1, \ldots, u_n) \mapsto (tu_1, \ldots, tu_n)$ defines a k-map $\mathbf{A}^1_k \to X$ carrying 1 to $y' \in X(k)$ and 0 to $y \in Y(k)$. □

The map $q_{X'}$ in Proposition A.5.7 has an interesting and useful surjectivity property in the smooth case:

Lemma A.5.10 *Let k be a field, k' a finite local k-algebra with residue field purely inseparable over k, and X' a smooth quasi-projective k'-scheme. The k'-map $q_{X'} : \mathrm{R}_{k'/k}(X')_{k'} \to X'$ is surjective on B'-points for any k'-algebra B'.*

Proof The residue field hypothesis implies that $A := k' \otimes_k k'$ is a local k-algebra with residue field k' via the multiplication map $m : A \to k'$ defined by $c \otimes c' \mapsto cc'$. The natural multiplication map $\pi : k' \otimes_k B' \to B'$ is induced by m via scalar extension. Since $\ker m$ has vanishing nth power for some $n > 0$, so does $\ker \pi$. Hence, $X'(\pi) : X'(k' \otimes_k B') \to X'(B')$ is surjective due to the k'-smoothness of X'. But $X'(\pi)$ is $q_{X'}$ on B'-points, by Proposition A.5.7, so we are done. □

Without smoothness conditions on X', the map $q_{X'}$ in Lemma A.5.10 can fail to be surjective. An example is $X' = \mathrm{Spec}(K')$ for a nontrivial finite extension K' of k' in characteristic $p > 0$ with $k' \subseteq k^{1/p}$ but $(K')^p \cap k'$ not contained in k, in which case $\mathrm{R}_{k'/k}(X')$ is empty. The following refinement in the smooth case is a generalization of [SGA3, XVII, App. III, Prop. 5.1(i)].

Proposition A.5.11 *Let $B \to B'$ be a finite flat map of noetherian rings and X' a smooth quasi-projective B'-scheme such that $X' \to \mathrm{Spec}\, B'$ is surjective.*

(1) *The B-scheme $X = \mathrm{R}_{B'/B}(X')$ is smooth, and the canonical map*

$$q : X_{B'} = \mathrm{R}_{B'/B}(X')_{B'} \to X'$$

(from (A.5.3)) is smooth and surjective. In particular, q is faithfully flat.

Assume $X' \to \operatorname{Spec} B'$ *has non-empty equidimensional fibers, and that for some* $y \in \operatorname{Spec} B$ *the local fiber-degrees of* $\operatorname{Spec} B' \to \operatorname{Spec} B$ *at all* y' *over* y *such that* $\dim X'_{y'} > 0$ *have sum* > 1. *Then the fibers of* $q_{y'}$ *have pure dimension* > 0 *for all* y' *over* y.

(2) *Assume that* X' *is a* B'*-group, and that* B *and* B' *are artin local rings for which the induced map on residue fields is purely inseparable. Then* $\ker q$ *has connected unipotent fibers over* $\operatorname{Spec} B'$.

(3) *If* $X' \to \operatorname{Spec} B'$ *has geometrically connected fibers and* $\operatorname{Spec} B' \to \operatorname{Spec} B$ *is surjective then* $X \to \operatorname{Spec} B$ *has geometrically connected fibers.*

Proof By Proposition A.5.2(4), the smoothness of X is inherited from that of X'; likewise, the formation of X is compatible with any scalar extension on B. Thus, for the rest we may and do assume that $B = k$ is an algebraically closed field. Passing to factor rings of B' is harmless for (1) and (3) apart from the analysis of fiber dimensions for q in (1), so now assume B' is local. Hence, B' is a finite local k-algebra with residue field k. Now the surjectivity in (1) follows from Lemma A.5.10, and (3) follows from Proposition A.5.9.

To prove the smoothness in (1), since the source and target are smooth it suffices to check smoothness for the induced map $q_0 : X \to X'_0 := X' \otimes_{B'} k$ between special fibers over the residue field k of B'. For smoothness of q_0 it suffices to check surjectivity of the induced map between tangent spaces at k-points. Choose $x \in X(k)$ and let $x' \in X'(B')$ be the corresponding section, with specialization $x'_0 \in X'_0(k)$ (so $x'_0 = q_0(x)$). Calculating with dual numbers shows that $dq_0(x)$ is the natural k-linear reduction map $\operatorname{Tan}_x(X) = \operatorname{Tan}_{x'}(X') \to \operatorname{Tan}_{x'_0}(X'_0)$. This is surjective, so q is smooth.

For the assertion on fiber dimensions in (1), we may continue to assume that B is an algebraically closed field k as above. Now B' is just a finite k-algebra (not necessarily local). Let $y'_1, \ldots, y'_m$ be the maximal ideals of B', $d_i = \dim X'_{y'_i} \geqslant 0$, and $e_i = \dim_k(B'_{y'_i}) \geqslant 1$. The hypothesis on fibral degrees implies that $\delta_i := (\sum_j e_j d_j) - d_i > 0$ for each i (as one sees by separately considering the cases that $d_i = 0$ or $d_i > 0$). Each fiber of $q_{y'_i}$ is already known to be smooth and non-empty; we will show that the tangent space at any of its k-points has dimension δ_i. The target of the smooth surjective map $q_{y'_i}$ is non-empty of pure dimension d_i, so we just have to check that the smooth y'_i-fiber of $X_{B'}$ has pure dimension $\sum e_j d_j$. More specifically, it is equivalent to show that all tangent spaces at k-points of the smooth X have dimension $\sum e_j d_j$. For $x \in X(k)$ and the corresponding section $x' \in X'(B') = \prod X'_{y'_j}(B'_{y'_j})$ with jth component $x'_j \in X'_{y'_j}(B'_{y'_j})$, the tangent space $\operatorname{Tan}_x(X)$ is identified with $\prod_j \operatorname{Tan}_{x'_j}(X'_j)$ (by consideration with dual numbers). But $\operatorname{Tan}_{x'_j}(X'_j)$ is a free $B'_{y'_j}$-module of rank d_j (by the smoothness of X'_j at x'_j), so its k-dimension is $e_j d_j$.

It remains to show that if X' is a B'-group then $\ker q$ has connected unipotent fibers over $\operatorname{Spec} B'$. By Proposition A.5.7, the canonical map $q_0 : X = \mathrm{R}_{B'/k}(X') \to X'_0$ is defined functorially on A-points (for a k-algebra A) by

$$\mathrm{R}_{B'/k}(X')(A) = X'(B' \otimes_k A) \to X'((B'/\mathfrak{m}) \otimes_k A) = X'_0(A')$$

where $\mathfrak{m}$ is the maximal ideal of B'. The map q_0 has now been expressed entirely in terms of the k-algebra B' with residue field k and the smooth quasi-projective B'-group X'. Letting $B'_n = B'/\mathfrak{m}^{n+1}$ ($n \geqslant 0$) and $X'_n = X' \otimes_{B'} B'_n$, we see that q_0 is the composition of the natural maps

$$\mathrm{R}_{B'_{n+1}/k}(X'_{n+1}) \to \mathrm{R}_{B'_n/k}(X'_n),$$

and it is precisely these latter maps that are studied in [Oes, A.3.5] (assuming X' is affine, which is never used in the proofs), and we record that result in Proposition A.5.12 below. From this it follows that $\ker q_0$ admits a composition series whose successive quotients are the vector groups $\mathrm{Lie}(X'_0) \otimes_k (\mathfrak{m}^i/\mathfrak{m}^{i+1})$. In particular, $\ker q_0$ is smooth, connected, and unipotent. □

Here is a key fact that was mentioned in the preceding proof.

Proposition A.5.12 *Let k be a field and $(B, \mathfrak{m})$ a finite local k-algebra with $B/\mathfrak{m} = k$. Let $B_i = B/\mathfrak{m}^{i+1}$ for $i \geqslant 0$. Let H be a smooth quasi-projective B-group, $H_i = H \otimes_B B_i$, and $G_i = \mathrm{R}_{B_i/k}(H_i)$. The natural map $G_{i+1} \to G_i$ is a smooth surjection and its kernel is the k-vector group corresponding to the finite-dimensional k-vector space* $\mathrm{Lie}(H_0) \otimes_k \mathfrak{m}^{i+1}/\mathfrak{m}^{i+2}$.

Proof This is [Oes, Prop. A.3.5]. It is assumed there that H is affine, but this hypothesis is never needed in the argument. (In fact, since $A \otimes_k B$ is local for any local k-algebra A, if S is any k-scheme then every B-map $S_B \to H$ lands in open affines of H Zariski-locally over S. Thus, the construction of $\mathrm{R}_{B/k}(Y)$ as a locally finite type k-scheme works for any locally finite type B-scheme Y, and likewise with each B_i in place of B, so the hypothesis that H is quasi-projective can be dropped from the proposition.) □

The preceding considerations with Weil restrictions lead to an interesting phenomenon, as follows. Let k'/k be a nontrivial purely inseparable finite extension of fields, and G' a pseudo-reductive k'-group (e.g., a connected reductive k'-group). By Proposition A.5.11 (applied to k'/k), the kernel U' of the natural map $q : \mathrm{R}_{k'/k}(G')_{k'} \to G'$ is the k'-unipotent radical. By [SGA3, XVII, App. III, Prop. 5.1(ii)], if a smooth closed k-subgroup H of $\mathrm{R}_{k'/k}(G')$ has the property that $U' \subseteq H_{k'}$ then $H = \mathrm{R}_{k'/k}(G')$. In particular, $\mathrm{R}_{k'/k}(G')$ does not contain a proper parabolic k-subgroup, even if $k = k_s$ (even though it does contain a split maximal k-torus when $k = k_s$, as does any smooth connected affine k-group). This is a striking contrast with the case of connected reductive groups.

When working with finite étale schemes over an imperfect field, the operation of purely inseparable Weil restriction is invisible. This is made precise by:

Proposition A.5.13 *Let k'/k be a purely inseparable finite extension of fields, X' a finite étale k'-scheme, and Y a finite étale k-scheme. Then $X := \mathrm{R}_{k'/k}(X')$ is a finite étale k-scheme and the natural map $q_{X'} : X_{k'} \to X'$ is an isomorphism, and likewise the natural map $j_Y : Y \to \mathrm{R}_{k'/k}(Y_{k'})$ is an isomorphism.*

In particular, the adjoint functors $\mathrm{R}_{k'/k}$ and $(\cdot) \otimes_k k'$ are naturally inverse equivalences between the categories of finite étale schemes over k' and over k, and if $f' : G' \to H'$ is an étale isogeny between smooth connected k'-groups then $\mathrm{R}_{k'/k}(f')$ is an étale isogeny with the same degree as f'.

Proof The assertion about étale isogenies is immediate from the rest via (A.5.2) with finite étale K' there. The first part of the proposition is a special case of a general fact from étale sheaf theory, but we can easily give a direct proof as follows. We can assume that $k = k_s$, so any X' is a finite disjoint union of copies of $\mathrm{Spec}\, k'$. But $q_{Y_{k'}} \circ (j_Y)_{k'}$ is always the identity map of $Y_{k'}$, so it suffices to prove that j_Y is an isomorphism for all Y. Since k'/k is purely inseparable, $\mathrm{R}_{k'/k}$ commutes with the formation of disjoint unions (Corollary A.5.4(2)). Thus, we may assume $Y = \mathrm{Spec}\, k$, and then the isomorphism property is a special case of the final claim in Proposition A.5.7. □

We conclude this section with several results concerning how Weil restriction interacts with interesting structures in the case of smooth groups of finite type over a field.

Proposition A.5.14 *Let G be a nontrivial connected semisimple group over a field k, and assume that G is simply connected. There exists a nonzero finite étale k-algebra k' and a semisimple k'-group G' with absolutely simple fibers over the factor fields of k' such that $G \simeq \mathrm{R}_{k'/k}(G')$ as k-groups.*

If $(G'', k''/k)$ is another such pair then every k-group isomorphism

$$\mathrm{R}_{k'/k}(G') \simeq \mathrm{R}_{k''/k}(G'')$$

arises via Weil restriction of a unique pair (f, α) where $\alpha : k'' \simeq k'$ is a k-algebra isomorphism and $f : G' \simeq G''$ is a group isomorphism over $\mathrm{Spec}(\alpha)$, so the pair $(G', k'/k)$ is unique up to unique isomorphism. Moreover, G is k-simple if and only if k' is a field.

This is a refinement of [BoTi1, 6.21(ii)].

Proof By the uniqueness claim concerning isomorphisms, Galois descent permits us to assume $k = k_s$ apart from the k-simplicity claim. Now assume $k = k_s$. Thus $G \simeq \mathrm{R}_{k'/k}(G')$ for $G' = \coprod G_i$ and $k' = k^I$ for a non-empty finite set I.

Due to the simply connectedness of G, it is uniquely (up to rearrangement of factors) a direct product $\prod G_i$ of connected semisimple groups G_i that are each absolutely simple and simply connected. (This rests on the canonical decomposition of a connected semisimple group over a field into an almost direct product of simple factors; see [Bo2, 22.9, 22.10].)

It remains to show for general k that $\mathrm{R}_{k'/k}(G')$ is k-simple when k' is a field, and that in general k-isomorphisms between such k-groups $\mathrm{R}_{k'/k}(G')$ arise uniquely from isomorphisms between such pairs $(G', k'/k)$. First we verify the k-simplicity when k' is a field. Consider the product decomposition

$$\mathrm{R}_{k'/k}(G')_{k_s} \simeq \prod_{\sigma} G' \otimes_{k',\sigma} k_s \tag{A.5.7}$$

in which the factors are indexed by the k-embeddings σ of k' into k_s. These factors are all k_s-simple, since G' is absolutely simple by hypothesis, so they are the simple factors of $\mathrm{R}_{k'/k}(G')_{k_s}$. Every smooth connected normal subgroup of a connected semisimple group over a field is a product among the simple factors (this is an immediate consequence of [Bo2, 22.10(i)]), so by transitivity of the $\mathrm{Gal}(k_s/k)$-action on the factors on the right side of (A.5.7) we see that there are no nontrivial proper $\mathrm{Gal}(k_s/k)$-stable smooth connected normal k_s-subgroups of $\mathrm{R}_{k'/k}(G')_{k_s}$. Hence, $\mathrm{R}_{k'/k}(G')$ is k-simple.

To prove the desired description of all k-group isomorphisms

$$\mathrm{R}_{k'/k}(G') \simeq \mathrm{R}_{k''/k}(G'')$$

it suffices (by Galois descent) to consider the case when k is separably closed. Then $k' = k^n$ and $k'' = k^m$ as k-algebras, and $G' = \coprod_{i=1}^n G'_i$ and $G'' = \coprod_{j=1}^m G''_j$ for connected semisimple k-groups G'_i and G''_j that are simply connected. In such cases $\mathrm{R}_{k'/k}(G') = \prod_i G'_i$, so the ordered n-tuple $\{G'_i\}$ is determined as the set of k-simple factors (up to rearrangement) of $\mathrm{R}_{k'/k}(G')$. In view of the uniqueness of the set of k-simple factors, we must have $n = m$, so the problem is to show that all k-group isomorphisms $\prod G'_i \simeq \prod G''_i$ arise from unique k-group isomorphisms $G'_i \simeq G''_{t(i)}$ for a unique permutation t of $\{1, \dots, n\}$. Since the k-simple factors are characterized as the minimal nontrivial smooth connected normal k-subgroups, we are done. □

Proposition A.5.15 *Let k be a field, k' a nonzero finite reduced k-algebra, G' a smooth affine k'-group, and $G = \mathrm{R}_{k'/k}(G')$.*

(1) *For any smooth closed subscheme V' of G' and associated smooth closed subscheme $V = \mathrm{R}_{k'/k}(V')$ in G we have $Z_G(V) = \mathrm{R}_{k'/k}(Z_{G'}(V'))$ as subschemes of G. In particular, $Z_G = \mathrm{R}_{k'/k}(Z_{G'})$.*

(2) *There is a natural bijection between the set of maximal k-tori T in G and the set of maximal k'-tori T' in G' by the requirement that $T \subseteq \mathrm{R}_{k'/k}(T')$. The maximal k-split torus T_0 in T is maximal as a k-split torus in $\mathrm{R}_{k'/k}(G')$ if and only if the maximal k'-split torus T'_0 in T' is maximal as a k'-split torus in G'.*

A bijection between the sets of maximal k-split tori and maximal k'-split tori is likewise defined by the requirement $T_0 \subseteq \mathrm{R}_{k'/k}(T'_0)$.

(3) *If T' is a maximal k'-torus in G' and T is the maximal k-torus in $\mathrm{R}_{k'/k}(T')$ then*

$$\mathrm{R}_{k'/k}(Z_{G'}(T')) = Z_G(\mathrm{R}_{k'/k}(T')) = Z_G(T).$$

In particular, if C' is a Cartan k'-subgroup of G' then $\mathrm{R}_{k'/k}(C')$ is a Cartan k-subgroup of G, and every Cartan k-subgroup of G arises in this way for a unique C'.

This proposition is true more generally for smooth groups of finite type (not necessarily affine). All that is missing to make the argument below work in this greater generality is enough foundational material concerning tori in such groups (e.g., preservation of maximality under extension of the ground field, behavior under smooth surjections, and rational conjugacy results in the maximal split case). Lemma C.4.4 and Proposition C.4.5 below provide the required ingredients beyond the affine case.

Proof We postpone the assertions concerning split tori in (2) to the end of the proof because for the proof of the other assertions we may assume that k is separably closed. (This rests on using Galois descent for the proof of the bijectivity among maximal tori in (2).) Passing to factor fields of k' lets us reduce to the case when k' is a field (purely inseparable of finite degree over k). We will assume that we are in this setting until we indicate otherwise.

First consider (1). To prove the equality $Z_G(V) = \mathrm{R}_{k'/k}(Z_{G'}(V'))$ as closed subschemes of $G = \mathrm{R}_{k'/k}(G')$ we compare A-valued points for any k-algebra A. Choose $g \in G(A)$, and let g' denote the corresponding point in $G'(A')$ with $A' = k' \otimes_k A$. We need to prove that g centralizes V_A if and only if g' centralizes $V'_{A'}$. The identification $G_A = \mathrm{R}_{A'/A}(G'_{A'})$ carries the conjugation automorphism $c_g : x \mapsto gxg^{-1}$ of G_A over to $\mathrm{R}_{A'/A}(c_{g'})$. The natural map $q_{G'} : G_{k'} \to G'$ corresponding to

$$G(B') = G'(k' \otimes_k B') \to G'(B')$$

for k'-algebras B' is surjective by Proposition A.5.11(1).

Extending scalars to A' gives a quotient map

$$(G_A) \otimes_A A' = (G_{k'}) \otimes_{k'} A' \twoheadrightarrow G'_{A'}$$

carrying $(V_A) \otimes_A A'$ onto $V'_{A'}$ (since V' is k'-smooth) and carrying $(c_g)_{A'}$ over to $c_{g'}$. The identification of c_g with $\mathrm{R}_{A'/A}(c_{g'})$ and of $c_{g'}$ with an automorphism induced by $(c_g)_{A'}$ imply that c_g is trivial on V_A if and only if $c_{g'}$ is trivial on $V'_{A'}$. This proves that $\mathrm{R}_{k'/k}(Z_{G'}(V')) = Z_G(V)$ inside of G. Taking $V' = G'$ then yields the relation $Z_G = \mathrm{R}_{k'/k}(Z_{G'})$ between scheme-theoretic centers.

Now consider the claim in (2) concerning maximal tori. For any maximal k-torus T in G, the natural surjective map $G_{k'} = \mathrm{R}_{k'/k}(G')_{k'} \twoheadrightarrow G'$ (which is even smooth, by Proposition A.5.11(1)) carries $T_{k'}$ onto a maximal k'-torus T' of G', so by the universal property of Weil restriction we have $T \subseteq \mathrm{R}_{k'/k}(T')$. Since k and k' are separably closed, we can apply Proposition A.2.10 to obtain the desired bijection among maximal tori.

To prove (3), let $V = \mathrm{R}_{k'/k}(T')$. Since T' is a torus, the centralizer $Z_{G'}(T')$ is k'-smooth (Proposition A.2.5). Thus, the centralizer $Z_G(V) = \mathrm{R}_{k'/k}(Z_{G'}(T'))$ is k-smooth. Letting T be the maximal k-torus in $V = \mathrm{R}_{k'/k}(T')$, we wish to prove that $Z_G(V) = Z_G(T)$. The containment $Z_G(V) \subseteq Z_G(T)$ is clear, and to prove the reverse containment it suffices to check equality on k-points since $k = k_s$ and both $Z_G(V)$ and $Z_G(T)$ are smooth. That is, we want any $g \in Z_{G(k)}(T)$ to lie in $Z_G(V)(k) = Z_{G'(k')}(T')$. Since $G_{k'} \to G'$ carries $T_{k'}$ onto T', this is clear.

Finally, we return to the situation over a general field k and address the claims in (2) concerning maximal split tori. We may again assume that k' is a field. Observe that if S' is a k'-anisotropic torus then the k-group $\mathrm{R}_{k'/k}(S')$ (which is not a k-torus if $S' \neq 1$ and k'/k is not separable) is k-anisotropic in the sense that there is no nontrivial k-homomorphism $\mathrm{GL}_1 \to \mathrm{R}_{k'/k}(S')$.

Consider a matching pair (T, T') of maximal tori, and the maximal k-split torus $T_0 \subseteq T$ and maximal k'-split torus $T'_0 \subseteq T'$. Since the quotient T'/T'_0 is a k'-anisotropic torus, so the k-group

$$\mathrm{R}_{k'/k}(T')/\mathrm{R}_{k'/k}(T'_0) \simeq \mathrm{R}_{k'/k}(T'/T'_0)$$

is k-anisotropic, necessarily $T_0 \subseteq \mathrm{R}_{k'/k}(T'_0)$. If T_0 is a maximal k-split torus in $\mathrm{R}_{k'/k}(G')$ then since $T_0 \subseteq \mathrm{R}_{k'/k}(T'_0)$ it follows that T'_0 is a maximal k'-split torus in G'. Indeed, assume T'_0 is contained in a k'-split torus $S' \subseteq G'$, and let S be the maximal k-split torus in $\mathrm{R}_{k'/k}(S')$. Then $T_0 \subseteq S$, so $T_0 = S$, but the maps $S_{k'} \to S'$ and $(T_0)_{k'} \to T'_0$ are surjective, so $T'_0 = S'$.

Conversely, if T'_0 is a maximal k'-split torus in G' then we wish to prove that T_0 is a maximal k-split torus in G. Obviously $(T_0)_{k'}$ maps onto T'_0 in G' since T_0 is the maximal k-split torus in $\mathrm{R}_{k'/k}(T'_0)$, so if S_0 is a k-split torus of G containing T_0 then the image of $(S_0)_{k'}$ under the canonical surjection $\mathrm{R}_{k'/k}(G')_{k'} \twoheadrightarrow G'$ is a k'-split torus containing T'_0. Maximality of T'_0 then forces

$(S_0)_{k'}$ to map into T'_0, so $S_0 \subseteq \mathrm{R}_{k'/k}(T'_0)$. But T_0 is the maximal k-split torus in $\mathrm{R}_{k'/k}(T'_0)$, so the containment $T_0 \subseteq S_0$ is an equality. That is, T_0 is maximal as a k-split torus in $\mathrm{R}_{k'/k}(G')$.

Since every maximal split torus in a smooth affine group over a field is also the maximal split torus in some maximal torus over the ground field, the preceding arguments show that for every maximal k'-split torus T'_0 in G' the maximal k-split torus T_0 in $\mathrm{R}_{k'/k}(T'_0)$ is maximal as a k-split torus in $\mathrm{R}_{k'/k}(G')$, and moreover that every maximal k-split torus T_0 in $\mathrm{R}_{k'/k}(G')$ arises in this way for some T'_0. Observe that T'_0 is uniquely determined by T_0, since we have already seen that it is the image of $(T_0)_{k'}$ under the canonical quotient map $\mathrm{R}_{k'/k}(G')_{k'} \twoheadrightarrow G'$. □

Recall that if G is a smooth connected affine group over a field k, a *Levi k-subgroup* of G is a smooth connected k-subgroup $L \subseteq G$ such that the natural map $L_{\overline{k}} \ltimes \mathscr{R}_u(G_{\overline{k}}) \to G_{\overline{k}}$ is an isomorphism.

Corollary A.5.16 *Let k'/k be a purely inseparable finite extension of fields, and G a connected reductive k-group. The geometric unipotent radical of the smooth connected affine k-group $\mathrm{R}_{k'/k}(G_{k'})$ is defined over k', and the closed k-subgroup $\iota : G \to \mathrm{R}_{k'/k}(G_{k'})$ is a Levi subgroup.*

Proof For any k'-algebra B there is a natural quotient map $k' \otimes_k B \twoheadrightarrow B$ as k'-algebras (using the left k'-structure on the source) and hence we naturally get a map of groups $\mathrm{R}_{k'/k}(G_{k'})_{k'}(B) = G_{k'}(k' \otimes_k B) \to G_{k'}(B)$. By Proposition A.5.7, this is the smooth surjective map of k'-groups $q_{(G_{k'})} : \mathrm{R}_{k'/k}(G_{k'})_{k'} \to G_{k'}$ for which $\iota_{k'}$ is a section. By Proposition A.5.11 the kernel of $q_{(G_{k'})}$ is smooth, connected, and unipotent. Hence, $\ker q_{(G_{k'})}$ is a k'-descent of the geometric unipotent radical of $\mathrm{R}_{k'/k}(G')$, and G is a Levi k-subgroup via ι. □

A.6 Groups without Levi subgroups

In this section we construct many smooth connected affine groups in nonzero characteristic not containing a Levi subgroup, even if the ground field is algebraically closed. We never need such a construction, but we record it here for three reasons: (i) no example of this sort seems to be in the literature, (ii) it is an interesting contrast with our result (Theorem 3.4.6) that pseudo-reductive groups over a separably closed field k always admit a Levi k-subgroup, and (iii) the construction involves the Greenberg functor, which is (roughly) a mixed characteristic analogue of the Weil restriction functor that we discussed in §A.5.

Let k be an algebraically closed field with $\mathrm{char}(k) = p > 0$. For any non-commutative connected reductive group scheme H over the artin local ring

$W_2(k)$ of length-2 Witt vectors for k, we claim that the group $H(W_2(k))$ viewed as a k-group (in a manner that is made rigorous below) has no Levi k-subgroup. For example, $\mathrm{SL}_2(W_2(k))$ viewed as a k-group has no Levi k-subgroup.

The idea is to identify $H(k)$ as the maximal reductive quotient of $H(W_2(k))$ and to show that if this quotient map has a k-group section (i.e., if there exists a Levi k-subgroup in $H(W_2(k))$) then by inspecting suitable "root groups" such a section induces a k-group section to the canonical quotient map $W_2 \twoheadrightarrow \mathbf{G}_\mathrm{a}$. No such section exists, since $\mathbf{G}_\mathrm{a}$ is p-torsion whereas $W_2[p]$ is the preimage of $\alpha_p \subseteq \mathbf{G}_\mathrm{a}$. Hence, $H(W_2(k))$ viewed as a k-group has no Levi subgroup.

In the instructive and interesting special case $H = \mathrm{SL}_2$, most of the functorial considerations below are not necessary and so the argument simplifies a lot. (The general case will essentially be reduced to this special case.) To analyze the general case rigorously without matrix calculations, we first briefly review the Greenberg functor that is analogous to Weil restriction.

For $n \geqslant 1$, let W_n denote the functor of length-n Witt vectors $(w_0, \ldots, w_{n-1})$ on k-algebras. For every k-algebra B, the ring $W_n(B)$ is naturally a $W_n(k)$-algebra. In view of the universal formulas for the ring scheme structure on W_n, for any affine $W_n(k)$-scheme X of finite type, the functor

$$B \rightsquigarrow X(W_n(B))$$

is represented by an affine k-scheme X_n of finite type. (The correspondence $X \rightsquigarrow X_n$ is the *Greenberg functor* on affine $W_n(k)$-schemes of finite type. Loosely speaking, the Greenberg functor is a mixed characteristic analogue of Weil restriction through the finite flat covering $\mathrm{Spec}(k[t]/(t^n)) \to \mathrm{Spec}(k)$.) The formation of X_n is compatible with products in X over $W_n(k)$, and $X_n = \mathrm{Spec}\, k$ when $X = \mathrm{Spec}\, W_n(k)$, so X_n is naturally a k-group scheme when X is a $W_n(k)$-group scheme.

Example A.6.1 Let $X = \mathbf{G}_a$ over $W_n(k)$. Then $X_n = W_n$ as a k-group.

In general, the k-scheme X_n is smooth when X is $W_n(k)$-smooth, due to the functorial criterion for smoothness and the fact that the map $W_n(B') \to W_n(B)$ is surjective with square-zero kernel when $B' \to B$ is surjective with square-zero kernel. The relationship between connectedness properties of X and X_n requires the following fibration construction.

Define $\overline{X} = X \otimes_{W_n(k)} k$, so there is a natural k-scheme map

$$\pi_n : X_n \to \overline{X}$$

defined functorially by

$$X(W_n(B)) \to X(B) = \overline{X}(B)$$

for k-algebras B. The map π_n is a smooth surjection when X is smooth, and it is a k-homomorphism when X is a $W_n(k)$-group scheme. We now apply this with $n = 2$.

Lemma A.6.2 *Let X be an affine $W_2(k)$-scheme of finite type, and choose $x \in X(W_2(k))$. Let $\overline{x} \in \overline{X}(k)$ be its reduction.*

There is a natural isomorphism of k-schemes $\theta_x : \pi_2^{-1}(\overline{x}) \simeq \mathrm{T}_{\overline{x}}(\overline{X})^{(p)}$ onto the Frobenius twist of the tangent space to $\overline{X}$ at $\overline{x}$, with $\theta_x(x) = 0$. In particular, if X is connected then X_2 is connected.

Proof The fiber scheme $\pi_2^{-1}(\overline{x})$ over k represents the functor of points in $X(W_2(B))$ reducing to the image of $\overline{x} \in \overline{X}(k)$ in $\overline{X}(B)$ for varying k-algebras B. We need to identify this with the functor $B \rightsquigarrow B \otimes_k \mathrm{T}_{\overline{x}}(\overline{X})^{(p)}$.

Let A be the coordinate ring of X, so B-points of $\pi_2^{-1}(\overline{x})$ are $W_2(k)$-algebra maps $A \to W_2(B)$ having the form

$$a \mapsto a(x) + (0, h(a))$$

for a map of sets $h : A \to B$ (where $a(x) \in W_2(k)$ is viewed in $W_2(B)$ in the evident manner).

Using the ring scheme structure on W_2, $a \mapsto a(x) + (0, h(a))$ is a $W_2(k)$-algebra map if and only if $h(a) = \overline{h}(a \bmod p)$ for a (unique) additive map

$$\overline{h} : \overline{A} = A/pA \to B$$

such that $\overline{h}(1) = 0$, $\overline{h}(c\overline{a}) = c^p\overline{h}(\overline{a})$ for all $c \in W_2(k)/(p) = k$ and $\overline{a} \in \overline{A}$, and

$$\overline{h}(\overline{a} \cdot \overline{a}') = \overline{a}(\overline{x})^p\overline{h}(\overline{a}') + \overline{a}'(\overline{x})^p\overline{h}(\overline{a})$$

for all $\overline{a}, \overline{a}' \in \overline{A}$. Such maps factor through $\overline{A}/\mathfrak{m}_{\overline{x}}^2$, vanish on k, and are k-linearized after applying a Frobenius scalar extension on the source. This is precisely the set

$$\mathrm{Hom}_k(\mathrm{Cot}_{\overline{x}}(\overline{X})^{(p)}, B) \simeq \mathrm{T}_{\overline{x}}(\overline{X})^{(p)} \otimes_k B. \qquad \square$$

Remark A.6.3 By functoriality of the formation of π_n and of tangent spaces with respect to products, θ_x is compatible with products in (X, x). Hence, if X is a $W_2(k)$-group and $x = 1$ then θ_x is a k-group isomorphism between $\pi_2^{-1}(1) = \ker \pi_2$ and $\mathrm{Lie}(\overline{X})^{(p)}$.

The functoriality of θ_x with respect to automorphisms of (X, x) can be made explicit: if $\sigma \in \mathrm{Aut}_{W_2(k)}(X, x)$ then the induced automorphism on $\mathrm{T}_{\overline{x}}(\overline{X})^{(p)}$ via θ_x is $\mathrm{d}\overline{\sigma}(\overline{x})^{(p)}$, with $\overline{\sigma} \in \mathrm{Aut}_k(\overline{X}, \overline{x})$ denoting $\sigma \bmod p$.

Now let H be a smooth affine $W_2(k)$-group (such as SL_2) and let $G = H_2$ be the associated smooth affine k-group. By Lemma A.6.2, G is connected when H is connected. By Remark A.6.3, there is an exact sequence of smooth k-groups

$$1 \to \mathrm{Lie}(\overline{H})^{(p)} \to G \overset{\pi}{\to} \overline{H} \to 1 \qquad \text{(A.6.1)}$$

where $\pi = \pi_2$.

Since the group scheme $\ker \pi = \mathrm{Lie}(\overline{H})^{(p)}$ is commutative, the conjugation action by G on $\ker \pi$ factors through an action by $\overline{H}$ on $\mathrm{Lie}(\overline{H})^{(p)}$. The description of functoriality of θ_x with respect to automorphisms in Remark A.6.3 implies that this action by $\overline{H}$ is the Frobenius scalar extension of the adjoint action of $\overline{H}$ on $\mathrm{Lie}(\overline{H})$.

Assume that H is a non-commutative connected reductive $W_2(k)$-group. Since $W_2(k)$ is an artin local ring with algebraically closed residue field, H is a split reductive group scheme over $W_2(k)$. (To avoid the theory of split reductive groups over artinian rings, the reader may prefer to work with a fixed nontrivial Chevalley group, such as SL_2.) The smooth surjective map $\pi : G \twoheadrightarrow \overline{H}$ is the maximal reductive quotient of G since $\ker \pi$ is a vector group.

Proposition A.6.4 *Let k be an algebraically closed field with* $\mathrm{char}(k) > 0$, *and let H be a non-commutative connected reductive $W_2(k)$-group. Let $G = H_2$ be the associated smooth connected affine k-group. This k-group has no Levi k-subgroup. That is, the maximal reductive quotient map $\pi : G \to \overline{H}$ has no k-group section.*

Proof We assume that there is a k-group section s to π, and we seek a contradiction. The idea is to modify s so it is compatible with $W_2(k)$-points of a root group in G and k-points of a root group in $\overline{H}$ (relative to suitable maximal tori that are compatible via π). The root groups are $\mathbf{G}_a$ (over $W_2(k)$ and k respectively), so by Example A.6.1 with $n = 2$, the map s would restrict to a section of the quotient map $W_2 \to \mathbf{G}_a$. But no such section of the latter sort exists.

Step 1. To get started, we will arrange compatible choices of maximal tori in G and $\overline{H}$. Let T be a maximal $W_2(k)$-torus in H. Then T_2 is a smooth commutative subgroup of $G = H_2$. It is an extension of the maximal k-torus $\overline{T} = T \bmod p$ of $\overline{H}$ by the vector group corresponding to the k-vector space $\mathrm{Lie}(\overline{T})^{(p)}$. Hence, the maximal k-torus $\mathscr{T}$ in T_2 is a maximal k-torus of G.

All maximal k-tori in G are $G(k)$-conjugate [Bo2, Cor. 11.3], so for a suitable $g \in G(k)$ we may replace the section s with the section $c_{g^{-1}} \circ s \circ c_{\pi(g)}$ (where c_γ denotes conjugation by γ) to arrange that the Levi k-subgroup $s(\overline{H})$ contains $\mathscr{T}$.

Under the reduction map π, the k-torus $\mathscr{T}$ is carried isomorphically onto the maximal torus $\overline{T}$ in $\overline{H}$, so $s(\overline{T}) = \mathscr{T}$. For any cocharacter $\lambda : \mathrm{GL}_1 \to T$ over $W_2(k)$, let $[\lambda] : \mathrm{GL}_1 \to \mathscr{T}$ be the map between maximal tori defined by the k-group map $\lambda_2 : (\mathrm{GL}_1)_2 \to T_2$ arising from λ by functoriality. Concretely, $[\lambda](t) = \lambda([t])$ in $T_2(W_2(k))$ where $[t] \in W_2(k)^\times$ is the Teichmüller lift of $t \in k^\times$. In particular, for each coroot $\overline{c} : \mathrm{GL}_1 \to \overline{T}$ we get a 1-parameter subgroup $[\overline{c}] : \mathrm{GL}_1 \to \mathscr{T}$ in G defined by $[\overline{c}] = s \circ \overline{c}$.

For the associated cocharacter $\overline{\lambda} : \mathrm{GL}_1 \to \overline{T}$ over k, the restriction of s to $\overline{T}$ satisfies $s(\overline{\lambda}(t)) = [\lambda](t)$ for $t \in \mathrm{GL}_1$.

STEP 2. Now we use root groups. The set of roots $\Phi(H,T)$ is non-empty since H is non-commutative. Choose $a \in \Phi(H,T)$, so the root group U_a in H is isomorphic to $\mathbf{G}_\mathrm{a}$ over $W_2(k)$. Let $\overline{a} \in \Phi(\overline{H},\overline{T})$ denote the reduction of a over k, and let

$$U_{\overline{a}} := U_a \bmod p \subseteq \overline{H}$$

be the $\overline{a}$-root group (k-isomorphic to $\mathbf{G}_\mathrm{a}$).

Consider $s(U_{\overline{a}}) \subseteq G$. We claim that this subgroup is contained in the unipotent subgroup $(U_a)_2 \subseteq H_2 = G$ over k. Letting $\overline{T}_{\overline{a}}$ denote the unique codimension-1 torus in $\overline{T}$ killed by $\overline{a}$, $s(U_{\overline{a}})$ is centralized by $s(\overline{T}_{\overline{a}})$, and the subgroup

$$s(\overline{T}_{\overline{a}})(k) \subseteq \mathscr{T}(k) \subseteq T_2(k) = T(W_2(k))$$

is schematically dense in the unique codimension-1 $W_2(k)$-torus T_a in T that is killed by a. Thus,

$$s(U_{\overline{a}}) \subseteq Z_G((T_a)_2)$$

inside of G by comparison of k-points (since $k = \overline{k}$).

To get finer control on $s(U_{\overline{a}})$, we need to use the 1-parameter subgroup

$$[\overline{a}^\vee] : \mathrm{GL}_1 \to Z_G((T_a)_2).$$

The maximal k-torus $\mathscr{T}$ in $Z_G((T_a)_2)$ is the almost direct product of the central torus $s(\overline{T}_{\overline{a}})$ and the 1-dimensional torus $[\overline{a}^\vee](\mathrm{GL}_1)$, so we will study the conjugation action of $[\overline{a}^\vee](\mathrm{GL}_1)$ on $Z_G((T_a)_2)$.

First, we need to give an alternative description of $Z_G((T_a)_2)$. By the self-contained Proposition A.8.10(1),(2) below, applied over $W_2(k)$, the centralizer functor $Z_H(T_a)$ inside of H is represented by a smooth closed $W_2(k)$-subgroup. By functorial considerations, we get

$$Z_G((T_a)_2) = Z_H(T_a)_2.$$

Thus, the affine finite type k-group $Z_G((T_a)_2)$ is k-smooth since $Z_H(T_a)$ is $W_2(k)$-smooth, and it is connected as well because the $W_2(k)$-smooth $Z_H(T_a)$ is connected (as $Z_H(T_a) \bmod p$ is the torus centralizer $Z_{\overline{H}}(\overline{T}_{\overline{a}})$ over a field).

Applying Lemma 2.1.5 to the 1-parameter subgroup $[\overline{a}^\vee] : \mathrm{GL}_1 \to Z_G((T_a)_2)$ over k provides a smooth connected unipotent k-subgroup

$$U^+ := U_{Z_G((T_a)_2)}([\overline{a}^\vee]) \subseteq Z_G((T_a)_2)$$

that represents the functor of points $g \in Z_G((T_a)_2)(B) = Z_H(T_a)(W_2(B))$ such that

$$[\overline{a}^\vee](t) \cdot g \cdot [\overline{a}^\vee](t)^{-1} \to 1$$

in $Z_G((T_a)_2)_B$ as $t \to 0$, with B any k-algebra. Also, $\mathrm{Lie}(U^+)$ is the span of the $[\overline{a}^\vee](\mathrm{GL}_1)$-weight spaces on $\mathrm{Lie}(Z_G((T_a)_2))$ for the weights whose pairing against $[\overline{a}^\vee]$ is > 0.

Step 3. Since $\mathscr{T}(k) \subseteq T(W_2(k))$, and the Teichmüller lift $\mathrm{GL}_1 \to W_2^\times$ extends to a k-scheme map $\mathbf{A}^1 \to W_2$ carrying 0 to 0, the subgroup $U^+(k)$ ($\subseteq G(k) = H(W_2(k))$) contains $(U_a)_2(k) = U_a(W_2(k))$. Likewise, $s(U_{\overline{a}}) \subseteq U^+$ since $[\overline{a}^\vee] = s \circ \overline{a}^\vee$ and $\overline{a}^\vee(t) \cdot u \cdot \overline{a}^\vee(t)^{-1} \to 1$ as $t \to 0$ for $u \in U_{\overline{a}}(k)$. Hence, $s(U_{\overline{a}})$ and $(U_a)_2$ are contained in U^+.

We claim that $s(U_{\overline{a}}) \subseteq (U_a)_2$. It suffices to show that the inclusion $(U_a)_2 \subseteq U^+$ is an equality. By smoothness and connectedness, such an equality amounts to a dimension calculation: proving $\dim U^+ = 2$. We shall describe $\mathrm{Lie}(U^+)$ inside of $\mathrm{Lie}(G)$ to see that it is 2-dimensional.

Passing to Lie algebras on (A.6.1), $\mathrm{Lie}(G)$ is an extension of $\mathrm{Lie}(\overline{H})$ by $\mathrm{Lie}(\overline{H})^{(p)}$. This is compatible with an action of $\mathscr{T} \simeq \overline{T}$ that induces the adjoint action of $\overline{T}$ on the quotient $\mathrm{Lie}(\overline{H})$ and the natural scalar extension of this action on the subspace $\mathrm{Lie}(\overline{H})^{(p)}$.

The root system associated to $(\overline{H}, \overline{T})$ is reduced, and the roots occurring on $\mathrm{Lie}(\overline{H})^{(p)}$ are the p-powers of the ones on $\mathrm{Lie}(\overline{H})$, so each root line in $\mathrm{Lie}(\overline{H})$ admits a unique $\overline{T}$-equivariant lift into $\mathrm{Lie}(G)$. Hence, $\mathrm{Lie}(U^+)$ is spanned by a 1-dimensional weight space for $\overline{a}^p$ on $\mathrm{Lie}(\overline{H})^{(p)}$ and a 1-dimensional weight space for $\overline{a}$. It follows that $\dim U^+ = 2$, so indeed $s(U_{\overline{a}}) \subseteq (U_a)_2$.

We conclude that s restricts to a section of the reduction map $(U_a)_2 \to U_{\overline{a}}$. But this reduction map is the natural map $W_2 \twoheadrightarrow \mathbf{G}_\mathrm{a}$ for which there is no k-group section. □

A.7 Lie algebras and Weil restriction

In the development of the structure theory of pseudo-reductive groups, Weil restriction and infinitesimal group schemes are useful notions. For this reason, it is sometimes necessary to do computations with Lie algebras of Weil restrictions of group schemes of finite type over a field, so we need to describe such Lie algebras in concrete terms. Also, in our study of exotic pseudo-reductive groups it will be useful to work with Lie algebras of infinitesimal group schemes. Both of these topics are taken up in this section.

As a preliminary step, we first review how a Lie algebra is associated to any group scheme G over any ring k. An elegant treatment of the formalism of Lie algebras of group schemes (and even group functors) is given in [DG, II, §4, §7]. For the convenience of the reader who is more familiar with algebraic groups than algebraic geometry, we begin with a self-contained development of that formalism. (Restricting to the affine finite type case below will be sufficient for our needs, and it makes some of our arguments more concrete.)

A natural Lie algebra structure on $\mathfrak{g} = \mathrm{Tan}_e(G)$ can be defined in a couple of different (but equivalent) ways. The most classical approach is via left-invariant global vector fields (or rather, left-invariant derivations of the structure sheaf, to allow non-smoothness), but for some functorial arguments it will be convenient to have a more direct group-theoretic formulation near the identity. To avoid confusion, we now review both viewpoints and prove that they agree.

Let us begin with the global approach through vector fields. A k-linear derivation D of the structure sheaf $\mathscr{O}_G$ is *left-invariant* if, for all k-algebras B and $g \in G(B)$, the B-linear derivation D_B of $\mathscr{O}_{G_B}$ commutes with pullback along the left multiplication map $\ell_g : h \mapsto gh$. This condition can also be expressed in terms of the $k[\epsilon]$-scheme automorphism $\widetilde{D}$ of $G_{k[\epsilon]}$ that is the identity on topological spaces and the $k[\epsilon]$-algebra automorphism $1 + \epsilon D$ on the sheaf of $k[\epsilon]$-algebras $\mathscr{O}_{G_{k[\epsilon]}}$:

Lemma A.7.1 *The global derivation D is left-invariant if and only if the diagram*

$$\begin{array}{ccc} G_{k[\epsilon]} \times_{\mathrm{Spec}\, k[\epsilon]} G_{k[\epsilon]} & \xrightarrow{m} & G_{k[\epsilon]} \\ {\scriptstyle 1\times\widetilde{D}}\downarrow & & \downarrow{\scriptstyle \widetilde{D}} \\ G_{k[\epsilon]} \times_{\mathrm{Spec}\, k[\epsilon]} G_{k[\epsilon]} & \xrightarrow[m]{} & G_{k[\epsilon]} \end{array} \tag{A.7.1}$$

commutes. In such cases, D is uniquely determined by its specialization $D(e) \in \mathfrak{g}$ at e.

Proof The key point is the observation that to check such commutativity it suffices to do so against all points $g \in G(B) \subseteq G_{k[\epsilon]}(B[\epsilon])$ in the left factor $G_{k[\epsilon]}$ along the left side of (A.7.1) where $\mathrm{Spec}\, B$ varies through an open affine cover of G and g is the corresponding inclusion map (as then the corresponding points of $G_{k[\epsilon]}$ constitute an open affine covering).

Working more generally with any k-algebra B and any $g \in G(B)$, the $B[\epsilon]$-scheme automorphisms $\widetilde{D}_{B[\epsilon]}$ and $\ell_{g_{B[\epsilon]}}$ of $G_{B[\epsilon]}$ trivially commute modulo ϵ, and at the level of structure sheaves over $B[\epsilon]$ they commute precisely when D_B and ℓ_g commute on $\mathscr{O}_{G_B}$.

If an automorphism ϕ of a group functor commutes with all left translations then it must be right multiplication by $\phi(e)$ (so ϕ is determined by $\phi(e)$). Hence, if $\widetilde{D}$ is left-invariant then it is right multiplication by

$$\widetilde{D}(e) = 1 + \epsilon D(e) \in \ker(G(k[\epsilon]) \to G(k)) = 1 + \epsilon\mathfrak{g}.$$

(Conversely, right multiplication on $G_{k[\epsilon]}$ by $1 + \epsilon v$ for $v \in \mathfrak{g}$ is a $k[\epsilon]$-scheme automorphism of $G_{k[\epsilon]}$ that reduces to the identity modulo ϵ and so has the form $\widetilde{D}$ for a unique D that is necessarily left-invariant.) □

In concrete terms, $\widetilde{D}(e)$ corresponds to the augmented k-algebra map

$$k \oplus \mathscr{I}_G/\mathscr{I}_G^2 = \mathscr{O}_G/\mathscr{I}_G^2 \to k[\epsilon]$$

given by $a + b\epsilon \mapsto a + (D(\widetilde{b}))(e)\epsilon$ for a local section $\widetilde{b}$ of $\mathscr{I}_G$ lifting b. In other words, $\widetilde{D}(e)$ is the "tangent vector" along e induced by the "vector field" D. (Keep in mind that we do not assume G is k-smooth, so the classical language of vector fields is not entirely appropriate.) This proves:

Lemma A.7.2 *Every element of $\mathfrak{g}$ uniquely extends to a k-linear left-invariant derivation of the structure sheaf $\mathscr{O}_G$. Explicitly, if $v \in \mathfrak{g}$ corresponds to D_v then right multiplication by $1 + \epsilon v$ on $G_{k[\epsilon]}$ is the identity on the topological space and $1 + \epsilon D_v$ on the structure sheaf.*

In view of this lemma, we can define a Lie algebra structure $[\cdot, \cdot]_G$ on $\mathfrak{g}$ over k by forming the commutator of left-invariant derivations, exactly as in the classical smooth case. By construction, this Lie algebra structure is compatible with change of the base ring in the sense that the base change map (2.1.3) is a map of Lie algebras over k'.

To analyze further properties of the Lie bracket, such as its functoriality in G, we now express the bracket in another way via infinitesimal methods, following [DG, II, §4, 4.1–4.2]. (A related technique is used in [Ser2] to define a Lie algebra structure on smooth formal groups.) For any k-algebra B and $g \in G(B)$, consider the conjugation automorphism c_g of G_B defined by $h \mapsto ghg^{-1}$. This induces a B-linear automorphism $\mathrm{Ad}_G(g)$ of $\mathrm{Tan}_{e_B}(G_B)$ by functoriality. Explicitly, for $X \in \mathrm{Tan}_{e_B}(G_B)$,

$$1 + \epsilon' \mathrm{Ad}_G(g)(X) = g(1 + \epsilon' X)g^{-1} \qquad \text{(A.7.2)}$$

in $G_B(B[\epsilon']) = G(B[\epsilon'])$. Using $B[\epsilon]$ in place of B, if $g \in G(B[\epsilon])$ then $\mathrm{Ad}_G(g)$ is a $B[\epsilon]$-linear automorphism of

$$\mathrm{Tan}_{e_{B[\epsilon]}}(G_{B[\epsilon]}) = B[\epsilon] \otimes_B \mathrm{Tan}_{e_B}(G_B)$$

(equality since $B[\epsilon]$ is finite free over B) whose reduction modulo ϵ is the adjoint action of g mod ϵ. Thus, if g mod $\epsilon = 1$ then $\mathrm{Ad}_G(g) = \mathrm{id} + \epsilon \cdot \mathrm{ad}_G(g)$ for a unique B-linear endomorphism $\mathrm{ad}_G(g)$ of $\mathrm{Tan}_{e_B}(G_B)$. Writing $g = 1 + \epsilon v$ for $v \in \mathrm{Tan}_{e_B}(G_B)$, we have made a construction

$$\mathrm{ad}_G : \mathrm{Tan}_{e_B}(G_B) \to \mathrm{End}_B(\mathrm{Tan}_{e_B}(G_B))$$

via $\mathrm{Ad}_G(1+\epsilon v) = \mathrm{id} + \epsilon \cdot \mathrm{ad}_G(v)$, so $(v, w) \mapsto (\mathrm{ad}_G(v))(w)$ is B-bilinear and natural in B and G.

Example A.7.3 The construction of ad_G can be described in more classical terms when G is locally of finite presentation and $\mathrm{Cot}_e(G)$ is k-flat, such as when k is a field (with G not necessarily smooth). Assume we are in such a case, so (2.1.3) is always an isomorphism and hence we may use Yoneda-style arguments with $\mathfrak{g}$ (i.e., $\mathrm{Tan}_{e_B}(G_B) = \mathfrak{g}_B$ for any k-algebra B). Thus, $g \mapsto \mathrm{Ad}_G(g) = \mathrm{d}(c_g)(e)$ defines a k-group map $G \to \mathrm{GL}(\mathfrak{g})$, and $\mathrm{d}(\mathrm{Ad}_G)(e) : \mathfrak{g} \to \mathfrak{gl}(\mathfrak{g})$ is ad_G.

Before we relate ad_G and the Lie bracket, we relate ad_G more directly to the infinitesimal group structure. This requires introducing an additional infinitesimal parameter.

Lemma A.7.4 *For the k-algebra $k[\epsilon, \epsilon'] = k[\epsilon] \otimes_k k[\epsilon]$ and any $v, w \in \mathfrak{g}$, the commutator*

$$(1+\epsilon v)(1+\epsilon' w)(1+\epsilon v)^{-1}(1+\epsilon' w)^{-1} \in G(k[\epsilon, \epsilon'])$$

is equal to $1 + \epsilon\epsilon'(\mathrm{ad}_G(v))(w)$.

Specializing this lemma along $\epsilon' = \epsilon$ gives nothing interesting, so it is essential that we have introduced two "independent" infinitesimal parameters.

Proof Identifying $k[\epsilon, \epsilon']$ with $k[\epsilon][\epsilon']$ and using scalar extension from k to $k[\epsilon']$, the point $1 + \epsilon v \in G_{k[\epsilon]}(k[\epsilon][\epsilon'])$ comes from $G_{k[\epsilon]}(k[\epsilon])$. Thus, the first three terms of the commutator have product $1 + \epsilon' \mathrm{Ad}_G(1+\epsilon v)(w)$ by (A.7.2) (with $B = k[\epsilon]$, $g = 1 + \epsilon v$, and $X = w$). But $\mathrm{Ad}_G(1+\epsilon v)(w)$ as a $k[\epsilon, \epsilon']$-point of G may be computed as an element of $\mathrm{Tan}_{e_{k[\epsilon]}}(G_{k[\epsilon]})$. By definition of ad_G, we have $\mathrm{Ad}_G(1+\epsilon v) = \mathrm{id} + \epsilon \cdot \mathrm{ad}_G(v)$. Applying this to w, we get

$$\begin{aligned}(1+\epsilon v)(1+\epsilon' w)(1+\epsilon v)^{-1} &= 1 + \epsilon' w + \epsilon\epsilon'(\mathrm{ad}_G(v))(w) \\ &= 1 + \epsilon'(w + \epsilon(\mathrm{ad}_G(v))(w)).\end{aligned}$$

Since $\epsilon'^2 = 0$, this final term is $(1 + \epsilon\epsilon'(\mathrm{ad}_G(v))(w))(1 + \epsilon' w)$ by computing the effect of both sides on the structure sheaf of $G_{k[\epsilon,\epsilon']}$ via right translation. □

Now we can prove that ad_G computes the Lie algebra structure defined via global derivations:

Proposition A.7.5 *The map* $\mathrm{ad}_G : \mathfrak{g} \to \mathfrak{gl}(\mathfrak{g})$ *is the adjoint representation* $v \mapsto [v, \cdot]_G$ *for the Lie algebra structure on* $\mathfrak{g}$.

We only need this result for G affine over a field, but we must avoid smoothness hypotheses.

Proof Choose $v, w \in \mathfrak{g}$ and consider the $k[\epsilon]$-scheme automorphism $\widetilde{D}_v$ of $G_{k[\epsilon]}$ and the $k[\epsilon']$-scheme automorphism $\widetilde{D}_w$ of $G_{k[\epsilon']}$. Extending scalars from $k[\epsilon]$ and $k[\epsilon']$ to $k[\epsilon, \epsilon']$ via the natural maps gives the respective $k[\epsilon, \epsilon']$-algebra automorphisms $1 + \epsilon D_v$ and $1 + \epsilon' D_w$ of $\mathscr{O}_{G_{k[\epsilon,\epsilon']}}$ (lifting the identity modulo (ϵ, ϵ')). The commutator of these latter automorphisms is

$$(1 + \epsilon D_v) \circ (1 + \epsilon' D_w) \circ (1 + \epsilon D_v)^{-1} \circ (1 + \epsilon' D_w)^{-1} = 1 + \epsilon\epsilon'[D_v, D_w].$$

Hence, it arises from right translation by $1 + \epsilon\epsilon'[D_v, D_w](e)$ (Lemma A.7.2). Passage from scheme morphisms to pullback on structure sheaves is contravariant, so this right translation computes the commutator automorphism $(\widetilde{D}_w^{-1}, \widetilde{D}_v^{-1})$ of $G_{k[\epsilon,\epsilon']}$.

For any k-algebra B and $g \in G(B)$, let ρ_g denote right translation by g on G_B. Thus, $\widetilde{D}_v = \rho_{1+\epsilon v}$ on $G_{k[\epsilon]}$, and $\widetilde{D}_w = \rho_{1+\epsilon' w}$ on $G_{k[\epsilon']}$. It follows that on $G_{k[\epsilon,\epsilon']}$ we have

$$(\widetilde{D}_w^{-1}, \widetilde{D}_v^{-1}) = \rho_{1+\epsilon' w}^{-1} \circ \rho_{1+\epsilon v}^{-1} \circ \rho_{1+\epsilon' w} \circ \rho_{1+\epsilon v}.$$

But $\rho_g \circ \rho_h = \rho_{hg}$, so the 4-fold composite is right translation by

$$\begin{aligned}&(1 + \epsilon v)(1 + \epsilon' w)(1 + \epsilon v)^{-1}(1 + \epsilon' w)^{-1} \\ &= 1 + \epsilon\epsilon'(\mathrm{ad}_G(v))(w) \in G(k[\epsilon, \epsilon'])\end{aligned}$$

due to Lemma A.7.4.

We conclude that for $x = (\mathrm{ad}_G(v))(w)$ and $y = [D_v, D_w](e)$ in $\mathfrak{g}$, right translations on $G_{k[\epsilon\epsilon']}$ by $1 + \epsilon\epsilon' x$ and $1 + \epsilon\epsilon' y$ coincide after scalar extension to $k[\epsilon, \epsilon']$. But $G(k[\epsilon\epsilon']) \to G(k[\epsilon, \epsilon'])$ is injective since $k[\epsilon\epsilon'] \to k[\epsilon, \epsilon']$ is an injective map between local rings, so $x = y$. □

As a consequence of Proposition A.7.5 and the obvious functoriality of ad_G in G, it follows that the Lie bracket is functorial in G. Moreover, we immediately obtain the following general compatibility of Lie algebras and Weil restriction:

Corollary A.7.6 *Let $k \to k'$ be a finite flat map of noetherian rings, G' a quasi-projective k'-group scheme of finite type, and $G = \mathrm{R}_{k'/k}(G')$.*

The natural isomorphism of groups $G(k[\epsilon]) \simeq G'(k'[\epsilon])$ induces an isomorphism of subgroups $\mathrm{Lie}(G) \simeq \mathrm{Lie}(G')$ that is compatible with the structure of Lie algebra over k on both sides and is functorial in G'.

Note that the quasi-projectivity hypothesis in this corollary always holds when k is a field, k' is a finite product of fields, and G' is a k'-group of finite type. We will only need the corollary for such k and k'. (If we work more generally with group functors then the hypotheses on the groups and rings can be weakened considerably.)

A useful application of Corollary A.7.6 is in the determination of fields of definition of geometric unipotent radicals. The idea is to reduce nontrivial problems concerning fields of definition for subgroups to easier problems in linear algebra concerning fields of definition for tangent spaces. For example, consider a purely inseparable finite extension of fields k'/k, a smooth connected k'-group G', and the smooth connected k-group $G = \mathrm{R}_{k'/k}(G')$. By Proposition A.5.11, the natural k'-group map $q : \mathrm{R}_{k'/k}(G')_{k'} \to G'$ is a smooth surjection with connected unipotent kernel. Thus, $\ker q$ is a k'-descent of $\mathscr{R}_u(\mathrm{R}_{k'/k}(G')_{\overline{k}})$ when G' is reductive. In [SGA3, XVII, App. III, (i) in Cor. to Prop. 5.1] it is shown that if $k' \neq k$ and G' is not unipotent then the field of definition over k for $\ker q$ is not k. We will prove a more precise statement in such cases: the field of definition is equal to k'. In fact, even the field of definition over k for the k'-subspace

$$\begin{aligned}\ker(\mathrm{Lie}(q)) = \mathrm{Lie}(\ker q) &\subseteq \mathrm{Lie}(\mathrm{R}_{k'/k}(G')_{k'})\\ &= k' \otimes_k \mathrm{Lie}(\mathrm{R}_{k'/k}(G')) = k' \otimes_k \mathrm{Lie}(G')\end{aligned}$$

turns out to be k'. (It is rather amazing that in these cases the field of definition of the Lie algebra cannot be strictly smaller than the field of definition for the group in nonzero characteristic.) The determination of this minimal field of definition over k is of vital importance in our work, due to its role in the proof of Theorem 1.6.2, so here is a self-contained development.

We begin with a lemma concerning fields of definition in linear algebra (in the sense of Remark 1.1.7).

Lemma A.7.7 *Let $K'/k'/k$ be a tower of field extensions, and W' a k'-vector space. Consider an intermediate K'-subspace V as in:*

$$\ker(K' \otimes_k W' \to K' \otimes_{k'} W') \subseteq V \subseteq K' \otimes_k W'.$$

If the field of definition F/k for V inside of K' does not contain k' then $V = K' \otimes_k W'$.

Proof Proceeding by contradiction, suppose V is a proper K'-subspace (so $W' \neq 0$). Choose a collection of k'-lines $\{L_i'\}$ that directly span W', and let $V_i = V \cap (K' \otimes_{k'} L_i')$. For every i, clearly

$$\ker(K' \otimes_k L_i' \to K' \otimes_{k'} L_i') \subseteq V_i \subseteq K' \otimes_k L_i'$$

and the field of definition over k for V_i inside of $K' \otimes_k L_i'$ is contained in F (and so does not contain k'). There must be some i such that V_i is a proper K'-subspace of $K' \otimes_k L_i'$, so to get a contradiction we may rename such an L_i' as W' to reduce to the case $\dim_{k'} W' = 1$. But then $\ker(K' \otimes_k W' \twoheadrightarrow K' \otimes_{k'} W')$ is a K'-hyperplane, so its containment in the proper K'-subspace V forces it to equal V.

Now we can replace the k'-line W' with k' to reduce to a problem in ring theory: if $\mu : K' \otimes_k k' \to K'$ is the natural map defined by multiplication and the ideal $\ker \mu$ viewed as a K'-subspace is defined over a subfield $F \subseteq K'$ over k then we claim that F contains k'. The map μ is the scalar extension by $K' \otimes_{k'} (\cdot)$ applied to the natural map $m : k' \otimes_k k' \to k'$. Hence, $\ker \mu$ is defined over the subfield $k' \subseteq K'$ over k, so by minimality $F \subseteq k'$. We can therefore replace K' with k' to reduce to showing that if $F \subseteq k'$ is a subfield over k such that $\ker(m) = k' \otimes_F I$ for an F-subspace $I \subseteq F \otimes_k k'$ then necessarily $F = k'$.

If such an I exists then

$$k' = (k' \otimes_k k')/\ker(m) = k' \otimes_F ((F \otimes_k k')/I),$$

so $(F \otimes_k k')/I$ is 1-dimensional over F yet its (F-isomorphic) image in k' is k'. Thus, $[k' : F] = 1$, so $F = k'$. □

Proposition A.7.8 *Let k'/k be a finite extension of fields, K'/k' any extension field, and G' a smooth connected k'-group. Let $G = \mathrm{R}_{k'/k}(G')$, and let $q : G_{k'} \to G'$ be the natural surjective k'-homomorphism. Consider a smooth K'-subgroup scheme $H \subseteq G_{K'}$ that contains* $(\ker q) \otimes_{k'} K'$.

(1) *Let $F \subseteq K'$ be the field of definition over k for H. If $H \neq G_{K'}$ then F contains k'.*
(2) *If G' is affine and not unipotent and k'/k is purely inseparable then the field of definition over k for $\mathscr{R}_u(G_{\overline{k}})$ coincides with the field of definition over k' for $\mathscr{R}_u(G'_{\overline{k}})$ inside of $\overline{k}$. The same holds for geometric radicals if G' is affine and not solvable.*

Proof We first deduce (2) from (1), so assume k'/k is purely inseparable. In particular, k' has a unique k-embedding into $\overline{k}$. By Proposition A.5.11, $\ker q$ is a smooth connected unipotent k'-group. Hence, $(\ker q) \otimes_{k'} \overline{k}$ is contained inside of $\mathscr{R}_u(G_{\overline{k}})$. The exact sequence

$$1 \to \ker q \to G_{k'} \to G' \to 1 \tag{A.7.3}$$

implies that if G' is not unipotent then the smooth connected affine k-group G cannot be unipotent. Thus, when G' is affine and non-unipotent we can apply (1) to $K' = \overline{k}$ and $H = \mathscr{R}_u(G_{\overline{k}})$. It follows that in such cases the field of definition F/k for H contains k'. In other words, F is also the field of definition over k' for H. But $G_{k'}/\ker q = G'$ and over $\overline{k}$ this carries $H/(\ker q)_{\overline{k}}$ over to $\mathscr{R}_u(G'_{\overline{k}})$, so for any subfield L of $\overline{k}$ over k' we see that H is defined over L (as a subgroup of G_L) if and only if $\mathscr{R}_u(G'_{\overline{k}})$ is defined over L (as a subgroup of G'_L). This settles (2) for unipotent radicals, and the assertion for radicals in the non-solvable case goes the same way.

To prove (1) we will show that if F does not contain k' then $H = G_{K'}$. Since G is k-smooth and connected and H is K'-smooth, it suffices to show that $\mathrm{Tan}_e(H) = \mathrm{Tan}_e(G_{K'})$. Observe that $V := \mathrm{Tan}_e(H)$ is a K'-subspace of $\mathrm{Tan}_e(G)_{K'}$, and its field of definition over k is certainly contained in F. Since V contains $\mathrm{Lie}(\ker q)_{K'} = \ker(\mathrm{Lie}(q)_{K'})$, and (by Corollary A.7.6) $\mathrm{Lie}(q)$ is identified with the kernel of the natural multiplication map

$$k' \otimes_k \mathrm{Lie}(G') \to \mathrm{Lie}(G'),$$

we are reduced to a problem in linear algebra that is settled by Lemma A.7.7. □

The next issue to address is a criterion for $\mathrm{R}_{k'/k}(G')$ to be its own derived group, with G' a connected semisimple k'-group and k'/k a finite extension of fields that may not be separable. It is instructive to first give an example in which $\mathrm{R}_{k'/k}(G')$ admits a nontrivial commutative quotient.

Example A.7.9 Let k'/k be a purely inseparable finite extension of fields and $G'_1 \to G'_2$ a central isogeny between connected semisimple k'-groups with kernel μ'. The exact sequence of fppf group sheaves (1.3.3) shows that the k-group $\mathrm{R}_{k'/k}(G'_1)/\mathrm{R}_{k'/k}(\mu')$ is a normal subgroup of $\mathrm{R}_{k'/k}(G'_2)$, and that the cokernel Q by this subgroup is commutative.

Thus, $\mathrm{R}_{k'/k}(G'_2)$ is not its own derived group if $\dim Q > 0$, which is to say that $\dim \mathrm{R}_{k'/k}(\mu') > 0$. This occurs when $k' \neq k$ and μ' is not k'-étale, by Example 1.3.2. For example, $\mathrm{R}_{k'/k}(\mathrm{PGL}_p)$ is not its own derived group when $\mathrm{char}(k) = p > 0$ and k'/k is a nontrivial purely inseparable finite extension.

We now establish a general criterion for a Weil restriction to be its own derived group, and then apply it in the simply connected semisimple case.

Proposition A.7.10 *Let k be a field and A a nonzero finite k-algebra. Let G be a smooth affine A-group such that each geometric fiber of G over* $\operatorname{Spec} A$ *is its own derived group and acts on its Lie algebra with vanishing space of coinvariants. The smooth affine k-group $\mathrm{R}_{A/k}(G)$ is its own derived group, or equivalently it has no nontrivial commutative quotient over k.*

In our study of exotic pseudo-reductive groups, this result is used in the generality in which it is stated (without reductivity hypotheses on G and with A possibly non-reduced). Before we prove Proposition A.7.10, we make some remarks on its hypotheses. First, the vanishing hypothesis on fibral coinvariants implies the fibral perfectness hypothesis when the fibers are connected. Second, the affine hypothesis on G can be relaxed to finite type because the only role of affineness is to ensure that $\mathrm{R}_{A/k}(G)$ exists as a scheme. Existence only requires that any finite set of points of G lies in an affine open subscheme, and this latter property holds in the finite type case since G_{red} is quasi-projective over the finite product of fields A_{red} (due to Proposition A.3.5).

Proof By extending scalars from k to $\overline{k}$ and from A to $A \otimes_k \overline{k}$, we may assume that k is algebraically closed. In this case the problem is to show that the abstract group $G(A) = \mathrm{R}_{A/k}(G)(k)$ is its own derived group. Passing to the local factor rings of A allows us to assume that A is local. Let G_0 be the special fiber of G.

We may induct on the length of A, the case of length 1 being obvious. In the general case with A of length larger than 1, there is an ideal I of A that is killed by the maximal ideal of A and is of dimension 1 over k. Thus, the result holds for $\mathrm{R}_{(A/I)/k}(G \bmod I)$ by induction, so $G(A/I)$ has no nontrivial commutative quotient.

Observe that the reduction map $\pi : G(A) \to G(A/I)$ is surjective by A-smoothness of G. We claim that the image of $\ker \pi$ in any commutative quotient of $G(A)$ is trivial, so any commutative quotient of $G(A)$ is also a commutative quotient of $G(A/I)$ and thus is trivial. To prove the triviality of the image of $\ker \pi$ in any commutative quotient of $G(A)$, first observe that $\ker \pi$ is naturally identified with the underlying additive group of $I \otimes_k \mathrm{Lie}(G_0)$ in such a way that the natural conjugation action by $G(A/I)$ on $\ker \pi$ factors through the natural adjoint action by $G_0(k)$ on $I \otimes_k \mathrm{Lie}(G_0)$. Recall also that $\dim_k I = 1$. Hence, in any commutative quotient of $G(A)$ the image of $\ker \pi$ is a quotient of the additive group $\mathrm{Lie}(G_0)$ in which the image of every element of the form $g_0.v - v$ vanishes (for $g_0 \in G_0(k)$ acting on $v \in \mathrm{Lie}(G_0)$ by the adjoint action).

The differences $g_0.v - v$ depend k-linearly on v and so generate $\mathrm{Lie}(G_0)$ as an additive group since we assume $\mathrm{Lie}(G_0)$ has vanishing space of coinvariants for the adjoint action. Thus, $\ker \pi$ has vanishing image in every commutative quotient of $G(A)$, as desired. □

Corollary A.7.11 *Let k be a field and k' a nonzero finite reduced k-algebra. Let G' be a semisimple k'-group whose fibers over the factor fields of k' are connected and simply connected. Then $\mathrm{R}_{k'/k}(G')$ is its own derived group.*

Proof We may and do assume that k' is a field. By Proposition A.7.10, it suffices to prove that if $\widetilde{G}$ is a connected and simply connected semisimple group over a field K then the adjoint representation of $\widetilde{G}$ on $\mathrm{Lie}(\widetilde{G})$ has vanishing space of coinvariants.

We can assume that K is separably closed, so $\widetilde{G}$ contains a maximal K-torus T that is K-split. Since $\widetilde{G}$ is simply connected, any pair of opposite root groups relative to T generates an SL_2 (rather than PGL_2). These copies of SL_2 generate $\widetilde{G}$, and their Lie algebras span $\mathrm{Lie}(\widetilde{G})$ since $\widetilde{G}$ is simply connected, so it suffices to check the vanishing of the coinvariants for SL_2 acting on $\mathfrak{sl}_2$ by the adjoint action. This is a straightforward calculation (even in characteristic 2, where the analogue for PGL_2 acting on $\mathfrak{pgl}_2 = \mathfrak{gl}_2/\mathfrak{z}$ is false). □

Remark A.7.12 By Corollary A.7.11, the k'-group G'_2 in Example A.7.9 must violate the vanishing hypothesis on coinvariants in Proposition A.7.10 when $k' \neq k$ and μ' is not k'-étale. To verify the non-vanishing directly, consider the G'_2-equivariant quotient $\mathfrak{g} = \mathrm{coker}(\mathrm{Lie}(G'_1) \to \mathrm{Lie}(G'_2))$ of $\mathrm{Lie}(G'_2)$. This is nonzero for dimension reasons, since $\mathrm{Lie}(\mu') \neq 0$.

To check that $\mathfrak{g}$ has trivial G'_2-action, we may assume that k is separably closed. Hence, $G'_1(k')$ has Zariski-dense image in G'_2, so it is enough to show that the $G'_1(k')$-action on $\mathfrak{g}$ is trivial. But there is a $G'_1(k')$-equivariant identification of $\mathfrak{g}$ with a subgroup of the degree-1 fppf cohomology $\mathrm{H}^1(\mathrm{Spec}(k'[\epsilon]), \mu'_{k'[\epsilon]})$ over the dual numbers $k'[\epsilon]$, and the $G'_1(k')$-action on this cohomology is trivial due to the centrality of μ' in G'_1.

Now we shift our attention from the Lie bracket to a Lie algebra structure that is specific to prime characteristic. Let p be a prime. For any Lie algebra $\mathfrak{g}$ over an $\mathbf{F}_p$-algebra k, a *p-Lie algebra* structure on $\mathfrak{g}$ is a set-theoretic map $X \mapsto X^{[p]}$ from $\mathfrak{g}$ to itself such that $\mathrm{ad}(X^{[p]}) = \mathrm{ad}(X)^p$, $(cX)^{[p]} = c^p X^{[p]}$ for all $c \in k$, and $(X+Y)^{[p]}$ satisfies an identity whose precise formulation we omit here; see [Bo2, 3.1] or [DG, II, §7, 3.1–3.3] for the complete definition. One example of such a structure is the p-power operation in any associative k-algebra (equipped with its standard Lie algebra structure via commutators). Thus, for any group scheme G over k, $\mathrm{Lie}(G)$ admits a p-Lie algebra structure via the p-power operation on left-invariant global k-linear derivations of $\mathscr{O}_G$ (within the associative algebra of all k-linear endomorphisms of $\mathscr{O}_G$). We will write $\mathrm{Lie}_p(G)$ when we wish to think of $\mathrm{Lie}(G)$ equipped with its p-Lie algebra structure arising from G.

Lemma A.7.13 *The p-Lie algebra structure is functorial in G. Moreover, if* $G = \mathrm{R}_{k'/k}(G')$ *for a finite flat map* $k \to k'$ *of noetherian rings and a quasi-projective* k'*-group* G' *then the natural identification* $\mathrm{Lie}(G) \simeq \mathrm{Lie}(G')$ *of Lie algebras over k respects the p-Lie algebra structures.*

Proof In the spirit of the proof of Proposition A.7.5, we give an infinitesimal description of the p-operation, from which functoriality and the compatibility with Weil restriction will be immediate.

Consider the p-fold tensor power

$$k[\epsilon_1, \ldots, \epsilon_p] = k[\epsilon]^{\otimes p}.$$

Since each ϵ_i^2 vanishes, for $s = \sum \epsilon_i$ it is clear that $1, s, s^2, \ldots, s^{p-1}$ and $e := \prod \epsilon_i$ linearly span the k-subalgebra of $\mathfrak{S}_p$-invariant elements. As a k-algebra, this is $k[s, e]$ with relations $s^p = se = e^2 = 0$. In particular, there is a unique k-algebra map $k[s, e] \twoheadrightarrow k[\epsilon]$ satisfying $s \mapsto 0$ and $e \mapsto \epsilon$. For any $v \in \mathfrak{g}$, the elements $1 + \epsilon_i v \in G(k[\epsilon_1, \ldots, \epsilon_p])$ pairwise commute (set $w = v$ in Lemma A.7.4 and use Proposition A.7.5). Thus,

$$\prod (1 + \epsilon_i v) \in G(k[\epsilon_1, \ldots, \epsilon_p])^{\mathfrak{S}_p} = G(k[s, e]),$$

so specialization along $s \mapsto 0$ and $e \mapsto \epsilon$ yields an element $1 + \epsilon[v]$ for some $[v] \in \mathfrak{g}$.

It suffices to prove $v \mapsto [v]$ is the p-operation. Right translation by $1 + \epsilon[v]$ on $G_{k[\epsilon]}$ is specialization along $k[s, e] \twoheadrightarrow k[\epsilon]$ of right translation by the point $\prod(1 + \epsilon_i v) \in G_{k[s,e]}(k[s, e])$. Thus, on structure sheaves, $1 + \epsilon D_{[v]}$ is the specialization of $\prod(1 + \epsilon_i D_v)$. But this latter specialization is clearly $1 + \epsilon D_v^p$, so $D_{[v]} = D_v^p$. By definition, $D_{v^{[p]}} = D_v^p$, so we are done. □

When G is affine, Lemma A.7.13 allows us to compute the p-operation on $\mathrm{Lie}_p(G)$ using a k-subgroup inclusion of G into GL_n (since the p-operation is the p-power map $L \mapsto L^p$ on $\mathrm{Mat}_n(k) = \mathfrak{gl}_n$, as follows from applying the proof of Lemma A.7.13 to $G = \mathrm{GL}_n$). The usefulness of the p-operation for our purposes is due to the following result.

Proposition A.7.14 *Let k be a field with* $\mathrm{char}(k) = p > 0$ *and consider the functor* $G \rightsquigarrow \mathrm{Lie}_p(G)$ *from locally finite type k-groups to p-Lie algebras over k. This restricts to an equivalence on the full subcategory of finite k-group schemes* $\mathscr{G}$ *for which the relative Frobenius morphism* $F_{\mathscr{G}/k} : \mathscr{G} \to \mathscr{G}^{(p)}$ *is trivial.*

Moreover, if B is any k-algebra and $\mathscr{G}$ is a finite k-group with trivial relative Frobenius morphism then the natural map

$$\mathrm{Hom}_B(\mathscr{G}_B, G_B) \to \mathrm{Hom}(\mathrm{Lie}_p(\mathscr{G})_B, \mathrm{Lie}_p(G_B)) \qquad \text{(A.7.4)}$$

is bijective for any locally finite type k-group G.

Proof This is a special case of [SGA3, $\mathrm{VII_A}$, 7.2, 7.4] (which works more generally with group schemes that may not arise via base change from a field). □

Definition A.7.15 Let k be a field with $\mathrm{char}(k) = p > 0$. A k-group scheme $\mathscr{G}$ has *height* $\leqslant 1$ if $\mathscr{G}$ is k-finite and the relative Frobenius morphism $F_{\mathscr{G}/k}$ is trivial.

Example A.7.16 Let G be a k-group locally of finite type and $\mathfrak{n}$ a p-Lie subalgebra of $\mathfrak{g} = \mathrm{Lie}(G)$. Let N be the unique k-group of height $\leqslant 1$ such that $\mathrm{Lie}(N) = \mathfrak{n}$ as p-Lie algebras. By (A.7.4), the inclusion $\mathfrak{n} \hookrightarrow \mathfrak{g}$ of p-Lie algebras arises from a unique k-homomorphism $f : N \to G$. The k-group $\ker f$ has height $\leqslant 1$ (as this is inherited from N), and $\mathrm{Lie}(\ker f) = \ker(\mathfrak{n} \to \mathfrak{g}) = 0$. Hence, $\ker f = 1$, so f is a closed immersion (Proposition A.2.1). In other words, $\mathfrak{n}$ uniquely "exponentiates" to a closed k-subgroup $N \subseteq G$ of height $\leqslant 1$.

We claim that N is (scheme-theoretically) normal in G if and only if $\mathfrak{n}$ is stable under the adjoint action of G on $\mathfrak{g}$. Indeed, normality says that N_B is carried back into itself under $G(B)$-conjugation for all k-algebras B, whereas stability under the adjoint action says that $\mathfrak{n}_B$ is stable under the adjoint action of $G(B)$ on $\mathfrak{g}_B$ for all k-algebras B. By the bijectivity of (A.7.4), these two conditions are equivalent. In the special case that G also has height $\leqslant 1$, N is normal in G if and only if $\mathfrak{n}$ is an ideal in $\mathfrak{g}$. This amounts to verifying that $\mathfrak{n}$ is an ideal in $\mathfrak{g}$ if and only if $\mathfrak{n}$ is stable under the adjoint action of G on $\mathfrak{g}$. Consider the closed k-subgroup scheme $H \subseteq G$ defined as the preimage under $\mathrm{Ad}_G : G \to \mathrm{GL}(\mathfrak{g})$ of the subgroup that stabilizes $\mathfrak{n}$. Since G has height $\leqslant 1$, by Proposition A.7.14 we have $H = G$ if and only if the inclusion of Lie algebras $\mathfrak{h} \hookrightarrow \mathfrak{g}$ is an equality. By Proposition A.7.5, $\mathfrak{h}$ is the Lie-theoretic normalizer of $\mathfrak{n}$ in $\mathfrak{g}$. Hence, $\mathfrak{h} = \mathfrak{g}$ if and only if $\mathfrak{n}$ is an ideal in $\mathfrak{g}$. This proves that for G of height $\leqslant 1$, normality of N in G is equivalent to $\mathfrak{n}$ being an ideal in $\mathfrak{g}$.

In case $\mathfrak{n}$ is G-stable and G is a smooth affine k-group, there are two quotients of G naturally associated to $\mathfrak{n}$: the quotient G/N in the sense of group schemes, and the quotient $G/\mathfrak{n}$ in the sense of Borel [Bo2, §17]. We never use this second point of view, but since it has the advantage of enabling one to study quotients by normal infinitesimal subgroup schemes entirely in the language of smooth groups and Lie algebras, we now prove that these two quotients coincide.

To prove the equality of G/N and $G/\mathfrak{n}$ as quotients of G, we first note that (by construction) the quotient map $G \to G/\mathfrak{n}$ is a purely inseparable isogeny whose Lie algebra kernel is $\mathfrak{n}$ and it is universal among k-homomorphisms $G \to H$ to smooth affine k-groups such that the tangent map at the identity kills $\mathfrak{n}$. In particular, $G/\mathfrak{n}$ uniquely dominates G/N as quotients of G. By Proposition A.7.14, the quotient map $G \to G/\mathfrak{n}$ kills N since it kills $\mathfrak{n} = \mathrm{Lie}(N)$ on Lie algebras, so G/N uniquely dominates $G/\mathfrak{n}$ as quotients of G. This proves that $G/N = G/\mathfrak{n}$ as quotients of G.

Suppose G has height $\leqslant 1$ and that N is normal in G, or equivalently (as we have seen above) that $\mathfrak{n}$ is an ideal in $\mathfrak{g}$. In such cases $\mathfrak{g}' := \mathfrak{g}/\mathfrak{n}$ is a p-Lie algebra, and we claim that the natural map $\mathfrak{g}' \to \mathrm{Lie}_p(G/N)$ of p-Lie algebras is an isomorphism. This follows from Proposition A.7.14 because these p-Lie algebras (compatibly) represent the same covariant functor. (More specifically, a k-homomorphism $G \to H$ to a finite k-group scheme of height $\leqslant 1$ factors through G/N if and only if the induced map of p-Lie algebras $\mathfrak{g} \to \mathfrak{h}$ factors through $\mathfrak{g}'$.) In particular, the equivalence $\mathscr{G} \rightsquigarrow \mathrm{Lie}_p(\mathscr{G})$ from the category of finite k-group schemes of height $\leqslant 1$ to the category of p-Lie algebras carries short exact sequences of such k-group schemes to short exact sequences of p-Lie algebras.

A.8 Lie algebras and groups of multiplicative type

We now turn to the properties of centralizers under the action of a torus, and more generally under the action of a group scheme of multiplicative type. For example, the study of exotic pseudo-reductive groups involves working with centralizers of a μ_p-action in characteristic $p \in \{2, 3\}$. Also, centralizers under a torus action on a possibly non-smooth group are useful in our work with pseudo-reductive groups in all nonzero characteristics.

Before we prove results concerning centralizers for the action of a group scheme of multiplicative type over a field, it will be convenient to review some basic results concerning the classification of such group schemes and their finite-dimensional linear representations. The first key fact gives a useful interpretation of character groups:

Proposition A.8.1 *Let k be a field. The functors $\Lambda \rightsquigarrow \underline{\mathrm{Hom}}(\Lambda, \mathrm{GL}_1)$ and $M \rightsquigarrow \mathrm{X}(M) := \underline{\mathrm{Hom}}(M, \mathrm{GL}_1)$ are quasi-inverse anti-equivalences between the category of k-groups M of multiplicative type and the category of étale k-groups Λ whose group of k_s-points is a finitely generated $\mathbf{Z}$-module.*

Proof By Galois equivariance, we may and do assume k is separably closed. If H is a $\overline{k}$-subgroup scheme of GL_1^n, then by [Bo2, 8.2] we have $H^0_{\mathrm{red}} \simeq \mathrm{GL}_1^r$ for some $0 \leqslant r \leqslant n$. The action on GL_1^n by some element of $\mathrm{GL}_n(\mathbf{Z})$ identifies H^0_{red} with the direct factor of GL_1^n given by the first product of the first r factors, so the action of the same matrix carries H over to $\mathrm{GL}_1^r \times \mu$ for some finite $\overline{k}$-subgroup scheme $\mu \subseteq \mathrm{GL}_1^{n-r}$. In particular, $\mu \subseteq \mu_d^{n-r}$ for an integer $d > 0$ killing μ.

Cartier duality for finite commutative $\overline{k}$-group schemes thereby implies that the $\overline{k}$-subgroup schemes of GL_1^n are in natural bijection with quotients of $\mathrm{X}(\mathrm{GL}_1^n) = \mathbf{Z}^n$, where a quotient $\mathbf{Z}^n/L$ corresponds to the $\overline{k}$-subgroup scheme defined by intersecting the kernels of all characters $\chi \in L$. By inspection, each such subgroup scheme over $\overline{k}$ arises from a k-subgroup scheme of GL_1^n. Applying these considerations to graphs of homomorphisms, we see that the functors in question give an anti-equivalence between the category of k-group schemes M that arise as a closed k-subgroup of some GL_1^n and the category of finitely generated $\mathbf{Z}$-modules.

It remains to show (for $k = k_s$) that any k-group scheme M of multiplicative type is a k-subgroup of some GL_1^n. We have seen above that $M_{\overline{k}} \simeq \mathrm{GL}_1^r \times \mu$ for some $r \geqslant 0$ and the Cartier dual μ of a commutative finite constant $\overline{k}$-group. For an integer $d > 0$ that kills μ, by calculating over $\overline{k}$ we see that the d-torsion $M[d]$ is a finite k-group scheme with étale Cartier dual and $M/M[d]$ is a k-torus. But $k = k_s$, so finite étale k-schemes are constant and $M/M[d] \simeq \mathrm{GL}_1^r$ for some $r \geqslant 0$. Presenting the Cartier dual of $M[d]$ as a quotient of $(\mathbf{Z}/d\mathbf{Z})^e$ for some $e > 0$ exhibits $M[d]$ as a k-subgroup of μ_d^e. This provides an inclusion of k-group schemes $j : M[d] \hookrightarrow \mathrm{GL}_1^e$, so we may form a pushout diagram of k-group schemes

$$\begin{array}{ccccccccc} 0 & \longrightarrow & M[d] & \longrightarrow & M & \longrightarrow & \mathrm{GL}_1^r & \longrightarrow & 0 \\ & & {\scriptstyle j}\downarrow & & \downarrow{\scriptstyle f} & & \| & & \\ 0 & \longrightarrow & \mathrm{GL}_1^e & \longrightarrow & E & \longrightarrow & \mathrm{GL}_1^r & \longrightarrow & 0 \end{array}$$

in which the rows are exact for the fppf topology. In particular, $\ker f = 0$ and the smooth group $(E_{\overline{k}})_{\mathrm{red}}$ is compatibly an extension of GL_1^r by GL_1^e over $\overline{k}$, so the inclusion $(E_{\overline{k}})_{\mathrm{red}} \hookrightarrow E_{\overline{k}}$ is an equality. Hence, E is smooth and connected, so E is a k-torus (apply [Bo2, 10.6(2)] over $\overline{k}$) and thus $E \simeq \mathrm{GL}_1^{e+r}$ since $k = k_s$. The middle vertical arrow thereby exhibits M as a k-subgroup of some GL_1^n. □

Corollary A.8.2 *Let M be a group scheme of multiplicative type over a field. Every k-subgroup scheme of M is of multiplicative type, and the closed subscheme $M_{\mathrm{red}} \subseteq M$ is a smooth k-subgroup.*

In particular, M_{red}^0 is a k-torus.

Proof Since $M_{\overline{k}}$ is a subgroup scheme of a torus, every k-subgroup scheme of M is certainly of multiplicative type (by definition). To analyze M_{red}, we may assume that k is separably closed. By applying the structure theorem for finitely generated $\mathbf{Z}$-modules to $\mathrm{X}(M)$ and dualizing back, we conclude that $M \simeq \mathrm{GL}_1^r \times \mu$ for a finite multiplicative group μ (dual to an ordinary finite commutative group). Hence, $M_{\mathrm{red}} = \mathrm{GL}_1^r \times \mu_{\mathrm{red}}$. This reduces the problem to the case $M = \mu$, so by passing to primary parts we can separately treat the cases when M is étale or infinitesimal, each of which is clear. □

Example A.8.3 The final claim in Corollary A.8.2 is specific to the case of groups of multiplicative type because over any imperfect field k:

(i) the underlying reduced scheme of an affine k-group scheme G of finite type need not be a k-subgroup scheme,
(ii) even when G_{red} is a k-subgroup, it may not be k-smooth.

The main issue is that for schemes X and Y of finite type over an imperfect field, the inclusion $(X \times Y)_{\mathrm{red}} \subseteq X_{\mathrm{red}} \times Y_{\mathrm{red}}$ may not be an equality.

For an example of (i), assume k is imperfect of characteristic $p > 0$ and consider the isomorphism $\mathrm{H}^1(k, \mu_p) \simeq k^\times/(k^\times)^p$ provided by fppf Kummer theory. This cohomology group classifies isomorphism classes of commutative p-torsion k-group extensions of $\mathbf{Z}/p\mathbf{Z}$ by μ_p. (For the convenience of the interested reader, here is a proof. The δ-functor $\mathrm{H}^\bullet(k, \cdot)$ on p-torsion fppf abelian sheaves over $\mathrm{Spec}\, k$ is the derived functor of $\Gamma(k, \cdot) = \mathrm{Hom}_k(\mathbf{Z}/p\mathbf{Z}, \cdot)$. Thus, for p-torsion abelian sheaves $\mathscr{F}$, the group $\mathrm{H}^1(k, \mathscr{F})$ is naturally isomorphic to the group $\mathrm{Ext}_k^1(\mathbf{Z}/p\mathbf{Z}, \mathscr{F})$ classifying isomorphism classes of extensions $\mathscr{E}$ of $\mathbf{Z}/p\mathbf{Z}$ by $\mathscr{F}$ in the category of p-torsion abelian sheaves over $\mathrm{Spec}\, k$. Since we can view $\mathscr{E}$ as an $\mathscr{F}$-torsor over $\mathbf{Z}/p\mathbf{Z}$, by descent theory for affine maps it follows that $\mathscr{E}$ is represented by a finite k-group scheme when $\mathscr{F}$ is, such as for $\mathscr{F} = \mu_p$. The notion of short-exactness for finite commutative k-groups matches that for fppf abelian sheaves over $\mathrm{Spec}\, k$.)

Consider such an extension G of $\mathbf{Z}/p\mathbf{Z}$ by μ_p classified by $a \bmod (k^\times)^p$ for some $a \in k^\times$ that is not a pth power in k. The fiber of $G \twoheadrightarrow \mathbf{Z}/p\mathbf{Z}$ over $i \in \mathbf{Z}/p\mathbf{Z}$ is $\mathrm{Spec}\, k[x]/(x^p - a^{[i]})$ for any representative $[i] \in \mathbf{Z}$ of i. Hence, $G^0 = \mu_p$ and $G - G^0$ is reduced, so G_{red} is the disjoint union of $G - G^0$ and the identity point. This is étale at the identity but is not étale over k, so G_{red} does not admit a k-group scheme structure.

The preceding example rests on an inhomogeneity among distinct connected components, so here is a less elementary example of (i) that has the virtue of being connected. Choose $c \in k - k^p$, and let

$$H = \{(x, y) \in \mathbf{G}_{\mathrm{a}}^2 \mid x^p = cy^p\}.$$

This is a reduced k-subgroup scheme of the affine k-plane (as an additive group) and it is not k-smooth. Clearly $(H_{\overline{k}})_{\mathrm{red}}$ is the line $x = c^{1/p}y$ in the affine $\overline{k}$-plane. In particular, H is geometrically irreducible over k. The k-group scheme

$$G = H \times H = \{(x, y, x', y') \in \mathbf{G}_a^4 \mid x^p = cy^p,\ x'^p = cy'^p\}$$

is geometrically irreducible over k, and it is non-reduced since in $k[G]$ clearly

$$(xy' - yx')^p = x^p y'^p - y^p x'^p = cy^p y'^p - cy^p y'^p = 0$$

yet $xy' - yx' \neq 0$ in $k[G]$ (as we see by using the evident $k[y, y']$-basis $\{x^i x'^j\}_{0 \leqslant i,j < p}$ of $k[G]$). In [DeGr, VI$_{\rm A}$, Ex. 1.3.2(2)] there is an elegant argument due to Raynaud showing that $G_{\rm red}$ is not a k-subgroup scheme of G (and in fact that $G_{\rm red}$ has a 4-dimensional tangent space at the identity but a tangent space of smaller dimension at the geometric point $(c^{1/p}, 1, c^{1/p}, 1)$, so $(G_{\rm red})_{\overline{k}}$ does not admit a group scheme structure).

To give an example of (ii), let k'/k be a purely inseparable extension of degree $p = \mathrm{char}(k) > 0$ and let $H = \mathrm{R}_{k'/k}(\mathrm{GL}_1)$. Consider $G = H[p]$, so $G_{\overline{k}} = \mu_p \times \mathscr{R}_u(H_{\overline{k}})$. Thus, G is not smooth. Writing $k' = k(\alpha)$ with $\alpha^p = a \in k - k^p$, we have

$$G(B) = \{x_0 + \alpha \otimes x_1 + \cdots + \alpha^{p-1} \otimes x_{p-1} \in k' \otimes_k B \mid \\ x_i \in B, x_0^p + a x_1^p + \cdots + a^{p-1} x_{p-1}^p = 1\}$$

functorially in k-algebras B. In other words, G is the hypersurface in affine p-space over k defined by the vanishing of the polynomial

$$x_0^p + a x_1^p + \cdots + a^{p-1} x_{p-1}^p - 1.$$

This polynomial is irreducible over k because $a \in k - k^p$. We conclude that $G_{\rm red} = G$. A more elementary example is $G = \{x^p + ay^p = 0\}$ with $a \in k - k^p$.

Remark A.8.4 Instances of (i) (i.e., examples of affine k-group schemes of finite type whose underlying reduced scheme is not a subgroup scheme) naturally arise as the group scheme of p-torsion in a Tate elliptic curve when k is a Laurent series field of characteristic p.

Example A.8.5 Let k be a nonzero commutative ring with connected spectrum. For any $m \geqslant 1$ we have

$$\mathrm{Hom}_k(\mu_m, \mathrm{GL}_1) = \mathrm{End}_k(\mu_m) = \mathbf{Z}/m\mathbf{Z}$$

by Cartier duality. Likewise, we claim that $\mathrm{End}_k(\mathrm{GL}_1) = \mathbf{Z}$. Connectedness reduces this to a problem over the local rings of k, for which we can reduce to the noetherian case. Scalar extension to the completion and comparison against the special fiber reduces the problem to the case of artin local k, in which case an elementary induction on the length of k does the job.

We conclude that it is reasonable to define a *k-split group of multiplicative type* to be a k-group that is isomorphic to a finite product of copies of GL_1 and k-groups μ_m for various $m \geqslant 1$. The interest in this notion is that it is "dual" to the category of finitely generated $\mathbf{Z}$-modules via the functors

$$M \rightsquigarrow \mathrm{X}(M) := \mathrm{Hom}_k(M, \mathrm{GL}_1), \quad \Lambda \rightsquigarrow \mathrm{D}_k(\Lambda) := \mathbf{Hom}(\Lambda, \mathrm{GL}_1)$$

(where $\mathbf{Hom}(\Lambda, \mathrm{GL}_1)$ denotes the affine k-group representing the functor on k-algebras defined by $B \rightsquigarrow \mathrm{Hom}_{\mathbf{Z}}(\Lambda, B^{\times})$). This duality is verified by using product decompositions to reduce to the special cases $M = \mathrm{GL}_1$ and $M = \mu_m$, as well as $\Lambda = \mathbf{Z}$ and $\Lambda = \mathbf{Z}/m\mathbf{Z}$, with $m \geqslant 1$.

If we drop the connectedness hypothesis on $\mathrm{Spec}\, k$ in Example A.8.5 then we can go in the reverse direction. That is, given a finitely generated $\mathbf{Z}$-module Λ we can construct the affine k-group

$$\mathrm{D}_k(\Lambda) := \mathbf{Hom}(\Lambda, \mathrm{GL}_1).$$

However, $\mathrm{X}(\mathrm{D}_k(\Lambda))$ is the set of global sections of the constant sheaf Λ for the Zariski topology on $\mathrm{Spec}\, k$, so this is not Λ if $\mathrm{Spec}\, k$ is not (non-empty and) connected. Nonetheless, for our purposes the following definition is sufficient.

Definition A.8.6 Let k be a ring. A *k-split group of multiplicative type* is a k-group M that is k-isomorphic to $\mathrm{D}_k(\Lambda)$ for some finitely generated $\mathbf{Z}$-module Λ. (Equivalently, M is a finite product of copies of GL_1 and k-groups μ_m for various $m \geqslant 1$.)

Example A.8.7 Let k be a commutative ring with connected (non-empty) spectrum, M a split k-group of multiplicative type, and $X = \mathrm{X}(M)$. For every $\chi \in X$, the k-homomorphism $\chi : M \to \mathrm{GL}_1$ is identified with an element e_χ in the coordinate ring $k[M]$ of M. Since $(e_\chi e_{\chi'})(m) = \chi(m)\chi'(m) = (\chi\chi')(m)$ for all points m of M, we have $e_\chi e_{\chi'} = e_{\chi\chi'}$. Likewise, e_1 is the constant function 1, and the comultiplication is $e_\chi \mapsto e_\chi \otimes e_\chi$ since $\chi(mm') = \chi(m)\chi(m')$.

The group algebra $k[X]$ over k has a natural structure of commutative and co-commutative Hopf algebra, and we have just constructed a natural map $k[X] \to k[M]$. Since $\mathrm{X}(M \times M') = \mathrm{X}(M) \oplus \mathrm{X}(M')$ and M is a product of copies of GL_1 and k-groups μ_m, we verify that $k[X] \to k[M]$ is an isomorphism by reducing to a direct calculation when $M = \mathrm{GL}_1$ and $M = \mu_m$ ($m \geqslant 1$).

This procedure works without connectedness hypotheses on $\mathrm{Spec}\, k$ provided that we begin with the $\mathbf{Z}$-module rather than the k-group. That is, if Λ is a finitely generated $\mathbf{Z}$-module and $M = \mathrm{D}_k(\Lambda)$ is the associated k-group of multiplicative type then the natural map from Λ to $\mathrm{X}(M)$ defines a map of Hopf algebras $k[\Lambda] \to k[M]$ over k. This is seen to be an isomorphism by reducing to the special cases $\Lambda = \mathbf{Z}$ and $\Lambda = \mathbf{Z}/m\mathbf{Z}$.

It is an important fact in the theory of split tori over a field that their finite-dimensional linear representations are direct sums of characters (and so are completely reducible). The following lemma records the generalization for split groups M of multiplicative type over any ring k, using *functorial linear representations* of M on a k-module V; this means a B-linear representation of $M(B)$ on $B \otimes_k V$ functorially in k-algebras B.

Lemma A.8.8 *Let k be a commutative ring, and Λ a finitely generated $\mathbf{Z}$-module. Let $M = \mathrm{D}_k(\Lambda)$ be the corresponding k-split k-group of multiplicative type. There is a natural bijection between the set of functorial linear representations ρ of M on a k-module V and the set of k-linear Λ-gradings on V. The correspondence is that if $V = \bigoplus_{\lambda \in \Lambda} V_\lambda$ then*

$$\rho(t)\left(\sum v_\lambda\right) = \sum \lambda(t) v_\lambda.$$

In particular, all such ρ are semisimple when k is a field.

Proof Let $\underline{\mathrm{End}}(V)$ denote the linear endomorphism functor of V; this associates to any k-algebra B the B-algebra $\mathrm{End}_B(B \otimes_k V)$ of B-linear endomorphisms of $B \otimes_k V$. (This is generally not representable, unless V is locally free of finite rank over k.) For any affine k-group scheme G, to give a functorial linear representation ρ of G on V is to give a $k[G]$-valued point of $\underline{\mathrm{End}}(V)$ that specializes to the identity automorphism of V at the identity of G and in general is multiplicative.

That is, we get a linear endomorphism of the $k[G]$-module $V \otimes_k k[G]$, or equivalently a k-linear map $\alpha_\rho : V \to V \otimes_k k[G]$, satisfying a multiplicative action condition and an identity element condition. If $\alpha_\rho(v) = \sum v_i \otimes f_i$ then for any k-algebra B and $g \in G(B)$ we have

$$\rho(g)(v) = \sum f_i(g) v_i$$

in $V_B := V \otimes_k B$. Likewise, if $\alpha_\rho(v_i) = \sum_{j \in J_i} v_{ij} \otimes h_{ij}$ then for any B and $g, g' \in G(B)$ we compute

$$\rho(gg')(v) = \sum f_i(gg') v_i,$$

$$\rho(g)(\rho(g')(v)) = \rho(g)\left(\sum f_i(g') v_i\right) = \sum_i \sum_{j \in J_i} h_{ij}(g) f_i(g') v_{ij}.$$

Thus, if $m_G^* : k[G] \to k[G] \otimes_k k[G]$ denotes the comultiplication then the multiplicative action condition $\rho(gg') = \rho(g) \circ \rho(g')$ says exactly that the elements

$$(\mathrm{id}_V \otimes m_G^*)\left(\sum v_i \otimes f_i\right) = \sum v_i \otimes m_G^*(f_i),$$

$$(\alpha_\rho \otimes \mathrm{id}_{k[G]})\left(\sum v_i \otimes f_i\right) = \sum_i \left(\sum_{j \in J_i} v_{ij} \otimes h_{ij}\right) \otimes f_i$$

in $V \otimes_k k[G] \otimes_k k[G]$ have the same pullback to V_B under all pairs $(g, g') \in G(B) \times G(B)$ for all B, which is to say that they coincide in $V \otimes_k k[G] \otimes_k k[G]$.

Since $\sum v_i \otimes f_i = \alpha_\rho(v)$ with $v \in V$ arbitrary (and v_i and f_i depending on v), the action condition says exactly that the diagram of k-linear maps

$$\begin{array}{ccc} V & \xrightarrow{\alpha_\rho} & V \otimes_k k[G] \\ {\scriptstyle \alpha_\rho}\downarrow & & \downarrow{\scriptstyle \mathrm{id}_V \otimes m_G^*} \\ V \otimes_k k[G] & \xrightarrow[\alpha_\rho \otimes \mathrm{id}_{k[G]}]{} & V \otimes_k k[G] \otimes_k k[G] \end{array} \tag{A.8.1}$$

commutes.

Now consider the special case when $G = M = \mathrm{D}_k(\Lambda)$ is k-split of multiplicative type, so the k-algebra $k[M]$ is naturally identified with $\bigoplus_{\lambda \in \Lambda} ke_\lambda$ where $e_\lambda(t) = \lambda(t)$ for all points t of M, so $e_\lambda e_{\lambda'} = e_{\lambda\lambda'}$. The comultiplication is $m_M^* : e_\lambda \mapsto e_\lambda \otimes e_\lambda$. The map α_ρ can be uniquely written as $v \mapsto \sum f_\lambda(v) \otimes e_\lambda$ for k-linear maps $f_\lambda : V \to V$ (a finite sum for each $v \in V$), so $\rho(g)(v) = \sum_\lambda \lambda(g) f_\lambda(v)$ for points g of G valued in any k-algebra. The diagram (A.8.1) encoding the action condition $t.(t'.v) = (tt').v$ says that

$$\sum_\lambda e_\lambda \otimes e_\lambda \otimes f_\lambda(v) = \sum_{\lambda, \lambda'} e_\lambda \otimes e_{\lambda'} \otimes f_{\lambda'}(f_\lambda(v)),$$

or in other words that $f_{\lambda'} \circ f_\lambda$ vanishes if $\lambda' \neq \lambda$ and is equal to f_λ if $\lambda' = \lambda$. That is, the linear maps f_λ are projectors onto mutually independent submodules $V_\lambda \subseteq V$.

Since $e_\lambda(1) = 1$ for all λ, the condition $1.v = v$ says that $v = \sum_\lambda f_\lambda(v)$ for all v (a finite sum), so $V = \bigoplus V_\lambda$. Since $\rho(g)(v) = \sum_\lambda \lambda(g) f_\lambda(v)$ (a finite sum for any specific $v \in V$), the M-action on V restricts to the action on each V_λ through λ. □

Remark A.8.9 If V is a finite free k-module of rank $n \geqslant 1$ and $k \neq 0$ then we can interpret the map α_ρ in the preceding proof very concretely in terms of a

k-basis $\{v_1, \ldots, v_n\}$ of V by taking $J_i = \{1, \ldots, n\}$ and $v_{ij} = v_j$ in the calculations (as we may).

The action condition gives the identity $f_i(g'g) = \sum_j h_{ji}(g')f_j(g)$ for all $g', g \in G(B)$ with a k-algebra B. This says that $\alpha_\rho(gv) = \sum v_i \otimes f_i(xg)$ for all $v \in V_B$ (where $\alpha_\rho(v) = \sum v_i \otimes f_i$), or in other words that the map $\alpha_\rho : V \to V \otimes k[G]$ is G-equivariant if we use ρ on the source and $\mathrm{id}_V \otimes r_G$ on the target, with r_G the right regular representation of G on $k[G]$. That is, upon fixing an ordered k-basis of V, α_ρ is a G-equivariant embedding of V into $k[G]^n$ with each factor $k[G]$ endowed with the right regular G-action. This embedding encodes the representation space V in terms of its "matrix coefficients".

For any μ_p-action on a group scheme G locally of finite type over a field of characteristic $p > 0$, the centralizer of μ_p on G can be studied via Lie algebra centralizers. This viewpoint is used by Borel in [Bo2, 9.1–9.2], where infinitesimal group schemes are avoided, but for our purposes it is useful to have both viewpoints available. The third part of Proposition A.8.10 below unifies them. Before stating the result, we recall that an affine group scheme H of finite type over an algebraically closed field F is *linearly reductive* if its finite-dimensional algebraic representations are completely reducible. For example, by Lemma A.8.8 and an averaging argument, linear reductivity holds if the identity component H^0 is of multiplicative type and the component group H/H^0 has order not divisible by $\mathrm{char}(F)$.

We also observe that if H is a flat affine group scheme over a ring k and M is a finitely presented k-module equipped with a functorial linear left H-action (i.e., an R-linear left action of $H(R)$ on M_R naturally in k-algebras R) then the linear dual $M^\vee = \mathrm{Hom}(M, k)$ inherits a natural linear left H-action. Indeed, to define an $H(R)$-action on $(M^\vee)_R$ naturally in all k-algebras R it suffices (by the proof of Yoneda's Lemma) to treat flat R since $k[H]$ is k-flat, and for such R we have $(M^\vee)_R = \mathrm{Hom}_R(M_R, R)$ on which there is an evident R-linear left $H(R)$-action. The submodule $(M^\vee)^H$ of elements $\ell \in M^\vee$ such that the element $\ell_R \in (M^\vee)_R$ is $H(R)$-invariant for all R consists of those linear forms $\ell : M \to k$ that are H-equivariant (in the functorial sense), again since it suffices to consider flat R (such as $R = k[H]$). A case of much interest is $M = \mathrm{Cot}_y(Y)$ for a k-scheme Y locally of finite presentation that is equipped with a left H-action, with $y \in Y(k)$ fixed by the H-action; in such cases $M^\vee = \mathrm{Tan}_y(Y)$.

Proposition A.8.10 *Let Y be a separated scheme over a ring k, and let H be an affine flat k-group scheme equipped with a left action on Y. Assume that for some faithfully flat extension $k \to k'$ the coordinate ring $k'[H_{k'}] = k' \otimes_k k[H]$ is a projective k'-module.*

(1) *The invariant locus of the H-action on Y (i.e., the functor assigning to any k-algebra R the subset of points $y \in Y(R)$ such that H_R acts trivially*

on y) is represented by a closed subscheme Y^H in Y, with Y^H locally of finite type over k if Y is. If Y is locally of finite presentation over k then $\mathrm{Tan}_y(Y^H) = \mathrm{Tan}_y(Y)^H$ *for any* $y \in Y^H(k)$.

(2) *Assume k is noetherian, Y and H are locally of finite type, and the geometric fibers of* $H \to \mathrm{Spec}\, k$ *are linearly reductive. Let* Y' *be another separated k-scheme locally of finite type equipped with a left H-action. If* $f : Y \to Y'$ *is a smooth H-equivariant k-map then the induced map* $Y^H \to Y'^H$ *is smooth. In particular, if Y is smooth then* Y^H *is smooth.*

(3) *Assume k is a field with* $\mathrm{char}(k) = p > 0$, $M := H$ *is of multiplicative type and p-torsion, and* $G := Y$ *is equipped with a k-group structure respected by the M-action. Consider the induced action of M on* $\mathfrak{g} = \mathrm{Lie}(G)$ *and the induced Lie algebra map* $\rho : \mathfrak{m} = \mathrm{Lie}(M) \to \mathfrak{gl}(\mathfrak{g})$. *The representation* ρ *determines* G^M *and* $\mathrm{Lie}(G^M)$ *via the formulas*

$$G^M = \{g \in G \mid (\mathrm{Ad}(g))(\rho(X)) = \rho(X) \text{ for all } X \in \mathfrak{m}\},$$
$$\mathrm{Lie}(G^M) = \{v \in \mathfrak{g} \mid \rho(X)(v) = 0 \text{ for all } X \in \mathfrak{m}\}.$$

Before we prove this result, we make some remarks on its hypotheses. The separatedness hypothesis on Y is satisfied when k is a field and Y is a k-group scheme. Two interesting cases for which the projectivity hypothesis on $k[H]$ is satisfied after a faithfully flat scalar extension are when k is a field and when H is a group of multiplicative type in the sense of Definition A.8.6, or more generally a k-group of multiplicative type in the sense of [SGA3, IX, 1.1]. (By the deeper results in [RG, I, 3.3.3–3.3.5], the homogeneity arising from the k-group structure on H implies that $k[H]$ is projective over k whenever H is finitely presented and affine over k with connected fibers. We do not use this fact, but it shows that the projectivity hypothesis on H in Proposition A.8.10 is satisfied in most cases of "practical" interest.)

Finally, as a special case of (3), if $\dim \mathfrak{m} = 1$ (e.g., $M = \mu_p$) and M is a subgroup scheme of G acting via conjugation, then for a nonzero X in $\mathfrak{m}$ we get expressions for G^M and its Lie algebra in terms of the effect on X of the adjoint actions of G and $\mathfrak{g}$ on $\mathfrak{g}$. More specifically, if G is smooth then $G^M = Z_G(M)$ and the formulas in (3) show that in such cases $Z_G(M)$ and $\mathrm{Lie}(Z_G(M))$ respectively coincide with $Z_G(X)$ and $\mathfrak{z}_{\mathfrak{g}}(X)$ in the sense of [Bo2, 9.1].

Proof of Proposition A.8.10 To prove (1), by descent theory it is harmless to make a faithfully flat scalar extension on k, so we can assume that $k[H]$ is a projective k-module. Consider the map $\alpha : H \times Y \to Y \times Y$ defined by $(h, y) \mapsto (h.y, y)$. Since the diagonal $\Delta_{Y/k}$ is closed in $Y \times Y$ (as Y is separated), the locus $D := \alpha^{-1}(\Delta_{Y/k})$ of points (h, y) such that $h.y = y$ is a closed subscheme in $H \times Y$. If Y is locally of finite type over k then so is D.

Associated to D we wish to define an "ideal of coefficients" $\mathscr{J}$ on Y, finitely generated if Y is locally of finite type. This rests on the following general observation. If R is a commutative ring, P is a projective R-module, M is any R-module, and $f \in P \otimes M$ then for any expression $\sum x_j \otimes m_j$ for f we can consider the submodule of M generated by the m_j. We claim that among the submodules of M obtained in this way, there is a unique one that is contained in all others; this is the *module of coefficients* $C(f) \subseteq M$. To prove this, consider a free R-module of the form $F = P \oplus Q$ with an R-basis $\{e_i\}$. Identifying $P \otimes M$ as a submodule of $F \otimes M$, we can uniquely write $f = \sum e_i \otimes c_i$ with $c_i = 0$ for all but finitely many i. Letting $\pi : F \to P$ be the projection, we then have $f = \sum \pi(e_i) \otimes c_i$ in $P \otimes M$. But for any expression $\sum x_j \otimes m_j$ for f in $P \otimes M$ as a finite sum of elementary tensors, we can write $x_j = \sum r_{ij} e_i$ with $r_{ij} \in R$, so

$$f = \sum_i e_i \otimes \left(\sum_j r_{ij} m_j \right).$$

Thus, $c_i = \sum_j r_{ij} m_j$, so defining $C(f)$ to be the R-span of the c_i proves the existence of a module of coefficients.

For open affine U in Y, let I_U be the ideal of $D \cap (H \times U)$ in the ring $k[H \times U] = k[H] \otimes_k k[U]$ with $k[H]$ a projective k-module. Define J_U be the $k[U]$-linear span of the k-modules of coefficients of all elements $f \in I_U$. It is clear that J_U is an ideal and we claim its formation is compatible with localization on U, so the ideals J_U globalize to an ideal $\mathscr{J}$ on Y. This assertion only involves $k[H]$ through its underlying k-module structure, so by k-projectivity we can replace it with a free k-module. Then the claim is clear via a k-basis.

For any k-algebra R and $y \in Y(R)$, the condition that y is centralized by the H_R-action on Y_R is that the section $\sigma_y : H_R \to H_R \times Y_R$ defined by $h \mapsto (h, y)$ lies entirely in D_R. We claim that this is the same as $\mathscr{J}$ being killed by pullback along $y : \operatorname{Spec} R \to Y$. To prove it we can work Zariski-locally on R to reduce to the case when y factors through an affine open $U \subseteq Y$, so if $k[H] \hookrightarrow F$ is a k-linear split injection into a free k-module and $\{e_i\}$ is a k-basis of F then for any $f \in I_U$ we have uniquely $f = \sum e_i \otimes a_{i,f}$ with $\{a_{i,f}\}_{i,f}$ a generating set of J_U (by the construction of J_U using modules of coefficients). Hence, $\sigma_y^*(f) = \sum e_i \otimes y^*(a_{i,f})$ in $k[H] \otimes_k R = R[H_R]$, so σ_y factors through D precisely when all $y^*(a_{i,f})$ vanish in R, which is to say that y pulls J_U back to 0 in R. This proves that the zero scheme of $\mathscr{J}$ in Y represents Y^H. By evaluating on points valued in the flat algebra of dual numbers over k, for any $y \in Y^H(k)$ we thereby obtain the description of $\operatorname{Tan}_y(Y^H)$ as the H-invariants in $\operatorname{Tan}_y(Y)$.

To prove (2) we first note that taking $Y' = \operatorname{Spec} k$ yields the second part as a special case of the first. To prove the smoothness of $Y^H \to Y'^H$ in general, we may assume k is artin local with algebraically closed residue field κ and it suffices to check the infinitesimal lifting criterion for smoothness using points valued in finite local k-algebras. That is, if $R \twoheadrightarrow R_0$ is a surjection between finite local k-algebras and $y_0 \in Y^H(R_0)$ has image $y_0' = f(y_0) \in Y'^H(R_0)$ which is equipped with a lift $y' \in Y'^H(R)$ then we just have to construct a lift $y \in Y^H(R)$ of y_0 such that $f(y) = y'$. By induction on the length of R, we can assume that $I := \ker(R \twoheadrightarrow R_0)$ is killed by $\mathfrak{m}_R$.

Let $\overline{y}_0 = y_0 \bmod \mathfrak{m}_R \in Y^H(\kappa)$ and $\overline{y}_0' = f(\overline{y}_0) = y_0' \bmod \mathfrak{m}_R \in Y'^H(\kappa)$, so the fiber $Y_{\overline{y}_0'} = f^{-1}(\overline{y}_0')$ through $\overline{y}_0$ is smooth since f is smooth. Smoothness of f ensures that for any R-algebra A and the R_0-algebra $A_0 = A \otimes_R R_0 = A/IA$, the map

$$Y(A) \to Y(A_0) \times_{Y'(A_0)} Y'(A) \tag{A.8.2}$$

is surjective.

The fiber of (A.8.2) over $((y_0)_{A_0}, y_A')$ is a torsor over the A_0-module

$$F(A) = IA \otimes_\kappa \mathrm{Tan}_{\overline{y}_0}(Y_{\overline{y}_0'}) = IA \otimes_{A_0} (A_0 \otimes_\kappa \mathrm{Tan}_{\overline{y}_0}(Y_{\overline{y}_0'}))$$

naturally in A; this torsor structure is denoted $v + \xi$ for $v \in F(A)$. For any $h \in H(A)$ with reduction $h_0 \in H(A_0) = H_\kappa(A_0)$ and any $v \in F(A)$, we have $h.(v+\xi) = h_0.v + h.\xi$. Thus, for a point $y \in Y(R)$ over (y_0, y') the $H(A)$-action on y_A for varying A satisfies $h.y_A = c(h) + y_A$ for a map of set-valued functors $c : H_R \to F$ (depending on the initial choice of y) that satisfies the 1-cocycle condition $c(h'h) = h'.c(h) + c(h')$. The notion of a 1-coboundary is defined in the evident manner, namely maps of the form $h \mapsto h.v - v$ for some $v \in F(R)$, and c modulo 1-coboundaries is independent of the initial choice of y.

The problem of finding an H_R-invariant y is precisely the condition that c is a 1-coboundary. Letting $\underline{V}$ denote the vector group over κ corresponding to the finite-dimensional κ-vector space $F(R) = I \otimes_\kappa \mathrm{Tan}_{\overline{y}_0}(Y_{\overline{y}_0'})$, since $IR[H] = I \otimes_R R[H]$ (due to the R-flatness of $R[H]$) we get the equality

$$F(R[H]) = IR[H] \otimes_\kappa \mathrm{Tan}_{\overline{y}_0}(Y_{\overline{y}_0'}) = I \otimes_\kappa \kappa[H] \otimes_\kappa \mathrm{Tan}_{\overline{y}_0}(Y_{\overline{y}_0'}) = \underline{V}(\kappa[H])$$

which carries c over to a 1-cocycle $c_0 : H_\kappa \to \underline{V}$. This respects the notion of 1-coboundary on both sides, so the proof of (2) is reduced to verifying the triviality of the quotient set $\mathrm{H}^1(H_\kappa, \underline{V})$. Such cohomology classes correspond exactly to isomorphism classes of H_κ-equivariant linear extensions of $\mathbf{G}_\mathrm{a}$ by $\underline{V}$, as is readily checked by a computation with a linear splitting. By hypothesis all linear representations of H_κ over κ are completely reducible, so any H_κ-equivariant linear extension of $\mathbf{G}_\mathrm{a}$ by $\underline{V}$ is a split extension and hence (2) is proved.

Finally, consider (3). We can assume $k = \overline{k}$, so M is a power of μ_p. Since the formation of Lie algebras commutes with scheme-theoretic intersections and

$G^{M' \times M''} = G^{M'} \cap G^{M''}$ for any direct product decomposition $M = M' \times M''$, we may reduce to the case $M = \mu_p$. Thus, we assume for the rest of the argument that $M = \mu_p$ and $k = \overline{k}$ with characteristic $p > 0$. Also, if we let $G' = G \rtimes M$ be the semidirect product defined by the action then $G'^M = G^M \rtimes M$ inside of G' and $\mathrm{Lie}(G'^M) = \mathrm{Lie}(G^M) \oplus \mathfrak{m}$ inside of $\mathrm{Lie}(G') = \mathfrak{g} \oplus \mathfrak{m}$, so the proposed formulas for G^M and its Lie algebra would follow from those for G'^M and its Lie algebra. Hence, we may and do rename G' as G to reduce to the case when M is a k-subgroup of G acting via conjugation.

For a k-algebra R and $g \in G(R)$, the M_R-action is trivial on $\mathfrak{g}$ precisely when the R-group map $M_R \to G_R$ defined by $t \mapsto gtg^{-1}$ is the canonical inclusion. Since $M = \mu_p$, by Proposition A.7.14 any R-group map $M_R \to G_R$ is uniquely determined by its associated map on Lie algebras. In particular, the R-group map $M_R \to G_R$ defined by g-conjugation on M_R is the canonical inclusion if and only if the induced map on Lie algebras over R is the canonical inclusion. But the map on Lie algebras is $X \mapsto (\mathrm{Ad}(g))(X)$. Setting such a map equal to the inclusion map $X \mapsto X$ is a Zariski-closed condition on $g \in G$, and by using a single X that spans $\mathfrak{m}$, we get a closed k-subgroup scheme of G that must be G^M and is given by the desired recipe in (3).

To compute $\mathrm{Lie}(G^M)$ in terms of the Lie-theoretic description of G^M, consider the k-subgroup $\mathrm{Fix}(\mathfrak{m}) \subseteq \mathrm{GL}(\mathfrak{g})$ of linear automorphisms of $\mathfrak{g}$ that restrict to the identity on the subspace $\mathfrak{m}$. The preceding alternative description of G^M gives a cartesian square

$$\begin{array}{ccc} G^M & \longrightarrow & \mathrm{Fix}(\mathfrak{m}) \\ \downarrow & & \downarrow \\ G & \xrightarrow[\mathrm{Ad}]{} & \mathrm{GL}(\mathfrak{g}) \end{array}$$

Since the formation of Lie algebras is compatible with fiber products, we therefore get a cartesian square of Lie algebras

$$\begin{array}{ccc} \mathrm{Lie}(G^M) & \longrightarrow & \mathfrak{z}_{\mathfrak{g}}(\mathfrak{m}) \\ \downarrow & & \downarrow \\ \mathfrak{g} & \xrightarrow[\mathrm{ad}]{} & \mathfrak{gl}(\mathfrak{g}) \end{array}$$

from which we see that $\mathrm{Lie}(G^M)$ is as desired. □

Corollary A.8.11 *Let G be a connected group scheme of finite type over a field k, and let H be an affine k-group scheme of finite type equipped with a left* *action on G over k. If $H_{\overline{k}}$ is linearly reductive then the H-action on G is trivial if and only if the induced action of H on* Lie(G) *is trivial.*

Proof The action is trivial if and only if $G^H = G$. Since $\mathrm{Lie}((G^H)^0) = \mathrm{Lie}(G^H) = \mathrm{Lie}(G)^H$, it suffices to prove that $(G^H)^0 = G$ when their Lie algebras coincide. We may also assume $k = \overline{k}$, so G_{red} is smooth and connected, hence it is irreducible and therefore G is irreducible. It suffices to prove that G^H and G coincide formally at the identity; i.e., their complete local rings at the identity coincide. Indeed, in such cases the closed G^H would contain a (dense) open subscheme around the identity in the irreducible G, so $G(k)$-translations would imply $G^H = G$.

Since the H-action on G preserves the identity section, it acts on the infinitesimal neighborhoods of the identity (even when H is not smooth). In particular, if R is the finite local k-algebra of such a neighborhood then H acts algebraically on R viewed as a k-algebra with finite dimension as a k-vector space and it suffices to prove that this action is trivial for each such R (as then G^H would contain all infinitesimal neighborhoods of the identity in G, so G^H and G would coincide formally at the identity). Let $\mathfrak{m}$ be the nilpotent maximal ideal of R, so $k = R/\mathfrak{m}$. The Lie algebra hypothesis implies that H acts trivially on $\mathfrak{m}/\mathfrak{m}^2$, so it acts trivially on each successive quotient $\mathfrak{m}^i/\mathfrak{m}^{i+1}$ for $i \geqslant 1$. Hence, H acts on the finite-dimensional k-vector space R through a k-group map $H \to \mathrm{GL}(R)$ factoring through a smooth unipotent subgroup of $\mathrm{GL}(R)$. But H is linearly reductive, so the k-group map $H \to \mathrm{GL}(R)$ is trivial. This gives the triviality of the H-action on $\mathrm{Spec}(R)$. □

The following result is of independent interest, and it will be used in the proof of Proposition A.8.14 below.

Proposition A.8.12 *Let G be a connected reductive group over a field k, and H an affine k-group scheme of finite type equipped with a left action on G. Assume that $H_{\overline{k}}$ is linearly reductive. Then the identity component of the scheme-theoretic fixed locus G^H is a connected reductive k-group.*

This result was originally proved in [PY1, 2.4] for finite étale H with $\mathrm{char}(k) \nmid \#H$. Over any field k of characteristic $p > 0$ there are counterexamples for more general H, such as $H = \mathrm{PGL}_n$ acting via conjugation through its natural inclusion into $G = \mathrm{GL}(\mathfrak{gl}_n)$ with $p|n$.

Proof We may and do assume that $k = \overline{k}$. Proposition A.8.10(1),(2) ensures that G^H is a smooth closed k-subgroup of G. In general, the identity component of a smooth closed k-subgroup G' of G is reductive if G/G' is affine. Indeed, we can assume G' is connected, and then the reductivity of G' follows from [Bo1, Thm. 1.1(i)] (whose proof via étale cohomology works for connected G') or

[Ri, Thm. A] (whose proof uses Mumford's conjecture, proved by Haboush). Thus, it suffices to prove that G/G^H is affine.

Define $G \rtimes H$ to be the semidirect product against the H-action on G, and let $\pi : G \rtimes H \to H$ be the natural projection. Let $\mathscr{R}$ be the functor whose value on a k-algebra R is the set of R-group sections ρ to π_R. There is a natural left action of G on $\mathscr{R}$ via conjugation, and for the canonical point $\rho_0 \in \mathscr{R}(k)$ corresponding to the natural section of π the G-stabilizer of ρ_0 is represented by G^H. More generally, if $\rho \in \mathscr{R}(k)$ is any k-group section to π then its functorial stabilizer G_ρ in G is represented by a closed subscheme (as we can use ρ to rewrite $G \rtimes H$ as a semidirect product for the action of H on G through ρ-conjugation, and G_ρ is the H-centralizer for this new action), so it suffices to prove that the quotients G/G_ρ are affine for all ρ.

A mild complication is that $\mathscr{R}$ is generally not representable (apart from special cases, such as when H is étale), but we can approximate it by representable subfunctors as follows.

Lemma A.8.13 *There is a directed system of G-stable subfunctors $\mathscr{R}_i$ in $\mathscr{R}$ such that each $\mathscr{R}_i$ is represented by an affine k-scheme of finite type and $\mathscr{R} = \bigcup \mathscr{R}_i$ as functors on k-algebras.*

Proof In general, for any pair of affine k-group schemes $\mathscr{G}$ and $\mathscr{G}'$ of finite type, consider the functor $\mathscr{H} = \underline{\mathrm{Hom}}(\mathscr{G}', \mathscr{G})$ of group scheme homomorphisms (i.e., its value on any k-algebra R is the set of R-group maps $\mathscr{G}'_R \to \mathscr{G}_R$). The coordinate ring A of $\mathscr{G}$ is the directed union of finite-dimensional k-subspaces that generate A as a k-algebra and are stable under the left and right action of $\mathscr{G}$ on A (by viewing A as a $\mathscr{G} \times \mathscr{G}$-representation space via $(g_1, g_2).a : g \mapsto a(g_1^{-1} g g_2)$); the same holds for $\mathscr{G}'$ and its coordinate ring A'. For any such pair of subspaces $V \subseteq A$ and $V' \subseteq A'$ we define the subfunctor $\mathscr{H}_{V',V} \subseteq \mathscr{H}$ to have as R-points the R-homomorphisms $f : \mathscr{G}'_R \to \mathscr{G}_R$ such that the pullback map f^* carries V_R into V'_R. It is clear that $\mathscr{H}(R)$ is the directed union of its subsets $\mathscr{H}_{V',V}(R)$ for all R, and that each subfunctor $\mathscr{H}_{V',V}$ is stable under the natural $\mathscr{G}$-action on $\mathscr{H}$ via conjugation. We claim that each $\mathscr{H}_{V',V}$ is represented by an affine k-scheme of finite type.

Since A is a quotient of the symmetric algebra $\mathrm{Sym}(V)$ by a finitely generated ideal $J = (h_1, \ldots, h_n)$, if d is the maximal degree of the h_i and we denote by $V'_{\leqslant d} \subseteq A'$ the k-linear span of monomials in V' of degree at most d then for any R-linear map $T : V_R \to V'_R$ the unique R-algebra extension $[T] : \mathrm{Sym}(V)_R \to A'_R$ carries the h_i into $(V'_{\leqslant d})_R$. Thus, it is easy to check that the condition $[T](J) = 0$ is a Zariski-closed condition on the scheme of linear maps $\underline{\mathrm{Hom}}(V, V')$, and for such T the condition that the resulting R-algebra map $A_R \to A'_R$ corresponds to an R-group homomorphism is an additional Zariski-closed condition. This proves the asserted representability of each $\mathscr{H}_{V',V}$.

In the same way, the functor of group scheme sections to a fixed k-homomorphism $\mathscr{G}' \to \mathscr{G}$ is a directed union of subfunctors that are represented by affine k-schemes of finite type. Applying this to $\pi : G \rtimes H \to H$, we get the desired G-stable subfunctors $\mathscr{R}_i$ in $\mathscr{R}$. □

Returning to the proof of Proposition A.8.12, since each $\rho_0 \in \mathscr{R}(k)$ lies in some $\mathscr{R}_{i_0}(k)$, the G-orbit of ρ_0 lies in $\mathscr{R}_{i_0}$. Thus, to prove G/G_{ρ_0} is affine for each ρ_0 it suffices to work with each $\mathscr{R}_i$ separately. Fix i and consider the left action of G on $\mathscr{R}_i$. We claim the action map $\alpha : G \times \mathscr{R}_i \to \mathscr{R}_i \times \mathscr{R}_i$ defined by $(g, \rho) \mapsto (g.\rho, \rho)$ is smooth.

Consider the smooth projection map

$$q : G \times \mathscr{R}_i \times \mathscr{R}_i \to \mathscr{R}_i \times \mathscr{R}_i$$

and the H-action on the source by $h.(g, \rho, \rho') = (\rho(h)g\rho'(h)^{-1}, \rho, \rho')$. This makes q an H-equivariant map (using trivial action on the target), so by Proposition A.8.10(2) the induced map between subschemes of H-invariants is smooth. But the H-invariants of the source consist of triples (g, ρ, ρ') such that $g.\rho' = \rho$, so smoothness implies that there is no obstruction to infinitesimal lifting of points of G conjugating one point of $\mathscr{R}_i$ into another. This is precisely the functorial criterion for smoothness of α.

For each $\rho_0 \in \mathscr{R}_i(k)$, the pullback of α to ρ_0-fibers under second projection to $\mathscr{R}_i$ is the orbit map $G \to \mathscr{R}_i$ defined by $g \mapsto g.\rho_0$, so this map is smooth since α is smooth. Thus, all G-orbits in $\mathscr{R}_i$ are open and smooth. But the orbits are pairwise disjoint, so it follows that the affine k-scheme $\mathscr{R}_i$ is smooth with only finitely many orbits, and hence all orbits are affine. For each $\rho_0 \in \mathscr{R}_i(k)$, smoothness of the orbit map for ρ_0 implies that the affine orbit is identified with G/G_{ρ_0}, so G/G_{ρ_0} is affine. □

Proposition A.8.14 *Let G be a smooth connected affine group over a field k and H an affine k-group of finite type equipped with a left action on G. Assume that $H_{\overline{k}}$ is linearly reductive.*

(1) *If a closed normal k-subgroup N in G is preserved by H and either N or H is smooth then $(G^H)^0 \to ((G/N)^H)^0$ is surjective.*
(2) *Assume that the H-action preserves $\mathscr{R}_{u,k}(G)$ and the $H_{\overline{k}}$-action on $G_{\overline{k}}$ preserves $\mathscr{R}_u(G_{\overline{k}})$. Then*

$$(\mathscr{R}_{u,k}(G) \cap (G^H)^0)^0 = \mathscr{R}_{u,k}((G^H)^0).$$

In particular, the natural map $(G^H)^0 \to G/\mathscr{R}_{u,k}(G)$ factors as

$$(G^H)^0 \to (G^H)^0/\mathscr{R}_{u,k}((G^H)^0) \twoheadrightarrow ((G/\mathscr{R}_{u,k}(G))^H)^0 \qquad \text{(A.8.2)}$$

with the second map an étale unipotent isogeny (which is an isomorphism when k is perfect or the H-action on G is induced by the identification of H with a closed k-subgroup of a k-torus T acting on G, and in the latter case $\mathscr{R}_{u,k}(G)^H = \mathscr{R}_{u,k}((G^H)^0)$). Also, if G is pseudo-reductive over k then $(G^H)^0$ is pseudo-reductive over k.

(3) *If G is pseudo-reductive over k, T is a maximal k-torus in G, and H is a k-subgroup of T acting on G via conjugation, then $\Phi((G^H)^0_{k_s}, T_{k_s})$ is the set of roots in $\Phi(G_{k_s}, T_{k_s})$ that are trivial on H_{k_s}.*

Note that the hypotheses in (2) that H preserves $\mathscr{R}_{u,k}(G)$ and $H_{\overline{k}}$ preserves $\mathscr{R}_u(G_{\overline{k}})$ are satisfied whenever the H-action is the restriction of an action on G by a smooth k-group H' containing H as a closed k-subgroup (since $H'(k_s)$ is schematically dense in H'_{k_s}). This applies in particular when $H' = G$ (i.e., H is a k-subgroup scheme of G acting via conjugation) and when H' is a k-torus. Also, in (1) the necessity of assuming either N or H is smooth is seen by considering the Frobenius isogeny $G \to G^{(p)}$ of a split connected semisimple group G in characteristic $p > 0$ containing a non-central μ_p; let $N = H = \mu_p$ act by conjugation on G and act trivially on $G^{(p)}$.

Proof We first prove (1) for smooth N (in which case our argument will apply to any k-group scheme G locally of finite type, without smoothness, connectedness, or affineness hypotheses). The quotient map $G \to G/N$ is smooth since N is smooth, so by Proposition A.8.10(2) the induced homomorphism $G^H \to (G/N)^H$ is smooth. Thus, this map restricts to a surjection between identity components. To prove (1) when H is smooth, we may and do assume $k = \overline{k}$, so N_{red} is a smooth closed k-subgroup of G that is normal and preserved by H (as we can check by computing with the actions of $G(k)$ and $H(k)$, due to the smoothness of G and H). The settled case of smooth N therefore implies that the natural map $(G^H)^0 \to ((G/N_{\text{red}})^H)^0$ is surjective, so it suffices to check that the natural map $((G/N_{\text{red}})^H)^0 \to ((G/N)^H)^0$ is surjective. This is the map between identity components induced by the natural map $(G/N_{\text{red}})^H \to (G/N)^H$ between affine k-group schemes of finite type, so it is enough to show that this latter map is surjective. Even better, it is bijective on k-points with $k = \overline{k}$ because $N(k) = N_{\text{red}}(k)$ and the formation of H-invariants coincides with the formation of (scheme-theoretic) $H(k)$-invariants (due to the smoothness of H).

To prove (2), first note that since (by hypothesis) H preserves $\mathscr{R}_{u,k}(G)$,

$$\mathscr{R}_{u,k}(G) \cap G^H = \mathscr{R}_{u,k}(G)^H.$$

In particular, using Proposition A.8.10(2) when $\text{char}(k) > 0$, this intersection is smooth. It is visibly unipotent and normal in G^H, so

$$(\mathscr{R}_{u,k}(G) \cap G^H)^0 \subseteq \mathscr{R}_{u,k}((G^H)^0).$$

To prove the reverse inclusion (which will imply that the second map in (A.8.2) is an étale unipotent isogeny, with surjectivity following from (1)), it suffices to prove that $\mathscr{R}_{u,k}((G^H)^0)_{\overline{k}} \subseteq \mathscr{R}_u(G_{\overline{k}})$ (as then the smooth connected normal k-subgroup of G generated by $\mathscr{R}_{u,k}((G^H)^0)$ would be unipotent and hence would be contained in $\mathscr{R}_{u,k}(G)$).

Since $\mathscr{R}_{u,k}(G')_{\overline{k}} \subseteq \mathscr{R}_u(G'_{\overline{k}})$ for any smooth connected affine k-group G', and by hypothesis $H_{\overline{k}}$ preserves $\mathscr{R}_u(G_{\overline{k}})$, by taking $G' = (G^H)^0$ we see that it suffices to consider the case when k is algebraically closed. In other words, to prove (2) apart from the isomorphism claim in (A.8.2) for perfect k or for H acting through a k-torus action on G, we just need to show that if k is algebraically closed then the image of $(G^H)^0$ in G^{red} is reductive. But we can apply (1) to the quotient map $G \to G^{\text{red}}$ to identify the image of $(G^H)^0$ in G^{red} with $((G^{\text{red}})^H)^0$, so we can assume that G is reductive. Then the reductivity of $(G^H)^0$ is Proposition A.8.12.

When k is perfect, the final map in (A.8.2) must be an isomorphism. Indeed, in this case it is an étale unipotent isogeny between smooth connected affine k-groups, and the target group is reductive, so both groups are reductive. But in a connected reductive group over any field there are no nontrivial étale unipotent normal subgroups (as an étale normal closed subgroup must be central, due to connectedness considerations, and the center of a connected reductive group is of multiplicative type).

Suppose instead that H is a closed k-subgroup scheme of a k-torus T equipped with an action on G (from which the H-action is defined via restriction), and allow k to be arbitrary. To prove that the second map in (A.8.2) is an isomorphism in such cases we just have to prove that the étale unipotent quotient

$$(\mathscr{R}_{u,k}(G) \cap (G^H)^0)/\mathscr{R}_{u,k}((G^H)^0)$$

is trivial, which is to say that the intersection $\mathscr{R}_{u,k}(G) \cap (G^H)^0$ is connected. This is a normal open and closed k-subgroup of $\mathscr{R}_{u,k}(G) \cap G^H = \mathscr{R}_{u,k}(G)^H$, so it suffices to note that this latter k-group is connected. This connectedness follows from Proposition 2.1.12(3) applied to the smooth connected unipotent k-group $\mathscr{R}_{u,k}(G)$ equipped with the induced T-action. (To see that the T-action on G preserves $\mathscr{R}_{u,k}(G)$ it suffices to observe that $T(k_s)$ clearly keeps $\mathscr{R}_{u,k_s}(G_{k_s})$ stable inside G_{k_s}.) This method establishes the equality

$$\mathscr{R}_{u,k}(G)^H = \mathscr{R}_{u,k}((G^H)^0)$$

when the H-action on G is induced by a T-action as above. (Since we never enlarged the constant field k in the argument apart from passage to k_s, the same reasoning proves that $\mathscr{R}_{us,k}(G)^H = \mathscr{R}_{us,k}((G^H)^0)$ when the H-action on G is induced by a T-action as above; note that $\mathscr{R}_{us,k}(G)^H$ is smooth by Proposition A.8.10, connected by Proposition 2.1.12(3), and k-split by Proposition B.4.5.) This completes the proof of (2).

By (2), the k-group $(G^H)^0$ in (3) is pseudo-reductive over k and contains T as a maximal k-torus. Thus, the identification of $\mathrm{Lie}((G^H)^0)$ with $\mathrm{Lie}(G)^H$ proves (3). □

Remark A.8.15 In Proposition A.8.14(1), when N is smooth and unipotent (and we allow G to be any k-group scheme locally of finite type) we claim that $G^H \to (G/N)^H$ is surjective. For example, if H is a k-subgroup of a torus acting on a smooth connected affine k-group G then the left-exact sequence

$$1 \to \mathscr{R}_u(G_{\overline{k}})^{H_{\overline{k}}} \to (G_{\overline{k}})^{H_{\overline{k}}} \to (G_{\overline{k}}^{\mathrm{red}})^{H_{\overline{k}}} \to 1$$

is short exact (choose $N = \mathscr{R}_u(G_{\overline{k}})$ inside $G_{\overline{k}}$). Since the kernel term is connected (by Proposition 2.1.12(3)), it follows that G^H is connected if and only if $(G_{\overline{k}}^{\mathrm{red}})^{H_{\overline{k}}}$ is connected. A notable example where this applies is when $\mathscr{D}(G_{\overline{k}}^{\mathrm{red}}) = \mathrm{SL}_2$ and H is a k-subgroup of a k-torus in G acting by conjugation.

To prove the surjectivity of $G^H \to (G/N)^H$ we may assume $k = \overline{k}$, so H is linearly reductive. Our problem is to show that for any $x \in (G/N)^H(k)$ the x-fiber Y of $G \to G/N$ has non-empty H-fixed locus. This fiber is an N-torsor and there are evident left actions of N and H on Y that are compatible in the sense that $h.(n.y) = (h.n)(h.y)$ (using the H-action on N induced from its action on G). Rather generally, if Y is a finite type k-scheme equipped with compatible actions of N and H that make Y a homogeneous space for the smooth unipotent N (in the sense that the action map for N through some $y \in Y(k)$ identifies Y with $N/\mathrm{Stab}_N(y)$, in which case Y is smooth and the same holds for any $y \in Y(k)$) then we claim that Y^H is non-empty (equivalently, $Y(k)$ contains an H-fixed point).

If N' is an H-stable smooth unipotent normal subgroup of N then the quotient Y/N' clearly exists in the category of k-schemes (as a coset space for N modulo a closed subgroup scheme). This quotient is a homogeneous space for N/N' admitting a compatible action by H, with each fiber of $Y \to Y/N'$ a homogeneous space for N'. Thus, by induction on $\dim N$ using the descending central series (which is stable under the automorphism functor of N, and hence under H), we may assume N is commutative, and even p-torsion when $\mathrm{char}(k) = p > 0$. The commutativity of N ensures that the stabilizer scheme S in N at any $y \in Y(k)$ is equal to the functorial kernel of the action of N on Y, so S is stable under H. Thus, by replacing N with N/S we may assume Y is an N-torsor. Likewise, by working with N^0 and N/N^0 we may assume that N is either connected or finite étale.

Assume N is finite étale, so Y is also finite étale. The H-actions on both N and Y factor through H/H^0, so we can assume that the linearly reductive H is also finite étale. Thus, H is a finite constant k-group with order not divisible by $\mathrm{char}(k)$. We may assume $\mathrm{char}(k) = p > 0$, since in characteristic 0 the finite unipotent N must be trivial. The obstruction to finding an H-fixed point on the $N(k)$-torsor $Y(k)$ lies in $\mathrm{H}^1(H(k), N(k))$. But $N(k)$ is p-torsion and $H(k)$ is finite of order not divisible by p, so the obstruction vanishes.

Now suppose that N is connected, so it is a vector group (see Lemma B.1.10 when $\mathrm{char}(k) > 0$). The H-action on N might not respect a linear structure when $\mathrm{char}(k) > 0$, but this can be fixed via an H-equivariant refinement of Lemma B.1.10: there is an H-equivariant étale isogeny $N \to V$ onto a linear representation V of H. Indeed, since N is a vector group, we see via a choice of linear structure on N that the coordinate ring $k[N]$ is generated by the k-subspace $\mathrm{Hom}_{k\text{-gp}}(N, \mathbf{G}_{\mathrm{a}})$ of the augmentation ideal J. This k-subspace is visibly an H-subrepresentation of $k[N]$, and it maps onto J/J^2, so by the linear reductivity of H there is an H-submodule $V' \subseteq \mathrm{Hom}_{k\text{-gp}}(N, \mathbf{G}_{\mathrm{a}})$ that maps k-isomorphically onto J/J^2. Hence, for the dual space V of V' (viewed as a linear representation of H) we obtain an H-equivariant k-homomorphism $f : N \to \underline{V}$ that is an étale isogeny (since the induced map between Lie algebras is dual to the isomorphism $V' \simeq J/J^2$).

By filtering N via its finite étale subgroup $\ker f$ and its quotient $\underline{V}$, we can finally arrange that $N = \underline{V}$ is a linear representation for H. The obstruction to finding an H-invariant point in the N-torsor Y lies in $\mathrm{H}^1(H, \underline{V})$ by the same cocycle argument that arose in the proof of Proposition A.8.10(2). In that proof we also noted that this cohomology group vanishes because it is identified with the set of H-equivariant extensions of k by V as linear representations (so the vanishing follows from the linear reductivity of H).

Remark A.8.16 The hypothesis in Proposition A.8.14 that $H_{\overline{k}}$ preserves $\mathscr{R}_u(G_{\overline{k}})$ neither implies nor is implied by the property that the H-action preserves $\mathscr{R}_{u,k}(G)$. We illustrate this with examples over imperfect fields k of characteristic $p > 0$, using $H = \mu_p$.

Let $k' = k(a^{1/p})$ with $a \in k - k^p$, and define $G = \mathrm{R}_{k'/k}(\mathscr{G}_{k'})$ for a nontrivial connected reductive k-group $\mathscr{G}$. Let H act on G by the functorial rule that for any k-algebra R the action of $\zeta \in \mu_p(R)$ on $G(R) = \mathscr{G}(k' \otimes_k R)$ is induced by the R-algebra automorphism of $k' \otimes_k R = R[x]/(x^p - a)$ defined by $x \mapsto \zeta x$. Clearly H preserves $\mathscr{R}_{u,k}(G) = 1$, and to see that $H_{\overline{k}}$ does not preserve $\mathscr{R}_u(G_{\overline{k}})$ it suffices to verify this on Lie algebras. The maximal reductive quotient of $G_{\overline{k}}$ is identified with $\mathscr{G}_{\overline{k}}$ via the functorial map $G_{\overline{k}}(R) = \mathscr{G}(k' \otimes_k R) \to \mathscr{G}(R)$ (for $\overline{k}$-algebras R) defined by the natural k-algebra map $k' \otimes_k R \to R$ (using the unique k-embedding of k' into $\overline{k}$). We conclude that

$$\mathrm{Lie}(\mathscr{R}_u(G_{\overline{k}})) = \ker(k' \otimes \mathrm{Lie}(\mathscr{G})_{\overline{k}} \to \mathrm{Lie}(\mathscr{G})_{\overline{k}}),$$

using the natural k'-linear structure on $\mathrm{Lie}(\mathscr{G})_{\overline{k}}$. Thus, for any $v \in \mathrm{Lie}(\mathscr{G})_{\overline{k}}$ clearly $w := a^{1/p} \otimes v - 1 \otimes a^{1/p} v \in \mathrm{Lie}(\mathscr{R}_u(G_{\overline{k}}))$. But for a $\overline{k}$-algebra R and $\zeta \in \mu_p(R)$ we have $\zeta.w = a^{1/p} \otimes (\zeta - 1)v + w$, so by choosing $v \neq 0$ and $\zeta \neq 1$ we ensure that $\zeta.w \notin \mathrm{Lie}(\mathscr{R}_u(G_{\overline{k}}))_R$.

In the other direction, assume $[k : k^p] \geqslant p^2$ and choose $a, b \in k - k^p$ that are distinct elements of a p-basis, so the extensions $k_1 = k(a^{1/p})$, $k_2 = k(a^{1/p}, b^{1/p})$, and $k_3 = k(a^{1/p^2}, b^{1/p})$ have respective k-degrees p, p^2, and p^3. Note that k_3 has degree p over $k(a^{1/p^2} + b^{1/p})$ with primitive generator a^{1/p^2}, and $k^{1/p} \cap k_3 = k_2$. We define G to be the pushout quotient

$$G = (\mathrm{R}_{k_3/k}(\mathrm{GL}_1) \times \mathrm{GL}_1)/\mathrm{R}_{k_1/k}(\mathrm{GL}_1)$$

using the inclusion $j : \mathrm{R}_{k_1/k}(\mathrm{GL}_1) \hookrightarrow \mathrm{R}_{k_3/k}(\mathrm{GL}_1) \times \mathrm{GL}_1$ defined by $j(t) = (t, t^p)$ (which makes sense since $(k_1)^p \subseteq k$). To define an action of $H = \mu_p$ on the k-group G, for k-algebras R consider the action of $\zeta \in \mu_p(R)$ on the R-algebra $k_3 \otimes_k R$ defined by the trivial action on $k(a^{1/p^2} + b^{1/p})$ and the effect $a^{1/p^2} \mapsto \zeta a^{1/p^2}$ on a^{1/p^2} (so, for example, $\zeta.b^{1/p} = b^{1/p} + (1-\zeta)a^{1/p^2} \notin k_2 \otimes R$). This naturally defines an H-action on $\mathrm{R}_{k_3/k}(\mathrm{GL}_1)$ that is trivial on $\mathrm{R}_{k_1/k}(\mathrm{GL}_1)$ and does *not* preserve the k-subgroup $\mathrm{R}_{k_2/k}(\mathrm{GL}_1)$, so it defines an H-action on G that does not preserve the smooth connected unipotent k-subgroup

$$U := \mathrm{R}_{k_2/k}(\mathrm{GL}_1)/\mathrm{R}_{k_1/k}(\mathrm{GL}_1) = \mathrm{R}_{k_1/k}(\mathrm{R}_{k_2/k_1}(\mathrm{GL}_1)/\mathrm{GL}_1)$$

(embedded into G via the k-embedding $k_2 \hookrightarrow k_3$ and the p-power map $\mathrm{R}_{k_2/k}(\mathrm{GL}_1) \to \mathrm{GL}_1$).

We claim that the inclusion $U \subseteq \mathscr{R}_{u,k}(G)$ is an equality, or in other words that G/U is pseudo-reductive. In view of the short exact sequence

$$1 \to \mathrm{R}_{k_3/k}(\mathrm{GL}_1)/\mathrm{R}_{k_2/k}(\mu_p) \to G/U \to \mathrm{GL}_1 \to 1,$$

it suffices to prove that the smooth connected left term is pseudo-reductive. This affine k-group admits an evident map into the pseudo-reductive k-group $\mathrm{R}_{k_3/k}(\mathrm{GL}_1/\mu_p)$ with kernel $Q = \mathrm{R}_{k_3/k}(\mu_p)/\mathrm{R}_{k_2/k}(\mu_p)$ that satisfies $Q(k_s) = 1$ (see Example 5.3.7, noting that $k_2 = k^{1/p} \cap k_3$), so the pseudo-reductivity of G/U is proved.

To show that $H_{\overline{k}}$ preserves $\mathscr{R}_u(G_{\overline{k}})$, it is equivalent that the $H_{\overline{k}}$-action on $\mathrm{R}_{k_3/k}(\mathrm{GL}_1)_{\overline{k}}$ preserves the unipotent radical up to multiplication against μ_p in the unique GL_1 subgroup. As $\overline{k}$-algebras, clearly $k_3 \otimes_k \overline{k} = \overline{k}[x, y]/(x^{p^2}, y^p)$ for $x = a^{1/p^2} \otimes 1 - 1 \otimes a^{1/p^2}$ and $y = b^{1/p} \otimes 1 - 1 \otimes b^{1/p}$. For any $\overline{k}$-algebra R,

$$\mathrm{R}_{k_3/k}(\mathrm{GL}_1)_{\overline{k}}(R) = (k_3 \otimes_k R)^\times = ((k_3 \otimes_k \overline{k}) \otimes_{\overline{k}} R)^\times = (R[x, y]/(x^{p^2}, y^p))^\times$$

and $(\mathscr{R}_u(\mathrm{R}_{k_3/k}(\mathrm{GL}_1)))(R)$ is the subgroup of elements with constant term 1, so our problem is to show that the subgroup of elements with constant term in $\mu_p(R)$ is preserved under the R-algebra automorphisms of $R[x,y]/(x^{p^2},y^p)$ determined by

$$[\zeta]: x \mapsto x + a^{1/p^2}(\zeta - 1), \ \ [\zeta]: y \mapsto y - a^{1/p^2}(\zeta - 1)$$

for every $\zeta \in \mu_p(R)$. If $f \in R[x,y]$ has constant term in $\mu_p(R)$ then $f([\zeta](x),[\zeta](y))$ has constant term

$$f(a^{1/p^2}(\zeta-1), -a^{1/p^2}(\zeta-1)) \in f(0,0)+(\zeta-1)R \subseteq \mu_p(R)+\alpha_p(R) = \mu_p(R),$$

so we are done.

Appendix B

Tits' work on unipotent groups in nonzero characteristic

In our study of pseudo-reductive groups, we need to use some fundamental results of Tits (building on earlier work of M. Rosenlicht [Ros1], [Ros2]) concerning the structure of smooth connected unipotent groups and torus actions on such groups over an arbitrary (especially imperfect) ground field of nonzero characteristic. These results were presented by Tits in a course at Yale University in 1967, and lecture notes [Ti1] for that course were circulated but never published. Much of the course was concerned with general results on linear algebraic groups that are available now in many standard references (such as [Bo2], [Hum2], and [Spr]).

The only previously available written account (with proofs) of Tits' structure theory of unipotent groups is his unpublished Yale lecture notes, though a summary of the statements of his results is given in [Oes, Ch. V]. In this appendix we give a self-contained development of the theory, with complete proofs, and in some parts we have reproduced arguments from Tits' lecture notes whereas in other parts we provide new proofs that are shorter and simpler. We go beyond what we need in order that this appendix may serve as a useful general reference on Tits' work on this topic.

Throughout this appendix, k is an arbitrary field with characteristic $p > 0$, except in Proposition B.4.6.

B.1 Subgroups of vector groups

A smooth connected solvable k-group G is *k-split* if it admits a composition series whose successive quotients are k-isomorphic to $\mathbf{G}_{\mathrm{a}}$ or GL_1. In the case of tori this is a widely used notion, and it satisfies convenient properties, such as: (i) every subtorus or quotient torus (over k) of a k-split k-torus is k-split, (ii) every k-torus is an almost direct product of its maximal k-split subtorus and its maximal k-anisotropic subtorus. However, in contrast with the case of tori, it is not true for general smooth connected solvable G that the k-split property is inherited by smooth connected normal k-subgroups:

Example B.1.1 Suppose that k is imperfect and choose $a \in k - k^p$. The k-group

$$U := \{y^p = x - ax^p\}$$

is a k-subgroup of the k-split $G = \mathbf{G}_a^2$ and it becomes isomorphic to $\mathbf{G}_a$ over $k(a^{1/p})$ but it is not k-isomorphic to $\mathbf{G}_a$ as a k-scheme, let alone as a k-group. Indeed, the regular compactification of U has a unique point at infinity, but that point has residue field $k(a^{1/p})$ so it is not k-rational (in contrast with $\mathbf{G}_a$).

Tits introduced an analogue for unipotent k-groups of the notion of anisotropicity for tori over a field. This rests on a preliminary understanding of the properties of subgroups of vector groups, so we take up that study now. The main case of interest to us will be imperfect ground fields.

Definition B.1.2 A polynomial $f \in k[x_1, \ldots, x_n]$ is a *p-polynomial* if every monomial appearing in f has the form $c_{ij}x_i^{p^j}$ for some $c_{ij} \in k$; that is, $f = \sum f_i(x_i)$ with $f_i(x_i) = \sum_j c_{ij}x_i^{p^j} \in k[x_i]$. (In particular, $f_i(0) = 0$ for all i. Together with the identity $f = \sum f_i(x_i)$, this uniquely determines each f_i in terms of f. Note that $f(0) = 0$.)

Proposition B.1.3 *A polynomial $f \in k[x_1, \ldots, x_n]$ is a p-polynomial if and only if the associated map of k-schemes $\mathbf{G}_a^n \to \mathbf{G}_a$ is a k-homomorphism.*

Proof This is elementary and is left to the reader. □

A nonzero polynomial over k is *separable* if its zero scheme in affine space is generically k-smooth.

Proposition B.1.4 *Let $f \in k[x_1, \ldots, x_n]$ be a nonzero polynomial such that $f(0) = 0$. Then the subscheme $f^{-1}(0) \subseteq \mathbf{G}_a^n$ is a smooth k-subgroup if and only if f is a separable p-polynomial.*

Proof The "if" direction is clear. For the converse, we assume that $f^{-1}(0)$ is a smooth k-subgroup and we denote it as G. The smoothness implies that f is separable. To prove that f is a p-polynomial, by Proposition B.1.3 it suffices to prove that the associated map of k-schemes $\mathbf{G}_a^n \to \mathbf{G}_a$ is a k-homomorphism. Without loss of generality, we may assume that k is algebraically closed.

For any $\alpha \in G(k)$, $f(x + \alpha)$ and $f(x)$ have the same zero scheme (namely, G) inside of $\mathbf{G}_a^n$. Thus, $f(x + \alpha) = c(\alpha)f(x)$ for a unique $c(\alpha) \in k^\times$. Consideration of a highest-degree monomial term appearing in f implies that

$c = 1$. Pick $\beta \in k^n$, so $f(\beta + \alpha) - f(\beta) = 0$ for all $\alpha \in G(k)$. Thus $f(\beta + x) - f(\beta)$ vanishes on G, so $f(\beta + x) - f(\beta) = g(\beta)f(x)$ for a unique $g(\beta) \in k$. Consideration of a highest-degree monomial term in f forces $g(\beta) = 1$. Hence, $f(\beta + x) = f(\beta) + f(x)$ for all $\beta \in k^n$. This says that f is additive. □

Corollary B.1.5 *Let $G \subseteq \mathbf{G}_{\mathrm{a}}^n$ be a smooth k-subgroup of codimension 1. Then G is the zero scheme of a separable nonzero p-polynomial in $k[x_1, \dots, x_n]$.*

Proof Since G is smooth of codimension 1 in $\mathbf{G}_{\mathrm{a}}^n$, it is the zero scheme of a separable nonzero polynomial $f \in k[x_1, \dots, x_n]$. By Proposition B.1.4, f is a p-polynomial. □

Definition B.1.6 If $f = \sum_{i=1}^n f_i(x_i)$ is a p-polynomial over k in n variables with $f_i(0) = 0$ for all i, then the *principal part* of f is the sum of the leading terms of the f_i.

Lemma B.1.7 *Let V be a vector group of dimension $n \geqslant 1$ over k, and let $f : V \to \mathbf{G}_{\mathrm{a}}$ be a k-homomorphism. Then the following are equivalent:*

(1) *there exists a non-constant k-scheme morphism $f' : \mathbf{A}_k^1 \to V$ such that $f \circ f' = 0$;*
(2) *for every k-group isomorphism $h : \mathbf{G}_{\mathrm{a}}^n \simeq V$, the principal part of the p-polynomial $f \circ h \in k[x_1, \dots, x_n]$ has a nontrivial zero in k^n;*
(3) *there exists a k-group isomorphism $h : \mathbf{G}_{\mathrm{a}}^n \simeq V$ such that $f \circ h$ "only depends on the last $n - 1$ coordinates" (i.e.,* $\ker(f \circ h)$ *contains the first factor of $\mathbf{G}_{\mathrm{a}}^n$).*

In this lemma, it is not sufficient in (2) to consider just a single choice of h. For example, if k is imperfect and $a \in k - k^p$, then $f = y^p - (x + ax^p)$ has principal part $y^p - ax^p$ with no zeros on $k^2 - \{0\}$. Composing f with the k-automorphism $(x, y) \mapsto (x, y + x^p)$ yields $y^p + x^{p^2} - (x + ax^p)$ whose principal part is $y^p + x^{p^2}$, which has zeros on $k^2 - \{0\}$.

Proof We will show that (1) ⇒ (2) ⇒ (3) ⇒ (1).

For (1) ⇒ (2), assume that (1) holds and let $\varphi = h^{-1} \circ f'$. Let $\varphi_i : \mathbf{G}_{\mathrm{a}} \to \mathbf{G}_{\mathrm{a}}$ be the ith component of φ, and $a_i t^{s_i}$ denote the leading term of $\varphi_i(t)$, with $s_i = 0$ when $\varphi_i = 0$. For some i we have $s_i > 0$, since some φ_i is non-constant (as φ is non-constant, because of the same for f'). Let $\sum_{i=1}^n c_i x_i^{p^{m_i}}$ be the principal part of $f \circ h$, so

$$0 = f(h(\varphi(t))) = \sum_{i=1}^{n} c_i a_i^{p^{m_i}} t^{s_i p^{m_i}} + \cdots$$

since $f \circ h \circ \varphi = f \circ h \circ h^{-1} \circ f' = f \circ f' = 0$. Let $N = \max_i\{s_i p^{m_i}\} > 0$. Define $b_i = a_i$ if $s_i p^{m_i} = N$ (so $b_i \neq 0$), and $b_i = 0$ if $s_i p^{m_i} < N$. Since the coefficient of the term of degree N in $f(h(\varphi(t)))$ must be zero, we have $\sum_{i=1}^{n} c_i b_i^{p^{m_i}} = 0$ with $b_i \in k$ and some b_i is nonzero, so (2) holds.

To prove (2) $\Rightarrow$ (3), assume (2) holds and let $h : \mathbf{G}_{\mathrm{a}}^n \simeq V$ be any k-group isomorphism. We may assume $f \neq 0$, so the principal part of $f \circ h$ is nonzero. The proof will proceed by induction on the sum d of the degrees of nonzero terms of the principal part $\sum_{i=1}^{n} c_i x_i^{p^{m_i}}$ of $f \circ h$. If $c_r = 0$ for some r, we are done by interchanging x_r and x_1. So we may assume that all c_i are nonzero and, upon permuting the coordinates, that $m_1 \geqslant \cdots \geqslant m_n \geqslant 0$. By (2), there exist $(a_1, \ldots, a_n) \in k^n - \{0\}$ such that $\sum_{i=1}^{n} c_i a_i^{p^{m_i}} = 0$. Let $r \geqslant 0$ be minimal such that $a_r \neq 0$. Define the k-group isomorphism $h' : \mathbf{G}_{\mathrm{a}}^n \simeq \mathbf{G}_{\mathrm{a}}^n$ by $h'(y_1, \ldots, y_n) = (x_1, \ldots, x_n)$ with

$$x_1 = y_1, \quad \ldots, \quad x_{r-1} = y_{r-1},$$

$$x_r = a_r y_r, \quad x_{r+1} = y_{r+1} + a_{r+1} y_r^{p^{m_r - m_{r+1}}}, \quad \ldots, \quad x_n = y_n + a_n y_r^{p^{m_r - m_n}}.$$

Thus, $f \circ h \circ h'$ is a p-polynomial with principal part

$$\sum_{i \neq r} c_i y_i^{p^{m_i}} + \sum_{i=1}^{n} c_i a_i^{p^{m_i}} \cdot y_r^{p^{m_r}} = \sum_{i \neq r} c_i y_i^{p^{m_i}}$$

since $\sum_{i=1}^{n} c_i a_i^{p^{m_i}} = 0$. The sum of the degrees of the nonzero terms of the principal part of $f \circ h \circ h'$ is strictly smaller than d since $c_r \neq 0$, so the induction hypothesis applies.

Finally, we assume (3) and prove (1). Let $h : \mathbf{G}_{\mathrm{a}}^n \to V$ be a k-isomorphism such that $\ker(f \circ h)$ contains the first factor of $\mathbf{G}_{\mathrm{a}}^n$. Define $\varphi : \mathbf{G}_{\mathrm{a}} \to \mathbf{G}_{\mathrm{a}}^n$ by $\varphi(t) = (t, 0, 0, \ldots, 0)$. Finally, let $f' = h \circ \varphi$. Then $f \circ f' = f \circ h \circ \varphi = 0$. □

Lemma B.1.8 *If K/k is a Galois extension and a p-polynomial of the form $\sum_{i=1}^{n} c_i x_i^{p^{m_i}}$ over k has a zero in $K^n - \{0\}$ then it has a zero in $k^n - \{0\}$.*

Proof The proof is by induction on n. The terms may be ordered so that $m_1 \geqslant m_2 \geqslant \cdots$. If $n = 1$, then since $c_1 a_1^{p^{m_1}} = 0$ with $a_1 \in K^{\times}$ we see that $c_1 = 0$, so $c_1 x_1^{p^{m_1}}$ has a zero in $k^{\times}$.

Now suppose $n > 1$ and that $\sum_{i=1}^{n} c_i a_i^{p^{m_i}} = 0$ with $a_i \in K$ not all zero. Let $a = (a_1, \ldots, a_n)$. If $a_n = 0$ then the theorem is true by the induction hypothesis. If $a_n \neq 0$, we may assume $a_n = 1$ by replacing a_i with $a_i / a_n^{p^{m_n - m_i}}$ for all i. For all $\sigma \in \mathrm{Gal}(K/k)$, the point $a - \sigma(a)$ is a zero of $\sum c_i x_i^{p^{m_i}}$. If not all a_i belong to k then $a - \sigma(a) \neq 0$ for some σ, so since $a_n - \sigma(a_n) = 0$ we may again apply the inductive hypothesis. □

Lemma B.1.9 *Let V be a vector group over k, K/k a Galois extension, and $f : V \to \mathbf{G}_{\mathrm{a}}$ a k-homomorphism. The equivalent conditions* (1), (2), *and* (3) *of Lemma B.1.7 hold over K if and only if they hold over k.*

Proof It is clear that if (1) holds over k, it also holds over K. On the other hand, by Lemma B.1.8, (2) is true over k if it is true over K. □

Lemma B.1.10 *Every smooth p-torsion commutative affine k-group G embeds as a k-subgroup of a vector group over k. Moreover, G admits an étale isogeny onto a vector group over k, and if G is connected and $k = \overline{k}$ then G is a vector group over k.*

Proof We first construct the embedding into a vector group over k, and then at the end use this to make the étale isogeny. Consider the canonical k-subgroup inclusion $G \hookrightarrow \mathrm{R}_{k'/k}(G_{k'})$ for any finite extension field k'/k. Since $\mathrm{R}_{k'/k}(\mathbf{G}_{\mathrm{a}}) \simeq \mathbf{G}_{\mathrm{a}}^{[k':k]}$, it is harmless (for the purpose of finding an embedding into a vector group over k) to replace k with a finite extension. If $G_{\overline{k}}$ embeds as a subgroup of $\mathbf{G}_{\mathrm{a}}^N$ over $\overline{k}$, the embedding descends to a finite extension k'/k inside of $\overline{k}$. Hence, for the construction of the embedding into a vector group we can now assume that k is algebraically closed.

The component group G/G^0 is a power of $\mathbf{Z}/p\mathbf{Z}$. Thus, since G is commutative and p-torsion, the connected-étale sequence of G splits. That is, $G = G^0 \times (\mathbf{Z}/p\mathbf{Z})^n$ for some $n \geqslant 0$. The finite constant k-group $\mathbf{Z}/p\mathbf{Z}$ is a k-subgroup of $\mathbf{G}_{\mathrm{a}}$, so we can assume that G is connected. We shall prove that G is a vector group. By [Bo2, 10.6(2), 10.9], G has a composition series whose successive quotients are $\mathbf{G}_{\mathrm{a}}$. By induction on $\dim G$, it suffices to prove that a commutative extension U of $\mathbf{G}_{\mathrm{a}}$ by $\mathbf{G}_{\mathrm{a}}$ over k is a split extension if $p \cdot U = 0$.

Let W_2 be the additive k-group of Witt vectors of length 2, so there is a canonical exact sequence of k-groups

$$0 \to \mathbf{G}_{\mathrm{a}} \to W_2 \to \mathbf{G}_{\mathrm{a}} \to 0.$$

It is a classical fact (see [Ser1, Ch. VII, §2, Lemma 3]) that every commutative extension U of $\mathbf{G}_{\mathrm{a}}$ by $\mathbf{G}_{\mathrm{a}}$ over k is obtained by pullback of this Witt vector extension along a (unique) k-homomorphism $f : \mathbf{G}_{\mathrm{a}} \to \mathbf{G}_{\mathrm{a}}$. In other words, there is a unique pullback diagram

$$\begin{array}{ccccccccc} 0 & \longrightarrow & \mathbf{G}_{\mathrm{a}} & \longrightarrow & U & \longrightarrow & \mathbf{G}_{\mathrm{a}} & \longrightarrow & 0 \\ & & \| & & \downarrow{\scriptstyle f'} & & \downarrow{\scriptstyle f} & & \\ 0 & \longrightarrow & \mathbf{G}_{\mathrm{a}} & \longrightarrow & W_2 & \longrightarrow & \mathbf{G}_{\mathrm{a}} & \longrightarrow & 0 \end{array}$$

and we claim that if U is p-torsion then $f = 0$ (so the top row is a split sequence). Clearly $f'(U) \subseteq W_2[p]$, but the maximal smooth k-subgroup of

$W_2[p]$ is the kernel term $\mathbf{G}_a$ along the bottom row. Hence, $f'(U)$ is killed by the quotient map along the bottom row, so $f = 0$.

Now return to the setting of a general ground field k, and fix a k-subgroup inclusion of G into a vector group V, say with codimension c. Choose a linear structure on V (in the sense of Definition A.1.1). Then $W \mapsto \mathrm{Lie}(W)$ is a bijection between the set of linear subgroups of V and the set of linear subspaces of $\mathrm{Lie}(V)$. Hence, if we choose W so that $\mathrm{Lie}(W)$ is complementary to $\mathrm{Lie}(G)$ then the natural map $G \to V/W$ is an isomorphism on Lie algebras, so it is an étale isogeny. Since W is a linear subgroup of V, the quotient V/W is a vector group over k. □

Proposition B.1.11 *Let $V_1, \ldots, V_n$ be k-groups isomorphic to $\mathbf{G}_a$, and let $V = \prod_{i=1}^n V_i$. Let U be a smooth k-subgroup of V such that U_{k_s} is the k_s-subgroup of V_{k_s} generated by images of k_s-scheme morphisms $\mathbf{A}^1_{k_s} \to V_{k_s}$ that pass through* 0.

There exists a k-group automorphism $h : V \simeq V$ such that $h(U)$ is the direct product of some of the V_i inside of V. In particular, U is a vector group over k and is a k-group direct factor of V.

Proof The proof is by induction on n and is trivial for $n = 1$. Now consider $n > 1$. The case $U = V$ is trivial, so we can assume $\dim U \leqslant n-1$. First assume that $\dim U = n-1 > 0$. By Corollary B.1.5, U is the kernel of a k-homomorphism $f : V \to \mathbf{G}_a$. By hypothesis, there exists a non-constant k_s-scheme morphism $\mathbf{A}^1_{k_s} \to U_{k_s}$, so by Lemma B.1.7 (applied over k_s) and Lemma B.1.9 there exists a k-group automorphism h' of V such that $h'(U) \supseteq V_1$. But then $h'(U) = V_1 \times U'$, where U' denotes the projection of $h'(U)$ into $V' = \prod_{i=2}^n V_i$. Applying the induction hypothesis to V' and U', we are done.

Suppose now that $\dim U < n-1$, and let U' denote the projection of U into the product V' as defined above. By the inductive hypothesis, after relabeling $V_2, \ldots, V_n$ there exists a k-group automorphism $h_1 : V' \to V'$ such that $h_1(U') = \prod_{i=2}^r V_i$ for some $r < n$. Setting

$$h' = \mathrm{id}_{V_1} \times h_1 : V \simeq V,$$

we then have $h'(U) \subseteq \prod_{i=1}^r V_i$, and we can again apply induction. The proof is now complete. □

Corollary B.1.12 *In a smooth p-torsion commutative affine k-group G, every smooth k-subgroup that is a vector group is a k-group direct factor.*

Proof This is a consequence of Proposition B.1.11, provided that G is a k-subgroup of a vector group. Such an embedding is provided by Lemma B.1.10. □

The following proposition is a useful refinement of Lemma B.1.10.

Proposition B.1.13 *Let k be an infinite field of characteristic $p > 0$ and let U be a smooth p-torsion commutative affine k-group. Then U is k-isomorphic to a k-subgroup of codimension 1 in a k-vector group. In particular, U is isomorphic (as a k-group) to the zero scheme of a separable nonzero p-polynomial over k.*

This proposition is also true for finite k if U is connected since then it is a vector group; see Corollary B.2.7.

Proof By Lemma B.1.10, U can be identified with a k-subgroup of a k-vector group V. Let $m = \dim V - \dim U$. If $m \leqslant 1$ then we are done by Corollary B.1.5, so we assume $m > 1$. We will show that U can be embedded in a k-vector group W with $\dim W = \dim V - 1$, which will complete the argument via induction on m. The vector group W will arise as a quotient of V.

The k-linear subspace $\mathrm{Lie}(U)$ in $\mathrm{Lie}(V)$ has codimension m. Fix a linear structure on V (in the sense of Definition A.1.1). Since $m \geqslant 2$, the Zariski closure $\mathbf{G}_{\mathrm{a}}.U$ ($\subseteq V$) of the image of the multiplication map $\mathbf{G}_{\mathrm{a}} \times U \to V$ is a closed subscheme of V with nonzero codimension. By irreducibility of V, the union $\mathrm{Lie}(U) \cup (\mathbf{G}_{\mathrm{a}}.U)$ inside of V is a proper closed subscheme of V.

Since $V(k)$ is Zariski-dense in V (as k is infinite), there exists $v \in V(k)$ with

$$v \notin \mathrm{Lie}(U) \cup (\mathbf{G}_{\mathrm{a}}.U).$$

Let $L \subseteq V$ be the k-subgroup corresponding to the line $kv \subseteq V(k)$. Consider the canonical k-homomorphism $\phi : V \to W := V/L$, and let $\psi = \phi|_U$. We shall prove $\ker \psi = 1$, from which it follows that ψ identifies U with a k-subgroup of W.

It suffices to show that $\mathrm{Lie}(\psi)$ is injective (so $\ker \psi$ is étale) and that $\psi|_{U(\overline{k})}$ is injective. The map $\mathrm{Lie}(\psi)$ has kernel $L \cap \mathrm{Lie}(U) = \{0\}$, so it is indeed injective. If $\psi|_{U(\overline{k})}$ is not injective then the line L would lie in $\mathbf{G}_{\mathrm{a}}.U$ since $\mathbf{G}_{\mathrm{a}}.U$ is stable under the $\mathbf{G}_{\mathrm{a}}$-multiplication on V. But the point $v \in L(k)$ does not lie in $(\mathbf{G}_{\mathrm{a}}.U)(\overline{k})$, due to how we chose v, so indeed $\psi|_{U(\overline{k})}$ is injective. □

B.2 Wound unipotent groups

It is natural to consider a smooth connected unipotent k-group U to be analogous to an anisotropic torus if U does not contain $\mathbf{G}_{\mathrm{a}}$ as a k-subgroup. This concrete viewpoint is inconvenient for developing a general theory, but eventually we will prove that it gives the right concept. A more convenient definition to get the theory off the ground requires going beyond the category of k-groups, as follows.

Definition B.2.1 A smooth connected unipotent k-group U is *k-wound* if every map of k-schemes $\mathbf{A}^1_k \to U$ is a constant map to a point in $U(k)$.

Remark B.2.2 Note that the definition of k-wound also makes sense in characteristic 0, where it is only satisfied by $U = 1$ (since a nontrivial smooth connected unipotent group in characteristic 0 always contains $\mathbf{G}_a$ as a subgroup over the ground field). Thus, although in this appendix we only work with ground fields of nonzero characteristic, it is convenient in practice (for handling some trivialities) to make the convention that "wound" means "trivial" for smooth connected unipotent groups in characteristic 0.

Whereas anisotropicity for a torus over a field is insensitive to purely inseparable extension of the ground field but is often lost under a separable algebraic extension of the ground field, the k-wound property will behave in the opposite manner: it is insensitive to a separable extension on k (such as scalar extension from a global field to a completion) but is often lost under a purely inseparable extension on k.

Example B.2.3 Assume k is imperfect and choose $a \in k - k^p$. The k-group $\{y^p = x - ax^p\}$ becomes isomorphic to $\mathbf{G}_a$ over $k(a^{1/p})$ but it is k-wound; i.e., over k it cannot be the target of a non-constant k-scheme morphism $\mathbf{A}^1_k \to U$. Indeed, such a map would be dominant and so would extend to a finite surjective map between the regular compactifications, contradicting that the unique point at infinity for $\mathbf{A}^1_k$ is k-rational but the one for U is not (its residue field is $k(a^{1/p})$).

The k-group U is a k-wound subgroup of the k-split group $\mathbf{G}_a^2$. In [Oes, Ch. V, 3.5] there is an example over any imperfect field k of a 2-dimensional k-wound smooth connected p-torsion commutative affine group G admitting a 1-dimensional (necessarily k-wound) smooth connected k-subgroup G' such that $G/G' \simeq \mathbf{G}_a$ as k-groups. Thus, k-wound groups can admit smooth quotient maps onto nonzero k-split groups.

Example B.2.4 Assume k is infinite. By Corollary B.1.5 and Proposition B.1.13, smooth p-torsion commutative affine k-groups G are precisely the zero schemes of separable nonzero p-polynomials f over k. Since G is connected if and only if it is geometrically irreducible (as for any k-group scheme of finite type), we see that G is connected if and only if f is irreducible over k, as well as if and only if f is geometrically irreducible over k. Assume G is connected.

If the principal part $f_{\rm prin}$ of f does not have a zero on $k^n - \{0\}$ then by Lemma B.1.7 it follows that G is k-wound. The converse is false, as we saw following the statement of Lemma B.1.7. However, if G is k-wound and if $f_{\rm prin}$ has a zero on $k^n - \{0\}$ then the calculation in the proof of (2) $\Rightarrow$ (3) in

Lemma B.1.7 (taking h to be the identity map of $\mathbf{G}_a^n$) shows that we can find a p-polynomial $F \in k[x_1, \ldots, x_n]$ having zero scheme k-isomorphic to G as a k-group (so F is geometrically irreducible over k) with the sum of the degrees of the monomials appearing in F_{prin} strictly less than the corresponding sum for f_{prin}. Continuing in this way, we eventually arrive at a choice of f having zero scheme G (as a k-group) such that f_{prin} has no zeros on $k^n - \{0\}$. In this sense, the zero schemes of geometrically irreducible p-polynomials f over k for which f_{prin} has no nontrivial k-rational zero are precisely the p-torsion commutative k-wound groups (up to k-isomorphism).

Theorem B.2.5 *Every smooth connected p-torsion commutative affine k-group U is a direct product $U = V \times W$ of a vector group V and a smooth connected unipotent k-group W such that W_{k_s} is k_s-wound. In this decomposition, the subgroup V is uniquely determined: V_{k_s} is generated by the images of k_s-scheme morphisms $\varphi : \mathbf{A}^1_{k_s} \to U_{k_s}$ passing through the identity.*

Proof By Galois descent, there is a unique smooth connected k-subgroup V of U such that V_{k_s} is generated by the images of k_s-scheme morphisms $\varphi : \mathbf{A}^1_{k_s} \to U_{k_s}$ that pass through the identity. By Lemma B.1.10, we can identify U with a k-subgroup of a vector group over k. Thus, by Proposition B.1.11, V is a vector group over k and (by Corollary B.1.12) we have $U = V \times W$ as k-groups for some k-subgroup W of U. Since U is a smooth connected unipotent k-group, so is its direct factor W. Clearly, W_{k_s} is k_s-wound (due to the definition of V).

Now we prove that V in this decomposition is unique. Consider any decomposition of k-groups $U = V' \times W'$, where V' is a vector group over k and W' is a smooth connected unipotent k-subgroup of U such that W'_{k_s} is k_s-wound. The image of any k_s-scheme morphism $\varphi : \mathbf{A}^1_{k_s} \to U_{k_s}$ passing through the identity is contained in V'_{k_s} because otherwise the composite of φ and the canonical projection $U_{k_s} \to W'_{k_s}$ would be a non-constant k_s-scheme morphism from $\mathbf{A}^1_{k_s}$ to W'_{k_s} (contradicting that W'_{k_s} is assumed to be k_s-wound). Hence, $V \subseteq V'$, so $V' = V \times V'_1$ with V'_1 the image of the vector group V' under the projection $U \twoheadrightarrow W$. Since W_{k_s} is k_s-wound and V' is a vector group, $V'_1 = 0$. That is, $V' = V$. □

Corollary B.2.6 *A smooth connected p-torsion commutative affine k-group U is k-wound if and only if U_{k_s} is k_s-wound, and also if and only if there are no nontrivial k-homomorphisms $\mathbf{G}_a \to U$. The k-group U is a vector group over k if and only if U_{k_s} is a vector group over k_s.*

Proof This is immediate from Theorem B.2.5. □

Corollary B.2.7 *If k is perfect then a smooth connected p-torsion commutative affine k-group is a vector group.*

Proof By Corollary B.2.6, we may assume that k is algebraically closed. This case is part of Lemma B.1.10. □

Example B.2.8 Let k be a field and let G be a commutative pseudo-reductive k-group (e.g., $\mathrm{R}_{k'/k}(T')$ for a purely inseparable finite extension k'/k and a k'-torus T'). For the maximal k-torus T in G, consider the smooth connected commutative unipotent quotient $U = G/T$. We claim that U is k-wound. Since G_{k_s} is pseudo-reductive we may assume $k = k_s$, so T is k-split. By definition, we need to prove that any map of k-schemes $f : \mathbf{A}^1_k \to U$ is constant.

Consider the pullback $G \times_U \mathbf{A}^1_k$. This is a T-torsor over $\mathbf{A}^1_k$, so it is trivial since T is split and $\mathrm{Pic}(\mathbf{A}^1_k) = 1$. A choice of splitting defines a k-scheme morphism $\widetilde{f} : \mathbf{A}^1_k \to G$ over f, so it suffices to prove that $\widetilde{f}$ is constant. Using a translation, we may assume $\widetilde{f}(0) = 1$. We claim that for any smooth connected commutative k-group C and any k-scheme morphism $h : \mathbf{A}^1_k \to C$ satisfying $h(0) = 1$, the smooth connected k-subgroup of C generated by the image of h is unipotent. Applying this to the commutative pseudo-reductive G would then force $\widetilde{f} = 1$, as desired.

To prove our claim concerning C we may assume that k is algebraically closed, so C is a direct product of a torus and a unipotent group. Using projections to factors, it suffices to treat the case $C = \mathrm{GL}_1$. In this case h is a nowhere-vanishing polynomial in one variable with value 1 at the origin, so $h = 1$.

Example B.2.9 Here is an example of a 2-dimensional *non-commutative* wound smooth connected unipotent group U over an arbitrary imperfect field k of characteristic $p > 0$. Choose $a \in k - k^p$, and consider the smooth connected k-subgroups of $\mathbf{G}_{\mathrm{a}}^2$ defined by

$$G = \{x = x^{p^2} + ay^{p^2}\}, \quad C^{\pm} = \{x = \pm(x^p + ay^p)\}.$$

Their closures in $\mathbf{P}^2_k$ are regular with a unique point at infinity, and this point is not k-rational, so these groups are wound. We will construct a non-commutative central extension U of G by C^-, so U must be a k-wound smooth connected unipotent k-group. (The construction will work "universally" over the polynomial ring $\mathbf{F}_p[a]$, yielding the desired k-group via base change.)

Define the k-morphism $f : G \to C^+$ by $(x, y) \mapsto (x^{p+1}, xy^p)$ and consider the symmetric bi-additive 2-coboundary $b = -df : G \times G \to C^+$ defined by

$$b(g, g') = f(g + g') - f(g) - f(g') = (xx'^p + x^p x', xy'^p + x'y^p)$$

for points $g = (x, y)$ and $g' = (x', y')$ of G. The related map $b^- : G \times G \to C^-$ defined by

$$b^-((x, y), (x', y')) = (xx'^p - x^p x', xy' p - x'y^p)$$

is easily checked to be an alternating bi-additive 2-cocycle, and if $p \neq 2$ then b^- is *not* symmetric. Thus, if $p \neq 2$ then the associated k-group U with underlying scheme $C^- \times G$ and composition law

$$(c, g)(c', g') = (c + c' + b^-(g, g'), g + g')$$

is a non-commutative central extension of G by C^- (with identity $(0, 0)$ and inversion $-(c, g) = (-c, -g)$).

To handle the case $p = 2$, we consider a variant on this construction. For any p and $\zeta \in \mathbf{F}_{p^2} - \mathbf{F}_p$ consider the bi-additive map $b_\zeta : G \times G \to C^+$ over $\mathbf{F}_{p^2}[a]$ defined by $b_\zeta(g, g') = b(g, \zeta g') = b(\zeta^p g, g')$. This is easily seen to be a 2-cocycle that is not symmetric, so it defines a non-commutative central extension U_ζ of G by C^+ over $\mathbf{F}_{p^2}[a]$ ($U_\zeta = C^+ \times G$ as $\mathbf{F}_{p^2}[a]$-schemes, equipped with the composition law $(c, g)(c', g') = (c + c' + b_\zeta(g, g'), g + g')$, identity $(0, 0)$, and inversion $-(c, g) = (-c - b_\zeta(g, -g), -g)$). Taking $p = 2$, so $C^+ = C^-$ and each ζ is a primitive cube root of unity, we have $\zeta^{-1} = \zeta + 1$ and $b_{\zeta+1} = b_\zeta + b = b_\zeta - df$, so for each ζ we obtain an isomorphism of central extensions $U_\zeta \simeq U_{\zeta+1}$ via $(c, g) \mapsto (c + f(g), g)$. Letting σ be the nontrivial automorphism of $\mathbf{F}_4$, upon fixing ζ we get an $\mathbf{F}_4[a]$-isomorphism $[\sigma] : U_\zeta \simeq U_{\zeta+1} = \sigma^*(U_\zeta)$ corresponding to the automorphism $(c, g) \mapsto (c + f(g), g)$ of $C^- \times G$. By inspection the automorphism $\sigma^*([\sigma]) \circ [\sigma]$ of U_ζ is the identity map, so $[\sigma]$ defines a descent datum on the central extension U_ζ relative to the quadratic Galois covering $\mathrm{Spec}(\mathbf{F}_4[a]) \to \mathrm{Spec}(\mathbf{F}_2[a])$. The descent is a non-commutative central extension of G by C^- over $\mathbf{F}_2[a]$, so it yields the desired k-group by base change.

To get results on k-wound groups beyond the commutative p-torsion case, we need to study smooth connected p-torsion central k-subgroups in a general smooth connected unipotent k-group U. This is taken up in the next section.

B.3 The cckp-kernel

In a smooth connected unipotent k-group U, any two smooth connected p-torsion central k-subgroups generate a third such subgroup. Hence, the following definition makes sense.

Definition B.3.1 The *cckp-kernel* of U is its maximal smooth connected p-torsion central k-subgroup.

Note that if $U \neq 1$ then its cckp-kernel is nontrivial, since the latter contains the cckp-kernel of the last nontrivial term of the descending central series of U.

By Galois descent and specialization (as in the proof of Proposition 1.1.9(1)), the formation of the *cckp*-kernel commutes with any separable extension on k.

Proposition B.3.2 *Let U be a smooth connected unipotent k-group, and let k'/k be a separable extension. Let F denote the cckp-kernel of U. Then U is k-wound if and only if $U_{k'}$ is k'-wound, and the quotient U/F is k-wound whenever U is k-wound. Also, the following conditions are equivalent:*

(1) *U is k-wound,*
(2) *U does not have a central k-subgroup k-isomorphic to $\mathbf{G}_a$,*
(3) *the cckp-kernel F of U is k-wound.*

This proposition implies that U is k-wound if and only if U admits no nontrivial k-homomorphism from $\mathbf{G}_a$. Such a characterization of the k-wound property is analogous to the characterization of anisotropic tori over a field in terms of homomorphisms from GL_1 over the ground field.

Proof Obviously (1) $\Rightarrow$ (2). By Theorem B.2.5, (2) and (3) are equivalent. Also, by specialization (as in the proof of Proposition 1.1.9(1)), if U_K is not K-wound for some separable extension K/k then the same holds with K/k taken to be some finite separable extension. Thus, to prove the equivalence of (1), (2), and (3) and the fact that $U_{k'}$ is k'-wound whenever U is k-wound, it suffices to show that if U_{k_s} is not k_s-wound then the *cckp*-kernel F of U is not k-wound.

Thus let $\varphi : \mathbf{A}^1_{k_s} \to U_{k_s}$ be a non-constant k_s-scheme morphism. Composing with a $U(k_s)$-translation if necessary, we may assume $\varphi(0) = 1$. We may choose such a φ so that $\varphi(\mathbf{A}^1_{k_s})$ is central. Indeed, suppose $\varphi(\mathbf{A}^1_{k_s})$ is non-central, so U is not commutative and there exists $g \in U(k_s)$ not centralizing $\varphi(\mathbf{A}^1_{k_s})$. The k_s-scheme morphism $\varphi' : \mathbf{A}^1_{k_s} \to U_{k_s}$ defined by $\varphi'(x) = g^{-1}\varphi(x)^{-1}g\varphi(x)$ carries 0 to 1, so it is then non-constant, and its image lies in derived group $\mathscr{D}(U_{k_s}) = \mathscr{D}(U)_{k_s}$. The k-subgroup $\mathscr{D}(U)$ has smaller dimension than U and is nontrivial since the smooth connected k-group U is not commutative. Hence, by iteration with the descending central series, the required non-constant φ with $\varphi(\mathbf{A}^1_{k_s})$ central is eventually obtained. We may also assume that $\varphi(\mathbf{A}^1_{k_s})$ is p-torsion by replacing the original φ with $p^e \cdot \varphi$ for some $e \geqslant 0$.

The nontrivial k_s-subgroup generated by $\varphi(\mathbf{A}^1_{k_s})$ then lies in the *cck*$_s$ p-kernel of U_{k_s}; i.e., it lies in F_{k_s}. Thus F_{k_s} is not k_s-wound, so by Corollary B.2.6 the k-group F is not k-wound.

It remains to show that if U is k-wound then U/F is k-wound. For this we may, in view of the preceding conclusions, assume that $k = k_s$. Suppose that U is k-wound and U/F is not k-wound. Thus, there exists a central k-subgroup A of U/F that is k-isomorphic to $\mathbf{G}_a$. Let π denote the canonical homomorphism

$U \to U/F$. The k-subgroup scheme $\pi^{-1}(A)$ in U is an extension of A by F, so it is smooth, connected, and unipotent.

We claim that $\pi^{-1}(A)$ is central in U. If not, we get a non-constant k-scheme morphism $\varphi : \mathbf{A}^1_k \to U$ (contradicting that U is k-wound) as follows. Choose $g \in U(k)$ not centralizing $\pi^{-1}(A)$ (recall $k = k_s$), identify $\mathbf{G}_a$ with $A = \pi^{-1}(A)/F$, and define $\varphi : \pi^{-1}(A)/F \to U$ by $xF \mapsto gxg^{-1}x^{-1}$. Thus, $\pi^{-1}(A)$ is central in U. Similarly, $\pi^{-1}(A)$ is p-torsion because otherwise we would get a non-constant k-scheme morphism $\psi : \mathbf{A}^1_k \to U$ via $\psi(xF) = x^p$. We have shown that $\pi^{-1}(A)$ lies in the cckp-kernel F of U, so the given inclusion $F \subseteq \pi^{-1}(A)$ is an equality. Hence, $A = 1$, which is absurd since $A \simeq \mathbf{G}_a$. □

Corollary B.3.3 *Let U be a smooth connected unipotent k-group. Define the ascending chain of smooth connected normal k-subgroups $\{U_i\}_{i\geqslant 0}$ as follows: $U_0 = 1$ and U_{i+1}/U_i is the cckp-kernel of U/U_i for all $i \geqslant 0$. These subgroups are stable under k-group automorphisms of U, their formation commutes with any separable extension of k, and $U_i = U$ for sufficiently large i. If U is k-wound then so is every U/U_i.*

Moreover, if H is a smooth k-group equipped with an action on U then H carries each U_i back into itself.

Proof The well-posedness of the definition (e.g., that U/U_1 is k-wound when U is k-wound) and the compatibility with separable extension on k follow from Proposition B.3.2. By dimension considerations, $U_i = U$ for sufficiently large i since the cckp-kernel of a nontrivial smooth connected unipotent k-group is nontrivial.

Finally, if H is a smooth k-group acting on U then we need to prove that H carries each U_i into itself. For this we may extend scalars to k_s, so k is separably closed. Then the H-stability of U_i is equivalent to the $H(k)$-stability of U_i, and this latter property is a special case of each U_i being stable under all k-automorphisms of U. □

We will now prove the following structure theorem that is analogous to the canonical presentation of a torus over a field as an extension of an anisotropic torus by a split torus.

Theorem B.3.4 *Let U be a smooth connected unipotent k-group. Then there exists a unique smooth connected normal k-split k-subgroup $U_{\rm split}$ of U such that $U/U_{\rm split}$ is k-wound in the sense of Definition B.2.1.*

The subgroup $U_{\rm split}$ contains the image of every k-homomorphism from a k-split smooth connected unipotent k-group into U. Also, the kernel of every k-homomorphism from U into a k-wound smooth connected unipotent k-group contains $U_{\rm split}$, and the formation of the k-subgroup $U_{\rm split}$ is compatible with any separable extension of k.

Proof The proof is by induction on $\dim U$. If U is k-wound then $U_{\text{split}} = \{1\}$ satisfies the requirements and is unique as such. Assume that U is not k-wound, and let A be a smooth central k-subgroup isomorphic to $\mathbf{G}_a$ (Proposition B.3.2). Let $H = U/A$. By induction, there exists a smooth connected normal k-subgroup H_{split} in H with the desired properties in relation to H (in the role of U). Let U_{split} be the corresponding subgroup of U containing A. It is k-split, and $U/U_{\text{split}} \simeq H/H_{\text{split}}$ is k-wound.

Let U' be a smooth connected unipotent k-group having a composition series

$$U' = U'_0 \supseteq U'_1 \supseteq \cdots$$

with successive quotients k-isomorphic to $\mathbf{G}_a$, and let $\varphi : U' \to U$ be a k-homomorphism. There exists a minimal i such that $\varphi(U'_i) \subseteq U_{\text{split}}$. If $i > 0$ then there is induced a k-homomorphism $\mathbf{G}_a \simeq U'_{i-1}/U'_i \to U/U_{\text{split}}$ with nontrivial image. This contradicts that U/U_{split} is k-wound. Thus, $i = 0$; i.e., $\varphi(U') \subseteq U_{\text{split}}$. It follows in particular that U_{split} is unique. Also, for any k-homomorphism $\varphi : U \to U''$ into a k-wound smooth connected unipotent k-group U'' we have $\varphi(U_{\text{split}}) \subseteq U''_{\text{split}} = \{1\}$. This says that $\ker \varphi$ contains U_{split}.

The last assertion of the theorem follows from Proposition B.3.2. Indeed, if k'/k is a separable extension and $U' := U_{k'}$ then $(U_{\text{split}})_{k'} \subseteq U'_{\text{split}}$ and the k'-split quotient $U'_{\text{split}}/(U_{\text{split}})_{k'}$ is a k'-subgroup of the k'-group $(U/U_{\text{split}})_{k'}$ that is k'-wound (by Proposition B.3.2). This forces $U'_{\text{split}} = (U_{\text{split}})_{k'}$. □

Corollary B.3.5 *Let G be a smooth connected affine k-group. The maximal k-split smooth connected unipotent normal k-subgroup $\mathscr{R}_{us,k}(G)$ in G is equal to $\mathscr{R}_{u,k}(G)_{\text{split}}$. In particular, $\mathscr{R}_{u,k}(G)/\mathscr{R}_{us,k}(G)$ is k-wound and the formation of $\mathscr{R}_{us,k}(G)$ commutes with separable extension on k.*

Proof By Galois descent, $\mathscr{R}_{us,k_s}(G_{k_s})$ descends to a smooth connected unipotent normal k-subgroup of G. Such a descent is k-split, since the k-split property of smooth connected unipotent k-groups is insensitive to separable extension on k (due to Theorem B.3.4). Thus, the descent is contained in $\mathscr{R}_{us,k}(G)$, so the inclusion $\mathscr{R}_{us,k}(G)_{k_s} \subseteq \mathscr{R}_{us,k_s}(G_{k_s})$ is an equality. In other words, the formation of $\mathscr{R}_{us,k}(G)$ is compatible with separable algebraic extension on k. Hence, to prove the compatibility with general separable extension on k and the agreement with the maximal k-split smooth connected k-subgroup of $\mathscr{R}_{u,k}(G)$, we may assume that $k = k_s$. But $\mathscr{R}_{u,k}(G)_{\text{split}}$ is a characteristic k-subgroup of G, so it is normal due to the Zariski-density of $G(k)$ in G when $k = k_s$. This proves that $\mathscr{R}_{u,k}(G)_{\text{split}} \subseteq \mathscr{R}_{us,k}(G)$. The reverse inclusion is clear.

The compatibility of the formation of $\mathscr{R}_{us,k}(G)$ now follows from such a compatibility for two constructions: the formation of U_{split} in Theorem B.3.4, and the formation of $\mathscr{R}_{u,k}(G)$ (by Proposition 1.1.9(1)). □

B.4 Torus actions on unipotent groups

Consider the action of a k-torus T on a smooth connected unipotent k-group U. This induces a linear representation of T on $\mathrm{Lie}(U)$, so if T is k-split then we get a weight space decomposition of $\mathrm{Lie}(U)$. If U is a vector group then it is natural to wonder if this decomposition of $\mathrm{Lie}(U)$ can be lifted to the group U. When $\dim U > 1$, the T-action may not respect an initial choice of linear structure on U (in the sense of Definition A.1.1) since $\mathrm{char}(k) = p > 0$, so we first seek a T-equivariant linear structure.

For example, if $U = \mathbf{G}_{\mathrm{a}}^2$ with its usual linear structure and $T = \mathrm{GL}_1$ with the action $t.(x, y) = (tx, (t^p - t)x^p + ty)$ then the T-action is not linear and the action on $\mathrm{Lie}(U) = k^2$ has the single weight given by the identity character of T. But note that if we define the additive non-linear automorphism $f : U \simeq U$ by $f(x, y) = (x, y + x^p)$ then the f-twisted T-action $(t, (x, y)) \mapsto f^{-1}(t.f(x, y))$ is the usual linear action $(t, (x, y)) \mapsto (tx, ty)$.

Tits proved rather generally that if a k-split T acts on U with only nontrivial weights on $\mathrm{Lie}(U)$, then there are nontrivial constraints on the possibilities for U as a k-group and that (after passing to a suitable characteristic composition series for U) the action can always be described in terms of linear representations of T. To explain his results in this direction, we begin with the following proposition that generalizes Lemma B.1.10 by incorporating an action by an affine group scheme of finite type.

Proposition B.4.1 *Let U be a smooth p-torsion commutative affine k-group equipped with an action by an affine k-group scheme H of finite type. There exist a linear representation of H on a finite-dimensional k-vector space V and an H-equivariant isomorphism of U onto a k-subgroup of $\underline{V}$.*

Proof Let $\mathbf{Hom}(U, \mathbf{G}_{\mathrm{a}})$ be the covariant functor assigning to any k-algebra R the R-module $\mathrm{Hom}_R(U_R, \mathbf{G}_{\mathrm{a}})$ of R-group morphisms $\phi : U_R \to \mathbf{G}_{\mathrm{a}}$ (with R-module structure defined via the R-linear structure on the R-group $\mathbf{G}_{\mathrm{a}}$). There is a natural R-linear injection $\mathbf{Hom}(U, \mathbf{G}_{\mathrm{a}})(R) \hookrightarrow R[U_R] = R \otimes_k k[U]$ defined by $\phi \mapsto \phi^*(x)$ (where x is the standard coordinate on $\mathbf{G}_{\mathrm{a}}$), and its image is the R-submodule of "group-like" elements: those f satisfying $m_R^*(f) = f \otimes 1 + 1 \otimes f$ (where $m : U \times U \to U$ is the group law). This is an R-linear condition on f and is functorial in R, so by k-flatness the R-module of group-like elements over R is J_R where $J \subseteq k[U]$ is the k-subspace of group-like elements over k. In particular, the natural map $R \otimes_k \mathrm{Hom}(U, \mathbf{G}_{\mathrm{a}}) \to \mathrm{Hom}_R(U_R, \mathbf{G}_{\mathrm{a}})$ is an isomorphism.

The (left) H-action on U defines a left H-action on $\mathbf{Hom}(U, \mathbf{G}_{\mathrm{a}})$ (via $(h.\phi)(u) = \phi(h^{-1}.u)$) making the k-linear inclusion $\mathrm{Hom}(U, \mathbf{G}_{\mathrm{a}}) \hookrightarrow k[U]$ an H-equivariant map. Thus, $\mathrm{Hom}(U, \mathbf{G}_{\mathrm{a}})$ is the directed union of H-stable finite-dimensional k-subspaces, due to the same property for $k[U]$

(Proposition A.2.3). By Lemma B.1.10 there is a k-subgroup inclusion $j: U \hookrightarrow \mathbf{G}_a^n$ for some $n \geqslant 1$. Let $W \subseteq \mathrm{Hom}(U, \mathbf{G}_a)$ be an H-stable finite-dimensional k-subspace containing $j^*(x_1), \ldots, j^*(x_n)$. The canonical map $U \to W^* = \mathrm{Spec}(\mathrm{Sym}(W))$ is an H-equivariant closed immersion that is a k-homomorphism (since W consists of group-like elements in $k[U]$ that generate $k[U]$ as a k-algebra). □

We now apply our work with wound groups to analyze the structure of smooth connected unipotent k-groups equipped with a sufficiently nontrivial action by a k-torus.

Proposition B.4.2 *Let U and V be as in Proposition B.4.1, and replace H with a k-torus T there. Let $V = V_0 \times V'$ be the unique T-equivariant k-linear decomposition of V with $V_0 = V^T$ (so V' is the span of the isotypic k-subspaces for the nontrivial irreducible representations of T over k that occur in V). The product map*

$$\iota : (U \cap \underline{V}_0) \times (U \cap \underline{V}') \to U$$

is an isomorphism and there is a T-equivariant k-linear decomposition $V' = V_1' \times V_2'$ of V' and a T-equivariant k-automorphism α of the additive k-group $\underline{V}$ such that

$$\alpha(U) = (\alpha(U) \cap \underline{V}_0) \times \underline{V}_1'.$$

In particular, if $V^T = 0$ then the k-group U is a vector group admitting a T-equivariant linear structure.

Proof Clearly $\underline{V}_0 = Z_{\underline{V}}(T)$ as k-subgroups of $\underline{V}$, so $U_0 := U \cap \underline{V}_0$ is $Z_U(T)$. This is smooth since U is smooth. We will first prove that ι is an isomorphism, so $U \cap \underline{V}'$ is smooth.

Since the formation of V' clearly commutes with scalar extension on k, to establish that ι is an isomorphism we may assume k is algebraically closed. Choose $s \in T(k)$ such that for every weight χ of T in V', $\chi(s) \neq 1$. Consider the k-linear map $f : V \to V$ defined by $f(v) = s \cdot v - v$. It is obvious that f maps V onto V' with $\ker f = Z_V(s) = V_0$ and that the restriction of f to V' is a linear automorphism. The image $f(U)$ is a smooth k-subgroup of $\underline{V}'$, and it lies in U due to the T-stability of U inside of $\underline{V}$. By definition, $\underline{V}'$ has a T-equivariant composition series whose successive quotients are 1-dimensional vector groups with a nontrivial T-action. Hence, all T-stable k-subgroup schemes of $\underline{V}'$ are connected. In particular, $f(U)$ is connected.

Since $U_0 \cap f(U) = 0$ (as $\underline{V}_0 \cap \underline{V}' = 0$), under addition $U_0 \times f(U)$ is a k-subgroup of U. Thus, $f : U \to f(U)$ is a map onto a k-subgroup of U and the restriction of this map to $f(U)$ is therefore an endomorphism $f(U) \to f(U)$ with trivial kernel. But $f(U)$ is smooth and connected, so this endomorphism is an automorphism. In other words, $f : U \twoheadrightarrow f(U)$ is a projector up to an automorphism of $f(U)$. Since $U \cap \ker f = U \cap \underline{V}_0 = U_0$, this shows that the k-subgroup inclusion $U_0 \times f(U) \hookrightarrow U$ is an isomorphism, so $f(U) = U \cap \underline{V}'$. This completes the proof that ι is an isomorphism.

Let $U' = U \cap \underline{V}'$ and define $V_1' = \mathrm{Lie}(U')$. Then V_1' is a T-stable k-linear subspace of V'. Complete reducibility of k-linear representations of T provides a T-stable k-linear complement V_2' of V_1' in V'. Using the decomposition $\underline{V}' = \underline{V}_1' \times \underline{V}_2'$, the projection $U' \to \underline{V}_1'$ is an isomorphism on Lie algebras, so it is étale. By T-equivariance, the finite étale kernel is T-stable and therefore centralized by the connected T. But $Z_{\underline{V}'}(T) = 0$, so this kernel vanishes. In other words, $U' \to \underline{V}_1'$ is an isomorphism. It follows that the k-subgroup $U' \subseteq \underline{V}' = \underline{V}_1' \times \underline{V}_2'$ is the graph of a T-equivariant k-homomorphism $g : \underline{V}_1' \to \underline{V}_2'$. The T-equivariant k-automorphism α of $\underline{V}$ may be taken to be the automorphism that is the identity on $\underline{V}_0$ and is the inverse of the map $(v_1, v_2) \mapsto (v_1, g(v_1) + v_2)$ on $\underline{V}_1' \times \underline{V}_2'$. □

Theorem B.4.3 *Let T be a k-torus and U a smooth p-torsion commutative affine k-group. Suppose that there is given an action of T on U over k. Then $U = U_0 \times U'$ with $U_0 = Z_U(T)$ and U' a T-stable k-subgroup that is a vector group admitting a linear structure relative to which T acts linearly. Moreover, U' is uniquely determined and is functorial in U.*

Proof By Propositions B.4.1 and B.4.2 we get the existence of U'. To prove the uniqueness and functoriality of U', we may assume $k = k_s$. Under the decomposition of U' into weight spaces relative to a T-equivariant linear structure on U', all T-weights must be nontrivial due to the definition of U_0. Hence, the canonical map $T \times U \to U$ defined by $(t, u) \mapsto t.u - u$ has image U'. This proves the uniqueness and functoriality of U'. □

If U in Theorem B.4.3 is k-wound, then it must coincide with U_0 and so have trivial T-action. This is a special case of the following general consequence of invariance of the wound property with respect to separable extension of the ground field (Proposition B.3.2):

Proposition B.4.4 *Let T be a k-torus and U a k-wound smooth connected unipotent k-group. The only T-action on U is the trivial one.*

Proof Our aim is to prove that the k-subgroup scheme $Z_U(T)$ is equal to U. For the k-group $G = U \rtimes T$, we have that the torus centralizer $Z_G(T)$ is equal to $Z_U(T) \rtimes T$. But $Z_G(T)$ is smooth and connected, so the same holds for $Z_U(T)$. Since $Z_U(T)$ is a scheme-theoretic centralizer, $\mathrm{Lie}(Z_U(T))$ is the T-centralizer in Lie(U). Hence, to prove that $Z_U(T) = U$ it suffices (by smoothness and connectedness of U) to prove that T acts trivially on Lie(U).

By Proposition B.3.2, we may extend scalars to k_s, so T is k-split. Assume that the T-action on Lie(U) is nontrivial, so the action on Lie(U) by some GL_1 in T is nontrivial. We may replace T by this GL_1, so there is an action $\lambda : \mathrm{GL}_1 \times U \to U$ that induces a nontrivial action on Lie(U). Composing the action with inversion on GL_1 if necessary allows us to arrange that there is a nonzero weight space in Lie(U) with a positive weight. We now apply the theory of §2.1 to the semidirect product $G = U \rtimes \mathrm{GL}_1$ (see Remark 2.1.11). By Lemma 2.1.5 and Proposition 2.1.8(1), there is a smooth connected k-subgroup $U_G(\lambda) \subseteq U$ whose Lie algebra is the span in Lie(U) of the positive weight spaces. Hence, $U_G(\lambda) \neq 0$. But $U_G(\lambda)$ is k-split (Proposition 2.1.10), contradicting that U is k-wound. □

Proposition B.4.5 *Let T be a k-torus acting on a smooth connected unipotent k-group U. For any closed k-subgroup scheme S of T, U is k-split if and only if $U^S := Z_U(S)$ is k-split.*

Note that the unipotent k-group scheme U^S is connected by Proposition 2.1.12(3), and it is smooth by Proposition A.8.10. Thus, it makes sense to consider whether or not U^S is "k-split".

Proof The k-split property for smooth connected unipotent k-groups is insensitive to separable extension on k (Theorem B.3.4), so we may and do assume $k = k_s$. In particular, all k-tori are split. By Lie algebra considerations, the smooth connected U^S is generated by the k-subgroup U^T and the root groups of U^S with respect to T; these root groups are those of U with respect to T for the nontrivial T-weights on Lie(U) whose restriction to S is trivial. Likewise, U is generated by the k-subgroup U^S and the root groups of U with respect to T associated to the T-weights on Lie(U) whose restriction to S is nontrivial. Since the image of a k-split smooth connected unipotent k-group under a k-homomorphism is k-split, any smooth connected unipotent k-group $\mathscr{U}$ that is generated by k-split smooth connected k-subgroups cannot admit a nontrivial k-wound quotient. Thus, by Theorem B.3.4, any such $\mathscr{U}$ must be k-split, so the k-split property of all root groups $U_{(a)}$ as in Lemma 2.3.3 for U and U^S with respect to T (Proposition 2.1.10) implies that U^S is k-split when U^T is k-split and likewise that U is k-split when U^S is k-split.

To prove the converse, we assume that U is k-split and nontrivial, and we need to prove that U^S is k-split. As we have just noted, for this purpose it suffices to show that U^T is k-split, so we now forget about S. Choose a descending sequence of smooth connected k-split normal subgroups $\{U_i\}_{i \geqslant 0}$ of U that are each stable under the action of T and for which each (k-split smooth connected unipotent) successive quotient U_i/U_{i+1} is commutative and p-torsion, with $U_0 = U$ and $U_i = 1$ for sufficiently large i. (For example, if $\{C_j(U)\}$ is the descending central series of U then one can take the sequence $\{U_i\}$ to be the sequence obtained by arranging the smooth connected normal k-split subgroups $C_j(U)^{p^a}C_{j+1}(U)$, for $a, j \geqslant 0$, in a descending sequence relative to inclusion.) By dropping repetitions, we may and do assume $U_i/U_{i+1} \neq 1$ whenever $U_i \neq 1$, so $U_1 \neq 1$ provided that U is not a commutative p-torsion group.

We argue by induction on $\dim U > 0$. Suppose that U is not commutative and p-torsion, so U_1 and U/U_1 are nontrivial. The short exact sequence

$$1 \to U_1 \to U \to U/U_1 \to 1$$

among smooth connected k-groups induces a T-equivariant short exact sequence of Lie algebras, so it gives rise to the short exact sequence

$$1 \to U_1^T \to U^T \to (U/U_1)^T \to 1.$$

(The exactness follows from that of the Lie algebras and the fact that all terms are smooth and connected.) Since U_1 and U/U_1 have strictly smaller dimension than U (as each is nontrivial), by induction the outer terms of the second short exact sequence are k-split, so U^T is as well. It remains to treat the case that U is commutative and p-torsion. In such cases, by Theorem B.4.3 the k-group U^T is a direct factor of the k-split U, so U^T is k-split. □

Proposition B.4.6 *Let S be a split torus over a field k of characteristic exponent $p \geqslant 1$, and let $f: G \to G'$ be a surjective S-equivariant k-homomorphism between smooth affine k-groups equipped with S-actions. Every S-weight on* $\mathrm{Lie}(G')$ *is a p-power multiple of an S-weight on* $\mathrm{Lie}(G)$.

Proof By Proposition A.2.5 the centralizer G^S in G for the S-action is smooth and connected, since $G^S \times S$ is the centralizer of S in the smooth connected affine group $G \rtimes S$. Applying Proposition A.2.8 to the surjective homomorphism $G \rtimes S \twoheadrightarrow G' \rtimes S$ implies that $G^S \to G'^S$ is surjective, so by Proposition A.2.5 the case of the trivial S-weight is settled.

Consider a nontrivial S-weight $a \in \mathrm{X}(S)$ that occurs in $\mathrm{Lie}(G')$. Let S_a be the codimension-1 subtorus $(\ker a)^0_{\mathrm{red}}$. We may replace G with G^{S_a}, G' with G'^{S_a}, and S with S/S_a to reduce to the case that $S = \mathrm{GL}_1$. Choosing $\lambda \in \mathrm{X}_*(S)$ such that the composition $a \circ \lambda \in \mathbf{Z}$ is positive, the a-weight space in $\mathrm{Lie}(G')$ is supported inside of $\mathrm{Lie}(U_{G'}(\lambda))$ (using terminology as in Remark 2.1.11 with abstract torus actions). By Corollary 2.1.9 (and Remark 2.1.11), the natural map $U_G(\lambda) \to U_{G'}(\lambda)$ is surjective. Thus, by Proposition 2.1.10 (and Remark 2.1.11) we may rename $U_G(\lambda)$ as G and rename $U_{G'}(\lambda)$ as G' to reduce the general problem (without reference to a) to the case that G is k-split unipotent (and nontrivial) and all weights in $X(S) = \mathbf{Z}$ are > 0.

Since G is not k-wound, by Proposition B.3.2 it contains a nontrivial central vector group V. By Galois descent from k_s, there is such a V that is S-stable, so $V' := f(V)$ is an S-stable central vector group in G' (by Theorem B.2.5). Any S-weight on $\mathrm{Lie}(G')$ occurs in either $\mathrm{Lie}(V')$ or $\mathrm{Lie}(G'/V')$, and $G/V \to G'/V'$ is an S-equivariant surjection with G/V also k-split (Proposition A.1.4). Hence, by induction on $\dim G$ we reduce to the case that G and G' are vector groups. By Theorem B.4.3 there is a decomposition $G = \prod L_i$ where $L_i = \mathbf{G}_{\mathrm{a}}$ on which S acts via $t.x = t^{n_i}x$ for some $n_i > 0$. Running through the same filtration argument again, we may assume $G = \mathbf{G}_{\mathrm{a}}$ with S-action $t.x = t^n x$ for some $n > 0$. We can also assume $G' \neq 1$, so $G' = \mathbf{G}_{\mathrm{a}}$ with S-action $t.x = t^{n'}x$ for some $n' > 0$. The map $f : G \twoheadrightarrow G'$ is a nonzero additive polynomial in one variable such that $f(t^n x) = t^{n'} f(x)$. Thus, $f(x) = cx^{p^e}$ for some $c \in k^\times$ and $e \geqslant 0$ with $n' = p^e n$. □

Remark B.4.7 A proof of Proposition B.4.6 that avoids Theorem B.4.3 (and only requires G and G' to be smooth of finite type rather than affine) can be given by using Proposition A.7.14 and the exactness properties of p-Lie algebras of infinitesimal group schemes in characteristic $p > 0$ (as discussed at the end of Example A.7.16). The argument proceeds by reduction to the case when $H = \ker f$ is infinitesimal, for which one proves that each S-weight on $\mathrm{Lie}(G')$ is $p^i a$ for an S-weight a on $\mathrm{Lie}(G)$ and an integer $0 \leqslant i \leqslant h$ with h minimal so that $F_{H/k,h}$ vanishes. To prove this case, one reduces to when $h = 1$ and then factorizes $F_{G/k}$ through f and factorizes $F_{G'/k}$ through $f^{(p)}$.

Appendix C

Rational conjugacy and relative root systems

Let G be a smooth connected affine group over a field k. In [BoTi2], Borel and Tits announced (without proof) some remarkable results generalizing important theorems when G is reductive. Among these are the $G(k)$-conjugacy of maximal k-split k-tori, maximal k-split smooth connected unipotent k-subgroups, and minimal pseudo-parabolic k-subgroups, as well as the Bruhat decomposition for $G(k)$ (relative to a choice of minimal pseudo-parabolic k-subgroup). In this appendix we use §§2.1–3.5 and Appendix B to prove these results, following the ideas outlined in [Ti3, §§2 and 3] (with some scheme-theoretic improvements). We give some generalizations in §C.4 for group schemes locally of finite type over a field. We also develop a theory of k-root systems and associated root groups in smooth connected affine groups over any field k (with results that are most satisfactory in the pseudo-reductive case, eliminating pseudo-split hypotheses from some results in §3.3).

Nothing in this appendix is used in the main text except for Theorem C.2.3 and Theorem C.2.29, which are used in Chapter 9, and the self-contained Lemma C.4.1, which is used in several places.

C.1 Pseudo-completeness

We shall prove that the coset space G/P modulo a pseudo-parabolic k-subgroup P satisfies the following variant of the valuative criterion for properness.

Definition C.1.1 A scheme X over a field k is *pseudo-complete* over k if it is of finite type and separated and $X(R) = X(K)$ for any discrete valuation ring R over k with fraction field K and residue field separable over k.

For any pseudo-complete X, if C is a smooth curve over k and $c \in C$ is a closed point such that $k(c)/k$ is *separable* then any k-morphism $C - \{c\} \to X$ uniquely extends to a k-morphism $C \to X$.

Proposition C.1.2 *Let X be a scheme over a field k.*

(1) *If X is pseudo-complete over k then so is any closed subscheme.*
(2) *If k'/k is an extension of fields then X is pseudo-complete over k if $X_{k'}$ is pseudo-complete over k'. The converse holds when k'/k is separable.*
(3) *In the definition of pseudo-completeness for X, it suffices to consider complete discrete valuation rings over k with residue field that is separably closed and separable over k.*

Proof If Z is closed in X then $Z(R) = Z(K) \cap X(R)$ for any domain R and $K = \mathrm{Frac}(R)$, so (1) follows. To prove (3), we may assume that X is separated and of finite type. In particular, $X(R) \subseteq X(K)$ for any discrete valuation ring R over k with $K = \mathrm{Frac}(R)$. If $R \to R'$ is a local extension of discrete valuation rings over k with $K \to K'$ the corresponding extension of fraction fields then $X(R) = X(R') \cap X(K)$ inside of $X(K')$ since $R = R' \cap K$ inside of K. Hence, $X(R) = X(K)$ if $X(R') = X(K')$. Taking R' to be the completion of a strict henselization of R, (3) follows.

Since the properties of being separated and of finite type descend through a ground field extension, to prove (2) we have to show that if X is separated and of finite type then it satisfies the valuative criterion defining pseudo-completeness over k if $X_{k'}$ satisfies the analogous criterion over k', and conversely when k'/k is separable. We first check preservation under separable extension on k, so let k'/k be separable and consider a discrete valuation ring R' over k' with fraction field K' and residue field separable over k'. By transitivity of separability in field extensions, this residue field is also separable over k, so by the assumed valuative criterion for X over k we have $X(R') = X(K')$. But $X(A) = X_{k'}(A)$ for any k'-algebra A, so $X_{k'}(R') = X_{k'}(K')$. This proves the "converse" part of (2).

To prove the first part of (2) (i.e., pseudo-completeness descends from $X_{k'}$ over k' to X over k for any k'/k), we shall adapt the argument used to prove (3). Consider a discrete valuation ring R over k whose residue field F is separable over k. We will construct a discrete valuation ring R' over k' with residue field F' separable over k' and a local injection $R \to R'$ over $k \to k'$. Granting this, if $K \to K'$ is the induced extension of fraction fields then $X(R') = X_{k'}(R') = X_{k'}(K') = X(K')$. Thus, as in the proof of (3), we deduce that

$$X(R) = X(R') \cap X(K) = X(K') \cap X(K) = X(K).$$

To construct such an R' over k', let t be a uniformizer of R. Consider the ring $A = R \otimes_k k'$. The quotient $A/tA = F \otimes_k k'$ is a reduced ring since F/k is separable. Thus, for a minimal prime ideal $\mathfrak{p}$ of $F \otimes_k k'$, the local ring $F' = (F \otimes_k k')_{\mathfrak{p}}$ is a field. This field is separable over k' since F/k is separable. Let P be the corresponding prime ideal of A, so $R' := A_P$ is a local k'-algebra in which t is not a zero divisor and such that $R'/tR' = F'$. Thus, t generates the maximal ideal of R', so if R' is noetherian (e.g., if k'/k is finitely generated) then it is a discrete valuation ring with residue field F' and uniformizer t such that $R \to R'$ is a local injection over $k \to k'$. This settles the case when k'/k is finitely generated. Since a direct limit of separable field extensions is separable and a direct limit of discrete valuation rings with a common uniformizer is a discrete valuation ring with the same uniformizer, the general case follows by expressing k' as a direct limit of subfields k'_i finitely generated over k and using that a minimal prime of $F \otimes_k k'$ contracts to a minimal prime of each $F \otimes_k k'_i$ (by flatness). □

It follows from Proposition C.1.2(3) that pseudo-completeness is equivalent to properness when k is perfect, since properness can be checked using the valuative criterion with complete discrete valuation rings having an algebraically closed residue field. Hence, pseudo-completeness is of interest primarily for imperfect k. Also, the separability hypothesis on k'/k in Proposition C.1.2(2) is essential, as the following example illustrates.

Example C.1.3 Let k'/k be any finite extension of fields, and X' a projective k'-scheme. The Weil restriction $X = \mathrm{R}_{k'/k}(X)$ exists as a k-scheme of finite type, and it is separated since the functor $\mathrm{R}_{k'/k}$ preserves separatedness (as follows by using consideration of diagonal maps and closed immersions). However, such an X is *never* proper when k'/k is not separable and X' is non-empty and smooth with pure dimension $d' > 0$, as we saw in Example A.5.6. In particular, $X_{\overline{k}}$ is not proper over $\overline{k}$ in such cases.

But X is always pseudo-complete over k. Indeed, if B is a discrete valuation ring over k with residue field F and fraction field K then by definition $X(B) = X'(B')$ for the finite flat B-algebra $B' = B \otimes_k k'$, and likewise $X(K) = X'(K')$ for $K' = K \otimes_k k'$. The quotient of B' by a uniformizer of B is the ring $F \otimes_k k'$, and if F/k is separable then this tensor product is a finite product of fields. In such cases the localization of B' at each maximal ideal is therefore a discrete valuation ring. Hence, if F/k is separable then B' is regular of pure dimension 1 (and so B' is a finite product of Dedekind domains) with total ring of fractions K', so by the usual valuative criterion of properness for the projective X' we deduce that $X'(B') = X'(K')$. This proves that X is pseudo-complete over k.

Remark C.1.4 Let X be a separated k-scheme of finite type. In [Ti3, §2.3], Tits uses only $R = k_s[\![t]\!]$ in his definition of pseudo-completeness for X (which he calls *relative completeness*), so his definition may initially seem weaker than Definition C.1.1. In fact, these two definitions are equivalent, and they are also equivalent to the geometric condition that k-morphisms to X from punctured smooth curves over k_s extend across k_s-rational punctures. Below we prove the equivalence of these definitions, using Chow's Lemma and approximation arguments, but we will never use the equivalences. One advantage of Definition C.1.1 is its insensitivity to arbitrary separable extension on k (which is not obvious for either of the other two definitions). But such strength of the definition is precisely because it requires the consideration of many discrete valuation rings R. In Proposition C.1.6 we will directly verify this stronger-looking definition for some natural examples.

First we prove the equivalence between Tits' definition and the definition using punctured smooth curves over k_s. We can assume $k = k_s$, so the problem is to show that if the valuative criterion using $k[\![t]\!]$ fails then it also fails using the local ring at a k-rational point on a smooth curve over k. The henselization $B = k\{t\}$ of $k[t]_{(t)}$ is a direct limit of such "algebraic" local rings, and $\widehat{B} = k[\![t]\!]$. Let $K = B[1/t]$ denote the fraction field of B and $\widehat{K} = \widehat{B}[1/t] = k(\!(t)\!)$. It suffices to show that if $X(\widehat{B}) \neq X(\widehat{K})$ then $X(B) \neq X(K)$. Consider any $x \in X(\widehat{K})$ not in $X(\widehat{B})$. We seek to make an analogous point with (B, K) replacing $(\widehat{B}, \widehat{K})$.

Without loss of generality we can replace X with the closure of the image of x, so X is irreducible and reduced with x mapping onto its generic point. Since X is separated over k, by Chow's Lemma there is a proper birational map $\pi : X' \to X$ with X' a quasi-projective integral k-scheme, so x uniquely lifts to $x' \in X'(\widehat{K})$ and clearly $x' \notin X'(\widehat{B})$. By quasi-projectivity, X' is open in a projective k-scheme $\overline{X}'$, so by the valuative criterion for properness x' does extend to a point $\overline{x}' \in \overline{X}'(\widehat{B})$. The map $\overline{x}' : \mathrm{Spec}\,\widehat{B} \to \overline{X}'$ carries the closed point to a k-rational point $\overline{x}'_0 \in \overline{X}'(k) - X'(k)$. By Artin approximation (or an earlier version due to Greenberg [Gre]) applied to the excellent henselian discrete valuation ring B (and $\overline{X}'_B$), there is a point $\overline{x} \in \overline{X}'(B)$ lifting $\overline{x}'_0 \in \overline{X}'(k)$ that is as "close" as we wish to $\overline{x}'$ (using the t-adic topology). Since $\overline{x}'$ carries $\mathrm{Spec}\,\widehat{K}$ into X', the radical coherent ideal of $\overline{X}' - X'$ in $\overline{X}'$ has $\overline{x}'$-pullback in $\widehat{B}$ equal to (t^r) for some $r \geqslant 1$. Hence, by choosing $\overline{x}$ sufficiently close to $\overline{x}'$ we can arrange that the restriction $\overline{x}_\eta : \mathrm{Spec}\,K \to \overline{X}'$ of $\overline{x}$ lands in X'. The composite k-map $\mathrm{Spec}\,K \to X' \xrightarrow{\pi} X$ cannot extend to a B-point, for if it did then this would uniquely lift to a B-point of X' extending $\overline{x}_\eta$ (due to properness of π), contradicting that the B-point $\overline{x} \in \overline{X}'(B) - X'(B)$ extending $\overline{x}_\eta$ is unique in $\overline{X}'(B)$ (due to the separatedness of $\overline{X}'$ over k).

In a similar manner we can show that Tits' notion of relative completeness implies the valuative criterion for pseudo-completeness (the converse being trivial). In other words, if this valuative criterion fails for X using some discrete valuation ring B over k with residue field F separable over k then we claim that it already fails for $k_s[\![t]\!]$. We can replace B with the completion of a strict henselization, so B is complete and F is separably closed. Let k_s be the separable closure of k in F. By Proposition C.1.2(2), we can rename k_s as k. Since F/k is a separable extension of fields, there is a k-algebra isomorphism $B \simeq F[\![t]\!]$. (Indeed, by separability, F/k is formally smooth in the sense of commutative algebra [Mat, Thm. 26.9]. Hence, the quotient map $B/\mathfrak{m}^2 \twoheadrightarrow F$ admits a k-algebra section, and continuing inductively there is a k-algebra map $F \to B$ lifting the identity on residue fields. Any uniformizer of B may be taken as t.) Thus, we have a point $x \in X(F(\!(t)\!))$ not in $X(F[\![t]\!])$. As above, we may replace X with a closed subscheme so that X is integral and x dominates the generic point of X, and after applying Chow's Lemma we may also assume that X is quasi-projective over k.

Let $\overline{X}$ be a projective compactification of X, so $x \in X(F(\!(t)\!))$ uniquely extends to a point $\overline{x} \in \overline{X}(F[\![t]\!])$. Using the Artin/Greenberg approximation as above, we get a point $\overline{y} \in \overline{X}(F\{t\})$ whose generic point lies in X but whose special point does not. In other words, the radical ideal of $\overline{X} - X$ has $\overline{y}$-pullback in $F\{t\}$ equal to (t^r) for some $r \geqslant 1$. By direct limit considerations, we can assume that F is finitely generated over k, so $F = \mathrm{Frac}(A)$ for a smooth k-algebra domain A. Since F is a direct limit of such algebras A, for some such A there is an étale map $\mathrm{Spec}\, R \to \mathrm{Spec}(A[t])$ with $R \subseteq F\{t\}$ such that $\overline{y} \in \overline{X}(R)$ with the radical ideal of $\overline{X} - X$ pulling back to (t^r) in R. Specialization at a suitable k-point of A then gives a point in $\overline{X}(k\{t\})$ whose generic point lies in X and special point does not. Passing to the completion, we have constructed a point in $X(k(\!(t)\!))$ not lying in $X(k[\![t]\!])$. This completes the proof of the equivalence of the three definitions of pseudo-completeness over k.

Pseudo-complete k-schemes satisfy an analogue of the Borel fixed point theorem:

Proposition C.1.5 *Let k be a field and X a pseudo-complete k-scheme such that $X(k) \neq \emptyset$. Assume that X is endowed with a left action by a smooth connected affine k-group H that is k-split solvable. There is a point in $X(k)$ fixed by H.*

At the beginning of §B.1 we reviewed the notion of a k-split solvable smooth connected affine k-group.

Proof We may assume $H \neq 1$, so by definition of being k-split there is a k-split solvable smooth connected normal k-subgroup N of H such that H/N is k-isomorphic to $\mathbf{G}_{\mathrm{a}}$ or GL_1. By induction on $\dim H$, there is some $x \in X(k)$ fixed by N. Consider the subfunctor of X whose B-points for any k-algebra B consist of those $x \in X(B)$ fixed by N_B. This is represented by a closed subscheme X' in X. Indeed, by Galois descent we can assume $k = k_s$, and then

$N(k)$ is Zariski-dense in N. Thus, this subfunctor of X is represented by

$$\bigcap_{n\in N(k)} \{x \in X \mid nx = x\}.$$

Since X' is closed in X, it inherits the pseudo-completeness property from X. Also, induction gives that $X'(k)$ is non-empty. Thus, we may replace X with X' and replace H with H/N to reduce to the case where H is either $\mathbf{G}_a$ or GL_1. Choose $x \in X(k)$ and consider the orbit map $H \to X$ defined by $g \mapsto gx$. By pseudo-completeness, this extends to a map $f : \mathbf{P}^1_k \to X$. The natural action of H on itself extends to an action on $\mathbf{P}^1_k$ fixing the point ∞ at infinity, and f is H-equivariant. Hence, $f(\infty) \in X(k)$ is fixed by H. □

Proposition C.1.6 *Let G be a smooth connected affine group over a field k, and P a pseudo-parabolic k-subgroup of G. The coset space G/P is pseudo-complete over k.*

Proof By Proposition 2.2.10, we may assume that G is pseudo-reductive over k. To prove the pseudo-completeness of G/P it is harmless to make a separable extension on k, so we can assume that $k = k_s$. Hence, $G(k)$ is Zariski-dense in G. Let $G' = G^{\mathrm{red}}_{\overline{k}}$. By Proposition 3.5.4, the image P' of $P_{\overline{k}}$ in G' is a parabolic subgroup, so G'/P' is proper over $\overline{k}$.

To check pseudo-completeness of G/P over k, choose a discrete valuation ring A over k with residue field F that is separable over k. Let $K = \mathrm{Frac}(A)$ and let $t \in A$ be a uniformizer. Since $k = k_s$, $F' := F \otimes_k \overline{k}$ is a field. Thus, $A' := A \otimes_k \overline{k}$ is a discrete valuation ring with uniformizer t, residue field F', and fraction field $K' = K \otimes_k \overline{k}$. Pick $x \in (G/P)(K)$, and consider its image $x' \in (G'/P')(K') = (G'/P')(A')$. This specializes to a point $x'_0 \in (G'/P')(F')$. Since $G(k)$ is Zariski-dense in G, it has Zariski-dense image in $(G'/P')_{F'}$. Hence, by applying a suitable left $G(k)$-translation to x (which is harmless for the purpose of checking if $x \in (G/P)(A)$) we can arrange that x'_0 lies in $U'(F')$ for any desired non-empty open subset $U' \subseteq G'/P'$. It would then follow that $x' \in U'(A')$, since an open subset of $\mathrm{Spec}\, A'$ containing the closed point is the entire space. We will make a useful choice of such a U' shortly.

Since G is pseudo-reductive, there is a 1-parameter k-subgroup $\lambda : \mathrm{GL}_1 \to G$ such that $P = P_G(\lambda)$. Let $\lambda' : \mathrm{GL}_1 \to G'$ be the induced homomorphism over $\overline{k}$, so $P_{G'}(\lambda')$ is the image of $P_{\overline{k}}$ (i.e., $P' = P_{G'}(\lambda')$) and $U_{G'}(-\lambda')$ is the image of $U_G(-\lambda)_{\overline{k}}$ (Corollary 2.1.9). The natural maps $U_G(-\lambda) \times P \to G$ and $U_{G'}(-\lambda') \times P' \to G'$ are open immersions (Proposition 2.1.8(3)), so $U_G(-\lambda)$ and $U_{G'}(-\lambda')$ are naturally identified with dense open subschemes U and U' of G/P and G'/P' respectively.

Using the Zariski topology on $G(k)$, we claim that there is a dense open locus of $g \in G(k)$ such that $gx' \in U'(A')$. This follows from considering the closed point of $\operatorname{Spec} A'$ and using the general fact that if X is a k-scheme in which $X(k)$ is Zariski-dense and if $\mathscr{U}$ is an open subscheme of X_K for an extension field K/k then $X(k)$ is Zariski-dense in X_K and the set

$$X(k) \cap \mathscr{U} := \{x \in X(k) \mid x_K \in \mathscr{U}(K) \text{ inside } X(K)\}$$

is open in $X(k)$. These assertions concerning $X(k)$ are consequences of the more general fact that $X_K \to X$ is open (see [EGA, IV$_2$, 2.4.10]). We likewise see that there is a dense open locus of $g \in G(k)$ such that $gx \in U(K)$. Choosing g in the overlap of these two dense open subsets, we may arrange that $x \in U(K)$ and $x' \in U'(A')$.

It now suffices to prove that under the map $U(K) \to U'(K')$, the preimage of $U'(A')$ is $U(A)$. Pick a maximal k-torus T of G containing $\lambda(\mathrm{GL}_1)$, so $T_{\overline{k}}$ is carried isomorphically onto a maximal torus T' in G' containing $\lambda'(\mathrm{GL}_1)$. Since $U(A) = U_{\overline{k}}(A') \cap U(K)$ inside of $U_{\overline{k}}(K')$, we are led to focus attention on the quotient map of $\overline{k}$-groups $U_{\overline{k}} \to U'$. The kernel of this map has no nontrivial points of $U(k)$. Indeed, this kernel is contained in $\mathscr{R}_u(G_{\overline{k}})$, so if it contains a nontrivial element of $U(k)$ then it would contain infinitely many such elements due to $T(k)$-equivariance (since $U = U_G(-\lambda)$), contradicting pseudo-reductivity of G over k.

By Corollary 3.3.12 (applied with $-\lambda$ in place of λ), U is directly spanned by the subgroups $U_a = U_{(a)}$ for the non-divisible $a \in \Phi(G, T)$ such that $\langle a, \lambda \rangle < 0$, and the quotient map $G_{\overline{k}} \twoheadrightarrow G'$ carries each $(U_{(a)})_{\overline{k}}$ onto the corresponding 1-dimensional root group $U'_{(a)}$ in G'. If we choose a T-equivariant linear structure on $U_{(a)}$, as we may always do, then restricting the quotient map $(U_{(a)})_{\overline{k}} \twoheadrightarrow U'_{(a)}$ to a basis of lines arising from T-stable lines of $U_{(a)}$ yields endomorphisms of $\mathbf{G}_\mathrm{a}$ over $\overline{k}$ given by monomials (due to the T-equivariance). Recalling that $U(k) \to U'(\overline{k})$ is injective, we are reduced to the following special situation.

Let f be a p-polynomial in $n \geqslant 1$ variables $x_1, \ldots, x_n$ over $\overline{k}$ (in the sense of Definition B.1.2, so f has vanishing constant term) and assume that f has no nontrivial k-rational zero. In particular, every x_i occurs in f or else we could construct a nontrivial k-rational zero of f. Assume that in fact each variable occurs in a single monomial: $f = \sum c_i x_i^{q_i}$ for some $c_i \in \overline{k}^\times$ and some $q_i \geqslant 1$ that is a power of the characteristic exponent of k (so $f \neq 0$). Our problem is to show that if $\xi_1, \ldots, \xi_n \in K$ and $f(\xi) \in A'$ then $\xi_i \in A$ for all i. Since each variable x_i intervenes in f only in a term of the form $c_i x_i^{q_i}$ with $q_i \geqslant 1$, if some ξ_i vanishes then we can drop the corresponding x_i term from f without affecting the hypothesis of no nontrivial k-rational zeros and preserving the condition that $f(\xi) = 0$. Hence, we may assume that $\xi_i \neq 0$ for all i.

Let $m_i = -\mathrm{ord}_A(\xi_i) \in \mathbf{Z}$, and assume that $m_i \geqslant 1$ for some i (i.e., $\xi_i \notin A$ for some i). Thus, $N := \max_i m_i q_i \geqslant 1$. Let $a_i = t^{m_i}\xi_i \bmod t \in A/tA = F$, so $a_i \neq 0$ for all i and the element $t^N f(\xi) = 0$ in $A/tA = F$ is equal to $\sum c_i a_i^{q_i} t^{N-m_i q_i} \bmod t$. Hence, a nontrivial F-rational zero of f is given by the point $(a'_1, \ldots, a'_n) \in F^n$ with $a'_i = a_i$ when $m_i q_i = N$ and $a'_i = 0$ otherwise. We conclude that the p-polynomial $f = \sum c_i x_i^{q_i}$ over $\overline{k}$ acquires a nontrivial zero in the separable extension F/k when f is viewed as a polynomial over the field $F \otimes_k \overline{k}$. Such a nontrivial zero occurs in a subfield L of F that is finitely generated over k, and hence it occurs in $R_{\overline{k}}$ for a smooth finitely generated k-subalgebra R of F. Since the k-rational points of $\mathrm{Spec}\, R$ are Zariski-dense (by smoothness, as $k = k_s$), by viewing f as a polynomial over $R_{\overline{k}}$ we can specialize at some k-rational point of $\mathrm{Spec}\, R$ to get a nontrivial zero of f over k. This is a contradiction. □

We warn the reader that the converse to Proposition C.1.6 is false. To explain this, it is convenient to use the following terminology:

Definition C.1.7 For a smooth connected affine group G over a field k, a smooth k-subgroup Q is *quasi-parabolic* if the coset space G/Q is pseudo-complete over k.

By Proposition C.1.6, pseudo-parabolic k-subgroups Q in G are quasi-parabolic. Since pseudo-completeness is insensitive to separable extension on k (Proposition C.1.2(2)), the same holds for quasi-parabolicity. Also, any quasi-parabolic k-subgroup Q in G contains every k-split solvable smooth connected *normal* k-subgroup H of G. (To prove this, we can assume $k = k_s$, so then by Proposition C.1.5, some $G(k)$-conjugate of H is contained in Q. By normality, $H \subseteq Q$, as desired.) It follows that a k-split smooth connected solvable k-group has no proper quasi-parabolic k-subgroup, and that any quasi-parabolic k-subgroup of a smooth connected unipotent k-group U contains the (normal) maximal k-split smooth connected k-subgroup U_{split}.

In contrast with pseudo-parabolicity, a smooth k-subgroup of G containing a quasi-parabolic k-subgroup can fail to be quasi-parabolic. For example, in Lemma C.1.8 below we will show that if $\mathrm{char}(k) > 0$ then $\{1\}$ is quasi-parabolic in any k-wound smooth connected unipotent k-group U, yet over any imperfect field k there are examples (see [Oes, Ch. V, 3.5]) of 2-dimensional k-wound U admitting $\mathbf{G}_{\mathrm{a}}$ as a quotient by a smooth connected normal k-subgroup U', so U' is not quasi-parabolic in U even though $\{1\}$ is. Such examples are of limited interest, since we are largely interested in pseudo-reductive G.

It is natural to wonder: is every quasi-parabolic k-subgroup actually pseudo-parabolic when G is pseudo-reductive? For example, in the reductive case this amounts to asking if the quasi-parabolic k-subgroups are parabolic (by Proposition 2.2.9).

The answer to the question is negative in general over imperfect fields, though affirmative in the reductive case (over any ground field). First we give a result in the unipotent case.

Lemma C.1.8 *Let U be a smooth connected unipotent group over a field k. There exist proper quasi-parabolic k-subgroups in U if and only if U is not k-split, and if U is k-wound then it is pseudo-complete over k.*

Proof We have seen above that U_{split} is contained in any quasi-parabolic k-subgroup of U, so by passing to U/U_{split} it suffices to treat the case $p = \text{char}(k) > 0$ and k-wound U. That is, we aim to prove that U is pseudo-complete when it is nontrivial and k-wound. By Proposition B.3.2 and Proposition C.1.2(2), we may assume $k = k_s$.

Applying Proposition B.3.2 again, there is a composition series $\{U_i\}$ for U consisting of smooth connected normal k-subgroups such that each U_i/U_{i+1} is k-wound, commutative, and p-torsion. We will induct on the length of this composition series. Granting the case of length 1 for a moment, we may assume that there is an exact sequence

$$1 \to U' \to U \to V \to 1$$

with U' pseudo-complete and V a p-torsion commutative k-wound smooth connected unipotent k-group. By the base case for the induction (that we are assuming for now), V is also pseudo-complete over k. To prove that U is pseudo-complete over k, it suffices (by Proposition C.1.2(3)) to check the valuative criterion for complete discrete valuation rings A over k with residue field that is separably closed and separable over k. Let K be the fraction field of A and consider the commutative diagram

$$\begin{array}{ccccccccc} 1 & \longrightarrow & U'(A) & \longrightarrow & U(A) & \longrightarrow & V(A) & \longrightarrow & 1 \\ & & \downarrow & & \downarrow & & \downarrow & & \\ 1 & \longrightarrow & U'(K) & \longrightarrow & U(K) & \longrightarrow & V(K) & & \end{array}$$

The bottom row is exact, and the top row is also exact because A is complete with separably closed residue field and $U \to V$ is a smooth surjection. The outer vertical maps are isomorphisms due to the pseudo-completeness of U' and V over k, so the middle vertical map is also an isomorphism.

It only remains now to treat the case where U is commutative and p-torsion. In this case it is the zero locus in $\mathbf{A}_k^n$ of a nonzero separable p-polynomial $f \in k[x_1, \ldots, x_n]$ (Proposition B.1.13). We need to show that if $\{f = 0\}$ is k-wound then any solution to $f = 0$ in K^n actually lies in A^n. By Lemma B.1.7, the k-wound property implies that after a suitable additive k-automorphism of $\mathbf{A}_k^n$ (which we may certainly apply without loss of generality) it can be arranged that the principal part $f_{\rm prin}$ of f has no nontrivial zero in k^n. This is the only property of $f_{\rm prin}$ that we shall use (together with the fact that $f_{\rm prin}(0) = 0$). In particular, $f_{\rm prin}$ involves each variable, so $f_{\rm prin} = \sum c_i x_i^{q_i}$ for some $c_i \in k^\times$ and p-power $q_i \geqslant 1$.

We now adapt the calculation at the end of the proof of Proposition C.1.6 to get a contradiction if f vanishes at a point $\xi \in K^n$ not in A^n. For any i_0, dropping the x_{i_0} terms from f yields a polynomial in $n - 1$ variables having no nontrivial zero on k^{n-1}. Thus, we can assume $\xi_i \neq 0$ for all i. Let t be a uniformizer of A and $F = A/tA$ the residue field. Define $m_i = -{\rm ord}_A(\xi_i) \in \mathbf{Z}$, so $m_i \geqslant 1$ for some i. Hence, $N = \max_i m_i q_i \geqslant 1$. The residue class $z_i := t^{m_i}\xi_i \bmod t \in F$ is nonzero for all i, and the vanishing residue class $t^N f(\xi) \bmod t \in A/tA = F$ is also equal to $\sum c_i z_i^{q_i} t^{N - m_i q_i} \bmod t$. This is precisely $f_{\rm prin}(z_1', \ldots, z_n')$ with $z_i' = z_i$ when $m_i q_i = N$ and $z_i' = 0$ otherwise. Hence, $f_{\rm prin}$ acquires a nontrivial zero in the separable extension F/k, so by a specialization argument as at the end of the proof of Proposition C.1.6, it has a nontrivial zero in k^n. □

The preceding lemma is a key ingredient in the following general result concerning quasi-parabolic subgroups of pseudo-reductive groups.

Theorem C.1.9 *Let G be a pseudo-reductive group over a separably closed field k. The quasi-parabolic k-subgroups of G are precisely the pseudo-parabolic k-subgroups if and only if the Cartan k-subgroups of G are tori. In particular:*

(1) *if G is reductive then its quasi-parabolic k-subgroups are precisely its parabolic k-subgroups,*
(2) *if G is not reductive (so k is imperfect) and* ${\rm char}(k) \neq 2$ *then G contains a quasi-parabolic k-subgroup that is not pseudo-parabolic,*
(3) *if k is imperfect of characteristic 2 then there exist non-reductive standard pseudo-reductive k-groups G in which all quasi-parabolic k-subgroups are pseudo-parabolic.*

Since quasi-parabolicity and pseudo-parabolicity are insensitive to separable extension on k, part (1) of the theorem is valid over any field k.

Proof Let P be a minimal pseudo-parabolic k-subgroup of G, and T a maximal k-torus in P. Let $U = \mathscr{R}_{u,k}(P)$, so U is k-split because G is pseudo-reductive (Corollary 2.2.5). Since P is minimal and $k = k_s$, it follows from Proposition 3.3.15 that $P = Z_G(T) \cdot U$. In particular, $H = T \ltimes U$ is a normal k-subgroup of P. Consider the smooth fibration $G/H \to G/P$ that is a torsor for the group $P/H = Z_G(T)/T$. This quotient is unipotent, and it is commutative since $Z_G(T)$ is a Cartan k-subgroup of the pseudo-reductive G (Proposition 1.2.4). Since $Z_G(T)$ is pseudo-reductive and T is k-split, it follows that P/H is k-wound (as any commutative k-group extension of $\mathbf{G}_{\mathrm{a}}$ by GL_1 is a split extension). Thus, by Lemma C.1.8, P/H is pseudo-complete over k.

The quotient space G/P is pseudo-complete since P is pseudo-parabolic, and the structure group P/H for the fibration $\pi : G/H \to G/P$ is pseudo-complete. We claim that this implies that G/H is also pseudo-complete (i.e., H is quasi-parabolic in G). To check this, it suffices to verify the valuative criterion using a complete discrete valuation ring A over k with residue field that is separably closed and separable over k. Let K denote the fraction field of A, and choose any $x \in (G/H)(K)$. Let $y = \pi(x) \in (G/P)(K)$. Since $(G/P)(K) = (G/P)(A)$, we view y as an A-point and pull back the fibration π along y to get a P/H-torsor E over $\mathrm{Spec}\, A$. The k-group P/H is smooth and A is complete with separably closed residue field, so such a torsor must be trivial. That is, $E \simeq (P/H)_A$ as A-schemes. In particular, $E(A) = E(K)$. But $x \in E(K)$, so x extends to an A-point of E, which in turn defines an A-point of G/H that extends x. Thus, H is indeed quasi-parabolic in G.

Now assume that the quasi-parabolic k-subgroups of G are precisely the pseudo-parabolic k-subgroups. By construction, H is contained in the minimal pseudo-parabolic k-subgroup P, so $H = P$. Since $Z_P(T) = Z_G(T)$, we conclude that $Z_H(T) = Z_G(T)$. But $H = T \ltimes U$ with T acting on $U = \mathscr{R}_{u,k}(P)$ with only nontrivial weights on $\mathrm{Lie}(U)$. Thus, the inclusion $T \subseteq Z_H(T)$ between smooth connected k-groups is an equality on Lie algebras, so it is an equality of groups. This proves that the Cartan k-subgroups of G are tori.

Conversely, assume that the Cartan k-subgroups of G are tori and let H be any quasi-parabolic k-subgroup of G. The minimal pseudo-parabolic k-subgroups of G have the form $Q = T \ltimes U$ for a maximal k-torus T and a k-split smooth connected unipotent k-group U. Such a Q is a k-split solvable smooth connected k-group since $k = k_s$, so by pseudo-completeness of G/H and Proposition C.1.5 there exists $g \in G(k)$ such that $gQg^{-1} \subseteq H$. But gQg^{-1} is pseudo-parabolic, so H is also pseudo-parabolic (Proposition 3.5.8).

In the reductive case (for which Cartan k-subgroups are always tori), we conclude that the quasi-parabolic k-subgroups are the pseudo-parabolic k-subgroups. These are precisely the parabolic k-subgroups, by Proposition 2.2.9. Now suppose that G is not reductive, so k is imperfect. In particular, $p = \mathrm{char}(k) > 0$. If $p \neq 2$, then by Theorem 11.1.1 the Cartan k-subgroups of G are not tori. Hence, by what we have just proved above, G contains quasi-parabolic k-subgroups that are not pseudo-parabolic. Finally, assume k is imperfect with characteristic 2. In this case, Example 11.1.2 gives examples of non-reductive standard pseudo-reductive k-groups in which all Cartan

k-subgroups are tori, and so in such k-groups the quasi-parabolic k-subgroups are precisely the pseudo-parabolic k-subgroups. □

C.2 Conjugacy results in the smooth affine case

As an application of Proposition C.1.6, we can prove $G(k)$-conjugacy results that are well known in the connected reductive case but not so widely known in the general case. The proofs rest on the following lemma that generalizes a known result in the connected reductive case [Bo2, 20.5].

Lemma C.2.1 *Let G be a smooth connected affine group over a field k, and P a pseudo-parabolic k-subgroup of G. The map $G \to G/P$ has sections Zariski-locally on G/P if k is infinite, and in general $G(k) \to (G/P)(k)$ is surjective.*

Our proof is a variation on the argument in [Spr, 15.1.3]. The main difficulty, in comparison with the connected reductive case, is that $G(k)$ may not be Zariski-dense in G when k is imperfect.

Proof If k is finite then the desired surjectivity is a well-known consequence of Lang's Theorem on the vanishing of degree-1 Galois cohomology for smooth connected k-groups (such as P); see [Bo2, 16.5(ii)]. Thus, we may assume k is infinite. We shall argue by induction on $\dim G$, the case of dimension 0 being clear. Now assume $\dim G > 0$ and that the result is known for all smooth connected affine k-groups of strictly smaller dimension. Let $U = \mathscr{R}_{us,k}(G)$ be the maximal k-split smooth connected unipotent normal k-subgroup of G, so $U \subseteq P$. The map $G \to G/U$ is a U-torsor over G/U, so it has sections over every affine open subset of G/U because the pointed étale cohomology set $\mathrm{H}^1(X, U)$ is trivial for every affine k-scheme X (as we may verify by using the k-split property of U to reduce to the case $U = \mathbf{G}_a$, for which we can apply the equality of Zariski and étale cohomology for quasi-coherent sheaves). The quotient group P/U is a pseudo-parabolic k-subgroup of G/U with $(G/U)/(P/U) = G/P$, so if $U \neq 1$ then the settled case of G/U gives the result for G. We can therefore assume $U = 1$.

By Proposition 2.2.4, the triviality of U implies that $P = P_G(\lambda)$ for some k-homomorphism $\lambda : \mathrm{GL}_1 \to G$. The multiplication map $U_G(-\lambda) \times P \to G$ is an open immersion, so $U_G(-\lambda)$ maps isomorphically onto a dense open subscheme $V \subseteq G/P$. If $G(k)$ is Zariski-dense in G then its image in G/P is Zariski-dense, so G/P is covered by the translates $g.V$ for $g \in G(k)$. Since $g.U_G(-\lambda)$ maps isomorphically onto $g.V$, we get the desired sections Zariski-locally on G/P when $G(k)$ is Zariski-dense in G.

In general, let G_t be the smooth connected k-subgroup of G generated by the k-tori. By Proposition A.2.11, this is normal in G and unirational. In particular,

$G_t(k)$ is Zariski-dense in G_t since k is infinite. Note that λ factors through G_t, so $P_t := P \cap G_t = P_{G_t}(\lambda)$. The normality of G_t in G implies that $\mathscr{R}_{us,k}(G_t) = 1$ since $\mathscr{R}_{us,k}(G) = 1$. In particular, P_t is a pseudo-parabolic k-subgroup of G_t. Since either $\dim G_t < \dim G$ or $G_t = G$ (in which case $G(k)$ is Zariski-dense in G), the map $G_t \to G_t/P_t$ has Zariski-local sections. We shall prove that the natural map $G_t/P_t \to G/P$ is an isomorphism of k-schemes, so we will then be done.

The quotient G/G_t makes sense as a k-group since G_t is normal in G, and since λ has trivial image in this quotient we see via Corollary 2.1.9 that the k-subgroup $P = P_G(\lambda)$ in G maps onto G/G_t. It follows that the natural k-homomorphism $\theta : G_t \rtimes P \to G$ is surjective, and its kernel is isomorphic to $G_t \cap P = P_t$ (which is smooth). Since $P_t \rtimes P$ is a k-subgroup of $G_t \rtimes P$ containing $\ker \theta$, we get the composite isomorphism

$$G_t/P_t \simeq (G_t \rtimes P)/(P_t \rtimes P) \simeq ((G_t \rtimes P)/\ker\theta)/((P_t \rtimes P)/\ker\theta) = G/P$$

that is the natural map. □

Lemma C.2.2 *Let G be a smooth connected affine group over a field k, and H a k-split solvable smooth connected k-subgroup. For any pseudo-parabolic k-subgroup $P \subseteq G$, there exists $g \in G(k)$ such that $g^{-1}Hg \subseteq P$.*

Proof Consider the left translation action of H on G/P. By Propositions C.1.6 and C.1.5, there exists $x \in (G/P)(k)$ fixed by H. This lifts to some $g \in G(k)$ by Lemma C.2.1, so $HgP \subseteq gP$. In other words, $g^{-1}Hg \subseteq P$. □

Theorem C.2.3 (Borel–Tits) *Let G be a smooth connected affine group over a field k. All maximal k-split k-tori in G are $G(k)$-conjugate.*

See Proposition A.2.10 for a simpler proof when $k = k_s$. The common dimension of the maximal k-split tori in G is called its *k-rank*.

Proof We argue by induction on $\dim G$, the case of dimension 0 being clear. Now assume $\dim G > 0$ and that the result holds in all strictly smaller dimensions. Let $U = \mathscr{R}_{u,k}(G)$. By Lemma 2.2.3, there is a proper pseudo-parabolic k-subgroup P in G if and only if G/U contains a non-central k-split k-torus.

First suppose that G does not contain any proper pseudo-parabolic k-subgroup, so all k-split tori in G map into the maximal central k-split k-torus S in G/U. Thus, the preimage H of S in G is a smooth connected solvable k-subgroup of G that contains all k-split tori of G. If $H \neq G$ then we are done since $\dim H < \dim G$, so we can assume G is solvable. In this case, by [Bo2,

19.2] the maximal k-tori in G are $G(k)$-conjugate, so the maximal k-split k-tori in G are also $G(k)$-conjugate.

Assume instead that there is a proper pseudo-parabolic k-subgroup P in G, and let S be a maximal k-split k-torus in G. Since S is k-split solvable, by Lemma C.2.2 there exists $g \in G(k)$ such that $g^{-1}Sg \subseteq P$. Hence, we can again conclude by induction on dimension since $\dim P < \dim G$. □

As an application of Theorem C.2.3, for any smooth connected affine group G over a field k and any pseudo-parabolic k-subgroup P of G, the maximal k-split tori S of P are maximal k-split tori in G. More importantly, if P is minimal for the property of containing S then it is minimal as a pseudo-parabolic k-subgroup of G. (In particular, every k-split k-torus in G lies in a minimal pseudo-parabolic k-subgroup, though this also follows from Lemma C.2.2.) Indeed, in the case that S is a maximal k-torus in G this was proved in the discussion preceding Proposition 3.3.15 by passing up to k_s so as to apply the conjugacy result in Proposition A.2.10, and we can carry out the same argument using S directly over k by using Theorem C.2.3.

Proposition C.2.4 *Let P be a pseudo-parabolic k-subgroup of a smooth connected affine group G over a field k. Let S be a maximal k-split k-torus in P. Then $Z_G(S) \subseteq P$, and P is a minimal pseudo-parabolic k-subgroup of G if and only if $P = Z_G(S) \cdot \mathscr{R}_{us,k}(P)$.*

Proof By Remark 3.5.6 and Proposition 3.5.8, P is minimal in G if and only if P has no proper pseudo-parabolic k-subgroup of its own. Applying Lemma 2.2.3(1) to P, we conclude that P is minimal in G if and only if all k-split k-tori in $\overline{P} := P/\mathscr{R}_{u,k}(P)$ are central. It is equivalent to say that one maximal k-split k-torus in $\overline{P}$ is central.

Consider the image $\overline{S}$ of S in $\overline{P}$. For any k-torus $\overline{T}$ in $\overline{P}$ containing $\overline{S}$, the preimage H of $\overline{T}$ in P is a smooth connected solvable k-subgroup of P that is an extension of $\overline{T}$ by the unipotent group $\mathscr{R}_{u,k}(P)$. Hence, any maximal k-torus of H maps isomorphically onto $\overline{T}$. Applying this to a maximal k-torus of H that contains S, it follows from the maximality of S as a k-split k-torus in P that $\overline{S}$ is maximal as a k-split k-torus in $\overline{P}$. We conclude that P is minimal in G if and only if $\overline{S}$ is central in $\overline{P}$. But $Z_P(S) \to Z_{\overline{P}}(\overline{S})$ is surjective (Proposition A.2.5), so P is minimal in G if and only if $P = Z_P(S) \cdot \mathscr{R}_{u,k}(P)$.

We will now prove that $P = Z_P(S) \cdot \mathscr{R}_{us,k}(P)$ when P is minimal. For this, we can replace P by $P/\mathscr{R}_{us,k}(P)$ to reduce to the case where (by Corollary B.3.5) $\mathscr{R}_{u,k}(P)$ is k-wound. But according to Proposition B.4.4, a k-wound smooth connected unipotent k-group does not admit a nontrivial action by a k-torus. Hence, $\mathscr{R}_{u,k}(P) \subseteq Z_P(S)$. This proves that $P = Z_P(S) \cdot \mathscr{R}_{us,k}(P)$.

Finally we will show that $Z_P(S) = Z_G(S)$ for all pseudo-parabolic k-subgroups P containing S. Since $P = P_G(\lambda)\mathscr{R}_{u,k}(G)$ for some $\lambda : \mathrm{GL}_1 \to G$, by Theorem C.2.3 applied to P we can replace λ with a suitable $P(k)$-conjugate to arrange that λ takes values in S. Then $Z_G(S) \subseteq Z_G(\lambda) \subseteq P$, so indeed $Z_G(S) = Z_P(S)$. □

In the pseudo-reductive case, there is an alternative minimality criterion in the setting of Proposition C.2.4: by Theorem C.2.3 applied to P we can find $\lambda \in X_*(S)$ such that $P = P_G(\lambda)$, so $Z_G(S) \subseteq Z_G(\lambda)$, and then P is minimal in G if and only if $Z_G(S) = Z_G(\lambda)$. Indeed, since $P = Z_G(\lambda) \ltimes U_G(\lambda)$ we have $P/\mathscr{R}_{us,k}(P) = Z_G(\lambda)$, so the minimality criterion in Proposition C.2.4 yields the desired criterion in the pseudo-reductive case.

Theorem C.2.5 (Borel–Tits) *Let G be a smooth connected affine group over a field k. All minimal pseudo-parabolic k-subgroups of G are $G(k)$-conjugate.*

Proof Let P and Q be minimal pseudo-parabolic k-subgroups of G, and let S be a maximal k-split k-torus in P. The smooth connected solvable k-subgroup $H := S \ltimes \mathscr{R}_{us,k}(P) \subseteq P$ is k-split. By Lemma C.2.2 there exists $g \in G(k)$ such that $g^{-1}Hg \subseteq Q$. Replacing P with $g^{-1}Pg$ and S with $g^{-1}Sg$ reduces the problem to the case where $\mathscr{R}_{us,k}(P) \subseteq Q$ and $S \subseteq Q$. As noted above Proposition C.2.4, S is a maximal k-split k-torus of G and hence of Q. The above and Proposition C.2.4 for P and Q imply easily that $P \subseteq Q$, so $P = Q$ by minimality of Q. □

Corollary C.2.6 *Let G be a smooth connected affine k-group, and let P and Q be pseudo-parabolic k-subgroups of G. Let K be a separable extension of k. If P_K and Q_K are $G(K)$-conjugate then P and Q are $G(k)$-conjugate.*

Proof Choose minimal pseudo-parabolic k-subgroups P' and Q' of G with $P' \subseteq P$ and $Q' \subseteq Q$. The $G(k)$-conjugacy of P' and Q' (Theorem C.2.5) allows us to replace Q with a suitable $G(k)$-conjugate so that $P \cap Q$ contains the minimal pseudo-parabolic subgroup P'. By Proposition 3.5.12(1), the k-subgroup $P \cap Q$ is smooth and connected, so it is a pseudo-parabolic k-subgroup of G (Proposition 3.5.8). Thus, if P_K and Q_K are $G(K)$-conjugate then Proposition 3.5.12(3) implies that $P_K = Q_K$, so $P = Q$. □

Proposition C.2.7 *Let G be a smooth connected affine group over a field k and P and Q be two pseudo-parabolic k-subgroups. Then there exists a maximal k-split k-torus S of G such that $P \cap Q$ contains the centralizer $Z_G(S)$ of S.*

Proof We may and do assume that P is a minimal pseudo-parabolic k-subgroup. By Proposition 3.5.12, $P \cap Q$ is a smooth connected k-subgroup and $(P \cap Q)_{k_s}$ contains a maximal k_s-torus of G_{k_s}. Hence, any maximal k-torus T of $P \cap Q$ is maximal in G. Fix such a T.

Under the natural surjective homomorphism $\pi : P \to P/\mathscr{R}_{u,k}(P) =: H$, the maximal k-torus T maps isomorphically onto a maximal k-torus of H. In particular, $\pi(T)$ contains the maximal central k-torus of H. Since P is a minimal pseudo-parabolic k-subgroup of G, by Proposition C.2.4 the center of H contains a k-torus isomorphic to a maximal k-split k-torus of G. We conclude from this that the maximal k-split k-subtorus S of T is isomorphic to a maximal k-split k-torus of G. Hence, S is a maximal k-split k-torus of G (Theorem C.2.3). Now, by conjugacy of maximal k-split k-tori of P (resp. Q) under $P(k)$ (resp. $Q(k)$), we see that there exist 1-parameter k-subgroups λ, μ : $\mathrm{GL}_1 \rightrightarrows S$ such that P contains $P_G(\lambda)$ and Q contains $P_G(\mu)$. But since both $P_G(\lambda)$ and $P_G(\mu)$ clearly contain $Z_G(S)$, so does $P \cap Q$. □

The following gives the Bruhat decomposition of $G(k)$ for an arbitrary smooth connected affine group G over any field k.

Theorem C.2.8 (Borel–Tits) *Let G be a smooth connected affine group over a field k, S a maximal k-split k-torus, and P a pseudo-parabolic k-subgroup of G containing S. Let $N = N_G(S)$ and $Z = Z_G(S) \subseteq P$. Then $G(k) = P(k)N(k)P(k)$ and if P is minimal then the surjective map $N(k)/Z(k) \to P(k)\backslash G(k)/P(k)$ is bijective.*

Proof Let $g \in G(k)$. We consider the pseudo-parabolic k-subgroup gPg^{-1}. According to Proposition C.2.7, $P \cap gPg^{-1}$ contains a maximal k-split k-torus T of G. Thus, P contains the maximal k-split k-tori S, T, and $g^{-1}Tg$. By conjugacy of such tori under $P(k)$ (Theorem C.2.3) we deduce that there exist $p_1, p_2 \in P(k)$ such that $p_1^{-1}Tp_1 = S = p_2g^{-1}Tgp_2^{-1}$. Then $n := p_1^{-1}gp_2^{-1}$ lies in $N_G(S)(k)$. Thus, $g = p_1np_2 \in P(k)N(k)P(k)$.

Now we prove that $N(k)/Z(k) \to P(k)\backslash G(k)/P(k)$ is bijective when P is minimal. Let $\overline{P} = P/\mathscr{R}_{u,k}(P)$, so minimality of P says exactly that the map $Z \to \overline{P}$ is surjective (Proposition C.2.4). The natural k-homomorphism $S \to \overline{P}$ has trivial kernel (as $S \cap \mathscr{R}_{u,k}(P)$ is both unipotent and of multiplicative type), so S embeds as a central k-torus in $\overline{P}$. In particular, any $g \in P(\overline{k}) \cap N(\overline{k})$ must centralize $S_{\overline{k}}$, so $P(\overline{k}) \cap N(\overline{k}) = Z(\overline{k})$. Thus for any $n \in N(\overline{k}) - Z(\overline{k})$ we must have $P_{\overline{k}} \cap (P_{\overline{k}}nP_{\overline{k}}) = \emptyset$. In particular, $P(k)nP(k)$ is disjoint from $P(k)$ for any $n \in N(k) - Z(k)$. This is a special case of what we wish to prove, and we will use it to deduce the general case.

Consider any $n, n' \in N(k)$ such that $P(k)nP(k) = P(k)n'P(k)$. We will prove $n^{-1}n' \in Z(k)$. The equality of double cosets implies that $nq = pn'$ for some $p, q \in P(k)$, so

$$pSp^{-1} = (pn')S(pn')^{-1} = n(qSq^{-1})n^{-1} \subseteq nPn^{-1}.$$

Letting H be the smooth connected k-subgroup of P and nPn^{-1} generated by S and pSp^{-1}, it follows that S and pSp^{-1} are maximal k-split k-tori in H. Thus, by

Theorem C.2.3 (applied to H), there exists $h \in H(k)$ such that $pSp^{-1} = hSh^{-1}$, so $p = hz$ for some $z \in N(k)$. But $z = h^{-1}p \in P(k)$, so $z \in P(k) \cap N(k) = Z(k)$. Since $H \subseteq nPn^{-1}$ we can write $h = np'n^{-1}$ for some $p' \in P(k)$, so

$$nqn'^{-1} = p = hz = np'n^{-1}z.$$

It follows that $p'^{-1}q = n^{-1}zn' = n^{-1}n'(n'^{-1}zn')$. This implies that $n^{-1}n'Z(k)$ meets $P(k)$, so $P(k)n^{-1}n'P(k)$ meets $P(k)$. Hence, $n^{-1}n' \in Z(k)$. □

Remark C.2.9 In the setting of Theorem C.2.8, when P is minimal in G we can use the second assertion in Proposition 2.1.12 to prove that if $n, n' \in N(k)$ are distinct modulo $Z(k)$ then PnP and $Pn'P$ are disjoint as locally closed subvarieties of G; this refines the fact that n and n' have distinct images in $P(k)\backslash G(k)/P(k)$. To prove this refinement, we first reduce to the pseudo-reductive case as follows. Let $\pi : G \to G' = G/\mathscr{R}_{u,k}(G)$ be the maximal pseudo-reductive quotient, so $S' := \pi(S)$ is a maximal k-split k-torus in G' and $Z' := Z_{G'}(S') = \pi(Z)$. Let $N' = N_{G'}(S')$, so $\pi(N) \subseteq N'$. Since $\pi : S \to S'$ is an isomorphism, the map $N(k)/Z(k) \to N'(k)/Z'(k)$ is injective. The image $P' = \pi(P) = P/\mathscr{R}_{u,k}(G)$ is a pseudo-parabolic k-subgroup of G' which is minimal (by Proposition 2.2.10 and the minimality hypothesis for P), and $PnP = Pn'P$ if and only if $P'\pi(n)P' = P'\pi(n')P'$, so we can replace (G, S, P) with (G', S', P') and replace n and n' with their images in $N'(k)$ to reduce to the pseudo-reductive case. In particular, $P = P_G(\lambda)$ and by Theorem C.2.3 (applied to P) we can arrange that $\lambda \in \mathrm{X}_*(S)$. Note that $Z_G(\lambda) = Z_G(S)$ due to the minimality of P in G (and the minimality criterion in the pseudo-reductive case, as noted following the proof of Proposition C.2.4). This equality of centralizers is what will be used; neither the pseudo-reductivity of G nor the pseudo-parabolicity of P will be needed below.

Consider the following general assertion for an arbitrary smooth connected affine k-group G, k-torus S in G, and k-homomorphism $\lambda : \mathrm{GL}_1 \to S$ such that $Z_G(S) = Z_G(\lambda)$: for $N = N_G(S)$, $Z = Z_G(S)$, $P = P_G(\lambda)$, $U = U_G(\lambda)$, and distinct elements $n, n' \in N(k)$, UnU is disjoint from $Un'U$. Once this is proved, we apply it to the preceding setup after scalar extension to $\overline{k}$ (noting that if $n, n' \in N(k)$ are distinct modulo $Z(k)$ then they are distinct modulo $Z(\overline{k})$) to deduce that $(PnP)(\overline{k})$ is disjoint from $(Pn'P)(\overline{k})$ since $P = Z \ltimes U$. So now we may assume $k = \overline{k}$ and just have to show that if $nu = u'n'$ for some $u, u' \in U(k)$ then $n = n'$. Since n' normalizes S it is easy to check that the solvable smooth connected group $V := n'^{-1}Un'$ is normalized by S (since U is normalized by $S \subseteq Z$), so by Proposition 2.1.12(2) we have $(V \cap U^-) \times (V \cap Z) \times (V \cap U) = V$ via multiplication, with $U^- := U_G(-\lambda)$. But $V \cap Z = n'^{-1}(U \cap Z)n' = 1$. Hence, $n'^{-1}u'n' \in V(k) = (V \cap U^-)(k) \times (V \cap U)(k)$, so $n'^{-1}nu = n'^{-1}u'n' = vu''$ with $v \in U^-(k)$ and $u'' \in U(k)$. Thus, $n'^{-1}n = vu''u^{-1} \in U^-(k) \times U(k)$.

This element lies in $N(k)$, and we claim that this forces $v, u''u^{-1} \in Z(k) = Z_G(\lambda)(k)$, from which it follows that $v = u''u^{-1} = 1$, so $n' = n$ as desired.

To prove the above claim it will suffice to show that if H is any smooth subgroup scheme of Z (such as S) then

$$N_G(H) \cap (U^- \times Z \times U) = (U^- \cap Z_G(H)) \times N_Z(H) \times (U \cap Z_G(H)).$$

Working functorially, consider a k-algebra R and points $u' \in U^-(R)$, $z \in Z(R)$, and $u \in U(R)$ such that $u'zu$ normalizes H_R in G_R. We aim to show that u and u' centralize H_R and z normalizes H_R. To check this we may work with points of H valued in an R-algebra R', so by renaming R' as R we aim to show that u and u' centralize $H(R)$ and z normalizes $H(R)$. Let ϕ be the automorphism of H_R defined by conjugation by $u'zu$, so for any $h \in H(R)$ we have

$$u' \cdot zh \cdot h^{-1}uh = (\phi(h)u'\phi(h)^{-1}) \cdot \phi(h)z \cdot u.$$

But $U^- \times Z \times U$ is a subfunctor of G via multiplication, so $u' = \phi(h)u'\phi(h)^{-1}$, $zh = \phi(h)z$, and $h^{-1}uh = u$. Thus, u and u' indeed centralize $H(R)$ and z normalizes $H(R)$. This completes the proof that UnU and $Un'U$ are disjoint if $n' \neq n$.

Proposition C.2.10 *Let G be a smooth connected affine group over a field k, S a k-split torus in G, and $W(G,S) = N_G(S)/Z_G(S)$ the associated finite étale quotient group.*

The k-group $W(G,S)$ is constant, and if S is a maximal k-split torus then the natural homomorphism $N_G(S)(k) \to W(G,S)(k)$ is surjective. In particular, $N_G(S)(k)/Z_G(S)(k) \to W(G,S)(k)$ is an isomorphism if S is a maximal k-split torus in G.

Proof To prove that $W(G,S)$ is a constant k-group, it suffices to prove that all of its k_s-points are invariant under $\Gamma = \mathrm{Gal}(k_s/k)$; i.e., the natural action of Γ on $W(G,S)(k_s) = N_G(S)(k_s)/Z_G(S)(k_s)$ is trivial. Choose any $g \in N_G(S)(k_s)$, so $x \mapsto gxg^{-1}$ is a k_s-automorphism of S_{k_s}. But any k_s-automorphism of S_{k_s} is defined over k since S is k-split, so for any $\gamma \in \Gamma$ we have $\gamma(gxg^{-1}) = g\gamma(x)g^{-1}$ for $x \in S(k_s)$. This says that $\gamma(g)^{-1}g \in Z_G(S)(k_s)$, so the image of g in $W(G,S)(k_s)$ coincides with that of $\gamma(g)$. Hence, the Γ-action is indeed trivial.

Now assume that S is a maximal k-split torus, and let $W = W(G,S)$. There is a natural action of $W(k)$ on the set $\mathscr{P}$ of minimal pseudo-parabolic k-subgroups of G containing S, described as follows. For a given $P \in \mathscr{P}$ and $w \in W(k)$ let us choose an $n \in N_G(S)(k_s)$ which maps onto w. Then $nP_{k_s}n^{-1}$ is a pseudo-parabolic k_s-subgroup of G_{k_s} which by the Galois criterion is defined over k and hence descends to a k-subgroup of G containing $Z_G(S)$.

By Proposition 3.5.2(1) this k-subgroup of G is a pseudo-parabolic subgroup, and as its dimension equals that of P, it is in fact a minimal pseudo-parabolic k-subgroup (by Theorem C.2.5); we will denote this minimal pseudo-parabolic k-subgroup by $w \circ P$ since it is easily seen to depend only on w and not on a particular choice of n. The action of $W(k)$ on $\mathscr{P}$ is then $(w, P) \mapsto w \circ P$. It readily follows from Proposition 3.5.7 and the fact that $N_G(S) \cap P = Z_G(S)$ for any $P \in \mathscr{P}$ (see the proof of Theorem C.2.8) that $W(k)$ acts freely on $\mathscr{P}$.

On the other hand, given P, $P' \in \mathscr{P}$, by applying Theorems C.2.5 and C.2.3 we can find an $n \in N_G(S)(k)$ such that $nPn^{-1} = P'$. Thus, under the natural action of $W(k)$ on $\mathscr{P}$ described above, the image W' of $N_G(S)(k)$ in $W(k)$ acts transitively on $\mathscr{P}$. As the action of $W(k)$ on $\mathscr{P}$ is free, we conclude that $W' = W(k)$; i.e., $N_G(S)(k)$ maps onto $W(k)$. □

Definition C.2.11 A *quasi-reductive* k-group is a smooth connected affine k-group G such that $\mathscr{R}_{us,k}(G)$ is trivial, or, equivalently, by Corollary B.3.5, $\mathscr{R}_{u,k}(G)$ is k-wound. (By Proposition B.4.4 any k-torus of a quasi-reductive G commutes with $\mathscr{R}_{u,k}(G)$.)

Remark C.2.12 (1) Given a smooth connected affine k-group G, a smooth connected normal k-subgroup N, and a k-torus T in G, the images of $N_{\overline{k}}$ and $Z_G(T)_{\overline{k}}$ in the reductive quotient $G^{\rm ss}_{\overline{k}} := G_{\overline{k}}/\mathscr{R}_u(G_{\overline{k}})$ are reductive and hence $\mathscr{R}_u(N_{\overline{k}})$ and $\mathscr{R}_u(Z_G(T)_{\overline{k}})$ are contained in $\mathscr{R}_u(G_{\overline{k}})$. This implies that both $\mathscr{R}_{us,k}(N)$ and $\mathscr{R}_{us,k}(Z_G(T))$ are contained in $\mathscr{R}_{us,k}(G)$, so if G is quasi-reductive then so are N and $Z_G(T)$.

(2) Parts (ii) and (iii) of Lemma 1.2.5 hold if the pseudo-reductivity hypothesis on G there is relaxed to quasi-reductivity. To explain this, we shall freely use the notation introduced in the proof of assertion (ii) in that lemma (e.g., Z is as in that proof, not as in the statement of the lemma). The only role of pseudo-reductivity was to show that Z commutes with N. We will show this commutation holds whenever $\mathscr{R}_{u,k}(G)$ is k-wound. Since Z' commutes with N', clearly $(Z, N)_{\overline{k}} \subseteq \mathscr{R}_u(G_{\overline{k}})$. But $(Z, N) \subseteq N$ due to the normality of N in G, and it is normal in N, so $(Z, N) \subseteq \mathscr{R}_{u,k}(N)$. Thus, the natural action of Z on $N/\mathscr{R}_{u,k}(N)$ is trivial. By applying Proposition A.2.8 to the quotient map $Z \ltimes N \twoheadrightarrow Z \times (N/\mathscr{R}_{u,k}(N))$ and the torus $Z \subseteq Z \ltimes N$, we see that the smooth connected subgroup N^Z of Z-fixed points in N maps onto $N/\mathscr{R}_{u,k}(N)$, so $N = N^Z \mathscr{R}_{u,k}(N)$. But the smooth connected unipotent $\mathscr{R}_{u,k}(N)$ is k-wound because it is contained in the k-wound $\mathscr{R}_{u,k}(G)$ (due to normality of N in G), so the action of Z on it is trivial. We conclude that $N = N^Z$, as desired.

C.2.13 The k-root system of G and a Tits system in $G(k)$. Let G be a smooth connected affine k-group and $\pi : G \to \overline{G} := G/\mathscr{R}_{u,k}(G)$ its maximal pseudo-reductive quotient. We fix a maximal k-split torus S of G, and a minimal pseudo-parabolic k-subgroup P of G containing S. Let $N = N_G(S)$ and

$Z = Z_G(S)$, so $Z = N \cap P$ (by Theorem C.2.8). Our aim in the remainder of §C.2 is to construct a root system ${}_k\overline{\Phi}$ associated to (G, S) such that (i) $W({}_k\overline{\Phi})$ is naturally identified with $N(k)/Z(k)$, (ii) the set of parabolic subsets of ${}_k\overline{\Phi}$ is in natural bijective correspondence with the set of pseudo-parabolic k-subgroups of G containing S, and (iii) the hypothesis that G is pseudo-split in Corollary 3.3.12 to Corollary 3.3.16, Proposition 3.4.2, and Proposition 3.5.1 can be removed by using maximal k-split tori. We will also exhibit a Tits system associated to $(G(k), P(k), N(k))$; the definition of a Tits system will be recalled later.

Let T be a maximal k-torus of G containing S. The map π carries the tori S and T isomorphically onto their respective images $\overline{S}$ and $\overline{T}$ in $\overline{G}$ (which are respectively a maximal k-split torus and a maximal k-torus of $\overline{G}$). In this way, we identify $\mathrm{X}(\overline{S})$ and $\mathrm{X}(\overline{T})$ with $\mathrm{X}(S)$ and $\mathrm{X}(T)$ respectively. By definition the *root system* of G_{k_s} with respect to T_{k_s} is the (possibly non-reduced) root system $\Phi(\overline{G}_{k_s}, \overline{T}_{k_s})$ of $\overline{G}_{k_s}$ with respect to $\overline{T}_{k_s}$ viewed as a root system in $\mathrm{X}(T_{k_s})_{\mathbf{Q}}$ via the identification of $\mathrm{X}(\overline{T}_{k_s})$ with $\mathrm{X}(T_{k_s})$. (The discussion immediately following Definition 3.2.5 shows that $\Phi(\overline{G}_{k_s}, \overline{T}_{k_s})$ is a root system in its $\mathbf{Q}$-linear span.) We will often denote the root system $\Phi(\overline{G}_{k_s}, \overline{T}_{k_s})$ by $\overline{\Phi}$ in the sequel.

The k-groups $W := W(G, T) = N_G(T)/Z_G(T)$ and $\overline{W} := W(\overline{G}, \overline{T}) = N_{\overline{G}}(\overline{T})/Z_{\overline{G}}(\overline{T})$ are finite étale (Lemma A.2.9), and by Lemma 3.2.1 the natural homomorphism $W \to \overline{W}$ is a k-isomorphism. Since $Z_G(T)$ is smooth, the homomorphism $N_G(T)(k_s)/Z_G(T)(k_s) \to W(k_s)$ is an isomorphism. By Proposition 3.2.7, $W(k_s)$ maps isomorphically onto the Weyl group of $\overline{\Phi}$.

According to Proposition C.2.10, the finite étale k-group $W(G, S) = N/Z$ is constant and $N(k)/Z(k) \to W(G, S)(k)$ is an isomorphism. Let $\Phi(\overline{G}, \overline{S}) = \overline{\Phi}|_{\overline{S}_{k_s}} - \{0\}$. Via the identification $\mathrm{X}(\overline{S}) = \mathrm{X}(\overline{S}_{k_s})$, $\Phi(\overline{G}, \overline{S})$ is the set of nonzero weights for the adjoint action of the k-split $\overline{S}$ on the Lie algebra of $\overline{G}$. Define ${}_kW = W(G, S)(k)$ and call it the *k-Weyl group* (or *relative Weyl group*) of (G, S) or of G (cf. Theorem C.2.15).

Denote by ${}_k\overline{\Phi}$ the subset of $\mathrm{X}(S)$ corresponding to $\Phi(\overline{G}, \overline{S}) \subseteq \mathrm{X}(\overline{S})$ under the identification of $\mathrm{X}(\overline{S})$ with $\mathrm{X}(S)$, so ${}_k\overline{\Phi} \subseteq \Phi(G, S)$; in general this latter inclusion need not be an equality. When $\mathscr{R}_{u,k}(G) \subseteq Z_G(S)$ (for example, if G is quasi-reductive) then ${}_k\overline{\Phi}$ equals the set $\Phi(G, S)$ of nontrivial S-weights on $\mathrm{Lie}(G)$, and in such cases we will often denote ${}_k\overline{\Phi}$ by ${}_k\Phi$.

By Lemma 2.2.3(1) and Theorem C.2.3, ${}_k\overline{\Phi}$ is empty if and only if G does not contain proper pseudo-parabolic k-subgroups. It is clear that the natural action of ${}_kW$ on $\mathrm{X}(S)$ carries ${}_k\overline{\Phi}$ into itself. We will prove in Theorem C.2.15 that ${}_k\overline{\Phi}$ is a root system in its $\mathbf{Q}$-span in $\mathrm{X}(S)_{\mathbf{Q}}$, with ${}_kW$ mapping isomorphically onto $W({}_k\overline{\Phi})$. We call ${}_k\overline{\Phi}$ the *k-root system of G with respect to S* (or *relative root system*), and we call the elements of ${}_k\overline{\Phi}$ the *k-roots* of G with respect to S.

For $a \in {}_k\overline{\Phi}$, let $\mathscr{S}_a = \ker a$ and $S_a = (\ker a)^0_{\mathrm{red}}$, so $\mathscr{S}_a$ is a k-group scheme of multiplicative type and S_a is a torus. Beware that $\mathscr{S}_a$ may not be smooth. By Proposition A.8.10(1), there exists a closed k-subgroup scheme $Z_G(\mathscr{S}_a)$ in G representing the centralizer of $\mathscr{S}_a$ under the conjugation action on G. The k-group $Z_G(\mathscr{S}_a)$ is smooth by Proposition A.8.10(2), and we define G_a to be its identity component. Note that the k-subgroups G_a, $\mathscr{S}_a$, and S_a in G are unaffected by replacing a with $-a$. (In the main text with pseudo-reductive groups, notation such as G_a generally denotes the derived group of the centralizer of a suitable torus. In the remainder of §C.2 we consider rather general G, possibly not pseudo-reductive, and we use the identity component of a torus centralizer – such as G_a above – without passing to its derived subgroup.)

The following assertions are immediate from Proposition A.8.14. The image $\overline{G}_a$ of G_a in $\overline{G} = G/\mathscr{R}_{u,k}(G)$ is pseudo-reductive and equal to the identity component of the functorial centralizer of the natural $\mathscr{S}_a$-action on $\overline{G}$ (again, Proposition A.8.10(1),(2) provides the existence and smoothness of such a centralizer scheme). Moreover, the natural map $G_a/\mathscr{R}_{u,k}(G_a) \to \overline{G}_a$ is an isomorphism, so it induces an isomorphism between the Lie algebras and an isomorphism $T \simeq \overline{T}$ between maximal k-tori. This identifies $\Phi((G_a/\mathscr{R}_{u,k}(G_a))_{k_s}, T_{k_s})$ and $\Phi((\overline{G}_a)_{k_s}, \overline{T}_{k_s})$ in $\mathrm{X}(T_{k_s}) = \mathrm{X}(\overline{T}_{k_s})$, and this common subset consists of those $\alpha \in \overline{\Phi}$ whose restriction to $S_{k_s} \simeq \overline{S}_{k_s}$ is an integral multiple of a (in $\mathrm{X}(S_{k_s}) = \mathrm{X}(S)$).

Lemma C.2.14 *The smooth connected k-group G_a has proper pseudo-parabolic k-subgroups, and all such pseudo-parabolic k-subgroups are minimal. There are exactly two such k-subgroups P_a and P'_a that contain S.*

Proof The natural map $q : G_a^{\mathrm{pred}} := G_a/\mathscr{R}_{u,k}(G_a) \to \overline{G}_a$ is an isomorphism (Proposition A.8.14(2)). Thus, by Proposition 2.2.10, forming images and scheme-theoretic preimages under $G_a \to \overline{G}_a$ defines an inclusion-preserving bijective correspondence between the sets of pseudo-parabolic k-subgroups of G_a and $\overline{G}_a$. Hence, to prove the existence and minimality of proper pseudo-parabolic k-subgroups in G_a it is equivalent to prove the same for $\overline{G}_a$.

Since $\overline{S}$ is a maximal k-split torus in $\overline{G}_a$, by Theorem C.2.3 every pseudo-parabolic k-subgroup of $\overline{G}_a$ containing $\overline{S}$ has the form $P_{\overline{G}_a}(\overline{\mu})$ for some $\overline{\mu} \in \mathrm{X}_*(\overline{S})$. But $\overline{S}_a := \pi(S_a)$ is a central torus of $\overline{G}_a$ with codimension 1 in $\overline{S}$, so it is the maximal central k-split torus in $\overline{G}_a$ (as $a \in {}_k\overline{\Phi}$) and $P_{\overline{G}_a}(\overline{\mu})$ only depends on the image of $\overline{\mu}$ in $\mathrm{X}_*(\overline{S}/\overline{S}_a) \simeq \mathbf{Z}$. For example, this image vanishes if and only if $\overline{\mu}$ is central in $\overline{G}_a$, which is equivalent to the condition $P_{\overline{G}_a}(\overline{\mu}) = \overline{G}_a$ (Lemma 2.2.3(2)), so $\overline{G}_a$ admits a proper pseudo-parabolic k-subgroup. By Remark 2.1.7 we have $P_{\overline{G}_a}(\overline{\mu}) = P_{\overline{G}_a}(\overline{\mu}^n)$ for any integer $n \geqslant 1$, so all proper pseudo-parabolic k-subgroups of $\overline{G}_a$ (and hence of G_a) are minimal.

Fix a 1-parameter subgroup $\lambda_a : \mathrm{GL}_1 \to S$ such that $\langle a, \lambda_a \rangle > 0$, and let $\overline{\lambda}_a \in \mathrm{X}_*(\overline{S})$ correspond to λ_a. The pseudo-parabolic k-subgroup $P_a = P_{G_a}(\lambda_a)\mathscr{R}_{u,k}(G_a)$ of G_a contains S and has image $P_{\overline{G}_a}(\overline{\lambda}_a)$ in $\overline{G}_a$ (Corollary 2.1.9), so it is a proper k-subgroup due to the non-centrality of $\overline{\lambda}_a$ in $\overline{G}_a$ (Lemma 2.2.3(2)). Hence, P_a is a minimal pseudo-parabolic k-subgroup of G_a. Likewise, $P'_a := P_{G_a}(-\lambda_a)\mathscr{R}_{u,k}(G_a)$ is a minimal pseudo-parabolic k-subgroup of G_a containing S, and $P'_a \neq P_a$ (as we see by comparing the Lie algebras of their images $P_{\overline{G}_a}(\pm\overline{\lambda}_a)$ in $\overline{G}_a$). The preceding arguments show that $P_{\overline{G}_a}(\pm\overline{\lambda}_a)$ are the only proper pseudo-parabolic k-subgroups of $\overline{G}_a$ that contain $\overline{S}$, so P_a and P'_a are the only proper pseudo-parabolic k-subgroups of G_a that contain S. □

By Theorems C.2.5 and C.2.3, there exists $n_a \in N_{G_a}(S)(k)$ such that $P'_a = n_a P_a n_a^{-1}$. Note that n_a does not centralize S, since via its natural action on $\mathrm{X}_*(S)$ we have

$$P_{G_a}(n_a.\lambda_a)\mathscr{R}_{u,k}(G_a) = n_a P_a n_a^{-1} = P'_a \neq P_a = P_{G_a}(\lambda_a)\mathscr{R}_{u,k}(G_a).$$

Let $r_a \in {}_kW = N(k)/Z(k)$ denote the nontrivial image of n_a. Under the natural faithful linear action of the finite group ${}_kW$ on $\mathrm{X}(S)_{\mathbf{Q}}$, the element r_a acts nontrivially and restricts to the identity on a hyperplane (since n_a centralizes the codimension-1 subtorus S_a in S), so r_a acts as a reflection. Since the codimension-1 subtorus S_a in S is central in G_a, it follows that the subgroup $W(G_a, S)(k)$ of ${}_kW$ has order 2 and is generated by the reflection r_a. In particular, r_a is independent of the choice of n_a.

Recall that every pseudo-parabolic k-subgroup of G containing S also contains $Z_G(S)$ (Proposition C.2.4) and hence contains T. Now let Q be such a pseudo-parabolic k-subgroup of G and $\overline{Q}$ be its image in $\overline{G}$. Then $\overline{Q}$ is a pseudo-parabolic k-subgroup of $\overline{G}$ containing $\overline{S}$. We will denote by $\Phi(\overline{Q}, \overline{S})$ the set of nontrivial weights of $\overline{S}$ on $\mathrm{Lie}(\overline{Q})$ ($\subseteq \mathrm{Lie}(\overline{G})$).

Let P be a minimal pseudo-parabolic k-subgroup of G containing S. The image $\overline{P}$ of P in $\overline{G}$ is a minimal pseudo-parabolic k-subgroup of $\overline{G}$ (by Proposition 2.2.10); it contains $\overline{S}$ and $\overline{T}$. Since the subset $\Phi(\overline{P}, \overline{S}) \subseteq \mathrm{X}(\overline{S}) = \mathrm{X}(\overline{S}_{k_s})$ consists of the nontrivial restrictions to $\overline{S}_{k_s}$ of the parabolic subset $\Phi(\overline{P}_{k_s}, \overline{T}_{k_s})$ in $\Phi = \Phi(\overline{G}_{k_s}, \overline{T}_{k_s})$, it is clear that $\Phi(\overline{G}, \overline{S}) = \Phi(\overline{P}, \overline{S}) \cup -\Phi(\overline{P}, \overline{S})$. The minimality of $\overline{P}$ in $\overline{G}$ gives more: this union is *disjoint*. Indeed, by Proposition C.2.4 we have $\overline{P} = Z_{\overline{G}}(\overline{S}) \cdot \mathscr{R}_{us,k}(\overline{P})$ with $Z_{\overline{G}}(\overline{S})$ pseudo-reductive (Proposition 1.2.4) and $\mathscr{R}_{us,k}(\overline{P}) \cap Z_{\overline{G}}(\overline{S}) = 1$ (because this intersection is smooth and connected, by Proposition A.2.5), so $\overline{P} = Z_{\overline{G}}(\overline{S}) \ltimes \mathscr{R}_{us,k}(\overline{P})$. Hence, it suffices to show that a pair of opposite nontrivial $\overline{S}$-weights cannot both arise on the Lie algebra of $\mathscr{R}_{us,k}(\overline{P})$. Theorem C.2.3 applied to $\overline{P}$ implies $\overline{P} = P_{\overline{G}}(\overline{\mu}) = Z_{\overline{G}}(\overline{\mu}) \ltimes U_{\overline{G}}(\overline{\mu})$ for some $\overline{\mu} : \mathrm{GL}_1 \to \overline{S}$, so $\mathscr{R}_{us,k}(\overline{P}) = U_{\overline{G}}(\overline{\mu})$

and hence $\Phi(\overline{P}, \overline{S})$ is the set of $\overline{a} \in \Phi(\overline{G}, \overline{S})$ such that $\langle \overline{a}, \overline{\mu} \rangle > 0$. Thus, no such opposite nontrivial weights exist.

Since $\Phi(\overline{P}, \overline{S})$ and $-\Phi(\overline{P}, \overline{S})$ are disjoint, and $Z_{\overline{G}}(\overline{S})$ is contained in $\overline{P}$ (Proposition C.2.4), the intersection of $\Phi(\overline{P}_{k_s}, \overline{T}_{k_s})$ and $-\Phi(\overline{P}_{k_s}, \overline{T}_{k_s})$ is the root system $\Phi(Z_{\overline{G}}(\overline{S})_{k_s}, \overline{T}_{k_s})$ of $Z_{\overline{G}}(\overline{S})_{k_s}$ with respect to $\overline{T}_{k_s}$. On the other hand, since the subset $\Phi(\overline{P}_{k_s}, \overline{T}_{k_s})$ of $\overline{\Phi}$ is parabolic, by Proposition 2.2.8 there is a positive system of roots $\overline{\Phi}^+$ of $\overline{\Phi}$ contained in $\Phi(\overline{P}_{k_s}, \overline{T}_{k_s})$ and a subset $\overline{\Delta}_0$ of the basis $\overline{\Delta}$ of $\overline{\Phi}^+$ such that $\Phi(\overline{P}_{k_s}, \overline{T}_{k_s})$ is the union of $\overline{\Phi}^+$ and the set $[\overline{\Delta}_0]$ of roots that are integral linear combinations of elements in $\overline{\Delta}_0$. Then the intersection $\Phi(\overline{P}_{k_s}, \overline{T}_{k_s}) \cap -\Phi(\overline{P}_{k_s}, \overline{T}_{k_s})$ is the subset $[\overline{\Delta}_0]$, and we conclude that the root system of $Z_{\overline{G}}(\overline{S})_{k_s}$ with respect to $\overline{T}_{k_s}$ is $[\overline{\Delta}_0]$ and $\overline{\Delta}_0$ is a basis of this root system. Therefore, $\overline{\Delta}_0$ is precisely the subset of $\overline{\Delta}$ consisting of positive simple roots whose restriction to $\overline{S}_{k_s}$ is trivial. Let ${}_k\overline{\Delta}$ be the subset of $\mathrm{X}(S)$ obtained from $(\overline{\Delta} - \overline{\Delta}_0)|_{\overline{S}_{k_s}}$ using the identification of $\mathrm{X}(S)$ with $\mathrm{X}(\overline{S}_{k_s})$.

For $w \in {}_kW = W(G,S)(k) = N(k)/Z(k)$ and a pseudo-parabolic k-subgroup Q of G containing S, define $w \circ Q := nQn^{-1}$ for a representative $n \in N(k)$ of w (the choice of which does not matter, since $Z \subseteq Q$ by Proposition C.2.4). This is a left action of ${}_kW$ on the set of such Q. Recall that ${}_k\overline{\Phi} = \Phi(\overline{G}, \overline{S})$.

Theorem C.2.15 *There is a canonical subset ${}_k\overline{\Phi}^\vee \subseteq \mathrm{X}_*(S)$ such that $({}_k\overline{\Phi}, \mathrm{X}(S), {}_k\overline{\Phi}^\vee, \mathrm{X}_*(S))$ is a root datum. A basis of the root system ${}_k\overline{\Phi}$ (in its* **Q***-span inside $\mathrm{X}(S)_{\mathbf{Q}}$) is given by ${}_k\overline{\Delta}$, and the natural action of ${}_kW$ on $\mathrm{X}(S)_{\mathbf{Q}}$ identifies ${}_kW$ with the Weyl group of this root datum.*

If G is quasi-reductive then: the root system ${}_k\overline{\Phi}$ is the set of nontrivial S-weights on Lie(G), *$S' := (S \cap \mathscr{D}(G))^0_{\mathrm{red}}$ is a maximal k-split torus of $\mathscr{D}(G)$, S is an almost direct product of S' and the maximal central split k-torus S_0 of G, ${}_k\overline{\Phi}^\vee \subseteq \mathrm{X}_*(S')$, and the* **Q***-span of ${}_k\overline{\Phi}$ inside $\mathrm{X}(S)_{\mathbf{Q}} = \mathrm{X}(S')_{\mathbf{Q}} \oplus \mathrm{X}(S_0)_{\mathbf{Q}}$ is $\mathrm{X}(S')_{\mathbf{Q}}$. In particular, if G is quasi-reductive then the root system ${}_k\overline{\Phi}$ has rank equal to* $\dim S'$.

The assignment $Q \mapsto {}_k\overline{\Phi}_Q := \Phi(Q/\mathscr{R}_{u,k}(G), \overline{S})$ is a bijection from the set of pseudo-parabolic k-subgroups of G containing S onto the set of parabolic subsets of ${}_k\overline{\Phi}$, with $Q \subseteq Q'$ if and only if ${}_k\overline{\Phi}_Q \subseteq {}_k\overline{\Phi}_{Q'}$. The positive system of roots ${}_k\overline{\Phi}_P$ has basis ${}_k\overline{\Delta}$.

This generalizes Proposition 3.3.15(1),(2) and Proposition 3.5.1(1),(3) (and by Lemma 3.2.4 the bijection $a \mapsto a^\vee$ satisfying root datum axioms is uniquely determined). It also implies that the natural action of ${}_kW$ on the set of minimal pseudo-parabolic subgroups of G containing S is simply transitive, since $W({}_k\overline{\Phi})$ acts simply transitively on the set of minimal parabolic subsets of ${}_k\overline{\Phi}$ by [Bou2, VI, §1.5, Thm. 2(i); §1.7, Prop. 20, Cor. 1]. (A pseudo-parabolic

k-subgroup $P \subseteq G$ minimal among those containing S is also minimal as a pseudo-parabolic k-subgroup of G without reference to S, since all maximal k-split k-tori in P are $P(k)$-conjugate, by Theorem C.2.3.) Note also that $\mathscr{R}_{us,k}(G) = 1$ if and only if $\mathscr{R}_{u,k}(G)$ is k-wound (Corollary B.3.5), and in such cases $\mathscr{R}_{u,k}(G) \subseteq Z_G(S)$ (Proposition B.4.4); the standard semidirect product $G = \mathbf{G}_a \rtimes \mathrm{GL}_1$ shows that if $\mathscr{R}_{us,k}(G) \neq 1$ then $(S \cap \mathscr{D}(G))^0_{\mathrm{red}}$ and the maximal central split k-torus can be trivial even when $S \neq 1$.

Proof of Theorem C.2.15 We first treat the quasi-reductive case, and then the general case.

Step 1. Assume that G is quasi-reductive, so the k-torus S centralizes the k-wound $\mathscr{R}_{u,k}(G)$ (Proposition B.4.4). Thus, the set ${}_k\overline{\Phi}$ of nonzero S-weights on $\mathrm{Lie}(G/\mathscr{R}_{u,k}(G))$ coincides with the set of nonzero S-weights on $\mathrm{Lie}(G)$. To simplify the notation, we will denote ${}_k\overline{\Phi}$ by ${}_k\Phi$ when G is quasi-reductive.

Let S_0 be the maximal split central k-torus in G (or equivalently, the maximal k-subtorus of S that is central in G). By applying Remark C.2.12(2) to a maximal k-torus of G containing S, we see that $S' := (S \cap \mathscr{D}(G))^0_{\mathrm{red}}$ is a maximal k-split torus of $\mathscr{D}(G)$ and S is an almost direct product of S_0 and S', so $S' \to S/S_0$ is an isogeny and $\dim(S/S_0)$ is equal to the k-rank of $\mathscr{D}(G)$.

Since S_0 is central in G, all S-weights on $\mathrm{Lie}(G)$ are trivial on S_0 and so they can and will be viewed as characters on S'. Consider the subtorus $Z = (\bigcap_{a \in {}_k\Phi} \ker a)^0_{\mathrm{red}}$ of S'. By Lie algebra considerations, $\mathscr{D}(G)$ is generated by $Z_{\mathscr{D}(G)}(S')$ and the k-root groups $\{U_a\}_{a \in {}_k\Phi}$, so Z is central in $\mathscr{D}(G)$. Hence, $Z_{\overline{k}}$ maps isomorphically onto a central k-torus in the semisimple derived group of the maximal reductive quotient $G^{\mathrm{red}}_{\overline{k}}$ of $G_{\overline{k}}$, so $Z = 1$. This implies that ${}_k\Phi$ spans $\mathrm{X}(S')_{\mathbf{Q}}$.

Continuing to assume that G is quasi-reductive, if we apply the preceding conclusions for G to $\overline{G}$ as well, using $\overline{S}' := (\overline{S} \cap \mathscr{D}(\overline{G}))^0_{\mathrm{red}}$ and the maximal central split k-torus $\overline{S}_0$ of $\overline{G}$, the fact that the natural map $S \to \overline{S}$ is an isomorphism implies that the homomorphisms $S' \to \overline{S}'$ and $S_0 \to \overline{S}_0$ are isomorphisms.

Step 2. We now analyze the root datum properties for any G, and then pass to the pseudo-reductive quotient to verify the additional assertions. For $a \in {}_k\overline{\Phi}$, the element $r_a \in N_{G_a}(S)(k)/Z_{G_a}(S)(k)$ as in the discussion following Lemma C.2.14 induces an involution of S that is given by conjugation against an element $n_a \in N_{G_a}(S)(k)$. The element n_a centralizes the k-subgroup $\mathscr{S}_a := \ker(a) \subseteq S$ due to the definition of G_a, so the endomorphism $S \to S$ defined by $t \mapsto t/r_a(t)$ kills $\ker a$. By identifying $S/(\ker a)$ with GL_1 via a, we see that there is a unique k-homomorphism $a^\vee : \mathrm{GL}_1 \to S$ (the *coroot* associated to a) such that $t/r_a(t) = a^\vee(a(t))$ for all points t of S, and $a^\vee$ is nontrivial. Clearly $r_a(t) = t/a^\vee(a(t))$, but $a(r_a(t)) = a(t)^{-1}$ (since inversion is the only nontrivial automorphism of $S/(\ker a) = \mathrm{GL}_1$), so $a(t)/a(t)^{\langle a,a^\vee\rangle} = a(t)^{-1}$ for all $t \in S$. This proves that $\langle a, a^\vee\rangle = 2$ for all a, and the reflection $x \mapsto x - \langle x, a^\vee\rangle a$ on $\mathrm{X}(S) = \mathrm{X}(\overline{S})$ induced by each r_a preserves ${}_k\overline{\Phi}$ (since the same holds for the action induced by any $w \in {}_kW$). In particular, $r_a(b) = b - \langle b, a^\vee\rangle a$ for all $a, b \in {}_k\overline{\Phi}$, with $\langle b, a^\vee\rangle \in \mathbf{Z}$, so ${}_k\overline{\Phi}$ is a root system (in its $\mathbf{Q}$-span in $\mathrm{X}(S)_{\mathbf{Q}}$) and $W({}_k\overline{\Phi}) \subseteq {}_kW$ inside $\mathrm{Aut}(\mathrm{X}(S))$ ($W({}_k\overline{\Phi})$ is the subgroup generated by the reflections r_a).

Define ${}_k\overline{\Phi}^\vee$ to be the set of (nontrivial) cocharacters of S having the form $a^\vee$ for $a \in {}_k\overline{\Phi}$. The unique characterization of the reflection r_a implies that $r_{w(a)} = w \circ r_a \circ w^{-1}$ as endomorphisms of S for all $w \in {}_kW$, so the unique characterization of coroots implies $w(a^\vee) = w(a)^\vee$ in $\mathrm{X}_*(S)$. In particular, the reflection $r_a^\vee$ of $\mathrm{X}_*(S)$ dual to r_a on $\mathrm{X}(S)$ carries $b^\vee$ to $r_a(b)^\vee$ for all $b \in {}_k\overline{\Phi}$, so each $r_a^\vee$ preserves ${}_k\overline{\Phi}^\vee$. To conclude that we have a root datum, it remains to show that the assignment $a \mapsto a^\vee$ is injective. This will be established in Step 4 below via calculations in the pseudo-reductive case.

Step 3. By Lemma 3.2.1 the natural maps $W(G,T) \to W(\overline{G},\overline{T})$ and $W(G,S) \to W(\overline{G},\overline{S})$ are isomorphisms, as is $W(G_a,S) \to W(\overline{G}_a,\overline{S})$. Also, essentially by definition, the identification $\mathrm{X}(S) = \mathrm{X}(\overline{S})$ carries ${}_k\Phi$ onto $\Phi(\overline{G},\overline{S})$ and moreover the identification $\mathrm{X}_*(S) = \mathrm{X}_*(\overline{S})$ clearly respects the formation of coroots. We have also seen that if G is quasi-reductive then $S' \to \overline{S}'$ and $S_0 \to \overline{S}_0$ are isomorphisms. In view of the bijective correspondence in Proposition 2.2.10 between the sets of pseudo-parabolic k-subgroups of G and of $\overline{G}$ (respecting containment of T and $\overline{T}$), and the fact that a pseudo-parabolic k-subgroup containing S necessarily contains $Z_G(S) \supseteq T$, we may replace (G,P,T,S) with $(\overline{G},\overline{P},\overline{T},\overline{S})$ to reduce to the case of pseudo-reductive G. With G now pseudo-reductive, we shall write Δ, Δ_0, Φ, Φ^+, ${}_k\Delta$, ${}_k\Phi$, and ${}_k\Phi_Q$ rather than $\overline{\Delta}$, $\overline{\Delta}_0$, $\overline{\Phi}$, $\overline{\Phi}^+$, ${}_k\overline{\Delta}$, ${}_k\overline{\Phi}$, and ${}_k\overline{\Phi}_Q$ respectively.

We next reduce to using $G' = \mathscr{D}(G)$ in place of G. By Lemma 1.2.5(iii), the maximal k-torus T is an almost direct product of the maximal central k-torus T_0 in G and the maximal k-torus $T' = T \cap G'$ in G'. Semisimplicity of the isogeny category of k-tori then implies that the k-subtorus S is an almost direct product of the maximal k-split central k-torus S_0 and the k-torus $S' = (S \cap G')^0_{\mathrm{red}}$ that is necessarily a maximal k-split k-torus in G'. The commutativity of G/G' implies that the root spaces in $\mathrm{Lie}(G)_{k_s}$ for the action of T_{k_s} lie in $\mathrm{Lie}(G')_{k_s}$. The finite-index inclusion $\mathrm{X}(T_{k_s}) \hookrightarrow \mathrm{X}(T'_{k_s}) \oplus \mathrm{X}(T_{0k_s})$ carries $\Phi := \Phi(G_{k_s}, T_{k_s})$ into $\mathrm{X}(T'_{k_s})$ with image $\Phi' := \Phi(G'_{k_s}, T'_{k_s})$, and it is equivariant with respect to the natural homomorphism $W(G'_{k_s}, T'_{k_s}) \to W(G_{k_s}, T_{k_s})$. In particular, there is a natural correspondence between parabolic subsets of Φ and Φ', so Δ corresponds to a basis Δ' for a positive system of roots in Φ' (carrying Δ_0 to the analogous Δ'_0). Also, if $\alpha \in \Phi$ restricts to $\alpha' \in \Phi'$ then the natural map $W((G'_{k_s})_{\alpha'}, T'_{k_s}) \to W((G_{k_s})_\alpha, T_{k_s})$ between groups of order 2 is injective and hence an equality, so r_α restricts to $r_{\alpha'}$.

By Proposition 1.2.6 we have $G = Z_G(T)\cdot G'$, so $G = Z_G(S)\cdot G'$. Therefore, $N_G(S) = Z_G(S)\cdot N_{G'}(S)$. As $S' = (S\cap G')^0_{\mathrm{red}}$ and S_0 is central, we see that $N_G(S) = N_G(S')$ and $Z_G(S) = Z_G(S')$. Hence, $N_G(S) = Z_G(S)\cdot N_{G'}(S) = Z_G(S)\cdot N_{G'}(S')$, so the natural homomorphism

$$W(G',S') = N_{G'}(S')/Z_{G'}(S') \longrightarrow N_G(S)/Z_G(S) = W(G,S)$$

is an isomorphism. Replacing T_{k_s} with S in the argument in the preceding paragraph, we see that the inclusion $\mathrm{X}(S) \hookrightarrow \mathrm{X}(S')\oplus \mathrm{X}(S_0)$ of finite index carries ${}_k\Phi$ onto ${}_k\Phi'$ (and ${}_k\Delta$ onto ${}_k\Delta'$) equivariantly with respect to the natural isomorphism $W(G',S') \to W(G,S)$, and if $a \in {}_k\Phi$ restricts to $a' \in {}_k\Phi'$ then the reflection $r_a \in {}_kW$ restricts to the reflection $r_{a'} \in {}_kW'$ on $\mathrm{X}(S')$ and restricts to the identity on $\mathrm{X}(S_0)$, so (by uniqueness) $a^\vee$ is the composition of $a'^\vee$ with the inclusion of S' into S. Any pseudo-parabolic k-subgroup of G containing the almost direct product $S = S'\cdot S_0$ must contain $Z = Z_G(S)$ (Proposition C.2.4) and has the form $P_G(\lambda)$ for some $\lambda \in \mathrm{X}_*(S')$ (due to Remark 2.1.7 and Theorem C.2.3). Thus, the equality $G = Z\cdot G'$ (which follows from Proposition 1.2.6) implies that $Q \mapsto Q\cap G'$ and $Q' \mapsto ZQ'$ are inverse bijections between the set of pseudo-parabolic k-subgroups of G containing S and the set of pseudo-parabolic k-subgroups of G' containing S'. We may now replace (G,S,T,P,Δ) with $(G',S',T',P\cap G',\Delta')$, so G is perfect (Proposition 1.2.6).

Step 4. Since G is perfect, $G^{\mathrm{red}}_{\overline{k}}$ is semisimple, so there is no nontrivial central torus in $G_{\overline{k}}$. Thus, Φ spans $\mathrm{X}(T_{k_s})_{\mathbf{Q}}$ and ${}_k\Phi$ spans $\mathrm{X}(S)_{\mathbf{Q}}$ (by Corollary A.8.11, applied to a torus in the intersection of the kernels of all elements of Φ, or of ${}_k\Phi$). Thus, for each $a \in {}_k\Phi$ the reflection r_a is uniquely determined by the property that it negates $a \neq 0$ and preserves ${}_k\Phi$ (see [Bou2, VI, §1.1, Lemma 1]). Since $(\mathrm{X}(S)_{\mathbf{Q}}, {}_k\Phi)$ is a root system, it follows from [Bou2, VI, §1.1, Lemma 2] that the equality $a^\vee = b^\vee$ for roots a and b forces $a = b$. Hence, $a \mapsto a^\vee$ is injective, so the root datum axioms are satisfied.

Step 5. Now we turn to the analysis of ${}_k\Delta$ and ${}_kW$. Every nontrivial S-weight on $\mathrm{Lie}(G)$ is the restriction of a nontrivial T_{k_s}-weight on $\mathrm{Lie}(G)_{k_s}$, and each $\beta \in \Phi$ has the form $\sum_{\alpha\in\Delta} n_\alpha \alpha$ with integers n_α that are either all $\geqslant 0$ or all $\leqslant 0$. Hence, each $b \in {}_k\Phi$ has the form $\sum_{a\in {}_k\Delta} n_a a$ with *integers* n_a that are either all $\geqslant 0$ or all $\leqslant 0$. In particular, ${}_k\Delta$ spans $\mathrm{X}(S)_{\mathbf{Q}}$ (so $\#_k\Delta \geqslant \dim S$) and if ${}_k\Delta$ is linearly independent then its elements must be non-divisible in ${}_k\Phi$. Thus, linear independence of ${}_k\Delta$ would imply that it is a basis for a positive system of roots in ${}_k\Phi$ (by [Bou2, VI, §1.7, Cor. 3]), so the subgroup $W({}_k\Phi)$ of ${}_kW$ would be generated by the reflections r_a for $a \in {}_k\Delta$. It remains to prove: $\#_k\Delta = \dim S$, the inclusion $W({}_k\Phi) \subseteq {}_kW$ is an equality, $Q \mapsto \Phi(Q,S)$ is a bijection from the set of pseudo-parabolic k-subgroups of G containing S to the set of parabolic subsets of ${}_k\Phi$, and $Q \subseteq Q'$ if $\Phi(Q, S) \subseteq \Phi(Q', S)$.

To determine $\#_k\Delta$, we first describe the "$*$-action" of $\Gamma := \mathrm{Gal}(k_s/k)$ on Δ. For $\gamma \in \Gamma$, the subset $\gamma(\Phi^+)$ is a positive system of roots in Φ, so there is an element $w_\gamma \in W(k_s)$ such that $w_\gamma(\gamma(\Phi^+)) = \Phi^+$. (Note that since the Weyl group $W(k_s)$ acts simply transitively on the set of positive systems of roots in Φ, w_γ is unique.) Then, clearly, $w_\gamma(\gamma(\Delta)) = \Delta$. The correspondence $\alpha \mapsto \gamma * \alpha := w_\gamma(\gamma(\alpha))$ defines a continuous action of Γ on Δ which is called the $*$-*action*.

In the sequel, for simplicity, we will denote $\mathscr{R}_{us,k}(P)$ by U and denote $Z_G(S)$ by Z. Since P is minimal, Proposition C.2.4 gives that $P = Z \ltimes U$. We will now show that for every $\gamma \in \Gamma$, there exists a representative n_γ of w_γ in $N_Z(T)(k_s)$. Since $T \subseteq P$, clearly P_{k_s} is a pseudo-parabolic k_s-subgroup of G_{k_s} containing T_{k_s}. The pseudo-parabolic k_s-subgroups of P_{k_s} are precisely the pseudo-parabolic k_s-subgroups of G_{k_s} which are contained in P_{k_s} (Proposition 3.5.8), and each of them contains U_{k_s}. The quotient map $P_{k_s} \to P_{k_s}/U_{k_s} \simeq Z_{k_s}$ induces a bijective correspondence between the set of pseudo-parabolic k_s-subgroups of P_{k_s} and the set of pseudo-parabolic k_s-subgroups of Z_{k_s} (Proposition 2.2.10). Under this correspondence, the minimal pseudo-parabolic k_s-subgroups of P_{k_s} containing T_{k_s} correspond to the minimal pseudo-parabolic k_s-subgroups of Z_{k_s} containing T_{k_s}.

On the other hand, there is a natural bijective correspondence between the set of minimal pseudo-parabolic k_s-subgroups of Z_{k_s} containing T_{k_s} and the set of positive systems of roots in the root system $\Phi(Z_{k_s}, T_{k_s})$ of Z_{k_s}, and the group $N_Z(T)(k_s)$ acts transitively on both of these sets (Proposition 3.3.15). There is a similar bijective correspondence between the set of minimal pseudo-parabolic k_s-subgroups of P_{k_s} containing T_{k_s} and the set of positive systems of roots in the root system Φ which are contained in the parabolic set $\Phi_P := \Phi(P_{k_s}, T_{k_s}) = \Phi^+ \cup [\Delta_0]$ (Propositions 3.5.1(3) and 3.3.15(1)). We conclude that $N_Z(T)(k_s)$ ($\subseteq N_G(T)(k_s)$) acts transitively on the set of positive systems of roots in Φ which are contained in Φ_P. The parabolic set Φ_P is stable under Γ since P is defined over k. Hence, for $\gamma \in \Gamma$, $\gamma(\Phi^+)$ is a positive system of roots contained in Φ_P, so there is an $n_\gamma \in N_Z(T)(k_s)$ that carries $\gamma(\Phi^+)$ onto Φ^+. From the uniqueness of w_γ, we see that n_γ represents it.

The subset Δ_0 can be described as the set of simple roots α such that $-\alpha$ lies in Φ_P. From this description it is obvious that Δ_0 is stable under the $*$-action. As n_γ lies in $N_Z(T)(k_s)$, and Δ_0 is a basis of the root system $\Phi(Z_{k_s}, T_{k_s})$, w_γ lies in the group generated by the reflections r_α for $\alpha \in \Delta_0$. Hence, for any $\alpha \in \Delta$ and $\gamma \in \Gamma$ the difference $\gamma * \alpha - \gamma(\alpha) = w_\gamma(\gamma(\alpha)) - \gamma(\alpha)$ belongs to the integral span of Δ_0. This implies that the restrictions of $\gamma * \alpha$ and $\gamma(\alpha)$ to S_{k_s} are equal. But the restriction of $\gamma(\alpha)$ to S_{k_s} coincides with the restriction of α. Thus all roots belonging to a Γ-orbit in Δ under the $*$-action restrict to the same element in $X(S_{k_s}) = X(S)$. We will prove that the number of Γ-orbits in $\Delta - \Delta_0$ under the $*$-action is $\dim S$, so the inequality $\#_k\Delta \geqslant \dim S$ is an equality. It will then follow that ${}_k\Delta$ is a basis of the root system ${}_k\Phi$ and that a pair of elements of $\Delta - \Delta_0$ have the same restriction to S_{k_s} if and only if they lie in the same Γ-orbit under the $*$-action.

To prove that the number of Γ-orbits in $\Delta - \Delta_0$ under the $*$-action is $\dim S$, we will relate it to the set of pseudo-parabolic k-subgroups of G containing the minimal pseudo-parabolic k-subgroup P. This set can be parameterized in two different ways. First, we have:

Lemma C.2.16 *The map $Q \mapsto \Phi(Q, S)$ is a bijection from the set of pseudo-parabolic k-subgroups of G containing S onto the set of parabolic subsets of ${}_k\Phi$. For any such Q and Q', $Q \subseteq Q'$ if and only if $\Phi(Q, S) \subseteq \Phi(Q', S)$. In particular, $\Phi(P, S)$ is a positive system of roots in ${}_k\Phi$ and the set of Q that contain P is parameterized by the set of subsets of the basis of $\Phi(P, S)$.*

Proof By Theorem C.2.3, any such Q has the form $Q = P_G(\lambda)$ for some $\lambda \in X_*(S)$, so $\Phi(Q, S) = ({}_k\Phi)_{\lambda \geqslant 0}$. Thus, $\Phi(Q, S)$ is a parabolic subset of ${}_k\Phi$ and every parabolic subset of ${}_k\Phi$ arises from some Q (by Proposition 2.2.8). To prove that $Q \subseteq Q'$ if $\Phi(Q, S) \subseteq \Phi(Q', S)$ (the converse being obvious), fix a maximal k-torus T of G containing S. Since Q and Q' contain $Z_G(S)$ (Proposition C.2.4), they contain T. Thus, $Q_{k_s} \subseteq Q'_{k_s}$ if and only if $\Phi(Q_{k_s}, T_{k_s}) \subseteq \Phi(Q'_{k_s}, T_{k_s})$ inside Φ (Proposition 3.5.1(3)). As $Q = P_G(\lambda)$, $\Phi(Q_{k_s}, T_{k_s}) = \Phi_{\lambda_{k_s} \geqslant 0}$. Since λ_{k_s} is valued in S_{k_s}, for any $\alpha \in \Phi$ with restriction $a = \alpha|_{S_{k_s}} \in X(S_{k_s}) = X(S)$ we have $\langle \alpha, \lambda_{k_s} \rangle = \langle a, \lambda \rangle$. Hence, $\Phi(Q_{k_s}, T_{k_s})$ is the set of nonzero T_{k_s}-weights on $\mathrm{Lie}(Q_{k_s})$ contained in the preimage of $\{0\} \cup ({}_k\Phi)_{\lambda \geqslant 0} = \{0\} \cup \Phi(Q, S)$ under the restriction map $X(T_{k_s}) \to X(S_{k_s}) = X(S)$. Thus, if $\Phi(Q, S) \subseteq \Phi(Q', S)$ then $Q_{k_s} \subseteq Q'_{k_s}$, so $Q \subseteq Q'$.

To prove the last assertion, let ${}_k\Phi^+ = \Phi(P, S)$ be the positive system of roots in ${}_k\Phi$ determined by the minimal pseudo-parabolic k-subgroup P. The description of parabolic sets containing ${}_k\Phi^+$ given by Proposition 2.2.8(2) implies that the set of such parabolic sets is in bijective correspondence with the set of subsets of the basis of ${}_k\Phi^+$. □

On the other hand, given any pseudo-parabolic k-subgroup Q of G containing P (and hence T), the set $\Phi_Q := \Phi(Q_{k_s}, T_{k_s})$ is a parabolic subset of Φ such that: (i) Φ_Q contains Φ_P, and (ii) Φ_Q is stable under the natural action of Γ on $X(T_{k_s})$. By Proposition 3.5.2, $Q \mapsto \Phi_Q$ is a bijection from the set of such Q onto the set of parabolic subsets of Φ satisfying (i) and (ii). A parabolic subset of Φ containing $\Phi_P = \Phi^+ \cup [\Delta_0]$ is of the form $\Phi^+ \cup [\Delta_0 \cup \Delta']$, for a unique subset Δ' of $\Delta - \Delta_0$. Such a parabolic subset is stable under the action of w_γ, for every $\gamma \in \Gamma$, since n_γ lies in $N_Z(T)(k_s)$ $(\subseteq Z(k_s) \subseteq P(k_s))$. Hence,

$\Phi^+ \cup [\Delta_0 \cup \Delta']$ is stable under the natural action of Γ on $X(T_{k_s})$ if and only if for every $\gamma \in \Gamma$, $w_\gamma(\gamma(\Phi^+ \cup [\Delta_0 \cup \Delta'])) = \Phi^+ \cup [\Delta_0 \cup \Delta']$. But from the definition of the $*$-action it is clear that $w_\gamma(\gamma(\Phi^+ \cup [\Delta_0 \cup \Delta'])) = \Phi^+ \cup [\Delta_0 \cup \gamma * \Delta']$. Therefore, $\Phi^+ \cup [\Delta_0 \cup \Delta']$ is stable under the natural action of Γ on $X(T_{k_s})$ if and only if Δ' is stable under the $*$-action of Γ.

Putting everything together, the set of pseudo-parabolic k-subgroups of G containing P is parameterized in two ways: by the set of subsets of the basis of ${}_k\Phi^+ = \Phi(P, S)$ (there are $2^{\dim S}$ such subsets), and by the set of subsets of $\Delta - \Delta_0$ which are stable under the $*$-action of Γ. If r denotes the number of Γ-orbits in $\Delta - \Delta_0$ (under the $*$-action), then the number of Γ-stable subsets is clearly 2^r. Thus, $2^r = 2^{\dim S}$, so $r = \dim S$, and hence, $\dim S = \#_k\Delta$.

STEP 6. Since ${}_k\Delta$ is the basis of a positive system of roots in ${}_k\Phi$, the subset

$$\overline{C} = \{\lambda \in X_*(S)_{\mathbf{R}} \mid \langle a, \lambda\rangle \geqslant 0 \text{ for all } a \in {}_k\Delta\}$$

is a fundamental domain for the action of $W({}_k\Phi)$ on $X_*(S)_{\mathbf{R}}$ [Bou2, V, §3.3, Thm. 2].

For $\lambda \in X_*(S) \cap \overline{C}$, we claim that the pseudo-parabolic k-subgroup $P_G(\lambda)$ contains P. By Propositions 2.2.10 and 3.5.1(3) it is equivalent to show that every $\beta \in \Phi(P_{k_s}, T_{k_s}) = \Phi^+ \cup [\Delta_0]$ satisfies $\langle \beta, \lambda_{k_s}\rangle \geqslant 0$. Since $\langle \beta, \lambda_{k_s}\rangle$ only depends on the restriction of β to S_{k_s}, and more specifically this pairing vanishes for $\beta \in [\Delta_0]$, we may and do restrict our attention to $\beta \in \Phi^+$, and even to $\beta \in \Delta - \Delta_0$. Thus, it suffices to prove $\langle a, \lambda\rangle \geqslant 0$ for all $a \in {}_k\Delta$. This in turn is immediate from the condition $\lambda \in \overline{C}$.

Any pseudo-parabolic k-subgroup Q of G containing S is of the form $Q = P_G(\mu)$ for some $\mu \in X_*(S)$ (Theorem C.2.3), and since μ can be carried into $\overline{C}$ by the action of some element of $W({}_k\Phi)$ we see that: (i) some member Q' of the $W({}_k\Phi)$-orbit of Q contains P and (ii) if Q is a minimal pseudo-parabolic k-subgroup then $Q' = P$. This proves that $W({}_k\Phi)$ acts transitively on the set $\mathscr{P}$ of minimal pseudo-parabolic k-subgroups of G containing S. Since ${}_kW$ acts simply transitively on $\mathscr{P}$ (see the proof of Proposition C.2.10), and $W({}_k\Phi) \subseteq {}_kW$, we conclude that $W({}_k\Phi) = {}_kW$. This completes the proof of Theorem C.2.15. □

In the setting of Theorem C.2.15 it is natural to compare the set $\Phi(G_{\overline{k}}^{\rm red}, S_{\overline{k}})$ of nonzero $S_{\overline{k}}$-weights on the Lie algebra of the maximal reductive quotient of $G_{\overline{k}}$ with the root system ${}_k\overline{\Phi}$ of nonzero S-weights on $\mathrm{Lie}(G/\mathscr{R}_{u,k}(G))$ (or equivalently the set of nonzero S-weights on $\mathrm{Lie}(G)$, when $\mathscr{R}_{us,k}(G) = 1$). Via the equality $X(S) = X(S_{\overline{k}})$, clearly $\Phi(G_{\overline{k}}^{\rm red}, S_{\overline{k}}) \subseteq {}_k\overline{\Phi}$.

Theorem C.2.17 (Borel–Tits) *The set $\Phi := \Phi(G_{\overline{k}}^{\rm red}, S_{\overline{k}}) \subseteq X(S_{\overline{k}})$ is a root system in its $\mathbf{Q}$-linear span, and the inclusion $\Phi \subseteq {}_k\overline{\Phi}$ is an equality if*

$\mathrm{char}(k) \neq 2$ or k is perfect. In general, the respective sets Φ_{nm} and ${}_k\overline{\Phi}_{\mathrm{nm}}$ of non-multipliable elements in Φ and ${}_k\overline{\Phi}$ coincide and

$$\Phi \subseteq {}_k\overline{\Phi} \subseteq \Phi_{\mathrm{nm}} \coprod \frac{1}{2}\cdot \Phi_{\mathrm{nm}},$$

so the natural map ${}_kW \to W(\Phi) = W(\Phi_{\mathrm{nm}})$ is an isomorphism.

This result is announced in [BoTi2, Thm. 3] (with some details given in [Ti3, §4]). In the special case that G is pseudo-reductive and pseudo-split (i.e., S is a maximal k-torus of G), the root system Φ is reduced and so this result recovers Theorem 2.3.10. In §9.8 we give many examples of the failure of Φ to exhaust ${}_k\overline{\Phi}$ in the pseudo-split absolutely pseudo-simple case over any imperfect field k of characteristic 2.

Proof We may and do assume that G is pseudo-reductive, so we write ${}_k\Phi$ rather than ${}_k\overline{\Phi}$. The equality of Φ and ${}_k\Phi$ for perfect k (e.g., $\mathrm{char}(k) = 0$) is obvious. Apart from the asserted equality of Φ and ${}_k\Phi$ when $\mathrm{char}(k) \neq 2$, it suffices to show that the non-multipliable elements a in the root system ${}_k\Phi$ lie in Φ. Indeed, once this is shown we see that Φ and ${}_k\Phi$ have the same $\mathbf{Q}$-span, and the reflections of this span arising from $W({}_k\Phi) = {}_kW = N_G(S)(k)/Z_G(S)(k)$ establish the root system axioms for Φ. Any $a \in \Phi_{\mathrm{nm}}$ must be non-multipliable in ${}_k\Phi$ (as otherwise $2a$ would be a non-multipliable element of ${}_k\Phi$, so it would lie in Φ, contrary to the non-multipliability of a in Φ).

Now choose a non-multipliable $a \in {}_k\Phi$, so ${}_k\Phi \cap \mathbf{Q}_{>0}\cdot a$ is equal to either $\{a\}$ or $\{a, a/2\}$. We shall prove that $a \in \Phi$. Since $\langle a\rangle \cap {}_k\Phi = \{a\}$, by Lemma 3.3.8 the k-group $U_{\langle a\rangle}(G)$ has Lie algebra that is the a-weight space $\mathrm{Lie}(G)_a$ in $\mathrm{Lie}(G)$. By Lemma 3.3.9 we have

$$U_{\langle a\rangle}(\mathscr{R}_u(G_{\overline{k}})) = (\mathscr{R}_u(G_{\overline{k}}) \cap U_{\langle a\rangle}(G)_{\overline{k}})^0,$$

so $\mathrm{Lie}(U_{\langle a\rangle}(\mathscr{R}_u(G_{\overline{k}}))) = \mathrm{Lie}(\mathscr{R}_u(G_{\overline{k}})) \cap \mathrm{Lie}(G_{\overline{k}})_a$. But if $a \notin \Phi$ then this intersection is $\mathrm{Lie}(G_{\overline{k}})_a$, so the containment $U_{\langle a\rangle}(\mathscr{R}_u(G_{\overline{k}})) \subseteq U_{\langle a\rangle}(G)_{\overline{k}}$ between smooth connected groups is an equality. This would force $U_{\langle a\rangle}(G) \subseteq \mathscr{R}_{u,k}(G)$ by Lemma 1.2.1, so it is trivial since G has been assumed to be pseudo-reductive, contradicting that $a \in {}_k\Phi$. Hence, $a \in \Phi$. In particular, it follows (as we have already explained) that Φ is a root system.

It remains to prove that if $\mathrm{char}(k) > 2$ and $a \in {}_k\Phi$ is non-multipliable but divisible then $a/2 \in \Phi$. Since $U_{\langle a/2\rangle}(G)$ has Lie algebra that is the span of the (nonzero) weight spaces for a and $a/2$ on $\mathrm{Lie}(G)$, to exhibit $a/2$ as an $S_{\overline{k}}$-weight on $\mathrm{Lie}(G_{\overline{k}}^{\mathrm{red}})$ it suffices to prove that

$$\mathrm{Lie}(\mathscr{R}_u(G_{\overline{k}})) \cap \mathrm{Lie}(U_{\langle a/2\rangle}(G)) = 0$$

inside $\mathrm{Lie}(G)_{\overline{k}}$. This intersection is an S-stable Lie subalgebra of $\mathrm{Lie}(U_{\langle a/2\rangle}(G))$, so it suffices to show that this Lie subalgebra has the form $\mathrm{Lie}(U)$ for a smooth connected k-subgroup $U \subseteq U_{\langle a/2\rangle}(G)$ such that $U_{\overline{k}} \subseteq \mathscr{R}_u(G_{\overline{k}})$ (as then $U \subseteq \mathscr{R}_{u,k}(G) = 1$). Note that the set of S-weights on $\mathrm{Lie}(U_{\langle a/2\rangle}(G))$ induced by the S-action on $U_{\langle a/2\rangle}(G)$ has the form $\{b, 2b\}$ for $b = a/2$, so we may conclude by applying the final part of the following proposition to $U_{\langle a/2\rangle}(G)$ and $U_{\langle a/2\rangle}(\mathscr{R}_u(G_{\overline{k}})) = (\mathscr{R}_u(G_{\overline{k}}) \cap U_{\langle a/2\rangle}(G)_{\overline{k}})^0$. □

Proposition C.2.18 *Let k be a field of characteristic $p > 0$ and S a split k-torus. Let $b \in \mathrm{X}(S)$ be a nontrivial character, and n a positive integer. Let $\mathscr{U}_S$ be the category of smooth connected unipotent k-groups U equipped with an S-action such that the set of S-weights on* $\mathrm{Lie}(U)$ *is contained in $\Lambda := \{nb, \dots, (np-1)b\}$, using S-equivariant homomorphisms as morphisms. Let $\mathscr{L}_S$ be the category of finite-dimensional Lie algebras over k equipped with an S-action whose set of weights is contained in Λ, using S-equivariant maps as morphisms.*

(1) *For each $\mathfrak{u}$ in $\mathscr{L}_S$ and $v_1, \dots, v_{p-1} \in \mathfrak{u}$,* $\mathrm{ad}(v_1) \circ \cdots \circ \mathrm{ad}(v_{p-1}) = 0$. *The S-equivariant k-group structure defined by the Baker–Campbell–Hausdorff law* $\mathrm{BCH}_{<p}$ *in degrees $< p$ on the affine space $\underline{\mathfrak{u}}$ covariantly associated to $\mathfrak{u}$ is in $\mathscr{U}_S$ and its Lie algebra is canonically S-equivariantly isomorphic to $\mathfrak{u}$.*
(2) *The functor $U \mapsto$* $\mathrm{Lie}(U)$ *from $\mathscr{U}_S$ to $\mathscr{L}_S$ is an equivalence. The unique isomorphism* $\underline{\mathrm{Lie}(U)} \simeq U$ *in $\mathscr{U}_S$ inducing the identity on Lie algebras is functorial in U.*
(3) *A map $f : U' \to U$ in $\mathscr{U}_S$ has trivial kernel if and only if* $\mathrm{Lie}(f)$ *is injective. For each U in $\mathscr{U}_S$ the S-stable smooth closed k-subgroups of U are connected, so $U' \mapsto$* $\mathrm{Lie}(U')$ *is a bijection from the set of S-stable smooth closed k-subgroups of U onto the set of S-stable Lie subalgebras of* $\mathrm{Lie}(U)$.

It makes sense to define a k-group structure in (1) via the $\mathrm{BCH}_{<p}$ law (see [Bou1, Ch. II, §6.4, Thm. 2], using the origin of $\mathfrak{u}$ as the identity and negation on $\mathfrak{u}$ as inversion) because the BCH law in characteristic 0 involves a denominator divisible by p only for terms involving a composition of at least $p-1$ operators of the form $\mathrm{ad}(v_i)$ (see [Bou1, Ch. II, §8.1, Prop. 1]) and such compositions vanish on $\mathfrak{u}$. Also, the requirement for the S-weights to lie in the fixed interval Λ cannot be relaxed too much, since $U^{(p)}$ with its natural S-action (through composition with the Frobenius isogeny $S \to S^{(p)}$ and the natural $S^{(p)}$-action on $U^{(p)}$) has weights lying in $p\Lambda$ and the Frobenius isogeny $U \to U^{(p)}$ is S-equivariant but vanishes under the Lie functor.

Proof To prove that $v := [v_1, [v_2, \ldots, [v_{p-1}, v_p] \ldots]]$ vanishes for all $v_1, \ldots, v_p \in \mathfrak{u}$, by multilinearity we may assume that each v_i is an S-eigenvector, say $t.v_i = b(t)^{m_i} v_i$ for all $t \in S$ and some $m_i \geqslant n$. The S-equivariance of the Lie bracket implies that $t.v = b(t)^{\sum m_i} v$. But $\sum m_i \geqslant np > np - 1$, so the S-eigenspace in which v lies must vanish. Now consider the affine space $\underline{\mathfrak{u}}$ equipped with the k-group law defined by the $\mathrm{BCH}_{<p}$ formula (using the origin as the identity). This is S-equivariant since the S-action on $\mathfrak{u}$ respects the Lie bracket. To finish the proof of (1) we have to show that the canonical S-equivariant identification of $\mathrm{Lie}(\underline{\mathfrak{u}})$ with $\mathfrak{u}$ as k-vector spaces recovers the given Lie algebra structure on $\mathfrak{u}$. If $p = 2$ then $[v_1, v_2] = 0$ and moreover $\underline{\mathfrak{u}}$ is commutative since $\mathrm{BCH}_{<2}$ is addition, so this case is clear. If $p > 2$ then the degree-2 part of the $\mathrm{BCH}_{<p}$ group law on $\underline{\mathfrak{u}}$ is $B(v, w) = (1/2)[v, w]$, so by Lemma A.7.4 the Lie bracket for U is

$$B(v, w) - B(w, v) = [v, w].$$

This completes the proof of (1).

The first assertion in (3) is reduced to showing that a finite étale S-stable subgroup of U must vanish, or more generally that all S-stable smooth closed k-subgroups of U are connected. We may assume $k = \overline{k}$, and by hypothesis $U = U(\lambda)$ for $\lambda : \mathrm{GL}_1 \to S$ such that $\langle b, \lambda \rangle > 0$, so for each $u \in U(k)$ the path $t \mapsto \lambda(t) u \lambda(t)^{-1}$ extended across $t = 0$ connects u to 1. Hence, U is connected. The second assertion in (3) is a formal consequence of (2), so it remains to prove (2). In view of the $\mathrm{BCH}_{<p}$ construction in (1), the only problem is to show that the Lie functor is fully faithful since the functoriality of the resulting unique isomorphism $\underline{\mathrm{Lie}(U)} \simeq U$ lifting the identity on Lie algebras is then immediate by the faithfulness of the Lie functor on the category $\mathscr{U}_S$.

For any U in $\mathscr{U}_S$ we claim that the p-operation $v \mapsto v^{[p]}$ on $\mathrm{Lie}(U)$ (as reviewed above Lemma A.7.13) must vanish. By (1), the algebraic map $v \mapsto \mathrm{ad}(v)^{p-1}$ vanishes. Thus, the general formula for the interaction of the p-operation with addition implies that $(v + w)^{[p]} = v^{[p]} + w^{[p]}$, so it suffices to show that $v^{[p]} = 0$ when v is an eigenvector for S. We have $t.v = b(t)^m v$ for $t \in S$ and some $m \in [n, np - 1]$. The S-action on $\mathrm{Lie}(U)$ respects the p-operation, so $t.v^{[p]} = (t.v)^{[p]} = b(t)^{mp} v^{[p]}$ with $mp \geqslant np$. Thus, $v^{[p]}$ lies in the S-eigenspace of $\mathrm{Lie}(U)$ for a character that is not an S-weight, so $v^{[p]} = 0$.

It follows from Proposition A.7.14 that for any U and U' in $\mathscr{U}_S$, every map $\mathrm{Lie}(U') \to \mathrm{Lie}(U)$ in $\mathscr{L}_S$ arises from a unique S-equivariant k-homomorphism $\ker F_{U'/k} \to \ker F_{U/k}$ between the Frobenius kernels. Hence, it suffices to show that every S-equivariant k-homomorphism $\ker F_{U'/k} \to \ker F_{U/k}$ uniquely extends to an S-equivariant k-homomorphism $U' \to U$. (Indeed, the functoriality of the isomorphism in (2) then holds because the equality of a pair of maps that it asserts can be checked after applying the Lie functor.)

Choose a set of elements $\{x_1, \ldots, x_d\}$ of the augmentation ideal I of $k[U]$ that are S-eigenvectors lifting an S-eigenbasis of $I/I^2 = \mathrm{Lie}(U)^*$. We claim that the k-algebra map $k[X_1, \ldots, X_d] \to k[U]$ defined by $X_i \mapsto x_i$ is an isomorphism. Injectivity holds because the induced map between the respective completions at the origin of affine n-space and the identity point of U is an isomorphism. For surjectivity, by complete reducibility of the S-action on I we just have to show that each S-eigenspace in I is spanned by monomials in the x_i. Each S-weight on I occurs on some I^j/I^{j+1}, and the hypothesis concerning the S-weights that occur on $\mathrm{Lie}(U)$ shows that the S-weights on the quotient I^j/I^{j+1} of $(I/I^2)^{\otimes j}$ are multiples of b against integers that tend to infinity with j. Thus, every S-eigenspace in I is finite-dimensional and so maps isomorphically onto the S-eigenspace for the same weight in I/I^j for some large j. Since I/I^j is spanned by monomials in the x_i, all of which are S-eigenvectors, the desired surjectivity is proved.

We transport the S-action and Hopf algebra structure from $k[U]$ to $k[X_1, \ldots, X_d]$, so for each i and all $t \in S$ we have $t.X_i = b(t)^{m_i} X_i$ for an integer m_i between m and mp for some nonzero integer m, with $m_i \neq mp$ for all i. Each monomial in the X_j is an S-eigenvector. For any integer μ between m and mp with $\mu \neq mp$, we claim that the S-eigenspace in $k[X_1, \ldots, X_d]$ for the character $\chi : t \mapsto b(t)^{\mu}$ *injects* into the χ-eigenspace of $k[X_1, \ldots, X_d]/(X_1^p, \ldots, X_n^p)$. Indeed, any nonzero S-eigenvector in $(X_1^p, \ldots, X_n^p)$ must have S-weight outside the given interval of possibilities for μ (due to the contraints on the m_i), so there is no such eigenvector.

Choose analogous S-eigenvectors $\{x'_1, \ldots, x'_{d'}\}$ in the augmentation ideal I' of $k[U']$, so the k-algebra map $k[X'_1, \ldots, X'_{d'}] \to k[U']$ defined by $X'_j \mapsto x'_j$ is also an isomorphism. The hypothesis that Lie algebra weights lie in the fixed interval Λ for both U and U' implies that the S-eigenspaces in $k[X'_1, \ldots, X'_{d'}]$ for the eigencharacter χ_i of any X_i injects into the χ_i-eigenspace of $k[X'_1, \ldots, X'_{d'}]/(X_1'^p, \ldots, X_{d'}'^p)$. An S-equivariant map of pointed k-schemes $U' \to U$ corresponds to an S-equivariant map $f : k[X_1, \ldots, X_d] \to k[X'_1, \ldots, X'_{d'}]$ respecting the augmentation ideals, and by eigenspace considerations we see that f is uniquely determined by the induced map

$$\bar{f} : k[X_1, \ldots, X_d]/(X_1^p, \ldots, X_d^p) \to k[X'_1, \ldots, X'_{d'}]/(X_1'^p, \ldots, X_{d'}'^p).$$

We conclude that every S-equivariant pointed map $\ker F_{U'/k} \to \ker F_{U/k}$ uniquely extends to an S-equivariant pointed map $U' \to U$. So far we have not used the group law structure, just the pointed k-scheme structure and the S-weight hypothesis on the tangent space at the marked point. Since $U' \times U'$ satisfies the same hypotheses as a pointed k-scheme (ignoring any k-group structure on it), if $h : U' \to U$ is a pointed S-equivariant map then to check the equality of the pointed S-equivariant maps

$$h \circ m_{U'}, m_U \circ (h \times h) : U' \times U' \rightrightarrows U$$

it is equivalent to do so on the Frobenius kernels at the identity points of each side. This says exactly that h is a k-homomorphism if and only if its restriction between Frobenius kernels is a k-homomorphism. □

Remark C.2.19 The faithfulness part of (2) above can be generalized as follows. Let S be a split torus over a field k of characteristic $p > 0$, and let $f_1, f_2 : G \rightrightarrows G'$ be *distinct* S-equivariant k-homomorphisms between connected k-group schemes of finite type such that $\mathrm{Lie}(f_1) = \mathrm{Lie}(f_2)$. Then we claim that there exists a (possibly trivial) S-weight a on $\mathrm{Lie}(G)$ and an integer $n > 0$ such that $p^n a$ is an S-weight on $\mathrm{Lie}(G')$. In particular, if no S-weight on $\mathrm{Lie}(G')$ has the form $p^n a$ for an S-weight a on $\mathrm{Lie}(G)$ and an integer $n > .0$ then every S-equivariant k-homomorphism $f : G \to G'$ is uniquely determined by $\mathrm{Lie}(f)$.

To prove this claim, let $G(n) = \ker F_{G/k,n}$ be the kernel of the n-fold relative Frobenius morphism $F_{G/k,n} : G \to G^{(p^n)}$ for integers $n > 0$. By Proposition A.7.14 and the assumed equality $\mathrm{Lie}(f_1) = \mathrm{Lie}(f_2)$, f_1 and f_2 coincide on $G(1)$. By connectedess and translation considerations over $\overline{k}$, the maps induced by f_1 and f_2 between formal groups (i.e., completions at the identity) must be distinct. Thus, there exists an integer $n > 0$ such $f_1|_{G(n)} = f_2|_{G(n)}$ but $f_1|_{G(n+1)} \neq f_2|_{G(n+1)}$. We will construct an S-equivariant morphism of pointed k-schemes $c : G(n+1)/G(n) \to G'$ such that the induced map between tangent spaces at the identity is nonzero. Then some S-weight on $\mathrm{Lie}(G')$ occurs on $\mathrm{Lie}(G(n+1)/G(n)) \subseteq \mathrm{Lie}(G^{(p^n)}) = \mathrm{Lie}(G)^{(p^n)}$, and the S-weights on $\mathrm{Lie}(G)^{(p^n)}$ are p^n times the S-weights on $\mathrm{Lie}(G)$.

Consider the k-scheme morphism $z : G/G(n) \to G'$ defined by $gG(n) \mapsto f_1(g)f_2(g)^{-1}$. Since f_1 and f_2 coincide on $G(n)$, and the k-subgroup $G(n)$ in G is normal, clearly

$$z(\overline{g})f_1(x)z(\overline{g})^{-1} = f_1(x)$$

for points $\overline{g}$ of $G/G(n)$ and x of $G(n)$. Thus, z is valued in the closed k-subgroup scheme $G'' = G'^{G(n)}$ of points of G' centralized under the common left conjugation action of $G(n)$ through f_1 and through f_2 (Proposition A.8.10(1) provides the existence of G''). Since $f_1|_{G(n+1)} \neq f_2|_{G(n+1)}$, the restriction c of z to $G(n+1)/G(n)$ is nontrivial. Let $\rho : G(n+1)/G(n) \to \underline{\mathrm{Aut}}_{G''/k}$ denote the S-equivariant left action of $G(n+1)/G(n)$ on the $G(n)$-centralizer G'' through the conjugation action of $G(n+1)$ via f_2. For the nontrivial S-equivariant map of pointed k-schemes $c : G(n+1)/G(n) \to G''$ clearly $c(y_1 y_2) = c(y_1) \cdot \rho(y_1)(c(y_2))$, so $s : y \mapsto (c(y), y)$ is an S-equivariant homomorphic section to the natural k-homomorphism

$$G'' \rtimes (G(n+1)/G(n)) \to G(n+1)/G(n).$$

But the infinitesimal k-group scheme $G(n+1)/G(n)$ has vanishing relative Frobenius morphism, so (by Proposition A.7.14) s is uniquely determined by $\mathrm{Lie}(s)$. Since s is *not* the section $y \mapsto (1, y)$, it follows that the S-equivariant map induced by c between tangent spaces at the identity points is nonzero, so we are done.

A *Tits system* is a 4-tuple $(\mathscr{G}, B, N, \Sigma)$ where $\mathscr{G}$ is an abstract group, B and N are subgroups, and $\Sigma \subseteq N/(B \cap N)$ is a subset such that the following four axioms are satisfied: (T1) $B \cup N$ generates $\mathscr{G}$ and $B \cap N$ is normal in N, (T2) the elements of Σ have order 2 in the *Weyl group* $W := N/(B \cap N)$ and generate W, (T3) for all $\sigma \in \Sigma$ and $w \in W$, $\sigma B w \subseteq BwB \cup B\sigma wB$ (using any representatives for σ and w in N, the choices of which do not matter), and (T4) $\sigma B \sigma \not\subseteq B$ for all $\sigma \in \Sigma$ (which is equivalent to $\sigma B \sigma \neq B$, since $\sigma^2 = 1$ in W). We refer the reader to [Bou2, IV, §2] for the basic properties of Tits systems. By [Bou2, IV, §2.5, Rem. (1)], Σ is uniquely determined by $(\mathscr{G}, B, N)$.

The following theorem was announced by Borel and Tits in [BoTi2] without proof.

Theorem C.2.20 *Let G be a smooth connected affine group over a field k, S a maximal k-split torus in G, $N = N_G(S)$, $Z = Z_G(S)$, and P a minimal pseudo-parabolic k-subgroup of G containing S. Let ${}_k\overline{\Delta}$ be the basis of the positive system of roots ${}_k\overline{\Phi}^+$ in ${}_k\overline{\Phi}$ associated to P, and let R denote the set $\{r_a \mid a \in {}_k\overline{\Delta}\}$. The 4-tuple $(G(k), P(k), N(k), R)$ is a Tits system with Weyl group ${}_kW$ $(= N(k)/Z(k))$.*

This is the *standard Tits system* associated to (G, S, P). We will provide a proof of Theorem C.2.20 after some general preparations involving a notion of "root group" associated to any $a \in \Phi(G, S)$. Note that by Theorem C.2.15, ${}_kW$ naturally coincides with the combinatorial Weyl group $W({}_k\overline{\Phi})$.

C.2.21 Root groups in G. For any $a \in \mathrm{X}(S) - \{0\}$, let $\langle a \rangle$ be the semigroup consisting of positive integral multiples of a. By Proposition 3.3.6 (applied to the semigroup $A = \langle a \rangle$) there is a unique smooth connected unipotent k-subgroup $U_a = H_{\langle a \rangle}(G)$ in G that is normalized by S and for which $\mathrm{Lie}(U_a)$ is the span of the weight spaces in $\mathrm{Lie}(G)$ for the weights in $\Phi(G, S)$ that are positive integral multiples of a. Moreover, U_a contains every smooth connected k-subgroup $H \subseteq G$ normalized by S such that S acts on $\mathrm{Lie}(H)$ only with weights that are positive integral multiples of a. Let $\mathscr{S}_a = \ker a$ and $G_a = Z_G(\mathscr{S}_a)^0$ be as in Remark C.2.12(2). For $\lambda_a \in \mathrm{X}_*(S)$ such that $\langle a, \lambda_a \rangle > 0$ we have $U_{G_a}(\lambda_a) \subseteq U_a$. This is an equality because $U_a \subseteq G_a$ by Corollary A.8.11 (forcing $\mathrm{Lie}(U_{G_a}(\lambda_a)) = \mathrm{Lie}(U_a)$). In particular, if $H \subseteq G$ is a smooth closed k-subgroup normalized by S then $U_a \cap H$ is smooth and connected (by Remark 2.1.11 and Proposition 2.1.8).

We call U_a the *root group of G associated to* $a \in \mathrm{X}(S) - \{0\}$ (even if $a \notin {}_k\overline{\Phi}$, or $a \notin \Phi(G,S)$!). The U_a of most interest are for $a \in \Phi(G,S)$, but in some later considerations with general connected smooth affine k-groups it is convenient to allow the possibility $a \notin \Phi(G,S)$. The above dynamic description of U_a is compatible with any extension of the ground field, so by using scalar extension to k_s we see that $(U_a)_{k_s}$ is normalized by $Z_G(S)(k_s)$ and hence U_a is normalized by $Z_G(S)$ for any $a \in \mathrm{X}(S) - \{0\}$. Since $U_a = U_{G_a}(\lambda_a)$, the root group U_a is a k-split smooth connected unipotent group (Proposition 2.1.10); this also follows from Lemma 3.3.8. When G is pseudo-reductive and pseudo-split with S a k-split maximal k-torus in G, the above recovers Definition 2.3.13 for $a \in \Phi(G,S)$.

Proof of Theorem C.2.20 The Bruhat decomposition (Theorem C.2.8) implies $P(k) \cap N(k) = Z(k)$ and also axiom (T1), so the Weyl group $N(k)/(P(k) \cap N(k))$ is indeed ${}_kW$. By Theorem C.2.15, the quotient group $N(k)/Z(k) = {}_kW$ is generated by R. This is axiom (T2). To prove axiom (T4), for $r = r_a \in R$ (with $a \in {}_k\overline{\Delta} \subseteq {}_k\overline{\Phi} \subseteq \Phi(G,S)$) it suffices to prove that $rP(k)r \neq P(k)$. Clearly rPr contains $rU_ar = U_{r(a)} = U_{-a}$, so it suffices to prove that $U_{-a}(k)$ is not contained in $P(k)$.

By applying Proposition 3.3.10 to the semigroup A consisting of negative integral multiples of a we see that $U_{-a} \cap P$ is a smooth connected k-subgroup of the k-split unipotent group U_{-a}. By Lemma 3.3.8, $U_{-a} \cap P$ is also k-split. To show that it is a proper k-subgroup of U_{-a}, note that $-a$ is a root of the maximal pseudo-reductive quotient $\overline{G} = G/\mathscr{R}_{u,k}(G)$ but not of the image $\overline{P}$ of P in $\overline{G}$. Thus, consideration of the S-equivariant surjection $\mathrm{Lie}(G) \to \mathrm{Lie}(\overline{G})$ implies that $\mathrm{Lie}(U_{-a})$ is not contained in $\mathrm{Lie}(P)$, so indeed $U_{-a} \cap P \neq U_{-a}$.

For the verification of (T4), it now suffices to show that if $U \hookrightarrow U'$ is an inclusion between k-split smooth connected unipotent k-groups and $U(k) = U'(k)$ then $U = U'$. Note that U and U' are affine spaces since both of them are connected k-split unipotent groups. Thus, the case of infinite k is settled by Zariski-density considerations. If k is finite of cardinality q then we may instead argue by dimension, since $\#U(k) = q^{\dim U}$ and $\#U'(k) = q^{\dim U'}$.

It remains to prove axiom (T3). That is, for $r := r_a$ (with $a \in {}_k\overline{\Delta}$) and $w \in {}_kW = W(G,S)(k)$, we claim that

$$rP(k)\{w, rw\}P(k) \subseteq P(k)\{w, rw\}P(k). \qquad (*)$$

We first reduce the proof of $(*)$ to the case where G is pseudo-reductive. The natural homomorphism $W(G,S) \to W(\overline{G}, \overline{S})$ is a k-isomorphism (Lemma 3.2.1), so from the Bruhat decomposition (Theorem C.2.8) we see that $\pi : G \to \overline{G}$ maps any two distinct double cosets of $P(k)$ in $G(k)$ into two distinct double cosets of $\overline{P}(k)$ in $\overline{G}(k)$. Hence, it suffices to prove $(*)$ with G replaced by the maximal pseudo-reductive quotient $\overline{G}$. Accordingly, we assume now that G is pseudo-reductive, so ${}_k\overline{\Phi} = \Phi(G,S)$. We shall denote this root system by ${}_k\Phi$ and its basis corresponding to ${}_k\overline{\Delta}$ by ${}_k\Delta$. As we will see below, the proof of (∗) for reductive groups given in [Bo2, p. 237] works for pseudo-reductive groups as well.

Let $\lambda: \mathrm{GL}_1 \to S$ be a 1-parameter k-subgroup such that $P = P_G(\lambda) = Z_G(\lambda) \ltimes U_G(\lambda)$, so by Proposition C.2.4 and the minimality of P we have $Z_G(\lambda) = Z_G(S) = Z$ and $U_G(\lambda) = \mathscr{R}_{us,k}(P)$ (since $Z_G(\lambda)$ is pseudo-reductive and $U_G(\lambda)$ is k-split). Thus, $P = Z \ltimes \mathscr{R}_{us,k}(P)$ and ${}_k\Phi^+ = \Phi(P,S) = ({}_k\Phi)_{\lambda \geqslant 0} = ({}_k\Phi)_{\lambda > 0}$, so $\langle a, \lambda \rangle > 0$. More generally, a k-root $b \in {}_k\Phi$ is *positive* (i.e., it lies in ${}_k\Phi^+$) if and only if $\langle b, \lambda \rangle > 0$. If necessary, replacing w with rw, we may and will assume that $w^{-1}(a)$ is positive.

Now let $U = \mathscr{R}_{us,k}(P) = U_G(\lambda)$, so for $G_a := Z_G(\ker a)^0$ the root group U_a is equal to $U_{G_a}(\lambda) = G_a \cap U$. Let $S_a = (\ker a)^0_{\mathrm{red}}$. As a is non-divisible in $\Phi(G,S)$ (since $a \in {}_k\Delta$), an element of $\Phi(G,S)$ is a rational multiple of a if and only if it is an integral multiple of a. Thus, the containment $G_a \subseteq Z_G(S_a)$ between smooth connected k-groups is an equality via comparison of their Lie algebras. Choose a 1-parameter k-subgroup $\lambda_a : \mathrm{GL}_1 \to S_a$ such that $\langle b, \lambda_a \rangle > 0$ for all $b \in {}_k\Delta - \{a\}$. Note that $\langle a, \lambda_a \rangle = 0$, so $G_a = Z_G(\lambda_a)$. Let $V = U_G(\lambda_a)$, so $P_G(\lambda_a) = Z_G(\lambda_a) \ltimes U_G(\lambda_a) = G_a \ltimes V$ and hence $U = U_a \ltimes V$. We also noted above that $P = Z \ltimes U$. Since G_a contains Z as well as a representative of $r = r_a$ (by definition of r_a), it follows that V is normalized by any representative of r and hence,

$$rP(k)\{w, rw\}P(k) = Z(k)rU(k)\{w, rw\}P(k) = Z(k)V(k)rU_a(k)\{w, rw\}P(k).$$

Thus,

$$rP(k)\{w, rw\}P(k) \subseteq P(k)G_a(k)wP(k).$$

For the pseudo-parabolic k-subgroup $P_a := P_{G_a}(\lambda) = G_a \cap P = Z \ltimes U_a$ of G_a, the Bruhat decomposition (Theorem C.2.8) for $G_a(k)$ with respect to $P_a(k)$ implies that

$$G_a(k) = Z(k)U_a(k)\{1, r\}U_a(k),$$

so

$$rP(k)\{w, rw\}P(k) \subseteq P(k)\{1, r\}U_a(k)wP(k).$$

But the positivity of $w^{-1}(a)$ implies that $w^{-1}U_a w \subseteq P$, so

$$rP(k)\{w, rw\}P(k) \subseteq P(k)\{w, rw\}P(k). \qquad \square$$

Remark C.2.22 Since ${}_kW = W({}_k\overline{\Phi})$ acts simply transitively on the set of minimal pseudo-parabolic k-subgroups of G containing S (as we explained immediately after the statement of Theorem C.2.15), if P is such a k-subgroup corresponding to a positive system of roots ${}_k\overline{\Phi}^+$ then for a suitable $w_0 \in {}_kW$ the k-subgroup $w_0 \circ P$ corresponds to the positive system of roots $-{}_k\overline{\Phi}^+$. Now assume that S commutes with $\mathscr{R}_{u,k}(G)$; i.e., $\mathscr{R}_{u,k}(G) \subseteq Z_G(S)$ (e.g., G is quasi-reductive). Then $P = P_G(\lambda) = Z_G(\lambda) \ltimes U_G(\lambda)$ for some $\lambda \in X_*(S)$ (by Theorem C.2.3), and $w_0 \circ P = P_G(-\lambda)$. Thus, $P \cap w_0 \circ P = Z_G(\lambda) = Z_G(S) =: Z$ (by considering the weights of the S-action on Lie algebras), so $Z(k) \subseteq \bigcap_{w \in {}_kW} wP(k)w^{-1} \subseteq P(k) \cap w_0P(k)w_0^{-1} = Z(k)$; i.e., $Z(k) = \bigcap_{w \in {}_kW} wP(k)w^{-1}$. For $N := N_G(S)$ we have $Z(k) = P(k) \cap N(k)$ by the Bruhat decomposition of $G(k)$ relative to the minimal P (Theorem C.2.8), so if $Z_G(S)$ contains $\mathscr{R}_{u,k}(G)$ then $\bigcap_{n \in N(k)} nP(k)n^{-1} = P(k) \cap N(k)$; i.e., in such cases the Tits system in Theorem C.2.20 is saturated (in the sense of [Bou2, IV, §2, Exer. 5(a)]). Since $P(k)$ is a semidirect product of its subgroup $Z(k) = P(k) \cap N(k)$ against the nilpotent subgroup $U_G(\lambda)(k)$, the Tits system is split. Also recall from Corollary 2.2.5 that $\mathscr{R}_{us,k}(P) = U_G(\lambda)\mathscr{R}_{us,k}(G)$.

Theorem C.2.23 *Let G be a smooth connected affine group over a field k, P a minimal pseudo-parabolic k-subgroup of G, and S a maximal k-split torus of G contained in P. Let $N = N_G(S)$, $Z = Z_G(S)$, and ${}_kW = N(k)/Z(k)$. Let ${}_k\overline{\Delta}$ and R be as in Theorem C.2.20.*

(i) *There are $2^{\#R}$ subgroups of $G(k)$ containing $P(k)$, and any such subgroup equals $Q(k)$ for a unique pseudo-parabolic k-subgroup Q containing P.*
(ii) *For pseudo-parabolic k-subgroups Q and Q' of G, $Q \subseteq Q'$ if and only if $Q(k) \subseteq Q'(k)$. In particular, $Q = Q'$ if and only if $Q(k) = Q'(k)$.*

Proof Since $(G(k), P(k), N(k), R)$ is a Tits system (with Weyl group ${}_kW$), by [Bou2, IV, §2.5, Thm. 3(b)] there is a natural bijection $X \mapsto G(k)_X$ from the set of subsets $X \subseteq R$ to the set of subgroups of $G(k)$ containing $P(k)$, so there are exactly $2^{\#R}$ subgroups of $G(k)$ containing $P(k)$. Explicitly, relative to the Bruhat decomposition $G(k) = \coprod_{w \in {}_kW} P(k)n_wP(k)$ for a choice of representative $n_w \in N(k)$ of each $w \in {}_kW$ (Theorem C.2.8), $G(k)_X$ is the union of the $P(k)$-double cosets indexed by the elements of the subgroup ${}_kW_X$ of ${}_kW$ generated by X. Note that the subgroup $G(k)_X$ is the unique subgroup of $G(k)$ containing $P(k)$ such that the image of $N(k) \cap G(k)_X$ in ${}_kW$ is ${}_kW_X$. (In particular, ${}_kW_X$ determines X.)

Any smooth closed k-subgroup Q of G containing P is pseudo-parabolic (Proposition 3.5.8), and there are exactly $2^{\#R}$ such k-subgroups (as we saw at the end of Step 5 in the proof of Theorem C.2.15). We will now display them all. Let $\overline{G} = G/\mathscr{R}_{u,k}(G)$ be the maximal pseudo-reductive quotient of G and $\pi : G \to \overline{G}$ be the quotient map. As usual, we identify S with its maximal k-split torus image $\overline{S}$ in $\overline{G}$. For each subset $I \subseteq {}_k\overline{\Delta}$, we fix a cocharacter $\lambda_I \in X_*(S)$ such that $\langle a, \lambda_I \rangle \geqslant 0$ for all $a \in {}_k\overline{\Delta}$, with equality if and only if $a \in I$.

Define $P_I = P_G(\lambda_I)\mathscr{R}_{u,k}(G) = Z_G(\lambda_I)U_G(\lambda_I)\mathscr{R}_{u,k}(G)$, so $\pi(P_I) = P_{\overline{G}}(\lambda_I)$. In view of how ${}_k\overline{\Delta}$ is defined in terms of P, by Theorem C.2.15 and weight space computations in $\mathrm{Lie}(\overline{G})$ we see that $P = P_\emptyset$ and $P_J \subseteq P_{J'}$ when $J \subseteq J'$, so $P \subseteq P_I$. We will prove that if $J \neq J'$ then $P_J(k) \neq P_{J'}(k)$ (so $P_J \neq P_{J'}$). It would then follow that the collection $\{P_I \mid I \subseteq {}_k\overline{\Delta}\}$ consists of $2^{\#_k\overline{\Delta}} = 2^{\#R}$ pseudo-parabolic k-subgroups of G containing P, and the collection $\{P_I(k) \mid I \subseteq {}_k\overline{\Delta}\}$ consists of $2^{\#R}$ subgroups of $G(k)$ containing $P(k)$. Hence, the former would be the collection of all pseudo-parabolic k-subgroups of G containing P and the latter would be the collection of all subgroups of $G(k)$ containing $P(k)$, so (i) would be proved.

Fix an I, and let $N_I = N \cap P_I = N_{P_I}(S)$. We have noted above that I is determined by the subgroup ${}_kW_I \subseteq {}_kW$ generated by $\{r_a\}_{a\in I} \subseteq R$, so it suffices to show that the image of $N(k) \cap P_I(k) = N_I(k)$ in ${}_kW$ is ${}_kW_I$. This image $N_I(k)/Z(k)$ is the k-Weyl group of (P_I, S). By Lemma 3.2.1 and Proposition C.2.10, the natural map ${}_kW \to {}_k\overline{W} := W(\overline{G}, \overline{S})(k)$ is an isomorphism, and similarly with $(G, \overline{G})$ replaced by $(P_I, \overline{P}_I)$ (where $\overline{P}_I$ is the maximal pseudo-reductive quotient of P_I); we will use this to reduce our k-Weyl group assertion for (G, P_I, S) to an analogue for $(\overline{G}, Z_{\overline{G}}(\lambda_I), S)$.

For a k-torus T in G that normalizes a smooth connected k-subgroup H of G, since the k-subgroup $Z_{T\ltimes H}(T) = T \ltimes (Z_G(T) \cap H)$ is smooth and connected (Proposition A.2.5) we see that $Z_G(T) \cap H$ is smooth and connected. Thus, $Z_G(\lambda_I) \cap \mathscr{R}_{u,k}(G)$ and $Z_G(\lambda_I) \cap \mathscr{R}_{u,k}(P_I)$ are smooth and connected (and visibly unipotent). But $P_I = Z_G(\lambda_I) \cdot U_G(\lambda_I)\mathscr{R}_{u,k}(G)$ and the unipotent smooth normal k-subgroup $U_G(\lambda_I)\mathscr{R}_{u,k}(G)$ of P_I is connected, so the maximal pseudo-reductive quotients of P_I and $Z_G(\lambda_I)$ coincide. The pseudo-reductivity of

$$Z_{\overline{G}}(\lambda_I) = \pi(Z_G(\lambda_I)) = Z_G(\lambda_I)/(Z_G(\lambda_I) \cap \mathscr{R}_{u,k}(G))$$

implies that the maximal pseudo-reductive quotient of $Z_G(\lambda_I)$ is identified with $Z_{\overline{G}}(\lambda_I)$, so the maximal pseudo-reductive quotient of P_I is identified with $Z_{\overline{G}}(\lambda_I)$. Assertion (i) is now reduced to showing that the subgroup $W(Z_{\overline{G}}(\lambda_I), \overline{S})(k) \subseteq {}_k\overline{W}$ is generated by $\{r_a\}_{a\in I}$. The root system ${}_k\overline{\Phi}_I := \Phi(Z_{\overline{G}}(\lambda_I), \overline{S})$ is obviously the span of I in the root system ${}_k\overline{\Phi}$ of $(\overline{G}, \overline{S})$, with I moreover as a basis, so $W(Z_{\overline{G}}(\lambda_I), \overline{S})(k)$ is generated by reflections $\{r'_{a,I}\}_{a\in I}$. It remains to note that the natural inclusion $W(Z_{\overline{G}}(\lambda_I), \overline{S})(k) \hookrightarrow {}_k\overline{W}$ inside $\mathrm{GL}(\mathrm{X}(S)_{\mathbf{Q}})$ carries $r'_{a,I}$ to r_a for each $a \in I$, due to the well-known fact that for any nonzero finite-dimensional vector space V over a field of characteristic 0 and any finite subgroup $F \subseteq \mathrm{GL}(V)$, a reflection $r \in F$ is uniquely determined by the line that it negates.

To prove assertion (ii), consider arbitrary pseudo-parabolic k-subgroups Q and Q' of G such that $Q(k) \subseteq Q'(k)$. Fix a minimal pseudo-parabolic k-subgroup P contained in Q. By the conjugacy of minimal pseudo-parabolic

k-subgroups of G (Theorem C.2.5), there exists $g \in G(k)$ such that $gPg^{-1} \subseteq Q'$. In the terminology of [Bou2, IV, §2.6], the "parabolic" subgroups of $G(k)$ (relative to the Tits system $(G(k), P(k), N(k), R)$) are the subgroups that contain a $G(k)$-conjugate of $P(k)$. Hence, $Q'(k)$ is such a subgroup. But $P(k)$ and $gP(k)g^{-1}$ are contained in $Q'(k) \supseteq Q(k)$, so $g \in Q'(k)$ by [Bou2, IV, §2.6, Thm. 4(i)]. Thus, $P \subseteq Q'$, so Q and Q' are pseudo-parabolic k-subgroups containing the common minimal pseudo-parabolic k-subgroup P. Then $Q \cap Q'$ is another such k-subgroup yet its k-points coincide with those of Q, so (i) implies that $Q = Q \cap Q' \subseteq Q'$ and (ii) is proved. □

We will now prove the following generalization of Proposition 3.4.2.

Proposition C.2.24 *Let G be a smooth connected affine k-group, S a maximal k-split torus in G, $N = N_G(S)$, and $Z = Z_G(S)$. Let ${}_k\overline{\Phi}$ be the k-root system of G with respect to S. Assume that no nonzero weight of S on* $\mathrm{Lie}(\mathscr{R}_{u,k}(G))$ *is an integral multiple of a root in ${}_k\overline{\Phi}$. Choose $a \in {}_k\overline{\Phi}$ and let $U_{\pm a}$ be the root groups associated to $\pm a$.*

If $2a \notin {}_k\overline{\Phi}$ then U_a is commutative and if moreover $\mathrm{char}(k) = p > 0$ *then U_a is p-torsion. If $2a \in {}_k\overline{\Phi}$ then $(U_a, U_a) \subseteq U_{2a}$ and U_{2a} is contained in the center of U_a.*

For a nontrivial element $u \in U_a(k)$, the following properties are satisfied:

(i) *There exist unique $u', u'' \in U_{-a}(k)$ such that $m(u) := u'uu''$ normalizes S. The action of $m(u)$ on $\mathrm{X}(S)$ is the reflection r_a, and moreover $u', u'' \neq 1$.*
(ii) *For any field extension K of k and any $z \in Z_G(S)(K)$, if zuz^{-1} lies in $U_a(k)$ then $zu'z^{-1}, zu''z^{-1} \in U_{-a}(k)$ and $m(zuz^{-1}) = zm(u)z^{-1}$.*
(iii) *If a is a non-multipliable root then $u' = u''$, $u' = m(u)^{-1} \cdot u \cdot m(u)$, and $m(u)^2 \in S(k)$.*

It is not generally true, even when G is reductive (and non-split), that $u' = u''$.

Proof Let F_a be the identity component of the centralizer of $\mathscr{S}_a = \ker a$ in G, so F_a contains $U_{\pm a}$ and its k-unipotent radical $\mathscr{R}_{u,k}(F_a)$ is equal to $\mathscr{R}_{u,k}(G)^{\mathscr{S}_a}$ (Proposition A.8.14(2)). Since no nonzero weight of S on the Lie algebra of $\mathscr{R}_{u,k}(G)$ is an integral multiple of a, we conclude that $\mathscr{R}_{u,k}(F_a) \subseteq Z_{F_a}(S)$. But $Z = Z_G(S) \subseteq F_a$, so we may and do replace G with F_a. Now ${}_k\Delta = \{a\}$, ${}_k\overline{\Phi} = \Phi(G, S)$, the Weyl group ${}_kW$ is a group of order 2 generated by r_a, and $P := Z \ltimes U_{-a}$ is a minimal pseudo-parabolic k-subgroup of G.

Since ${}_k\overline{\Phi}$ is a root system, the only positive integral multiple of a, other than a, which can be in ${}_k\overline{\Phi}$ is $2a$. By Proposition 3.3.5, if $2a \in {}_k\overline{\Phi}$ then $(U_a, U_a) \subseteq U_{2a}$ and U_{2a} is contained in the center of U_a; likewise, if $2a \notin {}_k\overline{\Phi}$ then U_a is commutative. Furthermore, if $\mathrm{char}(k) = p > 0$ and $2a \notin {}_k\overline{\Phi}$ then the commutative group U_a is p-torsion by Example 3.3.3.

Now consider a nontrivial element $u \in U_a(k)$. Let $P' = r_a P r_a = Z \ltimes U_a$ be the pseudo-parabolic k-subgroup "opposite" to P. By the Bruhat decomposition (Theorem C.2.8),

$$G(k) = P(k) \cup P(k) r_a P(k) = P(k) \cup U_{-a}(k)(N(k) - Z(k))U_{-a}(k).$$

Since $U_a(k) \cap P(k)$ is trivial, u must lie in $U_{-a}(k)(N(k) - Z(k))U_{-a}(k)$, and therefore we can find $u', u'' \in U_{-a}(k)$ and $n \in N(k) - Z(k)$ such that $u = u'^{-1} n u''^{-1}$. Then $m(u) := u'uu'' = n \in N(k)$ and it maps onto r_a in the Weyl group ${}_kW$.

To see that u' and u'' are unique, we first note that for any $u', u'' \in U_{-a}(k)$ such that $n' := u'uu'' \in N(k)$, necessarily $n' \notin Z(k)$ since otherwise we would have $u = u'^{-1} n' u''^{-1} \in U_{-a}(k)Z(k) = P(k)$, a contradiction because $u \in U_a(k)$ and $U_a \cap P = 1$. We have arranged that $G = F_a$, so n' must represent r_a. Since any two representatives of r_a in $N(k)$ coincide modulo $Z(k)$, the uniqueness of u' and u'' is reduced to checking that if $v'n'v'' = n'z$ with $v', v'' \in U_{-a}(k)$ and $z \in Z(k)$ then $v' = v'' = 1$ and $z = 1$. Since

$$v'(n'v''n'^{-1}) = n'zn'^{-1} \in Z(k)$$

with $n'v''n'^{-1} \in U_a(k)$, it suffices to prove that $U_{-a}(k)Z(k)U_a(k)$ is a direct product set (via multiplication) inside of $G(k)$. This in turn is a special case of Proposition 2.1.8(3). To prove that $u', u'' \neq 1$, we just have to show that no $n \in N(k) - Z(k)$ can have the form $u'u$ or uu' for $u' \in U_{-a}(k)$. If $u'u = n$ then

$$u' = nu^{-1} = (nu^{-1}n^{-1})n \in U_{-a}(k)n,$$

forcing $n \in U_{-a}(k)$, an absurdity since $u \notin U_{-a}(k)$; the case $uu' = n$ is likewise ruled out, so we have proved assertion (i).

To prove assertion (ii), consider $z \in Z_G(S)(K)$ such that zuz^{-1} lies in $U_a(k)$. Then $m(zuz^{-1}) = u'_z(zuz^{-1})u''_z$ for some $u'_z, u''_z \in U_{-a}(k)$, so

$$z^{-1}m(zuz^{-1})z = z^{-1}u'_z z \cdot u \cdot z^{-1}u''_z z = (z^{-1}u'_z z \cdot u'^{-1})m(u)(u''^{-1} \cdot z^{-1}u''_z z).$$

Hence,

$$(m(u)^{-1} \cdot z^{-1}m(zuz^{-1})z)(u''^{-1} \cdot z^{-1}u''_z z)^{-1} = m(u)^{-1}(z^{-1}u'_z z \cdot u'^{-1})m(u).$$

The element on the left side of this equality lies in $P(K)$, whereas the element on the right side lies in $U_a(K)$. Therefore, the elements on both sides are 1. This implies that $zu'z^{-1} = u'_z$, $zu''z^{-1} = u''_z$, and $m(zuz^{-1}) = zm(u)z^{-1}$, so we have proved (ii).

We will now prove assertion (iii), so consider a non-multipliable root a. The root group U_a is commutative for such a, and if $\mathrm{char}(k) = p > 0$ then U_a is also p-torsion. Choose $s \in S(\overline{k})$ such that $a(s) = -1$. Conjugating the equation

$u'uu'' = m(u)$ by s, we obtain $u'^{-1}u^{-1}u''^{-1} = sm(u)s^{-1}$, which in turn implies that $u''uu' = sm(u)^{-1}s^{-1}$. From the uniqueness of u' and u'' we deduce that $u' = u''$, so $sm(u)^{-1}s^{-1} = m(u)$. Therefore, $m(u)^2 = m(u)sm(u)^{-1}\cdot s^{-1} \in S(k)$. Since $m(u) = u'uu'$, we see that

$$u' \cdot (m(u)^{-1}u'm(u)) \cdot (m(u)^{-1}um(u)) = m(u)$$

with $m(u)^{-1}u'm(u) \in U_a(k)$ and $m(u)^{-1}um(u) \in U_{-a}(k)$ because $m(u)$ normalizes S and acts as r_a on $\mathrm{X}(S)$. Thus, $m(m(u)^{-1}u'm(u)) = m(u)$ and $u' = m(u)^{-1}um(u)$. □

A subset Ψ of $\Phi(G,S)$ is *saturated* if the subsemigroup A of $\mathrm{X}(S)$ spanned by Ψ does not contain 0 and $\Psi = A \cap \Phi(G,S)$. If $\Phi := \Phi(G,S)$ is a root system (for example, if $\mathscr{R}_{u,k}(G) \subseteq Z_G(S)$), Φ^+ a positive system of roots in Φ, and $\Psi \subseteq \Phi^+$ a closed subset, then Ψ is saturated by Proposition 2.2.7. However, we will be interested in allowing for the possibility that $\Phi(G,S)$ is not a root system, so saturatedness will be a more useful concept below than closedness.

Particularly interesting examples of saturated subsets of $\Phi(G,S)$ are obtained as follows. Let $\lambda : \mathrm{GL}_1 \to S$ be a 1-parameter k-subgroup. Then

$$\Phi(G,S)_{\lambda>0} := \{a \in \Phi(G,S) \mid \langle a, \lambda\rangle > 0\}$$

is a saturated subset of $\Phi(G,S)$. Another example of a saturated subset which will be of interest to us is the following. Let $a, b \in \Phi(G,S)$ be two linearly independent elements, and let (a,b) denote the set of elements of $\Phi(G,S)$ of the form $ma + nb$ for positive integers m and n. Clearly (a,b) is a saturated subset.

We seek to construct a smooth connected unipotent k-subgroup U_Ψ attached to any saturated Ψ, and to describe it as a direct span (in any order) of suitable root groups. However, if $\Phi(G,S)$ is not a root system then we cannot expect a direct spanning result to hold when using only root groups U_c attached to $c \in \Psi$. The problem is that there may be distinct elements in a saturated Ψ that are not divisible in Ψ but share a common positive integral multiple in Ψ. For example, perhaps $c \in \mathrm{X}(S) - \{0\}$ with $2c, 3c, 6c \in \Psi$ but $c \notin \Psi$. (This problem cannot arise if $\Phi(G,S)$ is a root system, as then the only possible nontrivial linear dependence relations in Ψ are of the form $c' = 2c''$.) In such an example, the root group U_c contains U_{nc} for $n = 2, 3, 6$ where U_{2c} and U_{3c} each contain U_{6c}. To account for this phenomenon, we introduce the following terminology.

Definition C.2.25 A *half-line* in $\mathrm{X}(S)_{\mathbf{Q}}$ is the set of $\mathbf{Q}_{>0}$-multiples of a nonzero element. If $\{c_i\}$ is a non-empty subset of $\Phi(G,S)$ lying in a half-line L of $\mathrm{X}(S)_{\mathbf{Q}}$ (so the codimension-1 tori $(\ker c_i)^0_{\mathrm{red}}$ coincide) then their *greatest* *common divisor* is the unique $c \in \mathrm{X}(S)$ such that $c_i = n_i c$ with integers $n_i > 0$ satisfying $\gcd(n_i) = 1$.

Proposition C.2.26 *Let G be a smooth connected affine group over a field k, and S a maximal k-split k-torus in G.*

(1) *For any saturated subset Ψ of $\Phi(G,S)$, there is a unique S-stable smooth connected unipotent k-subgroup U_Ψ such that $\mathrm{Lie}(U_\Psi)$ is the span of the subspaces $\mathrm{Lie}(U_c)$ for $c \in \Psi$. It is normalized by $Z_G(S)$, and directly spanned in any order by the subgroups $U_c \cap U_\Psi$ for the greatest common divisors c of the non-empty intersections of Ψ with half-lines in $\mathrm{X}(S)_{\mathbf{Q}}$.*
(2) *For any pseudo-parabolic k-subgroup P of G containing S, the subset Ψ of ${}_k\overline{\Phi} \subseteq \Phi(G,S)$ (via the equality $\mathrm{X}(S) = \mathrm{X}(\overline{S})$) consisting of $a \in {}_k\overline{\Phi}$ such that $-a \notin {}_k\overline{\Phi}_P := \Phi(P/\mathscr{R}_{u,k}(G), \overline{S})$ is saturated, and $\mathscr{R}_{u,k}(P) = U_\Psi \mathscr{R}_{u,k}(G)$.*
(3) *A pseudo-parabolic k-subgroup P of G is minimal if and only if all split k-tori in $P/\mathscr{R}_{u,k}(P)$ are central.*

Before we prove this result, we record its relation with earlier results in the pseudo-split pseudo-reductive case. Part (1) applied to pseudo-split pseudo-reductive G recovers Corollaries 3.3.12 and 3.3.13(1), and if $\Phi(G,S)$ is a root system then the greatest common divisors that occur in the direct spanning description are precisely the non-divisible elements of Ψ. Part (2) applied to pseudo-split pseudo-reductive G recovers Proposition 3.5.1(2) and the direct spanning description in Corollary 3.3.16 for k-unipotent radicals of pseudo-parabolic k-subgroups of such G. Finally, part (3) for pseudo-split pseudo-reductive G recovers Propositions 3.3.15(3) and 3.5.1(4) because if H is a pseudo-reductive k-group with a central maximal k-torus then it is commutative (Proposition 1.2.3).

Proof To prove assertion (1), let A be the subsemigroup of $\mathrm{X}(S)$ generated by Ψ. Then as Ψ is saturated, $0 \notin A$ and $A \cap \Phi(G,S) = \Psi$. Let $U_\Psi = H_A(G)$, where the subgroup $H_A(G)$ is as in Proposition 3.3.6. Since $0 \notin A$, U_Ψ is a smooth connected unipotent k-subgroup (Proposition 3.3.6). To prove that $Z_G(S)$ normalizes U_Ψ, we note that the description of $H_A(G)$ makes sense without reference to the maximality of S as a k-split torus in G, and is compatible with any extension of the ground field. Thus, using scalar extension to k_s implies that $(U_\Psi)_{k_s}$ is normalized by $Z_G(S)(k_s)$, and hence U_Ψ is normalized by $Z_G(S)$. The rest of assertion (1) follows easily from Theorem 3.3.11 applied to the smooth connected solvable k-group $U_\Psi = H_A(G)$ equipped with its natural action by S (taking as the disjoint union decomposition $\Psi = \coprod \Psi_i$ where the Ψ_i are the non-empty intersections of Ψ with half-lines in $\mathrm{X}(S)_{\mathbf{Q}}$).

To prove assertion (2), we fix a $\lambda \in X_*(S)$ such that $P = P_G(\lambda)\mathscr{R}_{u,k}(G)$. (Such a λ exists by Theorem C.2.3 applied to P.) By Corollary 2.1.9 the overlap $\Phi(G,S)_{\lambda>0} \cap {}_k\overline{\Phi}$ equals Ψ, which shows that Ψ is saturated. To prove the rest of (2), we can and do replace G with the pseudo-reductive quotient $G/\mathscr{R}_{u,k}(G)$ and replace Ψ with the set of $a \in \Phi(G,S)$ such that $-a \notin \Phi(P,S)$. Now $P = P_G(\lambda)$ and $\mathscr{R}_{u,k}(P) = U_G(\lambda)$. Since $U_\Psi = U_G(\lambda)$ for $\Psi = \Phi(G,S)_{\lambda>0}$, this completes the proof of (2).

By Remark 3.5.6 and Proposition 3.5.8 (and Proposition 2.2.10), to prove assertion (3) we can rename $P/\mathscr{R}_{u,k}(P)$ as G to reduce to showing that if G is pseudo-reductive then it does not contain any proper pseudo-parabolic k-subgroups if and only if its k-split k-tori are all central. This equivalence is Lemma 2.2.3(1). □

If we impose the additional hypothesis on the pair (G,S) in Proposition C.2.26 that $\mathscr{R}_{u,k}(G) \subseteq Z_G(S)$ (equivalently, S acts trivially on $\mathrm{Lie}(\mathscr{R}_{u,k}(G))$), so $\Phi(G,S) = {}_k\overline{\Phi}$ and hence $\Phi(G,S)$ is a root system, then there is an "open cell" description of G (recovering Corollary 3.3.16 when G is pseudo-split and pseudo-reductive): for any positive system of roots ${}_k\overline{\Phi}^+$ in the k-root system, the multiplication map

$$U_{{}_k\overline{\Phi}^+} \times Z_G(S) \times U_{-{}_k\overline{\Phi}^+} \to G$$

is an open immersion. Indeed, the equality $\Phi(G,S) = {}_k\overline{\Phi}$ allows us to carry over the proof of Corollary 3.3.16 by using Proposition C.2.26(1) in place of Corollary 3.3.12.

The next proposition generalizes Corollary 3.3.13(2) and is immediate from the following ingredients: Proposition 3.3.5, the definition of the root groups U_a ($a \in X(S) - \{0\}$), and the construction of U_Ψ in Proposition C.2.26(1) for the saturated subset $\Psi = (a,b)$ in $\Phi(G,S)$ with independent $a, b \in \Phi(G,S)$.

Proposition C.2.27 *Let G be a smooth connected affine group over a field k, and S a maximal k-split torus. For any linearly independent $a, b \in \Phi(G,S)$, let (a,b) be the set of elements of $\Phi(G,S)$ of the form $ma + nb$, with m, n positive integers. Then*

$$(U_a, U_b) \subseteq U_{(a,b)} = \prod_i (U_{c_i} \cap U_{(a,b)}),$$

where the direct spanning is taken with respect to an arbitrary enumeration $\{c_i\}$ of the greatest common divisors (in the sense of Definition C.2.26) of the non-empty intersections of (a,b) with half-lines in $X(S)_{\mathbf{Q}}$. In particular, if (a,b) is empty, then U_a and U_b commute.

When $\Phi(G, S)$ is a root system, the c_i in Proposition C.2.27 are precisely the non-divisible elements of (a, b).

Remark C.2.28 Let G be a smooth connected affine k-group, and let S be a maximal k-split torus in G. Assume that S commutes with $\mathscr{R}_{u,k}(G)$ (i.e., $\mathscr{R}_{u,k}(G) \subseteq Z_G(S)$), so the set $\Phi := \Phi(G, S)$ of nonzero S-weights on $\mathrm{Lie}(G)$ is the root system of G with respect to S (in the sense of Remark C.2.13). For $a \in \Phi$, let U_a be the corresponding root group (in the sense of C.2.21), and let $M_a = m(u)Z_G(S)$, where u is any nontrivial element of $U_a(k)$ and the element $m(u) \in N_G(S)(k)$ is as in Proposition C.2.24. It is easy to see using Propositions C.2.24 and C.2.27 that $(Z_G(S)(k), (U_a(k), M_a(k))_{a\in\Phi})$ is a generating root datum (*donnée radicielle génératrice*) in $G(k)$ (in the sense reviewed in the proof of Theorem 3.4.5).

Let G be a smooth connected affine group over an arbitrary field k, and let S be a nontrivial k-split torus in G. We now present a technique for constructing pseudo-reductive k-subgroups of G with S as a maximal k-split torus and having a specified linearly independent subset of $\mathrm{X}(S)$ as a basis of the associated root system. Let Δ be a non-empty linearly independent subset of $\mathrm{X}(S)$, and let C be a smooth connected k-subgroup of $Z_G(S)$ containing S as a maximal k-split torus. For each $a \in \Delta$ suppose there is given a smooth connected k-subgroup F_a of G containing C such that $Z_{F_a}(S) = C$, and assume that the set $\Phi(F_a, S)$ of nonzero S-weights on $\mathrm{Lie}(F_a)$ contains $\pm a$ and is contained in $\mathbf{Z} \cdot a$. Since $Z_{F_a}(S) = C$, the maximal k-split torus S of C is also a maximal k-split torus of F_a for all $a \in \Delta$. For $a \in \Delta$, let $E_{\pm a}$ be the root groups of F_a corresponding to $\pm a$ (cf. C.2.21). With this notation, we will prove the following theorem that was suggested by [St, Thm. 5.4]. This theorem is used in the proof of Theorem 9.8.1 and in 9.8.3, and will be used in the proof of Theorem C.2.30 below. It can also be used to give a different proof of Theorem 3.4.6.

Theorem C.2.29 *Assume that for every $a, b \in \Delta$ with $a \neq b$, E_a commutes with E_{-b}. For the smooth connected k-subgroup F of G generated by the k-subgroups $\{F_a\}_{a\in\Delta}$, the following hold.*

(i) *The centralizer $Z_F(S)$ is equal to C, so S is a maximal k-split torus of F and if S is a maximal torus of C then it is a maximal torus of F.*
(ii) *Any nonzero weight of S on* $\mathrm{Lie}(F)$ *is either a nonnegative or a nonpositive integral linear combination of elements in Δ.*
(iii) *For each $a \in \Delta$, the $\pm a$-root groups in F are $E_{\pm a}$.*
(iv) *If for every $a \in \Delta$, F_a is quasi-reductive* (*resp. pseudo-reductive*) *then F is quasi-reductive* (*resp. pseudo-reductive*), *and in such cases Δ is a basis of the root system of F with respect to S.*

(v) *If the k-groups F_a are reductive for every $a \in \Delta$ then so is F.*

(vi) *If F_a is quasi-reductive for all $a \in \Delta$ then F is functorial with respect to isomorphisms in $(S, \Delta, \{F_a\}_{a\in\Delta})$ in the following sense: if $(G', S', C', \Delta', \{F'_{a'}\}_{a'\in\Delta'})$ is a second 5-tuple such that $E'_{a'}$ commutes with $E'_{-b'}$ for distinct $a', b' \in \Delta'$ and there are given a k-isomorphism $f_C : C \simeq C'$ restricting to $f_S : S \simeq S'$ satisfying $\mathrm{X}(f_S)(\Delta') = \Delta$ and k-isomorphisms $f_{a'} : F_{a' \circ f_S} \simeq F'_{a'}$ extending f_C for all $a' \in \Delta'$ (so every $F'_{a'}$ is quasi-reductive, and hence F' is quasi-reductive by* (iv)*), then there is a unique k-isomorphism $f : F \simeq F'$ extending $f_{a'}$ for all $a' \in \Delta'$.*

The main issue in applications of Theorem C.2.29 to the construction of pseudo-reductive groups is finding an appropriate $C \subseteq Z_G(S)$. (Away from the pseudo-split pseudo-reductive case, C is generally not commutative.)

Proof As an initial step, we give a concrete "open cell" in F and analyze the nontrivial S-weights on Lie(F). Choose a 1-parameter k-subgroup $\lambda : \mathrm{GL}_1 \to S$ such that $\langle a, \lambda \rangle > 0$ for all $a \in \Delta$ (as we may do since Δ is linearly independent in $\mathrm{X}(S)$). For each $a \in \Delta$ we have $E_{\pm a} = U_{F_a}(\pm\lambda)$, and likewise $C = Z_{F_a}(S) = Z_{F_a}(\lambda)$, so the multiplication map $E_{-a} \times C \times E_a \to F_a$ is an open immersion by Proposition 2.1.8(3). By definition F is generated by the F_a for $a \in \Delta$, so F is generated by C and the k-subgroups $\{E_{\pm a}\}_{a\in\Delta}$.

Step 1. For a smooth locally closed subscheme X of G, we denote its geometrically reduced closure by $\overline{X}$. Let U (resp. V) be the smooth connected k-subgroup of F generated by the E_a (resp. E_{-a}) for $a \in \Delta$. The subgroups U and V are normalized by C, and by Proposition 3.3.5 the weights of S on Lie(U) (resp. Lie(V)) are contained in the semigroup A^+ (resp. A^-) generated by Δ (resp. $-\Delta$). We are expecting $V \times C \times U$ to be an open subscheme of F via multiplication, so we are led to analyze the image I of the multiplication map $V \times CU \to F$. This image is an orbit of the group $V \times CU$ (as well as of the group $VC \times U$), so it is open in its closure $\overline{I}$.

Now we will show that $\overline{I} = F$, so I is open in F. We first show by induction on $r \geqslant 0$ that for any sequence of elements $a, a_1, \dots, a_r \in \Delta$,

$$E_a E_{-a_1} \cdots E_{-a_r} \subseteq \overline{I}. \qquad (*)$$

For $r = 0$, this is obvious. Now consider $r > 0$. If $a \neq a_1$ then $E_a E_{-a_1} = E_{-a_1} E_a$, whereas if $a = a_1$ then $E_a E_{-a_1} \subseteq F_a = \overline{E_{-a} C E_a}$. Thus, in both cases $(*)$ follows from the induction assumption. We have $E_a V \subseteq \overline{I}$ by $(*)$ since $V = E_{-a_1} \cdots E_{-a_r}$ for some sequence $a_1, \dots, a_r \in \Delta$. It follows that $E_a \cdot \overline{I} \subseteq \overline{I}$, and clearly $E_{-a} \cdot \overline{I} \subseteq \overline{I}$ and $C \cdot \overline{I} \subseteq \overline{I}$. As F is generated by C and the subgroups $\{E_{\pm a}\}_{a\in\Delta}$, we conclude that $F \cdot \overline{I} \subseteq \overline{I}$. It follows that $\overline{I} = F$, so I is an open subscheme of F.

Choose λ as above so that in addition $Z_G(\lambda) = Z_G(S)$; this just involves making sure that no nonzero S-weight on $\mathrm{Lie}(G)$ is trivial on λ. In particular, $Z_F(\lambda) = Z_F(S)$. Clearly $U \subseteq U_F(\lambda)$ and $V \subseteq U_F(-\lambda)$, and the multiplication map

$$U_F(-\lambda) \times Z_F(S) \times U_F(\lambda) \longrightarrow F$$

is an open immersion by Proposition 2.1.8(3); let Ω_F denote its image. Since $I = VCU$ ($\subseteq \Omega_F$) is a dense open subscheme of F, I is dense in Ω_F. But U, C, and V are closed in $U_F(\lambda)$, $Z_F(S)$, and $U_F(-\lambda)$ respectively, so $I = VCU$ is closed in Ω_F. We conclude that $VCU = I = \Omega_F = U_F(-\lambda)Z_F(S)U_F(\lambda)$, so $U = U_F(\lambda)$, $V = U_F(-\lambda)$, and $C = Z_F(S)$ (thereby proving (i)). Since the multiplication map

$$V \times C \times U \longrightarrow F$$

is an open immersion, $\mathrm{Lie}(F) = \mathrm{Lie}(V) + \mathrm{Lie}(C) + \mathrm{Lie}(U)$. Therefore, the nonzero weights of S on $\mathrm{Lie}(F)$ are all contained in $A^+ \cup A^-$. This proves (ii).

STEP 2. We wish now to prove (iii): for $a \in \Delta$, the $\pm a$-root groups of F are $E_{\pm a}$. For a subsemigroup Σ of $\mathrm{X}(S)$, we will let $H_\Sigma(F)$ denote the smooth connected k-subgroup given by Proposition 3.3.6 for F in place of G. For a nonzero $a \in \mathrm{X}(S)$, $\langle a \rangle$ will denote the subsemigroup $\mathbf{Z}_{\geqslant 1} \cdot a$. The a-root group of F is $H_{\langle a \rangle}(F)$ by definition. We shall prove that $H_{\langle a \rangle}(F) = E_a$ for every $a \in \Delta$. The inclusion $H_{\langle a \rangle}(U) \subseteq H_{\langle a \rangle}(F)$ is an equality on Lie algebras, hence an equality of k-groups, so it suffices to show that $H_{\langle a \rangle}(U) = E_a$. The set of S-weights on U is contained in the semigroup A^+ generated by Δ, and clearly $A^+ - \langle a \rangle$ is a subsemigroup of A^+, so $H_{A^+ - \langle a \rangle}(U)$ makes sense. This k-subgroup of U is normalized by $H_{\langle a \rangle}(U)$, due to Proposition 3.3.5. The multiplication map $\varphi : H_{\langle a \rangle}(U) \ltimes H_{A^+ - \langle a \rangle}(U) \to U$ is an isomorphism by Theorem 3.3.11. Since $H_{A^+ - \langle a \rangle}(U)$ contains E_b for all $b \in \Delta - \{a\}$, the restriction of φ to the subgroup $E_a \ltimes H_{A^+ - \langle a \rangle}(U)$ has image containing E_c for all $c \in \Delta$, so it is surjective since U is generated by these E_c. From this, and the fact that φ is an isomorphism, we conclude that $E_a = H_{\langle a \rangle}(U) = H_{\langle a \rangle}(F)$, as desired. A similar argument shows that $E_{-a} = H_{\langle -a \rangle}(F)$ for all $a \in \Delta$.

In the rest of the proof, we may and do assume that F_a is quasi-reductive for all $a \in \Delta$, so C is also quasi-reductive (since $\Delta \neq \emptyset$ and $C = Z_{F_a}(S)$ for any $a \in \Delta$).

STEP 3. We next prove (iv). The smooth connected unipotent k-subgroup $H_{\langle a \rangle}(\mathscr{R}_{us,k}(F)) \subseteq H_{\langle a \rangle}(F) = E_a \subseteq F_a$ must be contained in $\mathscr{R}_{us,k}(F_a)$ (as we can check on k_s-points), but $\mathscr{R}_{us,k}(F_a) = 1$ since F_a is quasi-reductive. Thus, $H_{\langle a \rangle}(\mathscr{R}_{us,k}(F)) = 1$, so consideration of its Lie algebra shows that no positive integral multiple of a is an S-weight on $\mathrm{Lie}(\mathscr{R}_{us,k}(F))$. The Lie algebra map induced by the smooth quotient map $F \twoheadrightarrow F^{\text{qred}} := F/\mathscr{R}_{us,k}(F)$ therefore induces an isomorphism between na-weight spaces for any $n \geqslant 1$. Taking $n = 1$, we see that a is a root of (F^{qred}, S). Since the nonzero weights of S on $\text{Lie}(F)$ are either nonnegative or nonpositive integral linear combinations of elements in Δ, we conclude that Δ is a basis of the root system of F^{qred} with respect to S. Recall that by definition, the root system ${}_k\overline{\Phi}$ of F with respect to S is that of F^{qred}, so Δ is a basis of ${}_k\overline{\Phi}$. Let W be the Weyl group of this root system.

For each $a \in \Delta$, we fix $n_a \in F_a(k)$ that normalizes S and for which the induced action of n_a on $\text{X}(S)$ is the reflection r_a in a. The subgroup $\mathscr{N}$ of $F(k)$ generated by the elements n_a ($a \in \Delta$) maps onto the Weyl group W. Let $w = r_{a_s} \cdots r_{a_1}$ (with $a_1, \ldots, a_s \in \Delta$) be the longest element of W, so $n = n_{a_s} \cdots n_{a_1}$ is a representative of w in $\mathscr{N}$. The element w is of order 2 and $w(\Delta) = -\Delta$, so $w(A^+) = A^-$ and $nUn^{-1} = V$. Now we observe (using the definition of r_a and the fact that Δ is a linearly independent set) that for $a \in \Delta$, $r_a(A^+) \cap A^-$ is the set of negative integral multiples of a in ${}_k\overline{\Phi}$.

Assume F is not quasi-reductive; i.e., $\mathscr{R}_{us,k}(F) \neq 1$. We seek a contradiction. Since C is a quasi-reductive k-subgroup of F, the normal smooth connected unipotent k-subgroup $\mathscr{R}_{us,k}(F)$ of F cannot be contained in C. As $C = Z_F(S)$, we conclude (by Corollary A.8.11, or Remark 2.1.11 applied to the action on $\mathscr{R}_{us,k}(F)$ by a suitable GL_1 inside S) that there is a nonzero weight of S on the Lie algebra of $\mathscr{R}_{us,k}(F)$. Such a weight belongs to $A^+ \cup A^-$. But the set of S-weights on $\text{Lie}(\mathscr{R}_{us,k}(F))$ is stable under the Weyl group W, so there is an S-weight on this Lie algebra, say b, which belongs to A^+. Let $w_0 = 1$ and for all positive integers $i \leqslant s$ define $w_i = r_{a_i} \cdots r_{a_1}$, so $w = w_s$. For all $i \leqslant s$, $w_i(b)$ is a nonzero S-weight on $\text{Lie}(\mathscr{R}_{us,k}(F))$. Moreover, $w_0(b) = b \in A^+$ and $w_s(b) \in A^-$. Let $j \leqslant s$ be the smallest positive integer such that $w_j(b)$ lies in A^-, and let $a_0 = w_{j-1}(b)$. Then $a_0 \in A^+$ and $r_{a_j}(a_0) \in A^-$. Therefore, a_0 is a positive integral multiple of the simple root a_j, and it (i.e., a_0) is a weight of S on $\text{Lie}(\mathscr{R}_{us,k}(F))$. But we showed above that no such multiple can be a weight of S on $\text{Lie}(\mathscr{R}_{us,k}(F))$. Thus, we have proved that F is quasi-reductive.

Suppose instead that F_a is pseudo-reductive for each $a \in \Delta$. The above argument carries over verbatim upon replacing $\mathscr{R}_{us,k}(\cdot)$ with $\mathscr{R}_{u,k}(\cdot)$ and replacing F^{qred} with $F^{\text{pred}} := F/\mathscr{R}_{u,k}(F)$ to show that F is pseudo-reductive in such cases.

Step 4. Now assuming that F_a is reductive for every $a \in \Delta$, we need to show that F is reductive. Applying the settled (iv) over $\overline{k}$ does the job.

Step 5. It remains to show (in the quasi-reductive case) F is functorial in the sense described in the theorem. Consider a 5-tuple $(G', S', C', \Delta', \{F'_{a'}\}_{a' \in \Delta'})$ given with compatible isomorphisms $f_C : C \simeq C'$ and $f_{a'} : F_{a' \circ f_S} \simeq F'_{a'}$ for all $a' \in \Delta'$. We claim these uniquely glue to an isomorphism $f : F \simeq F'$.

The uniqueness is clear since F is generated by the F_a for $a \in \Delta$ (and $a' \mapsto a' \circ f_S$ is a bijection $\Delta' \simeq \Delta$). The uniqueness argument uses that f is a k-homomorphism rather than a k-isomorphism, so for the proof of existence of f it suffices to construct f as a k-homomorphism (as we can then apply the same reasoning to the inverses of the given isomorphisms f_C and $\{f_{a'}\}$ to build a k-homomorphism $F' \to F$ that is inverse to f). Our method, following Steinberg's idea in the reductive case, is to use the conclusions from Steps 1–4 to construct a k-subgroup of $G \times G'$ that is the graph of a k-homomorphism $F \to F'$ which is the f that we seek.

Inside the group $\mathscr{G} := G \times G'$, let $\mathscr{S} \subseteq S \times S'$ be the graph of $f_S : S \simeq S'$ and let $\mathscr{C} \subseteq Z_G(S) \times Z_{G'}(S') = Z_{\mathscr{G}}(\mathscr{S})$ be the graph of $f_C : C \simeq C'$, so $\mathscr{S}$ is a maximal split k-torus in $\mathscr{C}$. Define $\Delta_0 \subseteq \mathrm{X}(\mathscr{S})$ to be the subset corresponding to Δ via the natural isomorphism $\mathrm{X}(S) \simeq \mathrm{X}(\mathscr{S})$, so the natural isomorphism $\mathrm{X}(S') \simeq \mathrm{X}(\mathscr{S})$ identifies Δ' with Δ_0 via $\mathrm{X}(f_S) : \Delta' \simeq \Delta$.

For $a_0 \in \Delta_0$ corresponding to $a' \in \Delta'$ and $a \in \Delta$, let $\mathscr{F}_{a_0} \subseteq \mathscr{G}$ be the smooth connected graph of $f_{a'} : F_a \simeq F'_{a'}$, so $\mathscr{C} \subseteq \mathscr{F}_{a_0}$ and $Z_{\mathscr{F}_{a_0}}(\mathscr{S})$ is visibly equal to the graph $\mathscr{C}$ of f_C. It is clear that the projection isomorphisms $\mathscr{F}_{a_0} \simeq F_a$ and $\mathscr{F}_{a_0} \simeq F'_{a'}$ respectively identify $\Phi(\mathscr{F}_{a_0}, \mathscr{S})$ with $\Phi(F_a, S)$ and $\Phi(F'_{a'}, S')$, so as a subset of $\mathrm{X}(\mathscr{S})$ we see that $\Phi(\mathscr{F}_{a_0}, \mathscr{S})$ contains $\pm a_0$ and is contained in $\mathbf{Z} \cdot a_0$. Since $f_{a'}$ restricts to f_S on S, it restricts to isomorphisms $E_{\pm a} \simeq E'_{\pm a'}$ between corresponding root groups. It is clear that the $\pm a_0$-root groups $\mathscr{U}_{\pm a_0}$ of $(\mathscr{F}_{a_0}, \mathscr{S})$ are the graphs of these root group isomorphisms. Hence, for distinct $a_0, b_0 \in \Delta_0$, $\mathscr{U}_{a_0}$ commutes with $\mathscr{U}_{-b_0}$ since E_a commutes with E_{-b} and likewise $E'_{a'}$ commutes with $E'_{-b'}$ (where b_0 corresponds to the elements $b \in \Delta$ and $b' \in \Delta'$).

Step 6. Let $\mathscr{F} \subseteq \mathscr{G} = G \times G'$ be the smooth connected k-subgroup generated by the $\mathscr{F}_{a_0}$ for $a_0 \in \Delta_0$. By parts (i), (ii), and (iii), $Z_{\mathscr{F}}(\mathscr{S}) = \mathscr{C}$ (so $\mathscr{S}$ is a maximal split k-torus in $\mathscr{F}$) and every nontrivial $\mathscr{S}$-weight on $\mathrm{Lie}(\mathscr{F})$ is a nonnegative or nonpositive integral linear combination of elements of Δ_0, with $\mathscr{F}$ having $\pm a_0$-root group $\mathscr{U}_{\pm a_0}$ for all $a_0 \in \Delta_0$. Moreover, by (iv), $\mathscr{F}$ is quasi-reductive and Δ_0 is a basis of the relative root system $\Phi(\mathscr{F}, \mathscr{S})$. Consider the first projection $p_1 : \mathscr{F} \to G$. This restricts to the natural isomorphisms $\mathscr{C} \simeq C$ and $\mathscr{F}_{a_0} \simeq F_a$ for all $a_0 \in \Delta_0$, so $p_1(\mathscr{F}) = F$. Likewise $p_2(\mathscr{F}) = F'$. We shall prove that the restriction $\pi : \mathscr{F} \twoheadrightarrow F$ of p_1 has trivial kernel and hence is an isomorphism, so the map $p_2 \circ \pi^{-1} : F \to F'$ is a k-homomorphism extending every $f_{a'}$; in other words, this would be the map f that we want.

Clearly π carries $\mathscr{S}$ isomorphically onto S via the natural map and carries $\mathscr{C} = Z_{\mathscr{F}}(\mathscr{S})$ isomorphically onto $C = Z_F(S)$ via the natural map. By design, π restricts to an isomorphism $\mathscr{U}_{\pm a_0} \simeq E_{\pm a}$ between corresponding root groups for all $a_0 \in \Delta_0$. Let $\Phi_0 = \Phi(\mathscr{F}, \mathscr{S})$, so every non-divisible root is in the orbit of Δ_0 under the action of the Weyl group W_0 of Φ_0. The surjectivity of $N_{\mathscr{F}}(\mathscr{S})(k) \to W_0$ implies that every non-divisible element of Φ_0 is in the

$N_{\mathscr{F}}(\mathscr{S})(k)$-orbit of an element of Δ_0, so for every non-divisible $a_0 \in \Phi_0$ the corresponding nontrivial character $a \in \mathrm{X}(S)$ lies in $\Phi = \Phi(G,S)$ and π carries $\mathscr{U}_{a_0}$ isomorphically onto the a-root group E_a of (G,S).

It is clear that π carries the open cell Ω_0 of $(\mathscr{F},\mathscr{S},\Delta_0)$ into the open cell Ω of (G,S,Δ) respecting the direct product structures for each open cell (indexed by non-divisible roots) and having trivial kernel on each root group and Cartan subgroup factor of Ω_0. Thus, $(\ker\pi)\cap\Omega_0 = 1$, so $\ker\pi$ is finite étale and therefore is central in $\mathscr{F}$. This forces $\ker\pi \subseteq Z_{\mathscr{F}}(\mathscr{S}) = \mathscr{C}$, yet $\pi : \mathscr{C} \to C$ is an isomorphism, so $\ker\pi = 1$. □

It is natural to wonder if the reduced root system ${}_k\overline{\Phi}'$ consisting of non-multipliable roots in the k-root system ${}_k\overline{\Phi}$ coincides with the root system of a split connected reductive k-subgroup of G containing S as a maximal torus. The following theorem provides an affirmative answer under some hypotheses on the S-action on $\mathrm{Lie}(\mathscr{R}_{u,k}(G))$. For connected reductive G it recovers [BoTi1, Thm. 7.2], and for pseudo-split pseudo-reductive G it recovers Theorem 3.4.6. For reductive G, the proof given below is simpler than the proof of [BoTi1, Thm. 7.2].

Theorem C.2.30 *Let G be a smooth connected affine group over an infinite field k, and S a maximal k-split torus in G. Assume $S \neq 1$. Let ${}_k\overline{\Phi}$ be the root system of G with respect to S, and let ${}_k\overline{\Phi}'$ be the root system consisting of non-multipliable roots in ${}_k\overline{\Phi}$. Let Δ be a basis of ${}_k\overline{\Phi}'$. Assume that no nonzero weight of S on* $\mathrm{Lie}(\mathscr{R}_{u,k}(G))$ *is an integral multiple of a root in ${}_k\overline{\Phi}'$, and the root groups U_a and U_{-b} commute for all distinct $a, b \in \Delta$.*

For each $a \in \Delta$, let E_a be a smooth connected 1*-dimensional k-subgroup of U_a that is normalized by S (i.e., E_a is a k-linear subspace of U_a of dimension* 1*). There is a unique k-split connected reductive k-subgroup F of G containing S and every E_a. The root system of F with respect to S is ${}_k\overline{\Phi}'$, and the root datum of (F,S) is obtained from the one for (G,S) in the sense of Theorem C.2.15 by removing the multipliable roots and their associated coroots.*

In the formulation of the theorem, we are invoking the consequence of Proposition C.2.24 that (since $2a \notin {}_k\overline{\Phi}$, as $a \in {}_k\overline{\Phi}'$) U_a is commutative and moreover is p-torsion when $\mathrm{char}(k) = p > 0$. The k-group U_a admits a unique linear structure that linearizes the S-action (use Theorem B.4.3 when $\mathrm{char}(k) > 0$). It is also helpful to remember that if G is not pseudo-reductive, ${}_k\overline{\Phi}$ is generally smaller than the set $\Phi(G,S)$ of nontrivial S-weights on $\mathrm{Lie}(G)$. We also note that by Proposition C.2.27, for distinct $a, b \in \Delta$ the subgroups U_a and U_{-b} commute when $\Phi(G,S)$ does not contain any elements of the form $ma - nb$, with positive integers m and n. This occurs, for example, when $\Phi(G,S) = {}_k\overline{\Phi}$.

Proof Let $\mathscr{S}$ ($\subseteq S$) be the intersection of the kernels of roots belonging to ${}_k\overline{\Phi}'$ and let G' be the identity component of the centralizer of $\mathscr{S}$ in G. Thus,

$$\mathscr{R}_{u,k}(G') = G' \cap \mathscr{R}_{u,k}(G)$$

by Proposition A.8.14(2). The root system of G' with respect to S is the reduced root system ${}_k\overline{\Phi}'$. After replacing G with G', we may and do assume that the root system ${}_k\overline{\Phi}$ of G is reduced.

For each $a \in \Delta$, we fix $u_a \in E_a(k) - \{1\}$. By Proposition C.2.24(iii), there is a unique $v_a \in U_{-a}(k)$ such that $n_a := m(u_a) = v_a u_a v_a$ normalizes S, and the induced action of n_a on $\mathrm{X}(S)$ is the reflection r_a in a. This proposition also gives that $n_a^2 \in S(k)$ and $v_a = n_a^{-1} u_a n_a$. The 1-dimensional smooth connected k-subgroup $E_{-a} := n_a^{-1} E_a n_a$ of U_{-a} is stable under the conjugation action of S, so it is a 1-dimensional k-linear subgroup. Clearly $v_a \in E_{-a}(k)$.

Step 1. We claim that the k-subgroup E_{-a} does not depend on the choice of u_a. Since S acts on $E_a \simeq \mathbf{G}_\mathrm{a}$ through $a \neq 1$, if u'_a is another choice then $u'_a = s u_a s^{-1}$ for some $s \in S(\overline{k})$. Proposition C.2.24(ii) yields that $n'_a := m(u'_a) = s n_a s^{-1}$, so

$$\begin{aligned}(E'_{-a})_{\overline{k}} &:= n_a'^{-1} (E_a)_{\overline{k}} n'_a \\ &= s n_a^{-1} s^{-1} (E_a)_{\overline{k}} s n_a s^{-1} \\ &= n_a^{-1} (n_a s n_a^{-1} s^{-1}) (E_a)_{\overline{k}} (n_a s n_a^{-1} s^{-1})^{-1} n_a.\end{aligned}$$

Since $n_a s n_a^{-1} s^{-1} \in S(\overline{k})$ and E_a is normalized by S, the right side is $n_a^{-1} (E_a)_{\overline{k}} n_a = (E_{-a})_{\overline{k}}$. Thus, the k-subgroups E'_{-a} and E_{-a} coincide over $\overline{k}$, so they are equal over k.

Step 2. Let F_a be the smooth connected k-subgroup of G generated by S and $E_{\pm a}$. We will first prove that F_a is a k-split connected reductive k-subgroup of G whose root system with respect to the maximal torus S is $\{\pm a\}$. The key point is to give a concrete subgroup of $F_a(k)$ that is Zariski-dense in F_a. Since $n_a = v_a u_a v_a \in F_a(k)$, the subset

$$\Gamma_a = E_a(k)\{1, n_a\} S(k) E_a(k)$$

is contained in $F_a(k)$. We will now prove that Γ_a is a subgroup of $F_a(k)$ that is Zariski-dense in F_a. It is clear that Γ_a contains 1 and is stable under inversion in $G(k)$ (as n_a normalizes S and $n_a^2 \in S(k)$), so we just have to show that Γ_a is stable under multiplication. Since it is stable under left and right multiplication by $E_a(k)$ and $S(k)$, and $n_a^2 \in S(k)$, to prove that Γ_a is a subgroup it suffices to show that $n_a E_a(k) n_a^{-1}$ is contained in Γ_a.

By the transitivity of the conjugation action of $S(\overline{k})$ on $E_a(\overline{k}) - \{1\}$, every nontrivial $u \in E_a(k)$ has the form $u = su_as^{-1}$ for some $s \in S(\overline{k})$. For such an s, $a(s) \in k^\times$ since $u, u_a \in E_a(k)$. Thus, the conjugation action of s on $(E_{\pm a})_{\overline{k}}$, and so also of $n_asn_a^{-1}$ on $(E_{\pm a})_{\overline{k}}$, is k-rational. Now

$$n_aun_a^{-1} = n_asu_as^{-1}n_a^{-1} = n_asn_a^{-1} \cdot n_au_an_a^{-1} \cdot (n_asn_a^{-1})^{-1},$$

so it suffices to prove that $n_au_an_a^{-1} \in E_a(k)n_aE_a(k)$. Indeed, the conjugation action of $n_asn_a^{-1}$ keeps $E_a(k)$ stable (since it is a k-rational action), and the conjugate

$$c_s := n_asn_a^{-1} \cdot n_a \cdot (n_asn_a^{-1})^{-1}$$

of n_a under $n_asn_a^{-1}$ is equal to the product $n_a \cdot sn_as^{-1}n_a^{-1}$ that lies in $n_aS(k)$ (because, by Proposition C.2.24(ii), $sn_as^{-1} = m(su_as^{-1}) = m(u) \in G(k)$, forcing $sn_as^{-1}n_a^{-1} \in S(k)$).

The formula $n_a = v_au_av_a$ with $v_a = n_a^{-1}u_an_a$ yields

$$n_a = n_av_an_a^{-1} \cdot n_au_an_a^{-1} \cdot n_av_an_a^{-1} = u_a \cdot n_au_an_a^{-1} \cdot u_a,$$

so $n_au_an_a^{-1} = u_a^{-1}n_au_a^{-1} \in E_a(k)n_aE_a(k) \subseteq \Gamma_a$. This proves that Γ_a is a subgroup. Since $S(k)$, $E_a(k)$, and $n_a^{-1}E_a(k)n_a = E_{-a}(k)$ are Zariski-dense in S, E_a, and E_{-a} respectively, we conclude that Γ_a is Zariski-dense in F_a.

By Proposition A.8.10(1),(2), the scheme-theoretic centralizer of $\ker a$ in G is smooth with Lie algebra spanned by S-weight spaces in $\mathrm{Lie}(G)$ for the trivial weight and $\pm a$ (as a is non-multipliable in $\Phi(G, S)$, due to our hypothesis on the S-weights of $\mathrm{Lie}(\mathscr{R}_{u,k}(G))$). Thus, the identity component G_a of this centralizer is generated by $Z_G(S)$ and $U_{\pm a}$. The multiplication map

$$U_{-a} \times Z_G(S) \times U_a \longrightarrow G_a$$

is an open immersion (by Proposition 2.1.8(3), since $U_a = U_{G_a}(\lambda)$ for $\lambda \in X_*(S)$ satisfying $\langle a, \lambda\rangle > 0$). Since $n_a^{-1}U_an_a = U_{-a}$, the map

$$U_a \times Z_G(S) \times U_a \longrightarrow G_a$$

defined by $(u', z, u'') \mapsto u'n_azu''$ is an isomorphism onto the open subscheme

$$\Omega_a := U_an_aZ_G(S)U_a$$

of G_a. But $P_a := Z_G(S) \ltimes U_a$ is a minimal pseudo-parabolic k-subgroup of G_a containing S, so the Bruhat decomposition (Theorem C.2.8) gives that $\Omega_a(k) \cap$

$P_a(k)$ is empty. In particular, the set $E_a(k)S(k)E_a(k)$ ($\subseteq P_a(k)$) is disjoint from $\Omega_a(k)$, so $\Gamma_a \cap \Omega_a(k) = E_a(k)n_aS(k)E_a(k)$.

The formation of closures is of local nature in any topological space, so by the Zariski-density of Γ_a in F_a we conclude that the subset

$$\Gamma_a \cap \Omega_a(k) = E_a(k)n_aS(k)E_a(k)$$

is Zariski-dense in $F_a \cap \Omega_a$. The Zariski closure of $E_a(k)n_aS(k)E_a(k)$ in Ω_a is clearly $E_an_aSE_a$, so the open subscheme $F_a \cap \Omega_a$ of F_a is equal to $E_an_aSE_a$. In particular, $\dim F_a = 2 + \dim S$. But $n_a \in F_a(k)$ and $E_{-a} = n_a^{-1}E_an_a$, so

$$F_a \cap U_{-a}Z_G(S)U_a = F_a \cap n_a^{-1}\Omega_a = n_a^{-1}(F_a \cap \Omega_a) = n_a^{-1}E_an_aSE_a = E_{-a}SE_a.$$

Since $U_{-a}Z_G(S)U_a$ is a direct product scheme, we see that $F_a \cap Z_G(S) = S$ (hence S is a maximal torus of F_a), $F_a \cap U_a = E_a$, and $F_a \cap U_{-a} = E_{-a}$.

The derived group of any solvable smooth connected affine group H is unipotent, so in any such H the normalizer of a maximal torus is equal to its centralizer. Since $n_a \in F_a(k)$ normalizes S but does not centralize S, it follows that F_a is not solvable. Thus, the connected solvable codimension-1 subgroups $B_a := S \ltimes E_a$ and $B'_a := S \ltimes E_{-a}$ of F_a are Borel k-subgroups of F_a. Since $B_a \cap B'_a = S$ is a torus, we conclude that F_a is reductive. The root system of F_a with respect to S is clearly $\{\pm a\}$.

STEP 3. Let F be the smooth connected k-subgroup generated by the F_a for varying $a \in \Delta$. By our hypothesis, for distinct $a, b \in \Delta$ the root groups U_a and U_{-b} commute with each other, so E_a and E_{-b} commute with each other. Also, the split k-torus S is maximal in each F_a. Now Theorem C.2.29 (with $C = S$, $\Phi = {}_k\overline{\Phi}'$) implies that S is a maximal torus in F and that F is reductive (hence split).

To prove that the root system of F with respect to the k-split maximal k-torus S is ${}_k\overline{\Phi}'$, first note that $\Phi(F, S)$ clearly contains Δ, so it contains ${}_kW \cdot \Delta = {}_k\overline{\Phi}'$. To establish the reverse containment, choose $a \in \Phi(F, S)$ and let $\mathscr{U}_a \simeq \mathbf{G}_\mathrm{a}$ be the corresponding root group in F. The map $h : F \to \overline{G} := G/\mathscr{R}_{u,k}(G)$ has unipotent kernel that must be infinitesimal (due to the reductivity of F), and it restricts to the natural isomorphism $S \simeq \overline{S}$ onto the maximal k-split torus $\overline{S}$ in $\overline{G}$. Thus, h carries $\mathscr{U}_a = \mathbf{G}_\mathrm{a}$ onto an $\overline{S}$-stable $\mathbf{G}_\mathrm{a}$ in $\overline{G}$. Since this isogeny between the $\mathbf{G}_\mathrm{a}$ is equivariant with respect to the isomorphism $S \simeq \overline{S}$, if $p \geqslant 1$ is the characteristic exponent of k then $p^ra \in \Phi(\overline{G}, \overline{S}) = {}_k\overline{\Phi}$ for some $r > 0$. Thus, a is a rational multiple of an element of ${}_k\overline{\Phi}'$. Since $\Phi(F, S)$ is reduced (as F is reductive and k-split), the containment ${}_k\overline{\Phi}' \subseteq \Phi(F, S)$ must therefore be an equality. It is clear that the root datum for (F, S) is obtained from the one for (G, S) in the asserted manner.

STEP 4. Now only the uniqueness of F remains to be proved. Suppose H is a split connected reductive k-subgroup of G that contains S and E_a for every $a \in \Delta$. Note that S must be a maximal k-torus in H, since H is k-split and S is maximal as a k-split torus in G. Choose $a \in \Delta$. From the uniqueness assertion in Proposition C.2.24(i) (along with the existence assertion applied to (H, S, a)), the element $n_a = m(u_a)$ introduced at the beginning of this proof must lie in H. Hence, the open subscheme $E_a n_a S E_a$ of F_a is contained in H, so F_a is contained in H for all $a \in \Delta$. Therefore, $F \subseteq H$, so ${}_k\overline{\Phi}' \subseteq \Phi(H,S)$. The above proof that $\Phi(F,S) = {}_k\overline{\Phi}'$ may now be applied with H in place of F to establish that $\Phi(H,S) = {}_k\overline{\Phi}'$. Hence, F and H have common dimension $\#_k\overline{\Phi}' + \dim S$, so $H = F$ by connectedness. □

In the following proposition and its proof, we will use the fact (from Proposition B.4.4) that an action by a k-torus on a k-wound smooth connected unipotent k-group is necessarily trivial. We will also use the following lemma.

Lemma C.2.31 *Let G and $\overline{G}$ be smooth connected affine k-groups and $f : G \to \overline{G}$ a smooth surjective k-homomorphism. Let N be a smooth connected normal k-subgroup of G and let $\overline{N} = f(N)$. Let S be a maximal k-split torus of G. Then $S_N := (S \cap N)^0_{\mathrm{red}}$ is a maximal k-split torus of N and the image of S_N in $\overline{N}$ is a maximal k-split torus of $\overline{N}$.*

Proof The first assertion is an immediate consequence of Theorem C.2.3. To prove the second assertion, let T be a maximal k-torus of G containing S. As N is a normal subgroup of G, $T_N := T \cap N$ is a maximal torus of N (Corollary A.2.7), and as f is surjective, $f(T)$ and $f(T_N)$ are maximal tori in $\overline{G}$ and $\overline{N}$ respectively. Thus, $f(T_N) = f(T) \cap f(N)$. On the other hand, as f is smooth (and surjective), $f(S)$ ($\subseteq f(T)$) is a maximal k-split torus of $\overline{G}$ and hence it contains a maximal k-split torus of $\overline{N}$. This implies that the maximal k-split subtorus of $f(T_N) = f(T) \cap f(N)$ is a maximal k-split torus of $f(N) = \overline{N}$. But f carries the maximal k-split torus S_N of T_N onto the maximal k-split torus of $f(T_N)$. This proves the lemma. □

Proposition C.2.32 *Let G be a quasi-reductive k-group, S a maximal k-split torus in G (so $\mathscr{R}_{u,k}(G) \subseteq Z_G(S)$), and Φ the k-root system of G with respect to S. Let S' be the maximal k-split torus $(S \cap \mathscr{D}(G))^0_{\mathrm{red}}$ of $\mathscr{D}(G)$. (Since $\mathscr{R}_{u,k}(G) \subseteq Z_G(S)$, Φ is the set of nonzero S-weights on* Lie(G) *and is the root system of $\overline{G} = G/\mathscr{R}_{u,k}(G)$ with respect to the isomorphic image of S that is a maximal k-split torus of $\overline{G}$.) For each irreducible component Ψ of Φ, let N_Ψ be the k-subgroup generated by the root groups U_a, $a \in \Psi$.*

(1) *The k-group G is generated by $Z_G(S)$ and the N_Ψ.*
(2) *Each N_Ψ is perfect, nontrivial, and normal in G.*

(3) *The maximal k-split torus* $S_\Psi := (S \cap N_\Psi)^0_{\mathrm{red}}$ *of* N_Ψ *is nontrivial,* $\prod_\Psi S_\Psi \to S'$ *is an isogeny, and the k-root system of* N_Ψ *with respect to* S_Ψ *is naturally identified with* Ψ *via the natural isomorphism* $\mathrm{X}(S')_{\mathbf{Q}} \simeq \prod_\Psi \mathrm{X}(S_\Psi)_{\mathbf{Q}}$.
(4) *Every k-isotropic perfect smooth connected normal k-subgroup of G contains some* N_Ψ*, and no* N_Ψ *has a nontrivial perfect smooth connected proper normal k-subgroup.*
(5) *If G is pseudo-reductive then no* N_Ψ *has a nontrivial smooth connected proper normal k-subgroup.*

Before we prove the proposition, we make some preliminary remarks. A smooth connected affine k-group is k*-isotropic* when it contains GL_1 as a k-subgroup, and k*-anisotropic* otherwise. Part (3) implies that if $\mathscr{D}(G)$ is k-isotropic then Φ is non-empty and each N_Ψ is k-isotropic. Also, the normal subgroup N_Ψ inherits quasi-reductivity from G.

Finally, the minimality of the N_Ψ as nontrivial smooth connected normal k-subgroups of G in the pseudo-reductive case (as in part (5)) fails in the general case. For example, if k is imperfect of characteristic p and k'/k is a nontrivial finite extension with $k'^p \subseteq k$ then the smooth connected k-group $G := \mathrm{R}_{k'/k}(\mathrm{SL}_p)/\mu_p$ is perfect with a split maximal k-torus S and the k-wound unipotent smooth connected central k-subgroup $\mathrm{R}_{k'/k}(\mu_p)/\mu_p = \mathrm{R}_{k'/k}(\mathrm{GL}_1)/\mathrm{GL}_1$ is equal to $\mathscr{R}_{u,k}(G)$ (see Example 1.3.5), so $\mathscr{R}_{u,k}(G)$ is k-wound and $\Psi := \Phi(G,S)$ is irreducible of type A_{p-1}. Clearly $G = N_\Psi$ and $\mathscr{R}_{u,k}(G)$ is a nontrivial smooth connected proper normal k-subgroup.

Proof Each k-subgroup N_Ψ is clearly normalized by $Z_G(S)$, and Lie algebra considerations imply that G is generated by $Z_G(S)$ and the N_Ψ. For a choice of Ψ and roots $a \in \Psi$ and $b \in \Phi - \Psi$, U_a commutes with U_b due to Proposition 3.3.5, since $ma + nb \notin \Phi$ for any positive integers m and n. Since G is generated by $Z_G(S)$ and the root groups $\{U_b\}_{b \in \Phi}$, we see that N_Ψ is a (smooth connected) nontrivial normal k-subgroup of G. By Remark C.2.12(2) (applied to a maximal k-torus of G containing S), S is an almost direct product of the maximal k-split torus $(S_\Psi \cap \mathscr{D}(N_\Psi))^0_{\mathrm{red}}$ of $\mathscr{D}(N_\Psi)$ and the maximal k-subtorus of S that centralizes N_Ψ. Thus, $(S, U_a) = (S_\Psi, U_a) \subseteq \mathscr{D}(N_\Psi)$ for all $a \in \Psi$, yet $(S, U_a) = U_a$ by Lemma 3.3.8, so N_Ψ is perfect and $S_\Psi \neq 1$. We have established (1), (2), and the first assertion in (3).

We now treat the case where G is pseudo-reductive, and at the end we will deduce the general case from this case. Note that every smooth connected normal k-subgroup of G is also pseudo-reductive, and the derived group of a pseudo-reductive group is perfect (Proposition 1.2.6). We also recall (Remark 3.1.10) that a smooth connected normal k-subgroup of a smooth connected normal k-subgroup of G is normal in G. If a k-subgroup is minimal among the non-commutative smooth connected normal k-subgroups of G then it is perfect and hence is contained in $\mathscr{D}(G)$ (which is itself perfect). By

Remark C.2.12(2) (applied to a maximal k-torus of G containing S), S is an almost direct product of S' and the maximal k-split central torus of G. Each N_Ψ lies in $\mathscr{D}(G)$ by perfectness, so the k-root system of G with respect to S is identified with that of $\mathscr{D}(G)$ with respect to S'. We may therefore replace G with $\mathscr{D}(G)$ (and S with S'), so G is perfect and $S' = S$. We may and do assume that $S \neq 1$.

Let $\{N_i\}_{i\in I}$ be the set of minimal non-commutative smooth connected normal k-subgroups of $G = \mathscr{D}(G)$, so every N_i is perfect. According to Proposition 3.1.8, the subgroups N_i satisfy the following properties: they pairwise commute, the product homomorphism

$$\pi : N := \prod_{i\in I} N_i \longrightarrow G$$

is surjective, every perfect smooth connected normal k-subgroup of G is generated by $\{N_j\}_{j\in J}$ for a unique subset $J \subseteq I$, and $\ker \pi$ is central in N and contains no nontrivial smooth connected k-subgroup (e.g., $\ker \pi$ does not contain a nontrivial k-torus). No N_i contains a nontrivial smooth connected proper normal k-subgroup. Indeed, if N' is such a k-subgroup of some N_{i_0} then N' is normal in G and due to minimality of N_{i_0} among non-commutative smooth connected normal k-subgroups of G, N' must be commutative. The image of the commutative normal $N'_{\overline{k}}$ in the connected semisimple $(N_{i_0})^{\mathrm{red}}_{\overline{k}}$ must be trivial. Hence, N' is unipotent, contradicting the pseudo-reductivity of G.

For $i \in I$, let $S_i = (S \cap N_i)^0_{\mathrm{red}}$, a maximal k-split torus of N_i. Let $I^\sharp$ be the set of $i \in I$ such that N_i is k-anisotropic. Thus, $S_i = 1$ precisely for $i \in I^\sharp$, and the pair (N_i, S_i) has a non-empty k-root system for $i \in I - I^\sharp$ (since otherwise S_i is central in N_i, so the torus $(S_i)_{\overline{k}}$ has nontrivial central image in the semisimple $(N_i)^{\mathrm{red}}_{\overline{k}}$, an absurdity). We claim that S is an isogenous quotient of the maximal k-split torus $\prod_{i\in I-I^\sharp} S_i$ of N. Choose a maximal k-torus T of G containing S, so (by Corollary A.2.7) $T_i := T \cap N_i$ is a maximal k-torus of N_i that contains S_i. The product $\prod T_i$ is a maximal k-torus of N, so its image in $G = \pi(N)$ is a maximal k-torus of G. This forces $\prod T_i \to T$ to be surjective, so the maximal split k-subtorus $\prod_{i\in I} S_i = \prod_{i\in I-I^\sharp} S_i$ of $\prod T_i$ maps onto the maximal split k-subtorus S of T.

The isogeny $\prod_{i\in I-I^\sharp} S_i \to S$ implies that: $N_i \subseteq Z_G(S)$ for every $i \in I^\sharp$, π induces an injective homomorphism of $\mathrm{X}_*(\prod_{i\in I-I^\sharp} S_i) = \prod_{i\in I-I^\sharp} \mathrm{X}_*(S_i)$ into $\mathrm{X}_*(S)$ whose image is a subgroup of finite index, and $I - I^\sharp$ is non-empty. Thus, for any subset $J \subseteq I^\sharp$, the k-subgroup $N_J := \langle N_j\rangle_{j\in J}$ is k-anisotropic (because we can apply the preceding considerations to $N_{I^\sharp}$ in the role of G to deduce that $N_{I^\sharp}$ is k-anisotropic) and Φ is not empty.

For any $\lambda \in \prod_{i\in I-I^\sharp} \mathrm{X}_*(S_i)$, the restriction of π to $U_N(\lambda)$ is an isomorphism onto $U_G(\pi\circ\lambda)$ since the kernel of π is central (forcing $\ker \pi \subseteq Z_N(\lambda)$). Hence, the set of root groups of G relative to S is the (disjoint) union of the set of root groups of N_i relative to S_i over all $i \in I - I^\sharp$. Thus, Φ is the direct sum of the non-empty k-root systems $\Phi_i := \Phi(N_i, S_i)$ for $i \in I - I^\sharp$ compatibly with the decomposition $X(S)_{\mathbf{Q}} = \prod_{i \in I - I^\sharp} X(S_i)_{\mathbf{Q}}$.

We claim that for each $i \in I - I^\sharp$, the k-root system Φ_i in $X(S_i)_{\mathbf{Q}} \neq 0$ is irreducible. Granting this, $\{\Phi_i\}_{i \in I - I^\sharp}$ is the set of irreducible components of the root system Φ in the $\mathbf{Q}$-vector space $X(S)_{\mathbf{Q}} = \prod_{i \in I - I^\sharp} X(S_i)_{\mathbf{Q}}$ and (by minimality) the nontrivial perfect normal k-subgroup N_{Φ_i} of N_i must equal N_i for each $i \in I - I^\sharp$. This would establish (3) and (4) for pseudo-reductive groups, as well as (5). Thus, for the purpose of settling the pseudo-reductive case we can replace (G, S) with the k-isotropic (N_{i_0}, S_{i_0}) for each $i_0 \in I - I^\sharp$ separately. Now the k-isotropic perfect pseudo-reductive G has no nontrivial smooth connected normal k-subgroup apart from itself. It would suffice to prove that the non-empty root system Φ is irreducible.

Let Ψ be an irreducible component of Φ; our aim is to prove that $\Psi = \Phi$. Since N_Ψ is a nontrivial smooth connected normal k-subgroup of G, $N_\Psi = G$. Suppose there exists $b \in \Phi - \Psi$, so the nontrivial smooth connected unipotent k-group U_b centralizes $N_\Psi = G$ and hence is central in G. This contradicts that G is pseudo-reductive, so $\Psi = \Phi$, and we have completely proved the proposition in case G is pseudo-reductive.

We now consider general quasi-reductive G. Let $\overline{G} = G/\mathscr{R}_{u,k}(G)$ denote the maximal pseudo-reductive quotient and let $\overline{S}$ denote the isomorphic image of S in $\overline{G}$ (so $\overline{S}$ is a maximal k-split torus of $\overline{G}$). Under the natural smooth quotient map $G \to \overline{G}$, each root group U_a of G with respect to S is mapped isomorphically onto the corresponding root group $\overline{U}_a$ of $\overline{G}$ (as we see by consideration of Lie algebras and the triviality of the S-action on $\mathscr{R}_{u,k}(G)$). In particular, each N_Ψ maps onto the corresponding k-subgroup $\overline{N}_\Psi \subseteq \overline{G}$.

Let N be a k-isotropic perfect smooth connected normal k-subgroup of G. Its image $\overline{N}$ in $\overline{G}$ is k-isotropic and perfect. As we saw in the proof in the pseudo-reductive case, $\overline{N} = \langle \overline{N}_j \rangle_{j \in J}$ for a unique (non-empty) subset $J \subseteq I$ and the maximal k-split torus of $\overline{N}$ contained in $\overline{S}$ is an almost direct product of the maximal k-split tori $(\overline{S}_j \cap \overline{N}_j)^0_{\mathrm{red}}$ of $\overline{N}_j$ for $j \in J$. The k-isotropicity of $\overline{N}$ therefore forces $\overline{N}_{j_0}$ to be k-isotropic for some $j_0 \in J$, so $\overline{N}_{j_0} = \overline{N}_\Psi$ for some Ψ. Since $\overline{N}_\Psi \subseteq \overline{N}$ and $\overline{N}_\Psi$ is perfect, the smooth connected normal k-subgroup $\mathscr{N} := (N, N_\Psi)$ of G that is contained in both N and N_Ψ maps onto $\overline{N}_\Psi$. Although $\mathscr{N} \twoheadrightarrow \overline{N}_\Psi$ might not be smooth, for each *non-divisible* $a \in \Psi$ the a-root group of $\mathscr{N}$ maps onto the a-root group of $\overline{N}_\Psi$, due to C.2.21, Corollary 2.1.9, and Remark 2.1.11 (since the homomorphism $Z_{\mathscr{N}}((\ker a)^0_{\mathrm{red}}) \to Z_{\overline{N}_\Psi}((\ker a)^0_{\mathrm{red}})$ is surjective). By dimension considerations for the a-root groups of G and $\overline{G}$, it follows that $\mathscr{N}$ contains the entire a-root group of G, and since this holds for all non-divisible $a \in \Psi$ we conclude that $\mathscr{N}$ contains N_Ψ. This establishes the first part of (4).

Since $N_{\Psi'}$ is not contained in N_Ψ for $\Psi' \neq \Psi$ (as we may compare their images in $\overline{G}$), it follows that the k-subgroups N_Ψ are precisely the k-isotropic minimal perfect smooth connected normal k-subgroups of G if for each Ψ the only nontrivial perfect smooth connected normal k-subgroup N' of N_Ψ is

N_Ψ itself. By (5) for $\overline{G}$ the (visibly nontrivial) image $\overline{N}'$ of N' in $\overline{G}$ is $\overline{N}_\Psi$, so $N' \to \overline{N}_\Psi$ is surjective. The same argument with root groups as in the proof of the first part of (4) now implies that N' contains the a-root group of (G, S) for all non-divisible $a \in \Psi$, so N' contains N_Ψ and hence $N' = N_\Psi$.

It remains to prove (3) for quasi-reductive G. By Lemma C.2.31, under the isomorphism $S \to \overline{S}$, the maximal k-split torus S_Ψ of N_Ψ is mapped onto the maximal k-split torus $\overline{S}_\Psi$ of $\overline{N}_\Psi$. Now let $\overline{S}_0$ be the maximal k-split central torus of $\overline{G}$. The k-torus $\overline{S}_0$ is killed by all $a \in \Phi(\overline{G}, \overline{S}) = \Phi(G, S)$, so under the isomorphism $S \to \overline{S}$ it lifts to a central k-torus S_0 of G contained in S. (It is easily seen that S_0 is the maximal k-split central torus of G.) The natural map $S_0 \times \prod_\Psi S_\Psi \to \overline{S}_0 \times \prod_\Psi \overline{S}_\Psi$ is clearly an isomorphism, and as the product map $\overline{S}_0 \times \prod_\Psi \overline{S}_\Psi \to \overline{S}$ is an isogeny and the natural map $S \to \overline{S}$ is an isomorphism, we conclude that the product map $S_0 \times \prod_\Psi S_\Psi \to S$ is an isogeny. Since $S_\Psi \subseteq N_\Psi \subseteq \mathscr{D}(G)$ and $\prod_\Psi \overline{S}_\Psi \to \overline{S}' := (\overline{S} \cap \mathscr{D}(\overline{G}))^0_{\rm red}$ is an isogeny, it follows that $\prod S_\Psi \to S'$ is an isogeny too. Thus, under the resulting isomorphism $\mathrm{X}(S')_\mathbf{Q} \simeq \prod_\Psi \mathrm{X}(S_\Psi)_\mathbf{Q}$, the k-root system for N_Ψ with respect to S_Ψ is identified with Ψ due to the settled pseudo-reductive case. □

Remark C.2.33 Let G be a quasi-reductive k-group, S a maximal k-split torus of G, and P a minimal pseudo-parabolic k-subgroup of G containing S. Let $N = N_G(S)$ and $Z = Z_G(S)$. As $\mathscr{R}_{u,k}(G)$ is k-wound, it commutes with S (Proposition B.4.4) and hence $Z \supseteq \mathscr{R}_{u,k}(G)$. By Corollary 2.2.5, Theorem C.2.3 and Proposition C.2.4, there exists a $\lambda \in \mathrm{X}_*(S)$ such that $P = P_G(\lambda) = U_G(\lambda) \rtimes Z_G(\lambda)$ with $Z_G(\lambda) = Z$ and $\mathscr{R}_{us,k}(P) = U_G(\lambda)$.

Let $G(k)^+$ be the normal subgroup of $G(k)$ generated by the k-rational points of the k-split unipotent radicals of pseudo-parabolic k-subgroups of G. Using the conjugacy of minimal pseudo-parabolic k-subgroups of G under $G(k)$ (Theorem C.2.5), it is easily seen that $G(k)^+$ is the normal subgroup of $G(k)$ generated by $\mathscr{R}_{us,k}(P)(k)$.

It is immediate from Proposition C.2.24 that $G(k)^+ \cap N(k)$ maps onto the k-Weyl group ${}_kW = N(k)/Z(k)$. Now let $\mathscr{G}$ be a subgroup of $G(k)$ that contains $G(k)^+$, $\mathscr{N} := N(k) \cap \mathscr{G}$, and $\mathscr{B} := P(k) \cap \mathscr{G}$. Then $\mathscr{N}$ maps onto ${}_kW$. As $U_G(\lambda)(k) \subseteq \mathscr{B}$, we see that $P(k) = U_G(\lambda)(k) \rtimes Z(k) = \mathscr{B}Z(k)$. Moreover, for all $n \in \mathscr{N}$ we have $P(k)nP(k) = \mathscr{B}n\mathscr{B}Z(k)$ and $\mathscr{G} \cap (P(k)nP(k)) = \mathscr{B}n\mathscr{B}$. It is seen from this, using the Bruhat decomposition of $G(k)$ (Theorem C.2.8), that $G(k) = \mathscr{G}Z(k)$. Now it is easily deduced from Theorem C.2.20 and Remark C.2.22 that the 4-tuple $(\mathscr{G}, \mathscr{B}, \mathscr{N}, R)$, with R as in Theorem C.2.20, is a saturated and split Tits system with Weyl group ${}_kW$. We shall call this Tits system a *standard* Tits system in $\mathscr{G}$. Its rank (i.e., the cardinality of R) is equal to the k-rank of $\mathscr{D}(G)$; see Theorem C.2.15.

The equality $G(k) = \mathscr{G}Z(k)$, together with Theorem C.2.3 (resp., Theorem C.2.5), also implies that all maximal k-split tori (resp., all minimal pseudo-parabolic k-subgroups) of G are conjugate to each other under $\mathscr{G}$.

We close this section with the following theorem.

Theorem C.2.34 *Let k be an infinite field, G a k-isotropic perfect quasi-reductive k-group, S a maximal k-split torus of G, and P a minimal pseudo-parabolic k-subgroup of G containing S. Let $U = \mathscr{R}_{us,k}(P)$, and let $G(k)^+$ be the normal subgroup of $G(k)$ generated by the $G(k)$-conjugates of $U(k)$. We assume that G does not contain any nontrivial perfect smooth connected proper normal k-subgroup. The following hold.*

(1) *The abstract group $G(k)^+$ is perfect.*
(2) *The group $G(k)^+$ is Zariski-dense in G.*
(3) *A subgroup of $G(k)$ that is normalized by $G(k)^+$ either contains $G(k)^+$ or is contained in the center of $G(k)$. Any non-central normal subgroup of $G(k)$ is Zariski-dense in G.*

Proof Let ${}_k\Phi$ be the k-root system of G with respect to S; this is non-empty by Proposition C.2.32(3). As G is quasi-reductive, ${}_k\Phi$ is simply the set of nontrivial S-weights on $\mathrm{Lie}(G)$. Let ${}_k\Phi^+$ $(\subseteq {}_k\Phi)$ be the positive system of k-roots determined by the minimal pseudo-parabolic k-subgroup P, so $U_a \subseteq U$ for all $a \in {}_k\Phi^+$. (In particular, $U \neq 1$.) Let P' be the minimal pseudo-parabolic k-subgroup containing S that corresponds to the positive system of k-roots $-{}_k\Phi^+$, and let $U' = \mathscr{R}_{us,k}(P')$. There exists $g \in N_G(S)(k)$ such that $P' = gPg^{-1}$, so $U' = gUg^{-1}$ and hence $U'(k) \subseteq G(k)^+$.

Fix $a \in {}_k\Phi^+$, and define $S_a = (\ker a)^0_{\mathrm{red}}$. For a nontrivial $u \in U_a(k)$, let $m(u) = u'uu''$ be as in Proposition C.2.24(i). The action of $m(u)$ on $\mathrm{X}(S)$ is the reflection r_a in a, so the conjugation action of $m(u)$ on S is trivial on S_a and the induced action on S/S_a is by inversion. For $s \in S(k)$, clearly $m(sus^{-1}) = sm(u)s^{-1}$, so the conjugation action of $m(sus^{-1})m(u)^{-1} = s \cdot m(u)s^{-1}m(u)^{-1}$ on U_a is the conjugation action of s^2. Since u' and u'' lie in $U'(k)$, $m(u)$ and $m(sus^{-1}) = sm(u)s^{-1}$ belong to $G(k)^+$. Thus, for $s \in S(k)$ the conjugation action of s^2 on U_a equals the conjugation action of an element of $G(k)^+$. From this it follows (via the infinitude of k, and the description of the S-action on suitable vector group subquotients of U_a over k) that the derived group of $G(k)^+$ contains $U_a(k)$ for all $a \in {}_k\Phi^+$. But these $U_a(k)$ generate $U(k)$ (due to the direct spanning property in Proposition C.2.26(1), where the greatest common divisors are actually roots since $\Phi(G, S)$ is a root system), so the derived group of $G(k)^+$ contains $U(k)$. Since $G(k)^+$ is the normal subgroup of $G(k)$ generated by the conjugates of $U(k)$, $G(k)^+$ is perfect. Thus, we have proved (1).

Since G is perfect by hypothesis, $G(k)$ is Zariski-dense in G (by Proposition A.2.11). Now to prove that $G(k)^+$ is Zariski-dense in G, we note that $U(k)$ is Zariski-dense in the k-split unipotent smooth connected k-subgroup $U \neq 1$, so the Zariski closure H of $G(k)^+$ in G is the subgroup generated by the k-subgroups gUg^{-1} for $g \in G(k)$. Hence, H is connected and nontrivial, and the Zariski-density of $G(k)$ in G implies that H is normal in G. As $G(k)^+$ is perfect, H is perfect. But we have assumed that G does not contain any nontrivial perfect smooth connected proper normal k-subgroups, so $H = G$; i.e., $G(k)^+$ is Zariski-dense in G.

The subgroup M generated by the k-subgroups gSg^{-1} for $g \in G(k)$ is a smooth connected k-subgroup of G that is normal (since $G(k)$ is Zariski-dense in G). We assert that $M = G$. The image of $M_{\overline{k}}$ in the maximal reductive quotient $G^{\rm red}_{\overline{k}}$ (this quotient is semisimple since G is perfect) is a nontrivial normal (and hence semisimple) subgroup. Thus, M cannot be solvable. But since G does not contain any nontrivial perfect smooth connected proper normal k-subgroup, any smooth connected proper normal k-subgroup of G is necessarily solvable, so $M = G$.

We will now show that the normal subgroup $C := \bigcap_{g \in G(k)} gP(k)g^{-1}$ in $G(k)$ is the center $\mathscr{C}$ of $G(k)$. Since the opposite pseudo-parabolic k-subgroup P' is $G(k)$-conjugate to P, C lies in $P(k) \cap P'(k) = Z(k)$, so C is a normal subgroup of $G(k)$ that commutes with S, and hence with every conjugate of S. As we saw above, the $G(k)$-conjugates of S generate G, so we conclude that C is central. On the other hand, $\mathscr{C}$ lies in $Z(k)$ and hence in $P(k)$. Thus, $\mathscr{C}$ lies in every $gP(k)g^{-1}$ ($g \in G(k)$) and hence in C. This proves that $C = \mathscr{C}$.

Since G does not contain any nontrivial perfect smooth connected proper normal k-subgroup, by Proposition C.2.32 the non-empty k-root system ${}_k\Phi$ is irreducible. Thus, when the k-Weyl group $({}_kW)(k) = W({}_k\Phi)$ is viewed as a Coxeter group in the natural way, its associated Coxeter graph is connected (due to the relationship between the Dynkin diagram of a root system and the Coxeter graph of the corresponding Weyl group; see [Bou2, VI, §4.2]). Also, $P = Z \ltimes U$ and U is unipotent, so $P(k) = Z(k) \ltimes U(k)$ and $U(k)$ is nilpotent. Now the first assertion of (3) follows from [Bou2, IV, §2.7, Thm. 5]. Thus, any non-central normal subgroup of $G(k)$ must contain $G(k)^+$, so the Zariski-density of the latter group in G implies the last assertion. □

C.3 Split unipotent subgroups of pseudo-reductive groups

Let G be a smooth connected affine group over a field k. If there is a non-central k-split k-torus T in G then there is a non-central 1-parameter k-subgroup $\lambda : \mathrm{GL}_1 \to G$. Hence, $Z_G(\lambda)$ is a proper subgroup of G, yet $U_G(-\lambda)\cdot Z_G(\lambda)\cdot U_G(\lambda)$ is open in G (Proposition 2.1.8(3)), so at least one of the subgroups $U_G(-\lambda)$ and $U_G(\lambda)$ is nontrivial. Since both $U_G(-\lambda)$ and $U_G(\lambda)$ are k-split smooth connected unipotent k-groups (Proposition 2.1.10), if G contains a non-central GL_1 then G contains $\mathbf{G}_{\rm a}$.

In this section we will prove a converse to the preceding conclusion for pseudo-reductive groups: if a pseudo-reductive G contains $\mathbf{G}_{\rm a}$ then it contains a non-central GL_1. In fact, we will go further and show that any such $\mathbf{G}_a$ must occur inside of $\mathscr{R}_{us,k}(P)$ for a proper pseudo-parabolic k-subgroup P of G. (Recall from Lemma 2.2.3 that the existence of such a P is equivalent to the existence of a non-central GL_1 in G.) We will also prove a unipotent analogue of Theorem C.2.3: all maximal k-split smooth connected unipotent k-subgroups U in a pseudo-reductive k-group G are $G(k)$-conjugate. This is closely related to the $G(k)$-conjugacy of minimal pseudo-parabolic k-subgroups P of G (Theorem C.2.5) because we will prove that all such U have the form $U = \mathscr{R}_{us,k}(P)$ for some minimal P.

The preceding facts were announced by Borel and Tits without proof. In [Ti3], Tits sketched proofs of some of them. We will provide proofs over a general ground field, based on a variant of an argument of Kempf in [Kem]. The key fact to be shown is that if G is pseudo-reductive over k then any nontrivial k-split smooth connected unipotent k-subgroup $U \subseteq G$ is contained in $\mathscr{R}_{us,k}(P)$ for some pseudo-parabolic k-subgroup P of G. (By pseudo-reductivity of G, such a P must be a proper pseudo-parabolic k-subgroup of G.)

We now fix a *pseudo-reductive* group G over a *separably closed* field k. (This restriction on k will be eliminated later by using uniqueness results to carry out Galois descent from k_s to k.) We also fix a faithful representation $\rho : G \hookrightarrow \mathrm{GL}_n$ over k, and let Mat_n denote the k-algebra scheme of $n \times n$ matrices, in which GL_n is the open unit group. Endow Mat_n with a structure of finite-dimensional G-module over k via the conjugation action $g.x = gxg^{-1}$ for $g \in G$ and $x \in \mathrm{Mat}_n$. Finally, define $\mathrm{X}_*(G) := \mathrm{Hom}_k(\mathrm{GL}_1, G)$ to be the set of all 1-parameter k-subgroups $\lambda : \mathrm{GL}_1 \to G$. There is a natural left action of $G(k)$ on the set $\mathrm{X}_*(G)$ via conjugation: for $g \in G(k)$ and $\lambda \in \mathrm{X}_*(G)$, define $(g.\lambda)(t) = g\lambda(t)g^{-1}$.

Let U be a nontrivial k-split smooth connected unipotent k-subgroup of G. For each $\lambda \in \mathrm{X}_*(G)$, consider the set $\mathscr{S}_\lambda$ of integers m such that for all $u \in U(k)$, the k-scheme map $\mathrm{GL}_1 \to \mathrm{Mat}_n$ defined by

$$t \mapsto t^{-m}(\lambda(t).(u-1))$$

(using the linear structure on Mat_n to define the scaling action by $t^{-m} \in \mathrm{GL}_1$) extends to a k-map $\mathbf{A}^1_k \to \mathrm{Mat}_n$. By considering the weights for $\lambda(\mathrm{GL}_1)$ acting on Mat_n, we see that all sufficiently negative m lie in $\mathscr{S}_\lambda$. Likewise, by choosing $u \in U(k) - \{1\}$ and considering the nontrivial components of $u - 1$ along some weight spaces for $\lambda(\mathrm{GL}_1)$ in Mat_n we see that $\mathscr{S}_\lambda$ is bounded above. Hence, it is legitimate to define

$$f(\lambda) = \max \mathscr{S}_\lambda \in \mathbf{Z}.$$

Note that $f(\lambda)$ depends on U (as well as on the fixed faithful representation ρ of G), but U will be fixed throughout our discussion.

Lemma C.3.1 *For $\lambda \in X_*(G)$, let $P = P_G(\lambda)$ be the associated pseudo-parabolic k-subgroup of G. For all $p \in P(k)$, we have $f(p.\lambda) = f(\lambda)$.*

Proof Let $p \in P(k)$. For all $u \in U(k)$ and $m \in \mathbf{Z}$, since the G-action on Mat_n is k-linear we see that

$$t^{-m}((p\lambda(t)p^{-1}).(u-1)) = (p\lambda(t)p^{-1}\lambda(t)^{-1}).(t^{-m}(\lambda(t).(u-1))).$$

By definition of $P = P_G(\lambda)$, the k-scheme map $\mathrm{GL}_1 \to G$ defined by $t \mapsto \lambda(t)p^{-1}\lambda(t)^{-1}$ extends to a k-scheme map $\mathbf{A}^1_k \to G$, so it is obvious from the above equation that $\mathscr{S}_\lambda \subseteq \mathscr{S}_{p.\lambda}$. Since $P_G(p.\lambda) = P_G(\lambda)$ and $p^{-1}.(p.\lambda) = \lambda$, we conclude that $\mathscr{S}_\lambda = \mathscr{S}_{p.\lambda}$. Hence, $f(p.\lambda) = f(\lambda)$. □

Now let S be any maximal k-torus of G. Any character χ of S gives rise to a $\mathbf{Z}$-linear functional $X_*(S) \to \mathbf{Z}$ via $\lambda \mapsto \langle \chi, \lambda \rangle$. Define

$$V_S = \mathbf{R} \otimes_{\mathbf{Z}} X_*(S).$$

Extending scalars to $\mathbf{R}$ defines an $\mathbf{R}$-linear functional $\ell^S_\chi : V_S \to \mathbf{R}$ that carries the lattice $X_*(S)$ into $\mathbf{Z}$ via $\langle \chi, \cdot \rangle$.

Using the left S-action on Mat_n via conjugation, consider the S-submodule $S.(U-1)$ of Mat_n generated by the points $u-1$ for $u \in U(k)$.

Definition C.3.2 Let Ξ_S be the set of S-weights occurring in the linear representation space $S.(U-1)$ for S. For $v \in V_S$, define $\ell^S(v) = \min_{\chi \in \Xi_S} \ell^S_\chi(v)$.

Since each ℓ^S_χ is a linear form on V_S, ℓ^S is a continuous function. It is easy to check that it satisfies the following three properties: (i) for $\lambda \in X_*(S)$, we have $f(\lambda) = \ell^S(\lambda)$, (ii) $\ell^S(sv) = s\ell^S(v)$ for any nonnegative real number s, and (iii) ℓ^S is a *concave function*; i.e., for all $0 \leqslant s \leqslant 1$ and $v, v' \in V_S$, we have $\ell^S(sv + (1-s)v') \geqslant s\ell^S(v) + (1-s)\ell^S(v')$. By (iii), for any $c \in \mathbf{R}$ the set $\{v \in V_S \mid \ell^S(v) \geqslant c\}$ is convex in V_S.

Definition C.3.3 A maximal k-torus S in G is *special* with respect to the nontrivial k-split smooth connected unipotent k-subgroup $U \subseteq G$ if the set $\{v \in V_S \mid \ell^S(v) \geqslant 1\}$ is non-empty, or equivalently there exists $\lambda \in X_*(S)$ such that $f(\lambda) = \ell^S(\lambda) > 0$. (This definition does not depend on the faithful representation ρ of G, as it says $U \subseteq U_G(\lambda)$.)

Lemma C.3.4 *There exist maximal k-tori in G that are special with respect to U.*

Proof By Lemma C.2.2, there exists a minimal pseudo-parabolic k-subgroup P of G that contains U. Since $k = k_s$ by our hypothesis, the minimality of P forces the pseudo-reductive k-group $P/\mathscr{R}_{us,k}(P)$ to be commutative (Corollary 2.2.5 and Proposition 3.3.15(3)), so it contains no nontrivial smooth connected unipotent k-subgroup. In particular, the image of U in this quotient is trivial, so $U \subseteq \mathscr{R}_{us,k}(P)$. Let S be a maximal k-torus of P, so S is also a maximal k-torus of G. We may and do choose $\lambda \in \mathrm{X}_*(S)$ such that $P = P_G(\lambda)$. Since $U \subseteq \mathscr{R}_{us,k}(P) = U_G(\lambda)$, we have $\ell^S(\lambda) = f(\lambda) \geqslant 1$. Thus, S is special with respect to U. □

It will be convenient to introduce a $\mathbf{Z}$-valued positive-definite inner product on $\mathrm{X}_*(S)$ for each maximal k-torus S of G (though only special S will be of interest to us) such that the isomorphism $\mathrm{X}_*(S) \simeq \mathrm{X}_*(gSg^{-1})$ defined by g-conjugation for any $g \in G(k)$ is an isometry. To make such a construction, fix one maximal k-torus T in G and let $W = N_G(T)(k)/Z_G(T)(k)$ be the associated finite Weyl group. There is a natural left W-action on $\mathrm{X}_*(T)$, akin to the left $G(k)$-action on $\mathrm{X}_*(G)$ via conjugation.

Lemma C.3.5 *The natural map $W\backslash \mathrm{X}_*(T) \to G(k)\backslash \mathrm{X}_*(G)$ is bijective.*

Proof In view of the $G(k)$-conjugacy of maximal k-tori of G (Proposition A.2.10), for any $\lambda \in \mathrm{X}_*(G)$ there is a $g \in G(k)$ such that $g.\lambda \in \mathrm{X}_*(T)$. This proves surjectivity.

To prove injectivity, consider $g \in G(k)$ and $\lambda \in \mathrm{X}_*(T)$ such that $g.\lambda$ also belongs to $\mathrm{X}_*(T)$. Hence, the smooth connected affine k-group $Z_G(\lambda)$ contains the maximal k-tori $g^{-1}Tg$ and T, so by $Z_G(\lambda)(k)$-conjugacy of maximal k-tori of $Z_G(\lambda)$ there exists $z \in Z_G(\lambda)(k)$ such that $z^{-1}g^{-1}Tgz = T$. This implies $gz \in N_G(T)(k)$, so gz represents a class $w \in W$ such that $g.\lambda = w.\lambda$. □

Now fix a W-invariant $\mathbf{Z}$-valued positive-definite inner product on $\mathrm{X}_*(T)$. For any maximal k-torus S of G, there exists $g \in G(k)$ such that $gTg^{-1} = S$. The identification of S with T via g-conjugation is independent of g up to the W-action on T. Hence, if we identify $\mathrm{X}_*(S)$ with $\mathrm{X}_*(T)$ via g-conjugation and thereby transfer the W-invariant positive-definite inner product on $\mathrm{X}_*(T)$ over to a $\mathbf{Z}$-valued inner product on $\mathrm{X}_*(S)$ then this latter inner product is independent of g. In this way, for every S we get a $\mathbf{Z}$-valued positive-definite inner product on the lattice $\mathrm{X}_*(S)$ such that for every $g \in G(k)$ the identification of $\mathrm{X}_*(S)$ with $\mathrm{X}_*(gSg^{-1})$ via g-conjugation is compatible with the inner product. Applying extension of scalars to $\mathbf{R}$ gives an inner product on each $V_S = \mathbf{R} \otimes_{\mathbf{Z}} \mathrm{X}_*(S)$ with an analogous compatibility property. We will denote the norm of $v \in V_S$ under this inner product by $\|v\|_S$.

Consider any maximal k-torus S in G that is special with respect to U (Lemma C.3.4). The non-empty convex closed set $\{v \in V_S \mid \ell^S(v) \geqslant 1\}$ contains a point v_S with minimal distance to the origin relative to the inner product we have constructed on V_S. By convexity, v_S is unique. We must have $\ell^S(v_S) = 1$, for if $\ell^S(v_S) > 1$ then by the homogeneity of ℓ^S we could scale v_S by some $c \in (0, 1)$ such that $\ell^S(cv_S) \geqslant 1$, contradicting the minimality of the distance from v_S to the origin. The points $c \cdot v_S$ with $c > 0$ can obviously be characterized as the nonzero elements of V_S where the function $v \mapsto \ell^S(v)/\|v\|_S$ takes the maximum value.

Lemma C.3.6 *There exists $c > 0$ such that $c \cdot v_S$ lies in the lattice $X_*(S)$ in V_S.*

Proof Since $\ell^S(v_S) = 1$, from the definition of ℓ^S we see that the set $\Xi'_S := \{\chi \in \Xi_S \mid \ell^S_\chi(v_S) = 1\}$ is non-empty. Consider the saturated **Z**-submodule

$$L' = \{\lambda \in X_*(S) \mid \ell^S_\chi(\lambda) = \ell^S_{\chi'}(\lambda) \text{ for all } \chi, \chi' \in \Xi'_S\}$$

of $X_*(S)$. Since each ℓ^S_χ is **Z**-valued on $X_*(S)$, the corresponding **R**-subspace

$$V' = \{v \in V_S \mid \ell^S_\chi(v) = \ell^S_{\chi'}(v) \text{ for all } \chi, \chi' \in \Xi'_S\}$$

of V_S is identified with $\mathbf{R} \otimes_{\mathbf{Z}} L'$, so $L' = V' \cap X_*(S)$ and L' is a lattice in V'. By definition of Ξ'_S we have $v_S \in V'$, so $V' \neq 0$.

The functions ℓ^S_χ for $\chi \in \Xi'_S$ restrict to a common linear form ϕ on V', and this linear form is **Z**-valued on the lattice L'. Since $\phi(v_S) = 1$ by definition of Ξ'_S, we see that $\phi \neq 0$. Hence, $H := \ker \phi$ is a hyperplane in V'. We claim that the line $\mathbf{R}v_S$ in V' through 0 is the orthogonal complement $H^\perp$ of H (in V') with respect to the inner product we have constructed on V_S at the outset, or in other words that v_S is perpendicular to H. Near v_S in V_S, the linear forms ℓ^S_χ for $\chi \in \Xi_S - \Xi'_S$ are all strictly larger than 1, so near v_S in V' the function ϕ coincides with ℓ^S. Hence, the point v_S in the half-space $\{\phi \geqslant 1\}$ is a local minimum for distance to the origin, and so it is the unique global minimum. It follows by elementary Euclidean geometry that v_S is perpendicular to the hyperplane $H = \ker \phi$ parallel to the half-space boundary $\phi^{-1}(1)$, as claimed.

Since ϕ and the inner product on $V_S = \mathbf{R} \otimes_{\mathbf{Z}} X_*(S)$ respectively arise from a **Z**-valued linear form on $X_*(S)$ and a **Z**-valued inner product on $X_*(S)$, orthogonal projection from $H = \ker \phi$ is defined over **Q**. Hence, the orthogonal complement $H^\perp = \mathbf{R}v_S$ meets $X_*(S)_{\mathbf{Q}}$ in a **Q**-line and so meets $X_*(S)$ in an infinite cyclic group. □

By Lemma C.3.6, there is a unique positive multiple λ_S of v_S that lies in $X_*(S)$ and generates $X_*(S) \cap \mathbf{R}v_S$. We call it the *optimal* 1-parameter k-subgroup of S. Of course, λ_S depends on U. Observe that λ_S also depends on the **Z**-valued inner product on $X_*(S)$ (constructed from a choice of W-invariant **Z**-valued inner product on $X_*(T)$ for an initial choice of maximal k-torus T in G). For $\lambda \in X_*(S)$ we have $\ell^S(\lambda) = f(\lambda)$, so the earlier characterization of the points $c \cdot v_S$ with $c > 0$ (as the nonzero elements of V_S where the function $v \mapsto \ell^S(v)/\|v\|_S$ takes the maximum value) leads to the following useful alternative description of λ_S:

$$\begin{array}{l}\lambda_S \text{ is the unique non-divisible element of the lattice } \mathrm{X}_*(S) \\ \text{where the function } \lambda \mapsto f(\lambda)/\|\lambda\|_S \text{ takes the maximum value.}\end{array} \quad (*)$$

Our interest in working with special maximal k-tori (with respect to U) is due to the following remarkable property of their associated optimal 1-parameter k-subgroups:

Lemma C.3.7 *For each special maximal k-torus S in G with respect to U, consider the pseudo-parabolic k-subgroup $P_S = P_G(\lambda_S)$ associated to the optimal 1-parameter k-subgroup λ_S. These k-subgroups all coincide; i.e., they are independent of S.*

Proof Let S' and S'' be two special maximal k-tori. Consider the intersection $P_{S'} \cap P_{S''}$. This intersection contains a maximal k-torus S of G (Proposition 3.5.12(1)), and $P_{S'}(k)$-conjugacy of maximal k-tori in $P_{S'}$ provides $p' \in P_{S'}(k)$ such that $p'S'p'^{-1} = S$. Likewise, there exists $p'' \in P_{S''}(k)$ such that $p''S''p''^{-1} = S$.

Since $f(\lambda') = f(p'.\lambda')$ and $\|\lambda'\|_{S'} = \|p'.\lambda'\|_S$ for all $\lambda' \in \mathrm{X}_*(S')$, by Lemma C.3.1 and the characterization $(*)$ of optimal 1-parameter subgroups given above, it follows that the maximal k-torus S in G is special and that its optimal 1-parameter k-subgroup λ_S is equal to $p'.\lambda_{S'}$. Hence, as $p' \in P_{S'}(k)$, we conclude that $P := P_G(\lambda_S)$ is equal to $P_{S'}$. In a similar manner, $\lambda_S = p''.\lambda_{S''}$ and $P = P_{S''}$. Thus, $P_{S'} = P_{S''}$. □

Now we are in position to prove the first main result, dropping the preceding hypothesis that the ground field is separably closed.

Theorem C.3.8 *Let G be a pseudo-reductive group over a field k, H a smooth closed k-subgroup, and U a k-split smooth connected unipotent k-subgroup of G normalized by H. There exists a pseudo-parabolic k-subgroup P of G containing H such that $U \subseteq \mathscr{R}_{us,k}(P)$.*

Note that we do not require H to be connected.

Proof We may assume that U is nontrivial, so the discussion above is applicable to U_{k_s} inside of G_{k_s}. Fix a faithful representation $\rho : G \hookrightarrow \mathrm{GL}_n$ over k. Let T be a maximal k-torus in G, so $W := W(G,T)(k_s)$ and $\mathrm{X}_*(T_{k_s})$ are equipped with natural actions by $\Gamma := \mathrm{Gal}(k_s/k)$ factoring through a finite discrete quotient. Hence, we can construct a W-invariant $\mathbf{Z}$-valued inner product on $\mathrm{X}_*(T_{k_s})$ that is also Γ-invariant.

Consider the natural left action by Γ on the set of closed subschemes Z of G_{k_s} by using the k-structure G on G_{k_s}. This action is defined via extension of scalars with respect to the natural action of Γ on k_s over k via automorphisms, which is to say

$$^{\gamma}Z := k_s \otimes_{\gamma,k_s} Z \subseteq k_s \otimes_{\gamma,k_s} G_{k_s} \simeq G_{k_s}$$

as k_s-schemes. For any maximal k_s-torus S in G_{k_s} and $\gamma \in \Gamma$ we can also apply such extension of scalars to 1-parameter subgroups of S, viewed as k_s-homomorphisms from the k_s-group GL_1 (where GL_1 is equipped with its natural k-structure) to S, and we can do similarly for characters. This identifies $\mathrm{X}_*(S)$ with $\mathrm{X}_*(^{\gamma}S)$ and identifies $\mathrm{X}(S)$ with $\mathrm{X}(^{\gamma}S)$ as $\mathbf{Z}$-modules. We claim that this identification of cocharacter groups respects the inner products defined by transport from $\mathrm{X}_*(T_{k_s})$.

To check the compatibility with inner products, pick $g \in G(k_s)$ such that $S = gT_{k_s}g^{-1}$. The inner product on $\mathrm{X}_*(S)$ is obtained from that on $\mathrm{X}_*(T_{k_s})$ via g-conjugation. Clearly $^{\gamma}S = \gamma(g)T_{k_s}\gamma(g)^{-1}$, so the inner product on $\mathrm{X}_*(^{\gamma}S)$ is obtained from that on $\mathrm{X}_*(T_{k_s})$ via $\gamma(g)$-conjugation. But the composite of g-conjugation $\mathrm{X}_*(T_{k_s}) \simeq \mathrm{X}_*(S)$ with γ-twisting $\mathrm{X}_*(S) \simeq \mathrm{X}_*(^{\gamma}S)$ is the same as the composite of the γ-action on $\mathrm{X}_*(T_{k_s})$ and $\gamma(g)$-conjugation $\mathrm{X}_*(T_{k_s}) \simeq \mathrm{X}_*(^{\gamma}S)$. Indeed, for any $\lambda \in \mathrm{X}_*(T_{k_s})$ the γ-twist of the map $k_s^{\times} \to S(k_s)$ defined by $t \mapsto g\lambda(t)g^{-1}$ is $t \mapsto \gamma(g) \cdot \gamma(\lambda)(t) \cdot \gamma(g)^{-1}$ into $^{\gamma}S(k_s)$. Thus, we are reduced to checking the Γ-invariance of the inner product chosen on $\mathrm{X}_*(T_{k_s})$. We arranged this invariance property when constructing the inner product on $\mathrm{X}_*(T_{k_s})$.

In addition to the γ-twisting identification of $\mathrm{X}_*(S)$ and $\mathrm{X}_*(^{\gamma}S)$ respecting inner products for $\gamma \in \Gamma$, the action of γ on the set of closed subschemes of $(\mathrm{Mat}_n)_{k_s}$ takes $S.(U_{k_s} - 1)$ to $^{\gamma}S.(U_{k_s} - 1)$. Hence, the isomorphism $\mathrm{X}(S) \to \mathrm{X}(^{\gamma}S)$ obtained via γ-twisting carries Ξ_S to $\Xi_{^{\gamma}S}$. It follows that the isometry $V_S \simeq V_{^{\gamma}S}$ defined by scalar extension to $\mathbf{R}$ carries ℓ_{χ}^S to $\ell_{\gamma(\chi)}^{^{\gamma}S}$ and ℓ^S to $\ell^{^{\gamma}S}$, so by definition we see that S is special if and only if $^{\gamma}S$ is special, with $\gamma(v_S) \in V_{^{\gamma}S}$ necessarily equal to $v_{^{\gamma}S}$. We conclude from the definitions that when S is special, $\gamma(\lambda_S) = \lambda_{^{\gamma}S}$ for any $\gamma \in \Gamma$.

Consider the common pseudo-parabolic k_s-subgroup $Q = P_{G_{k_s}}(\lambda_S)$ for all special maximal k_s-tori S in G_{k_s} (Lemma C.3.7). As H normalizes U, if S is a special torus then for every $h \in H(k_s)$ the torus $S_h := hSh^{-1}$ is also special, and the optimal 1-parameter k_s-subgroup λ_S is carried to the optimal 1-parameter subgroup λ_{S_h} under conjugation by h. Therefore, Q is normalized by $H(k_s)$, so the Zariski-density of $H(k_s)$ in H_{k_s} implies that H_{k_s} normalizes Q. Then by Proposition 3.5.7, $H_{k_s} \subseteq Q$. For any $\gamma \in \Gamma$, we have

$$^{\gamma}Q = P_{G_{k_s}}(\gamma(\lambda_S)) = P_{G_{k_s}}(\lambda_{^{\gamma}S}) = Q.$$

Thus, by Galois descent, Q descends to a k-subgroup P in G; this is a pseudo-parabolic k-subgroup by Proposition 3.5.2(1). The containments $H_{k_s} \subseteq Q = P_{k_s}$ and $U_{k_s} \subseteq U_{G_{k_s}}(\lambda_S)\,\mathscr{R}_{us,k_s}(Q) = \mathscr{R}_{us,k}(P)_{k_s}$ then imply that H is contained in P and U is contained in $\mathscr{R}_{us,k}(P)$. □

Corollary C.3.9 *Let G be a smooth connected affine group over a field k, and assume that the maximal pseudo-reductive quotient $G' = G/\mathscr{R}_{u,k}(G)$ does not contain a non-central k-split k-torus. Then every k-split smooth connected unipotent k-subgroup of G is contained in $\mathscr{R}_{u,k}(G)$ (and hence in $\mathscr{R}_{us,k}(G)$).*

Proof The formation of images of smooth connected k-subgroups under the quotient map $G \twoheadrightarrow G'$ preserves the property of being k-split unipotent, so we can replace G with G' so that G is pseudo-reductive. Assume that G contains a nontrivial k-split smooth connected unipotent k-subgroup U. By Theorem C.3.8, G contains a pseudo-parabolic k-subgroup P such that $U \subseteq \mathscr{R}_{us,k}(P)$. This forces $P \neq G$ (as $U \neq 1$), so G contains a proper pseudo-parabolic k-subgroup. Since G is pseudo-reductive over k, it then follows from Lemma 2.2.3 that G contains a non-central k-split k-torus, contrary to the hypotheses. □

Remark C.3.10 For a pseudo-reductive k-group G, the following are equivalent:

(i) every k-split torus is central,
(i′) $\mathscr{D}(G)$ is k-anisotropic (i.e., it has no nontrivial k-split torus),
(ii) $\mathbf{G}_a$ is not a k-subgroup of G,
(ii′) every smooth connected unipotent k-subgroup of G is k-wound (in the sense of Definition B.2.1).

Indeed, the equivalence between (i) and (i′) follows from Lemma 1.2.5(iii), the equivalence of (ii) and (ii′) follows from Proposition B.3.2, and Corollary C.3.9 gives that (i) implies (ii). Finally, (ii) implies (i) by Lemma 2.2.3 and Proposition 2.1.10.

Over an imperfect field, a k-anisotropic connected semisimple group can contain a nontrivial k-wound smooth connected unipotent k-subgroup. For example, let k be a local function field of characteristic $p > 0$ and D a central division algebra over k of rank p^2. By local class field theory such D exist and the field $k' = k^{1/p}$ embeds into D over k. This identifies $\mathrm{R}_{k'/k}(\mathrm{GL}_1)$ as a k-subgroup of the algebraic unit group $\underline{D}^{\times}$ over k, so the k-anisotropic adjoint connected semisimple quotient $\underline{D}^{\times}/\mathrm{GL}_1$ contains the $(p-1)$-dimensional commutative smooth connected k-group $U := \mathrm{R}_{k'/k}(\mathrm{GL}_1)/\mathrm{GL}_1$ as a k-subgroup. The k-group U is unipotent since it is p-torsion, and Example B.2.8 gives a direct proof that U is k-wound.

Even with the additional requirement that G is simply connected (in addition to being connected, semisimple, and k-anisotropic) it can happen that G contains a nontrivial smooth connected (k-wound) unipotent k-subgroup. Examples of such G are given in [GQ].

Proposition C.3.11 *Let G be a smooth connected affine group over a field k, and P a minimal pseudo-parabolic k-subgroup of G. Then $\mathscr{R}_{us,k}(P)$ contains every k-split smooth connected unipotent k-subgroup of P.*

Proof By Corollary B.3.5, it suffices to prove that $\mathscr{R}_{u,k}(P)$ contains every k-split smooth connected unipotent k-subgroup of P. By definition of pseudo-parabolicity we have $\mathscr{R}_{u,k}(G) \subseteq \mathscr{R}_{u,k}(P)$, so we can replace G with the pseudo-reductive group $G/\mathscr{R}_{u,k}(G)$ and replace P with $P/\mathscr{R}_{u,k}(G)$ to reduce to the case that G is pseudo-reductive.

Let $\lambda : \mathrm{GL}_1 \to G$ be a 1-parameter k-subgroup such that $P = P_G(\lambda) = Z_G(\lambda) \ltimes U_G(\lambda)$. Since $U_G(\lambda) = \mathscr{R}_{u,k}(P) = \mathscr{R}_{us,k}(P)$ (Corollary 2.2.5), we see that $Z_G(\lambda) \simeq P/\mathscr{R}_{us,k}(P)$ is pseudo-reductive over k, and by minimality of P in G the k-group $Z_G(\lambda)$ contains no proper pseudo-parabolic k-subgroup (due to Proposition 2.2.10 and Remark 3.5.6). Hence, by Lemma 2.2.3, $Z_G(\lambda)$ does not contain any non-central k-split tori. It then follows from Corollary C.3.9 that $Z_G(\lambda)$ contains no nontrivial k-split smooth connected unipotent k-subgroup.

Consider any k-split smooth connected unipotent k-subgroup U of P. The image of U in $P/\mathscr{R}_{us,k}(P)$ is a k-split smooth connected unipotent k-subgroup, but we have just seen that the only such k-subgroup is the trivial one. Hence, $U \subseteq \mathscr{R}_{us,k}(P)$. □

Theorem C.3.12 *Let G be a smooth connected affine group over a field k. All maximal k-split unipotent smooth connected k-subgroups (resp. all maximal k-split solvable smooth connected k-subgroups) of G are $G(k)$-conjugate to each other.*

Proof By Lemma C.2.2 and Proposition C.3.11, any maximal k-split unipotent smooth connected k-subgroup U of G has the form $U = \mathscr{R}_{us,k}(P)$ for some minimal pseudo-parabolic k-subgroup P of G. The $G(k)$-conjugacy of minimal pseudo-parabolic k-subgroups (Theorem C.2.5) then gives the desired $G(k)$-conjugacy of all such U.

Now let H be a maximal k-split solvable smooth connected k-subgroup of G. Then by Lemma C.2.2 there is a minimal pseudo-parabolic k-subgroup P which contains H. Since $H \cdot \mathscr{R}_{us,k}(P)$ is a k-split solvable smooth connected k-subgroup, maximality of H implies that it contains $\mathscr{R}_{us,k}(P)$. The structure of k-split solvable smooth connected affine k-groups [Bo2, 15.4(i)] implies that $H = S \ltimes \mathscr{R}_{us,k}(H)$ for a k-split k-torus S. According to Proposition C.3.11,

$\mathscr{R}_{us,k}(P)$ contains every k-split unipotent smooth connected k-subgroup of P. Hence, $\mathscr{R}_{us,k}(H) \subseteq \mathscr{R}_{us,k}(P)$, so $H = S \ltimes \mathscr{R}_{us,k}(P)$.

Since H is maximal among k-split solvable smooth connected k-subgroups, S is a maximal k-split torus of P. Proposition C.2.4 implies that $P = Z_G(S) \ltimes \mathscr{R}_{us,k}(P)$, and as S is central in $Z_G(S)$ we conclude that the k-subgroup $H = S \ltimes \mathscr{R}_{us,k}(P)$ is the unique maximal k-split solvable smooth connected k-subgroup of P. The $G(k)$-conjugacy of minimal pseudo-parabolic k-subgroups (Theorem C.2.5) now implies the $G(k)$-conjugacy of the maximal k-split solvable smooth connected k-subgroups of G. □

C.4 Beyond the smooth affine case

The rational conjugacy of maximal split tori in smooth connected affine groups is valid more generally in all groups of finite type over a field: it is not necessary to assume smoothness or affineness. To bypass affineness we will use general structure theorems as recorded in Theorem A.3.7 and Theorem A.3.9. The following lemma is useful for avoiding smoothness hypotheses more generally in situations that involve properties of rational points over separable extensions of the ground field, as we illustrate in Example C.4.3.

Lemma C.4.1 *Let X be a scheme locally of finite type over a field k. There is a unique geometrically reduced closed subscheme $X^{\natural} \subseteq X$ such that $X^{\natural}(k') = X(k')$ for all separable extension fields k'/k. The formation of $X^{\natural}$ is functorial in X and commutes with the formation of products over k and separable extension of the ground field. In particular, if X is a k-group scheme then $X^{\natural}$ is a smooth k-subgroup scheme.*

In this lemma we permit non-algebraic separable extensions. This is important for applications with k a global function field (over a finite field) and k' its completion at some place.

Proof First we check uniqueness. Let k_s/k be a separable closure. If $X^{\natural}$ is such a closed subscheme then $X^{\natural}(k_s) = X(k_s)$ is Zariski-dense in $X^{\natural}$ by geometric reducedness (which implies generic smoothness), so the closed subscheme $X^{\natural}_{k_s} \subseteq X_{k_s}$ has to be the schematic closure of the subset $X(k_s)$. This settles uniqueness, and for existence consider the Zariski closure Z of $X(k_s)$ in X_{k_s}. This is a closed subscheme, and Z is geometrically reduced over k_s (as we may see by reducing to the case of affine X and using injectivity of the natural map $K' \otimes_K \prod V_i \to \prod(K' \otimes_K V_i)$ for any field extension K'/K and any indexed collection of K-vector spaces V_i). By Galois descent, Z descends uniquely to a closed subscheme $X^{\natural} \subseteq X$ that is necessarily geometrically reduced over k. Thus, we just have to check that $X^{\natural}(k') = X(k')$ for all separable extensions k'/k.

If k'_s/k' denotes a separable closure then $X^\natural(k') \subseteq X^\natural(k'_s)$ is the subset of $\mathrm{Gal}(k'_s/k')$-fixed points, and likewise for $X(k') \subseteq X(k'_s)$. Thus, if the functorial inclusion $X^\natural(k'_s) \subseteq X(k'_s)$ is an equality then the same holds for k'-points. We may therefore restrict our attention to the case when k' is separably closed. Fixing a k-embedding of k_s into k' thereby reduces us to the case when k is separably closed (in view of how $X^\natural$ was defined via Galois descent). Hence, we now have that $X^\natural$ is (by definition) the Zariski closure $\overline{X(k)}$ of $X(k)$ in X.

Our aim is to prove $X(k') \subseteq \overline{X(k)}(k')$ for all separable extensions k'/k (when k is separably closed). Since k' is the direct limit of its k-smooth subalgebras, if $\mathscr{I}$ is the coherent (radical) ideal sheaf of $\overline{X(k)}$ in X then it suffices to show that for all geometrically reduced k-algebras A of finite type and $x \in X(A)$ the x-pullback of $\mathscr{I}$ to an ideal in A vanishes. Since A is reduced it suffices to check such vanishing in $A/\mathfrak{p}$ for a Zariski-dense set of primes $\mathfrak{p}$ of A. But A is a generically smooth k-algebra and k is separably closed, so such a Zariski-dense set is given by the k-rational points of $\mathrm{Spec}(A)$. That is, it suffices to treat the case $A = k$, and this case is a tautology by the definition of $\mathscr{I}$ in terms of $X(k)$. This settles the existence, and functoriality is clear by the construction. Uniqueness and the defining property of $X^\natural$ imply that its formation commutes with any separable extension on k.

It remains to prove that if X and Y are two k-schemes locally of finite type then the geometrically reduced closed subscheme $X^\natural \times Y^\natural \subseteq X \times Y$ is $(X \times Y)^\natural$. By functoriality with respect to the projections, we have $(X \times Y)^\natural \subseteq X^\natural \times Y^\natural$. Likewise, by functoriality the inclusion $X^\natural \times Y^\natural \hookrightarrow X \times Y$ factors through $(X \times Y)^\natural$ because $(X^\natural \times Y^\natural)^\natural = X^\natural \times Y^\natural$. Hence, the desired equality holds. □

Remark C.4.2 If G is a k-group scheme locally of finite type over k then $G^\natural$ is the maximal smooth closed k-subgroup scheme of G, so it is also denoted as G^{sm}. For any G-torsor E, $E^\natural$ is either empty or a G^{sm}-torsor, hence smooth, so we denote it as E^{sm}. When using Lemma C.4.1, it is difficult to control connectedness properties. More specifically, for any imperfect field k with $\mathrm{char}(k) = p > 0$ there exists a connected affine k-group scheme H of finite type such that H^{sm} is disconnected. In fact, there exist H such that $(H_{\overline{k}})_{\mathrm{red}} \simeq \mathbf{G}_{\mathrm{a}}$ but $H^{\mathrm{sm}} = \mathbf{Z}/p\mathbf{Z}$. For example, we can use $H = \{(y^p - y)^p = ax^p\}$ for $a \in k - k^p$.

Example C.4.3 Let G be a group scheme of finite type over a global field k, and let S be a finite set of places of k. Let $\text{Ш}^1_S(k, G)$ denote the pointed set of isomorphism classes of right G-torsors E for the fppf topology over $\mathrm{Spec}\, k$ that are locally trivial away from S (i.e., admit a k_v-point for all places $v \notin S$). By using Lemma C.4.1, the study of such torsors can be carried out with a smooth k-group of finite type in place of G.

To explain how this goes, we first make some preliminary observations. Any separable extension of k (such as k_v/k for any $v \notin S$) is a directed union of fraction fields of smooth k-algebras, so the specialization method as in Proposition 1.1.9(1) shows that such E are automatically torsors for the étale topology on $\operatorname{Spec} k$. In other words, E can be described as a descent over k of the trivial G_K-torsor G_K for some finite Galois extension K/k. Since G is quasi-projective over k (Proposition A.3.5), this description via Galois descent implies that E is quasi-projective over k.

The k-group G^{sm} is affine when G is affine, but it may be disconnected even if G is connected. The pointed set $Ш^1_S(k, G^{\mathrm{sm}})$ is more accessible than $Ш^1_S(k, G)$ in general because it can be studied in the framework of Galois cohomology (since G^{sm} is smooth). Thus, it is a useful fact that the natural map $Ш^1_S(k, G^{\mathrm{sm}}) \to Ш^1_S(k, G)$ defined via pushout of torsors along $G^{\mathrm{sm}} \to G$ is bijective. (Such a pushout applied to a G^{sm}-torsor E' is the quotient E of $E' \times G$ by the equivalence relation $(xg', g) \sim (x, g'g)$ for points g' of G^{sm}. This quotient exists as a trivial G_K-torsor for a finite Galois extension K/k trivializing the G^{sm}-torsor E', so it exists over k due to the effectivity of Galois descent for quasi-projective schemes over a field and $E^{\mathrm{sm}} = E'$ inside E) To prove bijectivity of the pushout map $Ш^1_S(k, G^{\mathrm{sm}}) \to Ш^1_S(k, G)$, we now describe an inverse in terms of a functor on torsors.

Let E be a right G-torsor over k such that $E(k_v)$ is non-empty for all places $v \notin S$. The compatibility of the formation of E^{sm} with respect to any separable extension of the ground field can be applied to the extension $k \to k_{v_0}$ for a single $v_0 \notin S$, so E^{sm} is a right G^{sm}-torsor inside the right G-torsor E because the analogous claim is obvious over k_{v_0}. Also, $E^{\mathrm{sm}}(k_v) = E(k_v)$ for all v since k_v/k is separable, and this is non-empty for $v \notin S$. Hence, E^{sm} represents a class in $Ш^1_S(k, G^{\mathrm{sm}})$. The inclusion $E^{\mathrm{sm}} \hookrightarrow E$ is equivariant with respect to the inclusion of G^{sm} into G and identifies E^{sm} as the maximal smooth closed subscheme of E, so pushout along $G^{\mathrm{sm}} \to G$ carries the class $[E^{\mathrm{sm}}]$ of E^{sm} in $Ш^1_S(k, G^{\mathrm{sm}})$ to the class $[E]$ of E in $Ш^1_S(k, G)$. A map $Ш^1_S(k, G) \to Ш^1_S(k, G^{\mathrm{sm}})$ in the other direction is induced by $[E] \mapsto [E^{\mathrm{sm}}]$, and we have just shown that it is an inverse to the map $Ш^1_S(k, G^{\mathrm{sm}}) \to Ш^1_S(k, G)$.

As an application of Lemma C.4.1 over general fields, we can analyze conjugacy of maximal split tori in possibly non-smooth groups and prove some other related facts without hypotheses of smoothness or affineness. We first check that the notion of "maximal k-torus" interacts well with extension on k in many situations going beyond the familiar smooth affine case.

Lemma C.4.4 *Let G be a group scheme locally of finite type over a field k, T a maximal k-torus in G, and K/k an extension field. If G is smooth or commutative, or if K/k is separable, then T_K is a maximal K-torus in G_K.*

Proof The formations of $G^{\mathrm{sm}} \subseteq G$ commutes with any separable extension on k, and this k-subgroup contains every k-torus of G, so the case when K/k

is separable is reduced to the case when G is smooth. In both the smooth and commutative cases we can replace G with G^0 so that G is connected and hence of finite type [SGA3, $\mathrm{VI_A}$, 2.4].

Suppose G is connected and commutative. Any k_s-torus in G_{k_s} has a $\mathrm{Gal}(k_s/k)$-orbit consisting of finitely many k_s-tori, so by commutativity they collectively generate a $\mathrm{Gal}(k_s/k)$-stable k_s-subtorus. Thus, by Galois descent, T_{k_s} is maximal in G_{k_s}. In general we can replace K with K_s, so we may assume k and K are separably closed. For any integer n not divisible by $\mathrm{char}(k)$, $G[n]$ is finite étale, hence a constant group. Every K-torus S in G_K contains as a Zariski-dense subset the union of the $S[n](K)$ for n not divisible by $\mathrm{char}(K)$. But $S[n](K) \subseteq G[n](K) = G[n](k)$ and so each $S[n]$ uniquely descends to a k-subgroup of G. The Zariski closure of these descents is a smooth k-subgroup H in G such that $H_K = S$, so H is a k-torus. In particular, if S contains T_K then H contains T. By maximality of T, this forces $S = T_K$ and so settles the commutative case.

Finally, assume G is connected and smooth. In this case, the result for affine G is well known; see Corollary A.2.6. In general, by Proposition A.8.10(1) we can replace G with $Z_G(T)^0$ so that T is central, and then pass to G/T so that $T = 1$. In this case we need to show that G_K has no nontrivial K-tori. By Theorem A.3.9 there is a short exact sequence

$$1 \to Z \to G \to \overline{G} \to 1$$

with $\overline{G}$ a smooth connected affine k-group and Z a smooth connected central k-subgroup. Since Z contains no nontrivial k-tori, the settled commutative case shows that this property persists after any ground field extension.

Let $\overline{T} \subseteq \overline{G}$ be a maximal k-torus, so $\overline{T}_K \subseteq \overline{G}_K$ is a maximal K-torus (as $\overline{G}$ is affine). We need to show that G_K has no nontrivial K-torus, so we can assume K is algebraically closed. Thus by conjugacy of maximal K-tori in $\overline{G}_K$, any K-torus in G_K can be conjugated into H_K for the preimage H of $\overline{T}$ in G. Hence we may assume $\overline{G}$ is a k-torus. With $\overline{G}$ a k-torus, G is necessarily commutative since its commutator map defines a pairing $\overline{G} \times \overline{G} \to Z$ that is easily checked to be bi-additive and hence trivial (since a bi-additive pairing between k-tori valued in any commutative group of finite type must be valued in a torus and hence vanish, due to the rigidity of tori). Thus, the settled commutative case implies that G_K contains no nontrivial K-tori. □

The hypotheses in Lemma C.4.4 are optimal in the sense that over any imperfect field k of characteristic $p > 0$ there are "many" counterexamples to the conclusion if we do not assume smoothness or commutativity of G or separability of K/k. To be precise, for any affine k-group scheme H of finite type we claim that there exists a finite flat surjective homomorphism $f : G \to H$ with infinitesimal kernel such that $G(k_s) = 1$. Granting this, for smooth connected H that is not unipotent (e.g., GL_n) the underlying reduced scheme $(G_{k_p})_{\mathrm{red}}$ of G_{k_p} over the perfect closure k_p of k is k_p-smooth and connected but not unipotent (as it admits an isogeny onto H_{k_p}), so it contains a nontrivial k_p-torus. Such a torus descends to a nontrivial K-torus inside G_K for a purely inseparable finite extension K/k, yet G contains no nontrivial k-torus (as $G(k_s) = 1$); any such G must be non-smooth and non-commutative (due to Lemma C.4.4).

To construct such a G for a given H, by consideration of fiber products we see that a k-subgroup inclusion of H into some GL_n reduces the problem to the case $H = \mathrm{GL}_n$. Let $j : \mathrm{GL}_n \hookrightarrow \mathrm{GL}_{2n}$ be the inclusion $h \mapsto \left(\begin{smallmatrix} h & 0 \\ 0 & 1_n \end{smallmatrix}\right)$ in terms of $n \times n$ block matrices. Pick $c \in k - k^p$ and let $u = \left(\begin{smallmatrix} 1_n & c\cdot 1_n \\ 0 & 1_n \end{smallmatrix}\right)$. Define the k-group scheme

$$G = \{(h, M) \in \mathrm{GL}_n \times \mathrm{GL}_{2n} \mid uj(h)u^{-1} = \phi(M)\},$$

where $\phi : \mathrm{GL}_{2n} \to \mathrm{GL}_{2n}$ is the Frobenius isogeny over k (i.e., p-power on matrix entries), and let $f : G \to \mathrm{GL}_n = H$ be projection to the first factor. The map f is obviously bijective on $\overline{k}$-points, so $\ker f$ is infinitesimal, and the k-homomorphism $G/(\ker f) \to H$ must be a closed immersion (Proposition A.2.1). This closed immersion is surjective on $\overline{k}$-points, so it is an isomorphism (as H is reduced). It follows that f is a finite flat surjection, and $G(k_s)$ consists of pairs $(\phi(A), M)$ with $M = \left(\begin{smallmatrix} A & B \\ 0 & 1_n \end{smallmatrix}\right)$ for $A \in \mathrm{GL}_n(k_s)$ and $B \in \mathrm{Mat}_n(k_s)$ satisfying $c\phi(1_n - A) = \phi(B)$. If $1_n - A$ has non-vanishing ij entry for some (i, j) then the same holds for B and $b_{ij}/(\delta_{ij} - a_{ij}) \in k_s^\times$ is a pth root of c, contradicting that c has no pth root in k (and so cannot acquire a pth root in a separable extension of k). Thus, $G(k_s) = 1$ as desired.

Proposition C.4.5 *Let G be a group scheme locally of finite type over an arbitrary field k.*

(1) *All maximal k-split k-tori in G are $G(k)$-conjugate.*
(2) *Let $f : G \twoheadrightarrow H$ be a surjective homomorphism onto a k-group H locally of finite type. If f is smooth then it carries maximal k-tori onto maximal k-tori, and likewise for maximal k-split k-tori. Moreover, if T is a maximal k-torus (resp. maximal k-split k-torus) in H then it lifts to one in G provided that $f^{-1}(T)^0_{\mathrm{red}}$ is smooth (e.g., f is smooth).*

Proof To prove (1), the same argument as at the beginning of the proof of Lemma C.4.4 allows us to reduce to the case when G is smooth and connected. By [BLR, 9.2/1], G is an extension of an abelian variety A by a possibly non-smooth connected affine k-group H. Every k-torus in G is contained in H, so we may replace G with H to reduce to the case when G is affine but possibly not smooth. Now going through the reduction to the smooth connected case again brings us to the smooth connected affine case. This case is Theorem C.2.3.

To prove (2), we first treat the assertion concerning images of maximal k-tori, so we may assume f is smooth. The case when G is affine (so H is also affine) is Proposition A.2.8. In general, let S be a maximal k-torus in G and let T be a maximal k-torus in H containing $f(S)$. To show $T = f(S)$ we may replace H with T and G with $f^{-1}(T)$ so that $H = T$ is a k-torus and G is smooth. In this case we want $f(S) = T$. We note for later purposes that smoothness of f will no longer be used.

Since G is smooth, by Lemma C.4.4 we may assume that k is algebraically closed. If the subtorus $f(G^0)$ in T is distinct from T then each connected component of G maps onto a translate of this subtorus and so f would not hit the generic point of T, contrary to the surjectivity hypothesis. Hence, $f(G^0) = T$, so we can replace G with G^0 so that G is smooth and connected.

By Theorem A.3.9, there is a central extension

$$1 \to Z \to G \to \overline{G} \to 1$$

with Z a smooth connected commutative k-group and $\overline{G}$ a smooth connected affine k-group. The natural map $\overline{G} = G/Z \to T/f(Z)$ induced by f between smooth connected affine k-groups is surjective, so by the known case of smooth connected affine groups there is a k-torus $\overline{S}$ in $\overline{G}$ containing the image of S such that $\overline{S}$ maps onto $T/f(Z)$.

We may replace G with the preimage of $\overline{S}$ in G (as this maps onto T under f), so G is a central extension of the torus $\overline{S}$ by the smooth connected Z. This forces G to be commutative, as we saw in the proof of Lemma C.4.4 (using rigidity of tori and the commutativity of Z). Thus, the quotient G/S makes sense and the map $G/S \to T/f(S)$ is surjective. By commutativity of G, the smooth connected quotient G/S contains no nontrivial torus. Hence, by Chevalley's structure theorem (Theorem A.3.7), G/S is an extension of an abelian variety by a smooth connected unipotent group. In particular, G has no nontrivial torus quotient, so $T/f(S)$ is trivial. That is, $T = f(S)$ as desired.

Having settled (2) for the formation of images of maximal k-tori over any k, now consider the image under f of a maximal k-split k-torus S_0 in G. The k-split k-torus $T_0 = f(S_0)$ in H lies in a maximal one, say Z_0, and we want $T_0 = Z_0$. We may replace H with Z_0 and G with $f^{-1}(Z_0)$ to reduce to the case where H is a k-split torus. In this case if S is a maximal k-torus in G containing S_0 then $f(S) = H$, so $f(S_0) = H$ since H is a k-split torus.

Consider the problem of lifting a maximal k-torus T in H to one in G, assuming that $G' := f^{-1}(T)^0_{\text{red}}$ is smooth. If a maximal k-torus S' in G' maps onto T then any k-torus S in G containing S' also maps onto T, so $S \subseteq G'$ and hence $S = S'$. That is, such an S' would be maximal in G and so would do the job. We can therefore replace G with G' so that $H = T$ is a k-torus and G is smooth and connected. The same argument works if we replace "maximal k-torus" with "maximal k-split k-torus" throughout. The preceding argument for images of k-tori may now be applied to get the required lift, since the earlier argument with smooth connected G and a k-torus H did not use the smoothness of f. □

Remark C.4.6 The conjugacy in Proposition C.4.5(1) allows us to generalize the surjectivity result in Lemma 3.2.1 as follows. Let $f: H \to H'$ be a surjective homomorphism between group schemes locally of finite type over a field F, and let T be an F-torus in H, with $T' := f(T)$ its F-torus image in H'. Assume that the F-torus $(T \cap \ker f)^0_{\text{red}}$ is maximal in $\ker f$ and remains maximal after scalar extension to $\overline{F}$ (e.g., this holds if $(\ker f_{\overline{F}})^0_{\text{red}}$ is unipotent, as in Lemma 3.2.1). In such cases we claim that the map $N_H(T) \to N_{H'}(T')$ between scheme-theoretic normalizers is surjective.

It suffices to prove surjectivity on F'-points for all algebraically closed F'/F, renamed as F. By conjugacy of maximal F-tori in $\ker f$ (Proposition C.4.5(1) applied to $\ker f$ over $F = \overline{F}$), the hypothesis on $T \cap \ker f$ implies that T is a maximal torus in $f^{-1}(f(T))$. Consider $h' \in N_{H'}(T')(F)$, and choose $h \in H(F)$ such that $f(h) = h'$. Then hTh^{-1} is contained in $f^{-1}(f(T))$. By dimension considerations and the conjugacy of maximal F-tori in $f^{-1}(f(T))$ (Proposition C.4.5(1)), it follows that $hTh^{-1} = gTg^{-1}$ for some F-point g in $f^{-1}(f(T))$.

In other words, $h^{-1}g \in N_H(T)(F)$, so $h'^{-1}f(g) = f(h^{-1}g) \in N_{H'}(T')(F)$. Replacing h' with $h' \cdot h'^{-1}f(g) = f(g)$ is therefore harmless. But $f(g) = f(t)$ for some $t \in T(F)$, so we are done.

References

[Be] H. Behr, *Arithmetic groups over function fields I. A complete characterization of finitely generated and finitely presented arithmetic subgroups of reductive algebraic groups*, J. Reine Angew. Math. **495**(1998), 79–118.

[Bo1] A. Borel, *On affine algebraic homogeneous spaces*, Arch. Math. **45**(1985), 74–78.

[Bo2] A. Borel, *Linear algebraic groups* (2nd ed.), Springer-Verlag, New York, 1991.

[BoSe] A. Borel, J.-P. Serre, *Théorèmes de finitude en cohomologie galoisienne*, Comment. Math. Helv. **39**(1964), 111–164.

[BoSp] A. Borel, T. Springer, *Rationality properties of linear algebraic groups* II, Tohoku Math. J. (2) **20**(1968), 443–497.

[BoTi1] A. Borel, J. Tits, *Groupes réductifs*, Publ. Math. IHES **27**(1965), 55–151.

[BoTi2] A. Borel, J. Tits, *Théorèmes de structure et de conjugaison pour les groupes algébriques linéaires*, C. R. Acad. Sci. Paris **287**(1978), 55–57.

[BLR] S. Bosch, W. Lütkebohmert, M. Raynaud, *Néron models*, Springer-Verlag, New York, 1990.

[Bou1] N. Bourbaki, *Lie groups and Lie algebras* (Ch. 1–3), Springer-Verlag, New York, 1989.

[Bou2] N. Bourbaki, *Lie groups and Lie algebras* (Ch. 4–6), Springer-Verlag, New York, 2002.

[Bri] M. Brion, *Anti-affine algebraic groups*, J. Algebra **321**(2009), 934–952.

[BrTi] F. Bruhat, J. Tits, *Groupes réductifs sur un corps local I*, Publ. Math. IHES **41**(1972), 5–251.

[Chev] C. Chevalley, *Une démonstration d'un théorème sur les groupes algébriques*, J. Math. Pures Appl., **39**(1960), 307–317.

[Chow] W.-L. Chow, "On the projective embedding of homogeneous varieties" in *Algebraic geometry and topology. A symposium in honor of S. Lefschetz*, pp. 122–128, Princeton University Press, Princeton, NJ, 1957.

[Con1] B. Conrad, *A modern proof of Chevalley's theorem on algebraic groups*, J. Ramanujan Math. Soc., **17**(2002), 1–18.

[Con2] B. Conrad, *Finiteness theorems for algebraic groups over function fields*, Compos. Math. **148**(2012), 555–639.

[CP] B. Conrad, G. Prasad, *Classification of pseudo-reductive groups*, Annals of Mathematics Studies, Princeton University Press, Princeton, NJ, to appear.

[DG] M. Demazure, P. Gabriel, *Groupes algébriques*, Masson, Paris, 1970.

[SGA3] M. Demazure, A. Grothendieck, *Schémas en groupes* I, II, III, Lecture Notes in Mathematics 151, 152, 153, Springer-Verlag, New York, 1970.

[DeGr] M. Demazure, A. Grothendieck, *Schémas en groupes* (new edition), Documents Mathématiques 7, 8, Société Mathématique de France, 2011.

[EKM] R. Elman, N. Karpenko, A. Merkurjev, *The algebraic and geometric theory of quadratic forms*, AMS Colloq. Publ. 56, American Mathematical Society, Providence, RI, 2008.

[GQ] P. Gille, A. Quéguiner-Mathieu, *Exemples de groupes semi-simples simplement connexes anisotropes contenant un sous-groupe unipotent*, Pure Appl. Math. Q. **9**(2013), 487–492.

[Gre] M. Greenberg, *Rational points in henselian discrete valuation rings*, Publ. Math. IHES **31**(1966), 59–64.

[EGA] A. Grothendieck, *Eléments de géométrie algébrique*, Publ. Math. IHES **4**, **8**, **11**, **17**, **20**, **24**, **28**, **32**, 1960–7.

[SGA1] A. Grothendieck, *Revêtements étales et groupe fondamental*, Lecture Notes in Mathematics 224, Springer-Verlag, New York, 1971.

[Hum1] J. Humphreys, *Introduction to Lie algebras and representation theory*, Springer-Verlag, New York, 1972.

[Hum2] J. Humphreys, *Linear algebraic groups* (2nd ed.), Springer-Verlag, New York, 1987.

[KMT] T. Kambayashi, M. Miyanishi, M. Takeuchi, *Unipotent algebraic groups*, Lecture Notes in Mathematics 414, Springer-Verlag, New York, 1974.

[Kem] G. Kempf, *Instability in invariant theory*, Ann. Math. **108**(1978), 299–316.

[Mat] H. Matsumura, *Commutative ring theory*, Cambridge University Press, 1990.

[Mum] D. Mumford, *Abelian varieties*, Oxford University Press, 1970.

[Oes] J. Oesterlé, *Nombres de Tamagawa et groupes unipotents en caractéristique p*, Invent Math. **78**(1984), 13–88.

[Pink] R. Pink, *On Weil restriction of reductive groups and a theorem of Prasad*, Math. Z. **248**(2004), 449–457.

[PR] V. P. Platonov, A. S. Rapinchuk, *Algebraic groups and number theory*, Academic Press, New York, 1994.

[PY1] G. Prasad, J.-K. Yu, *On finite group actions on reductive groups and buildings*, Invent. Math. **147**(2002), 545–560.

[PY2] G. Prasad, J.-K. Yu, *On quasi-reductive group schemes*, J. Algebraic Geom. **15**(2006), 507–549.

[RG] M. Raynaud, L. Gruson, *Critères de platitude et de projectivité. Techniques de "platification" d'un module*, Invent. Math. **13**(1971), 1–89.

[Ri] R. W. Richardson, *Affine coset spaces of reductive algebraic groups*, Bull. London Math. Soc. **9**(1977), 38–41.

[Ros1] M. Rosenlicht, *Some rationality questions on algebraic groups*, Ann. Mat. Pura Appl. **43**(1957), 25–50.

[Ros2] M. Rosenlicht, *Questions of rationality for solvable algebraic groups over nonperfect fields*, Ann. Mat. Pura Appl. **61**(1963), 97–120.

[Ru] P. Russell, *Forms of the affine line and its additive group*, Pacific J. Math. **32**(1970), 527–539.

[SS] C. Sancho de Salas, F. Sancho de Salas, *Principal bundles, quasi-abelian varieties, and structure of algebraic groups*, J. Algebra **322**(2009), 2751–2772.

[Ser1] J.-P. Serre, *Algebraic groups and class fields*, Springer-Verlag, New York, 1988.

[Ser2] J.-P. Serre, *Lie groups and Lie algebras*, Lecture Notes in Mathematics 1500, Springer-Verlag, New York, 1992.

[Ser3] J.-P. Serre, *Galois cohomology*, Springer-Verlag, New York, 1997.

[Spr] T. A. Springer, *Linear algebraic groups* (2nd ed.), Birkhäuser, New York, 1998.

[St] R. Steinberg, *The isomorphism and isogeny theorems for reductive algebraic groups*, J. Algebra **216**(1999), 366–383.

[Ti0] J. Tits, "Classification of algebraic semisimple groups" in *Algebraic groups and discontinuous groups*, Proc. Symp. Pure Math., vol. 9, American Mathematical Society, Providence, RI, 1966.

[Ti1] J. Tits, *Lectures on algebraic groups*, Yale University Press, New Haven, CT, 1967.

[Ti2] J. Tits, *Théorie des groupes*, Annuaire du Collège de France, 1991–92.

[Ti3] J. Tits, *Théorie des groupes*, Annuaire du Collège de France, 1992–93.

Index

Printed in the United States
By Bookmasters